Handbook of Formal Argumentation

Handbook of Formal Argumentation

Edited by

Pietro Baroni
Dov Gabbay
Massimiliano Giacomin
Leendert van der Torre

© Individual authors and College Publications 2018. All rights reserved.

ISBN 978-1-84890-275-6

College Publications
Scientific Director: Dov Gabbay
Managing Director: Jane Spurr

http://www.collegepublications.co.uk

Cover produced by Laraine Welch
Printed by Lightning Source, Milton Keynes, UK

All rights reserved. No part of this publication may be reproduced, stored in a retrieval system or transmitted in any form, or by any means, electronic, mechanical, photocopying, recording or otherwise without prior permission, in writing, from the publisher.

CONTENTS

Preface ... vii

PART A INTRODUCTION ... 1

FRANS H. VAN EEMEREN, BART VERHEIJ
Argumentation Theory in Formal and Computational Perspective ... 3

HENRY PRAKKEN
Historical Overview of Formal Argumentation ... 75

THOMAS F. GORDON
Towards Requirements Analysis for Formal Argumentation ... 145

PART B ARGUMENTATION FORMALISMS ... 157

PIETRO BARONI, MARTIN CAMINADA,
MASSIMILIANO GIACOMIN
Abstract Argumentation Frameworks and Their Semantics ... 159

GERHARD BREWKA, STEFAN ELLMAUTHALER,
HANNES STRASS, JOHANNES P. WALLNER, STEFAN WOLTRAN
Abstract Dialectical Frameworks ... 237

SANJAY MODGIL, HENRY PRAKKEN
Abstract Rule-Based Argumentation ... 287

KRISTIJONAS ČYRAS, XIUYI FAN,
CLAUDIA SCHULZ, FRANCESCA TONI
Assumption-Based Argumentation: Disputes, Explanations, Preferences ... 365

ALEJANDRO J. GARCÍA, GUILLERMO R. SIMARI
Argumentation Based on Logic Programming ... 409

PHILIPPE BESNARD, ANTHONY HUNTER
A Review of Argumentation Based on Deductive Arguments ... 437

PART C ARGUMENTATION AND DIALOGUES ... 485

MARTIN CAMINADA
Argumentation Semantics as Formal Discussion ... 487

FABRIZIO MACAGNO, DOUGLAS WALTON, CHRIS REED
Argumentation Schemes   519

KATARZYNA BUDZYNSKA, SERENA VILLATA
Processing Natural Language Argumentation   577

PART D ALGORITHMS AND IMPLEMENTATIONS   **629**

WOLFGANG DVOŘÁK, PAUL E. DUNNE
Computational Problems in Formal Argumentation and their Complexity 631

FEDERICO CERUTTI, SARAH A. GAGGL,
MATTHIAS THIMM, JOHANNES P. WALLNER
Foundations of Implementations for Formal Argumentation   689

PART E ANALYSIS   **769**

MARTIN CAMINADA
Rationality Postulates: Applying Argumentation Theory
for Non-monotonic Reasoning   771

LEENDERT VAN DER TORRE, SRDJAN VESIC
The Principle-Based Approach to Abstract Argumentation Semantics 797

RINGO BAUMANN
On the Nature of Argumentation Semantics: Existence and Uniqueness,
Expressibility, and Replaceability   839

PIETRO BARONI, MASSIMILIANO GIACOMIN, BEISHUI LIAO
Locality and Modularity in Abstract Argumentation   937

ALEXANDER BOCHMAN
Argumentation, Nonmonotonic Reasoning and Logic   981

Preface

This handbook is a community effort aimed at providing a comprehensive and up-to-date view of the state of the art and current trends in the lively research field of formal argumentation.

The Handbook of Formal Argumentation is an open-ended initiative of which the present volume is the first outcome and a foundation stone. Further volumes are planned to cover topics not included in the present one and the initiative is conceived to grow by the support and feeding it receives from the community members. Contributions of ideas, contents, and, last but not least, availability to share the coordination and organisation activities are more than welcome.

The experience with the first volume is particularly encouraging. After some initial discussions among the editors and a few preliminary interactions with some colleagues representative of various areas in the field, the initiative was first presented at a meeting held during the COMMA 2014 conference, aimed at verifying the interest and availability of the community with respect to the challenging goal of producing a handbook.

Very positive feedback, many suggestions, and general support collected in that occasion boosted the process. The handbook website was set up (www.formalargumentation.org) and an intense exchange of ideas at large was started, in order to shape the whole initiative and the contents of the first volume in particular. This preparatory effort culminated in the Dagstuhl Perspectives Workshop "Present and Future of Formal Argumentation" held at Schloss Dagstuhl from August 30th to September 4th, 2015.

The presentations and lively discussions at the workshop cemented the initiative and opened the way to the actual implementation of the first volume.

First versions of the chapters were produced and submitted in the first part of 2016. Each chapter was examined by two reviewers, whose feedback has been crucial to improve the quality of the final versions of the chapters produced by the authors by the summer of 2017.

Some of the chapters, suitably modified to stand as journal papers, have appeared in a special issue of the *IfCoLog Journal of Logics and their Applications* (Volume 4, Issue 8) published in September 2017.

Authors' proofreading finally completed the long effort resulting in the present volume, whose organization and contents are briefly summarised below.

The volume is organised into five parts: *A. Introduction, B. Argumentation formalisms, C. Argumentation and dialogues, D. Algorithms and implementations, E. Analysis.*

Chapters in Part A provide an overall perspective on the field before entering into more technical matters. Chapter 1 provides a general overview by discussing argumentation and argumentation theory before and after the 'formal and computational turn' witnessed in the last decades. Chapter 2 surveys the historical development of formal argumentation covering two main classes of formal systems, devoted to inference and dialogue respectively. Chapter 3 analyses the issue of identifying suitable requirements for formal argumentation models in the view of supporting the design and development of suitable software tools for real-world applications.

Chapters in Part B provide a comprehensive coverage of the argumentation formalisms available in the literature at various levels of abstraction. Chapter 4 introduces the formalism of abstract argumentation frameworks and their semantics, where arguments are abstract entities attacking each other and attention is focused on the problem of identifying the possible solutions to this conflict situation. Chapter 5 describes abstract dialectical frameworks generalising the argumentation frameworks presented in Chapter 4 by allowing arbitrary relationships among arguments to be expressed. At a different level of abstraction, the subsequent chapters deal with so-called *structured* argumentation formalisms, explicitly encompassing the internal structure of the arguments and the identification of the relation of attack among them. Chapter 6 reviews rule-based approaches to argumentation and in particular the $ASPIC^+$ framework. Chapter 7 presents assumption-based argumentation and the relevant computational tools, while Chapter 8 deals with argumentation based on logic programming and in particular describes the $DeLP$ formalism, combining techniques from both areas. Chapter 9 describes deductive argumentation, where arguments are built from a knowledge-base using the consequence relation of some logic formalism. Attacks between the arguments are expressed as logical relationships involving their premises and claims.

While chapters in Part B are prevalently, but not exclusively, oriented to the inferential aspects of formal argumentation, Part C is devoted to cover some of its many dialogical aspects. Chapter 10 provides an interpretation of argumentation semantics as formal discussion, analysing how the notions of acceptance embedded in different semantics correspond to different types of discussions. Chapter 11 is devoted to argumentation schemes—stereotypical patterns for argument construction—accompanied by a set of critical questions that the interlocutor can use to question the argument and evaluate its strength. Chapter 12 deals with the problem of processing natural language texts including arguments and argument exchanges, with a special focus on the growing research area of argument mining.

Part D is devoted to algorithmic, computational and implementation issues. Chapter 13 gives an overview of the core computational problems arising in formal argumentation and develops the relevant computational complexity analysis, with main reference to three formalisms, namely abstract argumentation frameworks, abstract dialectical frameworks, and assumption-based argumentation, as they allow to highlight different sources of complexity in different formal contexts. Chapter 14 deals with implementation issues by surveying general techniques as well as specific software systems in three main areas: abstract argumentation frameworks, structured argumentation frameworks, and approaches related to visualisation and analysis of argumentation processes.

Part E completes this volume by providing deeper analyses on some of the previously introduced topics. Chapter 15 discusses the interplay between argument construction at the structured level and argument evaluation at the abstract level and introduces some rationality postulates aimed at ensuring some desirable properties of the conclusions drawn at the end of the overall process. Chapter 16 provides a comprehensive overview of the general principles identified in the literature for analysis and comparison of abstract argumentation semantics, a systematic characterisation of existing semantics proposals based on these principles, and lays down the foundations for the use of a principle-based approach in other areas of formal argumentation. Chapter 17 focuses on some intrinsic properties related to abstract argumentation semantics, namely existence, uniqueness, expressibility, and replaceability, and provides a full technical treatment of these aspects for the main semantics considered in the literature, including the case of infinite sets of arguments. Chapter 18 surveys locality and modularity properties in abstract argumentation and provides some examples of their practical applicability. Finally, Chapter 19 discusses the links between argumentation, nonmonotonic reasoning, and logic and provides new perspectives on their relationships based on the formalism of abstract collective argumentation.

Altogether we believe that the chapters included in this volume achieve two goals. On the one hand they provide enough introductory material so that a newcomer can get acquainted with the essentials of the field. On the other hand, they cover more advanced issues so that anyone interested in the field may have a comprehensive, though of course not exhaustive, reference on the state of the art and get useful insights for future developments.

Looking forward to covering further aspects of formal argumentation in the next volumes of this handbook, we are pleased to conclude with some dutiful expressions of gratitude. We thankfully acknowledge the contribution of all the authors and the reviewers who made this volume possible and the help of all the colleagues who provided comments, suggestions, critiques, and encouragements during the development of the initiative. Last but not least, special thanks go to College Publications and in particular to Jane Spurr for her invaluable continued support.

Pietro Baroni
Dov Gabbay
Massimiliano Giacomin
Leendert van der Torre

PART A

INTRODUCTION

1
Argumentation Theory in Formal and Computational Perspective

FRANS H. VAN EEMEREN, BART VERHEIJ

ABSTRACT. Argumentation has been studied since Antiquity. Modern argumentation theory took inspiration from these classical roots, with Toulmin's 'The Uses of Argument' (1958) and Perelman and Olbrechts-Tyteca's 'The New Rhetoric' (1969) as representants of a neo-classical development. In the 1970s, a significant rise of the study of argumentation started, often in opposition to the logical formalisms of those days that lacked the tools to be of much relevance for the study of argumentation as it appears in the wild. In this period, argumentation theory, rhetoric, dialectics, informal logic, and critical thinking became the subject of productive academic study. Since the 1990s, innovations in artificial intelligence supported a formal and computational turn in argumentation theory, with ever stronger interaction with non-formal and non-computational scholars. The present chapter sketches argumentation and argumentation theory as it goes back to classical times, following the developments before and during the currently ongoing formal and computational turn.

1 Introduction

Argumentation has been studied since Antiquity. Several 20th century developments in the study of argumentation (in particular since the 1950s) were initiated by concerns that the formal methods of the time, especially classical formal logic, were not fully adequate for the study of argumentation. In recent years, such concerns have been addressed, and partially answered, using innovations in formal and computational methods, in particular in computer science and in artificial intelligence. We can speak of a formal and computational turn in the study of argumentation. This chapter sketches argumentation and argumentation theory as it goes back to classical times, following the developments before and during the currently ongoing formal and computational turn. While doing so, we explain what the study of argumentation, generally known as *argumentation theory*, involves. Our exposé is based on the *Handbook of Argumentation Theory* that we recently co-authored with Bart Garssen, Erik C.W. Krabbe, A. Francisca Snoeck Henkemans and Jean Wagemans (van Eemeren et al., 2014; in particular Chapters 1 and 11).[1]

---
[1]Relevant journals include: Artificial Intelligence, Artificial Intelligence and Law, Autonomous Agents and Multi-Agent Systems, Computational Intelligence, International Jour-

In Section 2, 'Argumentation and argumentation theory before the formal and computational turn', we define argumentation in the way this concept has been used in argumentation theory before the formal and computational turn; starting from this definition we explain what argumentation theory is about and describe its main aims. We introduce crucial concepts that play a major role in argumentation theory, and give an overview of prominent theoretical approaches. In Section 3, 'Formal and computational argumentation theory: precursors and first steps', we start the discussion of formal and computational approaches to argumentation by addressing precursors and first steps made, in particular in non-monotonic logic and defeasible reasoning. Section 4, 'Argumentation and the structure of arguments in formal and computational perspective', is about the formalization of argument attack, the structure of arguments, argument schemes and dialogue. In Section 5, 'Specific kinds of argumentation in formal and computational perspective', we discuss argumentation with rules, cases, values and evidence. We conclude the chapter by looking back at the formal and computational turn in argumentation theory using the crucial concepts of argumentation theory before that turn, and by an outlook into the future of argumentation theory.

2 Argumentation and argumentation theory before the formal and computational turn

Argumentation, a phenomenon we are all familiar with, arises in response to, or in anticipation of, a real or imagined difference of opinion. It comes into play in cases when people start defending a view they assume not to be shared by others. Not only the need for argumentation, but also the requirements argumentation has to fulfil and the structure of argumentation are connected with a context in which doubt, potential opposition, and perhaps also objections and counterclaims arise.

A definition of argumentation suitable to be used in argumentation theory should connect with commonly recognized characteristics of argumentation. It is important to realize however that there are striking differences between the meaning of the pivotal word 'argumentation' in English usage and the meaning of its lexical counterparts in other languages.[2] A first relevant difference is

---

nal of Cooperative Information Systems, International Journal of Human-Computer Studies, Journal of Logic and Computation, and The Knowledge Engineering Review. Contributions have also been made to journals that deal primarily with argumentation, such as Argumentation and Informal Logic. A journal devoted explicitly to the interdisciplinary area of AI is Argument and Computation. The biennial conference series COMMA is devoted to the study of computational models of argument. The first was held in Liverpool in 2006, followed by conferences in Toulouse (2008), Desenzano del Garda (2010), Vienna (2012), Pitlochry (2014), and Potsdam (2016). See http://www.comma-conf.org/. ArgMAS (Argumentation in Multi-Agent Systems) and CMNA (Computational Models of Natural Argument) are related workshops.

[2] For instance, in French 'argumentation,' in German 'Argumentation,' in Italian 'argomentazione,' in Portuguese 'argumentação,' in Spanish 'argumentación,' in Dutch 'argumentatie,' and in Swedish 'argumentation.'

that the meaning of argumentation in the latter naturally includes both argumentation as a process and argumentation as a product. Second, unlike the English word 'argumentation', its non-English counterparts pertain exclusively to a constructive effort to convince the addressee of the acceptability of one's standpoint, so that argumentation is immediately associated with reasonableness.[3] Third, in the non-English counterparts 'argumentation' is taken to refer only to the constellation of propositions put forward in defence of a standpoint without including the standpoint,[4] so that standpoint and argumentation are viewed as separate entities, which facilitates the study of their relationship (van Eemeren & Grootendorst, 1984, p. 18). Note that—as we will see below—since the formal and computational turn discussed below, attention for argumentation that goes against a standpoint has increased.

Next to the meaning of the non-English counterparts, which captures some vital characteristics, there are also some general characteristics of argumentation that are independent of any specific language that are taken into account in defining the term argumentation in argumentation theory. To begin with, argumentation is a *communicative act complex*,[5] whose structural design reflects the functional intent of the communicative moves that are made. Next, argumentation is an *interactional act complex* directed at eliciting a response that indicates acceptance of the standpoint that is defended, so that it is always part of an explicit or implicit dialogue with the addressee. Further, as a rational activity of reason, argumentation involves putting forward a constellation of propositions the arguer can be held accountable for, so that it is not just an expressive but creates commitments. Finally, in making an appeal to common critical standards of reasonableness in trying to convince the addressee, the arguer approaches the addressee as a rational judge who judges reasonably.[6]

Based on these starting points, defining argumentation starts from ordinary usage and is next made more precise and explicit in order to adequately serve its purpose in argumentation theory:

Argumentation is a communicative and interactional act complex aimed at resolving a difference of opinion with the addressee by putting forward a constellation of propositions *the arguer can be held accountable for* to make the standpoint at issue acceptable to *a rational judge who judges reasonably*.[7]

---

[3] This does not mean, of course, that in practice argumentation cannot be abused, so that there is no matter of acting reasonably.

[4] According to Tindale (1999, p. 45), it is 'the European fashion' to refer to the premises of an argument as the argumentation and to the conclusion by using another term, such as standpoint.

[5] Because argumentation can also be non-verbal, for instance, visual, it is defined here—more generally—as a 'communicative' rather than a 'verbal' ('linguistic') act complex.

[6] Although the terms rational and reasonable are often used interchangeably, we think that it is useful to make a distinction between acting 'rationally' in the sense of using one's faculty of reason and acting 'reasonably' in the sense of utilizing one's faculty of reason in an appropriate way.

[7] The term argumentation refers to the whole constellation of propositions put forward in defence of the standpoint. Because each of the propositions constituting the constellation has its own share in providing grounds for accepting the standpoint at issue, in principle,

Argumentation theory is the umbrella term used to denote the study of argumentation in all its manifestations and varieties, irrespective of the intellectual backgrounds, primary research interests and angles of approach of the theorists. Other general labels, such as informal logic and rhetoric, refer to specific theoretical perspectives on the study of argumentation (and usually also include other research interests than argumentation).

Because the standpoints at issue in a difference of opinion and the argumentation advanced to support them can pertain to all walks of life and all kinds of subjects, the scope of argumentation theory is very broad. It ranges from argumentative discourse in the public and the professional sphere to argumentative discourse in the personal or private sphere. The types of standpoints supported by argumentation may vary from descriptive standpoints to evaluative and prescriptive standpoints. It is in particular worth noting that argumentation is certainly not used only for truth-finding and truth-preservation.[8]

Scholars are often drawn to studying argumentation by their practical interest in improving the quality of argumentative discourse where this is called for. In order to be able to realize this ambition, they have to combine an empirical orientation towards how argumentative discourse is conducted with a critical orientation towards how it should be conducted. To give substance to this challenging combination, they need to carry out a comprehensive research programme that ensures that argumentative discourse will not only be examined descriptively as a specimen of verbal communication and interaction ('pragmatics') but also be measured against normative standards of reasonableness ('normative pragmatics') (van Eemeren, 1990).

In order to combine critical and empirical insights systematically, in argumentation theory argumentation scholars make it their business to bridge the gap between the normative dimension and the descriptive dimension of argumentative discourse. The complex problems that are at stake are to be solved with the help of a research programme with five interrelated components (van Eemeren & Grootendorst, 2004, pp. 9–41).[9] On the one hand, the programme has a *philosophical* component, in which a philosophy of reasonableness is developed, and a *theoretical* component, in which, starting from this philosophy, a model for argumentative discourse is designed. On the other hand, the programme has an *empirical* component, in which argumentative reality as it manifests itself in communicative and interactional exchanges is investigated. Next, in the pivotal *analytical* component of the research programme, the normative and the descriptive dimensions are systematically linked together by a theoretically motivated and empirically justified reconstruction of argumentative discourse. Finally, in the *practical* component the problems that occur in

---

these propositions by themselves also have an argumentative function. This is expressed terminologically by calling them the reasons that make up the argumentation as a whole.

[8]Generally, in discussing a claim to acceptance, argumentation has in fact no major role to play when a decisive solution can readily be offered by other means.

[9]The five components of a fully-fledged research programme in argumentation theory were introduced in van Eemeren (1987).

the various kinds of argumentative practices are identified, and methods are developed to tackle these problems.

In developing a philosophy of reasonableness argumentation theorists reflect in the philosophical component upon the rationale for the view of reasonableness that is to underlie their theoretical approach. Depending on the conception of reasonableness they favour, in the theoretical component standards for the validity, soundness or appropriateness of argumentation are adopted and theoretical models are developed based on these conceptions. Because the model of argumentation is in this case a normative instrument for assessing the quality of argumentation put forward in argumentative reality, the model constitutes a point of orientation for the empirical research that is to be carried out in argumentation theory but does not constitute a test of the model. The model indicates which factors and processes are worth investigating and to what extent the norms prevailing in argumentative reality agree with the theoretical standards, but deviations are not necessarily an indication of any wrongness in the model.[10]

Analytical research in argumentation theory is aimed at the reconstruction of argumentative discourse as it occurs in argumentative reality from the perspective of the model of argumentation that is chosen as the theoretical starting point. Whichever theoretical background they may have, argumentation theorists engaging in analytical research need to develop appropriate tools and methods for reconstructing argumentative discourse. Practical research in argumentation theory, finally, is aimed at analyzing the (spoken and written) argumentative practices that can be distinguished in the various communicative domains from the perspective of argumentation theory and developing instruments for intervention in argumentative discourse where this is due. The instruments for enhancing the quality of argumentative practices may consist of designs for the formats of communicative activity types or of methods for improving arguers' skills in analysing, evaluating and producing argumentative discourse.

In the end, the general objective of argumentation theory is a practical one: to provide adequate instruments for analysing, evaluating and producing argumentative discourse. Ultimately the *raison d'être* of the other components of the research programme carried out in argumentation theory is that they enable the systematic development of such instruments. When taken together, philosophical and theoretical insights into argumentative discourse, analytically connected with empirical insights, are to lead to methodical applications of argumentation theory to the various kinds of argumentative practices.

In pursuing their objective of improving the analysis, evaluation and production of argumentative discourse, argumentation theorists take account of the point of departure of argumentation, consisting of the explicit and implicit ma-

---

[10] Only in case of a purely descriptive theory the empirical research could be aimed at testing the model, but so far no fully-fledged argumentation theory without a critical dimension has been developed.

terial and procedural premises that serve as the starting point, and the layout of the argumentation displayed in the constellation of propositions explicitly or implicitly advanced in support of the standpoints at issue. Both the point of departure and the layout of argumentation are to be judged by appropriate standards of evaluation that are in agreement with all requirements a rational judge who judges reasonably should comply with. This means that the descriptive and normative aims of argumentation theory as a discipline can be specified as follows:[11]

1. Giving a descriptive account of the components of argumentative discourse that constitute together the point of departure of argumentation;

2. Giving a normative account of the standards for evaluating the point of departure of argumentation that are appropriate to a rational judge who judges reasonably;

3. Giving a descriptive account of the components of argumentative discourse that constitute together the layout of argumentation;

4. Giving a normative account of the standards for evaluating argumentation as it is laid out in argumentative discourse that are appropriate to a rational judge who judges reasonably.

2.1 Crucial concepts

Certain theoretical concepts are indispensable in developing instruments for methodically improving the quality of the analysis, evaluation and production of argumentative discourse. Among them are the notions of 'standpoint,' 'unexpressed premise,' 'argument scheme,' 'argumentation structure,' and 'fallacy.' All of them are immediately connected with central problem areas in argumentation theory.

Standpoints We use the term standpoint (or point of view) to refer to what is at issue in argumentative discourse in the sense of what is being argued about.[12] In advancing a standpoint the speaker or writer assumes a positive or negative position regarding a proposition. Because advancing a standpoint implies undertaking a positive or negative commitment, in view of the aim of resolving a difference of opinion, whoever advances a standpoint is obliged to defend their standpoint if challenged to do so by the listener or reader. The standpoints at issue in a difference of opinion can be descriptive, evaluative or

---

[11]The descriptive aims of argumentation theory are often associated with the 'emic' study of what is involved in justifying claims and what are good reasons for accepting a claim viewed from the 'internal' perspective of the arguers while the normative aims are associated with the 'etic' study of these matters viewed from the 'external' perspective of a critical theorist.

[12]The terms claim, conclusion, thesis and debate proposition are used to refer from different theoretical angles to virtually the same concept as the term standpoint. Terms such as belief, opinion and attitude usually refer to related concepts that are in relevant ways different from a standpoint.

prescriptive, but in all cases they can be reconstructed as a claim to acceptability (in case of a positive standpoint) or unacceptability (in case of a negative standpoint) regarding the proposition the standpoint pertains to.[13]

Unexpressed premises Unexpressed premises are often pivotal missing links in transferring acceptance from the premises that are explicitly put forward in the argumentation to the standpoint that is defended.[14] Such partly implicit argumentation, which is quite usual in ordinary argumentative discourse, is called enthymematic. The identification of elements left implicit in enthymematic argumentation is in practice usually unproblematic, but in some cases it can be a problem. According to most argumentation theorists, then carrying out a logical analysis does not suffice. Starting from a logical analysis, a pragmatic analysis needs to be carried out in which the analyst tries to identify the unexpressed premise by determining on the basis of the available contextual and background information to which implicit proposition the arguer can be held committed to.[15]

Argument schemes An argument(ation) scheme is an abstract characterization of the way in which in a particular type of argumentation a reason used in support of a standpoint is related to that standpoint in order to bring about a transfer of acceptance from that reason to the standpoint. Depending on the kind of relationship established in the argument scheme, specific kinds of evaluative questions—usually referred to as critical questions—are to be answered in evaluating the argumentation. These critical questions capture the specific pragmatic rationale for bringing about the transition of acceptance.[16]

Argumentation structures The argumentation structure of a piece of argumentative discourse characterizes the 'external' organization of the argumentation that is advanced: how do the reasons put forward in a particular argumentation hang together and in what way exactly do they relate to the standpoint at issue? In argumentation theory, various ways of combining reasons have been distinguished that characterize the different kinds of argumentation structures that can be instrumental in defending a standpoint.[17]

---

[13] For an overview of the various approaches to standpoints see Houtlosser (2001).

[14] Depending on the theoretical background of the theorists, other terms are used to refer to an unexpressed premise: implicit, suppressed, tacit, and missing premise, reason or argument, but also warrant, implicature, supposition, and even assumption, inference and implication.

[15] For an approach in which a logical analysis is used as a heuristic tool in carrying out a pragmatic analysis see van Eemeren and Grootendorst (1992, pp. 64–67; 2004, pp. 117–118). For the various kinds of resources that can be used in accounting for the reconstruction see van Eemeren (2010, pp. 16–19).

[16] For an overview of the study of argument schemes, see Garssen (2001); for attempts at formalization and the computational implications, see Walton, Reed and Macagno (2008, Ch. 11 and 12). A recent development is the study of what have been called prototypical argumentative patterns. These consist of constellations of argumentative moves in which a particular argument scheme or combination of argument schemes is used (van Eemeren, 2017).

[17] Different terminological conventions have been developed for naming the combinations of reasons and the divisions of the various types of structures are not always exactly the same. For an overview of the study of argumentation structures see Snoeck Henkemans (2001).

Fallacies The difference of opinion at issue in argumentative discourse will not be resolved satisfactorily if contaminators of the argumentative exchange enter the discourse that are not detected. Such contaminators, which may be so treacherous that they go unobserved in the argumentative exchange, are known as fallacies. Virtually every normative theory of argumentation includes a treatment of the fallacies. The degree to which a theory of argumentation makes it possible to give an adequate treatment of the fallacies can even be considered as a litmus test of the quality of the theory.[18]

2.2 Prominent theoretical approaches

Ancient dialectic and rhetoric—in combination with syllogistic logicare the forbears of modern argumentation theory.[19] The Aristotelian concept of dialectic is best understood as the art of inquiry through critical dialogue. In a dialogue that is dialectical in the Aristotelian sense the adequacy of any particular claim is supposed to be cooperatively assessed by eliciting premises that might serve as commonly accepted starting points, then drawing out implications from those starting points and determining their compatibility with the claim in question. Where contradictions emerge, revised claims might be put forward to avoid such problems. This method of regimented opposition amounts to a pragmatic application of logic, a collaborative method of putting logic into use so as to move from conjecture and opinion to more secure belief. Aristotle's rhetoric deals with the principles of effective persuasion leading to assent or consensus. It bears little resemblance to modern-day persuasion theories heavily oriented to the analysis of attitude formation and attitude change but largely indifferent to the problem of the invention of persuasive messages (Eagly & Chaiken, 1993; O'Keefe, 2002). In Aristotle's rhetoric, the emphasis is on the production of effective argumentation for an audience when the subject matter does not lend itself to a logical demonstration of certainty. When it comes to logical demonstration, the syllogism is the most prominent form; the enthymeme, thought of as an incomplete syllogism whose premises are acceptable to the audience, is its rhetorical counterpart. As yet, there is no unitary theory of argumentation available that encompasses the dialectical and rhetorical dimensions of argumentation and is universally accepted. The current state of the art in the argumentation theory (as it developed before the recent formal and computational turn) is characterized by the co-existence of a variety of theoretical perspectives and approaches, which differ considerably from each other in conceptualization, scope and theoretical refinement. Every fully-fledged theoretical approach to argumentation represents in fact a particular specification of what it means for a rational judge to judge reasonably and provides a definition of (crucial aspects of) the type of validity favoured by the

---

[18]For a more detailed overview of the study of fallacies see van Eemeren (2001).

[19]Although ancient dialectic and rhetoric are often discussed as if both of them were unified wholes, contributions to their development have been made by various scholars and their views were by no means always in harmony. In order to be accurate, we must therefore always indicate precisely to whose views exactly we are referring.

theorist.

Some argumentation theorists, especially those having a background in linguistics, discourse analysis or rhetoric, have a goal that is primarily (and sometimes even exclusively) descriptive. They are interested in finding out how in argumentative discourse speakers and writers try to convince or persuade others. Other argumentation theorists, often inspired by logic, philosophy or insights from law, study argumentation primarily for normative purposes. They are interested in developing validity or soundness criteria that argumentation must satisfy in order to qualify as rational or reasonable. Currently, however, most argumentation theorists seem to recognize that argumentation research has a descriptive as well as a normative dimension and that in argumentation theory both dimensions must be combined.[20]

Most modern approaches to argumentation are strongly affected by the perspectives on argumentation developed in Antiquity. Both the dialectical perspective (which nowadays usually incorporates the logical dimension) and the rhetorical perspective are represented prominently. Approaches to argumentation that are dialectically oriented tend to focus primarily on the quality of argumentation in defending standpoints in regulated critical dialogues. They put an emphasis on guarding the reasonableness of argumentation by means of regimentation. It is noteworthy that in the rhetorically oriented approaches to argumentation putting an emphasis on factors influencing the effectiveness of argumentation, effectiveness is usually viewed as a 'right to acceptance' that speakers or writers are, as it were, entitled to on the basis of the qualities of their argumentation rather than in terms of actual persuasive effects.[21]

In modern argumentation theory a remarkable revival has taken place of both dialectic and rhetoric. Unlike in Aristotle's approach, however, there is a wide conceptual gap between the two perspectives on argumentation, going together with a communicative gap between their protagonists. In recent times, some argumentation scholars have come to the conclusion that the dialectical and rhetorical views on argumentation are not per se incompatible. It has even been argued that re-establishing the link between dialectic and rhetoric will

---

[20]The infrastructure of the field of argumentation theory in terms of academic associations, journals and book series reflects to some extent the existing division in theoretical perspectives. The American Forensic Association (AFA), associated with the National Communication Association, and its journal *Argumentation & Advocacy* concentrate on argumentation, communication and debate. The Ontario Society for the Study of Argumentation (OSSA), the Association of Informal Logic and Critical Thinking (AILACT) and the electronic journal *Informal Logic* focus on informal logic. The International Society for the Study of Argumentation (ISSA), the journals *Argumentation* and *Journal of Argumentation in Context*, and the accompanying book series Argumentation Library and Argumentation in Context aim to cover the whole spectrum of argumentation theory. Other international journals relevant to argumentation theory are *Philosophy and Rhetoric*, *Logique et Analyse*, *Controversia*, *Pragmatics and Cognition*, *Argument and Computation*, and *Cogency*.

[21]Research aimed at examining the actual effectiveness of argumentation is usually called persuasion research. In practice, it generally amounts to quantitative empirical testing of the ways in which argumentation and other means of persuasion lead to changes of attitude in the recipients (O'Keefe, 2002).

enrich the analysis and evaluation of argumentative discourse (van Eemeren, 2010, especially Ch. 3).

In giving a brief overview of the current theoretical approaches, we first turn to two 'neo-classical' proposals developed in the 1950s: the Toulmin model and the 'new rhetoric'. In dealing with argumentation both aim to counterbalance the formal approach that modern logic provides for dealing with analytic reasoning.

In *The uses of argument*, first published in 1958, Toulmin (2003) reacted against the then dominant logical view that argumentation is just another specimen of the reasoning that the formal approach is qualified to deal with. As an alternative, he presented a model of the 'procedural form' of argumentation aimed at capturing the functional steps that can be distinguished in the defence of a standpoint by means of argumentation. The procedural form of argumentation is, according to Toulmin, 'field-independent', meaning that the steps that are taken are always the same, irrespective of the subject that is being discussed.[22]

In judging the validity of argumentation, Toulmin gives the term validity a different meaning than it has in formal logic. The validity of argumentation is in his view primarily determined by the degree to which the (usually implicit) warrant that connects the data advanced in the argumentation with the claim at issue is acceptable—or, if challenged, can be made acceptable by a backing. What kind of backing may be required in a particular case depends on the field to which the standpoint at issue belongs. This means that the criteria used in evaluating the validity of argumentation are in Toulmin's view 'field-dependent'. Thus, Toulmin puts the validity criteria for argumentation in an empirical and historical context.

In their monograph *The new rhetoric*, also first published in 1958, Perelman and Olbrechts-Tyteca (1969) regard argumentation—in line with classical rhetoric—as sound if it adduces or reinforces assent among the audience to the standpoint at issue. The audience addressed may be a 'particular' audience consisting of a specific person or group of people, but it can also be the 'universal' audience—the (real or imagined) audience that, in the arguer's view, embodies reasonableness.

Besides an overview of the elements of agreement that can in argumentation serve as points of departure (facts, truths, presumptions, values, value hierarchies and topoi[23]), Perelman and Olbrechts-Tyteca provide an overview of the argument schemes that in the layout of argumentation can be used to convince or persuade an audience. The argument schemes they distinguish remain for the most part close to the classical topical tradition. Apart from argumentative techniques of 'association', in which these argument schemes are employed, Perelman and Olbrechts-Tyteca also distinguish an argumentative

---

[22]It is noteworthy that Toulmin's model of the argumentative procedure is in fact conceptually equivalent to the extended syllogism known in Roman-Hellenistic rhetoric as epicheirema.

[23]Perelman and Olbrechts-Tyteca use the Latin equivalent loci.

technique of 'dissociation.' Dissociation divides an existing conceptual unity into two separate conceptual unities.

In spite of obvious differences between Toulmin's approach to argumentation and that of Perelman and Olbrechts-Tyteca, there are also some striking commonalities. Starting from an interest in the justification of views by means of argumentative discourse, both emphasize that values play a part in argumentation, both reject formal logic as a theoretical tool, and both turn for an alternative model to juridical procedures. A theoretical connection between the Toulmin model and the new rhetoric could be made by viewing the various points of departure distinguished in the new rhetoric as representing different types of data in the Toulmin model and its argument schemes as different types of warrants or backings.

Of the approaches to argumentation that have been developed more recently, formal dialectic, coined and instigated by Hamblin (1970), remains closest to formal logic, albeit logic in a dialectical garb. The scholars responsible for the revival of dialectic in the second part of the twentieth century treat argumentation as part of a formal discussion procedure for resolving a difference of opinion by testing the tenability of the 'thesis' at issue against challenges. Apart from the ideas about formal dialectic articulated by Hamblin, in designing such a procedure they make use of the 'dialogue logic' of the Erlangen School (Lorenzen & Lorenz, 1978), but also from insights advanced by Crawshay-Williams (1957) Næss (1966). The most complete proposal was presented by Barth and Krabbe (1982) in *From axiom to dialogue.* Their formal dialectic describes systems for determining by means of a regimented dialogue game between the proponent and the opponent of the thesis whether the proponent's thesis can be maintained given the premises allowed as 'concessions' by the opponent.

Building on the proposals for a dialogue logic made by the Erlangen School, Barth and Krabbe's formal dialectic offers a translation of formal logical systems into formal rules of dialogue. In *Commitment in dialogue,* Walton and Krabbe (1995) integrate the proposals of the Erlangen School with the more permissive kind of dialogues promoted in Hamblin's (1970) dialectical systems. After having provided a classification of the main types of dialogue, they discuss the conditions under which in argumentation commitments should be maintained or may be retracted without violating any of the rules of the type of dialogue concerned.

Related approaches can be found in some of the proposals made by formal and informal logicians. Out of dissatisfaction with the treatment of argumentation in logical textbooks, and inspired by the Toulmin model (and to a much lesser extent the new rhetoric), a group of Canadian and American philosophers have propagated since the 1970s an approach known as informal logic. The label informal logic refers in fact to a collection of logic-oriented normative approaches to the study of reasoning in ordinary language which remain closer to the practice of argumentation than is usually the case in formal logic. Informal logicians aim in the first place at developing adequate norms for interpreting, assessing and construing argumentation.

Since 1978, the journal *Informal Logic*,[24] started and edited by Blair and Johnson (later joined by others), has been the speaking voice of informal logic and the connected educational reform movement dedicated to 'critical thinking'. In their textbook *Logical self-defense*, Johnson and Blair (2006) have indicated what they have in mind when they speak of an informal logical alternative to formal logic. They explain that the premises of an argument have to meet the criteria of 'acceptability', 'relevance' and 'sufficiency'. Other informal logicians have adopted these three criteria, albeit sometimes under slightly different names (e.g., Govier, 1987).

Freeman (2005) provides, from an epistemological perspective on informal logic, a comprehensive theory of premise acceptability. Generally, however, informal logicians remain in the first place interested in the premise-conclusion relations in arguments (e.g., Walton, 1989). Most of them maintain that argumentation should be valid in some logical sense, but generally they do not stick to the formal criterion of deductive validity. Woods and Walton (1989) claim that each fallacy requires its own theoretical treatment, which leads them to applying a variety of logical systems in their theoretical treatment of the fallacies. Johnson (2000) also takes a predominantly logical approach, but he complements this approach with a 'dialectical tier', where the arguer discharges his or her dialectical obligations, for instance, by anticipating objections, and dealing with alternative positions. In Finocchiaro's contributions to informal logic, too, the logical and the dialectical approach are combined, albeit that the emphasis is more strongly on the dialectical dimension, and historical and empirical dimensions are added (e.g., Finocchiaro, 2005). The rhetorical perspective has received less attention from informal logicians. A notable exception is Christopher Tindale (1999, 2004).

In modern times, the study of rhetoric has fared considerably better in the United States than in Europe. Not only has classical rhetoric from the nineteenth century onwards been represented in the academic curriculum, but also has the development of modern rhetorical approaches been more prolific. In the last decades of the twentieth century, the image that rhetoric had acquired of being irrational and even anti-rational has been revised. Paying tribute to Perelman and Olbrechts-Tyteca's new rhetoric, in various countries various scholars have argued for a rehabilitation of the rhetorical approach. In spite of the unlimited extension in the United States in the 1960s of the scope of Big Rhetoric 'to the point that everything, or virtually everything, can be described as 'rhetorical' ' (Swearingen & Schiappa, 2009, p. 2), Wenzel (1987) emphasized the rational qualities of rhetoric. In Europe, Reboul (1990) and Kopperschmidt (1989a) argued at about the same time for giving rhetoric its rightful position in the study of argumentation beside dialectic.

Although all of them may be described as rhetoricians in the broad sense, the American scholars from the field of (speech) communication currently engaged

---

[24] At first named Informal Logic Newsletter.

in argumentation theory do not share a clearly articulated joint perspective. Their most obvious common feature is a concern with the connection between claims and the people engaged in some kind of argumentative practice. The American debate tradition in particular has had an enormous influence on American argumentation studies. More or less outside the immediate debate tradition, Zarefsky (2006, 2009), Leff (2003) and Schiappa (2002) have contributed profound historical rhetorical analyses. Fahnestock (1999, 2009) dealt theoretically with rhetorical figures and stylistics.

Concentrating on the public features of communicative acts, Jackson and Jacobs (1982) initiated a research programme for studying argumentation in informal conversations. Their joint research is aimed at understanding the reasoning processes by which individuals make inferences and resolve disputes in ordinary conversation. A related empirical angle in American argumentation research consists in the study of argument in natural settings, such as school board meetings, counseling sessions and public relations campaigns, to produce 'grounded theory'—a theory of the specific case.

A Toulminian concept that has strongly influenced American argumentation scholarship is the notion of 'field'. Toulmin (1972) describes fields as 'rational enterprises', which he equates with intellectual disciplines, and explores how the nature of reasoning differs from field to field. This treatment led to vigorous discussion about what defines a 'field of argument': subject matter, general perspective, world-view, or the arguer's purpose—to mention just a few of the possibilities. The concept of fields of argument encouraged recognition that the soundness of arguments is not something universal and necessary, but context-specific and contingent. Instead of the term fields, Goodnight prefers the term spheres, referring to 'the grounds upon which arguments are built and the authorities to which arguers appeal' (1982, p. 216). He uses 'argument' to mean interaction based on dissensus, so that the grounds of arguments lie in doubts and uncertainties. In a similar vein as Habermas (1984), Goodnight (2012) distinguishes between three spheres of argument: the 'personal' (or 'private') sphere, the 'public' sphere, and the 'technical' sphere.

Meanwhile, starting in the 1970s, in Europe a descriptive approach has developed in which argumentation is viewed as a linguistic phenomenon that not only manifests itself in language use, but is also inherent in most language use. In a number of publications (almost exclusively in French), the protagonists of this approach, Ducrot and Anscombre, have presented a linguistic analysis to show that almost all verbal utterances lead the listener or reader—often implicitly—to certain conclusions, so that their meaning is crucially argumentative. In *L'argumentation dans la langue* (Anscombre & Ducrot, 1983) they refer to the theoretical position they adopt as radical argumentativism. Their approach is characterized by a strong interest in words that can serve as argumentative 'operators' or 'connectors', giving linguistic utterances a specific argumentative force and argumentative direction (e.g., 'only', 'no less than', 'but', 'even', 'still', 'because', 'so'). Anscombre (1994) observes that the argumentative principles that are at issue here are on a par with the topoi from classical rhetoric.

It has become a tradition among a substantial group of European researchers, primarily based in the French-speaking world, to approach argumentation from a descriptive linguistic angle. Some of them continue the approach started by Ducrot and Anscombre. Others, such as Plantin (1996) and Doury (1997), build on this approach but are also—and often more strongly—influenced by conversation analysis and discourse analysis. Other researchers, based in Switzerland, who favour a linguistic approach, but allow also for normativity, are Rigotti (2009), Rocci (2009), and Greco Morasso (2011). They combine their linguistic approach with insights from other approaches, such as pragma-dialectics.

The pragma-dialectical theory of argumentation developed in Amsterdam combines a dialectical and a rhetorical perspective on argumentation and is both normative and descriptive. As van Eemeren and Grootendorst (1984) explain, pragma-dialecticians view argumentation as part of a discourse aimed at resolving a difference of opinion on the merits by methodically testing the acceptability of the standpoints at issue. The dialectical dimension of the approach is inspired by normative insights from critical rationalism and formal dialectics, the pragmatic dimension by descriptive insights from speech act theory, Gricean pragmatics and discourse analysis.

The various stages argumentative discourse must pass through to resolve a difference of opinion on the merits by a critical exchange of speech acts are in the pragma-dialectical theory laid down in an ideal model of a critical discussion (van Eemeren & Grootendorst, 2004). Viewed analytically, there should be a 'confrontation stage', in which the difference of opinion comes about, an 'opening stage', in which the point of departure of the discussion is determined, an 'argumentation stage', in which the standpoints at issue are defended against criticism, and a 'concluding stage', in which it is determined what the result of the discussion is. The model of a critical discussion defines the nature and the distribution of the speech acts that have a constructive role in the various stages of the resolution process. In addition, the standards of reasonableness authorizing the performance of particular speech acts in the various stages of a critical discussion are laid down in a set of dialectical rules for critical discussion. Any violation of any of the rules amounts to making an argumentative move that is an impediment to the resolution of a difference of opinion on the merits and is therefore fallacious (van Eemeren & Grootendorst, 1992).[25]

Because argumentative discourse generally diverges for various reasons from the ideal of a critical discussion, in the analysis of the discourse a reconstruction is required to achieve an analytic overview of all those, and only those,

---

[25]The extent to which the rules for critical discussion are capable of dealing with the defective argumentative moves traditionally designated as fallacies is viewed as a test of their 'problem-solving validity'. For experimental empirical research of the 'intersubjective acceptability' of the rules for critical discussion that lends them 'conventional validity' see van Eemeren, Garssen and Meuffels (2009).

speech acts that play a potential part in resolving a difference of opinion on the merits. Van Eemeren, Grootendorst, Jackson and Jacobs (1993) emphasize that the reconstruction should be guided by the theoretical model of a critical discussion and faithful to the commitments that may be ascribed to the arguers on the basis of their contributions to the discourse. Because the reconstruction of argumentative discourse as well as its evaluation can be made more pertinent, more precise, and also better accounted for if, next to the maintenance of dialectical reasonableness, the simultaneous pursuit of rhetorical effectiveness is taken into account, van Eemeren and Houtlosser (2002) developed the notion of strategic manoeuvring. This notion makes it possible to integrate relevant rhetorical insights systematically in the pragma-dialectical analysis and evaluation (van Eemeren, 2010).

3 Formal and computational argumentation theory: precursors and first steps

Today much research addresses argumentation using formal and computational methods. Precursors can be found in the fields of non-monotonic logic and logic programming, and first steps were made by philosophers addressing defeasible reasoning.

3.1 Non-monotonic logic

A relevant field predating the formal and computational study of argumentation is non-monotonic logic (Antonelli, 2010). A logic is non-monotonic when a conclusion that, according to the logic, follows from certain premises need not always follow when premises are added. In contrast, classical logic is monotonic. For instance, in a standard classical analysis, from premises 'Edith goes to Vienna or Rome' and 'Edith does not go to Rome', it follows that 'Edith goes to Vienna', irrespective of possible additional premises. The standard example of non-monotonicity used in the literature of the 1980s concerns the flying of birds. Typically, birds fly, so if you hear about a bird, you will conclude that it can fly. However, when you next learn that the bird is a penguin, you retract your conclusion. In a non-monotonic logic, a balance can be sought between the advantage of drawing a tentative conclusion, which is usually correct, and the risk of having to withdraw the conclusion in light of new information.

A prominent proposal in non-monotonic logic is Raymond Reiter's (1980) logic for default reasoning, using default rules. Reiter's first example of a default rule expresses that birds typically fly:

$BIRD(x) : M\ FLY(x)\ /\ FLY(x)$

The default rule expresses that, if x is a bird, and it is consistent to assume that x can fly, then by default one can conclude that x can fly. Other influential logical systems for non-monotonic reasoning include circumscription, auto-epistemic logic, and non-monotonic inheritance; each of them discussed in the representative overview of the study of non-monotonic logic at its heyday by Gabbay, Hogger and Robinson (1994).

3.2 Logic programming

A development related to non-monotonic logic is logic programming. The general idea underlying logic programming is that a computer can be programmed using logical techniques. In this view, computer programs are not only considered procedurally as recipes for how to achieve the program's aims, but also declaratively, in the sense that the program can be read like a text, for instance, as the rule-like knowledge needed to answer a question. In the logic programming language Prolog (the result of a collaboration between Colmerauer and Kowalski; see Kowalski, 2011), these are examples of facts and a rule (Bratko, 2001):

```
parent(pam, bob)
female(pam)
mother(X, Y) :- parent(X, Y), female(X)
```

This small logic program represents the facts that Pam is Bob's parent, and that Pam is female, and the rule that someone's mother is a female parent. Given this Prolog program, a computer can as expected derive that Pam is Bob's mother. In the interpretation of logic programs, the closed world assumption plays a key role: a logic program is assumed to describe all facts and rules about the world. For instance, in the program above it is assumed that all parent relations are given, so 'parent(tom, bob)' cannot be derived. By what is called negation as failure, it will be considered false that Tom is Bob's parent. If we add 'parent(tom, bob)' it becomes derivable that Tom is Bob's parent, showing the connection between logic programming's negation as failure and non-monotonic logic.

3.3 Themes and impact of non-monotonic logics

The study of non-monotonic logics gave hope that logical tools would become more relevant for the study of natural reasoning. To some extent this hope has been fulfilled, since certain themes that before were at the boundaries of logic, were now placed in the centre of attention. Examples of such themes are defeasible inference, consistency preservation, and uncertainty. In the handbook edited by Gabbay, Hogger and Robinson (1994), Donald Nute discusses defeasible inference that can be blocked or defeated in some way (Nute, 1994, p. 354). Interestingly, Donald Nute speaks of the presentation of sets of beliefs as reasons for holding other beliefs as advancing arguments. David Makinson (1994, p. 51) describes consistency preservation as the property that the conclusions drawn on the basis of certain premises can only be inconsistent in case the premises are inconsistent. Henry Kyburg (1994, p. 400) distinguishes three kinds of inference involving uncertainty: classical, deductive, valid inference about uncertainty; an 'inductive' kind where a conclusion can be false even when the premises are true (hence distinct from the idea of induction as going from the specific to the general, and closer to what today is often called 'defeasible'); and a kind of inference with uncertainty that gives probabilities of particular statements.

The study of non-monotonic logic has been very successful as a research enterprise, and coincided with innovations in computer programming in the form of logic-based languages such as Prolog, and to commercial applications: today's knowledge-based expert systems—in wide-spread use—often include some elementary form of non-monotonic reasoning.

At the same time, non-monotonic logic did not fulfil all expectations of the artificial intelligence community in which it was initiated. Matthew Ginsberg (1994), for instance, notes—somewhat disappointedly—that the field put itself "in a position where it is almost impossible for our work to be validated by anyone other than a member of our small subcommunity of Artificial Intelligence as a whole" (1994, p. 28–29) His diagnosis of this issue is that attention shifted from the key objective of building an intelligent artefact to the study of simple examples and mathematics. This leads him to plead for a more experimental, scientific attitude as opposed to a theoretical, mathematical focus.

3.4 Defeasible reasoning

In 1987, the publication of John Pollock's paper 'Defeasible reasoning' in *Cognitive Science* marked a turning point. The paper emphasized that the philosophical notion of 'defeasible reasoning' coincides with what in AI is called 'non-monotonic reasoning.' As philosophical heritage for the study of defeasible reasoning, Pollock (1987) refers to works by Roderick Chisholm (going back to 1957) and himself (earliest reference in 1967). Ronald Loui (1995) places the origins of the notion of 'defeasibility' a decade earlier, namely in 1948 when the legal positivist H. L. A. Hart presented the paper 'The ascription of responsibility and rights' at the Aristotelian Society (Hart, 1951). Although Toulmin (1958/2003) rarely uses the term defeasible in *The uses of argument*, he is obviously an early adopter of the idea of defeasible reasoning, but he is not mentioned by Pollock (1987). Like Pollock, he mentions Hart, but also another philosopher, David Ross, who applied the idea to ethics, recognizing that moral rules may hold prima facie, but can have exceptions.

In Pollock's approach (1987), 'reasoning' is conceived as a process that proceeds in terms of reasons. Pollock's reasons correspond to the constellations of premises and a conclusion which argumentation theorists and logicians call (elementary) arguments. Pollock distinguishes two kinds of reasons:

1. A reason is *non-defeasible* when it logically implies its conclusion;

2. A reason P for Q is *prima facie* when there is a circumstance R such that P ∧ R [where '∧' denotes logical conjunction] is not a reason for the reasoner to believe Q. R is then a *defeater* of P as a reason for Q.

Note how closely related the idea of a prima facie reason is to non-monotonic inference: Q can be concluded from P, but not when there is additional information R.

Pollock's standard example is about an object that looks red. 'X looks red to John' is a reason for John to believe that X is red, but there can be defeating

Figure 1. Pollock's red light example

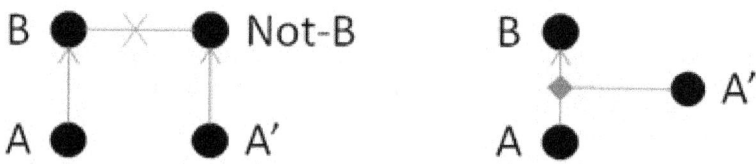

Figure 2. A rebutting defeater and an undercutting defeater

circumstances, for instance, when there is a red light illuminating the object. See Figure 1.

Pollock argues for the existence of two kinds of defeaters: 'rebutting' and 'undercutting defeaters.' A defeater is rebutting when it is a reason for the opposite conclusion (Figure 2, left). Undercutting defeaters attack the connection between the reason and the conclusion, and not the conclusion itself (Figure 2, right). The example about looking red concerns an undercutting defeater since when there is a red light it is not attacked that the object is red, but merely that the object's looking red is a reason for its being red.

A key element in Pollock's work on defeasible reasoning is the development of a theory of warrant. Pollock uses the term warrant as follows: a proposition is warranted in an epistemic situation if and only if an ideal reasoner starting in that situation would be justified in believing the proposition. Here justification is based on the existence of an undefeated argument with the proposition as conclusion. Pollock has developed his theory of warrant in a series of publications which formed the basis of his 1995 book *Cognitive Carpentry*. As a background for his approach to the structure of defeasible reasoning, Pollock provides a list of important classes of specific reasons: reasons based on logical deduction, perception, memory, statistics, or induction. Pollock's theory is embedded in what he called the OSCAR project (Pollock, 1995). This project aims at the implementation of a rational agent. In the project Pollock addresses both theoretical (epistemic) and practical reasoning.[26]

---

[26] See Hitchcock (2001, 2002) for a survey and a discussion of the OSCAR project for those interested in argumentation. Hitchcock also gives further information about Pollock's work

In a theory of defeasible reasoning based on arguments that can defeat each other, such as Pollock's, the question needs to be considered which arguments can defeat which other arguments. Different forms of argument defeat can be distinguished:

1. An argument can be *undermined*. In this form of defeat, the premises or assumptions of an argument are attacked.[27] Cf. the denial of the premises of an argument.

2. An argument can be *undercut*. In this form of defeat, the connection between a (set of) reason(s) and a conclusion in an argument is attacked. Cf. Pollock's undercutting defeaters.

3. An argument can be *rebutted*. In this form of defeat, an argument is attacked by giving an argument for an opposite conclusion. Cf. Pollock's rebutting defeaters.

4. An argument can be defeated by *sequential weakening*. Then each step in an argument is correct, but the argument breaks down when the steps are chained. An example is an argument based on the sorites paradox (Verheij 1996a, p. 122f.):

   > This body of grains of sand is a heap.
   > So, this body of grains of sand minus 1 grain is a heap.
   > So, this body of grains of sand minus 2 grains is a heap.
   > ...
   > So, this body of grains of sand minus n grains is a heap.

5. An argument can be defeated by parallel strengthening. This kind of defeat is associated with what has been called the 'accrual of reasons.' When reasons can accrue, it is possible that different reasons for a conclusion are together stronger than each reason separately. For instance, having robbed someone and having injured someone can be separate reasons for convicting someone. But when the suspect is a minor first offender, these reasons may each by itself be rebutted. On the other hand when a suspect has both robbed someone and also injured that person, the reasons may accrue and outweigh the fact that the suspect is a minor first offender. The argument for not punishing the suspect based on the reason that he is a minor first offender is defeated by the 'parallel strengthening' of the two arguments for punishing him.

Building on experiences in the ASPIC project,[28] the recent state-of-the-art

---

on practical reasoning, i.e., reasoning concerning what to do.

[27] This form of defeat is the basis of Bondarenko et al. (1997). We shall here not elaborate on the distinction between premises and assumptions. One way of thinking about assumptions is to see them as defeasible premises.

[28] The ASPIC project (full name: Argumentation Service Platform with Integrated Components) was supported by the EU 6th Framework Programme and ran from January 2004 to September 2007. In the project, academic and industry partners cooperated in developing argumentation-based software systems.

ASPIC+ system for the formal modelling of defeasible argumentation (Prakken, 2010)[29] uses the first three kinds of defeat. The final two kinds of defeat are distinguished by Verheij (1996a, p. 122f.). Pollock considered the accrual of reasons to be a natural idea, but argued against it (1995, p. 101f.). More recent discussions of the accrual of reasons are to be found in Prakken (2005), Gómez Lucero et al. (2009, 2013), and D'Avila Garcez et al. (2009, p. 155f.).

4 Argumentation and the structure of arguments in formal and computational perspective

4.1 Abstract argumentation

Phan Minh Dung's 1995 paper 'On the acceptability of arguments and its fundamental role in non-monotonic reasoning, logic programming and n-person games' in the journal *Artificial Intelligence* (Dung, 1995) reformed the formal study of non-monotonic logic and defeasible reasoning. By his focus on argument attack as an abstract formal relation, Dung gave the field of study a mathematical basis that inspired many new insights. Dung's approach and the work inspired by it are generally referred to as abstract argumentation.

Dung's paper is strongly mathematically oriented, and has led to intricate formal studies. However, the mathematical tools used by Dung are elementary, hence various concepts studied by Dung can be explained without going into much formal detail.

The central innovation of Dung's 1995 paper is that he started the formal study of the attack relation between arguments, thereby separating the properties depending exclusively on argument attack from any concerns related to the structure of the arguments. Mathematically speaking, the argument attack relation is a directed graph, the nodes of which are the arguments, whereas the edges represent that one argument attacks another. Such a directed graph is called an argumentation framework. Figure 3 shows an example of an argumentation framework, with the dots representing arguments, and the arrows (ending in a cross to emphasize the attacking nature of the connection[30]) representing argument attack.

In Figure 3, the argument α attacks the argument β, which in turn attacks both γ and δ, etc.

Dung's paper consists of two parts, corresponding to two steps in what he refers to as an 'analysis of the nature of human argumentation in its full generality' (Dung, 1995, p. 324). In the first step, Dung develops the theory of argument attack and how argument attack determines argument acceptability. In the second part, he evaluates his theory by two applications, one consisting of a study of the logical structure of human economic and social problems, the other comprising a reconstruction of a number of approaches to non-monotonic reasoning, among them Reiter's and Pollock's. Notwithstanding the relevance

---

[29] Prakken (2010) speaks of ways of attack, where argument defeat is the result of argument attack.

[30] This is especially helpful when also supporting connections are considered; see Section 4.2.

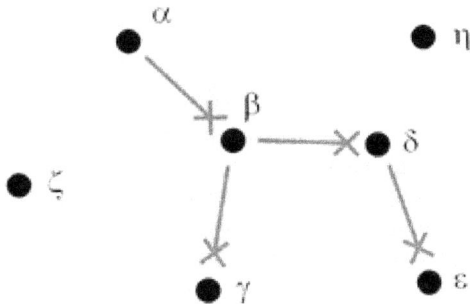

Figure 3. An argumentation framework representing attack between arguments

of the second part of the paper, the paper's influence is largely based on the first part about argument attack and acceptability.

In Dung's approach, the notion of an 'admissible set of arguments' is central. A set of arguments is admissible if two conditions obtain: (1) The set of arguments is conflict-free, i.e., does not contain an argument that attacks another argument in the set (nor self-attacking arguments). (2) Each argument in the set is acceptable with respect to the set, i.e., when an argument in the set is attacked by an argument (which by (1) cannot be in the set itself), the set contains an argument that attacks the attacker.

In other words, a set of arguments is admissible if it contains no conflicts and if the set also can defend itself against all attacks. An example of an admissible set of arguments for the framework in Figure 3 is $\{\alpha, \gamma\}$. Since α and γ do not attack one another the set is conflict-free. The argument α is acceptable with respect to the set since it is not attacked, so that it needs no defence. The argument γ is also acceptable with respect to $\{\alpha, \gamma\}$: the argument γ needs a defence against the attack by β, which defence is provided by the argument α, α being in the set. The set $\{\alpha, \beta\}$ is not admissible since it is not conflict-free. The set $\{\gamma\}$ is not admissible since it does not contain a defence against the argument β, which attacks argument γ.

Admissible sets of arguments can be used to define argumentation notions of what counts as a proof or a refutation.[31] An argument is '(admissibly) provable' when there is an admissible set of arguments that contains the argument. A minimal such set can be regarded as a kind of 'proof' of the argument, in the sense that the arguments in such a set are just enough to successfully defend the argument against counterarguments. An argument is '(admissibly) refutable' when there is an admissible set of arguments that contains an argument that attacks the former argument. A minimal such set can be regarded as a kind of 'refutation' of the attacked argument.

Dung speaks of the basic principle of argument acceptability using an infor-

---

[31] In the following, we make use of terminology proposed by Verheij (2007).

mal slogan: the one who has the last word laughs best. The argumentative meaning of this slogan can be explained as follows. When someone makes a claim, and that is the end of the discussion, the claim stands. But when there is an opponent raising a counterargument attacking the claim, the claim is no longer accepted—unless the proponent of the claim provides a counterattack in the form of an argument attacking the counterargument raised by the opponent. Whoever has raised the last argument in a sequence of arguments, counterarguments, counter-counterarguments, etc., is the one who has won the argumentative discussion.

Formally, Dung's argumentation principle 'the one who has the last word laughs best' can be illustrated using the notion of an 'admissible set of arguments'. In Figure 3, a proponent of the argument γ has the last word and laughs best, since the only counterargument β is attacked by the counter-counterargument α. Formally, this is captured by the admissibility of the set $\{\alpha, \gamma\}$.

Although the principle of argument acceptability and the concept of an admissible set of arguments seem straightforward enough, it turns out that intricate formal puzzles loom. This has to do with two important formal facts:

1. It can happen that an argument is both admissibly provable and refutable.

2. It can happen that an argument is neither admissibly provable nor refutable.

The two argumentation frameworks shown in Figure 4 provide examples of these two facts. In the cycle of attacks on the left, consisting of two arguments α and β, each of the arguments is both admissibly provable and admissibly refutable. This is a consequence of the fact that the two sets $\{\alpha\}$ and $\{\beta\}$ are each admissible. For instance, $\{\alpha\}$ is admissible since it is conflict-free and can defend itself against attacks: the argument α itself defends against its attacker α. By the admissibility of the set $\{\alpha\}$, the argument α is admissibly probable, and the argument β admissibly refutable.

The cycle of attacks on the right containing three arguments α_1, α_2 and α_3 is an example of the second fact above, the fact that it can happen that an argument is neither admissibly provable nor refutable. This follows from the fact that there is no admissible set that contains (at least) one of the arguments α_1, α_2 or α_3. Suppose that the argument α_3 is in an admissible set. Then the set should defend α_3 against the argument α_2, which attacks α_3. This means that α_1 should also be in the set, since it is the only argument that can defend α_3 against α_2. But this is not possible, because then α_1 and α_3 are both in the set, introducing a conflict in the set. As a result, there is only one admissible set: the empty set, which contains no arguments at all. We conclude that no argument is admissibly provable or admissibly refutable.

A related formal issue is that when two sets of arguments are admissible, it need not be the case that their union is admissible. The framework on the left in Figure 4 is an example. As we saw, the two sets $\{\alpha\}$ and $\{\beta\}$ are both admissible, but their union $\{\alpha, \beta\}$ is not, since it contains a conflict. This has

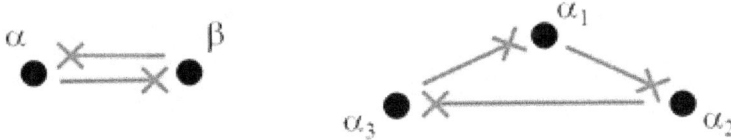

Figure 4. Arguments attacking each other in cycles

led Dung to propose the notion of a preferred extension of an argumentation framework, which is an admissible set that is as large as possible, in the sense that adding elements to the set makes it not admissible. The framework in Figure 3 has one preferred extension: the set $\{\alpha, \gamma, \delta, \zeta, \eta\}$. The framework in Figure 4 on the left has two preferred extensions $\{\alpha\}$ and $\{\beta\}$, the one on the right has one: the empty set.

Some preferred extensions have a special property, namely that each argument that is not in the set is attacked by an argument in the set. Such an extension is called a stable extension. Stable extensions are formally defined as conflict-free sets that attack each argument not in the set. It follows from this definition that a stable extension is also a preferred extension.

The preferred extension $\{\alpha, \gamma, \delta, \zeta, \eta\}$ of the framework in Figure 3, for instance, is stable, since the arguments β and ϵ, which are the only ones that are not in the set, are attacked by arguments in the set, α and δ, respectively. The preferred extensions $\{\alpha\}$ and $\{\beta\}$ of Figure 4 (left) are also stable. The preferred extension of Figure 4 (right), the empty set, is not stable, since none of the arguments α_1, α_2 and α_3 is attacked by an argument in the set. This example shows that there exist preferred extensions that are not stable. It also shows that there are argumentation frameworks that do not have a stable extension. In contrast, every argumentation framework has at least one preferred extension (which can be the empty set).

The concepts of preferred and stable extension of an argumentation framework can be regarded as different ways to interpret a framework, and therefore they are often referred to as 'preferred semantics' and 'stable semantics.' Dung (1995) proposed two other kinds of semantics: 'grounded semantics' and 'complete semantics,' and following his paper several additional kinds of semantics have been proposed (see Baroni et al., 2011, for an overview). By the abstract nature of argumentation frameworks, formal questions about the computational complexity of related algorithms and formal connections with other theoretical paradigms came within reach (see, e.g., Dunne & Bench-Capon, 2003, Dunne, 2007, and Egly, Gaggl & Woltran, 2010).

Dung's original definitions are in terms of mathematical sets. An alternative way of studying argument attack is in terms of labelling. Arguments are marked with a label, such as 'Justified' or 'Defeated' (or IN/OUT, +/-, 1/0,

'Warranted'/ 'Unwarranted,' etc.), and the properties of different kinds of labelling are studied in the field. For instance, the notion of a stable extension corresponds to the following notion in terms of labelling:

> A *stable labelling* is a function that assigns one label 'Justified' or 'Defeated' to each argument in the argumentation framework such that the following property holds: an argument α is labelled 'Defeated' if and only if there is an argument β that attacks α and that is labelled 'Justified.'

A stable extension gives rise to a stable labelling by labelling all arguments in the extension 'Justified' and all other arguments 'Defeated.' A stable labelling gives rise to a stable extension by considering the set of arguments labelled 'Justified.'

The idea of labelling arguments can be thought of in analogy with the truth functions of propositional logic, where propositions are labelled with truth-values 'true' and 'false' (or 1/0, t/f, etc.). In the formal study of argumentation, labelling techniques predate Dung's abstract argumentation (1995). Pollock (1994) uses labelling techniques in order to develop a new version of a criterion that determines warrant.

Verheij (1996b) applied the labelling approach to Dung's abstract argumentation frameworks. He uses argument labelling also as a technique to formally model which arguments are taken into account: in an interpretation of an abstract argumentation framework, the arguments that are assigned a label can be regarded as the ones taken into account, whereas the unlabelled arguments are not considered. Using this idea, Verheij defines two new kinds of semantics: the 'stage semantics' and the 'semi-stable semantics.'[32] Other authors using a labelling approach are Jakobovits and Vermeir (1999) and Caminada (2006). The latter author translated each of Dung's extension types into a mode of labelling.

As an illustration of the labelling approach, we give a labelling treatment of the grounded extension of an argumentation framework as defined by Dung.[33] Consider the following procedure in which gradually labels are assigned to the arguments of an argumentation framework:

1. Apply the following to each unlabelled argument α in the framework: if the argument α is only attacked by arguments that have been labelled 'Defeated' (or perhaps is not attacked at all), label the argument α as 'Justified.'

2. Apply the following to each unlabelled argument α in the framework: if the argument α is attacked by an argument that has been labelled 'Justified,' label the argument α as 'Defeated.'

---

[32] In establishing the concept Verheij (1996b) used the term admissible stage extensions. The now standard term semi-stable extension was proposed by Caminada (2006).

[33] Dung's own definition of grounded extension, which does not use labelling, is not discussed here.

3. If step 1 and/or step 2 have led to new labelling, go back to step 1; otherwise stop.

When this procedure is completed (which always happens after a finite number of steps when the argumentation framework is finite), the arguments labelled 'Justified' constitute the grounded extension of the argumentation framework. Consider, for instance, the framework of Figure 3. In the first step, the arguments α, ζ and η are labelled 'Justified.' The condition that all arguments attacking them have been 'Defeated' is vacuously fulfilled, since there are no arguments attacking them. In the second step the argument β is labelled 'Defeated', since α has been labelled 'Justified.' Then a second pass of step 1 occurs and the arguments γ and δ are labelled 'Justified,' since their only attacker β has been labelled 'Defeated.' Finally, the argument ϵ is labelled 'Defeated,' since δ has been labelled 'Justified.' The arguments α, γ, δ, ζ and η (i.e., those labelled 'Justified') together form the grounded extension of the framework. Every argumentation framework has a unique grounded extension. In the framework of Figure 3, the grounded extension coincides with the unique preferred extension that is also the unique stable extension. The framework in Figure 4 (left) shows that the grounded extension is not always a stable or preferred extension. Its grounded extension is here the empty set, but its two preferred and stable extensions are not empty.

4.2 Arguments with structure

Abstract argumentation, discussed in the previous subsection, focuses on the attack relation between arguments, abstracting from the structure of arguments. We now discuss various themes related to the structure of arguments for and against conclusions, and how it has been studied: arguments and specificity, the comparison of conclusive force, arguments with prima facie assumptions, arguments and classical logic, and the combination of support and attack.

Argument specificity An early theme in the formal study of argumentation was that of 'argument specificity' in relation to the resolution of a conflict between arguments. The key idea connecting arguments and specificity is that when two arguments are conflicting, with one of them being based on more specific information, the more specific argument wins the conflict, and defeats the more general argument.

Guillermo Simari and Ronald Loui (1992) have provided a mathematical formalization of this connection between arguments and specificity, taking inspiration from Poole's (1985) work in non-monotonic logic, and connecting to Pollock's work on argumentative warrant. In their proposal, an argument is a pair (T, h), with T being a set of defeasible rules that are applied to arrive at the argument's conclusion h given the argument's premises (formalized in the background knowledge). Arguments are assumed to be consistent, in the sense that no contradiction can be derived (not even defeasibly). Also arguments are assumed to be minimal, in the sense that all rules are needed to arrive at the conclusion. Formally, for an argument (T, h), it holds that when T' is the result of omitting one or more rules in T, the pair (T', h) is not an argument. Two arguments (T, h) and (T', h') disagree when h and h' are logically incompatible, given the background knowledge. An argument (T, h) counter-argues an argument (T', h') if (T, h) disagrees with an argument (T'', h'') that is a sub-argument of (T', h'), i.e., T'' is a subset of T'. An argument (T, h) defeats an argument (T', h') when (T, h) disagrees with a sub-argument of (T', h') that is strictly less specific. Simari and Loui's approach has been developed further—with applications in artificial intelligence, multi-agent systems, and logic by the Bahia Blanca group, led by Simari (e.g., García & Simari, 2004; Chesñevar et al., 2004; Falappa et al., 2002). García and Simari (2004) show the close connection between argumentation and logic programming that was also an inspiration for Dung (1995).

Conclusive force A second theme connected to arguments and their structure is conclusive force. Arguments that have more conclusive force will survive a conflict more easily than arguments with less conclusive force. One idea that connects conclusive force with argument defeat is the weakest link principle, which Pollock characterizes as follows:

The degree of support of the conclusion of a deductive argument is the minimum of the degrees of support of its premises (1995, p. 99).

Pollock presents the weakest link principle as an alternative to a Bayesian approach, which he rejects. Gerard Vreeswijk (1997) has proposed an abstract model of argumentation with defeasible arguments that focuses on the comparison of the conclusive force of arguments. In his model, conclusive force is not modelled directly but as an abstract comparison relation that expresses which arguments have more conclusive force than which other arguments. Vreeswijk defines an abstract argumentation system as a triple $(L, R, \leq)$, where L is a set of sentences expressing the claims made in an argument, R is a set of defeasible rules allowing the construction of arguments, and $\leq$ represents the conclusive force relation between arguments. The rules come in two flavours: strict and defeasible. Arguments are constructed by chaining rules. A set of arguments Σ is a defeater of an argument α if Σ and α are incompatible (i.e., imply an inconsistency), and α is not an underminer of Σ. An argument α is an underminer of a set of arguments Σ if Σ contains an argument β that has strictly lower conclusive force than α. Whereas Dung's (1995) system is abstract by its focus on argument attack, Vreeswijk's proposal is abstract in particular also because the conclusive force relation is left unspecified. Vreeswijk gives the following examples of conclusive force relations:

1. *Basic order*. In this order, a strict argument has more conclusive force than a defeasible argument. In a strict argument, no defeasible rule is used.

2. *Number of defeasible steps*. An argument has more conclusive force than another argument if it uses less defeasible steps. Vreeswijk remarks that this is not a very natural criterion, but it can be used to give formal

examples and counterexamples.

3. *Weakest link.* Here the conclusive force relation on arguments is derived from an ordering relation on the rules. An argument has more conclusive force than another if its weakest link is stronger than the weakest link of the other.

4. *Preferring the most specific argument.* Of two defeasible arguments, one has more conclusive force than the other if the first has the premises of the second among its conclusions.

Prima facie assumptions A third theme related to arguments and their structure is arguments with prima facie assumptions. In particular, the defeat of arguments can be the result of prima facie assumptions that are successfully attacked. In their abstract, argumentation-theoretic approach to default reasoning, Bondarenko, Dung, Kowalski, and Toni (1997) use such an approach. Using a given deductive system (L, R) that consists of a language L and a set of rules R, so-called 'deductions' are built by the application of rules. Given a deductive system (L, R), an assumption-based framework is then a triple (T, Ab, Contrary), where T is a set of sentences expressing the current beliefs, Ab expresses assumptions that can be used to extend T, and Contrary is a mapping from the language to itself that expresses which sentences are contraries of which other sentences. Bondarenko and colleagues define a number of semantics (similar to Dung's 1995 in the context of abstract argumentation). For instance, a stable extension is a set of assumptions Δ such that the following properties hold:

1. Δ is closed, meaning that Δ contains all assumptions that are logical consequences of the beliefs in T and Δ itself.

2. Δ does not attack itself, meaning that there is no deduction from the beliefs in T and Δ with a contrary of an element of Δ as conclusion.

3. Δ attacks each assumption not in Δ, meaning that, for every assumption outside Δ, there is a deduction from T and Δ with a contrary of that assumption as conclusion.

Verheij (2003a) has also developed an assumption-based model of defeasible argumentation. In contrast with Bondarenko et al. (1997), in Verheij's system, the rules from which arguments are constructed are part of the prima facie assumptions. Technically, the rules have become conditionals of the underlying language. As a result, it can be the issue of an argument whether some proposition supports another proposition. In this way, Pollock's undercutting defeaters can be modelled as an attack on a conditional. Pollock's example of an object that looks red (Section 3.4) is formalized using two conditional sentences:

```
looks_red ⤳ is_red
red_light ⤳ ×(looks_red ⤳ is_red)
```

The first expresses the conditional prima facie assumption that if something looks red, it is red. The second expresses an attack on this prima facie assumption: when there is a red light illuminating the object, it no longer holds that if the object looks red, it is red. The sentences illustrate the two connectives of the language: one to express the conditional (⤳), the other to express what is called dialectical negation (×). The two conditional sentences correspond exactly to two graphical elements in Figure 1: the first to the arrow connecting the reason and the conclusion, the second, nested, conditional to the arrow (ending in a diamond) that expresses the attack on the first conditional. This isomorphism between formal structures of the language and graphical elements has been used for the diagrams supported by the argumentation software ArguMed (Verheij, 2005b; see Section 4.5).

The use of assumptions raises the question how they are related to an argument's ordinary premises. Assumptions can be thought of as the defeasible premises of an argument, and as such they are akin to defeasible rules[34] with an empty antecedent. The Carneades framework (Gordon et al., 2007) distinguishes three kinds of argument premises: ordinary premises, presumptions (much like the prima facie assumptions discussed here) and exceptions (which are like the contraries of assumptions).

Arguments and classical logic A fourth theme connected to arguments and their structure is how they are related to classical logic. In particular, the relation between classical logic and defeasible argumentation remains a puzzle. Above we already saw different attempts at combining elements of classical logic and defeasible argumentation. In Pollock's system, classical logic is one source of reasons. Often conditional sentences ('rules') are used to construct arguments by chaining them (e.g., Vreeswijk, 1997). Chaining rule applications is closely related to the inference rule modus ponens of classical logic. Verheij's (2003a) system gives conditionals which validate modus ponens a central place. Bondarenko et al. (1997) allow generalized rules of inference by their use of a contingent deductive system as starting point.

Besnard and Hunter (2008) have proposed to formalize arguments in classical logic entirely. For them, an argument is a pair (Φ, α), such that Φ is a set of sentences and α is a sentence, and such that Φ is logically consistent, Φ logically entails α (in the classical sense), and Φ is a minimal such set. (Note the analogy with the proposal by Simari and Loui, 1992, discussed earlier.)

---

[34]Some would object to the use of the term rules here. Rules are here thought of in analogy with the inference rules of classical logic. An issue is then that, as such, they are not expressed in the logical object language, but in a meta-language. In the context of defeasible reasoning and argumentation (and also in non-monotonic logic), this distinction becomes less clear. Often there is one logical language to express ordinary sentences, a second formal language (with less structure and/or less semantics, and therefore not usually referred to as 'logical') used to express the rules, and the actual meta-language that is used to define the formal system.

Φ is the support of the argument, and α the claim. They define defeaters as arguments that refute the support of another argument. More formally, a defeater for an argument (Φ, α) is an argument (Ψ, β), such that β logically entails the negation of the conjunction of some of the elements of Φ. An undercut for an argument (Φ, α) is an argument (Ψ, β) where β is equal to (and not just entails) the negation of the conjunction of some of the elements of Φ. A rebuttal for an argument (Φ, α) is an argument (Ψ, β) such that $\beta \leftrightarrow \neg\alpha$ is a tautology. Besnard and Hunter give the following example (p. 46):

p	Simon Jones is a Member of Parliament.
p → ¬q	If Simon Jones is a Member of Parliament, then we need not keep quiet about details of his private life.
r	Simon Jones just resigned from the House of Commons.
r → ¬p	If Simon Jones just resigned from the House of Commons, then he is not a Member of Parliament.
¬p → q	If Simon Jones is not a Member of Parliament, then we need to keep quiet about details of his private life.

Then $(\{p, p \rightarrow \neg q\}, \neg q)$ is an argument with the argument $(r, r \rightarrow \neg p, \neg p)$ as an undercut and the argument $(r, r \rightarrow \neg p, \neg p \rightarrow q, q)$ as a rebuttal.

Besnard and Hunter focus on structural properties of arguments, in part because of the diversity of proposals for semantics (see Section 4.1). For instance, when they discuss these systems, they note that the semantic conceptualization of such systems is not as clear as the semantics of classical logic, which is the basis of their framework (p. 221, also p. 226). At the same time, they note that knowledge representation can be simpler in systems based on defeasible logic (see below) or inference rules.

Combining support and attack A fifth and final theme discussed here in connection with arguments and their structure is how support and attack are combined. In several proposals, support and attack are combined in separated steps. In the first step, argumentative support is established by constructing arguments for conclusions from a given set of possible reasons or rules (of inference). The second step determines argumentative attack. Attack is, for instance, based on defeaters or on the structure of the supporting arguments in combination with a preference relation on arguments. In the third and final step, it is determined which arguments are warranted or undefeated. We already saw that several criteria have been proposed (e.g., Pollock's gradual development of criteria for argumentative warrant, and Dung's abstract argumentation semantics).

An example of this modelling style is depicted in Figure 5. Three supporting arguments are shown. The first on the left shows that A supports B, which in turn supports C. In the middle of the figure, this argument is attacked by a second argument, which reasons from A' for Not-B (hence against B). This argument is in turn attacked by a third argument, which reasons from A"

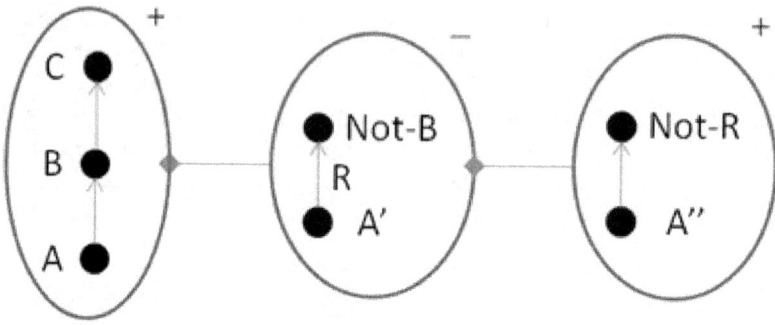

Figure 5. Supporting arguments that attack each other

Figure 6. The abstract argumentation framework associated with the example of Figure 5

against the support relation R between A' and Not-B. Using the terminology of Section 3.4, the first subargument of the first argument is rebutted by the second, which is undercut by the third. The arguments are marked with a + sign when they are warranted, and a − sign when they are defeated (which can be thought of as a variant of the labelling approaches of Section 4.1). The argument on the right is warranted, since it is not attacked. As a result, the middle argument is defeated, since it is attacked by a warranted argument. The left argument is then also warranted, since its only attacker is defeated. (See the procedure for computing the grounded extension of an argumentation framework discussed in Section 4.1.)

In this approach, the relation with Dung's abstract argumentation is that we can abstract from the structure of the supporting arguments resulting in an abstract argumentation framework. For the three arguments in Figure 5, we obtain the abstract framework shown in Figure 6. In this example, the argumentation semantics is unproblematic at the abstract argument attack level since the grounded extension coincides with the unique preferred extension that is also stable. Special care is needed to handle parts of arguments. For instance, the middle argument has the premise A', which is not attacked, and should therefore remain undefeated.

This type of combining support and attack is used in the ASPIC+ model (Prakken, 2010). A second approach does not separate support and attack when combining them. Arguments are constructed from reasons for and against

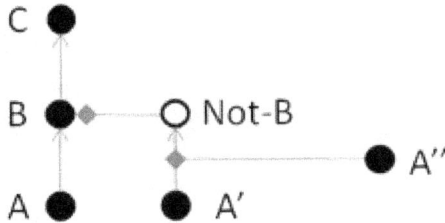

Figure 7. Arguments supporting and attacking conclusions

conclusions, which in turn determine whether a conclusion follows or not. Figure 7 models the same argumentative information as Figure 5, but now using this second approach.

Here the reason A" undercuts the argument from A' to Not-B, so Not-B is not supported (indicated by the open circle). As a result, Not-B does not actually attack B, which is therefore justified by A and in turn justifies C.

In this approach, for instance, conditional sentences are used to express which reasons support or attack which conclusions. An example is Nute's defeasible logic (Nute, 1994; Antoniou et al., 2001), which uses conditional sentences for the representation of strict rules and defeasible rules, and for defeater rules, which can block an inference based on a defeasible rule. Algorithms for defeasible logic have been designed with good computational properties. Another example of the approach is Verheij's DefLog (2003a), in which a conditional for the representation of support is combined with a negation operator for the representation of attack. A related proposal extending Dung's abstract argumentation frameworks by expressing both support and attack is bipolar argumentation (Cayrol & Lagasquie-Schiex, 2005; Amgoud et al., 2008). For DefLog and bipolar argumentation, generalisations of Dung's stable and preferred semantics are presented. DefLog has been used to formalize Toulmin's argument model (Verheij, 2005a).

A special case of the combination of support and attack occurs when the support and attack relations can themselves be supported or attacked. Indeed it can be at issue whether a reason supports or attacks a conclusion. The four ways of arguing about support and attack are illustrated in Figure 8, from left to right: support of a support relation, attack of a support relation, support of an attack relation, and attack of an attack relation, respectively.

For instance, Pollock's undercutting defeaters can be thought of as attacks of a support relation (second from the left in Figure 8). In Verheij's DefLog (2003a; 2005b), the four ways are expressed using nested conditional sentences, in a way that extends the expressiveness of Dung's frameworks. Modgil (2009) has studied attacks of attacks (rightmost in 11) in a system that also extends Dung's expressiveness.

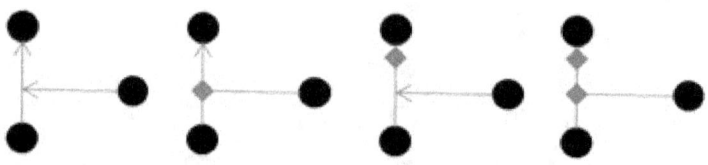

Figure 8. The four ways of arguing about support and attack

4.3 Formalizing argument schemes

Argumentation formalisms can only come to life when arguments are built from meaningful reasons. We already mentioned that Pollock made explicit which kinds of reasons he considered: deductive reasons, perception, memory, statistical syllogism, and induction.

An approach to the specification of meaningful kinds of reasons to construct arguments from is that of argument schemes, as they have been studied in argumentation theory. Argument schemes were already distinguished by Perelman and Olbrechts-Tyteca (1969).[35] In today's artificial intelligence research on argumentation, Douglas Walton's approach to argumentation schemes (his terminology) has been widely adopted (e.g., Walton et al., 2008).

Argument schemes can be thought of as analogues of the rules of inference of classical logic. An example of a rule of inference is, for instance, the following version of modus ponens:

P
If P, then Q
Therefore: Q

Whereas logical rules of inference, such as modus ponens, are abstract, strict, and (usually) considered to have universal validity, argumentation schemes are concrete, defeasible, and context-dependent. An example is the following scheme for witness testimony:

Witness A has testified that P.
Therefore: P

The use of this scheme is defeasible, as can be made explicit by asking critical questions, for instance:

Wasn't A mistaken?
Wasn't A lying?

---

[35] Although the term schème argumentative [argumentative scheme] was already used by Perelman and Olbrechts-Tyteca, according to Garssen (2001), van Eemeren, Grootendorst and Kruiger (1978, 1984) used the notion of argument(ation) scheme for the first time in its present sense. See also van Eemeren and Grootendorst (1992), Kienpointner (1992), Walton (1996), and Walton, Reed and Macagno (2008).

A key reason why argument schemes have been taken up in artificial intelligence is that the critical questions associated with them correspond to defeating circumstances. For instance, the question whether A was mistaken gives rise to the defeater 'A was mistaken'.

Bex, Prakken, Reed and Walton (2003) applied the concept of 'argumentation schemes' to the formalization of legal reasoning from evidence. An example of a scheme in that paper (taken from Walton, 1996) is the following.

> Argument from expert opinion
> Source E is an expert in domain D.
> E asserts that proposition A is known to be true (false).
> A is within D.
> Therefore, A may plausibly be taken to be true (false).

This scheme has the following critical questions:

1. Expertise question: How credible is E as an expert source?
2. Field question: Is E an expert in D?
3. Opinion question: What did E assert that implies A?
4. Trustworthiness question: Is E personally reliable as a source?
5. Consistency question: Is A consistent with what other experts assert?
6. Backup evidence question: Is E's assertion based on evidence?

The authors elaborate on how these and other argumentation schemes related to evidential reasoning can be formalized.

From the perspective of artificial intelligence, the work on argumentation schemes of Walton and his colleagues can be regarded as contributions to the theory of knowledge representation. Gradually, a collection of argumentation schemes is being developed. When appropriate, a scheme is added, and existing schemes are adapted, for instance, by refining the scheme's premises or critical questions. This knowledge representation point of view is developed by Verheij (2003b), who like Bex, Prakken, Reed and Walton (2003) formalizes argumentation schemes as defeasible rules of inference. He notes that in Walton's work argumentation schemes sometimes take the form of small derivations, or sequences of argumentation schemes; or even of a small prototypical dialogue. To streamline the work on knowledge representation, Verheij proposes to treat argumentation schemes as consisting of four elements: Conclusion, Premises, Conditions of use, and Exceptions. The Exceptions correspond to answers to the critical questions of an argumentation scheme. By this representation format, it is also possible to consider different roles of critical questions: critical questions concerning a conclusion, a premise, a condition of use, or an exception.

Reed and Rowe (2004) have incorporated argumentation schemes in their Araucaria tool for the analysis of argumentative texts. Rahwan, Zablith, and Reed (2007) have proposed formats for the integration of argumentation schemes in what is called the Semantic Web. The vision underlying the Semantic Web is that, when information on the Internet is properly tagged, it becomes possible to add meaning to such information that can be handled by a machine. For instance, when the Conclusion, Premises, Conditions of use, and Exceptions of an argumentation scheme are marked as such, software can be built that can handle these different elements of a scheme appropriately. Gordon, Prakken and Walton (2007) have integrated argumentation schemes in their Carneades model.

A fundamental issue concerning argumentation schemes is how to evaluate a scheme or set of schemes. When is a scheme good, under which circumstances? When is an adaptation appropriate? This issue is, for instance, discussed in Reed and Tindale (2010).

4.4 Formalizing argumentation dialogues

One reason why Toulmin's (2003) *The uses of argument* remains a thought-provoking study is his starting point that argument should be considered in its natural, critical, and procedural context. This starting point led him to propose that logic, in the sense of the theory of good argument, should be treated as 'generalized jurisprudence,' where a critical and procedural perspective on good argument is the norm. The critical and procedural sides of arguments come together in the study of argumentation dialogues.

The following is a fragment, taken from McBurney and Parsons (2002a), of an argumentation dialogue concerning the sale of a used car between a buyer (B) and seller (S), illustrating the study of argumentative dialogue in a computational setting:

S: BEGIN(PERSUASION(Make); PERSUASION(Condition_of_Engine); PERSUASION(Number_of_Owners))
S requests a sequence of three Persuasion dialogues over the purchase criteria Make, Condition of the Engine, and Number of Owners.
B: AGREE(PERSUASION(Make);PERSUASION(Condition_of_Engine); PERSUASION(Number_of_Owners)) PERSUASION Dialogue 1 in the sequence of three opens.
S: Argues that 'Make' is the most important purchase criterion, within any budget, because a typical car of one Make may remain in better condition than a typical car of another Make, even though older.
B: Accepts this argument.
PERSUASION Dialogue 1 closes upon acceptance of the proposition by B.
PERSUASION Dialogue 2 opens.
S: Argues that that 'Condition_of_Engine' is the next most important purchase criterion.
B: Does not accept this. Argues that he cannot tell the engine condition of any car without pulling it apart. Only S, as the Seller, is able to tell this. Hence, B must use 'Mileage' as a surrogate for 'Condition_of_Engine.'
PERSUASION Dialogue 2 closes with neither side changing its views: B does not accept 'Condition_of_Engine' as the second criterion, and S does not accept 'Mileage' as the second criterion. PERSUASION Dialogue 3 opens.

The fragment shows how dialogues about certain topics are opened and closed in relation to the arguments provided.

The formal and computational study of argumentation dialogues has primarily been performed in the fields of AI and law and of multi-agent systems, as addressed below.

In the field of AI and law, argumentation dialogues have been studied extensively (see Bench-Capon et al, 2004, 2009). Ashley's (1990) HYPO, to be discussed more extensively in Section 5.2, takes a 3–ply dialogue model as starting point, in which a proponent makes a claim, which can be attacked by an opponent, and then defended by the proponent. An early AI and law conception of argumentation dialogue is Thomas Gordon's (1993, 1995) *Pleadings game*. Gordon formalizes the pleading in a US-style civil law process, which is aimed at determining the legal and factual issues of a case. In the Pleadings Game, a proponent and opponent (in this setting referred to as 'plaintiff' and 'defendant') can concede, deny and defend claims, and also declare defeasible rules. Players can discuss the validity of a defeasible rule. Players are committed to the consequences of their claims, as prescribed by a non-monotonic logic underlying the Pleadings Game.

Other dialogue models of argumentation in AI and law have been proposed by Prakken and Sartor (1996, 1998), Hage, Leenes and Lodder (1993), and Lodder (1999). In Prakken and Sartor's approach (1996, 1998), dialogue models are presented as a kind of proof theory for their argumentation model. Prakken and Sartor interpret a proof as a dialogue between a proponent and opponent. An argument is justified when there is a winning strategy for the proponent of the argument. Hage, Leenes and Lodder (1993) and Lodder (1999) propose a model of argumentation dialogues with the purpose of establishing the law in a concrete case. They are inspired by the idea of law as a pure procedure (though not endorsing it): when the law is purely procedural, there is no criterion for a good outcome of a legal procedure other than the procedure itself.

Some models emphasize that the rules of argumentative dialogue can themselves be the subject of debate. An actual example is a parliamentary discussion about the way in which legislation is to be discussed. In philosophy, Suber has taken the idea of self-amending games to its extreme by proposing the game of Nomic, in which the players can gradually change the rules.[36] Proposals to formalize such meta-argumentation include Vreeswijk (2000) and Brewka (2001), who have proposed formal models of argumentative dialogues allowing self-amendments.[37]

In an attempt to clarify how logic, defeasibility, dialogue and procedure are related, Henry Prakken (1997, p. 270f.) proposed to distinguish four layers of argumentation models. The first is the logical layer, which determines contradiction and support. The second layer is dialectical, which defines what counts as attack, counterargument, and also when an argument is defeated. The third layer is procedural and contains the rules constraining a dialogue, for instance, which moves parties can make, when parties can make a move, and when the dialogue is finished. The fourth and final layer is strategic. At this layer, one finds the strategies and heuristics used by a good, effective arguer.

Jaap Hage (2000) addresses the question of why dialogue models of argu-

---

[36] http://en.wikipedia.org/wiki/Nomic. See also Hofstadter (1996, chapter 4).
[37] See also the study of Nomic by Vreeswijk (1995a).

mentation became popular in the field of AI and law. He gives two reasons. The first is that legal reasoning is defeasible, and dialogue models are a good tool to study defeasibility. The second reason is that dialogue models are useful when investigating the process of establishing the law in a concrete case. Hage recalls the legal theoretic discussion about the law as an open system, in the sense that there can be disagreement about the starting points of legal arguments. As a result, the outcome of a legal procedure is indeterminate. A better understanding of this predicament can be achieved by considering the legal procedure as an argumentative dialogue.

Hage (2000) then discusses three functions of dialogue models of argumentation in AI and law. The first function is to define argument justification, in analogy with dialogical definitions of logical validity as can be found in the work by Lorenzen and Lorenz (1978). In this connection, Hage refers to Barth and Krabbe's notion of the 'dialectical garb' of a logic as opposed to an axiomatic, inferential or model-theoretic garb (Barth & Krabbe, 1982, pp. 7–8). Hage generalizes the idea of dialectical garb to what he refers to as battle of argument models of defeasible reasoning in which arguments attack each other, such as Loui's (1987), Pollock's (1987, 1994), Vreeswijk's (1993), Dung's (1995), and Prakken and Sartor's (1996). Battle of argument models can or cannot be presented in a dialectical garb. In their dialectical garb, such models define the justification of an argument in terms of the existence of a winning strategy in an argumentative dialogue game.

The second function of dialogue models of argumentation that is distinguished by Hage is to establish shared premises. Proponent and opponent enter into a dialogue that leads to a shared set of premises. The conclusions that follow from these shared premises can be regarded as justified. In this category, Hage discusses Gordon's Pleadings Game, which we discussed above. Hage makes connections to legal theory, in particular to Alexy's (1978) procedural approach to legal justification, and the philosophy of truth and justification, in particular Habermas's (1973) consensus theory of truth, and Schwemmer's approach to justification, in which the basis of justification is only assumed as long as it is not actually questioned (Schwemmer & Lorenzen, 1973).

As a third and final function of dialogue models of argumentation in AI and law, Hage discusses the procedural establishment of law in a concrete case. In this connection, he discusses mediating systems, which are systems that support dialogues, instead of evaluating them. He uses Zeno (Gordon & Karacapilidis, 1997), Room 5 (Loui et al., 1997) (see also Section 4.5) and DiaLaw (Lodder, 1999) as examples. Hage argues that regarding the law as purely procedural is somewhat counterintuitive, since there exist cases in which there is a clear answer, which can be known even without actually going through the whole procedure. Hage speaks therefore of the law as an imperfect procedure, in which the correctness of the outcome is not guaranteed.

Outside the field of AI and law, one further function of dialogue models of argumentation has been emphasized, namely that a dialogue perspective on argumentation can have computational advantages. For instance, argumentative dialogue can be used to optimize search, for instance, by cutting off dead ends or focusing on the most relevant issues. Vreeswijk (1995b) takes this assumption as the starting point of a paper:

> If dialectical concepts like argument, debate, and resolution of dispute are seemingly so important in practical reasoning, there must be some reason as to why these techniques survived as rulers of commonsense argument. Perhaps the reason is that they are just most suited for the job (Vreeswijk, 1995b, p. 307).

Vreeswijk takes inspiration from a paper by Loui (1998), which circulated in an earlier version since 1992. Loui emphasises the relevance of protocol, the assignment of burdens to parties, termination conditions, and strategy. A key idea is that argumentation dialogues are well-suited for reasoning in a setting of bounded resources (see also Loui & Norman, 1995).

Inspired by the computational perspective on argumentation, approaches to argumentative dialogue have been taken up in the field of multi-agent systems.[38] The focus in that field is on the interaction between autonomous software agents that pursue their own goals or goals shared with other agents. Since the actions of one agent can affect those of another, beyond control of an individual agent or the system as a whole, the kinds of problems when designing multi-agent software systems are of a different nature than those in the design of software where control can be assumed to be centralized. Computational models of argumentation have inspired the development of interaction protocols for the resolution of conflicts among agents and for belief formation. The typology of argumentative dialogue that has been proposed by Douglas Walton and Erik Krabbe (1995) has been especially influential.[39] In this typology, seven dialogue types are distinguished:

1. Persuasion, aimed at resolving or clarifying an issue;

2. Inquiry, aimed at proving (or disproving) a hypothesis;

3. Discovery, aimed at choosing the best hypothesis for testing;

4. Negotiation, aimed at a reasonable settlement all parties can live with;

5. Information-seeking, aimed at the exchange of information;

6. Deliberation, aimed at deciding the best available course of action;

7. Eristic, aimed at revealing a deeper basis of conflict.

---

[38] For an overview of the field of multi-agent systems see the textbook by Wooldridge (2009), which contains a chapter entitled 'Arguing.'

[39] The 2000 Symposium on Argument and Computation at Bonskeid House Perthshire, Scotland, organized by Reed and Norman, has been a causal factor. See Reed and Norman (2004).

In particular, the persuasion dialogue, starting with a conflict of opinion and aimed at resolving the issue by persuading a participant, has been extensively studied. An early persuasion system—focusing on persuasion in a negotition setting—is Sycara's Persuader system (1989). Persuader, developed in the field of what was then called Distributed AI, uses the domain of labour negotiation as an illustration. An agent forms a model of another agent's beliefs and goals, and determines its actions in such a way that it influences the other agent. For instance, agents can choose a so-called 'threatening argument,' i.e., an argument that is aimed at persuading another agent to give up a goal. Here it is notable that in Walton and Krabbe's typology negotiation is a dialogue type different from persuasion.

Prakken (2006, 2009) gives an overview and analysis of dialogue models of persuasion. In a dialogue system, dialogues have a goal and participants. It is specified which kinds of moves participants can make, for instance, making claims or conceding. Participants can have specific roles, for instance, Proponent or Opponent. The actual flow of a dialogue is constrained by a protocol, consisting of rules for turn-taking and termination. Effect rules determine how the commitments of participants change after each dialogue move. Outcome rules define the outcome of the dialogue, by determining, for instance, in persuasion dialogues who wins the dialogue. These elements are common to all dialogue types. By specifying or constraining the elements, one generates a system of persuasion dialogue. In particular, the dialogue goal of persuasion dialogue consists of a set of propositions that are at issue and need to be resolved. Prakken formalizes these elements and then uses his analytic model to discuss several extant persuasion systems, among them Mackenzie's (1979) proposals, and Walton and Krabbe's (1995) model of what they call Permissive persuasion dialogue.

Sycara's Persuader system (1989) is a persuasion system applied to labour negotiation. Parsons, Sierra and Jennings (1998) also speak of negotiation as involving persuasion. Their model uses the Belief-Desire-Intention model of agents (Rao & Georgeff, 1995) and specifies logically how the beliefs, desires and intentions of the agents influence the process of negotiation.[40] Dignum, Dunin-Kęplicz and Verbrugge (2001) have studied the role of argumentative dialogue for the forming of coalitions of agents that create collective intentions. Argumentation about what to do rather than about what is the case has been studied in a dialogue setting by Atkinson and colleagues (Atkinson et al., 2005, 2006; Atkinson & Bench-Capon, 2007). In this connection, it is noteworthy that Pollock's OSCAR model (1995) is an attempt to combine theoretical reasoning—about what to believe—with practical reasoning—about what to do—, though in a single agent, non-dialogical setting. Amgoud (2009) discusses the application of dialogical argumentation to decision making (see also Girle et al., 2004). Deliberation has been studied by McBurney, Hitchcock

---

[40] A systematic overview of argumentation dialogue models of negotiation has been provided by Rahwan et al. (2003).

and Parsons (2007).

Several attempts have been made to systematize the extensive work on argumentation dialogue. Bench-Capon, Geldard and Leng (2000), for instance, propose a formal method for modelling argumentation dialogue. Prakken (2005b) provides a formal framework that can be used to study argumentation dialogue models with different choices of underlying argument model and reply structures. McBurney and Parsons (2002a, 2002b, 2009) have developed an abstract theory of argumentative dialogue in which syntactic, semantic, and pragmatic elements are considered.

4.5 Argumentation support software

When studying argumentation from an artificial intelligence perspective, it can be investigated how software tools can perform or support argumentative tasks. Some researchers in the field of argumentation in AI have openly addressed themselves to building an artificial arguer. The most prominent among them is John Pollock (see also Section 3.4), who titled one of his books about his OSCAR project ambitiously *How to build a person* (Pollock, 1989).[41] Most researchers however have not aimed at realizing the grand task of addressing the so-called 'strong AI' problem of building an intelligent artefact that can perform any intellectual task a human being can. Instead of building software mimicking human argumentative behaviour, the more modest aim of supporting humans performing argumentative tasks was chosen. A great deal of research has been aimed at the construction of argumentation support software. Here we discuss three recurring themes: argument diagramming in software, the integration of rules and argument schemes, and argument evaluation.[42]

Argument diagramming in software The first theme discussed is argument diagramming in software. In the literature on argumentation support software, much attention has been paid to argument diagramming. Different kinds of argument diagramming styles have been proposed, many inspired by non-computational research on argument diagrams. We shall discuss three styles: boxes and arrows, boxes and lines, and nested boxes.

The first style of argument diagramming uses boxes and arrows. Argumentative statements are enclosed in boxes, and their relations indicated by arrows. A common use of arrows is to indicate the support relation between a reason and a conclusion. An example of a software tool that uses boxes and arrows diagrams is the Araucaria tool by Chris Reed and Glenn Rowe (2004) (Figure 9[43]). The Araucaria tool has been designed for the analysis of written arguments. Vertical arrows indicate reasons and their conclusions, and horizontal bi-directional arrows indicate conflicts between statements. The Araucaria software was one step in the development by the Dundee Argumentation Research Group, led by Reed, of open source argumentation software. For this purpose, a repre-

---

[41] The book's subtitle adds modestly: A Prolegomenon.
[42] The reviews by Kirschner et al. (2003), Verheij (2005b), and Scheuer et al. (2010) provide further detail about argumentation support software.
[43] Source: http://staff.computing.dundee.ac.uk/creed/araucaria/.

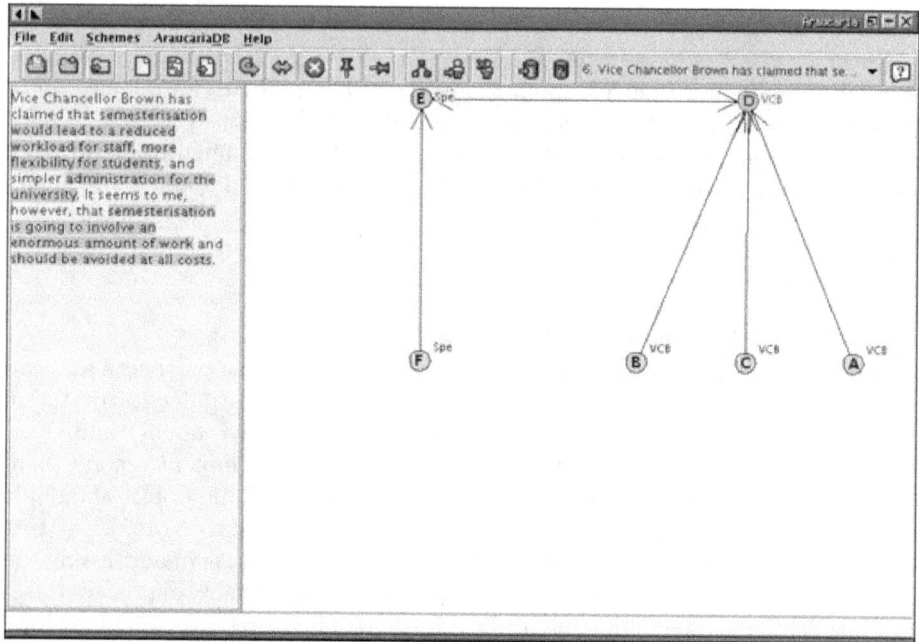

Figure 9. Boxes and arrows diagramming: The Araucaria system

sentation format, called the Argument mark-up language (AML), has been developed that allows for the exchange of arguments and their analyses using contemporary Internet technology. The format also allows for the exchange of sets of argument schemes (see Section 4.3) that can be used for argument analysis. Connected developments concerning machine-readable argument representation formats are the Argument interchange format (Chesñevar et al., 2006) and ArgDF, a proposal for a language allowing for a World wide argument web (Rahwan et al., 2007). One aim of the latter work is to develop classification systems for arguments, using ontology development techniques in Artificial Intelligence. In AI, an 'ontology' is a systematic conceptualization of a domain, often taking the form of a hierarchical system of concepts and their relations.

Another example of a system using boxes and arrows is the Hermes system (Karacapilidis & Papadias, 2001), an extension of the Zeno system (Gordon & Karacapilidis, 1997). Both Hermes and Zeno have been inspired by the IBIS approach. In IBIS, an abbreviation of Issue-Based Information Systems (Kunz & Rittel, 1970), problems are analysed in terms of issues, questions of fact, positions, and arguments. The focus is on what Rittel and Webber (1973) call wicked problems: problems with no definitive formulation, and no definitive solutions. Hence a goal of IBIS and systems such as Hermes and Zeno is to support the identification, structuring and settling of issues.

The second style of argument diagramming uses boxes and lines. In a boxes and lines style of argument diagramming, argumentative statements are depicted in boxes and their relations are indicated by (undirected) lines between them. This diagramming style abstracts from the directionality between statements, for instance, from a reason to a conclusion, or from a cause to an event. An example of a tool using the boxes and lines style is the Belvedere system (Suthers et al., 1995; Suthers, 1999). A goal of the system was to stimulate the critical discussion of science and public policy issues by middle school and high-school students, taking the cognitive limitations of the intended users into account. Such limitations include difficulty in focusing attention, lack of domain knowledge, and lack of motivation. In early versions, the diagrams were richly structured: there were links for support, explanation, causation, conjunction, conflict, justification, and undercutting. Link types could be distinguished graphically and by label. To prevent unproductive discussions about which structure to use, the graphical representation was significantly simplified in later versions (Suthers, 1999). Two types of statements were distinguished: data and hypotheses; and two link types: expressing a consistency and an inconsistency relation between statements. Figure 10[44] shows an example of a Belvedere screen using an even further simplified format with one statement type and one link type.

The third style of argument diagramming uses nested boxes. In this style, too, the argumentative statements are enclosed in boxes, but their relationships are indicated by the use of nesting. An example of the use of nested boxes is the Room 5 tool designed by Loui, Norman and a group of students (Loui et al., 1997). The Room 5 system aimed at the collaborative public discussion of pending Supreme Court cases. It was web-based, which is noteworthy as the proposal predates Google and Wikipedia. In its argument-diagramming format, a box inside a box expresses support, and a box next to a box indicates attack. In the argument depicted in the Room 5 screen shown in Figure 11[45], for instance, the punishability of John is supported by the reason that he has stolen a CD, and attacked by the reason that he is a minor first offender.

The integration of rules and argument schemes A second theme concerning the design of argumentation support software is the integration of rules and argument schemes. The integration of rules and argument schemes in argument diagramming software has been addressed in different ways: by the use of schematic arguments, conditional sentences, nested arrows and rule nodes. Consider, for instance, the elementary argument that Harry is a British subject because he is born in Bermuda (borrowed from Toulmin), and its underlying rule (or 'warrant' in Toulmin's terminology) that people born in Bermuda are British subjects.

A first approach is to consider such an argument as an instance of a scheme that abstracts from the person Harry in the argument. In Figure 12, an associ-

---

[44]Source: http://belvedere.sourceforge.net/.
[45]Screenshot of Room 5, as shown in Verheij (2005b). See also Bench-Capon et al. (2012).

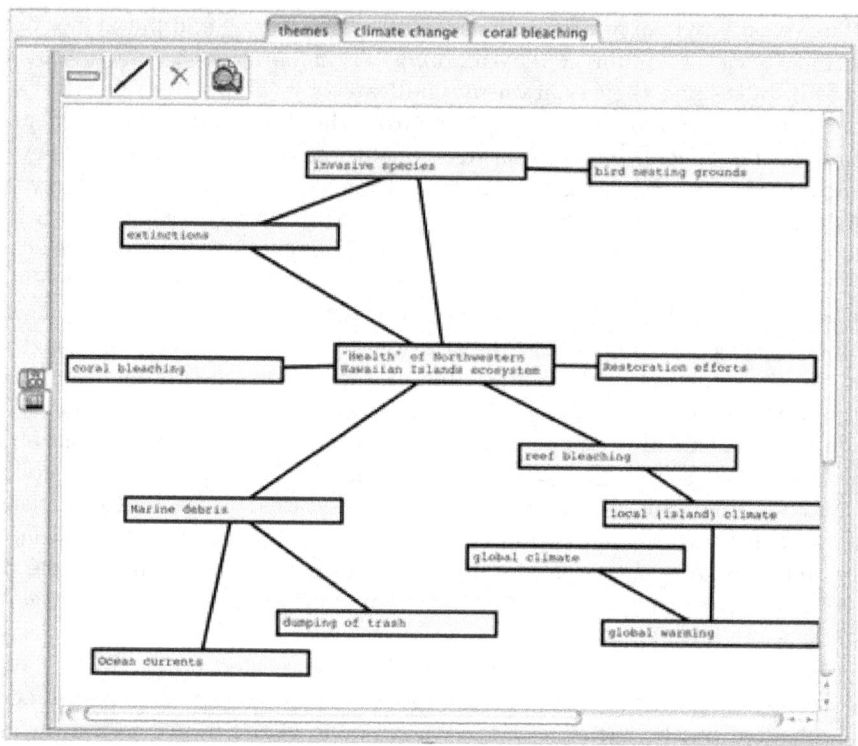

Figure 10. Boxes and lines diagramming: the Belvedere 4.1 system

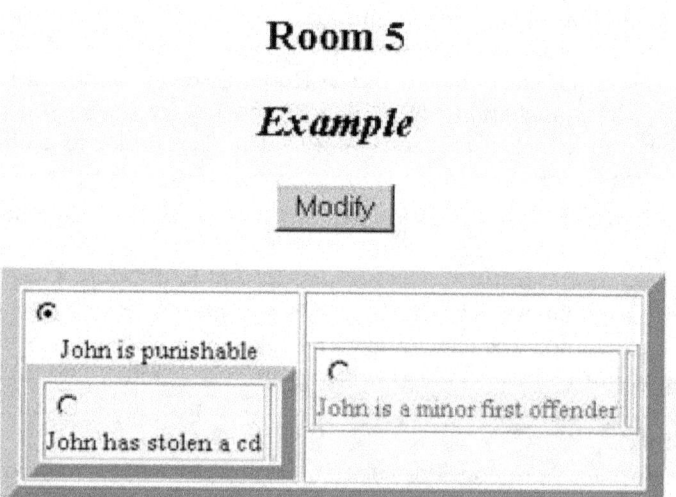

Figure 11. Nested boxes diagramming: the Room 5 system

Argumentation Theory in Formal and Computational Perspective   45

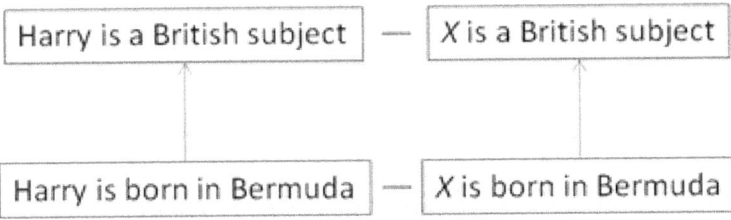

Figure 12. An elementary argument step as an instance of a schematic argument

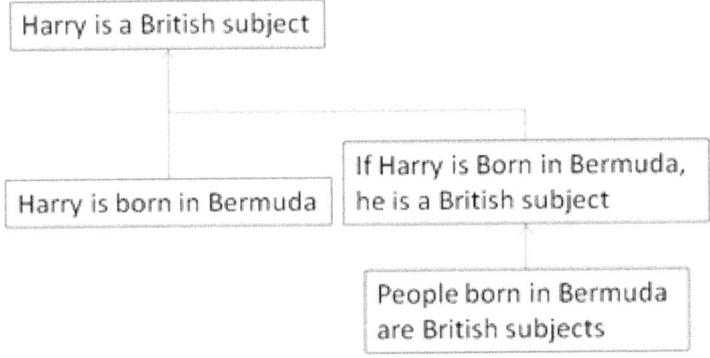

Figure 13. Using a conditional sentence

ated schematic argument is shown to the right of the argument about Harry. In the schematic argument, X appears as a variable that serves as the placeholder of someone's name. In software, the schematic argument is normally not shown graphically.

A second approach uses conditional sentences. The conditional sentence that expresses the connection between reason and conclusion is made explicit as an auxiliary premise. This conditional sentence can then be supported by further arguments, such as a warrant (as in Figure 13) or a backing. This approach is, for instance, proposed in the user-friendly Rationale[46] tool developed by van Gelder and his collaborators (van Gelder, 2007).

A third approach uses nested arrows. The arrows are treated as graphical expressions of the connection between the reason and conclusion, and can hence be argued about. In Figure 14, for instance, the warrant has been supplied as support for the connection between reason and conclusion. This approach has a straightforward generalisation when support and attack are combined (Section 4.2). The ArguMed tool developed by Verheij (2005b) uses this approach.

---

[46] http://rationale.austhink.com/.

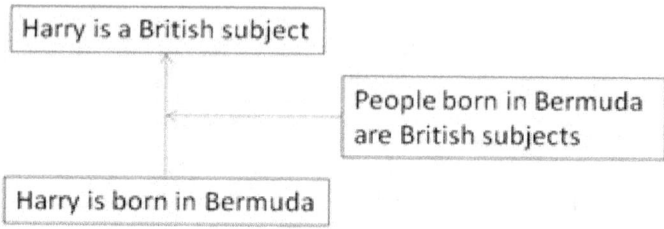

Figure 14. Nested arrows

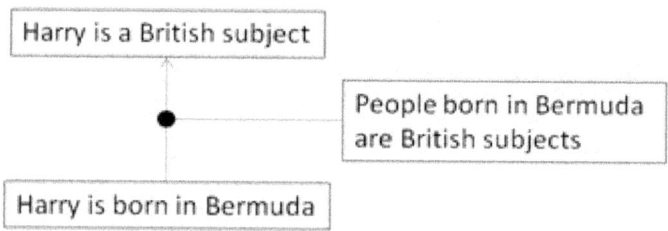

Figure 15. Rule nodes

A variation of the nested arrows approach uses rule nodes (Figure 15), instead of nested arrows. The AVERs tool (van den Braak et al., 2007) uses this approach.

Argument evaluation The third and final theme that we discuss in connection with the design of argumentation support software is argument evaluation. In argumentation software, different strategies for argument evaluation have been implemented. Some tools choose to leave argument evaluation as a task for the user of the system. For instance, in the Rationale system (van Gelder, 2007) a user can indicate which claims follow or do not follow given the reasons in the diagram. Specific graphical elements are used to show the user's evaluative actions.

In several other systems, some form of automatic evaluation has been implemented. Automatic evaluation algorithms can be logical, or numeric.

Logical evaluation algorithms in argumentation support tools have been grounded in versions of argumentation semantics (see Section 4.1). For instance, ArguMed (Verheij, 2005b) computes a version of stable semantics. Consider, for instance, Pollock's example of an undercutting defeater about red lights (see Section 3.4). ArguMed's evaluation algorithm behaves as expected: when the reason that the object looks red is assumed, the conclusion that the object is red will be justified, but that will no longer be the case when the defeater is added that the object is illuminated by a red light. A typical property of logical evaluation algorithms is reinstatement: when a defeating attacker of an initial argument is successfully attacked, the initial argument will no longer count as defeated and therefore be reinstated.

Numeric evaluation algorithms have been based on the numeric weights of the reasons supporting and attacking conclusions. A weight-based numeric evaluation algorithm has, for instance, been implemented in the Hermes system (Karacapilidis & Papadias, 2001). In Hermes, positions can be assigned a numeric score by adding the weights of active pro-positions and subtracting the weights of active con-positions. A proof standard can be used to determine an activation label of a position. In the proof standard called Preponderance of evidence, for instance, a position is active when the active pro-positions outweigh the active con-positions.

A numeric evaluation algorithm of a different kind has been implemented in the so-called 'Convince me' system (Schank, 1995). It uses ECHO, which is a connectionist version of Thagard's (1992) theory of explanatory coherence. In Convince me, statements are assigned numerical values by a step-wise constraint satisfaction algorithm. In the algorithm, incremental changes of the default weights of a statement are made by considering the excitatory and inhibitory links connected to a statement. When changes become too small to be taken into account (or computation is taking too long), the algorithm stops.

5 Specific kinds of argumentation in formal and computational perspective

In this section, we discuss specific kinds of argumentation using rules, cases, values and evidence. We end the section with applications and case studies.

5.1 Reasoning with rules

We already saw examples showing the close connections between argumentation research in artificial intelligence and legal applications. Since argumentation is an everyday task of professional lawyers this is not unexpected. An institutional reason however is that there exists an interdisciplinary research field, called artificial intelligence and law,[47] in which because of the nature of law the topic of argumentation has been given a great deal of attention. Early work in that field (e.g., McCarty, 1977; Gardner, 1987) already showed the intricacies and special characteristics of legal argumentation. Thorne McCarty (1977) attempted to formalize the detailed reasoning underlying a US Supreme Court case. Anne Gardner (1987) proposed a system aimed at what she called issue spotting. In a legal case, there is an issue when no rule applies or when conflicting rules apply and the conflict cannot be resolved. In this section, we pay special attention to the work inspired by developments in non-monotonic logic that has been carried out, mostly in the mid-1990s, regarding reasoning with (legal) rules.

Henry Prakken's (1997) book Logical tools for modelling legal argument provides an extensive and careful treatment of the contributions of techniques

---

[47]The primary journal of the field of AI & law is *Artifical Intelligence and Law*, with the biennial ICAIL and annual JURIX as the main conferences.

from non-monotonic logic to the formal modelling of legal reasoning.[48] The formal tools presented by Prakken have gradually evolved into the ASPIC+ model already mentioned (Prakken, 2010). Parts of the material were developed in close collaboration with Sartor (e.g., Prakken & Sartor, 1996, 1998; see also the excellent resource Sartor, 2005).

The following example shows how Prakken models a case in contract law (1997, p. 171). The example concerns the defeasible rule that contracts only bind the contracting parties (d_1), and a defeasible, possibly contravening, rule specifically for contracts that concern the lease of a house, saying that such contracts also bind future owners of the house (d_2). Another exception is added by a defeasible rule saying that, even in the case of a house lease, when a tenant agrees to make such a stipulation only the contracting parties are bound (d_3). The factual statements f_1 and f_2 say respectively (1) that a house lease is a special kind of contract and (2) that binding only the contracting parties and binding also future owners of a house do not go together.

d_1: x is a contract $\Rightarrow$ x only binds its parties
d_2: x is a lease of house y $\Rightarrow$ x binds all owners of y
d_3: x is a lease of house y $\wedge$ tenant has agreed in x that x only binds its parties $\Rightarrow$ x only binds its parties
f_1: $\forall x \forall y$(x is a lease of a house y $\rightarrow$ x is a contract)[49]
f_2: $\forall x \forall y$ $\neg$(x only binds its parties $\wedge$ x binds all owners of y)

When there is a contract about the lease of a house, there is an apparent conflict, since both d_1 and d_2 seem to apply. In the system, the application of d_2 blocks the application of d_2, using a mechanism of specificity defeat (see Section 4.2). In a case where also the condition of d_3 is fulfilled, namely when the tenant has agreed that the lease contract only binds the contracting parties, the application of rule d_3 blocks the application of rule d_2, which in that case does no longer block the application of d_1.

Prakken uses elements from classical logic (for instance, classical connectives and quantifiers) and non-monotonic logic (defeasible rules and their names), and shows how they can be used to model rules with exceptions, as they occur prominently in the law. He treats, for instance, the handling of explicit exceptions, preferring the most specific argument, reasoning with inconsistent information, and reasoning about priority relations.

In the same period, Hage developed Reason-based logic (Hage, 1997; see also Hage, 2005).[50] Hage presents Reason-based logic as an extension of first-order predicate logic in which reasons play a central role. Reasons are the result of the application of rules.[51] Treating them as individuals allows the expression of properties of rules. Whether a rule applies depends on the rule's conditions

---

[48]The book is based on Prakken's (1993) doctoral dissertation.

[49]'$\forall x$ ...' stands for 'for every entity x it holds that ...'. Similarly, for '$\forall y$ ...'

[50]Reason-based logic exists in a series of versions, some introduced in collaboration with Verheij (e.g., Verheij, 1996a).

[51]We shall simplify Hage's formalism a bit by omitting the explicit distinction between rules and principles.

being satisfied, but also on possible other reasons for or against applying the rule. Consider, for instance, the rule that thieves are punishable:

punishable: thief(x) ⇒ punishable(x)

Here 'punishable' before the colon is the rule's name. When John is a thief (expressed as thief(john)), the rule's applicability can follow:

Applicable(thief(john) ⇒ punishable(john))

This gives a reason that the rule ought to be applied. If there are no reasons against the rule's application, this leads to the obligation to apply the rule. From this it will follow that John is punishable.

A characteristic aspect of Reason-based logic is that it models the weighing of reasons. In this system, there is no numerical mechanism for weighing; rather it can be explicitly represented that certain reasons for a conclusion outweigh the reasons against the conclusion. When there is no weighing information the conflict remains unresolved and no conclusion follows.

Like Prakken, Hage uses elements from classical logic and non-monotonic logic. In his theory, because of the emphasis on philosophical and legal considerations, the flavour of Reason-based logic is less that of pure logic, but comes closer to representing the ways of reasoning in the domain of law. Where Prakken's book remains closer to the field of AI, Hage's book reads more like a theoretical essay in philosophy or law.

Reason-based logic has been applied, for instance, to a well-known distinction made by the legal theorist Dworkin (1978): whereas legal rules seem to lead directly to their conclusion when they are applied, legal principles are not as direct, and merely give rise to a reason for their conclusion. Only a subsequent weighing of possibly competing reasons leads to a conclusion. Different models of the distinction between rules and principles in Reason-based logic have been proposed. Hage (1997) follows Dworkin and makes a strict formal distinction, whereas Verheij et al. (1998) show how the distinction can be softened by presenting a model in which rules and principles are the extremes of a spectrum.

Loui and Norman (1995) have argued that there is a calculus associated with what they call the compression of rationales, i.e., the combination and adaptation of the rules underlying arguments which are akin to Toulmin's warrants. They give the following example of a compression of rules (rationales). When there is a rule 'vehicles used for private transportation are not allowed in the park' and also a rule 'vehicles are normally for private transportation,' then a two-step argument based on these two rules can be shortened when the so-called compression rationale 'no vehicles in the park,' based on these two rules, is used.

5.2 Case-based reasoning

Reasoning with rules (Section 5.1) is often contrasted with case-based reasoning. Whereas the former is about following rules that describe existing conditional patterns, the latter is about finding relevantly similar examples that, by analogy, can suggest possible conclusions in new situations. In the domain of law, rule-based reasoning is associated with the application of legal statutes, and case-based reasoning with the following of precedents. The contrast can be appreciated by looking at the following two examples.

Art. 300 of the Dutch Criminal Code
1. Inflicting bodily harm is punishable with up to two years of imprisonment or a fine of the fourth category.
2. When the fact causes grievous bodily harm, the accused is punished with up to four years of imprisonment or a fine of the fourth category.
3. [...]

Dutch Supreme Court July 9, 2002, NJ 2002, 499
Theft requires the taking away of a good. Can one steal an already stolen car? The Supreme Court's answer is: yes.

The first example is an excerpt from a statutory article expressing a material rule of Dutch criminal law, stating the kinds of punishment associated with inflicting bodily harm. The levels of punishment depend on specific conditions, with more severe bodily harm being punishable with longer imprisonment. The second example is a (very) brief summary of a Supreme Court decision. In this case, an already stolen car was stolen from the thief. One of the statutory requirements of the crime theft is that a good is taken away, and here the car was already taken away from the original owner of the car. The new legal question was addressed whether stealing from the original thief can count as theft from the car's owner. In other words, can an already stolen car still be taken away from the original owner? Here the Supreme Court decided that stealing a stolen car can count as theft since the original ownership is the deciding criterion; it does not matter whether a good is actually in the control of the owner at the time of theft. When used as a precedent, this Supreme Court decision has the effect that similar cases are decided alike.

In case-based reasoning, the stare decisis doctrine is leading: when deciding a new case one should not depart from an earlier, relevantly similar decision, but decide analogously. In the field of AI and law, Kevin Ashley's HYPO system (1990) counts as a milestone in the study of case-based reasoning.[52] In HYPO, cases are treated as sets of factors, where factors are generalised facts pleading for or against a case. Consider the following example about an employee who has been dismissed by his employer, and aims to void (i.e., cancel) the dismissal.[53]

Issue:
    Can a dismissal be voided?

Precedent case:
+   The employee's behaviour was always good.
−   There was a serious act of violence.

---

[52] See also Rissland and Ashley (1987), Ashley (1989), and Rissland and Ashley (2002).
[53] The example is inspired by the case material used by Roth (2003).

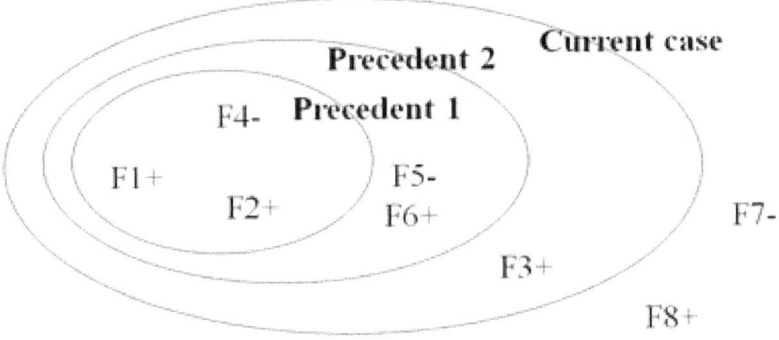

Figure 16. Factors in two precedent cases and the current case

Outcome:
+ (voided)

Current case:
+ The employee's behaviour was always good.
- There was a serious act of violence.
+ The working atmosphere was not affected.
Outcome:
?

There is a precedent case with one factor pleading for voidance (the good behaviour), and one pleading against voidance (the violence). In this precedent case, it was decided that voidance was in place. In the current case, the same factors apply, but there is also one additional factor pleading for voidance, namely that the working atmosphere was not affected. One could say that the decision taken in the precedent case is even more strongly supported in the current case. As a result, in HYPO and similar systems the suggested conclusion is that also in the current case voidance of the dismissal would be called for.

The example in Figure 16 shows that factors can be handled formally without knowing what they are about. There is a first precedent with pro-factors F1 and F2 and a con-factor F4. The second precedent has as additional factors a con-factor F5 and a pro-factor F6. The current case has all these factors and one more pro-factor F3. The domain also contains con-factor F7 and pro-factor F8 which do not apply to these cases.

Assume now that the first precedent was decided negatively, and the second positively. The second precedent is more on point, in the sense that it shares more factors with the current case than the first precedent. Since the current case even has an additional pro-factor, it is suggested that the current case should be decided positively, in analogy with precedent 2. Precedents do not always determine the outcome of the current case. For instance, if the second precedent had been decided negatively, there would be no suggested outcome

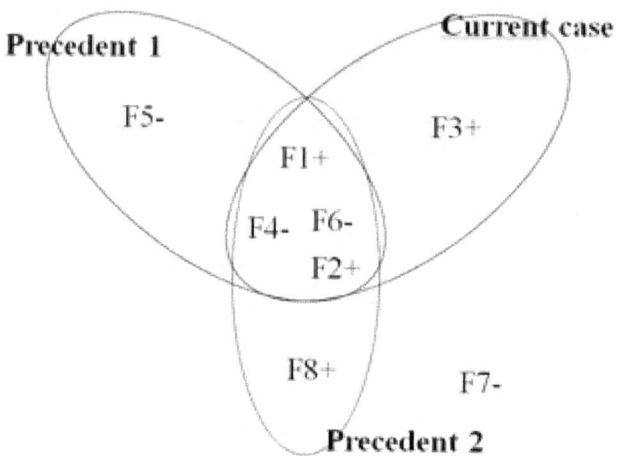

Figure 17. A different constellation of precedents

for the current case, since pro-factor F3 may be or may not be strong enough to turn the case.

Another formal example is shown in Figure 17. When both precedents have been decided positively, the suggested outcome for the current case is also positive. Precedent 1 can be followed because its support for a positive decision is weaker than that of the current case: the precedent has an additional con-factor, and the current case an additional pro-factor. Precedent 2 cannot be followed since F8 may be or may not be a stronger pro-factor than F3.

HYPO's aim is to form arguments about the current case, without determining a decision. This is made explicit in its model of 3–ply arguments. In HYPO's 3–ply model, the first argument move ('ply'), by the Proponent, is the citing of a precedent case in analogy with the current case. The analogy is based on the shared factors. The second argument move, by the Opponent, responds to the analogy, for instance, by distinguishing between the cited precedent case and the current case, pointing out differences in relevant factors, or by citing counterexamples. The third argument move, again by the Proponent, responds to the counterexamples, for instance, by making further distinctions.

HYPO's factors not only have a side (pro or con) associated with them, but can also come with a dimension pertaining in some way to the strength of the factor. This allows the citation of cases that share a certain factor, but have this factor with a different strength. For instance, by the use of dimensions, the good behaviour of the employee (of the first informal example) can come in gradations, say from good, via very good to excellent.

Vincent Aleven extended the HYPO model by the use of a factor hierarchy that allowed modelling of factors with hierarchical dependencies (Aleven, 1997; Aleven & Ashley, 1997a, 1997b). For instance, the factor that one has a family

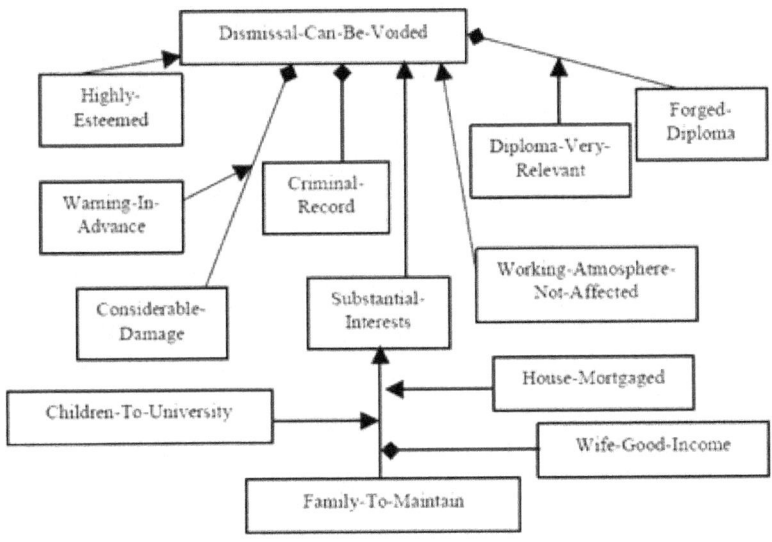

Figure 18. An entangled factor hierarchy (Roth, 2003)

to maintain is a special case of the factor that one has a substantial interest in keeping one's job. Inspired by Verheij's DefLog model (2003a), which allowed for reasoning about support and attack (Section 4.2), Roth (2003) developed case-based reasoning based on what he referred to as an entangled factor hierarchy, in order to expand the possible argumentative moves (Figure 18). For instance, the relevance of the factor that one has a family to maintain is strengthened by one's having children that go to university and weakened by one's having a wife with a good income. A factor hierarchy allows new kinds of argument moves by making it possible to downplay or emphasize a distinction. For instance, the factor of having a family to maintain can be downplayed by pointing out that one has a partner with a good income, or emphasized by mentioning that one has children going to university.

Proposals have been made to combine case-based and rule-based reasoning. For instance, Branting's GREBE model (1991, 2000) aims to generate explanations of decisions in terms of rules and cases. Both rules and cases can serve as warrants for a decision. Branting extends Toulmin's approach to warrants by using a so-called warrant reduction graph, in which warrants can be special cases of other warrants. Prakken and Sartor (1998) have applied their model of rule-based reasoning (Prakken & Sartor, 1996; see also Section 5.1) to the setting of case-based reasoning. Analogizing and distinguishing are connected to the deletion and addition of rule conditions that describe past decisions.

5.3 Values and audiences

Trevor Bench-Capon (2003) has developed a model of the values underlying arguments.[54] In this endeavour he refers to Perelman and Olbrechts-Tyteca's new rhetoric:

> If men oppose each other concerning a decision to be taken, it is not because they commit some error of logic or calculation. They discuss apropos the applicable rule, the ends to be considered, the meaning to be given to values, the interpretation and characterisation of facts (Perelman & Olbrechts-Tyteca, 1969, p. 150).

Because of the character of real-life argumentation, it is not to be expected that cases will be conclusively decided. Bench-Capon therefore aims to extend formal argumentation models by the inclusion of the values of the audiences addressed. This allows him to model the persuasion of an audience by means of argument.

Bench-Capon (2003) uses Dung's (1995) abstract argumentation frameworks as a starting point. He defines a value-based argumentation framework as a framework in which each argument has an associated (abstract) value. The idea is that values associated with an argument are promoted by accepting the argument. For instance, in a parliamentary debate about a tax raise it can be argued that accepting the raise will promote the value of social equality, while the value of enterprise is demoted. In an audience-specific argumentation framework, the preference ordering of the values can depend on an audience. For instance, the Labour Party may prefer the value of social equality, and the Conservative Party that of enterprise.

Bench-Capon continues to model defeat for an audience: an argument A defeats an argument B for audience a if A attacks B and the value associated with B is not preferred to the value associated with A for audience a. In his model, an attack succeeds, for instance, when the arguments promote the same value, or when there is no preference between the values. Dung's notions of argument acceptability, admissibility and preferred extension are then redefined relative to audience attack.

Bench-Capon uses a value-based argumentation framework with two values 'red' and 'blue' as an example (Figure 19). The underlying abstract argumentation framework is the same as that in Figure 6. In its unique preferred extension (which is also grounded and stable), A and C are accepted and B is rejected. For an audience preferring 'red,' defeat for the audience coincides with the underlying attack relation. In the preferred extension for an audience preferring 'red,' therefore, A and C are accepted and B is rejected. However, for an audience preferring 'blue,' A does not defeat B. But for such an audience B still defeats C. For a 'blue'-preferring audience, A and B are accepted and C is not.

Bench-Capon illustrates value-based argumentation by considering the case of a diabetic who almost collapses into a coma by lack of insulin, and therefore

---

[54]In AI and law, the importance of the modelling of the values and goals underlying legal decisions was already acknowledged by Berman and Hafner (1993).

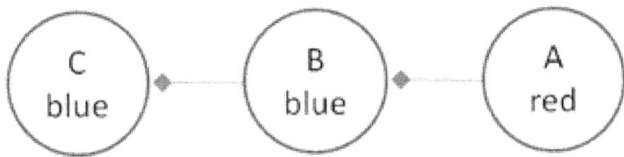

Figure 19. A value-based argumentation framework with two values (adapted from Bench-Capon, 2003)

takes another diabetic's insulin after entering her house. He analyses the case by discussing the roles of the value of property right infringement as opposed to that of saving one's life.

Bench-Capon and Sartor (2003) have used the value-based perspective in a treatment of legal reasoning that combines rule-based and case-based reasoning (see Sections 5.1 and 5.2). Legal reasoning takes the form of constructing and using a theory that explains a decision in terms of the values promoted and demoted by the decision. Precedent decisions have the role of revealing preferences holding between factors. This is similar to the role of precedents in HYPO that reveal how the factors in a precedent case are weighed. In Bench-Capon and Sartor's approach, the factor preferences in turn reveal preferences between values. The resulting preferences can then be used to decide new cases.

5.4 Burden of proof, evidence, and argument strength

Some arguments are more successful than others. An argument can meet or not meet the burden of proof fitting the circumstances of the debate. An argument can be founded on better evidence than another. An argument can also be stronger than another. In this section, we address the topics of burden of proof, evidence and argument strength.

Burden of proof and evidence The topic of burden of proof is strongly connected to the dialogical setting of argumentation. A burden of proof is assigned to a party in an argumentative dialogue when the quality of the arguments produced in the dialogue depends in part on whether the arguments produced by that party during the dialogue meet certain constraints. Such constraints can be procedural, for instance, requiring that a counterargument is met by a counterattack, or material, for instance, requiring that an argument is sufficiently strong in the light of the other arguments. Constraints of the latter, material, non-procedural type are also referred to as proof standards.

The topic of burden of proof is especially relevant in the law, as argumentation in court is often constrained by burden of proof constraints. As a result, in legal theory the topic has been studied extensively. The topic has also been addressed in AI approaches to argumentation, in particular by researchers connected to the field of AI and law (see also Section 4.4). In the Carneades argumentation model (Gordon et al., 2007), for instance, statements are categorized using three proof standards:

> SE (Scintilla of Evidence). A statement meets this standard if and only if it is supported by at least one defensible pro argument.
> BA (Best Argument). A statement meets this standard if and only if it is supported by some defensible pro argument with priority over all defensible con arguments.
> DV (Dialectical Validity). A statement meets this standard if and only if it is supported by at least one defensible pro argument and none of its con arguments are defensible.

A theme related to proof standards is argument accrual. What happens when there are several arguments for a conclusion? See Section 4.2, where research addressing the relation between argument defeat and accrual is discussed.

AI models of argumentation have been helpful in clarifying distinctions made in legal theory. Prakken and Sartor in particular have in a series of articles (Prakken & Sartor, 2007, 2009) contributed to the explication of different forms of burden of proof. They distinguish a burden of persuasion, a burden of production, and a tactical burden. A burden of persuasion requires that a party proves a statement to a specified degree (the standard of proof) or runs the risk of losing on the issue at the end of the debate. A burden of production has been assigned to a party when the party is required by law to provide evidence for a certain claim. Burdens of persuasion and burdens of production are assigned by the applicable law. The tactical burden of proof depends on a party's own assessment of whether sufficient grounds have been adduced about a claim made by the party. Prakken and Sartor connect these different notions to a formal dialogue model of argumentation.

Probability and other quantitative approaches to argument strength
Argument strength can be considered by using quantitative approaches. For instance, a conditional probability p(H|E), expressing the probability of a hypothesis H given the evidence E, can be interpreted as a measure of the strength of the argument for the hypothesis based on the evidence. The idea is that higher values of p(H|E) make H more strongly supported when given E. This interpretation of argument strength is associated with what is called Bayesian epistemology (Talbott, 2011). Bayesian epistemology provides in the following way an interpretation of the relevance of additional evidence, say E': additional evidence E' strengthens the argument E for H when $p(H|E \wedge E') > p(H|E)$. In this interpretation, Bayes' theorem:

$$p(H|E) = p(E|H) \times p(H)/p(E)$$

connects the strength of the argument from E to H and that of the argument from H to E, thereby reversing the direction of the arrow. This relation is helpful, when the values of p(E|H), p(H) and p(E) are available, or when they are more easily established than p(H|E) itself. Bayesian epistemology also provides a perspective on the comparison of hypotheses given additional evidence. When there are two hypotheses H and H', the odds form of Bayes' theorem can be used to update the odds of the hypotheses in light of new evidence E. The following relation shows how the prior odds p(H)/p(H') is connected to the posterior odds p(H|E)/p(H'|E):

$$p(H|E)/p(H'|E) = (p(H)/p(H')) \times (p(E|H)/p(E|H'))$$

This formal relation is helpful when the prior odds p(H)/p(H'), and the values of p(E|H) and p(E|H') are available.

Pollock has argued against a probabilistic account of argument strength (e.g., Pollock, 1995, 2006, 2010), referring to this position as 'generic Bayesianism' or 'probabilism.' Pollock argues that in a probabilistic account we would be justified in believing a mathematical theorem even before it is proven. This is especially absurd in cases such as Fermat's last theorem, which remained a conjecture for centuries before Wiles finally could complete a proof in the 1990s. Fitelson (2010) defends a probabilistic account against this and other criticisms advanced by Pollock.

Zukerman, McConachy and Korb (1998) have discussed the possibility of generating arguments from Bayesian networks, which are a widely studied tool for the representation of probabilistic information. Riveret et al. (2007) consider success in argument games in connection with probability. Dung and Thang (2010) have presented an approach to probabilistic argumentation in the setting of dispute resolution. Verheij (2012, 2017) has proposed a formal theory of defeasible argumentation in which logical and probabilistic properties are connected. Hunter (2013) discusses a model of deductive argumentation with uncertain premises. Verheij et al. (2016) discuss connections between arguments, scenarios and probabilities as normative tools in forensic reasoning with evidence.

Evidence and inference to the best explanation When an argument is aimed at establishing the truth, empirical evidence can be used to support alleged facts. For instance, a witness's testimony can provide evidence for the claim that the suspect was at the scene of a crime, a clinical test can provide evidence against a medical diagnosis, and the outcome of a laboratory experiment can be evidence confirming (or falsifying) a psychological phenomenon. The conclusions based on the available evidence can be regarded as hypothetical explanations for the occurrence of the evidence. As a result, reasoning on the basis of evidence is a specimen of what Peirce referred to as abductive reasoning, or inference to the best explanation: reasoning that goes from data describing something to a hypothesis that best explains or accounts for the data (Josephson & Josephson, 1996, p. 5). Josephson and Josephson conceive of inference to the best explanation as a kind of argument scheme (see Section 4.3):

> D is a collection of data (facts, observations, givens).
> H explains D (would, if true, explain D).
> No other hypothesis can explain D as well as H does.
> Therefore, H is probably true.
> (Josephson & Josephson, 1996, p. 5)

The explanatory connection between D and H is often regarded as going against the causal direction. For instance, a causal, expectation-evoking rule 'If there is a fire, then there is smoke' can be used to infer, or argue for, the effect 'there is smoke' after observing the cause 'there is fire.' The causal rule has an evidential, explanation-evoking counterpart, 'If there is smoke, then there is a fire,' that can be used to infer (argue for) the explanation 'there is a fire' after observing 'there is smoke.' Arguments based on causal or evidential rules are typically defeasible: not all fires generate smoke, and not all smoke stems from a fire.

In artificial intelligence, the distinction between causal and evidential rules has been emphasized by Pearl (1988, p. 499f.). He argues that special care is needed when mixing causal and evidential reasoning. To make his point, Pearl uses the following examples:

> Bill showed slight difficulties standing up, so I believed he was injured.
> Harry seemed injured, so I believed he would be unable to stand up.

The former uses the evidential pathway from the observation of Bill's difficulties in standing up to the explanation that he is injured, and the latter the reverse causal pathway from the observation of Harry's injuries to the effect that he is unable to stand up. The question is then addressed whether it is likely that Bill or Harry are likely to be drunk, drunkenness being a second cause for difficulties in standing up, independent from injury. Both Bill's and Harry's intoxicated state could be argued for using the evidential rule 'If someone has difficulties standing up, then he may be drunk.' However, for Bill the conclusion that he may be drunk seems more likely than for Harry, since for Bill both explanations for his difficulties in standing up, namely injury or being drunk, seem to be reasonable, whereas for Harry drunkenness is a less likely hypothesis now that an injury has been observed. The distinction between causal and evidential rules has played a central role in Pearl's thinking about causality (Pearl, 2000/2009), which relates to the probabilistic modelling tool of Bayesian Networks (see Jensen & Nielsen, 2007; Kjaerulff & Madsen, 2008). Bayesian Networks have been connected to the modelling of argumentation with legal evidence by Hepler, Dawid and Leucari (2007) and by Fenton, Neil and Lagnado (2012) (see also Taroni et al., 2006). Vlek et al. (2014, 2016) discuss the design and understanding of Bayesian Networks for evidential reasoning using scenarios. Timmer et al. (2017) discuss an algorithm to extract argumentative information from a Bayesian Network modeling hypotheses and evidence. Verheij (2017) investigates connections between arguments, scenarios and probabilities in one formal model.

The distinction between causal and evidential rules has also been used in the formalized hybrid argumentative-narrative model of reasoning with evidence developed by Bex and his colleagues (Bex et al., 2010, Bex, 2011). In this model, the elements of a scenario, or narrative, describing how a crime may have been committed, can be supported by arguments grounded in the available evidence. Causal connections between the elements of a scenario contribute to its coherence. It is possible that more than one scenario is available, each scenario with different evidential support and a different kind of coherence. Bex and Verheij (2012) have developed the argumentative-narrative model in terms of argument schemes and their associated critical questions (see Section 4.3).

5.5 Applications and case studies

A first reason for the popularity of argumentation research in the field of artificial intelligence is that it has led to theoretical advances. A second reason is that the theoretical advances have been corroborated by a variety of interesting applications and case studies, including advances in natural language processing. We give some examples.

Fox and Das (2000) provided a book-length study of AI technology in medical diagnosis and decision making, with much emphasis on the argumentative aspects (see also Fox and Modgil, 2006, where argumentation-based decision making is used to extend the Toulmin model). Aleven and Ashley (1997a, 1997b) developed a case-based argumentation tool that was empirically tested for its effects on learning. Buckingham Shum and Hammond (1994) approached the design of artefacts such as software as an argumentation problem. Grasso, Cawsey and Jones (2000) worked on argumentative conflict resolution in the context of health promotion. Teufel (1999) has worked on the problem of automatically estimating a sentence's role in argumentation, using a model of seven text categories called argumentative zones. Mochales Palau and Moens (2009) developed software for the mining of argumentative elements in legal texts. Hunter and Williams (2010) investigated the aggregation of evidence in a healthcare setting. Grasso (2002) and Crosswhite et al. (2004) have worked on the computational modelling of rhetorical aspects of argument. Reed and Grasso (2007) have collected argumentation-oriented research using natural language techniques. They discuss, for instance, the generation of argumentative texts as studied by Elhadad (1995), Reed (1999), Zukerman et al. (1998), and Green (2007).

Rahwan and McBurney (2007) edited a special issue on argumentation technology of the journal *IEEE Intelligent Systems*. Application areas addressed in the issue are medical decision-making, emotional strategies to persuade people to follow a healthy diet, ontology engineering, discussion mediation, and web services. In the 2012 edition of the COMMA conference proceedings series on the computational modelling of argument, a separate section was devoted to innovative applications. The topics included: automatic mining of arguments in opinions, a learning environment for scientific argumentation, semi-automatic analysis of online product reviews, argumentation with preferences in the setting of eco-efficient biodegradable packaging, hypothesis generation from cancer databases, sense making in policy deliberation, music recommendation, and argumentation about firewall policy. For applications focusing on argumentation support and facilitation, the reader is referred to Section 4.5.

In the domain of AI and law theories and systems were developed and tested by the use of case studies. For instance, McCarty (1977, 1995) analysed a seminal case in US tax law (Eisner v. Macomber, 252 U.S. 189 (1920)). In that case, the US Supreme Court decided that a federal rule of tax law was invalid.

McCarty's aims were set high, namely to build a software implementation that could handle a number of elusive, argumentative aspects of legal reasoning, illustrated in the majority opinion and dissenting opinions concerning the issues in this case. Quoting McCarty (1995):

1. Legal concepts cannot be adequately represented by definitions that state necessary and sufficient conditions. Instead, legal concepts are incurably 'open-textured'.

2. Legal rules are not static, but dynamic. As they are applied to new situations, they are constantly modified to 'fit' the new 'facts'. Thus the important process in legal reasoning is not theory application, but theory construction.

3. In this process of theory construction, there is no single 'right answer'. However, there are plausible arguments, of varying degrees of persuasiveness, for each alternative version of the rule in each new factual situation.

Berman and Hafner (1993) studied the 1805 Pierson v. Post case concerning the ownership of a dead fox chased by Post, but killed and taken by Pierson. They emphasize the teleological aspects of legal argumentation, in which the goals of legal rules and decisions are taken into account. Bex (2011) used the Anjum case, a Dutch high media profile murder case, to test his proposal for a hybrid argumentative-narrative model of reasoning with evidence. Atkinson (2012) edited an issue of the journal *Artificial Intelligence and Law* on the modelling of a 2002 case about the ownership of a baseball, representing possibly value in the order of a million dollars, being the one that Barry Bonds hit when he broke the record of home-runs in one season (Popov v Hayashi).

6 Conclusion

In the previous sections, we have introduced argumentation and argumentation theory as a field of study that goes back to classical times, passing through a neo-classical and anti-formal period in the second half of the 20nd century, and since the final decade of the 2nd millenium going through a formal and computational turn.

In Section 2, we discussed crucial concepts that have been indispensable in the study of argumentation before the recent formal and computational turn: standpoints, unexpressed premises, argument schemes, argumentation structures, and fallacies. All of these also played—and still play—a significant role in current formal and computational approaches to argumentation.

Standpoints occur in formal and computational work as the conclusions of arguments—possibly intermediate—, and as the commitments of the players in a computational dialogue game. Recently we see a move towards standpoints with a complex structure, in work that allows a complex hypothesis (such as a plan or a scenario) as the conclusion of an argument.

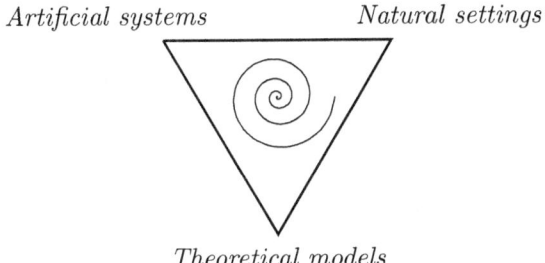

Figure 20. Perspectives on argumentation

Unexpressed premises have been studied in the context of manually analyzing argumentative texts in software tools. In today's research on argument mining, attempts are made to automatically understand argumentative texts, and we see that the ubiquity of unexpressed elements in argumentative discourse provides a significant hurdle.

Argument schemes have been the source of much interaction between the non-formal and formal/computational research communities. This is not a coincidence as argument schemes can be regarded as being intermediate between non-formal and the formal: argument schemes are formal in the sense that they have a well-organized structure, including elements such as premises, conclusions and critical questions; and argument schemes are non-formal in the sense that they handle just about every area of human reasoning, whether legal, medical, or common-sense. Because of their intermediate position, argument schemes have been referred to as semi-formal.

Argumentation structures have been extensively studied both in non-formal and in formal research into argumentation theory. Today's argumentation logics and argumentation diagramming tools provide carefully designed structuring tools that fit the non-formal theory well, and that have been applied to argument analysis and design. In the study of argumentation structures, we see perhaps most convincingly that the anti-logical period in argumentation theory of the second half of the 20nd century is now superseded by a fruitful interaction between formal and non-formal methods.

Fallacies have received mostly indirect attention in the formal and computational study of argumentation, in particular because the mirror image of fallacies—correct argumentation—is and always has been in the center of formal attention. Much progress has been made in the characterization of typically argumentative versions of validity, initially distancing from classical formal theories, and nowadays gradually returning to an integration with classical logic and standard probability theory, this time while engaging with the needs of actual human argumentation as uncovered in argumentation theory.

We hope that it has become clear that there are a great many issues that can be fruitfully researched if argumentation and artificial intelligence scholars cooperate (cf. the research programme initiated by Reed & Norman, Eds., 2004). The distinction between non-formal and formal argumentation theory becomes ever more blurred, and argumentation theory is ever further turning into an interdisciplinary enterprise, integrating insights from different perspectives (see Figure 20).

In the *theoretical models* perspective, the focus is on theoretical (possibly non-formal) and formal models of argumentation, for instance, extending the long tradition of philosophical and formal logic. In the *artificial systems* perspective, the aim is to build computer programmes that model or support argumentative tasks, for instance, in online dialogue games or in knowledge-based systems (computer programmes that reproduce the reasoning of an expert, for instance, in the law or in medicine). The *natural settings* perspective helps to ground research by concentrating on argumentation in its natural form, for instance, in the human mind or in an actual debate. We are curious where the continuing synergy between these perspectives will bring our understanding of argumentation, this utterly human characteristic of civilized coexistence.

BIBLIOGRAPHY

[Aleven, 1997] Aleven, V. (1997). *Teaching case-based reasoning through a model and examples.* Doctoral dissertation University of Pittsburgh.

[Aleven & Ashley, 1997a] Aleven, V., & Ashley, K. D. (1997a). Evaluating a learning environment for case-based argumentation skills. In *Proceedings of the sixth international conference on artificial intelligence and law* (pp. 170–179). New York: ACM Press.

[Aleven & Ashley, 1997b] Aleven, V., & Ashley, K. D. (1997b). Teaching case-based argumentation through a model and examples. Empirical evaluation of an intelligent learning environment. In B. du Boulay & R. Mizoguchi (Eds.), *Artificial intelligence in education. Proceedings of AI-ED 97 world conference* (pp. 87–94). Amsterdam: IOS Press.

[Alexy, 1978] Alexy, R. (1978). *Theorie der juristischen Argumentation. Die Theorie des rationale Diskurses as Theorie der juristischen Begründung* (A theory of legal argumentation. The theory of rational discourse as theory of juridical justification). Frankfurt am Main: Suhrkamp. (Spanish transl. by M. Atienza & I. Espejo as *Teoría de la argumentación jurídica*. Madrid: Centro de Estudios Constitucionales, 1989).

[Amgoud, 2009] Amgoud, L. (2009). Argumentation for decision making. In I. Rahwan & G. R. Simari (Eds.), *Argumentation in artificial intelligence* (pp. 301–320). Dordrecht: Springer.

[Amgoud et al., 2008] Amgoud, L., Cayrol, C., Lagasquie-Schiex, M. C., & Livet, P. (2008). On bipolarity in argumentation frameworks. *International Journal of Intelligent Systems*, 23(10), 1062–1093.

[Anscombre, 1994] Anscombre, J. C. (1994). La nature des topoï [The nature of the topoi]. In J. C. Anscombre (Ed.), *La théorie des topoï* [The theory of the topoi] (pp. 49–84). Paris: KimÃl'.

[Anscombre & Ducrot, 1983] Anscombre, J. C., & Ducrot, O. (1983). *L'argumentation dans la langue* [Argumentation in language]. Brussels: Pierre Mardaga.

[Antonelli, 2010] Antonelli, G. A. (2010). Non-monotonic logic. In E. N. Zalta (Ed.), *The Stanford encyclopedia of philosophy.* Summer 2010 ed. http://plato.stanford.edu/archives/sum2010/entries/logic-non-monotonic/.

[Antoniou et al., 2001] Antoniou, G., Billington, D., Governatori, G., & Maher, M. (2001). Representation results for defeasible logic. *ACM Transactions on Computational Logic*, 2(2), 255–287.

[Aristotle, 1984] Aristotle (1984). *The complete works of Aristotle. The revised Oxford translation.* 2 volumes. J. Barnes (Ed.). Transl. a.o. by W. A. Pickard-Cambridge (*Topics* and *Sophistical refutations*, 1928), J. L. Ackrill (*Categories* and *De interpretatione*, 1963),

A. J. Jenkinson (*Prior analytics*), and W. Rhys Roberts (*Rhetoric*, 1924). Princeton, NJ: Princeton University Press.
[Ashley, 1989] Ashley, K. D. (1989). Toward a computational theory of arguing with precedents. Accommodating multiple interpretations of cases. In *Proceedings of the second international conference on artificial intelligence and law* (pp. 93–102). New York: ACM Press.
[Ashley, 1990] Ashley, K. D. (1990). *Modeling legal argument. Reasoning with cases and hypotheticals*. Cambridge, MA: The MIT Press.
[Atkinson, 2012] Atkinson, K. (2012). Introduction to special issue on modelling Popov v. Hayashi. *Artificial Intelligence and Law*, 20, 1–14.
[Atkinson & Bench-Capon, 2007] Atkinson, K., & Bench-Capon, T. J. M. (2007). Practical reasoning as presumptive argumentation using action based alternating transition systems. *Artificial Intelligence*, 171, 855–874.
[Atkinson et al., 2005] Atkinson, K., Bench-Capon, T. J. M., & McBurney, P. (2005). A dialogue game protocol for multi-agent argument over proposals for action. *Autonomous Agents and Multi-Agent Systems*, 11, 153–171.
[Atkinson et al., 2006] Atkinson, K., Bench-Capon, T. J. M., & McBurney, P. (2006). Computational representation of practical argument. *Synthese*, 152, 157–206.
[Baroni et al., 2011] Baroni, P., Caminada, M., & Giacomin, M. (2011). An introduction to argumentation semantics. *Knowledge Engineering Review*, 26(4), 365–410.
[Barth, 1972] Barth, E. M. (1972). *Evaluaties. Rede uitgesproken bij de aanvaarding van het ambt van gewoon lector in de logica met inbegrip van haar geschiedenis en de wijsbegeerte van de logica in haar relatie tot de wijsbegeerte in het algemeen aan de Rijksuniversiteit te Utrecht op vrijdag 2 juni 1972* (Evaluations. Address given at the assumption of duties as profssor of logic including its history and philosophy of logic in relation to philosophy in general at the University of Utrecht on Friday, 2 June 1972). Assen: van Gorcum.
[Barth & Krabbe, 1982] Barth, E. M., & Krabbe, E. C. W. (1982). *From axiom to dialogue. A philosophical study of logics and argumentation*. Berlin/New York: Walter de Gruyter.
[Bench-Capon, 2003] Bench-Capon, T. J. M. (2003). Persuasion in practical argument using value-based argumentation frameworks. *Journal of Logic and Computation*, 13(3), 429–448.
[Bench-Capon & Dunne, 2007] Bench-Capon, T. J. M., & Dunne, P. E. (2007). Argumentation in artificial intelligence. *Artificial Intelligence*, 171, 619–641.
[Bench-Capon & Sartor, 2003] Bench-Capon, T. J. M., & Sartor, G. (2003). A model of legal reasoning with cases incorporating theories and values. *Artificial Intelligence*, 150, 97–143.
[Bench-Capon et al., 2012] Bench-Capon, T. J. M., Araszkiewicz, M., Ashley, K., Atkinson, K., Bex, F., Borges, F., Bourcier, D., Bourgine, D., Conrad, J. G., Francesconi, E., Gordon, T. F., Governatori, G., Leidner, J. L., Lewis, D. D., Loui, R. P., McCarty, L. T., Prakken, H., Schilder, F., Schweighofer, E., Thompson, P., Tyrrell, A., Verheij, B., Walton, D. N., & Wyner, A. Z. (2012). A history of AI and Law in 50 papers. 25 years of the international conference on AI and Law. *Artificial Intelligence and Law*, 20(3), 215–319.
[Bench-Capon et al., 2000] Bench-Capon, T. J. M., Geldard, T., & Leng, P. H. (2000). A method for the computational modelling of dialectical argument with dialogue games. *Artificial Intelligence and Law*, 8, 233–254.
[Bench-Capon et al., 2004] Bench-Capon, T. J. M., Freeman, J. B., Hohmann, H., & Prakken, H. (2004). Computational models, argumentation theories and legal practice. In Chr. A. Reed & T.·J. Norman (Eds.), *Argumentation machines. New frontiers in argument and computation* (pp. 85–120). Dordrecht: Kluwer.
[Bench-Capon et al., 2009] Bench-Capon, T. J. M., Prakken, H., & Sartor, G. (2009). Argumentation in legal reasoning. In I. Rahwan & G. R. Simari (Eds.), *Argumentation in artificial intelligence* (pp. 363–382). Dordrecht: Springer.
[Berger, 1977] Berger, F. R. (1977). *Studying deductive logic*. London: Prentice-Hall.
[Berman & Hafner, 1993] Berman, D., & Hafner, C. (1993). Representing teleological structure in case-based legal reasoning. The missing link. In *Proceedings of the fourth international conference on artificial intelligence and law*, 50–59. New York: ACM Press.
[Besnard & Hunter, 2008] Besnard, P., & Hunter, A. (2008). *Elements of argumentation*. Cambridge, MA: The MIT Press.

[Bex, 2011] Bex, F. J. (2011). *Arguments, stories and criminal evidence. A formal hybrid theory*. Dordrecht: Springer.
[Bex & Verheij, 2012] Bex, F. J., & Verheij, B. (2012). Solving a murder case by asking critical questions. An approach to fact-finding in terms of argumentation and story schemes. *Argumentation*, 26(3), 325–353.
[Bex et al., 2010] Bex, F. J., Koppen, P. van, Prakken, H., & Verheij, B. (2010). A hybrid formal theory of arguments, stories and criminal evidence. *Artificial Intelligence and Law*, 18(2), 123–152.
[Bex et al., 2003] Bex, F. J., Prakken, H., Reed, Chr., & Walton, D. N. (2003). Towards a formal account of reasoning about evidence. Argumentation schemes and generalisations. *Artificial Intelligence and Law*, 11, 125–165.
[Bondarenko, et al., 1997] Bondarenko, A., Dung, P. M., Kowalski, R. A., & Toni, F. (1997). An abstract, argumentation-theoretic approach to default reasoning. *Artificial Intelligence*, 93, pp. 63–101.
[Braak van den et al., 2007] Braak, S. W. van den, Vreeswijk, G., & Prakken, H. (2007). AVERs. An argument visualization tool for representing stories about evidence. In *Proceedings of the 11th international conference on artificial intelligence and law* (pp. 11–15). New York: ACM Press.
[Branting, 1991] Branting, L. K. (1991). Building explanations from rules and structured cases. *International Journal of Man-Machine Studies*, 34, 797–837.
[Branting, 2000] Branting, L. K. (2000). *Reasoning with rules and precedents. A computational model of legal analysis*. Dordrecht: Kluwer Academic.
[Bratko, 2001] Bratko, I. (2001). *PROLOG. Programming for artificial intelligence*. 3rd ed. Harlow: Pearson. (1st ed. 1986).
[Brewka, 2001] Brewka, G. (2001). Dynamic argument systems. A formal model of argumentation processes based on situation calculus. *Journal of Logic and Computation*, 11, 257–282.
[Buckingham Shum & Hammond, 1994] Buckingham Shum, S., & Hammond, N. (1994). Argumentation-based design rationale. What use at what cost? *International Journal of Human-Computer Studies*, 40(4), 603–652.
[Caminada, 2006] Caminada, M. (2006). Semi-stable semantics. In Dunne, P. E., Bench-Capon, T. & J. M. (Eds.), *Computational models of argument. Proceedings of COMMA 2006, September 11–12, 2006, Liverpool, UK. Frontiers in artificial intelligence and applications*, 144 (pp. 121–130). Amsterdam: IOS Press.
[Cayrol & Lagasquie-Schiex, 2005] Cayrol, C., & Lagasquie-Schiex, M. C. (2005). On the acceptability of arguments in bipolar argumentation frameworks. In L. Godo (Ed.), *Symbolic and quantitative approaches to reasoning with uncertainty. 8th European conference, ECSQARU 2005* (pp. 378–389). Berlin: Springer.
[Chesñevar et al., 2004] Chesñevar, C. I., Simari, G. R., Alsinet, T., & Godo, L. (2004). A logic programming framework for possibilistic argumentation with vague knowledge. In *Proceedings of the 20th conference on uncertainty in artificial intelligence* (pp. 76–84).
[Chesñevar et al., 2006] Chesnevar, J. McGinnis, Modgil, S., Rahwan, I., Reed, Chr., Simari, G., South, M., Vreeswijk, G., & Willmott, S. (2006). Towards an argument interchange format. *The Knowledge Engineering Review*, 21(4), 293–316.
[Crawshay-Williams, 1957] Crawshay-Williams, R. (1957). *Methods and criteria of reasoning. An inquiry into the structure of controversy*. London: Routledge & Kegan Paul.
[Crosswhite et al., 2004] Crosswhite, J., Fox, J., Reed, Chr. A., Scaltsas, T., & Stumpf, S. (2004). Computational models of rhetorical argument. In Chr. A. Reed & T. J. Norman (Eds.), *Argumentation machines. New frontiers in argument and computation* (pp.175–209). Dordrecht: Kluwer.
[d'Avila Garcez et al., 2009] d'Avila Garcez, A. S., Lamb, L. C., & Gabbay, D. M. (2009). *Neural-symbolic cognitive reasoning*. Springer: Berlin.
[Dignum et al., 2001] Dignum, F., Dunin-Kęplicz, B., & Verbrugge, R. (2001). Creating collective intention through dialogue. *Logic Journal of the IGPL* 9(2), 305–319.
[Doury, 1997] Doury, M. (1997). *Le débat immobile. L'Argumentation dans le débat médiatique surles parasciences*. [The immobile debate. Argumentation in the media debate on the parasciences]. Paris: KimÃĺ.

[Dung, 1995] Dung, P. M. (1995). On the acceptability of arguments and its fundamental role in non-monotonic reasoning, logic programming and n-person games. *Artificial Intelligence*, 77, 321–357.

[Dung & Thang, 2010] Dung, P. M., & Thang, P. M. (2010). Towards (probabilistic) argumentation for jury-based dispute resolution. In P. Baroni, F. Cerutti, M. Giacomin & G. R. Simari (Eds.), *Computational models of argument – Proceedings of COMMA 2010* (pp. 171–182). Amsterdam: IOS Press.

[Dunne, 2007] Dunne, P. E. (2007). Computational properties of argument systems satisfying graph-theoretic constraints. *Artificial Intelligence*, 171(10), 701–729.

[Dunne & Bench-Capon, 2003] Dunne, P. E., & Bench-Capon, T. J. M. (2003). Two party immediate response disputes. Properties and efficiency. *Artificial Intelligence*, 149(2), 221–250.

[Dworkin, 1978] Dworkin, R. (1978). *Taking rights seriously. New impression with a reply to critics.* London: Duckworth.

[Eagly & Chaiken, 1993] Eagly, A. H., & Chaiken, S. (1993). *The psychology of attitudes.* Fort Worth, TX: Harcourt Brace Jovanovich.

[van Eemeren, 1987] Eemeren, F. H. van (1987). Argumentation studies' five estates. In J. W. Wenzel (Ed.), *Argument and critical practices. Proceedings of the fifth SCA/AFA conference on argumentation* (pp. 9–24). Annandale, Virginia: Speech Communication Association.

[van Eemeren, 1990] Eemeren, F. H. van (1990). The study of argumentation as normative pragmatics. *Text. An Interdisciplinary Journal for the Study of Discourse*, 10(1/2), 37–44.

[van Eemeren, 2001] Eemeren, F. H. van (2001). Fallacies. In F. H. van Eemeren (Ed.), *Crucial concepts in argumentation theory* (pp. 135–164). Amsterdam: Amsterdam University Press.

[van Eemeren, 2010] Eemeren, F. H. van (2010). *Strategic maneuvering in argumentative discourse. Extending the pragma-dialectical theory of argumentation.* Amsterdam/Philadelphia: John Benjamins.

[van Eemeren, 2017] Eemeren, F. H. van (Ed.) (2017). *Prototypical argumentative patterns. Exploring the relationship between argumentative discourse and institutional context.* Amsterdam/Philadelphia: John Benjamins.

[van Eemeren et al., 2009] Eemeren, F. H. van, Garssen, B., & Meuffels, B. (2009). *Fallacies and judgments of reasonableness. Empirical research concerning the pragma-dialectical discussion rules.* Dordrecht etc.: Springer.

[van Eemeren & Grootendorst, 1984] Eemeren, F. H. van, & Grootendorst, R. (1984). *Speech acts in argumentative discussions. A theoretical model for the analysis of discussions directed towards solving conflicts of opinion.* Dordrecht/Cinnaminson: Foris & Berlin: de Gruyter. (transl. into Russian (1994c), Spanish (2013)).

[van Eemeren & Grootendorst, 1992] Eemeren, F. H. van, & Grootendorst, R. (1992). *Argumentation, communication, and fallacies. A pragma-dialectical perspective.* Hillsdale, NJ: Lawrence Erlbaum.

[van Eemeren & Grootendorst, 2004] Eemeren, F. H. van, & Grootendorst, R. (2004). *A systematic theory of argumentation. The pragma-dialectical approach.* Cambridge: Cambridge University Press.

[van Eemeren et al., 1978] Eemeren, F. H. van, Grootendorst, R., & Kruiger, T. (1978). *Argumentatietheorie* (Argumentation theory). Utrecht: Het Spectrum. (2nd enlarged ed. 1981; 3rd ed. 1986).

[van Eemeren et al., 1984] Eemeren, F. H. van, Grootendorst, R., & Kruiger, T. (1984). *The study of argumentation.* New York: Irvington. (Engl. transl. by H. Lake of F. H. van Eemeren, R. Grootendorst & T. Kruiger (1981). *Argumentatietheorie.* 2nd ed. Utrecht: Het Spectrum). (1 st ed. 1978).

[van Eemeren & Houtlosser, 2002] Eemeren, F. H. van, & Houtlosser, P. (2002a). Strategic maneuvering in argumentative discourse. Maintaining a delicate balance. In F. H. van Eemeren & P. Houtlosser (Eds.), *Dialectic and rhetoric. The warp and woof of argumentation analysis* (pp. 131–159). Dordrecht: Kluwer Academic.

[van Eemeren et al., 2010] Eemeren, F. H. van, Grootendorst, R., Jackson, S., & Jacobs, S. (1993). *Reconstructing argumentative discourse.* Tuscaloosa, AL: University of Alabama Press.

[van Eemeren et al., 2014] Eemeren, F.H. van, Garssen, B., Krabbe, E.C.W., Snoeck Henkemans, A.F., Verheij, B., & Wagemans, J.H.M. (2014). *Handbook of Argumentation Theory*. Dordrecht: Springer.
[Egly, Gaggl & Woltran, 2010] Egly, U., Gaggl, S. A., & Woltran, S. (2010). Answer-set programming encodings for argumentation frameworks. *Argument and Computation*, 1(2), 147–177.
[Elhadad, 1995] Elhadad, M. (1995). Using argumentation in text generation. *Journal of Pragmatics*, 24, 189–220.
[Fahnestock, 1999] Fahnestock, J. (1999). *Rhetorical figures in science*. New York, NY: Oxford University Press.
[Fahnestock, 2009] Fahnestock, J. (2009). Quid pro nobis. Rhetorical stylistics for argument analysis. In F. H. van Eemeren (Ed.), *Examining argumentation in context. Fifteen studies on strategic maneuvering* (pp. 131–152). Amsterdam: John Benjamins.
[Falappa et al., 2002] Falappa, M. A., Kern-Isberner, G., & Simari, G. R. (2002). Explanations, belief revision and defeasible reasoning. *Artificial Intelligence*, 141(1–2), 1–28.
[Fenton et al., 2012] Fenton, N. E., Neil, M., & Lagnado, D. A. (2012). A general structure for legal arguments using Bayesian networks. *Cognitive Science*, 37: 61–102.
[Finocchiaro, 1987] Finocchiaro, M. A. (1987). Six types of fallaciousness. Toward a realistic theory of logical criticism. *Argumentation*, 1, 263–282.
[Fitelson, 2010] Fitelson, B. (2010). Pollock on probability in epistemology. *Philosophical Studies*, 148, 455–465.
[Fox & Das, 2000] Fox, J., & Das, S. (2000). *Safe and sound. Artificial intelligence in hazardous applications*. Cambridge MA: The MIT Press.
[Fox & Modgil, 2006] Fox, J., & Modgil, S. (2006). From arguments to decisions. Extending the Toulmin view. In D. L. Hitchcock & B. Verheij (Eds.), *Arguing on the Toulmin model. New essays in argument analysis and evaluation* (pp. 273–287). Dordrecht: Springer.
[Freeman, 2005] Freeman, J. B. (2005). *Acceptable premises. An epistemic approach to an informal logic problem*. Cambridge: Cambridge University Press.
[Gabbay, et al., 1994] Gabbay, D. M., Hogger, C. J., & Robinson, J. A. (Eds., 1994). *Handbook of logic in artificial intelligence and logic programming, 3. Non-monotonic reasoning and uncertain reasoning*. Oxford: Clarendon Press.
[García, & Simari, 2004] García, A. J., & Simari, G. R. (2004). Defeasible logic programming. An argumentative approach. *Theory and practice of logic programming*, 4(2), 95–138.
[Gardner, 1987] Gardner, A. (1987). *An artificial intelligence approach to legal reasoning*. Cambridge.
[Garssen, 2001] Garssen, B. J. (2001). Argument schemes. In F. H. van Eemeren (Ed.), *Crucial concepts in argumentation theory* (pp. 81–99). Amsterdam: Amsterdam University Press.
[Gelder van, 2007] Gelder, T. van (2007). The rationale for Rationale. *Law, Probability and Risk*, 6, 23–42.
[Gelfond & Lifschitz, 1988] Gelfond, M., & Lifschitz, V. (1988). The stable model semantics for logic programming. In R. A. Kowalski & K. A. Bowen (Eds.), *Logic programming. Proceedings of the fifth international conference and symposium* (pp. 1070–1080). Cambridge MA: The MIT Press.
[Ginsberg, 1994] Ginsberg, M. L. (1994). AI and non-monotonic reasoning. In D. M. Gabbay, C. J. Hogger & J. A. Robinson (Eds.), *Handbook of logic in artificial intelligence and logic programming, 3. Non-monotonic reasoning and uncertain reasoning* (pp. 1–33). Oxford: Clarendon Press.
[Girle et al., 2004] Girle, R., Hitchcock, D. L., McBurney, P., & Verheij, B. (2004). Decision support for practical reasoning. A theoretical and computational perspective. In Chr. A. Reed & T. J. Norman (Eds.), *Argumentation machines. New frontiers in argument and computation* (pp. 55–83). Dordrecht: Kluwer Academic.
[Gómez Lucero et al., 2009] Gómez Lucero, M., Chesñevar, C., & Simari, G. (2009). Modelling argument accrual in possibilistic defeasible logic programming. In *ECSQARU âĂŸ09 Proceedings of the 10th European conference on symbolic and quantitative approaches to reasoning with uncertainty* (pp. 131–143). Springer: Berlin.

[Goodnight, 1982] Goodnight, G. Th. (1982). The personal, technical, and public spheres of argument. A speculative inquiry into the art of public deliberation. *Journal of the American Forensic Association*, 18, 214–227.

[Goodnight, 2012] Goodnight, G. T. (2012). The personal, technical, and public spheres. A note on 21st century critical communication inquiry. *Argumentation and Advocacy*, 48(4), 258–267.

[Gordon, 1993] Gordon, T. F. (1993). The Pleadings Game. *Artificial Intelligence and Law*, 2(4), 239–292.

[Gordon, 1995] Gordon, T. F. (1995). *The Pleadings Game. An artificial intelligence model of procedural justice.* Dordrecht: Kluwer.

[Gordon & Karacapilidis, 1997] Gordon, T. F., & Karacapilidis, N. (1997). The Zeno argumentation framework. In *Proceedings of the ICAIL 1997 conference* (pp. 10–18). New York: ACM Press.

[Gordon et al., 2007] Gordon, T. F., Prakken, H., & Walton, D. N. (2007). The Carneades model of argument and burden of proof. *Artificial Intelligence*, 171, 875–896.

[Govier, 1987] Govier, T. (1987). *Problems in argument analysis and evaluation.* Dordrecht/Providence, RI: Foris.

[Grasso, 2002] Grasso, F. (2002). Towards computational rhetoric. *Informal Logic*, 22, 195–229.

[Grasso, Cawsey & Jones, 2000] Grasso, F., Cawsey, A., & Jones, R. (2000). Dialectical argumentation to solve conflicts in advice giving. A case study in the promotion of healthy nutrition. *International Journal of Human-Computer Studies*, 53(6), 1077–1115.

[Greco Morasso, 2011] Greco Morasso, S. (2011). *Argumentation in dispute mediation. A reasonable way to handle conflict.* Amsterdam/Philadelphia: John Benjamins.

[Green, 2007] Green, N. (2007). A study of argumentation in a causal probabilistic humanistic domain. Genetic counseling. *International Journal of Intelligent Systems*, 22, 71–93.

[Habermas, 1973] Habermas, J. (1973). Wahrheitstheorien (Theories of truth). In H. Fahrenbach (Ed.), *Wirklichkeit und Reflexion. Festschrift für Walter Schulz zum 60. Geburtstag* (Reality and reflection. Festschrift for Walter Schulz in celebration of his 60th birthday) (pp. 211–265). Pfullingen: Günther Neske.

[Habermas, 1984] Habermas, J. (1984). *The theory of communicative action.* Vol. 1, *Reason and the rationalization of society.* Boston: Beacon. (English transl.; original work in German 1981).

[Hage, 1997] Hage, J. C. (1997). *Reasoning with rules. An essay on legal reasoning and its underlying logic.* Dordrecht: Kluwer Academic.

[Hage, 2000] Hage, J. C. (2000). Dialectical models in artificial intelligence and law. *Artificial Intelligence and Law*, 8, 137–172.

[Hage, 2005] Hage, J. C. (2005). *Studies in legal logic.* Berlin: Springer.

[Hage, Leenes & Lodder, 1993] Hage, J. C., Leenes, R., & Lodder, A. R. (1993). Hard cases. A procedural approach. *Artificial Intelligence and Law*, 2(2), 113–167.

[Hamblin, 1970] Hamblin, C. L. (1970). *Fallacies.* London: Methuen. Reprinted in 1986, with a preface by J. Plecnik & J. Hoaglund. Newport News, VA: Vale Press.

[Hart, 1951] Hart, H. L. A. (1951). The ascription of responsibility and rights. In A. Flew (Ed.), *Logic and language* (pp. 171–194). Oxford: Blackwell. (Originally Proceedings of the Aristotelian Society, 1948–1949).

[Hepler et al., 2007] Hepler, A. B., Dawid, A. P., & Leucari, V. (2007). Object-oriented graphical representations of complex patterns of evidence. *Law, Probability & Risk*, 6, 275–293.

[Hitchcock, 2001] Hitchcock, D. L. (2001). John L. Pollock's theory of rationality. In Chr. W. Tindale, H. V. Hansen & E. Sveda (Eds.), *Argumentation at the Century's turn.* Ontario Society for the Study of Argumentation. (Proceedings of the 3rd OSSA conference, 1999). (CD ROM).

[Hitchcock, 2002] Hitchcock, D. L. (2002). Pollock on practical reasoning. *Informal Logic*, 22, 247–256.

[Hofstadter, 1996] Hofstadter, D. (1996). *Metamagical themas. Questing for the essence of mind and pattern.* New York: Basic Books.

[Houtlosser, 2001] Houtlosser, P. (2001). Points of view. In F. H. van Eemeren (Ed.), *Crucial concepts in argumentation theory* (pp. 27–50). Amsterdam: Amsterdam University Press.

[Hunter, 2013] Hunter, A. (2013). A probabilistic approach to modelling uncertain logical arguments. *International Journal of Approximate Reasoning*, 54(1), 47–81.
[Hunter & Williams, 2010] Hunter, A., & Williams, M. (2010). Qualitative evidence aggregation using argumentation. In P. Baroni, F. Cerutti, M. Giacomin, & G. R. Simari (Eds.), *Computational models of argument – Proceedings of COMMA 2010* (pp. 287–298). Amsterdam: IOS Press.
[Jackson & Jacobs, 1982] Jackson, S., & Jacobs, S. (1982). The collaborative production of proposals in conversational argument and persuasion. A study of disagreement regulation. *Journal of the American Forensic Association*, 18, 77–90.
[Jakobovits & Vermeir, 1999] Jakobovits, H., & Vermeir, D. (1999). Robust semantics for argumentation frameworks. *Journal of Logic and Computation*, 9(2), 215–261.
[Jensen & Nielsen, 2007] Jensen, F. V., & Nielsen, T. D. (2007). *Bayesian networks and decision graphs*. New York: Springer.
[Johnson, 2000] Johnson, R. H. (2000). *Manifest rationality. A pragmatic theory of argument*. Mahwah, NJ: Lawrence Erlbaum.
[Johnson & Blair, 2006] Johnson, R. H., & Blair, J. A. (2006). *Logical self-defense* (reprint of Johnson & Blair, 1994). New York: International Debate Education Association. (1st ed. 1977).
[Josephson & Josephson, 1996] Josephson, J. R., & Josephson, S. G. (Eds., 1996). *Abductive inference. Computation, philosophy, technology*. Cambridge: Cambridge University Press.
[Karacapilidis & Papadias, 2001] Karacapilidis, N., & Papadias, D. (2001). Computer supported argumentation and collaborative decision making. The HERMES system. *Information Systems*, 26, 259–277.
[Kienpointner, 1992] Kienpointner, M. (1992). *Alltagslogik. Struktur and Funktion von Argumentationsmustern* (Everyday logic. Structure and functions of prototypes of argumentation). Stuttgart: Frommann-Holzboog.
[Kirschner, et al., 2003] Kirschner, P. A., Buckingham Shum, S. J., & Carr, C. S. (Eds., 2003). *Visualizing argumentation. Software tools for collaborative and educational sensemaking*. London: Springer.
[Kjaerulff & Madsen, 2008] Kjaerulff, U. B., & Madsen, A. L. (2008). *Bayesian networks and influence diagrams*. New York: Springer.
[Kopperschmidt, 1989] Kopperschmidt, J. (1989). *Methodik der Argumentationsanalyse* (Methodology of argumentation analysis). Stuttgart: Frommann-Holzboog.
[Kowalski, 2011] Kowalski, R. A. (2011). *Computational logic and human thinking. How to be artificially intelligent*. Cambridge: Cambridge University Press.
[Kunz & Rittel, 1970] Kunz, W., & Rittel, H. (1970). Issues as elements of information systems, Technical Report 0131, Universität Stuttgart, Institut für Grundlagen der Planung.
[Kyburg, 1994] Kyburg, H. E. (1994). Uncertainty logics. In D. M. Gabbay, C. J. Hogger & J. A. Robinson (Eds.), *Handbook of logic in artificial intelligence and logic programming, 3. Non-monotonic reasoning and uncertain reasoning* (pp. 397–438). Oxford: Clarendon Press.
[Leff, 2003] Leff, M. (2003). Rhetoric and dialectic in Martin Luther King's 'Letter from Birmingham Jail'. In F. H. van Eemeren, J. A. Blair, Ch. A. Willard & A. F. Snoeck Henkemans (Eds.), *Anyone who has a view. Theoretical contributions to the study of argumentation* (pp. 255–268). Dordrecht: Kluwer Academic.
[Lodder, 1999] Lodder, A. R. (1999). *DiaLaw. On legal justification and dialogical models of argumentation*. Dordrecht: Kluwer.
[Lorenzen & Lorenz, 1978] Lorenzen, P., & Lorenz, K. (1978). *Dialogische Logik* (Dialogic logic). Darmstadt: Wissenschaftliche Buchgesellschaft.
[Loui, 1987] Loui, R. P. (1987). Defeat among arguments. A system of defeasible inference. *Computational Intelligence*, 2, 100–106.
[Loui, 1995] Loui, R. P. (1995). Hart's critics on defeasible concepts and ascriptivism. *The fifth international conference on artificial intelligence and law. Proceedings of the conference* (pp. 21–30). New York: ACM. Extended report available at http://www1.cse.wustl.edu/~loui/ail2.pdf.
[Loui, 1998] Loui, R. P. (1998). Process and policy. Resource-bounded nondemonstrative reasoning. *Computational Intelligence*, 14, 1–38.
[Loui & Norman, 1995] Loui, R., & Norman, J. (1995). Rationales and argument moves. *Artificial Intelligence and Law*, 3, 159–189.

[Loui et al., 1997] Loui, R., Norman, J., Altepeter, J., Pinkard, D., Craven, D., Linsday, J., & Foltz, M. A. (1997). Progress on room 5. A testbed for public interactive semi-formal legal argumentation. In: *Proceedings of the sixth international conference on artificial intelligence and law* (pp. 207–214). New York: ACM Press.

[Mackenzie, 1979] Mackenzie, J. D. (1979). Question-begging in non-cumulative systems. *Journal of Philosophical Logic*, 8, 117–133.

[Makinson, 1994] Makinson, D. (1994). General patterns in non-monotonic reasoning. In D. M. Gabbay, C. J. Hogger & J. A. Robinson (Eds.), *Handbook of logic in artificial intelligence and logic programming, 3. Non-monotonic reasoning and uncertain reasoning* (pp. 35–110). Oxford: Clarendon Press.

[McBurney & Parsons, 2002a] McBurney, P. & Parsons, S. (2002a). Games that agents play. A formal framework for dialogues between autonomous agents. *Journal for Logic, Language and Information*, 11, 315–334.

[McBurney & Parsons, 2002b] McBurney, P. & Parsons, S. (2002b). Dialogue games in multi-agent systems. *Informal Logic*, 22, 257–274.

[McBurney & Parsons, 2009] McBurney, P., & Parsons, S. (2009). Dialogue games for agent argumentation. In I. Rahwan & G. R. Simari (Eds.), *Argumentation in artificial intelligence* (pp. 261–280). Dordrecht: Springer.

[McBurney, Hitchcock & Parsons, 2007] McBurney, P., Hitchcock, D. L., & Parsons, S. (2007). The eightfold way of deliberation dialogue. *International Journal of Intelligent Systems*, 22, 95–132.

[McCarty, 1977] McCarty, L. (1977). Reflections on TAXMAN. An experiment in artificial intelligence and legal reasoning. *Harvard Law Review*, 90, 89–116.

[McCarty, 1995] McCarty, L. (1995). An implementation of Eisner v. Macomber. In *Proceedings of the fifth international conference on artificial intelligence and law* (pp. 276–286). New York: ACM Press.

[Mochales Palau & Moens, 2009] Mochales Palau, R., & Moens, S. (2009). Argumentation mining. The detection, classification and structure of arguments in text. In *Proceedings of the 12th international conference on artificial intelligence and law (ICAIL 2009)* (pp. 98–107). New York: ACM Press.

[Modgil, 2009] Modgil, S. (2009). Reasoning about preferences in argumentation frameworks. *Artificial Intelligence*, 173(9–10), 901–934.

[Næss, 1966] Næss, A. (1966). *Communication and argument. Elements of applied semantics.* (A. Hannay, transl.). London: Allen and Unwin. (English transl. of *En del elementære logiske emner.* Oslo: Universitetsforlaget, 1947).

[Nute, 1994] Nute, D. (1994). Defeasible logic. In D. M. Gabbay, C. J. Hogger & J. A. Robinson (Eds.), *Handbook of logic in artificial intelligence and logic programming, 3. Non-monotonic reasoning and uncertain reasoning* (pp. 353–395). Oxford: Clarendon Press.

[O'Keefe, 2002] O'Keefe, D. J. (2002). *Persuasion. Theory and research.* 2nd ed. Thousand Oaks, CA: Sage. (1st ed. 1990).

[Parsons et al., 1998] Parsons, S., Sierra, C., & Jennings, N. R. (1998). Agents that reason and negotiate by arguing. *Journal of Logic and Computation*, 8, 261–292.

[Pearl, 1988] Pearl, J. (1988). *Probabilistic reasoning in intelligent systems. Networks of plausible inference.* San Francisco: Morgan Kaufmann Publishers.

[Pearl, 2009] Pearl, J. (2000/2009). *Causality. Models, reasoning, and inference.* 2nd ed. Cambridge: Cambridge University Press. (1st ed. 2000).

[Perelman & Olbrechts-Tyteca, 1969] Perelman, Ch., & Olbrechts-Tyteca, L. (1969). *The new rhetoric. A treatise on argumentation.* Notre Dame, IN: University of Notre Dame Press. (English transl. by J. Wilkinson and P. Weaver of Ch. Perelman & L. Olbrechts-Tyteca (1958). *La nouvelle rhétorique. Traité de l'argumentation.* Paris: Presses Universitaires de France. (3rd ed. Brussels: Éditions de l'Université de Bruxelles)).

[Plantin, 1996] Plantin, Chr. (1996). *L'argumentation* [Argumentation]. Paris: Le Seuil.

[Pollock, 1987] Pollock, J. L. (1987). Defeasible reasoning. *Cognitive Science*, 11, 481–518.

[Pollock, 1989] Pollock, J. L. (1989). *How to build a person. A prolegomenon.* Cambridge, MA: The MIT Press.

[Pollock, 1994] Pollock, J. L. (1994). Justification and defeat. *Artificial Intelligence*, 67, 377–407.

[Pollock, 1995] Pollock, J. L. (1995). *Cognitive Carpentry. A blueprint for how to build a person.* Cambridge, MA: The MIT Press.
[Pollock, 2006] Pollock, J. L. (2006). *Thinking about acting. Logical foundations for rational decision making.* New York: Oxford University Press.
[Pollock, 2010] Pollock, J. L. (2010). Defeasible reasoning and degrees of justification. *Argument & Computation*, 1(1), 7–22.
[Poole, 1985] Poole, D. L. (1985). On the comparison of theories. Preferring the most specific explanation. In *Proceedings of the ninth international joint conference on artificial intelligence* (pp. 144–147). San Francisco: Morgan Kaufmann.
[Prakken, 1993] Prakken, H. (1993). *Logical tools for modelling legal argument.* Doctoral dissertation Free University Amsterdam.
[Prakken, 1997] Prakken, H. (1997). *Logical tools for modelling legal argument. A study of defeasible reasoning in law.* Dordrecht: Kluwer.
[Prakken, 2005a] Prakken, H. (2005a). A study of accrual of arguments, with applications to evidential reasoning. In *Proceedings of the tenth international conference on artificial intelligence and law* (pp. 85–94). New York: ACM Press.
[Prakken, 2005b] Prakken, H. (2005b). Coherence and flexibility in dialogue games for argumentation. *Journal of Logic and Computation*, 15, 1009–1040.
[Prakken, 2006] Prakken, H. (2006). Formal systems for persuasion dialogue. *The Knowledge Engineering Review*, 21(2), 163–188.
[Prakken, 2009] Prakken, H. (2009). Models of persuasion dialogue. In I. Rahwan & G. R. Simari (Eds.), *Argumentation in artificial intelligence* (pp. 281–300). Dordrecht: Springer.
[Prakken, 2010] Prakken, H. (2010). An abstract framework for argumentation with structured arguments. *Argument and Computation*, 1, 93–124.
[Prakken & Sartor, 1996] Prakken, H., & Sartor, G. (1996). A dialectical model of assessing conflicting arguments in legal reasoning. *Artificial Intelligence and Law*, 4, 331–368.
[Prakken & Sartor, 1998] Prakken, H., & Sartor, G. (1998). Modelling reasoning with precedents in a formal dialogue game. *Artificial Intelligence and Law*, 6, 231–287.
[Prakken & Sartor, 2007] Prakken, H., & Sartor, G. (2007). Formalising arguments about the burden of persuasion. In *Proceedings of the eleventh international conference on artificial intelligence and law* (pp. 97–106). New York: ACM Press.
[Prakken & Sartor, 2009] Prakken, H., & Sartor, G. (2009). A logical analysis of burdens of proof. In H. Kaptein, H. Prakken & B. Verheij (Eds.), *Legal evidence and proof. Statistics, stories, logic* (pp. 223–253). Farnham: Ashgate.
[Prakken & Vreeswijk, 2002] Prakken, H., & Vreeswijk, G. A. W. (2002). Logics for defeasible argumentation. In D. Gabbay & F. Guenthner (Eds.), *Handbook of philosophical logic.* 2nd ed., IV (pp. 219–318). Dordrecht: Kluwer.
[Rahwan & McBurney, 2007] Rahwan, I., & McBurney, P. (2007). Argumentation technology. Guest editors' introduction. *IEEE Intelligent Systems*, 22(6), 21–23.
[Rahwan & Simari, 2009] Rahwan, I., & Simari, G. R. (Eds., 2009). *Argumentation in artificial intelligence.* Dordrecht: Springer.
[Rahwan et al., 2003] Rahwan, I., Ramchurn, S. D., Jennings, N. R., McBurney, P., Parsons, S., & Sonenberg, E. (2003). Argumentation-based negotiation. *Knowledge Engineering Review*, 18(4), 343–375.
[Rahwan et al., 2007] Rahwan, I., Zablith, F., & Reed, Chr. (2007). Laying the foundations for a world wide argument web. *Artificial Intelligence*, 171(10–15), 897–921.
[Rao & Georgeff, 1995] Rao, A., & Georgeff, M. (1995). BDI agents. From theory to practice. In *Proceedings of the 1st international conference on multi-agent systems* (pp. 312–319).
[Reboul, 1990] Reboul, O. (1990). Rhétorique et dialectique chez Aristote [Aristotle's views on rhetoric and dialectic]. *Argumentation.* 4, 35–52.
[Reed, 1997] Reed, Chr. A. (1997). Representing and applying knowledge for argumentation in a social context. *AI and Society*, 11(3–4), 138–154.
[Reed, 1999] Reed, Chr. A. (1999). The role of saliency in generating natural language arguments. In: *Proceedings of the 16th international joint conference on AI (IJCAI'99)* (pp. 876–881). San Francisco: Morgan Kaufmann.
[Reed & Grasso, 2007] Reed, Chr. A., & Grasso, F. (2007). Recent advances in computational models of natural argument. *International Journal of Intelligent Systems*, 22, 1–15.
[Reed & Norman et al., 2004] Reed, Chr. A., & Norman, T. J. (Eds., 2004). *Argumentation machines. New frontiers in argument and computation.* Dordrecht: Kluwer.

[Reed & Rowe, 2004] Reed, Chr. A., & Rowe, G. W. A. (2004). Araucaria. Software for argument analysis, diagramming and representation. *International Journal on Artificial Intelligence Tools*, 13(4), 961–979.

[Reed & Tindale, 2010] Reed, Chr. A., & Tindale, Chr. W. (Eds., 2010). *Dialectics, dialogue and argumentation. An examination of Douglas Walton's theories of reasoning*. London: College Publications.

[Reiter, 1980] Reiter, R. (1980). A logic for default reasoning. *Artificial Intelligence*, 13, 81–132.

[Rigotti, 2009] Rigotti, E. (2009). Whether and how classical topics can be revived within contemporary argumentation theory. In F. H. van Eemeren & B. Garssen (Eds.), *Pondering on problems of argumentation* (pp. 157–178). New York: Springer.

[Rissland & Ashley, 2002] Rissland, E. L., & Ashley, K. D. (1987). A case-based system for trade secrets law. In *Proceedings of the first international conference on artificial intelligence and law* (pp. 60–66). New York: ACM Press.

[Rissland & Ashley, 2002] Rissland, E. L., & Ashley, K. D. (2002). A note on dimensions and factors. *Artificial Intelligence and Law*, 10, 65–77.

[Rittel & Webber, 1973] Rittel, H., & Webber, M. (1973). Dilemmas in a general theory of planning. *Policy Sciences*, 155–169.

[Riveret et al., 2007] Riveret, R., Rotolo, A., Sartor, G., Prakken, H., & Roth, B. (2007). Success chances in argument games. A probabilistic approach to legal disputes. In A. R. Lodder & L. Mommers (Eds.), *Legal knowledge and information systems (JURIX 2007)* (pp. 99–108). Amsterdam: IOS Press.

[Rocci, 2008] Rocci, A. (2008). Modality and its conversational backgrounds in the reconstruction of argumentation. *Argumentation*, 22, 165–189.

[Roth, 2003] Roth, B. (2003). *Case-based reasoning in the law. A formal theory of reasoning by case comparison*. Doctoral dissertation University of Maastricht.

[Sartor, 2005] Sartor, G. (2005). *Legal reasoning. A cognitive approach to the law, 5. Treatise on legal philosophy and general jurisprudence*. Berlin: Springer.

[Schank, 1995] Schank, P. (1995). *Computational tools for modeling and aiding reasoning: Assessing and applying the Theory of Explanatory Coherence*. Doctoral dissertation University of California, Berkeley.

[Scheuer et al., 2010] Scheuer, O., Loll, F., Pinkwart, N., & McLaren, B. M. (2010). Computer-supported argumentation. A review of the state of the art. *Computer-Supported Collaborative Learning*, 5, 43–102.

[Schiappa, 2002] Schiappa, E. (2002). Evaluating argumentative discourse from a rhetorical perspective. Defining 'person' and 'human life' in constitutional disputes over abortion. In F. H. van Eemeren & P. Houtlosser (Eds.), *Dialectic and rhetoric. The warp and woof of argumentation analysis* (pp. 65–80). Dordrecht etc.: Kluwer.

[Schwemmer & Lorenzen, 1973] Schwemmer, O., & Lorenzen, P. (1973). *Konstruktive Logik, Ethik und Wissenschaftstheorie* (Constructive logic, ethics and theory of science). Mannheim: Bibliographisches Institut.

[Simari & Loui, 1992] Simari, G. R., & Loui, R. P. (1992). A mathematical treatment of defeasible reasoning and its applications. *Artificial Intelligence*, 53, 125–157.

[Snoeck Henkemans, 2001] Snoeck Henkemans, A. F. (2001). Argumentation structures. In F. H. van Eemeren (Ed.), *Crucial concepts in argumentation theory* (pp. 101–134). Amsterdam: Amsterdam University Press.

[Suthers, 1999] Suthers, D. (1999). Representational support for collaborative inquiry. In *Proceedings of the 32nd Hawaii international conference on the system sciences (HICSS-32)*. Institute of Electrical and Electronics Engineers (IEEE).

[Suthers et al., 1995] Suthers, D., Weiner, A., Connelly, J., & Paolucci, M. (1995). Belvedere. Engaging students in critical discussion of science and public policy issues. In *Proceedings of the 7th world conference on artificial intelligence in education (AIED âĂŸ95)*, 266–273. Washington.

[Swearngen & Schiappa, 2009] Swearngen, C. J., & Schiappa, E. (2009). Historical studies in rhetoric. Revisionist methods and new directions. In A. A. Lunsford, K. H. Wilson & R. A. Eberly (Eds.), *The Sage handbook of rhetorical studies* (pp. 1–12). Los Angeles, CA: Sage.

[Sycara, 1989] Sycara, K. (1989). Argumentation. Planning other agents' plans. In *Proceedings of the eleventh international joint conference on artificial intelligence* (pp. 517–523).

[Talbott, 2011] Talbott, W. (2011). Bayesian epistemology. In E. N. Zalta (Ed.), *The Stanford encyclopedia of philosophy*. Summer 2011 ed. https://plato.stanford.edu/entries/epistemology-bayesian/.

[Taroni et al., 2006] Taroni, F., Aitken, C., Garbolino, P., & Biedermann, A. (2006). *Bayesian networks and probabilistic inference in forensic science*. Chichester: Wiley.

[Teufel, 1999] Teufel, S. (1999). *Argumentative zoning. Information extraction from scientific articles*. Doctoral dissertation University of Edinburgh.

[Thagard, 1992] Thagard, P. (1992). *Conceptual revolutions*. Princeton, NJ: Princeton University Press.

[Timmer et al., 2017] Timmer, S., Meyer, J.J., Prakken, H., Renooij, S., & Verheij, B. (2017). A two-phase method for extracting explanatory arguments from Bayesian Networks. *International Journal of Approximate Reasoning*, 80, 475–494.

[Tindale, 1999] Tindale, Chr. W. (1996). From syllogisms to audiences. The prospects for logic in a rhetorical model of argumentation. In D. M. Gabbay & H. J. Ohlbach, *Practical reasoning. Proceedings of FAPR 1996* (pp. 596–605). Berlin: Springer.

[Tindale, 2004] Tindale, Chr. W. (1999), *Acts of arguing. A rhetorical model of argument*. Albany, NY: State University of New York Press.

[Toulmin, 2003] Toulmin, S. E. (1958). *The uses of argument*. Cambridge, England: Cambridge University Press. (updated ed. 2003).

[Verheij, 1996a] Verheij, B. (1996a). *Rules, reasons, arguments. Formal studies of argumentation and defeat*. Doctoral dissertation University of Maastricht.

[Verheij, 1996b] Verheij, B. (1996b). Two approaches to dialectical argumentation. Admissible sets and argumentation stages. In J.-J. Ch. Meyer & L. C. van der Gaag (Eds.), *NAIC'96. Proceedings of the eighth Dutch conference on artificial intelligence* (pp. 357–368). Utrecht: Utrecht University.

[Verheij, 2003a] Verheij, B. (2003a). DefLog. On the logical interpretation of prima facie justified assumptions. *Journal of Logic and Computation*, 13(3), 319–346.

[Verheij, 2003b] Verheij, B. (2003b). Dialectical argumentation with argumentation schemes. An approach to legal logic. *Artificial Intelligence and Law*, 11(1–2), 167–195.

[Verheij, 2005a] Verheij, B. (2005a). Evaluating arguments based on Toulmin's scheme. *Argumentation*, 19, 347–371. (Reprinted in D. L. Hitchcock & B. Verheij (Eds.) (2006), *Arguing on the Toulmin model. New essays in argument analysis and evaluation* (pp. 181–202). Dordrecht: Springer).

[Verheij, 2005b] Verheij, B. (2005b). *Virtual arguments. On the design of argument assistants for lawyers and other arguers*. The Hague: T. M. C. Asser Press.

[Verheij, 2007] Verheij, B. (2007). A labeling approach to the computation of credulous acceptance in argumentation. In M. M. Veloso (Ed.), *IJCAI 2007, Proceedings of the 20th international joint conference on artificial intelligence* (pp. 623–628). Hyderabad, India.

[Verheij, 2012] Verheij, B. (2012). Jumping to conclusions. A logico-probabilistic foundation for defeasible rule-based arguments. In L. Fariñas del Cerro, A. Herzig & J. Mengin (Eds.), *Logics in artificial intelligence. 13th European conference, JELIA 2012. Toulouse, France, September 2012. Proceedings (LNAI 7519)* (pp. 411–423). Springer, Berlin.

[Verheij, 2017] 205 Verheij, B. (2017). Proof with and without probabilities. Correct evidential reasoning with presumptive arguments, coherent hypotheses and degrees of uncertainty. *Artificial Intelligence and Law*, 25 (1), 127–154.

[Verheij et al., 2016] Verheij, B., Bex, F.J., Timmer, S., Vlek, C., Meyer, J.J., Renooij, S., & Prakken, H. (2016). Arguments, scenarios and probabilities: connections between three normative frameworks for evidential reasoning. *Law, Probability & Risk*, 15, 35–70.

[Verheij et al., 1998] Verheij, B., Hage, J. C., & Herik, H. J. van den (1998). An integrated view on rules and principles. *Artificial Intelligence and Law*, 6(1), 3–26.

[Vlek et al., 2014] Vlek, C., Prakken, H., Renooij, S., & Verheij, B. (2014). Building Bayesian Networks for legal evidence with narratives: a case study evaluation. *Artificial Intelligence and Law*, 22 (4), 375–421.

[Vlek et al., 2016] Vlek, C., Prakken, H., Renooij, S., & Verheij, B. (2016). A method for explaining Bayesian Networks for legal evidence with scenarios. *Artificial Intelligence and Law*, 24 (3), 285–324.

[Vreeswijk, 1993] Vreeswijk, G. A. W. (1993). *Studies in defeasible argumentation*. Doctoral dissertation Free University Amsterdam.

[Vreeswijk, 1995a] Vreeswijk, G. A. W. (1995a). Formalizing nomic. Working on a theory of communication with modifiable rules of procedure, Technical Report CS 95–02, Vakgroep Informatica (FdAW), Rijksuniversiteit Limburg, Maastricht.
[Vreeswijk, 1995b] Vreeswijk, G. A. W. (1995b). The computational value of debate in defeasible reasoning. *Argumentation*, 9, 305–342.
[Vreeswijk, 1997] Vreeswijk, G. A. W. (1997). Abstract argumentation systems. *Artificial Intelligence*, 90, 225–279.
[Vreeswijk, 2000] Vreeswijk, G. A. W. (2000). Representation of formal dispute with a standing order. *Artificial Intelligence and Law*, 8, 205–231.
[Walton, 1987] Walton, D. N. (1987). *Informal fallacies. Towards a theory of argument criticisms*. Amsterdam: John Benjamins.
[Walton, 1989] Walton, D. N. (1989). *Informal logic. A handbook for critical argumentation*. Cambridge: Cambridge University Press.
[Walton, 1996] Walton, D. N. (1996). *Argumentation schemes for presumptive reasoning*. Mahwah, NJ: Lawrence Erlbaum.
[Walton & Krabbe, 1995] Walton, D. N., & Krabbe, E. C. W. (1995). *Commitment in dialogue. Basic concepts of interpersonal reasoning*. Albany, NY: State University of New York Press.
[Walton et al., 2008] Walton, D. N., Reed, Chr. A., & Macagno, F. (2008). *Argumentation schemes*. Cambridge: Cambridge University Press.
[Wenzel, 1987] Wenzel, J. W. (1987). The rhetorical perspective on argument. In F. H. van Eemeren, R. Grootendorst, J. A. Blair & Ch. A. Willard (Eds.), *Argumentation. Across the lines of discipline. Proceedings of the conference on argumentation 1986* (pp. 101–109). Dordrecht/Providence: Foris.
[Wigmore, 1931] Wigmore, J. H. (1931). *The principles of judicial proof*. 2nd ed. Boston: Little Brown & Company. (1st ed. 1913).
[Woods & Walton, 1989] Woods, J., & Walton, D. N. (1989). *Fallacies. Selected papers 1972–1982*. Berlin/Dordrecht/Providence: de Gruyter/Foris.
[Wooldridge, 2009] Wooldridge, M. (2009). *An introduction to multiagent systems*. Chichester: Wiley.
[Zarefsky, 2006] Zarefsky, D. (2006). Strategic maneuvering through persuasive definitions. Implications for dialectic and rhetoric. *Argumentation*, 20(4), 399–416.
[Zarefsky, 2009] Zarefsky, D. (2009). Strategic maneuvering in political argumentation. In F. H. van Eemeren (Ed., 2009), *Examining argumentation in context. Fifteen studies on strategic maneuvering* (pp. 115–130). Amsterdam: John Benjamins.
[Zukerman et al., 1998] Zukerman, I., McConachy, R., & Korb, K. (1998). Bayesian reasoning in an abductive mechanism for argument generation and analysis. In *Proceedings of the fifteenth national conference on artificial intelligence (AAAI-98, Madison)* (pp. 833–838). Menlo Park: AAAI Press.

Frans H. van Eemeren
University of Amsterdam
Email: f.h.vaneemeren@uva.nl

Bart Verheij
University of Groningen
Email: bart.verheij@rug.nl

2
Historical Overview of Formal Argumentation

HENRY PRAKKEN

ABSTRACT. This chapter gives an overview of the history of formal argumentation in terms of a distinction between argumentation-based *inference* and argumentation-based *dialogue*. Systems for argumentation-based *inference* are about which conclusions can be drawn from a given body of possibly incomplete, inconsistent of uncertain information. They ultimately define a nonmonotonic notion of logical consequence, in terms of the intermediate notions of argument construction, argument attack and argument evaluation, where arguments are seen as constellations of premises, conclusions and inferences. Systems for argumentation-based *dialogue* model argumentation as a kind of verbal interaction aimed at resolving conflicts of opinion. They define argumentation protocols, that is, the rules of the argumentation game, and address matters of strategy, that is, how to play the game well. For both aspects of argumentation the main formal and computational models are reviewed and their main historical influences are sketched. Then some main applications areas are briefly discussed.

1 Introduction

This chapter gives an overview of the history of formal argumentation. There are two ways to write such an overview. One is to describe all significant research that has been done, while another is to give insight into the historical developments underlying the current state of the art. In this chapter I will do the latter. This will inevitably lead to a stronger focus on the early developments and a less detailed description of later research. Those who want more detail about the later research can consult the other chapters of this handbook.

The historical overview is given in terms of a distinction between argumentation-based *inference* and argumentation-based *dialogue*. Systems for argumentation-based *inference* are about which conclusions can be drawn from a given body of possibly incomplete, inconsistent of uncertain information. They ultimately define a nonmonotonic notion of logical consequence, in terms of the intermediate notions of argument construction, argument attack and argument evaluation, where arguments are seen as constellations of premises, conclusions and inferences. Systems for argumentation-based *dialogue* model argumentation as a kind of verbal interaction aimed at resolving conflicts of opinion. They define argumentation protocols (the rules of the argumentation game) and address matters of strategy (how to play the game well). While accounts of argumentation as inference assume a single static and global body of information from which the arguments and attacks are constructed, in studies of argumentation as dialogue this information is dynamic (it can change during a dialogue) and distributed over the dialogue's participants. Models of argumentation as inference can be embedded in models of argumentation as dialogue in two complementary ways: at each stage of a dialogue they can be 'globally' applied to the 'current' body of information; and within each dialogue participant they can be 'locally' applied as the participant's internal reasoning model.

Like all informal distinctions, the distinction between argumentation as inference and argumentation as dialogue breaks down at some point, and therefore I will also discuss work that cannot easily be classified as belonging to either inference or dialogue, especially work on argumentation dynamics that abstracts from agent-related and dialogical aspects. Another way in which a strict distinction between inference and dialogue causes problems for a historical overview is that some historical influences cannot clearly be described as influencing just models of inference or just models of dialogue. Some work has instead more generally promoted the idea of dialectics as constructing, criticising and comparing arguments, whether in an inferential or in a dialogical setting. One such historical influence was the development of dialogue logic [Lorenzen and Lorenz, 1978], which gives a game-theoretic formulation of the semantics of logical constants in terms of a dispute between a proponent and an opponent of a claim, plus a game-theoretic notion of logical consequence as the existence of a winning strategy for the proponent. This predates modern argument games for argumentation-based inference and also influenced the development of formal dialogue systems for argumentation. Having said so, in dialogue logic these ideas were only used to reformulate existing monotonic notions of logical consequence, so dialogue logic cannot be said to model genuine argumentation.

Another historical influence that is not confined to either inference or dialogue is early AI & Law work on the computational modelling of legal argument. Among the earliest work in AI and law on legal argument was the TAXMAN II project [McCarty, 1977; McCarty, 1995]). According to McCarty [1995], p. 285 "The task for a lawyer or a judge in a "hard case" is to construct a theory of the disputed rules that produces the desired legal result, and then to persuade the relevant audience that this theory is preferable to any theories offered by an opponent". Other influential early systems were the HYPO system [Rissland and Ashley, 1987; Ashley, 1990] and its successor the CATO system [Aleven and Ashley, 1991; Aleven, 2003]. These systems were meant to model how lawyers in common-law jurisdictions make use of past decisions when arguing a case. They did not compute an 'outcome' or 'winner' of a dispute; instead they were meant to generate debates as they could take place between 'good' common-law lawyers. Several researchers who later contributed to the general formal study of argumentation originate from AI & Law, such as Trevor Bench-Capon, Tom Gordon, Giovanni Sartor, Bart Verheij and myself.

The remainder of this chapter is divided into two main sections on, respectively, argumentation-based inference (Section 2) and dialogue (Section 3). Then some main applications areas are briefly discussed in Section 4 and some concluding remarks are made in Section 5.

2 Formal and computational models of argumentation-based inference

Nowadays, many systematic introductions to argumentation start with [Dung, 1995]'s theory of abstract argumentation frameworks, which takes the notions of argument and attack as primitive, i.e., nothing is assumed about about the structure of arguments or the nature of attack. Yet there had been quite some formal work on argumentation-based inference before Dung's landmark 1995 paper, and all this early work specified the structure of arguments and the nature of attack. The seminal paper in this respect was [Pollock, 1987]. Many ideas developed in this early body of work are still important today. The focus in this early work on structured argumentation agrees with the usual approaches in informal argumentation, which do not have arguments as the primitive notion but concepts like claims, reasons and grounds. For example, Walton [2006a], p. 285 defines the term 'argument' as "the giving of reasons to support or criticize a claim that is questionable, or open to doubt".

In this section first the three main historical sources of influence are sketched, namely, philosophy, nonmonotonic logic & logic programming, and informal logic & argumentation theory. Then the two seminal bodies of work are discussed in more more detail, John Pollock's argumentation-based system for defeasible reasoning and Phan Minh Dung's theory of abstract argumentation frameworks. Their works have inspired much research on, respectively, structured and abstract approaches to argumentation-based inference, which will subsequently be discussed.

2.1 Main historical influences

The formal and computational study of argumentation-based inference is generally regarded as a subfield of AI, originating from the study of nonmonotonic logic. However, there are two main other historical influences.

2.1.1 Philosophy

Arguably, the first mature formal system for argumentation-based inference was proposed by Pollock [1987][1]. John Pollock (1940-2009) was an influential American philosopher who made important contributions to various fields, including epistemology and cognitive science. In the last 25 years of his life he also contributed to artificial intelligence, starting with his classic 1987 paper on defeasible reasoning. Many important topics in the formal study of argumentation-based inference were first studied by Pollock, or first studied in detail, such as argument structure, the nature of defeasible reasons, the

---

[1]Several paragraphs in this subsection are, some with minor modifications, taken from Prakken and Horty [2012].

interplay between deductive and defeasible reasons, rebutting versus undercutting defeat, argument strength, argument labellings, self-defeat, and resource-bounded argumentation.

Pollock's work on formal argumentation was heavily influenced by the idea of defeasible reasons as developed in moral philosophy by Ross [1930] in his notion of *prima facie* moral rules, in epistemology by Chisholm [1957], Rescher [1977] and Pollock himself [1970, 1974], and as applied to practical reasoning by Raz [1975]. The term 'defeasibility' originates from legal philosophy, in particular from Hart [1949] (see the historical discussion in Loui [1995]). Hart observed that legal concepts are defeasible in that the conditions for when a fact situation classifies as an instance of a legal concept (such as 'contract'), are only ordinarily, or presumptively, sufficient. If a party in a law suit succeeds in proving these conditions, this does not have the effect that the case is settled; instead, legal procedure is such that the burden of proof shifts to the opponent, whose turn it then is to prove exceptional facts which, despite the facts proven by the proponent, nevertheless prevent the claim from being granted. For instance, insanity of one of the contracting parties is an exception to the legal rule that an offer and an acceptance constitute a binding contract. The notion of burden of proof was also studied by [Rescher, 1977], in the context of epistemology. Among other things, Rescher claimed that a dialectical model of scientific reasoning can explain the rational force of inductive arguments: they must be accepted if they cannot be successfully challenged in a properly conducted scientific dispute.

Pollock's work on formal argumentation originated as an attempt to make formal sense of the intuitive notion of defeasible reasoning that seemed to be at work in these papers and books. In fact, the task had been attempted before. There is an early paper by Chisholm [1974], a heroic effort whose failure is no surprise given the limited tools available at the time. Still, in spite of the blossoming of philosophical logic in the 1960's and 1970's, the logical study of defeasible reasoning had received almost no attention at all. It is fair to say that Pollock, working in isolation, was the first philosopher working in the field of philosophy, as opposed to computer science, to outline an adequate framework for defeasible reasoning.

2.1.2 Nonmonotonic logic and logic programming

The first AI systems for argumentation-based inference were not influenced by the above-discussed philosophical developments. Instead, they were presented as new ways to do nonmonotonic logic. Nonmonotonic logic had become fashionable around 1980 and a variety of approaches was being pursued. By the late 1980's, the field of nonmonotonic logic had been recognized as an important subfield of artificial intelligence. The field was motivated by the fact that commonsense reasoning often involves incomplete or inconsistent information, in which cases logical deduction is not a useful reasoning model. If information is incomplete, then nothing useful can be deductively derived, while if it is inconsistent, then anything is deductively implied. Nonmonotonic logics allow 'jumping to conclusions' in the absence of information to the contrary. The canonical example is 'birds typically fly, Tweety is bird, therefore (presumably) Tweety can fly'. This inference holds as long as no information is available that Tweety is not a typical bird with respect to flying, such as a penguin. Nonmonotonic logic can also model the derivation of useful conclusions from inconsistent information, namely, by focusing on consistent subsets of the inconsistent information. Several years after the first nonmonotonic logics were proposed in the now famous special issue on nonmonotonic logic of the *Artificial Intelligence* journal [Bobrow, 1980], the idea arose in this field that nonmonotonic inference can be modelled as the competition between arguments.

The earliest nonmonotonic reasoning systems with an argumentation flavour include the work of Touretzky [1984; 1986] on inheritance systems, later developed along with several collaborators [Horty et al., 1990]. Inheritance systems model reasoning about how objects inherit properties from the classes to which they belong. They are nonmonotonic since the inheritance of properties of classes by subclasses can be blocked by exceptions. For example, penguins do not inherit from birds the property of being able to fly. Although the work on inheritance systems did not use argumentation terms, such systems still have all the characteristics of argumentation systems. To start with, inheritance paths effectively are arguments. For example, the conclusion that Tweety the penguin can fly can be drawn via the path 'Penguins are birds and birds can fly' while the conclusion that Tweety the Penguin cannot fly can be drawn via the inheritance path 'Penguins cannot fly'. Inheritance systems also have various notions of conflict between inheritance plus definitions of whether a path is 'permitted' given its conflict relations with other paths. While the technical solutions devised in this work are now somewhat outdated, the work on inheritance paths has clearly influenced the development of the first AI argumentation systems. Among other things, the publications in inheritance are great sources of relevant examples.

An influential figure in the early days was Ron Loui. His [1987] paper was, although technically still preliminary, influential in promoting the idea of formulating nonmonotonic logic as argumentation. With Guillermo Simari he developed a technically mature version of his ideas [Simari and Loui, 1992]. Several other of his papers more generally promoted the idea of computational dialectics and were thus also relevant for dialogue models of argumentation. The fullest exposé of these ideas is [Loui, 1998], which circulated among researchers for several years until it was finally published in 1998.

Other relevant early work was the work of Nute [1988], later developed into so-called Defeasible Logic [Nute, 1994]. This approach is in spirit very close to argumentation but while in argumentation approaches conflict and defeat happen between arguments, in Defeasible Logic they happen between rules. For this reason the work on Defeasible Logic has diverged somewhat from the field of computational argument, although some work on the former has studied the formal relation with argumentation approaches. In particular, [Governatori *et al.*, 2004] studied to which extent defeasible logics can be reformulated in terms of Dung's theory of abstract argumentation frameworks.

Finally, the field of logic programming was influential since the idea arose to give semantics to negation as failure in argumentation-theoretic terms. If *not P* is assumed to hold because of the failure to derive *P*, then a derivation of *P* can be regarded as an attack on any derivation using *not P*. In other words, a logic-programming derivation can be regarded as a competition between arguments and counterarguments. Work on this idea of e.g. Geffner [1991] and Kakas *et al.* [1992] was a main source of inspiration of Dung's landmark [1995] paper on abstract argumentation frameworks.

2.1.3 Informal logic and informal argumentation theory

One would expect that the fields of informal logic and argumentation theory (which are often regarded as a single field) were also important historical influences on argumentation-based models of inference. However, in fact their influence has been relatively modest. In particular, the work of Toulmin [1958] and the resulting work on argumentation schemes was until around 2000 hardly linked to computational argument. An important event here was the 2000 Bonskeid Symposium on Argument and Computation in the Scottish mountains, organised by Tim Norman and Chris Reed, at which researchers from various formal and informal fields met in an informal setting. Various interdisciplinary collaborations resulted from this event, partly reported in [Reed and Norman, 2003].

Yet these fields originated from similar concerns about deductive logic as those that gave rise to the field of nonmonotonic logic in AI, namely, the inadequacy of deductive logic as a model of 'ordinary' reasoning. Stephen Toulmin, whose 1958 book *The Uses of Argument* is generally regarded as the origin of informal logic and argumentation theory, criticised the logicians of his days for neglecting many features of ordinary reasoning. In his well-known pictorial scheme for arguments (see Figure 1) he left room for "rebuttals" of an argument on the basis of exceptions to the "warrant" connecting the arguments "data" to its "claim". The idea of rebuttals is clearly related to Hart's [1949] ideas on exceptional circumstances that can defeat the application of a legal concept.

Toulmin's notion of a warrant was in informal logic and argumentation theory generalised into rich classifications of argument schemes for presumptive forms of reasoning, while his notion of a rebuttal was generalised into lists of critical questions attached to argument schemes [Walton, 1996]. The idea of argumentation schemes with critical questions has since the above-mentioned Bonskeid 2000 event often been used in formal and computational models of argumentation-based inference and dialogue.

Toulmin also argued that outside mathematics the validity of an argument does not depend on its syntactic form but on whether it can be defended in a properly conducted dispute, and that the task of logicians is to study the criteria for properly conducted disputes. This became an important and very

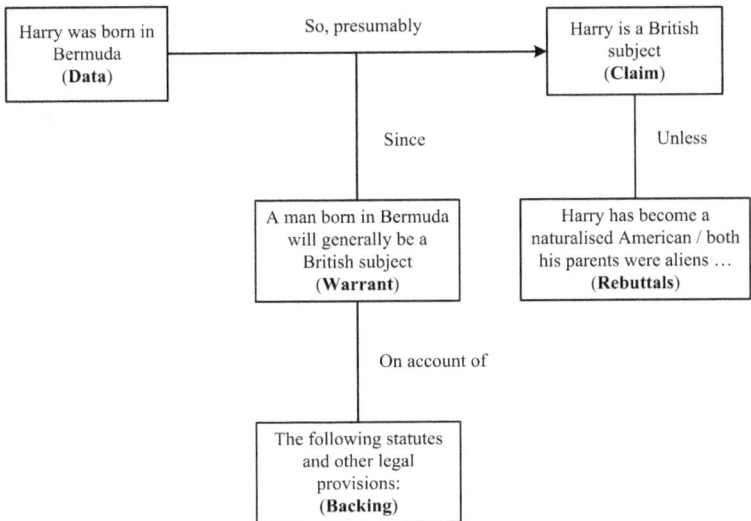

Figure 1. Toulmin argument scheme and an instance

influential idea, as further discussed below in Section 3 on argumentation-based dialogue. However, it also had an unfortunate effect. For decades, informal logic and argumentation theory rejected any use of formal methods in the study of ordinary reasoning, based on a mistaken equation of formal methods with deductive logic. As we now know after more than 35 years of research on nonmonotonic logic, belief revision and computational argument, many features of non-mathematical reasoning that Toulmin and his successors analysed can be formalised. For example, the AI work on argumentation schemes since 2000 has shown that reasoning with such schemes can to a large extent be formalised in modern argumentation logics.

2.2 Seminal work

I now discuss the two seminal contributions in the field, the ones of Pollock [1987] and Dung [1995]. These two papers successively introduced the two key ideas of the formal study of argumentation-based inference. Pollock introduced the notion of a defeasible reason, while Dung showed that argument evaluation can be formalised by assuming just two primitive notions of argument and attack. Neither of these ideas on their own define the field; it is their combination that makes the argumentation way of doing nonmonotonic logic so powerful.

2.2.1 Pollock's work

As said above, arguably, the first mature formal system for argumentation-based inference was proposed by Pollock [1987][2]. In fact, this work became

---

[2]Several parts of this subsection are reused or adapted from Prakken and Horty [2012].

close to being one of the first nonmonotonic logics at all. Concerning his 1987 paper, Pollock later wrote that he first developed the idea in 1979, but that he did not initially publish it because, as he says, "being ignorant of AI, I did not think anyone would be interested." [Pollock, 2007b, p. 469]. If Pollock had published this idea when it first occurred to him, the result would have been not only the first argument-based theory of defeasible reasoning, but one of the first systems of any kind for nonmonotonic reasoning.

I now discuss Pollock's system in some more detail, to illustrate that it introduced several fundamental ideas into our field. As usual in logic, arguments in Pollock's approach are inference graphs, in which a final conclusion is inferred from the premises via intermediate conclusions. Note that when an argument uses no premise more than once, the graph is a tree. What is unusual is Pollock's ideas on how conclusions can be supported by premises. The 'classic' logicians' view attacked by Toulmin [1958] had been that all arguments should be deductively valid, that is, the truth of their premises should guarantee the truth of their conclusion, and that the only source of fallibility of good arguments is their premises. Influenced by Toulmin, the fields of informal logic and argumentation theory had already questioned this view and argued that arguments that fail to meet this standard of inferential perfection can still be good, as long as they withstand critical scrutiny. Pollock [1987] gave us the tools to formalise this new account, with his notion of a defeasible reason.

In Pollock's approach, the inference rules (in his terminology "reasons") used to construct arguments come in two kinds: *deductive* and *defeasible* reasons (in his early work called "conclusive' and "prima facie" reasons). An argument can be defeated on its applications of defeasible reasons, which can happen in two ways. *Rebutting* defeaters attack the conclusion of a defeasible inference by supporting a conflicting conclusion. For example, 'Tweety can fly since it is a bird and birds typically fly' can be attacked by 'Tweety cannot fly since Tweety is a penguin and penguins cannot fly'. *Undercutting* defeaters instead attack the defeasible inference itself, without supporting a conflicting conclusion. For example: if the object looks red, this is a reason for concluding, defeasibly, that the object is red; but the presence of red illumination interrupts the reason relation without suggesting any conflicting conclusion. Pollock formalized several defeasible reasons that he found important in human cognition, such as reasons for perception, memory, induction, the statistical syllogism and temporal persistence, as well as undercutting defeaters for these reasons.

Pollock's notion of a defeasible reason is clearly related to argumentation theory's notion of an argumentation scheme: such schemes are defeasible reasons while many of their critical questions can be regarded as pointers to undercutting defeaters and other questions as pointers to rebutting defeaters or premise attacks.

Consider by way of example of Pollock's notions of reason, argument and conflict the following version of the Tweety example. Figure 2 contains two rebutting arguments for the conclusions that Tweety flies, respectively, does not fly, and an undercutting argument defeating the argument that Tweety flies. In this figure, deductive, respectively defeasible inferences are visualized with,

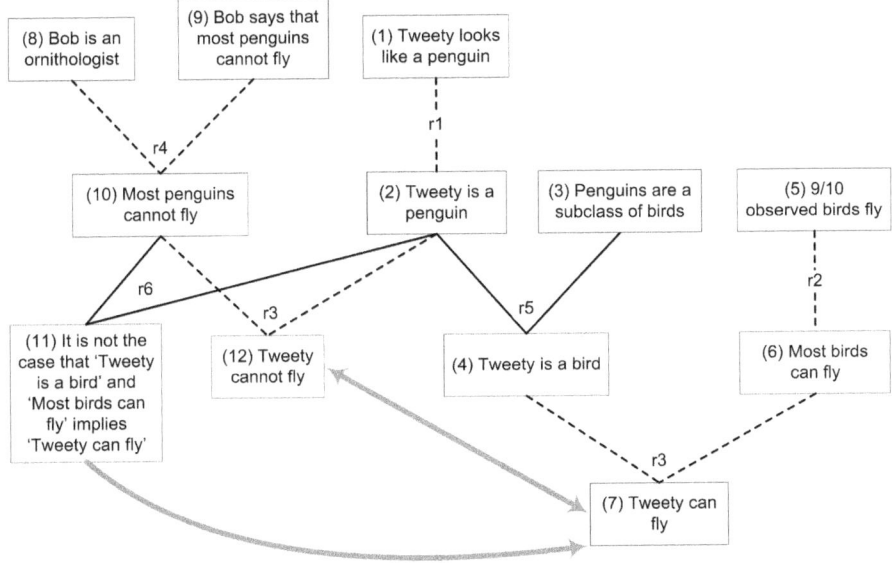

Figure 2. An example

respectively, solid and dotted lines without arrow heads, while defeat relations are displayed with arrows. The figure assumes four defeasible inference rules, informally paraphrased as follows:

r_1: That an object looks like having property P is a defeasible reason for believing that the object has property P

r_2: That n/m observed P's are Q's (where $n/m > 0,5$) is a defeasible reason for believing that most P's are Q's

r_3: That most P's are Q's and x is a P is a defeasible reason for believing that x is a Q

r_4: That an ornithologist says φ about birds is a defeasible reason for believing φ

Rule r_1 expresses that perceptions yield a defeasible reason for believing that what is perceived to be the case is indeed the case, rule r_2 captures enumerative induction, while r_3 expresses the statistical syllogism. Rule r_4 can be seen as a special case of the argumentation scheme from expert testimony; cf. [Walton, 1996].

Moreover, the figure assumes an obvious strict inference rule plus an undercutting defeater for r_3:

r_5: That P's are a subclass of Q's and a is a P is a deductive reason for believing that a is a Q

r_6: That x is an R, most R's are not Q's and R's are a subclass of P's is a deductive reason for believing $\neg r_3$

Rule r_6 is a special case of Pollock's "subproperty defeater" of the statistical syllogism, which says that conflicting statistical information about a subclass undercuts the statistical syllogism for the superclass.

Defeasible reasons should not be confused with nonmonotonic consequence notions. It is possible to design argumentation logics with nonmonotonic consequence notions in which nevertheless all arguments have to be deductively valid. For example, in classical argumentation arguments are classical implication relations from consistent subsets of a possibly inconsistent body of information and the only source of fallibility of arguments is their premises. Recent portrayals of Pollock's approach as 'deductive' [Hunter and Woltran, 2013] do no justice to his approach, given that Pollock strongly emphasised that "It is logically impossible to reason successfully about the world around us using only deductive reasoning. All interesting reasoning outside mathematics involves defeasible steps." [Pollock, 1995, p.41]. Pollock thus clearly rejected the conventional view that all arguments have to be deductively valid.

Defeasible reasons should also not be confused with deductive inference rules with assumption-type premises. Thinking otherwise would have the odd consequence that even the classically valid rules of inference become defeasible when applied to assumptions.

Once arguments can employ defeasible reasons, the support relation between their premises and conclusion can have varying strength. Pollock's 1987 system did not yet include a notion of strength but Pollock later took the notion of strength of arguments very seriously. Since his systems were meant for epistemic reasoning, he always formulated strength of reasons in terms of numerical degrees of belief. In his 1994 system, rebutting and undercutting arguments only succeed in defeating their target if the degree of belief of their conclusions is not lower than that of the attacked argument.

Finally, Pollock was well aware that just defining notions of argument and defeat are not enough and he spent much effort in designing well-behaved notions of argument acceptability. His two earliest definitions predate much current work on argumentation-based semantics. His 1987 proposal was by Dung [1995] proven to be an instance of Dung's grounded semantics, while his 1994 labelling definition predates the currently popular labeling approach to abstract argumentation and was by Jakobovits [2000] proven to be an instance of Dung's preferred semantics.

2.2.2 Dung's abstract argumentation frameworks

Dung's landmark 1995 paper is the origin of the second main idea of our field, namely, that argument evaluation can be formalised by assuming just two primitive notions of argument and attack. With just these two notions, Dung was able to develop an extremely rich and elegant abstract theory of argument evaluation. As apparent from this historic overview, Dung was not the first to study argument evaluation nor the first to provide well-behaved definitions. His great contribution was twofold: he showed that particular definitions of argument evaluation conformed to simple abstract patterns, and he showed that the same patterns are also implicit in other nonmonotonic logics, in logic programming and even in cooperative game theory. Exaggerating a little, one could say that while Pollock arguably was the father of argumentation in AI, Dung was the midwife, who smoothened its delivery into mainstream AI. His 1995 AI Journal paper was not the first work on argumentation-based inference, but its influence has been enormous, now being the de facto standard in the field. It is fair to say that Dung [1995] has made argumentation respectable in mainstream AI.

Nevertheless, the historic roots of Dung's 1995 paper should not be forgotten. As mentioned in the introduction to Section 2, all early work on argumentation-based inference specified the structure of arguments and the nature of attack (often called 'defeat'). Even Dung in his landmark 1995 paper stood in this tradition. Dung did two things: he developed the new idea of abstract argumentation frameworks, and he used this idea to reconstruct and compare a number of then mainstream nonmonotonic logics and logic-programming formalisms, namely, default logic [Reiter, 1980], Pollock's [1987] argumentation system and several logic-programming semantics. However, these days the second part of his paper, and also the third part on relations with cooperative game theory, is largely forgotten and his paper is almost exclusively cited for its general theory of abstract argumentation frameworks.

A historic overview of work on argumentation-based inference would not be complete without listing Dung's simple and elegant basic notions. An *abstract argumentation framework* (AF) is a pair $\langle AR, attacks \rangle$, where AR is a set arguments and $attacks \subseteq AR \times AR$ is a binary relation. The theory of AFs then addresses how sets of arguments (called *extensions*) can be identified which are internally coherent and defend themselves against attack. A key notion here is that of an argument being *acceptable with respect to* a set of arguments: $A \in AR$ is acceptable with respect to $S \subseteq AR$ if for all $A \in S$: if $B \in AR$ attacks A, then some $C \in S$ attacks B (nowadays it is more usual to say that $A \in AR$ is defended by $S \subseteq AR$). Then relative to a given AF various types of extensions can be defined as follows (here E is *conflict-free* if no argument in E attacks an argument in E):

- E is *admissible* if E is conflict-free and each argument in E is acceptable with respect to E;

- E is a *complete extension* if E is admissible and each argument that is acceptable with respect to E belongs to E;

- E is a *preferred extension* if E is a maximal (with respect to set inclusion) admissible set;

- E is a *stable extension* if E is conflict-free and attacks all arguments outside it;

- E is a *grounded extension* if E is the least fixpoint of operator F, where $F(S)$ returns all arguments acceptable to S.

Dung showed that the grounded extension is always unique but that there can be multiple extensions of the other types. Dung also showed that every stable extension is preferred but not vice versa, that the grounded extension is contained in every other extension, and that all extensions of any type are complete.

To illustrate how abstract argumentation frameworks can be instantiated, consider again Figure 2. There are three arguments. In fact, there are more arguments, since each of the three arguments we consider has several subarguments. However, none of these is attacked, so they can be ignored for simplicity. The two rebutting arguments for the conclusions that Tweety can fly, respectively, cannot fly attack each other, while the undercutting argument attacks the argument that Tweety flies. The resulting argumentation framework is shown in Figure 3. In this case the four semantics coincide: the set with the undercutting argument and the argument that Tweety cannot fly is the grounded extension, while it is also the unique complete, stable and preferred extension (the grey colourings indicate extension membership). To see why it is preferred, observe that the undercutting argument defends the argument that Tweety cannot fly against its rebutting attacker that Tweety can fly.

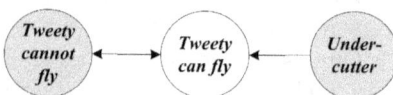

Figure 3. An abstract argumentation framework

To illustrate that argumentation frameworks can have multiple extensions, consider the simpler example in Figure 4 where the undercutting argument has been deleted from the AF of Figure 3. In grounded semantics the extension is empty (case a) but in preferred and stable semantics there are two extensions, depending on whether the argument that Tweety can (case b) or cannot fly (case c) is accepted. Finally, all three extensions are complete.

These examples point at a minor source of terminological confusion, since they use Dung's term 'attack' while Pollock always used 'defeat'. When Dung's 1995 paper appeared, 'defeat' was the standard term, not just in Pollock's work but essentially in all early work on argumentation-based inference. Current work on the $ASPIC^+$ framework [Prakken, 2010; Modgil and Prakken, 2013; Modgil and Prakken, 2014] also uses 'defeat' and reserves the term 'attack' for more basic, purely syntactical forms of conflicts between arguments. Defeat is

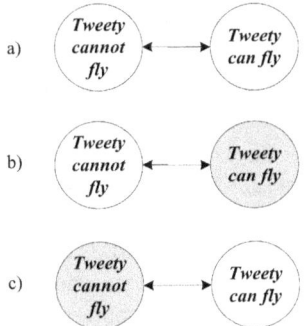

Figure 4. A simpler abstract argumentation framework and three extensions

then successful attack according to some notion of argument strength or preference, an idea present in much early work on argumentation-based inference, although usually not employing the term 'attack'. Thus it is not $ASPIC^+$'s attack relation but its defeat relation which instantiates Dung's notion of attack.

2.3 Other early work

Initial ideas In the same year in which Pollock published his seminal paper, Loui [1987] appeared as arguably the first AI paper that explicitly proposed to design nonmonotonic logics in the argumentation way. In 1992, Simari and Loui fully formalized Loui's [1987]'s initial ideas, which work in turn led to the development of Defeasible Logic Programming [Garcia et al., 1998; Garcia and Simari, 2004]. One year later, Konolige [1988] proposed an argumentation approach as a solution to the famous Yale Shooting problem in logic-based specifications of dynamic systems [Hanks and McDermott, 1986]. Although his formalism was still rather rudimentary, Konolige's discussion anticipates many issues and distinctions of later work, so that his paper can be regarded as one of the forerunners of the study of argumentation-based inference.

Argumentation as a proof theory for preferential entailment Around 1990, some papers proposed argumentation as a proof theory for model-theoretic notions of nonmonotonic consequence (preferential entailment). Baker and Ginsberg [1989] did this for a minimal-model semantics of prioritised circumscription, while Geffner [1992] and Geffner and Pearl [1992] did the same for their 'conditional entailment' semantics for default reasoning. The basic idea is twofold. First, given a propositional or first-order theory, an argument is a set or conjunction of assumptions consistent with the theory and that combined with the theory yields conclusions; and second, arguments can be attacked by arguments for the negation of the attacked argument or one of its assumptions. This idea later became the basis for assumption-based argumentation [Bondarenko et al., 1997], to be discussed in Section 2.4. Although the idea to found argumentation-based inference on preferential entailment is very interesting, it has since then not been further pursued.

Abstract argumentation systems Lin and Shoham [1989] were the first to propose the idea of abstraction in structured argumentation. They developed the notion of abstract argumentation structures with strict and defeasible rules and they showed how a number of existing nonmonotonic logics could be reconstructed as such structures. Gerard Vreeswijk further developed these ideas into his abstract argumentation systems [Vreeswijk, 1991; Vreeswijk, 1993b; Vreeswijk, 1997]. Since several of Vreeswijk's ideas are included in today's *ASPIC$^+$* framework, it is worthwhile summarising some of his definitions. Like Lin & Shoham, Vreeswijk defined arguments in terms of an unspecified logical language $\mathcal{L}$, only assumed to contain the symbol $\bot$, denoting 'falsum' or 'contradiction,' and two unspecified sets of strict ($\rightarrow$) and defeasible ($\Rightarrow$) inference rules defined over $\mathcal{L}$. In addition, he defined the main elements that are missing in Lin & Shoham's system, namely, notions of conflict and defeat between arguments. Vreeswijk defined arguments as follows:

Definition 2.1 *An argument σ is:*

1. *φ if $\varphi \in \mathcal{L}$; in that case:* $\text{Prem}(\sigma) = \{\varphi\}$, $\text{Conc}(\sigma) = \varphi$, $\text{Sent}(\sigma) = \{\varphi\}$;

2. *$\sigma_1, \ldots \sigma_n \rightarrow \varphi$ where $\sigma_1, \ldots, \sigma_n$ is a finite, possibly empty sequence of arguments such that* $\text{Conc}(\sigma_1) = \varphi_1, \ldots, \text{Conc}(\sigma_n) = \varphi_n$ *for some strict rule* $\varphi_1, \ldots, \varphi_n \rightarrow \varphi$, *and* $\varphi \notin \text{Sent}(\sigma_1) \cup \ldots \cup \text{Sent}(\sigma_n)$; *in this case:* $\text{Prem}(\sigma) = \text{Prem}(\sigma_1) \cup \ldots \cup \text{Prem}(\sigma_n)$, $\text{Conc}(\sigma) = \psi$, $\text{Sent}(\sigma) = \text{Sent}(\sigma_1) \cup \ldots \cup \text{Sent}(\sigma_n) \cup \{\varphi\}$;

3. *$\sigma_1, \ldots \sigma_n \Rightarrow \varphi$ where $\sigma_1, \ldots, \sigma_n$ is a finite, possibly empty sequence of arguments such that* $\text{Conc}(\sigma_1) = \varphi_1, \ldots, \text{Conc}(\sigma_n) = \varphi_n$ *for some defeasible rule* $\varphi_1, \ldots, \varphi_n \Rightarrow \varphi$, *and* $\varphi \notin \text{Sent}(\sigma_1) \cup \ldots \cup \text{Sent}(\sigma_n)$; *with the further attributes defined as in (2).*

Note that this definition, unlike most other definitions of arguments in the formal literature, excludes circular arguments.

Vreeswijk's notion of conflicts between arguments is unusual in that a counterargument is a *set* of arguments: a set Σ of arguments is *incompatible* with an argument τ iff the conclusions of $\Sigma \cup \{\tau\}$ give rise to a strict argument for $\bot$. While unusual, there is nothing obviously wrong with this kind of definition. The reason why currently conflict is usually defined as a relation between individual arguments is probably that such definitions better fit with Dung's theory of abstract argumentation frameworks. Vreeswijk's approach might fit better with generalisations of Dung's theory that allow attacks from sets of arguments to arguments [Bochman, 2003; Nielsen and Parsons, 2007b]. Recently, Baroni *et al.* [2015] have combined the *ASPIC$^+$* framework with a Vreeswijk-style definition of conflict.

Conflicts can in Vreeswijk's approach be resolved with any reflexive and transitive ordering on arguments that the user likes to adopt. A set of arguments

Σ is *undermined* by an argument τ if $\sigma < \tau$ for some $\sigma \in \Sigma$. Then a set of arguments Σ is a *defeater* of σ if Σ is incompatible with σ and not undermined by it.

Finally, Vreeswijk defined argument acceptability ("warrant") with a definition that is close but not equivalent to Dung's [1995] stable semantics. In light of the modern theory of abstract argumentation frameworks, Vreeswijk's definition of warrant is, unlike the rest of his approach, somewhat premature. This is understandable, since Vreeswijk developed his approach before 1995.

Logic-programming approaches The work on argumentation semantics for logic-programming's negation as failure did not only inspire Dung to develop his theory of abstract argumentation frameworks but also gave rise to logic-programming systems for argumentation with explicit negation. Two early papers here were Dung [1993] and Dimopoulos and Kakas [1995]. The first of these papers was in turn a source of inspiration for Prakken and Sartor's [Prakken and Sartor, 1997] argument-based logic programming system with defeasible priorities. Theirs was arguably the first system that was explicitly designed as an instance of Dung's [1995] approach. Strictly speaking, it was technically based not on Dung [1995] but on Dung [1993], but a reformulation in terms of abstract argumentation is trivial. Like all other work reviewed so far, it distinguished between strict and defeasible inference rules. Unlike Dimopoulos and Kakas [1995] but like Dung [1995], its language had both explicit negation and negation as failure, with corresponding "rebutting" attacks on defeasibly derived conclusions and "undercutting" attacks on negation-as-failure premises. One innovative feature was that it allowed argumentation about preferences inside the argumentation system, while another innovative feature was that the system had the first published argument game meant as a proof theory for the semantics of abstract argumentation frameworks (for more on argument games see Section 2.5.2 below).

Defeasible vs. plausible reasoning As apparent from the overview so far, until 1993 almost all accounts of argumentation-based inference made a distinction between deductive (or 'strict') and defeasible inference rules, introduced in philosophy by Pollock [1970; 1974] and in AI by Pollock [1987] and Touretzky [1984]. This approach is still being pursued today, notably in Defeasible Logic Programming, Defeasible Logic and the *ASPIC*$^+$ framework. In this approach a special definition of arguments is needed that regulates the interplay between strict and defeasible reasons (such as the above one of Vreeswijk [1993b; 1997]), since with two kinds of inference rules one cannot rely on a single given logical consequence notion to specify how conclusions are supported by premises. Around 1993 an alternative approach to structured argumentation emerged, according to which arguments are constructed in a single given deductive logic, obviating the need of a separate definition of an argument beyond being a premises-conclusion pair. In understanding and relating the two approaches, the philosophical distinction between *plausible* and *defeasible* reasoning is relevant; cf. Rescher [1976; 1977] and Vreeswijk [1993b], Ch. 8. Following Rescher,

Vreeswijk described plausible reasoning as sound (i.e., deductive) reasoning on an uncertain basis and defeasible reasoning as unsound (but still rational) reasoning on a solid basis. In other words, argumentation models of plausible reasoning locate all fallibility of an argument in its premises, while argumentation models of defeasible reasoning locate all fallibility in its defeasible inferences. Thus plausible-reasoning approaches effectively view argumentation as a kind of inconsistency handling, since in these approaches conflicts between arguments can only arise if the knowledge base is inconsistent. By contrast, in defeasible-reasoning approaches conflicts can arise from consistent knowledge bases, since in those approaches it is the application of defeasible rules that makes an argument fallible.

Two groups in particular initiated the plausible-reasoning approach to argumentation, respectively at Queen Mary's University in London and at INRIA in Toulouse. Elvang-Göransson et al. [1993] conceived of arguments as premise-conclusion pairs (δ, p) where δ is a subset of a possibly inconsistent database Δ and there exists a natural-deduction proof of p from δ. Arguments can be attacked in two ways: an argument (δ', q) *rebuts* (δ, p) if q is logically equivalent to $\neg p$ and it *undercuts* it if q is logically equivalent to $\neg r$ for some $r \in \delta$. Note that Elvang-Göransson et al. thus introduced a terminological confusion into the literature that exists until today. While they fully adopted Pollock's [1974; 1987]'s terminology, they only partly adopted its meaning, since Pollock used the term 'undercutter' not for premise attack but for attack on the application of a defeasible inference rule. Today, Pollock's meaning of the term 'undercutter' is adopted in the $ASPIC^+$ framework and Dung's recent work on structured argumentation frameworks, while Elvang-Göransson et al.'s meaning is fashionable in work on classical and Tarskian argumentation.

Elvang-Göransson et al. classified arguments into five classes of increasing degrees of acceptability: arguments, consistent arguments (i.e., arguments with consistent premises), non-rebutted consistent arguments, non-rebutted and non-undercut consistent arguments, and "tautological" arguments (i.e., arguments with an empty set of premises). In light of modern work this definition of argument acceptability seems somewhat ad-hoc. Among other things, it does not model the notions of defense and admissibility that are so beautifully modelled by Dung [1995]. The ideas of Elvang-Göransson et al. were further developed by Krause et al. [1995], replacing classical logic by intuitionistic logic as the underlying logic and adding notions of argument structure and argument strength.

Around the same time as Elvang-Göransson et al., Benferhat et al. [1993] proposed a similar system, containing what now is the standard definition of an argument in this approach, adding to Elvang-Göransson et al.'s definition the requirements that the set of premises is consistent and subset-minimal:

Definition 2.2 *Given a database* Σ, *a set* $\Sigma_i \subseteq \Sigma$ *is an argument for a formula* φ *iff:*

1. $\Sigma_i \not\vdash \bot$; and

2. $\Sigma_i \vdash \varphi$; and

3. for all $\psi \in \Sigma_i$: $\Sigma_i \setminus \{\psi\} \not\vdash \varphi$

Here, $\vdash$ denotes classical propositional consequence. Benferhat et al. did not define explicit notions of attack. Instead they defined φ to be an *argumentative consequence* of Σ if given Σ there exists an argument for φ but not for $\neg\varphi$. They also studied alternative consequence notions and their relations, and refined their system with a preference relation on the database. Their approach was related to abstract argumentation by Cayrol [1995], who among other things proved that with Elvang-Göransson et al.'s undercutting relation as the attack relation, the stable extensions given a database are in a one-to-one correspondence with the database's maximal consistent subsets. This result was later generalised by Amgoud and Besnard [2013] for any abstract Tarskian logic and by Modgil and Prakken [2013] in the context of the $ASPIC^+$ framework.

The ideas of Elvang-Göransson et al. and Benferhat et al. were picked up by e.g. Amgoud and Cayrol [1998] and Besnard and Hunter [2001] and evolved into classical, or classical-logic argumentation [Besnard and Hunter, 2008; Gorogiannis and Hunter, 2011, e.g.] and its generalisations to deductive [Besnard and Hunter, 2014] and abstract Tarskian argumentation [Amgoud and Besnard, 2013], to be further discussed below.

2.4 Structured argumentation: developments until now

While until 1995 work on structured argumentation had specific and sometimes ad-hoc definitions of argument evaluation, since 1995 most work on structured argumentation adopts Dung's approach or at least explicates the relation with it. Work that adopts Dung's approach does so by giving definitions of the structure of arguments and the nature of attack. Thus abstract argumentation frameworks are generated, so that arguments can be evaluated according to one of the abstract argumentation semantics and their acceptability status can be used to define nonmonotonic consequence notions for their statements. However, there is also work that deviates from Dung's approach. In this section I will give an overview of these research strands.

2.4.1 Argumentation models of plausible reasoning

Current argumentation models of plausible reasoning are essentially of two kinds.

Assumption-based argumentation Around the same time as argumentation was proposed as a way of inconsistency handling in classical logic, assumption-based argumentation (ABA) emerged from attempts to give an argumentation-theoretic semantics to logic-programming's negation as failure [Bondarenko et al., 1993; Bondarenko et al., 1997]. Like the classical-logic approaches, ABA also assumes a unique 'base logic', which in ABA is called a "deductive system", consisting of set of inference rules defined over some logical language. Given a set of so-called 'assumptions' formulated in the logical language, arguments are then deductions of claims using rules and supported by sets of assumptions. Contrary to in classical and abstract argumentation, the premises of ABA arguments, i.e., its assumptions, do not have to be consistent. ABA leaves both the logical language and set of inference rules unspecified in general, so it is like Vreeswijk's [1993b; 1997] approach and the later $ASPIC^+$ framework, an abstract framework for structured argumentation. However, unlike these approaches, ABA only allows attacks on an argument's assumptions, so that ABA's rules are effectively equivalent to Vreeswijk's and $ASPIC^+$'s strict inference rules (as formally confirmed in [Prakken, 2010]).

In order to express conflicts between arguments, ABA makes like Vreeswijk a minimum assumption on the logical language, which in ABA is that each assumption in the logical language has a *contrary*. That b is a contrary of a, written as $b = \overline{a}$, informally means that b contradicts a. An argument using an assumption a is then attacked by any argument for conclusion $\overline{a}$. Contrary relations do not have to be symmetric. This feature allows an argumentation-theoretic semantics for negation as failure (*not*) by for every formula *not p* letting $p = \overline{not\ p}$ but not vice versa. However, ABA's application is not limited to logic programming; in the landmark ABA paper [Bondarenko et al., 1997], it is instantiated with various nonmonotonic logics, including default logic, circumscription and Poole's [1989] Theorist system.

Although ABA and Dung's approach clearly have commonalities, ABA as originally formulated by Bondarenko *et al.* [1997] does not generate abstract argumentation frameworks. Instead, its extensions are (in some sense maximal) sets of assumptions, induced by transforming attack relations between arguments to attack relations between sets of assumptions. Only ten years later was ABA given an explicit Dungean formulation by Dung *et al.* [2007]. Currently, there is some controversy about whether the correspondence holds for all current abstract argumentation semantics or not; cf. Gabbay [2015] and Caminada [2015].

ABA was originally used theoretically as a framework for nonmonotonic logic. Over the years, the focus has shifted somewhat to developing algorithms and implementations and to applying these to a wide range of reasoning and decision problems. For more details the reader is referred to the other chapters in this handbook.

An interesting variant of assumption-based argumentation is Verheij's [2003] DefLog system. Verheij assumes a logical language with just two connectives, a unary connective × which informally stands for 'it is defeated that' and a binary connective $\rightsquigarrow$ for expressing defeasible conditionals. Verheij then assumes a single inference scheme for this language, namely, modus ponens for $\rightsquigarrow$. A set of sentences T is said to *support* a sentence φ if φ is in T or follows from T by repeated application of $\rightsquigarrow$-modus ponens. Moreover, T is said to *attack* φ if T supports $\times\varphi$. Verheij then considers partitions (J, D) of sets of sentences Δ such that J (the "justified" sentences) is conflict-free and attacks every sentence in D (the "defeated" sentences). As observed by Verheij, DefLog can be encoded as an ABA instance with stable semantics by setting ABA's assumptions to Δ, defining the ABA ABF contrary mapping as $\times \varphi = \overline{\varphi}$ for any φ and letting ABA's set of rules be generated by the modus scheme for $\leadsto$.

Classical, deductive and Tarskian argumentation The initial work of Elvang-Göransson *et al.* [1993] and Benferhat *et al.* [1993] led to a family of approaches usually called 'classical' or 'deductive' argumentation [Amgoud and Cayrol, 2002; Besnard and Hunter, 2001; Kaci *et al.*, 2007; Besnard and Hunter, 2008; Amgoud and Vesic, 2010; Kaci, 2010]. The first name refers to instances with as base logic classical propositional or first-order logic, while the term 'deductive argumentation' is used for approaches that abstract from particular base logics, as long as they are "deductive". Often the term 'deductive' is here used in an informal sense. For example, Besnard and Hunter [2014] describe a deductive inference as an inference that is "infallible in the sense that it does not introduce uncertainty". This agrees with Pollock's notion of a deductive reason. Recently Amgoud and Besnard [2010; 2013] gave a precise interpretation by assuming that the base logic satisfies the properties of a so-called Tarskian abstract logic.

In all these approaches arguments are, as in Benferhat *et al.* [1993] for the special case of classical propositional logic, premises-conclusion pairs such that the premises are, according to the base logic, consistent and subset-minimal sets logically implying their conclusion. Unlike in many other approaches, these approaches do not commit to specific definitions of argument attack but explore the consequences of various definitions, all exhibiting some form of premise- and/or conclusion attack. Given that these approaches locate all fallibility of arguments in their premises, one might expect that definitions that only allow premise attack are the best-behaved. This was formally confirmed by Gorogiannis and Hunter [2011] and Amgoud and Besnard [2013] who, for respectively classical and Tarskian argumentation, showed that when abstract argumentation frameworks are generated, only particular forms of premise attack fully guarantee the consistency of the conclusion sets of extensions of abstract argumentation frameworks.

Until these investigations, research in this strand was not much concerned with argument evaluation. Instead, other properties were studied, such as relations between kinds of attack, and the formalisms were used as a tool for investigating dialogue-related questions, such as enthymemes [Black and Hunter, 2012] and persuasive force of arguments [Hunter, 2004]. See for further details Besnard and Hunter [2008] and other chapters in this handbook.

2.4.2 Argumentation models of defeasible reasoning

Defeasible Logic Programming Defeasible Logic Programming, or DeLP [Garcia *et al.*, 1998; Garcia and Simari, 2004] is a further development of Simari and Loui's [1992] argumentation system with strict and defeasible rules. While Simari and Loui only allowed specificity as a source of preferences, DeLP allows any preference ordering. DeLP's logic-programming rules can contain both explicit negation and negation as failure. It is noteworthy that while the consequence notion of Simari and Loui's system is equivalent to Dung's [1995] grounded semantics, DeLP as described by Garcia et al. [1998] and Garcia and Simari [2004] does not conform to any of Dung's semantics. Instead, it is based on the notion of a dialectical tree, which essentially captures all ways in which a proponent and an opponent of a claim can have a debate about the claim by defeating each other's arguments. This notion is very similar to the notion of an argument game as a proof theory for the semantics of abstract argumentation frameworks (see further Section 2.5.2). However, while the constraints on argument games are based on the semantics for abstract argumentation frameworks, DeLP's constraints on dialectical trees are based on intuitions concerning concrete examples.

A unifying approach: the $ASPIC^+$ framework The $ASPIC^+$ framework [Prakken, 2010; Modgil and Prakken, 2013; Modgil and Prakken, 2014] unifies plausible and defeasible reasoning. Its main sources of inspiration are the systems of Pollock [1987; 1994; 1995] and Vreeswijk [1993b; 1997], which model defeasible reasoning. However, $ASPIC^+$ adds to these systems the possibility to attack an argument's premises, which makes it also suitable for modelling plausible reasoning. Apart from this, $ASPIC^+$ adopts Pollock's distinction between deductive (strict) and defeasible inference rules, Vreeswijk's definition of an argument and Pollock's notions of rebutting and undercutting attack, with the exception that in $ASPIC^+$, unlike in Pollock's systems, undercutting attack succeeds as defeat irrespective of preferences. Also, like Vreeswijk, $ASPIC^+$ abstracts from particular logical languages, sets of inference rules and argument orderings. Unlike Vreeswijk's particular method of argument evaluation, $ASPIC^+$ generates abstract argumentation frameworks, so that any semantics for such frameworks can be used to evaluate arguments.

A preliminary version of $ASPIC^+$ was developed during the EC-sponsored ASPIC project, which ran from 2004 to 2007. This version was used by Caminada and Amgoud [2007] as a vehicle for proposing the idea of rationality postulates for structured argumentation. The first publication focusing on $ASPIC^+$ as a framework for structured argumentation was Prakken [2010]. Modgil and Prakken [2013] proposed some small modifications and variations and proved further results on the framework and its relation with other work. Recently, several other variations of the $ASPIC^+$ framework have been studied, which are further described in this handbook's chapter on rule-based argumentation.

Its abstract nature makes that $ASPIC^+$ can be instantiated in many different ways and captures a number of other approaches as special cases. For example, Prakken [2010] proves that Dung et al.'s [2007]'s version of assumption-based argumentation can be reconstructed as a special case of $ASPIC^+$ with only strict inference rules, no unattackable premises and no preferences. And Modgil and Prakken [2013] reconstruct two forms of classical argumentation as studied by Gorogiannis and Hunter [2011] as the special case with only strict rules, being all valid classical inferences from finite sets, no unattackable premises, no preferences and the constraint that an argument's premises are classically consistent and subset-minimal. They then generalise this reconstruction with a preference relation on the knowledge base and prove that the resulting stable extensions are in a one-to-one correspondence with Brewka's [1989] preferred subtheories. Thus they also extend Cayrol's [1995] similar result without preferences for maximal consistent subsets.

Not only $ASPIC^+$ but also assumption-based argumentation is an abstract model of structured argumentation. Compared to ABA, $ASPIC^+$ is more complex, with its two kinds of inference rules, its three kinds of attack and its explicit preferences to distinguish between attack and defeat. As stated by Toni [2014], the philosophy behind ABA is instead to translate preferences and defeasible rules into ABA rules plus ABA assumptions, so that rebutting and undercutting attack and the application of preferences all reduce to premise attack. This approach has its merits but it is an open question whether $ASPIC^+$ can in its full generality be translated into ABA. Currently there are only partial answers to this question. Dung and Thang [2014] prove for the case without preferences that defeasible $ASPIC^+$ rules can be translated to ABA rules with assumption premises. Moreover, in an early paper, Kowalski and Toni [1996] give a partial method for encoding rule preferences with explicit assumption premises. However, it remains to be seen whether this can be done for any argument ordering. Moreover, $ASPIC^+$ representations of examples are often arguably closer to natural-language than ABA presentations, in which every conflict has to be translated to premise attack and every preference statement to explicit exceptions. If the aim is to formalise modes of reasoning in a way that corresponds with human modes of reasoning and debate, then there is some merit in having a theory with explicit notions of rebutting and undercutting attack and preference application.

2.4.3 The study of rationality postulates

An important recent development is the introduction by Caminada and Amgoud [2005; 2007] of the idea of rationality postulates for structured argumentation. According to Caminada and Amgoud, all systems of structured argumentation that have notions of negation, strict rules and subarguments should satisfy the following properties:

Sub-argument Closure: For any argument A in E, all sub-arguments of A are in E.

Closure under Strict Rules: If E contains arguments with conclusions $\alpha_1, \ldots .\alpha_n$, then any arguments obtained by applying only strict inference rules to these conclusions, are in E.

Direct Consistency: The set of conclusions of all arguments in E are directly consistent, i.e., it contains no pair of formulas φ and $\neg\varphi$.

Indirect Consistency: The set of conclusions of all arguments in E are indirectly consistent, i.e., its closure under strict rules is directly consistent.

$ASPIC^+$ unconditionally satisfies closure under subarguments. Whether $ASPIC^+$ satisfies closure under strict rules and the consistency postulates depends on whether the non-attackable premises are consistent, on structural properties of the strict rules and on properties of the argument ordering [Caminada and Amgoud, 2007; Prakken, 2010; Modgil and Prakken, 2013]. These results on $ASPIC^+$ directly generalise to systems that can be reconstructed within $ASPIC^+$, such as assumption-based argumentation and several forms of classical and deductive argumentation with preferences. Recently, Dung and Thang [2014] identified alternative and partly weaker sufficient conditions for satisfying strict closure and consistency.

Three further rationality postulates were proposed by Caminada *et al.* [2012] and are about the extent to which contradictions can trivialise the set of conclusions. These postulates have been further studied by Wu and Podlaszewski [2015].

Although Caminada and Amgoud defined their postulates for rule-based systems, they can be straightforwardly adapted to systems that define argument structure in terms of consequence notions instead of inference rules, such as classical and deductive argumentation. In particular the consistency postulates have been studied for these approaches [Gorogiannis and Hunter, 2011; Amgoud and Besnard, 2013]. One insight here (of which the core is already in Caminada and Amgoud [2007]) is that satisfaction of the consistency postulates partly depends on the definitions of attack and defeat. Building on this idea, Dung [2014; 2016] proposes several desirable properties for defeat relations (which in line with his 1995 paper he calls 'attack' relations) and studies their effect on satisfaction of the consistency postulates.

Finally, the recent research on rationality postulates is reminiscent of work in other areas of nonmonotonic logic on general properties of nonmonotonic consequence notions [Gabbay, 1985; Kraus *et al.*, 1990; Makinson, 1994]. One much discussed property in that body of work is *cautious monotony*. Informally, this property is that if φ and ψ are implied by a knowledge base and φ is added to the knowledge base, then ψ is still implied by the new knowledge base. Recently, Dung [2014; 2016] has argued that this property should hold for credulous argumentation-based inference, i.e., for membership of at least one extension. By contrast, Prakken and Vreeswijk [2002], Section 4.4 argue that satisfaction of this property is not desirable in general, since strengthening a nonmonotonic conclusion to an indisputable fact can give arguments using the fact the power to defeat other arguments that they did not have before; and this may well result in the loss of the extension from which the conclusion was promoted to an indisputable fact.

2.4.4 Preferences and argument strength

An important element in many argumentation systems is the use of some notion of preference or strength to resolve conflicts between arguments. In Dungean terms, this boils down to defining his attack relation in terms of a more basic, non-evaluative notion of conflict between arguments and some binary preference relation on arguments. As noted above, most work before Dung [1995] used the term 'defeat' instead of 'attack' while much work after 1995 explicitly renamed Dung's attack relation to 'defeat' in order be able to call the more basic, non-evaluative notion of conflict 'attack'. This is what I will also do in this section. The use of preferences then amounts to checking which attacks succeed as defeats.

Arguably the first systems embodying some form of argument preference were the inheritance systems of Touretzky [1984] and Horty et al. [1990], which used syntactic specificity checks on inheritance paths to let inheritance paths from more specific classes defeat conflicting inheritance paths form more general classes. Loui [1987] and Simari and Loui [1992] also used specificity for conflict resolution.

Although Pollock's earliest system, from 1987, did not yet include a notion of strength, Pollock later took the notion of strength of arguments very seriously. Since his systems were meant for epistemic reasoning, he always formulated strength in terms of numerical degrees of belief. His approach here was non-standard. Against Bayesian approaches, he argued that degrees of belief and justification do not conform to the laws of probability theory. In his [1994, 1995], Pollock used a weakest-link approach to compute the strength of arguments: given numerical strengths of reasons (where deductive reasons have infinite strength), the strength of an argument's conclusion is the minimum of the strengths of the reason with which the conclusion is derived and the strengths of the intermediate conclusions to which this reason is applied. While thus arguments can have various strengths, defeat is still an all-or-nothing matter in that defeaters that are weaker than their target cannot affect the status of their target at all. This allows a reconstruction of Pollock's [1994, 1995] approach in terms of Dung's theory of abstract argumentation frameworks. Later, in his [2002, 2007a, 2010] Pollock explored the idea that weaker defeaters can still weaken the justification status of their stronger targets. To formalize this, he now made the justification status of statements a matter of numerical degree, being a function of the strengths of both supporting and defeating arguments. Thus in his latest work he deviated from a Dungean approach.

Similar to Pollock's [1994; 1995] way to use degrees of belief is Chesñevar et al.'s [2004] use of possibilistic logic in the context of Defeasible Logic Programming. In this paper, possibilistic strengths are added to rules, which are propagated through arguments according to possibilistic logic. Then the propagated strengths are used to resolve attacks into defeats.

Other early work resolved attacks with qualitative preference relations on premises or inference rules. One of the first argumentation models of defeasible reasoning with rule preferences from arbitrary sources was Prakken [1993], developed into Prakken and Sartor [1997]. One of the first argumentation models of plausible reasoning with prioritized knowledge bases was Benferhat *et al.* [1993]. Amgoud and Cayrol [1998; 2002] combined Benferhat *et al.*'s idea of prioritised knowledge bases and Cayrol's [1995] Dungean modelling of classical argumentation with Prakken and Sartor's way to distinguish between attack and defeat in Dung's grounded semantics and their argument game for it. Later papers included preferences in classical argumentation in other ways; e.g. Amgoud and Vesic [2010] and Kaci [2010].

Vreeswijk [1993a; 1997] was the first to include a binary argument ordering as primitive in his approach. The $ASPIC^+$ framework adopts this idea and several papers on $ASPIC^+$ study instantiations with qualitative preference relations on defeasible rules and attackable premises, building on the work of Benferhat *et al.* [1993], Prakken and Sartor [1997] and their successors. Recently, Dung [2014; 2016] has also contributed to this study.

Since there is not a unique kind of content of arguments, there is also not a unique kind of argument preference. In epistemic reasoning, argument preferences are often based on probabilistic considerations, degrees of belief, or on credibility estimates of information sources. In argumentation as decision making they have been based on preferences for decision outcomes. In normative (legal or moral reasoning) they have been derived from hierarchical relations between elements of normative systems. In addition, some have modelled argumentation *about* preference relations within argumentation logics. One of the first proposals of this kind was made by Prakken and Sartor [1997]. Modgil [2009] extended abstract argumentation frameworks with the possibility to attack attacks. Modgil then, among other things, showed that Prakken and Sartor's proposal can be reconstructed as an instance of his 'extended argumentation frameworks'.

One question here is whether preference relations logically behave the same regardless of their source. Dung [2016] seems to answer this question affirmatively, while Modgil and Prakken [2014] suggest that the right way to use preferences may depend on the kind of content of arguments, for example, on whether the reasoning is epistemic, normative or about decision making.

2.5 Abstract argumentation: developments into now

In the first years after publication of Dung's landmark paper it gave rise to two kinds of follow-up work. Some continued to use *AFs* as Dung did in his paper, namely, to reconstruct and compare existing systems for structured argumentation as instances of *AFs*. In line with this was work on developing new systems for structured argumentation as instances of *AFs*. Others further developed the theory of abstract argumentation frameworks in the form of proof of properties (such as complexity results), reformulations (e.g. in terms of labellings), argument games as a proof theory, and algorithms. Somewhat later a third kind of follow-up work emerged, namely, extending AFs with new elements without specifying the structure of arguments. I now briefly review these three bodies of work.

2.5.1 Instantiating abstract argumentation frameworks

Some continued Dung's work on reconstructing and comparing existing systems for structured argumentation as instances of AFs. For example, Jakobovits [2000] and Jakobovits and Vermeir [1999b] showed that Pollock's [1994; 1995] system for defeasible reasoning has preferred semantics and Cayrol [1995] related various forms of classical argumentation to Dung's stable semantics and (with Amgoud in [Amgoud and Cayrol, 2002]) to Dung's grounded semantics for AFs. More recent work in this vein is Gorogiannis and Hunter [2011] and Amgoud and Besnard [2013].

Others developed new systems for structured argumentation as an instantiation of abstract argumentation frameworks. As described above, possibly the first system developed in this way was Prakken and Sartor's [1997] system for argumentation-based logic programming. More recently, the $ASPIC^+$ framework was designed in this way.

2.5.2 Developing the theory of abstract argumentation frameworks

Labellings A few years after Dung introduced his extension-based approach to abstract argumentation, an alternative labelling-based approach became popular, based on the following definition:

> A *labelling* of an $AF = \langle AR, attacks \rangle$ assigns to zero or more members of AR either the status *in* or *out* (but not both) such that:
>
> 1. an argument is *in* iff all arguments attacking it are *out*.
>
> 2. an argument is *out* iff it is attacked by an argument that is *in*.
>
> Let $In = \{A \in AR \mid A \text{ is } in\}$ and $Out = \{A \in AR \mid A \text{ is } out\}$ and $Undecided = AR \setminus (In \cup Out)$. Then
>
> 1. A labelling is *stable* if $Undecided = \emptyset$.
>
> 2. A labelling is *preferred* if $Undecided$ is minimal (wrt set inclusion)
>
> 3. A labelling is *grounded* if $Undecided$ is maximal (wrt set inclusion)
>
> 4. Any labelling is *complete*.

These notions coincide with Dung's extension-based definitions as follows. Let $S \in \{\text{stable, preferred, grounded, complete}\}$. Then (In, Out) is an S-labelling iff In is an S-extension.

To illustrate the labelling definition, in Figure 3 the grey-white colourings correspond to the *in-out* labels in the unique stable/preferred/grounded/complete labelling. In Figure 4(b,c) the grey-white colourings correspond to the *in-out* labels of the two stable-and-preferred labellings but in Figure 4(b,c) both arguments are undecided.

Actually, Pollock was a source of inspiration here too, since he used a labelling definition in his [1994; 1995] system. Pollock was possibly in turn inspired by Doyle's [1979] justification-based truth maintenance systems. Pollock's 1994 system was, as just noted, by Jakobovits [2000] proved to be an instance of Dung's preferred semantics. Jakobovits' PhD thesis contains an in-depth investigation of the labelling approach, summarised by Jakobovits and Vermeir [1999b]. Other early work on labellings was done by Verheij [1996] and the labelling approach was finally popularised by Caminada [2006].

Argument games Both the extension- and the labelling-based approach can be regarded as a semantics of argumentation-based inference in that the main focus is on characterising properties of *sets* of arguments, without specifying procedures for determining whether a given argument is a member of the set. The proof theory of argumentation-based inference amounts to specifying such procedures. An elegant form of such a proof theory is that of an *argument game* between a proponent and an opponent of an argument. The precise rules of the game depend on the semantics the game is meant to capture. The rules should be chosen such that the existence of a winning strategy (in the usual game-theoretic sense) for the proponent of an argument corresponds to the investigated semantic status of the argument, for example, 'being in the grounded' or 'being in at least one (or in all) preferred extensions'.

To give an idea, the following game is sound and complete for grounded semantics in that the proponent of argument A has a winning strategy just in case A is in the grounded extension. The proponent starts a game with an argument and then the players take turns, trying to defeat the previous move of the other player. In doing so, the proponent must strictly defeat the opponent's arguments while he is not allowed to repeat his own arguments. A game is terminated if it cannot be extended with further moves. The player who moves last in a terminated game wins the game. Thus the proponent has a winning strategy if he has a way to make the opponent run out of moves (from the implicitly assumed AF) whatever choice the opponent makes.

The idea of argument games had been around since the beginning of the formal study of argumentation (see e.g. Vreeswijk [1993a]) but they were not formally linked to argumentation-based semantics until the mid 1990s. Dung [1995] refers to a technical report [Dung, 1992] that was never formally published and in which he proposed argument games for two logic-programming semantics. Prakken and Sartor [1997] proposed an argument game for their logic-programming instantiation of Dung's grounded semantics. Arguably the first publication on argument games for abstract argumentation semantics was Prakken [1999], who proposed the above game for grounded semantics as an abstraction of the game of Prakken and Sartor. Vreeswijk and Prakken [2000] proposed argument games for preferred semantics, which were further developed and studied by Dunne and Bench-Capon [2003].

New semantics and general study of semantics While Dung [1995] originally proposed four semantics for abstract argumentation frameworks, in later years several alternative semantics were proposed; cf. Baroni *et al.* [2011a]. A related development is the study of general characterisations of types of semantics and their properties and relations, initiated by Baroni and Giacomin [2007] and further pursued by e.g. Dvorak and Woltran [2011] and Baroni *et al.* [2014]. Baroni and Giacomin [2007] also had a normative aim, namely, to propose a set of principles for the evaluation of semantics for abstract argumentation frameworks. Thus their work can be seen as an abstract counterpart of Caminada and Amgoud's [2007] introduction of rationality postulates for structured argumentation formalisms (see Section 2.4.3 above). Part E of Volume 1 of this handbook reviews this line of research in detail.

Complexity results and algorithms The graph-based format of abstract argumentation frameworks naturally lends itself to studies of computational complexity. A leading figure here has been Paul Dunne [Dunne and Bench-Capon, 2002; Dunne and Bench-Capon, 2003; Dunne, 2007].

Algorithms for proof theories for abstract argumentation frameworks were proposed by e.g. Cayrol *et al.* [2003], Vreeswijk [2006] and Verheij [2007]. Early work on algorithms for enumerating extensions or labellings is reviewed by Modgil and Caminada [2009]. An interesting strategy for developing algorithms is encoding argumentation frameworks in some other formalism and to utilise algorithms for the other formalism. For example, Besnard and Doutre [2004] encoded abstract argumentation frameworks in propositional logic in order to apply model-checking and SAT solver techniques. They also proposed an equation checking approach, which was later further developed by Gabbay [2011]. Some other examples of this approach are Grossi's [2010] encoding of abstract argumentation frameworks in modal logic and Egli *et al.*'s [2010] encoding in answer set programming.

2.5.3 Adding new elements to abstract argumentation frameworks

A third research strand in the abstract approach to argumentation is to extend AFs with new elements without specifying the structure of arguments. In this subsection I briefly discuss various ways in which this has been done.

Adding preferences or values Amgoud and Cayrol [1998] added to abstract argumentation frameworks a a preference relation on AR, resulting in *preference-based argumentation frameworks* ($PAFs$), which are a triple $\langle AR, attacks, \preceq \rangle$. An argument A then *defeats* an argument B if A attacks B and $A \not\prec B$. Thus each PAF generates an AF of the form $\langle AR, defeats \rangle$, to which Dung's theory of abstract argumentation frameworks can be applied.

Bench-Capon [2003] proposed a variant of idea called *value-based argumentation frameworks* ($VAFs$), in which each argument is said to promote some value. The notion of value should be taken here not in a numerical sense but in the sense of, for example, legal, moral or societal values, such as welfare, equality, fairness, certainty of the law, freedom of speech, privacy, and so on. Attacks are in $VAFs$ resolved in terms of one or more orderings on the values. These value orderings are assumed to be provided by an audience evaluating the arguments.

Adding abstract support relations There have been several recent proposals to extend Dung's [1995] well-known abstract argumentation frameworks (AFs) with abstract support relations, such as Cayrol and Lagasquie-Schiex's [2005b; 2009; 2013] Bipolar Argumentation Frameworks (BAFs), the work of Martinez et al. [2006] and Oren and Norman's [2008] Evidential Argumentation Systems (EASs). Various semantics for such frameworks have been defined, claimed to capture different notions of support. For example, Martinez et al. want to abstract from subargument relations in systems for structured argumentation. Boella et al. [2010a] study semantics of what they call "deductive" support, which satisfies the constraint that if A is acceptable and A is a deductive support of B, then B is acceptable. Nouioua and Risch [2011] consider "necessary support", which satisfies the constraint that if B is acceptable and A is a necessary support of B, then A is acceptable.

Other additions Both Bochman [2003] and Nielsen and Parsons [2007b] generalised Dung's attack relation to a relation from *sets* of arguments to arguments. As noted above, Modgil [2009] extended abstract argumentation frameworks with attacks on attacks, as an abstraction of earlier proposals to model reasoning about priorities in nonmonotonic logics. Coste-Marquis et al. [2006] added constraints to argumentation frameworks in the form of propositional encodings of properties of extensions. Finally, Dunne et al. [2011] added weights to attacks, the idea being that attacks that are of insufficient weight (modelled by a "weight budget") can be ignored.

A word of caution Although it is tempting to extend abstract argumentation frameworks with additional elements, a word of caution is in order. One should resist the temptation to think that for any given argumentation phenomenon the most principled analysis is at the level of abstract argumentation frameworks. In fact, it often is the other way around, since at the abstract level crucial notions like claims, reasons and grounds are abstracted away.

An example where this leads to problems is the way preferences are used in *PAFs* and *VAFs* to resolve attacks. As shown in work on structured argumentation with preferences (e.g. Pollock's or Vreeswijk's system, $ASPIC^+$ or DeLP), the structure of arguments is crucial in determining how preferences must be applied to attacks. Consider the following semi-formal example adapted from Prakken [2012] and Modgil and Prakken [2013], which can easily be formalised in any of the above-mentioned systems for structured argumentation.

$A = p$
$B_1 = \neg p$
$B_2 = \neg p$, therefore, *presumably*, q

Here p and $\neg p$ are default assumptions. Note that B_1 is a subargument of B_2, so B_2 includes B_1 as part of itself. The arguments with their internal structure and their direct attack relations are displayed in Figure 5. In any of the above systems for structured argumentation we then have that A and

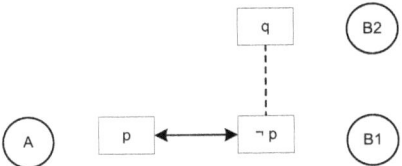

Figure 5. Argument structure and direct attack

B_1 directly attack each other while, moreover, A indirectly attacks B_2, since it directly attacks B_2's subargument B_1. So we have the abstract argumentation framework displayed in Figure 6(a).

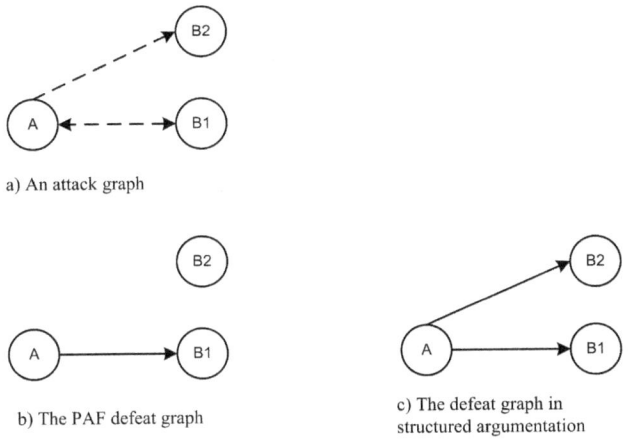

Figure 6. The abstract attack and defeat graphs

Assume next that A is preferred over B_1 and B_2 is preferred over A. Such an ordering could, for instance, be the result of comparing arguments according to their last fallible elements. A *PAF* modelling then generates the following single defeat relation: A defeats B_1; see Figure 6(b). Then we have a single extension (in whatever semantics), namely, $\{A, B_2\}$. So not only A but also B_2 is justified.

However, this violates Caminada and Amgoud's [2007] rationality postulate of subargument closure of extensions, since B_2 is in the extension while its subargument B_1 is not. (Prakken [2012] also discusses examples in which the postulate of indirect consistency is violated.) The cause of the problem is that the *PAF* modelling of this example cannot recognise that the reason why A attacks B_2 is that A directly attacks B_1, which is a subargument of B_2. So the *PAF* modelling fails to capture that in order to check whether A's attack on B_2 succeeds, we should compare A not with B_2 but with B_1. Now since $B_2 \prec A$

we also have that A defeats B_2; see Figure 6(c). So the single extension (in whatever semantics) is $\{A\}$, so closure under subarguments is respected.

This shows that *PAFs* (and also *VAFs*) only behave correctly under the assumption that all attacks are direct. We can conclude that for a principled analysis of the use of preferences to resolve attacks, the structure of arguments must be made explicit.

More generally, this analysis shows that in proposing an abstract model of argumentation, it is important to be aware what is abstracted from. Yet in the study of abstract support relations there is, unlike with Dung's original abstract frameworks, hardly any formal study of the relation between the abstract and the structured level. In consequence, it remains unclear what exactly is being modelled. One of the few studies in this vein is my [Prakken, 2014], in which I studied to what extent bipolar and evidential abstract frameworks can be interpreted as abstracting from the inferential relations in structured argumentation, as captured in $ASPIC^+$ by its subargument relations. I obtained mixed results. A form of BAFs that by Boella *et al.* [2010a] was claimed to be suitable for "deductive support" turned out to have no relation with classical-logic approaches to structured argumentation but Oren and Norman's [2008] evidential frameworks turned (for preferred semantics) out to be a suitable abstraction of $ASPIC^+$'s subargument relation. The same holds (for all four of Dung's [1995] semantics) for Dung and Thang's [2014] proposal. They add a binary support relation to abstract argumentation frameworks with the sole additional constraint that if B supports C and A attacks B then A also attacks C, and they then evaluate arguments as in Dung [1995] by only taking the thus constrained attack relation into account. The resulting system conforms to Nouioua and Risch's [2011] notion of "necessary support". Apart from these results it is still an open question what abstract models of argumentation with support relations abstract from.

These discussions lead me to propose a (to some readers possibly controversial) methodological guideline that every new proposal for extending abstract argumentation frameworks should in the same paper be accompanied by at least one non-trivial instantiation in order to demonstrate the significance of the new extension. Work that respects this guideline, respects the historic origins of the abstract study of argumentation, since the prime example of how this guideline can be applied is Dung [1995], who instantiated his frameworks with four nonmonotonic logics. A more recent example is Modgil [2009], who showed that his 'extended argumentation frameworks', which extend abstract argumentation frameworks with attacks on attacks, can be instantiated with Prakken and Sartor's [1997] modelling of reasoning about preferences.

2.6 Further developments

I now briefly sketch some important further developments in the formal study of argumentation as inference.

2.6.1 More recent graph-based approaches

Since 2007 several graph-based approaches have been proposed, in which not arguments and their relations but statements and their relations are the main focus of attention. This idea also goes back to the work of Pollock, since the system of Pollock [1994] is strictly speaking not formalised in terms of arguments but in terms of so-called 'inference graphs', in which nodes are connected either by inference links (applications of inference rules) or by defeat links. The nodes are 'lines of argument', which are propositions plus an encoding of the argument lines from which they are derived. Nodes are evaluated in terms of the recursive structure of the graph. As noted above, Jakobovits [2000] proved that Pollock's system can be given an equivalent formulation as an instance of Dung's abstract argumentation frameworks with preferred semantics.

Gordon et al. [2007] proposed the Carneades framework 'of argument and burden of proof'. Carneades' main structure is that of an argument graph, which, despite its name, is similar to Pollock's inference graphs. Statement nodes are linked to each other via argument nodes, which record the inferences from one or more nodes to another. This notion of an argument does not have the usual recursive structure in systems for structured argumentation but instead stands for a single inference step. Unlike Pollock, Carneades does not express conflicts as a special type of link between statement nodes. Instead, inferences (i.e., arguments) can be either pro or con a statement. The evaluation of statements in an argument graph is, as with Pollock's inference graphs, defined in terms of the recursive structure of the graph. Statements are acceptable if they satisfy their 'proof standard'. The general framework abstracts from their nature but Gordon et al. give several examples of proof standards.

Inspired by Carneades, Brewka and Woltran [2010] proposed their *Abstract Dialectical Frameworks*, which are directed graphs in which nodes are arguments, statements or positions which can be accepted or not and the links represent dependencies between arguments. The dialectical status (accepted or rejected) of a node depends on the status of its parents as specified in an acceptance condition for the node. Brewka and Woltranpresent ADFs as generalisations of abstract argumentation frameworks. In a purely technical sense they are, but so are assumption-based argumentation, Deflog and $ASPIC^+$, which can all represent AFs as a special case. For example, in assumption-based argumentation arguments from the AF can be made assumptions and an assumption can be said to be a contrary of another assumption if it attacks it in the AF. So far, applications of ADFs have instead interpreted the nodes as statements, e.g. Strass [2013], thus making ADFs more similar to Pollock's inference graphs. Future research should shed more light on the potential of ADFs as generalisations of abstract argumentation frameworks in a conceptual sense also.

2.6.2 Decision making as argumentation

While most early work on argumentation-based inference was on epistemic reasoning (what is the case?), in recent years there has been much attention for practical reasoning (what should we do?). Among the first papers on this topic was Fox and Parsons [1997], motivated by medical decision making.

One strand of work was initiated by Grasso *et al.*'s [2001] design for a nutrition advice system and Bench-Capon's [2003] formal work on value-based argumentation frameworks. Both works were influenced by Perelman and Olbrechts-Tyteca's [1969] idea (further discussed in Section 3.1) that whether an argument in ordinary discourse is good does not depend on its logical form but on whether it is capable of persuading the addressed audience, which in turn depends on the extent to which it takes the audience's "values" into account. The work on value-based argumentation frameworks was further developed by e.g. Atkinson [2005] and Atkinson and Bench-Capon [2007], who instantiated value-based argumentation frameworks with an argumentation scheme approach inspired by Walton's [2006b] schemes for practical reasoning. This work has among other things been applied to legal interpretation [Atkinson *et al.*, 2005a], seen as a decision problem in which the various interpretation options promote or demote various legal or societal values.

Another strand of work is the work of Amgoud and others, e.g. Amgoud *et al.* [2005; 2009], Amgoud and Prade [2009], which combines argumentation models of the inferential aspects of argumentation with models of qualitative decision theory for the choice aspects of decision making.

To compare and contrast the various bodies of work, note that decision making has various aspects: identifying possible decision options in the form of possible actions, identifying the decision criteria (preferences, desires, goals, values), determining the consequences of actions with respect to these criteria and choosing between the decision options. There is consensus that all these aspects up to the choice problem can be modelled as argumentation as inference, but there is no consensus whether they can be modelled as instantiating the above-discussed general models of structured argumentation or whether special argumentation formats should be developed. Examples of the former approach are Kakas and Moraitis [2003], who model decision-making arguments in Dimopoulos and Kakas's [1995] logic-programming system for argumentation, van der Weide *et al.* [2011; 2011], who model practical-reasoning arguments in a combination of $ASPIC^+$ with Wooldridge *et al.*'s [2006] formal model of meta-argumentation, and Fan and Toni [2013], who model decision making in assumption-based argumentation. Examples of the latter approach are Amgoud [2009], who proposes specific formats for decision-making arguments, and Atkinson [2005] and Atkinson and Bench-Capon [2007], who define special argument-schemes for practical-reasoning.

Another issue is whether argumentation-based decision-making can be fully modelled as argumentation-based inference. Amgoud [2009], p. 318 claims that decision making goes beyond inference when a choice has to be made between the decision options; all argumentation can do according to her is generating the decision options that have justified epistemic subarguments or assumptions. Accordingly, Amgoud and Prade [2009] model the choice between epistemically justified decision options outside their argumentation model in terms of models of qualitative decision theory. By contrast, the above-mentioned work in $ASPIC^+$ and assumption-based argumentation tries to model choice through the general conflict-resolution mechanisms of argumentation-based inference, such as $ASPIC^+$'s argument ordering.

2.6.3 Argumentation combined with probability theory

A recent trend is the combination of argumentation-based inference with probability theory. This is not surprising, since argumentation has from the early days on been proposed as a model for reasoning under uncertainty. Yet systematic studies of the combination of argumentation with probability were sparse until recently.

Argumentation has been combined with probability theory for three different kinds of purposes. First, there has been some work in which probabilistic models are the object of argumentative discourse, such as Nielsen and Parsons [2007a], who model how Bayesian networks can be jointly constructed in an argumentation process. In all other work the uncertainty does not concern the probabilistic but the argumentation model. Two approaches can be distinguished, depending on whether the uncertainty is *in* or *about* the arguments. When the uncertainty is in the arguments, probabilities are *intrinsic* to an argument in that they are used for capturing the strength of an argument given uncertainty concerning the truth of its premises or the reliability of its inferences. An example is default reasoning with probabilistic generalisations, as in *The large majority of Belgian people speak French, Mathieu is Belgian, therefore (presumably) Mathieu speaks French*. Clearly, if all premises of an argument are certain and it only makes deductive inferences, the argument should be given maximum probabilistic strength. Hunter [2013] calls this use of probability the *epistemic* approach.

When the uncertainty is about the arguments, probabilities are *extrinsic* to an argument in that they are used for expressing uncertainty about whether arguments are accepted as existing by some arguing agent. Hunter [2014] gives the example of a dialogue participant who utters an enthymeme and where the listener can imagine two reasonable premises that the speaker had in mind: the listener can then assign probabilities to these options, which translate into probabilities on which argument the speaker meant to construct. This uncertainty has nothing to do with the intrinsic strengths of the two candidate completed arguments: one might be stronger than the other while yet the other is more likely the argument that the speaker had in mind. Hunter [2013] calls this use of probability the *constellations* approach. Note that in this approach even deductive arguments from certain premises can have less than maximal strength.

The intrinsic, or epistemic approach can be applied in two ways: by simply computing probability values of conclusions or by using such probabilities to resolve attacks into defeats. Computing probability values of conclusions is done in early work by Haenni et al. [2000]. Their argumentation model is a rather specific one for diagnosis and has no clear relations with more general structured and abstract models of argumentation. More recent work in this vein is Dung and Thang [2010], who within assumption-based argumentation allow rules to be labelled with probabilities. As noted above in Section 2.4.4, Pollock [2002; 2007a; 2010] (using a non-standard account of probability) made the justification status of statements a matter of numerical degree, being a function of the strengths of both supporting and defeating arguments.

Other examples of the intrinsic/epistemic approach are methods for extracting arguments from (qualitative or quantitative) Bayesian networks. Older work in this vein is Parsons [1998a; 1998b], using a logic similar to the one of Krause et al. [1995] and Williams and Williamson [2006], using the logic of Prakken and Sartor [1997]. In this work no probability values of conclusions are computed: Parsons just generates arguments while Williams & Williamson generate abstract argumentation frameworks using rebutting attack without preferences. Recent work is of Timmer et al. [2017], who generates arguments in $ASPIC^+$ and resolves their rebutting attacks with probabilistic strengths of arguments. The latter is also done in two pieces of work using alternatives for standard probability theory, viz. Pollock's [1994; 1995] use of degrees of belief and Chesñevar et al.'s [2004] use of possibilistic logic (both discussed above in Section 2.4.4).

The extrinsic/constellations approach has been largely applied to abstract argumentation frameworks, as in Li et al. [2012] and Hunter [2014] (but see the early work of Riveret et al. [2007; 2008] using the logic of Prakken and Sartor [1997]). For a recent overview see Hunter and Thimm [2016]. In this approach, probabilities can unlike in the intrinsic approach, also be attached to, for instance, legal rules or moral value judgements. Another difference is that the extrinsic use of probability is defined on top of a separate model of argumentation existing independently of the probabilistic model, while in the second use probability is part of the argumentation model itself.

Assigning probabilities to arguments in the abstract is problematic, since in probability theory probabilities are assigned to the truth of statements or to outcomes of events, and an argument is neither a statement nor an event. What is required here is a precise specification of what the probability of an argument means. If it corresponds to the degree of justification of the argument's justification, then this should arguably be specified at the level of structured argumentation. For a preliminary attempt to do so in the context of classical-logic argumentation see Hunter [2013]. If the probability of an argument corresponds to the probability of a statement about the argument, then the nature of that statement should be made clear. More generally, here too the need arises to be explicit about what is abstracted from, in this case in abstract models of probabilistic argumentation.

2.6.4 Argumentation dynamics

A development that is in the border area of inference and dialogue is the logical study of the dynamics of argumentation, insofar as it abstracts from agent aspects and the dialogical setting. For example, Coste-Marquis *et al.* [2007] study the merging of abstract argumentation frameworks, with attention for the resolution of conflicts between the merged frameworks. Also, much work has recently been done on the nature and effects of change operations on a given argumentation state [Modgil, 2006; Rotstein *et al.*, 2008; Baumann and Brewka, 2010; Baroni *et al.*, 2011b]. Among other things, enforcing and preservation properties are studied. Enforcement concerns the extent to which desirable outcomes can or will be obtained by changing an argumentation state, while preservation is about the extent to which the current status of arguments is preserved under change. Quite recently, the revision of argumentation frameworks has been studied analogously to revising belief sets or bases in belief revision, i.e. as incorporating new or deleting old elements while keeping the changes minimal [Coste-Marquis *et al.*, 2014].

Almost all current work on argumentation dynamics concerns abstract argumentation frameworks. In particular the following operations have been studied: addition or deletion of (sets of) arguments (e.g. [Baumann and Brewka, 2010; Cayrol *et al.*, 2010; Baumann, 2012b; Baumann, 2012a]) and addition or deletion of (sets of) attack relations (e.g. [Modgil, 2006; Boella *et al.*, 2010b; Baroni *et al.*, 2011b; Bisquert *et al.*, 2013]). Deleting attacks can be seen as an abstraction from the use of preferences to resolve attacks into defeats.

This current abstract work on argumentation dynamics abstracts from the structure of arguments and the nature of their conflicts, which is a significant limitation. See e.g. Modgil and Prakken [2012], who for this reason propose a model of preference dynamics in $ASPIC^+$. For example, abstract models of argumentation dynamics do not recognise that some arguments are not attackable (such as deductive arguments with certain premises) or that some attacks cannot be deleted (for example between arguments that were determined to be equally strong), or that the deletion of one argument implies the deletion of other arguments (when the deleted argument is a subargument of another, as in Figure 6 above), or that the deletion or addition of one attack implies the deletion or addition of other attacks (for example, attacking an argument implies that all arguments of which the attacked argument is a subargument are also attacked; in Figure 6 above attacking B_1 implies attacking B_2). All this means that formal results about the abstract model may only be relevant for specific cases and may fail to cover many realistic situations in argumentation. To give a very simple example, in models that allow the addition of arguments and attacks, any non-selfattacking argument can be made a member of every extension by simply adding non-attacked attackers of all its attackers. However, this result at the abstract level does not carry over to instantiations in which not all arguments are attackable. Here, too, the importance shows of being aware what the model abstracts from.

2.6.5 Other work

I end this section on argumentation-based inference with a very brief review of some other relevant work (without any hope of being complete).

Wooldridge et al. [2006] proposed a formalism for meta-argumentation, supporting a hierarchical formalisation of logic-based arguments. At each level of the hierarchy, arguments, statements and positions can refer to arguments, statements and positions at lower levels. This is achieved by using a hierarchical first-order meta-logic, a type of first-order logic in which individual terms in the logic can refer to terms in another language. One application of this formalism is van der Weide's [2011] model of reasoning about preferences in argumentation about decision making.

Finally, a recent trend is to develop gradual notions of argument acceptability in terms of structural properties of abstract argumentation frameworks [Cayrol and Lagasquie-Schiex, 2005a; Grossi and Modgil, 2015].

3 Formal and computational models of argumentation-based dialogue

So far we have discussed argumentation as a form of (nonmonotonic) inference. However, argumentation can also be seen as a form of dialogue, in which two or more agents aim to resolve a conflict of opinion by verbal means. When argumentation is viewed as a kind of dialogue between 'real' agents (whether human or artificial), new issues arise, namely, the *distributed* nature of information (over the agents), the *dynamic* nature of information, since agents do not reveal everything they believe from the start and since they can learn from each other, and *strategic* issues, since agents will have their internal preferences, desires and goals. At first sight, it might be thought that the argument games for argumentation-based semantics discussed above in Section 2.5.2 are dialogical models of argumentation. However, this is not the case, since they are not meant for discussions between real agents but as a proof theory. There is no dynamics, no distributed information and the notions of a proponent and an opponent are just proof-theoretic metaphors, not real agents with preferences, desires and goals.

Research on argumentation-based dialogue divides into research on communication languages and protocols (their formal definition and study of their properties) and research on agent behaviour in argumentation dialogues (strategies, tactics, heuristics). Some work studies the combination of a protocol and agent behaviour within that protocol. The main idea of work on argumentation protocols is that such protocols should promote fair and effective resolution of conflicts of opinion. In work on argumentative agent design the agents are assumed to adhere to this purpose of the dialogue but within the rules of the protocol they can pursue their own interests and objectives.

Research on argumentation-based dialogue is often done against the background of Walton's [1984] classification of dialogues into six types according to their goal (see also e.g. [Walton and Krabbe, 1995]). *Persuasion* aims to resolve a difference of opinion, *negotiation* tries to resolve a conflict of interest by reaching a deal, *information seeking* aims at transferring information, *deliberation* wants to reach a decision on a course of action, *inquiry* is aimed at "growth of knowledge and agreement" and *quarrel* is the verbal substitute of a fight. This classification is not meant to be exhaustive and leaves room for dialogues of mixed type. Persuasion can, given its purpose, be seen as 'pure' argumentation and is often embedded in other dialogue types in that dialogues of other types may shift to persuasion if a conflict of opinion arises. For example, in information-seeking a conflict of opinion could arise on the credibility of a source of information, in deliberation the participants may disagree about likely effects of plans or actions and in negotiation they may disagree about the reasons why a proposal is in one's interest; also, in all three cases the participants may disagree about relevant factual matters.

The formal study of argumentation-based dialogue is less substantial and less advanced than the formal study of argumentation-based inference. Unlike with inference, it largely consists of a variety of different approaches and individual systems, with few unifying accounts or general frameworks. For these reasons this section is shorter than Section 2 on argumentation-based inference.

3.1 Main historical influences

As noted in Section 2.1.3, Toulmin [1958] claimed that outside mathematics the validity of an argument does not depend on its syntactic form but on whether it can be defended in a properly conducted dispute. It might be argued that Toulmin thus anticipated argumentation-based inference, especially in argument-game form. However, more importantly, he thus planted the seed of an idea that later became prominent in informal logic and argumentation theory, namely, that arguments can only be evaluated in the context of a dialogue. Toulmin's call to logicians of his days to study the criteria for properly conducted disputes can be regarded as a call to study dialogue protocols for argumentation.

The Belgian philosopher Chaïm Perelman also emphasised the dialogical nature of argument evaluation. However, he did not address protocol but strategy, in arguing that arguments in ordinary discourse should not be evaluated in terms of their syntactic form but on their rhetorical potential to persuade an audience [Perelman and Olbrechts-Tyteca, 1969]. In particular, an argument is more persuasive the more it takes the audience's "values" into account. For example, an argument that governments should not tap internet communications of their citizens since this infringes on their privacy is not very persuasive to an audience that values security over privacy. While initially Perelman's work was only influential in informal logic and argumentation theory, it was around 2000 taken up by AI researchers in both inferential and dialogical models of argumentation about action selection, starting with Grasso *et al.*'s [2001] design of a nutrition advice system and Bench-Capon's [2003] formal work on value-based argumentation frameworks (the latter was discussed above in Section 2.6.2). Other work more generally aimed at characterising the persuasive force of arguments in terms of the similarity of an argument with the beliefs of a typical audience [Hunter, 2004].

While argumentation logics define notions of consequence from a given body of information, dialogue systems for argumentation regulate disputes between real agents. Systems for persuasion dialogues were already studied in medieval times [Angelelli, 1970]. The modern study of formal dialogue systems for persuasion probably started with two publications by Charles Hamblin [1970, 1971], who coined the term 'formal dialectic', and was also inspired by speech act theory in philosophy [Searle, 1969] and dialogue logic [Lorenzen and Lorenz, 1978]. It should be noted that formal systems for persuasion dialogue differ from dialogue logics in one crucial respect. Dialogue logic aims to define the semantics of logical operators in terms of rules of attack and defence. Accordingly the purpose of a dialogue is to determine whether a proposition is implied by a given set of propositions and the roles of proponent and opponent are just logical metaphors, just as in the logical argument games discussed above in Section 2.5.2. By contrast, the purpose of a persuasion dialogue is to resolve a conflict of opinion between real agents, who can ask for and provide substantive reasons for their claims.

Initially, formal systems for argumentation as dialogue were studied only within philosophical logic and argumentation theory; see, for example, [Mackenzie, 1979; Mackenzie, 1990; Woods and Walton, 1978; Walton and Krabbe, 1995]. From the early nineties the study of argumentation dialogues was taken up in several fields of computer science. In Artificial Intelligence logical models of commonsense reasoning have been extended with formal models of persuasion dialogue as a way to deal with resource-bounded reasoning [Loui, 1998; Brewka, 2001]. Persuasion dialogues have also been used in the design of intelligent tutoring systems [Moore, 1993; Yuan, 2004] and were proposed as an element of computer-supported collaborative argumentation [Maudet and Moore, 1999]. In AI & law formal dialogue systems for persuasion were developed as a model of procedural justice in the sense of e.g. Alexy [1978]. See, for example, [Gordon, 1994; Hage et al., 1993; Bench-Capon, 1998; Bench-Capon et al., 2000; Lodder, 1999; Prakken, 2001a; Prakken, 2008]. Finally, in the field of multi-agent systems dialogue systems have been incorporated into models of rational agent interaction based on the observation that many kinds of agent interaction (such as negotiation and group decision making) involve argumentation. Accordingly, interaction protocols for various dialogue types involving argumentation have been designed [Parsons and Jennings, 1996; Kraus et al., 1998; Parsons et al., 1998; Amgoud et al., 2000a; McBurney and Parsons, 2002; Parsons et al., 2002; Parsons et al., 2003].

Most dialogue systems for argumentation are formulated in an informal mathematical metalanguage, but some have studied the full formalisation of protocols in logical action languages, such as Brewka [2001] in the situation calculus, Bodenstaff et al. [2006] in the event calculus and Artikis et al. [2007] in C^{++}.

3.2 General remarks on dialogue systems for argumentation

Persuasion is usually modelled as a two-party dialogue between a proponent and an opponent of an initial claim. Essentially, dialogue systems define a *communication language* (the well-formed utterances) and a *protocol* (when a well-formed utterance may be made and when the dialogue terminates). The communication language consists of a set of locutions applied to statements or arguments expressed in a logical language according to some adopted monotonic or nonmonotonic logic. If this logic is nonmonotonic, it can but need not be an argumentation logic.

Dialogue systems define the principles of coherent dialogue. Carlson [1983] defined coherence in terms of the purpose of a dialogue. According to him, whereas logic defines the conditions under which a proposition is true, dialogue systems define the conditions under which an utterance is appropriate, and this is the case if the utterance furthers the purpose of the dialogue in which it is made. Thus according to Carlson the principles governing the meaning and use of utterances should not be defined at the level of individual speech acts but at the level of the dialogue in which the utterance is made. This justifies why most work on argumentation dialogues, like Carlson, takes a game-theoretic approach to dialogues, where speech acts are viewed as moves in a game and rules for their appropriateness are formulated as rules of the game. Loui [1998] distinguished between *effectiveness* and *fairness* of dialogue systems. Effectiveness means that the protocol furthers the purpose of the dialogue (in the case of persuasion that the conflict of opinion is resolved). Some aspects of effectiveness are efficiency (how long are dialogues and is there a guarantee of termination?) and relevance (is every move relevant to the dialogue topic?). Fairness means that the participants have a fair opportunity to argue their case. Some aspects of fairness are that the participants always have the opportunity to move relevant moves and that the outcome of a dialogue agrees with the parties' commitments.

Communication language Here are some common speech acts that can be found in the literature on persuasion dialogues, with their informal meaning and the various terms with which they have been denoted in the literature.

- *claim* φ (assert, statement, ...). The speaker asserts that φ is the case.

- *why* φ (challenge, deny, question, ...) The speaker challenges that φ is the case and asks for reasons why it would be the case.

- *concede* φ (accept, admit, ...). The speaker admits that φ is the case.

- *retract* φ (withdraw, no commitment, ..) The speaker declares that he is not committed (any more) to φ. Retractions are 'really' retractions if the speaker is committed to the retracted proposition, otherwise it is a mere declaration of non-commitment (e.g. in reply to a question).

- φ *since* S (argue, argument, ...) The speaker provides reasons why φ is the case. Some protocols do not have this move but require instead that reasons be provided by a *claim* φ or *claim* S move in reply to a *why* ψ move (where S is a set of propositions). Also, in some systems the reasons provided for φ can have structure, for example, of a proof three or a deduction.

- *question* φ (...) The speaker asks another participant's opinion on whether φ is the case.

Structural degrees of freedom Dialogue systems can vary in their structural properties in several ways [Loui, 1998]: whether players can reply just once to the other player's moves or may try alternative replies (*unique- vs. multi-reply protocols*); whether players can make just one or may make several moves before the turn shifts (*unique- vs. multi-move protocols*); and whether the turn shifts as soon as the player-to-move has made himself the winning side or may shift later (*immediate- vs. non-immediate-reply protocols*). According to Loui [1998], the desired degree of structural 'strictness' of a dialogue system depends on the context of a dialogue. In contexts with little time and resources a unique-move, unique- and immediate reply protocol may be best, to force the participants not to waste resources, while in other contexts with more time and resources it is better to allow the participants more freedom to explore alternatives and return to earlier choices.

Commitments An important notion in systems for argumentation dialogue is that of propositional *commitments* [Walton and Krabbe, 1995]. Commitments are an agent's publicly declared points of view about a proposition, which may or may not agree or coincide with the agent's internal beliefs. An example of where they often do not agree is criminal trial, where the accused may very well publicly defend his innocence while he knows he is guilty. Commitments are typically incurred by stating claims or arguments, while they are typically lost by retracting a claim or argument. Commitments can serve several purposes in dialogue systems. One role is in enforcing a participant's dialogical consistency, for instance, by requiring him to keep his commitments consistent at all times or to make them consistent upon demand, or to defend one's commitments when challenged or else give them up. Another role of commitments is to determine termination and outcome of a dialogue. For example, persuasion dialogues can be defined to terminate if the opponent is committed to the proponent's main claim or the proponent is not committed any more to the main claim.

3.3 Some work on systems for persuasion dialogue

Since persuasion is 'pure' argumentation, I now review some historically important work on systems for persuasion dialogue in more detail. Then I will more briefly review work that embeds argumentation in systems for other kinds of dialogues.

3.3.1 Mackenzie [1979]

Mackenzie's [1979] system has been historically influential especially for its set of locutions. His system has the *claim, why, concede* and *retract* locutions.

The logical language is that of propositional logic but the logic is not full PL but instead a restricted notion of "immediate consequence", to capture resource-bounded reasoning (e.g. p, $p \to q$ and $q \to r$ immediately imply q but not r). Arguments are moved implicitly, by replying to a *why* move with a *claim*. An argument may be incomplete but its mover becomes committed to the material implication *premises $\to$ conclusion*. In addition, Mackenzie has a *question* speech act, which asks the hearer to declare a standpoint with respect to a proposition, and a *resolve* speech act for demanding resolution of conflicts in or logical implication by commitments. Mackenzie does not define outcomes or termination of dialogues. This makes his system underspecified as to the dialogue purpose, so that it can be extended to various types of dialogues. The protocol is unique-move and unique-reply but it nevertheless hardly enforces coherence of dialogues. Only the moves required after *why* and *question* and the use of the *resolve* move are constrained; the participants may freely exchange unrelated claims, and may freely challenge, retract or question. For instance, the following dialogue is legal:

P: *claim* p, O: *claim* q, P: *question* r, O: *claim* $\neg r$, P: *retract* s.

3.3.2 Walton & Krabbe [1995]

Walton and Krabbe [1995] developed the ideas of Mackenzie [1979] and also Woods and Walton [1978] into a full system for persuasion dialogues. To Mackenzie's locutions they added an explicit *since* locution for moving arguments. In their system, the only way to attack an argument is by challenging its premises, so the underlying logic is monotonic. The dialogues allowed by Walton and Krabbe are much more focused than Mackenzie's, since moves in a new turn must reply to a move in the previous turn of the other player. So, for instance, in the just-given example dialogue in Mackenzie's system, O's *claim* q move is not allowed and O must instead either concede or challenge p. This constraint also makes backtracking and postponement of replies impossible. Apart from this, the protocol allows that more than one move is made in one turn and alternative arguments for the same challenged proposition are moved. However, each move from the last turn must be replied-to (though other moves may be made as well).

Commitments are used by the protocol to enforce a participant's dialogical coherence. For example, if a participant's commitments logically imply an assertion of the other participant but do not contain that assertion, then the initial participant must either concede the assertion or retract one of the implying commitments.

The following example illustrates how the system deals with implicit premises:

P_1: *claim* this car is safe O_1: *why* is this car safe?; P_2: this car is safe *since* it has an airbag; P_2: safe *since* airbag.

Now the opponent must either challenge or concede both the explicit premise that the car has an airbag and the implicit premise 'if the car has an airbag, then it is safe'.

3.3.3 Gordon's Pleadings Game

Gordon's [1995] work on the *Pleadings Game* is seminal AI & Law work on the modelling of legal procedures as dialogue games. The game was intended as a normative model of civil pleading in Anglo-American legal systems, where the participants aim to identify the issues to be decided in court. The underlying logic is a nonmonotonic one, viz. conditional entailment [Geffner and Pearl, 1992], which as discussed above in Section 2.3 has a model-theoretic semantics and an argument-based proof theory. The game contains speech acts for conceding and challenging a claim, for stating and conceding arguments, and for challenging challenges of a claim. The latter has the effect of leaving the claim for trial. The Pleadings Game can be argued to have an implicit distinction between attacking and surrendering replies (as later made explicit in [Prakken, 2005]) in its distinction between three kinds of moves that have been made during a dialogue: the *open moves*, which have not yet been replied to, the *conceded moves*, which are the arguments and claims that have been conceded, and the *denied moves*, which are the claims and challenges that have been challenged and the arguments that have been attacked with counterarguments. The protocol is multi-move but unique-reply. At each turn a player must respond in some allowed way to every open move of the other player that is still 'relevant' (in a sense similar but not identical to that of Prakken [2005]), and may reply to any other open move. If no allowed move can be made, the turn shifts to the other player, except when this situation occurs at the beginning of a turn, in which case the game terminates. Move legality is further defined by specific rules for the various speech acts, which are mostly standard.

The result of a terminated game is twofold: a list of issues identified during the game (i.e., the claims on which the players disagree), and a winner, if there is one. Winning is defined relative to the background theory constructed during a game. If issues remain, there is no winner and the case must be decided by the court. If no issues remain, then the plaintiff wins iff his main claim is defeasibly implied by the final background theory, while the defendant wins otherwise.

3.3.4 Deriving locutions from argument schemes

The Toulmin Diagram Game (TDG) [Bench-Capon, 1998; Bench-Capon et al., 2000] was intended to produce more natural dialogues than the "stilted" ones produced by systems such as those reviewed thus far. To this end, its speech acts are based on an adapted version of Toulmin's [1958] well-known argument scheme. In this scheme a *claim* is supported by *data*, which support is *warranted* by an inference license, which possibly has *presuppositions*, and which is *backed* by grounds for its acceptance; finally, a claim can be attacked with a *rebuttal*, which itself is a claim and thus the starting point of a counterargument. Arguments can be chained by regarding data also as claims, for which further data can be provided.

The locutions of TDG's communication language correspond to the elements of this scheme, as shown in Table 1. For ease of comparison, this table has an

Table 1. Attackers and surrenders in TDG

Locutions	Attacks	Surrenders
$claim\ \varphi$	$why\ \varphi$	$concede\ \varphi$
$why\ \varphi$	$supply\ data_\varphi\ \psi$	$retract\ \varphi$
$concede\ \varphi$		
$supply\ data_\psi\ \varphi$	$so_\psi\ \varphi$	$concede\ \varphi$
	$why\ \varphi$	
$so_\psi\ \varphi$	$supply\ warrant\ \psi \Rightarrow \varphi$	
$supply\ warrant\ w$	$presupposing\ w$	$OK\ w$
	$on\ account\ of\ w$	
$presupposing\ w$	$supply\ presupposition_w\ \varphi$	$retract\ w$
$on\ account\ of\ w$	$supply\ backing_w\ b$	$retract\ w$
$supply\ backing_w\ b$		

explicit reply structure as in [Prakken, 2005], to be discussed below, although the original TDG system leaves this structure implicit in its protocol.

The idea to generate natural dialogues by defining the communication language in terms of some argumentation scheme was later applied to practical reasoning by Atkinson et al. [2005b; 2006], who embedded Atkinson's [2005] argumentation scheme for practical reasoning in a dialogue system for persuasion over action.

3.4 Later formal work

All systems reviewed so far are either philosophically motivated or geared towards application domains, and none of them were formally investigated on their properties. This changed in later AI work on dialogue systems for argumentation, some of which I will now discuss.

3.4.1 Parsons, Wooldridge & Amgoud [2003]

Parsons et al. [2002; 2003] were among the first to undertake a systematic formal study of argumentation as dialogue. They proposed dialogue systems for various types of dialogues involving argumentation and formally investigated them on various kinds of properties. In all of them the underlying logic is nonmonotonic, namely, Amgoud and Cayrol's [2002] system for classical-logic argumentation with grounded semantics. In this section I discuss their system for persuasion dialogues. Its communication language consists of claims, challenges, concessions and questions. Arguments are moved implicitly as *claim* replies to *why* moves (where sets of propositions may be claimed). The protocol has a rigid, unique-move and unique-reply nature, except that each premise of an argument may be responded to in turn. Unlike the above work, Parsons et al. make several assumptions on agent behaviour. Participants have their own, possibly inconsistent belief base and they are assumed to adopt an assertion and acceptance attitude, which they must respect throughout the dialogue. Moreover, claims moved in support of other claims must be from the participant's internal belief base.

Parsons et al. distinguish the following assertion attitudes: a *confident* agent can assert any proposition for which he can construct an argument, a *careful* agent can do so only if he can construct such an argument and cannot construct a stronger counterargument and a *thoughtful* agent can do so only if he can construct an acceptable argument for the proposition (according to grounded semantics). The corresponding acceptance attitudes also exist: a *credulous* agent concedes a proposition if he can construct an argument for it, a *cautious* agent does so only if in addition he cannot construct a stronger counterargument and a *skeptical* agent does so only if he can construct an acceptable argument for the proposition. In verifying these attitudes, each player must reason with its own beliefs and the commitments of the other side.

Consider the following example, where the proponent P believes p and $p \to q$, the opponent believes r and $r \to \neg q$ and all formulas are of equal preference. If P starts with *claim* q, then O must, depending on its dialogical attitudes, concede q if possible, otherwise claim $\neg q$ if possible, otherwise challenge q. If O is credulous or cautious, then perhaps surprisingly she must concede, since she has to reason with P's commitment p so she can construct a trivial argument for q, namely, $\{q\} \vdash q$. In both cases the dialogue terminates with agreement. By contrast, if O is skeptical, she has to challenge q. Then P has to move *claim* $\{p, p \to q\}$. Then O, being skeptical, must challenge both p and $p \to q$. The proponent then has to reply with *claim* $\{p\}$ and *claim* $\{p \to q\}$, after which the dialogue terminates without agreement, because the players are not allowed to repeat their moves, while O's acceptance attitude tells her to repeat her last two challenges.

Parsons et al. [2002; 2003] were among the first to undertake a systematic formal study of argumentation as dialogue. They proposed dialogue systems for various types of dialogues involving argumentation and formally investigated them on various kinds of properties. In all of them the underlying logic is nonmonotonic, namely, Amgoud and Cayrol's [2002] system for classical-logic argumentation with grounded semantics. In this section I discuss their system for persuasion dialogues. Its communication language consists of claims, challenges, concessions and questions. Arguments are moved implicitly as *claim* replies to *why* moves (where sets of propositions may be claimed). The protocol has a rigid, unique-move and unique-reply nature, except that each premise of an argument may be responded to in turn. Unlike the above work, Parsons et al. make several assumptions on agent behaviour. Participants have their own, possibly inconsistent belief base and they are assumed to adopt an assertion and acceptance attitude, which they must respect throughout the dialogue. Moreover, claims moved in support of other claims must be from the participant's internal belief base.

Parsons et al. investigate various properties of the protocols and their outcomes. Some results are on whether termination of dialogues is guaranteed. Other results are on the computational complexity of the various aspects. Yet

Table 2. An example communication language in Prakken's framework

Locutions	Attacks	Surrenders
claim φ	why φ	concede φ
φ since S	why $\psi(\psi \in S)$	concede ψ ($\psi \in S$)
	φ' since S' (φ' since S' defeats φ since S)	concede φ
why φ	φ since S	retract φ
concede φ		
retract φ		

other results concern possible agent behaviours. For example, they studied the extent to which one agent can mislead the other agent by making her concede a proposition he himself does not believe. They thus were among the first to address issues of trust in argumentation dialogue.

A very interesting aspect of this work is the definition of the various dialogical attitudes. However, these notions are perhaps better seen as aspects of strategy than of protocol, since if they are referred to by the protocol, an outside observer cannot verify protocol compliance, which is often regarded as a drawback of communication protocols.

3.4.2 Prakken [2005]

In [Prakken, 2005] a framework for specifying two-party persuasion dialogues is presented, which is then instantiated with some example protocols. The aim of this work was to allow a more general study of properties of dialogue systems of argumentation than the work reviewed so far. To this end, the framework largely abstracts from the logical language, the logic and the communication language, except that the communication language has to have an explicit reply structure and that underlying logic is assumed to be a system that is much like a preliminary version of $ASPIC^+$. Moreover, different protocols were defined, all extending a partial core protocol.

A main motivation of the framework was to ensure focus of dialogues while yet allowing for freedom to move alternative replies and to postpone replies. This was achieved with two main features of the framework. Firstly, an explicit reply structure on the communication language is assumed (implicit in several other systems), where each move either *attacks* or *surrenders to* its target. An example $\mathcal{L}_c$ of this format is displayed in Table 2. Secondly, winning is defined for each dialogue, whether terminated or not, and it is defined in terms of a notion of *dialogical status* of moves. The *dialogical status* of a move is recursively defined as follows, exploiting the tree structure of dialogues. A move is *in* if it is surrendered or else if all its attacking replies are *out*. This implies that a move without replies is *in*. And a move is *out* if it has a reply that is *in*. Actually, this has to be refined to allow that some premises of an argument are conceded while others are challenged; see [Prakken, 2005] for the details. Then a dialogue is (currently) won by the proponent if its initial move is *in* while it is (currently) won by the opponent otherwise.

Together, these two features of the framework allow for a notion of relevance that ensures focus while yet leaving the desired degree of freedom (generalised from [Prakken, 2001b]): a move is *relevant* just in case making its target *out* would make the speaker the current winner. Termination is defined as the situation that a player is to move but has no legal moves. The players can also agree to terminate a dialogue.

Consider by way of example the following dialogue in a protocol that allows replies to all moves of the other player but only if the move is relevant.

P_1:   *claim p*
O_1:   *why p*   (replying to P_1)
P_2:   *p since q*   (replying to O_1)
O_2:   *why q*   (replying to P_2)
P_3:   *p since r*   (replying to O_1)

At this point a reply to P_2 is irrelevant, since P_2 is *out*, so replying to it cannot change the status of P_1. Note that the dialogue can only terminate after either P has replied to O_1 with *retract p* or O has replied to P_1 with *concede p*. In all other cases, legal moves can always be made.

3.4.3 Argument games as dialogue systems

Argument games for abstract argumentation semantics were above in Section 2.5.2 discussed as a proof theory for abstract argumentation semantics. However, they have also been studied as genuine dialogue games for disagreeing agents, by dropping the assumption that all arguments are taken from a fixed and globally known argumentation framework [Loui, 1998; Jakobovits and Vermeir, 1999a; Jakobovits, 2000; Prakken, 2001b]. If this assumption is dropped, the properties of the game can change. A positive change is proven by Jakobovits, viz. that certain dynamic argument-game protocols prevent the construction of AFs containing odd loops (it is well known that such theories may have no extensions). A negative result is proven in [Prakken, 2001b], viz. that the dynamified game for grounded semantics loses soundness with respect to the joint framework constructed during a dialogue. However, if the game is changed by allowing any relevant reply (in the sense of [Prakken, 2005]) to any earlier move of the other side, then soundness is restored.

While the study of argument games as dialogue systems is theoretically very interesting, their very simple logic and communication language make that they cannot be a realistic model of persuasion dialogue.

3.5 Persuasion embedded in other types of dialogues

I now briefly review work that embeds argumentation in a dialogue system for other types of dialogues.

3.5.1 Negotiation

Much work on embedding argumentation into negotiation protocols is motivated by the claim that argumentation can be beneficial to negotiation. From the point of view of the negotiating agents, adding reasons for a proposal could increase the chance of acceptance. This was the idea of Sycara's [1985; 1990] early work on modelling threats and reward in labour negotiation. For example 'if you do not accept our offer, we will go on strike' (a threat) or 'if you accept that you have to work during the weekends, you will receive an increase in salary' (a reward). This idea was generalised by Parsons *et al.* [1998] and Kraus *et al.* [1998] for BDI-style agents, that is, agents that form their intentions to act according to their beliefs and, possibly prioritised, desires [Rao and Georgeff, 1991]. The general idea is that the other agent should be made to change its beliefs or preferences in such a way that it will form the intention to accept or make an offer that the initial agent wants.

From the perspective of protocol design the idea is that if negotiating agents exchange and discuss reasons for their proposals and rejections, the negotiation process may become more efficient and the negotiation outcome may be of higher quality. If an agent explains why he rejects a proposal, the other agent knows which of her future proposals will certainly be rejected so she will not waste effort at such proposals. Thus efficiency is promoted. In such exchanges, reasons are not only exchanged, they can also become the subject of debate. Suppose a car seller offers a Peugeot to the customer but the customer rejects the offer on the grounds that French cars are not safe enough. The car seller might then try to persuade the customer that he is mistaken about the safety of French cars. If she succeeds in persuading the customer that he was wrong, she can still offer her Peugeot. Thus the quality of the negotiation is promoted, since the buyer has revised his preferences to bring them in agreement with reality. This example illustrates that a negotiation dialogue (where the aim is to reach a deal) sometimes contains an embedded persuasion dialogue (where the aim is to resolve a conflict of opinion).

Since all this is about giving reasons for or against acting in a certain way, the kind of argumentation that is involved is, inferentially speaking, argumentation about decision options (see Section 2.6.2 above), although it can, as the car sales example shows, also shift to epistemic argumentation about the underlying facts. The early work of Sycara [1985; 1990] and Kraus *et al.* [1998] applied informal rhetorical models of argumentation. Later work incorporated formal inferential models of argumentation in negotiation protocols. For example, Parsons *et al.* [1998] embed Krause *et al.*'s [1995] logic of argumentation, Amgoud *et al.* [2000b] embed Amgoud and Cayrol's [1998] model of classical argumentation with preferences, Amgoud and Prade [2004] incorporate the model summarised by Amgoud [2009], and van Veenen and Prakken [2006] combine Wooldridge and Parson's [2000] negotiation protocol with one of Prakken's [2005] persuasion protocols, thus also embedding its preliminary version of *ASPIC*$^+$.

3.5.2 Deliberation

The purpose of deliberation is to agree on a course of action. It differs from persuasion over action, as modelled by e.g. Atkinson et al. [2005b; 2006] in that at the start of a deliberation dialogue there typically just is a problem and no proposed solutions yet. It differs from negotiation in that deliberating agents are assumed not to be self-interested but collaborative, sharing the goals of the group or community they are part of. The group may be small, such as a few people choosing a restaurant for dinner, it may be big, such as in parliamentary debate, and it may be huge, such as in public debate about political or societal issues. Clearly, different settings require different kinds of protocols.

Embedding argumentation in deliberation has much the same benefits as embedding it in negotiation: for the agents it may increase the chance of acceptance of their proposals, and for the dialogue it may increase the quality of the outcome. Research on deliberation with argumentation started later than research on argumentation-based negotiation and is not as extensive. Here is brief overview of some work.

Tang and Parsons [2005] proposed a rather specific dialogue system for argumentation about means-end planning, not based on a formal model of argumentation.

McBurney et al. [2007] proposed a framework for multi-agent deliberation dialogues. The protocol is intended to allow for the open nature of deliberation, giving the agents much freedom for establishing goals, constraints, perspectives, facts, actions and evaluations. Accordingly, the dialogue cyclicly moves through various stages. After initial inform and propose stages, the agents evaluate and decide on actions in the consideration, revision, recommendation and confirmation stages. The framework does not assume a specific argumentation logic.

Black and Atkinson et al. [2011] proposed a much more rigid system for two-agent deliberation based on Atkinson's [2005] embedding of an argument scheme for practical reasoning in value-based argumentation frameworks. The rigidness of the system allows them to show that if the agents adhere to the dialogue protocol and construct their arguments on the basis of their own belief bases, then any agreed proposal is also acceptable to both agents individually.

Finally, Kok et al. [2011] proposed a dialogue system for multi-agent deliberation dialogues as part of an experimental setup for testing the usefulness of argumentation in such dialogues. The system is an instance of Prakken's [2005] framework for persuasion but adapted to deliberation. It incorporates $ASPIC^+$ as the underlying logic.

3.5.3 Inquiry

Only little work has been done on embedding argumentation in inquiry. Early work is McBurney and Parsons' [2001] model of scientific inquiry. More recently, Black and Hunter [2007; 2009] embedded Garcia and Simari's [2004] DeLP argumentation system in a protocol for inquiry dialogue. They combined the protocol with a strategy that selects exactly one of the legal moves to make. This allowed them to prove soundness and completeness properties with respect to the participants' belief bases, provided the agents construct their arguments from their own belief base.

3.6 Work on strategic aspects of argumentation

Dialogue systems for argumentation only cover the rules of the game, i.e., which moves are allowed; they do not cover principles for playing the game well, i.e., strategies, tactics and heuristics for the individual players. Above we already discussed some work that studies the combination of a protocol with strategies, such as [Black and Atkinson, 2011] for deliberation and [Black and Hunter, 2007; Black and Hunter, 2009] for inquiry. Moreover, as remarked above, the assertion and acceptance policies studied by Parsons *et al.* [2002; 2003] could be seen as heuristics for move selection (although Parsons *et al.* make them part of their protocols).

Other early work on strategic aspects of argumentation is of Amgoud and Maudet [2002], who, building on the even earlier work of Moore [1993] on argumentation dialogues for intelligent tutoring, formulated move selection strategies and tactics based on human strategies in natural dialogues. One example is that agents have to choose between a *build* or *destroy* attitude, i.e., whether they want to support their own or to attack their opponent's position. This idea was later also used by Kok [2013] in his simulation experiments on whether argumentation is beneficial to deliberating agents.

In the context of dialogue games for abstract argumentation, Paul Dunne studied issues arising from the mismatch between the purpose of persuasion dialogues and the arguing agent's own objectives. In [Dunne, 2003] he studied the use of delay tactics and in [Dunne, 2006] he studied situations where agents have a 'hidden agenda'.

More recently, there is an emerging research strand on opponent modelling for strategic purposes, for example in terms of probability distributions or expected-utility distributions over the possible actions of the opponent [Matt and Toni, 2008; Thimm and Garcia, 2010; Oren and Norman, 2010; Hadjinikolis *et al.*, 2013; Rienstra *et al.*, 2013]. Somewhat earlier, Riveret *et al.* [2008] probabilistically modelled not an opponent but an impartial adjudicator who has the power to accept or reject premises of arguments put forward by the adversaries. In this work, probabilistic game theory can be used to determine optimal strategies.

Other recent work that uses game theory is that on mechanism design for argumentation [Rahwan and Larson, 2008; Rahwan *et al.*, 2009]. The goal here is to develop protocols that make unwanted behaviour (such as lying or withholding information) suboptimal.

All this recent work on strategic aspects of argumentation is still preliminary and therefore I leave a further description to other chapters in this handbook. Until these chapters are available, the reader can consult [Thimm, 2014] for a recent overview. I confine myself to one concluding observation. On the one hand, the recent work on strategy, heuristics and tactics is a natural continuation of the earlier work on communication languages and protocols. However, in one respect it is a step backwards, since it generally assumes much simpler dialogue systems than were developed before, with, for example, much recent work assuming simple dialogue games for abstract argumentation semantics.

4 Application areas

Formal and computational models of argumentation have been applied in several areas. Although a comprehensive review is beyond the scope of this chapter, a brief overview is in order. For more detailed overviews the reader can consult [Modgil et al., 2013] and some references given below. I will mainly focus on three main application areas, viz. medicine, law and debating technologies. In addition, in the literature many specific applications can be found, such as to recommender systems, trust and reputation management, robot soccer, waste management, licensing policy management, the internet of things, and so on.

Below I will only discuss applications of formal models of argumentation. In several areas there is much applied research based on informal or ad-hoc models of argumentation. For example, argumentation has been used in work on risk assessment and design rationale in software engineering for explaining why a design meets a design requirement or avoids a risk [Haley et al., 2008; Franqueira et al., 2011] . Moreover, there is quite some work on support tools for argument visualisation [Reed et al., 2007; ter Berg et al., 2009] and collaborative argumentation and decision making [Conklin et al., 2001; Scheuer et al., 2010; Kirschner et al., 2003], sometimes in educational contexts [Pinkwart and McLaren, 2012], and with applications for the social web [Schneider et al., 2013]. Finally, recently research in argument mining [Palau and Moens, 2009; Lippi and Torroni, 2016] has become popular, which aims to recognise (elements of) arguments and their relations in natural-language texts.

As for the nature of the applications mentioned below, theoretical, user-oriented and fielded applications can be distinguished. *Theoretical applications* use a non-trivial domain example to demonstrate the adequacy or motivate design features of the model. In *user-oriented applications* (which usually are of computational architectures) the usefulness of the architecture for designated types of users or tasks is an essential aspect. *Fielded applications* have actually been used by the intended user group in a realistic context, either experimentally or in actual use.

4.1 Medical applications

Medicine has been an important application field of argumentation, with John Fox as a historically influential figure. Several systems developed by him and his colleagues have been experimentally tested or are even in actual use [Fox et al., 2007], so these count as fielded applications. While their underlying argumentation model is rather simple, this group also studied formal foundations of their systems, e.g. in [Elvang-Göransson et al., 1993] and [Krause et al., 1995]. Moreover, Fox and Parsons [1997] proposed one of the first formal argumentation-based models of decision making, using arguments for expressing and comparing the positive and negative effects of medical treatments. This idea was combined with an argument-scheme approach by Tolchinsky et al. [2006; 2012], who present a model for multi-agent deliberation about safety-critical medical actions, such as donor organ selection for patients. The intended system plays the role of a mediating agent whose task is to inform the participants about their valid move options, to decide whether an argument is relevant enough to be admitted into the process, and to evaluate the admitted arguments in order to assess whether the proposed action should be undertaken. Since this system was tested experimentally with medical doctors, it counts as a fielded application.

More recently, Hunter and Williams [2012] have applied argumentation in a user-oriented way to the problem of aggregating evidence-based arguments for and against treatment options from clinical trials. They use preference-based abstract argumentation frameworks instantiated with one-steps applications of domain-specific inference rules, and express argument preferences in terms of outcome indicators of the treatments. The approach was evaluated by comparison with recommendations made in published healthcare guidelines.

4.2 Legal applications

There has been much cross-pollination with the field of AI & Law [Prakken and Sartor, 2015]. This is understandable, given the inherently adversarial nature of the law and the importance of written justifications of legal decisions. Rule-based argumentation formalisms such as assumption-based argumentation and the system of Prakken and Sartor [1997] have been applied to preference-based reasoning with conflicting rules [Kowalski and Toni, 1996; Prakken and Sartor, 1996]. Prakken & Sartor [2009] also used their logic to formalise notions of burdens of proof, as was done by Gordon and Walton [2009] with their Carneades system. Work on applying dialogue systems to the formalisation of legal procedure was discussed above in Sections 3.1 and 3.3.3.

An important contribution of AI & Law to the formal study of argumentation is the study of the role of cases in argumentation; for a recent detailed overview see [Bench-Capon, 2017]. In Section 1 above the still influential HYPO system [Ashley, 1990] and its successor CATO [Aleven, 2003] were mentioned. Their underlying argumentation model is for 'factor'- or 'dimension'-based reasoning, where cases are collections of abstract fact patterns that favour or oppose a conclusion, either in an all-or nothing fashion (factors) or to varying degrees (dimensions). This work inspired subsequent formal work using the tools of formal argumentation, e.g. [Hage, 1993; Loui et al., 1993; Prakken and Sartor, 1998; Bench-Capon and Sartor, 2003]. A key idea in this work is that case decisions give rise to conflicting rules (or conflicting sets of reasons) plus a preference expressing how the court resolved this conflict. In the notation of Prakken and Sartor [1998]:

$r_1:$ Pro-factors $\Rightarrow$ Decision
$r_2:$ Con-factors $\Rightarrow$ Not Decision
$r_1 > r_2$

The rule preference expresses the court's decision that the pro factors in the body of rule r_1 together outweigh the con factors in the body of rule r_2. This approach allows for 'a fortiori' reasoning in that adding factors to a pro-decision rule or removing factors from a con-decision rule does not affect the rule priority. Horty [2011], using a non-argumentation-based nonmonotonic logic, formalises the conditions under which a decision is allowed or forced by body of precedents and then uses this to also formalise the concepts of following, distinguishing and overruling a precedent.

A related line of research is to compare cases not in terms of their factors but in terms of underlying legal and social values. Berman and Hafner [1993] argued that often a factor can be said to favour a decision by virtue of the purposes served or values promoted by taking that decision because of the factor. A choice in case of conflicting factors is then explained in terms of a preference ordering on the purposes, or values, promoted or demoted by the decisions suggested by the factors. Cases can then be compared in terms of the values at stake rather than on the factors they contain. Bench-Capon [2002] first computationally modelled this approach, leading to a series of papers culminating in [Prakken et al., 2015] and using argument schemes for practical reasoning of the kinds also used in argumentation-based models of decision making (see Section 2.6.2 above).

All the AI & law applications mentioned so far are theoretical applications. User-oriented legal applications of argumentation are rare, with most applications in the field of e-democracy, e.g. [Cartwright and Atkinson, 2009; Gordon, 2011]. Finally, to my knowledge only one fielded application exists, namely, the CATO system, which was experimentally tested for teaching case-based argumentation skills to American law students.

4.3 Debating technologies

Most work on debating technologies is based on informal or ad-hoc models of argumentation; for overviews see the references given above. An exception is the work of the Arg-tech group at the University of Dundee, Scotland, led by Chris Reed. This group has developed various user-oriented web-based argumentation tools partly based on formal foundations [Bex et al., 2013a]. For example, they have been using the so-called Argument Interchange Format [Chesñevar et al., 2006], which was given a logical foundation in $ASPIC^+$ by Bex et al. [2013b] and they have an online implementation of an instance of $ASPIC^+$ called TOAST [Snaid and Reed, 2012]. Several tools developed by the Arg-tech group have been experimentally tested with intended users, so these count as fielded applications.

5 Conclusion

Looking back on the history of formal research on argumentation, there is a marked difference between the study of argumentation as inference and that of argumentation as dialogue. The theory of argumentation-based inference is mature, with an almost universally accepted formal foundation in Dung's theory of abstract argumentation frameworks and its extensions and with a converging study of structured argumentation, with just a small number of general frameworks and increasing knowledge about their relations. By contrast, the study of argumentation-based dialogue consists of a variety of different approaches and individual systems, all exciting work but with few unifying accounts or general frameworks. There are a few exceptions, such as a series of papers just after 2000 by Peter McBurney, Simon Parsons and others on principles for the design of dialogue systems e.g. [McBurney et al., 2002; McBurney and Parsons, 2002], and my own formal framework for persuasion dialogue in [Prakken, 2005]. However, this work is still far from being foundational.

In my own personal opinion, the following are the four main main theoretical contributions of the field.

1. The idea that dialectical evaluation of arguments can be formalised. While logic textbooks routinely write that a valid argument does not dictate the acceptance of its conclusion since it can always be attacked on its premises, formal argumentation has shown that attack relations between arguments conform to patterns that can be formally studied. In its purest form this is captured in Dung's [1995] theory of abstract argumentation frameworks.

2. The idea of defeasible rules. Dogma has it that all arguments should be deductively valid, that is, the truth of their premises should guarantee the truth of their conclusion. The fields of informal logic, argumentation theory and epistemology have questioned this dogma and argued that arguments that fail to meet this standard of perfection can still be good, as long as they withstand critical scrutiny. The field of formal argumentation has shown that this idea can be formalised.

3. The idea that the principles for evaluating arguments in the context of a dialogue can be formalised. Toulmin [1958] first proposed that arguments should be evaluated not on their syntactic form but on whether they can be defended in a properly conducted dispute. He urged logicians of his day to study the principles of proper dispute. The formal study of dialogue studies has met this challenge and thus also opened the prospects for precise formal studies of strategy and tactics for persuasion.

4. The idea that reasoning under uncertainty can be formalised in a qualitative way. There is an increasing trend of advocating quantitative (especially Bayesian) models of uncertainty as the only way to reason about

uncertainty. Likewise with quantitative models of decision making. However, for humans such quantitative theories are often hard to grasp, while they largely ignore the dialogical and procedural aspects of reasoning. This is especially a problem for applications with humans in the loop, such as support tools for human argumentation and decision making. Our field has shown that a natural qualitative theory of reasoning under uncertainty can be formalised.

However, there is, in my opinion, also an unfortunate recent development. While Dung's [1995] idea of abstract argumentation frameworks was a major breakthrough and is deservedly a key element in the formal study of argumentation-based inference, not all follow-up work is of the same generality. We have seen that several proposals for extending abstract argumentation frameworks with new elements implicitly make assumptions that are not in general satisfied. The same holds for work on the dynamics of abstract argumentation and for some work on probabilistic abstract argumentation. The resulting formalisms are thus abstract but not general in that they model special cases, such as the case in which all arguments, or all attacks, are independent of each other, or the special case in which all arguments are attackable.

It is worth noting that the word 'abstract' in Dung's [1995] notion of abstract argumentation frameworks does not qualify 'argumentation' but 'frameworks'. In Dung's terminology, it is the framework that is abstract, not the argumentation. Strictly speaking there is no such thing as abstract argumentation, just as there is no such thing as structured argumentation. All there is is argumentation, which can be studied at various levels of abstraction. And in real argumentation not arguments but things like claims, reasons and grounds are the most basic elements. There is nothing wrong in principle with abstract studies of argumentation: abstraction is an indispensable tool in any kind of research. However, one should not forget that we all study the same phenomenon, so that the various levels of abstraction should be connected. I remind the reader of my (perhaps controversial) proposal in Section 2.5.3 of a methodological guideline that every new proposal for extending abstract argumentation frameworks with new elements should in the same paper be accompanied by at least one non-trivial instantiation, in order to demonstrate the significance of the new extension. In doing so, we would respect the historic roots of the abstract study of argumentation, since in his original 1995 paper Dung respected this guideline in a way that has since never been equalled.

It is time to conclude. The formal and computational study of argumentation has established itself as a mature field of research. Argumentation is a key word or topic in all main AI conferences, papers on argumentation are published in the major AI journals, and the field has its own COMMA conference plus several workshops (CNMA, ArgMas, TAFA). Theoretically, the field is in a healthy state with much exciting research. With respect to applications this is less so, but this holds for all theoretically interesting fields of research. There is every hope to be optimistic here too, as long as a too strong focus on abstract argumentation is avoided. Unlike, for example, constraint satisfaction or model checking, argumentation is not just a technique but an important aspect of human life. There will therefore always be the need for support tools for argumentation, and our field is arguably in an excellent position to provide these tools. In any case, it provides their formal foundations.

BIBLIOGRAPHY

[Aleven and Ashley, 1991] V. Aleven and K.D. Ashley. Toward an intelligent tutoring system for teaching law students to argue with cases. In *Proceedings of the Third International Conference on Artificial Intelligence and Law*, pages 42–52, New York, 1991. ACM Press.

[Aleven, 2003] V. Aleven. Using background knowledge in case-based legal reasoning: a computational model and an intelligent learning environment. *Artificial Intelligence*, 150:183–237, 2003.

[Alexy, 1978] R. Alexy. *Theorie der juristischen Argumentation. Die Theorie des rationalen Diskurses als eine Theorie der juristischen Begründung*. Suhrkamp Verlag, Frankfurt am Main, 1978.

[Amgoud and Besnard, 2010] L. Amgoud and Ph. Besnard. A formal analysis of logic-based argumentation systems. In *Proceedings of the 4th International Conference on Scalable Uncertainty Management (SUM'10)*, number 6379 in Springer Lecture Notes in AI, pages 42–55, Berlin, 2010. Springer Verlag.

[Amgoud and Besnard, 2013] L. Amgoud and Ph. Besnard. Logical limits of abstract argumentation frameworks. *Journal of Applied Non-classical Logics*, 23:229–267, 2013.

[Amgoud and Cayrol, 1998] L. Amgoud and C. Cayrol. On the acceptability of arguments in preference-based argumentation. In *Proceedings of the 14th Conference on Uncertainty in Artificial Intelligence*, pages 1–7, 1998.

[Amgoud and Cayrol, 2002] L. Amgoud and C. Cayrol. A model of reasoning based on the production of acceptable arguments. *Annals of Mathematics and Artificial Intelligence*, 34:197–215, 2002.

[Amgoud and Maudet, 2002] L. Amgoud and N. Maudet. Strategical considerations for argumentative agents (preliminary report). In *Proceedings of the Ninth International Workshop on Nonmonotonic Reasoning*, pages 399–407, Toulouse, France, 2002.

[Amgoud and Prade, 2004] L. Amgoud and H. Prade. Reaching agreement through argumentation: A possibilistic approach. In *Principles of Knowledge Representation and Reasoning: Proceedings of the Ninth International Conference*, pages 175–182. AAAI Press, 2004.

[Amgoud and Prade, 2009] L. Amgoud and H. Prade. Using arguments for making and explaining decisions. *Artificial Intelligence*, 173:413–436, 2009.

[Amgoud and Vesic, 2010] L. Amgoud and S. Vesic. Handling inconsistency with preference-based argumentation. In *Proceedings of the 4th International Conference on Scalable Uncertainty Management (SUM'10)*, number 6379 in Springer Lecture Notes in AI, pages 56–69, Berlin, 2010. Springer Verlag.

[Amgoud et al., 2000a] L. Amgoud, N. Maudet, and S. Parsons. Modelling dialogues using argumentation. In *Proceedings of the Fourth International Conference on MultiAgent Systems*, pages 31–38, 2000.

[Amgoud et al., 2000b] L. Amgoud, S. Parsons, and N. Maudet. Arguments, dialogue, and negotiation. In *Proceedings of the Fourteenth European Conference on Artificial Intelligence*, pages 338–342, 2000.

[Amgoud et al., 2005] L. Amgoud, J.-F. Bonnefon, and H. Prade. An argumentation-based approach to multiple criteria decision. In *Proceedings of the 8th European Conference on Symbolic and Quantitative Approaches to Reasoning with Uncertainty (ECSQARU 05)*, number 3571 in Springer Lecture Notes in AI, pages 269–280, Berlin, 2005. Springer Verlag.

[Amgoud, 2009] L. Amgoud. Argumentation for decision making. In I. Rahwan and G.R. Simari, editors, *Argumentation in Artificial Intelligence*, pages 301–320. Springer, Berlin, 2009.

[Angelelli, 1970] I. Angelelli. The techniques of disputation in the history of logic. *The Journal of Philosophy*, 67:800–815, 1970.

[Artikis et al., 2007] A. Artikis, M.J. Sergot, and J. Pitt. An executable specification of a formal argumentation protocol. *Artificial Intelligence*, 171:776–804, 2007.

[Ashley, 1990] K.D. Ashley. *Modeling Legal Argument: Reasoning with Cases and Hypotheticals*. MIT Press, Cambridge, MA, 1990.

[Atkinson and Bench-Capon, 2007] K. Atkinson and T.J.M. Bench-Capon. Practical reasoning as presumptive argumentation using action based alternating transition systems. *Artificial Intelligence*, 171:855–874, 2007.

[Atkinson et al., 2005a] K. Atkinson, T.J.M. Bench-Capon, and P. McBurney. Arguing about cases as practical reasoning. In *Proceedings of the Tenth International Conference on Artificial Intelligence and Law*, pages 35–44, New York, 2005. ACM Press.

[Atkinson et al., 2005b] K. Atkinson, T.J.M. Bench-Capon, and P. McBurney. A dialogue game protocol for multi-agent argument over proposals for action. *Journal of Autonomous Agents and Multi-Agent Systems*, 11:153–171, 2005.

[Atkinson et al., 2006] K. Atkinson, T.J.M. Bench-Capon, and P. McBurney. Computational representation of persuasive argument. *Synthese*, 152:157–206, 2006.

[Atkinson, 2005] K. Atkinson. *What Should We Do?: Computational Representation of Persuasive Argument in Practical Reasoning*. PhD Thesis, Department of Computer Science, University of Liverpool, Liverpool, UK, 2005.

[Baker and Ginsberg, 1989] A.B. Baker and M.L. Ginsberg. A theorem prover for prioritized circumscription. In *Proceedings of the 11th International Joint Conference on Artificial Intelligence*, pages 463–467, 1989.

[Baroni and Giacomin, 2007] P. Baroni and M. Giacomin. On principle-based evaluation of extension-based argumentation semantics. *Artificial Intelligence*, 171:675–700, 2007.

[Baroni et al., 2011a] P. Baroni, M. Caminada, and M. Giacomin. An introduction to argumentation semantics. *The Knowledge Engineering Review*, 26:365–410, 2011.

[Baroni et al., 2011b] P. Baroni, P.E. Dunne, and M. Giacomin. On the resolution-based family of abstract argumentation semantics and its grounded instance. *Artificial Intelligence*, 175:791–813, 2011.

[Baroni et al., 2014] P. Baroni, G. Boella, F. Cerutti, M. Giacomin, L. van der Torre, and S. Villata. On the input/output behavior of argumentation frameworks. *Artificial Intelligence*, 217:144–197, 2014.

[Baroni et al., 2015] P. Baroni, M. Giacomin, and B. Lao. Dealing with generic contrariness in structured argumentation. In *Proceedings of the 24th International Joint Conference on Artificial Intelligence*, pages 2727–2733, 2015.

[Baumann and Brewka, 2010] R. Baumann and G. Brewka. Expanding argumentation frameworks: Enforcing and monotonicity results. In P. Baroni, F. Cerutti, M. Giacomin, and G.R. Simari, editors, *Computational Models of Argument. Proceedings of COMMA 2010*, pages 75–86. IOS Press, Amsterdam etc, 2010.

[Baumann, 2012a] R. Baumann. Normal and strong expansion equivalence for argumentation frameworks. *Artificial Intelligence*, 193:18–44, 2012.

[Baumann, 2012b] R. Baumann. What does it take to enforce an argument? minimal change in abstract argumentation. In *Proceedings of the 20th European Conference on Artificial Intelligence*, pages 127–132, 2012.

[Bench-Capon and Sartor, 2003] T.J.M. Bench-Capon and G. Sartor. A model of legal reasoning with cases incorporating theories and values. *Artificial Intelligence*, 150:97–143, 2003.

[Bench-Capon et al., 2000] T.J.M. Bench-Capon, T. Geldard, and P.H. Leng. A method for the computational modelling of dialectical argument with dialogue games. *Artificial Intelligence and Law*, 8:233–254, 2000.

[Bench-Capon, 1998] T.J.M. Bench-Capon. Specification and implementation of Toulmin dialogue game. In *Legal Knowledge-Based Systems. JURIX: The Eleventh Conference*, pages 5–19, Nijmegen, 1998. Gerard Noodt Instituut.

[Bench-Capon, 2002] T.J.M. Bench-Capon. The missing link revisited: the role of teleology in representing legal argument. *Artificial Intelligence and Law*, 10:79–94, 2002.

[Bench-Capon, 2003] T.J.M. Bench-Capon. Persuasion in practical argument using value-based argumentation frameworks. *Journal of Logic and Computation*, 13:429–448, 2003.

[Bench-Capon, 2017] T.J.M. Bench-Capon. HYPOs legacy: introduction to the virtual special issue. *Artificial Intelligence and Law*, 25:205–250, 2017.

[Benferhat et al., 1993] S. Benferhat, D. Dubois, and H. Prade. Argumentative inference in uncertain and inconsistent knowledge bases. In *Proceedings of the 9th International Conference on Uncertainty in Artificial Intelligence*, pages 411–419, 1993.

[Berman and Hafner, 1993] D.H. Berman and C.D. Hafner. Representing teleological structure in case-based legal reasoning: the missing link. In *Proceedings of the Fourth International Conference on Artificial Intelligence and Law*, pages 50–59, New York, 1993. ACM Press.

[Besnard and Doutre, 2004] Ph. Besnard and S. Doutre. Checking the acceptability of a set of arguments. In *Proceedings of the Tenth International Workshop on Nonmonotonic Reasoning*, pages 59–64, Chistler, Canada, 2004.

[Besnard and Hunter, 2001] Ph. Besnard and A. Hunter. A logic-based theory of deductive arguments. *Artificial Intelligence*, 128:203–235, 2001.

[Besnard and Hunter, 2008] Ph. Besnard and A. Hunter. *Elements of Argumentation*. MIT Press, Cambridge, MA, 2008.

[Besnard and Hunter, 2014] Ph. Besnard and H. Hunter. Constructing argument graphs with deductive arguments: a tutorial. *Argument and Computation*, 5:5–30, 2014.

[Bex et al., 2013a] F.J. Bex, J. Lawrence, M. Snaith, and C. Reed. Implementing the argument web. *Communications of the ACM*, 56:66–73, 2013.

[Bex et al., 2013b] F.J. Bex, S.J. Modgil, H. Prakken, and C.A. Reed. On logical specifications of the argument interchange format. *Journal of Logic and Computation*, 23:951–989, 2013.

[Bisquert et al., 2013] P. Bisquert, C. Cayrol, F. Dupin de Saint-Cyr, and M-C Lagasquie-Schiex. Goal-driven changes in argumentation: a theoretical framework and a tool. In *Proceedings of the 25th International Conference on Tools with Artificial Intelligence (ICTAI 2013)*, pages 610–617, 2013.

[Black and Atkinson, 2011] E. Black and K.D. Atkinson. Agreeing what to do. In P. McBurney, I. Rahwan, and S. Parsons, editors, *Argumentation in Multi-Agent Systems, 7th International Workshop, ArgMAS 2010, Toronto, Canada, May 2010. Revised Selected and Invited Papers*, number 6614 in Springer Lecture Notes in AI, pages 1–88. Springer Verlag, Berlin, 2011.

[Black and Hunter, 2007] E. Black and A. Hunter. A generative inquiry dialogue system. In *Proceedings of the Sixth International Conference on Autonomous Agents and Multiagent Systems*, pages 1010–1017, 2007.

[Black and Hunter, 2009] E. Black and A. Hunter. An inquiry dialogue system. *Journal of Autonomous Agents and Multi-Agent Systems*, 19:173–209, 2009.

[Black and Hunter, 2012] E. Black and A. Hunter. A relevance-theoretic framework for constructing and deconstructing enthymemes. *Journal of Logic and Computation*, 22:55–78, 2012.

[Bobrow, 1980] D.G. Bobrow, editor. *Artificial Intelligence*, volume 13. 1980. Special issue on Non-Monotonic Logic.

[Bochman, 2003] A. Bochman. Collective argumentation and disjunctive logic programming. *Journal of Logic and Computation*, 13:405–428, 2003.

[Bodenstaff et al., 2006] L. Bodenstaff, H. Prakken, and G. Vreeswijk. On formalising dialogue systems for argumentation in the event calculus. In *Proceedings of the Eleventh International Workshop on Nonmonotonic Reasoning*, pages 374–382, Windermere, UK, 2006.

[Boella et al., 2010a] G. Boella, D.M. Gabbay, L. van der Torre, and S. Villata. Support in abstract argumentation. In P. Baroni, F. Cerutti, M. Giacomin, and G.R. Simari, editors, *Computational Models of Argument. Proceedings of COMMA 2010*, pages 111–122. IOS Press, Amsterdam etc, 2010.

[Boella et al., 2010b] G. Boella, S. Kaci, and L. van der Torre. Dynamics in argumentation with single extensions: attack refinement and the grounded extension (extended version). In P. McBurney, I. Rahwan, S. Parsons, and N. Maudet, editors, *Argumentation in Multi-Agent Systems, 6th International Workshop, ArgMAS 2009, Budapest, Hungary, May 12, 2009. Revised Selected and Invited Papers*, number 6057 in Springer Lecture Notes in AI, pages 150–159. Springer Verlag, Berlin, 2010.

[Bondarenko et al., 1993] A. Bondarenko, R.A. Kowalski, and F. Toni. An assumption-based framework for non-monotonic reasoning. In *Proceedings of the second International*

Workshop on Logic Programming and Nonmonotonic Logic, pages 171–189, Lisbon (Portugal), 1993.
[Bondarenko et al., 1997] A. Bondarenko, P.M. Dung, R.A. Kowalski, and F. Toni. An abstract, argumentation-theoretic approach to default reasoning. *Artificial Intelligence*, 93:63–101, 1997.
[Brewka and Woltran, 2010] G. Brewka and S. Woltran. Abstract dialectical frameworks. In *Principles of Knowledge Representation and Reasoning: Proceedings of the Twelfth International Conference*, pages 102–111. AAAI Press, 2010.
[Brewka, 1989] G. Brewka. Preferred subtheories: An extended logical framework for default reasoning. In *Proceedings of the 11th International Joint Conference on Artificial Intelligence*, pages 1043–1048, 1989.
[Brewka, 2001] G. Brewka. Dynamic argument systems: a formal model of argumentation processes based on situation calculus. *Journal of Logic and Computation*, 11:257–282, 2001.
[Caminada and Amgoud, 2005] M. Caminada and L. Amgoud. An axiomatic account of formal argumentation. In *Proceedings of the 20th National Conference on Artificial Intelligence*, Pittsburgh, PA, 2005.
[Caminada and Amgoud, 2007] M. Caminada and L. Amgoud. On the evaluation of argumentation formalisms. *Artificial Intelligence*, 171:286–310, 2007.
[Caminada et al., 2012] M. Caminada, W.A. Carnielli, and P.E. Dunne. Semi-stable semantics. *Journal of Logic and Computation*, 22:1207–1254, 2012.
[Caminada et al., 2015] M. Caminada, S. Sa, J. Alcantara, and W. Dvorak. On the difference between assumption-based argumentation and abstract argumentation. *IFCoLog Journal of Logic and its Applications*, 2:15–34, 2015.
[Caminada, 2006] M. Caminada. On the issue of reinstatement in argumentation. In M. Fischer, W. van der Hoek, B. Konev, and A. Lisitsa, editors, *Logics in Artificial Intelligence. Proceedings of JELIA 2006*, number 4160 in Springer Lecture Notes in AI, pages 111–123, Berlin, 2006. Springer Verlag.
[Carlson, 1983] L. Carlson. *Dialogue Games: an Approach to Discourse Analysis*. Reidel Publishing Company, Dordrecht, 1983.
[Cartwright and Atkinson, 2009] D. Cartwright and K. Atkinson. Using computational argumentation to support e-participation. *IEEE Intelligent Systems*, 24:42–52, 2009. Special Issue on Transforming E-government and E-participation through IT.
[Cayrol and Lagasquie-Schiex, 2005a] C. Cayrol and M.-C. Lagasquie-Schiex. Graduality in argumentation. *Journal of Artificial Intelligence Research*, 23:245–297, 2005.
[Cayrol and Lagasquie-Schiex, 2005b] C. Cayrol and M.-C. Lagasquie-Schiex. On the acceptability of arguments in bipolar argumentation. In L. Godo, editor, *Proceedings of the 8nd European Conference on Symbolic and Quantitative Approaches to Reasoning with Uncertainty (ECSQARU 05)*, number 3571 in Springer Lecture Notes in AI, pages 378–389, Berlin, 2005. Springer Verlag.
[Cayrol and Lagasquie-Schiex, 2009] C. Cayrol and M.-C. Lagasquie-Schiex. Bipolar abstract argumentation systems. In I. Rahwan and G.R. Simari, editors, *Argumentation in Artificial Intelligence*, pages 65–84. Springer, Berlin, 2009.
[Cayrol and Lagasquie-Schiex, 2013] C. Cayrol and M.-C. Lagasquie-Schiex. Bipolarity in argumentation graphs: Towards a better understanding. *International Journal of Approximate Reasoning*, 54:876–899, 2013.
[Cayrol et al., 2003] C. Cayrol, S. Doutre, and J. Mengin. On decision problems related to the preferred semantics for argumentation frameworks. *Journal of Logic and Computation*, 13, 2003. 377-403.
[Cayrol et al., 2010] C. Cayrol, F Dupin de Saint-Cyr, and M.-C. Lagasquie-Schiex. Change in abstract argumentation frameworks: adding an argument. *Journal of Artificial Intelligence Research*, 38:49–84, 2010.
[Cayrol, 1995] C. Cayrol. On the relation between argumentation and non-monotonic coherence-based entailment. In *Proceedings of the 14th International Joint Conference on Artificial Intelligence*, pages 1443–1448, 1995.
[Chesñevar et al., 2004] C.I. Chesñevar, G. Simari, T. Alsinet, and L. Godo. A logic programming framework for possibilistic argumentation with vague knowledge. In *Proceedings of the 18th Conference on Uncertainty in Artificial Intelligence*, pages 76–84, 2004.

[Chesñevar et al., 2006] C.I. Chesñevar, J. McGinnis, S. Modgil, I. Rahwan, C. Reed, G. Simari, M. South, G. Vreeswijk, and S. Willmott. Towards an argument interchange format. *The Knowledge Engineering Review*, 21:293–316, 2006.

[Chisholm, 1957] R. Chisholm. *Perceiving: A Philosophical Study*. Cornell University Press, Ithaca, 1957.

[Chisholm, 1974] Roderick Chisholm. Practical reason and the logic of requirement. In Stephan Körner, editor, *Practical Reason*, pages 2–13. Blackwell Publishing Company, 1974.

[Conklin et al., 2001] J. Conklin, A. Selvin, S. Buckingham Shum, and M. Sierhuis. Facilitating hypertext for collective sensemaking: 15 years on from gIBIS. In *Proceedings of the The Twelfth ACM Conference on Hypertext and Hypermedia (Hypertext 2001)*, New York, 2001. ACM Press. Also available as Technical Report KMI-TR-112, Knowledge Media Institute, The Open University, UK.

[Coste-Marquis et al., 2006] S. Coste-Marquis, C. Devred, and P. Marquis. Constrained argumentation frameworks. In *Principles of Knowledge Representation and Reasoning: Proceedings of the Tenth International Conference*, pages 112–122. AAAI Press, 2006.

[Coste-Marquis et al., 2007] S. Coste-Marquis, C. Devred, S. Konieczny, and P. Marquis. On the merging of dung's argumentation systems. *Artificial Intelligence*, 171:730–753, 2007.

[Coste-Marquis et al., 2014] S. Coste-Marquis, S. Konieczny, J.-G. Mailly, and P. Marquis. On the revision of argumentation systems: minimal change of arguments statuses. In *Principles of Knowledge Representation and Reasoning: Proceedings of the Fourteenth International Conference*, pages 72–81. AAAI Press, 2014.

[Dimopoulos and Kakas, 1995] Y. Dimopoulos and A.C. Kakas. Logic programming without negation as failure. In *Proceedings of the Fifth International Logic Programming Symposium*, pages 369–384, Cambridge, MA, 1995. MIT Press.

[Doyle, 1979] J. Doyle. Truth maintenance systems. *Artificial Intelligence*, 12:231–272, 1979.

[Dung and Thang, 2010] P.M. Dung and P.M. Thang. Towards (probabilistic) argumentation for jury-based dispute resolution. In P. Baroni, F. Cerutti, M. Giacomin, and G.R. Simari, editors, *Computational Models of Argument. Proceedings of COMMA 2010*, pages 171–182. IOS Press, Amsterdam etc, 2010.

[Dung and Thang, 2014] P.M. Dung and P.M. Thang. Closure and consistency in logic-associated argumentation. *Journal of Artificial Intelligence Research*, 49:79–109, 2014.

[Dung et al., 2007] P.M. Dung, P. Mancarella, and F. Toni. Computing ideal sceptical argumentation. *Artificial Intelligence*, 171:642–674, 2007.

[Dung, 1992] P.M. Dung. Logic programming as dialog game. Technical Report, Division of Computer Science, Asian Institute of Technology, Bangkok, 1992.

[Dung, 1993] P.M. Dung. An argumentation semantics for logic programming with explicit negation. In *Proceedings of the Tenth Logic Programming Conference*, pages 616–630, Cambridge, MA, 1993. MIT Press.

[Dung, 1995] P.M. Dung. On the acceptability of arguments and its fundamental role in nonmonotonic reasoning, logic programming, and n–person games. *Artificial Intelligence*, 77:321–357, 1995.

[Dung, 2014] P.M. Dung. An axiomatic analysis of structured argumentation for prioritized default reasoning. In *Proceedings of the 21st European Conference on Artificial Intelligence*, pages 267–272, 2014.

[Dung, 2016] P.M. Dung. An axiomatic analysis of structured argumentation with priorities. *Artificial Intelligence*, 231:107–150, 2016.

[Dunne and Bench-Capon, 2002] P.E. Dunne and T.J. Bench-Capon. Coherence in finite argumentation systems. *Artificial Intelligence*, 141:187–203, 2002.

[Dunne and Bench-Capon, 2003] P.E. Dunne and T.J. Bench-Capon. Two party immediate response disputes: Properties and efficiency. *Artificial Intelligence*, 149:221–250, 2003.

[Dunne et al., 2011] P.E. Dunne, A. Hunter, P. McBurney, S. Parsons, and M. Wooldridge. Weighted argument systems: Basic definitions, algorithms, and complexity results. *Artificial Intelligence*, 175:457–486, 2011.

[Dunne, 2003] P.E. Dunne. Prevarication in dispute protocols. In *Proceedings of the Ninth International Conference on Artificial Intelligence and Law*, pages 12–21, New York, 2003. ACM Press.

[Dunne, 2006] P.E. Dunne. Suspicion of hidden agenda in persuasive argument. In P.E. Dunne and T.B.C. Bench-Capon, editors, *Computational Models of Argument. Proceedings of COMMA 2006*, pages 329–340. IOS Press, Amsterdam etc, 2006.

[Dunne, 2007] P.E. Dunne. Computational properties of argument systems satisfying graph-theoretical constraints. *Artificial Intelligence*, 171:701–729, 2007.

[Dvorak and Woltran, 2011] W. Dvorak and S. Woltran. On the intertranslatability of argumentation semantics. *Journal of Artificial Intelligence Research*, 41:445–475, 2011.

[Egly et al., 2010] U. Egly, S. Gaggl, and S. Woltran. Answer-set programming encodings for argumentation frameworks. *Argument and Computation*, 1:147–177, 2010.

[Elvang-Göransson et al., 1993] M. Elvang-Göransson, J. Fox, and P. Krause. Acceptability of arguments as logical uncertainty. In *Proceedings of the 2nd European Conference on Symbolic and Quantitative Approaches to Reasoning with Uncertainty (ECSQARU 93)*, pages 85–90, Berlin, 1993. Springer Verlag.

[Fan and Toni, 2013] X. Fan and F. Toni. Decision making with assumption-based argumentation. In E. Black, S. Modgil, and N. Oren, editors, *Theories and Applications of Formal Argumentation - Second International Workshop, TAFA 2013*, number 8306 in Springer Lecture Notes in AI, pages 127–142, Berlin, 2013. Springer Verlag.

[Fox and Parsons, 1997] J. Fox and S. Parsons. On using arguments for reasoning about actions and values. In *Proceedings of the AAAI Spring Symposium on Qualitative Preferences in Deliberation and Practical Reasoning*, Stanford, CA, 1997.

[Fox et al., 2007] J. Fox, D. Glasspool, D. Greca, S. Modgil, M. South, and V. Patkar. Argumentation-based inference and decision making – a medical perspective. *IEEE Intelligent Systems*, 22:34–41, 2007.

[Franqueira et al., 2011] V.N.L. Franqueira, T.T. Tun, Y.Yu, R. Wieringa, and B. Nuseibeh. Risk and argument: a risk-based argumentation method for practical security. In *Proceedings of the 19th IEEE International Requirements Engineering Conference*, pages 239–248, Trento, Italy, 2011.

[Gabbay, 1985] D. Gabbay. Theoretical foundations for non-monotonic reasoning in expert systems. In Krzysztof R. Apt, editor, *Logics and Models of Concurrent Systems*, pages 439–457. Springer-Verlag New York, Inc., New York, NY, USA, 1985.

[Gabbay, 2011] D. Gabbay. Introducing equational semantics for argumentation networks. In *Proceedings of the 11th European Conference on Symbolic and Quantitative Approaches to Reasoning with Uncertainty (ECSQARU 2011)*, number 6717 in Springer Lecture Notes in Computer Science, pages 19–35, Berlin, 2011. Springer Verlag.

[Gabbay, 2015] D. Gabbay. Editorial comment about "on the difference between ABA and AA. *IFCoLog Journal of Logic and its Applications*, 2:1–14, 2015.

[Garcia and Simari, 2004] A.J. Garcia and G.R. Simari. Defeasible logic programming: An argumentative approach. *Theory and Practice of Logic Programming*, 4:95–138, 2004.

[Garcia et al., 1998] A.J. Garcia, G.R. Simari, and C.I. Chesñevar. An argumentative framework for reasoning with inconsistent and incomplete information. In *Proceedings of the ECAI'98 Workshop on Practical Reasoning and Rationality*, Brighton, UK, 1998.

[Geffner and Pearl, 1992] H. Geffner and J. Pearl. Conditional entailment: bridging two approaches to default reasoning. *Artificial Intelligence*, 53:209–244, 1992.

[Geffner, 1991] H. Geffner. Beyond negation as failure. In *Principles of Knowledge Representation and Reasoning: Proceedings of the second International Conference*, pages 218–229. AAAI Press, 1991.

[Geffner, 1992] H. Geffner. *Default reasoning: causal and conditional theories*. MIT Press, Cambridge, MA, 1992.

[Gordon and Walton, 2009] T.F. Gordon and D.N. Walton. Proof burdens and standards. In I. Rahwan and G.R. Simari, editors, *Argumentation in Artificial Intelligence*, pages 239–258. Springer, Berlin, 2009.

[Gordon et al., 2007] T.F. Gordon, H. Prakken, and D.N. Walton. The Carneades model of argument and burden of proof. *Artificial Intelligence*, 171:875–896, 2007.

[Gordon, 1994] T.F. Gordon. The Pleadings Game: an exercise in computational dialectics. *Artificial Intelligence and Law*, 2:239–292, 1994.

[Gordon, 1995] T.F. Gordon. *The Pleadings Game. An Artificial Intelligence Model of Procedural Justice*. Kluwer Academic Publishers, Dordrecht/Boston/London, 1995.

[Gordon, 2011] T.F. Gordon. The policy making tool of the IMPACT argumentation toolbox. In *Proceedings of the Jurix Workshop on Modelling Policy-Making (MPM 2011)*, pages 29–38, 2011.
[Gorogiannis and Hunter, 2011] N. Gorogiannis and A. Hunter. Instantiating abstract argumentation with classical-logic arguments: postulates and properties. *Artificial Intelligence*, 175:1479–1497, 2011.
[Governatori et al., 2004] G. Governatori, M.J. Maher, G. Antoniou, and D. Billington. Argumentation semantics for defeasible logic. *Journal of Logic and Computation*, 14:675–702, 2004.
[Grasso et al., 2001] F. Grasso, A. Cawsey, and R. Jones. Dialectical argumentation to solve conflicts in advice giving: a case study in the promotion of healthy nutrition. *International Journal of Human-Computer Studies*, 53:1077–1115, 2001.
[Grossi and Modgil, 2015] D. Grossi and S. Modgil. On the graded acceptability of arguments. In *Proceedings of the 24th International Joint Conference on Artificial Intelligence*, pages 868–874, 2015.
[Grossi, 2010] D. Grossi. On the logic of argumentation theory. In *Proceedings of the 9th International Conference on Autonomous Agents and Multiagent Systems*, pages 1147–1154, 2010.
[Hadjinikolis et al., 2013] C. Hadjinikolis, S. Modgil, E. Black, P. McBurney, and Y. Siantos. Opponent modelling in persuasion dialogues. In *Proceedings of the 23rd International Joint Conference on Artificial Intelligence*, pages 164–170, 2013.
[Haenni et al., 2000] R. Haenni, J. Kohlas, and N. Lehmann. Probabilistic argumentation systems. In J. Kohlas and S. Moral, editors, *Defeasible reasoning and uncertainty management systems: algorithms*. Oxford University Press, Oxford, 2000.
[Hage et al., 1993] J.C. Hage, R.E. Leenes, and A.R. Lodder. Hard cases: a procedural approach. *Artificial Intelligence and Law*, 2:113–166, 1993.
[Hage, 1993] J.C. Hage. Monological reason-based logic: a low-level integration of rule-based reasoning and case-based reasoning. In *Proceedings of the Fourth International Conference on Artificial Intelligence and Law*, pages 30–39, New York, 1993. ACM Press.
[Haley et al., 2008] C. Haley, R. Laney, J. Moffett, and B. Nuseibeh. Security requirements engineering: A framework for representation and analysis. *IEEE Transactions on Software Engineering*, 34(1):133–153, 2008.
[Hamblin, 1970] C.L. Hamblin. *Fallacies*. Methuen, London, 1970.
[Hamblin, 1971] C.L. Hamblin. Mathematical models of dialogue. *Theoria*, 37:130–155, 1971.
[Hanks and McDermott, 1986] S. Hanks and D. McDermott. Default reasoning, nonmonotonic logics and the frame problem. In *Proceedings of the 5thth National Conference on Artificial Intelligence*, pages 328–333, 1986.
[Hart, 1949] H.L.A. Hart. The ascription of responsibility and rights. In *Proceedings of the Aristotelean Society*, pages 99–117, 1949. Reprinted in *Logic and Language. First Series*, ed. A.G.N. Flew, 145–166. Oxford: Basil Blackwell.
[Horty et al., 1990] J. Horty, R.H. Thomason, and D.S. Touretzky. A skeptical theory of inheritance in nonmonotonic semantic networks. *Artificial Intelligence*, 42:311–348, 1990.
[Horty, 2011] J. Horty. Rules and reasons in the theory of precedent. *Legal Theory*, 17:1–33, 2011.
[Hunter and Thimm, 2016] A. Hunter and M. Thimm. On partial information and contradictions in probabilistic abstract argumentation. In *Principles of Knowledge Representation and Reasoning: Proceedings of the Fifteenth International Conference*, pages 53–62. AAAI Press, 2016.
[Hunter and Williams, 2012] A. Hunter and M. Williams. Aggregating evidence about the positive and negative effects of treatments. *Artificial Intelligence in Medicine*, 56:173–190, 2012.
[Hunter and Woltran, 2013] A. Hunter and S. Woltran. Structural properties for deductive argument systems. In *Proceedings of the 12th European Conference on Symbolic and Quantitative Approaches to Reasoning with Uncertainty (ECSQARU 2013)*, number 7958 in Springer Lecture Notes in Computer Science, pages 278–289, Berlin, 2013. Springer Verlag.
[Hunter, 2004] A. Hunter. Making arguments more believable. In *Proceedings of the 19th National Conference on Artificial Intelligence*, pages 6269–274, 2004.

[Hunter, 2013] A. Hunter. A probabilistic approach to modelling uncertain logical arguments. *International Journal of Approximate Reasoning*, 54:47–81, 2013.
[Hunter, 2014] A. Hunter. Probabilistic qualification of attack in abstract argumentation. *International Journal of Approximate Reasoning*, 55:607–638, 2014.
[Jakobovits and Vermeir, 1999a] H. Jakobovits and D. Vermeir. Dialectic semantics for argumentation frameworks. In *Proceedings of the Seventh International Conference on Artificial Intelligence and Law*, pages 53–62, New York, 1999. ACM Press.
[Jakobovits and Vermeir, 1999b] H. Jakobovits and D. Vermeir. Robust semantics for argumentation frameworks. *Journal of Logic and Computation*, 9:215–261, 1999.
[Jakobovits, 2000] H. Jakobovits. *On the Theory of Argumentation Frameworks*. Doctoral dissertation Free University Brussels, 2000.
[Kaci et al., 2007] S. Kaci, L. van der Torre, and E. Weydert. On the acceptability of incompatible arguments. In *Proceedings of the 9th European Conference on Symbolic and Quantitative Approaches to Reasoning with Uncertainty (ECSQARU 07)*, number 4724 in Springer Lecture Notes in AI, page 247258, Berlin, 2007. Springer Verlag.
[Kaci, 2010] S. Kaci. Refined preference-based argumentation frameworks. In P. Baroni, F. Cerutti, M. Giacomin, and G.R. Simari, editors, *Computational Models of Argument. Proceedings of COMMA 2010*. IOS Press, Amsterdam etc, 2010.
[Kakas and Moraitis, 2003] A.C. Kakas and P. Moraitis. Argumentation based decision making for autonomous agents. In *Proceedings of the Second International Conference on Autonomous Agents and Multiagent Systems*, pages 883–890, 2003.
[Kakas et al., 1992] A.C. Kakas, R.A. Kowalski, and F. Toni. Abductive logic programming. *Journal of Logic and Computation*, 2:719–770, 1992.
[Kirschner et al., 2003] P.A. Kirschner, S.J. Buckingham Schum, and C.S. Carr. *Visualizing Argumentation. Software Tools for Collaborative and Educational Sense-Making*. Springer, London etc, 2003.
[Kok et al., 2011] E. Kok, J.-J.Ch. Meyer, H. Prakken, and G. Vreeswijk. A formal argumentation framework for deliberation dialogues. In P. McBurney, I. Rahwan, and S. Parsons, editors, *Argumentation in Multi-Agent Systems, 7th International Workshop, ArgMAS 2010, Toronto, Canada, May 2010. Revised Selected and Invited Papers*, number 6614 in Springer Lecture Notes in AI, pages 31–48. Springer Verlag, Berlin, 2011.
[Kok, 2013] E.M. Kok. *Exploring the Practical Benefits of Argumentation in Multi-Agent Deliberation*. Doctoral dissertation Department of Information and Computing Sciences, Utrecht University, 2013.
[Konolige, 1988] K. Konolige. Defeasible argumentation in reasoning about events. In Z.W. Ras and L. Saitta, editors, *Methodologies for Intelligent Systems*, page 380390. Elsevier, Amsterdam, 1988.
[Kowalski and Toni, 1996] R.A. Kowalski and F. Toni. Abstract argumentation. *Artificial Intelligence and Law*, 4:275–296, 1996.
[Kraus et al., 1990] S. Kraus, D.J. Lehmann, and M. Magidor. Nonmonotonic reasoning, preferential models and cumulative logics. *Artificial Intelligence*, 44:167–207, 1990.
[Kraus et al., 1998] S. Kraus, K. Sycara, and A. Evenchik. Reaching agreements through argumentation: a logical model and implementation. *Artificial Intelligence*, 104:1–69, 1998.
[Krause et al., 1995] P. Krause, S. Ambler, M. Elvang-Gøransson, and J. Fox. A logic of argumentation for reasoning under uncertainty. *Computational Intelligence*, 11(1):113–131, 1995.
[Li et al., 2012] H. Li, N. Oren, and T. Norman. Probabilistic argumentation frameworks. In S. Modgil, N. Oren, and F. Toni, editors, *Theorie and Applications of Formal Argumentation. First International Workshop, TAFA 2011. Barcelona, Spain, July 16-17, 2011, Revised Selected Papers*, number 7132 in Springer Lecture Notes in AI, pages 1–16, Berlin, 2012. Springer Verlag.
[Lin and Shoham, 1989] F. Lin and Y. Shoham. Argument systems. A uniform basis for nonmonotonic reasoning. In *Principles of Knowledge Representation and Reasoning: Proceedings of the First International Conference*, pages 245–255, San Mateo, CA, 1989. Morgan Kaufmann Publishers.
[Lippi and Torroni, 2016] M. Lippi and P. Torroni. Argument mining: state of the art and emerging trends. *ACM Transactions on Internet Technology*, 10(2), 2016.

[Lodder, 1999] A.R. Lodder. *DiaLaw. On Legal Justification and Dialogical Models of Argumentation*. Law and Philosophy Library. Kluwer Academic Publishers, Dordrecht/Boston/London, 1999.
[Lorenzen and Lorenz, 1978] P. Lorenzen and K. Lorenz. *Dialogische Logik*. Wissenschaftliche Buchgesellschaft, Darmstadt, 1978.
[Loui et al., 1993] R.P. Loui, J. Norman, J. Olson, and A. Merrill. A design for reasoning with policies, precedents, and rationales. In *Proceedings of the Fourth International Conference on Artificial Intelligence and Law*, pages 202–211, New York, 1993. ACM Press.
[Loui, 1987] R.P. Loui. Defeat among arguments: a system of defeasible inference. *Computational Intelligence*, 2:100–106, 1987.
[Loui, 1995] R.P. Loui. Hart's critics on defeasible concepts and ascriptivism. In *Proceedings of the Fifth International Conference on Artificial Intelligence and Law*, pages 21–30, New York, 1995. ACM Press.
[Loui, 1998] R.P. Loui. Process and policy: resource-bounded non-demonstrative reasoning. *Computational Intelligence*, 14:1–38, 1998.
[Mackenzie, 1979] J.D. Mackenzie. Question-begging in non-cumulative systems. *Journal of Philosophical Logic*, 8:117–133, 1979.
[Mackenzie, 1990] J.D. Mackenzie. Four dialogue systems. *Studia Logica*, 51:567–583, 1990.
[Makinson, 1994] D. Makinson. General patterns in nonmonotonic reasoning. In D. Gabbay, C.J. Hogger, and J.A. Robinson, editors, *Handbook of Logic in Artificial Intelligence and Logic Programming*, pages 35–110. Clarendon Press, Oxford, 1994.
[Martinez et al., 2006] D.C. Martinez, A.J. Garcia, and G.R. Simari. On acceptability in abstract argumentation frameworks with an extended defeat relation. In P.E. Dunne and T.B.C. Bench-Capon, editors, *Computational Models of Argument. Proceedings of COMMA 2006*, pages 273–278. IOS Press, Amsterdam etc, 2006.
[Matt and Toni, 2008] P.-A. Matt and F. Toni. A game-theoretic measure of argument strength for abstract argumentation. In S. Hoelldobler, C. Lutz, and H. Wansing, editors, *Proceedings of the 11th European Conference on Logics in Artificial Intelligence (JELIA 2008)*, number 5293 in Springer Lecture Notes in AI, pages 285–297, Berlin, 2008. Springer Verlag.
[Maudet and Moore, 1999] N. Maudet and D. Moore. Dialogue games for computer-supported collaborative argumentation. In *Proceedings of the Workshop on Computer-Supported Collaborative Argumentation for Learning Communities*, Stanford, 1999.
[McBurney and Parsons, 2001] P. McBurney and S. Parsons. Representing epistemic uncertainty by means of dialectical argumentation. *Annals of Mathematics and Artificial Intelligence*, 32:125–169, 2001.
[McBurney and Parsons, 2002] P. McBurney and S. Parsons. Games that agents play: A formal framework for dialogues between autonomous agents. *Journal of Logic, Language and Information*, 13:315–343, 2002.
[McBurney et al., 2002] P. McBurney, S. Parsons, and M. Wooldridge. Desiderata for agent argumentation protocols. In *Proceedings of the First International Conference on Autonomous Agents and Multiagent Systems*, pages 402–409, 2002.
[McBurney et al., 2007] P. McBurney, D. Hitchcock, and S. Parsons. The eightfold way of deliberation dialogue. *International Journal of Intelligent Systems*, 22:95–132, 2007.
[McCarty, 1977] L.T. McCarty. Reflections on TAXMAN: An experiment in artificial intelligence and legal reasoning. *Harvard Law Review*, 90:89–116, 1977.
[McCarty, 1995] L.T. McCarty. An implementation of Eisner v. Macomber. In *Proceedings of the Fifth International Conference on Artificial Intelligence and Law*, pages 276–286, New York, 1995. ACM Press.
[Modgil and Caminada, 2009] S. Modgil and M. Caminada. Proof theories and algorithms for abstract argumentation frameworks. In I. Rahwan and G.R. Simari, editors, *Argumentation in Artificial Intelligence*, pages 105–129. Springer, Berlin, 2009.
[Modgil and Prakken, 2012] S. Modgil and H. Prakken. Resolutions in structured argumentation. In B. Verheij, S. Woltran, and S. Szeider, editors, *Computational Models of Argument. Proceedings of COMMA 2012*, pages 310–321. IOS Press, Amsterdam etc, 2012.
[Modgil and Prakken, 2013] S. Modgil and H. Prakken. A general account of argumentation with preferences. *Artificial Intelligence*, 195:361–397, 2013.

[Modgil and Prakken, 2014] S. Modgil and H. Prakken. The ASPIC+ framework for structured argumentation: a tutorial. *Argument and Computation*, 5:31–62, 2014.
[Modgil et al., 2013] S. Modgil, F. Toni, F. Bex, I. Bratko, C. Chesñevar, W. Dvorak, M. Falappa, X. Fan, S. Gaggl, A. Garcia, M. Gonzalez, T. Gordon, J. Leite, M. Mozina, C. Reed, G. Simari, S. Szeider, P. Torroni, and S. Woltran. The added value of argumentation. In S. Ossowski, editor, *Agreement Technologies*, pages 357–403. Springer, 2013.
[Modgil, 2006] S. Modgil. Hierarchical argumentation. In M. Fischer, W. van der Hoek, B. Konev, and A. Lisitsa, editors, *Logics in Artificial Intelligence. Proceedings of JELIA 2006*, number 4160 in Springer Lecture Notes in AI, pages 319–332, Berlin, 2006. Springer Verlag.
[Modgil, 2009] S. Modgil. Reasoning about preferences in argumentation frameworks. *Artificial Intelligence*, 173:901–934, 2009.
[Moore, 1993] D. Moore. *Dialogue game theory for intelligent tutoring systems*. PhD Thesis, Leeds Metropolitan University, 1993.
[Nielsen and Parsons, 2007a] S.H. Nielsen and S. Parsons. An application of formal argumentation: fusing Bayesian networks in multi-agent systems. *Artificial Intelligence*, 171:754–775, 2007.
[Nielsen and Parsons, 2007b] S.H. Nielsen and S. Parsons. A generalization of Dung's abstract framework for argumentation: Arguing with sets of attacking arguments. In S. Parsons, N. Maudet, and I. Rahwan, editors, *Argumentation in Multi-Agent Systems. Third International Workshop, ArgMAS 2006, Hakodate, Japan, May 8, 2006, Revised Selected and Invited Papers*, number 4766 in Springer Lecture Notes in AI, pages 54–73. Springer Verlag, Berlin, 2007.
[Nouioua and Risch, 2011] F. Nouioua and V. Risch. Argumentation frameworks with necessities. In *Proceedings of the 4th International Conference on Scalable Uncertainty Management (SUM'11)*, number 6929 in Springer Lecture Notes in AI, pages 163–176, Berlin, 2011. Springer Verlag.
[Nute, 1988] Donald Nute. Defeasible reasoning: A philosophical analysis in prolog. In J. Fetzer, editor, *Aspects of Artificial Intelligence*, pages 251–288. Kluwer Academic Publishers, 1988.
[Nute, 1994] D. Nute. Defeasible logic. In D. Gabbay, C.J. Hogger, and J.A. Robinson, editors, *Handbook of Logic in Artificial Intelligence and Logic Programming*, pages 253–395. Clarendon Press, Oxford, 1994.
[Oren and Norman, 2008] N. Oren and T.J. Norman. Semantics for evidence-based argumentation. In Ph. Besnard, S. Doutre, and A. Hunter, editors, *Computational Models of Argument. Proceedings of COMMA 2008*, pages 276–284, Amsterdam etc, 2008. IOS Press.
[Oren and Norman, 2010] N. Oren and T. Norman. Arguing using opponent models. In P. McBurney, I. Rahwan, S. Parsons, and N. Maudet, editors, *Argumentation in Multi-Agent Systems, 6th International Workshop, ArgMAS 2009, Budapest, Hungary, May 12, 2009. Revised Selected and Invited Papers*, number 6057 in Springer Lecture Notes in AI, pages 160–174. Springer Verlag, Berlin, 2010.
[Palau and Moens, 2009] R. Mochales Palau and M.-F. Moens. Argumentation mining: the detection, classification and structure of arguments in text. In *Proceedings of the Tenth International Conference on Artificial Intelligence and Law*, pages 98–107, New York, 2009. ACM Press.
[Parsons and Jennings, 1996] S. Parsons and N. Jennings. Negotiation through argumentation - a preliminary report. In *Proceedings of the 2nd International Conference in Multiagent Systems*, pages 267–274, Kyoto, Japan, 1996.
[Parsons et al., 1998] S. Parsons, C. Sierra, and N.R. Jennings. Agents that reason and negotiate by arguing. *Journal of Logic and Computation*, 8:261–292, 1998.
[Parsons et al., 2002] S. Parsons, M. Wooldridge, and L. Amgoud. An analysis of formal interagent dialogues. In *Proceedings of the First International Conference on Autonomous Agents and Multiagent Systems*, pages 394–401, 2002.
[Parsons et al., 2003] S. Parsons, M. Wooldridge, and L. Amgoud. Properties and complexity of some formal inter-agent dialogues. *Journal of Logic and Computation*, 13, 2003. 347-376.

[Parsons, 1998a] S. Parsons. On precise and correct qualitative probabilistic reasoning. *International Journal of Approximate Reasoning*, 35:111–135, 1998.

[Parsons, 1998b] S. Parsons. A proof-theoretic approach to qualitative probabilistic reasoning. *International Journal of Approximate Reasoning*, 19:265–297, 1998.

[Perelman and Olbrechts-Tyteca, 1969] Ch. Perelman and L. Olbrechts-Tyteca. *The New Rhetoric. A Treatise on Argumentation*. University of Notre Dame Press, Notre Dame, Indiana, 1969.

[Pinkwart and McLaren, 2012] N. Pinkwart and B.M. McLaren, editors. *Educational Technologies for Teaching Argumentation Skills*. Bentham Science Publishers, Sharjah, UAE, 2012.

[Pollock, 1970] John Pollock. The structure of epistemic justification. In *Studies in the Theory of Knowledge*, American Philosophical Quarterly Monograph Series, number 4, pages 62–78. Basil Blackwell Publisher, Inc., 1970.

[Pollock, 1974] J.L. Pollock. *Knowledge and Justification*. Princeton University Press, Princeton, 1974.

[Pollock, 1987] J.L. Pollock. Defeasible reasoning. *Cognitive Science*, 11:481–518, 1987.

[Pollock, 1994] J.L. Pollock. Justification and defeat. *Artificial Intelligence*, 67:377–408, 1994.

[Pollock, 1995] J.L. Pollock. *Cognitive Carpentry. A Blueprint for How to Build a Person*. MIT Press, Cambridge, MA, 1995.

[Pollock, 2002] J.L. Pollock. Defeasible reasoning with variable degrees of justification. *Artificial Intelligence*, 133:233–282, 2002.

[Pollock, 2007a] J.L. Pollock. Reasoning and probability. *Law, Probability and Risk*, 6:43–58, 2007a.

[Pollock, 2007b] J.L. Pollock. Defeasible reasoning. In J. Adler and L. Rips, editors, *Reasoning: Studies of Human Inference and its Foundations*, pages 451–470. Cambridge, Cambridge University Press, 2007b.

[Pollock, 2010] J.L. Pollock. Defeasible reasoning and degrees of justification. *Argument and Computation*, 1:7–22, 2010.

[Poole, 1989] D.L. Poole. Explanation and prediction: an architecture for default and abductive reasoning. *Computational Intelligence*, 5:97–110, 1989.

[Prakken and Horty, 2012] H. Prakken and J. Horty. An appreciation of John Pollock's work on the computational study of argument. *Argument and Computation*, 3:1–19, 2012.

[Prakken and Sartor, 1996] H. Prakken and G. Sartor. A dialectical model of assessing conflicting arguments in legal reasoning. *Artificial Intelligence and Law*, 4:331–368, 1996.

[Prakken and Sartor, 1997] H. Prakken and G. Sartor. Argument-based extended logic programming with defeasible priorities. *Journal of Applied Non-classical Logics*, 7:25–75, 1997.

[Prakken and Sartor, 1998] H. Prakken and G. Sartor. Modelling reasoning with precedents in a formal dialogue game. *Artificial Intelligence and Law*, 6:231–287, 1998.

[Prakken and Sartor, 2009] H. Prakken and G. Sartor. A logical analysis of burdens of proof. In H. Kaptein, H. Prakken, and B. Verheij, editors, *Legal Evidence and Proof: Statistics, Stories, Logic*, pages 223–253. Ashgate Publishing, Farnham, 2009.

[Prakken and Sartor, 2015] H. Prakken and G. Sartor. Law and logic: A review from an argumentation perspective. *Artificial Intelligence*, 227:214–225, 2015.

[Prakken and Vreeswijk, 2002] H. Prakken and G.A.W. Vreeswijk. Logics for defeasible argumentation. In D. Gabbay and F. Günthner, editors, *Handbook of Philosophical Logic*, volume 4, pages 219–318. Kluwer Academic Publishers, Dordrecht/Boston/London, second edition, 2002.

[Prakken et al., 2015] H. Prakken, A.Z. Wyner, T.J.M. Bench-Capon, and K. Atkinson. A formalisation of argumentation schemes for legal case-based reasoning in ASPIC+. *Journal of Logic and Computation*, 25:1141–1166, 2015.

[Prakken, 1993] H. Prakken. An argumentation framework in default logic. *Annals of Mathematics and Artificial Intelligence*, 9:91–132, 1993.

[Prakken, 1999] H. Prakken. Dialectical proof theory for defeasible argumentation with defeasible priorities (preliminary report). In J.-J.Ch. Meyer and P.-Y. Schobbens, editors, *Formal Models of Agents*, number 1760 in Springer Lecture Notes in AI, pages 202–215, Berlin, 1999. Springer Verlag.

[Prakken, 2001a] H. Prakken. Modelling reasoning about evidence in legal procedure. In *Proceedings of the Eighth International Conference on Artificial Intelligence and Law*, pages 119–128, New York, 2001. ACM Press.

[Prakken, 2001b] H. Prakken. Relating protocols for dynamic dispute with logics for defeasible argumentation. *Synthese*, 127:187–219, 2001.

[Prakken, 2005] H. Prakken. Coherence and flexibility in dialogue games for argumentation. *Journal of Logic and Computation*, 15:1009–1040, 2005.

[Prakken, 2008] H. Prakken. A formal model of adjudication dialogues. *Artificial Intelligence and Law*, 16:305–328, 2008.

[Prakken, 2010] H. Prakken. An abstract framework for argumentation with structured arguments. *Argument and Computation*, 1:93–124, 2010.

[Prakken, 2012] H. Prakken. Some reflections on two current trends in formal argumentation. In *Logic Programs, Norms and Action. Essays in Honour of Marek J. Sergot on the Occasion of his 60th Birthday*, pages 249–272. Springer, Berlin/Heidelberg, 2012.

[Prakken, 2014] H. Prakken. On support relations in abstract argumentation as abstractions of inferential relations. In *Proceedings of the 21st European Conference on Artificial Intelligence*, pages 735–740, 2014.

[Rahwan and Larson, 2008] I. Rahwan and K. Larson. Mechanism design for abstract argumentation. In *Proceedings of the Seventh International Conference on Autonomous Agents and Multiagent Systems*, pages 1031–1038, 2008.

[Rahwan et al., 2009] I. Rahwan, K. Larson, and F. Thomé. A characterisation of strategy-proofness for grounded argumentation semantics. In *Proceedings of the 21st International Joint Conference on Artificial Intelligence*, pages 251–256, 2009.

[Rao and Georgeff, 1991] A.S. Rao and M.P. Georgeff. Modelling rational agents within a BDI-architecture. In *Principles of Knowledge Representation and Reasoning: Proceedings of the second International Conference*, pages 473–484. AAAI Press, 1991.

[Raz, 1975] J. Raz. *Practical Reason and Norms*. Princeton University Press, Princeton, 1975.

[Reed and Norman, 2003] C. Reed and T.J. Norman, editors. *Argumentation Machines. New Frontiers in Argument and Computation*, volume 9 of *Argumentation Library*. Kluwer Academic Publishers, Boston/Dordrecht/London, 2003.

[Reed et al., 2007] C. Reed, D. Walton, and F. Macagno. Argument diagramming in logic, law and artificial intelligence. *The Knowledge Engineering Review*, 22:87–109, 2007.

[Reiter, 1980] R. Reiter. A logic for default reasoning. *Artificial Intelligence*, 13:81–132, 1980.

[Rescher, 1976] N. Rescher. *Plausible Reasoning*. Van Gorcum, Assen, 1976.

[Rescher, 1977] N. Rescher. *Dialectics: a Controversy-oriented Approach to the Theory of Knowledge*. State University of New York Press, Albany, N.Y., 1977.

[Rienstra et al., 2013] T. Rienstra, M. Thimm, and N. Oren. Opponent modelling with uncertainty for strategic argumentation. In *Proceedings of the 23rd International Joint Conference on Artificial Intelligence*, pages 332–338, 2013.

[Rissland and Ashley, 1987] E.L. Rissland and K.D. Ashley. A case-based system for trade secrets law. In *Proceedings of the First International Conference on Artificial Intelligence and Law*, pages 60–66, New York, 1987. ACM Press.

[Riveret et al., 2007] R. Riveret, A. Rotolo, G. Sartor, H. Prakken, and B. Roth. Success chances in argument games: a probabilistic approach to legal disputes. In A.R. Lodder and L. Mommers, editors, *Legal Knowledge and Information Systems. JURIX 2007: The Twentieth Annual Conference*, pages 99–108. IOS Press, Amsterdam etc., 2007.

[Riveret et al., 2008] R. Riveret, H. Prakken, A. Rotolo, and G. Sartor. Heuristics in argumentation: a game-theoretical investigation. In Ph. Besnard, S. Doutre, and A. Hunter, editors, *Computational Models of Argument. Proceedings of COMMA 2008*, pages 324–335. IOS Press, Amsterdam etc, 2008.

[Ross, 1930] W.D. Ross. *The Right and the Good*. Oxford University Press, Oxford, 1930.

[Rotstein et al., 2008] N.D. Rotstein, M.O. Moguillansky, M.A. Falappa, A.J. Garcia, and G.R. Simari. Argument theory change: Revision upon warrant. In Ph. Besnard, S. Doutre, and A. Hunter, editors, *Computational Models of Argument. Proceedings of COMMA 2008*, pages 336–347, Amsterdam etc, 2008. IOS Press.

[Scheuer et al., 2010] O. Scheuer, F. Loll, N. Pinkwart, and B.M. McLaren. Computer-supported argumentation: A review of the state-of-the-art. *International Journal of Computer-Supported Collaborative Learning*, 5:43–102, 2010.

[Schneider et al., 2013] J. Schneider, T. Groza, and A. Passant. A review of argumentation for the social semantic web. *Semantic Web*, 4:159–218, 2013.

[Searle, 1969] J.R. Searle. *Speech Acts. An Essay in the Philosophy of Language*. Cambridge University Press, Cambridge, UK, 1969.

[Simari and Loui, 1992] G.R. Simari and R.P. Loui. A mathematical treatment of defeasible argumentation and its implementation. *Artificial Intelligence*, 53:125–157, 1992.

[Snaid and Reed, 2012] M. Snaid and C. Reed. TOAST: Online ASPIC+ argumentation. In B. Verheij, S. Woltran, and S. Szeider, editors, *Computational Models of Argument. Proceedings of COMMA 2012*, pages 509–510. IOS Press, Amsterdam etc, 2012.

[Strass, 2013] H. Strass. Instantiating knowledge bases in abstract dialectical frameworks. In J. Leite, T.C. Son, P. Torroni, L. van der Torre, and S. Woltran, editors, *Proceedings of the 14th International Workshop on Computational Logic in Multi-Agent Systems (CLIMA XIV)*, number 8143 in Springer Lecture Notes in AI, pages 86–101, Berlin, 2013. Springer Verlag.

[Sycara, 1985] K.P. Sycara. Arguments of persuasion in labour mediation. In *Proceedings of the 9th International Joint Conference on Artificial Intelligence*, pages 294–296, 1985.

[Sycara, 1990] K.P. Sycara. Persuasive argumentation in negotiation. *Theory and Decision*, 28:203–242, 1990.

[Tang and Parsons, 2005] Y. Tang and S. Parsons. Argumentation-based dialogues for deliberation. In *Proceedings of the Fourth International Conference on Autonomous Agents and Multiagent Systems (AAMAS-05)*, pages 552–559, 2005.

[ter Berg et al., 2009] T. ter Berg, T. van Gelder, F. Patterson, and S. Teppema. *Critical Thinking: Reasoning and Communicating with Rationale*. Pearson Education Benelux, Amsterdam, 2009.

[Thimm and Garcia, 2010] M. Thimm and A.J. Garcia. Classification and strategical issues of argumentation games on structured argumentation frameworks. In *Proceedings of the 9th International Conference on Autonomous Agents and Multiagent Systems*, pages 1247–1254, 2010.

[Thimm, 2014] M. Thimm. Strategic argumentation in multi-agent systems. *Kuenstliche Intelligenz*, 28:159–168, 2014.

[Timmer et al., 2017] S. Timmer, J.-J.Ch. Meyer, H. Prakken, S. Renooij, and B. Verheij. A two-phase method for extracting explanatory arguments from Bayesian networks. *International Journal of Approximate Reasoning*, 80:475–494, 2017.

[Tolchinsky et al., 2006] P. Tolchinsky, U. Cortes, S. Modgil, F. Caballero, and A. Lopez-Navidad. Increasing human-organ transplant availability: argumentation-based agent deliberation. *IEEE Intelligent Systems*, 21:30–37, 2006.

[Tolchinsky et al., 2012] P. Tolchinsky, S. Modgil, K.D. Atkinson, P. McBurney, and U. Cortes. Deliberation dialogues for reasoning about safety critical actions. *Journal of Autonomous Agents and Multi-Agent Systems*, 25:209–259, 2012.

[Toni, 2014] F. Toni. A tutorial on assumption-based argumentation. *Argument and Computation*, 5:89–117, 2014.

[Toulmin, 1958] S.E. Toulmin. *The Uses of Argument*. Cambridge University Press, Cambridge, 1958.

[Touretzky, 1984] D.S. Touretzky. Implicit ordering of defaults in inheritance systems. In *Proceedings of the 4th National Conference on Artificial Intelligence*, pages 322–325, 1984.

[Touretzky, 1986] David Touretzky. *The Mathematics of Inheritance Systems*. Morgan Kaufmann, 1986.

[van der Weide et al., 2011] T. van der Weide, F. Dignum, J.-J.Ch. Meyer, H. Prakken, and G. Vreeswijk. Arguing about preferences and decisions. In P. McBurney, I. Rahwan, and S. Parsons, editors, *Argumentation in Multi-Agent Systems, 7th International Workshop, ArgMAS 2010, Toronto, Canada, May 2010. Revised Selected and Invited Papers*, number 6614 in Springer Lecture Notes in AI, pages 68–85. Springer Verlag, Berlin, 2011.

[van der Weide, 2011] T.L. van der Weide. *Arguing to Motivate Decisions*. Doctoral dissertation Department of Information and Computing Sciences, Utrecht University, 2011.

[van Veenen and Prakken, 2006] J. van Veenen and H. Prakken. A protocol for arguing about rejections in negotiation. In S. Parsons, N. Maudet, P. Moraitis, and I. Rahwan, editors, *Argumentation in Multi-Agent Systems: Second International Workshop, ArgMAS 2005, Utrecht, Netherlands, July 26, 2005, Revised Selected and Invited Papers*, number 4049 in Springer Lecture Notes in AI, pages 138–153. Springer Verlag, Berlin, 2006.

[Verheij, 1996] B. Verheij. Two approaches to dialectical argumentation: admissible sets and argumentation stages. In *Proceedings of the Eighth Dutch Conference on Artificial Intelligence (NAIC-96)*, pages 357–368, Utrecht, The Netherlands, 1996.

[Verheij, 2003] B. Verheij. DefLog: on the logical interpretation of prima facie justified assumptions. *Journal of Logic and Computation*, 13:319–346, 2003.

[Verheij, 2007] B. Verheij. A labeling approach to the computation of credulous acceptance in argumentation. In *Proceedings of the 2oth International Joint Conference on Artificial Intelligence (IJCAI-07)*, pages 623–628, 2007.

[Vreeswijk and Prakken, 2000] G.A.W. Vreeswijk and H. Prakken. Credulous and sceptical argument games for preferred semantics. In *Proceedings of the 7th European Workshop on Logics in Artificial Intelligence (JELIA'2000)*, number 1919 in Springer Lecture Notes in AI, pages 239–253, Berlin, 2000. Springer Verlag.

[Vreeswijk, 1991] G.A.W. Vreeswijk. The feasibility of defeat in defeasible reasoning. In *Principles of Knowledge Representation and Reasoning: Proceedings of the Second International Conference*, pages 526–534, San Mateo, CA, 1991. Morgan Kaufmann Publishers.

[Vreeswijk, 1993a] G.A.W. Vreeswijk. Defeasible dialectics: a controversy-oriented approach towards defeasible argumentation. *Journal of Logic and Computation*, 3:317–334, 1993.

[Vreeswijk, 1993b] G.A.W. Vreeswijk. *Studies in Defeasible Argumentation*. Doctoral dissertation Free University Amsterdam, 1993.

[Vreeswijk, 1997] G.A.W. Vreeswijk. Abstract argumentation systems. *Artificial Intelligence*, 90:225–279, 1997.

[Vreeswijk, 2006] G.A.W. Vreeswijk. An algorithm to compute minimally grounded and admissible defence sets in argumentation sytems. In P.E. Dunne and T.B.C. Bench-Capon, editors, *Computational Models of Argument. Proceedings of COMMA 2006*, pages 109–120. IOS Press, Amsterdam etc, 2006.

[Walton and Krabbe, 1995] D.N. Walton and E.C.W. Krabbe. *Commitment in Dialogue. Basic Concepts of Interpersonal Reasoning*. State University of New York Press, Albany, NY, 1995.

[Walton, 1984] D.N. Walton. *Logical dialogue-games and fallacies*. University Press of America, Inc., Lanham, MD, 1984.

[Walton, 1996] D.N. Walton. *Argumentation Schemes for Presumptive Reasoning*. Lawrence Erlbaum Associates, Mahwah, NJ, 1996.

[Walton, 2006a] D.N. Walton. *Fundamentals of Critical Argumentation*. Cambridge University Press, Cambridge, 2006.

[Walton, 2006b] D.N. Walton. Metadialogues for resolving burden of proof disputes. *Argumentation*, 2006. To appear.

[Williams and Williamson, 2006] M. Williams and J. Williamson. Combining argumentation and Bayesian Nets for breast cancer prognosis. *Journal of Logic, Language and Information*, 15:155–178, 2006.

[Woods and Walton, 1978] J. Woods and D.N. Walton. Arresting circles in formal dialogues. *Journal of Philosophical Logic*, 7:73–90, 1978.

[Wooldridge and Parsons, 2000] M. Wooldridge and S. Parsons. Languages for negotiation. In *Proceedings of the Fourteenth European Conference on Artificial Intelligence*, pages 393–400, 2000.

[Wooldridge et al., 2006] M. Wooldridge, P. McBurney, and S. Parsons. On the meta-logic of arguments. In S. Parsons, N. Maudet, P. Moraitis, and I. Rahwan, editors, *Argumentation in Multi-Agent Systems: Second International Workshop, ArgMAS 2005, Utrecht, Netherlands, July 26, 2005, Revised Selected and Invited Papers*, number 4049 in Springer Lecture Notes in AI, pages 42–56. Springer Verlag, Berlin, 2006.

[Wu and Podlaszewski, 2015] Y. Wu and M. Podlaszewski. Implementing crash-resistence and non-interference in logic-based argumentation. *Journal of Logic and Computation*, 25:303–333, 2015.

[Yuan, 2004] T. Yuan. *Human-computer debate, a computational dialectics approach*. PhD Thesis, Leeds Metropolitan University, 2004.

Henry Prakken
Department of Information and Computing Sciences, Utrecht University, The Netherlands
Faculty of Law, University of Groningen, The Netherlands
Email: H.Prakken@uu.nl

3
Towards Requirements Analysis for Formal Argumentation

THOMAS F. GORDON

ABSTRACT. We suggest applying software engineering requirements analysis methods to the development and evaluation of formal models of argumentation. Our aim and purpose is to help assure that formal argumentation *models* the full scope of argumentation as it is understood and studied in the humanities and social sciences, so as to provide a foundation for software tools supporting real argumentation tasks, in a wide variety of application domains.

1 Introduction

Argumentation is an application domain for mathematics, formal logic, and computer science. Thus the development of formal models of argumentation should be an interdisciplinary endeavor, with the participation of argumentation experts. Typically argumentation experts are humanities and social science scholars, in fields such as philosophy, languages, law and politics, religion, literature, speech communication or rhetoric. When developing formal models of argument, we need to engage with these experts to learn to speak their language, understand their conceptions of argumentation and take advantage of their expertise to help us identify the requirements which formal models of argument need to satisfy in order to serve as a foundation for tools, typically software systems, for supporting argumentation tasks in practice.

While argumentation is mostly *studied* in the humanities and social sciences, argumentation is *applied and used* in every field and in everyday life. Thus applications of formal models of argumentation can be found in fields such as law, medicine, natural science and engineering. When developing such formalizations, it is recommended to also collaborate with experts and potential users in the application domain.

The aim of this chapter is to stimulate a constructive debate about how best to develop and validate formal models of argumentation, not to criticize prior work. The chapter presents the views of the author, since there is as yet no consensus in the field on the issues raised here.

The view presented here is based on the assumption that formal models of argument are mostly intended to be a foundation for software systems for supporting argumentation tasks in real-world application scenarios. Such formal models can be viewed as software specifications. Indeed, some formalization methods make it is possible to automatically extract executable software programs from formalizations. (See, e.g., [Chlipala, 2013].) Thus, in this context it makes sense, we claim, to develop, evaluate and validate formal models of argumentation by applying methods from software engineering, including requirements analysis methods.

The chapter does not itself present such a requirements analysis, let alone the "authoritative" analysis. It would be quite presumptuous to pretend to be able to do so, without knowing more about the context of the particular application and without the participation of application domain experts or representatives of the intended stakeholders and users. The chapter does however include a high-level, domain-independent analysis of common argumentation tasks, which may be useful as a starting point.

The rest of this chapter is organized as follows. The next section proposes a scope and context for research on formal models of argumentation, with the aim of having the scope be sufficiently broad to cover argumentation as it is understood in the humanities and related fields. Next we present a particular requirements analysis method from software engineering. The aim is not so much to recommend this particular method, as to illustrate the basic idea and goals of requirements analysis. The following section presents a high-level overview of many argumentation tasks. The concluding section reiterates the main points.

2 Scope and Context

The aim of formal argumentation, we claim, is to develop formal models of argumentation which are useful as a foundation for developing software tools for supporting various argumentation tasks in practical applications. Our aim should be to avoid developing a separate technical understanding of *argument* and *argumentation* with only a weak connection to how these concepts are understood in the humanities and related fields, both by scholars and practitioners.

Good starting points for gaining an understanding of argumentation include the Handbook of Argumentation Theory [van Eemeren *et al.*, 2014b]. (Chapter 11 of the handbook [van Eemeren *et al.*, 2014a] summarizes prior work on computational models of argument in the field of Artificial Intelligence.) A briefer introduction is Walton's Foundations of Critical Argumentation [Walton, 2006]. See also Chapter 1 of this handbook (Argumentation Theory in Formal and Computational Perspective), by van Eemeren and Verheij.

The Handbook of Argumentation Theory aims to connect the definition of argumentation with "commonly recognized characteristics of argumentation as it is known from everyday practice" . It recognizes that "argument" is both a process, typically dialogues, and a "product", often conceived as a kind of structure linking premises to conclusions. After a thorough consideration of various aspects of the concept of argumentation, including differences between English and other European languages, the Handbook of Argumentation Theory proposes the following definition:

> Argumentation is a communicative and interactional act complex

aimed at resolving a difference of opinion with the addressee by putting forward a constellation of propositions the arguer can be held accountable for to make the standpoint at issue acceptable to a rational judge who judges reasonably. [van Eemeren *et al.*, 2014b, pg. 7]

While this definition may be general enough, it has the disadvantage of perhaps suggesting that argumentation is limited to or focused on persuasion dialogues, where the starting point is a claim by a proponent (arguer), which is then called into question by the respondent (addressee), and the goal of the dialogue is to resolve a conflict of opinion, by trying to put forward arguments which should be sufficient to persuade the respondent to accept the claim. However, argumentation takes place in other kinds of dialogues [Walton, 1998], in particular deliberation dialogues [Walton, 2015], where the starting point is an issue or problem to solve, and the goal of the dialogue is to balance the pros and cons of various options to find the best option, ideally, or at least an acceptable option. Let me propose a definition which more clearly covers argumentation in deliberation as well as persuasion dialogues:

Argumentation is a rational process, typically in dialogues, for making and justifying decisions of various kinds of issues, in which arguments pro and con alternative resolutions of the issues (options or positions) are put forward, evaluated, resolved and balanced.

This proposed definition more clearly emphasizes the importance of argumentation for making justified decisions, not only when resolving conflicts of opinion in persuasion dialogues, but also, e.g., when deciding courses of action in deliberation dialogues. In persuasion dialogues, the issue is whether or not to accept the claim put forward by the proponent. The respondent is the decider. If the respondent is not persuaded by the arguments, he or she will decide to reject the claim.

Here are some kinds of issues typically resolved via argumentation:

- Practical reasoning issues, about which course of action to take, considering multiple evaluation criteria, including soft and hard constraints, values, goals, preconditions and effects. Arguing about government policies is of this type [Fairclough and Fairclough, 2013].

- Theoretical issues, about which scientific theory to prefer, taking into consideration the experimental evidence and other evaluation criteria, such as simplicity, following Occam's razor.

- Text interpretation issues, about interpreting religious, legal and other texts, applying various methods of text interpretation (e.g. literal, historical, teleological) in accordance with hermeneutic principles, to find the best or most coherent interpretations.

- Factual issues, about what happened in some case, by weighing conflicting evidence, including testimonial evidence, or by constructing and evaluating alternative narratives (stories) to choose the most coherent narratives.

- Conceptual issues, about the meaning of concepts, or relationships among concepts. For example whether freedom means lack of government regulation or rather government by the people and for the people, no matter how much regulation the people choose to accept. Or whether one fact situation can or should be subsumed under some abstract concept. For example, whether the concept "vehicle" includes baby carriages or tanks (used in memorials), in a regulation prohibiting vehicles in parks [Hart, 1961].

Some of these kinds of issues may be resolved using methods other than argumentation, under special circumstances. Argumentation is a generally applicable method, appropriate when the special circumstances required for applying more specific methods are not satisfied. Conditions which make argumentation appropriate include:

- Facts and knowledge are not readily available, but must first be "found" via discovering and interpreting texts, evidence and data. There is no prior "knowledge base" to use for constructing arguments. Rather the knowledge base is constructed along with the arguments, incrementally and iteratively.

- There can be both too little and too much information (information overload).

- The issue or problem is not well-formed (semi-decidable), because the problem space is not enumerable. The problem space is incrementally defined during the problem solving process. There is no "stopping rule" for testing whether a solution has been found.

- Problem solving and reasoning resources are limited. It may be necessary to interrupt the decision-making process at any time and accept a potentially less than optimal solution (bounded rationality).

- There is no independent procedure by which the correctness of the decision reached via argumentation can be evaluated. (See Rawl's distinction between perfect, imperfect and pure procedural justice [Rawls, 1971].)

- There are multiple and conflicting opinions, goals, interests and values, from various stakeholders.

- The issues are not Boolean (yes/no). There is no definitive formulation of the issue. There is the possibility to *re-frame* the issue to admit further solutions. Solutions are not true or false, but rather good or bad (acceptable/justified or not).

- Arguments pro and con the options need to be weighed and balanced, taking into consideration such factors as interests, values and goals.

- Conclusions are defeasible and may need to be reconsidered in the light of further arguments, until a decision has been made.

Rittel, one of the developers of Issue-Based Information Systems (IBIS), advocated in an article with Weber the use of argumentation for city planning, after recognizing that many of the conditions listed above apply in the domain of city planning [Rittel and Webber, 1973].

Examples of further domains where argumentation is widely used and applied include:

- Engineering design rationales, e.g. [Jarczyk et al., 1992; Shum and Hammond, 1994]

- Medical diagnosis, e.g. [Fox et al., 2007; Williams and Hunter, 2007; Lu and Lajoie, 2008; Fan et al., 2013]

- Humanities research [Nelson et al., 1987]

- Education [Kirschner et al., 2003; Andrews, 2009; Mirza and Perret-Clermont, 2009]

- Policy analysis and discourse, including participation by citizens and civil society, e.g. [Hajer et al., 1993; Fairclough and Fairclough, 2013]

- Law, public administration, and alternative dispute resolution, e.g. [Prakken and Sartor, 2015; Walton and Lodder, 2005]

- Police and intelligence agency investigations, e.g. [Nissan, 2012] and [Heuer, 2010, Chapter 8].

- Scientific discourses, for example in the debate about climate change, e.g. [Betz and Cacean, 2011].

3 Requirements Analysis Methodology

The traditional software engineering methodology, still used by some, is the waterfall methodology. It is a process model consisting of several phases:

- Requirements Analysis
- Design
- Implementation
- Verification, and
- Maintenance

The process is mostly sequential, from one phase to the next, but the model does allow for some feedback from later phases to earlier phases. For example, problems arising in the design phase can cause requirements to be revised.

Currently fashionable is *agile* software development, such as Fowler's lightweight development process [Fowler and Scott, 2000] or the Scrum methodology [Sims and Johnson, 2011]. Rather than trying to design the whole system before beginning work on the implementation, the development process is broken down into smaller parts, in a cyclic process. In each iteration of the process, called a *sprint*, a small number of requirements are selected from a *backlog* of requirements, and system modifications are designed, implemented, tested, and documented to meet these requirements. Each sprint should produce a working, useful system, mitigating the risks of the waterfall model, where no system exists until much later in the process. The phases of the waterfall model remain important when applying agile development methods but occur repeatedly, in each iteration (sprint), rather than only once per project.

Here we want to focus on an agile method for requirements analysis [Leffingwell, 2010]. Requirements analysis takes place in collaboration with domain experts and potential users. In particular, when developing formal models of argumentation, one should collaborate with domain experts in argumentation, typically humanities scholars. Requirements are captured by defining roles, personas, user stories, scenarios and acceptance tests. Roles classify the different kinds of users of a system. For argumentation, these roles might include the participants or parties (e.g. proponent and respondent), some neutral third-parties (e.g. moderator, judge) and, last but not least, the audience (e.g. jury). Personas are more concrete descriptions of fictional persons or characters in each of the roles, to help the developers to have a better understanding of the skills, interests and attitudes of typical users. User stories describe particular requirements by instantiating a template of the following form:

> As a [role], I want to be able to perform [action], so that I can [goal].[1]

For example: As a party, I want to be able to put forward arguments so that I can support my claims.

Scenarios are more concrete illustrations of interactions among several user stories. In scenarios, user stories are instantiated with personas and more particular actions and goals and then linked together into sequences of actions. (This terminology may be somewhat confusing, because it is scenarios, not user stories, which read like narratives.) Finally, acceptance tests can be defined for each user story, with the aim to operationalize the evaluation of whether or not the user story has been correctly and completely implemented. We will come back to the topic of evaluation below, when discussing verification and validation.

---

[1] Interestingly, the template for user stories bears some similarity to an argumentation scheme for value-based practical reasoning [Atkinson and Bench-Capon, 2007].

The user stories collected during the requirements phase are entered into a database, called the *backlog*. At the beginning of each sprint, user stories to be implemented during the sprint are selected from the backlog. Work then begins by next designing the system to meet these selected requirements. A variety of design methods exist, ranging from informal, such as user-interface mock-ups, across semi-formal, such as Unified Modeling Language (UML) diagrams [Fowler and Scott, 2000], to formal methods, using mathematical models, simulation, formal logic and proof assistants [Chlipala, 2013; Paulson and Nipkow, 1990].

After the design is ready, implementation can begin. This is where computer programming, unit testing and debugging comes into play. It may be appropriate to quickly implement a first version of the system, using *rapid prototyping* with high-level programming languages. After the prototype has been validated the system may be reimplemented for efficiency reasons, if necessary, using a lower-level programming language. Often, however, this optimization effort is not worth the required effort.

The sprint is not complete until the new version of the system has been verified, validated and documented. Verification is the process of checking whether the system correctly implements its specifications. When proof assistants are used to design the system using formal methods, verification is assured, since these assistants provide tools for automatically generating correct implementations from formal specifications. Validation is the process of checking whether the implemented system in fact meets the requirements. This is a more open, empirical issue, which cannot be completely formalized or automated. Systems which are verified may fail validation tests, for example if the specifications fail to correctly or completely capture requirements. The acceptance tests developed during the requirements analysis phase facilitate a somewhat more objective evaluation, since the tests were developed *before* the system was designed or implemented. Without acceptance tests, it might be difficult to resist the temptation to develop validation criteria which one knows are satisfied by system. Thus acceptance tests play a role in software engineering similar to double-blind and other safeguards of experimental science methodologies.

It is not our purpose here to identify particular requirements or to promote a particular software engineering methodology, but rather only to encourage researchers in the field of formal models of argument to use *some* methodology for developing and validating models in a requirements-driven way, so that the models are more likely to be practically useful as a foundation for software tools for argumentation tasks.

4 Argumentation Tasks

Formal models are abstractions designed for particular tasks. Details of the domain irrelevant for the task are abstracted away in the model. A broad overview of the range and scope of argumentation tasks could be useful for classifying prior formal models of argumentation, understanding their scope and limitations, and contribute to a road-map suggesting topics for future research. Figure 1 presents such a broad overview of argumentation tasks, based on [Brewka and Gordon, 1994; Prakken, 1995; Bench-Capon and Sartor, 2003; Gordon, 2009], in the form of a high-level *use-case* diagram [Fowler and Scott, 2000].

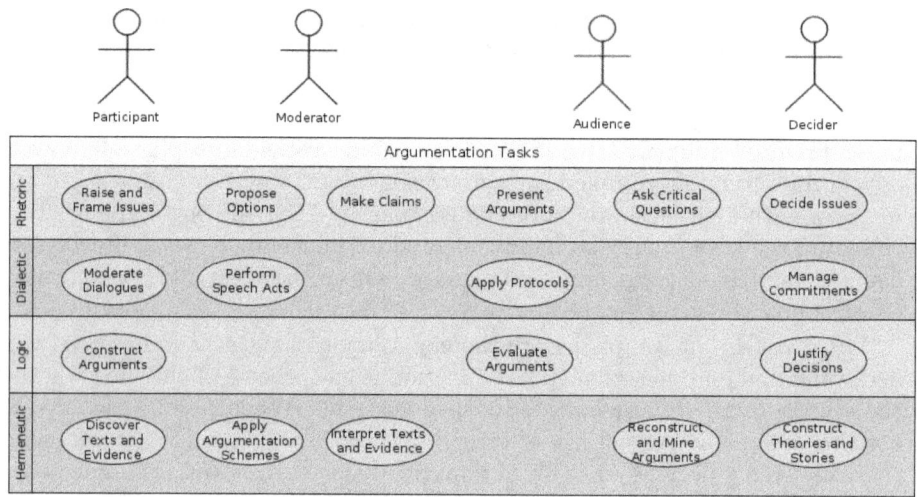

Figure 1. Argumentation Tasks

Four layers are distinguished in this diagram, inspired by classical Greek philosophy. In general, the layers build upon each other, from bottom to top. That is, tasks in higher levels can make use of tasks in lower levels. However, we do not want to interpret this diagram restrictively or preclude more complex interdependencies. For example, structures in the logic layer may be used to guide the interpretation of texts or the mining of arguments in the hermeneutic layer [Cocarascu and Toni, 2016]. The diagram does not define processes. The ordering of tasks is unspecified. That said, tasks on the left generally occur before and feed into tasks on the right. Also left unspecified, to avoid making the diagram overly complex, are dependencies (use relations) among the tasks and any indication of which user roles perform which tasks.

From bottom to top:

- The *hermeneutic layer* is responsible for interpreting structured and unstructured data, in particular natural language texts, to derive, reconstruct or "mine" arguments, as well as to construct candidate theories ("knowledge bases") and narratives ("stories"). Argumentation schemes can be used as heuristics to guide the search for interpretations.

- The *logical layer* is responsible for constructing and evaluating arguments and justifying conclusions. Arguments are constructed by applying argumentation schemes, in their function as inference rules, to axioms or assumptions of theories constructed in the hermeneutic layer. Argumentation schemes extend deductive inference rules of classical logic with domain-specific and contextual rules for deriving conclusions which are only presumptively or defeasibly true. Argument evaluation includes the resolution of conflicts and attack relations among arguments, but also the balancing of pro and con arguments. The logical layer is primarily normative: it specifies which conclusions or claims can be justifiably accepted by rational agents.

- The *dialectical layer* is responsible for regulating and supporting the process of argumentation, typically in dialogues, guiding and facilitating the process to help assure it achieves its normative goals. Moderators, mediators and other independent third parties have roles to play here. One of the tasks is to help participants to follow procedural rules, i.e. the argumentation protocol appropriate for the particular dialogue type. This task in turn can require keeping track of the commitments of the participants in the dialogue, from their claims, assertions and retractions.

- Finally, the *rhetorical layer* is responsible for helping participants to "play the game" well. Whereas the dialectical layer supports the normative goals of the dialogue as a whole, this layer consists of tasks helping actors to protect and promote their own interests. Tasks here include selecting among arguments which could be made and presenting these arguments clearly and persuasively, taking into consideration the intended audience, perhaps using argument visualization (mapping) techniques. The task of deciding issues has also been put in this layer, since even though the decision must be justified, there may be multiple justifiable options for the decider to choose from and the persuasive presentation of the decision and its justification has rhetorical aspects.

The four actors shown at the top of the diagram are abstract roles, not individuals. In particular procedures, a person may have more than one role, some roles may be combined or some roles may need to be distinguished further. For example, in lawsuits a judge may have the moderator role during the trial but share the decider role with a jury, with the judge responsible for deciding legal issues and the jury responsible for factual issues. Again, we have left unspecified which actor is responsible for performing each task. This level of detail may depend on the dialogue type and is outside the scope of this chapter.

There is surely room for discussion about which tasks to include in each of the layers, or whether still other layers might be needed. For example, [Prakken, 1995] included argument evaluation in the dialectic layer and had a separate procedure layer for managing dialogues.

Although the names of the layers have been inspired by ancient Greek philosophy, it is not necessary for our purposes here to try to perform a scholarly or historical analysis of how the meanings of these terms have been used in philosophy. According to the Stanford Encyclopedia of Philosophy[2], hermeneutics as a "systematic activity" goes back to the Homeric epics; logic as a "fully systematic discipline" began with Aristotle; dialectic was invented by Zeno of Elea; and rhetoric, although a much older topic, was highly influenced by Aristotle's *Rhetoric* [Aristotle, 2016].

We also do not claim that this model is comprehensive, covering all possible argumentation tasks. Rather, here the goal is only to try to present a broad overview of many argumentation tasks, to get a better understanding of the potential scope of formal argumentation.

5 Conclusions

This chapter encourages researchers in the field of formal argumentation to:

- take an interdisciplinary approach, working with experts from the humanities, social sciences and the application domain, as well as potential users;

- try to develop formalizations which model conceptions of argument and argumentation from the humanities;

- adapt and apply requirements analysis and other software engineering methods to develop and validate formal models of argument, helping to assure the models provide suitable foundations for building practically useful software tools supporting real argumentation tasks; and

- adopt a broad understanding of the scope of the field of formal argumentation, covering the hermeneutical, logical, dialectical, rhetorical and other aspects and tasks of argumentation in practice.

Acknowledgments

I would like to thank Doug Walton for his collaboration and support over the years, as well as my former employer, Fraunhofer FOKUS, for providing me the opportunity to conduct somewhat longer-term research than is typical for the organization.

BIBLIOGRAPHY

[Andrews, 2009] Richard Andrews. *Argumentation in higher education: Improving practice through theory and research.* Routledge, 2009.

[Aristotle, 2016] Aristotle. *Rhetoric.* Arcadia eBook, 2016.

[Atkinson and Bench-Capon, 2007] Katie Atkinson and Trevor J. M. Bench-Capon. Practical reasoning as presumptive argumentation using action based alternating transition systems. *Artificial Intelligence*, 171(10-15):855–874, 2007.

[Bench-Capon and Sartor, 2003] Trevor Bench-Capon and Giovanni Sartor. A model of legal reasoning with cases incorporating theories and values. *Artificial Intelligence*, 150(1–2):97–143, November 2003.

---

[2]https://plato.stanford.edu/

[Betz and Cacean, 2011] Gregor Betz and Sebastian Cacean. The moral controversy about Climate Engineering - an argument map. Technical report, Karlsruhe Institute of Technology, 2011.

[Brewka and Gordon, 1994] Gerhard Brewka and Thomas F Gordon. How to buy a Porsche: An approach to defeasible decision making. In *Working Notes of the AAAI-94 Workshop on Computational Dialectics*, pages 28–38, Seattle, Washington, 1994.

[Chlipala, 2013] Adam Chlipala. *Certified Programming with Dependent Types – A Pragmatic Introduction to the Coq Proof Assistant*. MIT Press, 2013.

[Cocarascu and Toni, 2016] Oana Cocarascu and Francesca Toni. Argumentation for machine learning: A survey. In *Proceeding of the 2016 conference on Computational Models of Argument (COMMA 2016)*, pages 219–230. IOS Press, 2016.

[Fairclough and Fairclough, 2013] Isabela Fairclough and Norman Fairclough. *Political discourse analysis: A method for advanced students*. Routledge, 2013.

[Fan et al., 2013] Xiuyi Fan, Robert Craven, Ramsay Singer, Francesca Toni, and Matthew Williams. Assumption-based argumentation for decision-making with preferences: A medical case study. In *International Workshop on Computational Logic in Multi-Agent Systems*, pages 374–390. Springer, 2013.

[Fowler and Scott, 2000] Martin Fowler and Kendall Scott. *UML Distilled – A Brief Guide to the Standard Object Modeling Language*. Addison Wesley Longman, Inc., 2nd edition, 2000.

[Fox et al., 2007] John Fox, David Glasspool, Dan Grecu, Sanjay Modgil, Matthew South, and Vivek Patkar. Argumentation-based inference and decision making–a medical perspective. *IEEE intelligent systems*, 22(6), 2007.

[Gordon, 2009] Thomas F Gordon. *Foundations of Argumentation Technology – Summary of Habilitation Thesis*. PhD thesis, Technical University of Berlin, 2009.

[Hajer et al., 1993] Maarten A Hajer, Robert Hoppe, Bruce Jennings, Frank Fischer, and John Forester. *The argumentative turn in policy analysis and planning*. Duke University Press, 1993.

[Hart, 1961] H L A Hart. *The Concept of Law*. Clarendon Press, Oxford, 1961.

[Heuer, 2010] Richards J. Heuer. *Psychology of intelligence analysis*. Central Intelligence Agency, 2010.

[Jarczyk et al., 1992] Alex P J Jarczyk, Peter Löffler, and Frank M Shipmann III. Design Rational for Software Engineering: A Survey. In *Proceedings of the 25th International Conference on System Sciences*, volume 2, pages 557–586, 1992.

[Kirschner et al., 2003] Paul A. Kirschner, Simon J. Buckingham-Shum, and Chad S. Carr. *Visualizing Argumentation: Software Tools for Collaborative Education and Sense-Making*. Springer-Verlag, London, 2003.

[Leffingwell, 2010] Dean Leffingwell. *Agile Software Requirements: Lean Requirements Practices for Teams, Programs and the Enterprise*. Addison-Wesley, 2010.

[Lu and Lajoie, 2008] Jingyan Lu and Susanne P Lajoie. Supporting medical decision making with argumentation tools. *Contemporary Educational Psychology*, 33(3):425–442, 2008.

[Mirza and Perret-Clermont, 2009] Nathalie Muller Mirza and Anne Nelly Perret-Clermont, editors. *Argumentation and Education: Theoretical Foundations and Practices*. Springer, 2009.

[Nelson et al., 1987] John S Nelson, Allan Megill, and Donald N McCloskey. *The rhetoric of the human sciences: Language and argument in scholarship and public affairs*. Univ of Wisconsin Press, 1987.

[Nissan, 2012] Ephraim Nissan. *Computer Applications for Handling Legal Evidence, Police Investigation and Case Argumentation*, volume 5. Springer Science & Business Media, 2012.

[Paulson and Nipkow, 1990] Lawrence C Paulson and Tobias Nipkow. Isabelle tutorial and user's manual. Technical Report Technical Report No. 189, University of Cambridge, Computer Laboratory, 1990.

[Prakken and Sartor, 2015] Henry Prakken and Giovanni Sartor. Law and logic: A review from an argumentation perspective. *Artificial Intelligence*, 227:214–245, 2015.

[Prakken, 1995] Henry Prakken. From logic to dialectic in legal argument. In *Proceedings of the Fifth International Conference on Artificial Intelligence and Law*, pages 165–174, Maryland, 1995.

[Rawls, 1971] John Rawls. *A Theory of Justice*. Belknap Press of Harvard University Press, 1971.

[Rittel and Webber, 1973] Horst W J Rittel and Melvin M Webber. Dilemmas in a General Theory of Planning. *Policy Science*, 4:155–169, 1973.

[Shum and Hammond, 1994] Simon Buckingham Shum and Nick Hammond. Argumentation-Based Design Rationale: What Use and What Cost? *International Journal of Human-Computer Studies*, 40(4):603–652, 1994.

[Sims and Johnson, 2011] Chris Sims and Hillary Louise Johnson. *The Elements of Scrum*. Dymaxicon, 2011.

[van Eemeren et al., 2014a] Frans van Eemeren, Bart Garssen, Erik C.W. Krabbe, A. Francisca Snoeck Henkemans, Bart Verheij, and Jean H.M. Wagemanns, editors. *Handbook of Argumentation Theory*, chapter Argumentation and Artificial Intelligence, pages 615–676. Springer, 2014.

[van Eemeren et al., 2014b] Frans H. van Eemeren, Bart Garssen, Erik C.W. Krabbe, A. Francisca Snoeck Henkemans, Bart Verheij, and Jean H. M. Wagemans. *Handbook of Argumentation Theory*. Springer, 2014.

[Walton and Lodder, 2005] Douglas N Walton and Arno R Lodder. What Role can Rationale Argument Play in ADR and Online Dispute Resolution? In *Second International ODR Workshop*, IAAIL Workshop Series, pages 69–76, Nijmegen, The Netherlands, 2005. Wolf Legal Publishers.

[Walton, 1998] Douglas Walton. *The New Dialectic: Conversational Contexts of Argument*. University of Toronto Press, Toronto; Buffalo, 1998.

[Walton, 2006] Douglas Walton. *Fundamentals of Critical Argumentation*. Cambridge University Press, Cambridge, UK, 2006.

[Walton, 2015] Douglas Walton. *Goal-Based Reasoning for Argumentation*. Cambridge University Press, 2015.

[Williams and Hunter, 2007] Matt Williams and Anthony Hunter. Harnessing ontologies for argument-based decision-making in breast cancer. In *19th IEEE International Conference on Tools with Artificial Intelligence, 2007 (ICTAI 2007)*, volume 2, pages 254–261. IEEE, 2007.

Thomas F. Gordon
Berlin, Germany
Web site: http://www.tfgordon.de

PART B

ARGUMENTATION FORMALISMS

4
Abstract Argumentation Frameworks and Their Semantics

PIETRO BARONI, MARTIN CAMINADA, MASSIMILIANO GIACOMIN

ABSTRACT. The current chapter presents an overview on the state of the art of Dung's abstract argumentation frameworks and their semantics, covering both some of the most influential literature proposals and some general issues concerning semantics definition and evaluation. As to the former point the chapter reviews Dung's original notions of complete, grounded, preferred, and stable semantics, as well as a variety of notions subsequently proposed in the literature namely, naïve, semi-stable, ideal, eager, stage, $\mathcal{CF}2$, and stage2 semantics, considering both the extension-based and the labelling-based approaches with respect to their definitions. As to the latter point the chapter analyzes the notions of argument justification and skepticism comparison and discusses semantics agreement.

1 Introduction

This chapter is devoted to the formalism of *abstract argumentation frameworks* introduced by Dung [1995]. This formalism is based on the idea that arguments are defeasible entities which may attack each other and whose acceptance is subject to evaluation. In presence of conflicts, an argument cannot be accepted just because it exists: its acceptance depends on the existence of possible counter arguments, that can then themselves be attacked by counter arguments, and so on. Formally, an argumentation framework is represented as a directed graph in which the arguments are represented as nodes and the attack relation is represented by the arrows. Given such a graph, one is naturally led to examine the question of which set(s) of arguments can be accepted: answering this question corresponds to defining an *argumentation semantics*. Various proposals have been formulated in this respect, and in the current chapter we will describe some of the mainstream approaches. Before entering the technical presentation, however, some important general considerations are worth introducing. As sketched above, the formalism of argumentation frameworks is exclusively centered on the notion of attack between arguments and on the evaluation of argument acceptability, based on the intuition that the existence of attacks prevents all arguments to be accepted together. In this formal context, arguments are deprived of all their features apart their identity: their origin, structure and any other characteristics differentiating them are abstracted away, leaving room only to the property of attacking (or being attacked by) their homogeneous (from the abstract point of view) peers. This extreme simplification of an otherwise complex and articulated phenomenon like argumentation is a key factor for understanding the huge interest that the study of this formalism has attracted, as well as its boundaries. Theoretical cleanness and wide applicability are among the formidable strengths following from the simplicity of the formalism. Argumentation frameworks allow the investigation of notions and properties (far from being trivially simple, by the way) which are purely related to the existence of attacks without being obfuscated or complicated by the many accidental or complementary properties of the entities involved in the attack themselves. In this sense Dung's theory can be regarded as an attack (or conflict) calculus, which has been initially formulated for the need of dealing with attacks between arguments but then stands on its own feet, even independently of the original interpretation in argumentative terms. This yields a powerful generality: as shown in Dung's original paper, several more specific formal settings can be regarded (and better understood) as special cases of argumentation frameworks, in areas ranging from nonmonotonic reasoning to game theory. Remarkably, some of these settings are only loosely related to the notion of argumentation. This has shown since the beginning that theoretical investigations of Dung's formalism enjoy a wide applicability across a variety of domains. It must also be observed, however, that dealing with attacks, while being crucial, is by no means sufficient to provide a formal counterpart of actual argumentation processes. In this sense, the temptation of considering abstract argumentation theory as a self-sufficient tool for formal argumentation should be regarded as an oversimplification. To avoid this risk, it is important to keep in mind that assessing argument acceptability in presence of attacks, which is the essence of Dung's theory, is only one specific (although important) aspect in formalizing argument-based reasoning and needs to be integrated and bridged with other formal components. As an example of how to apply Dung's theory within a broadened setting, consider the use of abstract argumentation theory for the purpose of nonmonotonic inference from a knowledge base.

In this context, one can distinguish three steps (see Figure 1). First of all, one would use an underlying knowledge base to generate a set of arguments and determine in which ways these arguments attack each other (step 1). The result is then an argumentation framework, to be represented as a directed graph in which the internal structure of the arguments, as well as the nature of the attack relation have been abstracted away. Based on this argumentation framework, the next step is to determine the sets of arguments that can be accepted, using a pre-defined criterion corresponding to an argumentation semantics (step 2). After the set(s) of accepted arguments have been identified, one then has to identify the set(s) of accepted conclusions (step 3), for which there exist various approaches.

As an example of how things work, suppose one applies the argumentation process in the context of logic programming. In particular, suppose the knowledge base consists of a logic program P of the following form.

Abstract Argumentation Frameworks and Their Semantics

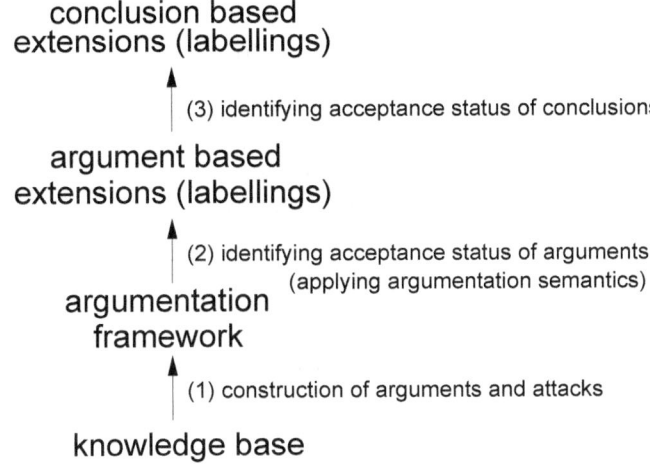

Figure 1. Argumentation for inference

$$b \leftarrow c, \text{not } a \qquad a \leftarrow \text{not } b$$
$$p \leftarrow c, d, \text{not } p \qquad p \leftarrow \text{not } a$$
$$c \leftarrow d \qquad d \leftarrow$$

In that case, following for instance the approach of [Wu et al., 2009] where arguments consist of trees of rules and attack is based on weak negation, one can construct the argumentation framework shown in Figure 2.

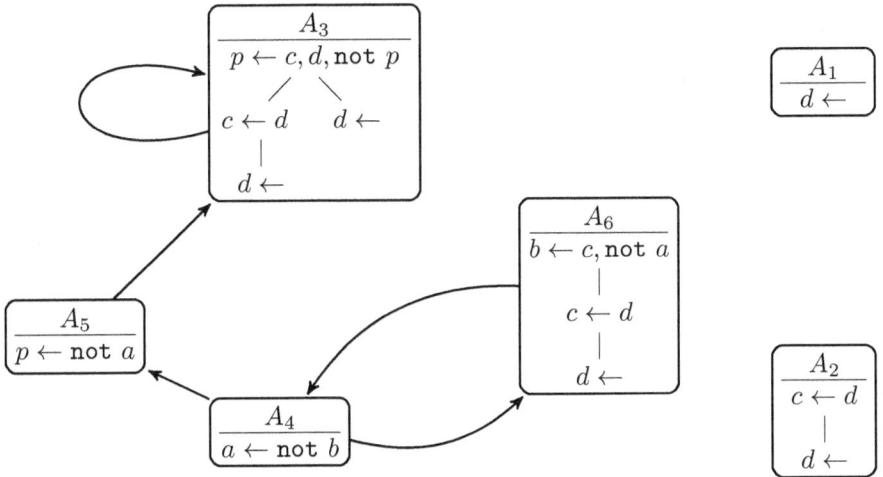

Figure 2. The argumentation framework built from the logic program P

This argumentation framework has exactly one stable extension[1] of arguments: $\{A_1, A_2, A_5, A_6\}$. This extension of arguments yields the following extension of conclusions: $\{d, c, p, b\}$. It has been proved [Dung, 1995] that when applying stable semantics at step 2, the overall result (extensions of conclusions) is precisely the same as when applying stable model semantics [Gelfond and Lifschitz, 1988; Gelfond and Lifschitz, 1991] to the original logic program. In a similar way, the three-step argumentation process can also be applied to simulate other logic programming semantics. We refer to [Dung, 1995; Wu et al., 2009; Caminada et al., 2015] for an overview.

As mentioned before, one of the strengths of the argumentation approach is that it turns out to be powerful enough to model not just logic programming but a whole range of formalisms, including Default Logic [Dung, 1995; Caminada et al., 2012] and Nute's Defeasible Logic [Governatori et al., 2004]. Other scholars have subsequently used the argumentation approach to specify their own formalisms for non-monotonic entailment, like ASPIC+ [Modgil and Prakken, 2014], ABA [Toni, 2014] and logic-based argumentation [Gorogiannis and Hunter, 2011].

Overall, the argumentation process described above leads to a number of questions:

1. *What is the content of the knowledge base with which arguments are constructed?* Different argument-based formalisms start with different types of knowledge bases. In the case of argument-based logic programming [Dung, 1995; Wu et al., 2009; Caminada et al., 2015], like described above, the knowledge base is simply a logic program. In the case of logic-based argumentation [Gorogiannis and Hunter, 2011] the knowledge base consists of propositions. In the case of ABA [Toni, 2014], it contains rules and assumptions. In the case of ASPIC+ [Modgil and Prakken, 2014], it contains rules and formulas, as well as a preference ordering, to determine argument strength. In spite of their differences, what unifies these formalisms is that each of them can be seen as applying the argumentation process of Figure 1. The reader can find details in chapters 6, 7, 8 and 9 of this volume.

2. *Given a knowledge base, how to precisely construct the associated argumentation framework?* Even for the same type of knowledge base, there can be several ways of constructing the associated argumentation framework, each with their own advantages and disadvantages. Details will be provided in chapters 6, 7, and 9.

3. *Once the argumentation framework has been constructed, how to determine which arguments to accept and reject?* This is the key question to be studied in the current chapter. Our aim is to provide an overview of

---

[1] A stable extension attacks precisely those arguments that are not in it. We refer to Section 3.6 for details.

criteria for argument acceptance, as have been stated in the literature. Hence, the current chapter focuses on step 2 of the argumentation process in Figure 1.

4. *How to make sure the overall outcome makes sense?* Abstract argumentation theory selects arguments purely on the basis of the topology of the graph, without looking at their actual contents. In particular, these contents can have a logical form (as is the case for instance in ABA [Toni, 2014], ASPIC+ [Modgil and Prakken, 2014] and logic-based argumentation [Gorogiannis and Hunter, 2011]). The question is how to make sure that the overall conclusions yielded by the formalism are consistent or satisfy any other desirable property. This is a crucial research issue: chapter 15 will present some key desirable properties as well as some of the approaches for satisfying them.

To complete this introduction, it is worth mentioning again that while inference from a knowledge base is an important domain for abstract argumentation theory, a wider range of applications can be found in the literature, ranging from decision making [Amgoud, 2009] to topics like coalition formation and the stable marriage problem [Dung, 1995]. It is also fair to mention that in the literature there are also formalisms for argument-based reasoning, like *DeLP* [García and Simari, 2004], which adopt alternative approaches with respect to abstract argumentation theory for the assessment of argument acceptance.

The remaining part of this chapter is structured as follows. First, in Section 2 we formally describe the notion of an argumentation framework and present some relevant basic concepts.

Then in Section 3 we present some relatively well-known and well-established argumentation semantics, both in terms of argument extensions and in terms of argument labellings. In Section 4 we provide a comprehensive treatment of the notions of argument justification and skepticism, including skepticism comparison between the reviewed semantics, while in Section 5 we discuss the issue of semantics agreement. Finally Section 6 provides a brief summary and concludes the chapter.

2 Basic concepts

Central to the theory of abstract argumentation is the notion of an *argumentation framework*, which, as mentioned in Section 1, is essentially a directed graph in which the arguments are represented by the nodes and the attack relation is represented by the arrows[2]. Given the tutorial nature of this chapter, we keep the presentation simple by restricting ourselves to finite argumentation frameworks, while briefly mentioning non-finite frameworks where appropriate. The reader may refer to chapter 17 for a coverage of properties of infinite argumentation frameworks.

---

[2]In Dung's theory, attack is a one-to-one relationship, which deviates from earlier work of, for instance, [Vreeswijk, 1993] which is centered around the notion of *collective attack*, meaning that a set of arguments is collectively attacking another argument.

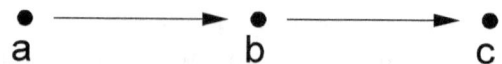

Figure 3. A simple argumentation framework

Definition 2.1 *An* argumentation framework *is a pair* $\langle Ar, att \rangle$ *in which* Ar *is a finite set of arguments and* $att \subseteq Ar \times Ar$.

We say that argument $a \in Ar$ attacks argument $b \in Ar$ (or that a is an *attacker* of b) iff $(a, b) \in att$. If $Args \subseteq Ar$ and $a \in Ar$ then we say that a attacks $Args$ iff there exists $b \in Args$ such that a attacks b. Likewise, we say that $Args$ attacks a iff there exists $b \in Args$ such that b attacks a. For $a \in Ar$ then we write a^- for $\{b \mid (b, a) \in att\}$ and a^+ for $\{b \mid (a, b) \in att\}$. Likewise, for $Args \subseteq Ar$ we write $Args^-$ for $\{b \mid \exists a \in Args : (b, a) \in att\}$ and $Args^+$ for $\{b \mid \exists a \in Args : (a, b) \in att\}$. All these notions refer to a given argumentation framework, which is left implicit in the relevant notation for the sake of simplicity and conciseness.

We will also need to consider the restriction of an argumentation framework to a subset of its arguments.

Definition 2.2 *Given an argumentation framework* $AF = \langle Ar, att \rangle$ *and a set of arguments* $Args \subseteq Ar$, *the* restriction *of* AF *to* $Args$, *denoted as* $AF{\downarrow}_{Args}$ *is the argumentation framework* $\langle Args, att \cap (Args \times Args) \rangle$.

An argumentation framework encodes, through the attack relation, the existing conflicts within a set of arguments. It is then interesting to identify the conflict outcomes, which, roughly speaking, means determining which arguments should be accepted (let's say, "survive the conflict") and which arguments should be rejected (let's say, "are defeated in the conflict"), according to some reasonable criterion.

Consider for instance the argumentation framework depicted in Figure 3. Which arguments are able to survive the conflict? Is there only one possibility or are there several solutions available? While the reader may resort to her/his personal intuition to devise a specific answer in this simple case, it appears that a well-defined systematic method is needed to deal with the case of arbitrarily complex argumentation frameworks: such a formal method to identify conflict outcomes for any argumentation framework is called *argumentation semantics*.

Two main approaches to the definition of argumentation semantics are available in the literature: the *labelling*-based approach and the *extension*-based approach.

The idea underlying the *labelling*-based approach is to give each argument a label. A sensible (though not the only possible) choice for the set of labels is: **in**, **out** or **undec**, where the label **in** means the argument is accepted, the label **out** means the argument is rejected and the label **undec** means one abstains from an opinion on whether the argument is accepted or rejected. Each argument then gets exactly one label. In Figure 3, one might start assigning the label in to argument a, as it does not receive attacks, then derive that the argument b should be out. Then the attack from b to c can be considered ineffective or, in other words, it can be said that a defends c against b and one can assume that c should be in in turn. While this labeling may sound reasonable, other choices are, at least in principle, available: e.g. one might assign all arguments the label in, but this seems incompatible with the existence of conflicts among them, or one might assign all arguments the label undec, but this seems excessively cautious at least as far as the unattacked argument a is concerned. Thus, a specific labelling-based argumentation semantics provides a way to select "reasonable" labellings among all the possible ones, according to some criterion embedded in its definition.

The idea underlying the *extension*-based approach is to identify sets of arguments, called extensions, which can survive the conflict together and thus represent collectively a reasonable position an autonomous reasoner might take. To illustrate how one could use an incremental procedure for extension construction, in Figure 3 one might start including the argument a, as it does not receive attacks, then exclude the argument b, and then assume that c should be included in turn, ending up with the extension $\{a, c\}$. Also in this case other choices are available, at least in principle: e.g. one might consider the extension $\{a, b, c\}$, but (again) this seems incompatible with the existing conflicts among arguments, or one might consider the empty set as extension, but this seems excessively cautious since at least a seems to deserve inclusion in any extension. Thus, a specific extension-based argumentation semantics provides a way to select "reasonable" sets of arguments among all the possible ones, according to some criterion embedded in its definition.

Let us now turn to the formal counterpart of the notions exemplified above.

A generic labelling assigns to each argument of an argumentation framework a label taken from a predefined set.

Definition 2.3 *Let $AF = \langle Ar, att \rangle$ be an argumentation framework and Λ a set of labels. A Λ–labelling is a total function $\mathcal{L}ab : Ar \longrightarrow \Lambda$. The set of all Λ–labellings of AF is denoted as $\mathfrak{L}(\Lambda, AF)$.*

A labelling-based semantics prescribes a set of labellings for any argumentation framework.

Definition 2.4 *Given an argumentation framework $AF = \langle Ar, att \rangle$ and a set of labels Λ, a labelling-based semantics σ associates with AF a subset of $\mathfrak{L}(\Lambda, AF)$, denoted as $\mathcal{L}_\sigma(AF)$.*

We will also need the notion of restriction of a labelling to a set of arguments.

Definition 2.5 *Given an argumentation framework $AF = \langle Ar, att \rangle$, a set of labels Λ, a Λ–labelling $\mathcal{L}ab$, and a set of arguments $Args \subseteq Ar$, the restriction of $\mathcal{L}ab$ to $Args$, denoted as $\mathcal{L}ab{\downarrow}_{Args}$ is defined as $\mathcal{L}ab \cap (Args \times \Lambda)$.*

In this chapter we focus on the case $\Lambda = \{\text{in}, \text{out}, \text{undec}\}$, a sensible choice for Λ which has received considerable attention in the literature [Caminada, 2006a; Caminada, 2007a; Rahwan and Larson, 2008; Caminada and Gabbay, 2009; Caminada and Pigozzi, 2011; Rahwan and Tohmé, 2010]. An alternative approach can be found in [Jakobovits and Vermeir, 1999], where a four-valued labelling is considered. The idea of labelling can also be put in correspondence with the notion of status assignment in inference graphs [Pollock, 1995]. A first investigation of the connections between defeat status assignments and extensions in Dung's argumentation frameworks was provided in [Verheij, 1996].

We will implicitly assume the use of $\Lambda = \{\text{in}, \text{out}, \text{undec}\}$, when the reference to the label set is omitted. In particular, given a labelling $\mathcal{L}ab$, we write $\text{in}(\mathcal{L}ab)$ for $\{a \mid \mathcal{L}ab(a) = \text{in}\}$, $\text{out}(\mathcal{L}ab)$ for $\{a \mid \mathcal{L}ab(a) = \text{out}\}$ and $\text{undec}(\mathcal{L}ab)$ for $\{a \mid \mathcal{L}ab(a) = \text{undec}\}$. A labelling can be represented as a set of pairs. For instance, the first labelling exemplified above for Figure 3 can be described as $\{(a, \text{in}), (b, \text{out}), (c, \text{in})\}$. Sometimes we will also represent a labelling $\mathcal{L}ab$ as the triple $(\text{in}(\mathcal{L}ab), \text{out}(\mathcal{L}ab), \text{undec}(\mathcal{L}ab))$. The same labelling for Figure 3 can thus be represented as $(\{a, c\}, \{b\}, \emptyset)$.

Turning to the extension-based approach, since an extension is just a set of arguments, the definition of extension-based semantics is quite simple and does not require preliminary notions.

Definition 2.6 *An extension-based semantics σ associates with any argumentation framework $AF = \langle Ar, att \rangle$ a subset of 2^{Ar}, denoted as $\mathcal{E}_\sigma(AF)$.*

Some observations about the relations between the labelling and extension-based approaches are worth remarking. First, as set membership can be formulated in terms of a simple binary labelling, e.g. with $\Lambda = \{\in, \notin\}$, the extension-based approach can be regarded as a special case of the general labelling-based approach. The latter is therefore more general, while the former, probably due to its simplicity, has received by far more attention in previous literature.

Considering the three-valued labelling we focus on in this chapter, a correspondence with the extension-based approach can be drawn, so that a semantics based on this labelling can be turned into an extension-based one through a simple mapping. In fact, given a labelling of an AF, the labels in can be understood as identifying the members of an extension. This kind of correspondence can be easily identified in the example concerning Figure 3 described above and is formally expressed by the following definitions.

Definition 2.7 *Given an argumentation framework $AF = \langle Ar, att \rangle$ and a labelling $\mathcal{L}ab$ the corresponding set of arguments $\text{Lab2Ext}(\mathcal{L}ab)$ is defined as $\text{Lab2Ext}(\mathcal{L}ab) = \text{in}(\mathcal{L}ab)$.*

Definition 2.8 *Given an argumentation framework $AF = \langle Ar, att \rangle$ and a labelling-based semantics σ, the set of extensions corresponding to $\mathcal{L}_\sigma(AF)$ is given by $\mathcal{E}_\sigma(AF) = \{\text{Lab2Ext}(\mathcal{L}ab) \mid \mathcal{L}ab \in \mathcal{L}_\sigma(AF)\}$.*

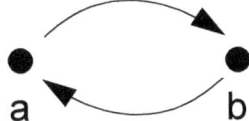

Figure 4. An argumentation framework with mutual attack

On the other hand, given a set of arguments E it is possible to define a corresponding three-valued labelling by distinguishing the arguments belonging to E, those attacked by some member of E, and those which neither belong to E nor are attacked by E. As this correspondence is well-defined only if E satisfies some basic conditions, we defer its formal definition to Section 3.1 (Definition 3.6).

We now introduce some notions which are common to both approaches.

First it can be noted that both approaches encompass (in general) the existence of a set of alternatives (either labellings or extensions) for a single argumentation framework. It may be the case, however, that a semantics σ is defined so that a univocal outcome is prescribed for each argumentation framework (formally for any argumentation framework AF, $|\mathcal{L}_\sigma(AF)| = 1$ or $|\mathcal{E}_\sigma(AF)| = 1$). In this case, the semantics is said to belong to the *unique-status* (or *single-status*) approach, while in the general case it is said to belong to the *multiple-status* approach.

Consider the argumentation framework of Figure 4 representing a mutual attack. A unique-status approach may prescribe the $\{(a, \text{undec}), (b, \text{undec})\}$ labelling or analogously a single empty extension, corresponding to an explicit abstention from decision. On the other hand, a multiple-status approach may encompass the two alternative labellings $\{(a, \text{in}), (b, \text{out})\}$ and $\{(a, \text{out}), (b, \text{in})\}$ or analogously the extensions $\{a\}$ and $\{b\}$, corresponding to two opposite ways of solving the conflict.

As evident from the previous example, a semantics σ does not provide, in general, the "last word" about the status of an argument a. In fact σ may prescribe both a labelling where a is labelled in and another where a is labelled out (or, analogously, an extension including a and another one not). In the view of producing a synthetic evaluation for each argument, one has then to consider questions like "Is being in in all labellings significantly different from being in only in some of them?" or "If an argument is not in in all labellings should it being labelled out or undec in the remaining labellings make some difference?". Analogous questions may arise for the extension-based approach. It emerges that the assessment of a synthetic *justification status* for each argument of an argumentation framework is a further distinct (and not trivial) step after the identification of labellings or extensions. This will be dealt with in Section 4.

3 An Overview of Argumentation Semantics

In this section we provide an overview of some well-known argumentation semantics, starting from the very basic notion of "naïve semantics" and then discussing Dung's original concepts of complete, stable, preferred and grounded semantics [Dung, 1995], as well as the subsequently introduced ideal [Dung et al., 2007], semi-stable [Verheij, 1996; Caminada, 2006b] and eager [Caminada, 2007b] semantics.[3] These semantics can be considered as mainstream, since they share a basic property called *admissibility* and have been subject to much study, including the specification of proof procedures and of properties regarding computational complexity. We also treat three additional semantics, namely stage [Verheij, 1996], $\mathcal{CF}2$ [Baroni et al., 2005b] and stage2 [Dvořák and Gaggl, 2012b; Dvořák and Gaggl, 2012a; Dvořák and Gaggl, 2016] semantics. Unlike the other semantics considered in this chapter, stage, $\mathcal{CF}2$ and stage2 semantics are not admissibility-based, but they have quite unique characteristics that make them worthwhile to examine.

The presentations of the various semantics roughly follow a common line. First, the underlying intuitive idea is introduced, then the semantics formal definition is given according to both the labelling and the extension-based approach, and finally the presentation is completed by discussing illustrative examples and examining additional important formal properties and inter-semantics relationships. We do not deal with algorithmic or implementation issues in this chapter, as this matter is extensively treated in chapter 14. However, without any claim of providing an adequate coverage of the state of the art, in some places we will mention some algorithms and the relevant literature sources which, in our opinion, represent useful readings to get a further insight on the nature and behaviour of the considered semantics. Each semantics is denoted by a short abbreviation for easy reference. As for examples, the relatively simple ones provided in Figures 5-7 will be used as a common reference throughout this section, adding other more specific and/or complex ones where necessary. We invite the reader to give a look to Figures 5-7 in order to set up a "personal view" on how the conflict they encode might be resolved, and then comparing this view with those emerging from the various semantics proposals analyzed in the following. Before dealing directly with semantics we need however to examine (in the next subsection) two general properties, which underlie most of them, namely admissibility[4] and conflict-freeness.

3.1 Admissibility and conflict-freeness

To introduce the notion of admissibility let us start from a very simple principle: for every argument a one accepts (or rejects) one should be able to explain why

---

[3]Please notice that terms like "preferred semantics" or "ideal semantics" correspond to existing terminology in the literature and do not imply any value judgements.

[4]Admissibility has been introduced as a semantic property, not as a semantics in [Dung, 1995]. In the subsequent literature, however, the term admissible semantics has often been used. We will also refer to admissible semantics later in the chapter where convenient for presentation purposes.

Figure 5. An argumentation framework with unidirectional and mutual attacks

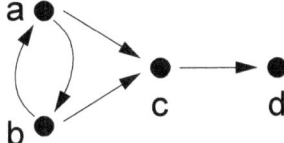

Figure 6. The case of "floating" acceptance

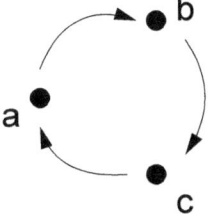

Figure 7. Cycle of three attacking arguments

it is accepted (or rejected), taking into account the acceptance or rejection of other arguments connected to a through the attack relation. This concept lends itself to slightly different, though converging, realisations in the labelling and in the extension-based approach.

In the labelling-based approach, assigning the **in** label to an argument a can be explained by having assigned the **out** label to all its attackers (or by a being attacked by no argument) so that a is not affected by any attack, while assigning the **out** label to a can be explained by having assigned the **in** label to one of its attackers, which enables a to be rejected.

This is expressed by the following definitions.

Definition 3.1 *Let $\mathcal{L}ab$ be a labelling of argumentation framework $\langle Ar, att \rangle$.*

- *An in-labelled argument is said to be* legally **in** *iff all its attackers are labelled* **out**.

- *An out-labelled argument is said to be* legally **out** *iff it has at least one attacker that is labelled* **in**.

Definition 3.2 *Let $AF = \langle Ar, att \rangle$ be an argumentation framework. An admissible labelling is a labelling $\mathcal{L}ab$ where each in-labelled argument is legally* **in** *and each out-labelled argument is legally* **out**.

Note that, according to this definition, for any argumentation framework a labelling where all arguments are **undec** is admissible. Let us now examine admissible labellings in the reference examples. Considering Figure 5, it is evident that a, having no attackers, can only be labelled legally **in** or **undec**. Considering the latter case, b can only be labelled **undec**, which implies that c cannot be legally **in**. If c is labelled **undec** then d is **undec** too, otherwise c is labelled **out** entailing that d is labelled **in**. This yields two admissible labellings, the trivial one $(\emptyset, \emptyset, \{a, b, c, d\})$ and $(\{d\}, \{c\}, \{a, b\})$. The case where a is labelled **in** leaves two alternatives for b. If b is labelled **undec** we have the same options as above for c and d yielding the two additional admissible labellings $(\{a\}, \emptyset, \{b, c, d\})$ and $(\{a, d\}, \{c\}, \{b\})$. Finally if b is labelled **out**, three alternatives are left open for c and d: they can be both labelled **undec** or c can be legally labelled **in** if d is labelled **out** and vice versa, yielding other three labellings: $(\{a\}, \{b\}, \{c, d\})$, $(\{a, c\}, \{b, d\}, \emptyset)$, $(\{a, d\}, \{b, c\}, \emptyset)$.

In Figure 6, with a similar reasoning as in the previous example it can be noted that a and b can be both labelled **undec** or one **in** and the other **out**. The first case yields only the trivial labelling $(\emptyset, \emptyset, \{a, b, c, d\})$, in the other cases c may be labelled **undec**, yielding d **undec**, or **out** leaving for d both the options **undec** and **in**. Altogether there are seven admissible labellings whose enumeration is left to the reader (see Table 2 later).

In Figure 7 no admissible labellings besides the trivial one $(\emptyset, \emptyset, \{a, b, c\})$ are possible, since no argument defends itself and every argument attacks the

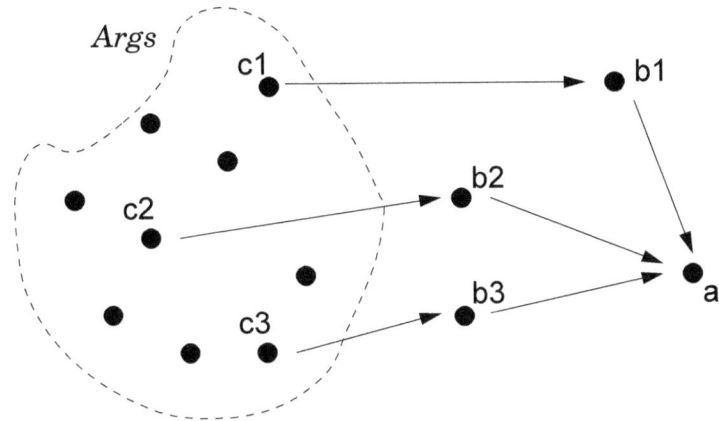

Figure 8. $Args$ defends argument a

argument which would defend it: for instance a would need a defense against c but attacks its potential defender b, and similarly for the other two arguments.

Turning now to the extension-based approach, the inclusion of an argument a in an extension E can be supported by the fact that E rules out all the attackers of a by in turn attacking them (if any). To put it in other words, E "defends" a. This is formalized in the following definitions.

Definition 3.3 Let $AF = \langle Ar, att \rangle$ be an argumentation framework and $Args \subseteq Ar$. The set $Args$ defends[5] $a \in Ar$ iff $\forall b \in a^- \exists c \in Args : c$ attacks b. The function $F_{AF} : 2^{Ar} \longrightarrow 2^{Ar}$ such that $F_{AF}(Args) = \{a \mid Args$ defends $a\}$ is called the characteristic function of AF.

An example of defense is given in Figure 8. Here we have an argument a that has three attackers: b_1, b_2 and b_3. $Args$ defends a because it attacks all these attackers.

Having introduced the notion of defense, a basic requirement for a set of arguments is the capability to defend all its elements. It is however natural to also require that the set of arguments features some sort of "internal coherence": no conflict should be allowed within a set of arguments which are considered able to survive the conflict *together*. This leads to the definition of conflict-free set.

Definition 3.4 Let $AF = \langle Ar, att \rangle$ be an argumentation framework and $Args \subseteq Ar$. The set $Args$ is conflict-free iff $\neg \exists a, b \in Args : a$ attacks b.

Note that this definition also rules out sets containing *self-attacking* (also called *self-defeating*) arguments (in the case $a = b$).

---

[5] The original terminology in [Dung, 1995] was that an argument a is *acceptable* w.r.t. a set of arguments $Args$. However, the more intuitive expression that an argument a is *defended* by a set of arguments $Args$ is commonly used in the literature and we prefer it.

Admissible labellings	Admissible sets
$\{(a, \text{undec}), (b, \text{undec}), (c, \text{undec}), (d, \text{undec})\}$	$\emptyset$
$\{(a, \text{undec}), (b, \text{undec}), (c, \text{out}), (d, \text{in})\}$	$\{d\}$
$\{(a, \text{in}), (b, \text{undec}), (c, \text{undec}), (d, \text{undec})\}$	$\{a\}$
$\{(a, \text{in}), (b, \text{undec}), (c, \text{out}), (d, \text{in})\}$	$\{a, d\}$
$\{(a, \text{in}), (b, \text{out}), (c, \text{undec}), (d, \text{undec})\}$	$\{a\}$
$\{(a, \text{in}), (b, \text{out}), (c, \text{in}), (d, \text{out})\}$	$\{a, c\}$
$\{(a, \text{in}), (b, \text{out}), (c, \text{out}), (d, \text{in})\}$	$\{a, d\}$

Table 1. Admissibile labellings and sets in the example of Figure 5

An admissible set [Dung, 1995] is required to be both internally coherent and able to defend its elements.

Definition 3.5 *Let* $AF = \langle Ar, att \rangle$ *be an argumentation framework. A set* $Args \subseteq Ar$ *is called an* admissible set *iff* $Args$ *is conflict-free and* $Args \subseteq F_{AF}(Args)$. *The set of admissible sets of AF is denoted as* $\mathcal{AS}(AF)$.

As evident from this definition, the empty set is admissible for any argumentation framework. Apart from this trivial case, let us examine conflict-free and admissible sets in the reference examples. Considering Figure 5, one can observe that the non empty conflict-free sets are $\{a\}$, $\{b\}$, $\{c\}$, $\{d\}$, $\{a, c\}$, $\{a, d\}$, $\{b, d\}$. Among them, $\{a\}$, having no attackers, is admissible (actually $F_{AF}(\{a\}) = \{a\}$). The sets $\{b\}$ and $\{c\}$ are not admissible (b does not defend itself from a and c does not defend itself from b), while $\{d\}$ is, as it defends itself from c (in particular $F_{AF}(\{d\}) = \{a, d\}$). Moreover the sets $\{a, c\}$ and $\{a, d\}$ are admissible (in the former case c defends itself from the attack by d and is defended by a against b, in the latter both a and d are able to defend themselves), while $\{b, d\}$ is not (a defense for b against a is lacking). Applying analogous considerations in Figure 6, it can be seen that the non empty admissible sets are $\{a\}$, $\{b\}$, $\{a, d\}$ and $\{b, d\}$. On the other hand, in Figure 7 only the empty set is admissible since the non empty conflict-free sets are just the singletons $\{a\}$, $\{b\}$, $\{c\}$ but no argument defends itself from the attack it receives.

As probably noticed by the reader, the above examples point out a correspondence between admissible labellings and admissible sets (see an overview in Tables 1 and 2 for the two more articulated examples).

Before stating this correspondence in the general case, we need to provide the mapping from sets of arguments to labellings that was not introduced in previous section since it is well-defined only for conflict-free sets of arguments[6].

---

[6]If a set $Args$ of arguments is not conflict-free $Args \cap Args^+$ is not empty, i.e. some argument would be labelled both in and out according to Ext2Lab($Args$).

Admissible labellings	Admissible sets
$\{(a, \text{undec}), (b, \text{undec}), (c, \text{undec}), (d, \text{undec})\}$	$\emptyset$
$\{(a, \text{in}), (b, \text{out}), (c, \text{undec}), (d, \text{undec})\}$	$\{a\}$
$\{(a, \text{in}), (b, \text{out}), (c, \text{out}), (d, \text{undec})\}$	$\{a\}$
$\{(a, \text{in}), (b, \text{out}), (c, \text{out}), (d, \text{in})\}$	$\{a, d\}$
$\{(a, \text{out}), (b, \text{in}), (c, \text{undec}), (d, \text{undec})\}$	$\{b\}$
$\{(a, \text{out}), (b, \text{in}), (c, \text{out}), (d, \text{undec})\}$	$\{b\}$
$\{(a, \text{out}), (b, \text{in}), (c, \text{out}), (d, \text{in})\}$	$\{b, d\}$

Table 2. Admissibile labellings and sets in the example of Figure 6

The idea is that the members of the set are labelled in, the arguments attacked by the set are labelled out and the remaining arguments are labelled undec.

Definition 3.6 *Given an argumentation framework $AF = \langle Ar, att \rangle$ and a conflict-free set $Args \subseteq Ar$ the corresponding labelling $\text{Ext2Lab}(Args)$ is defined as $\text{Ext2Lab}(Args) = (Args, Args^+, Ar \setminus (Args \cup Args^+))$.*

Let us call an extension-based semantics conflict-free if each of its extensions is a conflict-free set. We can then extend the above definition to sets of extensions.

Definition 3.7 *Given an argumentation framework $AF = \langle Ar, att \rangle$ and a conflict-free extension-based semantics σ, the set of labellings corresponding to $\mathcal{E}_\sigma(AF)$ is given by $\mathcal{L}_\sigma(AF) = \{\text{Ext2Lab}(E) \mid E \in \mathcal{E}_\sigma(AF)\}$.*

The correspondence between admissible labellings and admissible sets stated by Proposition 3.8 has been proved in [Caminada and Gabbay, 2009].

Proposition 3.8 *For any argumentation framework $AF = \langle Ar, att \rangle$*

- *if $Args$ is an admissible set then $\text{Ext2Lab}(Args)$ is an admissible labelling;*

- *if Lab is an admissible labelling then $\text{Lab2Ext}(Lab)$ is an admissible set.*

It can be noted that the correspondence is not bijective, since different admissible labellings may give rise to the same admissible set. For instance, in the argumentation framework of Figure 5 both $(\{a\}, \{b\}, \{c, d\})$ and $(\{a\}, \emptyset, \{b, c, d\})$ are admissible labellings, whose set of in-labelled arguments yields the same admissible set $\{a\}$.

To complete the correspondence it is also possible to define a notion of a conflict-free labelling which parallels the one of conflict-free set[7].

---
[7]We use the Definition of [Caminada, 2011]. Note that clause 2. is needed for defining stage labellings (see Section 3.9).

Definition 3.9 *Let $\mathcal{L}ab$ be a labelling of an argumentation framework $AF = \langle Ar, att \rangle$. $\mathcal{L}ab$ is* conflict-free *iff for each $a \in Ar$ it holds that:*

1. *if a is labelled* in *then it does not have an attacker that is labelled* in

2. *if a is labelled* out *then it has at least one attacker that is labelled* in

When comparing a conflict-free labelling to an admissible labelling it can be noticed that the condition on out labelled arguments (second bullet) is essentially the same. However, the condition for in-labelled arguments (first bullet) is weaker for conflict-free labellings than for admissible labellings. It then follows that every admissible labelling is also a conflict-free labelling (just like every admissible set is also a conflict-free set by definition).

Finally, it is worth recalling that admissibility and defense are related by a basic property. In terms of extensions, if an admissible sets defends an argument, it is possible to add the argument to the set while preserving its admissibility and its capability to defend any other argument. This was proved in the so called Dung's Fundamental Lemma [Dung, 1995] recalled below.

Lemma 3.10 *For any argumentation framework $AF = \langle Ar, att \rangle$, let $\mathcal{A}rgs$ be an admissible set and a, b be arguments defended by $\mathcal{A}rgs$. Then*

1. $\mathcal{A}rgs' = \mathcal{A}rgs \cup \{a\}$ *is an admissible set;*

2. *b is defended by $\mathcal{A}rgs'$.*

Apart from admissibility as commonly applied in the literature, there also exists the related concept of *strong admissibility* [Baroni and Giacomin, 2007b; Caminada, 2014]. In order to describe this, we first need to introduce the concept of a min-max numbering. Basically, what a min-max numbering does is to assign to each in or out-labelled argument a natural number (or ∞) such that the min-max number of each in-labelled argument becomes the maximal value of its out-labelled attackers, plus one, and the min-max value of each out-labelled argument becomes the minimal value of its in-labelled attackers, plus one.

Definition 3.11 *Let $\mathcal{L}ab$ be an admissible labelling of argumentation framework $\langle Ar, att \rangle$. A* min-max numbering *is a total function $\mathcal{MM}_{\mathcal{L}ab} : \text{in}(\mathcal{L}ab) \cup \text{out}(\mathcal{L}ab) \to \mathbb{N} \cup \{\infty\}$ such that for each $a \in \text{in}(\mathcal{L}ab) \cup \text{out}(\mathcal{L}ab)$ it holds that:*

- *if $\mathcal{L}ab(a) = \text{in}$ then $\mathcal{MM}_{\mathcal{L}ab}(a) = max(\{\mathcal{MM}_{\mathcal{L}ab}(b) \mid b$ attacks a and $\mathcal{L}ab(b) = \text{out}\}) + 1$ (with $max(\emptyset)$ defined as 0)*

- *if $\mathcal{L}ab(a) = \text{out}$ then $\mathcal{MM}_{\mathcal{L}ab}(a) = min(\{\mathcal{MM}_{\mathcal{L}ab}(b) \mid b$ attacks a and $\mathcal{L}ab(b) = \text{in}\}) + 1$ (with $min(\emptyset)$ defined as ∞)*

It can be proved that each admissible labelling $\mathcal{L}ab$ has a unique min-max labelling $\mathcal{MM}_{\mathcal{L}ab}$ [Caminada, 2014]. As an example of a min-max numbering, take the argumentation framework of Figure 5 and admissible labelling $\mathcal{L}ab = (\{a,c\}, \{b,d\}, \emptyset)$. Argument a is labelled in and does not have any attackers. Hence, $\mathcal{MM}_{\mathcal{L}ab}(a) = 1$ (as $max(\emptyset) = 0$). Argument b is labelled out and has only one attacker that is labelled in (a), whose min-max value we have just determined to be 1. Hence, $\mathcal{MM}_{\mathcal{L}ab}(b) = 2$. For argument c the situation is more complex, as it has two attackers (b and d) and we have determined the min-max value of only one of these ($\mathcal{MM}_{\mathcal{L}ab}(b) = 2$). This means we first need to "guess" the min-max number of d in order to determine the min-max number of c. If we would for instance guess the min-max number of d as 2, then c will be assigned the min-max number of 3, which then implies that d should actually have been assigned a min-max number of 4 (contradiction). It can be verified that whatever natural number we initially assign as a min-max value to d, the number will later turn out to be too small. The only solution is to assign both c and d not with a natural number but with ∞. In that case, $\mathcal{MM}_{\mathcal{L}ab}(c) = max(\{2, \infty\}) + 1 = \infty$ and $\mathcal{MM}_{\mathcal{L}ab}(d) = max(\{\infty\}) + 1 = \infty$, thus satisfying the constraints of a min-max numbering.

We are now ready to provide the definition of a strongly admissible labelling.

Definition 3.12 *A strongly admissible labelling is an admissible labelling whose min-max numbering yields natural numbers only (so no argument is numbered ∞).*

From Definition 3.12 it directly follows that every strongly admissible labelling is also an admissible labelling. Apart from applying the labelling-based approach, it is also possible to express strong admissibility using the extension-based approach [Baroni and Giacomin, 2007b; Caminada, 2014].

Definition 3.13 *Let $\langle Ar, att \rangle$ be an argumentation framework. $Args \subseteq Ar$ is strongly admissible iff every $a \in Args$ is defended by some $Args' \subseteq Args \setminus \{a\}$ which is strongly admissible.*

The basis of this recursive definition is given by the facts that the empty set is strongly admissible and that unattacked arguments are defended by the empty set. Intuitively, the defense of every argument in a strongly admissible set is "rooted" in an unattacked argument (a notion called *strong defense* in [Baroni and Giacomin, 2007b]). In case there are no unattacked arguments in a framework, the empty set is its only strongly admissible set.

It can be proved that each strongly admissible set is conflict free and admissible [Baroni and Giacomin, 2007b; Caminada, 2014]. As an example, consider again the argumentation framework of Figure 5. The set $\{a\}$ is strongly admissible, because a is defended by $\emptyset$ (and $\emptyset \subseteq \{a\} \setminus \{a\}$) which is trivially strongly admissible since it has no elements. Also, the set $\{a, c\}$, although admissible, is not strongly admissible. This is because no subset of $\{a, c\} \setminus \{c\}$ can defend c against d. Correspondence between strongly admissible sets and strongly admissible labellings can be established through the functions `Ext2Lab` and `Lab2Ext`.

Proposition 3.14 *For any argumentation framework $AF = \langle Ar, att \rangle$*

- *if $Args$ is a strongly admissible set then $\mathtt{Ext2Lab}(Args)$ is a strongly admissible labelling;*

- *if Lab is a strongly admissible labelling then $\mathtt{Lab2Ext}(Lab)$ is a strongly admissible set.*

It can be noted that the correspondence is not bijective, since different strongly admissible labellings may give rise to the same strongly admissible set. For instance, in the argumentation framework of Figure 5 both $(\{a\}, \{b\}, \{c, d\})$ and $(\{a\}, \emptyset, \{b, c, d\})$ are strongly admissible labellings, whose set of `in`-labelled arguments yields the same strongly admissible set $\{a\}$.

Finally, an equivalent non-recursive characterisation for finite strongly admissible sets has been provided in [Baumann et al., 2016] and is recalled in the Proposition 3.15.

Proposition 3.15 *Let $\langle Ar, att \rangle$ be an argumentation framework and $Args \subseteq Ar$ a finite set of arguments. $Args$ is strongly admissible iff there are finitely many and pairwise disjoint sets $Args_1, \ldots Args_n$ such that $Args = \bigcup_{1 \leq i \leq n} Args_i$, $Args_1 \subseteq \mathrm{F}_{AF}(\emptyset)$ and for every $1 \leq j \leq n-1$ $Args_{j+1} \subseteq \mathrm{F}_{AF}(\bigcup_{1 \leq i \leq j} Args_i)$.*

In words, a strongly admissible set can be constructed starting from a set of unattacked arguments $Args_1$ and then adding iteratively further arguments which are defended by those already included in the set.

3.2 Naïve semantics

Naïve semantics (denoted as $\mathcal{NA}$) corresponds to selecting as many arguments as possible, provided that there are no attacks among them. It is a sort of greedy strategy, driven by the only criterion of avoiding conflicts. Formally it corresponds to requiring conflict-freeness together with a maximality property and can be easily expressed in both the labelling-based and extension-based approach.

Definition 3.16 *Let Lab be a labelling of an argumentation framework $\langle Ar, att \rangle$. Lab is a naïve labelling iff it is a conflict-free labelling whose set of `in`-labelled arguments is maximal (w.r.t. set inclusion) with respect to all conflict-free labellings.*

Definition 3.17 *Let $\langle Ar, att \rangle$ be an argumentation framework. A set $Args \subseteq Ar$ is called a naïve extension iff $Args$ is a maximal conflict-free set.*

It can immediately be observed from the definitions above that naïve semantics ignores the direction of attacks, which makes it very simple but at the same time rather poor, since it overlooks an essential information carried by the formalism.

Let us illustrate naïve semantics with an example[8].

In Figure 5, there are nineteen conflict-free labellings, including for instance $(\emptyset, \emptyset, \{a, b, c, d\})$, $(\{a\}, \emptyset, \{b, c, d\})$, $(\{d\}, \{c\}, \{a, b\})$ and many others. Among them, there are four naïve labellings with set of in-labelled arguments $\{a, c\}$, namely $(\{a, c\}, \emptyset, \{b, d\})$, $(\{a, c\}, \{b\}, \{d\})$, $(\{a, c\}, \{d\}, \{b\})$, $(\{a, c\}, \{b, d\}, \emptyset)$, four naïve labellings where the set of in-labelled arguments is $\{a, d\}$, namely $(\{a, d\}, \emptyset, \{b, c\})$, $(\{a, d\}, \{b\}, \{c\})$, $(\{a, d\}, \{c\}, \{b\})$, $(\{a, d\}, \{b, c\}, \emptyset)$, and two naïve labellings with set of in-labelled arguments $\{b, d\}$, namely $(\{b, d\}, \emptyset, \{a, c\})$ and $(\{b, d\}, \{c\}, \{a\})$.

In the same example, there are eight conflict-free sets of arguments, namely $\emptyset, \{a\}, \{b\}, \{c\}, \{d\}, \{a, c\}, \{a, d\}, \{b, d\}$. Clearly only $\{a, c\}, \{a, d\}, \{b, d\}$ are maximal, i.e. naïve extensions.

From the example it emerges clearly that the relationships between naïve labellings and naïve extensions is in general many-to-one: many naïve labellings may correspond to the same naïve extension and each naïve extension corresponds to at least one, and in general many, naïve labellings.

Proposition 3.18 *For any argumentation framework $\langle Ar, att \rangle$, if $\mathcal{L}ab$ is a naïve labelling there is a naïve extension $\mathcal{A}rgs$ such that $\mathcal{A}rgs = \texttt{Lab2Ext}(\mathcal{L}ab)$. If $\mathcal{A}rgs$ is a naïve extension then $\texttt{Ext2Lab}(\mathcal{A}rgs)$ is a naïve labelling.*

3.3 Complete Semantics

Complete semantics ($\mathcal{CO}$) can be regarded as a strengthening of the basic requirements enforced by the idea of admissibility. Intuitively, while admissibility requires one to be able to give reasons for accepted and rejected arguments but leaves one free to abstain on any argument, complete semantics requires one to abstain only if there are no good reasons to do otherwise. That is, if one abstains from having an opinion on whether the argument is accepted or rejected, then one should have insufficient grounds to accept the argument (meaning that not all its attackers are rejected) and insufficient grounds to reject the argument (meaning that it does not have an attacker that is accepted). Note in particular that, while the trivial solution of leaving anything undecided is always admissible, it is not always complete since there can be arguments one has good reasons not to abstain about.

In the labelling-based approach, the intuition described above corresponds to extending Definition 3.1 in order to encompass a notion of an argument being *legally undecided*.

---

[8] A summary of the outcomes produced by all the semantics considered in this chapter on all the examples presented in this section is given in Tables 5 and 6 at the end of the section. For the sake of compactness, the summary is given in terms of extensions, the corresponding labellings being derivable with the `Ext2Lab` function.

Definition 3.19 *Let $\mathcal{L}ab$ be a labelling of an argumentation framework $\langle Ar, att \rangle$.*

- *An* **undec**-*labelled argument is said to be legally* **undec** *iff not all its attackers are labelled* **out** *and it doesn't have an attacker that is labelled* **in**.

Definition 3.20 *A* complete labelling *is a labelling where every* **in**-*labelled argument is legally* **in**, *every* **out**-*labelled argument is legally* **out** *and every* **undec** *labelled argument is legally* **undec**.

It is clear from Definitions 3.20 and 3.2 that every complete labelling is an admissible labelling (but the reverse does not hold in general).

An alternative characterisation of a complete labelling can be provided (a formal proof can be found in [Caminada and Gabbay, 2009]).

Proposition 3.21 *A labelling $\mathcal{L}ab$ of an argumentation framework (Ar, att) is a complete labelling iff for each argument $a \in Ar$ it holds that:*

1. *a is labelled* **in** *iff all its attackers are labelled* **out**, *and*

2. *a is labelled* **out** *iff it has at least one attacker that is labelled* **in**.

Although Proposition 3.21 does not explicitly mention **undec**, it follows that each argument that is labelled **undec** does not have all its attackers **out** (otherwise it would have to be labelled **in** by point 1) and it does not have an attacker that is labelled **in** (otherwise it would have to be labelled **out** by point 2). Therefore, each **undec**-labelled argument is legally **undec**. Comparing Proposition 3.21 with Definition 3.5 one can appreciate the difference between admissible and complete labellings from another perspective: in an admissible labelling an argument can be labelled **in** *only if* all its attackers are labelled **out**, but need not to be labelled **in** if this condition holds. In a complete labelling an argument is labelled **in** *if and only if* all its attackers are labelled **out**: thus complete labellings are more constrained and this corresponds to the lesser freedom of abstaining mentioned at the beginning of this section.

Turning to the extension-based approach, a complete extension is a conflict-free set which includes precisely those arguments it defends. That is, if an argument is defended by the set it should be in the set, and if an argument is not defended by the set, it should not be in the set. Technically this means that a complete extension is a conflict-free fixed point of the characteristic function, as stated in the following definition [Dung, 1995].

Definition 3.22 *Let $AF = \langle Ar, att \rangle$ be an argumentation framework. A set $Args \subseteq Ar$ is called a* complete extension *iff $Args$ is conflict-free and $Args = F_{AF}(Args)$.*

Fig. #	Members of $\mathcal{L}_{CO}(AF)$	Members of $\mathcal{E}_{CO}(AF)$
Fig. 5	$\{(a,\text{in}),(b,\text{out}),(c,\text{undec}),(d,\text{undec})\}$ $\{(a,\text{in}),(b,\text{out}),(c,\text{in}),(d,\text{out})\}$ $\{(a,\text{in}),(b,\text{out}),(c,\text{out}),(d,\text{in})\}$	$\{a\}$ $\{a,c\}$ $\{a,d\}$
Fig. 6	$\{(a,\text{undec}),(b,\text{undec}),(c,\text{undec}),(d,\text{undec})\}$ $\{(a,\text{in}),(b,\text{out}),(c,\text{out}),(d,\text{in})\}$ $\{(a,\text{out}),(b,\text{in}),(c,\text{out}),(d,\text{in})\}$	$\emptyset$ $\{a,d\}$ $\{b,d\}$
Fig. 7	$\{(a,\text{undec}),(b,\text{undec}),(c,\text{undec})\}$	$\{\emptyset\}$

Table 3. Complete labellings and extensions in the examples of Figures 5- 7

It is clear from Definitions 3.22 and 3.5 that every complete extension is an admissible set (but the reverse does not hold in general).

Let us now provide some examples to illustrate the notion of complete semantics. In Figure 5, one can observe that, among the seven admissible labellings, only $(\{a\},\{b\},\{c,d\})$, $(\{a,c\},\{b,d\},\emptyset)$, and $(\{a,d\},\{b,c\},\emptyset)$ are complete. In particular, note that a is legally in in all labellings because all its attackers are out (trivially, because it has no attackers). b is legally out in all labellings because it has an attacker (a) that is in. On the other hand, c and d can be both legally undec, or one legally in and the other legally out. Analogously, in the same figure it can be noted that $\{a\}$ is a complete extension (a has no attackers and is therefore trivially defended by any set, a defends c from b but not from d), and $\{a,c\}$ and $\{a,d\}$ are complete extensions too.

In Figure 6, the trivial labelling $(\emptyset, \emptyset, \{a,b,c,d\})$ is complete, as well as $(\{a,d\},\{b,c\},\emptyset)$ and $(\{b,d\},\{a,c\},\emptyset)$. Analogously, $\emptyset$ is a complete extension (no unattacked arguments exist, which would be the only arguments defended by the empty set) as well as $\{a,d\}$ and $\{b,d\}$, while $\{a\}$ and $\{b\}$ are not complete extensions since they both defend also argument d.

In Figure 7 the only complete labelling is the trivial one $(\emptyset,\emptyset,\{a,b,c\})$ and analogously the only complete extension is $\emptyset$ (as it was the case for admissible labellings/sets).

As the above examples also show, there is a direct mapping between complete labellings and complete extensions: it has been proved in [Caminada and Gabbay, 2009] that this correspondence is bijective, as stated in the following proposition.

Proposition 3.23 *For any argumentation framework* $AF = \langle Ar, att \rangle$, $\mathcal{L}ab$ *is a complete labelling iff there is a complete extension* $\mathcal{A}rgs$ *such that* $\mathcal{L}ab =$ Ext2Lab($\mathcal{A}rgs$).

Table 3 shows this correspondence on the examples discussed above.

3.4 Grounded Semantics

If one regards each complete labelling (or complete extension) as a reasonable position one can take in the presence of the conflicting information expressed in the argumentation framework, then a possible question is to examine what is the most "grounded" position one can take, namely the position which is least questionable. The idea is then to accept only the arguments that one cannot avoid to accept, to reject only the arguments that one cannot avoid to reject, and abstaining as much as possible. This gives rise to the most skeptical (or least committed) semantics among those based on complete extensions, namely the *grounded* semantics ($\mathcal{GR}$).

This idea has a straightforward formal counterpart in terms of a minimality requirement[9].

Definition 3.24 *Let $AF = \langle Ar, att \rangle$ be an argumentation framework. The grounded labelling of AF is a complete labelling $\mathcal{L}ab$ where $\text{in}(\mathcal{L}ab)$ is minimal (w.r.t. set inclusion), i.e. there is no complete labelling $\mathcal{L}ab'$ such that $\text{in}(\mathcal{L}ab') \subsetneq \text{in}(\mathcal{L}ab)$.*

Definition 3.25 *Let $AF = \langle Ar, att \rangle$ be an argumentation framework. The grounded extension of AF is a minimal (w.r.t. set inclusion) complete extension of AF (i.e. a minimal conflict-free fixed point of the characteristic function F_{AF}).*

As we have already seen complete labellings and extensions in the examples of Figures 5-7, one can easily identify those featuring the minimality property required by the above definitions. In the example of Figure 5, the grounded labelling is $(\{a\}, \{b\}, \{c,d\})$ while the grounded extension is $\{a\}$. In both Figures 6 and 7 the grounded labelling is the trivial one $((\emptyset, \emptyset, \{a,b,c,d\})$ and $(\emptyset, \emptyset, \{a,b,c\})$ respectively), and analogously the grounded extension is the empty set in both cases.

The uniqueness of the grounded labelling and extension in these examples is not accidental. Considering the grounded extension, since F_{AF} is monotonic it follows from the Knaster-Tarski theorem that F_{AF} has a unique smallest fixed point. It can then be proved that this fixed point is also conflict-free [Dung, 1995].

Proposition 3.26 *For any argumentation framework $AF = \langle Ar, att \rangle$, the following statements are equivalent:*

1. *Args is a minimal conflict-free fixed point of F_{AF}*

2. *Args is the smallest fixed point of F_{AF}*

---

[9]Definition 3.25 is not literally the same as the one originally given by [Dung, 1995]. We provide this equivalent version as it is more coherent with our presentation line.

It follows that:

- the grounded extension is unique (i.e. grounded semantics belongs to the unique-status approach);

- the grounded extension is the least complete extension, in particular it is included in any complete extension.

The grounded extension of an argumentation framework AF will be denoted as $GE(AF)$.

In virtue of the one-to-one correspondence between complete extensions and complete labellings established in Section 3.3, it can be proved that the grounded labelling is unique and coincides with Ext2Lab($Args$) where $Args$ is the grounded extension. Similarly, if Lab is the grounded labelling, then Lab2Ext(Lab) is the grounded extension.

As a confirmation of the intuitive meaning stated at the beginning of the section, it turns out that the grounded semantics can be described not only in terms of minimizing acceptance. In fact, the complete labelling where in(Lab) is minimal is also the complete labelling Lab where out(Lab) is minimal, and the complete labelling Lab where undec(Lab) is maximal. This is stated in Proposition 3.28, whose proof is based on Lemma 3.27 (see [Caminada, 2006a; Caminada and Gabbay, 2009] for details).

Lemma 3.27 *Given two complete labellings Lab_1 and Lab_2 of an argumentation framework $\langle Ar, att \rangle$, it holds that* in(Lab_1) $\subseteq$ in(Lab_2) *iff* out(Lab_1) $\subseteq$ out(Lab_2).

Proposition 3.28 *Let Lab be a complete labelling of an argumentation framework $\langle Ar, att \rangle$. The following statements are equivalent.*

1. *Lab is the complete labelling where* in(Lab) *is minimal (w.r.t. set inclusion)*

2. *Lab is the complete labelling where* out(Lab) *is minimal (w.r.t. set inclusion)*

3. *Lab is the complete labelling where* undec(Lab) *is maximal (w.r.t. set inclusion)*

Given the bijective correspondence between complete labellings and complete extensions, the above proposition can be equivalently formulated for the extension-based approach.

Proposition 3.29 *Let E be a complete extension of an argumentation framework $\langle Ar, att \rangle$. The following statements are equivalent.*

1. *E is the least (w.r.t. set inclusion) complete extension*

2. E is the complete extension such that E^+ is minimal (w.r.t. set inclusion)

3. E is the complete extension such that $Ar \setminus (E \cup E^+)$ is maximal (w.r.t. set inclusion)

There also exists a connection between grounded semantics and the concept of strong admissibility [Baroni and Giacomin, 2007b; Caminada, 2014].

Theorem 3.30 *Let $AF = \langle Ar, att \rangle$ be an argumentation framework. The grounded extension of AF is the unique maximal (w.r.t set inclusion) strongly admissible set of AF.*

Finally, an interesting property proved in [Dung, 1995] provides a useful "constructive" characterisation of grounded semantics[10] for finite (and more generally finitary[11]) argumentation frameworks.

Proposition 3.31 *The grounded extension of any finitary argumentation framework AF is equal to $\bigcup_{i=1,\ldots,\infty} F_{AF}^i(\emptyset)$, where $F_{AF}^1(\emptyset) = F_{AF}(\emptyset)$ and for $i > 1$ $F_{AF}^i(\emptyset) = F_{AF}(F_{AF}^{i-1}(\emptyset))$.*

On the basis of Proposition 3.31 the grounded labelling (or equivalently extension) can be obtained incrementally by first labelling **in** those arguments which do not receive attacks. Then the arguments attacked by those labelled **in** are labelled **out**. The same steps are iterated considering only those arguments which have not been labelled yet, namely repeating the procedure on an argumentation framework obtained by suppressing the already labelled arguments. In particular, this corresponds to labelling **in** those unlabelled arguments which only receive attacks from arguments labelled **out**, and then labelling **out** those attacked by the newly labelled **in** arguments. The procedure is then iterated until an iteration does not produce any newly **in** or **out** labelled argument. Then, any still unlabelled arguments are labelled **undec**.

It can be noted that the first iteration corresponds to labelling **in** the arguments in $F_{AF}^1(\emptyset)$ and **out** the arguments attacked by $F_{AF}^1(\emptyset)$, the second iteration to labelling **in** the arguments in $F_{AF}^2(\emptyset)$ and **out** the arguments attacked by $F_{AF}^2(\emptyset)$, and so on. This procedure can be applied to the examples and provides another way to see that the grounded extension includes those and only those arguments whose defense is "rooted" in unattacked arguments and is the maximal strongly admissible set.

If the aim is not so much to compute the entire grounded extension (labelling) but merely to examine whether or not an argument is in the grounded extension (labelled **in** by the grounded labelling) then one could also use the proof procedures described in [Modgil and Caminada, 2009; Caminada, 2015].

---

[10] Note that the characterisation of strongly admissible sets in Proposition 3.15 can be seen as a generalisation of the intuition underlying this traditional result in the case of finite frameworks.

[11] An argumentation framework is finitary if every argument receives a finite number of attacks.

3.5 Preferred Semantics

While grounded semantics takes a skeptical, or least-commitment, standpoint, one can also consider the alternative view oriented at accepting as many arguments as reasonably possible. This may give rise to mutually exclusive alternatives for acceptance: for instance a mutual attack can be reasonably resolved by accepting either of the conflicting arguments, but clearly not both (these alternatives are called non-skeptical solutions in the examples below).

The idea of maximizing accepted arguments is expressed by *preferred semantics* ($\mathcal{PR}$) whose description in the labelling-based and extension-based approaches is given in the following definitions.

Definition 3.32 *Let $AF = \langle Ar, att \rangle$ be an argumentation framework. A preferred labelling of AF is a complete labelling $\mathcal{L}ab$ where $\text{in}(\mathcal{L}ab)$ is maximal (w.r.t. set-inclusion) among all complete labellings, i.e. there is no complete labelling $\mathcal{L}ab'$ such that $\text{in}(\mathcal{L}ab') \supsetneq \text{in}(\mathcal{L}ab)$.*

Definition 3.33 *Let $AF = \langle Ar, att \rangle$ be an argumentation framework. A preferred extension is a maximal (w.r.t. set-inclusion) admissible set of AF.*

Considering the examples of Figures 5-7, the existence of multiple preferred labellings (or extensions) immediately emerges. For instance, in Figure 5 two non-skeptical solutions exist for the mutual attack between c and d, giving rise to the preferred labellings $(\{a,c\}, \{b,d\}, \emptyset)$ and $(\{a,d\}, \{b,c\}, \emptyset)$. Similarly, two preferred extensions exist, namely $\{a,c\}$ and $\{a,d\}$.

In Figure 6 again two alternative non-skeptical solutions exist for the mutual attack between a and b. In both cases, c is then rejected and d accepted. This intuitive description corresponds to the two preferred labellings $(\{a,d\}, \{b,c\}, \emptyset)$ and $(\{b,d\}, \{a,c\}, \emptyset)$ and, analogously, to the preferred extensions $\{a,d\}$ and $\{b,d\}$.

In Figure 7 instead, no non-trivial solutions to the conflict are available under the constraint of admissibility, as the reader may remember from previous subsections. It then follows that the unique preferred labelling in this case is $(\emptyset, \emptyset, \{a,b,c\})$ and, similarly, the only preferred extension is $\emptyset$.

As usual, the correspondences in the above examples are not accidental: it can be proved that an analogous version of Proposition 3.23 holds for preferred semantics, i.e. there is a bijective correspondence between preferred labellings and preferred extensions through the Ext2Lab (and Lab2Ext) functions.

It turns out that the complete labellings with maximal in are the same as the complete labellings with maximal out, as stated in Proposition 3.34 whose proof is based on Lemma 3.27.

Proposition 3.34 *Given an argumentation framework $AF = \langle Ar, att \rangle$ the following statements are equivalent.*

1. *$\mathcal{L}ab$ is a complete labelling where $\text{in}(\mathcal{L}ab)$ is maximal (w.r.t. set inclusion) among all complete labellings.*

2. $\mathcal{L}ab$ is a complete labelling where $\mathsf{out}(\mathcal{L}ab)$ is maximal (w.r.t. set inclusion) among all complete labellings.

An analogous formulation of Proposition 3.34 for the extension-based approach could be provided in a straightforward way.

Relationships of preferred extensions with other semantics notions have been analyzed in [Dung, 1995]. Preferred extensions can for instance equivalently be characterized as maximal complete extensions.

Proposition 3.35 *Let $AF = \langle Ar, att \rangle$ be an argumentation framework and let $\mathcal{A}rgs \subseteq Ar$. The following statements are equivalent.*

1. *$\mathcal{A}rgs$ is a maximal (w.r.t. set inclusion) admissible set of AF*

2. *$\mathcal{A}rgs$ is a maximal (w.r.t. set inclusion) complete extension of AF*

This in particular implies that the grounded extension is included in any preferred extension, as it is in any complete extension. By definition, the grounded extension coincides with the intersection of all complete extensions: one may then wonder whether this holds also for preferred extensions. The answer is negative, as shown for instance by the example of Figure 6 where the grounded extension is $\emptyset$ while the intersection of the preferred extensions is $\{d\}$. Again, this fact can be easily translated to the labelling-based approach referring to the in-labelled arguments.

An algorithm that produces all preferred labellings (and therefore also produces all preferred extensions) is described in [Caminada, 2007a; Modgil and Caminada, 2009; Nofal et al., 2014]. If the aim is merely to determine whether an argument is in at least one preferred extension (labelled in by at least one preferred labelling) then one could also use the proof procedures described in [Vreeswijk and Prakken, 2000; Vreeswijk, 2006; Verheij, 2007; Modgil and Caminada, 2009; Caminada et al., 2016]. Proof procedures for determining whether an argument is in every preferred extension (labelled in by every preferred labelling) are provided in [Cayrol et al., 2003; Modgil and Caminada, 2009].

3.6 Stable Semantics

So far we have discussed semantics according to the intuitive idea that an argument can be accepted, rejected or left undecided. One can however prefer more committed evaluations, in which there is no room for neutrality or shades of gray and everything is just black or white. This means that undecided arguments are simply "forbidden" as in statements like "you're either with us or against us."

This clear-and-strong view corresponds to *stable* semantics ($\mathcal{ST}$) and has a direct formulation in both the labelling-based and extension-based approach.

Definition 3.36 *Let $\mathcal{L}ab$ be a labelling of argumentation framework $AF = \langle Ar, att \rangle$. $\mathcal{L}ab$ is a stable labelling of AF iff it is a complete labelling with $\mathsf{undec}(\mathcal{L}ab) = \emptyset$.*

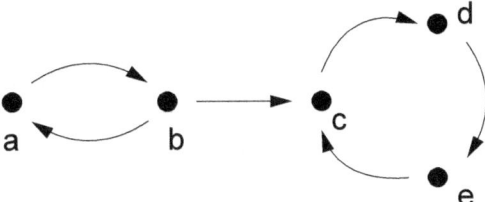

Figure 9. An argumentation framework where preferred and stable semantics differ

Definition 3.37 *Let $AF = \langle Ar, att \rangle$ be an argumentation framework. A stable extension of AF is a conflict-free set $Args$ such that $Args \cup Args^+ = Ar$.*

In the example shown in Figure 5 there are two stable labellings, namely $(\{a,c\}, \{b,d\}, \emptyset)$ and $(\{a,d\}, \{b,c\}, \emptyset)$. Similarly, two stable extensions exist, namely $\{a,c\}$ and $\{a,d\}$. In Figure 6 the labellings $(\{a,d\}, \{b,c\}, \emptyset)$ and $(\{b,d\}, \{a,c\}, \emptyset)$ are stable and, analogously, there are two stable extensions, namely $\{a,d\}$ and $\{b,d\}$.

Figure 7 shows that the strong view underlying stable semantics cannot be universally applied. In fact, no labelling nor extension complying with the definition can be identified (the requirements of conflict-freeness and ability to attack all other arguments are incompatible in this case). This can be regarded as a limitation of stable semantics as "stable extensions do not capture the intuitive semantics of every meaningful argumentation system" [Dung, 1995]. Looking at this fact from another perspective, differently from other semantics reviewed so far, in the case of stable semantics the trivial labelling (or extension) does not represent the "default" conflict resolution one can resort to when nothing else is reasonable. It follows that, using a terminology from [Baroni and Giacomin, 2009a], stable semantics is not *universally defined*, since there are argumentation frameworks where it is intrinsically impossible to apply its "in-or-out" view. No other argumentation semantics considered in the literature has this limitation.

Apart from this critical case, the reader may have noticed that the stable labellings (extensions) coincide with the preferred ones in the other two cases. One may then wonder whether stable semantics (leaving apart critical cases) coincides with preferred semantics in general. The answer is negative, as shown by the argumentation framework of Figure 9. Here one can verify that there are three complete labellings, namely $(\emptyset, \emptyset, \{a,b,c,d,e\})$, $(\{a\}, \{b\}, \{c,d,e\})$ and $(\{b,d\}, \{a,c,e\}, \emptyset)$, and, correspondingly, three complete extensions. Two of the three labellings (extensions) are preferred, namely $(\{a\}, \{b\}, \{c,d,e\})$ and $(\{b,d\}, \{a,c,e\}, \emptyset)$, but clearly only the last one is stable.

Let us now generalize this and possibly related observations, examining properties of stable semantics in general.

First it is possible to characterize the concept of a stable labelling in other terms. In particular note that the difference between a complete labelling and an admissible labelling is that a complete labelling has the additional requirement that every **undec**-labelled argument is legally **undec**. However, if, as in Definition 3.36, there are no **undec**-labelled arguments in the first place, then this extra requirement becomes superfluous. Moreover, the fact that anything that is not labelled **in** is labelled **out** ensures that every stable labelling is also preferred (but not viceversa, as we have already seen). These considerations are summarized in Proposition 3.38 (notice that point 3 of Proposition 3.38 coincides with Definition 3.36).

Proposition 3.38 *Let $\mathcal{L}ab$ be a labelling of an argumentation framework $AF = (Ar, att)$. The following statements are equivalent:*

1. *$\mathcal{L}ab$ is a conflict-free labelling with $\text{undec}(\mathcal{L}ab) = \emptyset$*

2. *$\mathcal{L}ab$ is an admissible labelling with $\text{undec}(\mathcal{L}ab) = \emptyset$*

3. *$\mathcal{L}ab$ is a complete labelling with $\text{undec}(\mathcal{L}ab) = \emptyset$*

4. *$\mathcal{L}ab$ is a preferred labelling with $\text{undec}(\mathcal{L}ab) = \emptyset$*

On the other hand, it is immediate to see that a stable extension is an admissible set, hence the equivalent characterisations given in Proposition 3.39 (again, note that point 1 of Proposition 3.39 coincides with Definition 3.37).

Proposition 3.39 *Let $AF = \langle Ar, att \rangle$ be an argumentation framework and $Args \subseteq Ar$ a set of arguments. The following statements are equivalent:*

1. *$Args$ is a conflict-free set with $Args \cup Args^+ = Ar$*

2. *$Args$ is an admissible set such that $Args \cup Args^+ = Ar$*

3. *$Args$ is a complete extension such that $Args \cup Args^+ = Ar$*

4. *$Args$ is a preferred extension such that $Args \cup Args^+ = Ar$*

5. *$Args^+ = Ar \backslash Args$*

As probably evident from above, the bijective labellings-extensions correspondence through `Ext2Lab` (and `Lab2Ext`) holds for stable semantics too as proved in [Caminada and Gabbay, 2009].

An algorithm that produces all stable labellings (and therefore also all stable extensions) is described in [Caminada, 2007a; Modgil and Caminada, 2009]. If the aim is merely to determine whether an argument is **in** at least one stable extension (labelled **in** by at least one stable labelling) then one could also use the proof procedures described in [Caminada and Wu, 2009]. Proof procedures for determining whether an argument is in every stable extension (labelled **in** by every stable labelling) are also provided in [Caminada and Wu, 2009].

3.7 Semi-Stable Semantics

As illustrated in the previous section, the requirement of "forbidding" undecided arguments turns out to yield no results in some cases. A more sophisticated idea consists in expressing a definite opinion on the largest possible set of arguments, while restricting as much as possible (but not necessarily avoiding) those which are left undecided. This intuition lies at the basis of *semi-stable* semantics ($\mathcal{SST}$), which can be defined as follows.

Definition 3.40 *Let $\mathcal{L}ab$ be a labelling of an argumentation framework $AF = \langle Ar, att \rangle$. $\mathcal{L}ab$ is a* semi-stable labelling *of AF iff $\mathcal{L}ab$ is a complete labelling where* undec($\mathcal{L}ab$) *is minimal (w.r.t. set inclusion) among all complete labellings, i.e. there is no complete labelling $\mathcal{L}ab'$ such that* undec($\mathcal{L}ab'$) $\subsetneq$ undec($\mathcal{L}ab$).

Definition 3.41 *Let $AF = \langle Ar, att \rangle$ be an argumentation framework. A* semi-stable extension *of AF is a complete extension $\mathcal{A}rgs$ where $\mathcal{A}rgs \cup \mathcal{A}rgs^+$ is maximal (w.r.t. set inclusion) among all complete extensions, i.e there is no complete extension $\mathcal{A}rgs'$ such that $(\mathcal{A}rgs' \cup \mathcal{A}rgs'^+) \supsetneq (\mathcal{A}rgs \cup \mathcal{A}rgs^+)$.*

It follows directly that each stable labelling is also a semi-stable labelling and that semi-stable labellings coincide with stable labellings when the latter exist. This is because a stable labelling is a complete labelling with an empty set of undec-labelled arguments. Hence, it is a complete labelling where the set of undec-labelled arguments is minimal (so a semi-stable labelling). Furthermore, if there exists at least one stable labelling then the set of undec-labelled arguments has to be empty in any complete labelling with a minimal set of undec-labelled arguments (semi-stable labelling) and hence any such a labelling has to be stable. The same relationship holds between stable and semi-stable extensions: each stable extension is a semi-stable extension, and semi-stable extensions coincide with stable extensions when the latter exist. Accordingly, we already know, from previous section, the behaviour of semi-stable semantics in the examples of Figures 5 and 6.

Even in situations where stable extensions/labellings do not exist, the existence of semi-stable labellings (or extensions) is anyway guaranteed, since they are selected among the (always existing) complete ones. In particular, in the example of Figure 7 the only semi-stable labelling (extension) is (again) the trivial one.

The maximization requirement imposed by semi-stable semantics is intuitively similar, but clearly different, from the maximization requirement in the definition of preferred semantics. One may wonder whether these different maximizations actually lead to the same results. The answer is negative, see also [Verheij, 2003], as shown by the example of Figure 9, where there are two preferred labellings (and then two corresponding extensions) namely ($\{a\}, \{b\}, \{c, d, e\}$) and ($\{b, d\}, \{a, c, e\}, \emptyset$), but only the latter is semi-stable (as well as stable).

Equivalent characterisations of semi-stable semantics in terms of admissible labellings/sets and of preferred labellings/extensions are available, see e.g. [Caminada and Gabbay, 2009], as summarized in the following propositions.

Proposition 3.42 *Let $\mathcal{L}ab$ be a labelling of an argumentation framework $AF = (Ar, att)$. The following statements are equivalent.*

1. *$\mathcal{L}ab$ is a complete labelling where $\text{undec}(\mathcal{L}ab)$ is minimal (w.r.t. set inclusion) among all complete labellings*

2. *$\mathcal{L}ab$ is an admissible labelling where $\text{undec}(\mathcal{L}ab)$ is minimal (w.r.t. set inclusion) among all admissible labellings*

3. *$\mathcal{L}ab$ is a preferred labelling where $\text{undec}(\mathcal{L}ab)$ is minimal (w.r.t. set inclusion) among all preferred labellings*

Proposition 3.43 *Let $AF = \langle Ar, att \rangle$ be an argumentation framework, and let $Args \subseteq Ar$. The following statements are equivalent.*

1. *$Args$ is a complete extension where $Args \cup Args^+$ is maximal (w.r.t. set inclusion) among all complete extensions*

2. *$Args$ is an admissible set where $Args \cup Args^+$ is maximal (w.r.t. set inclusion) among all admissible sets*

3. *$Args$ is a preferred extension where $Args \cup Args^+$ is maximal (w.r.t. set inclusion) among all preferred extensions*

Finally, the usual bijective labellings-extension correspondence holds for semi-stable semantics too, see [Caminada, 2007a; Caminada and Gabbay, 2009]. An algorithm that produces all semi-stable labellings (and therefore also all semi-stable extensions) is described in [Caminada, 2007a; Modgil and Caminada, 2009].

The concept of semi-stable semantics can be traced back to the notion of admissible stage extensions (see Section 3.9) introduced by [Verheij, 1996]. Although there are differences in the basic formalisation (Verheij for instance does not use the standard extension-based approach) it can be proved that Verheij's approach is equivalent to that of Caminada, who, independently from Verheij, rediscovered the same concept under the name of semi-stable semantics [Caminada, 2006b].

3.8 Ideal and Eager Semantics

The notion of *ideal* semantics ($\mathcal{ID}$) can perhaps be best explained using a description concerning a judgment aggregation context [Caminada and Pigozzi, 2011], where different people have different opinions on a set of arguments, each opinion being expressed as a labelling, and an aggregated opinion, namely an aggregated labelling, has to be produced. In particular the *ideal* labelling/extension results from the following assumption on the aggregation procedure:

- each participant tries to accept as many arguments as possible, that is, the opinions to be aggregated correspond to the set of the preferred labellings/extensions;
- each argument is (tentatively) accepted or rejected only if there is unanimity on it by all participants, otherwise it is regarded as undecided;
- the resulting labelling/extension may or may not correspond to a defensible position, namely may or may not be admissible: if it is not, water it down (by abstaining about some tentatively accepted or rejected arguments in the aggregated judgment) until it becomes defensible

In order to formally define the concept of the ideal labelling, according to the intuition outlined above, we first need to treat some preliminaries (see [Caminada and Pigozzi, 2011]).

Definition 3.44 *Let $\mathcal{L}ab_1$ and $\mathcal{L}ab_2$ be labellings of an argumentation framework $AF = \langle Ar, att\rangle$. We say that $\mathcal{L}ab_2$ is more or equally committed than $\mathcal{L}ab_1$ ($\mathcal{L}ab_1 \sqsubseteq \mathcal{L}ab_2$) iff $\text{in}(\mathcal{L}ab_1) \subseteq \text{in}(\mathcal{L}ab_2)$ and $\text{out}(\mathcal{L}ab_1) \subseteq \text{out}(\mathcal{L}ab_2)$. We say that $\mathcal{L}ab_2$ is compatible with $\mathcal{L}ab_1$ ($\mathcal{L}ab_1 \approx \mathcal{L}ab_2$) iff $\text{in}(\mathcal{L}ab_1) \cap \text{out}(\mathcal{L}ab_2) = \emptyset$ and $\text{out}(\mathcal{L}ab_1) \cap \text{in}(\mathcal{L}ab_2) = \emptyset$.*

It holds that "$\sqsubseteq$" defines a partial order (reflexive, anti-symmetric, transitive) on the labellings of an argumentation framework. We can therefore talk about a labelling being "bigger" or "smaller" than another labelling with respect to "$\sqsubseteq$". The relation "$\approx$", although reflexive and symmetric, is not an equivalence relation, since it does not satisfy transitivity.[12] It holds that "$\sqsubseteq$" is at least as strong as "$\approx$"; that is, if $\mathcal{L}ab_1 \sqsubseteq \mathcal{L}ab_2$ then $\mathcal{L}ab_1 \approx \mathcal{L}ab_2$.[13]

The idea of "$\sqsubseteq$" is to define what it means for a labelling to be more committed than another labelling (this is a special case of skepticism comparison, an issue which will be dealt with systematically in Section 4). For instance, the grounded labelling is the least committed labelling among all complete labellings. The idea of "$\approx$" is to define when a labelling of one person might still be acceptable to another person. To see this, first consider that by requiring that $\text{in}(\mathcal{L}ab_1) \cap \text{out}(\mathcal{L}ab_2) = \emptyset$ and $\text{out}(\mathcal{L}ab_1) \cap \text{in}(\mathcal{L}ab_2) = \emptyset$, the relation "$\approx$" does not allow for conflicts between in and out. That is, if there is an argument that is accepted by agent Ag_1 but rejected by agent Ag_2 (or vice versa) then their labellings are not compatible. However, it is less problematic to have conflicts between in and undec, or between out and undec. Thus, compatibility provides an indication of how easy or difficult it is to share a position that is not one's own. It is easier to do this for a labelling that is compatible than for a labelling that is not compatible. In the former case the worst that can happen

---

[12]As a counterexample, consider $AF = (\{a,b\}, \{(a,b),(b,a)\})$. Let $\mathcal{L}ab_1 = (\{a\},\{b\},\emptyset)$, $\mathcal{L}ab_2 = (\emptyset,\emptyset,\{a,b\})$ and $\mathcal{L}ab_3 = (\{b\},\{a\},\emptyset)$. It holds that $\mathcal{L}ab_1 \approx \mathcal{L}ab_2$ and $\mathcal{L}ab_2 \approx \mathcal{L}ab_3$ but $\mathcal{L}ab_1 \not\approx \mathcal{L}ab_3$.

[13]This is because $\mathcal{L}ab_1 \approx \mathcal{L}ab_2$ iff $\text{in}(\mathcal{L}ab_1) \subseteq \text{in}(\mathcal{L}ab_2) \cup \text{undec}(\mathcal{L}ab_2)$ and $\text{out}(\mathcal{L}ab_1) \subseteq \text{out}(\mathcal{L}ab_2) \cup \text{undec}(\mathcal{L}ab_2)$.

is that one has to abstain from something one accepts or rejects (or have to accept or reject something where one did not have an explicit opinion about). In the latter case, however, one has to make statements that go directly against one's private position.

To come back to the informal description of ideal semantics, we assume a meeting in which every preferred labelling is represented. The meeting then discusses each argument, one by one, with the aim to define an *initial labelling*. If everybody agrees that the argument is labelled **in** (that is, the argument is labelled **in** in every preferred labelling) then the argument is also labelled **in** in the initial tentative labelling. If everybody agrees that the argument is labelled **out** (that is, the argument is labelled **out** in every preferred labelling) then the argument is labelled **out** in the tentative labelling. In all other cases, the argument is labelled **undec** in the tentative labelling. After this process is over, and the tentative labelling has been finished, the meeting goes to the second phase, in which the initial labelling is "watered down" in order to become an admissible labelling. This is done by iteratively relabelling each argument that is illegally **in** or illegally **out** to **undec**. When there are no more arguments left that are illegally **in** or illegally **out**, the result is the *ideal labelling*. It was proved in [Caminada and Pigozzi, 2011] that this process results in constructing the most committed ("biggest") admissible labelling that is less or equally committed than each preferred labelling. This leads to the following definition of ideal semantics.

Definition 3.45 *Let $AF = \langle Ar, att \rangle$ be an argumentation framework. The ideal labelling of AF is the biggest admissible labelling that is smaller or equal to each preferred labelling.*

The uniqueness of the ideal labelling and the fact that the ideal labelling is a complete labelling have been proved in [Caminada and Pigozzi, 2011]. Since the grounded labelling is the smallest complete labelling (w.r.t. "$\sqsubseteq$") it directly follows that the ideal labelling is bigger or equal to the grounded labelling.

Proposition 3.46 *Let $AF = \langle Ar, att \rangle$ be an argumentation framework, let $\mathcal{L}ab_{grounded}$ be its grounded labelling and $\mathcal{L}ab_{ideal}$ be its ideal labelling. It holds that $\mathcal{L}ab_{grounded} \sqsubseteq \mathcal{L}ab_{ideal}$.*

There are several ways of describing the ideal labelling [Caminada, 2011].

Proposition 3.47 *Let $\mathcal{L}ab$ be a labelling of an argumentation framework $AF = \langle Ar, att \rangle$. The following statements are equivalent.*

1. *$\mathcal{L}ab$ is the biggest admissible labelling that is smaller or equal to each preferred labelling*

2. *$\mathcal{L}ab$ is the biggest admissible labelling that is compatible with each admissible labelling*

3. $\mathcal{L}ab$ is the biggest admissible labelling that is compatible with each complete labelling

4. $\mathcal{L}ab$ is the biggest admissible labelling that is compatible with each preferred labelling

The concept of ideal semantics was originally introduced in terms of extensions in [Dung et al., 2007], drawing inspiration from the analogous concept of ideal sceptical semantics in extended logic programs [Alferes et al., 1993].

Definition 3.48 *Let $AF = \langle Ar, att \rangle$ be an argumentation framework. An admissible set $\mathcal{A}rgs$ is called* ideal *iff it is a subset of each preferred extension. The* ideal extension *of AF is a maximal (w.r.t. set-inclusion) ideal set.*

It turns out that the ideal extension of any argumentation framework AF, denoted in the following as $ID(AF)$, is unique (which implies that it is also the biggest ideal set) and that it is also a complete extension [Dung et al., 2007]. It then follows directly that the ideal extension is a superset of the grounded extension.

Proposition 3.49 *Let $AF = \langle Ar, att \rangle$ be an argumentation framework. It holds that $GE(AF) \subseteq ID(AF)$.*

There are several ways of describing the ideal extension.

Proposition 3.50 *Let $AF = \langle Ar, att \rangle$ be an argumentation framework, and let $\mathcal{A}rgs \subseteq Ar$. The following statements are equivalent.*

1. *$\mathcal{A}rgs$ is the biggest admissible set that is a subset of each preferred extension*

2. *$\mathcal{A}rgs$ is the biggest admissible set that is not attacked by any admissible set*

3. *$\mathcal{A}rgs$ is the biggest admissible set that is not attacked by any complete extension*

4. *$\mathcal{A}rgs$ is the biggest admissible set that is not attacked by any preferred extension*

In Proposition 3.50 the equivalence between points 1 and 2 follows from [Dung et al., 2007, Theorem 3.3]. The equivalence between points 2, 3 and 4 follows from the fact that an argument (or set) is attacked by an admissible set iff it is attacked by a complete extension iff it is attacked by a preferred extension.

The bijective labellings-extensions correspondence through Ext2Lab (and Lab2Ext) also holds for ideal semantics [Caminada, 2011].

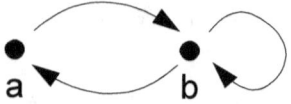

Figure 10. The ideal labelling can be less skeptical than the grounded labelling

Ideal semantics is similar to grounded semantics in the sense that it always yields a unique labelling (extension). Actually it can be seen that the ideal labelling (extension) coincides with the grounded labelling (extension) in the examples of Figures 5, 6 and 7. In particular, referring to extensions, in Figure 5 the intersection of preferred extensions $\{a\}$ coincides with the grounded extension; in Figure 6 the intersection of preferred extensions $\{d\}$ is not admissible and its only admissible subset is the empty set (coinciding with the grounded extension); in Figure 7 there is only one (empty) preferred extension, which coincides with the grounded and ideal extension.

However, as shown in Propositions 3.46 and 3.49, in general ideal semantics tends to be less skeptical than grounded semantics. As an example, in the argumentation framework of Figure 10 the grounded labelling is $(\emptyset, \emptyset, \{a, b\})$ (the grounded extension is $\emptyset$) whereas the ideal labeling is $(\{a\}, \{b\}, \emptyset)$ (the ideal extension is $\{a\}$).

To determine whether an argument is an element of the ideal extension, point 2 of Proposition 3.50 implies that it is sufficient to determine whether it is an element of an admissible set that is not attacked by any admissible set. Proof procedures for this are straightforward and have been described in [Dung et al., 2007].

An alternative approach that is very close to ideal semantics is that of *eager semantics* ($\mathcal{EAG}$) [Caminada, 2007b]. Where the ideal extension is the (unique) biggest admissible (and complete) subset of each preferred extension, the eager extension is the (unique[14]) biggest admissible (and complete) subset of each semi-stable extension. This of course admits an equivalent formulation in terms of labellings.

Definition 3.51 *Let $AF = \langle Ar, att \rangle$ be an argumentation framework. The eager extension of AF is the maximal (w.r.t. set-inclusion) admissible set which is included in every semi-stable extension.*

Definition 3.52 *Let $AF = \langle Ar, att \rangle$ be an argumentation framework. The eager labelling of AF is the biggest admissible labelling that is smaller or equal to each semi-stable labelling.*

The eager extension is a superset of the ideal extension, making eager semantics (to the best of our knowledge) the most credulous unique status semantics

---

[14]Uniqueness of the eager extension is guaranteed for finite and finitary frameworks, while it may not hold in general for infinite frameworks.

that has been proposed in the literature. The eager extension and the associated eager labelling can be computed by first calculating all semi-stable labellings (using for instance the algorithm of [Caminada, 2007a]) and subsequently applying the judgement aggregation operators specified in [Caminada and Pigozzi, 2011].

3.9 Stage Semantics

The concept of *stage* semantics ($\mathcal{STG}$) has been introduced in [Verheij, 1996] and further developed in [Verheij, 2003] in different formal settings with respect to the ones considered in this chapter. Precise (and rather straightforward) correspondences can nevertheless be drawn so that we can describe stage semantics in terms of labellings and extensions, as was done for all other semantics in this chapter. In essence, a stage labelling is a conflict-free labelling where undec is minimal, while a stage extension is a conflict-free set of arguments $Args$, where $Args \cup Args^+$ is maximal.

Definition 3.53 *Let $AF = \langle Ar, att \rangle$ be an argumentation framework. A labelling $\mathcal{L}ab$ is called a* stage labelling *of AF iff it is a conflict-free labelling where $\text{undec}(\mathcal{L}ab)$ is minimal (w.r.t. set-inclusion) among all conflict-free labellings, i.e. there is no conflict-free labelling $\mathcal{L}ab'$ such that $\text{undec}(\mathcal{L}ab') \subsetneq \text{undec}(\mathcal{L}ab)$.*

Definition 3.54 *Let $AF = \langle Ar, att \rangle$ be an argumentation framework. A* stage extension *of AF is a conflict-free set $Args \subseteq Ar$ where $Args \cup Args^+$ is maximal (w.r.t. set inclusion) among all conflict-free sets, i.e there is no conflict-free set $Args'$ such that $(Args' \cup Args'^+) \supsetneq (Args \cup Args^+)$.*

It holds that every stable labelling (extension) is also a stage labelling (extension).

Theorem 3.55 *Let $\mathcal{L}ab$ be a labelling of an argumentation framework $AF = \langle Ar, att \rangle$. If $\mathcal{L}ab$ is a stable labelling of AF then $\mathcal{L}ab$ is also a stage labelling of AF.*

Theorem 3.56 *Let $AF = \langle Ar, att \rangle$ be an argumentation framework and $Args \subseteq Ar$. If $Args$ is a stable extension of AF then $Args$ is also a stage extension of AF.*

If there exists at least one stable labelling (extension), then each stage labelling (extension) is also a stable labelling (extension).

Theorem 3.57 *Let $AF = \langle Ar, att \rangle$ be an argumentation framework. If there exists at least one stable labelling of AF then every stage labelling is also a stable labelling.*

Theorem 3.58 *Let $AF = \langle Ar, att \rangle$ be an argumentation framework. If there exists at least one stable extension of AF then every stage extension is also a stable extension.*

There also exists an alternative way to describe the concept of stage semantics. In essence, a stage labelling is a stable labelling of a maximal subgraph of the argumentation framework that has at least one stable labelling, augmented with undec labels for the arguments that did not make their way into the subgraph. Similarly, what a stage extension does is taking a maximal subgraph of the argumentation framework that has at least one stable extension. A stage extension is then a stable extension of such a maximal subgraph.

Theorem 3.59 *Let $\mathcal{L}ab$ be a labelling of an argumentation framework $AF = \langle Ar, att \rangle$. The following two statements are equivalent.*

1. *$\mathcal{L}ab$ is a conflict-free labelling where $\text{undec}(\mathcal{L}ab)$ is minimal (w.r.t. set inclusion) among all conflict-free labellings*

2. *$\mathcal{A}rgs = \text{in}(\mathcal{L}ab) \cup \text{out}(\mathcal{L}ab)$ is a maximal subset of Ar such that $AF\downarrow_{\mathcal{A}rgs}$ has a stable labelling, and $\mathcal{L}ab\downarrow_{\mathcal{A}rgs}$ is a stable labelling of $AF\downarrow_{\mathcal{A}rgs}$.*

Theorem 3.60 *Let $AF = \langle Ar, att \rangle$ be an argumentation framework and $\mathcal{A}rgs \subseteq Ar$. The following two statements are equivalent.*

1. *$\mathcal{A}rgs$ is a conflict-free set where $\mathcal{A}rgs \cup \mathcal{A}rgs^+$ is maximal (w.r.t. set inclusion) among all conflict-free sets.*

2. *$\mathcal{A}rgs \cup \mathcal{A}rgs^+$ is a maximal subset of Ar such that $AF\downarrow_{\mathcal{A}rgs\cup\mathcal{A}rgs^+}$ has a stable extension, and $\mathcal{A}rgs$ is a stable extension of $AF\downarrow_{\mathcal{A}rgs\cup\mathcal{A}rgs^+}$.*

The bijective labellings-extensions correspondence through Ext2Lab (and Lab2Ext) also holds for stage semantics, as proved in [Caminada, 2011]. An algorithm that produces all stage labellings (and therefore also all stage extensions) is described in [Caminada, 2010].

To exemplify stage labellings (extensions) let us refer as usual to the examples of Figures 5-7. Stage labellings (extensions) coincide with stable labellings (extensions), when the latter exist, as in the case of Figures 5 and 6. On the other hand, in the case of Figure 7, differently from all other semantics examined so far, stage semantics prescribes three non-trivial labellings, namely ($\{a\}, \{b\}, \{c\}$), ($\{b\}, \{c\}, \{a\}$), ($\{c\}, \{a\}, \{b\}$) (and of course the corresponding three non-empty extensions, $\{a\}, \{b\}$, and $\{c\}$).

Using the technical properties and the examples described above, we are now ready to describe the intuition behind stage semantics. In essence, stage semantics shares with stable semantics some sort of preference for strongly committed evaluations with respect to the undecided ones. As already seen, such an attitude is not universally applicable: the solution of stage semantics is to consider the maximal restrictions where this attitude is still applicable. In other terms, stage semantics can be interpreted as the attempt to identify and then ignore the minimal amounts of information that prevent the application of a black-and-white view of the world. Note that different information can

Figure 11. Stage semantics differs from semi-stable semantics

be ignored in different labellings (extensions), for instance in the example of Figure 7 arguments a, b, and c are alternatively ignored.

The idea of minimizing the set of undec-labelled arguments or, alternatively, of maximizing the range ($\mathcal{A}rgs \cup \mathcal{A}rgs^+$) of extensions is common to stage and semi-stable semantics. However, where semi-stable semantics aims to maximize the range under the condition of admissibility, stage semantics tries to maximize the range under the weaker condition of conflict-freeness. As shown above, this amounts to taking the stable labellings (extensions) of the biggest subframework that has at least one stable labelling (extension). Hence, the approach of stage semantics is comparable with the approach of handling inconsistent knowledge bases, where one can select maximal consistent subsets of the knowledge base, and then examine what holds in all of them (in the intersection of all their models). That is, it is as if stage semantics interprets the absence of stable labellings/extensions as some form of "inconsistency", which needs to be handled taking the "maximal consistent subframeworks". On the other hand, in semi-stable semantics as well as in most other semantics all arguments play a role in all extensions/labellings. In particular, an undecided argument keeps the capability to cause other arguments to be undecided, while this is not the case in stage semantics. An example is shown in Figure 11. Here, the other semantics considered up to now in the current chapter (with the exception of naïve semantics) yield a single labelling ($\emptyset, \emptyset, \{a,b\}$) corresponding to the extension $\emptyset$, whereas stage semantics yields a single labelling ($\{b\}, \emptyset, \{a\}$) corresponding to the extension $\{b\}$. In essence, what stage semantics does is to ignore argument a, since this argument causes the framework not to have any stable labelling/extension. $\mathcal{CF}2$ and stage2 semantics, examined in the next section, show the same behaviour in this example.

Another example to illustrate the difference between stage semantics and semi-stable semantics is given in Figure 12. Here, semi-stable semantics yields a single extension $\{a\}$, corresponding to a labelling ($\{a\}, \{b\}, \{c\}$). Stage semantics yields two extensions, the first one being equivalent to the one yielded by semi-stable semantics, the second one being $\{b\}$, corresponding to a labelling ($\{b\}, \{c\}, \{a\}$). The first stage extension (labelling) is the result of ignoring argument c, the second stage extension (labelling) is the result of ignoring argument a. For both possibilities, the remaining argumentation framework is a maximal one that has at least one stable extension (labelling). It can therefore be observed that under stage semantics, even an argument without any attackers (like argument a in Figure 12) is not always labelled in. With any

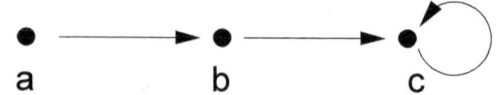

Figure 12. A peculiar case for stage semantics

other semantics considered in this chapter[15], however, an argument without any attackers is *always* labelled in.

3.10 $\mathcal{CF}2$ and stage2 semantics

With the exception of naïve and stage semantics, all semantics reviewed so far are admissibility-based, i.e. the labellings (extensions) they prescribe are admissible. Moreover they are compatible with the basic skeptical view represented by grounded semantics, in the sense that in any of their labellings (extensions) the accepted arguments are a superset of those accepted by the grounded semantics. Focusing now on those of these semantics which are multiple-status (namely complete, preferred, stable and semi-stable), one can notice that odd-length unidirectional attack cycles cause a sort of singularity in their behaviour. For instance, considering the example of Figure 7 only the trivial labelling (extension) is prescribed and, in the case of stable semantics, no labelling (extension) at all exists. This gives rise to a sort of unbalanced treatment of even-length and odd-length unidirectional attack cycles: non-trivial labellings (extensions) exist for the former ones, while they do not exist for the latter. This has been regarded as problematic by [Pollock, 2001], since in some contexts an "equal" treatment of cycles, independently of their length, can be more appropriate[16]. It is evident that this requires giving up the property of admissibility, as no non-trivial admissible labellings (extensions) exist for the example of Figure 7. In fact, the behaviour of stage semantics goes in that direction, since in the example of Figure 7 it prescribes three non-trivial labellings, namely $(\{a\},\{b\},\{c\})$, $(\{b\},\{c\},\{a\})$, $(\{c\},\{a\},\{b\})$, or, analogously, three non-empty extensions, namely $\{a\}$, $\{b\}$, $\{c\}$. Stage semantics however shows a peculiar behaviour and strongly departs from grounded semantics in some cases. As already commented in Section 3.9 a stage labelling (or extension) may even exclude from acceptance an unattacked argument (a in the example of Figure 12) while including an argument attacked by it (b in the same example). This kind of behaviour has no parallel in all other semantics considered in this chapter and, as such, appears rather hard to justify. Then the question arises as to whether it is possible to define a multiple-status semantics which is not admissibility-based, treats in an "equal" way odd and even-length

---

[15] With the exception of naïve semantics, which ignores the direction of attacks.

[16] [Pollock, 2001] discusses odd-length attack cycles in the context of a set of "reference" inference graphs for testing the intuitive validity of justification status assignments. Actually, the paper where the problem is raised [Pollock, 2001] is mainly focused on an approach to reasoning with variable degrees of justification and does not provide an explicit "solution" to this problematic example.

unidirectional attack cycles, while preserving compatibility with the grounded semantics in any case.

$\mathcal{CF}2$ [Baroni and Giacomin, 2003; Baroni et al., 2005b] and stage2 ($\mathcal{STG}2$) [Dvořák and Gaggl, 2012b; Dvořák and Gaggl, 2012a; Dvořák and Gaggl, 2016] semantics satisfy the above requirements. In fact, to achieve this objective a relatively sophisticated semantics definition scheme has been devised called *SCC-recursiveness*. The SCC-recursive scheme is based on the graph theoretical notion of a strongly connected component (SCC). In short, strongly connected components provide a unique partition of a directed graph into disjoint parts where all nodes are mutually reachable (it is assumed that reachability is a reflexive relation). Formally, strongly connected components are the equivalence classes induced by the path equivalence (i.e. mutual reachability) relation between nodes. To illustrate this notion, in the example of Figure 5 there are three SCCs, namely $\{a\}$, $\{b\}$, and $\{c,d\}$, in Figure 6 there are three SCCs too, namely $\{a,b\}$, $\{c\}$, and $\{d\}$, while the argumentation framework of Figure 7 consists of a unique SCC, namely $\{a,b,c\}$. As another example, in Figure 9 there are two SCCs, namely $\{a,b\}$ and $\{c,d,e\}$.

An important property of the SCC decomposition is that the graph obtained considering SCCs as single nodes is acyclic, i.e. the attack relation induces a partial order between the SCCs. The SCC-recursive scheme exploits this property and can be intuitively regarded as a constructive procedure to incrementally build extensions (or labellings) following the partial order of SCCs. In a nutshell, one "locally" applies some extension selection criterion to the *initial* SCCs, i.e. those not receiving attacks from other ones. Then, for each possible choice identified in the initial SCCs, one accordingly suppresses some arguments from the initial argumentation framework and the procedure is recursively applied to the new argumentation framework resulting from this modification, until no remaining arguments are left to process. In the case of $\mathcal{CF}2$ semantics, the "local" selection criterion[17] applied to SCCs is quite simple and corresponds to the intuition underlying naïve semantics: all maximal conflict free sets are selected. In the case of stage2 semantics the criterion corresponds to stage semantics, namely the conflict free sets with maximal range are selected. However embedding these criteria within the SCC recursive scheme gives rise to different results with respect to the original naïve and stage semantics.

We now provide a formal definition of $\mathcal{CF}2$ and stage2 semantics in terms of extensions (as this is their original and easier to follow formulation), exemplify their behaviour and review their properties. For further details and more extensive explanations of the SCC-recursive scheme the reader may refer to the original source [Baroni et al., 2005b] and to chapter 18 in this volume. A labelling-based formulation of $\mathcal{CF}2$ and stage2 semantics will be examined at the end of the section.

---

[17] It can be remarked that all Dung's original semantics can be equivalently characterized using SCC-recursive definitions similar to Definition 3.61, as proved in [Baroni et al., 2005b].

Definition 3.61 *Given an argumentation framework* $AF = \langle Ar, att \rangle$, *a set* $Args \subseteq Ar$ *is an extension of* $\mathcal{CF}2$ *semantics if and only if*

- $Args$ *is a naïve extension of* AF *if* $|\text{SCCS}_{AF}| = 1$
- $\forall S \in \text{SCCS}_{AF} \ (Args \cap S) \in \mathcal{E}_{\mathcal{CF}2}(AF{\downarrow}_{UP_{AF}(S,Args)})$ *otherwise*

where

- SCCS_{AF} *denotes the set of strongly connected components of* AF
- *for any* $Args, S \subseteq Ar$, $UP_{AF}(S, Args) = \{a \in S \mid \nexists b \in Args \setminus S : (b, a) \in att\}$.

Definition 3.62 *Given an argumentation framework* $AF = \langle Ar, att \rangle$, *a set* $Args \subseteq Ar$ *is an extension of stage2 semantics if and only if*

- $Args$ *is a stage extension of* AF *if* $|\text{SCCS}_{AF}| = 1$
- $\forall S \in \text{SCCS}_{AF} \ (Args \cap S) \in \mathcal{E}_{\mathcal{STG}2}(AF{\downarrow}_{UP_{AF}(S,Args)})$ *otherwise*

Definitions 3.61 and 3.62 are quite complicated and their detailed illustration is beyond the scope of the chapter. We remark only that the recursion is well-founded since, in their second branch, the semantics itself is applied to a set of restricted (and disjoint) argumentation frameworks, each including a strictly lesser number of arguments with respect to the original one. This ensures that the base case, namely the application of $\mathcal{CF}2$ or stage2 semantics to an argumentation framework consisting of a single SCC (first branch of Definitions 3.61 and 3.62) is reached in a finite number of steps. Note in particular that an argumentation framework including 0 or 1 arguments necessarily consists of a single SCC.

In spite of technical complications, the idea underlying $\mathcal{CF}2$ and stage2 semantics is relatively simple and can be illustrated with reference to our examples (since in these examples their extensions coincide, we refer to $\mathcal{CF}2$ only in the following description). In Figure 5 there is one initial SCC, namely $\{a\}$, and of course it contains only one maximal conflict-free set, namely $\{a\}$ itself, which is selected for extension building. The subsequent (according to the partial order induced by the attack relation) SCC, namely $\{b\}$, is suppressed as its only element is attacked by the already selected argument a. The last SCC, namely $\{c, d\}$, then remains unaffected by previously selected elements and we can select its maximal conflict-free subsets $\{c\}$ and $\{d\}$ to be combined with the previous selection, leading to the $\mathcal{CF}2$ extensions $\{a, c\}$ and $\{a, d\}$.

In Figure 6 there is one initial SCC, namely $\{a, b\}$, whose maximal conflict-free sets are $\{a\}$ and $\{b\}$, each representing a starting point for further extension construction. As a matter of fact, in both cases the subsequent SCC, namely $\{c\}$ is suppressed, leaving the remaining SCC, $\{d\}$, unaffected and providing

$\{d\}$ itself as maximal conflict-free subset. It turns out that there are two $\mathcal{CF}2$ extensions, namely $\{a,d\}$ and $\{b,d\}$.

The argumentation framework of Figure 7 consists of only one SCC and therefore its $\mathcal{CF}2$ extensions coincide with its maximal conflict-free subsets $\{a\}$, $\{b\}$ and $\{c\}$.

In the example of Figure 9, the application of $\mathcal{CF}2$ semantics definition is more articulated. The (again unique) initial SCC is $\{a,b\}$, which, as in the previous case, yields $\{a\}$ and $\{b\}$ as starting points for further extension construction. Considering $\{a\}$, we have that b is attacked by the extension and the subsequent SCC $\{c,d,e\}$ is left unaffected. As a consequence, all its maximal conflict-free subsets $\{c\}$, $\{d\}$ and $\{e\}$ are available, yielding the three $\mathcal{CF}2$ extensions $\{a,c\}$, $\{a,d\}$ and $\{a,e\}$. Considering $\{b\}$, both a and c are attacked by the extension and therefore suppressed. The restriction of the argumentation framework to the set $\{d,e\}$ then remains to be evaluated. As $\{d\}$ is the initial SCC of this restricted argumentation framework, it is selected and then the subsequent SCC $\{e\}$ is entirely suppressed, yielding a further $\mathcal{CF}2$ extension $\{b,d\}$.

Finally, in the example of Figure 12 a unique $\mathcal{CF}2$ extension is identified, namely $\{a\}$, yielding agreement with grounded semantics.

While in the examples above the extensions of $\mathcal{CF}2$ and stage2 semantics coincide, they can be different in some cases. A simple example is an argumentation framework consisting of six arguments $a_1,\ldots,a_6$ arranged into an attack cycle, i.e. such that a_i attacks a_{i+1} for $i=1,\ldots,5$ and a_6 attacks a_1. In this case, consisting of a unique SCC, there are five naïve and $\mathcal{CF}2$ extensions, namely $\{a_1,a_3,a_5\}$, $\{a_2,a_4,a_6\}$, $\{a_1,a_4\}$, $\{a_2,a_5\}$, $\{a_3,a_6\}$, while only two of them (clearly the first ones) are also stage and stage2 extensions.

Having exemplified the behaviour of $\mathcal{CF}2$ and stage2 semantics, we summarize in Proposition 3.63 some of their known properties in relation to other semantics notions (in particular naïve, grounded, stable and preferred).

Proposition 3.63 *For any argumentation framework $AF = \langle Ar, att \rangle$*

- $\mathcal{E}_{STG2}(AF) \subseteq \mathcal{E}_{CF2}(AF) \subseteq \mathcal{E}_{NA}(AF)$;

- *the grounded extension is included in any $\mathcal{CF}2$ and stage2 extension;*

- *any stable extension is also a $\mathcal{CF}2$ and stage2 extension;*

- *for any preferred extension E there is a $\mathcal{CF}2$ extension E' such that $E \subseteq E'$ (note that this property does not hold for stage2 semantics)*

As mentioned above, $\mathcal{CF}2$ and stage2 semantics have been conceived and defined in the extension-based setting. The same semantic notion can however be expressed using the SCC-recursive scheme in the labelling context.

Definition 3.64 *Given an argumentation framework $AF = \langle Ar, att \rangle$, a labelling $\mathcal{L}ab$ is a $\mathcal{CF}2$ labelling if and only if*

- if $|SCCS_{AF}| = 1$, $\mathcal{L}ab$ is a conflict-free labelling with maximal $\text{in}(\mathcal{L}ab)$ among conflict-free labellings and such that $a \in \text{in}(\mathcal{L}ab) \Rightarrow a^+ \subseteq \text{out}(\mathcal{L}ab)$;

- otherwise, $\forall S \in SCCS_{AF}$ $\mathcal{L}ab\!\downarrow_{UP_{AF}(S,Args)}$ is a $\mathcal{CF}2$ labelling of $AF\!\downarrow_{UP_{AF}(S,Args)}$ and all arguments in $S \setminus UP_{AF}(S,Args)$ are labelled out.

where all notations are as in Definition 3.61.

Definition 3.65 *Given an argumentation framework $AF = \langle Ar, att \rangle$, a labelling $\mathcal{L}ab$ is a stage2 labelling if and only if*

- if $|SCCS_{AF}| = 1$, $\mathcal{L}ab$ is a stage labelling;

- otherwise, $\forall S \in SCCS_{AF}$ $\mathcal{L}ab\!\downarrow_{UP_{AF}(S,Args)}$ is a stage2 labelling of $AF\!\downarrow_{UP_{AF}(S,Args)}$ and all arguments in $S \setminus UP_{AF}(S,Args)$ are labelled out.

where all notations are as in Definition 3.61.

By inspection of Definitions 3.61 vs. 3.64 and 3.62 vs. 3.65, it can be seen that the bijective labellings-extensions correspondence through `Ext2Lab` (and `Lab2Ext`) holds for $\mathcal{CF}2$ and stage2 semantics.

3.11 Roundup

We now provide an overview of how the semantics that have been treated until now are related. In Figure 13 we graphically depict what can be seen as an ontology of argumentation semantics. The figure shows for instance that every stable labelling is also a stage labelling, a semi-stable labelling and a $\mathcal{CF}2$ labelling, that every semi-stable labelling is also a preferred labelling, etc.

The same relations of Figure 13 also hold for the extension-based approach. In Table 4 we provide an overview of how the admissibility-based semantics can be expressed in terms of complete labellings.

As explicitly stated in Section 2, this chapter is focused on finite argumentation frameworks and the analysis of semantics properties we have carried out relies on this assumption. One may wonder what is the impact of this restriction and what would be the implications of considering also infinite frameworks. While providing a full answer to this question is beyond the scope of this chapter, we observe in particular that in infinite frameworks the notion of maximality w.r.t set inclusion is less immediate than in finite frameworks and the existence of maximal sets of arguments respecting some criterion, which is guaranteed in finite frameworks, may fail to be achieved in infinite ones. As an example of the consequences of this fact, a semantics which is universally defined or unique status in the context of finite frameworks may not be so when considering also infinite ones, implying (among other consequences) that the skepticism comparison of Section 4 does not extend directly to the infinite case. The reader may refer to chapter 17 for a treatment of these issues.

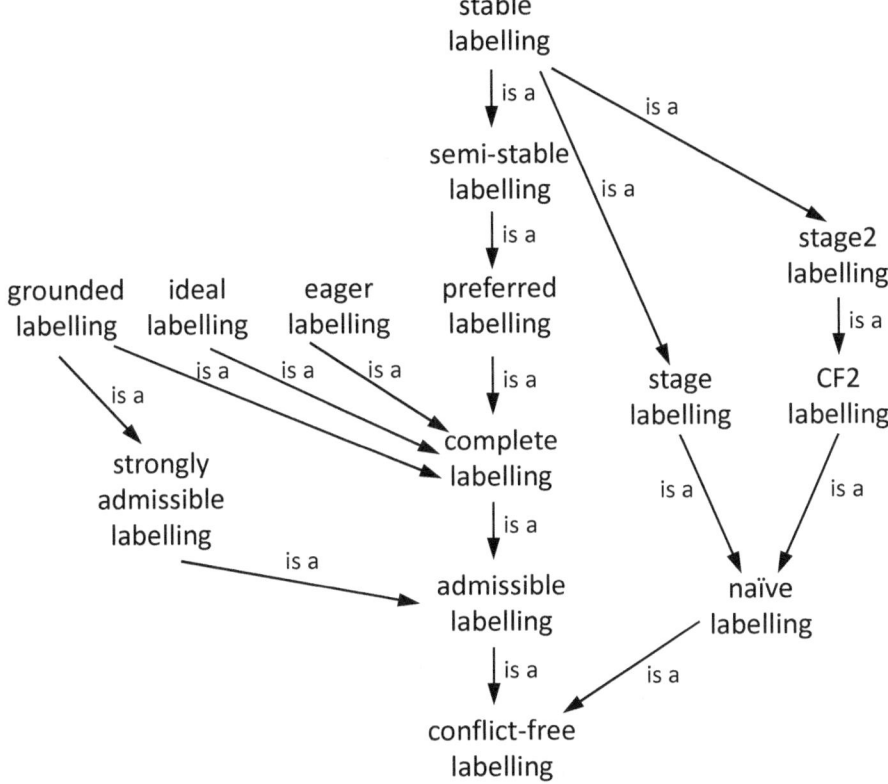

Figure 13. Relations among alternative labelling and extension notions

Table 4. Describing admissibility based semantics in terms of complete labellings

restriction on complete labelling	resulting semantics
no restrictions	complete semantics
empty undec	stable semantics
maximal in	preferred semantics
maximal out	preferred semantics
maximal undec	grounded semantics
minimal in	grounded semantics
minimal out	grounded semantics
minimal undec	semi-stable semantics
maximal w.r.t. $\sqsubseteq$ while compatible with each complete labelling	ideal semantics

Fig. #	$\mathcal{E}_{\mathcal{NA}}(AF)$	$\mathcal{E}_{CO}(AF)$	$\mathcal{E}_{\mathcal{GR}}(AF)$	$\mathcal{E}_{\mathcal{PR}}(AF)$	$\mathcal{E}_{ST}(AF)$
Fig. 3	$\{\{a,c\},\{b\}\}$	$\{\{a,c\}\}$	$\{\{a,c\}\}$	$\{\{a,c\}\}$	$\{\{a,c\}\}$
Fig. 4	$\{\{a\},\{b\}\}$	$\{\emptyset,\{a\},\{b\}\}$	$\{\emptyset\}$	$\{\{a\},\{b\}\}$	$\{\{a\},\{b\}\}$
Fig. 5	$\{\{a,c\},\{a,d\},\{b,d\}\}$	$\{\{a\},\{a,c\},\{a,d\}\}$	$\{\{a\}\}$	$\{\{a,c\},\{a,d\}\}$	$\{\{a,c\},\{a,d\}\}$
Fig. 6	$\{\{a,d\},\{b,d\},\{c\}\}$	$\{\emptyset,\{a,d\},\{b,d\}\}$	$\{\emptyset\}$	$\{\{a,d\},\{b,d\}\}$	$\{\{a,d\},\{b,d\}\}$
Fig. 7	$\{\{a\},\{b\},\{c\}\}$	$\{\emptyset\}$	$\{\emptyset\}$	$\{\emptyset\}$	$\emptyset$
Fig. 9	$\{\{a,c\},\{a,d\},\{a,e\},\{b,d\},\{b,e\}\}$	$\{\emptyset,\{a\},\{b,d\}\}$	$\{\emptyset\}$	$\{\{a\},\{b,d\}\}$	$\{\{b,d\}\}$
Fig. 10	$\{\{a\}\}$	$\{\emptyset,\{a\}\}$	$\{\emptyset\}$	$\{\{a\}\}$	$\{\{a\}\}$
Fig. 11	$\{\emptyset\}$	$\{\emptyset\}$	$\{\emptyset\}$	$\{\emptyset\}$	$\emptyset$
Fig. 12	$\{\{a\}\}$	$\{\{a\}\}$	$\{\{a\}\}$	$\{\{a\}\}$	$\{\{a\}\}$

Table 5. A summary of the examples: part I

Fig. #	$\mathcal{E}_{SST}(AF)$	$\mathcal{E}_{ID}(AF)$	$\mathcal{E}_{EAG}(AF)$	$\mathcal{E}_{STG}(AF)$	$\mathcal{E}_{CF2}(AF)$	$\mathcal{E}_{STG2}(AF)$
Fig. 3	$\{\{a,c\}\}$	$\{\{a,c\}\}$	$\{\{a,c\}\}$	$\{\{a,c\}\}$	$\{\{a,c\}\}$	$\{\{a,c\}\}$
Fig. 4	$\{\{a\};\{b\}\}$	$\{\emptyset\}$	$\{\emptyset\}$	$\{\{a\},\{b\}\}$	$\{\{a\},\{b\}\}$	$\{\{a\},\{b\}\}$
Fig. 5	$\{\{a,c\},\{a,d\}\}$	$\{\{a\}\}$	$\{\{a\}\}$	$\{\{a,c\},\{a,d\}\}$	$\{\{a,c\},\{a,d\}\}$	$\{\{a,c\},\{a,d\}\}$
Fig. 6	$\{\{a,d\},\{b,d\}\}$	$\{\emptyset\}$	$\{\emptyset\}$	$\{\{a,d\},\{b,d\}\}$	$\{\{a,d\},\{b,d\}\}$	$\{\{a,d\},\{b,d\}\}$
Fig. 7	$\{\emptyset\}$	$\{\emptyset\}$	$\{\emptyset\}$	$\{\{a\},\{b\},\{c\}\}$	$\{\{a\},\{b\},\{c\}\}$	$\{\{a\},\{b\},\{c\}\}$
Fig. 9	$\{\{b,d\}\}$	$\{\emptyset\}$	$\{\{b,d\}\}$	$\{\{b,d\}\}$	$\{\{a,c\},\{a,d\},\{a,e\},\{b,d\}\}$	$\{\{a,c\},\{a,d\},\{a,e\},\{b,d\}\}$
Fig. 10	$\{\{a\}\}$	$\{\{a\}\}$	$\{\{a\}\}$	$\{\{a\}\}$	$\{\{a\}\}$	$\{\{a\}\}$
Fig. 11	$\{\emptyset\}$	$\{\emptyset\}$	$\{\emptyset\}$	$\{\{b\}\}$	$\{\{b\}\}$	$\{\{b\}\}$
Fig. 12	$\{\{a\}\}$	$\{\{a\}\}$	$\{\{a\}\}$	$\{\{a\},\{b\}\}$	$\{\{a\}\}$	$\{\{a\}\}$

Table 6. A summary of the examples: part II

4 Argument Justification and Skepticism

4.1 The notion of justification status

Argumentation semantics can allow for the presence of more than one extension (or labelling) of arguments, following a tradition in nonmonotonic reasoning [Reiter, 1980; Pollock, 1995]. Hence, when one is interested in the overall status of a particular argument, one needs to have a way of taking the multiplicity of extensions (or labellings) into account. At a basic level, two very simple (and, in a sense, extreme) alternatives for the notion of justification status can be considered: *skeptical justification* requires that an argument is accepted in all labellings (or extensions), while *credulous justification* requires that an argument is accepted in at least one labelling (or extension). This is formalized in Definitions 4.2 and 4.3. Note that we assume, as in previous literature [Baroni and Giacomin, 2007b; Baroni and Giacomin, 2009b], that using these simple notions is meaningful only when the set of labellings or extensions is not empty, otherwise the basis for evaluation is lacking. To express this in a concise way, we introduce a specific notation in Definition 4.1.

Definition 4.1 *Given a labelling-based semantics σ, $\mathcal{DL}_\sigma = \{AF : \mathcal{L}_\sigma(AF) \neq \emptyset\}$. Given an extension-based semantics σ, $\mathcal{DE}_\sigma = \{AF : \mathcal{E}_\sigma(AF) \neq \emptyset\}$.*

Definition 4.2 (course-grained justification status, labelling-based)
Given a labelling-based semantics σ and an argumentation framework $AF \in \mathcal{DL}_\sigma$, an argument a is skeptically justified *(or* skeptically accepted*) if $\forall \mathcal{L}ab \in \mathcal{L}_\sigma(AF)$ $\mathcal{L}ab(a) = \mathtt{in}$; an argument a is* credulously justified *(or* credulously accepted*) if $\exists \mathcal{L}ab \in \mathcal{L}_\sigma(AF) : \mathcal{L}ab(a) = \mathtt{in}$.*

Definition 4.3 (course-grained justification status, extension-based)
Given an extension-based semantics σ and an argumentation framework $AF \in \mathcal{DE}_\sigma$, an argument a is skeptically justified *(or* skeptically accepted*) if $\forall E \in \mathcal{E}_\sigma(AF)$ $a \in E$; an argument a is* credulously justified *(or* credulously accepted*) if $\exists E \in \mathcal{E}_\sigma(AF) : a \in E$.*

Skeptical justification implies credulous justification, as long as the set of extensions (or labellings) is not empty. Also, a third justification status can be derived: an argument is *not justified* (or *rejected*) if it is not credulously justified (and hence also not skeptically justified, assuming that the set of extensions (or labellings) is not empty).

It can be noted that in any unique-status semantics skeptical and credulous acceptance coincide, so that an argument can only be accepted or rejected. In this context it is possible, however, to consider two levels of rejection, in fact a rejected argument can be attacked or not by the unique extension (or, analogously, can be labelled **out** or **undec** in the unique labelling). The former case corresponds to a stronger form of rejection (these arguments have been sometimes called *defeated outright* in the literature [Pollock, 1992]) while in the latter case rejection is clearly weaker (these arguments being called *provisionally defeated* according to the same terminology).

While the brief remarks above correspond to the prevailing approaches to the notion of justification status in the literature, one may observe that a more systematic treatment is possible, by combining the ideas concerning the status of an argument with respect to a single labelling (or extension) and those referring to a plurality of them. In fact, an argument can be in one of three possible states with respect to a single labelling (namely, in, out or undec) and correspondingly can be accepted, defeated outright or provisionally defeated with respect to a single extension. If a plurality of labellings (or extensions) is considered, we can ask ourselves the following three questions: is the argument accepted (labelled in) by at least one extension (labelling), is the argument rejected outright (labelled out) by at least one extension (labelling), and is the argument provisionally rejected (labelled undec) by at least one extension (labelling)? The answer to these three questions gives rise to a concept we will sometimes refer to as the *fine-grained justification status*, to distinguish it from the earlier introduced concepts of sceptical justification and credulous justification, which we will sometimes refer to as the *course-grained justification status*.

Fine-grained justification is also able to properly treat with a distinct status the case where no labellings or extensions at all exist. For the labellings approach, fine-grained justification status can be defined as follows.

Definition 4.4 (fine-grained justification status, labelling-based)
Given a labelling-based semantics σ and an argumentation framework $AF = \langle Ar, att \rangle$, the possible justification states of an argument a are defined by a function $\mathcal{JS}_\sigma^{AF} : Ar \to 2^{\{\text{in},\text{out},\text{undec}\}}$ such that $\mathcal{JS}_\sigma^{AF}(a) = \{\mathcal{L}ab(a) \mid \mathcal{L}ab \in \mathcal{L}_\sigma(AF)\}$.

If we assume a labelling-based semantics to specify the reasonable positions (labellings) one can take in the presence of the conflicting information specified in the argumentation framework, then one can give an intuitive interpretation of the concept of a justification status. For instance, the justification status of {in} means the argument has to be accepted (labelled in) in every reasonable position. Similarly, the justification status {in, undec} means that in every reasonable position the argument is either accepted (labelled in) or abstained from having an explicit opinion on (labelled undec), but the argument cannot be rejected (labelled out). Such an interpretation of the notion of justification status is for instance used in [Wu and Caminada, 2010; Dvořák, 2012].

It is also possible to define the notion of fine-grained justification status in terms of the extensions approach, as is done below.

Definition 4.5 (fine-grained justification status, extension-based)
Given an extension-based semantics σ and an argumentation framework $AF = \langle Ar, att \rangle$, the possible justification states of an argument a are defined by the following mutually exclusive conditions:

Argument	$\mathcal{JS}^{AF}_{\mathcal{CO}}$	$\mathcal{JS}^{AF}_{\mathcal{GR}}$	$\mathcal{JS}^{AF}_{\mathcal{PR}}$	$\mathcal{JS}^{AF}_{\mathcal{NA}}$
a	{in, out, undec}	{undec}	{in, out}	{in, out, undec}
b	{in, out, undec}	{undec}	{in, out}	{in, out, undec}
c	{out, undec}	{undec}	{out}	{in, out, undec}
d	{in, undec}	{undec}	{in}	{in, out, undec}

Table 7. Argument justification statuses in the example of Figure 6

- $\mathcal{E}_\sigma(AF) = \emptyset$

- $\mathcal{E}_\sigma(AF) \neq \emptyset$ and $\forall E \in \mathcal{E}_\sigma(AF)\ a \in E$;

- $\mathcal{E}_\sigma(AF) \neq \emptyset$ and $\forall E \in \mathcal{E}_\sigma(AF)\ a \in E^+$;

- $\mathcal{E}_\sigma(AF) \neq \emptyset$ and $\forall E \in \mathcal{E}_\sigma(AF)\ a \notin (E \cup E^+)$;

- $\exists E \in \mathcal{E}_\sigma(AF) : a \in E^+, \exists E \in \mathcal{E}_\sigma(AF) : a \notin (E \cup E^+)$, and $\nexists E \in \mathcal{E}_\sigma(AF) : a \in E$;

- $\exists E \in \mathcal{E}_\sigma(AF) : a \in E, \exists E \in \mathcal{E}_\sigma(AF) : a \notin (E \cup E^+)$, and $\nexists E \in \mathcal{E}_\sigma(AF) : a \in E^+$;

- $\exists E \in \mathcal{E}_\sigma(AF) : a \in E, \exists E \in \mathcal{E}_\sigma(AF) : a \in E^+$, and $\nexists E \in \mathcal{E}_\sigma(AF) : a \notin (E \cup E^+)$;

- $\exists E \in \mathcal{E}_\sigma(AF) : a \in E, \exists E \in \mathcal{E}_\sigma(AF) : a \in E^+$, and $\exists E \in \mathcal{E}_\sigma(AF) : a \notin (E \cup E^+)$.

Intuitively, for an argument a, each item of Definition 4.5 corresponds to a possible value $\mathcal{JS}^{AF}_\sigma(a)$ in Definition 4.4, i.e. to a subset of {in, out, undec}. For instance the first item corresponds to $\emptyset$, the following three items correspond to {in}, {out}, {undec} respectively, the fifth item corresponds to {out, undec} and so on.

As an example of how labelling-based justification status works, consider again the argumentation framework of Figure 6. If one applies complete semantics, three complete labellings are yielded: $(\{a,d\},\{b,c\},\emptyset)$, $(\{b,d\},\{a,c\},\emptyset)$, $(\emptyset,\emptyset,\{a,b,c,d\})$. This implies that $\mathcal{JS}^{AF}_{\mathcal{CO}}(a) = \{\text{in}, \text{out}, \text{undec}\}$, $\mathcal{JS}^{AF}_{\mathcal{CO}}(b) = \{\text{in}, \text{out}, \text{undec}\}$, $\mathcal{JS}^{AF}_{\mathcal{CO}}(c) = \{\text{out}, \text{undec}\}$ and $\mathcal{JS}^{AF}_{\mathcal{CO}}(d) = \{\text{in}, \text{undec}\}$.

Clearly, the justification status depends on the adopted semantics, for instance with grounded semantics only the third labelling is considered, yielding the status {undec} for all arguments, while with preferred semantics only the first two labellings are considered so that, for each argument, the label undec is "removed" from the statuses listed above for complete semantics. Table 7 summarizes the justification statuses of arguments for the various

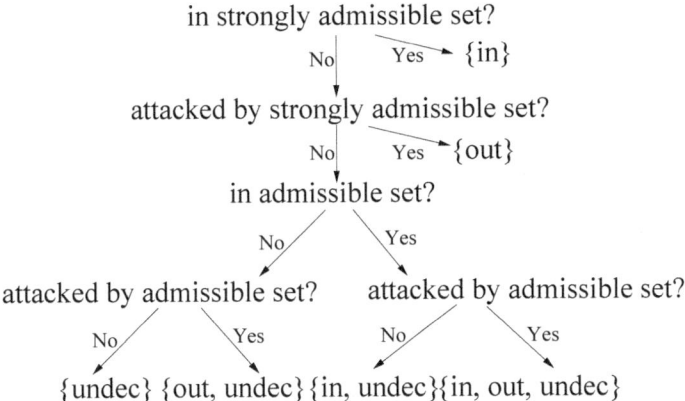

Figure 14. The justification status under complete semantics can be determined using admissibility and strong admissibility.

labelling-based[18] semantics in the example of Figure 6 (note that in this example $\mathcal{JS}^{AF}_{\mathcal{GR}} = \mathcal{JS}^{AF}_{\mathcal{ID}} = \mathcal{JS}^{AF}_{\mathcal{EAG}}$ and $\mathcal{JS}^{AF}_{\mathcal{PR}} = \mathcal{JS}^{AF}_{\mathcal{ST}} = \mathcal{JS}^{AF}_{\mathcal{SST}} = \mathcal{JS}^{AF}_{\mathcal{STG}} = \mathcal{JS}^{AF}_{\mathcal{CF}2} = \mathcal{JS}^{AF}_{\mathcal{STG}2}$).

In general, complete semantics gives rise to the following justification statuses [Caminada and Wu, 2010]:

- {in}, called *strong accept*
- {out}, called *strong reject*
- {in, undec}, called *weak accept*
- {out, undec}, called *weak reject*
- {undec}, called *determined borderline*
- {in, out, undec}, called *undetermined borderline*

For complete semantics, the justification status of an argument can also be determined using the concepts of admissible and strongly admissible sets, as is indicated in Figure 14.

The reader may have noticed that complete semantics yields only six different justification statuses, although theoretically it seems like eight (2^3) are possible. This is because some combinations of labels cannot occur under complete semantics. For instance, there cannot be an argumentation framework $AF = \langle Ar, att \rangle$ and argument $a \in Ar$ such that $\mathcal{JS}^{AF}_{\mathcal{CO}}(a) = \{\text{in}, \text{out}\}$. This is because if a is labelled in by a complete labelling and out by another, then

---

[18] These justification statuses have a direct correspondence in the extension-based approach, with the exception of naïve semantics, since in this case different labellings may correspond to the same extension.

there exists a third complete labelling (the grounded one) where a is labelled undec [Caminada and Wu, 2010] so the justification status should have been $\{\text{in}, \text{out}, \text{undec}\}$ instead. Similarly, $\mathcal{JS}_{\mathcal{CO}}^{AF}(a)$ cannot be $\emptyset$ because there always exists at least one complete labelling.

It can be observed that different semantics yield a different range of justification statuses [Dvořák, 2012].

Proposition 4.6 Let $AF = \langle Ar, att \rangle$ be an argumentation framework and let $a \in Ar$.

1. $\mathcal{JS}_{\mathcal{GR}}^{AF}(a) \in \{\{\text{in}\}, \{\text{out}\}, \{\text{undec}\}\}$
2. $\mathcal{JS}_{\mathcal{AD}}^{AF}(a) \in \{\{\text{undec}\}, \{\text{in}, \text{undec}\}, \{\text{out}, \text{undec}\}, \{\text{in}, \text{out}, \text{undec}\}\}$
3. $\mathcal{JS}_{\mathcal{CO}}^{AF}(a) \in 2^{\{\text{in},\text{out},\text{undec}\}} \setminus \{\{\text{in}, \text{out}\}, \emptyset\}$
4. $\mathcal{JS}_{\mathcal{ST}}^{AF}(a) \in \{\{\text{in}\}, \{\text{out}\}, \{\text{in}, \text{out}\}, \emptyset\}$
5. $\mathcal{JS}_{\mathcal{PR}}^{AF}(a) \in 2^{\{\text{in},\text{out},\text{undec}\}} \setminus \{\emptyset\}$
6. $\mathcal{JS}_{\mathcal{SST}}^{AF}(a) \in 2^{\{\text{in},\text{out},\text{undec}\}} \setminus \{\emptyset\}$
7. $\mathcal{JS}_{\mathcal{STG}}^{AF}(a) \in 2^{\{\text{in},\text{out},\text{undec}\}} \setminus \{\emptyset\}$
8. $\mathcal{JS}_{\mathcal{ID}}^{AF}(a) \in \{\{\text{in}\}, \{\text{out}\}, \{\text{undec}\}\}$
9. $\mathcal{JS}_{\mathcal{EAG}}^{AF}(a) \in \{\{\text{in}\}, \{\text{out}\}, \{\text{undec}\}\}$

As observed in [Dvořák, 2012], the justification status w.r.t. admissible, complete and preferred semantics only differs on the undec labels.

Proposition 4.7 Let $AF = \langle Ar, att \rangle$ be an argumentation framework and let $a \in Ar$.

1. $\mathcal{JS}_{\mathcal{AD}}^{AF}(a) = \mathcal{JS}_{\mathcal{CO}}^{AF}(a) \cup \{\text{undec}\}$
2. $\mathcal{JS}_{\mathcal{AD}}^{AF}(a) = \mathcal{JS}_{\mathcal{PR}}^{AF}(a) \cup \{\text{undec}\}$
3. $\mathcal{JS}_{\mathcal{CO}}^{AF}(a) = \begin{cases} \mathcal{JS}_{\mathcal{GR}}^{AF}(a) & \text{if } a \in GE(AF) \cup GE(AF)^+ \\ \mathcal{JS}_{\mathcal{AD}}^{AF}(a) & \text{otherwise} \end{cases}$
4. $\mathcal{JS}_{\mathcal{CO}}^{AF}(a) = \begin{cases} \mathcal{JS}_{\mathcal{GR}}^{AF}(a) & \text{if } a \in GE(AF) \cup GE(AF)^+ \\ \mathcal{JS}_{\mathcal{PR}}^{AF}(a) \cup \{\text{undec}\} & \text{otherwise} \end{cases}$

Fine-grained justification status, as described above, offers a number of advantages compared to the more traditional course-grained justification status that is based on sceptical and credulous acceptance. For instance, the computational complexity of *weak acceptance* under preferred semantics is lower than that of traditional sceptical preferred [Dvořák, 2012]. Furthermore, labelling-based justification status offers a subtle way of dealing with the notion of *floating conclusions* [Caminada and Wu, 2010]. Briefly, a conclusion is "floating" when it is supported by different arguments which are labelled in in different labellings, so that, even if there is no individual argument for the conclusion which is labelled in in all labellings, there is at least an argument labelled in supporting the conclusion in every labelling. In other words, the conclusion has different supports in different labellings, but has at least one in every labelling.

Still, labelling-based justification status has not yet received the same level of attention as the more traditional extension-based justification status, and some research questions are still open.[19]

4.2 Skepticism and skepticism relations

The term *skepticism* has been used in the literature (often in an informal way) to discuss argumentation semantics behaviour, e.g. by observing that a semantics is "more skeptical" than another one. Intuitively, a skeptical attitude tends to make less committed choices about the justification of the arguments, as well exemplified by the traditional notions of skeptical and credulous acceptance recalled in Section 4.1. In other words, a skeptical attitude tends to leave arguments in an "undecided" justification state and to accept (or reject) as least arguments as possible, while a less skeptical (or more credulous) attitude corresponds to more extensive acceptance (or rejection) of arguments. Note, in particular, that the notion of commitment (or decidedness) of a justification state must be clearly distinguished from the notion of acceptance: two justification states corresponding to definite acceptance and definite rejection, though reflecting antithetical choices about the state of an argument, have both the same highest level of commitment.

Which are the formal counterparts of these basic intuitions?

Starting from basic elements, we first need to define a criterion to compare extensions and labellings with respect to skepticism. As to extensions, this is quite simple: an extension E_1 is "more skeptical" than (to be precise, at least as skeptical as) an extension E_2 if $E_1 \subseteq E_2$, since then E_1 supports the acceptance of no more arguments than E_2. As to labellings, we have to consider both the in and out labels as being both more committed choices than undec. We can then state that a labelling $\mathcal{L}ab_1$ is at least as skeptical as a labelling $\mathcal{L}ab_2$ according to the inclusion of both the sets of in and out labelled arguments. These intuitions are formalized in Definition 4.8.

Definition 4.8 *Given two extensions E_1 and E_2 of an argumentation framework AF, E_1 is at least as skeptical as E_2, denoted as $E_1 \preceq E_2$ if and only if $E_1 \subseteq E_2$. Given two labellings $\mathcal{L}ab_1$ and $\mathcal{L}ab_2$ of an argumentation framework AF, $\mathcal{L}ab_1$ is at least as skeptical as $\mathcal{L}ab_2$, denoted as $\mathcal{L}ab_1 \preceq \mathcal{L}ab_2$, if and only if $\mathcal{L}ab_1 \sqsubseteq \mathcal{L}ab_2$ (see Definition 3.44).*

---

[19]For instance, how to find efficient proof procedures for determining the fine-grained justification status with respect to preferred, semi-stable and stage semantics.

While the above relations are sufficient to compare unique-status semantics, the next step is to introduce skepticism relations between non-empty[20] sets of extensions or labellings in order to compare multiple-status semantics. As more extensively discussed in [Baroni and Giacomin, 2009b], several alternatives can be considered for this issue.

As a first basic step, one can consider a comparison method based on inclusion of the sets of accepted arguments, either according to skeptical or credulous acceptance. This gives rise to the skepticism relations stated in the following definitions.

Definition 4.9 *Given two non-empty sets of extensions $\mathcal{E}_1$ and $\mathcal{E}_2$ of an argumentation framework AF, $\mathcal{E}_1 \preceq_\cap^E \mathcal{E}_2$ if and only if $\bigcap_{E_1 \in \mathcal{E}_1} E_1 \subseteq \bigcap_{E_2 \in \mathcal{E}_2} E_2$.*

Definition 4.10 *Given two non-empty sets of extensions $\mathcal{E}_1$ and $\mathcal{E}_2$ of an argumentation framework AF, $\mathcal{E}_1 \preceq_\cup^E \mathcal{E}_2$ if and only if $\bigcup_{E_1 \in \mathcal{E}_1} E_1 \subseteq \bigcup_{E_2 \in \mathcal{E}_2} E_2$.*

Definition 4.11 *Given two non-empty sets of labellings $\mathcal{L}_1$ and $\mathcal{L}_2$ of an argumentation framework AF, $\mathcal{L}_1 \preceq_\cap^L \mathcal{L}_2$ if and only if $\bigcap_{Lab_1 \in \mathcal{L}_1} \text{in}(Lab_1) \subseteq \bigcap_{Lab_2 \in \mathcal{L}_2} \text{in}(Lab_2)$.*

Definition 4.12 *Given two non-empty sets of labellings $\mathcal{L}_1$ and $\mathcal{L}_2$ of an argumentation framework AF, $\mathcal{L}_1 \preceq_\cup^L \mathcal{L}_2$ if and only if $\bigcup_{Lab_1 \in \mathcal{L}_1} \text{in}(Lab_1) \subseteq \bigcup_{Lab_2 \in \mathcal{L}_2} \text{in}(Lab_2)$.*

To exemplify the above notions, consider first the example of Figure 4. In the extension-based approach, the grounded and ideal semantics prescribe the set of extensions $\mathcal{E}_1 = \{\emptyset\}$ while all other semantics prescribe $\mathcal{E}_2 = \{\{a\}, \{b\}\}$. Clearly, $\mathcal{E}_1 \preceq_\cap^E \mathcal{E}_2$, $\mathcal{E}_2 \preceq_\cap^E \mathcal{E}_1$, $\mathcal{E}_1 \preceq_\cup^E \mathcal{E}_2$, while it is not the case that $\mathcal{E}_2 \preceq_\cup^E \mathcal{E}_1$ (denoted as $\mathcal{E}_2 \npreceq_\cup^E \mathcal{E}_1$). For the same example in the labelling-based approach grounded and ideal semantics prescribe the set of labellings $\mathcal{L}_1 = \{(\emptyset, \emptyset, \{a, b\})\}$ while all other semantics prescribe $\mathcal{L}_2 = \{(\{a\}, \{b\}, \emptyset), (\{b\}, \{a\}, \emptyset)\}$. Again, it can be seen that $\mathcal{L}_1 \preceq_\cap^L \mathcal{L}_2$, $\mathcal{L}_2 \preceq_\cap^L \mathcal{L}_1$, $\mathcal{L}_1 \preceq_\cup^L \mathcal{L}_2$, while $\mathcal{L}_2 \npreceq_\cup^L \mathcal{L}_1$.

Considering the example of Figure 6, in the extension-based approach the grounded and ideal semantics prescribe the set of extensions $\mathcal{E}_1 = \{\emptyset\}$ while all other semantics prescribe $\mathcal{E}_2 = \{\{a, d\}, \{b, d\}\}$. It turns out that $\mathcal{E}_1 \preceq_\cap^E \mathcal{E}_2$ and $\mathcal{E}_1 \preceq_\cup^E \mathcal{E}_2$, while $\mathcal{E}_2 \npreceq_\cap^E \mathcal{E}_1$ and $\mathcal{E}_2 \npreceq_\cup^E \mathcal{E}_1$. The case of labellings is perfectly analogous with $\mathcal{L}_1 = \{(\emptyset, \emptyset, \{a, b, c, d\})\}$ for grounded and ideal semantics and $\mathcal{L}_2 = \{(\{a, d\}, \{b, c\}, \emptyset), (\{b, d\}, \{a, c\}, \emptyset)\}$ for other semantics yielding $\mathcal{L}_1 \preceq_\cap^L \mathcal{L}_2$ and $\mathcal{L}_1 \preceq_\cup^L \mathcal{L}_2$, while $\mathcal{L}_2 \npreceq_\cap^L \mathcal{L}_1$ and $\mathcal{L}_2 \npreceq_\cup^L \mathcal{L}_1$.

Figure 9 provides a more articulated case for comparison.

In the extension-based approach grounded and ideal semantics prescribe $\mathcal{E}_1 = \{\emptyset\}$, preferred semantics prescribes $\mathcal{E}_2 = \{\{a\}, \{b, d\}\}$, $\mathcal{CF}2$ semantics prescribes $\mathcal{E}_3 = \{\{a, c\}, \{a, d\}, \{a, e\}, \{b, d\}\}$, while stable, semi-stable and

---

[20] We assume that an empty set of extensions/labellings corresponds to a peculiar case and cannot be involved in skepticism comparison.

stage semantics prescribe $\mathcal{E}_4 = \{\{b,d\}\}$. It follows that for any $i,j \in \{1,2,3\}$ $\mathcal{E}_i \preceq^E_\cap \mathcal{E}_j$, while for any $i \in \{1,2,3\}$ $\mathcal{E}_i \preceq^E_\cap \mathcal{E}_4$ and $\mathcal{E}_4 \not\preceq^E_\cap \mathcal{E}_i$. On the other hand, these sets are completely ordered according to $\preceq^E_U$ since $\mathcal{E}_1 \preceq^E_U \mathcal{E}_4 \preceq^E_U \mathcal{E}_2 \preceq^E_U \mathcal{E}_3$. Again, the case of labellings is perfectly analogous.

As a further step in the analysis of skepticism relations, one may observe that also explicitly rejected arguments should be taken into account in a similar way as accepted arguments: this gives rise to the following definitions.

Definition 4.13 *Given two non-empty sets of extensions $\mathcal{E}_1$ and $\mathcal{E}_2$ of an argumentation framework AF, $\mathcal{E}_1 \preceq^E_{\cap\rightarrow} \mathcal{E}_2$ if and only if $\mathcal{E}_1 \preceq^E_\cap \mathcal{E}_2$ and $\bigcap_{E_1 \in \mathcal{E}_1} E_1^+ \subseteq \bigcap_{E_2 \in \mathcal{E}_2} E_2^+$.*

Definition 4.14 *Given two non-empty sets of extensions $\mathcal{E}_1$ and $\mathcal{E}_2$ of an argumentation framework AF, $\mathcal{E}_1 \preceq^E_{U\rightarrow} \mathcal{E}_2$ if and only if $\mathcal{E}_1 \preceq^E_U \mathcal{E}_2$ and $\bigcup_{E_1 \in \mathcal{E}_1} E_1^+ \subseteq \bigcup_{E_2 \in \mathcal{E}_2} E_2^+$.*

Definition 4.15 *Given two non-empty sets of labellings $\mathcal{L}_1$ and $\mathcal{L}_2$ of an argumentation framework AF, $\mathcal{L}_1 \preceq^L_{\cap\rightarrow} \mathcal{L}_2$ if and only if $\mathcal{L}_1 \preceq^L_\cap \mathcal{L}_2$ and $\bigcap_{Lab_1 \in \mathcal{L}_1} \text{out}(Lab_1) \subseteq \bigcap_{Lab_2 \in \mathcal{L}_2} \text{out}(Lab_2)$.*

Definition 4.16 *Given two non-empty sets of labellings $\mathcal{L}_1$ and $\mathcal{L}_2$ of an argumentation framework AF, $\mathcal{L}_1 \preceq^L_{U\rightarrow} \mathcal{L}_2$ if and only if $\mathcal{L}_1 \preceq^L_U \mathcal{L}_2$ and $\bigcup_{Lab_1 \in \mathcal{L}_1} \text{out}(Lab_1) \subseteq \bigcup_{Lab_2 \in \mathcal{L}_2} \text{out}(Lab_2)$.*

Consider again the example of Figure 4, and for a set of extensions $\mathcal{E}$ let us denote in the following $\mathcal{E}^+ = \{E^+ \mid E \in \mathcal{E}\}$. Then, referring to the already mentioned sets of extensions $\mathcal{E}_1 = \{\emptyset\}$ and $\mathcal{E}_2 = \{\{a\},\{b\}\}$ we have $\mathcal{E}_1^+ = \{\emptyset\}$ and $\mathcal{E}_2^+ = \{\{b\},\{a\}\}$. Clearly, $\mathcal{E}_1 \preceq^E_{\cap\rightarrow} \mathcal{E}_2$, $\mathcal{E}_2 \preceq^E_{\cap\rightarrow} \mathcal{E}_1$, $\mathcal{E}_1 \preceq^E_{U\rightarrow} \mathcal{E}_2$ while $\mathcal{E}_2 \not\preceq^E_{U\rightarrow} \mathcal{E}_1$. For the same example in the labelling-based approach, it can analogously be seen that $\mathcal{L}_1 \preceq^L_{\cap\rightarrow} \mathcal{L}_2$, $\mathcal{L}_2 \preceq^L_{\cap\rightarrow} \mathcal{L}_1$, $\mathcal{L}_1 \preceq^L_{U\rightarrow} \mathcal{L}_2$ and $\mathcal{L}_2 \not\preceq^L_{U\rightarrow} \mathcal{L}_1$.

In the example of Figure 6, we refer again to $\mathcal{E}_1 = \{\emptyset\}$ and $\mathcal{E}_2 = \{\{a,d\},\{b,d\}\}$, yielding $\mathcal{E}_1^+ = \{\emptyset\}$ and $\mathcal{E}_2^+ = \{\{b,c\},\{a,c\}\}$. Then $\mathcal{E}_1 \preceq^E_{\cap\rightarrow} \mathcal{E}_2$ and $\mathcal{E}_1 \preceq^E_{U\rightarrow} \mathcal{E}_2$, while $\mathcal{E}_2 \not\preceq^E_{\cap\rightarrow} \mathcal{E}_1$ and $\mathcal{E}_2 \not\preceq^E_{U\rightarrow} \mathcal{E}_1$. The case of labellings is perfectly analogous with $\mathcal{L}_1 \preceq^L_{\cap\rightarrow} \mathcal{L}_2$ and $\mathcal{L}_1 \preceq^L_{U\rightarrow} \mathcal{L}_2$, while $\mathcal{L}_2 \not\preceq^L_{\cap\rightarrow} \mathcal{L}_1$ and $\mathcal{L}_2 \not\preceq^L_{U\rightarrow} \mathcal{L}_1$.

Figure 9 provides again a more articulated case.

Considering the sets of extensions $\mathcal{E}_1 = \{\emptyset\}$, $\mathcal{E}_2 = \{\{a\},\{b,d\}\}$, $\mathcal{E}_3 = \{\{a,c\},\{a,d\},\{a,e\},\{b,d\}\}$, and $\mathcal{E}_4 = \{\{b,d\}\}$ we have $\mathcal{E}_1^+ = \{\emptyset\}$, $\mathcal{E}_2^+ = \{\{b\},\{a,c,e\}\}$, $\mathcal{E}_3^+ = \{\{b,d\},\{b,e\},\{b,c\},\{a,c,e\}\}$, $\mathcal{E}_4^+ = \{\{a,c,e\}\}$. It follows that for any $i,j \in \{1,2,3\}$ $\mathcal{E}_i \preceq^E_{\cap\rightarrow} \mathcal{E}_j$, while for any $i \in \{1,2,3\}$ $\mathcal{E}_i \preceq^E_{\cap\rightarrow} \mathcal{E}_4$ and $\mathcal{E}_4 \not\preceq^E_{\cap\rightarrow} \mathcal{E}_i$. On the other hand, $\mathcal{E}_1 \preceq^E_{U\rightarrow} \mathcal{E}_4 \preceq^E_{U\rightarrow} \mathcal{E}_2 \preceq^E_{U\rightarrow} \mathcal{E}_3$. Again, the case of labellings is perfectly analogous.

To have an example where the relations of the $\preceq_\cap$ kind differ from those of the $\preceq_{\cap\rightarrow}$ kind consider the example of Figure 15. In the extension-based approach all semantics but stable[21], stage, $\mathcal{CF}2$, and stage2 semantics prescribe the set

---
[21]The set of stable extensions is empty in this case.

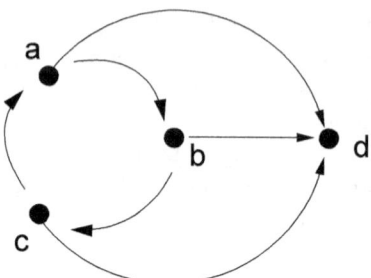

Figure 15. Cycle of three attacking arguments in turn attacking another argument

of extensions $\mathcal{E}_1 = \{\emptyset\}$ with $\mathcal{E}_1^+ = \{\emptyset\}$, while stage, $\mathcal{CF}2$, and stage2 semantics prescribe $\mathcal{E}_2 = \{\{a\}, \{b\}, \{c\}\}$ with $\mathcal{E}_2^+ = \{\{d\}\}$. It follows that $\mathcal{E}_1 \preceq_\cap^E \mathcal{E}_2$ and $\mathcal{E}_2 \preceq_\cap^E \mathcal{E}_1$, while $\mathcal{E}_1 \preceq_{\cap\rightarrow}^E \mathcal{E}_2$ but $\mathcal{E}_2 \not\preceq_{\cap\rightarrow}^E \mathcal{E}_1$. Similar considerations apply in the labelling-based approach.

Definitions 4.9-4.16 treat sets of extensions or labellings "as a whole" by simply considering their intersection or union: for instance, very different sets of extensions are treated in the same way if they have an empty intersection. In order to take account of how single extensions or labellings are defined, a different kind of definition is needed: the skepticism relation between two sets (let say $\mathcal{X}_1$ and $\mathcal{X}_2$) of extensions or labellings should be based on some comparison between their individual elements. In particular, according to a skeptical approach to argument justification, in order to state that $\mathcal{X}_1$ is at least as skeptical as $\mathcal{X}_2$, one may require that every element in $\mathcal{X}_2$ has a more skeptical counterpart in $\mathcal{X}_1$, while, according to a credulous approach, one may require dually that every element in $\mathcal{X}_1$ has a less skeptical counterpart in $\mathcal{X}_2$. This general idea is formalized by the following definitions, which resort to the basic comparisons between single extensions and labellings identified in Definition 4.8.

Definition 4.17 *Given two non-empty sets of extensions $\mathcal{E}_1$ and $\mathcal{E}_2$ of an argumentation framework AF, $\mathcal{E}_1 \preceq_{\cap+}^E \mathcal{E}_2$ if and only if $\forall E_2 \in \mathcal{E}_2 \; \exists E_1 \in \mathcal{E}_1 : E_1 \preceq E_2$.*

Definition 4.18 *Given two non-empty sets of extensions $\mathcal{E}_1$ and $\mathcal{E}_2$ of an argumentation framework AF, $\mathcal{E}_1 \preceq_{\cup+}^E \mathcal{E}_2$ if and only if $\forall E_1 \in \mathcal{E}_1 \; \exists E_2 \in \mathcal{E}_2 : E_1 \preceq E_2$.*

Definition 4.19 *Given two non-empty sets of labellings $\mathcal{L}_1$ and $\mathcal{L}_2$ of an argumentation framework AF, $\mathcal{L}_1 \preceq_{\cap+}^L \mathcal{L}_2$ if and only if $\forall Lab_2 \in \mathcal{L}_2 \; \exists Lab_1 \in \mathcal{L}_1 : Lab_1 \preceq Lab_2$.*

Definition 4.20 *Given two non-empty sets of labellings $\mathcal{L}_1$ and $\mathcal{L}_2$ of an argumentation framework AF, $\mathcal{L}_1 \preceq^L_{\cup+} \mathcal{L}_2$ if and only if $\forall Lab_1 \in \mathcal{L}_1 \; \exists Lab_2 \in \mathcal{L}_2 : Lab_1 \preceq Lab_2$.*

Let us exemplify these relations.

In the example of Figure 4, referring to the already mentioned sets of extensions $\mathcal{E}_1 = \{\emptyset\}$ and $\mathcal{E}_2 = \{\{a\}, \{b\}\}$ we have $\mathcal{E}_1 \preceq^E_{\cap+} \mathcal{E}_2$ but, differently from the previously considered relations, $\mathcal{E}_2 \not\preceq^E_{\cap+} \mathcal{E}_1$. On the other hand, $\mathcal{E}_1 \preceq^E_{\cup+} \mathcal{E}_2$ and $\mathcal{E}_2 \not\preceq^E_{\cup+} \mathcal{E}_1$. As usual, analogous relations hold for the labelling-based approach.

Similarly, in the example of Figure 6, with $\mathcal{E}_1 = \{\emptyset\}$ and $\mathcal{E}_2 = \{\{a,d\}, \{b,d\}\}$ it holds $\mathcal{E}_1 \preceq^E_{\cap+} \mathcal{E}_2$ and $\mathcal{E}_1 \preceq^E_{\cup+} \mathcal{E}_2$, while $\mathcal{E}_2 \not\preceq^E_{\cap+} \mathcal{E}_1$ and $\mathcal{E}_2 \not\preceq^E_{\cup+} \mathcal{E}_1$. It goes without saying that the same holds in the labelling-based approach.

Finally consider the case of Figure 9 with sets of extensions $\mathcal{E}_1 = \{\emptyset\}$, $\mathcal{E}_2 = \{\{a\}, \{b,d\}\}$, $\mathcal{E}_3 = \{\{a,c\}, \{a,d\}, \{a,e\}, \{b,d\}\}$, and $\mathcal{E}_4 = \{\{b,d\}\}$. We can first observe that for $i \in \{2,3,4\}$ $\mathcal{E}_1 \preceq^E_{\cap+} \mathcal{E}_i$ and (differently from previous relations) $\mathcal{E}_i \not\preceq^E_{\cap+} \mathcal{E}_1$. Then we can note that $\mathcal{E}_2 \preceq^E_{\cap+} \mathcal{E}_4$ and $\mathcal{E}_3 \preceq^E_{\cap+} \mathcal{E}_4$ since the only element of $\mathcal{E}_4$ (namely $\{b,d\}$) is a superset of (actually coincides with) an element of either $\mathcal{E}_2$ or $\mathcal{E}_3$. Also $\mathcal{E}_2 \preceq^E_{\cap+} \mathcal{E}_3$ since the elements $\{a,c\}, \{a,d\}$, and $\{a,e\}$ of $\mathcal{E}_3$ are supersets of $\{a\}$ in $\mathcal{E}_2$ and $\{b,d\}$ is present both in $\mathcal{E}_3$ and $\mathcal{E}_2$. With similar observations it can be seen that $\mathcal{E}_3 \not\preceq^E_{\cap+} \mathcal{E}_2$, $\mathcal{E}_4 \not\preceq^E_{\cap+} \mathcal{E}_2$, and $\mathcal{E}_4 \not\preceq^E_{\cap+} \mathcal{E}_3$. Turning to the relation corresponding to the credulous perspective, it can immediately be observed that for $i \in \{2,3,4\}$ $\mathcal{E}_1 \preceq^E_{\cup+} \mathcal{E}_i$ and $\mathcal{E}_i \not\preceq^E_{\cup+} \mathcal{E}_1$. Also, $\mathcal{E}_2 \preceq^E_{\cup+} \mathcal{E}_3$ since $\{a\}$ is included in some elements of $\mathcal{E}_3$ and $\{b,d\}$ is present both in $\mathcal{E}_2$ and $\mathcal{E}_3$. On the other hand, $\mathcal{E}_3 \not\preceq^E_{\cup+} \mathcal{E}_2$. Differently from the skeptical perspective, $\mathcal{E}_4 \preceq^E_{\cup+} \mathcal{E}_2$ and $\mathcal{E}_4 \preceq^E_{\cup+} \mathcal{E}_3$ (the only element of $\mathcal{E}_4$, namely $\{b,d\}$ is present both in $\mathcal{E}_2$ and $\mathcal{E}_3$) while it can be easily seen that $\mathcal{E}_2 \not\preceq^E_{\cup+} \mathcal{E}_4$ ($\{a\}$ is not included in any element of $\mathcal{E}_4$) and $\mathcal{E}_3 \not\preceq^E_{\cup+} \mathcal{E}_4$ (as above for sets $\{a,c\}, \{a,d\}, \{a,e\}$). Again, the case of labellings is perfectly analogous.

A stronger skepticism relation, unifying the skeptical and credulous perspectives, can be obtained by combining together the relations $\preceq_{\cap+}$ and $\preceq_{\cup+}$.

Definition 4.21 *Given two non-empty sets of extensions $\mathcal{E}_1$ and $\mathcal{E}_2$ of an argumentation framework AF, $\mathcal{E}_1 \preceq^E_\oplus \mathcal{E}_2$ if and only if $\mathcal{E}_1 \preceq^E_{\cap+} \mathcal{E}_2$ and $\mathcal{E}_1 \preceq^E_{\cup+} \mathcal{E}_2$.*

Definition 4.22 *Given two non-empty sets of labellings $\mathcal{L}_1$ and $\mathcal{L}_2$ of an argumentation framework AF, $\mathcal{L}_1 \preceq^L_\oplus \mathcal{L}_2$ if and only if $\mathcal{L}_1 \preceq^L_{\cap+} \mathcal{L}_2$ and $\mathcal{L}_1 \preceq^L_{\cup+} \mathcal{L}_2$.*

As also evident from their definitions, the various skepticism relations introduced above are related each other by implication. In particular, two implications chains can be identified in correspondence with the skeptical or credulous perspective. In fact, given two sets of extensions $\mathcal{E}_1$ and $\mathcal{E}_2$ of an argumentation framework AF, it holds that:

(1) $\qquad \mathcal{E}_1 \preceq^E_\oplus \mathcal{E}_2 \Rightarrow \mathcal{E}_1 \preceq^E_{\cap+} \mathcal{E}_2 \Rightarrow \mathcal{E}_1 \preceq^E_{\cap\rightarrow} \mathcal{E}_2 \Rightarrow \mathcal{E}_1 \preceq^E_\cap \mathcal{E}_2$

(2) $\qquad \mathcal{E}_1 \preceq^E_\oplus \mathcal{E}_2 \Rightarrow \mathcal{E}_1 \preceq^E_{\cup+} \mathcal{E}_2 \Rightarrow \mathcal{E}_1 \preceq^E_{\cup\rightarrow} \mathcal{E}_2 \Rightarrow \mathcal{E}_1 \preceq^E_\cup \mathcal{E}_2$

The only nontrivial implications in (1) and (2) concern the fact that $\preceq^E_{\cap+}$ implies $\preceq^E_{\cap\rightarrow}$, and, similarly, $\preceq^E_{\cup+}$ implies $\preceq^E_{\cup\rightarrow}$: they have been proved in [Baroni and Giacomin, 2009b].

Using Definitions 4.11, 4.12, 4.15, 4.16, 4.19, 4.20, 4.22, and the same kind of reasoning it is possible to prove that the analogous relations hold in the labelling based approach. In fact, given two sets of labellings $\mathcal{L}_1$ and $\mathcal{L}_2$ of an argumentation framework AF, it holds that:

(3) $\qquad \mathcal{L}_1 \preceq^L_\oplus \mathcal{L}_2 \Rightarrow \mathcal{L}_1 \preceq^L_{\cap+} \mathcal{L}_2 \Rightarrow \mathcal{L}_1 \preceq^L_{\cap\rightarrow} \mathcal{L}_2 \Rightarrow \mathcal{L}_1 \preceq^L_\cap \mathcal{L}_2$

(4) $\qquad \mathcal{L}_1 \preceq^L_\oplus \mathcal{L}_2 \Rightarrow \mathcal{L}_1 \preceq^L_{\cup+} \mathcal{L}_2 \Rightarrow \mathcal{L}_1 \preceq^L_{\cup\rightarrow} \mathcal{L}_2 \Rightarrow \mathcal{L}_1 \preceq^L_\cup \mathcal{L}_2$

Turning to the comparison between semantics, for a given generic relation $\preceq$ concerning either extensions or labellings it is quite natural to define an induced relation of skepticism between two semantics σ_1 and σ_2, by requiring that $\preceq$ holds for their sets of extensions or labellings. As it may happen that either σ_1 or σ_2 prescribes an empty set of extensions (or labellings) in some cases, the induced relation has to refer to a set of argumentation frameworks where both σ_1 and σ_2 prescribe non-empty sets of extensions (or labellings), namely to $\mathcal{DE}_{\sigma_1} \cap \mathcal{DE}_{\sigma_2}$ (or $\mathcal{DL}_{\sigma_1} \cap \mathcal{DL}_{\sigma_2}$) using the notation of Definition 4.1.

Definition 4.23 *Let $\preceq^E$ be a skepticism relation between sets of extensions, σ_1 and σ_2 be extension-based argumentation semantics, and $\mathcal{A}$ be a set of argumentation frameworks with $\mathcal{A} \subseteq (\mathcal{DE}_{\sigma_1} \cap \mathcal{DE}_{\sigma_2})$. The skepticism relation $\preceq^{SE}$ induced by $\preceq^E$ between σ_1 and σ_2 with reference to $\mathcal{A}$ is defined as follows: $\sigma_1 \preceq^{SE} \sigma_2$ if and only if $\forall AF \in \mathcal{A}\ \mathcal{E}_{\sigma_1}(AF) \preceq^E \mathcal{E}_{\sigma_2}(AF)$.*

Definition 4.24 *Let $\preceq^L$ be a skepticism relation between sets of labellings, σ_1 and σ_2 be labelling-based argumentation semantics, and $\mathcal{A}$ be a set of argumentation frameworks with $\mathcal{A} \subseteq (\mathcal{DL}_{\sigma_1} \cap \mathcal{DL}_{\sigma_2})$. The skepticism relation $\preceq^{SL}$ induced by $\preceq^L$ between σ_1 and σ_2 with reference to $\mathcal{A}$ is defined as follows: $\sigma_1 \preceq^{SL} \sigma_2$ if and only if $\forall AF \in \mathcal{A}\ \mathcal{L}_{\sigma_1}(AF) \preceq^L \mathcal{L}_{\sigma_2}(AF)$.*

Focusing on the extension-based approach, while Definition 4.23 considers a generic set of argumentation frameworks $\mathcal{A} \subseteq (\mathcal{DE}_{\sigma_1} \cap \mathcal{DE}_{\sigma_2})$ to define a skepticism relation, clearly the most interesting skepticism relation is the one corresponding to the case $\mathcal{A} = (\mathcal{DE}_{\sigma_1} \cap \mathcal{DE}_{\sigma_2})$. Then, when considering a skepticism comparison concerning more than two semantics $\sigma_1, \sigma_2, \ldots, \sigma_N$ it is reasonable to consider a common reference $\mathcal{A} = \bigcap_{i=1\ldots N} \mathcal{DE}_{\sigma_i}$. As to the semantics discussed in this chapter, only stable semantics may prescribe an empty set of extensions/labellings. Therefore two reference sets can be considered: the universe of all argumentation frameworks if stable semantics is not involved in the comparison, or $\mathcal{DE}_{ST}$ otherwise. Clearly the same considerations hold in the labelling-based approach by replacing $\mathcal{DE}$ with $\mathcal{DL}$.

It is worth noting that, in general, two semantics σ_1 and σ_2 may not be comparable with respect to skepticism. For instance, it may be the case that

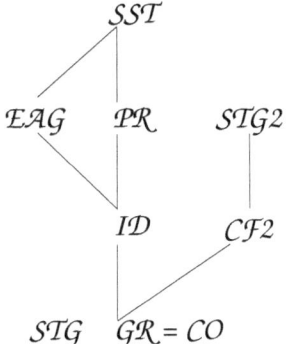

Figure 16. $\preceq^{SE}_{\cap+}$, $\preceq^{SE}_{\cap\to}$ and $\preceq^{SE}_{\cap}$ relations for any argumentation framework

there are two argumentation frameworks AF_1 and AF_2 such that σ_1 behaves more skeptically than σ_2 in the case of AF_1 but σ_2 behaves more skeptically than σ_1 in the case of AF_2, or that the two semantics yield incomparable sets of extensions for some given argumentation framework. Furthermore, the order between two semantics may be different according to the credulous or skeptical perspective.

A detailed analysis of skepticism relations between most extension-based semantics has been carried out in [Baroni and Giacomin, 2009b] to which the reader may refer for details: we report here only the resulting partial orders, graphically presented as Hasse diagrams (with the addition of eager, stage, and stage2 semantics[22] with respect to [Baroni and Giacomin, 2009b]). As mentioned above, distinct Hasse diagrams are presented for the case where stable extensions exist and for the general one.

The partial orders[23] induced by all the relations corresponding to the skeptical perspective, namely $\preceq^{SE}_{\cap+}$, $\preceq^{SE}_{\cap\to}$ and $\preceq^{SE}_{\cap}$ coincide.

The Hasse diagram corresponding to the general case is shown in Figure 16: grounded semantics is the most skeptical one and since the grounded extension is the least complete extension it turns out that $\mathcal{GR} \preceq^{SE}_{\cap+} \mathcal{CO}$ and $\mathcal{CO} \preceq^{SE}_{\cap+} \mathcal{GR}$. Ideal, preferred, and semi-stable semantics are all comparable among them and orderly less skeptical, with eager semantics in between ideal and semi-stable, while not comparable with preferred. Stage2 semantics is less skeptical than $\mathcal{CF}2$, which in turn is comparable with $\mathcal{GR}$ and $\mathcal{CO}$, while no other relation holds with the other semantics. Stage semantics is not comparable with any other, also due to its peculiar behaviour in some cases, exemplified in the

---

[22]The authors are grateful to Wolfgang Dvořák for suggesting the extension of these results to eager and stage2 semantics.

[23]The skepticism relations described in the following have been analyzed in [Baroni and Giacomin, 2009b] for the extension-based approach. Due to the one-to-one correspondence between extensions and labellings holding for all the semantics involved in the comparison, it is possible to prove that the skepticism relations hold also in the labelling-based approach.

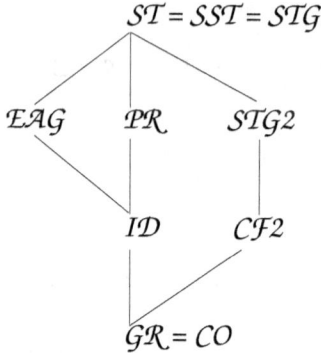

Figure 17. $\preceq^{SE}_{\cap+}$, $\preceq^{SE}_{\cap\rightarrow}$ and $\preceq^{SE}_{\cap}$ relations for argumentation frameworks in $\mathcal{DE}_{ST}$ ($\mathcal{DL}_{ST}$)

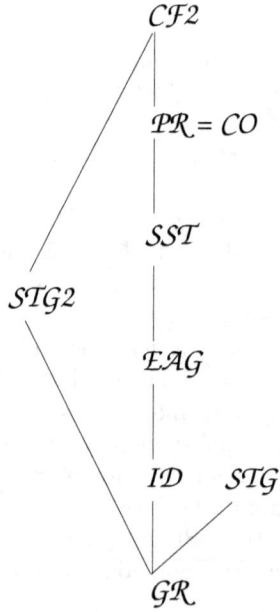

Figure 18. $\preceq^{SE}_{\cup+}$, $\preceq^{SE}_{\cup\rightarrow}$ and $\preceq^{SE}_{\cup}$ relations for any argumentation framework

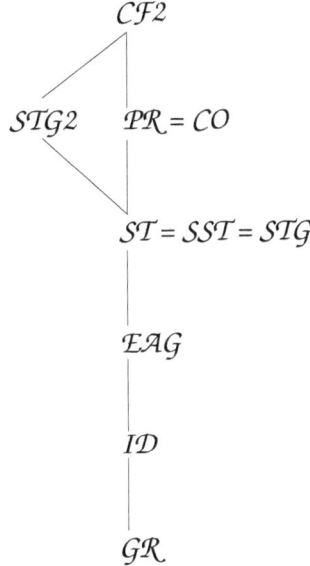

Figure 19. $\preceq^{SE}_{\cup+}$, $\preceq^{SE}_{\cup\to}$ and $\preceq^{SE}_{\cup}$ relations for argumentation frameworks in $\mathcal{DE}_{ST}$ ($\mathcal{DL}_{ST}$)

argumentation framework of Figure 12.

The Hasse diagram for $\preceq^{SE}_{\cap+}$, $\preceq^{SE}_{\cap\to}$ and $\preceq^{SE}_{\cap}$ considering only the argumentation frameworks where stable extensions exist (and then coincide with semi-stable and stage extensions) is shown in Figure 17. It can be noted that in this context stage2 semantics is comparable with (and more skeptical than) stable semantics.

Turning to skepticism relations based on the credulous perspective, namely $\preceq^{SE}_{\cup+}$, $\preceq^{SE}_{\cup\to}$ and $\preceq^{SE}_{\cup}$, the Hasse diagram corresponding to the general case is shown in Figure 18. An almost complete ordering is achieved due to the change of perspective. In particular, complete semantics is in mutual relation with preferred semantics: $\mathcal{PR} \preceq^{SE}_{\cup+} \mathcal{CO}$ and $\mathcal{CO} \preceq^{SE}_{\cup+} \mathcal{PR}$ since preferred extensions are maximal complete extensions. Moreover one can note that $\mathcal{CF}2$ is now comparable with any other one (and is actually the least skeptical semantics) and that the orderings between $\mathcal{PR}$ and $\mathcal{SST}$ and between $\mathcal{CF}2$ and stage2 are inverted with respect to Figure 16.

The Hasse diagram for $\preceq^{SE}_{\cup+}$, $\preceq^{SE}_{\cup\to}$ and $\preceq^{SE}_{\cup}$ considering only the argumentation frameworks where stable extensions exist is shown in Figure 19: here an almost total order is achieved, which obeys the same relations as the general case but where stable, semi-stable and stage semantics coincide and stage2 is between stable and $\mathcal{CF}2$ semantics.

Finally, the Hasse diagrams for the relations arising from the conjunction of

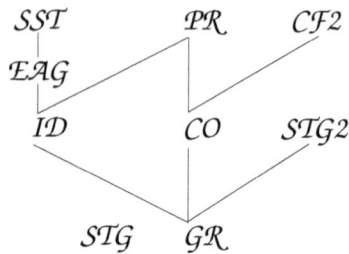

Figure 20. $\preceq_{\oplus}^{SE}$ relation for any argumentation framework.

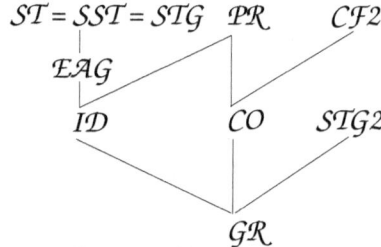

Figure 21. $\preceq_{\oplus}^{SE}$ relation for argumentation frameworks in $\mathcal{DE}_{ST}$ ($\mathcal{DL}_{ST}$).

the skeptical and credulous perspective are shown in Figures 20 and 21, for the general case and for argumentation frameworks where stable extensions exist respectively. As obvious, stronger relations entail lesser comparability between semantics, but one can note in particular that the role of $\mathcal{GR}$ as "bottom" skeptical reference with respect to all other semantics (but $\mathcal{STG}$) is confirmed.

5 Semantics agreement

The study of many different argumentation semantics in the literature indicates the richness and inherent complexity of the problem of "solving conflicts among arguments". In general terms, the introduction of a new semantics can be motivated by the need to achieve a particular desired outcome in some specific example considered particularly relevant and/or by the objective of satisfying some properties, at an intuitive or formal level.

While semantics differences obviously attract attention in analyses and comparisons, and have been discussed in Section 3, it is also important to characterize situations where argumentation semantics agree, i.e. exhibit the same behaviour. This is useful from several viewpoints. On one hand, situations where "most" (or even all) existing semantics agree can be regarded as providing a sort of reference behaviour against which further proposals should be confronted. On the other hand, it may be the case that in a specific application domain there are some restrictions on the structure of the argumentation frameworks that need to be considered. It is then surely interesting to know whether these restrictions lead to semantics agreement, since in this case the choice of the semantics to be adopted turns out to have no influence and, in a sense, the outcomes are universally supported in a semantics-independent way.

In this section we review the existing results about semantics agreement covering two distinct, though related, issues.

The first issue concerns the identification of the classes of argumentation frameworks where a given set of semantics agree. As will be shown in Section 5.1, using general properties of argumentation semantics only, it is possible to prove that there is a limited number of distinct agreement classes and identify them regardless of the topological properties of the argumentation frameworks belonging to them.

The second issue, dealt with in Section 5.2, concerns, in a complementary way, the analysis of argumentation frameworks where semantics agree from a topological viewpoint.

Before we enter the matter, some basic definitions and properties on semantics agreement need to be introduced.

Semantics agreement concerns comparing the set of extensions or, equivalently, of labellings prescribed by different semantics for the same framework. The analysis will be developed focusing on the extension-based approach, as this allows a more compact presentation. Furthermore, to avoid to deal with spurious situations, whenever we will discuss the properties of agreement of a given semantics σ we will implicitly refer, if not differently specified, only to argumentation frameworks belonging to $\mathcal{DE}_\sigma$, i.e. such that the set of extensions prescribed by σ is not empty.

The notion of agreement for a set of semantics on an argumentation framework is defined in the obvious way.

Definition 5.1 *Let $\mathbb{S}$ be a set of argumentation semantics and AF an argumentation framework such that $AF \in \bigcap_{\sigma \in \mathbb{S}} \mathcal{DE}_\sigma$. We say that AF is an agreement framework for $\mathbb{S}$ (or, equivalently, that the semantics in $\mathbb{S}$ agree on AF) iff $\forall \sigma_1, \sigma_2 \in \mathbb{S}$ it holds that $\mathcal{E}_{\sigma_1}(AF) = \mathcal{E}_{\sigma_2}(AF)$. The set of the agreement frameworks for $\mathbb{S}$ is denoted as $\mathcal{AGR}(\mathbb{S})$ and called the* agreement class *of $\mathbb{S}$.*

In general, it may be the case that $\mathcal{AGR}(\mathbb{S}_1) = \mathcal{AGR}(\mathbb{S}_2)$ for different sets of semantics $\mathbb{S}_1$ and $\mathbb{S}_2$: this in fact motivates the analysis carried out in Section 5.1. Moreover it is immediate to note that $\mathbb{S}_1 \subseteq \mathbb{S}_2 \Rightarrow \mathcal{AGR}(\mathbb{S}_2) \subseteq \mathcal{AGR}(\mathbb{S}_1)$.

5.1 Agreement classes

Systematic results on the identification of agreement classes have been obtained up to now for the semantics in the set $\Omega = \{\mathcal{GR}, \mathcal{ID}, \mathcal{CO}, \mathcal{PR}, \mathcal{ST}, \mathcal{CF}2, \mathcal{SST}\}$.

It can be noted that $\mathcal{AGR}(\mathbb{S}) \neq \emptyset$ for any set of semantics $\mathbb{S} \subseteq \Omega$ since all semantics belonging to Ω are obviously in agreement on the empty argumentation framework $AF_\emptyset = (\emptyset, \emptyset)$. Moreover, all semantics in Ω (and probably every

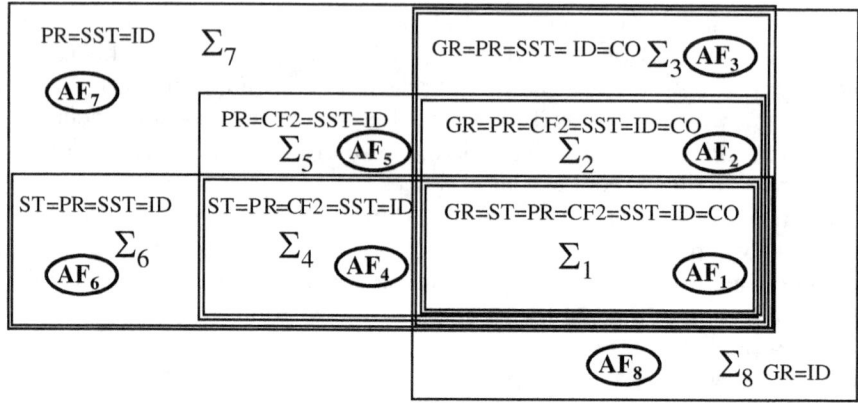

Figure 22. Venn diagram of unique-status agreement classes.

other reasonable argumentation semantics one can conceive) are in agreement also on attack-free argumentation frameworks. Namely, for an argumentation framework $AF = (\mathcal{A}rgs, \emptyset)$ for every $\sigma \in \Omega$ it holds that $\mathcal{E}_\sigma(AF) = \{\mathcal{A}rgs\}$.

As for the semantics not included in Ω, the following comments can be made. First, naïve semantics differs substantially from other semantics since, basically, it ignores the direction of attacks. Thus agreement with some other semantics can be achieved only in peculiar situations where the direction of attacks is not relevant (this is the case for instance of symmetric argumentation frameworks, mentioned in Section 5.2). As to stage semantics it must be recalled, first of all, that when stable extensions exist, they coincide with stage extensions. Thus agreement classes including stable semantics also implicitly include stage semantics. When stable extensions do not exist the peculiar behaviour of stage semantics in some cases (in particular the possibility that unattacked arguments are not included in an extension) implies that stage semantics may disagree with any other set of semantics. The identification and study of agreement classes involving eager or stage2 semantics appears to be potentially more interesting, but has not been developed in the literature yet.

Considering all subsets $\mathbb{S}$ of Ω such that $|\mathbb{S}| \geq 2$ gives rise, in principle, to 120 classes $\mathcal{AGR}(\mathbb{S})$ to be evaluated. Hovewer it has been proved in [Baroni and Giacomin, 2008] that the distinct agreement classes are only 14.

We will start by analyzing in subsection 5.1.1 unique-status agreement, namely the classes $\mathcal{AGR}(\mathbb{S})$ where $\mathbb{S}$ includes $\mathcal{GR}$ or $\mathcal{ID}$, and we will then examine agreement between multiple-status semantics in subsection 5.1.2.

5.1.1 Unique-status Agreement

The Venn diagram concerning unique-status agreement classes is shown in Figure 22 where rectangular boxes represent classes of agreement and small ellipses represent single argumentation frameworks to be used as specific examples.

The diagram will be illustrated in two main steps: first, we show that the set-theoretical relationships between the agreement classes $\Sigma_1,\ldots,\Sigma_8$ depicted in Figure 22 actually hold, then we explain why the classes $\Sigma_1,\ldots,\Sigma_8$ are the only meaningful ones in the context of unique-status agreement.

As to the first step, we proceed by following the partial order induced by inclusion (namely if $\Sigma_i \subsetneq \Sigma_j$ then $j > i$). While introducing each class Σ_i it will be necessary:

1. to identify which classes Σ_k, $k < i$, are included in Σ_i;

2. for each of these classes Σ_k to show that $\Sigma_i \setminus \Sigma_k \neq \emptyset$;

3. for any Σ_h such that $h < i$ and $\Sigma_h \not\subseteq \Sigma_i$ to examine $\Sigma_i \cap \Sigma_h$.

Point 1 will not be stressed since inclusion relationships can be directly derived from the inclusion of the relevant sets of semantics ($\mathbb{S}_1 \subseteq \mathbb{S}_2 \Rightarrow \mathcal{AGR}(\mathbb{S}_2) \subseteq \mathcal{AGR}(\mathbb{S}_1)$). As to point 2, examples of argumentation frameworks belonging to non-empty set differences will be given to prove that inclusion relationships are strict, while point 3 will be dealt with case by case.

Let us start from $\mathcal{AGR}(\{\mathcal{GR},\mathcal{ID},\mathcal{CO},\mathcal{PR},\mathcal{ST},\mathcal{CF}2,\mathcal{SST}\})$, denoted as Σ_1. This is the class of argumentation frameworks where all semantics show a uniform single status behaviour in agreement with grounded semantics. Σ_1 includes, for instance, attack-free argumentation frameworks like the very simple $AF_1 = \langle \{a\}, \emptyset \rangle$.

$\mathcal{AGR}(\{\mathcal{GR},\mathcal{ID},\mathcal{CO},\mathcal{PR},\mathcal{CF}2,\mathcal{SST}\})$, denoted as Σ_2, corresponds to a uniform single status behaviour in agreement with grounded semantics by all but stable semantics. As shown in Figure 22, Σ_2 strictly includes Σ_1 since $(\Sigma_2 \setminus \Sigma_1)$ includes in particular $AF_2 = \langle \{a,b\}, \{(b,b)\} \rangle$.

$\Sigma_3 \triangleq \mathcal{AGR}(\{\mathcal{GR},\mathcal{ID},\mathcal{CO},\mathcal{PR},\mathcal{SST}\})$ is the last class in Figure 22 concerning agreement with grounded semantics. $(\Sigma_3 \setminus \Sigma_2) \neq \emptyset$ since it includes for instance $AF_3 = \langle \{a,b,c\}, \{(a,b),(b,c),(c,a)\} \rangle$. It is now worth noting that $\Sigma_3 \cap \mathcal{DE}_{\mathcal{ST}} = \Sigma_2 \cap \mathcal{DE}_{\mathcal{ST}} = \Sigma_1$. In fact, since semi-stable extensions coincide with stable extensions when the latter exist, whenever $AF \in \mathcal{DE}_{\mathcal{ST}}$ it must be the case that $AF \in \mathcal{AGR}(\mathbb{S})$ where $\mathbb{S}$ includes both $\mathcal{ST}$ and $\mathcal{SST}$.

On the left of Σ_1, Σ_2 and Σ_3 the diagram of Figure 22 shows four classes where several multiple-status semantics exhibit a unique-status behaviour in agreement with ideal semantics, but not necessarily also with grounded semantics. The smallest of these classes is $\Sigma_4 \triangleq \mathcal{AGR}(\{\mathcal{ID},\mathcal{CF}2,\mathcal{ST},\mathcal{PR},\mathcal{SST}\})$. $(\Sigma_4 \setminus \Sigma_1) \neq \emptyset$ since it includes $AF_4 = \langle \{a,b\}, \{(a,b),(b,a),(b,b)\} \rangle$. Moreover, since $\Sigma_4 \subset \mathcal{DE}_{\mathcal{ST}}$, it is the case that $\Sigma_4 \cap (\Sigma_3 \setminus \Sigma_1) = \emptyset$. Not requiring agreement with stable semantics leads to $\Sigma_5 \triangleq \mathcal{AGR}(\{\mathcal{ID},\mathcal{CF}2,\mathcal{PR},\mathcal{SST}\})$. $(\Sigma_5 \setminus (\Sigma_4 \cup \Sigma_2)) \neq \emptyset$, since it includes $AF_5 = \langle \{a,b,c\}, \{(a,b),(b,a),(b,b),(c,c)\} \rangle$. Notice also that $\Sigma_5 \cap (\Sigma_3 \setminus \Sigma_2) = \emptyset$ since if $AF \in \Sigma_5 \cap \Sigma_3$ then, by definition of these classes, it also holds $AF \in \Sigma_2$. It is also clear that $\Sigma_5 \cap \mathcal{DE}_{\mathcal{ST}} = \Sigma_4$. Not requiring agreement with $\mathcal{CF}2$ semantics leads to $\Sigma_6 \triangleq \mathcal{AGR}(\{\mathcal{ID},\mathcal{ST},\mathcal{PR},\mathcal{SST}\})$. $(\Sigma_6 \setminus \Sigma_4) \neq \emptyset$ since it includes $AF_6 = \langle \{a,b,c\}, \{(a,b),(a,c),(b,c),(c,a)\} \rangle$.

Again, since $\Sigma_6 \subset \mathcal{DE}_{\mathcal{ST}}$ it holds $\Sigma_6 \cap (\Sigma_3 \setminus \Sigma_1) = \emptyset$ and $\Sigma_6 \cap (\Sigma_5 \setminus \Sigma_4) = \emptyset$. Finally, excluding both stable and $\mathcal{CF}2$ semantics from the required agreement corresponds to $\Sigma_7 \triangleq \mathcal{AGR}(\{\mathcal{ID}, \mathcal{PR}, \mathcal{SST}\})$. $(\Sigma_7 \setminus (\Sigma_6 \cup \Sigma_5 \cup \Sigma_3)) \neq \emptyset$, since it includes $AF_7 = \langle \{a,b,c,d,e\}, \{(a,b),(b,a),(b,b),(c,d),(d,e),(e,c)\} \rangle$.

The last class concerning unique-status agreement is $\Sigma_8 \triangleq \mathcal{AGR}(\{\mathcal{ID}, \mathcal{GR}\})$. By definition of the relevant sets of semantics (and since no other distinct agreement classes exist, as recalled below) it holds that $\Sigma_8 \cap \Sigma_7 = \Sigma_3$, moreover it is easy to see that $\Sigma_8 \setminus \Sigma_3 \neq \emptyset$, since it includes $AF_8 = \langle \{a,b\}, \{(a,b),(b,a)\} \rangle$.

In [Baroni and Giacomin, 2008] it is shown that no other classes than $\Sigma_1, \ldots, \Sigma_8$ are meaningful in the context of unique-status agreement. The proofs are based on a set of basic properties and relationships, which we recall together in Proposition 5.2 (see Lemmata 2-10 in [Baroni and Giacomin, 2008]) as they are of general interest in the analysis and understanding of argumentation semantics.

Proposition 5.2 *Given an argumentation framework AF the following statements hold:*

1. *if $|\mathcal{E}_{\mathcal{PR}}(AF)| = 1$, then*
   - $\mathcal{E}_{\mathcal{PR}}(AF) = \mathcal{E}_{\mathcal{ID}}(AF)$;
   - $\mathcal{E}_{\mathcal{PR}}(AF) = \mathcal{E}_{\mathcal{SST}}(AF)$;
   - *if $AF \in \mathcal{DE}_{\mathcal{ST}}$, $\mathcal{E}_{\mathcal{PR}}(AF) = \mathcal{E}_{\mathcal{ST}}(AF)$.*

2. *if $|\mathcal{E}_{\mathcal{CO}}(AF)| = 1$, then $\mathcal{E}_{\mathcal{CO}}(AF) = \mathcal{E}_{\mathcal{GR}}(AF) = \mathcal{E}_{\mathcal{PR}}(AF)$;*

3. *let $\sigma \in \{\mathcal{PR}, \mathcal{ST}, \mathcal{SST}, \mathcal{CF}2\}$, if $GE(AF) \in \mathcal{E}_\sigma(AF)$ then $\mathcal{E}_\sigma(AF) = \{GE(AF)\}$;*

4. *if $\mathcal{E}_{\mathcal{PR}}(AF) = \{GE(AF)\}$ then $\mathcal{E}_{\mathcal{CO}}(AF) = \{GE(AF)\}$;*

5. *if $\mathcal{E}_{\mathcal{ST}}(AF) = \{GE(AF)\}$ then $\mathcal{E}_{\mathcal{CF}2}(AF) = \{GE(AF)\}$;*

6. *if $\mathcal{E}_{\mathcal{CF}2}(AF) = \{GE(AF)\}$ then $\mathcal{E}_{\mathcal{CF}2}(AF) = \mathcal{E}_{\mathcal{PR}}(AF)$;*

7. *if $\mathcal{E}_{\mathcal{CF}2}(AF) = \{ID(AF)\}$ then $\mathcal{E}_{\mathcal{CF}2}(AF) = \mathcal{E}_{\mathcal{PR}}(AF)$;*

8. *if $\mathcal{E}_{\mathcal{SST}}(AF) = \{ID(AF)\}$ then $\mathcal{E}_{\mathcal{SST}}(AF) = \mathcal{E}_{\mathcal{PR}}(AF)$;*

9. *if $\mathcal{E}_{\mathcal{CF}2}(AF) \subseteq \mathcal{AS}(AF)$ then $\mathcal{E}_{\mathcal{CF}2}(AF) = \mathcal{E}_{\mathcal{PR}}(AF)$.*

These results can be briefly expressed in words as follows. Item 1 states that when there is a unique preferred extension, it is also the unique ideal, semi-stable, and (possibly) stable extension, while item 2 states that when there is a unique complete extension it is also the unique preferred extension and the grounded extension. Item 3 says that the set of preferred extensions can include the grounded extension only if it contains only the grounded extension (and the same holds for stable, semi-stable and $\mathcal{CF}2$ extensions). If the grounded extension is the only preferred extension it is also the only complete extension (item 4), while if it is the only stable extension it is also the only $\mathcal{CF}2$ extension (item 5) and the only preferred extension (item 6). Finally, if the set of $\mathcal{CF}2$ extensions or of semi-stable extensions contains only the ideal extension this is also the unique preferred extension (items 7 and 8) and if all $\mathcal{CF}2$ extensions are admissible they coincide with the preferred extensions (item 9).

On the basis of these results, it is shown in [Baroni and Giacomin, 2008] that any agreement class $\mathcal{AGR}(\mathbb{S})$ where $\mathbb{S}$ includes either $\mathcal{GR}$ or $\mathcal{ID}$ coincides with one of the classes $\Sigma_1, \ldots, \Sigma_8$. Note in particular that Σ_1, Σ_2 and Σ_3 are the only classes where $\mathcal{CO}$ appears: this is due to the fact that $GE(AF) \in \mathcal{E}_{\mathcal{CO}}(AF)$ which implies that agreement between complete semantics and other semantics can only occur when the set of extensions consists of the grounded extension only. For this reason $\mathcal{CO}$ will not need to be considered any more in next subsection concerning multiple-status agreement.

5.1.2 Multiple-status Agreement

The complete Venn diagram concerning all agreement classes is shown in Figure 23, where the rectangle with bold lines including all the others represents the universe of all finite argumentation frameworks and again rectangular boxes represent classes of agreement and small ellipses represent single argumentation frameworks to be used as specific examples. As in the previous subsection, the diagram will be illustrated by examining first the set-theoretical relationships between the agreement classes depicted in Figure 23 and then stating that no other meaningful classes exist.

The first step will encompass the same three points as in the previous subsection. In particular, as to point 3, it can be noticed from Figure 23 that most intersections between classes correspond to unions of previously identified classes and/or differences between classes and can be easily determined by considering the sets of semantics involved. For this reason, only intersections requiring specific explanations (in particular all those concerning Σ_8) will be explicitly discussed. Our analysis will now concern agreement classes involving only multiple-status semantics except $\mathcal{CO}$.

The smallest one includes all of them: $\Sigma_9 \triangleq \mathcal{AGR}(\{\mathcal{PR}, \mathcal{CF}2, \mathcal{ST}, \mathcal{SST}\})$. $(\Sigma_9 \setminus \Sigma_4) \neq \emptyset$ as it includes $AF_9 = \langle \{a,b,c,d\}, \{(a,b),(b,a),(a,c),(b,c),(c,d)\}\rangle$. Note that $AF_9 \in ((\Sigma_9 \setminus \Sigma_4) \cap \Sigma_8)$. Also $\Sigma_9 \setminus (\Sigma_4 \cup \Sigma_8)$ is not empty since it includes, for example, $AF_{9'} = \langle \{a,b,c,d\}, \{(a,b),(b,a),(a,c),(b,c),(c,d),(d,c)\}\rangle$.

$\Sigma_{10} \triangleq \mathcal{AGR}(\{\mathcal{PR}, \mathcal{CF}2, \mathcal{SST}\})$ covers the case where stable extensions do not exist, while all other multiple-status semantics agree. $\Sigma_{10} \setminus (\Sigma_9 \cup \Sigma_5) \neq \emptyset$ since it includes, for instance, $AF_{10} = \langle \{a,b,c\}, \{(a,b),(b,a),(c,c)\}\rangle$. Note that $AF_{10} \in ((\Sigma_{10} \setminus (\Sigma_9 \cup \Sigma_5)) \cap \Sigma_8)$. Also $(\Sigma_{10} \setminus (\Sigma_9 \cup \Sigma_5 \cup \Sigma_8))$ is not empty since it includes, for example, $AF_{10'} = AF_{10} \uplus AF_4$, where, given two argumentation frameworks $AF_1 = (Ar_1, att_1)$, $AF_2 = (Ar_2, att_2)$ such that $Ar_1 \cap Ar_2 = \emptyset$, we define[24] $AF_1 \uplus AF_2 \triangleq (Ar_1 \cup Ar_2, att_1 \cup att_2)$.

---

[24]While we have used the same labels $a, b, \ldots$ to denote arguments of our sample argumentation frameworks, we implicitly assume that arguments with the same label in different

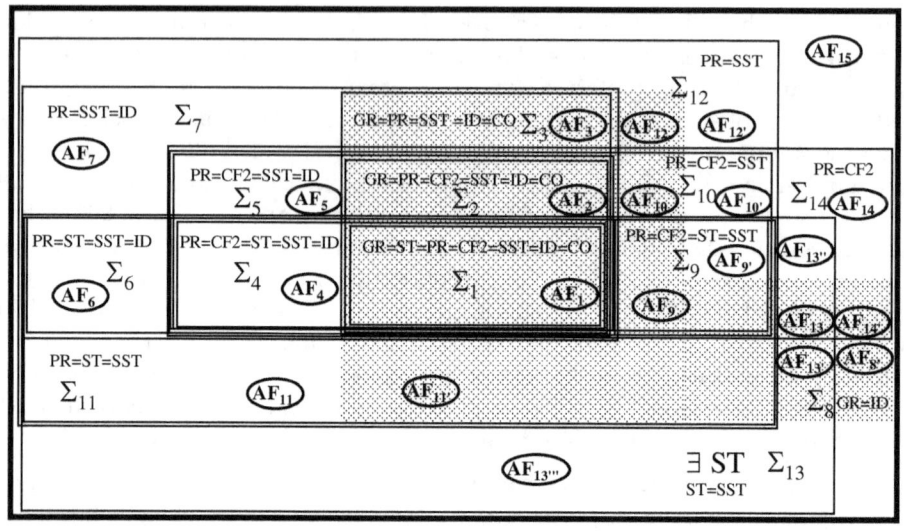

Figure 23. Venn diagram of agreement classes.

$\Sigma_{11} \triangleq \mathcal{AGR}(\{\mathcal{PR}, \mathcal{ST}, \mathcal{SST}\})$ coincides with the class of *coherent* argumentation frameworks considered in [Dung, 1995]. $(\Sigma_{11} \setminus (\Sigma_6 \cup \Sigma_9)) \neq \emptyset$ since it includes $AF_{11} = \langle \{a, b, c, d, e\}, \{(a, b), (a, c), (b, c), (c, a), (d, e), (e, d)\} \rangle$. It can be noted that $AF_{11} \notin \Sigma_8$. Also $(\Sigma_{11} \setminus (\Sigma_6 \cup \Sigma_9)) \cap \Sigma_8 \neq \emptyset$ since it includes $AF_{11'} = \langle \{a, b, c, d\}, \{(a, b), (a, c), (a, d), (b, c), (b, d), (c, a), (d, a), (d, b), (d, c)\} \rangle$.

We are now left with three classes where only a pair of multiple-status semantics are in agreement.

Let us start by considering $\Sigma_{12} \triangleq \mathcal{AGR}(\mathcal{PR}, \mathcal{SST})$. $\Sigma_{12} \setminus (\Sigma_7 \cup \Sigma_{10} \cup \Sigma_{11}) \neq \emptyset$ as it includes $AF_{12} = \langle \{a, b, c, d, e\}, \{(a, b), (b, a), (c, d), (d, e), (e, c)\} \rangle$. It can be noted that $AF_{12} \in \Sigma_8$. For an example of argumentation framework included in $\Sigma_{12} \setminus (\Sigma_7 \cup \Sigma_8 \cup \Sigma_{10} \cup \Sigma_{11})$ consider $AF_{12'} = AF_{12} \uplus AF_4$.

Finally, as already remarked, $\Sigma_{13} \triangleq \mathcal{AGR}(\mathcal{ST}, \mathcal{SST})$ coincides with the class $\mathcal{DE}_{\mathcal{ST}}$ of argumentation frameworks where stable extensions exist, while the last pair to be considered corresponds to $\Sigma_{14} \triangleq \mathcal{AGR}(\mathcal{PR}, \mathcal{CF}2)$. The part of the diagram still to be illustrated involves argumentation frameworks outside Σ_{12} and requires an articulated treatment, since the intersections $\Sigma_{13} \cap \Sigma_{14}$, $\Sigma_{13} \cap \Sigma_8$, and $\Sigma_{14} \cap \Sigma_8$ do not allow a simple characterisation in terms of the other identified classes. First the set difference $\Sigma_{13} \setminus \Sigma_{12}$ can be partitioned

---

argumentation frameworks are actually distinct. This assumption allows us to apply the combination operator $\uplus$ to any pair of sample argumentation frameworks while keeping a simple notation. To make the distinction explicit, argument a of AF_4 should be actually labeled a_4, argument a of AF_{10} should be labeled a_{10}, and so on, but we regard this as an unnecessary notational burden.

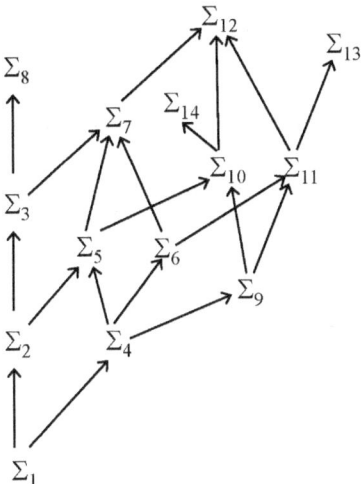

Figure 24. Inclusion relations between agreement classes.

into four non-empty subsets:

- $((\Sigma_{13} \setminus \Sigma_{12}) \cap \Sigma_{14} \cap \Sigma_8) \ni AF_{13} = \langle \{a,b,c\}, \{(a,b),(b,a),(a,c),(c,c)\} \rangle$;

- $((\Sigma_{13} \setminus \Sigma_{12}) \cap (\Sigma_8 \setminus \Sigma_{14})) \ni AF_{13'} = AF_{13} \uplus AF_{11'}$;

- $((\Sigma_{13} \setminus \Sigma_{12}) \cap (\Sigma_{14} \setminus \Sigma_8)) \ni AF_{13''} = AF_{13} \uplus AF_4$;

- $(\Sigma_{13} \setminus (\Sigma_{12} \cup \Sigma_{14} \cup \Sigma_8)) \ni AF_{13'''} = AF_{13} \uplus AF_4 \uplus AF_{11'}$.

Then, $\Sigma_{14} \setminus (\Sigma_{12} \cup \Sigma_{13})$ can be partitioned into two non-empty subsets:

- $(\Sigma_{14} \setminus (\Sigma_{12} \cup \Sigma_{13} \cup \Sigma_8)) \ni AF_{14} = AF_{13} \uplus AF_4 \uplus AF_2$;

- $((\Sigma_{14} \setminus (\Sigma_{12} \cup \Sigma_{13})) \cap \Sigma_8) \ni AF_{14'} = AF_{13} \uplus AF_2$.

Only the characterisation of Σ_8 remains to be completed, in fact $\Sigma_8 \setminus (\Sigma_{12} \cup \Sigma_{13} \cup \Sigma_{14})$ is not empty since it includes $AF_{8'} = AF_{13} \uplus AF_2 \uplus AF_{11'}$. Finally, argumentation frameworks where all semantics are in mutual disagreement also exist, like $AF_{15} = AF_{13} \uplus AF_2 \uplus AF_4 \uplus AF_{11'}$. The fact that no other agreement classes involving $\mathcal{CF}2$, $\mathcal{ST}$, $\mathcal{SST}$ and $\mathcal{PR}$ are meaningful is proved in [Baroni and Giacomin, 2008] on the basis of item 9 of Proposition 5.2.

The Hasse diagram corresponding to inclusion relationships between the agreement classes described above is shown in Figure 24 (where arrows point from subsets to supersets).

5.2 Topological properties and semantics agreement

Section 5.1 identifies the distinct agreement classes concerning a significant set of semantics and provides examples of argumentation frameworks belonging to them. It does not give any indication on how, given an argumentation framework, to answer the question of which agreement class(es) it belongs to (apart the obvious method of computing the sets of extensions prescribed by the various semantics considered). As a matter of fact, there are some significant relationships between agreement classes and some topological properties of argumentation frameworks: presenting them is the subject of this subsection.

5.2.1 Well-founded frameworks

The issue of single-status agreement has been considered as early as in the seminal paper by [Dung, 1995] where it is shown that a sufficient condition for agreement among grounded, preferred and stable semantics is that the argumentation framework is well-founded.

Definition 5.3 (Definition 29 of [Dung, 1995]) *An argumentation framework is well-founded iff there exists no infinite sequence $a_0, a_1, \ldots, a_n, \ldots$ of (not necessarily distinct) arguments such that for each i, a_{i+1} attacks a_i.*

In the case of a finite argumentation framework, well-foundedness coincides with acyclicity of the attack relation. In the light of the results presented in Section 5.1 acyclicity turns out to be a sufficient condition for membership in the agreement class Σ_1.

5.2.2 Determined argumentation frameworks

We consider now a more general conditions for agreement with grounded semantics. To this purpose we introduce the notion of *determined* argumentation framework.

Definition 5.4 *An argumentation framework $AF = \langle Ar, att \rangle$ is determined if and only if $\nexists a \in Ar : a \notin GE(AF) \wedge GE(AF) \cap a^- = \emptyset$.*

In words, an argumentation framework AF is determined if and only the grounded extension is also a stable extension. On the basis of the results of Section 5.1 the set of determined argumentation frameworks, denoted as $\mathcal{DET}$, coincides with the agreement class Σ_1.

Well-founded argumentation frameworks are a special case of determined argumentation frameworks: for finite frameworks, the absence of cycles is a sufficient but not necessary topological condition for $AF \in \mathcal{DET}$. A simple example of argumentation framework which is determined without being acyclic is the following: $\langle \{a, b, c\}, \{(a, b), (b, c), (c, b)\} \rangle$.

Actually, as shown in [Baroni and Giacomin, 2007a] the absence of cycles is necessary only in the initial SCCs, and then recursively in the initial SCCs of the restricted argumentation frameworks obtained by taking into account that the nodes corresponding to the initial SCCs are necessarily included in any extension. This observation gives rise to a full topological characterisation of determined argumentation frameworks, i.e. of Σ_1.

Definition 5.5 *An argumentation framework $AF = \langle Ar, att \rangle$ is initial-acyclic if $AF = \langle \emptyset, \emptyset \rangle$ or the following condition holds: $\forall S \in \mathcal{IS}(AF)$ S is monadic and $AF \downarrow_{UP_{AF}((Ar \setminus IN(AF)), IN(AF))}$ is initial-acyclic, where*

- *$\mathcal{IS}(AF)$ is the set of strongly connected components of AF which are not attacked by any other strongly connected component;*

- *$IN(AF) \triangleq \bigcup_{S \in \mathcal{IS}(AF)} S$;*

- *an argumentation framework is monadic iff it consists of a single non-self-defeating argument;*

- *for any $Args, S \subseteq Ar$, $UP_{AF}(S, Args) = \{a \in S \mid \nexists b \in Args \setminus S : (b, a) \in att\}$ (as in Definition 3.61).*

The base of this recursive definition is represented by the empty argumentation framework. The recursion is well-founded as the set $IN(AF)$ is non-empty for a non-empty argumentation framework, which means that at each recursive step an argumentation framework with a strictly lesser number of nodes is considered. The set of initial-acyclic argumentation frameworks is denoted by $\mathcal{IAA}$ and it is proved in [Baroni and Giacomin, 2007a] that $\mathcal{IAA} = \mathcal{DET}$.

Proposition 5.6 *For any argumentation framework $AF = \langle Ar, att \rangle$, $AF \in \mathcal{IAA}$ if and only if $AF \in \mathcal{DET}$.*

5.2.3 Almost determined argumentation frameworks

While determined argumentation frameworks ensure agreement among all the semantics belonging to the set Ω, it can be observed that there is a larger class of argumentation frameworks where an almost total agreement is reached. Consider for instance the case of an argumentation framework consisting just of a self-defeating argument, namely $AF = \langle \{a\}, \{(a,a)\} \rangle$. In this case we have that $\mathcal{E}_{\mathcal{GR}}(AF) = \{\emptyset\}$ and, in virtue of the conflict-free property, for any semantics σ which admits extensions on AF it must also hold that $\mathcal{E}_\sigma(AF) = \{\emptyset\}$. However, since stable semantics is unable to prescribe extensions in this case, $\mathcal{E}_{ST}(AF) = \emptyset \neq \{\emptyset\}$. In this case, disagreement arises from the non-existence of stable extensions rather than from the existence of extensions different from $GE(AF)$ and clearly AF belongs to Σ_2.

On the basis of this observation, it is useful to consider a further class of argumentation frameworks, called *almost determined*.

Definition 5.7 *An argumentation framework $AF = \langle Ar, att \rangle$ is almost determined if and only if for any $a \in Ar$, $(a \notin GE(AF) \wedge GE(AF) \cap a^- = \emptyset) \Rightarrow (a, a) \in att$.*

In words, an argumentation framework is almost determined if all the nodes which are not attacked nor included in the grounded extension are self-defeating. The set of almost determined argumentation frameworks is denoted as $\mathcal{AD}$. It is proved in [Baroni and Giacomin, 2007a] that frameworks in $\mathcal{AD}$ ensure agreement for all multiple status semantics in Ω but stable semantics (and actually for every SCC-recursive semantics satisfying some basic properties) and that outside $\mathcal{AD}$ there cannot be agreement with $\mathcal{CF}2$ semantics.

Proposition 5.8 *For any argumentation framework $AF = \langle Ar, att \rangle \in \mathcal{AD}$ it holds that $\mathcal{E}_\sigma(AF) = \{GE(AF)\}$ for $\sigma \in \{\mathcal{CO}, \mathcal{PR}, \mathcal{SST}, \mathcal{CF}2\}$.*

Proposition 5.9 *For any argumentation framework $AF = \langle Ar, att \rangle \notin \mathcal{AD}$ $\mathcal{E}_{\mathcal{CF}2}(AF) \neq \{GE(AF)\}$.*

The propositions above shows that $\mathcal{AD} = \Sigma_2$.

5.2.4 Limited controversial frameworks

Sections 5.2.1 - 5.2.3 deal with cases of single status agreement, turning now to multiple status agreement, a first basic result, concerning preferred and stable semantics, was introduced, again, in Dung's seminal paper with reference to the notion of *limited controversial* frameworks, based on the one of *controversial arguments*.

Definition 5.10 *Given an argumentation framework $AF = \langle Ar, att \rangle$ an argument a indirectly attacks an argument b iff there exists a finite sequence $a_0, \ldots a_{2n+1}$ such that $a = a_0$, $b = a_{2n+1}$, and $\forall i \in \{0, \ldots, 2n\}$ $a_i \in a_{i+1}^-$. An argument a indirectly defends an argument b iff there exists a finite sequence $a_0, \ldots a_{2n}$ such that $a = a_0$, $b = a_{2n}$, and $\forall i \in \{0, \ldots, 2n-1\}$ $a_i \in a_{i+1}^-$. An argument a is controversial with respect to an argument b if a indirectly attacks and indirectly defends b. An argument a is controversial if it is controversial with respect to an argument b.*

Definition 5.11 *An argumentation framework $AF = \langle Ar, att \rangle$ is limited controversial if and only if there is no infinite sequence of arguments $a_0, \ldots, a_n, \ldots$ such that a_{i+1} is controversial with respect to a_i.*

In the finite case, an argumentation framework is limited controversial if it does not include any odd-length cycle. In [Dung, 1995] it is shown that being limited controversial is a sufficient condition for being coherent, i.e. for membership in the agreement class Σ_{11}.

5.2.5 Agreement with stable semantics

The basic result on multiple status agreement recalled in Section 5.2.4 has been extended by [Baroni and Giacomin, 2007a] by characterizing a family of argumentation frameworks, called SCC-symmetric, where agreement is ensured for a class of multiple-status semantics including stable, preferred and $\mathcal{CF}2$ semantics.

First we need to introduce the notion of symmetric argumentation frameworks, noting also that symmetry is preserved in all the restrictions of a symmetric framework.

Definition 5.12 *An argumentation framework $AF = \langle Ar, att \rangle$ is symmetric if for any $a, b \in Ar$, $a \in b^- \Leftrightarrow b \in a^-$.*

Lemma 5.13 *Given a symmetric argumentation framework $AF = \langle Ar, att \rangle$ and a set $S \subseteq Ar$, $AF\downarrow_S$ is symmetric.*

As it will be more evident from Proposition 5.15, it is quite natural that extensions of a symmetric argumentation framework free of self-defeating arguments coincide with its maximal conflict free sets, if the multiple-status approach is adopted. Argumentation semantics satisfying this requirement will be called *-symmetric.

Definition 5.14 *An argumentation semantics σ is *-symmetric if for any argumentation framework AF which is symmetric and free of self-defeating arguments $\mathcal{E}_\sigma(AF) = \mathcal{E}_{\mathcal{NA}}(AF)$.*

Several significant multiple-status semantics, though their definition is based on quite different principles, share the property of being *-symmetric.

Proposition 5.15 *Stable semantics, preferred semantics, semi-stable semantics, and $\mathcal{CF}2$ semantics are *-symmetric.*

In symmetric argumentation frameworks non-mutual attacks cannot exist: this seriously limits their applicability for modeling practical situations. Their properties however provide the basis for analyzing a more interesting family of argumentation frameworks called *SCC-symmetric*.

Definition 5.16 *An argumentation framework AF is SCC-symmetric if $\forall S \in \text{SCCS}_{AF}$ $AF\downarrow_S$ is symmetric.*

Definition 5.16 is equivalent to impose that all attacks participating in an attack cycle are mutual, while non-mutual attacks are allowed outside cycles.

Proposition 5.17 *An argumentation framework AF is SCC-symmetric if and only if for every attack cycle, i.e. for every sequence $a_0, a_1, \ldots, a_n$ such that $a_0 = a_n$ and $\forall i \in \{0, \ldots, n-1\}$ $a_i \in a_{i+1}^-$, it holds that $\forall i \in \{1, \ldots, n\}$ $a_i \in a_{i-1}^-$.*

Theorem 5.18 provides the main result about agreement in SCC-symmetric argumentation frameworks.

Theorem 5.18 *In any argumentation framework which is SCC-symmetric and free of self-defeating arguments all SCC-recursive *-symmetric semantics are in agreement, i.e. they prescribe the same set of extensions.*

The following result immediately follows from the previous theorem and Proposition 5.15.

Corollary 5.19 *If an argumentation framework AF is SCC-symmetric and free of self-defeating arguments then $\mathcal{E}_{\mathcal{PR}}(AF) = \mathcal{E}_{CF2}(AF) = \mathcal{E}_{\mathcal{ST}}(AF) = \mathcal{E}_{\mathcal{SST}}(AF)$, thus in particular AF is coherent.*

Summing up, being SCC-symmetric is a sufficient condition for membership in the agreement class Σ_9.

It may be noted that the classes of SCC-symmetric and limited controversial argumentation frameworks are non-disjoint but distinct. In fact, an SCC-symmetric argumentation framework may contain cycles of any length, while a limited controversial argumentation framework may consist, for instance, of an even-length cycle which is not symmetric.

It is interesting to note that the property of SCC-symmetry may be recovered from assumptions on the underlying method of argument and attack construction, which have been considered in the literature on structured argumentation and are not directly related to decomposition into SCCs. For instance in [Baroni et al., 2005a] the case is considered where conflicts among arguments arise only from contradicting conclusions, namely only the *rebutting* kind of defeat is allowed while *undercutting* defeat is not (we follow here the terminology of [Pollock, 1992]). Briefly, rebutting defeats concern arguments with contradictory conclusions, while undercutting defeats concern questioning the applicability of the rule used to build an argument rather than its conclusion (see for instance [Modgil and Prakken, 2014]). This distinction is relevant because undercutting defeats can give rise to arbitrary attack cycles, while rebutting defeats cannot, since they are non mutual only if some preference ordering between arguments is in place and this ordering prevents the existence of attack cycles which are not symmetric. Formally, it is shown in Proposition 26 of [Baroni et al., 2005a] that if only rebutting defeats are allowed, the corresponding argumentation framework is SCC-symmetric (such a framework is called *r-type* in [Baroni et al., 2005a]). From another perspective, in [Kaci et al., 2006] it is shown that when the attack relation results from a symmetric conflict relation and a transitive preference relation between arguments the argumentation framework satisfies a property called *strict acyclicity*, which is actually equivalent to SCC-symmetry through the characterisation given in Proposition 5.17.

6 Conclusions

Starting from Dung's seminal paper [Dung, 1995] abstract argumentation has received a growing interest by the research community, witnessed by a large corpus of scientific literature where an increasing variety of alternative semantics proposals is complemented by studies on general principles and properties for their assessment and comparison. The current book chapter, which revises and updates a previous tutorial paper [Baroni et al., 2011a] is meant to provide a reasonably complete and up-to-date introductory survey on these aspects. In particular it provides a side-by-side treatment of the extension-based and labelling-based approaches and a coverage of skepticism-based semantics comparison and semantics agreement. While this can be sufficient to get a first impression of the subject, there are many other core aspects that are relevant for obtaining a comprehensive view, ranging from postulates and principles for argumentation semantics to notions of equivalence and locality and modularity issues: these additional aspects are covered in Part E of this volume.

Furthermore, some of the relevant aspects have only been briefly mentioned in the current chapter, for instance the procedures for computing labellings and extensions or for checking other semantics-related properties. The relevant theoretical and implementation issues are extensively treated in part D of this volume.

Many lines of further development have sprung up from Dung's core theory and are the subject of active investigation.

First, the basic model can be enriched considering other kinds of relationships between arguments in addition to attacks, like, for instance, support considered in bipolar argumentation frameworks [Cayrol and Lagasquie-Schiex, 2013; Cohen et al., 2014]. Abstract Dialectical Frameworks, presented in chapter 5, are a generalized graph-based abstract formalism able to capture different kinds of interactions between the graph nodes. Other extensions of the model include considering attacks to attacks [Modgil, 2009; Baroni et al., 2011b] and taking into account values and audiences in argument evaluation as in value-based argumentation frameworks [Bench-Capon, 2003].

The survey we presented in this chapter does not exhaustively treat the whole range of argumentation semantics in the literature. In particular, it is worth mentioning the notion of parametric semantics, namely semantics definition schemes which are generic with respect to the choice of an argumentation semantics, playing the role of a parameter in the context of a scheme.

Resolution-based semantics [Baroni et al., 2011c] is an example of parametric semantics. Here the idea is that, given an argumentation framework AF, a set of argumentation frameworks $\mathcal{RES}(AF)$ is generated, where each member of $\mathcal{RES}(AF)$ corresponds to a resolution of all the mutual attacks in AF. Then the semantics σ adopted as parameter is applied to the elements of $\mathcal{RES}(AF)$ and among the resulting extensions those which are minimal with respect to set inclusion are selected as extensions of the resolution-based semantics for σ. The reader is referred to [Baroni et al., 2011c] for all details, we only recall here that the instance of the resolution-based definition scheme based on grounded semantics shows unparalleled features in terms of principle-based evaluation, while also enjoying good computational complexity properties among multiple-status semantics [Baroni et al., 2009].

Another example of parametric semantics concerns the generalisation of the notion of ideal extension [Dunne et al., 2013]: basically this amounts to make Definitions 3.45 and 3.48 parametric by replacing the reference to preferred semantics with the reference to a generic semantics. This generalisation turns out to be suitably applicable to value-based argumentation frameworks.

While traditional abstract argumentation semantics produces qualitative assessments, a variety of quantitative approaches have been considered. Their study is quickly evolving, ranging from weighted argumentation frameworks [Dunne et al., 2011] to various flavors of probabilistic argumentation frameworks [Hunter, 2013; Hunter, 2014] to the equational approach to argumentation semantics [Gabbay, 2012], just to mention a few. The coverage of this kind of extensions is planned for the second volume of this handbook.

To conclude, we note that while the ongoing developments listed above promise to overcome some of the restrictive assumptions embedded in the original theory of abstract argumentation frameworks, they also witness the extraordinary interest and fertility of this formalism and ensure that it will continue to represent an active research subject for many years to come.

Acknowledgments

The authors are grateful to the reviewers for their careful analysis of the chapter and their insightful suggestions.

BIBLIOGRAPHY

[Alferes et al., 1993] José Júlio Alferes, Phan Minh Dung, and Luís Pereira. Scenario semantics of extended logic programs. In A. Nerode and L. Pereira, editors, *Proc. of the 2nd Int. Workshop on Logic Programming and Non-monotonic Reasoning (LPNMR 93)*, pages 334–348. MIT Press, 1993.

[Amgoud, 2009] Leila Amgoud. Argumentation for decision making. In I. Rahwan and G.R. Simari, editors, *Argumentation in Artificial Intelligence*, pages 301–320. Springer, 2009.

[Baroni and Giacomin, 2003] Pietro Baroni and Massimiliano Giacomin. Solving semantic problems with odd-length cycles in argumentation. In *Proc. of the 7th European Conf. on Symbolic and Quantitative Approaches to Reasoning with Uncertainty (ECSQARU 03)*, number 2711 in Springer Lecture Notes in AI, pages 440–451, Berlin, 2003. Springer Verlag.

[Baroni and Giacomin, 2007a] Pietro Baroni and Massimiliano Giacomin. Characterizing defeat graphs where argumentation semantics agree. In *Proc. of 1st Int. Workshop on Argumentation and Non-Monotonic Reasoning (ARGNMR07)*, pages 33–48, 2007.

[Baroni and Giacomin, 2007b] Pietro Baroni and Massimiliano Giacomin. On principle-based evaluation of extension-based argumentation semantics. *Artificial Intelligence*, 171(10-15):675–700, 2007.

[Baroni and Giacomin, 2008] Pietro Baroni and Massimiliano Giacomin. A systematic classification of argumentation frameworks where semantics agree. In *Proc. of the 2nd Int. Conf. on Computational Models of Argument (COMMA 2008)*, pages 37–48, 2008.

[Baroni and Giacomin, 2009a] Pietro Baroni and Massimiliano Giacomin. Semantics of abstract argument systems. In I. Rahwan and G.R. Simari, editors, *Argumentation in Artificial Intelligence*, pages 25–44. Springer, 2009.

[Baroni and Giacomin, 2009b] Pietro Baroni and Massimiliano Giacomin. Skepticism relations for comparing argumentation semantics. *Int. J. of Approximate Reasoning*, 50(6):854–866, 2009.

[Baroni et al., 2005a] Pietro Baroni, Massimiliano Giacomin, and Giovanni Guida. Self-stabilizing defeat status computation: dealing with conflict management in multi-agent systems. *Artificial Intelligence*, 165(2):187–259, 2005.

[Baroni et al., 2005b] Pietro Baroni, Massimiliano Giacomin, and Giovanni Guida. SCC-recursiveness: a general schema for argumentation semantics. *Artificial Intelligence*, 168(1-2):165–210, 2005.

[Baroni et al., 2009] Pietro Baroni, Paul E. Dunne, and Massimiliano Giacomin. Computational properties of resolution-based grounded semantics. In *Proc. of the 21st International Joint Conference on Artificial Intelligence (IJCAI 2009)*, pages 683–689, 2009.

[Baroni et al., 2011a] Pietro Baroni, Martin W. A. Caminada, and Massimiliano Giacomin. An introduction to argumentation semantics. *Knowledge Engineering Review*, 26(4):365–410, 2011.

[Baroni et al., 2011b] Pietro Baroni, Federico Cerutti, Massimiliano Giacomin, and Giovanni Guida. AFRA: argumentation framework with recursive attacks. *Int. J. of Approximate Reasoning*, 52(1):19–37, 2011.

[Baroni et al., 2011c] Pietro Baroni, Paul E. Dunne, and Massimiliano Giacomin. On the resolution-based family of abstract argumentation semantics and its grounded instance. *Artificial Intelligence*, 175(3-4):791–813, 2011.

[Baumann et al., 2016] Ringo Baumann, Thomas Linsbichler, and Stefan Woltran. Verifiability of argumentation semantics. In P. Baroni, T. F. Gordon, T. Scheffler, and M. Stede, editors, *Proc. of the 6th Int. Conf. on Computational Models of Argument (COMMA 2016)*, pages 83–94. IOS Press, 2016.

[Bench-Capon, 2003] Trevor J.M. Bench-Capon. Persuasion in practical argument using value-based argumentation frameworks. *J. of Logic and Computation*, 13(3):429–448, 2003.

[Caminada and Gabbay, 2009] Martin W. A. Caminada and Dov M. Gabbay. A logical account of formal argumentation. *Studia Logica*, 93(2-3):109–145, 2009. Special issue: new ideas in argumentation theory.

[Caminada and Pigozzi, 2011] Martin W. A. Caminada and Gabriella Pigozzi. On judgment aggregation in abstract argumentation. *Autonomous Agents and Multi-Agent Systems*, 22(1):64–102, 2011.

[Caminada and Wu, 2009] Martin W. A. Caminada and Yining Wu. An argument game of stable semantics. *Logic Journal of IGPL*, 17(1):77–90, 2009.

[Caminada and Wu, 2010] Martin W. A. Caminada and Yining Wu. On the justification status of arguments. In *Proc. of the 22nd Benelux Conference on Artificial Intelligence*, 2010.

[Caminada et al., 2012] Martin W. A. Caminada, Walter A. Carnielli, and Paul E. Dunne. Semi-stable semantics. *J. of Logic and Computation*, 22(5):1207–1254, 2012.

[Caminada et al., 2015] Martin W. A. Caminada, Samy Sá, João Alcântara, and Wolfgang Dvořák. On the equivalence between logic programming semantics and argumentation semantics. *Int. J. of Approximate Reasoning*, 58:87–111, 2015.

[Caminada et al., 2016] Martin W. A. Caminada, Wolfgang Dvořák, and Srdjan Vesic. Preferred semantics as socratic discussion. *J. of Logic and Computation*, 26(4):1257–1292, 2016.

[Caminada, 2006a] Martin W. A. Caminada. On the issue of reinstatement in argumentation. In M. Fischer, W. van der Hoek, B. Konev, and A. Lisitsa, editors, *Proc. of the 10th European Conf. on Logics in Artificial Intelligence (JELIA 2006)*, number 4160 in Lecture Notes in Computer Science, pages 111–123. Springer, 2006.

[Caminada, 2006b] Martin W. A. Caminada. Semi-stable semantics. In P. E. Dunne and T.J.M. Bench-Capon, editors, *Proc. of the 1st Int. Conf. on Computational Models of Argument (COMMA 2006)*, pages 121–130. IOS Press, 2006.

[Caminada, 2007a] Martin W. A. Caminada. An algorithm for computing semi-stable semantics. In *Proc. of the 9th European Conference on Symbolic and Quantitative Approaches to Reasoning with Uncertainty (ECSQARU 2007)*, number 4724 in Springer Lecture Notes in AI, pages 222–234, Berlin, 2007. Springer Verlag.

[Caminada, 2007b] Martin W. A. Caminada. Comparing two unique extension semantics for formal argumentation: ideal and eager. In M. M. Dastani and E. de Jong, editors, *Proc. of the 19th Belgian-Dutch Conference on Artificial Intelligence (BNAIC 2007)*, pages 81–87, 2007.

[Caminada, 2010] Martin W. A. Caminada. Preferred semantics as socratic discussion. In A. E. Gerevini and A. Saetti, editors, *Proceedings of the 11th AI*IA Symposium on Artificial Intelligence*, pages 209–216, 2010.

[Caminada, 2011] M. W. A. Caminada. A labelling approach for ideal and stage semantics. *Argument and Computation*, 2(1):1–21, 2011.

[Caminada, 2014] Martin W. A. Caminada. Strong admissibility revisited. In S. Parsons, N. Oren, C. Reed, and F. Cerutti, editors, *Proc. of the 5th Int. Conf. on Computational Models of Argument (COMMA 2014)*, pages 197–208. IOS Press, 2014.

[Caminada, 2015] Martin W. A. Caminada. A discussion game for grounded semantics. In E. Black, S. Modgil, and N. Oren, editors, *Proc. of the 3rd Int. Workshop on Theory and Applications of Formal Argumentation (TAFA 2015)*, volume 9524 of *Lecture Notes in Computer Science*, pages 59–73. Springer, 2015.

[Cayrol and Lagasquie-Schiex, 2013] Claudette Cayrol and Marie-Christine Lagasquie-Schiex. Bipolarity in argumentation graphs: Towards a better understanding. *Int. J. of Approximate Reasoning*, 54(7):876–899, 2013.

[Cayrol et al., 2003] Claudette Cayrol, Sylvie Doutre, and Jerome Mengin. On decision problems related to the preferred semantics for argumentation frameworks. *J. of Logic and Computation*, 13(3):377–403, 2003.

[Cohen et al., 2014] Andrea Cohen, Sebastian Gottifredi, Alejandro J. García, and Guillermo R. Simari. A survey of different approaches to support in argumentation systems. *Knowledge Engineering Review*, 29(5):513–550, 2014.

[Dung et al.; 2007] Phan Minh Dung, Paolo Mancarella, and Francesca Toni. Computing ideal sceptical argumentation. *Artificial Intelligence*, 171(10-15):642–674, 2007.

[Dung, 1995] Phan Minh Dung. On the acceptability of arguments and its fundamental role in nonmonotonic reasoning, logic programming and n-person games. *Artificial Intelligence*, 77:321–357, 1995.

[Dunne et al., 2011] Paul E. Dunne, Anthony Hunter, Peter McBurney, Simon Parsons, and Michael Wooldridge. Weighted argument systems: Basic definitions, algorithms, and complexity results. *Artificial Intelligence*, 175(2):457–486, 2011.

[Dunne et al., 2013] Paul E. Dunne, Wolfgang Dvořák, and Stefan Woltran. Parametric properties of ideal semantics. *Artificial Intelligence*, 202:1–28, 2013.

[Dvořák and Gaggl, 2012a] Wolfgang Dvořák and Sarah A. Gaggl. Computational aspects of cf2 and stage2 argumentation semantics. In *Proc. of the 4th Int. Conf. on Computational Models of Argument (COMMA 2012)*, pages 273–284, 2012.

[Dvořák and Gaggl, 2012b] Wolfgang Dvořák and Sarah A. Gaggl. Incorporating stage semantics in the SCC-recursive schema for argumentation semantics. In *Proc. of the 14th Int. Workshop on Non-Monotonic Reasoning (NMR 2012)*, pages 273–284, 2012.

[Dvořák and Gaggl, 2016] Wolfgang Dvořák and Sarah A. Gaggl. Stage semantics and the SCC-recursive schema for argumentation semantics. *J. of Logic and Computation*, 26(4):1149–1202, 2016.

[Dvořák, 2012] Wolfgang Dvořák. On the complexity of computing the justification status of an argument. In S. Modgil, N. Oren, and F. Toni, editors, *Proc. of the 1st Int. Workshop on Theory and Applications of Formal Argumentation (TAFA 2011)*, number 7132 in Lecture Notes in Computer Science, pages 32–49. Springer, 2012.

[Gabbay, 2012] Dov M. Gabbay. Equational approach to argumentation networks. *Argument & Computation*, 3(2-3):87–142, 2012.

[García and Simari, 2004] Alejandro J. García and Guillermo R. Simari. Defeasible logic programming: an argumentative approach. *Theory and Practice of Logic Programming*, 4(1):95–138, 2004.

[Gelfond and Lifschitz, 1988] Michael Gelfond and Vladimir Lifschitz. The stable model semantics for logic programming. In R.A. Kowalski and K. Bowen, editors, *Proceedings of the 5th Int. Conference/Symposium on Logic Programming*, pages 1070–1080. MIT Press, 1988.

[Gelfond and Lifschitz, 1991] Michael Gelfond and Vladimir Lifschitz. Classical negation in logic programs and disjunctive databases. *New Generation Computing*, 9(3/4):365–385, 1991.

[Gorogiannis and Hunter, 2011] Nikos Gorogiannis and Anthony Hunter. Instantiating abstract argumentation with classical logic arguments: Postulates and properties. *Artificial Intelligence*, 175(9-10):1479–1497, 2011.

[Governatori et al., 2004] Guido Governatori, Michael J. Maher, Grigoris Antoniou, and David Billington. Argumentation semantics for defeasible logic. *J. of Logic and Computation*, 14(5):675–702, 2004.

[Hunter, 2013] Anthony Hunter. A probabilistic approach to modelling uncertain logical arguments. *Int. J. of Approximate Reasoning*, 54(1):47–81, 2013.

[Hunter, 2014] Anthony Hunter. Probabilistic qualification of attack in abstract argumentation. *Int. J. of Approximate Reasoning*, 55(2):607–638, 2014.
[Jakobovits and Vermeir, 1999] Hadassa Jakobovits and Dirk Vermeir. Robust semantics for argumentation frameworks. *J. of Logic and Computation*, 9(2):215–261, 1999.
[Kaci et al., 2006] Souhila Kaci, Leendert W. N. van der Torre, and Emil Weydert. Acyclic argumentation: Attack = conflict + preference. In *Proc. of the 17th European Conference on Artificial Intelligence (ECAI 2006)*, pages 725–726, Riva del Garda, Italy, 2006. IOS Press.
[Modgil and Caminada, 2009] Sanjay Modgil and Martin W. A. Caminada. Proof theories and algorithms for abstract argumentation frameworks. In I. Rahwan and G.R. Simari, editors, *Argumentation in Artificial Intelligence*, pages 105–129. Springer, 2009.
[Modgil and Prakken, 2014] Sanjay Modgil and Henry Prakken. The ASPIC+ framework for structured argumentation: a tutorial. *Argument & Computation*, 5:31–62, 2014. Special Issue: Tutorials on Structured Argumentation.
[Modgil, 2009] Sanjay Modgil. Reasoning about preferences in argumentation frameworks. *Artificial Intelligence*, 173:901–1040, 2009.
[Nofal et al., 2014] Samir Nofal, Katie Atkinson, and Paul E. Dunne. Algorithms for decision problems in argumentation systems under preferred semantics. *Artificial Intelligence*, 207:23–51, 2014.
[Pollock, 1992] John L. Pollock. How to reason defeasibly. *Artificial Intelligence*, 57(1):1–42, 1992.
[Pollock, 1995] John L. Pollock. *Cognitive Carpentry. A Blueprint for How to Build a Person*. MIT Press, Cambridge, MA, 1995.
[Pollock, 2001] John L. Pollock. Defeasible reasoning with variable degrees of justification. *Artificial Intelligence*, 133:233–282, 2001.
[Rahwan and Larson, 2008] Iyad Rahwan and Kate Larson. Pareto optimality in abstract argumentation. In D. Fox and C. P. Gomes, editors, *Proc. of 23rd AAAI Conference on Artificial Intelligence (AAAI 2008)*, pages 150–155. AAAI Press, 2008.
[Rahwan and Tohmé, 2010] Iyad Rahwan and Fernando Tohmé. Collective argument evaluation as judgement aggregation. In W. van der Hoek, G. A. Kaminka, Y. Lespérance, M. Luck, and S. Sen, editors, *Proc. of the 9th Int. Conf. on Autonomous Agents and Multiagent Systems (AAMAS 2010)*, pages 417–424, 2010.
[Reiter, 1980] Raymond Reiter. A logic for default reasoning. *Artificial Intelligence*, 13:81–132, 1980.
[Toni, 2014] Francesca Toni. A tutorial on assumption-based argumentation. *Argument & Computation*, 5:89–117, 2014. Special Issue: Tutorials on Structured Argumentation.
[Verheij, 1996] Bart Verheij. Two approaches to dialectical argumentation: admissible sets and argumentation stages. In J.-J.Ch. Meyer and L.C. van der Gaag, editors, *Proc. of the Eighth Dutch Conf. on Artificial Intelligence (NAIC'96)*, pages 357–368, 1996.
[Verheij, 2003] Bart Verheij. DEFLOG: on the logical interpretation of prima facie justified assumptions. *J. of Logic and Computation*, 13:319–346, 2003.
[Verheij, 2007] Bart Verheij. A labeling approach to the computation of credulous acceptance in argumentation. In M. M. Veloso, editor, *Proc. of the 20th Int. Joint Conference on Artificial Intelligence (IJCAI 2007)*, pages 623–628, 2007.
[Vreeswijk and Prakken, 2000] Gerard A. W. Vreeswijk and Henry Prakken. Credulous and sceptical argument games for preferred semantics. In *Proc. of the 7th European Workshop on Logic for Artificial Intelligence (JELIA-00)*, number 1919 in Lecture Notes in Computer Science, pages 239–253, Berlin, 2000. Springer Verlag.
[Vreeswijk, 1993] Gerard A. W. Vreeswijk. Studies in defeasible argumentation. PhD thesis at Free University of Amsterdam, 1993.
[Vreeswijk, 2006] Gerard A. W. Vreeswijk. An algorithm to compute minimally grounded and admissible defence sets in argument systems. In P. E. Dunne and T. J. M. Bench-Capon, editors, *Proc. of the 1st Int. Conf. on Computational Models of Argument (COMMA 2006)*, pages 109–120. IOS, 2006.
[Wu and Caminada, 2010] Yining Wu and Martin W. A. Caminada. A labelling-based justification status of arguments. *Studies in Logic*, 3(4):12–29, 2010.
[Wu et al., 2009] Yining Wu, Martin W. A. Caminada, and Dov M. Gabbay. Complete extensions in argumentation coincide with 3-valued stable models in logic programming. *Studia Logica*, 93(1-2):383–403, 2009. Special issue: new ideas in argumentation theory.

Pietro Baroni
Dip. Ingegneria dell'Informazione
University of Brescia, Italy
Email: pietro.baroni@unibs.it

Martin Caminada
School of Computer Science & Informatics
Cardiff University, UK
Email: CaminadaM@cardiff.ac.uk

Massimiliano Giacomin
Dip. Ingegneria dell'Informazione
University of Brescia, Italy
Email: massimiliano.giacomin@unibs.it

5
Abstract Dialectical Frameworks

GERHARD BREWKA, STEFAN ELLMAUTHALER, HANNES STRASS,
JOHANNES P. WALLNER, STEFAN WOLTRAN

ABSTRACT. This handbook chapter describes abstract dialectical frameworks, or ADFs for short. ADFs are generalizations of the widely used Dung argumentation frameworks. Whereas the latter focus on a single relation among abstract arguments, namely attack, ADFs allow arbitrary relationships among arguments to be expressed. For instance, arguments may support each other, or a group of arguments may jointly attack another one while each single member of the group is not strong enough to do so. This additional expressiveness is achieved by handling acceptance conditions for each argument explicitly.

The semantics of ADFs are inspired by approximation fixpoint theory (AFT), a general algebraic theory for approximation based semantics developed by Denecker, Marek and Truszczyński. We briefly introduce AFT and discuss its role in argumentation. This puts us in a position to formally introduce ADFs and their semantics. In particular, we show how the most important Dung semantics can be generalized to ADFs. Furthermore, we illustrate the use of ADFs as semantical tool in various modelling scenarios, demonstrating how typical representations in argumentation can be equipped with precise semantics via translations to ADFs. We also present GRAPPA, a related approach where the semantics of arbitrary labelled argument graphs can be directly defined in an ADF-like manner, circumventing the need for explicit translations. Finally, we address various computational aspects of ADFs, like complexity, expressiveness and realizability, and present several implemented systems.

1 Introduction

This chapter is about abstract dialectical frameworks, or ADFs for short. ADFs are generalizations of Dung argumentation frameworks (AFs, see Chapter 4 of this Handbook). AFs are very popular tools in argumentation. They abstract away from the content of particular arguments and focus on conflicts among arguments, where each argument is viewed as an atomic item. The only information AFs take into account is whether an argument attacks another one or not. Based on a set of arguments and an attack relation, different AF semantics single out coherent subsets of arguments which "fit" together, according to specific criteria. More formally, an AF semantics takes an argumentation framework as input and produces as output a collection of sets of arguments, called extensions.

AFs are typically not used directly for knowledge representation purposes, but as semantical tools: given a knowledge base KB in some knowledge representation formalism, the set of arguments induced by KB is formally defined and the attack relation on these arguments is identified. This defines an AF that can be evaluated according to a chosen semantics. The KB formulas supported by accepted arguments are then the ones which are accepted. This stepwise evaluation is often referred to as the argumentation process [Caminada and Amgoud, 2007].

Given that AFs are in wide use, a natural question to ask is why a generalization of AFs is useful in the first place. There are at least two possible answers to this question:

- the generalization is more expressive than AFs,
- the generalization allows for easier modelling.

In fact, it turns out that both answers apply to ADFs. We will discuss the issue of expressiveness in detail in Section 6.2. For the time being let us focus on the modelling issue. AFs restrict their attention to the attack relation, and the basic intuition is the following: assume an argument b is attacked by argument c, then whenever c is accepted b is defeated. But how about more fine-grained – or entirely different – relations which could be of potential interest? What if c alone is not strong enough and a second argument, say d, is needed to jointly defeat b? And, maybe even more importantly, aren't there situations where accepting an argument can be a reason for accepting another one, in other words, where arguments are in support rather than in attack relation? We do not claim here that examples like the ones just discussed cannot be modelled at all with AFs. However, additional nodes in the AF argument graph will be needed which have the sole purpose of modelling other relations indirectly, via attack. These nodes will often be entirely unrelated to the original knowledge base and thus meaningless from the perspective of the application.

Indeed, for these reasons many authors have felt the need to extend the functionality available in AFs in one way or another. Examples of extensions described in the literature are preference or value-based AFs [Simari and Loui, 1992; Amgoud and Cayrol, 2002; Amgoud and Vesic, 2011; Bench-Capon, 2003], AFs with support relations [Cayrol and Lagasquie-Schiex, 2013; Oren and Norman, 2008; Polberg and Oren, 2014], necessities [Nouioua, 2013], set attacks [Nielsen and Parsons, 2007], attacks on attacks [Modgil, 2009], recursive attacks [Baroni et al., 2011] and AFs with weights [Martínez et al., 2008; Dunne et al., 2011; Coste-Marquis et al., 2012] or probabilities [Hunter, 2013; Thimm, 2012]. We refer the reader to [Brewka et al., 2014] for an overview of such extensions.

In a nutshell, ADFs are an attempt to unify several of these different approaches and to generalize AFs in a principled, systematic way. The basic idea is very simple. Consider again the conditions under which an argument, say b, with attackers c and d is accepted in an AF: b is accepted iff c is not accepted

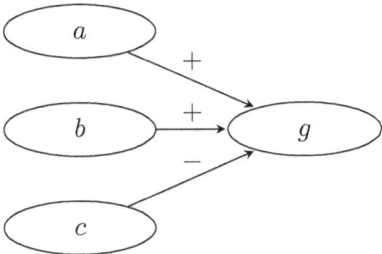

Figure 1: An argument with two supporters and one attacker.

and d is not accepted. This condition can easily be expressed as the propositional formula $\neg c \wedge \neg d$. The acceptance condition for each argument in an AF is obtained in exactly the same way, by constructing the conjunction of the negations of its attackers. Once the implicit acceptance conditions which are at work in AFs are made explicit this way, the generalization ADFs build upon are pretty straightforward: rather than using implicit acceptance conditions of the form we just saw, ADFs use explicit acceptance conditions which can conveniently be expressed as arbitrary propositional formulas.

Let us see how explicit acceptance conditions allow us to handle some of the examples discussed above. We start with joint attack. If b can only be defeated jointly by c and d, then all we have to do is change the acceptance condition accordingly: rather than a conjunction, we have to use the disjunction $\neg c \vee \neg d$ as acceptance condition for b. The effect is that b is only defeated when both c and d are accepted, as intended. As soon as one of them is not accepted, b is no longer defeated.

Support can be handled in a similar manner. Assume g has two supporting arguments a and b, and one attacking argument c, as illustrated in Figure 1. We use $+$ and $-$ to indicate support and attack, respectively.

Note that the information about supporting and attacking links in the graph does not sufficiently specify under what conditions g should be accepted. Let us call a link active if its source node is accepted. There are various options we may want to choose, all of them expressible as a particular acceptance condition for g:

- no negative and all positive links must be active: $\neg c \wedge (a \wedge b)$

- no negative and at least one positive link must be active: $\neg c \wedge (a \vee b)$

- no negative or both positive links must be active: $\neg c \vee (a \wedge b)$

- no negative or at least one positive link must be active: $\neg c \vee (a \vee b)$

- more positive than negative links must be active: $(\neg c \wedge (a \vee b)) \vee (a \wedge b)$

Note how it depends on the acceptance condition whether supporting links are "stronger than" attacking links (meaning that if all incoming links are active, the node is accepted), as in the last three items, or attacking links are "stronger than" supporting links (meaning that if all incoming links are active, the node is rejected), as for the first two items.

We hope these examples are sufficient to illustrate the additional modelling capabilities ADFs provide, and also the simplicity of the basic idea they rest upon. We will see, however, that generalizing the AF semantics to ADFs is far from being simple. This issue will be addressed in Section 3.

In spite of their additional expressiveness, we do not view ADFs primarily as a knowledge representation formalism. We rather consider them as "argumentation middleware", that is, as a framework which is particularly useful for providing semantics to other, maybe more user-friendly formalisms via translations [Brewka et al., 2014]. We will further illustrate this in Section 4.

The rest of this chapter is organized as follows. In Section 2 we recall some relevant background and in particular discuss some relationships between approximation fixpoint theory and AFs which will be useful later. Section 3 introduces ADFs and their semantics formally. The presentation of this section is based on [Brewka et al., 2013]. Section 4 illustrates the role of ADFs in argumentation, showing how they can be used for modelling. Section 5 describes GRAPPA (GRaph-based Argument Processing based on Patterns of Acceptance) along the lines of [Brewka and Woltran, 2014]; GRAPPA is an approach to graph-based argumentation which is closely related to ADFs and their underlying formal techniques. Section 6 discusses subclasses, computational aspects, and expressivity of ADFs. Section 6.1 focuses on an interesting special case of ADFs, so-called bipolar ADFs where each link in the ADF graph is attacking or supporting (or both). This rather expressive class is not only of practical interest, but also has nice computational properties. Expressiveness of ADFs and bipolar ADFs is investigated in Section 6.2, computational complexity in Section 6.3, and recent systems in Section 6.4. Section 7 concludes the chapter.

2 Approximation Fixpoint Theory in Abstract Argumentation

Denecker, Marek and Truszczyński [Denecker et al., 2000] (henceforth shortened to DMT) introduced an algebraic framework for studying semantics of knowledge representation formalisms. In this framework – approximation fixpoint theory (AFT) – knowledge bases are associated with operators (functions) on algebraic structures (for example lattices). The fixpoints of those operators are then studied in order to analyse the semantics of knowledge bases. While this technique is standard to define semantics of programming languages and has indeed been used in early works on logic programming [van Emden and Kowalski, 1976], the major invention of DMT has been the important concept of an approximation of an operator. In the study of semantics of knowledge representation formalisms, elements of lattices represent objects of interest.

Operators transform such objects into others according to the contents of a given knowledge base. Consequently, fixpoints of such operators are then objects that cannot be updated any more – informally speaking, the knowledge base can neither add information to a fixpoint nor remove information from it.

In classical approaches to fixpoint-based semantics, the underlying algebraic structure is the complete lattice of the set $\mathcal{V}_2 = \{v : A \to \{\mathbf{t}, \mathbf{f}\}\}$ of all two-valued interpretations over some vocabulary A ordered by the truth ordering $\leq_t$ with

$$v_1 \leq_t v_2 \text{ if and only if } \forall a \in A : v_1(a) = \mathbf{t} \implies v_2(a) = \mathbf{t}.[1]$$

Consequently, an operator O on this lattice $(\mathcal{V}_2, \leq_t)$ takes as input a two-valued interpretation $v \in \mathcal{V}_2$ and returns a revised interpretation $O(v) \in \mathcal{V}_2$. The intuition of the operator is that the revised interpretation $O(v)$ incorporates additional knowledge that is induced by the knowledge base associated to O from interpretation v. Based on this intuition, fixpoints of O correspond to the models of the knowledge base.

To study fixpoints of operators O, DMT investigate fixpoints of their *approximating operators* $\mathcal{O}$. When O operates on two-valued interpretations $\mathcal{V}_2$, its approximation $\mathcal{O}$ operates on *three-valued* interpretations $\mathcal{V}_3 = \{v : A \to \{\mathbf{t}, \mathbf{f}, \mathbf{u}\}\}$. The three truth values $\mathbf{t}$ (true), $\mathbf{f}$ (false), and $\mathbf{u}$ (undefined) can be ordered by the information ordering $\leq_i$. This ordering intuitively assigns a greater information content to the classical truth values $\{\mathbf{t}, \mathbf{f}\}$ than to undefined $\mathbf{u}$; more formally, we have $\mathbf{u} <_i \mathbf{t}$ and $\mathbf{u} <_i \mathbf{f}$ and $\leq_i$ is the reflexive transitive closure of $<_i$. The partially ordered set $(\{\mathbf{t}, \mathbf{f}, \mathbf{u}\}, \leq_i)$ forms a complete meet-semilattice with the meet operation $\sqcap_i$.[2] This meet can be read as *consensus* and assigns $\mathbf{t} \sqcap_i \mathbf{t} = \mathbf{t}$, $\mathbf{f} \sqcap_i \mathbf{f} = \mathbf{f}$, and returns $\mathbf{u}$ otherwise. The ordering $\leq_i$ can be generalized to three-valued interpretations in a pointwise fashion:

$$v_1 \leq_i v_2 \text{ if and only if } \forall a \in A : v_1(a) \in \{\mathbf{t}, \mathbf{f}\} \implies v_1(a) = v_2(a).[3]$$

Again, the resulting algebraic structure is a complete meet-semilattice; its $\leq_i$-maximal elements are exactly the two-valued interpretations $\mathcal{V}_2$, which form an $\leq_i$-antichain. Intuitively, in that complete meet-semilattice, a single three-valued interpretation

$$v : A \to \{\mathbf{t}, \mathbf{f}, \mathbf{u}\}$$

serves to approximate a set $[v]_2 = \{w \in \mathcal{V}_2 \mid v \leq_i w\}$ of two-valued interpretations. For example, for the vocabulary $A = \{a, b, c\}$, the three-valued interpretation $v = \{a \mapsto \mathbf{t}, b \mapsto \mathbf{u}, c \mapsto \mathbf{f}\}$ approximates the set $\{w_1, w_2\}$ of two-valued interpretations where $w_1 = \{a \mapsto \mathbf{t}, b \mapsto \mathbf{t}, c \mapsto \mathbf{f}\}$ and $w_2 = \{a \mapsto \mathbf{t}, b \mapsto \mathbf{f}, c \mapsto \mathbf{f}\}$.

---

[1] $(\mathcal{V}_2, \leq_t)$ is isomorphic to $(2^A, \subseteq)$ via $v \mapsto v^{-1}(\mathbf{t}) = \{a \in A \mid v(a) = \mathbf{t}\}$.
[2] A complete meet-semilattice is such that every non-empty finite subset has a greatest lower bound, the meet; and every non-empty directed subset has a least upper bound. A subset is directed iff any two of its elements have an upper bound in the set.
[3] $(\mathcal{V}_3, \leq_i)$ is isomorphic to $(\{M \subseteq A \cup \{\neg a \mid a \in A\} \mid a \in M \implies \neg a \notin M\}, \subseteq)$ via the mapping $v \mapsto \{a \in A \mid v(a) = \mathbf{t}\} \cup \{\neg a \mid a \in A, v(a) = \mathbf{f}\}$.

In a similar vein, a three-valued operator $\mathcal{O} : \mathcal{V}_3 \to \mathcal{V}_3$ *approximates* a two-valued operator $O : \mathcal{V}_2 \to \mathcal{V}_2$ if and only if

1. for all $v \in \mathcal{V}_2$, we have $\mathcal{O}(v) = O(v)$ ($\mathcal{O}$ agrees with O on two-valued v), and

2. for all $v_1, v_2 \in \mathcal{V}_3$, $v_1 \leq_i v_2 \implies \mathcal{O}(v_1) \leq_i \mathcal{O}(v_2)$ ($\mathcal{O}$ is $\leq_i$-monotone).

DMT [Denecker et al., 2000] showed that in this case fixpoints of $\mathcal{O}$ approximate fixpoints of O. More specifically, for every fixpoint v_2 of O, there is a fixpoint v_3 of $\mathcal{O}$ such that $v_2 \in [v_3]_2$. Moreover, an approximating operator $\mathcal{O}$ always has a fixpoint, which need not be the case for two-valued operators O. In particular, $\mathcal{O}$ has an $\leq_i$-least fixpoint, which approximates *all* fixpoints of O.

In subsequent work, DMT [Denecker et al., 2004] presented a general, abstract way to define the most precise approximation of a given operator $O : \mathcal{V}_2 \to \mathcal{V}_2$. Most precise here refers to a generalisation of $\leq_i$ to operators, where for $\mathcal{O}_1, \mathcal{O}_2 : \mathcal{V}_3 \to \mathcal{V}_3$, they define $\mathcal{O}_1 \leq_i \mathcal{O}_2$ iff for all $v \in \mathcal{V}_3$ it holds that $\mathcal{O}_1(v) \leq_i \mathcal{O}_2(v)$. Specifically, DMT then show that the most precise – called the *ultimate* – approximation of O is given by the operator $\mathcal{U}_O : \mathcal{V}_3 \to \mathcal{V}_3$ that maps a given $v \in \mathcal{V}_3$ to

$$\mathcal{U}_O(v) : A \to \{\mathbf{t}, \mathbf{f}, \mathbf{u}\} \quad \text{with} \quad a \mapsto \begin{cases} \mathbf{t} & \text{if } w(a) = \mathbf{t} \text{ for all } w \in \{O(x) \mid x \in [v]_2\} \\ \mathbf{f} & \text{if } w(a) = \mathbf{f} \text{ for all } w \in \{O(x) \mid x \in [v]_2\} \\ \mathbf{u} & \text{otherwise} \end{cases}$$

This definition is remarkable since previously, approximations of operators had to be devised by hand rather than automatically derived. DMT [Denecker et al., 2004] give additional definitions introducing stable semantics that are only of minor interest here and will be introduced in a special form later.

AFT on AFs

AFT can be used for defining semantics of AFs as follows [Strass, 2013a]. The stable semantics for AFs can be understood as a two-valued semantics given by the fixpoints of an operator (going back to Pollock [1987]) on two-valued interpretations.

Definition 2.1 *For each* AF $F = (A, R)$, *the operator* $U_F : \mathcal{V}_2 \to \mathcal{V}_2$ *yields – for a given interpretation* $v : A \to \{\mathbf{t}, \mathbf{f}\}$ *– a new interpretation*

$$U_F(v) : A \to \{\mathbf{t}, \mathbf{f}\} \quad \text{with} \quad a \mapsto \begin{cases} \mathbf{f} & \text{if } \exists b \in A : v(b) = \mathbf{t}, (b, a) \in R \\ \mathbf{t} & \text{otherwise} \end{cases}$$

Intuitively, all arguments that are attacked in F by some argument that is true in v are set to false in U_F and set to true otherwise, that is, if unattacked by all **t** arguments of v. (So the U is for "unattacked".) It is easy to see that the fixpoints of this operator exactly correspond to stable extensions [Strass, 2013a, Proposition 4.4].

Proposition 2.2 *Let $F = (A, R)$ be an AF and $v : A \to \{\mathbf{t}, \mathbf{f}\}$ be an interpretation. Then $v = U_F(v)$ iff the set $v^{-1}(\mathbf{t}) = \{a \in A \mid v(a) = \mathbf{t}\}$ is a stable extension of F.*

Example 2.3 *Consider the AF $F_1 = (A_1, R_1)$ with $A_1 = \{a, b\}$ and $R_1 = \{(a, b), (b, a)\}$:*

Below, we depict the complete lattice $(\{v : A_1 \to \{\mathbf{t}, \mathbf{f}\}\}, \leq_t)$ of two-valued interpretations over A_1 ordered by the truth ordering as a Hasse diagram (i.e. straight lines show direct $\leq_t$-neighbours), and how the operator U_{F_1} assigns its points to others (dashed arrows).

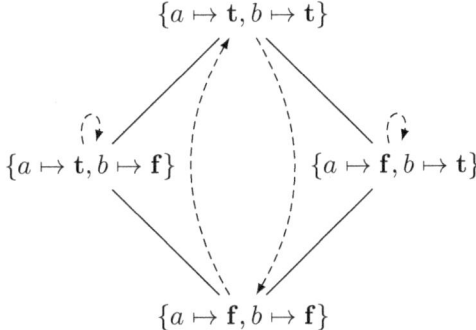

It can be seen from the diagram that the operator has two fixpoints, $\{a \mapsto \mathbf{t}, b \mapsto \mathbf{f}\}$ and $\{a \mapsto \mathbf{f}, b \mapsto \mathbf{t}\}$. They correspond one-to-one to the stable extensions $\{a\}$ and $\{b\}$ of the AF F_1.

Example 2.4 *In contrast, consider the AF $F_2 = (A_2, R_2)$ with $A_2 = \{a, b, c\}$ and $R_2 = \{(a, b), (b, c), (c, a)\}$:*

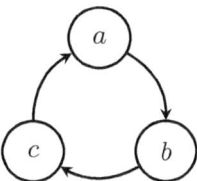

Again, we depict the complete lattice $(\{v : A_2 \to \{\mathbf{t}, \mathbf{f}\}\}, \leq_t)$ and how the operator U_{F_2} assigns its points to others.

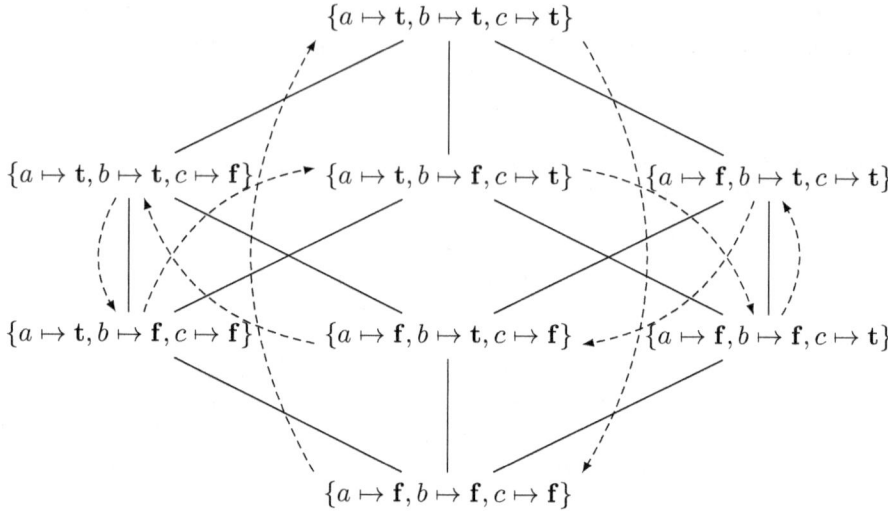

The picture makes it obvious that U_{F_2} has no fixpoint, in accordance with the fact that F_2 has no stable extension.

Using the definitions of Denecker, Marek and Truszczyński, it is easy to obtain the ultimate approximation of U_F. (See also [Strass, 2013a, Proposition 4.1].)

Corollary 2.5 *Given an interpretation $v : A \to \{\mathbf{t}, \mathbf{f}, \mathbf{u}\}$, the three-valued operator $\Upsilon_F : \mathcal{V}_3 \to \mathcal{V}_3$ yields a new interpretation*

$$\Upsilon_F(v) : A \to \{\mathbf{t}, \mathbf{f}, \mathbf{u}\} \quad \text{with} \quad a \mapsto \begin{cases} \mathbf{f} & \text{if } \exists b \in A : v(b) = \mathbf{t}, (b, a) \in R \\ \mathbf{t} & \text{if } \forall b \in A : (b, a) \in R \implies v(b) = \mathbf{f} \\ \mathbf{u} & \text{otherwise} \end{cases}$$

For any given AF F, the fixpoints of U_F constitute the stable semantics of F. The ultimate approximation Υ_F approximates U_F, thus the semantics induced by Υ_F then intuitively approximate AF stable semantics. More specifically, the following result is straightforward [Strass, 2013a, Section 4]:[4]

Proposition 2.6 *Let $F = (A, R)$ be an AF and $v : A \to \{\mathbf{t}, \mathbf{f}, \mathbf{u}\}$ be an interpretation.*

1. v is complete for F iff $v = \Upsilon_F(v)$.

---

[4]Given an AF $F = (A, R)$, an extension $E \subseteq A$ uniquely determines a three-valued interpretation v_E by letting $v_E(a) = \mathbf{t}$ if $a \in E$, $v_E(a) = \mathbf{f}$ if a is attacked by E in F, and $v_E(a) = \mathbf{u}$ otherwise. Similarly, a three-valued interpretation $v : A \to \{\mathbf{t}, \mathbf{f}, \mathbf{u}\}$ uniquely determines an extension $E_v = \{a \mid v(a) = \mathbf{t}\}$. This allows us to switch freely between extensions and interpretations.

2. v is admissible for F iff $v \leq_i \Upsilon_F(v)$.

3. v is preferred for F iff v is $\leq_i$-maximal admissible.

4. v is grounded for F iff v is the $\leq_i$-least fixpoint of Υ_F.

In the next section, we will use approximation fixpoint theory and this result to define the semantics of ADFs in a straightforward way.

Example 2.7 Consider the AF $F_3 = (A_3, R_3)$ with $A_3 = \{a, b\}$ and $R_3 = \{(a, b)\}$:

Below, we depict the associated meet-semilattice $(\{v : A_3 \to \{\mathbf{t}, \mathbf{f}, \mathbf{u}\}\}, \leq_i)$ of the set of all three-valued interpretations over A_3 ordered by the information ordering, and how the operator Υ_{F_3} maps those interpretations to others.

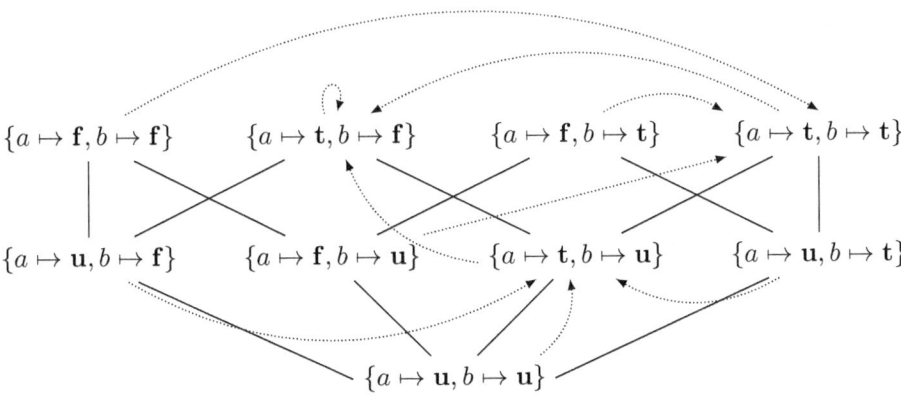

The picture shows how the grounded semantics can be obtained by following the dotted line starting in the $\leq_i$-least element up to the operator's single fixpoint. (In fact, it obviates that all (sufficiently long) sequences of operator applications lead to the fixpoint, showing that this interpretation really is the intended meaning of F_3.)

Example 2.8 Reconsider the AF $F_1 = (A_1, R_1)$ from Example 2.3 with $A_1 = \{a, b\}$ and $R_1 = \{(a, b), (b, a)\}$:

Again, we show the complete meet-semilattice $(\{v : A_1 \to \{\mathbf{t}, \mathbf{f}, \mathbf{u}\}\}, \leq_i)$ *along with the mappings of the operator* Υ_{F_1} *(dotted arrows).*

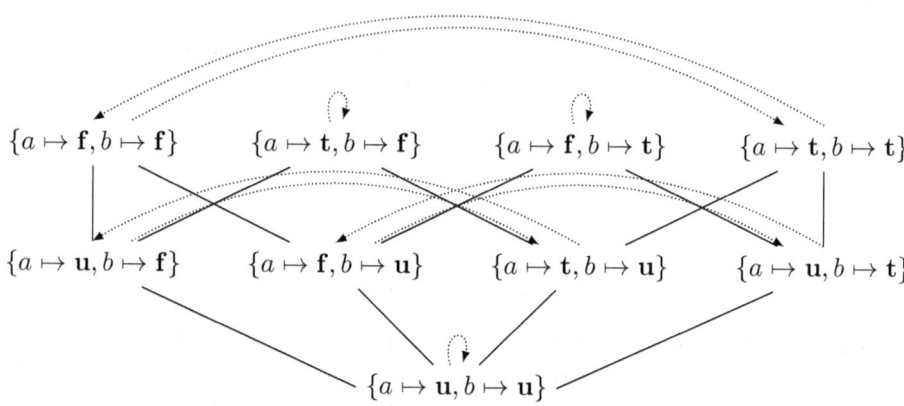

In this picture, the operator U_{F_1} re-appears in the top row of all two-valued interpretations. Those form a complete lattice with respect to $\leq_t$, but an antichain with respect to $\leq_i$. Likewise, the two fixpoints of U_{F_1} re-appear as fixpoints of Υ_{F_1} in the top row. The additional fixpoint of Υ_{F_1} consequently constitutes the grounded semantics of F_1.

As we have seen, the operator Υ_F arises naturally from a straightforward application of ultimate approximation [Denecker et al., 2004] to an operator proposed by Pollock [1987]. It is interesting to observe that the assignments of the operator correspond precisely to what has independently been defined as "legal argument labellings" [Caminada and Gabbay, 2009].

3 ADFs: Syntax and Semantics

Like an AF, an abstract dialectical framework (ADF) is a directed graph whose nodes represent arguments, statements or positions. One can think of the nodes as arbitrary items which can be accepted or not. The links represent dependencies. However, unlike a link in an AF, the meaning of an ADF link can vary. The status of a node s only depends on the status of its parents (denoted $par(s)$), that is, the nodes with a direct link to s. In addition, each node s has an associated acceptance condition C_s specifying the exact conditions under which s is accepted. C_s is a function assigning to each subset of $par(s)$ one of the truth values $\mathbf{t}, \mathbf{f}$.[5] Intuitively, if for some $R \subseteq par(s)$ we have $C_s(R) = \mathbf{t}$, then s will be accepted provided the nodes in R are accepted and those in $par(s) \setminus R$ are not accepted.

---

[5] In the original paper *in* and *out* were used. We prefer truth values here as they allow us to apply standard logical terminology.

Definition 3.1 *An abstract dialectical framework is a tuple* $D = (S, L, C)$ *where*

- *S is a set of statements (positions, nodes),*
- $L \subseteq S \times S$ *is a set of links,*
- $C = \{C_s\}_{s \in S}$ *is a set of total functions* $C_s : 2^{par(s)} \to \{\mathbf{t}, \mathbf{f}\}$, *one for each statement s.* C_s *is called acceptance condition of s.*

In many cases it is convenient to represent acceptance conditions as propositional formulas. For this reason we will frequently use a logical representation of ADFs (S, L, C) where C is a collection $\{\varphi_s\}_{s \in S}$ of propositional formulas.[6]

Example 3.2 *In the following* ADF, *which will act as running example throughout the chapter, we use formulas to specify acceptance conditions.*

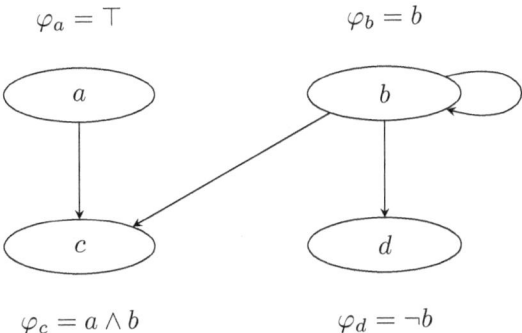

Intuitively, φ_a states that a should always be accepted. Condition φ_b expresses a kind of self-support, which can be utilized as a guess whether or not to accept b. Finally, c should be accepted if both a and b are, while d is attacked by statement b.

Unless specified differently we will tacitly assume that the acceptance formulas specify the parents a node depends on implicitly. It is then not necessary to give the links in the graph explicitly. We thus can represent an ADF D as a tuple (S, C) where S and C are as above and L is implicitly given as $(a, b) \in L$ iff a appears in φ_b.

The different semantics of ADFs over statements S are based (via approximation fixpoint theory) on the notion of a two-valued model. A two-valued interpretation $v : S \to \{\mathbf{t}, \mathbf{f}\}$ – a mapping from statements to the truth values true and false – is a *two-valued model* (*model*, if clear from the context) of an ADF (S, C) whenever for all statements $s \in S$ we have $v(s) = v(\varphi_s)$, that

---

[6] More precisely, each acceptance condition C_s will be represented as a propositional formula φ_s over the vocabulary $par(s)$.

is, v maps exactly those statements to true whose acceptance conditions are satisfied under v.[7]

Approximation Fixpoint Theory on ADFs

We now come back to AFT and illustrate its role to define semantics for ADFs [Strass, 2013a; Brewka et al., 2013]. As AFT deals with operator-based semantics and how to approximate them, the starting point is an operator for the two-valued semantics: the notion of an ADF model allows us to associate a two-valued operator to a given ADF.

Definition 3.3 *Let $D = (S, \{\varphi_s\}_{s \in S})$ be an ADF. The operator $G_D : \mathcal{V}_2 \to \mathcal{V}_2$ takes an input $v : S \to \{\mathbf{t}, \mathbf{f}\}$ and returns an updated interpretation*

$$G_D(v) : S \to \{\mathbf{t}, \mathbf{f}\} \quad \text{with} \quad s \mapsto v(\varphi_s)$$

In words, the operator takes a two-valued interpretation v and returns a two-valued interpretation $G_D(v)$ mapping each $s \in S$ to the truth value that is obtained by evaluating φ_s with v. It is easy to see that this operator characterises the ADF model semantics [Strass, 2013a, Proposition 3.4].

Proposition 3.4 *Let $D = (S, L, C)$ be an ADF and $v : S \to \{\mathbf{t}, \mathbf{f}\}$ be a two-valued interpretation. Then v is a (two-valued) model of D iff $v = G_D(v)$.*

Example 3.5 *For the ADF D from Example 3.2, Figure 2 depicts the complete lattice $(\{v : S \to \{\mathbf{t}, \mathbf{f}\}\}, \leq_t)$ and how the operator G_D assigns its points to others.*

Using the general operator-based definitions of Denecker, Marek and Truszczyński [Denecker et al., 2004], it is again straightforward to determine the ultimate approximation of G_D. Recall from the section on approximation fixpoint theory (Section 2) that the set $\mathcal{V}_3$ of all three-valued interpretations over S forms a complete meet-semilattice with respect to the information ordering $\leq_i$. The consensus meet operation $\sqcap_i$ of this semilattice is given by $(v_1 \sqcap_i v_2)(s) = v_1(s) \sqcap_i v_2(s)$ for all $s \in S$. The least element of this semilattice is the interpretation $v_\mathbf{u} : S \to \{\mathbf{u}\}$ mapping all statements to undefined – the least informative interpretation. The ultimate approximation of the two-valued ADF operator G_D is now obtained as follows [Strass, 2013a, Lemma 3.12]:

Corollary 3.6 *Let D be an ADF. The operator $\Gamma_D : \mathcal{V}_3 \to \mathcal{V}_3$ is the ultimate approximation of G_D and is defined as follows: for an ADF D and a three-valued interpretation v, the revised interpretation $\Gamma_D(v)$ is given by*

$$\Gamma_D(v) : S \to \{\mathbf{t}, \mathbf{f}, \mathbf{u}\} \quad \text{with} \quad s \mapsto \sqcap_i \{w(\varphi_s) \mid w \in [v]_2\}$$

---

[7] In an earlier paper [Brewka et al., 2013], there was the notion of a "three-valued model". The development and analysis of that concept has been discontinued.

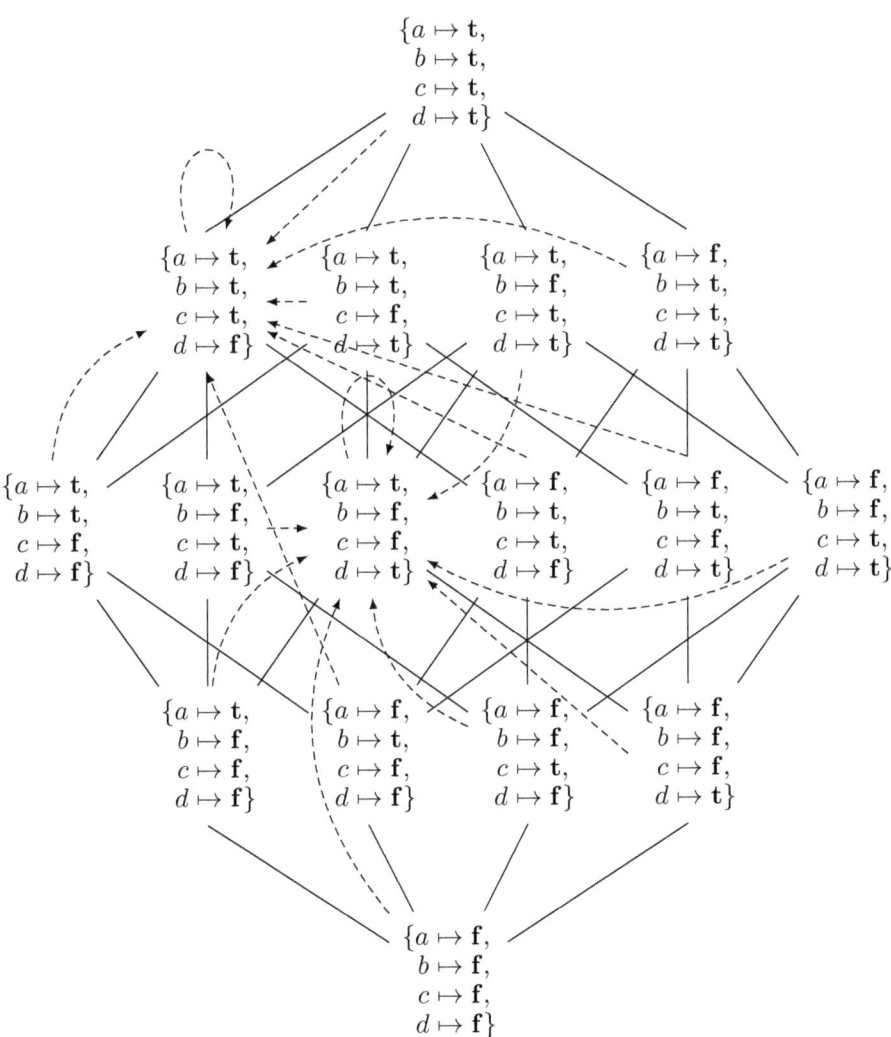

Figure 2: Complete lattice of two-valued interpretations for Example 3.2; dashed arrows visualise the assignments of the operator G_D. It can be readily seen that G_D has two fixpoints, whence D has two models (Proposition 3.4).

That is, for each statement s, the operator returns the consensus truth value for its acceptance formula φ_s, where the consensus takes into account all possible two-valued interpretations w that extend the input valuation v. If this v is two-valued, then $[v]_2 = \{v\}$, thus $\Gamma_D(v)(s) = v(\varphi_s) = G_D(v)(s)$ and Γ_D indeed approximates G_D.

Example 3.7 *Consider the* ADF $D_1 = (S_1, L_1, C_1)$ *given by* $S_1 = \{a, b, c\}$, *and* L_1 *and* C_1 *given as follows:*

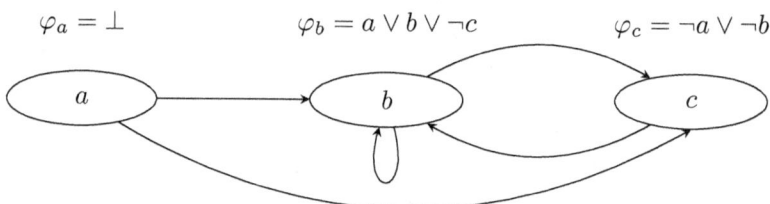

Roughly, a cannot be accepted. Statement b supports itself, and is furthermore supported by a and attacked by c – more precisely, b can be accepted if a can be accepted or b can be accepted or c can be rejected. In turn, c is jointly attacked by a and b – c can only be rejected if both a and b are accepted, otherwise c is accepted. Figure 3 shows the associated complete meet-semilattice $(\{v : S_1 \to \{\mathbf{t}, \mathbf{f}, \mathbf{u}\}\}, \leq_i)$ along with the mappings of the operator Γ_{D_1}.

It is now an easy corollary of Definition 2.6 to generalize the standard AF semantics to ADFs:

Definition 3.8 *Let* $D = (S, L, C)$ *be an* ADF *and* $v : S \to \{\mathbf{t}, \mathbf{f}, \mathbf{u}\}$ *be an interpretation.*

1. *v is* complete *for D iff $v = \Gamma_D(v)$.*

2. *v is* admissible *for D iff $v \leq_i \Gamma_D(v)$.*

3. *v is* preferred *for D iff v is $\leq_i$-maximal admissible.*

4. *v is* grounded *for D iff v is the $\leq_i$-least fixpoint of Γ_D.*

Incidentally, Brewka and Woltran [2010] already defined the operator Γ_D (manually) and used it to define the grounded semantics. Thus the grounded semantics can be seen as the greatest possible consensus between all acceptable ways of interpreting the ADF at hand. A three-valued interpretation is admissible for an ADF D iff it does not make an unjustified commitment that the operator Γ_D will subsequently revoke.

There is an alternative and perhaps slightly more accessible way of introducing the operator Γ_D. We will briefly pursue this way for illustration, and start

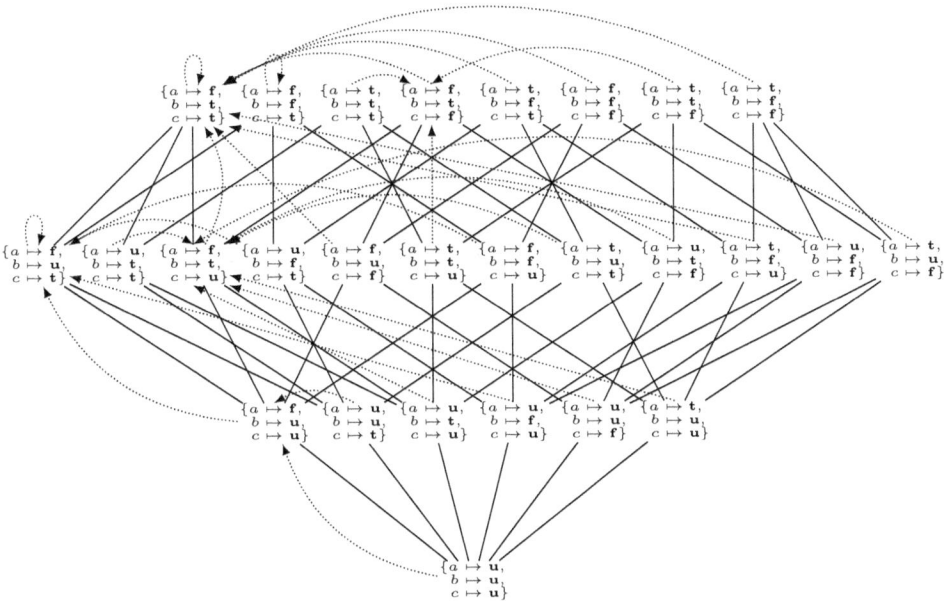

Figure 3: Complete meet-semilattice of three-valued interpretations over $S_1 = \{a, b, c\}$ under the information ordering for Example 3.7; dotted arrows visualise mappings of the operator Γ_{D_1}. It can be seen that Γ_{D_1} has a $\leq_i$-least fixpoint, which is situated right $\leq_i$-beneath its two-valued models, the other two fixpoints of Γ_{D_1}.

out with an additional definition. For a propositional formula φ over vocabulary S and a three-valued interpretation $v : S \to \{\mathbf{t}, \mathbf{f}, \mathbf{u}\}$, the *partial valuation of φ by v* is the formula

$$\varphi^v = \varphi[p/\top : v(p) = \mathbf{t}][p/\bot : v(p) = \mathbf{f}]$$

Intuitively, given a three-valued interpretation v and a formula φ, the partial evaluation of φ with v takes the two-valued part of v and replaces the evaluated variables with their truth values. For example, consider the propositional formula $\varphi = a \vee (b \wedge c)$ and the interpretation $v_1 = \{a \mapsto \mathbf{f}, b \mapsto \mathbf{t}, c \mapsto \mathbf{u}\}$. Statement c with $v_1(c) = \mathbf{u}$ will remain in φ, while a and b are replaced, and we get $\varphi^{v_1} = \bot \vee (\top \wedge c)$. Now assume that an ADF $D = (S, \{\varphi_s\}_{s \in S})$ is given via acceptance formulas; for this D and a three-valued interpretation v, the revised interpretation $\Gamma_D(v)$ is given by

$$\Gamma_D(v): S \to \{\mathbf{t}, \mathbf{f}, \mathbf{u}\} \quad \text{with} \quad s \mapsto \begin{cases} \mathbf{t} & \text{if } \varphi_s^v \text{ is irrefutable} \\ \mathbf{f} & \text{if } \varphi_s^v \text{ is unsatisfiable} \\ \mathbf{u} & \text{otherwise} \end{cases}$$

An irrefutable formula is a formula that is satisfied under any two-valued interpretation (i.e. the formula is a tautology).

For reasons of brevity, we will sometimes shorten the notation of a three-valued interpretation $v = \{a_1 \mapsto t_1, \ldots, a_n \mapsto t_n,\}$ with statements $a_1, \ldots, a_n$ and truth values $t_1, \ldots, t_n$ to $v \triangleq \{a_i \mid v(a_i) = \mathbf{t}\} \cup \{\neg a_i \mid v(a_i) = \mathbf{f}\}$. For instance, $v = \{a \mapsto \mathbf{t}, b \mapsto \mathbf{u}, c \mapsto \mathbf{f}\} \triangleq \{a, \neg c\}$.

We now show some concrete interpretations and semantics for an example.

Example 3.9 *As we have seen before, for the* ADF *D from Example 3.2 we obtain the following two-valued models:*

- $v_1 = \{a \mapsto \mathbf{t}, b \mapsto \mathbf{t}, c \mapsto \mathbf{t}, d \mapsto \mathbf{f}\} \triangleq \{a, b, c, \neg d\}$
- $v_2 = \{a \mapsto \mathbf{t}, b \mapsto \mathbf{f}, c \mapsto \mathbf{f}, d \mapsto \mathbf{t}\} \triangleq \{a, \neg b, \neg c, d\}$

Unfortunately, due to its sheer size ($3^4 = 81$ interpretations), we cannot depict the semi-lattice ($\{S \to \{\mathbf{t}, \mathbf{f}, \mathbf{u}\}\}, \leq_i$) and will henceforth resort to textual descriptions. The grounded interpretation of D is $v_3 = \{a \mapsto \mathbf{t}, b \mapsto \mathbf{u}, c \mapsto \mathbf{u}, d \mapsto \mathbf{u}\} \triangleq \{a\}$. The admissible interpretations (ordered by $\leq_i$) of our example ADF *are as follows:*

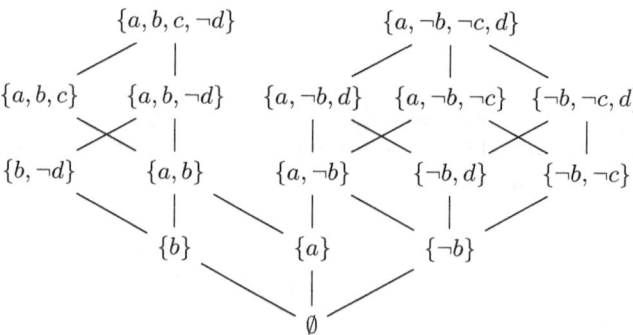

We verify that $v_4 \triangleq \{a, \neg b, \neg c\}$ is admissible in the example ADF. *Statement a's acceptance condition is a tautology. This means that under any three-valued interpretation v' it holds that $\Gamma_D(v')(a) = \mathbf{t}$, and, in particular, $\Gamma_D(v_4)(a) = v_4(a) = \mathbf{t}$. Acceptance condition of statement b is the formula b. Such an acceptance condition (a single unnegated variable) implies that for any three-valued interpretation v' that assigns a value to b, it holds that $\Gamma_D(v')(b) = v'(b)$. If b is assigned $\mathbf{t}$ by v', then $\varphi_b^{v'}$ is a tautology, if b is assigned*

f, then $\varphi_b^{v'}$ is unsatisfiable, and if b is assigned **u** by v', then $\varphi_b^{v'} = b$ is neither a tautology nor unsatisfiable. The acceptance condition of statement c is $a \wedge b$. Evaluating φ_c under v_4 gives $\varphi_c^{v_4} = \top \wedge \bot \equiv \bot$, and $\Gamma_D(v_4)(c) = \mathbf{f} = v_4(c)$. Finally, $v_4(d) = \mathbf{u}$ and $\varphi_d = \neg b$. Since for the undefined truth value it holds that $\mathbf{u} \leq_i \mathbf{t}$ and $\mathbf{u} \leq_i \mathbf{f}$, if a three-valued interpretation v' assigns undefined to a statement, then applying the operator Γ_D under v' cannot return a truth value with less information than **u** for that statement. For our example interpretation, we have $v_4(d) \leq_i \Gamma_D(v_4)(d) = \mathbf{t}$.

The complete interpretations of our example ADF are

$$v_3 \,\hat{=}\, \{a\}, \quad v_5 \,\hat{=}\, \{a,b,c,\neg d\}, \quad v_6 \,\hat{=}\, \{a, \neg b, \neg c, d\}.$$

The latter two, v_5 and v_6, are the preferred interpretations.

The definition of stable model semantics for ADFs [Brewka et al., 2013] is based on ideas from Logic Programming (LP) where stable models strengthen the notion of minimal models by excluding self-justifying cycles of atoms. In LP, this is achieved by a test which picks a candidate model M, uses M to reduce the original logic program to a program without negative literals, and then checks whether M coincides with the (typically unique) least model of the reduced program. This way self-justifying cycles cannot appear. What we do for an ADF D is very similar: to check whether a two-valued model v of D is stable we do the following:

- we eliminate in D all nodes with v-value **f** and corresponding links,

- we replace eliminated nodes in acceptance conditions by $\bot$,

- we check whether nodes that are **t** in v coincide with those that are **t** in the grounded interpretation of the reduced ADF.

This is captured in the following definition [Brewka et al., 2013, Definition 6]. (See also [Strass and Wallner, 2015, Proposition 2.4] for an alternative definition via AFT.)

Definition 3.10 Let $D = (S, L, C)$ be an ADF with $C = \{\varphi_s\}_{s \in S}$ and $v : S \to \{\mathbf{t}, \mathbf{f}\}$ be a two-valued model of D. Define the reduced ADF D^v with $D^v = (S^v, L^v, C^v)$, where

- $S^v = \{s \in S \mid v(s) = \mathbf{t}\}$

- $L^v = L \cap S^v \times S^v$

- $C^v = \{\varphi_s^v\}_{s \in S^v}$ where for each $s \in S^v$, we set $\varphi_s^v = \varphi_s[b/\bot : v(b) = \mathbf{f}]$.

Denote by w the unique grounded interpretation of D^v. Now the two-valued model v of D is a stable model of D if and only if for all $s \in S$, we find that $v(s) = \mathbf{t}$ implies $w(s) = \mathbf{t}$.

Note that a stable model of an ADF D is a model of D by definition (v is assumed to be a model). In the reduct for a model v, (i) only statements assigned to true by v are present, (ii) only links with both ends being statements assigned to true by v are considered, and (iii) in each acceptance formula of the remaining statements we replace statements $b \in S$ that v maps to false by their truth value, i.e., in these acceptance conditions variables assigned to false by v are replaced by $\bot$ (and the remaining statements/variables remain unmodified in the formulas). This definition straightforwardly expresses the intuition underlying stable models: if all statements the model v takes to be false are indeed false, we must find a constructive proof for all statements the model takes to be true.

Example 3.11 *Consider the ADF D given by*

$$\varphi_a = \top, \quad \varphi_b = \neg a \vee c, \quad \varphi_c = b.$$

It has two models: $v_1 = \{a \mapsto \mathbf{t}, b \mapsto \mathbf{t}, c \mapsto \mathbf{t}\}$ *and* $v_2 = \{a \mapsto \mathbf{t}, b \mapsto \mathbf{f}, c \mapsto \mathbf{f}\}$. *Let us check whether they are stable models. For v_1, the reduct, D^{v_1}, is equal to D (every statement is assigned to true by v_1, thus all statements and links remain in the reduct and no statement is replaced by $\bot$ in an acceptance condition). The grounded interpretation of D is $v_3 = \{a \mapsto \mathbf{t}, b \mapsto \mathbf{u}, c \mapsto \mathbf{u}\}$, implying that v_1 is not stable in D, since the grounded interpretation of D^{v_1} is not equal to v_1.*[8]

For the other model of D, the reduct $D^{v_2} = (S^{v_2}, L^{v_2}, C^{v_2})$ with $S^{v_2} = \{a\}$, $L^{v_2} = \emptyset$, and $\varphi_a = \top$. The grounded interpretation of D^{v_2} is $v_4 = \{a \mapsto \mathbf{t}\}$. The final condition of Definition 3.10, $v_2(a) = \mathbf{t}$ implies $v_4(a) = \mathbf{t}$, is satisfied, and, therefore, v_2 is a stable model of D. Further, v_2 is the only stable model of D, since we considered all models of D, only one being stable, and any other interpretation cannot be stable for D, since being a model is a prerequisite for being stable.

Next, we illustrate that there are cases where an ADF has a model, but no stable model.

Example 3.12 *Consider the ADF D given by*

$$\varphi_a = c, \quad \varphi_b = c, \quad \varphi_c = a \leftrightarrow b.$$

---

[8]The definition of stable models in this chapter, taken from [Brewka et al., 2013, Definition 6], supersedes the definition of stable models in the original paper on ADFs [Brewka and Woltran, 2010, Definition 6] in that the new definition corrects certain unintended results. For instance, v_1 in Example 3.11 is the only stable model according to the old definition, but this is not the case under the new definition. The model v_1 violates the basic intuition of stable semantics that all elements of a stable model should have a non-cyclic justification: in the model v_1 it holds that b is accepted because c is and vice versa (these two statements have supporting links to each other; see Section 6.1 for a formalization of attacking and supporting links between statements).

The only two-valued model of D is $v = \{a \mapsto \mathbf{t}, b \mapsto \mathbf{t}, c \mapsto \mathbf{t}\}$. Since c is true because a and b are and vice versa, the model contains unintended cyclic support and thus should not be stable. Indeed, for the reduct we get $D^v = D$. Let us compute the grounded semantics of D. We start with interpretation $w = \{a \mapsto \mathbf{u}, b \mapsto \mathbf{u}, c \mapsto \mathbf{u}\}$. Since none of the acceptance formulas is a tautology or an unsatisfiable formula, w is already a fixpoint of Γ_D and thus the grounded interpretation of D. Hence v is not a stable model and D has no stable models, just as intended. Since v is a minimal model of D the example illustrates that in Definition 3.10 we actually need the grounded semantics; requiring v to be among the (subset-inclusion or information) minimal two-valued models of the reduct is insufficient, in contrast to, e.g., stable semantics of logic programs.

For our running example, the concept of reduct is applied as follows.

Example 3.13 The ADF from Example 3.2 has two two-valued models, namely $v_1 = \{a \mapsto \mathbf{t}, b \mapsto \mathbf{t}, c \mapsto \mathbf{t}, d \mapsto \mathbf{f}\}$ and $v_2 = \{a \mapsto \mathbf{t}, b \mapsto \mathbf{f}, c \mapsto \mathbf{f}, d \mapsto \mathbf{t}\}$. We obtain the reducts for each model of D as follows:

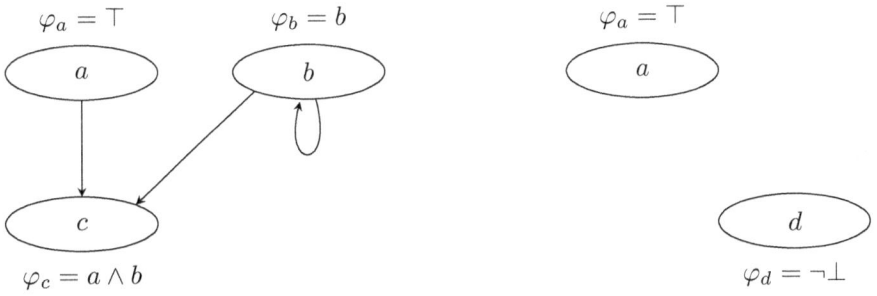

Reduct D^{v_1} Reduct D^{v_2}

The grounded interpretation of reduct D^{v_1} is $\{a\}$, v_1 is thus not a stable model of D. For v_2, the reduct D^{v_2} has the grounded interpretation $\{a \mapsto \mathbf{t}, d \mapsto \mathbf{t}\}$. The model v_2 of D is thus the single stable model of D.

Well-known relationships between semantics defined on Dung AFs carry over to ADFs. This is formalized in the next theorem [Brewka et al., 2013, Theorem 3].

Theorem 3.14 *Let D be an* ADF.

- *Each stable model of D is a two-valued model of D;*

- *each two-valued model of D is a preferred interpretation of D;*

- *each preferred interpretation of D is complete;*

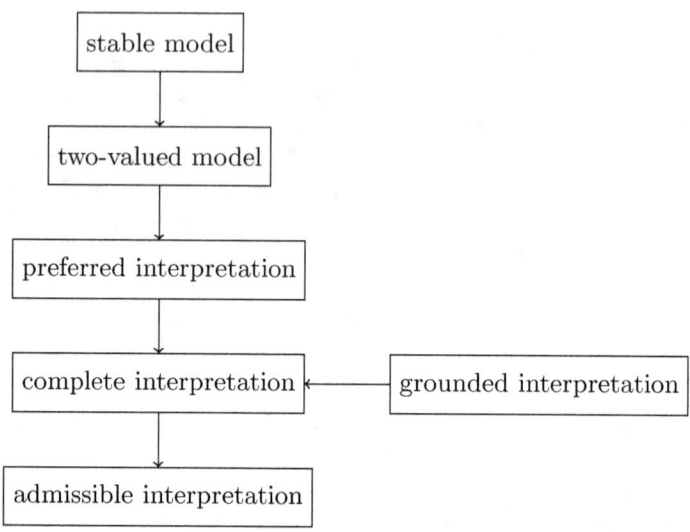

Figure 4: Relations between ADF semantics

- each complete interpretation of D is admissible;
- the grounded interpretation of D is complete.

We illustrate the relationships in Figure 4 where an arrow from a σ-interpretation to a τ-interpretation denotes that every σ-interpretation is a τ-interpretation. Further, again similarly as in AFs, any ADF possesses at least one admissible, complete, preferred, and grounded interpretation, while this is not guaranteed for models and stable models.

In addition to the semantical relationships generalizing those known from AFs, semantics on ADFs also directly generalize semantics for AFs. We first define for a given AF its associated ADF.

Definition 3.15 *For an AF $F = (A, R)$, define the ADF associated to F as $D_F = (A, R, C)$ with $C = \{\varphi_a\}_{a \in A}$ such that for each $a \in A$, the acceptance condition is given by*

$$\varphi_a = \bigwedge_{\substack{b \in A, \\ (b,a) \in R}} \neg b$$

Now we can formalize the way ADFs, and their semantics, generalize AFs in the next two theorems [Brewka et al., 2013].

Theorem 3.16 *Let $F = (A, R)$ be an AF and D_F its associated ADF. For any two-valued interpretation v for A, the following are equivalent:*

(A) the set $v^{-1}(\mathbf{t}) = \{a \in A \mid v(a) = \mathbf{t}\}$ is a stable extension of F,

(B) v is a stable model of D_F,

(C) v is a two-valued model of D_F.

Note that for AF-based ADFs, there is no distinction between models and stable models. The intuitive explanation for this is that stable semantics on ADFs breaks cyclic supports, which cannot arise in AFs because they cannot (directly) express support.

More generally, we can also show that our definitions are indeed proper generalizations of Dung's notions for AFs as given in Proposition 2.6. The result is due to [Brewka et al., 2013].

Theorem 3.17 *Let F be an AF and D_F its associated ADF. An interpretation is admissible, complete, preferred, grounded for F iff it is admissible, complete, preferred, grounded for D_F.*

On AFs, if v is a preferred interpretation (a stable model) for an AF F it holds that there is no preferred interpretation (stable model) $v' \neq v$ such that the set of statements assigned to true by v is a subset of the statements assigned to true by v', i.e., $\{s \mid v(s) = \mathbf{t}\} \not\subseteq \{s \mid v'(s) = \mathbf{t}\}$. On general ADFs, this property does not hold for preferred interpretations and two-valued models, i.e., there are ADFs with two preferred interpretations (models) v and v' such that $\{s \mid v(s) = \mathbf{t}\} \subseteq \{s \mid v'(s) = \mathbf{t}\}$.

Example 3.18 *Consider* ADF $D = (\{a\}, \{(a,a)\}, \{\varphi_a = a\})$. *Both* $v_1 = \{a \mapsto \mathbf{f}\}$ *and* $v_2 = \{a \mapsto \mathbf{t}\}$ *are models and preferred interpretations of D. It holds that* $\{s \mid v(s) = \mathbf{t}\} = \emptyset \subsetneq \{a\} = \{s \mid v'(s) = \mathbf{t}\}$.

On the other hand, for any ADF D with stable models v_1 and v_2, it holds that $v_1 \leq_t v_2$ implies $v_1 = v_2$ [Strass, 2013a, Proposition 3.8], that is, such strict relationships cannot occur between *stable* models. (This follows easily from AFT.)

4 ADFs as Modelling Tools

In this section we will provide various examples illustrating why – as we believe – ADFs are useful tools in formal argumentation. We discussed the term argumentation middleware in the introduction already. We now want to give a clearer picture what we actually mean by this. More precisely, we will discuss various graphical representations of argumentation scenarios users may find useful. In each case we define the semantics of the chosen representation by providing a formal translation to ADFs. The representation is thus equipped – via the translation – with the whole range of Dung semantics we have defined for ADFs. We also discuss how ADFs can serve as a tool for providing semantics to systems based on strict and defeasible inference rules, again via a translation.

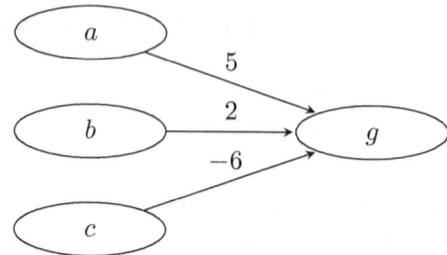

Figure 5: An argument graph with weighted links.

4.1 Weights and Preferences

In our informal discussion in the introduction we have already shown how graphical representations based on link types (+ for supporting, − for attacking) can be modeled using ADFs. The same is obviously true for links annotated with numerical weights. Throughout the chapter we will assume a positive weight represents support, a negative weight attack, in both cases with a given strength. An example can be found in Figure 5.

The figure uses a weighted graph to represent a simple argumentation scenario. We will provide the graph with a formal semantics based on translating it to an ADF. There are various ways of interpreting the numbers and of actually deriving specific ADF acceptance conditions from representations like this one. We first have to specify how the numbers should actually be used to decide whether a node is accepted or not. Recall that a link is active if its source node is accepted. A straightforward idea is to accept a node whenever the sum of the weights of all active links pointing to the node is positive. We will call this strategy *sum-of-weights* (sow). For node g in Figure 5 this amounts, as we will see, to the following acceptance condition: $(\neg c \wedge (a \vee b)) \vee (a \wedge b)$.

Secondly, we need to take care of those nodes which do not depend on other nodes, that is, nodes without incoming links. We will call these nodes *input nodes* and denote the input nodes of a graph G as $input(G)$. It is often useful to consider input nodes as parameters whose truth values can be chosen freely, with the aim to explore the consequences of a particular choice. Consequently, our translation will depend on the assignment of truth values to the input nodes.

Definition 4.1 *Let $G = (N, E, I)$ be a labelled graph with nodes N, edges E and (integer) labelling function $I : E \to \mathbb{Z}$. Let $A \subseteq input(G)$ be the subset of input nodes considered true (the other input nodes are considered false). The sum-of-weights translation of G under A is the ADF $D = (S, L, C)$ with $S = N$, $L = E$, and the acceptance condition C_s (represented as a formula ϕ_s) is defined as follows:*

$$\phi_s = \begin{cases} \top, & \text{if } s \in A \\ \bot, & \text{if } s \in input(G) \setminus A \\ \phi_{sow}(s), & \text{otherwise} \end{cases}$$

where the formula $\phi_{sow}(s)$ is the disjunction of all conjunctions of literals built from parent nodes of s which represent truth value assignments under which the sum of weights of active links is positive.

Let us check how the acceptance condition for node g in Figure 5 is obtained. The following table shows 8 possible assignments of truth values to g's parent nodes, together with the sum of values of active links:

a	b	c	
t	t	t	1
t	t	f	7
t	f	t	-1
t	f	f	5
f	t	t	-4
f	t	f	2
f	f	t	-6
f	f	f	0

The sum of weights of active links is positive in 4 of the 8 lines, the acceptance condition of g is the disjunction of the conjunctions corresponding to these lines, that is:

$$(a \wedge b \wedge c) \vee (a \wedge b \wedge \neg c) \vee (a \wedge \neg b \wedge \neg c) \vee (\neg a \wedge b \wedge \neg c)$$

which can be simplified to $(\neg c \wedge (a \vee b)) \vee (a \wedge b)$, the formula presented earlier.

Of course, there are many more strategies how to evaluate the numbers. One possibility is to check whether the maximal positive weight of an active link is higher than the maximal negative weight of an active link. This leads to a different definition of acceptance conditions for non-input nodes. We leave the details to the reader and just mention that in Figure 5 the acceptance condition for g under this new strategy becomes $(\neg c \wedge (a \vee b))$.

Qualitative preferences can be handled in a similar manner. Let us first introduce prioritized argument graphs.

Definition 4.2 *A prioritized argument graph is a tuple $G = (S, L^+, L^-, >)$ where S is the set of nodes, L^+ and L^- are subsets of $S \times S$, the supporting and attacking links, and $>$ is a strict partial order (irreflexive, transitive, antisymmetric) on S representing preferences among the nodes.*

As before, we will translate prioritized argument graphs to ADFs. We illustrate the translation using an example. Assume we are given the graph in Figure 6.

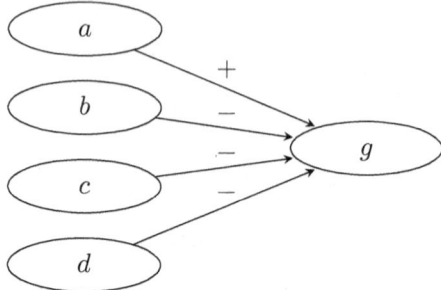

Figure 6: An argument graph with qualitative weights.

Assume further the preference ordering is $a > c$ and $g > d$, that is a is strictly preferred to c, g to d. We want to capture the following intuition: an attacker (represented by label $-$ in the graph) does not succeed if the attacked node is more preferred than the attacker, or if there is a more preferred supporting node (represented by label $+$ in the graph).

We treat input nodes as in Definition 4.1. The general scheme for deriving formulas expressing the corresponding acceptance condition ϕ_s for a node s with a non-empty set of parents is the following: we create a conjunction of implications, one for each attacker t of s which is not less preferred than s. The left side of the implication (the precondition) consists of the attacker t, the right side (conclusion) is the disjunction of all supporting nodes of s which are more preferred than t.

In the example above, the only attackers which are not less preferred than g are b and c. For b we obtain the implication $b \to \mathbf{f}$ (as there is no supporting node more preferred than b and the empty disjunction is equivalent to $\mathbf{f}$). For attacker c we obtain the implication $c \to a$, as a is more preferred than c. This yields the following acceptance condition for g: $(b \to \mathbf{f}) \wedge (c \to a)$ or, equivalently $\neg b \wedge (c \to a)$.

As a matter of fact, preferences are often not given in advance, as assumed in the example, but an issue of debate themselves. One way to model situations where the preference relation $>$ is established dynamically in the course of argumentation is the following. Let us assume some nodes represent (possibly conflicting) preference information, that is information about which pairs of nodes belong to $>$. The idea is to guess a (stable, preferred, grounded) interpretation M and then to verify whether M can be generated in a way satisfying the preference relation it contains. To do so we extract the preference information from the relevant nodes in M. We then check whether M can be reconstructed under this (now static) preference information using the techniques described above. We thus verify whether the preferences represented in the model itself were taken into account adequately.

Definition 4.3 *An argument graph with dynamic preferences is a tuple*

$$G = (S, L^+, L^-, P)$$

where S is the set of nodes, L^+ and L^- are subsets of $S \times S$, the supporting and attacking links, and $P : S \to S \times S$ is a partial function.

The function P assigns preference information to some of the nodes in S. If $P(a) = (b, c)$ then node a carries the information that b is preferred over c. For a three-valued interpretation M we use $>_M$ to denote the smallest strict partial order on S containing the set $\{(b, c) \mid P(a) = (b, c), M(a) = \mathbf{t}\}$. Note that $>_M$ may be undefined, e.g. if M contains two nodes with conflicting preference information. The semantics of argument graphs with dynamic preferences is now defined as follows:

Definition 4.4 *Let $G = (S, L^+, L^-, P)$ be an argument graph with dynamic preferences, A a subset of its input nodes. E is a (stable, preferred, grounded) interpretation of G under A iff $>_E$ is a strict partial order and E is a (stable, preferred, grounded) interpretation of the prioritized argument graph $D_E = (S, L^+, L^-, >_E)$ under A.*

We thus guess an interpretation E of the intended type, extract from E the corresponding strict partial order on S, and check whether E is among the intended interpretations of the (non-dynamic) prioritized argument graph which is based on the extracted preference information. The evaluation of the prioritized graph is based on the translation to ADFs described earlier in this section. For further details see [Brewka et al., 2013].

4.2 Proof Standards

Proof standards are well known and play an important role in legal reasoning. They are based on the intuitive idea that decisions or verdicts which have drastic consequences, say for a defendant, should be based on stronger, less doubtful criteria than decisions with limited consequences, say a small fine. Farley and Freeman [Farley and Freeman, 1995] introduced a model of legal argumentation which distinguishes four types of arguments (in decreasing order of strength):

- *valid* arguments based on deductive inference,
- *strong* arguments based on inference with defeasible rules,
- *credible* arguments where premises give some evidence,
- *weak* arguments based on abductive reasoning.

By using values $V = \{+v, +s, +c, +w, -v, -s, -c, -w\}$ we will distinguish pro and con links of the corresponding types in argument graphs, where the type of a link is inherited from the type of its source node.

Based on these argument types, Farley and Freeman define the following proof standards:

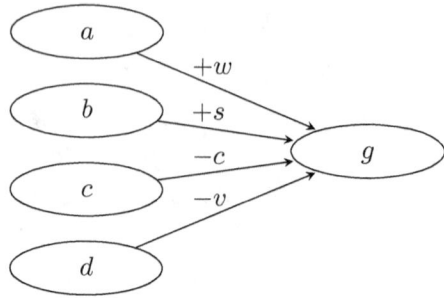

Figure 7: *A Farley/Freeman argument graph.*

- *Scintilla of Evidence*: at least one pro-argument is accepted.
- *Preponderance of Evidence*: at least one pro-argument is accepted, all accepted con arguments are outweighed by stronger accepted pro arguments.
- *Dialectical Validity*: there is at least one credible accepted pro-argument, none of the other side's arguments is accepted.
- *Beyond Reasonable Doubt*: there is at least one strong accepted pro-argument, none of the other side's arguments is accepted.
- *Beyond Doubt*: there is at least one valid active pro-argument, none of the other side's arguments is accepted.

Again we will show how these notions can be formalized using ADFs.

Consider the labelled graph in Figure 7. Let us focus on the acceptance condition for g, represented as a propositional formula. The condition obviously depends on g's proof standard. For scintilla of evidence it is sufficient that at least one pro-argument is accepted. There are two such arguments, a and b, the acceptance condition thus is $a \vee b$. For preponderance of evidence at least one pro-argument must be accepted, and in addition each accepted con-argument must be outweighed by a stronger pro-argument. In our case this means that if c is accepted, then the stronger pro-argument b must also be accepted, and d cannot be accepted, as there is no stronger pro-argument than the valid argument d. Taken together this yields the formula $(a \vee b) \wedge (c \rightarrow b) \wedge \neg d$. In a similar manner we obtain the formulas for g for the remaining proof standards, as shown in the following table:

Scintilla of evidence:	$a \vee b$
Preponderance of evidence:	$(a \vee b) \wedge (c \rightarrow b) \wedge \neg d$
Dialectical validity:	$b \wedge \neg c \wedge \neg d$
Beyond reasonable doubt:	$b \wedge \neg c \wedge \neg d$
Beyond doubt:	$\bot$

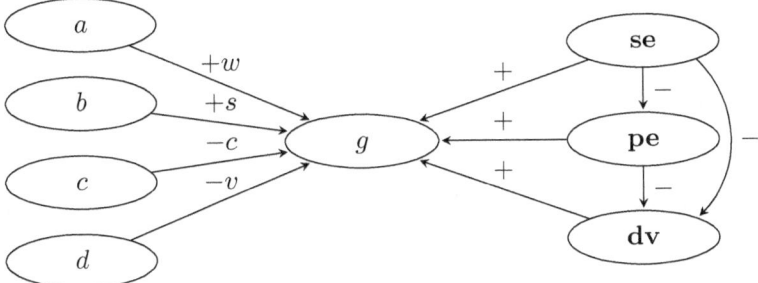

Figure 8: A graph with dynamic proof standards.

It is even possible to choose the proof standard dynamically. For ease of presentation let's focus on three proof standards, namely scintilla of evidence, preponderance of evidence and dialectical validity, represented as **se**, **pe** and **dv**, respectively.[9] Consider the graph in Figure 8 which should be viewed as part of a larger argument graph. The idea here is that scintilla of evidence is the default proof standard. If the corresponding node **se** is attacked from outside (e.g. since a crime was committed), then preponderance of evidence becomes the active proof standard. If also the corresponding node **pe** is attacked from outside (e.g. since the crime has serious consequences), then dialectical validity will be active. To model this intuition, the acceptance condition of node g becomes:

$$(\textbf{se} \wedge (a \vee b)) \vee (\textbf{pe} \wedge (a \vee b) \wedge (c \to b) \wedge \neg d) \vee (\textbf{dv} \wedge b \wedge \neg c \wedge \neg d).$$

4.3 Carneades

Carneades [Gordon et al., 2007] is an advanced model of argumentation based on a graphical representation of arguments and the propositions involved in them. Each proposition has an associated proof standard (scintilla of evidence, preponderance of evidence, clear and convincing evidence, beyond reasonable doubt, dialectical validity). There is some paraconsistency at work in the system as scintilla of evidence allows both a proposition and its negation to be accepted at the same time. The ADF graphs we will construct later will for this reason have separate nodes for each proposition p and its complement $\bar{p}$. A major restriction of Carneades is that cycles in the graph are not allowed (which means the system handles only cases where all Dung semantics coincide).

Let us start with some basic definitions underlying Carneades. Our presentation follows [Brewka and Gordon, 2010].

Definition 4.5 *An argument is a tuple* $\langle P, E, c \rangle$ *with premises P, exceptions E ($P \cap E = \emptyset$) and conclusion c. c and elements of P, E are literals.*

---
[9] The type of these nodes is irrelevant and thus left out.

An argument evaluation structure *(CAES) is a tuple* $\mathcal{S} = \langle \mathit{args}, \mathit{as}, \mathit{weight}, \mathit{standard} \rangle$, *where*

- *args is a set of arguments generating an acyclic argument graph,*
- *as is a consistent set of literals,*
- *weight assigns a real number to each argument, and*
- *standard maps propositions to a proof standard.*

The argument graph generated by a CAES is obtained as follows: each literal occurring in an argument *arg* becomes a node; each argument *arg* becomes a node; each premise of an argument *arg* is linked to the corresponding argument node *arg* via a link labelled with +, each exception via a link labelled with -; an additional link, labelled with weight(*arg*), connects *arg* and the conclusion of *arg*.

The central notions in Carneades are *applicability* of arguments and *acceptability* of propositions. These notions are defined via mutual recursion. Note that for the recursion to bottom out it is essential that Carneades is acyclic.

Definition 4.6 *We say an argument* $\langle P, E, c \rangle \in \mathit{args}$ *is* applicable *in* $\mathcal{S}$ *iff*

- $p \in P$ *implies* $p \in \mathit{as}$ *or* [$\overline{p} \notin \mathit{as}$ *and* p *acceptable in* $\mathcal{S}$], *and*
- $p \in E$ *implies* $p \notin \mathit{as}$ *and* [$\overline{p} \in \mathit{as}$ *or* p *is not acceptable in* $\mathcal{S}$].

Based on the applicability of arguments, we can define what it means for a proposition p to be *acceptable* in $\mathcal{S}$. As expected, acceptability depends on p's proof standard. The Carneades proof standards differ form those of Farley and Freeman. In particular, they depend on numerical values:

- *standard*(p) = *se*: there is an applicable argument for p,
- *standard*(p) = *pe*: p satisfies *se*, and the maximum weight assigned to an applicable argument pro p is greater than the maximum weight of an applicable argument con p,
- *standard*(p) = *ce*: p satisfies *pe*, and the maximum weight of an applicable pro argument exceeds a threshold α, and difference between the maximum weight of applicable pro arguments and the maximum weight of applicable con arguments exceeds a threshold β,
- *standard*(p) = *bd*: p satisfies *ce*, and the maximum weight of the applicable con arguments is less than a threshold γ,
- *standard*(p) = *dv*: there is an applicable argument pro p and no applicable argument con p.

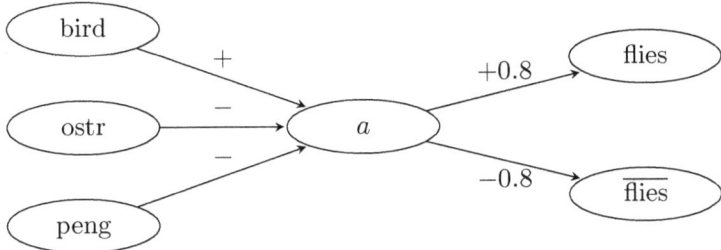

Figure 9: *A Carneades argument represented graphically.*

We now show how arguments and the generated argument graphs are represented using ADFs. The translation to ADFs is based on the techniques we have seen so far in this section. Consider the argument $a = \langle\{\text{bird}\}, \{\text{peng}, \text{ostr}\}, \text{flies}\rangle$ with $weight(a) = 0.8$. This argument is represented graphically as shown in Figure 9.

Apart from the duplication of propositions/complements the graphical representation corresponds to the original Carneades graph. Using techniques similar to the ones described earlier, we can properly define acceptance conditions such that an argument node is **t** in the ADF graph iff the argument is applicable, and a proposition node is **t** iff the proposition is acceptable. The acceptance condition of an argument node *arg* requires that all premises of *arg* are true, all exceptions false (assumptions can be handled by an easy preprocessing step). The acceptance condition of a proposition node depends on the proof standard and is modelled along the lines of what we have discussed earlier in this section. We leave the details to the reader. Note that we will resume our discussion of Carneades at the end of Section 5 where we show how the relevant acceptance conditions can be formalized in GRAPPA.

What has been gained by this reconstruction? Why is it useful? First of all, it shows the generality of ADFs. Secondly, it puts Carneades on safe formal ground. But in addition, and this is probably the main advantage, it allows us to give up the restriction of Carneades to acyclic argument graphs. Nothing in our translation rests on the assumption that Carneades is acyclic. The translation works perfectly well also for cyclic argument evaluation structures. The only difference is that the resulting ADF graph will have cycles as well. But handling cycles of this kind is part of the core functionality of ADFs, and they have a variety of different semantics to offer for this case, as we have seen in Section 3.

4.4 Rule-based Languages

A major strand of research in formal argumentation is concerned with using argumentation techniques to assign semantics to simple rule-based languages (see Chapter 6 of this handbook). Those languages are simple logic-inspired formalisms working with inference rules on a set of propositional literals. Inference rules can be strict, in which case the conclusion of the inference (a literal) must necessarily hold whenever all antecedents (also literals) hold. Inference rules can also be defeasible, which means that the conclusion *usually* holds whenever the antecedents hold. Here, the word "usually" suggests that there could be exceptional cases where a defeasible rule has not been applied [Pollock, 1987] (for example to avoid an imminent inconsistency).

Most of the existing works in this area translate rule-based languages to AFs by constructing arguments and identifying attacks. But this approach is not always without problems, as Caminada and Amgoud [Caminada and Amgoud, 2007] observed. (They even devised a set of *rationality postulates* for capturing the intended behavior of semantics for rule-based languages.) While there exist AF-based solutions to those problems [Wyner et al., 2013], we concentrate here on one approach using ADFs as target language [Strass, 2013b; Strass, 2015b]. Translating to ADFs instead of AFs has the additional benefit of tackling the problem of cyclic justifications amongst arguments on the semantic level instead of the syntactic one (like it is done in the ASPIC approach [Caminada and Amgoud, 2007] among others). We only give intuitions here and refer the reader to the original paper(s) for details [Strass, 2013b; Strass, 2015b].

Inspired by the approach of Wyner et al. [Wyner et al., 2013], Strass [Strass, 2013b; Strass, 2015b] directly uses the literals from the theory base as statements that express whether the literal holds. He also uses rule names as statements indicating that the rule is applicable. Additionally, for each rule r he creates a statement -r indicating that the rule has not been applied. Not applying a rule is acceptable for defeasible rules, but unacceptable for strict rules since it would violate the closure postulate. This is enforced via integrity constraints saying that it may not be the case in any model that the rule body holds but the head does not hold: Technically, for a strict rule r, he introduces a conditional self-attack of -r; this self-attack becomes active if (and only if) the body of r is satisfied but the head of r is not satisfied, thereby preventing this undesirable state of affairs from getting included in a model. Defeasible rules offer some degree of choice, whence it is left to the semantics whether or not to apply them. This choice is modelled by a mutual attack cycle between r and -r. The remaining acceptance conditions are equally straightforward:

- Opposite literals attack each other.

- A literal is accepted whenever some rule deriving it is applicable, that is, all rules with head ψ support statement ψ.

- A strict rule is applicable whenever all of its body literals hold, that is, the body literals of r are exactly the supporters of r.

- Likewise, a defeasible rule is applicable whenever all of its body literals hold, and additionally the negation of its head literal must not hold.

Strass [2013b, 2015b] showed that the approach satisfies the rationality postulates of Caminada and Amgoud [2007]. Furthermore, this method has a mild computational complexity (with an at most quadratic blowup from rule-based theory to ADF formalization, while there can be exponential to infinite blowup in other approaches).

5 Graph-based Argument Processing

We have seen in Section 4 how ADFs can be used to provide graphical representations of argumentation scenarios with semantics. The different approaches were based on translations from some graphical representation to ADFs. In a nutshell, the GRAPPA approach [Brewka and Woltran, 2014] described in this section addresses the opposite question: is it possible to extend the formal techniques underlying ADFs in such a way that the semantics of various kinds of graphical representations can be defined directly for these representations, without the detour of a translation? More specifically, we will consider arbitrary (edge) labelled graphs. Such graphs are highly popular for visualizing argumentation scenarios, and indeed this chapter (and the handbook) is full of such representations. The goal of this section is to define various semantics directly for such labelled graphs.

Another way of looking at the approach is the following: Dung AFs actually can be seen as graphs where all edges have the same label, which is left implicit for this reason. In addition, all nodes have the same type of acceptance condition. Dung's seminal contribution can thus be characterized as defining various semantics for specific graphs with a single label and uniform acceptance conditions. Our goal is to generalize this to arbitrary labelled graphs with flexible, user-defined acceptance conditions.

GRAPPA requires two major changes. First of all, the acceptance conditions can no longer be propositional formulas built from parent nodes, as in ADFs. We rather have to define them in terms of the labels of active links in the graph, that is links whose source nodes are accepted (true). More precisely, since it may be relevant whether there are multiple active links with the same label, we have to consider multisets of labels. An acceptance condition will thus be a function assigning a truth value to each multiset of labels. Secondly, we have to modify the operator Γ_D for ADFs D as defined in Section 3 in such a way that the new acceptance conditions are taken into account adequately.

In the following we describe multisets as functions into the natural numbers. Intuitively, the number assigned to an element describes the number of occurrences of the element in the multiset.

Definition 5.1 *An acceptance function over a set of labels L is a function* $c : (L \to \mathbb{N}) \to \{\mathbf{t}, \mathbf{f}\}$.

The set of all acceptance functions over L is denoted F^L.

Definition 5.2 *A labelled argument graph (LAG) is a tuple $G = (S, E, L, \lambda, \alpha)$*

where

- S is a set of nodes (statements),
- E is a set of edges (dependencies),
- L is a set of labels,
- $\lambda : E \to L$ assigns labels to edges,
- $\alpha : S \to F^L$ assigns L-acceptance-functions to nodes.

The characteristic operator Γ_G of a LAG G basically does what the corresponding operator does for ADFs: it takes a three-valued (or, equivalently, partial) interpretation v and produces a new one v'. In doing so, it checks which truth values of nodes in S can be justified by v. This is done by considering all possible completions of v, more precisely the multisets of active labels induced by completions of v. These multisets are obtained by including an occurrence of a particular label for each occurrence of that label in a link which is active in the completion. If the acceptance function of s yields $\mathbf{t}$ under all completions (more precisely, for all multisets induced by any completion), then v' assigns $\mathbf{t}$ to s. If the acceptance function of s yields $\mathbf{f}$ under all completions, then v' assigns $\mathbf{f}$ to s. In all other cases the value remains undefined.

Here are the formal details. Note that we represent here three-valued interpretations v as sets of literals: nodes true in v appear positively in the set, nodes assigned false appear negated, and undefined nodes are left out.

Definition 5.3 *Let $G = (S, E, L, \lambda, \alpha)$ be a LAG, v a three-valued interpretation of S. m_s^v, the multiset of active labels of $s \in S$ in G under v, is defined as*

$$m_s^v(l) = |\{(e, s) \in E \mid e \in v, \lambda((e, s)) = l\}|$$

for each $l \in L$.

The characteristic operator Γ_G of G takes a three-valued interpretation v of S and produces a revised three-valued interpretation $\Gamma_G(v)$ of S.

Definition 5.4 *Let $G = (S, E, L, \lambda, \alpha)$ be a LAG, v a three-valued interpretation of S. $\Gamma_G(v) = P_G(v) \cup N_G(v)$ with*

$$P_G(v) = \left\{ s \;\middle|\; \alpha(s)(m) = \mathbf{t} \text{ for each } m \in \{m_s^{v'} \mid v' \in [v]_c\} \right\}$$
$$N_G(v) = \left\{ \neg s \;\middle|\; \alpha(s)(m) = \mathbf{f} \text{ for each } m \in \{m_s^{v'} \mid v' \in [v]_c\} \right\}$$

With this new operator we can define the semantics of GRAPPA in exactly the same way as was done for ADFs:

Definition 5.5 Let $G = (S, E, L, \lambda, \alpha)$ be a LAG, v a three-valued interpretation of S.

- v is a model of G iff v is total and $v = \Gamma_G(v)$,
- v is grounded in G iff v is the least fixed point of Γ_G,
- v is admissible in G iff $v \subseteq \Gamma_G(v)$,
- v is preferred in G iff v is $\subseteq$-maximal admissible in G,
- v is complete in G iff $v = \Gamma_G(v)$.

Example 5.6 *This is a variation of Example 3.2. Consider the LAG with $S = \{a, b, c, d\}$ and $L = \{+, -\}$. The following graph shows the labels of each link.*

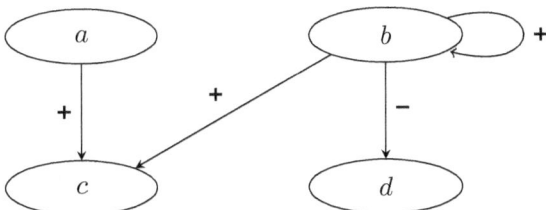

For simplicity, let us assume all nodes have the same acceptance condition requiring that all positive links must be active (that is the respective parents must be t) and no negative link is active.[10] *We obtain two models, namely $v_1 = \{a, b, c, \neg d\}$ and $v_2 = \{a, \neg b, \neg c, d\}$. The grounded interpretation is $v_3 = \{a\}$. The 16 admissible interpretations are exactly the same as for Example 3.9. Among the admissible interpretations $\{a, b, c, \neg d\}$ and $\{a, \neg b, \neg c, d\}$ are preferred. Complete interpretations are these two and in addition $\{a\}$.*

Now let us turn to stable semantics. The idea underlying stable semantics is to exclude self-justifying cycles. Again this semantics can be defined along the lines of the corresponding definition for ADFs in [Brewka et al., 2013]: take a model v, reduce the LAG based on v and check whether the grounded extension of the reduced LAG coincides with the nodes true in v. Here is the definition:

Definition 5.7 Let $G = (S, E, L, \lambda, \alpha)$ be a LAG, v a model of G, $S^v = v \cap S$. v is a stable model of G iff v restricted to S^v is the grounded interpretation of $G^v = (S^v, E^v, L, \lambda^v, \alpha^v)$, the v-reduct of G, where

- $E^v = E \cap (S^v \times S^v)$,

---

[10] In the pattern language developed later in this section this can be expressed as $(\#_t(+) - \#(+) = 0) \wedge (\#(-) = 0)$.

- λ^v is λ restricted to S^v,[11]

- α^v is α restricted to S^v.

Observe that in α^v we did not have to alter the values of the function, i.e. the true and false multisets remain the same (although some of them might become "unused" since the number of parents shrinked). We will see later that this exactly matches the stable semantics for ADFs from [Brewka et al., 2013]. For the moment, we continue our running example.

Example 5.8 For Example 5.6 we obtained two models, $v_1 = \{a, b, c, \neg d\}$ and $v_2 = \{a, \neg b, \neg c, d\}$. In v_1 the justification for b is obviously based on a cycle. The v_1-reduct of our graph is

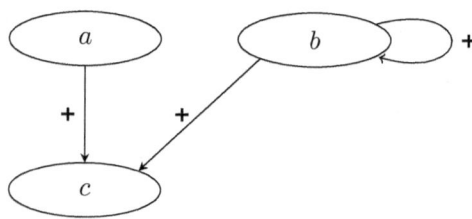

It is easy to see that the grounded interpretation of the reduced graph is $\{a\}$, v_1 is thus not a stable model, as intended. We leave it to the reader to verify that v_2 indeed is a stable model.

Results about the semantics carry over from ADFs [Brewka et al., 2013].

Proposition 5.9 Let G be a LAG. The following inclusions hold:

$$stb(G) \subseteq mod(G) \subseteq pref(G) \subseteq com(G) \subseteq adm(G),$$

where $stb(G), mod(G), pref(G), com(G)$ and $adm(G)$ denote the sets of stable models, models, preferred interpretations, complete interpretations and admissible interpretations of G, respectively. Moreover, $pref(G) \neq \emptyset$, whereas $mod(G') = \emptyset$ for some LAG G'.

A remaining question is how to actually specify acceptance functions for GRAPPA. In [Brewka and Woltran, 2014] a specific pattern language has been developed for this purpose. This pattern language allows for the specification of conditions on multisets of labels. In the patterns one can refer to the number of total and active labels of specific types, to minimal/maximal numerical labels of active links. It is also possible to use simple arithmetics and relations.

More precisely, GRAPPA acceptance functions are specified using *acceptance patterns* over a set of labels L defined as follows:

---

[11]Given a function $f : M \to N$ and $M' \subseteq M$, f restricted to M' is the function $f' : M' \to N$ such that $f'(m) = f(m)$ for all $m \in M'$.

- A *term* over L is of the form $\#(l)$, $\#_t(l)$ (with $l \in L$), or min, min_t, max, max_t, sum, sum_t, $count$, $count_t$.

- A *basic acceptance pattern* (over L) is of the form $a_1 t_1 + \cdots + a_n t_n \, R \, a$, where the t_i are terms over L, the a_is and a are integers and $R \in \{<, \leq, =, \neq, \geq, >\}$.

- An *acceptance pattern* (over L) is a basic acceptance pattern or a Boolean combination of acceptance patterns.

A GRAPPA instance then is a labelled argument graph with acceptance functions represented as acceptance patterns:

Definition 5.10 *A* GRAPPA *instance is a tuple* $G = (S, E, L, \lambda, \pi)$ *where S is a set of statements, E a set of edges, L a set of labels, λ an assignment of labels to edges, and π an assignment of acceptance patterns over L to all elements of S.*

We still need to specify what the acceptance function represented by a particular pattern assigned to a node s is. Recall that an acceptance function assigns a truth value in $\{\mathbf{t}, \mathbf{f}\}$ to a multiset of labels. We will define this function by specifying a satisfaction relation $\models$ between multisets and patterns: the basic idea is that a multiset receives value $\mathbf{t}$ iff it satisfies the corresponding pattern. The actual definition is slightly more complicated, though, as some of the terms (actually those indexed with t) are actually independent of the multiset, but depend on the node s, more precisely on the labels of links – active or not – with target s. For this reason, satisfaction of a pattern depends on both a multiset of labels and the node the pattern is assigned to via π. For a multiset of labels $m : L \to \mathbb{N}$ and $s \in S$ the value function val_s^m is:

$$\begin{aligned}
val_s^m(\#l) &= m(l) \\
val_s^m(\#_t l) &= |\{(e,s) \in E \mid \lambda((e,s)) = l\}| \\
val_s^m(min) &= \min\{l \in L \mid m(l) > 0\} \\
val_s^m(min_t) &= \min\{\lambda((e,s)) \mid (e,s) \in E\} \\
val_s^m(max) &= \max\{l \in L \mid m(l) > 0\} \\
val_s^m(max_t) &= \max\{\lambda((e,s)) \mid (e,s) \in E\} \\
val_s^m(sum) &= \sum_{l \in L} m(l) \\
val_s^m(sum_t) &= \sum_{(e,s) \in E} \lambda((e,s)) \\
val_s^m(count) &= |\{l \mid m(l) > 0\}| \\
val_s^m(count_t) &= |\{\lambda((e,s)) \mid (e,s) \in E\}|
\end{aligned}$$

$min_{(t)}$, $max_{(t)}$, $sum_{(t)}$ are undefined in case of non-numerical labels. For $\emptyset$ they yield the neutral element of the corresponding operation, i.e.

$$val_s^m(sum) = val_s^m(sum_t) = 0,$$
$$val_s^m(min) = val_s^m(min_t) = \infty,$$
$$val_s^m(max) = val_s^m(max_t) = -\infty.$$

Let m and s be as before. For basic acceptance patterns the *satisfaction relation* $\models$ is defined by

$$(m,s) \models a_1 t_1 + \cdots + a_n t_n R\, a \quad \text{iff} \quad \sum_{i=1}^{n}\left(a_i\, val_s^m(t_i)\right) R\, a.$$

The extension to Boolean combinations is as usual. The acceptance function represented by pattern p at node s then is the function assigning **t** to multiset m iff $(m,s) \models p$.

Example 5.11 *Let $L = \{\text{++},\text{+},\text{-},\text{--}\}$ be a set of labels representing strong support, support, attack and strong attack, respectively. Assume a node s is accepted if its (active) support is stronger than its attack, where we measure strength by counting the respective links, hereby multiplying strong support/attack with a factor of 2. This can be specified using the following pattern for s:*

$$2(\#\text{++}) + (\#\text{+}) - 2(\#\text{--}) - (\#\text{-}) > 0.$$

We conclude this section by showing how the necessary patterns for Carneades argument graphs, which we discussed in Section 4.3, can be defined in GRAPPA. Recall that these graphs have two kinds of nodes, argument nodes and propositions nodes. The pattern for all argument nodes is

$$((\#_t\text{+}) - (\#\text{+}) = 0) \wedge ((\#\text{-}) = 0).$$

which says that all premises and none of the exceptions must be accepted. The patterns for proposition nodes depend on their proof standard. Recall that some of these standards have additional numerical parameters α, β and γ. The terms max and min represent the maximal, respectively minimal, label of an active link:

- scintilla of evidence: $\max > 0$

- preponderance of evidence: $\max + \min > 0$

- clear and convincing evidence: $(\max > \alpha) \wedge (\max + \min > \beta)$

- beyond reasonable doubt: $(\max > \alpha) \wedge (\max + \min > \beta) \wedge (-\min < \gamma)$

- dialectical validity: $(\max > 0) \wedge (\min > 0)$

This representation of the acceptance conditions underlying Carneades is not only extremely simple. It has the big advantage that it is uniform: the patterns for all nodes with the same proof standard are actually the same. This is different from representations of proof standards and other notions we discussed in Section 4 in ADFs where the acceptance condition for each node depends on its specific parents.

6 Computational Aspects

In the introduction we discussed, in an informal manner, relationships between statements (arguments) that are supporting or attacking, in the sense that a statement can have a positive or negative influence on the acceptance of another statement. General ADFs have a generic notion of links (dependencies) between statements. However, such links can be formally categorized into 4 groups, depending on whether they have an attacking or supporting nature (or both or neither). This leads to the notion of so-called bipolar ADFs (BADFs for short) which contain only attacking or supporting dependencies. We will introduce them, based on the original definition of [Brewka and Woltran, 2010], in Section 6.1, together with the formalization of attacking and supporting links. Such BADFs are a subclass of general ADFs, yet have appealing computational properties. They generalize AFs in a direct manner, but are strictly "in-between" AFs and general ADFs w.r.t. their corresponding expressiveness. Results relating to expressiveness are presented in Section 6.2. Further, many frameworks arising in argumentation in AI, other than AFs, can be translated to BADFs [Polberg, 2016] (partially under semantics not discussed in this chapter).

From a computational perspective, BADFs have the following interesting properties: they have the same worst-time complexity as AFs for many semantics, while general ADFs typically exhibit higher computational complexity. We summarize these results in Section 6.3, followed by Section 6.4 that gives pointers to recent systems for computing reasoning tasks on ADFs and BADFs.

6.1 Bipolar ADFs

As we have seen in previous sections, the concept of acceptance condition is quite powerful. A natural question is to what extent different restrictions of acceptance conditions may form interesting subclasses of ADFs. One such subclass are bipolar ADFs, as already defined in [Brewka and Woltran, 2010]. This class relies on the concept of attacking and supporting links which are defined as follows.

Let $D = (S, L, C)$ be an ADF. Formally, a link $(r, s) \in L$ is

- supporting in D iff for all $R \subseteq par(s)$, we have $C_s(R) = \mathbf{t}$ implies $C_s(R \cup \{r\}) = \mathbf{t}$;

- attacking in D iff for all $R \subseteq par(s)$, we have $C_s(R \cup \{r\}) = \mathbf{t}$ implies $C_s(R) = \mathbf{t}$.

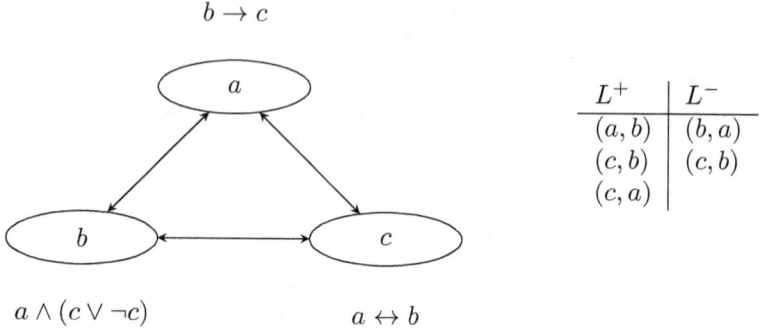

Figure 10: An ADF with link types.

We use $L^+ \subseteq L$ to denote all supporting and $L^- \subseteq L$ to denote all attacking links of L in an ADF $D = (S, L, C)$.

Example 6.1 *In Figure 10 we see an example ADF $D = (S, L, C)$ with $S = \{a, b, c\}$ and acceptance conditions $\varphi_a = b \to c$, $\varphi_b = a \land (c \lor \neg c)$, and $\varphi_c = a \leftrightarrow b$. On the right of that figure the link types are shown. Let us investigate why some of the links are supporting or attacking. Looking at the acceptance condition of a, φ_a, and the parents of a then we have the following relevant sets of statements (shown as two-valued interpretations):*

$$
\begin{aligned}
v_1 &\mathrel{\hat=} \{\neg b, \neg c\} & &\models \varphi_a \\
v_2 &\mathrel{\hat=} \{b, \neg c\} & &\not\models \varphi_a \\
v_3 &\mathrel{\hat=} \{\neg b, c\} & &\models \varphi_a \\
v_4 &\mathrel{\hat=} \{b, c\} & &\models \varphi_a
\end{aligned}
$$

We see, e.g., that the link (c, a) is supporting, because whenever c is added to a subset of parents that is mapped to $\mathbf{t}$ by C_a (switched to true in every model of φ_a) then the new set (interpretation) is again mapped to true by acceptance condition C_a (is a model of φ_a). More concretely, v_1, v_3, and v_4 are models of acceptance condition φ_a. Switching the truth value of c to true in each of them, results in v_3 and v_4 (assigning c to true in v_1 and v_3 results in both cases with v_3, and assigning c to true in v_4 is again equal to v_4). Both v_3 and v_4 are models of φ_a. This means (c, a) is a supporting link. Similarly, link (b, a) is attacking because whenever we remove b from a set of parents of a that is mapped to $\mathbf{t}$ by C_a we get a set that is likewise mapped to $\mathbf{t}$ by C_a.

Links (a, b) which are both attacking and supporting are so-called redundant links. The reason to call such a link redundant is that switching the truth value of a in any interpretation does not change the evaluation of acceptance condition φ_b w.r.t. the original interpretation and the modified interpretation. A link that is neither attacking nor supporting is called dependent.

Example 6.2 *Continuing Example 6.1, the link (c, b) is a redundant link. This link is both attacking and supporting. Redundancy means that the evaluation of φ_b is independent of the value of c (formula φ_b only depends on the truth value of a). In contrast, the links (b, c) and (a, c) are dependent links. For instance, $\{\neg a, \neg b\} \models \varphi_c$ and $\{a, \neg b\} \not\models \varphi_c$ taken together show that (a, c) is not supporting in this ADF. To see that (a, c) is not attacking, consider $\{a, b\} \models \varphi_c$ and $\{\neg a, b\} \not\models \varphi_c$.*

An ADF $D = (S, L, C)$ is bipolar (a BADF) if all links in L are supporting or attacking or both, i.e., $L = L^+ \cup L^-$. For example, our running example ADF from Example 3.2 is a BADF. Further, for any AF F its associated ADF D_F is bipolar, in fact each link in D_F is attacking.

Bipolar ADFs are still a quite expressible class; they allow acceptance conditions not only to express simply attack and support (for example $\neg a_1 \wedge \cdots \wedge \neg a_n \wedge s_1 \wedge \cdots \wedge s_m$ expressing that a statement is attacked by statements a_i and supported by statements s_j), but more advanced relations, like e.g. $((\neg a_1 \vee s_1) \wedge (\neg a_2 \vee s_2)) \vee \neg a_3$; in fact, all examples given in Section 4 are also bipolar ADFs. We would like to mention here that bipolar ADFs behave differently than the prominent class of bipolar AFs [Cayrol and Lagasquie-Schiex, 2013]. Indeed, several concepts of support relations have been discussed in the literature (abstract, deductive, necessary, and evidential support), thus a detailed discussion is beyond the scope of this chapter, and we refer the reader to works relating ADFs to formalisms including support [Polberg and Oren, 2014; Polberg, 2016]. However, what is important to state is that bipolar ADFs treat support and attack as equally strong concepts. Given the generality of bipolar ADFs which allow to "mix" support and attack as exemplified above, a distinct handling of support and attack in ADFs, e.g. as separated concepts in the language instead of a property of links and acceptance conditions, would require a lot of additional machinery.

Acceptance conditions in BADFs are, in fact, not only interesting for defining ADFs. The study of the concept of *bipolar Boolean functions* has meanwhile found applications outside of ADFs. Baumann and Strass (2016) have analyzed the integer sequence that arises when considering for each positive integer n the number of bipolar Boolean functions in n arguments. The resulting sequence is novel and has been added to the Online Encyclopedia of Number Sequences[12]. In further related work, Alviano, Faber, and Strass [Alviano et al., 2016] applied the concept of bipolar Boolean functions to aggregates in answer set programming and obtained a novel class of aggregates whose model checking problems (according to the semantics of Pelov et al. [Pelov et al., 2007] and Son and Pontelli [Son and Pontelli, 2007]) can be decided in deterministic polynomial time. They even identify a class that goes beyond bipolar Boolean functions but still retains polynomial-time decidability; this might constitute an interesting avenue for research that extends the bipolarity concept of ADFs.

---

[12] https://oeis.org/A245079

6.2 Expressiveness and Realizability

Expressiveness of a formalism $\mathcal{F}$ (i.e. the set of structures available in a formalism) with a semantics σ over a vocabulary A can be defined as the set of interpretation-sets over A that elements of $\mathcal{F}$ (the knowledge bases $\mathsf{kb} \in \mathcal{F}$ of that formalism) can produce. Formally, the *signature* of a formalism $\mathcal{F}$ w.r.t. semantics σ is the set

$$\Sigma^\sigma_\mathcal{F} = \{\sigma(\mathsf{kb}) \mid \mathsf{kb} \in \mathcal{F}\}$$

Intuitively, expressiveness is a basic measure of the capabilities of formalism $\mathcal{F}$ under σ, because it characterizes what "can and cannot be done" with $\mathcal{F}$ under semantics σ [Gogic et al., 1995]. Whenever we have two formalisms, say $\mathcal{F}_1$ and $\mathcal{F}_2$, that share a semantics σ and we find that $\Sigma^\sigma_{\mathcal{F}_1} \subsetneq \Sigma^\sigma_{\mathcal{F}_2}$, then this intuitively means that $\mathcal{F}_2$ is strictly more expressive than $\mathcal{F}_1$: all sets $V \subseteq \mathcal{V}_3$ that can be realized with $\mathcal{F}_1$ can be realized with $\mathcal{F}_2$, and there is at least one set $V \subseteq \mathcal{V}_3$ that can be realized with $\mathcal{F}_2$ but not with $\mathcal{F}_1$.

For AFs, BADFs and ADFs under various semantics, their relative expressiveness is summarized in the following result [Strass, 2015c; Strass, 2015a; Linsbichler et al., 2016a].

Theorem 6.3 *For* $\sigma \in \{adm, com, prf, mod\}$, *we find that*

$$\Sigma^\sigma_{AF} \subsetneq \Sigma^\sigma_{BADF} \subsetneq \Sigma^\sigma_{ADF}.$$

For the stable model semantics stb, we find that

$$\Sigma^{mod}_{AF} = \Sigma^{stb}_{AF} \subsetneq \Sigma^{stb}_{BADF} = \Sigma^{stb}_{ADF}.$$

Furthermore, for the model semantics we have

$$\Sigma^{mod}_{ADF} = \mathcal{V}_2 = \{v : A \to \{\mathbf{t}, \mathbf{f}\}\},$$

that is, ADFs *under the model semantics are universally expressive.*

Example 6.4 *We give example sets of interpretations that can be used to witness* $\Sigma^{prf}_{AF} \subsetneq \Sigma^{prf}_{BADF} \subsetneq \Sigma^{prf}_{ADF}$. *Consider* $S = \{a, b, c\}$ *and interpretations* $v_1 = \{a \mapsto \mathbf{t}, b \mapsto \mathbf{t}, c \mapsto \mathbf{f}\}$, $v_2 = \{a \mapsto \mathbf{t}, b \mapsto \mathbf{f}, c \mapsto \mathbf{t}\}$, *and* $v_3 = \{a \mapsto \mathbf{f}, b \mapsto \mathbf{t}, c \mapsto \mathbf{t}\}$. *To see that* $\{v_1, v_2, v_3\} \in \Sigma^{prf}_{BADF}$, *consider the* ADF *over* S *with acceptance conditions* $\varphi_a = \neg b \vee \neg c$, $\varphi_b = \neg a \vee \neg c$, *and* $\varphi_c = \neg a \vee \neg b$. *It is easy to verify that this* ADF *is bipolar and that* $\{v_1, v_2, v_3\}$ *constitute its preferred interpretations. On the other hand, from results in [Dunne et al., 2015] it follows that there is no* AF *with preferred extensions* $\{a, b\}$, $\{a, c\}$, *and* $\{b, c\}$. *In fact, this is quite easy to see: consider there would exist an* AF F *with those three preferred extensions. Then, there cannot be an attack in* F *between* a *and* b, *and moreover* $\{a, b\}$ *defends itself in* F; *the same holds for the pairs* a, c, *and* b, c. *But then,* $\{a, b, c\}$ *has to be conflict-free in* F *and defends itself, and thus* $\{a, b\}$ *(and likewise,* $\{a, c\}$ *and* $\{b, c\}$*) cannot be preferred in* F; *a contradiction.*

For $\Sigma^{prf}_{BADF} \subsetneq \Sigma^{prf}_{ADF}$, we use an example given in [Linsbichler et al., 2016b, Theorem 8]: consider $S' = \{a,b\}$ and interpretations $v_4 = \{a \mapsto \mathbf{t}, b \mapsto \mathbf{t}\}$, $v_5 = \{a \mapsto \mathbf{t}, b \mapsto \mathbf{f}\}$, and $v_6 = \{a \mapsto \mathbf{f}, b \mapsto \mathbf{u}\}$. For $X' = \{v_4, v_5, v_6\}$ we have $X' \in \Sigma^{prf}_{ADF}$, but $X' \notin \Sigma^{prf}_{BADF}$. For general ADFs, one example ADF is $D' = (S', L', \{\varphi_a = a, \varphi_b = a \leftrightarrow b\})$. All three interpretations v_4, v_5, and v_6 are preferred interpretations of D'. This ADF D' is not bipolar (due to φ_b, see Example 6.2). There is no BADF that has X' exactly as its preferred interpretations.[13]

While this shows that BADFs can do strictly more than AFs, and in turn ADFs can do strictly more than BADFs (with the exception of the stable model semantics), there is little information on *what exactly* these signatures look like. Work on precisely characterizing signatures has been carried out for AFs [Dunne et al., 2015]; the results can be found in Chapter 17 of this handbook. There has also been work on characterizing realizability for ADFs under two-valued [Strass, 2015a] and three-valued [Pührer, 2015; Linsbichler et al., 2016a] semantics.

Finally, initial results on characterizing the representational *succinctness* of these formalisms have recently been obtained. Succinctness not only takes into account *what* formalisms can realize, but also *to what representational cost*, that is, what amount of space is needed to represent the smallest knowledge base realizing some desired set of interpretations. Again, the capabilities of different formalisms can be compared with respect to this measure [Gogic et al., 1995]. As one promising result on ADFs, it turned out that even BADFs are exponentially more succinct than normal logic programs [Strass, 2015a].

6.3 Computational Complexity

The computational complexity of ADFs is well-studied [Strass and Wallner, 2014; Strass and Wallner, 2015; Gaggl et al., 2015; Brewka et al., 2013; Polberg and Wallner, 2017; Wallner, 2014]; for an overview we refer the reader to Chapter 13 of this volume. For the reader's convenience we repeat here the main results. For a specified semantics σ, the main reasoning tasks for ADFs to solve are:

- Credulous acceptance of a statement: is statement s assigned to true in at least one interpretation under semantics σ?

- Skeptical acceptance of a statement: is statement s assigned to true in all interpretations under semantics σ?

- Interpretation verification: is a given interpretation an interpretation under semantics σ?

- Interpretation existence: is there an interpretation under semantics σ?

---

[13]For an automated way to check whether for a given set of three-valued interpretations there is an ADF, BADF, or AF that has exactly this set as its σ-interpretations, one can use the system UNREAL [Linsbichler et al., 2016b], available at http://www.dbai.tuwien.ac.at/proj/adf/unreal/.

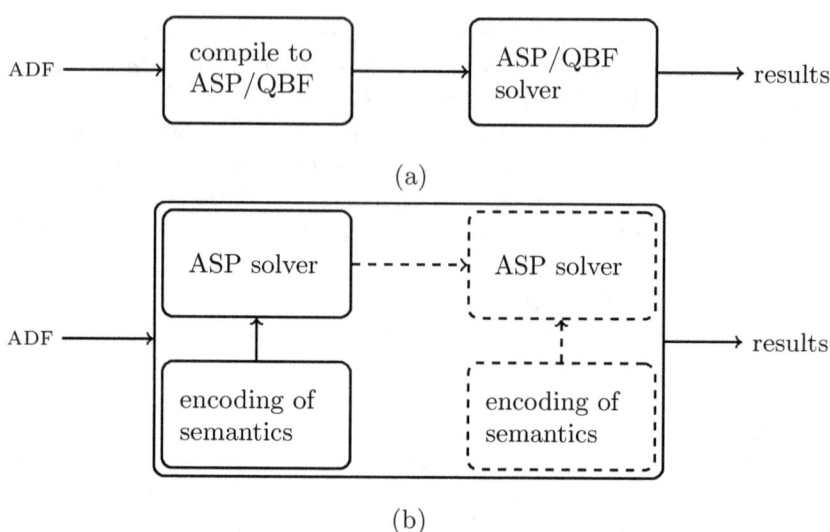

Figure 11: Workflow for systems based on (a) instance-based compilation (QADF, GRAPPAVIS, and YADF), and (b) static encodings (DIAMOND and GRAPPAVIS)

- Non-trivial interpretation existence: is there an interpretation under semantics σ assigning true or false to some statement?

Briefly put, complexity of reasoning tasks on general ADFs is situated one level higher in the polynomial hierarchy compared to the corresponding tasks on AFs. For BADFs complexity of reasoning stays at the same level as reasoning on AFs for most reasoning tasks, if the link type (attack or support) for each link is known (part of the input). Thus, BADFs offer more modeling capabilities than AFs while having the same (worst-case) computational cost as AFs for many reasoning tasks.

6.4 Systems

Systems for implementing reasoning on ADFs rely on declarative encodings in answer-set programming (ASP) [Brewka et al., 2011] or quantified Boolean satisfiability, and utilize available solvers for these languages [Gebser et al., 2011; Lonsing and Biere, 2010]. Most prominently, the DIAMOND family[14] [Strass and Ellmauthaler, 2017; Ellmauthaler and Strass, 2016; Ellmauthaler and Strass, 2014; Ellmauthaler and Strass, 2013] consists of ASP-based systems for reasoning on ADFs. In each DIAMOND version an ADF is encoded via ASP facts and, when augmented with static encodings for semantics, several reasoning tasks can be solved by computing answer-sets of the resulting ASP. Depending on the complexity of the reasoning task and used options in DIAMOND one call (in

---

[14] http://diamond-adf.sourceforge.net/

some family members two calls) to an ASP-solver are carried out to solve the given problem instance. DIAMOND includes dedicated BADF-specific encodings that make use of BADFs' upper complexity bounds.

The system QADF[15] [Diller et al., 2015] uses solvers for quantified Boolean formulas (QBFs) to perform reasoning on ADFs. In QADF, in contrast to DIAMOND, each ADF instance is compiled to a QBF incorporating both the input ADF and the chosen semantics, i.e., the encodings for the semantics are not static.

GRAPPAVIS[16] [Heißenberger, 2016] is a system implementing GRAPPA (see Section 5) and incorporates both instance-based compilation of GRAPPA input into declarative ASP encodings and static encodings for the semantics utilizing in both cases one ASP solver call.

The system YADF[17] [Brewka et al., 2017] is an ASP-based system for ADFs, based on the encodings for GRAPPA used in GRAPPAVIS. This system compiles ADF instances into one program to call an ASP solver (once).

The basic workflows for DIAMOND, QADF, GRAPPAVIS, and YADF are shown in Figure 11. With this figure we illustrate that QADF, GRAPPAVIS, and YADF implement algorithms that take an instance of an ADF, compile this instance, together with the chosen semantics and reasoning task, to one instance of an ASP or a QBF. On the other hand, DIAMOND and GRAPPAVIS implement algorithms that take an instance of an ADF, add to this instance a static encoding for the semantics and reasoning task, and give these to an ASP solver (with calling such a solver once or twice, depending on the task). The difference between (a) and (b) is that in (a) ADF and semantics have to be compiled together into one input for the solver, while for (b) semantics can be encoded separately (and modified separately).

A technique to cope with the high computational complexity of reasoning on ADFs was proposed by Linsbichler (2014). The technique is based on splitting the input ADF into partitions and solving one partition and transforming and solving the other partitions accordingly.

7 Conclusion

In this chapter, we have reviewed the argumentation formalism of abstract dialectical frameworks (ADFs). In contrast to Dung style frameworks, ADFs allow for a much more general specification of the interrelationship between the arguments. We have discussed how standard semantics like admissible, grounded, complete, preferred and stable can be generalized to ADFs by making use of the well known approximation fixpoint theory due to Denecker, Marek and Truszczyński [Denecker et al., 2004].

Alternative approaches to defining ADF semantics can be found in the works of Polberg and colleagues [Polberg et al., 2013; Polberg, 2014a; Polberg, 2014b;

---

[15]http://www.dbai.tuwien.ac.at/proj/adf/qadf/
[16]http://www.dbai.tuwien.ac.at/proj/adf/grappavis/
[17]http://www.dbai.tuwien.ac.at/proj/adf/yadf/

Polberg, 2015]. Likewise, further well-known semantics for ADFs have been generalized to ADFs, e.g. naive, stage, and the cf2 family of semantics [Gaggl and Strass, 2014] and an alternative, symmetric version of the naive semantics [Strass and Wallner, 2015].

A further subclass of ADFs, related to a certain notion of acyclicity and different from BADFs, is investigated in [Polberg, 2015; Polberg, 2016]. Other authors have analyzed the relationship of ADFs and logic programs [Strass, 2013a; Alviano and Faber, 2015] and in the course of that have defined new ADF semantics, like approximate stable models [Strass, 2013a], F-stable models [Alviano and Faber, 2015], and the grounded fixpoint semantics [Bogaerts et al., 2015]. The whole ADF formalism has even been lifted to the probabilistic case [Polberg and Doder, 2014].

We also addressed the modelling capabilities of ADFs; for a thorough discussion on the relation between ADFs and other argumentations frameworks, see also [Polberg, 2017]. A further application of ADFs in the context of legal reasoning can be found in [Al-Abdulkarim et al., 2014; Al-Abdulkarim et al., 2016]. The use of ADFs in text exploration has been investigated in [Cabrio and Villata, 2016]. Finally, we discussed the GRAPPA approach which makes use of ADF-like semantics in a flexible graph-based formalism. GRAPPA is the formal system underlying a mobile argumentation app developed by Pührer [2017].

Acknowledgements

We thank the anonymous reviewers for various comments which helped to significantly improve the chapter. This research has been supported by the German Research Foundation (DFG) (project BR 1817/7-2), the Austrian Science Fund (FWF) (projects I2854, P25521, and P30168-N31), and by Academy of Finland under grants 251170 COIN and 284591.

BIBLIOGRAPHY

[Al-Abdulkarim et al., 2014] Latifa Al-Abdulkarim, Katie Atkinson, and Trevor J. M. Bench-Capon. Abstract dialectical frameworks for legal reasoning. In Rinke Hoekstra, editor, *Proceedings of the 27th Conference on Legal Knowledge and Information Systems (JURIX)*, volume 271 of *Frontiers in Artificial Intelligence and Applications*, pages 61–70. IOS Press, 2014.

[Al-Abdulkarim et al., 2016] Latifa Al-Abdulkarim, Katie Atkinson, and Trevor J. M. Bench-Capon. A methodology for designing systems to reason with legal cases using abstract dialectical frameworks. *Artif. Intell. Law*, 24(1):1–49, 2016.

[Alviano and Faber, 2015] Mario Alviano and Wolfgang Faber. Stable model semantics of abstract dialectical frameworks revisited: A logic programming perspective. In Qiang Yang and Michael Wooldridge, editors, *Proceedings of the 24th International Joint Conference on Artificial Intelligence (IJCAI)*, pages 2684–2690. AAAI Press, 2015.

[Alviano et al., 2016] Mario Alviano, Wolfgang Faber, and Hannes Strass. Boolean functions with ordered domains in Answer Set Programming. In Dale Schuurmans and Michael P. Wellman, editors, *Proceedings of the Thirtieth AAAI Conference on Artificial Intelligence*, pages 879–885. AAAI Press, 2016.

[Amgoud and Cayrol, 2002] Leila Amgoud and Claudette Cayrol. A reasoning model based on the production of acceptable arguments. *Ann. Math. Artif. Intell*, 34(1-3):197–215, 2002.

[Amgoud and Vesic, 2011] Leila Amgoud and Srdjan Vesic. A new approach for preference-based argumentation frameworks. *Ann. Math. Artif. Intell*, 63:149–183, 2011.

[Baroni et al., 2011] Pietro Baroni, Federico Cerutti, Massimiliano Giacomin, and Giovanni Guida. AFRA: Argumentation framework with recursive attacks. *Int. J. Approx. Reasoning*, 52(1):19–37, 2011.

[Baumann and Strass, 2016] Ringo Baumann and Hannes Strass. On the number of bipolar Boolean functions. *International Journal of Integer Sequences*, 2016. Submitted.

[Bench-Capon, 2003] Trevor J. M. Bench-Capon. Persuasion in practical argument using value-based argumentation frameworks. *J. Log. Comput.*, 13(3):429–448, 2003.

[Bogaerts et al., 2015] Bart Bogaerts, Joost Vennekens, and Marc Denecker. Grounded fixpoints and their applications in knowledge representation. *Artif. Intell.*, 224:51–71, 2015.

[Brewka and Gordon, 2010] Gerhard Brewka and Thomas F. Gordon. Carneades and Abstract Dialectical Frameworks: A Reconstruction. In Pietro Baroni, Federico Cerutti, Massimiliano Giacomin, and Guillermo R. Simari, editors, *Proceedings of the 3rd International Conference on Computational Models of Argument (COMMA)*, volume 216 of *Frontiers in Artificial Intelligence and Applications*, pages 3–12. IOS Press, 2010.

[Brewka and Woltran, 2010] Gerhard Brewka and Stefan Woltran. Abstract Dialectical Frameworks. In Fangzhen Lin, Ulrike Sattler, and Miroslaw Truszczyński, editors, *Proceedings of the 12th International Conference on Principles of Knowledge Representation and Reasoning (KR)*, pages 102–111. AAAI Press, 2010.

[Brewka and Woltran, 2014] Gerhard Brewka and Stefan Woltran. GRAPPA: A semantical framework for graph-based argument processing. In Torsten Schaub, Gerhard Friedrich, and Barry O'Sullivan, editors, *Proceedings of the 21st European Conference on Artificial Intelligence (ECAI)*, volume 263 of *Frontiers in Artificial Intelligence and Applications*, pages 153–158. IOS Press, 2014.

[Brewka et al., 2011] Gerhard Brewka, Thomas Eiter, and Mirosław Truszczyński. Answer set programming at a glance. *Commun. ACM*, 54(12):92–103, 2011.

[Brewka et al., 2013] Gerhard Brewka, Stefan Ellmauthaler, Hannes Strass, Johannes P. Wallner, and Stefan Woltran. Abstract Dialectical Frameworks Revisited. In Francesca Rossi, editor, *Proceedings of the 23rd International Joint Conference on Artificial Intelligence (IJCAI)*, pages 803–809. AAAI Press / IJCAI, August 2013.

[Brewka et al., 2014] Gerhard Brewka, Sylwia Polberg, and Stefan Woltran. Generalizations of Dung Frameworks and Their Role in Formal Argumentation. *IEEE Intelligent Systems*, 29(1):30–38, 2014.

[Brewka et al., 2017] Gerhard Brewka, Martin Diller, Georg Heissenberger, Thomas Linsbichler, and Stefan Woltran. Solving advanced argumentation problems with answer-set programming. In Satinder P. Singh and Shaul Markovitch, editors, *Proceedings of the Thirty-First AAAI Conference on Artificial Intelligence*, pages 1077–1083. AAAI Press, 2017.

[Cabrio and Villata, 2016] Elena Cabrio and Serena Villata. Abstract dialectical frameworks for text exploration. In *Proceedings of the 8th International Conference on Agents and Artificial Intelligence (ICAART 2016)*, pages 85–95. SciTePress, 2016.

[Caminada and Amgoud, 2007] Martin Caminada and Leila Amgoud. On the evaluation of argumentation formalisms. *Artif. Intell.*, 171(5–6):286–310, 2007.

[Caminada and Gabbay, 2009] Martin W. A. Caminada and Dov M. Gabbay. A logical account of formal argumentation. *Studia Logica*, 93(2-3):109–145, 2009.

[Cayrol and Lagasquie-Schiex, 2013] Claudette Cayrol and Marie-Christine Lagasquie-Schiex. Bipolarity in argumentation graphs: Towards a better understanding. *Int. J. Approx. Reasoning*, 54(7):876–899, 2013.

[Coste-Marquis et al., 2012] Sylvie Coste-Marquis, Sébastien Konieczny, Pierre Marquis, and Mohand Akli Ouali. Weighted attacks in argumentation frameworks. In Gerhard Brewka, Thomas Eiter, and Sheila A. McIlraith, editors, *Proceedings of the 13th International Conference on Principles of Knowledge Representation and Reasoning (KR)*, pages 593–597. AAAI Press, 2012.

[Denecker et al., 2000] Marc Denecker, Victor W. Marek, and Miroslaw Truszczyński. Approximations, Stable Operators, Well-Founded Fixpoints and Applications in Nonmonotonic Reasoning. In *Logic-Based Artificial Intelligence*, pages 127–144. Kluwer Academic Publishers, 2000.

[Denecker et al., 2004] Marc Denecker, Victor W. Marek, and Miroslaw Truszczyński. Ultimate approximation and its application in nonmonotonic knowledge representation systems. *Inf. Comput.*, 192(1):84–121, 2004.

[Diller et al., 2015] Martin Diller, Johannes Peter Wallner, and Stefan Woltran. Reasoning in abstract dialectical frameworks using quantified boolean formulas. *Argument & Computation*, 6(2):149–177, 2015.

[Dunne et al., 2011] Paul E. Dunne, Anthony Hunter, Peter McBurney, Simon Parsons, and Michael Wooldridge. Weighted argument systems: Basic definitions, algorithms, and complexity results. *Artif. Intell.*, 175(2):457–486, 2011.

[Dunne et al., 2015] Paul E. Dunne, Wolfgang Dvořák, Thomas Linsbichler, and Stefan Woltran. Characteristics of multiple viewpoints in abstract argumentation. *Artif. Intell.*, 228:153–178, 2015.

[Ellmauthaler and Strass, 2013] Stefan Ellmauthaler and Hannes Strass. The DIAMOND system for argumentation: Preliminary report. In Michael Fink and Yuliya Lierler, editors, *Proceedings of the Sixth International Workshop on Answer Set Programming and Other Computing Paradigms (ASPOCP)*, September 2013.

[Ellmauthaler and Strass, 2014] Stefan Ellmauthaler and Hannes Strass. The diamond system for computing with abstract dialectical frameworks. In Simon Parsons, Nir Oren, Chris Reed, and Federico Cerutti, editors, *Proceedings of the 5th International Conference on Computational Models of Argument (COMMA)*, volume 266 of *Frontiers in Artificial Intelligence and Applications*, pages 233–240. IOS Press, 2014.

[Ellmauthaler and Strass, 2016] Stefan Ellmauthaler and Hannes Strass. DIAMOND 3.0 – A native C++ implementation of DIAMOND. In Pietro Baroni, editor, *Proceedings of the Sixth International Conference on Computational Models of Argument (COMMA)*, volume 287 of *Frontiers in Artificial Intelligence and Applications*, pages 471–472, Potsdam, Germany, September 2016. IOS Press.

[Farley and Freeman, 1995] Arthur M. Farley and Kathleen Freeman. Burden of proof in legal argumentation. In *Proceedings of the 5th International Conference on Artificial Intelligence and Law (ICAIL)*, pages 156–164, 1995.

[Gaggl and Strass, 2014] Sarah Alice Gaggl and Hannes Strass. Decomposing Abstract Dialectical Frameworks. In Simon Parsons, Nir Oren, and Chris Reed, editors, *Proceedings of the 5th International Conference on Computational Models of Argument (COMMA)*, volume 266 of *Frontiers in Artificial Intelligence and Applications*, pages 281–292. IOS Press, 2014.

[Gaggl et al., 2015] Sarah Alice Gaggl, Sebastian Rudolph, and Hannes Strass. On the computational complexity of naive-based semantics for abstract dialectical frameworks. In Qiang Yang and Michael Wooldridge, editors, *Proceedings of the 24th International Joint Conference on Artificial Intelligence (IJCAI)*, pages 2985–2991. IJCAI/AAAI, 2015.

[Gebser et al., 2011] Martin Gebser, Roland Kaminski, Benjamin Kaufmann, Max Ostrowski, Torsten Schaub, and Marius Schneider. Potassco: The Potsdam answer set solving collection. *AI Commun.*, 24(2):107–124, 2011.

[Gogic et al., 1995] Goran Gogic, Henry Kautz, Christos Papadimitriou, and Bart Selman. The comparative linguistics of knowledge representation. In *Proceedings of the 14th International Joint Conference on Artificial Intelligence (IJCAI)*, pages 862–869. Morgan Kaufmann, 1995.

[Gordon et al., 2007] Thomas F. Gordon, Henry Prakken, and Douglas Walton. The carneades model of argument and burden of proof. *Artif. Intell.*, 171(10-15):875–896, 2007.

[Heißenberger, 2016] G. Heißenberger. A system for advanced graphical argumentation formalisms. Master's thesis, TU Wien, 2016. Available at http://www.dbai.tuwien.ac.at/proj/adf/grappavis/.

[Hunter, 2013] Anthony Hunter. A probabilistic approach to modelling uncertain logical arguments. *Int. J. Approx. Reasoning*, 54(1):47 – 81, 2013.

[Linsbichler et al., 2016a] Thomas Linsbichler, Jörg Pührer, and Hannes Strass. Characterizing realizability in abstract argumentation. In Gabriele Kern-Isberner and Renata Wassermann, editors, *Proceedings of the 16th International Workshop on Non-Monotonic Reasoning (NMR)*, April 2016.

[Linsbichler et al., 2016b] Thomas Linsbichler, Jörg Pührer, and Hannes Strass. A uniform account of realizability in abstract argumentation. In Gal A. Kaminka, Maria Fox, Paolo Bouquet, Eyke Hüllermeier, Virginia Dignum, Frank Dignum, and Frank van Harmelen, editors, *ECAI 2016 - 22nd European Conference on Artificial Intelligence, 29 August-2 September 2016, The Hague, The Netherlands - Including Prestigious Applications of Artificial Intelligence (PAIS 2016)*, volume 285 of *Frontiers in Artificial Intelligence and Applications*, pages 252–260. IOS Press, 2016.

[Linsbichler, 2014] Thomas Linsbichler. Splitting abstract dialectical frameworks. In Simon Parsons, Nir Oren, Chris Reed, and Federico Cerutti, editors, *Proceedings of the 5th International Conference on Computational Models of Argument (COMMA)*, volume 266 of *Frontiers in Artificial Intelligence and Applications*, pages 357–368. IOS Press, 2014.

[Lonsing and Biere, 2010] Florian Lonsing and Armin Biere. DepQBF: A dependency-aware QBF solver. *JSAT*, 7(2-3):71–76, 2010.

[Martínez et al., 2008] Diego C. Martínez, Alejandro Javier García, and Guillermo Ricardo Simari. An abstract argumentation framework with varied-strength attacks. In Gerhard Brewka and Jérôme Lang, editors, *Proceedings of the 11th International Conference on Principles of Knowledge Representation and Reasoning (KR)*, pages 135–144. AAAI Press, 2008.

[Modgil, 2009] Sanjay Modgil. Reasoning about preferences in argumentation frameworks. *Artif. Intell.*, 173(9-10):901–934, 2009.

[Nielsen and Parsons, 2007] Søren Nielsen and Simon Parsons. A generalization of Dung's abstract framework for argumentation: Arguing with sets of attacking arguments. In Nicolas Maudet, Simon Parsons, and Iyad Rahwan, editors, *Proc. ArgMAS*, volume 4766 of *Lecture Notes in Computer Science*, pages 54–73. Springer, 2007.

[Nouioua, 2013] Farid Nouioua. AFs with necessities: Further semantics and labelling characterization. In Weiru Liu, V. S. Subrahmanian, and Jef Wijsen, editors, *Scalable Uncertainty Management - 7th International Conference, SUM 2013*, volume 8078 of *Lecture Notes in Computer Science*, pages 120–133. Springer, 2013.

[Oren and Norman, 2008] Nir Oren and Timothy J. Norman. Semantics for evidence-based argumentation. In Philippe Besnard, Sylvie Doutre, and Anthony Hunter, editors, *Proceedings of the 2nd International Conference on Computational Models of Argument (COMMA)*, volume 172 of *Frontiers in Artificial Intelligence and Applications*, pages 276–284. IOS Press, 2008.

[Pelov et al., 2007] Nikolay Pelov, Marc Denecker, and Maurice Bruynooghe. Well-founded and stable semantics of logic programs with aggregates. *Theory and Practice of Logic Programming*, 7(3):301–353, 2007.

[Polberg and Doder, 2014] Sylwia Polberg and Dragan Doder. Probabilistic abstract dialectical frameworks. In Eduardo Fermé and João Leite, editors, *Proceedings of the 14th European Conference on Logics in Artificial Intelligence (JELIA)*, volume 8761 of *Lecture Notes in Artificial Intelligence*, pages 591–599. Springer, 2014.

[Polberg and Oren, 2014] Sylwia Polberg and Nir Oren. Revisiting support in abstract argumentation systems. In Simon Parsons, Nir Oren, Chris Reed, and Federico Cerutti, editors, *Proceedings of the 5th International Conference on Computational Models of Argument (COMMA)*, volume 266 of *Frontiers in Artificial Intelligence and Applications*, pages 369–376, 2014.

[Polberg and Wallner, 2017] Sylwia Polberg and Johannes P. Wallner. Preliminary report on complexity analysis of extension-based semantics of abstract dialectical frameworks. Technical Report DBAI-TR-2017-103, TU Wien, 2017.

[Polberg et al., 2013] Sylwia Polberg, Johannes P. Wallner, and Stefan Woltran. Admissibility in the Abstract Dialectical Framework. In João Leite, Tran Cao Son, Paolo Torroni, Leon van der Torre, and Stefan Woltran, editors, *Proceedings of the 14th International Workshop on Computational Logic in Multi-Agent System (CLIMA)*, volume 8143 of *Lecture Notes in Artificial Intelligence*, pages 102–118. Springer, 2013.

[Polberg, 2014a] Sylwia Polberg. Extension-based semantics of abstract dialectical frameworks. In Sébastien Koniecny and Hans Tompits, editors, *Proceedings of the 15th International Workshop on Non-Monotonic Reasoning, NMR 2014*, pages 273–282, 2014.

[Polberg, 2014b] Sylwia Polberg. Extension-based semantics of abstract dialectical frameworks. In Ulle Endriss and João Leite, editors, *Proceedings of the 7th European Starting AI Researcher Symposium, STAIRS 2014*, pages 240–249, 2014.

[Polberg, 2015] Sylwia Polberg. Revisiting extension-based semantics of abstract dialectical frameworks. Technical Report DBAI-TR-2015-88, TU Wien, 2015.

[Polberg, 2016] Sylwia Polberg. Understanding the abstract dialectical framework. In Loizos Michael and Antonis C. Kakas, editors, *Logics in Artificial Intelligence - 15th European Conference, JELIA 2016*, volume 10021 of *Lecture Notes in Computer Science*, pages 430–446, 2016.

[Polberg, 2017] Sylwia Polberg. Intertranslatability of abstract argumentation frameworks. Technical Report DBAI-TR-2015-104, TU Wien, 2017.

[Pollock, 1987] John L Pollock. Defeasible reasoning. *Cognitive Science*, 11(4):481–518, 1987.

[Pührer, 2015] Jörg Pührer. Realizability of three-valued semantics for abstract dialectical frameworks. In Qiang Yang and Michael Wooldridge, editors, *Proceedings of the 24th International Joint Conference on Artificial Intelligence (IJCAI)*, pages 3171–3177. AAAI Press, 2015.

[Pührer, 2017] Jörg Pührer. ArgueApply: A mobile app for argumentation. In Marcello Balduccini and Tomi Janhunen, editors, *Proceedings of the 14th International Conference on Logic Programming and Nonmonotonic Reasoning (LPNMR 2017)*, volume 10377 of *Lecture Notes in Computer Science*, pages 250–262. Springer, 2017.

[Simari and Loui, 1992] Guillermo R. Simari and Ronald P. Loui. A mathematical treatment of defeasible reasoning and its implementation. *Artif. Intell.*, 53(2–3):125 – 157, 1992.

[Son and Pontelli, 2007] Tran Cao Son and Enrico Pontelli. A constructive semantic characterization of aggregates in answer set programming. *Theory and Practice of Logic Programming*, 7(3):355–375, 2007.

[Strass and Ellmauthaler, 2017] Hannes Strass and Stefan Ellmauthaler. goDIAMOND 0.6.6 – ICCMA 2017 system description, 2017. Second International Competition on Computational Models of Argumentation: http://www.dbai.tuwien.ac.at/iccma17/.

[Strass and Wallner, 2014] Hannes Strass and Johannes P. Wallner. Analyzing the Computational Complexity of Abstract Dialectical Frameworks via Approximation Fixpoint Theory. In Chitta Baral, Giuseppe De Giacomo, and Thomas Eiter, editors, *Proceedings of the 14th International Conference on Principles of Knowledge Representation and Reasoning (KR)*, pages 101–110. AAAI Press, 2014.

[Strass and Wallner, 2015] Hannes Strass and Johannes Peter Wallner. Analyzing the computational complexity of abstract dialectical frameworks via approximation fixpoint theory. *Artif. Intell.*, 226:34–74, 2015.

[Strass, 2013a] Hannes Strass. Approximating Operators and Semantics for Abstract Dialectical Frameworks. *Artificial Intelligence*, 205:39–70, 2013.

[Strass, 2013b] Hannes Strass. Instantiating Knowledge Bases in Abstract Dialectical Frameworks. In João Leite, Tran Cao Son, Paolo Torroni, Leon van der Torre, and Stefan Woltran, editors, *Proceedings of the 14th International Workshop on Computational Logic in Multi-Agent Systems (CLIMA)*, volume 8143 of *Lecture Notes in Artificial Intelligence*, pages 86–101. Springer, 2013.

[Strass, 2015a] Hannes Strass. Expressiveness of two-valued semantics for abstract dialectical frameworks. *J. Artif. Intell. Res. (JAIR)*, 54:193–231, 2015.

[Strass, 2015b] Hannes Strass. Instantiating rule-based defeasible theories in abstract dialectical frameworks and beyond. *J. Log. Comput.*, 2015.

[Strass, 2015c] Hannes Strass. The relative expressiveness of abstract argumentation and logic programming. In Sven Koenig and Blai Bonet, editors, *Proceedings of the 29th AAAI Conference on Artificial Intelligence (AAAI)*, pages 1625–1631. AAAI Press, 2015.

[Thimm, 2012] Matthias Thimm. A probabilistic semantics for abstract argumentation. In Luc De Raedt, Christian Bessière, Didier Dubois, Patrick Doherty, Paolo Frasconi, Fredrik Heintz, and Peter J. F. Lucas, editors, *Proceedings of the 20th European Conference on Artificial Intelligence (ECAI)*, volume 242 of *Frontiers in Artificial Intelligence and Applications*, pages 750–755. IOS Press, 2012.

[van Emden and Kowalski, 1976] Maarten H. van Emden and Robert A. Kowalski. The Semantics of Predicate Logic as a Programming Language. *J. ACM*, 23(4):733–742, 1976.

[Wallner, 2014] Johannes P. Wallner. *Complexity Results and Algorithms for Argumentation - Dung's Frameworks and Beyond*. PhD thesis, TU Vienna, Institute of Information Systems, 2014.

[Wyner et al., 2013] Adam Wyner, Trevor J. M. Bench-Capon, and Paul E. Dunne. On the instantiation of knowledge bases in abstract argumentation frameworks. In João Leite, Tran Cao Son, Paolo Torroni, Leon van der Torre, and Stefan Woltran, editors, *Proceedings of the 14th International Workshop on Computational Logic in Multi-Agent Systems (CLIMA)*, volume 8143 of *Lecture Notes in Artificial Intelligence*, pages 34–50. Springer, 2013.

Gerhard Brewka, Stefan Ellmauthaler, Hannes Strass
Leipzig University
Computer Science Institute
Intelligent Systems Department
Augustusplatz 10–11
04109 Leipzig, Germany

Johannes Wallner, Stefan Woltran
Vienna University of Technology
Institute of Information Systems
Database and Artificial Intelligence Group 184/2
Favoritenstraße 9–11
A-1040 Vienna, Austria

6
Abstract Rule-Based Argumentation
SANJAY MODGIL, HENRY PRAKKEN

ABSTRACT. This chapter reviews abstract rule-based approaches to argumentation, in particular the $ASPIC^+$ framework. In $ASPIC^+$ and its predecessors, going back to the seminal work of John Pollock, arguments can be formed by combining strict and defeasible inference rules and conflicts between arguments can be resolved in terms of a preference relation on arguments. This results in abstract argumentation frameworks (a set of arguments with a binary relation of defeat), so that arguments can be evaluated with the theory of abstract argumentation. First the basic $ASPIC^+$ framework is reviewed, possible ways to instantiate it are discussed and how these instantiations can satisfy closure and consistency properties. Then the relation between $ASPIC^+$ and other work in formal argumentation and nonmonotonic logic is discussed, including a review of how other approaches can be reconstructed as instantiations of $ASPIC^+$. Further developments and variants of the basic $ASPIC^+$ framework are also reviewed, including developments with alternative or generalised notions of attack and defeat and variants with further constraints on arguments. Finally, implementations and applications of $ASPIC^+$ are briefly reviewed and some open problems and avenues for further research are discussed.

1 Introduction

One of the oldest research strands in the logical study of argumentation is to allow for arguments that combine strict and defeasible inference rules. Strict inference rules are intended to capture deductively valid inferences, where the truth of the premises guarantees the truth of the conclusion. Defeasible inference rules are instead meant to capture presumptive inferences, where the premises only create a presumption in favour of the conclusion, which can be refuted by evidence to the contrary. This approach was introduced in AI by [Pollock, 1987; Pollock, 1990; Pollock, 1992; Pollock, 1994; Pollock, 1995], previously studied by e.g. [Lin and Shoham, 1989; Simari and Loui, 1992; Vreeswijk, 1997; Prakken and Sartor, 1997] and [Garcia and Simari, 2004] and currently studied by e.g. [Dung and Thang, 2014; Dung, 2014; Dung, 2016] and in work on the $ASPIC^+$ framework [Prakken, 2010; Modgil and Prakken, 2013; Modgil and Prakken, 2014; Caminada *et al.*, 2014; Li and Parsons, 2015; Grooters and Prakken, 2016].

While Dung's seminal theory of abstract argumentation frameworks [Dung, 1995] has proved to be extremely influential, it adopts a level of abstraction that precludes provision of guidelines for choosing how to define arguments and attacks from knowledge bases, and a study of how these choices should be made to ensure rational outcomes yielded by evaluation of the justified arguments under Dung's semantics. The above-mentioned work, which partly originates from before Dung's article, addresses these issues. This chapter presents the current consolidation of this research strand: the $ASPIC^+$ framework for structured argumentation. The $ASPIC$ framework was initially developed as an output of a European Union project on argumentation [Amgoud et al., 2006] and further developed into the $ASPIC^+$ framework, initially in [Prakken, 2010], and subsequently in [Modgil and Prakken, 2013]. The principal aims of $ASPIC^+$ were to: 1) generalise $ASPIC$ so as to provide a natural knowledge representation framework in which to formalise a wide variety of existing and novel instantiations of abstract argumentation frameworks, while; 2) providing guidelines for instantiations that use features typically incorporated at the abstract level of these frameworks; in particular the use of preferences, which were introduced at the abstract level to determine the success of attacks as defeats [Amgoud and Cayrol, 2002], but may violate rationality postulates unless one carefully accounts for their use when instantiating abstract argumentation frameworks.

Importantly, the strict and defeasible inference rules in $ASPIC^+$ are not part of the logical object language (in which the premises and conclusions of arguments are expressed), but are *metalevel* rules for encoding inference over well-formed formulas in some object level language. Also, the $ASPIC^+$ framework abstracts from the nature and origin of the inference rules and from the nature of the language over which they are defined. The resulting abstract nature[1] of $ASPIC^+$ means that it provides a framework enabling the study of various logical instantiations of abstract argumentation frameworks, and conditions under which the extensions of these frameworks (and hence the defined inference relation over the instantiating knowledge base of logical formulae, identified by the conclusions of justified arguments in extensions) satisfy the rationality postulates in [Caminada and Amgoud, 2007] (for example that the conclusions of arguments in an extension are mutually consistent). In fact, Assumption-Based Argumentation (ABA) ([Bondarenko et al., 1997]), which only has strict rules, can also be regarded as abstract rule-based argumentation, since ABA also abstracts from the nature and origin of its inference rules. However, we will (except for some brief comparisons) not discuss ABA in this chapter, as it is reviewed in another chapter of this handbook. The same holds for a particular instantiation of the rule-based approach: Defeasible Logic Programming.

In a rule-based approach, arguments are formed by chaining applications of inference rules into inference trees or graphs. This approach can be contrasted with approaches defined in terms of logical consequence notions, in which arguments are premises-conclusion pairs where the premises are consistent and imply the conclusion according to the consequence notion of some adopted 'base logic'. Examples of this approach are classical-logic argumen-

---

[1] The aforementioned features of $ASPIC^+$ are shared by earlier work in this tradition, such as the work of Pollock and [Vreeswijk, 1997], and justifies the title of this chapter.

tation [Cayrol, 1995; Besnard and Hunter, 2001; Besnard and Hunter, 2008; Gorogiannis and Hunter, 2011] and its generalisation into abstract Tarskian-logic argumentation [Amgoud and Besnard, 2013]. It is important to note that, unlike these logic-based approaches, rule-based approaches in general do not adopt a single base logic but two base logics, one for the strict and one for the defeasible rules. This issue will be discussed in detail in Section 3.1 of this chapter. Moreover, we will review how 'base logic' approaches [Hunter, 2010] can be formalised as instances of $ASPIC^+$ in which the logical language is a full propositional or first-order language and the only inference rules defined over this language are strict, and corresponds to the inference rules of the base logic.

This chapter is organised as follows. In Section 2 we incrementally introduce features of the $ASPIC^+$ framework. We first introduce the basic framework in which arguments are built from strict and or defeasible inference rules, and are grounded in fallible or infallible premises. Various notions of attacks as well as the use of preferences to determine defeats are defined. The basic framework can thus capture rule-based approaches to argumentation of the type dating back to John Pollock's work in formal epistemology, and formalisms for encoding the well-known schemes and critical questions approach to argumentation developed by the informal logic community (notably [Walton, 1996]), and widely used to accommodate more human orientated rather than formal logic based instantiations. We then define a version of $ASPIC^+$ that generalises the standard notion of negation used to identify when the claim of one argument is in conflict with an element in the attacked argument. In this way an asymmetric notion of conflict can be represented that allows for instantiations by logical languages with negation as failure, and the study of formalisms such as ABA as instances of $ASPIC^+$.

In Section 3 we provide guidance on how to choose and define the premises and strict and defeasible rules that comprise $ASPIC^+$ arguments, and the preference relations that are used to determine the success of attacks as defeats. We then specify formal guidelines as to how one should make the aforementioned choices to ensure satisfaction of the rationality postulates in [Caminada and Amgoud, 2007]. We also discuss the extent to which reasoning with defeasible rules and/or preferences can be reduced to reasoning in systems that do not distinguish between strict and defeasible rules, and/or do not use preferences. Finally, we discuss how argument schemes with critical questions can be reconstructed in $ASPIC^+$ as defeasible inference rules.

Section 4 then reviews the relation of $ASPIC^+$ with other works on argumentation and nonmonotonic logic. We show how some existing argumentation formalisms can be reconstructed in the $ASPIC^+$ framework; in particular, ABA as formulated in [Dung et al., 2007], the Carneades system [Gordon et al., 2007; Gordon and Walton, 2009a], and argumentation formalisms based on Tarskian abstract logics [Amgoud and Besnard, 2013] and in particular classical logic argumentation [Gorogiannis and Hunter, 2011]. We will also discuss how the inference relations of existing non-monotonic logics, in particular Preferred Subtheories [Brewka, 1989] and Prioritised Default Logic [Brewka, 1994a], can be endowed with argumentation semantics through instantiation of the $ASPIC^+$ framework. We conclude by reviewing how our structured approach to argumentation sheds light on developments of the theory of abstract argumentation frameworks, including the use of preferences and values, support relations, attacks on attacks, resolutions of attacks and the dynamics of abstract argumentation frameworks.

Further developments of the $ASPIC^+$ framework will be discussed in Section 5, in particular studies of alternative notions of attack, studies of generalised notions of attack and defeat, and studies of further consistency, minimality and chaining restrictions on arguments. Implementations and applications of $ASPIC^+$ are discussed in Section 6 and we conclude with a discussion of open problems and future research directions in Section 7.

2 $ASPIC^+$: Defining the Framework

2.1 The underlying ideas

People argue to remove doubt about a claim [Walton, 2006, p. 1], by giving reasons why one should accept the claim and by defending these reasons against criticism. The strongest way to remove doubt is to show that the claim deductively follows from indisputable grounds. A mathematical proof from the axioms of arithmetic is like this; its grounds are mathematical axioms, while its inferences are deductively sound. So such a proof cannot be attacked on its grounds or its inferences. However, in real life our grounds may not be indisputable and may provide less than conclusive support for their claim.

Suppose we believe that John was in Holland Park some morning and that Holland Park is in London. Then we can deductively reason from these beliefs, to conclude that John was in London that morning. While this reasoning cannot be attacked, the argument is still fallible since its grounds may turn out to be wrong. For instance, Jan may tell us that he met John in Amsterdam that morning around the same time, challenging our belief that John was in Holland Park that morning, since witnesses usually speak the truth. Maybe we have a supporting reason for our belief that John was in Holland Park; that we went jogging in Holland Park and saw John and that our senses are usually accurate. But given Jan's testimony, perhaps our senses betrayed us? But then we discover Jan has a reason to lie, since John is a suspect in a robbery in Holland Park that morning and Jan and John are friends. We then conclude that the basis for questioning our belief that John was in Holland Park that morning (namely, that witnesses usually speak the truth and Jan witnesses John in Amsterdam) does not apply to witnesses who have a reason to lie. So our reason in support of our belief is undefeated and we accept it.

This example is displayed in Figure 1, where the strict inference is visualised with solid lines, the defeasible inferences with dotted lines and the attack relations with arrow. The defeasible inferences within arguments are supposed to

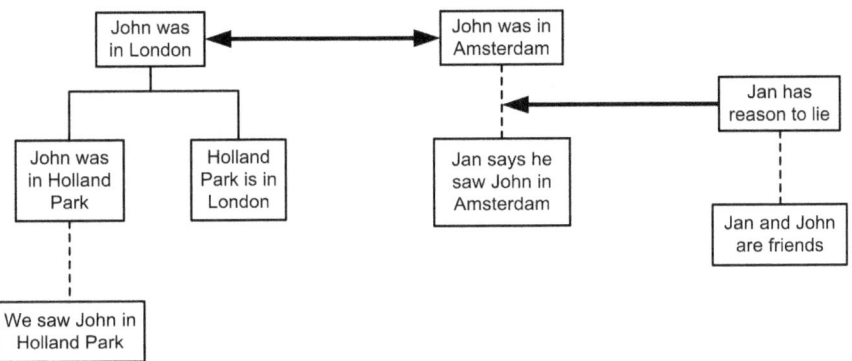

Figure 1. An informal example

be licensed by the generalisations in the example.

If we want to formalise a logic for argumentation, then this simple example already suggests a number of issues to be addressed. First, the claims and beliefs in our example were supported in various ways: in the first case we appealed to the principles of deductive inference when concluding that John was in London. $ASPIC^+$ is therefore designed so that arguments can be constructed using deductive or *strict* inference rules that license deductive inferences from premises to conclusions. However, in the other two cases the reasoning from grounds to claim appealed to the reliability of, respectively, our senses and witnesses as sources of information. Should these kinds of support (inferences) from grounds to claims be modelled as deductive?

To help answer this question, consider that our informal example contains three ways of attacking an argument: 1) Our initial argument that John was in London was attacked by the witness argument on its ground, or *premise*, that John was in Holland Park that morning; 2) The initial argument was then extended with an additional argument for the attacked premise, but the extended argument was still attacked (by the witness argument) on the (now) intermediate conclusion that John was in Holland Park that morning; 3) Finally, we counterattacked the witness argument not on a premise or conclusion but on the reasoning from the grounds to the claim: namely, the inference step from the premise that Jan said he met John in Amsterdam that morning to the claim that John was in Amsterdam that morning (note that here we regard the principle that witnesses usually speak the truth as an inference rule).

Now, returning to the question whether all kinds of inference should be deductive, the second type of attack would not be possible on the deductively inferred intermediate conclusion since the nature of deductive support is that if all antecedents of a deductively valid inference rule are true, then its consequent must also be true. So if we have reason to believe that the conclusion of a deductive inference is not true, then there must be something wrong with its premises (which may in turn be the conclusions of subarguments). It is for this very same reason that the third type of attack on a deductive inferential step is also not possible.

$ASPIC^+$ is therefore designed to comply with the common-sense and philosophically argued position ([Pollock, 1995, p.41]; [Pollock, 2009, p. 173]) advocating the rationality of supporting claims with grounds that do not deductively entail them. In other words, the fallibility of an argument need not only be located in its premises, but can also be located in the inference steps from premises to conclusion. Thus, arguments in $ASPIC^+$ can be constructed using *defeasible* inference rules, and arguments can be attacked on both the conclusions, and application of, such defeasible inference rules, in keeping with the interpretation that the premises of such a rule presumptively rather than deductively support their conclusions.

As well as *fallible* premises that can be attacked, $ASPIC^+$ also allows to distinguish premises that are axiomatic and so cannot be attacked. We discuss the uses of such premises in Section 2.2.1, but for the moment we can summarise by saying that $ASPIC^+$ arguments can be constructed from fallible and infallible premises (respectively called *ordinary* and *axiom* premises in Section 2.2.1), and strict and defeasible inference rules, and that arguments can be attacked on their ordinary premises, the conclusions of defeasible inference rules, and the defeasible inference steps themselves. Finally, a key feature of the $ASPIC^+$ framework is that it accommodates the use of preferences over arguments, so that an attack from one argument to another only succeeds (as a defeat) if the attacked argument is not stronger than (strictly preferred to) the attacking argument, according to some given preference relation. The justified $ASPIC^+$ arguments are then evaluated with respect to the abstract argumentation framework relating $ASPIC^+$ arguments by the defeat relation. Since requirements for use of preferences in argumentation (and more generally for conflict resolution in non-monotonic logics) are well established in the literature, we will here not justify the need for preferences. However, examples are given in the remainder of the chapter.

2.2 The basic framework with symmetric negation

2.2.1 Argumentation systems, knowledge bases, and arguments

$ASPIC^+$ is a general framework that allows one to choose a *logical language* $\mathcal{L}$ closed under negation $\neg$ (which we later replace with a more general notion of conflict) and two (possibly empty) sets of *strict* ($\mathcal{R}_s$) and *defeasible* ($\mathcal{R}_d$) inference rules. One also specifies well-formed formulas in $\mathcal{L}$ that correspond to (i.e., name) defeasible rules in $\mathcal{R}_d$ via a partial function n. These names can then be used when attacking arguments on defeasible inference steps. Informally, $n(r)$ is a well-formed formula (wff) in $\mathcal{L}$ which says that the defeasible rule $r \in \mathcal{R}$ is applicable, so that an argument claiming $\neg n(r)$ attacks the inference step in the corresponding rule[2].

---

[2]n is a partial function since you may want to enforce that some defeasible inference steps cannot be attacked.

Definition 2.1 *[Argumentation systems]* An argumentation system *is a triple* $AS = (\mathcal{L}, \mathcal{R}, n)$ *where:*

- $\mathcal{L}$ *is a logical language with a unary negation symbol* $\neg$.

- $\mathcal{R} = \mathcal{R}_s \cup \mathcal{R}_d$ *is a set of strict* $(\mathcal{R}_s)$ *and defeasible* $(\mathcal{R}_d)$ *inference rules of the form* $\varphi_1, \ldots, \varphi_n \to \varphi$ *and* $\varphi_1, \ldots, \varphi_n \Rightarrow \varphi$ *respectively (where* φ_i, φ *are meta-variables ranging over wff in* $\mathcal{L}$*), and* $\mathcal{R}_s \cap \mathcal{R}_d = \emptyset$.

- n *is a partial function such that* $n : \mathcal{R}_d \longrightarrow \mathcal{L}$.

We write $\psi = -\varphi$ *just in case* $\psi = \neg\varphi$ *or* $\varphi = \neg\psi$ *(we will sometimes informally say that formulas* φ *and* $-\varphi$ *are each other's negation).*

It is important to stress here that $ASPIC^+$'s strict and defeasible inference rules are *not* object-level formulae in the language $\mathcal{L}$, but are meta to the language, allowing one to deductively, respectively defeasibly, infer the rule's consequent from the rule's antecedents. Such inference rules may range over arbitrary formulae in the language, in which case they will, as usual in logic, be specified as *schemes*. For example, a scheme for strict inference rules capturing modus ponens for the material implication of classical logic can be written as $\alpha, \alpha \supset \beta \to \beta^3$, where α and β are metavariables for wff in $\mathcal{L}$. Alternatively, strict or defeasible inference rules may be domain-specific in that they reference specific formulae, as in the defeasible inference rule concluding that an individual flies if that individual is a bird: *Bird* $\Rightarrow$ *Flies*. We will further discuss these distinct uses of inference rules in Section 3.1.

$ASPIC^+$ also requires that one specify a knowledge base from which the premises of an argument can be taken, where one can distinguish between ordinary premises which are uncertain and so can be attacked, and axiom premises that are certain and so cannot be attacked.

Definition 2.2 *[Knowledge bases]* A knowledge base *in an* $AS = (\mathcal{L}, \mathcal{R}, n)$ *is a set* $\mathcal{K} \subseteq \mathcal{L}$ *consisting of two disjoint subsets* $\mathcal{K}_n$ *(the* axioms*) and* $\mathcal{K}_p$ *(the* ordinary premises*).*

An argumentation theory consists of an argumentation system and a knowledge base:

Definition 2.3 *[Argumentation theory]* An argumentation theory *is a tuple* $AT = (AS, \mathcal{K})$ *where* AS *is an argumentation system and* $\mathcal{K}$ *is a knowledge base in AS.*

$ASPIC^+$ arguments are now defined relative to an argumentation theory $AT = (AS, \mathcal{K})$, and chain applications of the inference rules from AS into inference graphs, starting with elements from the knowledge base $\mathcal{K}$. In what follows, for a given argument, the function Prem returns all the formulas of $\mathcal{K}$ (called *premises*) used to build the argument, Conc returns its conclusion, Sub

---

[3]In this chapter we use $\supset$ to denote the material implication connective of classical logic.

returns all its sub-arguments, `DefRules` returns all the defeasible rules of the argument and `TopRule` returns the last inference rule used in the argument.

Definition 2.4 *[Argument] An argument A on the basis of an argumentation theory with a knowledge base $\mathcal{K}$ and an argumentation system $(\mathcal{L}, \mathcal{R}, n)$ is any structure obtainable by applying one or more of the following steps finitely many times:*

1. φ *is an argument if* $\varphi \in \mathcal{K}$ *with:* $\text{Prem}(A) = \{\varphi\}$, $\text{Conc}(A) = \varphi$, $\text{Sub}(A) = \{\varphi\}$, $\text{DefRules}(A) = \emptyset$, $\text{TopRule}(A) = $ *undefined.*

2. $A_1, \ldots, A_n \to \psi$ *is an argument if* $A_1, \ldots, A_n$ *are arguments such that there exists a strict rule*
   $\text{Conc}(A_1), \ldots, \text{Conc}(A_n) \to \psi$ *in* $\mathcal{R}_s$.
   $\text{Prem}(A) = \text{Prem}(A_1) \cup \ldots \cup \text{Prem}(A_n)$,
   $\text{Conc}(A) = \psi$,
   $\text{Sub}(A) = \text{Sub}(A_1) \cup \ldots \cup \text{Sub}(A_n) \cup \{A\}$.
   $\text{DefRules}(A) = \text{DefRules}(A_1) \cup \ldots \cup \text{DefRules}(A_n)$.
   $\text{TopRule}(A) = \text{Conc}(A_1), \ldots, \text{Conc}(A_n) \to \psi$

3. $A_1, \ldots, A_n \Rightarrow \psi$ *is an argument if* $A_1, \ldots, A_n$ *are arguments such that there exists a defeasible rule* $\text{Conc}(A_1), \ldots, \text{Conc}(A_n) \Rightarrow \psi$ *in* $\mathcal{R}_d$.
   $\text{Prem}(A) = \text{Prem}(A_1) \cup \ldots \cup \text{Prem}(A_n)$,
   $\text{Conc}(A) = \psi$,
   $\text{Sub}(A) = \text{Sub}(A_1) \cup \ldots \cup \text{Sub}(A_n) \cup \{A\}$,
   $\text{DefRules}(A) = \text{DefRules}(A_1) \cup \ldots \cup \text{DefRules}(A_n) \cup \{\text{Conc}(A_1), \ldots, \text{Conc}(A_n) \Rightarrow \psi\}$,
   $\text{TopRule}(A) = \text{Conc}(A_1), \ldots, \text{Conc}(A_n) \Rightarrow \psi$.

The definition of each of these functions `Func` is extended to sets of arguments $S = \{A_1, \ldots, A_n\}$ as follows: $\text{Func}(S) = \text{Func}(A_1) \cup \ldots \cup \text{Func}(A_n)$. Moreover, for any argument A we define $\text{Prem}_n(A) = \text{Prem}(A) \cap \mathcal{K}_n$ and $\text{Prem}_p(A) = \text{Prem}(A) \cap \mathcal{K}_p$.

Example 2.5 *Consider a knowledge base in an argumentation system with* $\mathcal{L}$ *consisting of* $p, q, r, s, t, u, v, x, d_1, d_2, d_3, d_4, d_5$ *and their negations, with* $\mathcal{R}_s = \{s_1, s_2\}$ *and* $\mathcal{R}_d = \{d_1, d_2, d_3, d_4, d_5, d_6\}$*, where*[4]

d_1: $p \Rightarrow q$ d_4: $u \Rightarrow v$ s_1: $p, q \to r$
d_2: $s \Rightarrow t$ d_5: $v, x \Rightarrow \neg t$ s_2: $v \to \neg s$
d_3: $t \Rightarrow \neg d_1$

Let $\mathcal{K}_n = \{p\}$ *and* $\mathcal{K}_p = \{s, u, x\}$. *Note that in presenting the example, we have informally used names* d_i *to refer to defeasible inference rules. We now*

---

[4]In the examples that follow we may use terms of the form s_i, d_i or f_i, to identify strict or defeasible inference rules or items from the knowledge base. We will assume that the d_i names are those assigned by the n function of Definition 2.1.

define the n function that formally assigns wff d_i to such rules, i.e., for any rule informally referred to as d_i, we have that $n(d_i) = d_i$, so that '$n(d_1) = d_1$' is a shorthand for $n(p \Rightarrow q) = d_1$. In further examples we will often specify the n function in the same way.[5]

An argument for r (i.e., with conclusion r) is displayed in Figure 2, with the premises at the bottom and the conclusion at the top of the tree. In this and the next figure, the type of a premise is indicated with a superscript and defeasible inferences, underminable premises and rebuttable conclusions are displayed with dotted lines. The figure also displays the formal structure of the argument. We

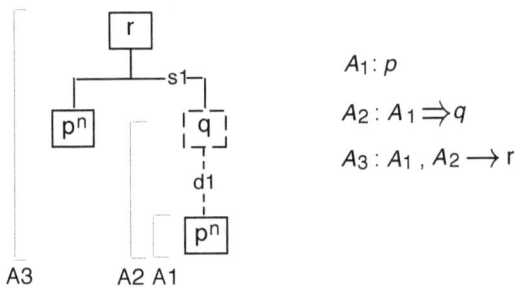

Figure 2. An argument

have that

$\text{Prem}(A_3) = \{p\}$ $\qquad$ $\text{DefRules}(A_3) = \{d_1\}$
$\text{Conc}(A_3) = r$ $\qquad$ $\text{TopRule}(A_3) = s_1$
$\text{Sub}(A_3) = \{A_1, A_2, A_3\}$

The distinction between two kinds of inference rules and two kinds of premises motivates a distinction into four kinds of arguments.

Definition 2.6 *[Argument properties]* An argument A is strict *if* $\text{DefRules}(A) = \emptyset$; defeasible *if* $\text{DefRules}(A) \neq \emptyset$; firm *if* $\text{Prem}(A) \subseteq \mathcal{K}_n$; plausible *if* $\text{Prem}(A) \cap \mathcal{K}_p \neq \emptyset$. An argument is fallible *if it is defeasible or plausible and infallible otherwise*. We write $S \vdash \varphi$ if there exists a strict argument for φ with all premises taken from S, and $S \mathrel{|\!\!\sim} \varphi$ if there exists a defeasible argument for φ with all premises taken from S.

Example 2.7 In Example 2.5 the argument A_1 is both strict and firm, while A_2 and A_3 are defeasible and firm. Furthermore,. we have that $\mathcal{K} \vdash p$, $\mathcal{K} \mathrel{|\!\!\sim} q$ and $\mathcal{K} \mathrel{|\!\!\sim} r$.

---

[5] In our further examples we will often leave the logical language $\mathcal{L}$ and the n function implicit, trusting that they will be obvious.

In logic-based approaches to argumentation [Besnard and Hunter, 2008; Amgoud and Besnard, 2013] arguments are often required to be minimal in that no proper subset of their premises should logically (according to the adopted base logic) imply the conclusion. In the $ASPIC^+$ context such a constraint would be fine for applications of strict rules and below we will review work that imposes such constraints on $ASPIC^+$ arguments (Sections 4.2 and 5.1). However, minimality cannot be required for application of defeasible inference rules, since defeasible rules that are based on more information may well make an argument stronger. For example, *Observations done in ideal circumstances are usually correct* is stronger than *Observations are usually correct*.

Another requirement of logic-based approaches, namely, that an argument's premises have to be consistent, can optionally be imposed in basic $ASPIC^+$, leading to two variants of the basic framework. We define a special class of arguments whose premises are 'c-consistent' (for 'contradictory-consistent'). In this way $ASPIC^+$ can be used as a framework for reconstructing logic-based argumentation formalisms, as we will further discuss in Section 4.2.

Definition 2.8 *[c-consistency]* A set $S \subseteq \mathcal{L}$ is c-consistent *if for no ϕ is it the case that $S \vdash \phi$ and $S \vdash -\phi$. Otherwise S is said to be c-inconsistent. We say that $S \subseteq \mathcal{L}$ is* minimally c-inconsistent *iff S is c-inconsistent and $\forall S' \subset S$, S' is c-consistent.*

Definition 2.9 *[c-consistent arguments]* An argument A is c-consistent *iff* $\mathrm{Prem}(A)$ *is c-consistent.*

2.2.2 Attack and defeat

$ASPIC^+$ generates abstract argumentation frameworks consisting of arguments related by binary defeats. Having defined arguments above, we now define the attack relation and then apply preferences to determine the defeat relation (in fact [Dung, 1995] called his relation "attack" but we reserve this term for the basic notion of conflict, to which we then apply preferences).

Attack We first present the three ways in which $ASPIC^+$ arguments can be in conflict (i.e., attack). Arguments can be attacked on a conclusion of a defeasible inference (rebutting attack), on a defeasible inference step itself (undercutting attack), or on an ordinary premise (undermining attack). In Section 2.1 we argued that arguments cannot be attacked on their strict inferences. In Section 3.3 we will also show that attacks on conclusions of strict inferences may result in violation of rationality postulates. In Section 5.3 we will discuss to what extent alternative definitions of rebutting attack still make sense.

To define undercutting attack, the function n of an AS is used, which assigns to elements of $\mathcal{R}_d$ a well-formed formula in $\mathcal{L}$. Recall that informally, $n(r)$ (where $r \in \mathcal{R}_d$) means that r is applicable. Then an argument using r is undercut by any argument with conclusion $-n(r)$.

Definition 2.10 *[Attacks]* A *attacks* B *iff* A *undercuts, rebuts or undermines* B*, where:*

- *A undercuts argument B (on B') iff* $\text{Conc}(A) = -n(r)$ *for some $B' \in \text{Sub}(B)$ such that B''s top rule r is defeasible.*

- *A rebuts argument B (on B') iff* $\text{Conc}(A) = -\varphi$ *for some $B' \in \text{Sub}(B)$ of the form $B''_1, \ldots, B''_n \Rightarrow \varphi$.*

- *Argument A undermines B (on φ) iff* $\text{Conc}(A) = -\varphi$ *for an ordinary premise φ of B.*

This definition allows for a distinction between direct and indirect attack: an argument can be indirectly attacked by directly attacking one of its proper subarguments. This distinction will turn out to be crucial for a proper application of preferences when determining whether attacks succeed as defeats.

Example 2.11 *In our running example argument A_3 cannot be undermined, since all its premises are axioms. A_3 can potentially be rebutted on A_2, with an argument for $\neg q$. However, the argumentation theory of our example does not allow the construction of such a rebuttal. Likewise, A_3 can potentially be undercut on A_2, with an argument for $\neg d_1$. Our example does allow the construction of such an undercutter, namely:*

$B_1: s$
$B_2: B_1 \Rightarrow t$
$B_3: B_2 \Rightarrow \neg d_1$

B_3 *has an ordinary premise s, and so can be undermined on B_1 with an argument for $\neg s$:*

$C_1: u$
$C_2: C_1 \Rightarrow v$
$C_3: C_2 \to \neg s$

Note that since C_3 has a strict top rule, argument B_1 does not in turn rebut C_3.

Argument B_3 can potentially be rebut or undercut on either B_2 or B_3, since both of these subarguments of B_3 have a defeasible top rule. Our argumentation theory only allows for a rebutting attack on B_2:

$C_1: u$
$C_2: C_1 \Rightarrow v$
$D_3: x$
$D_4: C_2, D_3 \Rightarrow \neg t$

All arguments and attacks in the example are displayed in Figure 3.

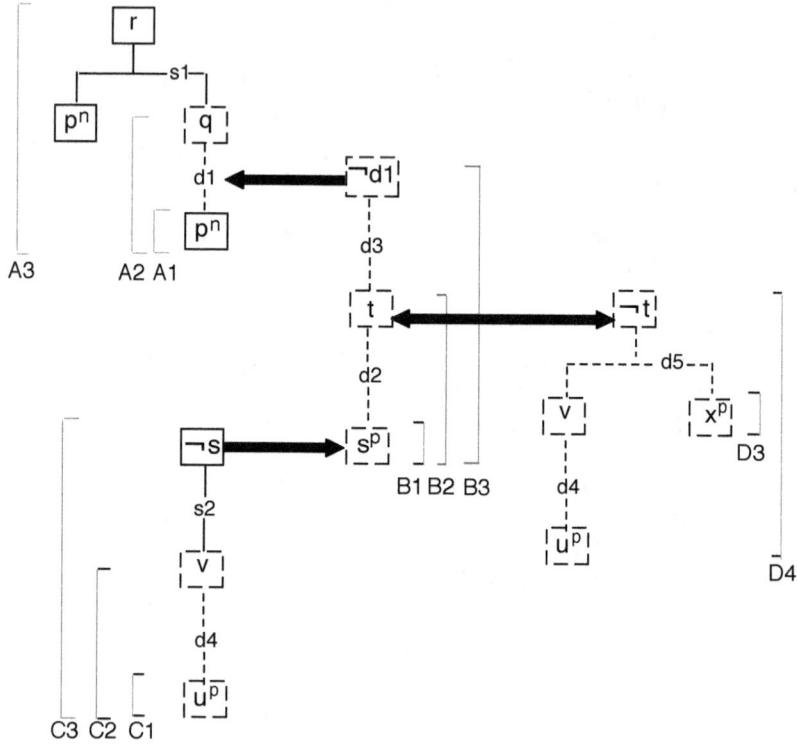

Figure 3. The arguments and attacks in the running example

Defeat The attack relation tells us which arguments are in conflict with each other. If an argument A *successfully attacks*, i.e., defeats, B, then A can be used as a counter-argument to B. Whether an attack from A to B (on its sub-argument B') succeeds as a defeat, may depend on the relative strength of A and B', i.e., whether B' is *strictly stronger than, or strictly preferred* to A. Only the success of undermining and rebutting attacks is contingent on preferences; undercutting attacks succeed as defeats independently of any preferences (see [Modgil and Prakken, 2013] for a discussion as to why this is the case). $ASPIC^+$ allows for any strict binary preference ordering $\prec$ on the set of all arguments that can be constructed on the basis of an argumentation theory. Note that in this chapter we formalise argument orderings not as they are defined in [Modgil and Prakken, 2013], but as they are defined in an erratum available online at https://nms.kcl.ac.uk/sanjay.modgil/AIJfinalErratum. The erratum essentially reverts to the directly defined strict partial ordering $\prec$ over arguments as employed in [Prakken, 2010]. Then (as illustrated in Section 3.2), the non-strict $\preceq$ is defined so that $A \preceq B$ iff $A \prec B$ or the fallible elements in

A and B that are used in deciding preferences, are the same. Moreover, [Modgil and Prakken, 2013] identify conditions under which argument orderings are well-behaved in that they ensure satisfaction of the rationality postulates. The erratum modifies these conditions, which in [Modgil and Prakken, 2013] are stated by reference to non-strict orderings over sets of defeasible rules (ordinary premises), but in the erratum are stated with respect to strict orderings over sets of defeasible rules (ordinary premises). This has been done in order to address a counterexample to rationality pointed out by Sjur Dyrkolbotn (personal communication), assuming the conditions as stated in [Modgil and Prakken, 2013][6]. We will review these conditions later in this chapter.

Definition 2.12 *[Successful rebuttal, undermining and defeat]*

- *A* successfully rebuts *B* if *A* rebuts *B* on B' and $A \not\prec B'$.

- *A* successfully undermines *B* if *A* undermines *B* on φ and $A \not\prec \varphi$.

- *A* defeats *B* iff *A* undercuts or successfully rebuts or successfully undermines *B*. (In general, we say *A* strictly defeats *B* if *A* defeats *B* and *B* does not defeat *A*).

The success of rebutting and undermining attacks thus involves comparing the conflicting arguments at the points where they conflict; that is, by comparing those arguments that are in a *direct* rebutting or undermining relation with each other. The definition of successful undermining exploits the fact that an argument premise is also a subargument, so the preference $A \not\prec \varphi$ is well defined.

Example 2.13 *In our running example, the undercutting attack of B_3 on A_2 (and thereby on A_3) succeeds as a defeat irrespective of the argument ordering between B_3 and A_2. The undermining attack of C_3 on B_1 succeeds if $C_3 \not\prec B_1$. If B_2 and D_4 are incomparable, then these two arguments defeat each other, while D_4 strictly defeats B_3. If $D_4 \prec B_2$ then B_2 strictly defeats D_4 while if $B_2 \prec D_4$ then D_4 strictly defeats both B_2 and B_3.*

Let us now put all these elements together; that is the arguments and attacks defined on the basis of an argumentation theory, and a preference ordering over the arguments (here we write '*(c-)SAF*' as meaning '*SAF* or *c-SAF*'):

Definition 2.14 *[c-SAFs] Let AT be an argumentation theory (AS, KB). A (c-)structured argumentation framework ((c-)SAF) defined by AT, is a triple $\langle \mathcal{A}, \mathcal{C}, \preceq \rangle$ where*

- *In a SAF, $\mathcal{A}$ is the set of all arguments constructed from KB in AS satisfying Definition 2.4;*

---

[6]Note that the erratum also addresses a counterexample to rationality in [Dung, 2016].

- In a c-SAF, $\mathcal{A}$ is the set of all c-consistent arguments constructed from KB in AS satisfying Definition 2.4;
- $\preceq$ is a preference ordering on $\mathcal{A}$;
- $(X, Y) \in \mathcal{C}$ iff X attacks Y.

Note that a c-SAF is a SAF in which all arguments are required to have a c-consistent set of premises.

Example 2.15 In our running example $\mathcal{A} = \{A_1, A_2, A_3, B_1, B_2, B_3, C_1, C_2, C_3, D_3, D_4\}$, while $\mathcal{C}$ is such that B_3 attacks both A_2 and A_3, argument C_3 attacks all of B_1, B_2, B_3, argument D_4 attacks both B_2 and B_3 and, finally, B_2 attacks D_4.

2.2.3 Generating abstract argumentation frameworks

We now instantiate abstract argumentation frameworks with $ASPIC^+$ arguments and defeats.

Definition 2.16 (Argumentation frameworks) An abstract argumentation framework (AF) corresponding to a (c-)SAF = $\langle \mathcal{A}, \mathcal{C}, \preceq \rangle$ is a pair $(\mathcal{A}, \mathcal{D})$ such that $\mathcal{D}$ is the defeat relation on $\mathcal{A}$ determined by $\langle \mathcal{A}, \mathcal{C}, \preceq \rangle$.

The justified arguments of the above defined abstract argumentation frameworks are then defined under various semantics, as in [Dung, 1995]:

Definition 2.17 [Dung Semantics] Let $(\mathcal{A}, \mathcal{D})$ be an AF and $S \subseteq \mathcal{A}$. Then:

- S is conflict free iff $\forall X, Y \in S$: $(X, Y) \notin \mathcal{D}$[7].

- $X \in \mathcal{A}$ is acceptable with respect to S iff $\forall Y \in \mathcal{A}$ such that $(Y, X) \in \mathcal{D}$: $\exists Z \in S$ such that $(Z, Y) \in \mathcal{D}$.

- S is an admissible set iff S is conflict free and $X \in S$ implies X is acceptable w.r.t. S.

- S is a complete extension iff S is admissible and if $X \in \mathcal{A}$ is acceptable w.r.t. S then $X \in S$;

- S is a preferred extension iff it is a set inclusion maximal complete extension;

- S is the grounded extension iff it is the set inclusion minimal complete extension;

---

[7]Note that in [Modgil and Prakken, 2013] we motivate the use of the $ASPIC^+$ attack relation to define conflict-free sets (a set of arguments is conflict-free if there does not exist an attack between any of its contained arguments), and then only use the $ASPIC^+$ defeat relation to determine the acceptability of arguments. It turns out that under certain conditions, this way of evaluating the status of arguments is equivalent to Definition 2.17's use of the defeat relation for *both* determining conflict freeness and acceptability of arguments.

- S is a stable *extension iff* S *is conflict free and* $\forall Y \notin S$, $\exists X \in S$ *s.t.* $(X, Y) \in \mathcal{D}$.

For $T \in \{complete,\ preferred,\ grounded,\ stable\}$, X is sceptically, *respectively* credulously *justified on the basis of AF under the T semantics if X belongs to all, respectively at least one, T extension of AF*.

It is now also possible to define a consequence notion for well-formed formulas. Several definitions are possible. One is:

Definition 2.18 *[Justified Formulae] A wff* $\varphi \in \mathcal{L}$ *is sceptically justified on the basis of a* (c-)SAF *under semantics T if* φ *is the conclusion of a sceptically justified argument on the basis of the AF corresponding to the* (c-)SAF *under semantics T, and* credulously justified *on the basis of a* (c-)SAF *under semantics T if* φ *is not sceptically justified and is the conclusion of a credulously justified argument on the basis of the AF corresponding to the* (c-)SAF *under semantics T*.

An alternative definition of skeptical justification is:

A wff $\varphi \in \mathcal{L}$ is *sceptically justified* on the basis of the *(c-)SAF* under semantics T if all T-extensions of the *AF* corresponding to the *(c-)SAF* contain an argument with conclusion φ.

While the original definition of skeptical justification requires that there is one argument for φ that is in all extensions, the alternative definition allows that different extensions contain different arguments for φ. In multiple-extension semantics this can make a difference in, for example, cases with so-called floating conclusions; cf. Example 25 of [Prakken and Vreeswijk, 2002].

Example 2.19 *In our running example, if D_4 strictly defeats B_2, then we get a unique extension in all semantics, namely, $E = \{A_1, A_2, A_3, C_1, C_2, C_3, D_3, D_4\}$. If in addition C_3 does not defeat B_1, then the extension also contains B_1. In both cases this yields that wff r is sceptically justified.*

Alternatively, if B_2 strictly defeats D_4, then the status of r depends on whether C_3 defeats B_1. If it does, then we again have a unique extension in all semantics consisting of the set S, so r is sceptically justified. By contrast, if C_3 does not defeat B_1, we obtain a unique extension with A_1, B_1, B_2, B_3, C_1, C_2, C_3 and D_3, so r is neither sceptically nor credulously justified.

Finally, if B_2 and D_4 defeat each other, then the outcome again depends on whether C_3 defeats B_1. If it does, then the situation is as in the previous case – a unique extension E – but if C_3 does not defeat B_1, then the grounded extension consists of A_1, B_1, C_1, C_2, C_3, D_3. So in the latter case, in grounded semantics r is neither sceptically nor credulously justified. However, in preferred and stable semantics we then obtain two alternative extensions: the first contains D_4, A_2 and A_3, while the second instead contains B_2 and B_3 and so excludes A_2 and A_3. So in the latter case r is credulously, but not sceptically justified under stable and preferred semantics.

2.3 The basic framework with possibly non-symmetric negation

The notion of an argumentation system in Section 2.2.1, assumed a language $\mathcal{L}$ with a unary negation symbol $\neg$, which was used in the definition of conflict-based attack. The standard classical interpretation of $\neg$ licenses a symmetric notion of conflict-based attack, so that an argument consisting of an ordinary premise ϕ or with a defeasible top rule concluding ϕ, *symmetrically* attacks an argument consisting of an ordinary premise $\neg\phi$ or with a defeasible top rule concluding $\neg\phi$. However, the $ASPIC^+$ framework as presented in [Prakken, 2010; Modgil and Prakken, 2013], accommodates a more general notion of conflict, by defining an argumentation system to additionally include a function $^-$ that, for any wff $\psi \in \mathcal{L}$, specifies the set of wff's that are in conflict with ψ, so that one can define both an asymmetric and symmetric notion of conflict-based attack. Formally:

Definition 2.20 *[$^-$ function]* $^-$ *is a function from $\mathcal{L}$ to $2^{\mathcal{L}}$, such that:*

- φ *is a* contrary *of ψ if $\varphi \in \overline{\psi}$, $\psi \notin \overline{\varphi}$;*
- φ *is a* contradictory *of ψ (denoted by '$\varphi = -\psi$'), if $\varphi \in \overline{\psi}$, $\psi \in \overline{\varphi}$.*

Now $\text{Conc}(A) \in \overline{\varphi}$ ($\text{Conc}(A) \in \overline{n(r)}$) replaces $\text{Conc}(A) = -\varphi$ ($\text{Conc}(A) = -n(r)$) in Definition 2.10's definition of attacks. This induces a generalised notion of an argumentation system as a four-tuple $AS = (\mathcal{L}, ^-, \mathcal{R}, n)$ where $\mathcal{L}$, $\mathcal{R}$ and n are defined as in Definition 2.1 and $^-$ is as just defined. The special case of Definition 2.1 can then be reformulated as the case where $^-$ is defined in terms of classical negation as $\alpha \in \overline{\beta}$ iff α is of the form $\neg\beta$ or β is of the form $\neg\alpha$ (i.e., for any wff α, α and $\neg\alpha$ are contradictories). Below we will continue to refer to the special case with $\neg$ as a triple, leaving the $^-$ function implicit.

The rationale for these more general notions of conflict and attack is twofold. Firstly, one can for pragmatic reasons state that two formulae are in conflict, rather than requiring that one implies the negation of another; for example, assuming a predicate language with the binary '$<$' relation, one can state that any two formulae of the form $\alpha < \beta$ and $\beta < \alpha$ are contradictories. Secondly, the $^-$ function allows for an asymmetric notion of negation. This enables reconstruction of assumption-based argumentation (ABA) in $ASPIC^+$ (indeed the idea of using a $^-$ function is taken from [Bondarenko et al., 1997]). We briefly review this reconstruction in Section 4.1. Closely related to its use in reconstructing ABA, the contrary function allows for the modelling of negation as failure (as in logic programming). Using the negation as failure symbol $\sim$ (also called 'weak' negation, in contrast to the 'strong' negation symbol $\neg$), then $\sim \alpha$ denotes the negation of α under the assumption that α is not provable (i.e., the negation of α is assumed in the absence of evidence for α). Given this intended reading of $\sim$ it is not meaningful to assert that such an assumption brings into question (and so initiates an attack on) the evidence whose very absence is required to make the assumption in the first place. In other words, if A is an argument consisting of the premise $\sim \alpha$, and B concludes α (the contrary of $\sim \alpha$), then B attacks A, but not vice versa. Furthermore, since the very construction of A is invalidated by evidence to the contrary, i.e., B, then such attacks succeed as defeats *independently* of preferences.

To accommodate the notion of contrary, and attacks on contraries succeeding as defeats independently of preferences, we further modify Definition 2.10 to distinguish the special cases where $\texttt{Conc}(A)$ is a contrary of φ, in which case we say that A *contrary rebuts* B and A *contrary undermines* B, and then modify Definition 2.12 so that:

- A successfully rebuts B if A contrary rebuts B, or A rebuts B on B' and $A \not\prec B'$.

- A successfully undermines B if A contrary undermines B, or A undermines B on ϕ and $A \not\prec \phi$.

The definition of undercutting attack does not need to be changed.

To illustrate the use of negation as failure, suppose one wants arguments to be built from a propositional language that includes both $\neg$ and $\sim$. One could then define $\mathcal{L}$ as a language of propositional literals, composed from a set of propositional atoms $\{a, b, c, \dots\}$ and the symbols $\neg$ and $\sim$. Then:

- α is a *strong literal* if α is a propositional atom or of the form $\neg \beta$ where β is a propositional atom (strong negation cannot be nested).

- α is a wff of $\mathcal{L}$, if α is a strong literal or of the form $\sim \beta$ where β is a strong literal (weak negation cannot be nested).

Then $\alpha \in \overline{\beta}$ iff (1) α is of the form $\neg \beta$ or β is of the form $\neg \alpha$; or (2) β is of the form $\sim \alpha$ (i.e., for any wff α, α and $\neg \alpha$ are contradictories and α is a contrary of $\sim \alpha$). Finally, for any $\sim \alpha$ that is in the antecedent of a strict or defeasible inference rule, one is required to include $\sim \alpha$ in the ordinary premises.

Consider now Example 2.5, where we now have that $u \in \overline{\sim u}$, and we replace the rule $d_4 : u \Rightarrow v$ with $d'_4: \sim u \Rightarrow v$, and add $\sim u$ to the ordinary premises: $\mathcal{K}_p = \{\sim u, s, u, x\}$. Then, the arguments C_3 and D_4 are now replaced by arguments C'_3 and D'_4 each of which contain the sub-argument $E : \sim u$ (instead of $C_1 : u$). Then $C_1 : u$ contrary undermines, and so defeats, C'_3 and D'_4 on $\sim u$.

3 Instantiating the *ASPIC*$^+$ Framework

ASPIC$^+$ is a framework for specifying systems, and so leaves one fully free to make choices as to the logical language, the strict and defeasible inference rules, the axioms and ordinary premises in a knowledge base, and the argument preference ordering. In this section we discuss various more or less principled ways to make these choices, and then show specific uses of *ASPIC*$^+$.

3.1 Choosing strict and defeasible rules

3.1.1 Domain specific strict inference rules

$ASPIC^+$ allows the specification of domain specific strict inference rules, as illustrated by the following example (based on Example 4 of [Caminada and Amgoud, 2007]) in which the strict inference rules capture definitional knowledge, namely, that bachelors are not married.

Example 3.1 Let $\mathcal{R}_d = \{d_1, d_2\}$ and $\mathcal{R}_s = \{s_1, s_2\}$, where:

$d_1 = $ WearsRing $\Rightarrow$ Married $\qquad s_1 = $ Married $\rightarrow \neg$Bachelor
$d_2 = $ PartyAnimal $\Rightarrow$ Bachelor $\qquad s_2 = $ Bachelor $\rightarrow \neg$Married

Finally, let $\mathcal{K}_p = \{WearsRing, PartyAnimal\}$. Consider the following arguments.

$A_1:$ WearsRing $\qquad B_1:$ PartyAnimal
$A_2:$ $A_1 \Rightarrow$ Married $\qquad B_2:$ $B_1 \Rightarrow$ Bachelor
$A_3:$ $A_2 \rightarrow \neg$Bachelor $\qquad B_3:$ $B_2 \rightarrow \neg$Married

We have that A_3 rebuts B_3 on its subargument B_2 while B_3 rebuts A_3 on its subargument A_2. Note that A_2 does not rebut B_3, since B_3 applies a strict rule; likewise for B_2 and A_3.

In Example 3.1, the rules s_1 and s_2 are 'transpositions' of each other, and $\mathcal{R}_s$ is 'closed under transposition', in the sense that:

Definition 3.2 (Closure under Transposition) Let $AT = (AS, \mathcal{K})$ be an argumentation theory. Then AT is closed under transposition iff if $\phi_1, \ldots, \phi_n \rightarrow \psi \in \mathcal{R}_s$, then for $i = 1 \ldots n$, $\phi_1, \ldots, \phi_{i-1}, -\psi, \phi_{i+1}, \ldots, \phi_n \rightarrow -\phi_i \in \mathcal{R}_s$.

In general it is a good idea to ensure that an argumentation theory is closed under transposition, since a strict (deductive) rule $q \rightarrow \neg s$ expresses that if q is true, then this guarantees the truth of $\neg s$, no matter what. Hence, if we have s, then q cannot hold, otherwise we would have $\neg s$. In general, if the negation of the consequent of a strict rule holds, then we cannot have all its antecedents, since if we had all of them, then its consequent would hold. This is the very meaning of a strict rule. So it is very reasonable to include in $\mathcal{R}_s$ the transposition of a strict rule that is in $\mathcal{R}_s$. A second reason for ensuring closure under transposition is that it ensures satisfaction of [Caminada and Amgoud, 2007]'s rationality postulates, as illustrated later in Section 3.3.

3.1.2 Strict inference rules and axioms based on deductive logics

Some find the use of domain-specific strict inference rules rather odd; why not instead express them as material implications in $\mathcal{L}$ and put them in the knowledge base as axiom premises? One then reserves the strict inference rules for general patterns of deductive inference, since one might argue that this is what inference rules are meant for in logic. $ASPIC^+$ therefore allows one to base your strict inference rules (and axioms) on a deductive logic of one's choice. One can do so by choosing a semantics for a particular choice of $\mathcal{L}$ with an associated monotonic notion of semantic consequence, and then letting $\mathcal{R}_s$ be rules that are sound with respect to that semantics. For example, suppose $\mathcal{R}_s$ should conform to classical logic, given a standard propositional (or first-order) language, such that arguments can contain any classically valid inference step over this language. This can be done in two ways: a crude way and a sophisticated way.

A crude way is to simply put all valid propositional (or first-order) inferences over your language of choice in $\mathcal{R}_s$. So if a propositional language has been chosen, then $\mathcal{R}_s$ can be defined as follows. (where $\vdash_{PL}$ denotes standard propositional-logic consequence). For any finite $S \subseteq \mathcal{L}$ and any $\varphi \in \mathcal{L}$:[8]

$$S \to \varphi \in \mathcal{R}_s \text{ if and only if } S \vdash_{PL} \varphi$$

In fact, with this choice of $\mathcal{R}_s$, strict parts of an argument don't need to be more than one step long. For example, if rules $S \to \varphi$ and $\varphi \to \psi$ are in $\mathcal{R}_s$, then $S \cup \{\varphi\} \to \psi$ will also be in $\mathcal{R}_s$. Note also that using this method, strict rules will be closed under transposition, because of the properties of classical logic.

It should be noted that this way of using a logic as the origin of the strict rule makes some implicit assumptions on the chosen logic, for example that it is compact (everything implied by an infinite set is implied by a finite subset) and satisfies the Cut rule (if S implies φ and $S \cup \{\varphi\}$ implies ψ then S implies ψ). In Section 5.1 we return to this issue.

Let us illustrate the crude approach with a variation of Example 3.1. We retain the defeasible rules d_1 and d_2 but we replace the domain-specific strict rules s_1 and s_2 with a single material implication $Married \supset \neg Bachelor$ in $\mathcal{K}_n$. Moreover, we put all propositionally valid inferences over our language in $\mathcal{R}_s$, including, for example, all inferences instantiating the modus ponens scheme $\varphi, \varphi \supset \psi \to \psi$. Then the arguments change as follows:

A_1: $WearsRing$
A_2: $A_1 \Rightarrow Married$
A_3: $Married \supset \neg Bachelor$
A_4: $A_2, A_3 \to \neg Bachelor$

B_1: $PartyAnimal$
B_2: $B_1 \Rightarrow Bachelor$
B_3: $Married \supset \neg Bachelor$
B_4: $B_2, B_3 \to \neg Married$

Now A_4 rebuts B_4 on B_2 while B_4 rebuts A_4 on A_2.

A sophisticated way to base the strict part of $ASPIC^+$ on a deductive logic of one's choice is to build an existing axiomatic system for the logic into $ASPIC^+$. Its axiom(s) (typically a handful) can be encoded in $\mathcal{K}_n$ and its inference rule(s) (typically just one or a few) in $\mathcal{R}_s$. For example, there are axiomatic systems for classical logic with just four axioms and just one inference rule, namely,

---

[8]Although antecedents of rules formally are sequences of formulas, we will sometimes abuse notation and write them as sets.

modus ponens (i.e, $\varphi \supset \psi, \varphi \rightarrow \psi)^9$. With this choice of $\mathcal{R}_s$, strict parts of an argument could be very long, since in logical axiomatic systems, proofs of even trivial validities might be long. However, this difference with the crude way is not very big, since if we want to be crude, we must, to know whether $S \rightarrow \varphi$ is in $\mathcal{R}_s$, first construct a propositional proof of φ from S.

With the sophisticated way of building classical logic into our argumentation system, argument A_4 in our example stays the same, since modus ponens is in $\mathcal{R}_s$. However, argument B_4 will change, since modus tollens is not in $\mathcal{R}_s$. In fact, B_4 will be replaced by a sequence of strict rule applications, together being an axiomatic proof of ¬*Married* from *Married* $\supset$ ¬*Bachelor* and *Bachelor*.

Note that in the sophisticated method, closure under transposition may not hold; our example above does not contain modus tollens (that is, $\varphi \supset \psi, -\psi \rightarrow -\varphi$). However, this desirable form of reasoning can also be enforced without explicitly transposing rules. Recall that $S \vdash \varphi$ was defined as 'there exists a strict argument for φ with all premises taken from S'. Now it turns out that if $\vdash$ contraposes, then this is just as good as closure of the strict rules under transposition. Contraposition of $\vdash$ means that if $S \vdash \varphi$, then if we replace one element s of S with $-\varphi$, then $-s$ is strictly implied (if $\vdash$ corresponds to classical provability, as enforced by our choice of axioms and inference rules, then $\vdash$ does indeed contrapose).

Definition 3.3 *[Closure under Contraposition]* Let $AT = (AS, \mathcal{K})$ be an argumentation theory. We say that AT is closed under contraposition *iff for all* $S \subseteq \mathcal{L}$, $s \in S$ and ϕ, if $S \vdash \phi$, then $S \backslash \{s\} \cup \{-\phi\} \vdash -s$.

Again, as will be discussed in Section 3.3, closure under contraposition also ensures satisfaction of rationality postulates.

3.1.3 Choosing defeasible inference rules

Regarding the choice of defeasible rules, the question as to whether these can be derived from a logic of our choice, just as with strict rules, is controversial. Some philosophers argue that all rule-like structures that we use in daily life are "inference licenses" and so cannot be expressed in the logical object language. In this view, all defeasible generalisations are inference rules, whether they are domain-specific or not, and are applied to formulas from $\mathcal{L}$ to support new formulas from $\mathcal{L}$.

Others (usually logicians) take a more standard-logic approach (e.g. [Kraus et al., 1990; Pearl, 1992]) whereby all contingent knowledge should be expressed in the object language, and so they reject the idea of domain-specific defeasible inference rules (for the same reason they don't like domain-specific strict rules). They introduce a new connective, e.g., $\leadsto$, into $\mathcal{L}$ where (informally) $p \leadsto q$ is read as "If p then normally/typically/usually q". They then want to give a model-theoretic semantics for this connective just as logicians give a

---

[9] As explained above, this strictly speaking is not a rule but a scheme, with meta variables ranging over $\mathcal{L}$.

model-theoretic semantics for all connectives, except that semantics for these defeasible conditionals focus on a *preferred class* of models (e.g., all models where things are as normal as possible) instead of *all* models of a theory as in semantics for deductive logics. Hence, the model-theoretic interpretation of $p \supset q$ is that q is true in *all* models of p, whereas the model theoretic interpretation of $p \rightsquigarrow q$ is that q is true in all *preferred* models of p.

What inference rules for $\rightsquigarrow$ could result from such an approach? On two things there is consensus: modus ponens for $\rightsquigarrow$ is defeasibly but not deductively valid, so the rule $\varphi \rightsquigarrow \psi, \varphi \Rightarrow \psi$ should go into $\mathcal{R}_d$. There is also consensus that contraposition for $\rightsquigarrow$ is deductively invalid, so the rule $\varphi \rightsquigarrow \psi \to -\psi \rightsquigarrow -\varphi$ should *not* go into $\mathcal{R}_s$. However, here the consensus ends. Should the defeasible analogue of this rule go into $\mathcal{R}_d$ or not? Opinions differ at this point[10].

Let us illustrate the difference between the two approaches, by including defeasible modus ponens for $\rightsquigarrow$ in $\mathcal{R}_d$, and replacing the defeasible inference rules d_1 and d_2 (in Example 3.1) with object-level conditionals expressed in $\mathcal{L}$ and included in $\mathcal{K}_p$:

$WearsRing \rightsquigarrow Married \in \mathcal{K}_p$ and $PartyAnimal \rightsquigarrow Bachelor \in \mathcal{K}_p$
$\mathcal{R}_d = \{\varphi \rightsquigarrow \psi, \varphi \Rightarrow \psi\}$

The arguments then change as follows (assuming the crude incorporation of classical logic):

A_1: $WearsRing$
A_2: $WearsRing \rightsquigarrow Married$
A_3: $A_1, A_2 \Rightarrow Married$
A_4: $Married \supset \neg Bachelor$
A_5: $A_3, A_4 \to \neg Bachelor$

B_1: $PartyAnimal$
B_2: $PartyAnimal \rightsquigarrow Bachelor$
B_3: $B_1, B_2 \Rightarrow Bachelor$
B_4: $Married \supset \neg Bachelor$
B_5: $B_3, B_4 \to \neg Married$

Now A_5 rebuts B_5 on B_3 while B_5 rebuts A_5 on A_3.

Concluding, if desired, at least some of the choices concerning defeasible inference rules can be based on model-theoretic semantics for nonmonotonic logics. However, it is an open question whether a model-theoretic semantics is the *only* criterion by which we can choose our defeasible rules. Some have based their choice on other criteria, since they do not primarily see defeasible rules as logical inference rules but as principles of human cognition or rational action, so that they should be based on foundations other than semantics. For example, John Pollock based his defeasible reasons on his account of epistemology. Others have based their choice of defeasible reasons on the study of argument schemes in informal argumentation theory. We give examples of both these approaches in Section 3.5.

---

[10] See Chapter 4 of [Caminada, 2004] for a very readable overview of the discussion.

3.2 Choosing argument preference orderings

A well studied use of preferences in the non-monotonic logic literature is based on the use of preference orderings over formulae in the language or defeasible inference rules. If $ASPIC^+$ is to be used as a framework for giving argumentation-based characterisations of non-monotonic formalisms augmented with preferences, then it needs to provide an account of how these preference orderings can be 'lifted' to preferences over arguments. Since $ASPIC^+$ uses defeasible inference rules and ordinary premises, both may come equipped with preference orderings $\leq$ on $\mathcal{R}_d$ and $\leq'$ on $\mathcal{K}_p$, which in general may be distinct, in keeping with the ontologically distinct nature of rules and premises. For example, the ordinary premises may represent the content of percepts from sensors or of witness testimonies, whose preference ordering reflects the relative reliability of the sensors, respectively witnesses. The defeasible rules may, for example, be ordered based on probabilistic strength, on temporal precedence (defeasible rules acquired later are preferred to those acquired earlier), on the basis of principles of legal precedence, and so on. The challenge is to then define a preference over arguments A and B based on the preferences over their constituent ordinary premises *and* defeasible rules.

We now define two argument preference orderings, called the weakest-link and last-link orderings. These orderings are in turn based on partial preorders $\leq$ on $\mathcal{R}_d$ and $\leq'$ on $\mathcal{K}_p$, where as usual, $X <^{(')} Y$ iff $X \leq^{(')} Y$ and $Y \not\leq^{(')} X$ (note that we may represent these orderings in terms of the strict counterpart they define). However, these preferences relate individual defeasible rules, respectively ordinary premises, whereas when comparing two arguments, we want to compare them on the (possibly non-singleton) *sets of* rules/premises that these arguments are constructed from. So, to define these argument preferences, we need to first define a strict set ordering $\triangleleft_\mathrm{s}$ over sets of rules/premises, where for any sets of defeasible rules/ordinary premises S and S', we intuitively want that:

1) if S is the empty set, it cannot be that $S \triangleleft_\mathrm{s} S'$;

2) if S' is the empty set, it must be that $S \triangleleft_\mathrm{s} S'$ for any non-empty S.

In other words, arguments that have no defeasible rules (ordinary premises) are, modulo the premises (rules), strictly stronger than (preferred to) arguments that have defeasible rules (ordinary premises). Hence the following definition explicitly imposes these constraints, and then gives two alternative ways of defining $\triangleleft_\mathrm{s}$; the so called Elitist and Democratic ways (i.e., s = Eli and Dem respectively). Eli compares sets on their minimal and Dem on their maximal elements.

Definition 3.4 *[Orderings $\triangleleft_\mathrm{s}$] Let Γ and Γ' be finite sets[11]. Then $\triangleleft_\mathrm{s}$ is defined as follows:*

---

[11] Notice that it suffices to restrict $\triangleleft$ to finite sets since $ASPIC^+$ arguments are assumed to be finite (in Definition 2.14) and so their sets of ordinary premises/defeasible rules must be finite.

1. If $\Gamma = \emptyset$ then $\Gamma \not\triangleleft_{\mathrm{s}} \Gamma'$;
2. If $\Gamma' = \emptyset$ and $\Gamma \neq \emptyset$ then $\Gamma \triangleleft_{\mathrm{s}} \Gamma'$;
   else, assuming a preordering $\leq$ over the elements in $\Gamma \cup \Gamma'$, then if :
3. $\mathrm{s} = \mathtt{Eli}$:
   $\Gamma \triangleleft_{\mathtt{Eli}} \Gamma'$ if $\exists X \in \Gamma$ s.t. $\forall Y \in \Gamma'$, $X < Y$.
   else, if:
4. $\mathrm{s} = \mathtt{Dem}$:
   $\Gamma \triangleleft_{\mathtt{Dem}} \Gamma'$ if $\forall X \in \Gamma$, $\exists Y \in \Gamma'$, $X < Y$.

For $\mathrm{s} = \mathtt{Eli}$ or $\mathrm{s} = \mathtt{Dem}$: $\Gamma \trianglelefteq_{\mathrm{s}} \Gamma'$ iff $\Gamma = \Gamma'$ or $\Gamma \triangleleft_{\mathrm{s}} \Gamma'$.

Now the **last-link principle** strictly prefers an argument A over another argument B if the last defeasible rules used in B are less preferred ($\triangleleft_{\mathrm{s}}$) than the last defeasible rules in A or, in case both arguments are strict, if the premises of B are less preferred than the premises of A. The concept of 'last defeasible rules' is defined as follows.

Definition 3.5 *[Last defeasible rules]* Let A be an argument.

- $\mathtt{LastDefRules}(A) = \emptyset$ iff $\mathtt{DefRules}(A) = \emptyset$.
- If $A = A_1, \ldots, A_n \Rightarrow \phi$, then $\mathtt{LastDefRules}(A) = \{\mathtt{Conc}(A_1), \ldots, \mathtt{Conc}(A_n) \Rightarrow \phi\}$, else $\mathtt{LastDefRules}(A) = \mathtt{LastDefRules}(A_1) \cup \ldots \cup \mathtt{LastDefRules}(A_n)$.

For example, letting $\mathcal{K} = \{p, q\}$, $\mathcal{R}_s = \{r, s \to t\}$ and $\mathcal{R}_d = \{p \Rightarrow r;\ q \Rightarrow s\}$, then
$\mathtt{LastDefRules}(A) = \{p \Rightarrow r;\ q \Rightarrow s\}$ where A is the argument for t.

The above definition is now used to compare pairs of arguments as follows:

Definition 3.6 *[Last link principle]* Let A and B be two arguments. Then $A \prec B$ iff:

1. $\mathtt{LastDefRules}(A) \triangleleft_{\mathrm{s}} \mathtt{LastDefRules}(B)$; or
2. $\mathtt{LastDefRules}(A)$ and $\mathtt{LastDefRules}(B)$ are empty and $\mathtt{Prem}_{\mathrm{p}}(A) \triangleleft_{\mathrm{s}} \mathtt{Prem}_{\mathrm{p}}(B)$.

Then $B \preceq A$ iff $B \prec A$ or, if $\mathtt{LastDefRules}(A) \neq \emptyset$ then $\mathtt{LastDefRules}(A) = \mathtt{LastDefRules}(B)$, else $\mathtt{Prem}_{\mathrm{p}}(A) = \mathtt{Prem}_{\mathrm{p}}(B)$.

Example 3.7 *Suppose in our running example that $u <' s$, $x <' s$, $d_2 < d_5$ and $d_4 < d_2$. Applying the last-link ordering to check whether C_3 defeats B_1, we compare $\mathtt{LastDefRules}(C_3) = \{d_4\}$ with $\mathtt{LastDefRules}(B_1) = \emptyset$. Clearly, $\{d_4\} \triangleleft_{\mathtt{Eli}} \emptyset$, so $C_3 \prec B_1$, so C_3 does not defeat B_1. Next, to check whether D_4*

defeats B_2, we compare $\text{LastDefRules}(B_2) = \{d_2\}$ with $\text{LastDefRules}(D_4) = \{d_5\}$. Since $d_2 < d_5$ we have that $\text{LastDefRules}(B_2) \vartriangleleft_{\text{Eli}} \text{LastDefRules}(D_4)$, so D_4 strictly defeats B_2.

The **weakest-link principle** considers not the last but *all* uncertain elements in an argument.

Definition 3.8 *[Weakest link principle]* Let A and B be two arguments. Then $A \prec B$ iff

1. If both B and A are strict, then $\text{Prem}_p(A) \vartriangleleft_s \text{Prem}_p(B)$, else;

2. If both B and A are firm, then $\text{DefRules}(A) \vartriangleleft_s \text{DefRules}(B)$, else;

3. $\text{Prem}_p(A) \vartriangleleft_s \text{Prem}_p(B)$ and $\text{DefRules}(A) \vartriangleleft_s \text{DefRules}(B)$

Then $B \preceq A$ iff $B \prec A$ or, $\text{DefRules}(A) = \text{DefRules}(B)$ and $\text{Prem}_p(A) = \text{Prem}_p(B)$.

Example 3.9 *In our running example to check whether C_3 defeats B_1 according to the weakest-link ordering, we first compare $\text{Prem}_p(C_3) = \{u\}$ with $\text{Prem}_p(B_1) = \{s\}$. Since $u <' s$ we have that $\text{Prem}_p(C_3) \vartriangleleft_{\text{Eli}} \text{Prem}_p(B_1)$. Also, $\text{DefRules}(C_3) = \{d_4\} \vartriangleleft_{\text{Eli}} \text{DefRules}(B_1) = \emptyset$, and so $C_3 \prec B_1$ and C_3 does not defeat B_1.*
For B_2 and D_4: $\text{Prem}_p(D_4) = \{u, x\} \vartriangleleft_{\text{Eli}} \text{Prem}_p(B_2) = \{s\}$ since $u <' s$ and $x <' s'$. Then since $d_4 < d_2$, $\text{DefRules}(D_4) = \{d_4, d_5\} \vartriangleleft_{\text{Eli}} \text{DefRules}(B_2)\{d_2\}$. So $D_4 \prec B_2$ and B_2 strictly defeats D_4.

We next present two examples illustrating the suitability of the last-, respectively, weakest-link orderings. Consider an example relating to whether people misbehaving in a university library may be denied access to the library.[12]

Example 3.10 *Let $\mathcal{K}_p = \{Snores, Professor\}$, $\mathcal{R}_d =$*

$\{Snores \Rightarrow_{d_1} Misbehaves;$
$Misbehaves \Rightarrow_{d_2} AccessDenied;$
$Professor \Rightarrow_{d_3} \neg AccessDenied\}$.

Assume that $Snores <' Professor$ and $d_1 < d_2$, $d_1 < d_3$, $d_3 < d_2$, and consider the following arguments.

A_1: $Snores$
A_2: $A_1 \Rightarrow Misbehaves$
A_3: $A_2 \Rightarrow AccessDenied$

B_1: $Professor$
B_2: $B_1 \Rightarrow \neg AccessDenied$

---

[12] In all examples below, sets that are not specified are assumed to be empty. Moreover, sometimes we will attach the rule names to the $\Rightarrow$ symbol. Note that these attached indices have no formal meaning and are for ease of reference only.

Let us apply the ordering to the arguments A_3 and B_2. The rule sets to be compared are $\texttt{LastDefRules}(A_3) = \{d_2\}$ and $\texttt{LastDefRules}(B_2) = \{d_3\}$. Since $d_3 < d_2$ we have that $\texttt{LastDefRules}(B_2) \vartriangleleft_{\texttt{Eli}} \texttt{LastDefRules}(A_3)$, hence $B_2 \prec A_3$. So A_3 strictly defeats B_2, hence A_3 is justified in any semantics, and we conclude AccessDenied.

With the weakest-link principle the ordering between A_3 and B_2 is different. Both A and B are plausible and defeasible so we are in case (3) of Definition 3.8. Since $Snores <'$ $Professor$, we have that $\texttt{Prem}_p(A_3) \vartriangleleft_{\texttt{Eli}} \texttt{Prem}_p(B_2)$. Furthermore, the rule sets to be compared are now $\texttt{DefRules}(A_3) = \{d_1, d_2\}$ and $\texttt{DefRules}(B_2) = \{d_3\}$. Since $d_1 < d_3$ we have that $\texttt{DefRules}(A_3) \vartriangleleft_{\texttt{Eli}} \texttt{DefRules}(B_2)$. So now we have that $A_3 \prec B_2$. Hence B_2 now strictly defeats A_3 and we conclude instead that $\neg AccessDenied$.

Which outcome is better? Some have argued that the last-link ordering gives the better outcome since the conflict really is between the two legal rules about whether someone may be denied access to the library, while d_1 just provides a sufficient condition for when a person can be said to misbehave. The existence of a conflict on whether someone may be denied access to the library is in no way relevant for the issue of whether a person misbehaves when snoring. More generally, it has been argued that for reasoning with legal (and other normative) rules the last-link ordering is appropriate. However, in an example of exactly the same form, with the legal rules replaced by empirical generalisations, intuitions seem to favour the weakest-link ordering:

Example 3.11 Let $\mathcal{K}_p = \{BornInScotland, FitnessLover\}$, $\mathcal{R}_d =$

$\{BornInScotland \Rightarrow_{d_1} Scottish;$
$Scottish \Rightarrow_{d_2} LikesWhisky;$
$FitnessLover \Rightarrow_{d_3} \neg LikesWhisky\}$.

Assume that $BornInScotland <' FitnessLover$ and $d_1 < d_2$, $d_1 < d_3$, $d_3 < d_2$, and consider the following arguments.

A_1:	$BornInScotland$	B_1:	$FitnessLover$
A_2:	$A_1 \Rightarrow Scottish$	B_2:	$B_1 \Rightarrow \neg LikesWhisky$
A_3:	$A_2 \Rightarrow LikesWhisky$		

This time it seems reasonable to conclude $\neg LikesWhisky$, since the epistemic uncertainty of the premise and d_1 of A_3 should propagate to weaken A_3. And this is the outcome given by the weakest-link ordering. So it could be argued that for epistemic reasoning the weakest-link ordering is appropriate.

3.3 The rationality postulates of Caminada and Amgoud (2007) and their satisfaction in $ASPIC^+$

$ASPIC^+$ leaves one fully free to choose a language, what is an axiom and what is an ordinary premise and how to specify strict and defeasible rules. However some care needs to be taken in making these choices, to ensure that the result of argumentation is guaranteed to be well-behaved in the sense that the desirable properties proposed by [Caminada and Amgoud, 2007] are satisfied. Before presenting these properties, we define required notions of direct and indirect consistency in terms of the contrary function (recall Definition 2.20).

Definition 3.12 *[Direct and Indirect Consistency] For any $S \subseteq \mathcal{L}$, let the closure of S under strict rules, denoted $Cl(S)$, be the smallest set containing S and the consequent of any strict rule in $\mathcal{R}_s$ whose antecedents are in $Cl(S)$. Then a set $S \subseteq \mathcal{L}$ is*

- *directly consistent iff $\nexists \, \psi, \varphi \in S$ such that $\psi \in \overline{\varphi}$*

- *indirectly consistent iff $Cl(S)$ is directly consistent.*

Let E be any complete extension of an abstract argumentation framework corresponding to a (c)-SAF as defined in Section 2.2.3.

Sub-argument Closure: For any argument A in E, all sub-arguments of A are in E, i.e., for all $A \in E$: if $A' \in \mathtt{Sub}(A)$ then $A' \in E$.

Closure under Strict Rules: If E contains arguments with conclusions $\alpha_1, \ldots . \alpha_n$, then any arguments obtained by applying only strict inference rules to these conclusions, are in E, i.e., $\{\mathtt{Conc}(A)|A \in E\} = Cl(\{\mathtt{Conc}(A)|A \in E\})$.

Direct Consistency: The conclusions of arguments in E are directly consistent, i.e., $\{\mathtt{Conc}(A)|A \in E\}$ is consistent.

Indirect Consistency: The conclusions of arguments in E are indirectly consistent, i.e., $Cl(\{\mathtt{Conc}(A)|A \in E\})$ is consistent.

We next review the work done on identifying sufficient conditions for AS-PIC^+ satisfying [Caminada and Amgoud, 2007]'s four rationality postulates.

3.3.1 The work of Caminada and Amgoud (2007), Prakken (2010) and Modgil and Prakken (2013)

The first relevant condition is that an argumentation theory is closed under transposition or contraposition. If neither is satisfied, then since strict rule applications cannot be attacked, direct consistency may be violated. Consider our first version of Example 3.1. Suppose we only have the strict rule s_1 so that B_3 cannot be constructed (given the absence of s_2). We still have that A_3 rebuts B_2. Suppose now that $d_1 < d_2$ and we apply the last-link argument ordering. Then A_3 does not defeat B_2. In fact, no argument in the example is defeated, so we end up with a single extension (under all semantics) which contains arguments for both *Bachelor* and $\neg Bachelor$ and so violates direct and indirect consistency. However, with transposition we also have s_2. Then B_3 can be constructed, which rebuts A_3 on A_2. Under the last-link ordering (assuming again that $d_1 < d_2$) we still have that A_3 does not defeat B_2, but now B_3 strictly defeats A_2. We have a unique extension in all semantics, containing all arguments except A_2 and A_3. This extension does not violate consistency.

One might argue that the above violation of consistency, before inclusion of the transposed rule s_2, arises because $ASPIC^+$ forbids attacks on strictly derived conclusions. Consistency would not be violated if B_2 was allowed to attack A_3. However, apart from the reasons discussed in Section 2.2.2, another reason for prohibiting attacks on strictly derived conclusions is that if allowed, extensions may not be strictly closed or indirectly consistent, even if the strict rules are closed under transposition. To see why, suppose we allow attacks on strict conclusions, so that B_2 attacks A_3, A_2 attacks B_3, and A_3 and B_3 attack each other in Example 3.1. Suppose also that all knowledge-base items and defeasible rules are of equal preference, and we apply the weakest- or last-link argument ordering. Then all rebutting attacks in the example succeed. But then the set $\{A_1, A_2, B_1, B_2\}$ is admissible and is in fact both a stable and preferred extension. But this violates strict closure and indirect consistency. The extension contains an argument for *Bachelor* but not for $\neg$*Married*, which strictly follows from it by rule s_2. Likewise, the extension contains an argument for *Married* but not for $\neg$*Bachelor*, which strictly follows from it by rule s_1. So the extension is not closed under strict rule application. Moreover, the extension is indirectly inconsistent, since its strict closure contains both *Married* and $\neg$*Married*, and both *Bachelor* and $\neg$*Bachelor*.

Other requirements for satisfying the postulates are expressed in the following definition of a 'well-defined' structured argumentation framework (recall Definition 2.14), which references the notion of a 'reasonable' preference relation that is subsequently explained and defined:

Definition 3.13 *[Well defined (c-)SAFs]* A *(c-)SAF* $(\mathcal{A}, \mathcal{C}, \preceq)$ *defined by an an argumentation theory* $AT = (AS, \mathcal{K})$, *where* $AS = (\mathcal{L}, ^-, \mathcal{R}, n)$ *and* $\mathcal{K} = \mathcal{K}_n \cup \mathcal{K}_p$, *is said to be* well defined *iff:*

- AT is closed under transposition *or* closed under contraposition.
- $Cl_{\mathcal{R}_s}(\mathcal{K}_n)$ is consistent (in which case $\mathcal{K}$ is said to be axiom consistent).
- If $\mathcal{A}$ is restricted to be the set of c-consistent arguments, then $\mathcal{A}$ is c-classical. That is to say, for any minimal c-inconsistent $S \subseteq \mathcal{L}$ and for any $\varphi \in S$, it holds that $S \setminus \{\varphi\} \vdash -\varphi$ (i.e., amongst all arguments defined there exists a strict argument with conclusion $-\varphi$ with all premises taken from $S \setminus \{\varphi\}$).
- well formed *if whenever* φ *is a contrary of* ψ *then:*
  - $\psi \notin \mathcal{K}_n$; and
  - ψ is not the consequent of a strict rule.
- $\preceq$ is reasonable.

The property of transposition (and the alternative contraposition) has been discussed above. That the axiom premises are required to be consistent when closed under strict rules is self-evident given that axiom premises represent indisputable information or axioms of a deductive logic. The c-classicality condition is only required to hold when using $ASPIC^+$ to reconstruct Tarskian logic, and in particular classical logic approaches to argumentation, where $\mathcal{A}$ is restricted to arguments with consistent premises. Intuitively, c-classicality says that for every minimally c-inconsistent set of wff and any of its elements the remaining maximally c-consistent subset gives rise to an argument against the element. The intuition underlying the well-formed property should be apparent given the motivation for use of the contrary function and preference independent attacks on contraries, as discussed in Section 2.3. We now elaborate on the notion of reasonable preference orderings.

Before doing so, we define the following notion of strict continuations of arguments, which we define in a slightly different way than [Modgil and Prakken, 2013]. The new definition is arguably simpler but does not affect the proofs of Modgil and Prakken. It identifies arguments that are formed by extending a set of arguments with only strict inferences into a new argument, so that the new argument can only be attacked on the arguments that it extends.

Definition 3.14 *[Strict continuations] The set of* strict continuations *of a set of arguments from $\mathcal{A}$ is the smallest set satisfying the following conditions:*

1. *Any argument A is a strict continuation of $\{A\}$.*

2. *If $A_1, \ldots, A_n$ and $S_1, \ldots, S_n$ are such that for each $i \in \{1, \ldots, n\}$, A_i is a strict continuation of S_i and $\{B_{n+1}, \ldots, B_m\}$ is a (possibly empty) set of strict-and-firm arguments, and $\text{Conc}(A_1), \ldots, \text{Conc}(A_n), \text{Conc}(B_{n+1}), \ldots, \text{Conc}(B_m) \to \varphi$ is a strict rule in R_s, then $A_1, \ldots, A_n, B_{n+1}, \ldots, B_m \to \varphi$ is a strict continuation of $S_1 \cup \ldots \cup S_n$.*

If argument A is a strict continuation of arguments $\{A_1, \ldots, A_n\}$, then A is a strict argument over $\{\text{Conc}(A_1), \ldots, \text{Conc}(A_n)\}$.

Example 3.15 In Example 2.5 (see Figure 3) all arguments are strict continuations of the singleton set containing themselves while A_3 is a strict continuation of $\{A_1, A_2\}$ and C_3 is a strict continuation of $\{C_2\}$.

Definition 3.16 *[Reasonable Argument Orderings] An argument ordering $\preceq$ is reasonable iff:*

1. *i) $\forall A, B$, if A is strict and firm and B is plausible or defeasible, then $B \prec A$;*
   ii) $\forall A, B$, if B is strict and firm then $B \not\prec A$;
   iii) $\forall A, A', B$ such that A' is a strict continuation of $\{A\}$, if $A \not\prec B$ then $A' \not\prec B$, and if $B \not\prec A$ then $B \not\prec A'$ (i.e., applying strict rules to a single argument's conclusion and possibly adding new axiom premises does not weaken, respectively strengthen, arguments).

2. Let $\{C_1, \ldots, C_n\}$ be a finite subset of $\mathcal{A}$, and for $i = 1 \ldots n$, let $C^{+\backslash i}$ be some strict continuation of $\{C_1, \ldots, C_{i-1}, C_{i+1}, \ldots, C_n\}$. Then it is not the case that: $\forall i, C^{+\backslash i} \triangleleft C_i$.

A reasonable argument ordering essentially amounts to requiring that: arguments that are both strict and firm are strictly preferred over all plausible or defeasible arguments, and no argument is strictly preferred to a strict and firm argument (1i) and 1ii)); the strength (and implied relative preference) of an argument is determined exclusively by the defeasible rules and/or ordinary premises (1iii)); the preference ordering is acyclic (2).

Indeed, a strict relation $\triangleleft$ (on sets of ordinary premises or defeasible rules) results in a preference ordering (under either weakest- or last-link) that is reasonable, if $\triangleleft$ satisfies the following conditions:

Definition 3.17 *[Inducing reasonable orderings]* $\triangleleft$ *is said to be* reasonable inducing *if $\triangleleft$ is a strict partial ordering (irreflexive and transitive) such that:*

for any $\text{kr} \in \{\text{LastDefRules}, \text{DefRules}, \text{Prem}_p\}$, *for all arguments* $B_1, \ldots, B_n, A$
such that $\bigcup_{i=1}^n \text{kr}(B_i) \triangleleft \text{kr}(A)$, *it holds that for some* $i = 1 \ldots n$, $\text{kr}(B_i) \triangleleft \text{kr}(A)$

It can be shown that both $\triangleleft_{\text{Eli}}$ and $\triangleleft_{\text{Dem}}$ (recall Definition 3.4) are reasonable inducing.

We are now in a position to state some important results proved in [Modgil and Prakken, 2013]. Any (c)-structured argumentation framework satisfies the rationality postulate of sub-argument closure. Moreover, if a (c-)structured argumentation framework is well-defined then the postulates of strict closure and direct and indirect consistency are also satisfied by the $ASPIC^+$ framework as defined with the contrary function in Section 2.3.

Theorem 3.18 *[Sub-argument Closure] Let $\Delta = (\mathcal{A}, \mathcal{C}, \preceq)$ be a (c-)SAF and E a complete extension of the AF corresponding to Δ. Then for all $A \in E$: if $A' \in \text{Sub}(A)$ then $A' \in E$.*

Theorem 3.19 *Let $\Delta = (\mathcal{A}, \mathcal{C}, \preceq)$ be a well-formed (c-)SAF and E a complete extension of the AF corresponding to Δ. Then*

Closure under Strict Rules	$\{\text{Conc}(A)|A \in E\} = Cl_{R_s}(\{\text{Conc}(A)|A \in E\})$;
Direct consistency	$\{\text{Conc}(A)|A \in E\}$ *is consistent;*
Indirect consistency	$Cl_{R_s}(\{\text{Conc}(A)|A \in E\})$ *is consistent.*

Finally, note that if no strict rules or axiom premises are included in the argumentation theory, then the preference ordering need *not* be reasonable in order for all four rationality postulates to be satisfied (indeed no assumptions as to the properties of the preference ordering are required in this case). Thus the requirement that the defined (c-)SAF be well-defined does not apply.

3.3.2 The work of Dung and Thang (2014) and Grooters (2014)

For the case without preferences and knowledge bases, [Dung and Thang, 2014] identify weaker conditions for satisfying the rationality postulates than those discussed above. [Dung and Thang, 2014] formulate their results in terms of an adaptation of [Amgoud and Besnard, 2013] abstract-logic approach to abstract argumentation with abstract attack and support relations between arguments. After defining their adaptation they apply it to what they call "rule-based systems", which are a pair of sets of strict and defeasible rules defined over a propositional literal language. Since they adopt the $ASPIC^+$ definitions of an argument and of defeat (which they call 'attack') they thus effectively study a class of $ASPIC^+$ instantiations with an empty knowledge base and with no preferences. Below we summarise their definitions and results as holding for this class of $ASPIC^+$ instantiations, adapting fragments of [Grooters, 2014] and [Grooters and Prakken, 2016]. In doing so, we implicitly assume a given $ASPIC^+$ structured argumentation framework generated by a rule-based instantiation in the sense of [Dung and Thang, 2014], which we will call a 'rule-based' $ASPIC^+$ SAF.

First, an argument A is a *basic defeasible argument* iff $\texttt{TopRule}(A) \in \mathcal{R}_d$, and a set X of arguments is called *inconsistent* if $\texttt{Conc}(X)$ is indirectly inconsistent.

Definition 3.20 *[Base of an argument] Let A be an argument and BA a finite set of subarguments of A. BA is a base of A if*

- $\texttt{Conc}(A) \in Cl_{\mathcal{R}_s}(\texttt{Conc}(BA))$;

- *For each argument C, C defeats A if and only if C defeats BA.*

The following example shows the intuitive idea of a base.

Example 3.21 Let $\mathcal{R}_s = \{c \to d\}$ and $\mathcal{R}_d = \{\Rightarrow a; \Rightarrow b; a, b \Rightarrow c\}$. Then the following arguments can be constructed: $A_1 :\Rightarrow a$, $A_2 :\Rightarrow b$, $A_3 : A_1, A_2 \Rightarrow c$ and $A_4 : A_3 \to d$. See Figure 4.

A_4 can only be attacked on its subarguments A_1, A_2, or A_3 because of the strict top rule. Every argument that attacks A_1 or A_2 also attacks A_3, so every argument that attacks A_4 also attacks A_3. It is easy to see that every argument that attacks A_3 also attacks A_4. $\texttt{Conc}(A_4) \subseteq Cl_{\mathcal{R}_s}(\texttt{Conc}(A_3))$, so $\{A_3\}$ is a base of A_4. The same kind of reasoning applies to the fact that the set $\{A_1, A_2, A_3\}$ is also a base of A_4.

However note that the set $\{A_1, A_2\}$ is not a base of A_4, because A_4 can be rebutted (on A_3) without A_1 or A_2 being attacked.

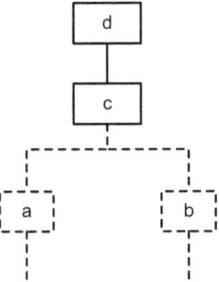

Figure 4. Arguments of Example 3.21

Definition 3.22 *[Generation of arguments] An argument A is said to be generated by a set of arguments S, if there is a base B of A such that $B \subseteq \mathtt{Sub}(S)$. The set of all arguments generated by S is denoted by $GN(S)$.*

[Dung and Thang, 2014] show that for every set of arguments S, $\mathtt{Sub}(S) \subseteq GN(S)$ and for every complete extension E, $GN(E) = E$. [Grooters, 2014] notes that these results immediately imply that each rule-based $ASPIC^+$ SAF satisfies the closure under subarguments postulate, since for every complete extension E: $\mathtt{Sub}(E) \subseteq GN(E) = E$ ([Dung and Thang, 2014] do not consider the subargument-closure postulate).

Definition 3.23 *[Compact] A rule-based $ASPIC^+$ SAF is compact if for each set of arguments S, $GN(S)$ is closed under strict rules.*

[Dung and Thang, 2014] show that each rule-based $ASPIC^+$ SAF is compact and that each compact rule-based $ASPIC^+$ SAF satisfies strict closure, so each rule-based $ASPIC^+$ SAF satisfies the closure under strict rules postulate.

Definition 3.24 *[Cohesive] A rule-based $ASPIC^+$ SAF is cohesive if for each inconsistent set of arguments S, $GN(S)$ is conflicting (attacks itself).*

Definition 3.25 *[Self-contradiction axiom] A rule-based $ASPIC^+$ SAF is said to satisfy the self-contradiction axiom if for each minimal inconsistent set $X \subseteq \mathcal{L}$: $\neg X \subseteq Cl_{\mathcal{R}_s}(X)$ (where $\neg X = \{\neg l \mid l \in L\}$).*

[Dung and Thang, 2014] then show that each cohesive rule-based $ASPIC^+$ SAF satisfies the indirect-consistency postulate and, moreover, that each rule-based $ASPIC^+$ SAF that satisfies the self-contradiction axiom is cohesive. Combining these two results, it follows that each rule-based $ASPIC^+$ SAF that satisfies the self-contradiction axiom, also satisfies indirect consistency. This result generalises the corresponding results discussed in the previous subsection, since satisfying the self-contradiction axiom is a weaker notion than closure under transposition. First, [Dung and Thang, 2014] prove that the latter implies the former in that each rule-based $ASPIC^+$ SAF that is closed under transposition satisfies the self-contradiction axiom. They then give the following counterexample to the converse implication.

Example 3.26 Let $\mathcal{L} = \{a, \neg a, b, \neg b\}$ and $\mathcal{R}_d = \emptyset$ and $\mathcal{R}_s = \{a \to b\} \cup \{x, \neg x \to y \mid x \in \{a, b\}$ and $y \in \mathcal{L}\}$. This system satisfies the self-contradiction axiom but is not closed under transposition.

It is worth noting that [Grooters, 2014] generalised all these results to the case with arbitrary logical languages with symmetric negation, c-consistent nonempty knowledge bases and reasonable argument orderings, and for both SAFs and for c-SAFs. Moreover, she did so alternatively for closure under transposition and closure under contraposition. In doing so, it was shown that the following weaker version of the self-contradiction axiom suffices:

Definition 3.27 *[Weak self-contradiction axiom]* A rule-based ASPIC$^+$ *(c-)SAF is said to satisfy the* weak self-contradiction axiom *if for each minimal inconsistent set $X \subseteq \mathcal{L}$ there is a $\sigma \in X$ such that $\neg \sigma \in Cl_{\mathcal{R}_s}(X)$.*

3.4 On the need for the various elements of $ASPIC^+$

$ASPIC^+$ as a general framework is quite expressive. The question therefore arises whether all these elements are really needed.

3.4.1 The need for knowledge bases

The $ASPIC$ system as presented in [Caminada and Amgoud, 2007] did not have knowledge bases. Instead, certain and uncertain premises were encoded as strict rules $\to \varphi$ and defeasible rules $\Rightarrow \varphi$. Others, such as [Dung and Thang, 2014], [Li and Parsons, 2015] and [Dung, 2016] also adopt this idea. Yet there are good reasons to retain knowledge bases. To start with, the distinction between knowledge (or beliefs) and inference rules is a natural one, widely adopted in logic. Furthermore, this distinction allows a systematic study of encodings of logical consequence notions in the set of strict rules, as we will see below. We therefore conclude that although dispensing with knowledge bases might have practical advantages in specific applications, a general theory of argumentation-based inference should retain the formal distinction between knowledge and inference rules.

3.4.2 The need for strict rules and axiom premises

[Li and Parsons, 2015] show that every $ASPIC^+$ SAF with a weakest-link ordering that satisfies the rationality postulates can be translated into a SAF with no strict rules and no axiom premises and that (for all of [Dung, 1995]'s semantics) validates exactly the same conclusions as the original SAF. Their basic idea is that each strict rule is translated to a corresponding defeasible rule and each axiom premise to an ordinary premise, and the argument ordering is then extended so as to give the new elements resulting from the translations of strict rules or axiom premises, precedence over all conflicting elements. While this result is theoretically interesting, we still believe that the distinction between strict and defeasible inference rules is a natural one and is philosophically grounded. For example, the observation that the inclusion of strict rules allows a systematic study of encodings of logical consequence notions also applies here. We also believe that the distinction between disputable (ordinary) and undisputable (axiom) premises is a practically useful one. For these reasons we claim that a general framework for structured argumentation should leave room for these distinctions.

3.4.3 The need for preferences

In the context of ABA, [Kowalski and Toni, 1996] proposed a way to encode preferences with a specific use of assumptions in strict rules with the effect that if a preferred rule applies, the assumption in a non-preferred conflicting rule is attacked. The same can in fact be done with defeasible rules. However, [Kowalski and Toni, 1996]'s proposal does not cover any of the argument orderings discussed in this chapter. Outside of argumentation, a systematic treatment for [Brewka, 1994b; Brewka, 1994a]'s prioritised default logic was given by [Delgrande and Schaub, 2000], who showed that prioritised default theories can be translated into equivalent ordinary default theories. In Section 4.5 we will discuss the relation between prioritised default logic and $ASPIC^+$.

In general, the question as to whether $ASPIC^+$ argument orderings can be encoded in $ASPIC^+$ rule sets or knowledge bases is still an open question. We conjecture that such translations may be very hard to give for argument orderings that depend on global properties of an argument, such as weakest-link orderings.

3.4.4 The need for defeasible rules

Perhaps the most controversial issue is whether defeasible inference rules are needed. In Section 2.1 we illustrated with an informal example that there are three ways to attack an argument: on its premises, on its defeasible inferences, and on the conclusions of its defeasible inferences. In Section 2.2.2 we saw that $ASPIC^+$ explicitly allows all three forms of attack. However, some would argue that the second and third type of attacks can be simulated using only deductive rules (specifically the deductive rules of classical logic) by augmenting the antecedents of these rules with normality premises. For example, with regard to the second type of attack, could we in our example of Section 2.1 not say that our argument claiming that John was in Holland Park that morning since we saw him there has an implicit premise *our senses functioned normally*, and that the argument that John was in Amsterdam that morning in fact attacks this implicit premise, rather than its claim, thus reducing attacks on conclusions to attacks on premises? With regard to the third type of attack, could we not say that instead of attacking the defeasible inference step from Jan's testimony to the claim that John was in Amsterdam, we could model this step as deductive, and then add the premise that normally witnesses speak the truth, and then direct the attack at this premise? In other words, can we reduce attacks on inferences to attacks on premises? These informal arguments have some formal backing since, as we will discuss in more detail in Section 5.2, [Dung and Thang, 2014] have shown that defeasible inference rules can in $ASPIC^+$ be reduced to strict rules.

In answer to these questions, we first claim that there is some merit in modelling the everyday practice of 'jumping to defeasible conclusions' and of considering arguments for contradictory conclusions. This is especially important given that one of the argumentation paradigm's key strengths is its characterisation of formal logical modes of reasoning in a way that corresponds with human modes of reasoning and debate.

We next note that some have argued that such deductive simulations are prone to yielding counterintuitive results. To illustrate, consider a instantiation of $ASPIC^+$ with no defeasible rules and in which the strict rules correspond to classical propositional logic as defined in Section 3.1.2, and assume that natural-language generalisations 'If P then normally Q' are formalised as material implications $P \supset Q$ in $\mathcal{K}_p$. The idea is that since $P \supset Q$ is an ordinary premise, its use as a premise can be undermined in exceptional cases. Observe that by classical reasoning we then have a strict argument for $\neg Q \supset \neg P$. Some say that this is problematic. Consider the following example: 'This alarm in this building usually does not give false alarms', so (strictly) 'false alarms in this building are usually not given by this alarm'. This strikes some as counterintuitive, since the first generalisation is consistent with the situation that this alarm is the only one in the building that gives false alarms, so the contraposition of 'If P then normally Q' cannot be deductively valid.

A more refined classical approach is to give the material implication an extra normality condition N, which informally reads as 'everything is normal as regards P implying Q', and which is also put in $\mathcal{K}_p$. The idea then is that exceptional cases give rise to underminers of N. However, $(P \wedge N) \supset Q$ also deductively contraposes, namely, as $(\neg Q \wedge N) \supset \neg P$, so we still have the controversial deductive validity of contraposition for generalisations. In the false-alarm example the contraposition of the rule with the added normality condition would read: 'any false alarm in this building which is usual with respect to false alarms in this building cannot be this alarm', which is clearly not deductively entailed by the initial generalisation given that it is consistent with the situation that this alarm is the only one in the building that gives false alarms.

One way to argue why classical simulations may give counter-intuitive results is to recall that a number of researchers provide statistical semantics for defeasible inference rules. These semantics regard a defeasible rule of the form $P \Rightarrow Q$ as a qualitative approximation of the statement that the conditional probability of Q, given P, is high. The laws of probability theory then tell us that this does not entail that the conditional probability of $\neg P$, given $\neg Q$, is high. The problem with the classical-logic approach is then that it conflates this distinction by turning the conditional probability of Q given P into the unconditional probability of $P \supset Q$, which then has to be equal to the unconditional probability of $\neg Q \supset \neg P$.

So far we have argued that contrapositive inferences with defeasible conditionals cannot be deductively valid (for a more detailed argument see [Modgil and Prakken, 2014, Section 4.5]). One way to respect this is to formalise defeasible natural-language conditionals as domain-specific defeasible inference rules in $ASPIC^+$ (see Section 3.1.3 above and in more detail [Modgil and Prakken, 2014]). However, this makes it hard to capture some logical properties of defeasible conditionals. For example, it might be argued that modus tollens and contraposition, although deductively invalid, are still defeasibly valid. For instance, in crime investigations the police often reason: if this person was at the crime, then we must be able to find his DNA at the crime scene; we have not been able to find his DNA at the crime scene, so presumably he was not at the crime scene. This seems a perfectly rational way of reasoning, provided that the modus-tollens inference is regarded as defeasible. Perhaps this can be captured by formalising generalisations with a defeasible object level connective $\leadsto$, as discussed above in Section 3.1.3 and by adding the appropriate strict and defeasible inference rules for $\leadsto$ to $\mathcal{R}_s$ and $\mathcal{R}_d$. For example, defeasible modus tollens could be added as follows:

$$\neg\psi, \varphi \leadsto \psi \Rightarrow \neg\varphi$$

However, doing so is not straightforward, since the above encoding of the defeasible modus pollens principle is in the form of an inference rule used in construction of $ASPIC^+$ arguments, while in contrast, the current nonmonotonic logics for defeasible conditionals model such principles at the level of the consequence relation (which in $ASPIC^+$ is defined in terms of the outcome of argument evaluation; cf. Definition 2.18 above). This suggests the following topic for future research: how to instantiate the sets of strict and defeasible rules in $ASPIC^+$ in such a way that the semantic insights on defeasible conditionals obtained in other areas of nonmonotonic logic are respected?

So far our discussion has focused on argumentation based reasoning as it applies to beliefs (i.e., reasoning about what is the case, often called *epistemic reasoning* by philosophers). However argumentation is often about what to do, prefer or value (what philosophers often call *practical reasoning*). Here too it has been argued on philosophical grounds that reasons for doing, preferring or valuing cannot be expressed in classical logic since they do not contrapose. This view can of course not be based on a statistical semantics, since statistics only applies to epistemic reasoning. Space limitations prevent us from giving more details about these philosophical arguments.

Finally, as further discussed in Section 4.1, [Dung and Thang, 2014] show for the case without preferences and knowledge bases that $ASPIC^+$ defeasible rules can be equivalently translated into theories of assumption-based argumentation (ABA). Since, as also discussed further in Section 4.1, ABA can be reconstructed as a special case of $ASPIC^+$ with no knowledge bases, defeasible rules or preferences, [Dung and Thang, 2014]'s result implies that the defeasible rules of $ASPIC^+$ SAFs with no knowledge bases or preferences can be translated into strict $ASPIC^+$ rules.

3.4.5 The value of translation results

Translation results like the ones of [Dung and Thang, 2014] and [Li and Parsons, 2015] on translating one type of rule into the other, and possible future results on encoding preferences in rules, are theoretically interesting and may have practical benefits. For example, [Dung and Thang, 2014]'s result makes it possible to use ABA tools for implementing fragments of $ASPIC^+$ without preferences. However, such translation results should be interpreted with care. Logic is full of such results and they do not necessarily mean that the translated system is less useful or less interesting. For example, nobody would say that the fact that all connectives of propositional logic can be translated into a single one means that presentations of propositional logic with the usual five or six connectives are unnecessarily complicated; on the contrary, versions with just one connective would lead to unnecessarily complex knowledge representations. Likewise, versions of $ASPIC^+$ with both strict and defeasible rules and with preferences may lead to more compact and more natural representations. Moreover, nobody would say that translations of modal logic into first-order predicate logic show that modal logic is superfluous. On the contrary, modal logics often provide systematic treatments of modalities in ways that their first-order translations do not. Likewise, $ASPIC^+$ provides a theory of reasoning with a combination of strict and defeasible rules and allows a general study of argumentation with preferences, something which formalisms with only strict or only defeasible rules or formalisms without preferences do not provide.

3.5 Argument schemes and critical questions

We concluded Section 3.1.3 by remarking on the use of defeasible inference rules as principles of cognition in John Pollock's work and as argument schemes in informal argumentation theory. We now illustrate how both approaches can be formalised in $ASPIC^+$ and how strict inference rules can also be accommodated when doing so.

John Pollock formalised defeasible rules for reasoning patterns involving perception, memory, induction, temporal persistence and the statistical syllogism, as well as undercutters for these reasons. In $ASPIC^+$ his principles of perception and memory can be written as follows:

$d_p(x, \varphi)$: $\text{Sees}(x, \varphi) \Rightarrow \varphi$
$d_m(x, \varphi)$: $\text{Recalls}(x, \varphi) \Rightarrow \varphi$

In fact, these defeasible inference rules are schemes for all their ground instances (that is, for any instance where x and φ are replaced by ground terms denoting a specific perceiving agent and a specific perceived state of affairs). Therefore, their names $d_p(x, \varphi)$ and $d_m(x, \varphi)$ as assigned by the n function are in fact also schemes for names. A proper name is obtained by instantiating these variables by the same ground terms as used to instantiate these variables in the scheme. Thus it becomes possible to formulate undercutters for one instance of the scheme (say for Jan who saw John in Amsterdam) while leaving another instance unattacked (say for Bob who saw John in Holland Park). Note, finally, that these schemes assume a naming convention for formulas in a first-order language, since φ is a term in the antecedent while it is a well-formed formula in the consequent. In the remainder we will leave this naming convention implicit.

Now undercutters for d_p state circumstances in which perceptions are unreliable, while undercutters of d_m state conditions under which memories may be flawed. For example, a well-known cause of false memories of events is that the memory is distorted by, for instance, seeing pictures in the newspaper or watching a TV programme about the remembered event. A general undercutter for distorted memories could be

$u_m(x, \varphi)$: $\texttt{DistortedMemory}(x, \varphi) \Rightarrow \neg d_m(x, \varphi)$

combined with information such as

$\forall x, \varphi(\texttt{SeesPicturesAbout}(x, \varphi) \supset \texttt{DistortedMemory}(x, \varphi))$

Pollock's epistemic inference schemes are in fact a subspecies of argument schemes. The notion of an argument scheme was developed in philosophy and is currently an important topic in the computational study of argumentation. Argument schemes are stereotypical non-deductive patterns of reasoning, consisting of a set of premises and a conclusion that is presumed to follow from them. Uses of argument schemes are evaluated in terms of critical questions specific to the scheme. An example of an epistemic argument scheme is the scheme from the position to know [Walton, 1996, pp. 61–63]:

> A is in the position to know whether P is true
> A asserts that P is true
> ―――――――――――――――――――
> P is true

Walton gives this scheme three critical questions:

1. Is A in the position to know whether P is true?
2. Did A assert that P is true?
3. Is A an honest (trustworty, reliable) source?

A natural way to formalise reasoning with argument schemes is to regard them as defeasible inference rules and to regard critical questions as pointers to counterarguments. For example, in the scheme from the position to know, questions (1) and (2) point to underminers (of, respectively, the first and second premise) while question (3) points to undercutters (the exception that the person is for some reason not credible).

Accordingly, we formalise the position to know scheme and its undercutter as follows:

$d_w(x,\varphi)$: $\texttt{PositionToKnow}(x,\varphi), \texttt{Says}(x,\varphi) \Rightarrow \varphi$
$u_w(x,\varphi)$: $\neg\texttt{Credible}(x) \Rightarrow \neg d_w(x,\varphi)$

We will now illustrate the modelling of both Pollock's defeasible reasons and Walton's argument schemes with our example from Section 2.1, focusing on a specific class of persons who are in the position to know, namely, witnesses. In fact, witnesses always report about what they observed in the past, so they will say something like "I remember that I saw that John was in Holland Park". Thus an appeal to a witness testimony involves the use of three schemes: first the position to know scheme is used to infer that the witness indeed remembers that he saw that John was in Holland Park, then the memory scheme is used to infer that he indeed saw that John was in Holland Park, and finally, the perception scheme is used to infer that John was indeed in Holland Park. Now recall that John was a suspect in a robbery in Holland Park, that Jan testified that he saw John in Amsterdam on the same morning, and that Jan is a friend of John. Suppose now we also receive information that Bob read newspaper reports about the robbery in which a picture of John was shown. One way to model this in $ASPIC^+$ is as follows.

The knowledge base consists of the following facts (since we don't want to dispute them, we put them in $\mathcal{K}_n$):

f_1: $\texttt{PositionToKnow}(Bob, \texttt{Recalls}(Bob, \texttt{Sees}(Bob, \texttt{InHollandPark}(John))))$
f_2: $\texttt{Says}(Bob, \texttt{Recalls}(Bob, \texttt{Sees}(Bob, \texttt{InHollandPark}(John))))$
f_3: $\texttt{SeesPicturesAbout}(Bob, \texttt{Sees}(Bob, \texttt{InHollandPark}(John)))$
f_4: $\forall x, \varphi.(\texttt{SeesPicturesAbout}(x,\varphi) \supset \texttt{DistortedMemory}(x,\varphi))$
f_5: $\forall x.\texttt{InHollandPark}(x) \supset \texttt{InLondon}(x)$
f_6: $\texttt{PositionToKnow}(Jan, \texttt{Recalls}(Jan, \texttt{Sees}(Jan, \texttt{InAmsterdam}(John))))$
f_7: $\texttt{Says}(Jan, \texttt{Recalls}(Jan, \texttt{Sees}(Jan, \texttt{InAmsterdam}(John))))$
f_8: $\texttt{Friends}(Jan, John)$
f_9: $\texttt{SuspectedRobber}(John)$
f_{10}: $\forall x, y, \varphi.\texttt{Friends}(x,y) \wedge \texttt{SuspectedRobber}(y) \wedge \texttt{InvolvedIn}(y,\varphi) \supset \neg\texttt{Credible}(x)$
f_{11}: $\texttt{InvolvedIn}(John, \texttt{Recalls}(Jan, \texttt{Sees}(Jan, \texttt{InAmsterdam}(John))))$
f_{12}: $\forall x \neg(\texttt{InAmsterdam}(x) \wedge \texttt{InLondon}(x))$

Combining this with the schemes from perception, memory and position to know, we obtain the following arguments (for reasons of space we don't list separate lines for arguments that just take an item from $\mathcal{K}$).

A_3: $f_1, f_2 \Rightarrow_{dw} \texttt{Recalls}(Bob, \texttt{Sees}(Bob, \texttt{InHollandPark}(John)))$
A_4: $A_3 \Rightarrow_{dm} \texttt{Sees}(Bob, \texttt{InHollandPark}(John))$
A_5: $A_4 \Rightarrow_{dp} \texttt{InHollandPark}(John)$
A_7: $A_5, f_5 \to \texttt{InLondon}(John)$

This argument is undercut (on A_4) by the following argument applying the undercutter for the memory scheme:

B_3: $f_3, f_4 \rightarrow$ DistortedMemory(Bob, Sees(Bob, InHollandPark($John$)))
B_4: $B_3 \Rightarrow_{um} \neg d_m(Bob,$ Sees(Bob, InHollandPark($John$)))

Moreover, A_7 is rebutted (on A_5) by the following argument:

C_3: $f_6, f_7 \Rightarrow_{dw}$ Recalls(Jan, Sees(Jan, InAmsterdam($John$)))
C_4: $C_3 \Rightarrow_{dm}$ Sees(Jan, InAmsterdam($John$))
C_5: $C_4 \Rightarrow_{dp}$ InAmsterdam($John$)
C_8: $C_5, f_5, f_{12} \rightarrow \neg$InHollandPark($John$)

This argument is also undercut, namely on C_3, based on the undercutter of the position to know scheme:

D_4: $f_8, f_9, f_{10}, f_{11} \rightarrow \neg$Credible($Jan$)
D_5: $D_4 \Rightarrow_{uw} \neg d_w(Jan,$ Recalls(Jan, Sees(Jan, InAmsterdam($John$))))

Finally, C_8 is rebutted on C_5 by the following continuation of argument A_7:

A_8: $A_7, f_5, f_{12} \Rightarrow \neg$InAmsterdam($John$)

A_8 is in turn undercut by B_4 (on A_4) and rebutted by C_8 (on A_5).

Because of the two undercutting arguments, neither of the testimony arguments are credulously or sceptically justified in any semantics. Let us now see what happens if we do not have the two undercutters. Then we must apply preferences to the rebutting attack of C_8 on A_5 and to the rebutting attack of A_8 on C_5. As it turns out, exactly the same preferences have to be applied in both cases, namely, those between the three defeasible-rule applications in the respective arguments. And this is what we intuitively want.

Finally, we note that counterarguments based on critical questions of argument schemes may themselves apply argument schemes. For example, we may believe that Jan and John are friends because another witness told us so. Or we may believe that Holland Park is in London because a London taxi driver told us so (an application of the so-called expert testimony scheme).

4 Relationship with other Argumentation Formalisms

As shown in various publications on $ASPIC^+$, its generality allows the reconstruction of various other systems and frameworks as special cases of $ASPIC^+$. In this section we review this work in some detail. We also discuss the relationship of $ASPIC^+$ with various developments of abstract argumentation frameworks.

4.1 Assumption-based argumentation

Assumption-based argumentation (ABA) emerged from attempts to give an argumentation-theoretic semantics to logic-programming's negation as failure, and has developed into a general framework for nonmonotonic logics [Bondarenko et al., 1993; Bondarenko et al., 1997; Toni, 2014]. ABA assumes a 'deductive system', consisting of a set of inference rules defined over some logical language. Given a set of so-called 'assumptions' formulated in the logical language, arguments are then deductions of claims using rules and supported by sets of assumptions. In general, ABA leaves both the logical language and set of inference rules unspecified, so that like $ASPIC^+$, it is an abstract framework for structured argumentation. However, unlike $ASPIC^+$, ABA only allows attacks on an argument's assumptions, so that ABA's rules are effectively strict inference rules. In order to express conflicts between arguments, ABA makes a minimum assumption on the logical language, namely, that each assumption has a contrary. That b is a contrary of a, written as $b = \overline{a}$, informally means that b contradicts a. An argument using an assumption a is then attacked by any argument for conclusion $\overline{a}$. In [Bondarenko et al., 1997] an argumentation-theoretic semantics is then given which is very much like [Dung, 1995]'s abstract approach, except that [Bondarenko et al., 1997] considers sets of assumptions rather than sets of arguments. However, [Dung et al., 2007] showed that an equivalent fully argument-based formulation can be given.

In this section we first discuss how ABA can be reconstructed in $ASPIC^+$ and then how some instantiations of $ASPIC^+$ can be reconstructed in ABA.

4.1.1 Reconstructing ABA in $ASPIC^+$

Above we remarked that [Bondarenko et al., 1997]'s version of ABA is strictly speaking not an instantiation of [Dung, 1995]'s abstract argumentation frameworks but that [Dung et al., 2007] gave an equivalent formulation of ABA in such frameworks. [Prakken, 2010] showed that this reconstructed version of ABA can in turn be reconstructed as a special case of $ASPIC^+$ extended with possibly non-symmetric negation (see Section 2.3 above). In $ASPIC^+$ as defined by [Prakken, 2010], the ordinary premises were further divided into 'really' ordinary premises and assumptions and the assumption premises were used to model ABA assumptions. However, as observed by [Modgil and Prakken, 2013, Section 3.1], one can do without such specialised premises and model assumptions as ordinary premises. ABA can then be reconstructed as the special case of $ASPIC^+$ with empty sets of defeasible rules and axiom premises and no preferences.

First the main definitions of ABA are recalled.

Definition 4.1 *(Def. 2.3 of [Dung et al., 2007].) A deductive system is a pair* $(\mathcal{L}, \mathcal{R})$ *where*

- $\mathcal{L}$ *is a formal language consisting of countably many sentences, and*

- $\mathcal{R}$ is a countable set of inference rules of the form $\alpha_1,\ldots,\alpha_n \to \alpha$.[13] $\alpha \in \mathcal{L}$ is called the conclusion of the inference rule, $\alpha_1,\ldots,\alpha_n \in \mathcal{L}$ are called the premises of the inference rule and $n \geq 0$.

Definition 4.2 *(Def. 2.5 of [Dung et al., 2007].)* An assumption-based argumentation framework *(ABF)* is a tuple $(\mathcal{L}, \mathcal{R}, \mathcal{A}, ^-)$ where

- $(\mathcal{L}, \mathcal{R})$ is a deductive system.
- $\mathcal{A} \subseteq \mathcal{L}$, $\mathcal{A} \neq \emptyset$. $\mathcal{A}$ is the set of assumptions.
- If $\alpha \in \mathcal{A}$, then there is no inference rule of the form $\alpha_1,\ldots,\alpha_n \to \alpha \in \mathcal{R}$.
- $^-$ is a total mapping from $\mathcal{A}$ into $\mathcal{L}$. $\overline{\alpha}$ is the contrary of α.

ABA arguments are then defined in terms of deductions. To remain as close as possible to $ASPIC^+$, we here give the tree-based definition of [Toni, 2014] (with some minor stylistic rephrasings). The proofs of [Prakken, 2010] instead use [Dung et al., 2007]'s sequence-based definition, which essentially presents one particular order in which a tree-style argument can be constructed.

Definition 4.3 *([Toni, 2014].)* A deduction *for a conclusion* α *supported by premises* $S \subseteq \mathcal{L}$ *is a finite tree with nodes labelled by sentences in* $\mathcal{L}$ *or by* τ[14]. *Each leaf is either* τ *or a sentence in* S. *each non-leave* α' *has, as children, the elements of the body of some rule in* $\mathcal{R}$ *with head* α'.

Then an assumption-based argument is defined as follows.

Definition 4.4 *(Def. 2.6 of [Dung et al., 2007].)* An argument *for a conclusion on the basis of an ABF is a deduction of that conclusion whose premises are all assumptions (in* $\mathcal{A}$*)*.

As for notation, the existence of an argument for a conclusion α supported by a set of assumptions A is denoted by $A \vdash \alpha$, or by $A \vdash_{ABF} \alpha$ if it has to be distinguished from the existence of a strict argument according to Definition 2.4 with the same premises and conclusion; the latter will below be denoted by $A \vdash_{AT} \alpha$.

Finally, Dung et al.'s notion of argument attack is defined as follows.

Definition 4.5 *(Def. 2.7 of [Dung et al., 2007].)*

- An argument $A \vdash \alpha$ attacks an argument $B \vdash \beta$ if and only if $A \vdash \alpha$ attacks an assumption in B;
- an argument $A \vdash \alpha$ attacks an assumption β if and only if α is the contrary $\overline{\beta}$ of β.

---

[13] In [Dung et al., 2007] the arrows are from right to left.
[14] τ represents 'true' and stands for the empty body of rules.

The $ASPIC^+$ argumentation theory corresponding to an assumption-based argumentation framework is then in [Prakken, 2010] defined as follows.[15]

Definition 4.6 *[Mapping ABFs to ATs] Given an assumption-based argumentation framework $ABF = (\mathcal{L}_{ABF}, \mathcal{R}_{ABF}, \mathcal{A}, ^-_{ABF})$, the corresponding argumentation theory $AT_{ABF} = (AS, \mathcal{K})$, where $AS = (\mathcal{L}_{AT}, ^-_{AT}, \mathcal{R}_{AT}, n)$ and $\mathcal{K} = \mathcal{K}_n \cup \mathcal{K}_p$, is defined as follows:*

- $\mathcal{L}_{AT} = \mathcal{L}_{ABF}$

- $\varphi \in \overline{\psi}_{AT}$ *iff* $\varphi = \overline{\psi}_{ABF}$

- $\mathcal{R}_{AT} = \mathcal{R}_s = \mathcal{R}_{ABF}$

- $\mathcal{K}_n = \emptyset$

- $\mathcal{K}_p = \mathcal{A}$

- *n is undefined.*

Then it can be shown that for all ABFs: there exists an argument $A \vdash_{ABF} \alpha$ if and only if there exists an argument $A \vdash_{AT} \alpha$. From this it follows for all ABFs and for every argument $A \vdash_{ABF} \alpha$ and every argument $A \vdash_{AT} \alpha$: $A \vdash_{ABF} \alpha$ is attacked by an argument $B \vdash_{ABF} \beta$ if and only if $A \vdash_{AT} \alpha$ is defeated by an argument $B \vdash_{AT} \beta$. Then the main correspondence result can be proven:

Theorem 4.7 (Thm. 8.8 of [Prakken, 2010]) *For all ABFs, and for any semantics S subsumed by complete semantics and any set E:*

1. *if E is an S-extension of ABF then E_{AT} is an S-extension of AT, where $E_{AT} = \{A \vdash_{AT} \alpha \mid A \vdash_{ABF} \alpha \in E\}$;*

2. *if E is an S-extension of AT then E_{ABF} is an S-extension of ABF, where $E_{ABF} = \{A \vdash_{ABF} \alpha \mid A \vdash_{AT} \alpha \in E\}$.*

Theorem 4.7 says that there is a one-to-one correspondence between the extensions of an ABF and those of its corresponding AT. Note also that the above results carry over to [Verheij, 2003]'s DefLog argumentation system since, as observed by Verheij, DefLog can be translated into ABA.

One virtue of this reconstruction of ABA in $ASPIC^+$ is that one can then identify conditions under which ABA satisfies rationality postulates (by requiring, for instance, that the strict rules are closed under transposition).

---

[15]In fact, in [Prakken, 2010] the ABA assumptions were translated into $ASPIC^+$ assumption-type premises, which in [Prakken, 2010] was an additional category of premises. However, as remarked by [Modgil and Prakken, 2013], the translation also succeeds when defined as below.

4.1.2 Reconstructing instantiations of $ASPIC^+$ in ABA

[Dung and Thang, 2014] have shown that their rule-based systems, which are a special case of $ASPIC^+$ with no knowledge base and no preferences, can be translated into ABA instantiations. They do this by translating every defeasible rule $p_1, \ldots, p_n \Rightarrow q$ as a strict rule $d_i, p_1, \ldots, p_n, not\neg q \to q$, where

- $d_i = n(p_1, \ldots, p_n \Rightarrow q)$ in $ASPIC^+$;
- $d_i, not\neg q \in \mathcal{A}$ (i.e., they are ABA assumptions);
- $q = \overline{not\neg q}$ and for all φ: $\varphi = \overline{\neg \varphi}$ and $\neg \varphi = \overline{\varphi}$

[Dung and Thang, 2014] then show (on the assumption that $ASPIC^+$ rule names do not occur as antecedents or consequents in $ASPIC^+$ rules), that for each semantics subsumed by complete semantics the resulting ABA framework validates the same conclusions as the original $ASPIC^+$ SAF. Generalising this result to cases with preferences is still an open question.

4.2 Tarskian abstract logics and classical-logic argumentation

[Amgoud and Besnard, 2013] present an abstract approach to defining the structure of arguments and attacks, based on Tarski's notion of an abstract logic that only assumes some unspecified logical language $\mathcal{L}$, and a consequence operator over this language, which to each subset of $\mathcal{L}$ assigns a subset of $\mathcal{L}$ (its logical consequences). Tarski then assumed a number of constraints on Cn (see [Amgoud and Besnard, 2013] for a more detailed account of these constraints). Finally, Tarski defined a set $S \subseteq \mathcal{L}$ as *consistent* iff $Cn(S) \neq \mathcal{L}$. In [Amgoud and Besnard, 2013], an argument is a pair (S, p) where $S \subseteq \mathcal{L}$ is consistent, $p \in Cn(S)$ and S is a minimal (under set inclusion) set satisfying these conditions. Then (S, p) attacks (T, q) iff $\{p, q'\}$ is inconsistent for some $q' \in T$.

[Modgil and Prakken, 2013, Section 5.2] show that $ASPIC^+$ can be used to reconstruct, and extend with preferences, the Tarskian logic approach. For the strict rules, they choose (for any finite $S \subseteq \mathcal{L}$):

$$S \to p \in \mathcal{R}_s \text{ iff } p \in Cn(S)$$

Then given any $\Sigma \subseteq \mathcal{L}$, they let $\mathcal{K}_p = \Sigma$, $\mathcal{R}_d = \emptyset$. Also, $\forall \phi \in \mathcal{L}$, ϕ has a contradictory ψ, and if $\phi = -\psi$ then $Cn(\{\phi, \psi\}) = \mathcal{L}$ and if $Cn(\{\phi, \psi\}) = \mathcal{L}$ then $\exists \phi' \in Cn(\{\phi\})$ s.t. $\phi' = -\psi$. They then show that given a reasonable argument preference ordering $\preceq$ (possibly defined on the basis of an ordering $\leq'$ over Σ), the c-SAF is well defined. Hence one obtains an account of [Amgoud and Besnard, 2013]'s Tarskian logic abstract argumentation approach that is extended with preferences and is well behaved with respect to rationality postulates. Two issues to note are that the reconstruction employs $ASPIC^+$ undermining attacks, which differ from the abstract logic attacks defined above which rely on showing that the claim and attacked premises are inconsistent. However, [Modgil and Prakken, 2013] show that the use of $ASPIC^+$ attacks does not change the outcome in the sense that the complete (and hence grounded, preferred and stable) extensions remain the same irrespective of whether we use the abstract logic notion of an attack instead. Moreover, $ASPIC^+$ imposes no subset minimality conditions on the premises of arguments. However, [Modgil and Prakken, 2013] show that if subset minimal arguments are not strengthened by adding 'irrelevant' premises – i.e., if A is subset minimal and $A \not\prec B$ then $A' \not\prec B$ where $\texttt{Prem}(A') \supset \texttt{Prem}(A)$ – then the conclusions of arguments in complete extensions remains the same whether or not we exclude arguments that are not subset minimal.

[Modgil and Prakken, 2013] then applied this to a reconstruction of so-called classical argumentation [Cayrol, 1995; Besnard and Hunter, 2001; Besnard and Hunter, 2008; Gorogiannis and Hunter, 2011], which formalises arguments as minimal classical consequences from consistent and finite premise sets in standard propositional or first-order logic. In particular, [Gorogiannis and Hunter, 2011] study classical logic instantiations of abstract argumentation frameworks. [Modgil and Prakken, 2013] reconstruct this as a specific instance of the above formulation of the Tarskian abstract logic approach, with Cn the classical consequence operator (below denoted as $\models$). This yields the following instantiation of $ASPIC^+$:

Definition 4.8 *[Classical argumentation with preferences reconstructed as an instance of* $ASPIC^+$*] Let $\mathcal{L}'$ be a classical-logic language, $\Sigma \subseteq \mathcal{L}'$ and $\leq'$ a partial preorder on Σ. A classical-logic argumentation theory based on $(\mathcal{L}', \Sigma, \leq')$ is a pair $(AS, \mathcal{K})$ such that AS is an argumentation system $(\mathcal{L}, ^-, \mathcal{R}, n)$ where:*

1. $\mathcal{L} = \mathcal{L}'$;
2. $\varphi \in \overline{\psi}$ iff $\varphi = \neg\psi$ or $\psi = \neg\varphi$;
3. $\mathcal{R}_d = \emptyset$, and for all finite $S \subseteq \mathcal{L}$ and $p \in \mathcal{L}$, $S \to p \in \mathcal{R}_s$ iff $S \models p$.

$\mathcal{K}$ *is a knowledge base such that* $\mathcal{K}_n = \emptyset$ *and* $\mathcal{K}_p = \Sigma$.
$(\mathcal{A}, \mathcal{C}, \preceq)$ *is the c-SAF based on* $(AS, \mathcal{K})$ *as defined in Definition 2.14 and where $\preceq$ is defined in terms of $\leq'$ as in Section 3.2.*

[Gorogiannis and Hunter, 2011] define seven attack relations and prove that only the following two ensure satisfaction of the rationality postulate of indirect consistency:

- Y *directly undercuts* X if $\texttt{Conc}(Y) \equiv \neg p$ for some $p \in \texttt{Prem}(X)$
- Y *directly defeats* X if $\texttt{Conc}(Y) \vdash_c \neg p$ for some $p \in \texttt{Prem}(X)$

Since classical logic can be specified as a Tarskian abstract logic, [Modgil and Prakken, 2013] can prove via their reconstruction of abstract-logic argumentation, that the $ASPIC^+$ notion of undermining attacks is equivalent to direct undercuts and defeats in that the complete extensions generated are the same. Moreover, from the results described above in Section 3.2 it follows that their extension of classical-logic argumentation with preferences satisfies the rationality postulates. Indeed, [Modgil and Prakken, 2013] argue that the extension to include preferences is needed if classical-logic argumentation is to be effectively used in arbitrating amongst conflicts, since as shown in ([Cayrol, 1995; Gorogiannis and Hunter, 2011; Amgoud and Besnard, 2013]), there is a one-to-one correspondence between the (premises of arguments in in) preferred/stable extensions of abstract argumentation frameworks instantiated by a classical-logic knowledge base and the maximal consistent subsets of the knowledge base. This is to be expected, given the monotonicity of classical logic (and thus the absence of logical mechanisms to withdraw previously derivable contradictory inferences).

4.3 Carneades

As shown by [Van Gijzel and Prakken, 2011; Van Gijzel and Prakken, 2012], the Carneades system of [Gordon et al., 2007; Gordon and Walton, 2009b] can be reconstructed as a special case of basic $ASPIC^+$ with a generalised contrariness relation. A Carneades argument is a triple $\langle P, E, c \rangle$ where P is a set of *premises*, E a set of *exceptions* and c the *conclusion*, which is either pro or con a *statement s*. Carneades does not assume that premises and conclusions are connected by inference rules. Also, all arguments are elementary, that is, they contain a single inference step; they are combined in recursive definitions of *applicability* of an argument and *acceptability* of its conclusion. In essence, an *argument* is *applicable* if (1) all its premises are given as facts or else are acceptable conclusions of other arguments, and (2) none of its exceptions are given as facts or as acceptable conclusions of other arguments. A *statement* is *acceptable* if it satisfies its *proof standard*. Facts are stated by an *audience*, which also provides numerical *weights* for each argument plus *thresholds* for argument weights and differences in argument weights. In the publications on Carneades five proof standards are defined. One is *preponderance of the evidence*:

> Statement p satisfies *preponderance of the evidence* iff there exists at least one applicable argument pro p for which the weight is greater than the weight of any applicable argument con p.

In the $ASPIC^+$ reconstruction of Carneades the facts are reconstructed as elements of $\mathcal{K}_n$, while the Carneades notions of applicability and acceptability are encoded in the $ASPIC^+$ defeasible inference rules. For every Carneades argument $a = \langle P, E, c \rangle$, a defeasible rule $P \Rightarrow_{app_a} arg_a$ is added, saying that if P then a is applicable[16]. Moreover, a defeasible rule $arg_a \Rightarrow_{acc_a} c$ is added, saying that if a is applicable, its conclusion is acceptable. Here, app_a and acc_a are the respective names of these rules in $\mathcal{L}$ according to the naming convention n. Thus a Carneades argument $\langle P, E, c \rangle$ pro statement s induces an $ASPIC^+$ argument:

---

[16]The idea to make the applicability step explicit by means of an argument node was adapted from [Brewka and Gordon, 2010].

A_1: P
A_2: $A_1 \Rightarrow_{app_a} arg_a$
A_3: $A_2 \Rightarrow_{acc_a} s$

It should be noted that effectively, a Carneades argument is analogous to a defeasible inference rule, since the representation (P, E, c) does not assume that the facts P are given as part of the argument; rather it is the *applicability* of the argument that depends on facts or arguments for P. This justifies the translation of Carneades arguments into $ASPIC^+$ defeasible rules.

Next, for each exception $e \in E$, a rule $e \Rightarrow \neg app_a$ is added to $\mathcal{R}_d$ and $\neg app_a = \overline{app_a}$ is added to the contrariness relation. So such rules can be used to undercut the $ASPIC^+$ version of an argument on its first step. Moreover, for each argument b with a conclusion c' that conflicts with s, we have that $arg_b = \overline{acc_a}$ if this is dictated by the proof standard for s. For example, if the standard for s is preponderance of the evidence, then $arg_b = \overline{acc_a}$ just in case $weight(a) \leq weight(b)$. Thus the Carneades proof standards and argument weights are not incorporated in the $ASPIC^+$ argument ordering but in the $ASPIC^+$ contrariness relation.

For example, a Carneades argument $b = \langle P', E', c' \rangle$ where c' is con s, induces an $ASPIC^+$ argument:

B_1: P'
B_2: $B_1 \Rightarrow_{app_b} arg_b$
B_3: $B_2 \Rightarrow_{acc_b} \neg s$

Then A_3 rebuts B_3 if $weight(b) < weight(a)$, B_3 rebuts A_3 if $weight(a) < weight(b)$ and both rebut each other if $weight(a) = weight(b)$. Since in the $ASPIC^+$ reconstruction all defeasible arguments are equally strong, all these rebutting attacks succeed as defeat.

[Van Gijzel and Prakken, 2011; Van Gijzel and Prakken, 2012] then prove that under this reconstruction, $ASPIC^+$ SAFs corresponding to a Carneades theory always have a unique extension, which is the same in all of [Dung, 1995]'s semantics. This perhaps surprising result is partly due to strong non-circularity assumptions made in Carneades on its 'inference graph', which contains all constructible arguments. [Van Gijzel and Prakken, 2011; Van Gijzel and Prakken, 2012] also prove that the conclusions of the justified arguments in $ASPIC^+$ correspond to the conclusions of the acceptable arguments in Carneades.

4.4 Defeasible Logic Programming

Defeasible logic programming (DeLP) is a logic-programming-based argumentation system originating from (but not equivalent to) [Simari and Loui, 1992]. The main publication on DeLP is [Garcia and Simari, 2004], which we will take as the basis for our discussion. Although DeLP is similar to $ASPIC^+$, it cannot be fully reconstructed as an instance. Elements of DeLP that instantiate $ASPIC^+$ are a predicate-logic literal language with ordinary negation, a set of indisputable facts, two sets of strict and defeasible rules, and a binary argument ordering. DeLP arguments can be reconstructed as $ASPIC^+$ arguments with the additional constraint that their sets of conclusions are consistent under application of strict rules in that for no φ it holds that $\text{Conc}(A) \vdash \varphi, \neg\varphi$.

DeLP's definition of attack is similar but not equivalent to $ASPIC^+$'s notion of rebutting attack. Instead (and translated to $ASPIC^+$ vocabulary), A attacks B at B's subargument B' if $\text{Conc}(A) \cup \text{Conc}(B') \vdash \varphi, \neg\varphi$ for some wff φ. Note that this allows an attack on a conclusion of a strict rule, but such an attack will never exist without an attack on a previous defeasible step in the argument as well. Apart from this difference, DeLP's notion of defeat is defined as in $ASPIC^+$: A defeats B if A attacks B on B' and $A \not\prec B'$. It remains to be investigated whether adopting DeLP's notion of rebutting attack in $ASPIC^+$ would lead to different outcomes.

A main difference with $ASPIC^+$ is that DeLP as defined in [Garcia and Simari, 2004] does not evaluate arguments by generating abstract argumentation frameworks. Instead, DeLP's notion of *warrant* is defined in a way that is similar to the argument game of grounded semantics [Prakken, 1999; Modgil and Caminada, 2009] but with some significant differences. Briefly, the argument game for grounded semantics is between a proponent and an opponent of an argument A, where the proponent begins with A and then the players take turns such that the opponent defeats or strictly defeats the proponent's previous argument while the proponent strictly defeats the opponent's previous argument; in addition, the proponent is not allowed to repeat his own arguments. An argument A is justified if the proponent has a winning strategy in a game starting with A. DeLP's notion of warrant is equivalent to this notion of justification but its game rules are different. First, if one player weakly defeats the previous argument then the next player must strictly defeat that argument, while if one player strictly defeats the previous argument then the next player may either weakly or strictly defeat it. Second, no player may reuse a subargument from one of its earlier moves.

It would be interesting to adopt the game rules of grounded semantics in DeLP's notion of warrant, which would then establish a clear link between DeLP and the theory of abstract argumentation. Among other things, this would facilitate the study of the satisfaction of rationality postulates in DeLP.

4.5 $ASPIC^+$ characterisations of non-monotonic reasoning formalisms

A key reason for the prominence of argumentation (in particular Dung's theory of abstract argumentation frameworks) in knowledge representation and reasoning, is its characterisation of non-monotonic reasoning in terms of the dialectical exchange of argument and counter-argument. Indeed, in [Dung, 1995], argumentation-based characterisations of logic programming, [Reiter, 1980]'s Default Logic and [Pollock, 1987]'s argumentation system are formalised. The theory thus provides foundations for reasoning by individual computational and human agents, and distributed non-monotonic reasoning ('dialogue') amongst agents.

$ASPIC^+$ continues in this tradition, formalising logic programming instantiations of abstract argumentation frameworks, whereby the defeasible rules are rules in a logic program, the strict rules and axiom premises are empty, the preference relation is empty, and (as described in Section 2.3) the ordinary premises are the negation as failure ($\sim$) assumptions in the antecedents of defeasible rules, and we define the contrary function $\forall \alpha \in \mathcal{L}: \alpha \in \overline{\sim \alpha}$.

Brewka's Preferred Subtheories [Brewka, 1989] can also be formalised as an instance of $ASPIC^+$'s formalisation of classical-logic argumentation (as outlined in Section 4.2). The arguments and attacks are defined by a base Σ of propositional classical wff equipped with a total ordering $\leq'$ which is used by the set comparison $\vartriangleleft_{Eli}$ to define weakest link preferences over arguments. One then obtains an argumentation-based characterisation of non-monotonic inference defined by Preferred Subtheories. The latter starts with a stratification $(\Sigma_1, \ldots, \Sigma_n)$ of the totally ordered Σ ($\alpha, \beta \in \Sigma_i$ iff $\alpha \equiv' \beta$ and $\alpha \in \Sigma_i, \beta \in \Sigma_j, i < j$ iff $\beta \in \Sigma <' \alpha \in \Sigma$). A 'preferred subtheory' (ps) is obtained by taking a maximal under set inclusion consistent subset of Σ_1, maximally extending this with a subset of Σ_2, and so on. Multiple individually consistent preferred subtheories may be constructed, and [Modgil and Prakken, 2013] show that each ps corresponds to the premises of arguments in a stable extension. Hence, α is classically entailed from a ps iff α is the conclusion of an argument in a stable extension. Then α is a sceptical (credulous) Preferred Subtheories inference iff α is entailed by all (respectively at least one) ps, iff α is sceptically (credulously) justified under the stable semantics (as defined in Definition 2.18).

More recently, $ASPIC^+$ has been used to provide an argumentative characterisation of Brewka's *Prioritised Default Logic (PDL)* [Brewka, 1994a]. PDL upgrades [Reiter, 1980]'s Default Logic to include a strict partial ordering $<_D$ on a finite set D of first order normal defaults of the form $\frac{\theta:\phi}{\phi}$. Then given a set W of first order formulae, and a linearisation $<^+$ of $<_D$, one iteratively applies the highest ordered default whose antecedent is in the first order closure of the result obtained in the previous iteration. Intuitively, one starts with the classical consequences E_0 of W, and then adds the consequent of the highest ordered default whose antecedent is contained in E_0. Then closure under classical consequence obtains E_1, to which one adds the consequent of the highest ordered default whose antecedent is contained in E_1, and so on, until $E_{n+1} = E_n$ is the unique extension of $(D, W, <)$. In [Young et al., 2016], an $ASPIC^+$ SAF is defined in which the contrary function is defined so as to formalise classical negation, $\mathcal{R}_s$ characterises inference in first order classical logic, the axiom premises $\mathcal{K}_n$ is defined as W ($\mathcal{K}_p = \emptyset$), $\mathcal{R}_d = \{\theta \Rightarrow \phi | \frac{\theta:\phi}{\phi} \in D\}$ (with the naming function n undefined), and $<_D$ the ordering on $\mathcal{R}_d$. A linear 'structure preference' ordering $<_{SP}$ is defined, which modifies $<_D$ so as to account for the dependency amongst rules in $\mathcal{R}_d$ (i.e., for any set of rules applicable given all rules thus far applied, $<_{SP}$ picks out the $<_D$ maximal rule, and the process is repeated for the set of rules that are subsequently applicable). Then the *disjoint* elitist ordering – $\Gamma \lhd_{\text{DEli}} \Gamma'$ iff $\exists r \in \Gamma \setminus \Gamma', \forall r' \in \Gamma' \setminus \Gamma : r <_{SP} r'$ – is used to define an ordering over arguments according to the weakest link principle. [Young et al., 2016] then show that the single extension E of $(D, W, <)$ corresponds to the conclusions of arguments in the (provably) unique stable extension of the corresponding $ASPIC^+$ SAF.

4.6 The relationship of $ASPIC^+$ with developments of the theory of abstract argumentation frameworks

$ASPIC^+$ is designed to generate abstract argumentation frameworks in the sense of [Dung, 1995]. Over the years, various extensions of abstract argumentation frameworks with further elements have been proposed, such as with preferences ([Amgoud and Cayrol, 1998]'s preference-based argumentation frameworks or *PAFs*), values ([Bench-Capon, 2003]'s value-based argumentation frameworks or *VAFs*), attacks on attacks ([Modgil, 2009]'s extended argumentation frameworks or *EAFs*) and abstract support relations between arguments (e.g. [Cayrol and Lagasquie-Schiex, 2009]'s bipolar argumentation frameworks or *BAFs*). The question arises as to what extent $ASPIC^+$ can be seen as instantiations of these frameworks. Moreover, work has recently been done on the dynamics of abstract argumentation frameworks, such as deleting or adding arguments or attacks; e.g. [Baroni and Giacomin, 2008; Baroni et al., 2011b; Baumann and Brewka, 2010]. The question also arises as to what extent can the dynamics of argumentation, as studied in these works, be applied to $ASPIC^+$. These questions are answered in this section.

4.6.1 E-$ASPIC^+$: Structuring Extended Argumentation Frameworks

[Modgil, 2009] extended abstract argumentation frameworks to accommodate arguments that attack attacks, and in so doing enabled integration of arguments that express preferences over other arguments. The essential idea is that given an attack from A to B, then if the argument C expresses a strict preference for B over A, C attacks (and so invalidates the success of) the attack from A to B. A modified definition of the acceptability of arguments was defined for these *Extended Argumentation Frameworks* (*EAFs*), and [Modgil, 2009] showed that one can reconstruct [Prakken and Sartor, 1997]'s logic-programming-based argumentation system with defeasible preferences as an instance of EAFs. In this reconstruction, arguments built from rules expressing preferences over other 'object level' rules, constitute arguments expressing preferences over the arguments built from the object level rules.

However, as with Dung's original abstract argumentation frameworks, the abstract *EAFs* can in principle yield extensions that violate the rationality postulates. Hence [Modgil and Prakken, 2010] define a version of $ASPIC^+$ – E-$ASPIC^+$ – that generate a special class of *bounded hierarchical EAFs* in which the finite arguments $\mathcal{A}$ can be stratified into $\mathcal{A}_1, \ldots, \mathcal{A}_n$, such that if $C \in \mathcal{A}_i$ ($i \neq 1$) expresses a preference for B over A, then $A, B \in \mathcal{A}_{i-1}$. As in $ASPIC^+$ arguments are constructed from strict and defeasible rules, and axiom and ordinary premises, and in addition to the usual notions of attack, E-$ASPIC^+$ defines a function over sets of arguments $\mathcal{A}' \subseteq \mathcal{A}$, that maps $\mathcal{A}'$ to a strict preference over some $B, A \in \mathcal{A}$. In this way, EAFs are conservatively modified to allow for attacks on attacks to originate from *sets of*, rather than *single*, arguments. As well as the notion of a well-defined SAF^{17} [Modgil and Prakken, 2010] additionally identify a condition that if $\mathcal{A}' \subseteq \mathcal{A}$ expresses that $A \prec B$ and $\mathcal{A}'' \subseteq \mathcal{A}$ expresses that $B \prec A$, then $\mathcal{A}'$ and $\mathcal{A}''$ respectively contain arguments X and Y that have contradictory conclusions, or some X and Y such that X can be extended by strict rules to an argument X^+ such that X^+ and Y have contradictory conclusions. [Modgil and Prakken, 2010] then show that the generated *bounded hierarchical EAF*s satisfy [Caminada and Amgoud, 2007]'s rationality postulates.

4.6.2 Abstract support relations

There have been several recent proposals to extend abstract argumentation frameworks with abstract support relations, such as [Cayrol and Lagasquie-Schiex, 2005; Cayrol and Lagasquie-Schiex, 2009; Cayrol and Lagasquie-Schiex, 2013]'s Bipolar Argumentation Frameworks ($BAFs$), the work of [Martinez et al., 2006] and [Oren and Norman, 2008]'s Evidential Argumentation Systems ($EASs$). Various semantics for such frameworks have been defined, claiming to capture different notions of support. For example, [Boella et al., 2010a] study semantics of what they call "deductive" support, which satisfies the constraint that if A is acceptable and A is a deductive support of B, then B is acceptable. [Nouioua and Risch, 2011] consider "necessary support", which satisfies the constraint that if B is acceptable and A is a necessary support of B, then A is acceptable.

One question is whether the $ASPIC^+$ notion of a subargument instantiates any of these notions. Here we first discuss [Dung and Thang, 2014]'s simple way of formalising [Nouioua and Risch, 2011] intuitions concerning necessary support, namely, by adding a binary support relation $\mathcal{S}$ on $\mathcal{A}$ to AFs with the sole additional constraint that if B supports C and A defeats B then A also defeats C. The semantics of the resulting abstract argumentation frameworks is simply defined by choosing one of the semantics for the corresponding pair $(\mathcal{A}, \mathcal{D})$. Thus the support relation $\mathcal{S}$ is only used to constrain the defeat relation $\mathcal{D}$. [Prakken, 2014] calls the resulting frameworks *SuppAFs* and notes that $ASPIC^+$ can be reconstructed as an instance of *SuppAFs* as follows. Take $\mathcal{D}$ to be $ASPIC^+$'s defeat relation and $\mathcal{S}$ to be $ASPIC^+$'s subargument relation between arguments. It is then immediate from Definitions 2.10 and 2.12 that $ASPIC^+$'s notion of defeat satisfies [Dung and Thang, 2014]'s constraint on $\mathcal{D}$ in terms of $\mathcal{S}$.

An equivalent reformulation of *SuppAFs* does make use of support relations in its semantics. In [Prakken, 2013] $ASPIC^+$ as presented above was reformu-

---

[17]Where the requirement that an argument ordering is reasonable is adapted to the setting of EAFs.

lated in terms of [Pollock, 1994]'s recursive labellings, and this reformulation was abstracted to *SuppAFs* in [Prakken, 2014]. First, [Prakken, 2013] defines a notion of p-defeat (for "Pollock-defeat"), which captures direct defeat between arguments:

Definition 4.9 *[p-Attack] A p-attacks B iff A p-undercuts, p-rebuts or p-undermines B, where:*

- *A p-undercuts argument B iff* $\mathrm{Conc}(A) = -n(r)$ *and B has a defeasible top rule r.*
- *A p-rebuts argument B iff* $\mathrm{Conc}(A) = -\mathrm{Conc}(B)$ *and B has a defeasible top rule.*
- *Argument A p-undermines B iff* $\mathrm{Conc}(A) = -\varphi$ *and* $B = \varphi$, $\varphi \notin \mathcal{K}_n$.

Definition 4.10 *[p-Defeat] A p-defeats B iff: A p-undercuts B, or; A p-rebuts/p-undermines B and* $A \not\prec B$.

Then [Prakken, 2013] proves that A defeats B according to Definition 2.12 iff A p-defeats B or A p-defeats a proper subargument B' of B. Now if the support relation of a SuppAF is taken to be $ASPIC^+$'s notion of an 'immediate' subargument and the defeat relation of a *SuppAF* is taken to be p-defeat, then the following definition is equivalent to [Dung, 1995]'s semantics for AFs (and so for *SuppAFs*).

Definition 4.11 *[p-labellings for* SuppAFs.*] Let* $(\mathcal{A}, \mathcal{D}, \mathcal{S})$ *be a* SuppAF *corresponding to a (c-)SAF* $= (\mathcal{A}, \mathcal{D})$ *where* $\mathcal{D}$ *is defined as p-defeat and where* $\mathcal{S}$ *is defined as* $(A, B) \in \mathcal{S}$ *iff* B *is of the form* $B_1, \ldots, B_n \to / \Rightarrow \varphi$ *and* $A = B_i$ *for some* $1 \leq i \leq n$. *Then* (In, Out) *is a p-labelling of* SuppAF *iff* $In \cap Out = \emptyset$ *and for all* $A \in \mathcal{A}$ *it holds that:*

1. *A is labelled in iff:*

    (a) *All arguments that p-defeat A are labelled out; and*

    (b) *All B that support A are labelled in.*

2. *A is labelled out iff:*

    (a) *A is p-defeated by an argument that is labelled in; or*

    (b) *Some B that supports A is labelled out.*

Exploiting the well-known correspondences between labelling- and extension-based semantics [Caminada, 2006], [Prakken, 2014] shows that the complete extensions defined thus for *SuppAFs* generated from $ASPIC^+$ with p-defeat are exactly the complete extensions of *SuppAFs* as generated above from $ASPIC^+$ with defeat.

[Prakken, 2014] also showed for preferred semantics that $ASPIC^+$ instantiates [Oren and Norman, 2008]'s evidential argumentation systems. One might expect that classical-logic instantiations of $ASPIC^+$ instantiate [Boella et al., 2010a]'s version of bipolar argumentation frameworks for "deductive support". However, [Prakken, 2014] showed that this is not the case. This raises the question as to how one might instantiate [Boella et al., 2010a]'s notion of deductive support.

More generally, the question arises as to the relation of the various accounts of abstract support relations with formalisms for structured argumentation. To the best of our knowledge, the only papers studying this question are [Prakken, 2014] and [Modgil, 2014]. [Modgil, 2014] discusses this issue under the assumption that arguments and their relations are constructed from a $ASPIC^+$ argumentation theory. He discusses how examples in the literature used to motivate the need for support relations essentially amount to the supporting argument A concluding some ϕ that is: 1) either a premise in the supported argument B; 2) the conclusion of a defeasible rule in B, or; 3) A provides the missing sub-argument for the *enthymeme* B (i.e., B is an incomplete argument). For example, letting A be an argument constructed from α and $\alpha \Rightarrow_{r_1} \beta$ then illustrating the three cases, B consists of: 1) β and $\beta \Rightarrow_{r_2} \delta$; 2) γ, $\gamma \Rightarrow_{r_3} \beta$ and $\beta \Rightarrow_{r_2} \delta$; 3) $\beta \Rightarrow_{r_2} \delta$.

Given this analysis, the underlying premises and rules can then be seen to generate additional arguments without the need for support relations; for example, in case 1) the additional argument $B' : A \Rightarrow_{r_2} \delta$. Hence, one would expect that the justification status of arguments obtained by the modified definitions of acceptability in abstract argumentation frameworks augmented by support relations, corresponds to their evaluation in a standard abstract argumentation framework of arguments and attacks, instantiated by the additional arguments generated by the same premises and rules. In case 1), this would mean that the status of B in the augmented framework in which B is supported by A, is the same as the status of B in the original framework consisting of A, B and B'. However, [Modgil, 2014] shows that this correspondence does not always hold[18]. He concludes from this that only when examining abstract concepts in a structured approach can one gain some insight into the appropriate use of these abstract level concepts in evaluating arguments. Indeed, [Modgil, 2014] provides a similar analysis of collective attacks [Nielsen and Parsons, 2007] and recursive attacks on attacks [Baroni et al., 2011a] that have been incorporated at the abstract level and that have led to modified definitions of acceptability.

4.6.3 Preference- and value-based argumentation frameworks

[Amgoud and Cayrol, 1998] added to abstract argumentation frameworks (AFs) a preference relation on $\mathcal{A}$, resulting in *preference-based argumentation frameworks* ($PAFs$), which are triples of the form $\langle \mathcal{A}, attacks, \preceq \rangle$. An argument A

---

[18] Note that [Modgil, 2014] is careful to acknowledge that these observations apply to the case where arguments and their relations are generated by instantiating sets of formulae, rather than by human authoring of arguments. He argues that in the latter context additional relations between arguments incorporated in abstract argumentation frameworks may well be warranted by human oriented uses of argument, and goes on to argue the need for complementary empirical studies of human argumentation.

then *defeats* an argument B if A attacks B and $A \not\prec B$. Thus each PAF generates an AF of the form $\langle \mathcal{A}, defeats \rangle$, to which Dung's theory of AFs can be applied. [Bench-Capon, 2003] proposed a variant called *value-based argumentation frameworks* ($VAFs$), in which each argument is said to promote some (legal, moral or societal) value. Attacks in an $VAFs$ succeed only if the value promoted by the attacked argument is strictly preferred to the value of the attacking argument, according to a given ordering on the values (an *audience*).

The question arises as to what happens if $ASPIC^+$ is reformulated so as to generate $PAFs$ instead of Dung's original AFs. This can be easily done, since $ASPIC^+$ instantiations already generate a set of arguments with an attack relation and define a binary argument ordering. However, this may lead to violation of rationality postulates, even in cases where $ASPIC^+$ satisfies them.

Consider the following example from [Prakken, 2012b; Modgil and Prakken, 2013].

$A : p$
$B_1 : \neg p$
$B_2 : B_1 \Rightarrow q$

Here p and $\neg p$ are ordinary premises. Note that B_1 is a subargument of B_2. In $ASPIC^+$ we then have that A and B_1 directly attack each other while, moreover, A indirectly attacks B_2, since it directly attacks B_2's subargument B_1. These attack relations are displayed in Figure 5(a).

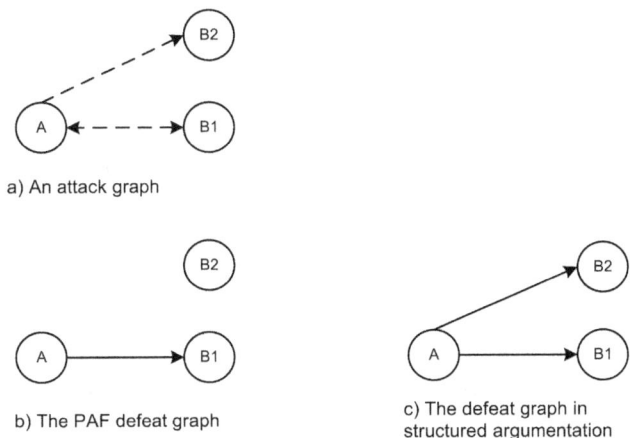

a) An attack graph

b) The PAF defeat graph

c) The defeat graph in structured argumentation

Figure 5. The attack graph

Assume next that $B_1 \prec A$ and $A \prec B_2$ (such an ordering could be the result of a last-link ordering). The PAF modelling then generates the following single defeat relation: A defeats B_1; see Figure 5(b). Then we have a single extension (in whatever semantics), namely, $\{A, B_2\}$. So not only A but also B_2 is justified.

However, this violates [Caminada and Amgoud, 2007]'s rationality postulate of subargument closure of extensions, since B_2 is in the extension while its subargument B_1 is not. This problem is not restricted to subargument closure; [Prakken, 2012b] also discusses examples in which the postulate of indirect consistency is violated.

The cause of the problem is that the PAF modelling of this example cannot recognise that the reason why A attacks B_2 is that A directly attacks B_1, which is a subargument of B_2. So the PAF modelling fails to capture that in order to check whether A's attack on B_2 succeeds, we should compare A not with B_2 but with B_1. Now since $B_1 \prec A$, then in $ASPIC^+$ we also have that A defeats B_2; see Figure 5(c). So the single extension (in whatever semantics) is $\{A\}$, and so closure under subarguments is respected.

This shows that under the assumption that $PAFs$ (and also $VAFs$) are instantiated by logical formulae, then these only behave correctly with respect to the rationality postulates, if all attacks are direct. We can conclude that for a principled analysis of the use of preferences to resolve attacks, the structure of arguments must be made explicit, since the structure of arguments is crucial in determining how preferences must be applied to attacks.

A more general word of caution is in order here. Although it is tempting to extend abstract argumentation frameworks with additional elements, one should resist the temptation to think that for any given argumentation phenomenon the most principled analysis is at the level of abstract argumentation. In fact, it often is the other way around, since at the abstract level crucial notions like claims, reasons and grounds are abstracted away.

4.6.4 Dynamics of argumentation

Recently much work has been done on the nature and effects of change operations on a given argumentation state, e.g. [Modgil, 2006; Baroni and Giacomin, 2008; Rotstein et al., 2008; Baumann and Brewka, 2010; Cayrol et al., 2010; Boella et al., 2010b; Baroni et al., 2011b]. Among other things, enforcing and preservation properties are studied. Enforcement concerns the extent to which desirable outcomes can or will be obtained by changing an argumentation state, while preservation is about the extent to which the current status of arguments is preserved under change. Almost all this work is done for abstract argumentation frameworks. In particular, the following operations on abstract argumentation frameworks have been studied: addition or deletion of (sets of) arguments and addition or deletion of (sets of) attack relations. Deleting attacks can here be seen as an abstraction from the use of preferences to resolve attacks into defeats.

The question arises as to what extent this work is relevant for $ASPIC^+$. Here too our above word of caution applies. At first sight, it would seem that the most principled analysis of argumentation dynamics is at the level of abstract argumentation frameworks. However, upon closer inspection it turns out that such analyses, because they ignore the structure of arguments, often implicitly make assumptions that are not in general satisfied by $ASPIC^+$

instantiations (and neither by other formalisms for structured argumentation). For example, abstract models of argumentation dynamics do not recognise that some arguments are not attackable (such as deductive arguments with certain premises) or that some attacks cannot be deleted (for example between arguments that were determined to be equally strong), or that the deletion of one argument implies the deletion of other arguments (when the deleted argument is a subargument of another, as in Figure 5 above), or that the deletion or addition of one attack implies the deletion or addition of other attacks (for example, attacking an argument implies that all arguments of which the attacked argument is a subargument are also attacked; in Figure 5 above attacking B_1 implies attacking B_2). These considerations imply that formal results pertaining to the abstract model are only relevant for specific cases, and fail to cover many realistic situations in argumentation that can be expressed in $ASPIC^+$. To give a very simple example, in models that allow the addition of arguments and attacks, any non-selfattacking argument A can be made a member of every extension by simply adding non-attacked attackers of all A's attackers. However, this result at the abstract level does not carry over to instantiations in which not all arguments are attackable. Here too, we see the importance of being aware of what the model abstracts from.

For these reasons we have in [Modgil and Prakken, 2012] proposed a model of preference dynamics in $ASPIC^+$, that arguably overcomes several limitations of [Baroni et al., 2011b]'s resolution-based semantics for abstract argumentation frameworks when applied to preference-based dynamics.[19] The latter allows that symmetric attacks are replaced by asymmetric attacks (i.e., the symmetric attacks are 'resolved'). We argued that from the perspective of instantiated abstract argumentation frameworks, it is the use of preferences that provides the clearest motivation for obtaining resolutions. But then studying the use of preferences at the structured $ASPIC^+$ level suggests that one must also account for the resolution of asymmetric attacks, that preferences may also result in removal of both attacks in a symmetric attack, and that certain resolutions may be impossible, because assuming a preference that removes one attack may necessarily imply removal of another attack, or because some attacks cannot be removed by preferences (e.g. undercut attacks and attacks on contraries). These subtleties can only be appreciated at the structured level, and are thus not addressed by the study of resolutions at the abstract level adopted by [Baroni et al., 2011b], in which only resolutions of symmetric attacks are considered, and all possible resolutions are considered possible.

5 Further Developments of ASPIC+

In Section 2 we presented what we called the 'basic' $ASPIC^+$ framework in two stages, first with symmetric negation and then generalising it with possibly asymmetric negation. As a matter of fact, this basic framework is the result

---

[19] We recognise that there may be other uses of resolution-based semantics to which our criticism does not apply.

of various revisions and incremental extensions [Amgoud et al., 2006; Prakken, 2010; Modgil and Prakken, 2013; Modgil and Prakken, 2014]. Also, in [Modgil and Prakken, 2013], the basic framework in fact comes in four variants, resulting from whether the premises of arguments are assumed to be c-consistent or not and whether conflict-freeness is defined with the attack or the defeat relation (recall footnote 7). So instead of a single $ASPIC^+$ framework there in fact exists a family of such frameworks. And this family is growing. In this section we discuss recent work that modifies the $ASPIC^+$ framework in some respects, especially with new constraints on arguments or with modified or generalised notions of attack. We consider this development of variants of $ASPIC^+$ a healthy situation, since it amounts to a systematic investigation of the effects of different design choices within a common approach, which may each be applicable to certain kinds of problems.

5.1 Consistency and chaining restrictions motivated by contamination problems

Some recent work on $ASPIC^+$ has studied further constraints on arguments in an attempt to address the so-called contamination problem originally discussed by [Pollock, 1994; Pollock, 1995].[20] This problem arises if the strict inference rules are chosen to correspond to classical logic and if they are then combined with defeasible rules. The problem is how the trivialising effect of the classical Ex Falso principle can be avoided when two arguments that use defeasible rules have contradictory conclusions. The problem is especially hard since any solution should arguably preserve satisfaction of the rationality postulates of [Caminada and Amgoud, 2007]. In addition, [Caminada et al., 2012] claim that any solution should also satisfy a new set of postulates that are meant to express the idea that information irrelevant to a part of the argumentation system should not affect the conclusions drawn from that part.

The following abstract example illustrates the problem. Assume that the strict rules of an argumentation system correspond to classical logic, i.e. $X \to \varphi \in \mathcal{R}_s$ if and only if $X \vdash \varphi$ and X is finite (where $\vdash$ denotes classical consequence).

Example 5.1 Let $\mathcal{R}_d = \{p \Rightarrow q;\ r \Rightarrow \neg q;\ t \Rightarrow s\}$, $\mathcal{K}_p = \emptyset$ and $\mathcal{K}_n = \{p, r, t\}$, while $\mathcal{R}_s$ corresponds to classical logic. Then the corresponding abstract argumentation framework includes the following arguments:

$A_1: p \quad A_2: A_1 \Rightarrow q$
$B_1: r \quad B_2: B_1 \Rightarrow \neg q \quad C: A_2, B_2 \to \neg s$
$D_1: t \quad D_2: D_1 \Rightarrow s$

Figure 6 displays these arguments and their attack relations. Argument C attacks D_2. Whether C defeats D_2 depends on the argument ordering but

---
[20]Some parts of this section have been taken or adapted from [Grooters and Prakken, 2016].

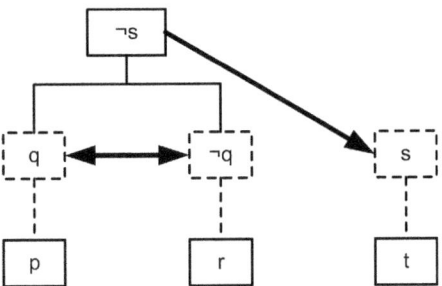

Figure 6. Illustrating trivialisation

plausible argument orderings are possible in which $C \not\prec D_2$ and so C defeats D_2. This is problematic, since s can be any formula, so any defeasible argument unrelated to A_2 or B_2, such as D_2, can, depending on the argument ordering, be defeated by C. Clearly, this is extremely harmful, since the existence of just a single case of mutual rebutting attack, which is very common, could trivialise the system. For instance, in this example neither of A_2 nor B_2 are in the grounded extension, since they defeat each other. But then the grounded extension does not defend D_2 against C and therefore does not contain D_2.

It should be noted that simply disallowing application of strict rules to inconsistent sets of formulas does not help, since then an argument for $\neg s$ can still be constructed as follows:

A_3: $\quad A_2 \to q \vee \neg s$
C': $\quad A_3, B_2 \to \neg s$

Note that argument C' does not apply any strict inference rule to an inconsistent set of formulas.

[Grooters and Prakken, 2016] propose the following formalisation of the property of trivialisation.

Definition 5.2 (Trivialising argumentation systems) *An argumentation system AS is trivialising iff for all $\varphi, \psi \in \mathcal{L}$ and all knowledge bases $\mathcal{K}$ such that $\{\varphi, \neg \varphi\} \subseteq \mathcal{K}$ a strict argument on the basis of $\mathcal{K}$ can be constructed in AS with conclusion ψ.*

The research problem then is identifying classes of non-trivialising argumentation systems. The argumentation system in our example is clearly trivialising since $\mathcal{R}_s$ contains strict rules $\varphi, \neg \varphi \to \psi$ for all $\varphi, \psi \in \mathcal{L}$.

Example 5.1 does not cause any problems for preferred or stable semantics, since A_2 and B_2 attack each other and at least one of these attacks will (with non-circular argument orderings) succeed as defeat. Therefore, all preferred or stable extensions contain either A_2 or B_2 but not both. Since both A_2 and B_2 attack C (by directly attacking one of its subarguments), C is for each preferred or stable extension defeated by at least one argument in the extension, so C is not in any of these extensions, so D_2 is in all these extensions. This is intuitively correct since there is no connection between D_2 and the arguments A_2 and B_2. [Pollock, 1994; Pollock, 1995] thought that this line of reasoning for preferred semantics suffices to show that his recursive-labelling approach (which was later in [Jakobovits and Vermeir, 1999] proved to be equivalent to preferred semantics) adequately deals with this problem. However, [Caminada, 2005] showed that the example can be extended in ways that also cause problems for preferred and stable semantics. Essentially, he replaced the facts p and r with defeasible arguments for p and r and let both these arguments be defeated by a self-defeating argument. On the one hand, such self-defeating arguments cannot be in any extension, since extensions are conflict free. However, if a self-defeating argument is not defeated by other arguments, it prevents any argument that it defeats from being acceptable with respect to an extension. In our example, if both A_2 and B_2 are defeated by a self-defeating argument that is otherwise undefeated, then neither A_2 not B_2 is in any extension, so no argument in an extension defends D_2 against C. To solve the problem, two approaches are possible. One is to change the definitions of the argumentation formalism, while the other is to derive the strict inference rules from a weaker logic than classical logic.

The first approach is taken by [Wu, 2012] and [Wu and Podlaszewski, 2015], who for the $ASPIC^+$ framework require that for each argument the set of conclusions of all its subarguments are classically consistent. They show that this solution partially works for a restricted version of $ASPIC^+$ without preferences, in that for complete semantics, both the original postulates of [Caminada and Amgoud, 2007] and the new ones of [Caminada et al., 2012] are satisfied. However, their results do not cover stable, preferred or grounded semantics, while they give counterexamples to the consistency postulates for the case with preferences.

A second approach to solve the problem is to replace classical logic as the source for strict rules with a weaker, monotonic paraconsistent logic, in order to invalidate the Ex Falso principle as a valid strict inference rule. [Grooters and Prakken, 2016] explored this possibility. They first showed that two well-known paraconsistent logics, the system C_ω of [Da Costa, 1974] and the Logic of Paradox of [Priest, 1979; Priest, 1989], cannot be used for these purposes, since they induce violation of the postulate of indirect consistency. They then investigated Rescher and Manor's 1970 paraconsistent consequence notion of *weak consequence*. A set S of wff's weakly' implies a wff φ just in case at least one consistent subset of S classically implies φ. While thus initially taking the second approach, [Grooters and Prakken, 2016] had to combine it with the first approach (changing the definitions). Chaining strict rules in arguments has to be disallowed since the notion of weak consequence does not satisfy the Cut rule. For a counterexample, consider the set $\Gamma = \{a, \neg a \wedge b\}$. Then $\Gamma \vdash_W b$ and $\Gamma, b \vdash_W a \wedge b$, while it is not the case that $\Gamma \vdash_W a \wedge b$.

[Grooters and Prakken, 2016] proved that this solution avoids trivialisation and for well-behaved c-$SAFs$ satisfies all closure and consistency postulates (where the strict-closure postulate has to be changed to closure under one-step application of strict rules). Illustrating their solution with the above example, we see that the contaminating argument C cannot be constructed since its conclusion $\neg s$ follows from no consistent subset of $\{q, \neg q\}$, while the contaminating argument C' cannot be constructed since it chains two strict rules.

[Grooters and Prakken, 2016] also showed that with [Wu and Podlaszewski, 2015]'s stronger condition that the set of all conclusions of all subarguments of an argument must be consistent, consistency and strict closure are not satisfied. [Grooters and Prakken, 2016] did not attempt to prove Caminada et al.'s 2012 'contamination' postulates, for two reasons. First, they wanted to obtain results for all of [Dung, 1995]'s semantics and, second, they argued that Caminada et al.'s postulates in fact capture a stronger intuitive notion than the notion of trivialisation.

The work of [Grooters and Prakken, 2016] gives rise to some more general observations on [Caminada and Amgoud, 2007]'s original postulate of closure under strict rules. Above we suggested that $\mathcal{R}_s$ can be chosen to correspond to any monotonic logic with consequence notion $\vdash$ by letting $S \to \varphi \in \mathcal{R}_s$ if and only if $S \vdash \varphi$ and S is finite. However, the fact that the weak-consequence notion $\vdash_W$ does not satisfy the Cut rule illustrates that when $\mathcal{R}_s$ is thus defined, a system that is closed under $\mathcal{R}_s$ as defined in Section 3.1.2, could allow for inferences that are invalid according to $\vdash$. For these reasons, [Grooters and Prakken, 2016] not only reformulated their definition of strict closure but also proposed a new rationality postulate of *logical closure* and showed that their adapted version of $ASPIC^+$ also satisfies this postulate for well-behaved c-$SAFs$.

We also briefly note that [Grooters and Prakken, 2016] also studied minimality constraints on strict-rule applications and the exclusion of circular arguments. They show that if these two constraints are combined with their adoption of weak consequence as the source of the strict rules, then if both the knowledge base and the set of defeasible rules is finite, then each argument has at most a finite number of attackers, i.e., their framework generates so-called finitary argumentation frameworks in the sense of [Dung, 1995], which is computationally beneficial.

Finally, [D'Agostino and Modgil, 2016] provide a formalisation of classical argumentation with preferences in which arguments are triples (Δ, Γ, α) such that α is classically entailed by $\Delta \cup \Gamma$[21], and where Δ are the premises assumed true, and Γ the premises supposed true 'for the sake of argument'. The idea is that if a trivialising argument $(\{q, \neg q\}, \emptyset, s)$ defeats $(\{s\}, \emptyset, s) \in E$ (where E is an extension under any semantics), then $Y = (\emptyset, \{q, \neg q\}, \bot)$ defeats

---

[21][D'Agostino and Modgil, 2016] allow for arguments with inconsistent premises, as they argue that arguments with inconsistent premises, and hence the trivialising effect of such arguments, should be excluded dialectically (as in real-world reasoning and debate), rather than checking for consistency prior to inclusion of the argument in an abstract argumentation framework.

$X = (\{q, \neg q\}, \emptyset, s)$ (Y supposes for the sake of argument the premises of X). Moreover, since the premises whose truth Y commits to are empty, Y cannot be defeated and so can be included in any E in order to defend $(\{s\}, \emptyset, s)$, thus negating the trivialising effect of X. [D'Agostino and Modgil, 2016] then show that under certain conditions, the consistency and closure postulates, as well as Caminada *et al.*'s additional contamination postulates are satisfied. As the authors note, an interesting direction for future research would be to see if their approach can be applied to the full $ASPIC^+$ framework.

5.2 Dung (2016) on rule-based argumentation systems

Recently, [Dung, 2016] has continued the formal study of [Dung and Thang, 2014]'s rule-based argumentation systems. Recall that these comprise of strict and defeasible inference rules over a propositional literal language, where axiom, respectively ordinary, premises p are simulated with rules $\rightarrow p$ and $\Rightarrow p$. [Dung, 2016] adds a transitive preference relation $\leq$ on $\mathcal{R}_d$, so that he defines rule-based systems as a triple $(\mathcal{R}_s, \mathcal{R}_d, \leq)$. In addition, he confines his study to knowledge bases with a consistent strict closure. Above we explained that [Dung and Thang, 2014] adopt the $ASPIC^+$ definitions of argument and defeat (which they call attack) and thus effectively study a class of $ASPIC^+$ instantiations. [Dung, 2016] also adopts the $ASPIC^+$ definition of an argument and still assumes that rule-based systems generate abstract argumentation frameworks in the sense of [Dung, 1995] (in our notation $(\mathcal{A}, \mathcal{D})$). However, Dung now abstracts from particular definitions of defeat ($\mathcal{D}$) and instead defines properties that defeat relations should have, thus effectively generalising $ASPIC^+$ on its notion of defeat. He then studies conditions under which defeat relations satisfy these properties.

Since this work is quite recent, we confine ourselves to a brief summary and discussion. In doing so, we will replace Dung's term 'attack' with 'defeat', in order to be consistent with the terminology in this chapter. This replacement is justified since in [Dung, 2016] it is the attack relation in terms of which arguments are evaluated, so it plays the role of $ASPIC^+$'s defeat relation.

Dung introduces two new rationality postulates. His postulate for *attack monotonicity* informally says that strengthening an argument cannot eliminate an attack of that argument on another. Let us illustrate this with Figure 2, interpreting the horizontal arrows as defeat relations. Then this postulate says, for instance, that if D_4's argument C_2 for v is replaced with a necessary premise v (or in [Dung, 2016]'s case a strict rule $\rightarrow v$) or with a strict and firm argument from u to v, then the new version of D_4 still defeats B_2. Next, Dung's postulate of *credulous cumulativity* informally means that changing a conclusion of an argument in some extension to a necessary fact cannot eliminate that extension.

Dung then identifies several sets of conditions under which one or both of these postulates and/or the original postulates of [Caminada and Amgoud, 2007] are satisfied. For the details of these very valuable results we refer the reader to his own publication. Dung then continues by investigating several definitions of defeat in terms of the preference relation $\leq$ on $\mathcal{R}_d$ on whether they satisfy these various postulates. Since he also assumes here that strict arguments cannot be defeated, this part of his study effectively concerns instantiations of $ASPIC^+$ as defined above in Section 2.2. Here Dung obtains both positive and negative results. For example, elitist orderings as defined in [Modgil and Prakken, 2013] are shown to satisfy attack monotonicity but not credulous cumulativity and indirect consistency, while democratic orderings as defined in [Modgil and Prakken, 2013] and Definition 3.4 above are shown to satisfy credulous cumulativity and indirect consistency but not attack monotonicity. As for Dung's results on consistency, these are a special case of [Modgil and Prakken, 2013]'s results for democratic orderings but they contain counterexamples to their results for the elitist orderings. However, these counterexamples do not apply to [Prakken, 2010]'s original way to define the elitist orderings, which has been incorporated in the above Definition 3.4, or to the erratum to [Modgil and Prakken, 2013] (which is available online at https://nms.kcl.ac.uk/sanjay.modgil/AIJfinalErratum).

The question arises as to whether Dung's two new postulates really are desirable in general. Our answer is positive for attack monotonicity but, following [Prakken and Vreeswijk, 2002, section 4.4], negative for credulous cumulativity. The point is that strengthening a defeasible conclusion to an indisputable fact may make arguments stronger than before, which can give them the power to defeat other arguments that they did not defeat before. This may in turn result in the loss of the extension from which the conclusion was promoted to an indisputable fact. We illustrate this with [Dung, 2016]'s own example. Informally: professors normally teach, administrators normally do not teach, deans are normally professors and all deans are administrators (so with transposition anyone who is not an administrator is not a dean). The question is whether some particular dean teaches. In rules:

$Dean \Rightarrow_{d1} Professor \quad Professor \Rightarrow_{d2} Teach \quad Administrator \Rightarrow_{d3} \neg Teach$
$Dean \rightarrow Administrator \quad \rightarrow Dean \quad \quad \quad \neg Administrator \rightarrow \neg Dean$

Assume further that $d_1 < d_3 < d_2$. We have the following arguments on whether the dean teaches:

$A_1: \quad \rightarrow Dean \quad\quad\quad B_1: \quad \rightarrow Dean$
$A_2: \quad A_1 \Rightarrow_{d1} Professor \quad\quad B_2: \quad B_1 \rightarrow Administrator$
$A_3: \quad A_2 \Rightarrow_{d2} Teach \quad\quad\quad B_3: \quad B_2 \Rightarrow_{d3} \neg Teach$

(A_1 and B_1 are, of course, the same argument; B_3 is called A_3 by [Dung, 2016], while he does not explicitly name A_1/B_1 and B_2.) With the elitist or democratic weakest-link ordering as defined in Definition 3.4 above, argument B_3 strictly defeats A_3, so in all semantics a unique extension is obtained in which the dean is a professor but does not teach.

Suppose now the defeasibly justified conclusion *Professor* is added as a fact. This gives rise to a new argument:

C_1: $\to Professor$
C_2: $C_1 \Rightarrow_{d2} Teach$

Now the elitist ordering yields that C_2 strictly defeats B_3, so again in all semantics a unique extension is obtained but now it contains that the dean teaches. So we have lost the original extension, which illustrates violation of credulous cumulativity.

In our opinion, this outcome is the intuitive one, since by adding *Professor* as a fact, we have promoted its status from a defeasibly justified conclusion to an indisputable fact; as a consequence, argument A_3 can be strengthened by replacing its defeasible subargument A_2 with the strict-and-firm subargument C_1; no wonder then that the thus strengthened argument C_2 has, unlike its weaker version A_3, the power to defeat B_3.

Despite this minor criticism, we believe that Dung's latest investigations are a very valuable addition to the study of rule-based argumentation.

5.3   Variants of rebutting attack

Several papers have considered alternative definitions of rebutting attack in which an argument can under specific conditions also be rebutted on the conclusions of strict inferences.

5.3.1   Unrestricted rebuts

In $ASPIC^+$ as presented so far, arguments can only be rebutted on conclusions of defeasible-rule applications. [Caminada and Amgoud, 2007] call this *restricted rebut*. They also study *unrestricted rebut*, which allows rebuttals on the conclusion of a strict inference provided that at least one of the argument's subarguments is defeasible. Their replacement of restricted with unrestricted rebut leads to a variant of their simplified version of $ASPIC^+$ (which is in fact equivalent to [Dung and Thang, 2014]'s rule-based systems). They prove that for grounded semantics the rationality postulates are (under the usual conditions) satisfied but they provide a counterexample for stable and preferred semantics, presented above in Section 3.3 with a modification of Example 3.1.

[Caminada et al., 2014] argue in favour of unrestricted rebut on the grounds that this would lead to more natural presentations of dialogues. They argue that when applying argumentation in dialogical settings, the notion of restricted rebuts sometimes forces agents to commit to statements they have insufficient reasons to believe. In abstract terms, suppose an agent Ag_1 submitting an argument A whose top rule is a strict rule $s_1 = \alpha_1, \ldots, \alpha_n \to \alpha$, where for $i = 1 \ldots n$, α_i is an ordinary premise in A or the head of a defeasible rule in A. Now suppose Ag_2 has an argument B that defeasibly concludes $\neg \alpha$. Since B does not rebut A on α, then to attack A requires that Ag_2 construct, for some $i = 1 \ldots n$, an argument B' that extends B and the arguments concluding α_j, $j \neq i$, with the transposition $s_1^i = \alpha_1, \ldots, \alpha_{i-1}, \neg \alpha, \alpha_{i+1}, \alpha_n \to \neg \alpha_i$. But then Ag_2 is forced to commit to her interlocutors' arguments concluding α_j, $j \neq i$, for which she has no reasons to believe.

[Caminada et al., 2014] give the following concrete example.

John: *"Bob will attend conferences AAMAS and IJCAI this year, as he has papers accepted at both conferences."*
Mary: *"That won't be possible, as his budget of £1000 only allows for one foreign trip."*

Formally, this discussion could be modelled using an argumentation theory with $\mathcal{R}_d \supseteq \{\text{accA} \Rightarrow \text{attA};\ \text{accI} \Rightarrow \text{attI};\ \text{budget} \Rightarrow \neg(\text{attA} \wedge \text{attI})\}$ and $\mathcal{R}_s \supseteq \{\rightarrow \text{accA};\ \rightarrow \text{accI};\ \rightarrow \text{budget};\ \text{attA}, \text{attI} \rightarrow \text{attA} \wedge \text{attI}\}$.

A direct formalisation of the above arguments is then:

J_1: $\rightarrow$ accA $\qquad\qquad$ M_1: $\rightarrow$ budget
J_2: $J_1 \Rightarrow$ attA $\qquad\qquad$ M_2: $M_1 \Rightarrow \neg(\text{attA} \wedge \text{attI})$
J_3: $\rightarrow$ accI
J_4: $J_3 \Rightarrow$ attI
J_5: $J_3, J_4 \rightarrow$ attA $\wedge$ attI

In $ASPIC^+$, Mary's argument does *not* attack John's argument, since the conclusion Mary wants to attack (attA $\wedge$ attI) is the consequent of a strict rule. Mary can only attack John's argument by attacking the consequent of one of the defeasible rules, that is, by uttering one of the following two statements.

Mary': *"Bob can't attend AAMAS because he will attend IJCAI, and his budget does not allow him to attend both."*
Mary'': *"Bob can't attend IJCAI because he will attend AAMAS, and his budget does not allow him to attend both."*

The associated formal counterarguments are as follows.[22]

M_1: $\rightarrow$ budget
M_2: $M_1 \Rightarrow \neg(\text{attA} \wedge \text{attI})$
J_3: $\rightarrow$ accI $\qquad$ J_1: $\rightarrow$ accA
J_4: $J_3 \Rightarrow$ attI $\qquad$ J_2: $J_1 \Rightarrow$ attA
M_5': $M_2, J_4 \rightarrow \neg$attA $\qquad$ M_5'': $M_2, J_2 \rightarrow \neg$attI

According to [Caminada et al., 2014] the problem with this is that Mary does not know which of the two conferences Bob will attend, but $ASPIC^+$ with restricted rebut forces her to assert that Bob will attend one or the other. They argue that from the perspective of commitment in dialogue [Walton and Krabbe, 1995], this is unnatural.

[Caminada et al., 2014] then define a restricted version of basic $ASPIC^+$ as presented above in Section 2.2 – which they call $ASPIC^-$ – that substitutes strict rules with empty antecedents for axiom premises, and defeasible rules

---

[22]Assuming $\mathcal{R}_s$ ito be closed under transposition, the fact that $\mathcal{R}_s$ contains attA, attI $\rightarrow$ attA $\wedge$ attI implies that $\mathcal{R}_s$ also contains $\neg(\text{attA} \wedge \text{attI}), \text{attI} \rightarrow \neg\text{attA}$ and attA, $\neg(\text{attA} \wedge \text{attI}) \rightarrow \neg\text{attI}$.

with empty antecedents for ordinary premises. Moreover, $ASPIC^-$ allows unrestricted rebuts on the conclusions of strict rules. They then show that under the assumption of a *total* ordering on the defeasible rules, and assuming either the `Elitist` or `Democratic` set comparisons used in defining weakest- or last-link preferences, all of [Caminada and Amgoud, 2007]'s rationality postulates are satisfied for well-behaved $SAFs$, but only for the grounded semantics. They have thus generalised [Caminada and Amgoud, 2007]'s results for some specific cases with preferences.

5.3.2 Weak rebuts and an alternative view on the rationality postulates

[Prakken, 2016] studies a weaker version of unrestricted rebut, motivated by the general observation that deductive inferences may weaken an argument. His argument is that when a deductive inference is made from the conclusions of at least two 'fallible' (defeasible or plausible) subarguments, the deductive inference can be said to aggregate the degrees of fallibility of the individual arguments to which it is applied. This in turn means that the deductive inference may be less preferred than either of these subarguments, so that a successful attack on the deductive inference does not necessarily imply a successful attack on one of its fallible subarguments. And this in turn means that there can be cases where it is rational to accept a set of arguments that is not strictly closed and that violate indirect consistency. Note that this line of reasoning does not apply to cases where a deductive inference is applied to at most one fallible subargument: then the amount of fallibility of the new argument is exactly the same as the amount of fallibility of the single fallible argument to which the deductive inference is applied. Accordingly, [Prakken, 2016] defines *weak rebut* as allowing rebuttals on the conclusion of a strict inference, provided that the strict inference is applied to at least two fallible subarguments. Moreover, he argues that there are cases where argument orderings cannot be required to satisfy all properties of a reasonable argument ordering as defined in Definition 3.16.

[Prakken, 2016] illustrates this with the lottery paradox, a well-known paradox from epistemology, first discussed by [Kyburg, 1961]. Imagine a fair lottery with one million tickets and just one prize. If the principle is accepted that it is rational to accept a proposition if its truth is highly probable, then for each ticket T_i it is rational to accept that T_i will not win while at the same time it is rational to accept that exactly one ticket will win. If we also accept that everything that deductively follows from a set of rationally acceptable propositions is rationally acceptable, then we have two rationally acceptable propositions that contradict each other: we can join all individual propositions $\neg T_i$ into a big conjunction $\neg T_1 \wedge \ldots \wedge \neg T_{1,000,000}$ with one million conjuncts, which contradicts the certain fact that exactly one ticket will win.

Many views on this paradox exist. [Prakken, 2016] wants to formalise the view that for each individual ticket it is rational to accept that it will not win while at the same time it is not rational to accept the conjunction of these acceptable beliefs. He considers the following modelling of the lottery paradox in $ASPIC^+$. Let $\mathcal{L}$ be a propositional language built from the set of atoms $\{T_i \mid 1 \leq i \leq 1,000,000\}$. Then let X denote a well-formed formula $X_1 \veebar \ldots \veebar X_{1,000,000}$ where $\veebar$ is exclusive or and where each X_i is of one of the following forms:

- If $i = 1$ then $X_i = T_1 \wedge \neg T_2 \wedge \ldots \wedge \neg T_n$
- If $i = n$ then $X_i = \neg T_1 \wedge \neg T_2 \wedge \ldots \wedge \neg T_{n-1} \wedge T_n$
- Otherwise $X_i = \neg T_1 \wedge \ldots \wedge \neg T_{i-1} \wedge T_i \wedge \neg T_{i+1} \wedge \ldots \wedge \neg T_n$

Next we choose $\mathcal{K}_p = \{\neg T_i \mid 1 \leq i \leq 1,000,000\}$, $\mathcal{K}_n = \{X\}$, $\mathcal{R}_s$ as consisting of all propositionally valid inferences from finite sets and $\mathcal{R}_d = \emptyset$.

The following arguments are relevant for any i such that $1 \leq i \leq 1,000,000$.

$$\neg T_i \quad \text{and} \quad \neg T_1, \ldots, \neg T_{i-1}, \neg T_{i+1}, \ldots, \neg T_{1,000,000}, X \to T_i \text{ (call it } A_i)$$

[Prakken, 2016] then equates rational acceptability with sceptical justification (see Definition 2.18 above). Making $\neg T_i$ sceptically justified for all i requires for all i that $A_i \prec \neg T_i$, to prevent A_i from defeating $\neg T_i$. Then we have a single extension in all semantics containing arguments for all conclusions $\neg T_i$ but not for their conjunction.

Note that adopting the above argument ordering requires that Condition (2) of Definition 3.16 of reasonable argument orderings is dropped, since it excludes such an argument ordering. On the other hand, Condition (1) of Definition 3.16 can be retained. In particular, Condition (1.iii) captures that applying a strict rule to the conclusion of a single argument A to obtain an argument A' does not change the 'preferedness' of A' compared to A. This is reasonable in general, since A and A' have exactly the same set of fallible elements (ordinary premises and/or defeasible inferences). [Prakken, 2016] calls argument orderings that satisfy Condition (1) of Definition 3.16 *weakly reasonable* argument orderings. Finally, he proposes weakened versions of the postulates of strict closure and indirect consistency, according to which these properties are only required to hold for subsets of extensions with at most one fallible argument. He then proves that if weak rebut is allowed in addition to restricted rebut and argument orderings are required to be weakly reasonable, then the original postulate of direct consistency plus the weakened postulates of strict closure and indirect consistency are satisfied if AT is closed under contraposition or transposition and $\text{Prem}(A) \cup \mathcal{K}_n$ is indirectly consistent.

[Prakken, 2016] concludes with some general observations on the relation between deduction and justification. He argues to have shown that preservation of truth (the definition of deductively valid arguments) does not imply preservation of rational acceptance, since truth and rational acceptance are different things. However, he also argues that deduction still plays an important role in argumentation. Deductive inference rules are still available as argument construction rules and if an argument with a strict top rule has no attackers or all its attackers are less preferred, then the argument may still be sceptically justified. The specifics of the adopted argument ordering are essential here. For instance, in the lottery paradox the argument ordering might allow that application of the conjunction rule to a small number of conclusions $\neg T_i$ is still sceptically justified.

5.4 Attacks from sets of arguments to arguments

[Baroni et al., 2015] consider a variant of $ASPIC^+$ by adapting an idea originally proposed by [Vreeswijk, 1997] in the context of his 'abstract argumentation systems', which are a predecessor of $ASPIC^+$. In Vreeswijk's systems a counterargument is in fact a *set* of arguments: a set Σ of arguments is *incompatible* with an argument τ iff the conclusions of $\Sigma \cup \{\tau\}$ give rise to a strict argument for $\bot$. [Baroni et al., 2015] adapt this idea to $ASPIC^+$, where the 'nodes' of the abstract argumentation frameworks generated by the modification are sets of arguments instead of individual arguments. They then prove satisfaction of [Caminada and Amgoud, 2007]'s rationality postulates under similar conditions as in [Modgil and Prakken, 2013].

[Baroni et al., 2015]'s proposal is motivated by criticism of the $ASPIC^+$ treatment of generalised contrariness relations. However, we believe that they just criticise specific uses of this generalised contrariness relation and that the problems they discuss can be avoided by proper definitions of contrariness. Nevertheless, their ideas are very interesting and also apply to basic $ASPIC^+$ with ordinary negation. For example, it would be interesting to see if their variant of $ASPIC^+$ provides an alternative way to model the examples discussed by [Caminada et al., 2014]. More generally, it would be interesting to see if their variant of $ASPIC^+$ can be reconstructed as generating AFs that allow attacks from sets of arguments to arguments as in e.g. [Bochman, 2003].

6 Implementations and applications

6.1 Implementations

Various implementations of instantiations of $ASPIC^+$ are available online, all with domain-specific inference rules defined over literal-like languages, and with argument orderings based on rule preferences.

The original ASPIC inference engine The original inference engine from the ASPIC project (designed by Matthew South on the basis of a prototype of Gerard Vreeswijk) is available online at http://aspic.cossac.org/, with a demonstrator with example inputs available at http://aspic.cossac.org/ArgumentationSystem/. Rules can be formulated over a language with predicate-logic literals with ordinary negation. The implementation allows for choosing between restricted and unrestricted rebut. The implementation of restricted rebut deviates from its formal definition in that it also allows rebuttals between two arguments that both have a strict top rule. Arguments can be evaluated alternatively with a last- and a weakest-link argument ordering and with sceptical grounded or credulous preferred semantics.

Visser's Epistemic and Practical Reasoner Wietske Visser took the ASPIC deliverable ([Amgoud et al., 2006]) as the basis for her Epistemic and Practical Reasoner (EPR), available at http://www.wietskevisser.nl/research/epr/. Rules can be formulated over a language of propositional literals with ordinary negation, optionally augmented with a 'desirable' modality for modelling practical reasoning. EPR implements argument games for sceptical grounded and credulous preferred semantics, as well as [Prakken, 2006]'s game for combined epistemic and practical reasoning. It also implements as an option [Prakken, 2005]'s mechanism for accrual of arguments.

ArgTech's TOAST Mark Snaith of ArgTech at the University of Dundee, Scotland, developed an implementation called TOAST ([Snaid and Reed, 2012]) based on [Prakken, 2010], available at www.arg-tech.org/index.php/toast-an-aspic-implementation/. Rules can be formulated over a language of propositional literals with ordinary negation plus optionally a user-specified contrariness relation. TOAST allows for argument evaluation with an elitist weakest- or last-link ordering and in grounded, preferred, stable and semi-stable semantics. Interestingly, TOAST can receive input specified in the AIF format, so that it can be connected to argumentation tools that can export to AIF ([Bex et al., 2013a]). More on this will be said in the following subsection.

6.2 Logical specifications of the Argument Interchange Format

There is substantial interest in the development of argumentation support tools enabling the structuring of individual arguments and the dialogical exchange of argument in offline and online tools supporting human reasoning and debate (for example see www.arg-tech.org). A key aim is to then organise human authored arguments into abstract argumentation frameworks, so ensuring that the assessment of arguments is formally and rationally grounded and enabling 'mixed initiative' argumentation integrating both machine and human authored arguments [Modgil et al., 2013]. These developments, as well as the burgeoning interest in logic-based models of argument, have motivated formulation of a standardised format – the *Argument Interchange Format (AIF)* [Chesñevar et al., 2006] – for representation of human authored arguments and arguments constructed in logic.

The *AIF* is an ontology that broadly speaking distinguishes between information (propositions and sentences) and schemes which are general patterns of reasoning such as applications of inference rules, or conflict or preferences between information. Instances of these information and schemes classes constitute nodes that can be organised into *AIF* graphs representing argumentation knowledge. In [Bex et al., 2013b], two-way translations are defined between *AIF* graphs and both $ASPIC^+$ and E-$ASPIC^+$ argumentation theories, and a number of information preserving properties are proved in both cases. The latter essentially prove that given certain assumptions on the given *AIF* graphs, the translation functions are identity-preserving (i.e. translating from the *AIF* graph to (E-)$ASPIC^+$ and back again yields the same graph as we started out with).

One can then translate *AIF* representations of human authored arguments and their interactions defined in the above-mentioned argumentation support tools, and translate these to instantiations of (E-)$ASPIC^+$ so enabling evaluation under Dung's semantics. This is explored in [Bex *et al.*, 2013b], in which arguments and their interactions authored in the *Rationale* tool [ter Berg *et al.*, 2009] are translated to the *AIF* and then to $ASPIC^+$ arguments, attacks and defeats. In this way, $ASPIC^+$ is placed in the wider spectrum of not just formal but also philosophical and linguistic approaches to argumentation.

6.3 Other applications of $ASPIC^+$

$ASPIC^+$ has been applied both in purely theoretical models and in implemented architectures.

6.3.1 Theoretical applications

Some theoretical applications of $ASPIC^+$ amount to the formulation of sets of argument schemes for specific forms of reasoning in $ASPIC^+$. [van der Weide *et al.*, 2011] and [van der Weide, 2011] use a combination of $ASPIC^+$ and [Wooldridge *et al.*, 2006]'s system for meta-argumentation for specifying argument schemes for reasoning about preferences in argumentation-based decision making. [Bench-Capon and Prakken, 2010] and [Bench-Capon *et al.*, 2011] formulate argument schemes for policy debates in E-$ASPIC^+$. [Prakken *et al.*, 2015] and [Bench-Capon *et al.*, 2013], inspired by earlier AI & Law work of e.g. [Ashley, 1990] and [Aleven, 2003], model factor-based legal reasoning with precedents in $ASPIC^+$, with argument schemes formalised as defeasible rules and auxiliary definitions concerning (sets of) factors, their origins, their relations and their preferences as first-order axioms. This allows the formalisation of arguments like the following:

> **Plaintiff** The current case and precedent *Bryce* share pro-plaintiff factors $\{f_1, f_2\}$ and pro-defendant factors $\{f_3\}$, the pro-plaintiff factors outweigh the pro-defendant factors since *Bryce* was decided for the plaintiff; therefore, the current case should be decided for me.
>
> **Defendant** But unlike the current case, Bryce also contained pro-plaintiff factor f_4, so it is relevantly different from the current case, so the outcome of *Bryce* does not control the current case.
>
> **Plaintiff** But the current case contains factor f_5 and both f_4 and f_5 are a special case of the more abstract factor f_6, so this difference between *Bryce* and the current case is not relevant.

Other theoretical applications of $ASPIC^+$ concern case studies. [Prakken, 2012a] modelled the legal and evidential reasoning in the American *Popov v.*

Hayashi case, an ownerships dispute between two baseball fans about a baseball hit in the 500th homerun of a famous American baseball player. [Prakken, 2015] modelled a legislative debate and an American labour law dispute as argumentation-based decision making involving goals, values and preferences.

Finally, some theoretical applications use $ASPIC^+$ as a component of a more general reasoning model. [Müller and Hunter, 2012] used a simple instantiation of $ASPIC^+$ with no knowledge base, only defeasible rules and no preferences as a reasoning component in a formal model of decision making. [Prakken et al., 2013] applied $ASPIC^+$ in a dialogue model of collaborative IT security risk assessment. Finally, [Timmer et al., 2017] used $ASPIC^+$ for generating explanations of forensic Bayesian networks.

6.3.2 Applications in implemented architectures

Some implemented architectures proposed in the literature have used implementations of $ASPIC^+$ as a component. [Kok, 2013] used $ASPIC^+$ as the agent reasoning mechanism in a testbed for inter-agent deliberation dialogue, meant for testing whether the use of argumentation is beneficial to the individual agents or to the group to which they belong. This testbed is available online at https://bitbucket.org/erickok/baidd. [Toniolo et al., 2015] used $ASPIC^+$ as a reasoning component in their *CISpaces* sensemaking tool for intelligence analysis. [Yun and Croitoru, 2016] used the original ASPIC inference engine for reasoning with possibly inconsistent ontologies in ontology-based data access. Finally, [van Zee et al., 2016] used the TOAST implementation of $ASPIC^+$ as a component of a framework for rationalising goal models using argument diagrams.

7 Open problems and avenues for future research

The study of abstract rule-based argumentation with both strict and defeasible rules has a long history, ultimately going back to the seminal work of [Pollock, 1987], passing through intermediate stages [Simari and Loui, 1992; Pollock, 1995; Vreeswijk, 1997; Prakken and Sartor, 1997; Garcia and Simari, 2004] and currently consolidated in the work on $ASPIC^+$. As this chapter has shown, the approach is a fruitful one, a mature metatheory is developing and there is a growing number of implementations and applications. Yet many open questions and avenues for future research remain. Here we list some of the (in our opinion) most important ones.

- The study of argument preference relations and their properties is relatively underdeveloped. More can be done here, for example, relating argument orderings to work in decision theory or to probability theory (see also the next point), or combining different preference criteria for different kinds of problems, such as for epistemic versus practical reasoning.

- A recent research trend in formal argumentation is the combination of argumentation-based inference with probability theory. This is not surprising, since argumentation has from the early days been proposed as a model for reasoning under uncertainty. One question that arises here is how characterisations of the strength or relative preference of arguments relate to probability theory. Much recent work on probabilistic argumentation assigns probabilities to arguments in abstract argumentation frameworks, as in [Li et al., 2012; Hunter and Thimm, 2014]. However, assigning probabilities to arguments is problematic, since in probability theory probabilities are assigned to the truth of statements or to outcomes of events, and an argument is neither a statement nor an event. What is required here is a precise specification of what the probability of an argument means in terms of its elements. How to do this in the context of abstract rule-based argumentation is still largely an open question. A preliminary answer is given by [Hunter, 2013] but only for the case of classical-logic argumentation.

- The contamination problems referred to in Section 5.1 remain to be solved for the fully general $ASPIC^+$ framework. As briefly discussed at the end of Section 5.1, the work of [D'Agostino and Modgil, 2016] suggests directions for future development of the $ASPIC^+$ framework such that one can establish conditions under which the additional rationality postulates of [Caminada et al., 2012] are satisfied.

- In contrast to abstract argumentation, the study of computational aspects of rule-based argumentation and the various ways it can be instantiated is seriously underdeveloped. Much work can still be done on algorithms and complexity results for rule-based argumentation involving defeasible rules and preferences.

- While there is a growing body of work on the dynamics of abstract argumentation, the work of [Modgil and Prakken, 2012] in $ASPIC^+$ is to our knowledge still the only account of the dynamics of structured argumentation. Much remains to be done here.

- Another important research topic is implementation of more expressive instantiations than those existing today. It would, for example, be interesting to integrate state-of-the art propositional, first-order or modal-logic theorem provers in $ASPIC^+$ implementations.

- Finally, with an eye to practical applications it is important to conduct comparative case studies involving various formalisms, such as $ASPIC^+$, assumption-based argumentation, Carneades or [Brewka and Woltran, 2010]'s abstract dialectical frameworks. It would be especially interesting to study issues like naturalness and conciseness of representations.

Acknowledgments

We would like to express our thanks to those who have worked on the precursors of the $ASPIC^+$ framework (reviewed at the beginning of this chapter), as well as those who have provided valuable feedback (including Peter Young and Leon van der Torre), and those who have taken the time and effort to critically and constructively study $ASPIC^+$ and as a consequence have helped us improve $ASPIC^+$; notably, Phan Minh Dung in [Dung, 2016] and especially Sjur Dyrkolbotn (personal communication), whose constructive observations motivated us to write the erratum (https://nms.kcl.ac.uk/sanjay.modgil/AIJfinalErratum) to [Modgil and Prakken, 2013].

BIBLIOGRAPHY

[Aleven, 2003] V. Aleven. Using background knowledge in case-based legal reasoning: a computational model and an intelligent learning environment. *Artificial Intelligence*, 150:183–237, 2003.

[Amgoud and Besnard, 2013] L. Amgoud and Ph. Besnard. Logical limits of abstract argumentation frameworks. *Journal of Applied Non-classical Logics*, 23:229–267, 2013.

[Amgoud and Cayrol, 1998] L. Amgoud and C. Cayrol. On the acceptability of arguments in preference-based argumentation. In *Proceedings of the 14th Conference on Uncertainty in Artificial Intelligence*, pages 1–7, 1998.

[Amgoud and Cayrol, 2002] L. Amgoud and C. Cayrol. A model of reasoning based on the production of acceptable arguments. *Annals of Mathematics and Artificial Intelligence*, 34:197–215, 2002.

[Amgoud et al., 2006] L. Amgoud, L. Bodenstaff, M. Caminada, P. McBurney, S. Parsons, H. Prakken, J. van Veenen, and G.A.W. Vreeswijk. Final review and report on formal argumentation system. Deliverable D2.6, ASPIC IST-FP6-002307, 2006.

[Ashley, 1990] K.D. Ashley. *Modeling Legal Argument: Reasoning with Cases and Hypotheticals*. MIT Press, Cambridge, MA, 1990.

[Baroni and Giacomin, 2008] P. Baroni and M. Giacomin. Resolution-based argumentation semantics. In Ph. Besnard, S. Doutre, and A. Hunter, editors, *Computational Models of Argument. Proceedings of COMMA 2008*, pages 25–36, Amsterdam etc, 2008. IOS Press.

[Baroni et al., 2011a] P. Baroni, F. Cerutti, M. Giacomin, and G. Guida. Afra: Argumentation framework with recursive attacks. *International Journal of Approximate Reasoning*, 52:19–37, 2011.

[Baroni et al., 2011b] P. Baroni, P.E. Dunne, and M. Giacomin. On the resolution-based family of abstract argumentation semantics and its grounded instance. *Artificial Intelligence*, 175:791–813, 2011.

[Baroni et al., 2015] P. Baroni, M. Giacomin, and B. Lao. Dealing with generic contrariness in structured argumentation. In *Proceedings of the 24th International Joint Conference on Artificial Intelligence*, pages 2727–2733, 2015.

[Baumann and Brewka, 2010] R. Baumann and G. Brewka. Expanding argumentation frameworks: Enforcing and monotonicity results. In P. Baroni, F. Cerutti, M. Giacomin, and G.R. Simari, editors, *Computational Models of Argument. Proceedings of COMMA 2010*, pages 75–86. IOS Press, Amsterdam etc, 2010.

[Bench-Capon and Prakken, 2010] T.J.M. Bench-Capon and H. Prakken. A lightweight formal model of two-phase democratic deliberation. In R.G.F. Winkels, editor, *Legal Knowledge and Information Systems. JURIX 2010: The Twenty-Third Annual Conference*, pages 27–36. IOS Press, Amsterdam etc., 2010.

[Bench-Capon et al., 2011] T.J.M. Bench-Capon, H. Prakken, and W. Visser. Argument schemes for two-phase democratic deliberation. In *Proceedings of the Thirteenth International Conference on Artificial Intelligence and Law*, pages 21–30, New York, 2011. ACM Press.

[Bench-Capon et al., 2013] T.J.M. Bench-Capon, H. Prakken, A.Z. Wyner, and K. Atkinson. Argument schemes for reasoning with legal cases using values. In *Proceedings of* the *Fourteenth International Conference on Artificial Intelligence and Law*, pages 13–22, New York, 2013. ACM Press.

[Bench-Capon, 2003] T.J.M. Bench-Capon. Persuasion in practical argument using value-based argumentation frameworks. *Journal of Logic and Computation*, 13:429–448, 2003.

[Besnard and Hunter, 2001] Ph. Besnard and A. Hunter. A logic-based theory of deductive arguments. *Artificial Intelligence*, 128:203–235, 2001.

[Besnard and Hunter, 2008] Ph. Besnard and A. Hunter. *Elements of Argumentation*. MIT Press, Cambridge, MA, 2008.

[Bex et al., 2013a] F.J. Bex, J. Lawrence, M. Snaith, and C. Reed. Implementing the argument web. *Communications of the ACM*, 56:66–73, 2013.

[Bex et al., 2013b] F.J. Bex, S.J. Modgil, H. Prakken, and C.A. Reed. On logical specifications of the argument interchange format. *Journal of Logic and Computation*, 23:951–989, 2013.

[Bochman, 2003] A. Bochman. Collective argumentation and disjunctive logic programming. *Journal of Logic and Computation*, 13:405–428, 2003.

[Boella et al., 2010a] G. Boella, D.M. Gabbay, L. van der Torre, and S. Villata. Support in abstract argumentation. In P. Baroni, F. Cerutti, M. Giacomin, and G.R. Simari, editors, *Computational Models of Argument. Proceedings of COMMA 2010*, pages 111–122. IOS Press, Amsterdam etc, 2010.

[Boella et al., 2010b] G. Boella, S. Kaci, and L. van der Torre. Dynamics in argumentation with single extensions: attack refinement and the grounded extension (extended version). In P. McBurney, I. Rahwan, S. Parsons, and N. Maudet, editors, *Argumentation in Multi-Agent Systems, 6th International Workshop, ArgMAS 2009, Budapest, Hungary, May 12, 2009. Revised Selected and Invited Papers*, number 6057 in Springer Lecture Notes in AI, pages 150–159. Springer Verlag, Berlin, 2010.

[Bondarenko et al., 1993] A. Bondarenko, R.A. Kowalski, and F. Toni. An assumption-based framework for non-monotonic reasoning. In *Proceedings of the second International Workshop on Logic Programming and Nonmonotonic Logic*, pages 171–189, Lisbon (Portugal), 1993.

[Bondarenko et al., 1997] A. Bondarenko, P.M. Dung, R.A. Kowalski, and F. Toni. An abstract, argumentation-theoretic approach to default reasoning. *Artificial Intelligence*, 93:63–101, 1997.

[Brewka and Gordon, 2010] G. Brewka and T.F. Gordon. Carneades and abstract dialectical frameworks: A reconstruction. In P. Baroni, F. Cerutti, M. Giacomin, and G.R. Simari, editors, *Computational Models of Argument. Proceedings of COMMA 2010*, pages 3–12. IOS Press, Amsterdam etc, 2010.

[Brewka and Woltran, 2010] G. Brewka and S. Woltran. Abstract dialectical frameworks. In *Principles of Knowledge Representation and Reasoning: Proceedings of the Twelfth International Conference*, pages 102–111. AAAI Press, 2010.

[Brewka, 1989] G. Brewka. Preferred subtheories: An extended logical framework for default reasoning. In *Proceedings of the 11th International Joint Conference on Artificial Intelligence*, pages 1043–1048, 1989.

[Brewka, 1994a] G. Brewka. Adding priorities and specificity to default logic. In C. MacNish, D. Pearce, and L. Moniz Pereira, editors, *Logics in Artificial Intelligence. Proceedings of JELIA 1994*, number 838 in Springer Lecture Notes in AI, pages 247–260, Berlin, 1994. Springer Verlag.

[Brewka, 1994b] G. Brewka. Reasoning about priorities in default logic. In *Proceedings of the 12th National Conference on Artificial Intelligence (AAAI-94)*, pages 247–260, 1994.

[Caminada and Amgoud, 2007] M. Caminada and L. Amgoud. On the evaluation of argumentation formalisms. *Artificial Intelligence*, 171:286–310, 2007.

[Caminada et al., 2012] M. Caminada, W.A. Carnielli, and P.E. Dunne. Semi-stable semantics. *Journal of Logic and Computation*, 22:1207–1254, 2012.

[Caminada et al., 2014] M. Caminada, S. Modgil, and N. Oren. Preferences and unrestricted rebut. In S. Parsons, N. Oren, C. Reed, and F. Cerutti, editors, *Computational Models of Argument. Proceedings of COMMA 2014*, pages 209–220. IOS Press, Amsterdam etc, 2014.

[Caminada, 2004] M. Caminada. *For the sake of the Argument. Explorations into argument-based reasoning*. Doctoral dissertation Free University Amsterdam, 2004.

[Caminada, 2005] M. Caminada. Contamination in formal argumentation systems. In *Proceedings of the Seventeenth Belgian-Dutch Conference on Artificial Intelligence (BNAIC-05)*, Brussels, Belgium, 2005.

[Caminada, 2006] M. Caminada. On the issue of reinstatement in argumentation. In M. Fischer, W. van der Hoek, B. Konev, and A. Lisitsa, editors, *Logics in Artificial Intelligence. Proceedings of JELIA 2006*, number 4160 in Springer Lecture Notes in AI, pages 111–123, Berlin, 2006. Springer Verlag.

[Cayrol and Lagasquie-Schiex, 2005] C. Cayrol and M.-C. Lagasquie-Schiex. On the acceptability of arguments in bipolar argumentation. In L. Godo, editor, *Proceedings of the 8nd European Conference on Symbolic and Quantitative Approaches to Reasoning with Uncertainty (ECSQARU 05)*, number 3571 in Springer Lecture Notes in AI, pages 378–389, Berlin, 2005. Springer Verlag.

[Cayrol and Lagasquie-Schiex, 2009] C. Cayrol and M.-C. Lagasquie-Schiex. Bipolar abstract argumentation systems. In I. Rahwan and G.R. Simari, editors, *Argumentation in Artificial Intelligence*, pages 65–84. Springer, Berlin, 2009.

[Cayrol and Lagasquie-Schiex, 2013] C. Cayrol and M.-C. Lagasquie-Schiex. Bipolarity in argumentation graphs: Towards a better understanding. *International Journal of Approximate Reasoning*, 54:876–899, 2013.

[Cayrol et al., 2010] C. Cayrol, F Dupin de Saint-Cyr, and M.-C. Lagasquie-Schiex. Change in abstract argumentation frameworks: adding an argument. *Journal of Artificial Intelligence Research*, 38:49–84, 2010.

[Cayrol, 1995] C. Cayrol. On the relation between argumentation and non-monotonic coherence-based entailment. In *Proceedings of the 14th International Joint Conference on Artificial Intelligence*, pages 1443–1448, 1995.

[Chesñevar et al., 2006] C.I. Chesñevar, J. McGinnis, S. Modgil, I. Rahwan, C. Reed, G. Simari, M. South, G. Vreeswijk, and S. Willmott. Towards an argument interchange format. *The Knowledge Engineering Review*, 21:293–316, 2006.

[Da Costa, 1974] N.C.A. Da Costa. On the theory of inconsistent formal systems. *Notre Dame Journal of Formal Logic*, 15(4):497–510, 1974.

[D'Agostino and Modgil, 2016] M. D'Agostino and S. Modgil. A rational account of classical logic argumentation for real-world agents. In *Proceedings of the Twenty Second European Conference on Artificial Intelligence (ECAI 2016)*, pages 141–149, 2016.

[Delgrande and Schaub, 2000] J. Delgrande and T. Schaub. Expressing preferences in default logic. *Artificial Intelligence*, 123:41–87, 2000.

[Dung and Thang, 2014] P.M. Dung and P.M. Thang. Closure and consistency in logic-associated argumentation. *Journal of Artificial Intelligence Research*, 49:79–109, 2014.

[Dung et al., 2007] P.M. Dung, P. Mancarella, and F. Toni. Computing ideal sceptical argumentation. *Artificial Intelligence*, 171:642–674, 2007.

[Dung, 1995] P.M. Dung. On the acceptability of arguments and its fundamental role in nonmonotonic reasoning, logic programming, and n–person games. *Artificial Intelligence*, 77:321–357, 1995.

[Dung, 2014] P.M. Dung. An axiomatic analysis of structured argumentation for prioritized default reasoning. In *Proceedings of the 21st European Conference on Artificial Intelligence*, pages 267–272, 2014.

[Dung, 2016] P.M. Dung. An axiomatic analysis of structured argumentation with priorities. *Artificial Intelligence*, 231:107–150, 2016.

[Garcia and Simari, 2004] A.J. Garcia and G.R. Simari. Defeasible logic programming: An argumentative approach. *Theory and Practice of Logic Programming*, 4:95–138, 2004.

[Gordon and Walton, 2009a] T.F. Gordon and D.N. Walton. Legal reasoning with argumentation schemes. In *Proceedings of the Twelfth International Conference on Artificial Intelligence and Law*, pages 137–146, New York, 2009. ACM Press.

[Gordon and Walton, 2009b] T.F. Gordon and D.N. Walton. Proof burdens and standards. In I. Rahwan and G.R. Simari, editors, *Argumentation in Artificial Intelligence*, pages 239–258. Springer, Berlin, 2009.

[Gordon et al., 2007] T.F. Gordon, H. Prakken, and D.N. Walton. The Carneades model of argument and burden of proof. *Artificial Intelligence*, 171:875–896, 2007.

[Gorogiannis and Hunter, 2011] N. Gorogiannis and A. Hunter. Instantiating abstract argumentation with classical-logic arguments: postulates and properties. *Artificial Intelligence*, 175:1479–1497, 2011.

[Grooters and Prakken, 2016] D. Grooters and H. Prakken. Two aspects of relevance in structured argumentation: minimality and paraconsistency. *Journal of Artificial Intelligence Research*, 56:197–245, 2016.

[Grooters, 2014] D. Grooters. Paraconsistent logics in argumentation systems. Master's thesis, Department of Information and Computing Sciences, Utrecht University, Utrecht, 2014.

[Hunter and Thimm, 2014] A. Hunter and M. Thimm. Probabilistic argumentation with incomplete information. In *Proceedings of the 21st European Conference on Artificial Intelligence*, pages 1033–1034, 2014.

[Hunter, 2010] A. Hunter. Base logics in argumentation. In P. Baroni, F. Cerutti, M. Giacomin, and G.R. Simari, editors, *Computational Models of Argument. Proceedings of COMMA 2010*, pages 275–286. IOS Press, Amsterdam etc, 2010.

[Hunter, 2013] A. Hunter. A probabilistic approach to modelling uncertain logical arguments. *International Journal of Approximate Reasoning*, 54:47–81, 2013.

[Jakobovits and Vermeir, 1999] H. Jakobovits and D. Vermeir. Robust semantics for argumentation frameworks. *Journal of Logic and Computation*, 9:215–261, 1999.

[Kok, 2013] E.M. Kok. *Exploring the Practical Benefits of Argumentation in Multi-Agent Deliberation*. Doctoral dissertation Department of Information and Computing Sciences, Utrecht University, 2013.

[Kowalski and Toni, 1996] R.A. Kowalski and F. Toni. Abstract argumentation. *Artificial Intelligence and Law*, 4:275–296, 1996.

[Kraus et al., 1990] S. Kraus, D.J. Lehmann, and M. Magidor. Nonmonotonic reasoning, preferential models and cumulative logics. *Artificial Intelligence*, 44:167–207, 1990.

[Kyburg, 1961] H. Kyburg. *Probability and the Logic of Rational Belief*. Wesleyan U. P. Middletown, CT, 1961.

[Li and Parsons, 2015] Z. Li and S. Parsons. On argumentation with purely defeasible rules. In C. Beierle and A. Dekhtyar, editors, *Proceedings of the 9th International Conference on Scalable Uncertainty Management (SUM'15)*, number 9310 in Springer Lecture Notes in AI, pages 30–43, Berlin, 2015. Springer Verlag.

[Li et al., 2012] H. Li, N. Oren, and T. Norman. Probabilistic argumentation frameworks. In S. Modgil, N. Oren, and F. Toni, editors, *Theorie and Applications of Formal Argumentation. First International Workshop, TAFA 2011. Barcelona, Spain, July 16-17, 2011, Revised Selected Papers*, number 7132 in Springer Lecture Notes in AI, pages 1–16, Berlin, 2012. Springer Verlag.

[Lin and Shoham, 1989] F. Lin and Y. Shoham. Argument systems. A uniform basis for nonmonotonic reasoning. In *Principles of Knowledge Representation and Reasoning: Proceedings of the First International Conference*, pages 245–255, San Mateo, CA, 1989. Morgan Kaufmann Publishers.

[Martinez et al., 2006] D.C. Martinez, A.J. Garcia, and G.R. Simari. On acceptability in abstract argumentation frameworks with an extended defeat relation. In P.E. Dunne and T.B.C. Bench-Capon, editors, *Computational Models of Argument. Proceedings of COMMA 2006*, pages 273–278. IOS Press, Amsterdam etc, 2006.

[Modgil and Caminada, 2009] S. Modgil and M. Caminada. Proof theories and algorithms for abstract argumentation frameworks. In I. Rahwan and G.R. Simari, editors, *Argumentation in Artificial Intelligence*, pages 105–129. Springer, Berlin, 2009.

[Modgil and Prakken, 2010] S. Modgil and H. Prakken. Reasoning about preferences in structured extended argumentation frameworks. In P. Baroni, F. Cerutti, M. Giacomin, and G.R. Simari, editors, *Computational Models of Argument. Proceedings of COMMA 2010*, pages 347–358. IOS Press, Amsterdam etc, 2010.

[Modgil and Prakken, 2012] S. Modgil and H. Prakken. Resolutions in structured argumentation. In B. Verheij, S. Woltran, and S. Szeider, editors, *Computational Models of Argument. Proceedings of COMMA 2012*, pages 310–321. IOS Press, Amsterdam etc, 2012.

[Modgil and Prakken, 2013] S. Modgil and H. Prakken. A general account of argumentation with preferences. *Artificial Intelligence*, 195:361–397, 2013.

[Modgil and Prakken, 2014] S. Modgil and H. Prakken. The ASPIC+ framework for structured argumentation: a tutorial. *Argument and Computation*, 5:31–62, 2014.

[Modgil et al., 2013] S. Modgil, F. Toni, F. Bex, I. Bratko, C. Chesñevar, W. Dvorak, M. Falappa, X. Fan, S. Gaggl, A. Garcia, M. Gonzalez, T. Gordon, J. Leite, M. Mozina, C. Reed, G. Simari, S. Szeider, P. Torroni, and S. Woltran. The added value of argumentation. In S. Ossowski, editor, *Agreement Technologies*, pages 357–403. Springer, 2013.

[Modgil, 2006] S. Modgil. Hierarchical argumentation. In M. Fischer, W. van der Hoek, B. Konev, and A. Lisitsa, editors, *Logics in Artificial Intelligence. Proceedings of JELIA 2006*, number 4160 in Springer Lecture Notes in AI, pages 319–332, Berlin, 2006. Springer Verlag.

[Modgil, 2009] S. Modgil. Reasoning about preferences in argumentation frameworks. *Artificial Intelligence*, 173:901–934, 2009.

[Modgil, 2014] S. Modgil. Revisiting abstract argumentation. In E. Black, S. Modgil, and N. Oren, editors, *Second International Workshop, TAFA 2013, Beijing, China, August 3-5, 2013, Revised Selected papers*, number 8306 in Springer Lecture Notes in AI, pages 1–15, Berlin, 2014. Springer Verlag.

[Müller and Hunter, 2012] J. Müller and A. Hunter. An argumentation-based approach for decision making. In *Proceedings of the 25th International Conference on Tools with Artificial Intelligence (ICTAI 2013)*, pages 564–571, 2012.

[Nielsen and Parsons, 2007] S.H. Nielsen and S. Parsons. A generalization of Dung's abstract framework for argumentation: Arguing with sets of attacking arguments. In S. Parsons, N. Maudet, and I. Rahwan, editors, *Argumentation in Multi-Agent Systems. Third International Workshop, ArgMAS 2006, Hakodate, Japan, May 8, 2006, Revised Selected and Invited Papers*, number 4766 in Springer Lecture Notes in AI, pages 54–73. Springer Verlag, Berlin, 2007.

[Nouioua and Risch, 2011] F. Nouioua and V. Risch. Argumentation frameworks with necessities. In *Proceedings of the 4th International Conference on Scalable Uncertainty Management (SUM'11)*, number 6929 in Springer Lecture Notes in AI, pages 163–176, Berlin, 2011. Springer Verlag.

[Oren and Norman, 2008] N. Oren and T.J. Norman. Semantics for evidence-based argumentation. In Ph. Besnard, S. Doutre, and A. Hunter, editors, *Computational Models of Argument. Proceedings of COMMA 2008*, pages 276–284, Amsterdam etc, 2008. IOS Press.

[Pearl, 1992] J. Pearl. Epsilon-semantics. In S.C. Shapiro, editor, *Encyclopedia of Artificial Intelligence*, pages 468–475. John Wiley & Sons, New York, 1992.

[Pollock, 1987] J.L. Pollock. Defeasible reasoning. *Cognitive Science*, 11:481–518, 1987.

[Pollock, 1990] J.L. Pollock. A theory of defeasible reasoning. *International Journal of Intelligent Systems*, 6:33–54, 1990.

[Pollock, 1992] J.L. Pollock. How to reason defeasibly. *Artificial Intelligence*, 57:1–42, 1992.

[Pollock, 1994] J.L. Pollock. Justification and defeat. *Artificial Intelligence*, 67:377–408, 1994.

[Pollock, 1995] J.L. Pollock. *Cognitive Carpentry. A Blueprint for How to Build a Person*. MIT Press, Cambridge, MA, 1995.

[Pollock, 2009] J.L. Pollock. A recursive semantics for defeasible reasoning. In I. Rahwan and G.R. Simari, editors, *Argumentation in Artificial Intelligence*, pages 173–197. Springer, Berlin, 2009.

[Prakken and Sartor, 1997] H. Prakken and G. Sartor. Argument-based extended logic programming with defeasible priorities. *Journal of Applied Non-classical Logics*, 7:25–75, 1997.

[Prakken and Vreeswijk, 2002] H. Prakken and G.A.W. Vreeswijk. Logics for defeasible argumentation. In D. Gabbay and F. Günthner, editors, *Handbook of Philosophical Logic*, volume 4, pages 219–318. Kluwer Academic Publishers, Dordrecht/Boston/London, second edition, 2002.

[Prakken et al., 2013] H. Prakken, D. Ionita, and R. Wieringa. Risk assessment as an argumentation game. In J. Leite, T.C. Son, P. Torroni, L. van der Torre, and S. Woltran, editors, *Proceedings of the 14th International Workshop on Computational Logic in Multi-Agent Systems (CLIMA XIV)*, number 8143 in Springer Lecture Notes in AI, pages 357–373, Berlin, 2013. Springer Verlag.

[Prakken et al., 2015] H. Prakken, A.Z. Wyner, T.J.M. Bench-Capon, and K. Atkinson. A formalisation of argumentation schemes for legal case-based reasoning in ASPIC+. *Journal of Logic and Computation*, 25:1141–1166, 2015.

[Prakken, 1999] H. Prakken. Dialectical proof theory for defeasible argumentation with defeasible priorities (preliminary report). In J.-J.Ch. Meyer and P.-Y. Schobbens, editors, *Formal Models of Agents*, number 1760 in Springer Lecture Notes in AI, pages 202–215, Berlin, 1999. Springer Verlag.

[Prakken, 2005] H. Prakken. A study of accrual of arguments, with applications to evidential reasoning. In *Proceedings of the Tenth International Conference on Artificial Intelligence and Law*, pages 85–94, New York, 2005. ACM Press.

[Prakken, 2006] H. Prakken. Combining sceptical epistemic reasoning with credulous practical reasoning. In P.E. Dunne and T.B.C. Bench-Capon, editors, *Computational Models of Argument. Proceedings of COMMA 2006*, pages 311–322. IOS Press, Amsterdam etc, 2006.

[Prakken, 2010] H. Prakken. An abstract framework for argumentation with structured arguments. *Argument and Computation*, 1:93–124, 2010.

[Prakken, 2012a] H. Prakken. Reconstructing Popov v. Hayashi in a framework for argumentation with structured arguments and Dungeon semantics. *Artificial Intelligence and Law*, 20:57–82, 2012.

[Prakken, 2012b] H. Prakken. Some reflections on two current trends in formal argumentation. In *Logic Programs, Norms and Action. Essays in Honour of Marek J. Sergot on the Occasion of his 60th Birthday*, pages 249–272. Springer, Berlin/Heidelberg, 2012.

[Prakken, 2013] H. Prakken. Relating ways to instantiate abstract argumentation frameworks. In K.D. Atkinson, H. Prakken, and A.Z. Wyner, editors, *From Knowledge Representation to Argumentation in AI, Law and Policy Making. A Festschrift in Honour of Trevor Bench-Capon on the Occasion of his 60th Birthday*, pages 167–189. College Publications, London, 2013.

[Prakken, 2014] H. Prakken. On support relations in abstract argumentation as abstractions of inferential relations. In *Proceedings of the 21st European Conference on Artificial Intelligence*, pages 735–740, 2014.

[Prakken, 2015] H. Prakken. Formalising debates about law-making proposals as practical reasoning. In M. Araszkiewicz and K. Płeszka, editors, *Logic in the Theory and Practice of Lawmaking*, pages 301–321. Springer, Berlin, 2015.

[Prakken, 2016] H. Prakken. Rethinking the rationality postulates for argumentation-based inference. In P. Baroni, T.F. Gordon, T. Scheffler, and M. Stede, editors, *Computational Models of Argument. Proceedings of COMMA 2016*, pages 419–430. IOS Press, Amsterdam etc, 2016.

[Priest, 1979] G. Priest. The logic of paradox. *Journal of Philosophical Logic*, 8(1):219–241, 1979.

[Priest, 1989] G. Priest. Reasoning about truth. *Artificial Intelligence*, 39(2):231–244, 1989.

[Reiter, 1980] R. Reiter. A logic for default reasoning. *Artificial Intelligence*, 13:81–132, 1980.

[Rescher and Manor, 1970] N. Rescher and R. Manor. On inference from inconsistent premises. *Journal of Theory and Decision*, 1:179–219, 1970.

[Rotstein et al., 2008] N.D. Rotstein, M.O. Moguillansky, M.A. Falappa, A.J. Garcia, and G.R. Simari. Argument theory change: Revision upon warrant. In Ph. Besnard, S. Doutre, and A. Hunter, editors, *Computational Models of Argument. Proceedings of COMMA 2008*, pages 336–347, Amsterdam etc, 2008. IOS Press.

[Simari and Loui, 1992] G.R. Simari and R.P. Loui. A mathematical treatment of defeasible argumentation and its implementation. *Artificial Intelligence*, 53:125–157, 1992.

[Snaid and Reed, 2012] M. Snaid and C. Reed. TOAST: Online ASPIC+ argumentation. In B. Verheij, S. Woltran, and S. Szeider, editors, *Computational Models of Argument. Proceedings of COMMA 2012*, pages 509–510. IOS Press, Amsterdam etc, 2012.

[ter Berg et al., 2009] T. ter Berg, T. van Gelder, F. Patterson, and S. Teppema. *Critical Thinking: Reasoning and Communicating with Rationale*. Pearson Education Benelux, Amsterdam, 2009.

[Timmer et al., 2017] S. Timmer, J.-J.Ch. Meyer, H. Prakken, S. Renooij, and B. Verheij. A two-phase method for extracting explanatory arguments from Bayesian networks. *International Journal of Approximate Reasoning*, 80:475–494, 2017.

[Toni, 2014] F. Toni. A tutorial on assumption-based argumentation. *Argument and Computation*, 5:89–117, 2014.
[Toniolo et al., 2015] A. Toniolo, T.J. Norman, A. Etuk, F. Cerutti, R. Wentao Ouyang, M. Srivastava, N. Oren, T. Dropps, J.A. Allen, and Paul Sullivan. Supporting reasoning with different types of evidence in intelligence analysis. In *Proceedings of the 14th International Conference on Autonomous Agents and Multiagent Systems (AAMAS'15)*, pages 781–789, 2015.
[van der Weide et al., 2011] T. van der Weide, F. Dignum, J.-J.Ch. Meyer, H. Prakken, and G. Vreeswijk. Arguing about preferences and decisions. In P. McBurney, I. Rahwan, and S. Parsons, editors, *Argumentation in Multi-Agent Systems, 7th International Workshop, ArgMAS 2010, Toronto, Canada, May 2010. Revised Selected and Invited Papers*, number 6614 in Springer Lecture Notes in AI, pages 68–85. Springer Verlag, Berlin, 2011.
[van der Weide, 2011] T.L. van der Weide. *Arguing to Motivate Decisions*. Doctoral dissertation Department of Information and Computing Sciences, Utrecht University, 2011.
[Van Gijzel and Prakken, 2011] B. Van Gijzel and H. Prakken. Relating Carneades with abstract argumentation. In *Proceedings of the 22nd International Joint Conference on Artificial Intelligence (IJCAI-11)*, pages 1113–1119, 2011.
[Van Gijzel and Prakken, 2012] B. Van Gijzel and H. Prakken. Relating Carneades with abstract argumentation via the ASPIC+ framework for structured argumentation. *Argument and Computation*, 3:21–47, 2012.
[van Zee et al., 2016] M. van Zee, D. Marosin, F.J. Bex, and S. Ghanavati. RationalGRL: A framework for rationalizing goal models using argument diagrams. In *Proceedings of the 35th International Conference on Conceptual Modeling (ER'2016)*, 2016.
[Verheij, 2003] B. Verheij. DefLog: on the logical interpretation of prima facie justified assumptions. *Journal of Logic and Computation*, 13:319–346, 2003.
[Vreeswijk, 1997] G.A.W. Vreeswijk. Abstract argumentation systems. *Artificial Intelligence*, 90:225–279, 1997.
[Walton and Krabbe, 1995] D.N. Walton and E.C.W. Krabbe. *Commitment in Dialogue. Basic Concepts of Interpersonal Reasoning*. State University of New York Press, Albany, NY, 1995.
[Walton, 1996] D.N. Walton. *Argumentation Schemes for Presumptive Reasoning*. Lawrence Erlbaum Associates, Mahwah, NJ, 1996.
[Walton, 2006] D.N. Walton. *Fundamentals of Critical Argumentation*. Cambridge University Press, Cambridge, 2006.
[Wooldridge et al., 2006] M. Wooldridge, P. McBurney, and S. Parsons. On the meta-logic of arguments. In S. Parsons, N. Maudet, P. Moraitis, and I. Rahwan, editors, *Argumentation in Multi-Agent Systems: Second International Workshop, ArgMAS 2005, Utrecht, Netherlands, July 26, 2005, Revised Selected and Invited Papers*, number 4049 in Springer Lecture Notes in AI, pages 42–56. Springer Verlag, Berlin, 2006.
[Wu and Podlaszewski, 2015] Y. Wu and M. Podlaszewski. Implementing crash-resistence and non-interference in logic-based argumentation. *Journal of Logic and Computation*, 25:303–333, 2015.
[Wu, 2012] Y. Wu. *Between Argument and Conclusion. Argument-based Approaches to Discussion, Inference and Uncertainty*. Doctoral Dissertation Faculty of Sciences, Technology and Communication, University of Luxembourg, 2012.
[Young et al., 2016] A.P. Young, S. Modgil, and O. Rodrigues. Prioritised default logic as rational argumentation. In *Proceedings of the 15th International Conference on Autonomous Agents and Multiagent Systems (AAMAS'16)*, pages 626–634, 2016.
[Yun and Croitoru, 2016] B. Yun and M. Croitoru. An argumentation workflow for reasoning in ontology based data access. In P. Baroni, T.F. Gordon, T. Scheffler, and M. Stede, editors, *Computational Models of Argument. Proceedings of COMMA 2016*, pages 61–68. IOS Press, Amsterdam etc, 2016.

Sanjay Modgil
Department of Informatics, King's College London, UK
Email: sanjay.modgil@kcl.ac.uk

Henry Prakken
Department of Information and Computing Sciences, Utrecht University, The Netherlands
Faculty of Law, University of Groningen, The Netherlands
Email: H.Prakken@uu.nl

7
Assumption-Based Argumentation: Disputes, Explanations, Preferences

KRISTIJONAS ČYRAS, XIUYI FAN, CLAUDIA SCHULZ, FRANCESCA TONI

ABSTRACT. Assumption-Based Argumentation (ABA) is a form of structured argumentation with roots in non-monotonic reasoning. As in other forms of structured argumentation, notions of argument and attack are not primitive in ABA, but are instead defined in terms of other notions. In the case of ABA these other notions are those of rules in a deductive system, assumptions, and contraries.

ABA is equipped with a range of computational tools, based on dispute trees and amounting to dispute derivations, and benefiting from equivalent views of the semantics of argumentation in ABA, in terms of sets of arguments and, equivalently, sets of assumptions. These computational tools can also provide the foundation for multi-agent argumentative dialogues and explanation of reasoning outputs, in various settings and senses.

ABA is a flexible modelling formalism, despite its simplicity, allowing to support, in particular, various forms of non-monotonic reasoning, and reasoning with some forms of preferences and defeasible rules without requiring any additional machinery. ABA can also be naturally extended to accommodate further reasoning with preferences.

1 Introduction

Assumption-Based Argumentation (ABA) [Bondarenko et al., 1993; Bondarenko et al., 1997; Dung et al., 2009; Toni, 2014] is a form of *structured argumentation* [Besnard et al., 2014] with roots in non-monotonic reasoning [Brewka et al., 1997]. Differently from abstract argumentation [Dung, 1995] but as in other forms of structured argumentation, e.g. DeLP [García and Simari, 2014] and deductive arguments [Besnard and Hunter, 2014], notions of argument and attack are not primitive in ABA, but are instead defined in terms of other notions. In the case of ABA these notions are those of *rules* in an underlying *deductive system*, *assumptions* and their *contraries*: arguments are supported by rules and assumptions and attacks are directed against (assumptions deducible from) assumptions supporting arguments, by building arguments for the contrary of these assumptions. Semantics of ABA frameworks can be characterised in terms of sets of assumptions (or *extensions*) [Bondarenko et al., 1993; Bondarenko et al., 1997; Dung et al., 2007] meeting desirable requirements, including, but not limited to, the two core requirements of *closedness* (where a set of assumptions is *closed* iff it consists of all the assumptions deducible from it) and *conflict-freeness* (where a set of assumptions is *conflict-free* iff it does not attack itself). The closedness requirement is guaranteed to be fulfilled automatically for all sets of assumptions for restricted kinds of *ABA frameworks*, referred to as *flat* [Bondarenko et al., 1997]. The ABA semantics of admissible, preferred, complete, well-founded, stable and ideal extensions [Bondarenko et al., 1997; Dung et al., 2007] differ in which additional desirable requirements they impose upon sets of assumptions, but can all be seen as providing argumentative counterparts of semantics that had previously been defined for non-monotonic reasoning, by appropriately instantiating (flat and non-flat) ABA frameworks [Bondarenko et al., 1993; Bondarenko et al., 1997] to "match" existing frameworks for non-monotonic reasoning.

Flat ABA is equipped with a range of computational tools, based on dispute trees [Dung et al., 2006; Dung et al., 2007] and amounting to dispute derivations [Dung et al., 2006; Dung et al., 2007; Toni, 2013], and benefiting from equivalent views of the semantics of argumentation in flat ABA, in terms of sets of arguments and, equivalently, sets of assumptions [Dung et al., 2007]. These computational tools can also provide the foundation for inter-agent *ABA dialogues* in various settings and senses [Fan and Toni, 2011b; Fan and Toni, 2011a; Fan and Toni, 2011c; Fan and Toni, 2012b; Fan and Toni, 2012a; Fan and Toni, 2012c; Fan et al., 2014; Fan and Toni, 2014b; Fan and Toni, 2016] and explanations of reasoning outputs, in various settings and senses, e.g. to explain (non-)membership in answer sets of logic programs [Schulz and Toni, 2016], to explain "goodness" of decisions [Fan and Toni, 2014a; Fan et al., 2013; Zhong et al., 2014] and, more generically, to explain admissibility of sentences in any flat instance of ABA [Fan and Toni, 2015c].

ABA is a flexible modelling formalism, despite its simplicity, allowing to support, in particular, reasoning with some forms of preferences and defeasible rules without requiring any additional machinery [Kowalski and Toni, 1996; Toni, 2008b; Thang and Luong, 2013; Fan et al., 2013], but accommodating preferences at the "object-level". ABA can also be naturally extended to accommodate further reasoning with preferences, e.g. as in [Wakaki, 2014] or as ABA^+ in [Čyras and Toni, 2016a; Čyras and Toni, 2016b].

This chapter is organised as follows. In Section 2 we recap the basic definitions of ABA frameworks and semantics, focusing on semantics that have been inspired by semantics for non-monotonic reasoning, and summarising properties of semantics, distinguishing amongst generic and flat ABA frameworks. In Section 3 we illustrate two instances of ABA, capturing autoepistemic logic and logic programming, and respectively requiring non-flat and flat ABA frameworks. From Section 4 to Section 7 we focus on flat ABA frameworks. In particular, in Section 4 we summarise how flat ABA frameworks can be equivalently understood, for all semantics considered in this chapter, as abstract argumentation frameworks [Dung, 1995], following the results in [Dung et al., 2009], and, vice versa, abstract argumentation frameworks can be equivalently understood, for all semantics considered in this chapter, as flat ABA frameworks, following the results in [Toni, 2012]. In Section 5 we provide an overview and illustration of the basis of all computational machinery for ABA, in the flat case, namely dispute trees [Dung et al., 2006; Dung et al., 2007] and dispute derivations [Dung et al., 2006; Dung et al., 2007; Toni, 2013]. In this section we also illustrate how this machinery can be adapted to provide a foundation for inter-agent ABA dialogues [Fan and Toni, 2014b]. In Section 6 we overview various uses of (flat) ABA to provide explanations of reasoning outputs [Schulz and Toni, 2016; Fan and Toni, 2014a; Fan et al., 2013; Zhong et al., 2014; Fan et al., 2014; Fan and Toni, 2015c]. In Section 7 we overview various existing approaches to accommodating preferences in (flat) ABA [Kowalski and Toni, 1996; Toni, 2008a; Toni, 2008b; Thang and Luong, 2013; Fan et al., 2013] or extending ABA to accommodate reasoning with preferences [Wakaki, 2014; Čyras and Toni, 2016a; Čyras and Toni, 2016b]. Finally, in Section 8 we conclude, emphasising, in particular, omissions and future work.

This chapter complements other earlier surveys of ABA [Dung et al., 2009; Toni, 2012; Toni, 2014]. In particular, all earlier surveys focused exclusively on flat ABA frameworks. These are powerful knowledge representation mechanisms, as, for example, they fully capture logic programming (see Section 3) and default logic [Reiter, 1980] (see [Bondarenko et al., 1997]), both widely used formalisms for non-monotonic reasoning and knowledge representation and reasoning, as well as, for instance, some forms of decision-making (see Section 7). However, non-flat frameworks allow to capture additional forms of reasoning, including the kind of non-monotonic reasoning encapsulated by autoepistemic logic (see Section 3), as well as circumscription [McCarthy, 1980], amongst others (see [Bondarenko et al., 1997]). For example, in non-flat ABA one can represent beliefs as assumptions that can be deduced via rules from other assumptions.

Moreover, differently from earlier surveys, this chapter summarises uses of ABA for non-monotonic reasoning (Section 3) and defeasible reasoning (Section 7) as well as the explanatory power of ABA (Section 6) afforded by its computational machinery. At the same time, this chapter ignores other aspects of ABA, emphasised instead in the earlier surveys, such as the equivalence between different presentations of ABA in the literature, e.g. alternative views of arguments (as trees [Dung et al., 2009] rather than as forward [Bondarenko et al., 1997] or backward [Dung et al., 2006] deductions).

This chapter is related to a number of other chapters in this handbook. In particular, it takes for granted notions from abstract argumentation, as overviewed in the chapter on *Abstract Argumentation Frameworks and Their Semantics* in this handbook. Moreover, the chapter on *Argumentation Based on Logic Programming* presents an approach to structured argumentation based on logic programming, but different from the logic programming instance of ABA, and the chapter on *Argumentation, Nonmonotonic Reasoning and Logic* overviews relationships between argumentation and non-monotonic reasoning more in general; the chapter on *Computational Problems in Formal Argumentation and Their Complexity* deals with computational complexity issues, that we neglect in this chapter for ABA (but we briefly consider in Section 8); finally, the chapter on *Foundations of Implementations for Formal Argumentation* overviews implementations of argumentation, including ones for ABA, that we ignore in this chapter (but again we briefly mention in Section 8).

2 ABA frameworks and semantics

In this section we introduce ABA frameworks [Bondarenko et al., 1993; Bondarenko et al., 1997; Dung et al., 2009; Toni, 2014] and their standard semantics of admissible, preferred, complete, well-founded (called *grounded* in the specific case of flat ABA frameworks), stable and ideal extensions [Bondarenko et al., 1993; Bondarenko et al., 1997; Dung et al., 2007] as sets of assumptions. All the semantics considered have counterparts in logic programming, in the sense that they correspond to semantics of logic programs in the logic programming instance of ABA (see Section 3).

Definition 2.1 *An* ABA framework *is a tuple* $\langle \mathcal{L}, \mathcal{R}, \mathcal{A}, \overline{} \rangle$ *where*

- $\langle \mathcal{L}, \mathcal{R} \rangle$ *is a deductive system, with $\mathcal{L}$ a language (a set of sentences) and $\mathcal{R}$ a set of (inference) rules, each with a head and a body, where the head is a sentence in $\mathcal{L}$, and the body consists of $m \geq 0$ sentences in $\mathcal{L}$;*

- $\mathcal{A} \subseteq \mathcal{L}$ *is a (non-empty) set, with elements referred to as* assumptions;

- $\overline{}$ *is a total mapping from $\mathcal{A}$ into $\mathcal{L}$; $\overline{a}$ is referred to as the* contrary *of a, for $a \in \mathcal{A}$.*

Rules in $\mathcal{R}$ can be written in different formats, e.g. a rule with head σ_0 and body $\sigma_1, \ldots, \sigma_m$ may be written as

$$\sigma_0 \leftarrow \sigma_1, \ldots, \sigma_m \qquad \text{or} \qquad \frac{\sigma_1, \ldots, \sigma_m}{\sigma_0}.$$

Note that $\leftarrow$ is not to be interpreted as logical implication, when used to represent rules in ABA as above. In the remainder of this chapter, we will use these two syntactic conventions for writing rules interchangeably. Moreover, unless specified otherwise, we will assume as given a generic ABA framework $\langle \mathcal{L}, \mathcal{R}, \mathcal{A}, \overline{} \rangle$. Note also that sentences have a contrary if, and only if, they are assumptions. This contrary is not to be confused with negation, which may or may not occur in $\mathcal{L}$.

Rules in ABA frameworks can be chained to form *deductions*. These can be defined in several ways, notably in a forward [Bondarenko et al., 1997], a backward [Dung et al., 2006] or a tree-style manner [Dung et al., 2009]. We use here the latter style, as follows:

Definition 2.2 *A deduction for* $\sigma \in \mathcal{L}$ *supported by* $S \subseteq \mathcal{L}$ *and* $R \subseteq \mathcal{R}$, *denoted* $S \vdash^{R} \sigma$ *(or simply* $S \vdash \sigma$*), is a (finite) tree with*

- *nodes labelled by sentences in* $\mathcal{L}$ *or by* τ,[1]
- *the root labelled by* σ,
- *leaves either* τ *or sentences in* S,
- *non-leaves* σ' *with, as children, the elements of the body of some rule in* $\mathcal{R}$ *with head* σ', *and* R *the set of all such rules.*

Example 2.3 *Consider an ABA framework* $\langle \mathcal{L}, \mathcal{R}, \mathcal{A}, \overline{} \rangle$ *with* $\mathcal{R} = \{x \leftarrow c, z \leftarrow y, b, \; y \leftarrow \; , \; a \leftarrow b\}$ *and* $\mathcal{A} = \{a, b, c\}$.[2] *The following are examples of deductions, denoted as indicated (first with the supporting rules and then without):*

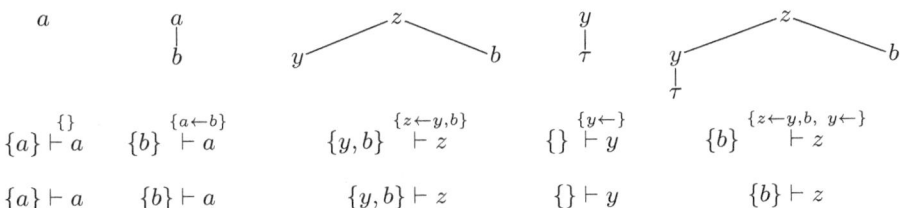

Note that deductions for assumptions have a non-empty rule support only if they occur as head of rules, and sentences occurring as head of rules with an empty body are always supported by an empty set of sentences (and a singleton set of rules).

Semantics of ABA frameworks are defined in terms of sets of assumptions meeting desirable requirements. One such requirement is being *closed* under deduction, defined as follows:

Definition 2.4 *The* closure *of a set of sentences* $S \subseteq \mathcal{L}$ *is*

$$Cl(S) = \{\sigma \in \mathcal{A} \mid \exists \, S' \vdash^{R} \sigma, \; S' \subseteq S, \; R \subseteq \mathcal{R}\}.$$

A set of assumptions $A \subseteq \mathcal{A}$ *is* closed *iff* $A = Cl(A)$.

In Example 2.3, $\{a, b\}$ is closed whereas $\{b\}$ is not.

---

[1] $\tau \notin \mathcal{L}$ represents "true" and stands for the empty body of rules. In other words, each rule with empty body can be interpreted as a rule with body τ for the purpose of presenting deductions as trees.

[2] Throughout, we often omit to specify the language $\mathcal{L}$, as it is implicit from the rules and assumptions. Also, if the contraries of assumptions are not explicitly defined, then they are assumed to be different from each other and any other explicitly mentioned sentences.

Note that, in some ABA frameworks, sets of assumptions are guaranteed to be closed. These ABA frameworks are referred to as *flat* and, as we will see later, exhibit additional properties than generic ABA frameworks.

Definition 2.5 *An ABA framework $\langle \mathcal{L}, \mathcal{R}, \mathcal{A}, \overline{} \rangle$ is flat iff for every $A \subseteq \mathcal{A}$, A is closed.*

The ABA framework in Example 2.3 is not flat, whereas the following is an example of a flat ABA framework.

Example 2.6 *An ABA framework with $\mathcal{R} = \{r \leftarrow b, c, \quad q \leftarrow \quad, \quad p \leftarrow q, a\}$ and $\mathcal{A} = \{a, b, c\}$ is guaranteed to be flat. Here, as in all flat ABA frameworks, deductions for assumptions can only be supported by an empty set of rules, e.g. there is a single deduction for a:*

$$\{a\} \vdash^{\{\}} a.$$

It is easy to see that if no assumption is the head of a rule, then an ABA framework is flat [Dung et al., 2006]. However, an ABA framework can be flat even if some assumptions are heads of rules. For instance, in an ABA framework with $\mathcal{R} = \{a \leftarrow x\}$ and $\mathcal{A} = \{a\}$, the assumption a appears as the head of the rule $a \leftarrow x$, but since x is not deducible from any set of assumptions, all sets of assumptions in this ABA framework are guaranteed to be closed, and so the framework is flat. Note, however, that "dummy" rules such as $a \leftarrow x$ above, whose body is not deducible from any set of assumptions, could without loss of generality be deleted from ABA frameworks, as they generate no conclusions. On the other hand, the ABA framework in Example 2.3 has no such "dummy" rules and is not flat (as, indeed, $\{b\}$ is not closed).

The remaining desirable requirements met by sets of assumptions, as semantics for ABA frameworks, are given in terms of a notion of *attack* between sets of assumptions, defined as follows:

Definition 2.7 *A set of assumptions $A \subseteq \mathcal{A}$ attacks a set of assumptions $B \subseteq \mathcal{A}$ iff there are $A' \subseteq A$ and $b \in B$ such that $A' \vdash \overline{b}$.*

The following definitions of semantics for ABA are adapted from [Bondarenko et al., 1993; Bondarenko et al., 1997; Dung et al., 2007].

Definition 2.8 *A set of assumptions (or* extension*) is* conflict-free *iff it does not attack itself. A set of assumptions/extension $A \subseteq \mathcal{A}$ is*

- admissible *iff it is closed, conflict-free and, for every $B \subseteq \mathcal{A}$, if B is closed and attacks A, then A attacks B;*

- preferred *iff it is maximally (w.r.t. $\subseteq$) admissible;*

- complete *iff it is admissible and contains all assumptions it defends, where A defends a iff for every $B \subseteq \mathcal{A}$, if B is closed and attacks $\{a\}$, then A attacks B;*
- stable *iff it is closed, conflict-free and, for every $a \notin A$, A attacks $\{a\}$;*
- well-founded *iff it is the intersection of all complete extensions;*
- ideal *iff A is maximal (w.r.t. $\subseteq$) such that*

  (i) it is admissible, and

  (ii) for all preferred extensions $P \subseteq \mathcal{A}$, $A \subseteq P$.

Note that ideal sets of assumptions were originally defined, in [Dung et al., 2007], in the context of flat ABA frameworks only. The original definition naturally generalises to general, possibly non-flat, ABA frameworks as given above. Note also that, in the case of flat ABA frameworks, the term *grounded* is conventionally used instead of *well-founded* (e.g. in [Dung et al., 2007]): we will adopt this convention too later in the chapter.

Example 2.9 *Consider a non-flat ABA framework with rules $\mathcal{R} = \{x \leftarrow c, z \leftarrow b, a \leftarrow b\}$, $\mathcal{A} = \{a, b, c\}$ and $\overline{a} = x$, $\overline{b} = y$, $\overline{c} = z$. Then, $\{c\}$ is closed and conflict-free. It is attacked by $\{b\}$, which cannot be counter-attacked but is not closed and thus can be disregarded; it is also attacked by the closed $\{a, b\}$, which is counter-attacked by $\{c\}$. Thus, $\{c\}$ is admissible, as well as preferred and complete. $\{\}$ is also admissible and complete, and thus well-founded, but not preferred. $\{b\}$ is not admissible, because it is not closed. Moreover, the closed $\{a, b\}$ is admissible because it is conflict-free and $\{b\}$ counter-attacks the closed $\{c\}$ which attacks $\{a, b\}$. Finally, $\{a, b\}$ is preferred and complete, and thus $\{\}$ is ideal.*

Note that a set of assumptions/extension can be seen as characterising the set of all sentences in the given ABA framework for which deductions exist supported by (subsets of) the extension:

Definition 2.10 *The* consequences *of an extension $A \subseteq \mathcal{A}$ is*

$$Cn(A) = \{\sigma \in \mathcal{L} \mid \exists\, A' \vdash \sigma,\ A' \subseteq A\}.$$

As an illustration, in Example 2.9, $Cn(\{c\}) = \{c, x\}$.

In the remainder of the chapter, when a sentence belongs to the consequences of an admissible / preferred / stable / complete / well-founded / ideal extension we will say that it is admissible / preferred / stable / complete / well-founded / ideal, respectively. Thus, in Example 2.9, x is admissible.

The following properties on relationships amongst extensions according to various semantics hold for generic (possibly non-flat) ABA frameworks:

Theorem 2.11 *Let $A \subseteq \mathcal{A}$ be a set of assumptions.*

(i) *If A is stable, then it is preferred.*

(ii) *If A is admissible, then there is some $P \subseteq \mathcal{A}$ such that P is preferred and $A \subseteq P$.*

(iii) *If A is stable, then it is complete.*

(iv) *If A is ideal and $S \subseteq \mathcal{A}$ is the intersection of all preferred extensions, then $A \subseteq S$.*

(v) *If A is the intersection of all preferred extensions and admissible, then it is ideal.*

(vi) *If A is ideal, then for each set of assumptions B attacking A there exists no admissible set of assumptions $B' \subseteq \mathcal{A}$ such that $B' \supseteq B$.*

(vii) *If A is well-founded, then for every $S \subseteq \mathcal{A}$, if S is stable, then $A \subseteq S$.*

Proof.

(i) See proof of Theorem 4.6 in [Bondarenko et al., 1997].

(ii) See proof of Theorem 4.9 in [Bondarenko et al., 1997].

(iii) See proof of Theorem 5.5 in [Bondarenko et al., 1997].

(iv) By definition, $A \subseteq P$ for every preferred $P \subseteq \mathcal{A}$, so $A \subseteq S$.

(v) The intersection of all preferred extensions A is a $\subseteq$-maximal set of assumptions that is contained in every preferred extension, so if A is in addition admissible, then it is by definition ideal.

(vi) Assume A is ideal and let B attack A. By contradiction, assume there exists an admissible $B' \supseteq B$. Then, by (ii) above, there is a preferred set of assumptions P such that $B' \subseteq P$. By definition of ideal extension, $A \subseteq P$, hence P is not conflict-free, contradicting its admissibility.

(vii) By definition, the well-founded extension is contained in every complete extension. Also, by (iii) above, every stable extension is complete. Therefore, the well-founded extension must be contained in every stable extension. ∎

Note that item (v) was given and proven in [Dung et al., 2007] (as Theorem 2.1(iv)) in the case of abstract argumentation frameworks [Dung, 1995].

The following properties on existence of extensions according to various semantics hold for generic (possibly non-flat) ABA frameworks.

Theorem 2.12

(i) *If there is an admissible extension, then there is at least one preferred extension.*

(ii) *If the empty set of assumptions is closed, then there is at least one preferred extension.*

(iii) *If the empty set of assumptions is closed, then there exists an ideal extension.*

Proof.

(i) Directly from Theorem 2.11(ii) (see also comments after the proof of Theorem 4.9 in [Bondarenko et al., 1997]).

(ii) Directly from (i) above, as the empty set, if closed, is necessarily admissible (see also comments after the proof of Theorem 4.9 in [Bondarenko et al., 1997]).

(iii) If {} is closed, then it is admissible. So by (i) above, there is a preferred extension. Hence, the intersection S of preferred extensions exists too. Given that {} is admissible, there must then be a $\subseteq$-maximal admissible subset of S, i.e. an ideal extension. ∎

For a simple example of a (necessarily non-flat) ABA framework in which the empty set is not closed, consider $\langle \mathcal{L}, \mathcal{R}, \mathcal{A}, \overline{} \rangle$ with $\mathcal{R} = \{a \leftarrow , \quad x \leftarrow a\}$, $\mathcal{A} = \{a\}$ and $\overline{a} = x$: here, $\{\} \vdash a$, so that $\{\}$ is not closed; note also that no set is admissible, because any admissible set needs to be a closed superset of the empty set, and since there are deductions $\{\} \vdash a$ as well as $\{\} \vdash x$, where x is the contrary of a, no closed superset of $\{\}$ is conflict-free.

Flat ABA frameworks fulfil the following property, often referred to as the *Fundamental Lemma* (see e.g. [Dung, 1995; Bondarenko et al., 1997]):

Theorem 2.13 *Let $\langle \mathcal{L}, \mathcal{R}, \mathcal{A}, \overline{} \rangle$ be a flat ABA framework, and let $A \subseteq \mathcal{A}$ be an admissible set of assumptions that defends assumptions $a, a' \in \mathcal{A}$. Then $A \cup \{a\}$ is admissible and defends a'.*

Proof. See proof of Theorem 5.7 in [Bondarenko et al., 1997]. ∎

Note that non-flat ABA frameworks need not in general fulfil the Fundamental Lemma: consider $\langle \mathcal{L}, \mathcal{R}, \mathcal{A}, \overline{} \rangle$ with $\mathcal{R} = \{c \leftarrow a, b\}$, $\mathcal{A} = \{a, b, c\}$ and $\overline{a} = x$, $\overline{b} = y$, $\overline{c} = z$; it is non-flat, because $\{a, b\} \vdash c$; observe that both $\{a\}$ and $\{b\}$ are closed and unattacked, so, for instance, $\{a\}$ is admissible and defends b; however, $\{a, b\}$ is not closed, and so not admissible.

Flat ABA frameworks also fulfil additional properties concerning relationships between semantics, as follows:

Theorem 2.14 Let $\langle \mathcal{L}, \mathcal{R}, \mathcal{A}, \bar{} \rangle$ be a flat ABA framework, and let $A \subseteq \mathcal{A}$ be a set of assumptions.

(i) If A is preferred, then it is complete.

(ii) If A is grounded, then it is minimally (w.r.t. $\subseteq$) complete.

(iii) If A is grounded, then for every $P \subseteq \mathcal{A}$, if P is preferred, then $A \subseteq P$.

(iv) If A is ideal, then it is complete.

(v) If A is ideal and $G \subseteq \mathcal{A}$ is grounded, then $A \supseteq G$.

(vi) If A is maximally (w.r.t. $\subseteq$) complete, then it is preferred.

(vii) If A is admissible, then it is ideal iff for each set of assumptions B attacking A there exists no admissible set of assumptions $B' \subseteq \mathcal{A}$ such that $B' \supseteq B$.

Proof.

(i) Directly from Theorem 5.7 in [Bondarenko et al., 1997], see Corollary 5.8 in [Bondarenko et al., 1997].

(ii) See proof of Theorem 6.2 in [Bondarenko et al., 1997].

(iii) See proof of Theorem 6.4 in [Bondarenko et al., 1997].

(iv) Let I be ideal and suppose it defends $a \in \mathcal{A}$. Due to flatness, $I \cup \{a\}$ is admissible, and hence contained in every preferred extension. So $a \in I$ by $\subseteq$-maximality of I.

(v) Directly from (iv) and (ii) above.

(vi) If A was $\subseteq$-maximally complete but not preferred, then, by Theorem 1(ii), there would be some preferred yet not complete P such that $A \subsetneq P \subseteq \mathcal{A}$, contrary to (i) above.

(vii) See Theorem 3.3 in [Dung et al., 2007].

∎

Note that items (iv) and (v) were given and proven in [Dung et al., 2007] (as items (ii) and (iii) respectively in Theorem 2.1), in the case of abstract argumentation frameworks. Also, (vii) was given and proven as Lemma 4(a) in [Dunne, 2009].

The following examples show that the properties in Theorem 2.14 may not hold, in general, for non-flat ABA frameworks.

Example 2.15 *Consider an ABA framework $\langle \mathcal{L}, \mathcal{R}, \mathcal{A}, \overline{} \rangle$ with $\mathcal{R} = \{d \leftarrow c\}$, $\mathcal{A} = \{a, b, c, d\}$ and $\overline{a} = p$, $\overline{b} = a$, $\overline{c} = b$, $\overline{d} = d$. Then $\{a\}$ is preferred and ideal, but not complete, as it defends c. (Cf. Theorem 2.14(i), (iv).) Note that $\{a, c\}$ is not admissible, as it is not closed whereas $\{a, c, d\}$ is closed, but not admissible as not conflict-free.*

In this example, there is no complete extension and thus no well-founded extension, and thus the ideal extension is not a superset of the well-founded extension.

Example 2.16 *Consider an ABA framework $\langle \mathcal{L}, \mathcal{R}, \mathcal{A}, \overline{} \rangle$ with $\mathcal{R} = \{p \leftarrow a, p \leftarrow b, c \leftarrow \}$, $\mathcal{A} = \{a, b, c, d\}$ and $\overline{a} = b$, $\overline{b} = a$, $\overline{c} = d$, $\overline{d} = p$. Here, the complete extensions are $\{a, c\}$ and $\{b, c\}$, and thus $\{c\}$ is well-founded, but it is not (minimally) complete, as it does not defend itself against (the attacking) $\{d\}$. (Cf. Theorem 2.14(ii).) This also shows that even if there is an admissible extension, there need not be an ideal extension.*

Example 2.17 *Consider an ABA framework $\langle \mathcal{L}, \mathcal{R}, \mathcal{A}, \overline{} \rangle$ with $\mathcal{R} = \{d \leftarrow c, f \leftarrow e, p \leftarrow d, p \leftarrow e\}$, $\mathcal{A} = \{a, b, c, d, e, f\}$ and $\overline{a} = f$, $\overline{b} = a$, $\overline{c} = b$, $\overline{d} = p$, $\overline{e} = q$, $\overline{f} = a$. Then $\{e, f, b\}$ is the only complete extension, and thus the well-founded extension. Moreover $\{a\}$ and $\{e, f, b\}$ are (the only) preferred extensions, and $\{e, f, b\} \not\subseteq \{a\}$. Therefore, there exists a preferred extension that does not contain the well-founded extension. (Cf. Theorem 2.14(iii).)*

Example 2.18 *Consider an ABA framework $\langle \mathcal{L}, \mathcal{R}, \mathcal{A}, \overline{} \rangle$ with $\mathcal{R} = \{q \leftarrow a, r \leftarrow b, c \leftarrow q, r, z \leftarrow a, z \leftarrow b, z \leftarrow c\}$, $\mathcal{A} = \{a, b, c\}$ and $\overline{a} = c$, $\overline{b} = c$, $\overline{c} = z$. Here, every $A \subseteq \mathcal{A}$ containing c is not conflict-free, so not admissible. Also, $\{a, b\}$ is not closed, so not admissible. However, $\{a\}$ is admissible, but not complete, as it defends b. Likewise $\{b\}$ is admissible, but not complete. Indeed, both $\{a\}$ and $\{b\}$ are preferred, yet not complete. Therefore, $\{\}$ is $\subseteq$-maximally complete, yet not preferred. (Cf. Theorem 2.14(vi).)*

Example 2.19 *Consider an ABA framework $\langle \mathcal{L}, \mathcal{R}, \mathcal{A}, \overline{} \rangle$ with $\mathcal{R} = \{z \leftarrow c, c \leftarrow a, b\}$, $\mathcal{A} = \{a, b, c\}$ and $\overline{a} = x$, $\overline{b} = y$, $\overline{c} = z$. Then $\{a\}$ is admissible (and preferred) and unattacked. Observe that $\{a, x\}$ is not closed, and $\{a, c\}$, $\{x, c\}$, $\{a, x, c\}$ are not conflict-free. So $\{b\}$ is preferred, yet $\{a\} \not\subseteq \{b\}$, so that A is not ideal. (Cf. Theorem 2.14(vii).)*

Flat ABA frameworks fulfil additional properties concerning existence of extensions w.r.t. various semantics, as follows:

Theorem 2.20 *Let $\langle \mathcal{L}, \mathcal{R}, \mathcal{A}, \overline{} \rangle$ be a flat ABA framework.*

(i) There is at least one preferred extension.

(ii) There is a unique ideal extension.

(iii) There is at least one complete extension.

(iv) There is a unique grounded extension and it is the least fixed point of Def, where, for $A \subseteq \mathcal{A}$, $Def(A) = \{a \in \mathcal{A} \mid A \text{ defends } a\}$.

Proof.

(i) Directly from the second item of Theorem 2.12, as, in the case of flat ABA frameworks, the empty set (like any other set of assumptions) is guaranteed to be closed (see also [Bondarenko et al., 1997]).

(ii) Follows from Theorem 2.12(iii).

(iii) Directly from (i) above and Theorem 2.14(i).

(iv) See proof of Theorem 6.2 in [Bondarenko et al., 1997].

∎

Note that (ii) above was given and proven in [Dung et al., 2007] in the case of abstract argumentation frameworks.

The following examples show that the properties in Theorem 2.20 may not hold, in general, for non-flat ABA frameworks.

Example 2.21 Consider an ABA framework $\langle \mathcal{L}, \mathcal{R}, \mathcal{A}, \overline{} \rangle$ with $\mathcal{R} = \{a \leftarrow \}$, $\mathcal{A} = \{a\}$ and $\overline{a} = a$. Here, $\{\}$ is not closed and $\{a\}$ is not conflict-free. Thus, no set of assumptions is admissible. Hence, there is no preferred, complete, ideal or well-founded extension.

Finally, consider an example which shows that, differently from flat ABA frameworks, in general, an ideal extension need not be unique for non-flat ABA frameworks.

Example 2.22 Consider an ABA framework $\langle \mathcal{L}, \mathcal{R}, \mathcal{A}, \overline{} \rangle$ with assumptions $\mathcal{A} = \{a, a', b, b', c, d\}$, rules $\mathcal{R} = \{c \leftarrow a, a', \ \overline{c} \leftarrow d, \ \overline{d} \leftarrow a, b, \ \overline{d} \leftarrow a', b', \ a' \leftarrow a, b, c, \ a \leftarrow a', b, c, \ a' \leftarrow a, b', c, \ a \leftarrow a', b', c\}$, and contraries $\overline{b} = b'$, $\overline{b'} = b$.[3] Here, $\{\}$ is closed, so admissible. There are two preferred extensions: $\{a, a', b, c\}$ and $\{a, a', b', c\}$. Their intersection $\{a, a', c\}$ is not admissible, because it cannot defend against (the closed attacking) $\{d\}$. Likewise, $\{a, c\}$ and $\{a', c\}$ are not admissible. Also, $\{a, a'\}$ is not closed. However, both $\{a\}$ and $\{a'\}$ are admissible, and hence ideal extensions.

Note that additional properties hold for (generic and/or flat) ABA frameworks of restricted kinds, for instance, where "cycles" are not allowed (e.g. *stratified* and *order-consistent* ABA frameworks, see [Bondarenko et al., 1997] for

---

[3] For readability, with an abuse of notation we may sometimes assume that the contraries (in this case, $\overline{c}$ and $\overline{d}$) of assumptions (in this case, c and d) are actually symbols in the language (different from other explicitly mentioned sentences).

details). Moreover, additional properties hold for other special classes of ABA frameworks, in addition to flat ABA frameworks, namely *normal* [Bondarenko et al., 1997] and *simple* [Dimopoulos et al., 2002] ABA frameworks (see [Bondarenko et al., 1997; Dimopoulos et al., 2002] for details).

3 ABA and non-monotonic reasoning

In this section we illustrate two instances of ABA for Non-Monotonic Reasoning, namely Autoepistemic Logic (AEL) [Moore, 1985] and Logic Programming (LP). The formal definitions of these instances, as well as correspondence results between the semantics for ABA as given in Definition 2.8 and their original semantics, can be found in [Bondarenko et al., 1997; Schulz and Toni, 2015]

For illustration, as well as a running example throughout the chapter, we will use the following extract from the Nationwide[4] building society's 2016 policy for UK/EU Breakdown Assistance:

COVERED FOR: UK/EU Breakdown Assistance for account holder(s) in any private car they are travelling in

NOT COVERED FOR: private cars not registered to the account holder(s) unless the account holder(s) are in the vehicle at the time of the breakdown

We consider a person, Mary (denoted simply as m), who is an account holder travelling in a friend's car (denoted as c) when the car breaks down somewhere in the EU. In the remainder of this section we show how the application of the policy above to Mary's case can be represented in the AEL and LP instances of ABA, as given in general in [Bondarenko et al., 1997]. In giving the concrete instantiations below we will use the following abbreviations: ah stands for "account holder"; tr stands for "travelling"; pr stands for "private vehicle"; cov stands for "covered"; reg stands for "registered"; cov' stands for "there is an exception to being not covered".

---

[4] www.nationwide.co.uk

3.1 Breakdown Assistance policy in the AEL instance of ABA

The application of the Breakdown Assistance policy to Mary's case can be represented in the AEL instance of ABA as follows:

$\mathcal{L} =$ a modal language containing a modal operator L
(where $L\sigma$ stands for "σ is believed") as well as atoms
$ah(m), tr(m,c), pr(c), cov(m,c), reg(c,m), cov'(m,c), in(m,c)$

$\mathcal{R} =$ a complete set of inference rules of classical logic for $\mathcal{L}$ together with the following inference rules (all with an empty body):

$$\frac{}{ah(m) \wedge tr(m,c) \wedge pr(c) \wedge \neg L \neg cov(m,c) \rightarrow cov(m,c)}$$

$$\frac{}{\neg reg(c,m) \wedge \neg Lcov'(m,c) \rightarrow \neg cov(m,c)}$$

$$\frac{}{in(m,c) \rightarrow cov'(m,c)} \quad \frac{}{ah(m)} \quad \frac{}{tr(m,c)}$$

$$\frac{}{pr(c)} \quad \frac{}{\neg reg(c,m)} \quad \frac{}{in(m,c)}$$

$\mathcal{A} = \{L\sigma, \neg L\sigma \mid \sigma \in \mathcal{L}\}$
$\overline{L\sigma} = \neg L\sigma$ and $\overline{\neg L\sigma} = \sigma$ for any $\sigma \in \mathcal{L}$

Note that, in this ABA framework, $\mathcal{R}$ includes domain-independent rules, e.g.

$$\frac{\sigma_1 \wedge \sigma_2}{\sigma_1} \text{ for any } \sigma_1, \sigma_2 \in \mathcal{L},$$

as well as domain-specific rules, e.g.

$$\frac{}{in(m,c)}.$$

Note also that this ABA framework (as well as any other AEL instance of ABA) is not flat [Bondarenko et al., 1997], as, for instance, the set of assumptions $\{Lcov(m), \neg Lcov(m)\}$ is not closed, because it is classically inconsistent. Nonetheless, for this instance, the empty set of assumptions is closed.

Given this representation in ABA, the problem of determining whether Mary should be covered or not amounts to determining whether $cov(m)$ is stable (following the conventional AEL approach of determining whether $cov(m)$ belongs to a consistent stable expansion [Moore, 1985] of the theory consisting of the heads of the domain-specific part of $\mathcal{R}$), or preferred, or well-founded etc. (by adopting any of the other ABA semantics). In this particular example, all ABA semantics agree that Mary should be covered, by assuming $\neg L \neg cov(m,c)$, in agreement with the original semantics of AEL, as predicted by the general correspondence Theorem 3.18 in [Bondarenko et al., 1997] and the fact that, in this example, all ABA semantics agree with the semantics of stable extensions. As an illustration, $\{\neg L \neg cov(m,c)\}$ is admissible, since it is conflict-free, closed and the (closed) set of assumptions $\{\neg Lcov'(m,c)\}$ attacking it, as well as all its (closed) supersets, are attacked by the (closed) empty set of assumptions.

Note that, in the AEL instance of ABA, beliefs, of the form $L\sigma$ or $\neg L\sigma$, are assumptions that may occur as heads of rules. For example, the earlier AEL instance of ABA may be extended so that $\mathcal{R}$ includes also

$$\overline{L\,ah(m)}$$

to represent that Mary is believed to be an account holder. This kind of knowledge cannot be directly represented in flat ABA.

3.2 Breakdown Assistance policy in the LP instance of ABA

The application of the Breakdown Assistance policy to Mary's case can be represented in the LP instance of ABA as follows:

$$\begin{aligned}
\mathcal{R} = \quad & \{cov(m,c) \leftarrow ah(m), tr(m,c), pr(c), not\ \neg cov(m,c), \\
& \neg cov(m,c) \leftarrow \neg reg(c,m), not\ cov'(m,c), \\
& cov'(m,c) \leftarrow in(m,c), \\
& ah(m) \leftarrow,\ tr(m,c) \leftarrow,\ pr(c) \leftarrow,\ \neg reg(c,m) \leftarrow,\ in(m,c) \leftarrow\} \\
\mathcal{A} = \quad & \{not\ p(t_1,t_2),\ not\ q(t) \mid p \in \{cov, tr, \neg cov, \neg reg, cov', in\}, \\
& \qquad\qquad q \in \{ah, pr\}, \\
& \qquad\qquad t_1, t_2, t \in \{m, c\}\} \\
\mathcal{L} = \quad & \mathcal{A} \cup \{x \mid not\ x \in \mathcal{A}\} \\
\overline{not\ x} = \quad & x\ \text{for any}\ not\ x \in \mathcal{A}
\end{aligned}$$

So, $\mathcal{L}$ is the Herbrand base of (the logic program) $\mathcal{R}$ together with all negation as failure (NAF) literals that can be built from this Herbrand base, and $\mathcal{A}$ is the set of all these NAF literals. Note that in this illustration we treat $\neg cov$ and $\neg reg$ as predicate symbols.

Given this representation in ABA, the problem of determining whether Mary should be covered or not amounts to determining, for instance, whether $cov(m,c)$ is stable (following the stable model semantics [Gelfond and Lifschitz, 1988] for $\mathcal{R}$, seen as a logic program, by virtue of the correspondence Theorem 3.13 in [Bondarenko et al., 1997]), or admissible/preferred (following the preferred extension semantics [Dung, 1991] for $\mathcal{R}$, by virtue of the correspondence Theorem 4.5 in [Bondarenko et al., 1997]), or grounded (following the well-founded model semantics [Gelder et al., 1991] for $\mathcal{R}$, by virtue of the correspondence Theorem 3.13 in [Bondarenko et al., 1997]), or ideal (following the scenario semantics [Alferes et al., 1993] for $\mathcal{R}$ In this particular example, all ABA semantics agree that Mary should be covered, by assuming $not\ cov(m,c)$. As an illustration, $\{not\ cov(m,c)\}$ is admissible, since it is conflict-free and the set of assumptions $\{not\ cov'(m,c)\}$ attacking it, as well as all its supersets, are attacked by the empty set of assumptions.

4 ABA versus abstract argumentation

In this section we focus on the relationship between flat ABA frameworks and Abstract Argumentation (AA) frameworks [Dung, 1995]. In particular, flat ABA is an instance of AA, under all semantics considered in this chapter and, conversely, AA is an instance of (flat) ABA.

Flat ABA frameworks are instances of AA frameworks where arguments are deductions supported by sets of assumptions and attacks are defined by appropriately lifting the notion of attack between sets of assumptions to a notion of attack between arguments [Dung et al., 2007; Toni, 2012].

Definition 4.1 Let $\mathcal{ABA} = \langle \mathcal{L}, \mathcal{R}, \mathcal{A}, ^- \rangle$ be a flat ABA framework.

- An argument for $\sigma \in \mathcal{L}$ supported by $A \subseteq \mathcal{A}$ and $R \subseteq \mathcal{R}$, denoted $A \vdash_{arg}^{R} \sigma$ (or simply $A \vdash_{arg} \sigma$), is such that there is a deduction $A \vdash^{R} \sigma$.

- An argument $A \vdash_{arg} \sigma$ attacks an argument $B \vdash_{arg} \pi$ iff there is $b \in B$ such that $\sigma = \overline{b}$.

Then $\mathcal{AA}(\mathcal{ABA}) = (Args, attack)$ is the corresponding AA framework of $\mathcal{ABA}$ with $Args$ the set of all arguments (as in the first bullet) and $attack$ the set of all pairs (a, b) such that $\mathsf{a}, \mathsf{b} \in Args$ and a attacks b (as in the second bullet).

Note that $Args$ contains an argument for every assumption in $\mathcal{A}$ as illustrated by the following example.

Example 4.2 Consider an ABA framework $\mathcal{ABA}$ with rules and assumptions as in Example 2.6 and $\overline{a} = r$, $\overline{b} = q$, $\overline{c} = p$. Then $\mathcal{AA}(\mathcal{ABA})$ is $(Args, attack)$ with $Args = \{\mathsf{a}, \mathsf{b}, \mathsf{c}, \mathsf{p}, \mathsf{q}, \mathsf{r}\}$ where $\mathsf{a} = \{a\} \vdash_{arg} a$, $\mathsf{b} = \{b\} \vdash_{arg} b$, $\mathsf{c} = \{c\} \vdash_{arg} c$, $\mathsf{p} = \{a\} \vdash_{arg} p$, $\mathsf{q} = \{\} \vdash_{arg} q$, $\mathsf{r} = \{b, c\} \vdash_{arg} r$, and $attack = \{(\mathsf{p}, \mathsf{c}), (\mathsf{p}, \mathsf{r}), (\mathsf{q}, \mathsf{b}), (\mathsf{q}, \mathsf{r}), (\mathsf{r}, \mathsf{a}), (\mathsf{r}, \mathsf{p})\}$.

The semantics of an AA framework corresponding to a flat ABA framework can be determined using the standard AA semantics [Dung, 1995; Dung et al., 2007]. For all ABA semantics considered in this chapter, the semantics of a flat ABA framework corresponds to the semantics of its corresponding AA framework, as follows:

Theorem 4.3 Let $\mathcal{ABA} = \langle \mathcal{L}, \mathcal{R}, \mathcal{A}, ^- \rangle$ be a flat ABA framework and let $\mathcal{AA}(\mathcal{ABA})$ be its corresponding AA framework.

(i) If a set of assumptions $A \subseteq \mathcal{A}$ is admissible / preferred / stable / complete / grounded / ideal in $\mathcal{ABA}$, then the union of all arguments supported by any $A' \subseteq A$ is admissible / preferred / stable / complete / grounded / ideal, respectively, in $\mathcal{AA}(\mathcal{ABA})$.

(ii) *The union of all sets of assumptions supporting the arguments in an admissible / preferred / stable / complete / grounded / ideal set of arguments in $\mathcal{AA}(\mathcal{ABA})$ is admissible / preferred / stable / complete / grounded / ideal, respectively, in $\mathcal{ABA}$.*

Proof. See the proof of Theorem 2.2 in [Dung et al., 2007] for admissible, grounded and ideal extensions, the proof of Theorem 1 in [Toni, 2012]for stable extensions[5], and the proof of Theorem 6.1 and 6.3 [Caminada et al., 2015] for complete and preferred extensions respectively. ∎

Note that for the preferred, stable, complete, grounded, and ideal semantics the correspondence between the extensions of a flat ABA framework and the extensions of the corresponding AA framework is one-to-one. For the admissible semantics, instead, the correspondence is one-to-many, i.e. the union of all sets of assumptions supporting the arguments in an admissible extension may be the same for various admissible extensions of the corresponding AA framework, as illustrated in the following example.

Example 4.4 *The ABA framework from Example 4.2 has two admissible extensions: {} and {a}. In contrast, the corresponding AA framework has five admissible extensions: $A_1 = \{\}$, $A_2 = \{q\}$, $A_3 = \{p\}$, $A_4 = \{p,q\}$, $A_5 = \{p,a\}$, $A_6 = \{q,a\}$, $A_7 = \{p,q,a\}$. However, the union of all sets of assumptions supporting the arguments in A_1 and A_2 is {}, so both correspond to the first admissible extension of the ABA framework. Similarly, the union of all sets of assumptions supporting arguments in the other admissible extensions (A_3 − − A_7) of the AA framework is {a}, so they all correspond to the second admissible extension of the ABA framework.*

Theorem 4.3 shows that, under the semantics considered therein, flat ABA frameworks are an instance of AA frameworks and the semantics of ABA can alternatively be defined in terms of extensions as sets of arguments, as in [Dung et al., 2007], rather than in terms of extensions as sets of assumptions, as in [Bondarenko et al., 1993; Bondarenko et al., 1997]. This implies, for example, that existing machinery for computing extensions of AA frameworks can be used to compute extensions of ABA frameworks whose corresponding AA frameworks are finite. Conversely, as shown below, AA frameworks are an instance of flat ABA frameworks, that is any AA framework can be translated into a corresponding flat ABA framework such that their respective extensions correspond [Toni, 2012]. This implies, in particular, that existing machinery for determining whether sentences are admissible / preferred / complete / grounded / ideal in flat ABA (see Section 5) can be used to determine whether arguments in an AA framework belong to an admissible / preferred / complete / grounded / ideal extension. The ABA framework corresponding to an AA

---

[5]The proof of Theorem 1 in [Toni, 2012] actually considers a different notion of stable extension, but can naturally be modified to prove the result indicated here.

framework has the arguments in the AA framework as (the only) assumptions and appropriate notions of contraries of these assumptions and rules to encode the attacks between the arguments in the original AA framework, as follows:

Definition 4.5 *Let* $\mathcal{AA} = (Args, attack)$ *be an AA framework. The corresponding ABA framework of* $\mathcal{AA}$ *is* $\mathcal{ABA}(\mathcal{AA}) = \langle \mathcal{L}, \mathcal{R}, \mathcal{A}, \overline{} \rangle$ *with*

- $\mathcal{A} = Args$;

- $\mathcal{L} = \mathcal{A} \cup \{\mathsf{a}^c \mid \mathsf{a} \in \mathcal{A}\}$;

- *for all* $\mathsf{a} \in \mathcal{A}$: $\overline{\mathsf{a}} = \mathsf{a}^c$;

- $\mathcal{R} = \{\mathsf{b}^c \leftarrow \mathsf{a} \mid (\mathsf{a}, \mathsf{b}) \in attack\}$.

Note that clearly the corresponding ABA framework of any AA framework is flat since assumptions never occur in the head of a rule, by construction.

Since the set of arguments in a given AA framework coincides with the set of assumptions of the corresponding ABA framework, there is a straightforward one-to-one correspondence between all semantics of the AA and ABA framework.

Theorem 4.6 *Let* $\mathcal{AA} = (Args, attack)$ *be an AA framework and let* $\mathcal{ABA}(\mathcal{AA})$ *be its corresponding ABA framework.*

(i) *If* $\mathsf{A} \subseteq Args$ *is admissible / preferred / stable / complete / grounded / ideal in* $\mathcal{AA}$, *then* A *is admissible / preferred / stable / complete / grounded / ideal, respectively, in* $\mathcal{ABA}(\mathcal{AA})$.

(ii) *If* $A \subseteq \mathcal{A}$ *is admissible / preferred / stable / complete / grounded / ideal set of arguments in* $\mathcal{ABA}(\mathcal{AA})$, *then* A *is admissible / preferred / stable / complete / grounded / ideal, respectively, in* $\mathcal{AA}$.

Proof. See proof of Theorem 2 in [Toni, 2012] for admissible. As noted in [Toni, 2012], the proof for other semantics is similar. ∎

Example 4.7 *Consider the AA framework* $\mathcal{AA}$ *with* $Args = \{\mathsf{a}, \mathsf{b}, \mathsf{c}\}$ *and* $attack = \{(\mathsf{a}, \mathsf{b}), (\mathsf{b}, \mathsf{a}), (\mathsf{b}, \mathsf{c})\}$. *The corresponding ABA framework is* $\mathcal{ABA}(\mathcal{AA})$ *with* $\mathcal{A} = \{\mathsf{a}, \mathsf{b}, \mathsf{c}\}$, $\overline{\mathsf{a}} = \mathsf{a}^c$, $\overline{\mathsf{b}} = \mathsf{b}^c$, $\overline{\mathsf{c}} = \mathsf{c}^c$, *and* $\mathcal{R} = \{\mathsf{b}^c \leftarrow \mathsf{a}, \; \mathsf{a}^c \leftarrow \mathsf{b}, \; \mathsf{c}^c \leftarrow \mathsf{b}\}$. *The admissible extensions of* $\mathcal{AA}$ *are* $\{\}$, $\{\mathsf{a}\}$, $\{\mathsf{b}\}$, *and* $\{\mathsf{a}, \mathsf{c}\}$, *which are exactly the admissible extensions of* $\mathcal{ABA}(\mathcal{AA})$. *Correspondence as dictated by Theorem 4.6 hold for the other semantics considered therein too.*

5 Dispute trees, dispute derivations and ABA dialogues

In this section we overview the main existing computational machinery for flat ABA frameworks, allowing to determine whether sentences are admissible (and therefore preferred, by Theorem 2.11 (ii), and complete, by Theorem 2.11 (ii) and Theorem 2.14 (i)), grounded, or ideal.[6] This machinery is based on the computation of *dispute trees* (overviewed in Section 5.1), using *dispute derivations* (illustrated in Section 5.2) that, in particular, can be executed amongst agents to form *ABA dialogues* (illustrated in Section 5.3).

5.1 Dispute trees

Dispute trees [Dung et al., 2006; Dung et al., 2007] provide an abstraction of the problem of determining whether arguments in AA frameworks belong to an admissible / grounded / ideal extension. Since flat ABA frameworks correspond to special instances of AA frameworks (see Section 4), dispute trees can be used to determine whether sentences are admissible / grounded / ideal, respectively, as well as identifying assumptions in admissible / grounded / ideal extensions for ABA, respectively, supporting arguments for these sentences. Dispute trees can be defined abstractly for any abstract argumentation framework as follows:

Definition 5.1 *Let* $(Args, attack)$ *be any abstract argumentation framework. A* dispute tree *for* $a \in Args$ *is a tree* $\mathcal{T}$ *such that:*

(i) *every node of* $\mathcal{T}$ *is of the form* $[\mathtt{L\!:\!x}]$, *with* $\mathtt{L} \in \{\mathtt{P}, \mathtt{O}\}$, $\mathtt{x} \in Args$: *the node is* labelled *by argument* $\mathtt{x}$ *and assigned the* status *of either* proponent $(\mathtt{P})$ *or* opponent $(\mathtt{O})$;

(ii) *the root of* $\mathcal{T}$ *is a* $\mathtt{P}$ *node labelled by* $\mathtt{a}$;

(iii) *for every* $\mathtt{P}$ *node* n, *labelled by some* $\mathtt{b} \in Args$, *and for every* $\mathtt{c} \in Args$ *such that* $\mathtt{c}$ *attacks* $\mathtt{b}$, *there exists a child of* n, *which is an* $\mathtt{O}$ *node labelled by* $\mathtt{c}$;

(iv) *for every* $\mathtt{O}$ *node* n, *labelled by some* $\mathtt{b} \in Args$, *there exists exactly one child of* n *which is a* $\mathtt{P}$ *node labelled by some* $\mathtt{c} \in Args$ *such that* $\mathtt{c}$ *attacks* $\mathtt{b}$;

(v) *there are no other nodes in* $\mathcal{T}$ *except those given by 1–4.*

The defence set *of a dispute tree* $\mathcal{T}$, *denoted by* $\mathcal{D}(\mathcal{T})$, *is the set of all arguments labelling* $\mathtt{P}$ *nodes in* $\mathcal{T}$.

---

[6]In general, this machinery cannot be used to determine whether a sentence is stable, as this requires the computation of a full extension, as discussed in [Dung et al., 2002]. However, for restricted types of flat ABA frameworks whose preferred extensions are guaranteed to be stable, determining whether a sentence is admissible amounts to determining whether it is stable, too.

Example 5.2 *Given the AA framework with $Args = \{a, b, c, d, e, f, g\}$ and $attack = \{(a, b), (b, c), (d, e), (d, f), (e, d), (e, f), (f, g), (g, f)\}$, consider the trees in the figure below. The tree on the left is not a dispute tree since an opponent node is a leaf node, thus violating condition (iv) in Definition 5.1. In contrast, the trees in the middle and on the right satisfy all conditions and are thus dispute trees for c and d, respectively.*

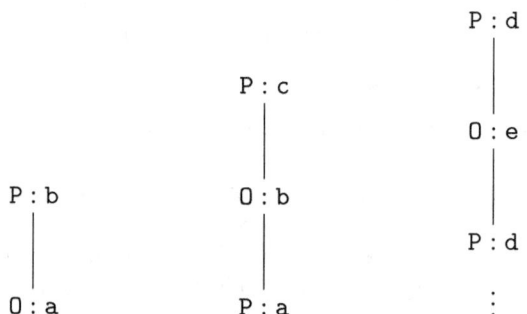

Figure 1. Only the middle and right of the three trees are dispute trees.

In order to help determine membership of arguments in admissible / grounded / ideal extensions of AA frameworks, dispute trees need to fulfil special requirements, as follows:

Definition 5.3 *Let $(Args, attack)$ be any abstract argumentation framework. A dispute tree $\mathcal{T}$ (for some argument in $Args$) is*

- admissible *iff no argument in $\mathcal{T}$ labels both P and O nodes;*
- grounded *iff it is finite;*
- ideal *iff for no argument a in $\mathcal{T}$ labelling an O node there exists an admissible dispute tree for a.*

Example 5.4 *Consider again the AA framework from Example 5.2. The dispute tree shown in the middle of Figure 1 is admissible since no argument labels both a proponent and an opponent node, as well as grounded since it is finite. Furthermore, the dispute tree is ideal since its only opponent node is labelled with b and there are no dispute trees for b, and thus there are no admissible dispute trees for b. The dispute tree for d on the right of Figure 1 is admissible, but not grounded since it is infinite. It is furthermore not ideal since there is an admissible dispute tree for e (obtained by exchanging d and e in the dispute tree for d on the right of Figure 1).*

The left of Figure 2 gives an example of a dispute tree which is ideal but not grounded. The opponent nodes of this dispute tree are all labelled by argument f.

Since the only dispute tree for f *is the one displayed in the middle of Figure 2, which is not an admissible dispute tree since argument* d *(as well as* e*) labels both an opponent and a proponent node, the dispute tree for* g *on the left of Figure 2 is ideal. Note that there are other admissible dispute trees for* g *which are not ideal. For example the one on the right of Figure 2 is not ideal since there exists an admissible dispute tree for* e.

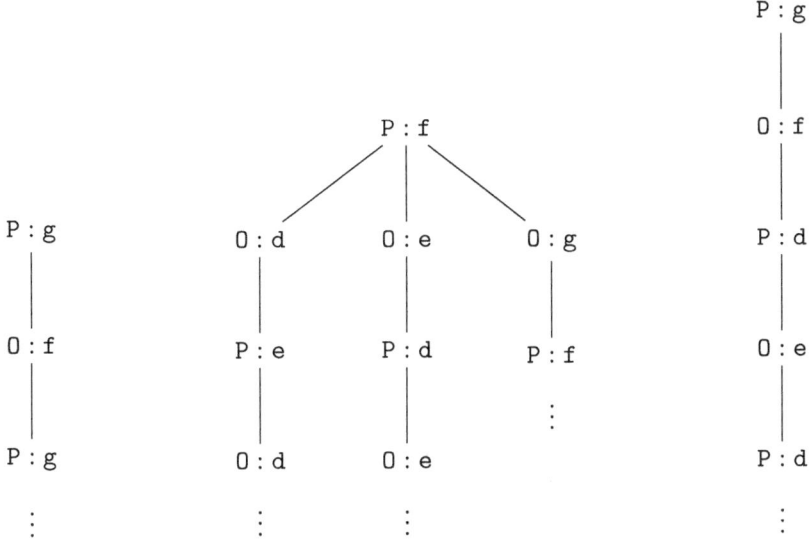

Figure 2. Three dispute trees constructed from the AA framework in Example 5.2.

Theorem 5.5 *Let* $(Args, attack)$ *be any abstract argumentation framework.*

(i) *If* $\mathcal{T}$ *is an admissible dispute tree for an argument* a *then the defence set of* $\mathcal{T}$ *is admissible.*

If a $\in$ A *for some admissible set of arguments* A $\subseteq Args$ *then there exists an admissible dispute tree for* a *with defence set* A' *such that* A' $\subseteq$ A *and* A' *is admissible.*

(ii) *If* $\mathcal{T}$ *is an ideal dispute tree for an argument* a *then the defence set* A *of* $\mathcal{T}$ *is such that* A *is admissible and* A $\subseteq$ I *with* I *the ideal extension of* $(Args, attack)$.

If a $\in$ I *with* I *the ideal extension of* $(Args, attack)$, *then there exists an ideal dispute tree for* a *with defence set* A *and* A $\subseteq$ I.

(iii) *If* $\mathcal{T}$ *is a grounded dispute tree for an argument* a *then the defence set* A *of* $\mathcal{T}$ *is such that* A *is admissible and* A $\subseteq$ G *with* G *the grounded extension of* $(Args, attack)$.

If a $\in G$ *with* G *the grounded extension of* $(Args, attack)$, *then there exists a grounded dispute tree for* a *with defence set* A *and* A $\subseteq G$.

Proof.

(i) See proof of Theorem 3.2 in [Dung et al., 2007].

(ii) See proof of Theorem 3.4 in [Dung et al., 2007].

(iii) Follows directly from Theorem 3.7 in [Kakas and Toni, 1999].

∎

Example 5.6 *As discussed in Example 5.4, the dispute tree in the middle of Figure 1 is admissible and grounded. As stated in Theorem 5.5 the defence set,* $\{a, c\}$, *is admissible and is a subset of the grounded extension of the AA framework from Example 5.2, in fact in this case it coincides with the grounded extension. The ideal extension of the AA framework is* $\{a, c, g\}$ *and we saw that there exists an ideal dispute tree for* g *(on the left of Figure 2) whose defence set is* $\{g\}$, *which is a subset of the ideal extension.*

In order to determine whether a sentence is admissible / grounded / ideal, given a flat ABA framework, a dispute tree for an argument for that sentence can be used, by virtue of the correspondence results overviewed in Section 4 and Theorem 5.5 above. For example, given the ABA framework in Section 3.2, the dispute tree in Figure 3 for argument $\{not \neg cov(m, c)\} \vdash_{arg} cov(m, c)$ can be used to determine that $cov(m, c)$ is admissible, grounded and ideal. Indeed, this is a dispute tree since the leaf node cannot be attacked and no other opponent node can attack the root. Moreover, it is trivially admissible and, since it is finite, it is grounded. Finally, it is ideal as no admissible dispute tree for its only opponent node exists (as $\{\} \vdash_{arg} cov'(m, c)$ cannot be attacked).

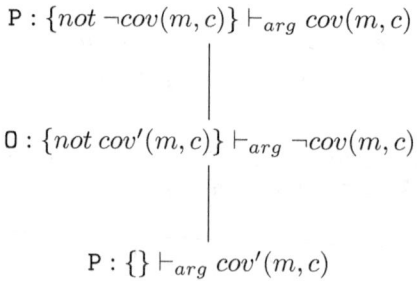

Figure 3. A dispute tree for $\{not \neg cov(m, c)\} \vdash_{arg} cov(m, c)$ for the flat ABA framework in Section 3.2.

5.2 Dispute derivations

Dispute derivations [Dung et al., 2006; Dung et al., 2007; Toni, 2013; Craven and Toni, 2016] are algorithms for determining whether a given sentence, in the language of a flat ABA framework, is admissible, grounded or ideal. Different kinds of dispute derivations can be defined for the different semantics, as in [Dung et al., 2006; Dung et al., 2007], or the same template of dispute derivations can be instantiated differently for the different semantics, as in [Toni, 2013; Craven and Toni, 2016] and, for the LP instance of ABA, in [Kakas and Toni, 1999]. Dispute derivations for determining whether a sentence is admissible can also be used to determine whether the sentence is complete or preferred [Toni, 2013]. All given notions of dispute derivations are defined as games between (fictional) *proponent* (P) and *opponent* (O) players, as for dispute trees. All given notions are sound and, for restricted types of flat ABA frameworks (referred to as *p-acyclic* [Dung et al., 2006]), complete [Dung et al., 2006; Dung et al., 2007; Toni, 2013]. The most recently defined types of dispute derivations are complete in general [Craven and Toni, 2016], for the admissible and grounded semantics. Different types of dispute derivations also differ in the data structures they deploy as well as their outputs:

- the dispute derivations of [Dung et al., 2006; Dung et al., 2007] deploy sets of assumptions and output admissible sets of assumptions in all cases, and, in the case of grounded/ideal semantics, these sets of assumptions are contained in the grounded/ideal extension, respectively;

- the dispute derivations of [Toni, 2013; Craven and Toni, 2016] deploy a mixture of sets of assumptions and sets of *potential arguments*, i.e. deductions supported by any sets of sentences (rather than assumptions) and with sentences in the support possibly marked as "seen", and output admissible sets of assumptions in all cases, as for the previous types of dispute derivations, as well as dialectical structures from which admissible / grounded / ideal dispute trees can be obtained.

We illustrate dispute derivations for the LP instance of ABA representation of the Breakdown Assistance policy, in Section 3.2, and refer to the original papers for formal definitions and results. In the illustration, we focus on the dispute derivations of [Toni, 2013], since they are generalisations of the earlier dispute derivations of [Dung et al., 2006; Dung et al., 2007] but still in the same spirit. Instead, the dispute derivations of [Craven and Toni, 2016] are based on a different conceptual model for ABA, where arguments and sets of arguments are defined as graphs instead (see [Craven and Toni, 2016] for details).

The (flat) ABA framework of Section 3.2 can be used to determine whether Mary should be covered, by determining whether $cov(m, c)$ is admissible (and thus, for this particular ABA framework, grounded, ideal etc.), i.e. if it belongs to an admissible extension. This can be determined in turn by means of a dispute derivation for $cov(m, c)$. This dispute derivation starts with a *potential argument* by P:[7]

$$\{\} \vdash^{p}_{\{cov(m,c)\}} cov(m,c),$$

namely a deduction $\{cov(m,c)\} \vdash cov(m,c)$ with no sentence in the support $\{cov(m,c)\}$ marked as "seen" (and the sentence $cov(m,c)$ in the support still "unseen"). In the first step of the derivation, then, P needs to "expand" its potential argument, while O is watching and can only put forward new potential arguments when P has sufficiently expanded its own potential arguments so as to have identified assumptions in their "unseen" support that O can attack (automatically rendering them "seen"). In this simple illustration, P will necessarily expand the initial potential argument to

$$\{\} \vdash^{p}_{\{ah(m),tr(m,c),pr(c),not\,\neg cov(m,c)\}} cov(m,c)$$

and identify the assumption $not\,\neg cov(m,c)$ as an element of the *defence set* of the dispute tree that the dispute derivation will output (if successful). At this stage O may opt to *eagerly* attack this assumption or *patiently* wait for P to carry on "expanding" its potential argument until it becomes an *actual argument*. This choice for O (and, in an analogous situation, for P) is dictated by the *selection function*, a parameter in the definition of dispute derivations. Whichever this selection function, at some later stage in the derivation the initial potential argument by P will become the *actual argument*

$$\{not\,\neg cov(m,c)\} \vdash^{p}_{\{\}} cov(m,c) \qquad (\text{P}_{cov})$$

attacked by a potential argument by O

$$\{not\,cov'(m,c)\} \vdash^{p}_{U} \neg cov(m,c) \qquad (\text{O}_{\neg cov}(U))$$

where, depending on the selection function, U may be as follows:

- $U = \{\neg reg(c,m)\}$, or
- $U = \{\}$.

In both cases, at some earlier stage, P will have chosen $not\,cov'(m,c)$, in the "unseen" support of a potential argument by O, as a *culprit*, causing that assumption to be marked as "seen" from that stage onwards. Note that O's potential argument $\text{O}_{\neg cov}(U)$, whichever U, is necessarily obtained by "expanding" the potential argument

$$\{\} \vdash^{p}_{\{\neg cov(m,c)\}} \neg cov(m,c)$$

---

[7]In general, a potential argument is of the form $A \vdash^{p}_{S} \sigma$, for $A \subseteq \mathcal{A}$, $S \subseteq \mathcal{L}$, and $\sigma \in \mathcal{L}$, where the superscript p stands for *"potential"*. Given $A \vdash^{p}_{S} \sigma$, there is a deduction for σ supported by $A \cup S$ (and some set of rules), with S the set of "unseen" sentences in this support and A the set of "seen" assumptions, as illustrated later.

put forward earlier by O to attack P's defence set element $not \neg cov(m,c)$. Also, when P "sees" $not\, cov'(m,c)$ and chooses it as a culprit in $\mathsf{O}_{\neg cov'}(U)$, it creates a potential argument

$$\{\} \vdash^p_{\{cov'(m,c)\}} cov'(m,c)$$

which is later "expanded" to

$$\{\} \vdash^p_{\{\}} cov'(m,c). \qquad (\mathrm{P}_{cov'})$$

Since O cannot possibly attack this argument, the derivation terminates successfully, returning, as output, the defence set $\{not\, \neg cov(m,c)\}$ as well as the dialectical structure

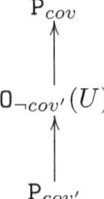

from which the dispute tree in Figure 3 is obtained.

In general, the defence set and the set of culprits are used to perform various kinds of *filtering* to save computation (to prevent players from attacking assumptions they have already attacked) as well as to guarantee that the computed defence set is conflict-free. Different semantics require different combinations of these filtering mechanisms. Moreover, the ideal semantics requires additional subcomputation to guarantee that the dispute tree is indeed ideal (namely that there exists no admissible dispute tree for the argument held at any opponent node).

5.3 ABA dialogues

ABA dialogues, as given in [Fan and Toni, 2014b; Fan and Toni, 2012a; Fan and Toni, 2011a], can be viewed as a distributed computation of dispute trees amongst agents, holding different ABA frameworks, but with the same underlying language $\mathcal{L}$.[8] An ABA dialogue is a sequence of *utterances*. The *content* of utterances may be a rule, an assumption, a contrary, or a claim whose "acceptability" (under admissible / grounded / ideal semantics) needs to be ascertained. The dialogue model can be used to support several dialogue types, e.g. information seeking and persuasion [Fan and Toni, 2011c; Fan and Toni, 2012a; Fan and Toni, 2012c; Fan et al., 2014].

---

[8]Here, as in [Gaertner and Toni, 2008], we (equivalently) define the contrary of an assumption as a total mapping from an assumption to a (non-empty) set of sentences, instead of a mapping from an assumption to a sentence as in the original ABA. This lends itself better to a dialogical setting, as agents may hold different sentences as contrary to the same assumption.

Syntactically, given two agents a_i and a_j, let $\mathcal{ID}$ be a (non-empty, possibly infinite) set that is totally ordered, with the ordering given by $<$, and contains a special element ID_0 which is the least element w.r.t. $<$. Then, utterances are denoted as tuples:

$$\langle a_i, a_j, T, C, \text{ID}\rangle,$$

where

- a_i is the agent making this utterance;
- a_j is the recipient;
- C (the *content*) is of one of the following forms:
  - $claim(\chi)$ for some $\chi \in \mathcal{L}$ (a *claim*),
  - $rl(\sigma_0 \leftarrow \sigma_1, \ldots, \sigma_m)$ for some $\sigma_0, \ldots, \sigma_m \in \mathcal{L}$ with $m \geq 0$ (a *rule*),
  - $asm(\alpha)$ for some $\alpha \in \mathcal{L}$ (an *assumption*),
  - $ctr(\alpha, \sigma)$ for some $\alpha, \sigma \in \mathcal{L}$ (a *contrary*),
  - a *pass* sentence π, such that $\pi \notin \mathcal{L}$.
- $\text{ID} \in \mathcal{ID} \setminus \{\text{ID}_0\}$ (the *identifier*).
- $T \in \mathcal{ID}$ (the *target*); we impose that $T < \text{ID}$.

Through a dialogue δ, the participating agents construct a joint *ABA framework* $\mathcal{F}_\delta$ *drawn from* δ. This $\mathcal{F}_\delta$ contains all information that the two agents have uttered in the dialogue and gives the context for examining the "acceptability" of the claim of the dialogue. Conceptually, a dialogue is "successful" if its claim is "acceptable" in $\mathcal{F}_\delta$. Note that the claim of a dialogue may be a belief, and acceptability thereof an indication that the agents may legitimately uphold the belief, or a course of actions, and acceptability thereof an indication that the agents may legitimately choose to adhere to it. Indeed, "acceptability" has so far shown to be an important criterion for assessing the outcome of various types of dialogues [Fan and Toni, 2011c; Fan and Toni, 2012a; Fan and Toni, 2012c; Fan et al., 2014], and thus "successful" dialogues can be seen as building blocks of a widely deployable framework for distributed interactions in multi-agent systems.

Rather than checking "success" retrospectively, this can be guaranteed constructively by means of *legal-move functions* (see [Fan and Toni, 2011a; Fan and Toni, 2014b] for details) guaranteed to generate "successful" dialogues if a limited form of retrospective checking by means of *outcome functions* succeeds [Fan and Toni, 2011a; Fan and Toni, 2014b]. Dialogue goals, e.g. information-seeking, inquiry or persuasion, can be modelled with *strategy-move functions* [Fan and Toni, 2012a]. Given a dialogue, a legal-move function returns a set of allowed utterances that can be uttered to extend the dialogue. Legal-move functions can thus be viewed as dialogue protocols. Outcome functions are mappings from dialogues to true / false. Given a dialogue, an outcome function returns true if a certain property holds for that dialogue. From utterances allowed by legal-move functions, strategy-move functions further select the ones advancing dialogues towards their goals.

We illustrate ABA dialogues for information seeking, persuasion and inquiry for the flat ABA framework in Section 3.2 again, and refer to the original papers for formal definitions and results.

Informally, information seeking dialogues are dialogues with the inquirer agent seeking some specific information from the inquiree agent. In an information seeking dialogue, the inquirer agent does nothing but posing its query, whereas the inquiree agent puts forward information it possesses in answering the query. With the breakdown assistance policy example, suppose that the inquirer agent a_1 asks the inquiree agent a_2 about the existence of argument for the sentence $cov'(m, c)$, as follows:

$$\langle a_1, a_2, 0, claim(cov'(m, c)), 1 \rangle$$
$$\langle a_2, a_1, 1, rl(cov'(m, c) \leftarrow in(m, c)), 2 \rangle$$
$$\langle a_2, a_1, 2, rl(in(m, c) \leftarrow), 3 \rangle$$

We can see that with a_1 and a_2 each using suitable strategy-move functions [Fan and Toni, 2012a], a_1 puts forward $cov'(m, c)$ as the claim of this dialogue and a_2 puts forward utterances 2 and 3 establishing the argument (in the ABA framework $\mathcal{F}_\delta$ drawn from the dialogue) for $cov'(m, c)$ supported by the empty set of assumptions and the two rules:

$$cov'(m, c) \leftarrow in(m, c) \text{ and } in(m, c) \leftarrow.$$

Persuasion dialogues are dialogues between two agents posing "incompatible" views towards some topic with the persuader trying to "prove" the topic and the persuadee trying to "disprove" it. Illustrating with the running example, we may have (for a_1 the persuader and a_2 the persuadee):

$$\langle a_1, a_2, 0, claim(not\ cov'(m, c)), 1 \rangle$$
$$\langle a_1, a_2, 1, asm(not\ cov'(m, c)), 2 \rangle$$
$$\langle a_2, a_1, 2, ctr(not\ cov'(m, c), cov'(m, c)), 3 \rangle$$
$$\langle a_2, a_1, 3, rl(cov'(m, c) \leftarrow in(m, c)), 4 \rangle$$
$$\langle a_2, a_1, 4, rl(in(m, c) \leftarrow), 5 \rangle$$

Here, a_1 tries to establish the acceptability of $not\ cov'(m, c)$ by claiming it as an assumption, thus forming the argument $\{not\ cov'(m, c)\} \vdash not\ cov'(m, c)$, whereas a_2 puts forward the attacking argument $\{\} \vdash cov'(m, c)$ with utterances 3, 4 and 5. The presented persuasion behaviours of both agents are formally defined with strategy-move functions in [Fan and Toni, 2012c].

Inquiry dialogues are about two agents jointly "proving" or "disproving" the acceptability of some claim. Both agents put forward information supporting or attacking the claim. Again illustrated with the breakdown assistance policy example, we may have:

$\langle a_1, a_2, 0, claim(cov(m,c)), 1 \rangle$
$\langle a_1, a_2, 1, rl(cov(m,c) \leftarrow ah(m), tr(m,c), pr(c), not\ \neg cov(m,c)), 2 \rangle$
$\langle a_1, a_2, 2, rl(ah(m) \leftarrow), 3 \rangle$
$\langle a_1, a_2, 2, rl(tr(m,c) \leftarrow), 4 \rangle$
$\langle a_1, a_2, 2, rl(pr(c) \leftarrow), 5 \rangle$
$\langle a_1, a_2, 2, asm(not\ \neg cov(m,c)), 6 \rangle$
$\langle a_2, a_1, 6, ctr(not\ \neg cov(m,c), \neg cov(m,c)), 7 \rangle$
$\langle a_2, a_1, 7, rl(\neg cov(m,c) \leftarrow \neg reg(c,m), not\ cov'(m,c)), 8 \rangle$
$\langle a_2, a_1, 8, rl(\neg reg(c,m) \leftarrow), 9 \rangle$
$\langle a_2, a_1, 8, asm(not\ cov'(m,c)), 10 \rangle$
$\langle a_2, a_1, 10, ctr(not\ cov'(m,c), cov'(m,c)), 11 \rangle$
$\langle a_2, a_1, 11, rl(cov'(m,c) \leftarrow in(m,c)), 12 \rangle$
$\langle a_2, a_1, 12, rl(in(m,c) \leftarrow), 13 \rangle$

With utterances 1-6, the argument $\{not\ \neg cov(m,c)\} \vdash cov(m,c)$ is formed. Utterances 7-10 form an attacking argument $\{not\ cov'(m,c)\} \vdash \neg cov(m,c)$, which is attacked by $\{\} \vdash cov'(m,c)$. The inquiry behaviour of agents is formally defined in [Fan and Toni, 2012a].

6 ABA and explanation

It is widely acknowledged that there is a strong interplay between argumentation and explanation, as for example discussed in [Seselja and Straßer, 2013]. In this section we overview existing proposals [Fan and Toni, 2015c; Schulz and Toni, 2016] using dispute trees in ABA (see Section 5) to provide (argumentative) explanations for why sentences should be concluded. Dispute trees for (flat) ABA can also serve as the basis for explanations in other settings, including various forms of decision-making [Fan and Toni, 2014a; Fan et al., 2014; Zhong et al., 2014; Fan et al., 2013] and case-based reasoning [Čyras et al., 2016] (see the original papers for details). In particular, natural language explanations can be drawn automatically from the dispute trees (see [Zhong et al., 2014; Mocanu et al., 2016] for details).

6.1 Dispute trees as explanations in flat ABA

We have seen (in Section 5) that dispute trees can be used to determine whether a sentence is admissible / grounded / ideal (and, as a consequence, preferred / complete). These dispute trees can also provide a computational counterpart for providing explanations for these sentences (being consequences of admissible / grounded / ideal / preferred / complete extensions, respectively). For example, the dispute tree in Figure 3 can be seen as providing an explanation for $cov(m,c)$, in the spirit of [Newton-Smith, 1981]:

> ...if I am asked to explain why I hold some general belief that p, I answer by giving my justification for the claim that p is true.

Hence, if a belief q does not contribute to the justification of p, q should not be in the explanation of p. Dispute trees are explanations for (the argument in their root supporting) a sentence in that everything in them contribute to justifying the sentence. This informal notion can be formalised in terms of a notion of *related admissibility* of ABA arguments [Fan and Toni, 2015c],[9] in turn defined using a notion of *r-defence* [Fan and Toni, 2015c], given as follows:

Definition 6.1 *Given an ABA framework* $\mathcal{ABA} = \langle \mathcal{L}, \mathcal{R}, \mathcal{A}, \bar{} \rangle$, *let* $\mathcal{AA}(\mathcal{ABA}) = (Args, attack)$ *be the corresponding AA framework of* $\mathcal{ABA}$.

- *Given* a, b $\in Args$, a *r-defends* b *iff:*

    (i) a = b, *or*

    (ii) there exists c $\in Args$ *such that* a *attacks* c *and* c *attacks* b, *or*

    (iii) there exists c $\in Args$ *such that* a *r-defends* c *and* c *r-defends* b.

- *Given* a $\in Args$ *and* $\sigma \in \mathcal{L}$, a *r-defends* σ *iff there exists* b $\in Args$ *such that* b *supports* σ *and* a *r-defends* b.

As an illustration, given the ABA framework in Section 3.2:

$\{\} \vdash_{arg} cov'(m,c)$ r-defends $\{\} \vdash_{arg} cov'(m,c)$,
$\{\} \vdash_{arg} cov'(m,c)$ r-defends $\{not \neg cov(m,c)\} \vdash_{arg} cov(m,c)$,
$\{\} \vdash_{arg} cov'(m,c)$ r-defends $cov'(m,c)$,
$\{not \neg cov(m,c)\} \vdash_{arg} cov(m,c)$ r-defends $cov(m,c)$,
$\{\} \vdash_{arg} cov'(m,c)$ r-defends $cov(m,c)$.

The notion of related admissibility is obtained by combining the r-defence relation and standard admissibility as follows:

Definition 6.2 *Given an ABA framework* $\mathcal{ABA}$, *let* $\mathcal{AA}(\mathcal{ABA}) = (Args, attack)$ *be the corresponding AA framework of* $\mathcal{ABA}$. *A set of arguments* A $\subseteq Args$ *is related admissible iff:*

(i) A *is admissible, and*

(ii) there exists a topic *sentence* σ *(of* A*) such that* σ *is supported by some argument in* A *and for all* b $\in$ A, b *defends* σ.

Intuitively, for a related admissible set of arguments A with topic sentence σ, no argument in A is "unrelated" to σ as all arguments in A r-defend σ. As an illustration, given the ABA framework in Section 3.2,

$$\{\{\} \vdash_{arg} cov'(m,c)\}$$

---

[9]The notions defined in this section can be defined trivially for any AA framework too, as in [Fan and Toni, 2015c]. The notions for AA frameworks corresponding to ABA frameworks, given below, are an instantiation of the notions for any AA frameworks.

is related admissible, with topic sentence $cov'(m,c)$, and

$$\{\{\} \vdash_{arg} cov'(m,c), \{not \neg cov(m,c)\} \vdash_{arg} cov(m,c)\}$$

is related admissible, with topic sentence $cov(m,c)$. Instead,

$$\{\{\} \vdash_{arg} cov'(m,c), \{not\ cov'(m,c)\} \vdash_{arg} \neg cov(m,c)\}$$

is not related admissible as it is not admissible; and

$$\{\{\} \vdash_{arg} ah(m), \{\} \vdash_{arg} pr(c)\}$$

is not related admissible as there does not exists a topic sentence σ such that it is defended by both $\{\} \vdash_{arg} ah(m)$ and $\{\} \vdash_{arg} pr(c)$.

Dispute trees correspond to explanations in that their defence sets are related admissible:

Theorem 6.3 *Given an ABA framework $\mathcal{ABA} = \langle \mathcal{L}, \mathcal{R}, \mathcal{A}, \overline{} \rangle$, let $\mathcal{AA}(\mathcal{ABA}) = (Args, attack)$ be the corresponding AA framework of $\mathcal{ABA}$. Let $\sigma \in \mathcal{L}$.*

(i) Let $\mathtt{a} = A \vdash_{arg} \sigma \in Args$ and $\mathcal{T}$ be a dispute tree for $\mathtt{a}$. If $\mathcal{T}$ is admissible / grounded / ideal, then $\mathcal{D}(\mathcal{T})$ is related admissible.

(ii) If $\mathtt{A} \subseteq Args$ is related admissible, with topic sentence σ, then there is an admissible dispute tree $\mathcal{T}$ such that $\mathtt{A}' = \mathcal{D}(\mathcal{T})$ and $\mathtt{A}' \subseteq \mathtt{A}$.

Proof.

(i) By definition 6.1, all arguments labelling P nodes ($\mathcal{D}(\mathcal{T})$) in a dispute tree r-defend the argument labelling the root note. By Theorem 5.5, all arguments labelling P nodes in an admissible / grounded / ideal dispute tree are admissible. Thus, by Definition 6.2, $\mathcal{D}(\mathcal{T})$ is related admissible.

(ii) If $\mathtt{A}$ is related admissible, by Definition 6.2, $\mathtt{A}$ is also admissible. By Theorem 5.5, there exists an admissible dispute tree $\mathcal{T}$ such that $\mathtt{A}' = \mathcal{D}(\mathcal{T})$ and $\mathtt{A}' \subseteq \mathtt{A}$.

∎

6.2 Explanations for answer set programming

We have seen in Section 3.2 that a logic program can be encoded as an (equivalent) ABA framework such that the semantics of the ABA framework coincide with the semantics of the underlying logic program [Bondarenko et al., 1997], for a wide range of semantics including the stable model (or *answer set*) semantics [Schulz and Toni, 2016; Schulz and Toni, 2015; Caminada and Schulz, 2015]. Logic programs under the answer set semantics (or *answer set programming*) can be applied in a wide range of scenarios [Baral and Uyan, 2001;

Lifschitz, 2002; Eiter *et al.*, 2008; Delgrande *et al.*, 2009; Ricca *et al.*, 2010; Gebser *et al.*, 2011b; Boenn *et al.*, 2011; Erdem, 2011; Ricca *et al.*, 2012; Terracina *et al.*, 2013], thanks also to the availability of efficient solvers for the computation of answer sets [Leone *et al.*, 2006; Gebser *et al.*, 2011a; Alviano *et al.*, 2015; Calimeri *et al.*, 2016]. These however do not provide any explanation of the answer sets computed. In particular, given one such answer set, there is no indication as to why a literal is or is not part of an answer set: this would instead be beneficial in human-computer interaction scenarios where logic programming is used for example to support human decision making.

As seen in Section 6.1, dispute trees do not only provide a way of determining whether or not a sentence is, for instance, admissible, but also an explanation as to *why* this is so.

Given that answer sets of a logic program correspond to stable extensions of the ABA framework encoding this logic program [Bondarenko *et al.*, 1997] and that if an answer set is guaranteed to exist then it is preferred (See Theorem 2.11 (i)), dispute trees can be used to determine, for a computed answer set and sentence in it, an explanation (in the form of a dispute tree) for why this is so. However, for the purpose of extracting explanations for literals in terms of other literals (rather than arguments, see [Schulz and Toni, 2016]), it is useful to single out, from the set of rules supporting ABA arguments, the rules with an empty body (referred to as *facts* in LP):

Definition 6.4 *Given a flat ABA framework* $\langle \mathcal{L}, \mathcal{R}, \mathcal{A}, \overline{} \rangle$, *we say that* $(A, F) \vdash_{arg}^{R} \sigma$ *is a fact-based-argument for* $\sigma \in \mathcal{L}$ *supported by* $A \subseteq \mathcal{A}$ *and* $F \subseteq \{\pi \leftarrow \mid \pi \leftarrow \in \mathcal{R}\}$, *if there is an argument* $A \vdash_{arg} \sigma$ *such that* $F = R \cap \{\pi \leftarrow \mid \pi \leftarrow \in \mathcal{R}\}$.

A generalisation of dispute trees, which we call *explanation trees* [Schulz and Toni, 2016], where nodes are labelled by fact-based-arguments[10] can be used to explain why a literal is contained in a given answer set.

As an example, consider the ABA framework in Section 3.2, and the logic program amounting to its rules. This logic program has only one answer set: $\{ah(m), tr(m, c), pr(c), \neg reg(c, m), in(m, c), cov(m, c), cov'(m, c)\}$.

The explanation tree in Figure 4 justifies why Mary is covered, i.e. why $cov(m, c)$ is contained in the answer set. It expresses that there is evidence that Mary is covered (given by the argument with conclusion $cov(m, c)$ in the root proponent node) since Mary is the account holder and she is travelling in a car which is a private vehicle (facts supporting the argument), and since it can be assumed that there is no evidence that Mary is not covered ($not \neg cov(m, c)$ is an assumption). Even though there is evidence against this assumption, i.e. there is evidence that that Mary is not covered (given by the argument with conclusion $\neg cov(m, c)$ in the opponent node) because she is not registered on

---

[10] For better readability we will omit the symbol $\leftarrow$ for all facts in the set F.

$$P : (\{not \neg cov(m,c)\}, \{ah(m), tr(m,c), pr(c)\}) \vdash_{arg} cov(m,c)$$

$$O : (\{not\, cov'(m,c)\}, \{\neg reg(c,m)\}) \vdash_{arg} \neg cov(m,c)$$

$$P : (\{\}, \{in(m,c)\}) \vdash_{arg} cov'(m,c)$$

Figure 4. An explanation tree justifying why Mary is covered in the running example

the car, this evidence can be disregarded since Mary was in the car at the time of the breakdown (given by the proponent argument with conclusion $cov'(m,c)$, which attacks the assumptions $not\, cov'(m,c)$ of the opponent node). Note that this explanation tree is the same as the dispute tree in Figure 3 except that it uses fact-based-arguments.

In contrast to dispute trees which are used to justify only the containment of an argument *in* an extension, explanation trees can also explain why a literal is *not in* an answer set. In that case, explanation trees have an opponent node as their root, as illustrated by the explanation tree below which justifies why it is not the case that Mary is not covered (why $\neg cov(m,c)$ is not part of the answer set)

$$O : (\{not\, cov'(m,c)\}, \{\neg reg(c,m)\}) \vdash_{arg} \neg cov(m,c)$$

$$P : (\{\}, \{in(m,c)\}) \vdash_{arg} cov'(m,c)$$

Note that this explanation tree is a sub-tree of the previous explanation tree in Figure 4 justifying why $cov(m,c)$ is contained in the answer set.

Since explanation trees whose root node is a proponent node are dispute trees and since arguments which are in a stable extension are also in an admissible extension (Theorem 2.11 (i)), it follows from the relationhip between admissible extensions and admissible dispute trees given in Theorem 5.5 (i) that explanation trees starting with proponent nodes are admissible dispute trees. Thus, for literals contained in the answer set, explanation trees illustrate that the literal is supported by an admissible subset of this answer set.

Explanation trees whose root node is an opponent node have an explanation tree for a literal contained in the answer set as its direct sub-tree. Thus, this direct sub-tree is an admissible dispute tree. This means that literals not contained in the answer set are justified by illustrating that they are attacked by an admissible subset of the answer set.

In summary, explanation trees provide justifications of literals with respect to an answer set in terms of *admissible subsets* of this answer set [Schulz and Toni, 2016].

7 ABA and reasoning with preferences

Argumentation and preferences come a long way, see e.g. [Simari and Loui, 1992]. In general, preferences can be used to express, for instance, agents' degrees of belief, imperatives (moral, legal, etc.), aims, wishes. There are numerous methods in knowledge representation and reasoning to account for preference information, see e.g. [Prakken and Sartor, 1999; Kakas and Moraitis, 2003; Delgrande *et al.*, 2004; Brewka *et al.*, 2010; Domshlak *et al.*, 2011], and, in particular, several argumentation formalisms handling preferences, see e.g. [Bench-Capon, 2003; Modgil, 2009; Modgil and Prakken, 2014; García and Simari, 2014; Besnard and Hunter, 2014; Amgoud and Vesic, 2014; Baroni *et al.*, 2011], where preferences help to discriminate amongst information such as extensions, arguments, assumptions, rules, decisions and goals [Wakaki, 2014; Besnard and Hunter, 2014; Čyras and Toni, 2016a; Modgil and Prakken, 2014; Fan *et al.*, 2013]. There are various ways to deal with preferences in ABA too [Kowalski and Toni, 1996; Toni, 2008b; Thang and Luong, 2013; Fan *et al.*, 2013; Wakaki, 2014; Čyras and Toni, 2016a; Čyras and Toni, 2016b]. In this section we illustrate (by way of examples) these latter approaches. At a high-level, they can be divided in two groups: *meta level* approaches ([Wakaki, 2014; Čyras and Toni, 2016a; Čyras and Toni, 2016b], see Section 7.1), which, roughly, account for preferences at the semantic level, and *object level* approaches ([Kowalski and Toni, 1996; Toni, 2008b; Thang and Luong, 2013; Fan *et al.*, 2013], see Section 7.2), which, roughly, encode preferences within the existing ABA components (e.g. rules and assumptions) and avoid the need to modify the semantics of ABA frameworks.

Note that the examples chosen for the illustrations in this section have been selected for their simplicity, to give a high-level idea of the various approaches overviewed, and may not convey the full sophistication and usefulness of these approaches: the interested reader can find details as well as formal results in the original papers.

7.1 Handling preferences in ABA at the meta-level

[Wakaki, 2014] follows the ideas of prioritized logic programming [Sakama and Inoue, 2000] and equips ABA with explicit preferences by introducing a binary preference relation $\leqslant$ over the language $\mathcal{L}$. (For $a, b \in \mathcal{L}$, $a \leqslant b$ expresses that 'a is less or equally preferred than b'.) This ordering $\leqslant$ is then used to compute, by comparing consequences of extensions, a preference ordering $\sqsubseteq$ over extensions so as to select the most "preferable" extensions (i.e. the $\sqsubseteq$-maximal ones) of the underlying ABA framework. Such meta-level preference treatment can be well illustrated via scenarios of decision making with preferences, as in the following example.

Example 7.1 *Mary needs to decide what insurance policy to buy. Following the approach of [Fan et al., 2013], information relevant to the decision making is represented via two tables, T_{DA} and T_{GA}, as illustrated in Table 1, where*

- T_{DA} *describes relations between* decision candidates *(Policy 1 (P_1), Policy 2 (P_2)) and* attributes *(£50, £70, no exceptions (no_ex));*

- T_{GA} *describes relations between* goals *(cheap and full coverage (full)) and* attributes.

	£50	£70	no_ex
P_1	0	1	1
P_2	1	0	0

	£50	£70	no_ex
cheap	1	0	0
full	0	0	1

Table 1. T_{DA}(left) and T_{GA}(right), for Example 7.1.

Intuitively, each decision candidate has certain attributes (P_1 has £70 and no_ex; P_2 has £50); and each goal can be met by certain attributes (cheap is met by £50; full is met by no_ex).

In addition, suppose that the goal full *is preferred over* cheap. *In p_ABA, we can represent this information as a framework $\langle \mathcal{L}, \mathcal{R}, \mathcal{A}, \bar{}, \leqslant \rangle$, with the underlying ABA framework $\langle \mathcal{L}, \mathcal{R}, \mathcal{A}, \bar{} \rangle$ with*

$$\mathcal{R} = \{£70 \leftarrow P_1, \quad no\_ex \leftarrow P_1, \quad £50 \leftarrow P_2, \quad cheap \leftarrow £50,$$
$$full \leftarrow no\_ex, \quad \overline{P_2} \leftarrow P_1, \quad \overline{P_1} \leftarrow P_2\},$$
$$\mathcal{A} = \{P_1, P_2\}, \text{ and}$$
$$cheap \leqslant full, \quad cheap \leqslant cheap, \quad full \leqslant full.$$

$\langle \mathcal{L}, \mathcal{R}, \mathcal{A}, \bar{} \rangle$ *has two preferred / stable extensions $\{P_1\}$ and $\{P_2\}$, with conclusions $Cn(\{P_1\}) = \{P_1, £70, no\_ex, full\}$ and $Cn(\{P_2\}) = \{P_2, £50, cheap\}$. We then find $\{P_2\} \sqsubseteq \{P_1\}$ and $\{P_1\} \not\sqsubseteq \{P_2\}$, so that $\{P_1\}$ is a $\sqsubseteq$-maximal extension, and is hence selected as the "preferable" one. Buying Policy 1 is thus deemed the better decision to take.*

Preferences in ABA can also be utilized to modify the attack relation between sets of assumptions, akin to approaches to argumentation with preferences such as [Bench-Capon, 2003; Modgil and Prakken, 2014; Amgoud and Vesic, 2014; Besnard and Hunter, 2014]. For instance, ABA$^+$ [Čyras and Toni, 2016a; Čyras and Toni, 2016b] equips ABA with a binary preference relation $\leqslant$ over assumptions, and incorporates preferences directly into the attack relation so as to *reverse* attacks that stem from sets containing assumptions less preferred than the one whose contrary is deduced, as illustrated next.

Example 7.2 *Suppose that Mary has decided to buy Policy 1, as suggested in Example 7.1. However, Mary has also found some information on the Internet* about the policy: source C says that under certain circumstances (c), the policy applies only to citizens of certain specified countries; source B says that sometimes (say, assuming c), the policy applies only to UK residents $(UK \leftarrow b,c)$; source A says that sometimes (assuming c) the policy applies only to non-UK residents $(non\_UK \leftarrow a,c)$. Mary trusts the source A the least (i.e. $a < b$, $a < c$). What is Mary justified believing in about the applicability of the policy, given certain circumstances?

We can formalize this in ABA^+ as follows: consider $\langle \mathcal{L}, \mathcal{R}, \mathcal{A}, ^{\overline{}}, \leqslant \rangle$ with

$$\mathcal{A} = \{a, b, c\},$$
$$\mathcal{R} = \{non\_UK \leftarrow a, c, \quad UK \leftarrow b, c\},$$
$$\overline{a} = UK, \quad \overline{b} = non\_UK,$$
$$a < b, \quad a < c,$$

where the assumptions stand for the possibility to trust the sources and preferences indicate their relative credibility, rules are drawn given that information from sources A and B is applicable under certain circumstances (c), also given that sources A and B are in conflict.

The underlying ABA framework $\langle \mathcal{L}, \mathcal{R}, \mathcal{A}, ^{\overline{}} \rangle$ admits both $\{a, c\}$ and $\{b, c\}$ as stable / preferred extensions. In ABA^+, attacks from $\{a, c\}$ to (any set of assumptions containing) b are reversed, due to the a's lower credibility in comparison with b. Hence, $\{b, c\}$ is a unique stable / preferred extension, arguably the desirable outcome. This can be seen clearly given the graph depicted below, omitting, for readability, assumption sets $\{\}$ and $\{a, b, c\}$, as well as attacks to and from them:

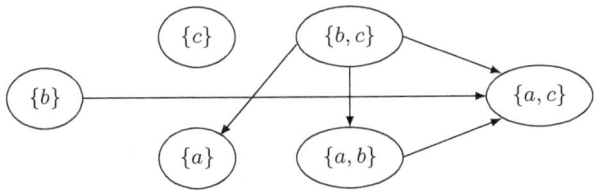

7.2 Handling preferences in ABA at the object-level

Instead of equipping ABA frameworks with explicit preference relations as in Section 7.1, and then modifying the semantics of ABA (by either comparing extensions or modifying the attack relation), preferences can be encoded within the existing components (rules, assumptions and contraries) without modifying the semantics.

For instance, [Kowalski and Toni, 1996; Toni, 2008b] deal with preferences between rules by adding conditions (i.e. assumptions) to the body of rules expressing that the rules are not attacked by other higher preference rules, by appropriately defining contraries of these assumptions. For illustration, consider the following example:

Example 7.3 *In our breakdown policy example of Section 3, the rules in the ABA instance for LP of section 3.2 can be modified by adding assumptions as follows:*

$$cov(m,c) \leftarrow ah(m), tr(m,c), pr(c), not \ \neg cov(m,c), a_{cov(m,c)},$$
$$\neg cov(m,c) \leftarrow \neg reg(c,m), not \ cov'(m,c), a_{\neg cov(m,c)}.$$

If a preference of the second rule over the first one is to be expressed, then one could assign contraries

$$\overline{a_{cov(m,c)}} = \neg cov(m,c), \quad \overline{a_{\neg cov(m,c)}} = a^c,$$

where a^c is new to $\mathcal{L}$.

More generally (as in [Toni, 2008b]), one can assume a naming function assigning distinguished names to elements (e.g. rules) of a given domain, and given preferences over the elements of the domain, consider a language that includes sentences expressing those preferences. For example, the two rules above can be given names r and r' respectively, and the language $\mathcal{L}$ would contain a "preference sentence" $r < r'$ expressing that the second rule is preferred over the first one. Then, when mapping the domain into an ABA framework, a rule

$$\neg cov(m,c) \leftarrow r < r', a_{\neg cov(m,c)},$$

could be added, so as to account for preferences, which could be stated e.g. via a rule $r < r' \leftarrow$. This way, ABA can also account for dynamic preferences (see e.g. [Prakken and Sartor, 1999]), i.e. preferences that are themselves deducible using rules, possibly from other assumptions.

Yet another way to deal with preferences in ABA on the object level is used in [Thang and Luong, 2013] when translating Brewka's preferred subtheories [Brewka, 1989] into ABA. To capture the interplay between classically inconsistent sentences and partial preference information among them, [Thang and Luong, 2013] introduce assumptions for representing sentences in the domain language as well as for determining their acceptance status in the construction of preferred subtheories, and further introduce rules for: deriving sentences from their corresponding assumptions; deriving contraries of the least preferred elements of minimally inconsistent subsets; enforcing (non-)acceptance of an assumption iff the statuses more preferred assumptions are determined. This is illustrated next.

Example 7.4 *Let us rewrite the rules from Example 7.2 as*

$$\alpha, \gamma \rightarrow \neg UK, \quad \beta, \gamma \rightarrow UK$$

(where $\rightarrow$ is material implication) to constitute the facts (world knowledge), and let $T = \{\alpha, \beta, \gamma\}$ be the theory representing the defeasible knowledge, with preferences $\alpha < \beta$ and $\alpha < \gamma$. This partial order $<$ admits two extensions to

total orders, namely $\alpha < \beta < \gamma$ and $\alpha < \gamma < \beta$, both of which result in the same preferred subtheory of T, namely $\{\beta, \gamma\}$.

The domain can be mapped into an ABA framework $\langle \mathcal{L}, \mathcal{R}, \mathcal{A}, \overline{} \rangle$ with (for readability treating contraries of assumptions as symbols in the language)

$$\mathcal{A} = \{a_\alpha, a_\beta, a_\gamma\} \cup \{b_\alpha, b_\beta, b_\gamma\},$$
$$\mathcal{R} = \{\alpha \leftarrow a_\alpha, \quad \beta \leftarrow a_\beta, \quad \gamma \leftarrow a_\gamma\} \cup \{\overline{a_\alpha} \leftarrow a_\beta, \overline{b_\beta}, a_\gamma, \overline{b_\gamma}\} \cup$$
$$\{\overline{a_\alpha} \leftarrow b_\alpha, \quad \overline{a_\beta} \leftarrow b_\beta, \quad \overline{a_\gamma} \leftarrow b_\gamma\} \cup$$
$$\{\overline{b_\beta} \leftarrow, \quad \overline{b_\gamma} \leftarrow, \quad \overline{b_\alpha} \leftarrow a_\beta, \overline{b_\beta}, a_\gamma, \overline{b_\gamma},$$
$$\overline{b_\alpha} \leftarrow \overline{a_\beta}, \overline{b_\beta}, a_\gamma, \overline{b_\gamma}, \quad \overline{b_\alpha} \leftarrow a_\beta, \overline{b_\beta}, \overline{a_\gamma}, \overline{b_\gamma}, \quad \overline{b_\alpha} \leftarrow \overline{a_\beta}, \overline{b_\beta}, \overline{a_\gamma}, \overline{b_\gamma}\}.$$

This $\langle \mathcal{L}, \mathcal{R}, \mathcal{A}, \overline{} \rangle$ has a unique stable extension $\{a_\beta, a_\gamma\}$, corresponding to the unique preferred subtheory of T.

Another example of preferences dealt with in ABA within the object-level is to support decision making with preferences over goals. Differently from the other approaches overviewed in this section, this method is specific to decision making settings, and uses preferences over sentences (the goals) within decision criteria (e.g. various kinds of "dominance", see [Fan et al., 2013]) for choosing "best" decisions. This can be illustrated in the context of the same decision making setting of Example 7.1:

Example 7.5 *Given the two tables, T_{DA} and T_{GA}, in Table 1, as well as the preference full $>$ cheap, the problem of identifying the "best" decisions, namely those "meeting the more preferred goals that no other decisions meet", can be represented in ABA with*

$$\mathcal{R} = \{ \quad has(P_1, £70) \leftarrow \quad , \quad has(P_1, no\_ex) \leftarrow \quad , \quad has(P_2, 50) \leftarrow \quad ,$$
$$satBy(cheap, £50) \leftarrow \quad , \quad satBy(full) \leftarrow no\_ex,$$
$$prefer(full, cheap) \leftarrow \quad \} \cup$$
$$\{ \quad met(X, Y) \leftarrow has(X, Z), satBy(Y, Z) \mid X \in \{P_1, P_2\},$$
$$Y \in \{cheap, full\}, \quad Z \in \{£50, £70, no\_ex\} \quad \} \cup$$
$$\{ \quad sel(X) \leftarrow met(X, Y), noBetterThan(X, Y) \mid X \in \{P_1, P_2\},$$
$$Y \in \{cheap, full\} \quad \} \cup$$
$$\{ \quad better(X, Y) \leftarrow met(X', Y'), prefer(Y', Y), X \neq X' \mid$$
$$X, X' \in \{P_1, P_2\}, \quad Y, Y' \in \{cheap, full\} \quad \}$$
$$\mathcal{A} = \{ \quad noBetterThan(X, Y) \mid X \in \{P_1, P_2\}, \quad Y \in \{cheap, full\} \quad \}$$
$$\overline{not\ x} = better(X, Y) \quad for\ any \quad x = noBetterThan(X, Y) \in \mathcal{A}$$

Then
$$\{\{noBetterThan(P_1, full)\} \vdash_{arg} sel(P_1)\}$$

is admissible whereas

$$\{\{noBetterThan(P_2, cheap)\} \vdash_{arg} sel(P_2)\}$$

is not, as the latter is attacked by $\{\} \vdash_{arg} better(P_2, cheap)$. *Indeed, Policy 1 is the "best" decision in this simple setting.*

8 Conclusion

This chapter overviews research, spanning over more than two decades (from [Bondarenko et al., 1993] onwards), on Assumption-Based Argumentation (ABA), a framework for structured argumentation motivated by and emerging from non-monotonic reasoning. We have focused on the semantic foundations of ABA, in the general case as well as for the special case of flat ABA frameworks, while also providing an overview of the computational machinery (flat) ABA is equipped with as well as its uses for explaining argumentative conclusions. Finally, we have overviewed, with the aid of examples, uses and generalisations of ABA to support reasoning with preferences.

This chapter is meant as a taster of ABA rather than a comprehensive technical presentation, and complements other earlier overviews [Dung et al., 2009; Toni, 2012; Toni, 2014]. In particular, it focuses on the case of general (possibly non-flat) frameworks rather than flat frameworks as in the earlier overviews, and provides a taster of explanation and the treatment of preferences in ABA.

We omitted to mention several aspects of ABA. For instance, there are several other instances of ABA for non-monotonic reasoning (see [Bondarenko et al., 1997]), and ABA has also been shown to admit Adaptive Logic and AS-PIC+ without preferences as instances [Heyninck and Straßer, 2016]. Other ABA semantics have been presented in the literature, e.g. the semi-stable semantics [Caminada et al., 2015]. Moreover, formulation of (some) ABA semantics in terms of labellings, in the spirit of those proposed for abstract argumentation [Caminada and Gabbay, 2009], have been proposed [Schulz and Toni, 2014; Schulz and Toni, 2015; Schulz and Toni, 2017]. Further, the computational complexity of several reasoning problems in several instances of ABA is known [Dimopoulos et al., 2002; Dunne, 2009], and several systems for (flat) ABA are publicly available (see robertcraven.org/proarg/ and www-abaplus.doc.ic.ac.uk). Recent work also shows that (sets of) arguments in ABA can be re-interpreted as graphs, with conceptual and computational advantages [Craven and Toni, 2016]. We have seen in Section 7 that ABA has been extended to accommodate reasoning with preferences: other extensions of ABA also exist, notably the probabilistic ABA of [Dung and Thang, 2010]. Finally, we have not delved into applications of ABA: these are overviewed in earlier surveys [Dung et al., 2009; Toni, 2012; Toni, 2014] or other papers [Gao et al., 2016; Fan and Toni, 2016]. In particular, [Gao et al., 2016] uses related admissibility in ABA (see Section 6.1) to coordinate and resolve conflicts amongst agents, while also guaranteeing that privacy is preserved, in some sense, whereas [Fan and Toni, 2016] reinterprets the problem of determining solutions in games in normal form in ABA, using ABA dialogues (as summarised in Section 5.3) to determine these solutions in a distributed fashion, without agents fully disclosing their preferences.

There are several open issues in ABA as well as several directions for future work. We have seen, in Section 6.2, that explanations as to why sentences are not "acceptable" may be useful [Schulz and Toni, 2016]. The concept of "not-explanations" can be defined, in general, in abstract argumentation [Fan and Toni, 2015b]: it would be useful to define this notion also for ABA. Other forms of explanations have been defined, notably for explaining inconsistencies in LP [Schulz et al., 2015]: it would be interesting to define a notion of explanation for the lack of (e.g. stable) extensions in generic ABA. Some preliminary work [Zhong et al., 2014; Mocanu et al., 2016] indicates that natural language explanations can be naturally drawn from dispute trees computed by dispute derivations: it would be interesting to develop this work further and test the usefulness of the generated explanations in practice. Further, in multi-agent settings, it would be interesting to further study strategic behaviour of agents using ABA as their language of interaction [Fan and Toni, 2012c; Fan and Toni, 2015a; Gao et al., 2016; Fan and Toni, 2016]. From a computational viewpoint, (flat) ABA is equipped with several (sound and complete) algorithms for determining the "acceptability" of sentences (and compute extensions "supporting" them): it would be interesting to see how these algorithms can be generalised to the case of any, possibly non-flat, ABA frameworks and/or deployed when preferences are given, e.g. in the spirit of Gorgias (see gorgiasb.tuc.gr/index.html) and dealt with at the meta-level (as in Section 7.1). Moreover, it would be interesting to identify (sound and complete) computational machinery for determining extensions of ABA, without having to resort to implementations of abstract argumentation by using the mapping described in Section 4.

BIBLIOGRAPHY

[Alferes et al., 1993] José Júlio Alferes, Phan Minh Dung, and Luís Moniz Pereira. Scenario semantics of extended logic programs. In A. Nerode and L. Pereira, editors, *Proceedings of the 2nd International Workshop on Logic Programming and Nonmonotonic Reasoning (LPNMR'93)*, pages 334–348. MIT Press, 1993.

[Alviano et al., 2015] Mario Alviano, Carmine Dodaro, Nicola Leone, and Francesco Ricca. Advances in WASP. In *Proceedings of the 13th International Conference on Logic Programming and Nonmonotonic Reasoning (LPNMR'15)*, pages 40–54, 2015.

[Amgoud and Vesic, 2014] Leila Amgoud and Srdjan Vesic. Rich preference-based argumentation frameworks. *International Journal of Approximate Reasoning*, 55(2):585–606, 2014.

[Baral and Uyan, 2001] Chitta Baral and Cenk Uyan. Declarative specification and solution of combinatorial auctions using logic programming. In *Proceedings of the 6th International Conference on Logic Programming and Nonmonotonic Reasoning (LPNMR'01)*, pages 186–199, 2001.

[Baroni et al., 2011] Pietro Baroni, Federico Cerutti, Massimiliano Giacomin, and Giovanni Guida. AFRA: argumentation framework with recursive attacks. *International Journal of Approximate Reasoning*, 52(1):19–37, 2011.

[Bench-Capon, 2003] Trevor J. M. Bench-Capon. Persuasion in practical argument using value based argumentation frameworks. *Journal of Logic and Computation*, 13(3):429–448, 2003.

[Besnard and Hunter, 2014] Philippe Besnard and Anthony Hunter. Constructing argument graphs with deductive arguments: A tutorial. *Argument & Computation*, 5(1):5–30, 2014.

[Besnard et al., 2014] Philippe Besnard, Alejandro Javier García, Anthony Hunter, Sanjay Modgil, Henry Prakken, Guillermo Ricardo Simari, and Francesca Toni. Introduction to structured argumentation. *Argument & Computation*, 5(1):1–4, 2014.

[Boenn et al., 2011] Georg Boenn, Martin Brain, Marina De Vos, and John P. Fitch. Automatic music composition using answer set programming. *Theory and Practice of Logic Programming*, 11(2-3):397–427, 2011.

[Bondarenko et al., 1993] Andrei Bondarenko, Francesca Toni, and Robert A. Kowalski. An assumption-based framework for mon-monotonic reasoning. In L. M. Pereira and A. Nerode, editors, *Proceedings of the 2nd International Workshop on Logic Programming and Non-monotonic Reasoning (LPNMR'93)*, pages 171–189, Lisbon, Portugal, June 1993. MIT Press.

[Bondarenko et al., 1997] Andrei Bondarenko, Phan Minh Dung, Robert A. Kowalski, and Francesca Toni. An abstract, argumentation-theoretic approach to default reasoning. *Artificial Intelligence*, 93(1-2):63–101, 1997.

[Brewka et al., 1997] Gerhard Brewka, Jürgen Dix, and Kurt Konolige. *Nonmonotonic Reasoning: An Overview*, volume 73 of *CSLI Lecture Notes*. CSLI Publications, Stanford, CA, 1997.

[Brewka et al., 2010] Gerhard Brewka, Mirosław Truszczyński, and Stefan Woltran. Representing preferences among sets. In Maria Fox and David Poole, editors, *24th AAAI Conference on Artificial Intelligence (AAAI'10)*, pages 273–278, Atlanta, Georgia, 2010. AAAI Press.

[Brewka, 1989] Gerhard Brewka. Preferred subtheories: An extended logical framework for default reasoning. In N. S. Sridharan, editor, *11th International Joint Conference on Artificial Intelligence (IJCAI'89)*, pages 1043–1048, Detroit, 1989. Morgan Kaufmann.

[Calimeri et al., 2016] Francesco Calimeri, Martin Gebser, Marco Maratea, and Francesco Ricca. Design and results of the fifth answer set programming competition. *Artificial Intelligence*, 231:151–181, 2016.

[Caminada and Gabbay, 2009] Martin Caminada and Dov M. Gabbay. A logical account of formal argumentation. *Studia Logica*, 93(2-3):109–145, 2009.

[Caminada and Schulz, 2015] Martin Caminada and Claudia Schulz. On the equivalence between assumption-based argumentation and logic programming. In *Proceedings of the 1st International Workshop on Argumentation and Logic Programming (ArgLP'15)*, 2015.

[Caminada et al., 2015] Martin Caminada, Samy Sá, João Alcântara, and Wolfgang Dvořák. On the difference between assumption-based argumentation and abstract argumentation. *The IfCoLog Journal of Logics and their Applications*, 2(1):16–34, 2015.

[Craven and Toni, 2016] Robert Craven and Francesca Toni. Argument graphs and assumption-based argumentation. *Artificial Intelligence*, 233:1–59, 2016.

[Čyras and Toni, 2016a] Kristijonas Čyras and Francesca Toni. ABA+: Assumption-based argumentation with preferences. In Chitta Baral, James P. Delgrande, and Frank Wolter, editors, *Principles of Knowledge Representation and Reasoning: Proceedings of the 15th International Conference, (KR'16)*, pages 553–556. AAAI Press, 2016.

[Čyras and Toni, 2016b] Kristijonas Čyras and Francesca Toni. Properties of ABA+ for non-monotonic reasoning. In *Proceedings of the 16th International Workshop on Non-Monotonic Reasoning (NMR'16), CoRR*, volume abs/1603.08714, 2016.

[Čyras et al., 2016] Kristijonas Čyras, Ken Satoh, and Francesca Toni. Abstract argumentation for case-based reasoning. In Chitta Baral, James P. Delgrande, and Frank Wolter, editors, *Principles of Knowledge Representation and Reasoning: Proceedings of the 15th International Conference, (KR'16)*, pages 549–552. AAAI Press, 2016.

[Delgrande et al., 2004] James P. Delgrande, Torsten H. Schaub, Hans Tompits, and Kewen Wang. A classification and survey of preference handling approaches in nonmonotonic reasoning. *Computational Intelligence*, 20(2):308–334, 2004.

[Delgrande et al., 2009] James P. Delgrande, Torsten Grote, and Aaron Hunter. A general approach to the verification of cryptographic protocols using answer set programming. In

Proceedings of the 10th International Conference on Logic Programming and Nonmonotonic Reasoning (LPNMR'09), pages 355–367, 2009.

[Dimopoulos et al., 2002] Yannis Dimopoulos, Bernhard Nebel, and Francesca Toni. On the computational complexity of assumption-based argumentation for default reasoning. *Artificial Intelligence*, 141(1/2):57–78, 2002.

[Domshlak et al., 2011] Carmel Domshlak, Eyke Hüllermeier, Souhila Kaci, and Henri Prade. Preferences in AI: an overview. *Artificial Intelligence*, 175(7-8):1037–1052, 2011.

[Dung and Thang, 2010] Phan Minh Dung and Phan Minh Thang. Towards (probabilistic) argumentation for jury-based dispute resolution. In *Proceedings of the 2010 Conference on Computational Models of Argument (COMMA'10)*, pages 171–182. IOS Press, 2010.

[Dung et al., 2002] Phan Minh Dung, Paolo Mancarella, and Francesca Toni. Argumentation-based proof procedures for credulous and sceptical non-monotonic reasoning. In A.C. Kakas and F. Sadri, editors, *Computational Logic: Logic Programming and Beyond - Essays in Honour of Robert A. Kowalski*, volume 2408 of *Lecture Notes in Artificial Intelligence*, pages 289–310. Springer Verlag, 2002.

[Dung et al., 2006] Phan Minh Dung, Robert A. Kowalski, and Francesca Toni. Dialectic proof procedures for assumption-based, admissible argumentation. *Artificial Intelligence*, 170:114–159, 2006.

[Dung et al., 2007] Phan Minh Dung, Paolo Mancarella, and Francesca Toni. Computing ideal sceptical argumentation. *Artificial Intelligence*, 171(10–15):642–674, 2007.

[Dung et al., 2009] Phan Minh Dung, Robert A. Kowalski, and Francesca Toni. Assumption-based argumentation. In Iyad Rahwan and Guillermo R. Simari, editors, *Argumentation in Artificial Intelligence*, pages 25–44. Springer, 2009.

[Dung, 1991] Phan Minh Dung. Negations as hypotheses: An abductive foundation for logic programming. In Koichi Furukawa, editor, *Logic Programming, Proceedings of the 8th International Conference (ICLP'91)*, pages 3–17. MIT Press, 1991.

[Dung, 1995] Phan Minh Dung. On the acceptability of arguments and its fundamental role in non-monotonic reasoning, logic programming and n-person games. *Artificial Intelligence*, 77:321–357, 1995.

[Dunne, 2009] Paul E. Dunne. The computational complexity of ideal semantics. *Artificial Intelligence*, 173(18):1559–1591, 2009.

[Eiter et al., 2008] Thomas Eiter, Giovambattista Ianni, Thomas Lukasiewicz, Roman Schindlauer, and Hans Tompits. Combining answer set programming with description logics for the semantic web. *Artificial Intelligence*, 172(12-13):1495–1539, 2008.

[Erdem, 2011] Esra Erdem. Applications of answer set programming in phylogenetic systematics. In *Logic Programming, Knowledge Representation, and Nonmonotonic Reasoning - Essays Dedicated to Michael Gelfond on the Occasion of His 65th Birthday*, pages 415–431, 2011.

[Fan and Toni, 2011a] Xiuyi Fan and Francesca Toni. Assumption-based argumentation dialogues. In *Proceedings of the 22nd International Joint Conference on Artificial Intelligence (IJCAI'11)*, pages 198–203, 2011.

[Fan and Toni, 2011b] Xiuyi Fan and Francesca Toni. Conflict resolution with argumentation dialogues. In *Proceedings of the 2011 International Conference on Autonomous Agents & Multiagent Systems (AAMAS'11)*, pages 1095–1096, 2011.

[Fan and Toni, 2011c] Xiuyi Fan and Francesca Toni. A first step towards argumentation dialogues for discovery. In *Theory and Applications of Formal Argumentation - 1st International Workshop, (TAFA'11)*, pages 263–279, 2011.

[Fan and Toni, 2012a] Xiuyi Fan and Francesca Toni. Agent strategies for ABA-based information-seeking and inquiry dialogues. In *20th European Conference on Artificial Intelligence (ECAI'12)*, pages 324–329, 2012.

[Fan and Toni, 2012b] Xiuyi Fan and Francesca Toni. Argumentation dialogues for two-agent conflict resolution. In *Proceedings of the 2012 Conference on Computational Models of Argument (COMMA'12)*, 2012.

[Fan and Toni, 2012c] Xiuyi Fan and Francesca Toni. Mechanism design for argumentation-based persuasion. In *Proceedings of the 2012 Conference on Computational Models of Argument (COMMA'12)*, pages 322–333, 2012.

[Fan and Toni, 2014a] Xiuyi Fan and Francesca Toni. Decision making with assumption-based argumentation. In Elizabeth Black, Sanjay Modgil, and Nir Oren, editors, *Theory* and *Applications of Formal Argumentation - 2nd International Workshop (TAFA'13)*, volume 8306 of *Lecture Notes in Computer Science*, pages 127–142. Springer, 2014.

[Fan and Toni, 2014b] Xiuyi Fan and Francesca Toni. A general framework for sound assumption-based argumentation dialogues. *Artificial Intelligence*, 216:20–54, 2014.

[Fan and Toni, 2015a] Xiuyi Fan and Francesca Toni. Mechanism design for argumentation-based information-seeking and inquiry. In Qingliang Chen, Paolo Torroni, Serena Villata, Jane Yung-jen Hsu, and Andrea Omicini, editors, *Principles and Practice of Multi-Agent Systems - 18th International Conference (PRIMA'15)*, volume 9387 of *Lecture Notes in Computer Science*, pages 519–527. Springer, 2015.

[Fan and Toni, 2015b] Xiuyi Fan and Francesca Toni. On computing explanations for non-acceptable arguments. In Elizabeth Black, Sanjay Modgil, and Nir Oren, editors, *Theory and Applications of Formal Argumentation - 3rd International Workshop (TAFA'15)*, volume 9524 of *Lecture Notes in Computer Science*, pages 112–127. Springer, 2015.

[Fan and Toni, 2015c] Xiuyi Fan and Francesca Toni. On computing explanations in argumentation. In Blai Bonet and Sven Koenig, editors, *Proceedings of the 29th AAAI Conference on Artificial Intelligence (AAAI'15)*, pages 1496–1502. AAAI Press, 2015.

[Fan and Toni, 2016] Xiuyi Fan and Francesca Toni. On the interplay between games, argumentation and dialogues. In Catholijn M. Jonker, Stacy Marsella, John Thangarajah, and Karl Tuyls, editors, *Proceedings of the 2016 International Conference on Autonomous Agents & Multiagent Systems (AAMAS'16)*, pages 260–268. ACM, 2016.

[Fan et al., 2013] Xiuyi Fan, Robert Craven, Ramsay Singer, Francesca Toni, and Matthew Williams. Assumption-based argumentation for decision-making with preferences: A medical case study. In João Leite, Tran Cao Son, Paolo Torroni, Leon van der Torre, and Stefan Woltran, editors, *Computational Logic in Multi-Agent Systems: 14th International Workshop (CLIMA'13)*, volume 8143 of *Lecture Notes in Computer Science*, pages 374–390. Springer, 2013.

[Fan et al., 2014] Xiuyi Fan, Francesca Toni, Andrei Mocanu, and Matthew Williams. Dialogical two-agent decision making with assumption-based argumentation. In *Proceedings of the 2014 International Conference on Autonomous Agents & Multiagent Systems (AAMAS'14)*, pages 533–540, 2014.

[Gaertner and Toni, 2008] Dorian Gaertner and Francesca Toni. Hybrid argumentation and its properties. In *Proceedings of the 2008 Conference on Computational Models of Argument (COMMA'08)*, pages 183–195, 2008.

[Gao et al., 2016] Yang Gao, Francesca Toni, Hao Wang, and Fanjiang Xu. Argumentation-based multi-agent decision making with privacy preserved. In Catholijn M. Jonker, Stacy Marsella, John Thangarajah, and Karl Tuyls, editors, *Proceedings of the 2016 International Conference on Autonomous Agents & Multiagent Systems (AAMAS'16)*, pages 1153–1161. ACM, 2016.

[García and Simari, 2014] Alejandro Javier García and Guillermo Ricardo Simari. Defeasible logic programming: DeLP-servers, contextual queries, and explanations for answers. *Argument & Computation*, 5(1):63–88, 2014.

[Gebser et al., 2011a] Martin Gebser, Benjamin Kaufmann, Roland Kaminski, Max Ostrowski, Torsten Schaub, and Marius Thomas Schneider. Potassco: The Potsdam answer set solving collection. *AI Communications*, 24(2):107–124, 2011.

[Gebser et al., 2011b] Martin Gebser, Torsten Schaub, Sven Thiele, and Philippe Veber. Detecting inconsistencies in large biological networks with answer set programming. *Theory and Practice of Logic Programming*, 11(2-3):323–360, 2011.

[Gelder et al., 1991] Allen Van Gelder, Kenneth A. Ross, and John S. Schlipf. The well-founded semantics for general logic programs. *Journal of the ACM*, 38(3):620–650, 1991.

[Gelfond and Lifschitz, 1988] Michael Gelfond and Vladimir Lifschitz. The stable model semantics for logic programming. In R. Kowalski and K. A. Bowen, editors, *Logic Programming, Proceedings of the 5th International Conference (ICLP'88)*, pages 1070–1080. MIT Press, 1988.

[Heyninck and Straßer, 2016] Jesse Heyninck and Christian Straßer. Relations between assumption-based approaches in nonmonotonic logic and formal argumentation. *CoRR*, abs/1604.00162, 2016.

[Kakas and Moraitis, 2003] Antonis C. Kakas and Pavlos Moraitis. Argumentation based decision making for autonomous agents. In *Proceedings of the 2003 International Conference on Autonomous Agents & Multiagent Systems (AAMAS'03)*, pages 883–890, Melbourne, 2003. ACM Press.

[Kakas and Toni, 1999] Antonis C. Kakas and Francesca Toni. Computing argumentation in logic programming. *Journal of Logic and Computation*, 9(4):515–562, 1999.

[Kowalski and Toni, 1996] Robert A. Kowalski and Francesca Toni. Abstract argumentation. *Artificial Intelligence and Law*, 4(3–4):275–296, 1996.

[Leone et al., 2006] Nicola Leone, Gerald Pfeifer, Wolfgang Faber, Thomas Eiter, Georg Gottlob, Simona Perri, and Francesco Scarcello. The DLV system for knowledge representation and reasoning. *ACM Transactions on Computational Logic*, 7(3):499–562, 2006.

[Lifschitz, 2002] Vladimir Lifschitz. Answer set programming and plan generation. *Artificial Intelligence*, 138(1-2):39–54, 2002.

[McCarthy, 1980] John McCarthy. Circumscription - a form of non-monotonic reasoning. *Artificial Intelligence*, 13(1-2):27–39, 1980.

[Mocanu et al., 2016] Andrei Mocanu, Xiuyi Fan, Francesca Toni, Matthew Williams, and Jiarong Chen. Online argumentation-based platform for recommending medical literature. In I. Hatzilygeroudis, V. Palade, and J. Prentzas, editors, *Combinations of Intelligent Methods and Applications*, volume 46 of *Smart Innovation, Systems and Technologies*, pages 97–115. Springer, 2016.

[Modgil and Prakken, 2014] Sanjay Modgil and Henry Prakken. The ASPIC+ framework for structured argumentation: A tutorial. *Argument & Computation*, 5(1):31–62, 2014.

[Modgil, 2009] Sanjay Modgil. Reasoning about preferences in argumentation frameworks. *Artificial Intelligence*, 173(9-10):901–934, 2009.

[Moore, 1985] Robert Moore. Semantical considerations on non-monotonic logic. *Artificial Intelligence*, 25(1):75–94, 1985.

[Newton-Smith, 1981] William H. Newton-Smith. *The Rationality of Science*. Routledge, 1981.

[Prakken and Sartor, 1999] Henry Prakken and Giovanni Sartor. A system for defeasible argumentation, with defeasible priorities. In Michael Wooldridge and Manuela Veloso, editors, *Artificial Intelligence Today*, volume 1600 of *Lecture Notes in Computer Science*, pages 365–379. Springer, 1999.

[Reiter, 1980] Raymond Reiter. A logic for default reasoning. *Artificial Intelligence*, 13(1-2):81–132, 1980.

[Ricca et al., 2010] Francesco Ricca, Antonella Dimasi, Giovanni Grasso, Salvatore Maria Ielpa, Salvatore Iiritano, Marco Manna, and Nicola Leone. A logic-based system for e-tourism. *Fundamenta Informaticae*, 105(1-2):35–55, 2010.

[Ricca et al., 2012] Francesco Ricca, Giovanni Grasso, Mario Alviano, Marco Manna, Vincenzino Lio, Salvatore Iiritano, and Nicola Leone. Team-building with answer set programming in the Gioia-Tauro Seaport. *Theory and Practice of Logic Programming*, 12(3):361–381, 2012.

[Sakama and Inoue, 2000] Chiaki Sakama and Katsumi Inoue. Prioritized logic programming and its application to commonsense reasoning. *Artificial Intelligence*, 123(1-2):185–222, 2000.

[Schulz and Toni, 2014] Claudia Schulz and Francesca Toni. Complete assumption labellings. In *Proceedings of the 2014 Conference on Computational Models of Argument (COMMA'14)*, pages 405–412, 2014.

[Schulz and Toni, 2015] Claudia Schulz and Francesca Toni. Logic programming in assumption-based argumentation revisited - semantics and graphical representation. In *Proceedings of the 29th AAAI Conference on Artificial Intelligence (AAAI'15)*, pages 1569–1575, 2015.

[Schulz and Toni, 2016] Claudia Schulz and Francesca Toni. Justifying answer sets using argumentation. *Theory and Practice of Logic Programming*, 16(1):59–110, 2016.

[Schulz and Toni, 2017] Claudia Schulz and Francesca Toni. Labellings for assumption-based and abstract argumentation. *International Journal of Approximate Reasoning*, 84:110–149, 2017.

[Schulz et al., 2015] Claudia Schulz, Ken Satoh, and Francesca Toni. Characterising and explaining inconsistency in logic programs. In Francesco Calimeri, Giovambattista Ianni, and Miroslaw Truszczynski, editors, *Proceedings of the 13th International Conference on*

Logic Programming and Nonmonotonic Reasoning (LPNMR'15), volume 9345 of *Lecture Notes in Computer Science*, pages 467–479. Springer, 2015.

[Seselja and Straßer, 2013] Dunja Seselja and Christian Straßer. Abstract argumentation and explanation applied to scientific debates. *Synthese*, 190(12):2195–2217, 2013.

[Simari and Loui, 1992] Guillermo Ricardo Simari and Ronald Prescott Loui. A mathematical treatment of defeasible reasoning and its implementation. *Artificial Intelligence*, 53(2-3):125–157, 1992.

[Terracina et al., 2013] Giorgio Terracina, Alessandra Martello, and Nicola Leone. Logic-based techniques for data cleaning: An application to the Italian national healthcare system. In *Proceedings of the 12th International Conference on Logic Programming and Nonmonotonic Reasoning (LPNMR'13)*, pages 524–529, 2013.

[Thang and Luong, 2013] Phan Minh Thang and H.T. Luong. Translating preferred subtheories into structured argumentation. *Journal of Logic and Computation*, 2013.

[Toni, 2008a] Francesca Toni. Assumption-based argumentation for closed and consistent defeasible reasoning. In Ken Satoh, Akihiro Inokuchi, Katashi Nagao, and Takahiro Kawamura, editors, *New Frontiers in Artificial Intelligence, JSAI 2007 Conference and Workshops, Miyazaki, Japan, June 18-22, 2007, Revised Selected Papers*, volume 4914 of *Lecture Notes in Computer Science*, pages 390–402. Springer, 2008.

[Toni, 2008b] Francesca Toni. Assumption-based argumentation for epistemic and practical reasoning. In Pompeu Casanovas, Giovanni Sartor, Nuria Casellas, and Rossella Rubino, editors, *Computable Models of the Law, Languages, Dialogues, Games, Ontologies*, volume 4884 of *Lecture Notes in Computer Science*, pages 185–202. Springer, 2008.

[Toni, 2012] Francesca Toni. Reasoning on the web with assumption-based argumentation. In Thomas Eiter and Thomas Krennwallner, editors, *Reasoning Web. Semantic Technologies for Advanced Query Answering. 8th International Summer School. Proceedings.*, volume 7487 of *LNCS*, pages 370–386. Springer, 2012.

[Toni, 2013] Francesca Toni. A generalised framework for dispute derivations in assumption-based argumentation. *Artificial Intelligence*, 195:1–43, 2013.

[Toni, 2014] Francesca Toni. A tutorial on assumption-based argumentation. *Argument & Computation*, 5(1):89–117, 2014.

[Wakaki, 2014] Toshiko Wakaki. Assumption-based argumentation equipped with preferences. In Hoa Khanh Dam, Jeremy V. Pitt, Yang Xu, Guido Governatori, and Takayuki Ito, editors, *Principles and Practice of Multi-Agent Systems - 17th International Conference (PRIMA'14)*, volume 8861 of *Lecture Notes in Computer Science*, pages 116–132. Springer, 2014.

[Zhong et al., 2014] Qiaoting Zhong, Xiuyi Fan, Francesca Toni, and Xudong Luo. Explaining best decisions via argumentation. In *Proceedings of the European Conference on Social Intelligence (ECSI'14)*, pages 224–237, 2014.

Kristijonas Čyras
Imperial College London, UK
Email: k.cyras@imperial.ac.uk

Xiuyi Fan
Nanyang Technological University, Singapore
Email: xyfan@ntu.edu.sg

Claudia Schulz
Imperial College London, UK
Email: claudia.schulz@imperial.ac.uk

Francesca Toni
Imperial College London, UK
Email: ft@imperial.ac.uk

8
Argumentation Based on Logic Programming

ALEJANDRO J. GARCÍA, GUILLERMO R. SIMARI

ABSTRACT. Among of the programming paradigms based on formal logic, Logic Programming has been a successful effort to create a declarative model of expressing computational processes producing significant theoretical and practical results; as such, the area has contributed computationally attractive systems with remarkable success in many applications. By blending concepts from the areas of Logic Programming and Argumentation, Defeasible Logic Programming (DeLP) proposes a computational reasoning system with an argumentation engine at its core capable of obtaining answers from a knowledge base which is represented with a language that uses logic programming constructs extended with defeasible rules. The careful integration of foundational intuitions and concepts from both areas has formulated a framework that inherits from the logic programming field its expressivity and computational efficiency and receives from argumentation theory a human-like reasoning model facilitating its use in applications. In this chapter, the basic elements of logic programming will be succinctly recalled, and the DeLP language will be formally introduced together with the warranting process that obtains the answers for queries. DeLP-Servers, which give possibly distributed client agents running on remote hosts the ability to consult different reasoning services are presented. Finally, some extensions and applications of DeLP are briefly described.

1 Introduction

From the earliest beginnings of Artificial Intelligence, researchers working in Knowledge Representation and Reasoning have been engaged in the problem of understanding how humans obtain useful conclusions from their store of beliefs using their cognitive abilities. The goal of these investigations was and is to discover the inner workings of the processes with the goal of getting a computationally effective way of simulating them [Turing, 1950]. Logic Programming is one of the programming paradigms based on formal logic that represents a successful effort to develop a declarative form of expressing computational processes and has produced significant theoretical and practical results. Logic Programming has developed computationally attractive systems with remarkable success in many applications. By combining ideas from the areas of Logic Programming and Argumentation, Defeasible Logic Programming (abbreviated DeLP) [García, 2000; García and Simari, 2014] proposes a computational reasoning system with an argumentation engine at its core to obtain answers from a knowledge base which is represented with a language that uses logic programming constructs extended with defeasible rules. The confluence of notions which are important by themselves produces a framework that inherits from the logic programming field its expressivity and computational efficiency and takes from argumentation theory a human-like reasoning model facilitating its use in applications.

2 Knowledge Representation and Reasoning

As the field of Knowledge Representation and Reasoning developed, different proposals have been advanced, offering interesting and highly valuable alternatives. Among these efforts were McCarthy's Circumscription [McCarthy, 1980], Reiter's Default Logic [Reiter, 1980], McDermott and Doyle's Nonmonotonic Logic [McDermott and Doyle, 1980], Moore's Autoepistemic Logic [Moore, 1984], the work of Pollock on defeasible reasoning [Pollock, 1987; Pollock, 1996], and Loui on argumentation [Loui, 1987]. In particular, Default Logic and Autoepistemic Logic found an interesting embedding in logic programming [Gelfond and Lifschitz, 1988] and argumentation was proposed as a general tool for Knowledge Representation and Reasoning [Lin and Shoham, 1989]. These lines of research, showing the importance and substance of the field, brought for many proposals exploring possible avenues to produce a computational approach to different aspects of commonsense reasoning. Focusing on argumentation community work, it is possible to mention the work of Loui [Loui, 1987], Nute [Nute, 1987; Nute, 1988], Lin and Shoham [Lin and Shoham, 1989], Simari and Loui [Simari, 1989; Simari and Loui, 1992] that considered defeasible reasoning and argumentation, Dung [Dung, 1993; Dung, 1995] on abstract argumentation frameworks, Bondarenko et al. [Bondarenko et al., 1993; Bondarenko et al., 1997] leading to Assumption-Based Argumentation, Prakken [Prakken, 2010] and Modgil and Prakken [Modgil and Prakken, 2013] on the development of ASPIC+, and Besnard and Hunter [Besnard and Hunter, 2001; Besnard and Hunter, 2008] on the introduction of an argumentation system based on classical logic. Several accounts of the research on argumentation systems have been produced, see for instance [Chesñevar et al., 2000; Prakken and Vreeswijk, 2002; Bench-Capon and Dunne, 2007; Besnard and Hunter, 2008; Rahwan and Simari, 2009].

Logic programming contributed to the area of Nonmonotonic and Defeasible Reasoning with efficient computational tools that are based on sound theoretical results attracting the attention of real-world practitioners wishing to implement intelligent applications. These characteristics made natural to choose logic programming as the substratum necessary to implement defeasible reasoning.

Defeasible Logic Programming, or DeLP, is an efficient computational system that recognize its origins in [Simari, 1989; Simari and Loui, 1992], which in turn extended the work Pollock [Pollock, 1987], Loui [Loui, 1987], and Nute [Nute, 1987; Nute, 1988]. The system employs an argumentation mechanism to obtain answers from a knowledge base which described using a logic programming language, but including defeasible rules as in [Simari, 1989; Simari and Loui, 1992]. The logic programming language of DeLP includes classical negation and default negation (see [Baral and Gelfond, 1994]), sharing the declarative capability of logic programming for representing knowledge, and adding the possibility of representing a weaker relation between the head and the body of rules as *defeasible rules*. The inferential mechanism will weight reasons for and against a potential conclusion deciding whether it can be obtained (warranted) from the knowledge base (see [García, 2000; García and Simari, 2004; García and Simari, 2014]).

A recent development, that is currently extended as the core inferential engine for agents in multi-agent systems (MAS), a Defeasible Logic Programming Server, or DeLP-Server for short [García et al., 2007b] has been introduced. The DeLP reasoning engine constitutes a central part of the DeLP-Server system (see Section 5 below for a more complete description). The DeLP-Servers aim at supporting the implementation of argumentative reasoning services working over knowledge bases represented as defeasible logic programs in a MAS environment, providing a cognitive infrastructure necessary for the autonomous agents. Multi-agent systems are implemented as part of a distributed environment, where the location of hosts for services and agents is irrelevant; some of the available services for the agents in the system could be DeLP-Servers that support different reasoning services which own specialized knowledge bases local to the server. These DeLP-Servers also provide a way of integrating agents' knowledge as part of query in the form of a defeasible logic program that will act as a complement of the stored knowledge base; several different ways of performing the integration are provided as local operators of the server. The integration will be temporary and the server will not be permanently modified. Also, in recent developments [Teze et al., 2014; Teze et al., 2015], the possibility of changing the comparison criteria for arguments has been analyzed.

3 The Elements of Logic Programming

The field of Logic Programming has evolved since its beginnings in the late 1960's and 1970's from the inception of PROLOG (*PROgramming in LOGic*) to full-fledged answer set solvers for Answer Set Programming (ASP) [Baral, 2003; Gelfond, 2008] which are computationally efficient and appealing for their capability of representing problems declaratively (see for instance the DLV system [Leone et al., 2006]). A recent recounting on the history of Logic Programming can be found in [Kowalski, 2014]. Given a logic programming language, a program is expressed as a set of sentences denoted as follows:

$$H \leftarrow B_1, B_2, \ldots, B_n$$

These sentences represent elements of the addressed application domain and are also called rules or clauses. Rules have both a declaratively read: "*If B_1 and*

$B_2 \ldots$ and B_n then H", and a procedurally read as a goal-reduction procedure: "*to show H, show B_1 and show B_1 ... and show B_n.*" The first part of the rule, H, is called the head of the rule and $B_1, B_2, \ldots, B_n$ is called the body. Facts are rules that have no body, and they are usually written in the form:

$$H \leftarrow \text{ or simply } H.$$

The simplest form of a rule is when $H, B_1, \ldots, B_n$ are all atomic formulæ, in these case the clauses are called *definite clauses* or *Horn clauses* and a program containing only this type of clauses is called a *definite logic program*. Nevertheless, there exist many extensions such as *normal logic programs, extended logic programs,* and *disjunctive logic programs* which in turn admit other categories (regarding the last mentioned extensions the interested reader is referred to [Minker, 1994]). Normal logic programs are built using normal clauses which are clauses where some atoms can be preceded by the default negation *not*, having the following structure:

$$H \leftarrow H_1, ..., H_n, not\ H_{n+1}, \ldots not\ H_m.$$

where $H, H_i, 1 \leq i \leq m$, are atoms.

Extended logic programs are defined through extended clauses, that are defined as:

$$L \leftarrow L_1, ..., L_n, not\ L_{n+1}, \ldots not\ L_m.$$

where $L, L_i, 1 \leq i \leq m$, are literals, *i.e.*, atoms possibly preceded by the strict (classical) negation $\sim$. Note that the introduction of the default negation *not* affecting the literals in the body of a normal or extended clause transforms logic programming in a nonmonotonic knowledge representation formalism which has been shown to possess great expressiveness.

In certain systems of logic programming, *e.g.*, ASP, logic programs just have a declarative reading, avoiding a procedural interpretation that could be used to change the behavior of the program. Clauses like the following:

$$mortal(X) \leftarrow human(X).$$

can be thought as a declarative definition of being *mortal* or, procedurally, as a form to test if a given individual substituted in lieu of the X is *human* and therefore *mortal*. Facts can also have a procedural interpretation, for example, the clause:

$$human(socrates).$$

could be used as a stipulation that socrates is *human*, or as a way of finding out through a query whether *socrates* is *human*. The declarative reading of logic programs can be used by a programmer to verify their correctness. Moreover, research has produced effective logic-based program transformation techniques which can be used to transform logic programs into logically equivalent programs that are more efficient.

As a logic-based system, logic programming shares with logical systems many desirable features. One that is relevant here is that logic programming is declarative in nature since tries to separate the "what" computation should be performed from the "how" that computation should be done.[1] This essential characteristic gives an excellent framework to represent knowledge independently of how the system is realized computationally with the advantage of favoring portability and the integration in different real-world scenarios. Thus, code optimization over the computational aspect of the problem can be accomplished separately without affecting the representation and systems themselves become extensible to a greater extent helping to address the representation of particular forms of knowledge. We will now summarily describe the field of knowledge representation to situate logic programming and defeasible logic programming in it.

4 Defeasible Logic Programming (DeLP): A Language for Argumentation

The computational framework of *Defeasible Logic Programming* (DeLP), as introduced in [García and Simari, 2004], is a formalism that combines techniques of both logic programming and defeasible argumentation. As in logic programming, in DeLP knowledge is represented using facts and rules; however, DeLP also provides the possibility of representing information in the form of *weak rules* in a declarative manner. These weak rules are the key element for introducing *defeasibility* [Pollock, 1995] and they are used to represent a relation between pieces of knowledge that could be *defeated* when all things have been considered. DeLP uses a defeasible argumentation inference mechanism for warranting the entailed conclusions.

A *defeasible logic program* (*DeLP-program* for short), is a set of facts, strict rules and defeasible rules, defined as follows.

Definition 4.1 (Fact)
A fact is a literal; that is, a ground atom or a negated ground atom.

Definition 4.2 (Strict Rule)
A Strict Rule is an ordered pair, denoted "Head $\leftarrow$ Body", whose first member, Head, is a literal, and whose second member, Body, is a finite non-empty set of literals. A strict rule with head L_0 and body $\{L_1, \ldots, L_n\}$ can also be written as: $L_0 \leftarrow L_1, \ldots, L_n, (n > 0)$.

The syntax of strict rules correspond to *basic rules* in Logic Programming [Lifschitz, 1996a], but we call them 'strict' to emphasize the difference with the 'defeasible' ones. Note that strong negation may be used in the head of the rules; some examples of strict rules are: "$\sim innocent \leftarrow guilty$", "$dead \leftarrow \sim alive$".

---

[1] The reader can find an interesting discussion about declarativeness in Robert Harper's blog https://goo.gl/MJVnS3, in particular we use the term "declarative" following Definition 4.

Definition 4.3 (Defeasible Rule)

A Defeasible Rule is an ordered pair, denoted "Head—≺ Body", whose first member, Head, is a literal, and whose second member, Body, is a finite non-empty set of literals. A defeasible rule with head L_0 and body $\{L_1, \ldots, L_n\}$ can also be written as: $L_0 \prec L_1, \ldots, L_n$, $(n > 0)$.

Syntactically, the symbol " —≺ " is all that distinguishes a defeasible rule from a strict one. Pragmatically, a defeasible rule is used to represent defeasible knowledge, *i.e.*, tentative information that may be used if nothing could be posed against it. Thus, whereas a strict rule is used to represent non-defeasible information such as: *"bird ← penguin"* which expresses that *"all penguins are birds"*, a defeasible rule is used to represent defeasible knowledge such as *"flies—≺ bird"* which conveys that: *"reasons to believe that it is a bird provide reasons to believe that it flies"*, or *"birds are presumed to fly"* or *"usually, a bird can fly."*

Notice that the symbols " —≺ " and " ← " denote meta-relations between sets of literals, and have no interaction with language symbols; thus, there is no contraposition for program rules. In our examples, we will follow standard typographic conventions of Logic Programming, conveniently extending them for representing DeLP rules.

It is important to remark that a defeasible rule is not a default rule; although both are meta-relations. In a default rule $\alpha : \beta / \gamma$, where α is the *Prerequisite*, β is the *Justification* and γ is the *Conclusion*. The justification β represents a consistency requirement that explicitly controls the applicability of the rule blocking that application when β is inconsistent with the current beliefs; meanwhile, in a defeasible rule the justification part is absent and the defeasible rule solely stands for a weak connection between head (conclusion) and body (prerequisite) of this rule and nothing else.

To obtain the conclusion of a defeasible rule, a global analysis that considers arguments and counter-arguments involving the rule is performed by a dialectical inference mechanism; therefore, in a defeasible rule there is no need of a programmed check such as the justification part in a default rule. This characteristic becomes important when the defeasible program changes since is not necessary to consider the possible interactions with other rules, making the representation more declarative; notice that when the beliefs supported by knowledge represented by default rules change it is possible that some of those rules might need to be modified in the justification part, possibly leading to a cascade effect in the changes, see [Brewka and Eiter, 2000]. Generally speaking, changes in the defeasible part of a defeasible program, will be effected with the sole addition or removal of defeasible rules to the representation; thus, if the knowledge base is dynamic, defeasible rules are a better alternative. We refer the interested reader to [Nute, 1994] where other nonmonotonic theories are compared.

The information that is represented by a strict rule is not tentative. For instance, given "$\sim innocent \leftarrow guilty$" and the fact *guilty*, it will not be possible to infer *innocent*; that is, if *guilty* has a strict derivation then ∼*innocent* will also have a strict derivation and *innocent* cannot be concluded. However, as it will be shown below, if *guilty* is a tentative conclusion, then ∼*innocent* will also be a tentative conclusion, and *innocent* may be concluded if there is pertinent information that supports that literal.

When required, a DeLP-program, which following the logic programming standard presentation is a finite set of facts and rules (strict and defeasible), is denoted by the pair (Π, Δ) distinguishing the subset Π of facts and strict rules (that represents non-defeasible knowledge), and the subset Δ of defeasible rules. Observe that strict and defeasible rules are ground; however, following the usual convention [Lifschitz, 1996b], some examples will make use of "schematic rules" with (schematic) variables[2]. To distinguish variables, they are written with an uppercase letter, and in the examples a period may be used as a delimiter between rules or facts. As in logic programming, queries can be posed to a program and a DeLP-*query* is a ground literal that DeLP will try to warrant.

Example 4.4 *Consider the DeLP-program* $(\Pi_{4.4}, \Delta_{4.4})$ *where:*

$$\Pi_{4.4} = \left\{ \begin{array}{l} \sim risky\_co(X) \leftarrow strong\_co(X). \\ in\_fusion(X,Y) \leftarrow declared\_in\_fusion(X,Y). \\ web\_good\_price(acme). \\ declared\_in\_fusion(acme, steel). \\ web\_strong\_co(steel). \end{array} \right\}$$

$$\Delta_{4.4} = \left\{ \begin{array}{l} buy\_stock(X) \prec good\_price(X). \\ \sim buy\_stock(X) \prec good\_price(X), risky\_co(X). \\ good\_price(X) \prec web\_good\_price(X). \\ \sim good\_price(X) \prec web\_downtrend(X). \\ strong\_co(X) \prec web\_strong\_co(X). \\ risky\_co(X) \prec in\_fusion(X,Y). \\ risky\_co(X) \prec rumors\_of\_debt(X). \\ in\_fusion(X) \prec rumors\_in\_funsion(X). \\ \sim risky\_co(X) \prec in\_fusion(X,Y), strong\_co(Y). \end{array} \right\}$$

All the literals with names prefixed *web* should be read as information obtained from a web page, *i.e.*, *web_good_price* means *a good price of stock reported on a web page* and *web_downtrend* should be understood as *downtrend information of a particular stock observed in a web site*; also, the ending *co* is intended as an abbreviation of *company*, *i.e.*, *strong_co* should be understood as *strong_company*.

Defeasible rules allow to obtain tentative conclusions. A *defeasible derivation* of a literal H from a DeLP-program (Π, Δ), denoted by $(\Pi, \Delta) \mathrel{|\!\sim} H$, is a finite

---
[2]A schematic rule represents a template that abbreviates all the possible ground rules that can be produced by consistently instantiating the schematic variables that appear on that rule with the individual constants appearing in the DeLP-program.

sequence of ground literals $L_1, L_2, \ldots, L_n = H$, where either:

1. L_i is a fact in Π, or

2. there exists a rule R_i in (Π, Δ) (strict or defeasible) with head L_i and body $B_1, B_2, \ldots, B_k$ and every literal of the body is an element L_j of the sequence appearing before L_i ($j < i$).

For instance, the sequence of ground literals:

$$declared\_in\_fusion(acme, steel), in\_fusion(acme, steel), risky\_co(acme)$$

is a defeasible derivation from $(\Pi_{4.4}, \Delta_{4.4})$ for the literal $risky\_co(acme)$, i.e., a defeasible derivation that *acme is a risky company*. In DeLP, a derivation from $(\Pi, \emptyset)$, i.e., a derivation that uses only strict rules and facts, is called a *strict derivation*; for instance,

$$declared\_in\_fusion(acme, steel), in\_fusion(acme, steel)$$

is a strict derivation for the literal $in\_fusion(acme, steel)$. Note that although $\sim risky\_co(steel)$ is the head of a strict rule, there is no strict derivation for it; the sequence:

$$web\_strong\_co(steel), strong\_co(steel), \sim risky\_co(steel)$$

corresponds to a defeasible derivation for $\sim risky\_co(steel)$ since it makes use of the defeasible rule $strong\_co(steel) \prec web\_strong\_co(steel)$. In order words, a derivation is defeasible if at least one defeasible rule is used, otherwise it is strict.

Two literals are said to be contradictory when one is the complement of the other with respect to strong negation; for instance, the literals $good\_price(acme)$ and $\sim good\_price(acme)$ are contradictory. Since Π represents non-defeasible information, it is natural to require certain internal coherence for this set. Consequently, in DeLP two contradictory literals cannot have a strict derivation from a valid program.

Nevertheless, from a valid DeLP-program, still it is possible to have defeasible derivations for contradictory literals (for example, $risky\_co(acme)$ and $\sim risky\_co(acme)$ from $(\Pi_{4.4}, \Delta_{4.4})$). In these situations, in order to decide which literal is accepted as warranted, DeLP incorporates a defeasible argumentation formalism that allows the identification of the pieces of knowledge that are in contradiction. Then, a dialectical process is used for deciding which information prevails as warranted. This process involves the construction, evaluation, and contraposition of arguments and counter-arguments, as described next.

Definition 4.5 (Argument Structure) *Let H be a ground literal, (Π, Δ) a DeLP-program, and $\mathcal{A} \subseteq \Delta$. The pair $\langle \mathcal{A}, H \rangle$ is an argument structure if:*

1. $(\Pi, \mathcal{A}) \mathrel{|\!\sim} H$, *i.e., there exists a defeasible derivation for H from $(\Pi, \mathcal{A})$,*

2. *there is no defeasible derivation from* $(\Pi, \mathcal{A})$ *of contradictory literals, and*
3. *there is no proper subset* $\mathcal{A}'$ *of* $\mathcal{A}$ *such that* $\mathcal{A}'$ *satisfies* (1) *and* (2).

In an argument structure $\langle \mathcal{A}, H \rangle$, H is the claim and $\mathcal{A}$ is the argument supporting that claim. Next, we show a few argument structures that can be obtained from $(\Pi_{4.4}, \Delta_{4.4})$ from Example 4.4 which will be used for other examples below.

$$\langle \mathcal{A}_1, buy\_stock(acme) \rangle,$$
$$\langle \mathcal{A}_2, \sim buy\_stock(acme) \rangle,$$
$$\langle \mathcal{A}_3, risky\_co(acme) \rangle,$$
$$\langle \mathcal{A}_4, \sim risky\_co(acme) \rangle$$

where:

$$\mathcal{A}_1 = \left\{ \begin{array}{l} buy\_stock(acme) \prec good\_price(acme). \\ good\_price(acme) \prec web\_good\_price(acme). \end{array} \right\}$$

$$\mathcal{A}_2 = \left\{ \begin{array}{l} \sim buy\_stock(acme) \prec good\_price(acme), risky\_co(acme). \\ good\_price(acme) \prec web\_good\_price(acme). \\ risky\_co(acme) \prec in\_fusion(acme, steel). \end{array} \right\}$$

$$\mathcal{A}_3 = \left\{ \ risky\_co(acme) \prec in\_fusion(acme, steel). \ \right\}$$

$$\mathcal{A}_4 = \left\{ \begin{array}{l} \sim risky\_co(acme) \prec in\_fusion(acme, steel), strong\_co(steel). \\ strong\_co(steel) \prec web\_strong\_co(steel). \end{array} \right\}$$

A DeLP-query H succeeds, *i.e.*, it is warranted from a DeLP-program, if it is possible to build an argument $\mathcal{A}$ that supports H and $\mathcal{A}$ is such that there is no undefeated argument obtained from the DeLP-program that defeats $\mathcal{A}$ (Definition 4.7). To verify this condition, the process implements an exhaustive dialectical analysis that involves the construction and evaluation of arguments that either support or interfere with the query under analysis. That is, given an argument $\mathcal{A}$ that supports H, the warrant procedure will evaluate if there are other arguments that counter-argue or attack $\mathcal{A}$ or a sub-argument of $\mathcal{A}$. An argument $\langle \mathcal{C}, P \rangle$ is a *sub-argument* of $\langle \mathcal{A}, H \rangle$ if $\mathcal{C} \subseteq \mathcal{A}$ (e.g., $\langle \mathcal{A}_3, risky\_co(acme) \rangle$ is a sub-argument of $\langle \mathcal{A}_2, \sim buy\_stock(acme) \rangle$).

Definition 4.6 (Counter-Argument)

An argument $\langle \mathcal{B}, S \rangle$ *is a counter-argument for* $\langle \mathcal{A}, H \rangle$ *at literal* P, *if there exists a sub-argument* $\langle \mathcal{C}, P \rangle$ *of* $\langle \mathcal{A}, H \rangle$ *such that* P *and* S *disagree, that is, there exist two contradictory literals that have a strict derivation from* $\Pi \cup \{S, P\}$. *The literal* P *is referred to as the counter-argument point and* $\langle \mathcal{C}, P \rangle$ *as the disagreement sub-argument.*

Consider the arguments introduced above. In that scenario, the argument $\langle \mathcal{A}_1, buy\_stock(acme) \rangle$ is a counter-argument for $\langle \mathcal{A}_2, \sim buy\_stock(acme) \rangle$ and vice versa; and the argument $\langle \mathcal{A}_3, risky\_co(acme) \rangle$ is a counter-argument for

$\langle \mathcal{A}_4, \sim risky\_co(acme) \rangle$ and vice versa. Also note since $\langle \mathcal{A}_3, risky\_co(acme) \rangle$ is a sub-argument of $\langle \mathcal{A}_2, \sim buy\_stock(acme) \rangle$ then, $\langle \mathcal{A}_4, \sim risky\_co(acme) \rangle$ is a counter-argument for $\langle \mathcal{A}_2, \sim buy\_stock(acme) \rangle$.

Consider the program (Π_2, Δ_2) below:

$$\Pi_2 = \left\{ \begin{array}{l} h \leftarrow a \\ \sim h \leftarrow c \\ b \\ d \end{array} \right\} \qquad \Delta_2 = \left\{ \begin{array}{l} a \prec b \\ c \prec d \end{array} \right\}$$

Observe that from (Π_2, Δ_2) the contradictory literals h and $\sim h$ can be derived. Hence, $\langle \{a \prec b\}, a \rangle$ is a counter-argument for $\langle \{c \prec d\}, c \rangle$ and vice versa.

Given an argument $\langle \mathcal{A}, H \rangle$, there could be several counter-arguments attacking different points in $\mathcal{A}$, or different counter-arguments attacking the same point in $\mathcal{A}$. In DeLP, in order to verify whether an argument is non-defeated, all of its associated counter-arguments $\mathcal{B}_1$, $\mathcal{B}_2$, ..., $\mathcal{B}_k$ have to be examined, each of them being a potential reason for rejecting $\mathcal{A}$. If any $\mathcal{B}_i$ is (somehow) "better" than, or unrelated to, $\mathcal{A}$, then $\mathcal{B}_i$ is a candidate for defeating $\mathcal{A}$. However, if for some $\mathcal{B}_i$, the argument $\mathcal{A}$ is "better" than $\mathcal{B}_i$, then $\mathcal{B}_i$ will not be a defeater for $\mathcal{A}$. The process requires a preference relation to compare arguments and counter-arguments.

In DeLP the *argument comparison criterion* is modular, and thus the most appropriate criterion for the domain that is being represented can be introduced. In the literature of DeLP different criteria have been defined: in [García and Simari, 2004] a criterion that uses rule priorities was introduced; in [García and Simari, 2004; Stolzenburg et al., 2003], a syntactic criterion called *generalized specificity* was described; and more recently, in [Ferretti et al., 2008] a comparison criterion based on priorities among selected literals of the program was defined. For the following definitions we will abstract away from the argument comparison criterion, assuming there exists one that will be denoted "$\succ$".

Definition 4.7 (Defeaters: Proper and Blocking) *Let $\langle \mathcal{B}, S \rangle$ be a counter-argument for $\langle \mathcal{A}, H \rangle$ at point P, and $\langle \mathcal{C}, P \rangle$ the disagreement sub-argument. If $\langle \mathcal{B}, S \rangle \succ \langle \mathcal{C}, P \rangle$ (i.e., $\langle \mathcal{B}, S \rangle$ is "better" than $\langle \mathcal{C}, P \rangle$) then $\langle \mathcal{B}, S \rangle$ is a proper defeater for $\langle \mathcal{A}, H \rangle$. If $\langle \mathcal{B}, S \rangle$ is unrelated by the preference relation to $\langle \mathcal{C}, P \rangle$, (i.e., $\langle \mathcal{B}, S \rangle \not\succ \langle \mathcal{C}, P \rangle$, and $\langle \mathcal{C}, P \rangle \not\succ \langle \mathcal{B}, S \rangle$) then $\langle \mathcal{B}, S \rangle$ is a blocking defeater for $\langle \mathcal{A}, H \rangle$. Finally, $\langle \mathcal{B}, S \rangle$ is a defeater for $\langle \mathcal{A}, H \rangle$, if $\langle \mathcal{B}, S \rangle$ is either a proper or blocking defeater for $\langle \mathcal{A}, H \rangle$.*

In the examples below, we will use a comparison criterion referred to as *rule's priorities* that was introduced in [García and Simari, 2004]. This argument preference relation uses a partial order defined over defeasible rules that is denoted "$<_R$". The intuitive meaning of "$R_1 <_R R_2$" is that: *the rule R_2 is preferred over R_1 in the application domain described by the program*. Using *rule's priorities*, an argument $\mathcal{A}_1$ is preferred to $\mathcal{A}_2$ ($\mathcal{A}_1 \succ \mathcal{A}_2$) if the following two conditions hold:

(1) there exists at least a rule $r_a \in \mathcal{A}_1$ and a rule $r_b \in \mathcal{A}_2$ such that $r_b <_R r_a$,

(2) and there is no rule $r'_b \in \mathcal{A}_2$, and $r'_a \in \mathcal{A}_1$ such that $r'_a <_R r'_b$.

With the following partial order defined over defeasible rules from $(\Pi_{4.4}, \Delta_{4.4})$:

$(buy\_stock(X)\prec good\_price(X)) <_R$
$(\sim buy\_stock(X)\prec good\_price(X), risky\_co(X))$

$(risky\_co(X)\prec in\_fusion(X,Y)) <_R$
$(\sim risky\_co(X)\prec in\_fusion(X,Y), strong\_co(Y))$

and using the criterion *rule's priorities* it holds that $\langle \mathcal{A}_4, \sim risky\_co(acme) \rangle$ is a proper defeater for $\langle \mathcal{A}_3, risky\_co(acme) \rangle$. Therefore, since the argument $\mathcal{A}_3$ is a subargument of $\mathcal{A}_2$ then, $\langle \mathcal{A}_4, \sim risky\_co(acme) \rangle$ is a proper defeater for $\langle \mathcal{A}_2, \sim buy\_stock(acme) \rangle$. Also note that $\langle \mathcal{A}_2, \sim buy\_stock(acme) \rangle$ is a proper defeater for $\langle \mathcal{A}_1, buy\_stock(acme) \rangle$.

If an argument $\langle \mathcal{A}_1, H_1 \rangle$ is defeated by $\langle \mathcal{A}_2, H_2 \rangle$, then $\langle \mathcal{A}_2, H_2 \rangle$ represents a reason for rejecting $\langle \mathcal{A}_1, H_1 \rangle$. Nevertheless, a defeater $\langle \mathcal{A}_3, H_3 \rangle$ for $\langle \mathcal{A}_2, H_2 \rangle$ may also exist, rejecting $\langle \mathcal{A}_2, H_2 \rangle$ and reinstating $\langle \mathcal{A}_1, H_1 \rangle$. Note that the argument $\langle \mathcal{A}_3, H_3 \rangle$ may itself become defeated, reinstating $\langle \mathcal{A}_2, H_2 \rangle$, and so on. In this manner, a sequence of arguments (called argumentation line) arises, where each element after the first is a defeater of its predecessor.

Definition 4.8 (Argumentation Line) *An* argumentation line *for* $\langle \mathcal{A}_1, H_1 \rangle$ *is a sequence of argument structures,* $\Lambda = [\langle \mathcal{A}_1, H_1 \rangle, \langle \mathcal{A}_2, H_2 \rangle, \ldots \langle \mathcal{A}_n, H_n \rangle]$, *where each element of the sequence* $\langle \mathcal{A}_i, H_i \rangle$, $1 < i \leq n$, *is a defeater of its predecessor* $\langle \mathcal{A}_{i-1}, H_{i-1} \rangle$.

In an argumentation line $\Lambda = [\langle \mathcal{A}_1, H_1 \rangle, \langle \mathcal{A}_2, H_2 \rangle, \ldots \langle \mathcal{A}_n, H_n \rangle]$, the first element $\langle \mathcal{A}_1, H_1 \rangle$, becomes a *supporting* argument for H_1, $\langle \mathcal{A}_2, H_2 \rangle$ an *interfering* argument, $\langle \mathcal{A}_3, H_3 \rangle$ a supporting argument, $\langle \mathcal{A}_4, H_4 \rangle$ an interfering one, and so on. Thus, an argumentation line can be split into two disjoint sets: $\Lambda_S = \{\langle \mathcal{A}_1, H_1 \rangle, \langle \mathcal{A}_3, H_3 \rangle, \langle \mathcal{A}_5, H_5 \rangle, \ldots\}$ of supporting arguments, and $\Lambda_I = \{\langle \mathcal{A}_2, H_2 \rangle, \langle \mathcal{A}_4, H_4 \rangle, \ldots\}$ of interfering arguments. Considering the arguments generated above from $(\Pi_{4.4}, \Delta_{4.4})$ of Example 4.4, we have that

$\Lambda = [\langle \mathcal{A}_1, buy\_stock(acme) \rangle, \langle \mathcal{A}_2, \sim buy\_stock(acme) \rangle, \langle \mathcal{A}_4, \sim risky\_co(acme) \rangle]$

is an argumentation line, with two supporting arguments and an interfering one.

Note that an infinite argumentation line may arise if an argument structure is reintroduced in the sequence. Clearly, this situation is undesirable as it leads to the construction of an infinite sequence of arguments. Circular argumentation and other forms of *fallacious argumentation* were studied in detail in [Simari et al., 1994; García and Simari, 2004]. In DeLP, argumentation lines have to be *acceptable*, that is, (1) they have to be finite, (2) supporting (resp. interfering) arguments have to be non-contradictory, (3) an argument cannot appear twice, and (4) every argument introduced after the first must be a defeater of the previous one in the sequence that must be undefeated up to that point; these conditions are reflected in the following definition.

Definition 4.9 (Acceptable Argumentation Line)
An argumentation line $\Lambda = [\langle \mathcal{A}_1, H_1 \rangle, \ldots \langle \mathcal{A}_n, H_n \rangle]$ is acceptable iff:

(1) Λ is a finite sequence.

(2) The set Λ_S of supporting arguments (resp. Λ_I), is concordant (we say that a set of arguments $\{\langle \mathcal{A}_i, H_i \rangle\}_{i=1}^n$ is concordant iff $\Pi \cup \bigcup_{i=1}^n \mathcal{A}_i$ is non-contradictory).

(3) No argument $\langle \mathcal{A}_k, H_k \rangle$ in Λ is a disagreement sub-argument of an argument $\langle \mathcal{A}_i, H_i \rangle$ appearing earlier in Λ ($i < k$).

(4) For all i, such that $\langle \mathcal{A}_i, H_i \rangle$ is a blocking defeater for $\langle \mathcal{A}_{i-1}, H_{i-1} \rangle$, if $\langle \mathcal{A}_{i+1}, H_{i+1} \rangle$ exists, then $\langle \mathcal{A}_{i+1}, H_{i+1} \rangle$ is a proper defeater for $\langle \mathcal{A}_i, H_i \rangle$.

A single acceptable argumentation line for $\langle \mathcal{A}_1, H_1 \rangle$ may not be enough to establish whether $\langle \mathcal{A}_1, H_1 \rangle$ is an undefeated argument. The reason is that, potentially, there may exist several defeaters, say $\langle \mathcal{B}_1, Q_1 \rangle$, $\langle \mathcal{B}_2, Q_2 \rangle$, ..., $\langle \mathcal{B}_k, Q_k \rangle$ for $\langle \mathcal{A}_1, H_1 \rangle$, and therefore k argumentation lines for $\langle \mathcal{A}_1, H_1 \rangle$ need to be considered; since for each defeater $\langle \mathcal{B}_i, Q_i \rangle$ there can be several defeaters, the exhaustive process of considering all the possibilities will produce a set of argumentation lines. This set of argumentation lines which start with the same argument can be conceptually thought as a path from the root of a tree to a leaf in that tree. The tree structure will be referred to as a *dialectical tree*, and this tree will integrate into one structure all the argumentation lines that start with the same argument. Notice that since different arguments may share the same literal as a point of attack, the same counterargument which exploits that attack opportunity could appear in several argumentation lines. In this structure, the root will be labeled with $\langle \mathcal{A}_1, H_1 \rangle$ and every node (except the root) represents a defeater (proper or blocking) of its parent. As we remarked before, each path from the root to a leaf will correspond to a different acceptable argumentation line.

Definition 4.10 (Dialectical Tree) *A dialectical tree $\mathcal{T}_{\langle \mathcal{A}_1, H_1 \rangle}$ for $\langle \mathcal{A}_1, H_1 \rangle$, is defined as follows:*

1. *The root of the tree is labeled with $\langle \mathcal{A}_1, H_1 \rangle$.*

2. *Let N be a non-root node labeled $\langle \mathcal{A}_n, H_n \rangle$, and $[\langle \mathcal{A}_1, H_1 \rangle, \ldots, \langle \mathcal{A}_n, H_n \rangle]$ be the sequence of labels of the path from the root to N. Let $\{\langle \mathcal{B}_1, Q_1 \rangle, \langle \mathcal{B}_2, Q_2 \rangle, \ldots, \langle \mathcal{B}_k, Q_k \rangle\}$ be the set of all the defeaters for $\langle \mathcal{A}_n, H_n \rangle$. For each defeater $\langle \mathcal{B}_i, Q_i \rangle$ ($1 \leq i \leq k$), such that the argumentation line $\Lambda' = [\langle \mathcal{A}_1, H_1 \rangle, \ldots, \langle \mathcal{A}_n, H_n \rangle, \langle \mathcal{B}_i, Q_i \rangle]$ is acceptable, the node N has a child N_i labeled $\langle \mathcal{B}_i, Q_i \rangle$. If there is no defeater for $\langle \mathcal{A}_n, H_n \rangle$ or there is no $\langle \mathcal{B}_i, Q_i \rangle$ such that Λ' is acceptable, then N is a leaf.*

A dialectical tree provides a structure integrating all the possible acceptable argumentation lines that can be generated for deciding whether an argument is undefeated. In order to compute whether the status of the root is defeated or undefeated, a recursive marking process is introduced next. Let $\mathcal{T}$ be a dialectical tree, a *marked dialectical tree*, denoted $\mathcal{T}^*$, can be obtained marking every node in $\mathcal{T}$ as follows: (1) each leaf in $\mathcal{T}$ is marked as **U** in $\mathcal{T}^*$; (2) an inner node N of $\mathcal{T}$ is marked as **D** in $\mathcal{T}^*$ iff N has at least a child marked as **U**, otherwise N is marked as **U**. Thus, the root $\langle \mathcal{A}, H \rangle$ of a dialectical tree will be marked **U** if and only if all its children are marked as **D** (that is to say, all the defeaters for $\langle \mathcal{A}, H \rangle$ are defeated).

Figure 1 shows three different marked dialectical trees each one for a different argument ($\mathcal{B}_1$, $\mathcal{B}_2$, $\mathcal{B}_2$), all of them supporting the same conclusion H. Nodes (arguments) are depicted as triangles with the mark (**D** or **U**) inside. **D**-nodes are also colored darker than **U**-nodes. An arrow is used for representing defeat. Following the procedure described above on the trees depicted in Figure 1, we have that the root $\langle \mathcal{B}_1, H \rangle$ of the tree on the left is marked **D**, the root $\langle \mathcal{B}_2, H \rangle$ of the tree at the center is marked **D**, but the root $\langle \mathcal{B}_2, H \rangle$ of the tree on the right is marked **U**. A ground literal H is considered to be *warranted*

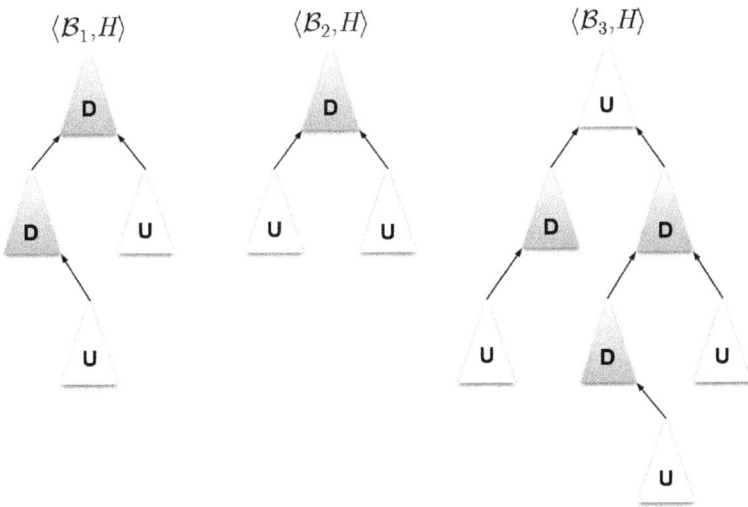

Figure 1. Marked dialectical trees

when there exists at least an argument $\mathcal{A}$ for H such that all the defeaters for $\langle \mathcal{A}, H \rangle$ are defeated; therefore, this marking procedure provides an effective way of determining if a DeLP-query H is warranted. Note that given a DeLP-query H, there might exist several arguments supporting H; therefore, H is warranted if there is at least one argument $\mathcal{A}$ for H such that the root of a dialectical tree for $\langle \mathcal{A}, H \rangle$ is marked as **U**.

Definition 4.11 (Warranted Literals) *Let $\mathcal{P}$ be a DeLP-program, H a ground literal, $\langle \mathcal{A}, H \rangle$ an argument from $\mathcal{P}$ and $\mathcal{T}^*$ a marked dialectical tree for $\langle \mathcal{A}, H \rangle$. Literal H is warranted from $\mathcal{P}$ iff the root of $\mathcal{T}^*$ is marked as **U**.*

Going back to the DeLP-program $(\Pi_{4.4}, \Delta_{4.4})$ of Example 4.4 we observe that the ground literal $buy\_stock(acme)$ is warranted because there is a dialectical tree for $\langle \mathcal{A}_1, buy\_stock(acme) \rangle$ and its root is marked **U** (see Figure 2, left), whereas $\sim buy\_stock(acme)$ is not warranted (see Figure 2, right). Note that from $(\Pi_{4.4}, \Delta_{4.4})$, the literal $\sim risky\_co(acme)$ is warranted since $\langle \mathcal{A}_4, \sim risky\_co(acme) \rangle$ has no defeaters and hence, its marked dialectical tree will have only one node (the root) marked **U**.

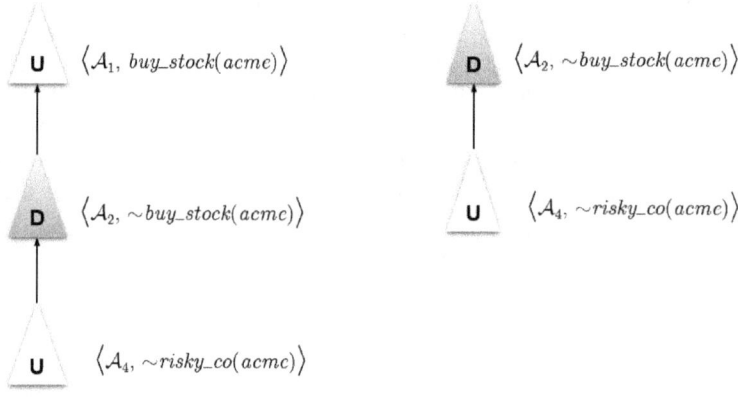

Figure 2. Two marked dialectical trees from $(\Pi_{4.4}, \Delta_{4.4})$

In [García, 2000; García and Simari, 2004] it was proved that in DeLP any claim H that has a strict derivation is warranted. For instance, from $(\Pi_{4.4}, \Delta_{4.4})$ $declared\_in\_fusion(acme, steel)$ is warranted. This is so because any literal H that has a strict derivation from a program (Π, Δ) has no possible counterargument, thus precluding the possibility that a defeater could exist. The reason for this is that no argument that disagrees with H could be built because the definition of argument requires it to be non-contradictory with Π. Also note that in DeLP self-defeating arguments cannot be built by definition, and if a literal H is warranted then the complement with respect to strong negation ($\overline{H}$) cannot be warranted, since it constitute a defeater of the argument for H.

An interpreter of DeLP takes a program $\mathcal{P}$, and a DeLP-query H as input, returning one of the following four possible answers: YES, if H is warranted from $\mathcal{P}$; NO, if the complement of H is warranted from $\mathcal{P}$; UNDECIDED, if neither H nor its complement are warranted from $\mathcal{P}$; or UNKNOWN, if H is not in the language of the program $\mathcal{P}$.

Following our running example, from $(\Pi_{4.4}, \Delta_{4.4})$ the answer for the query

$buy\_stock(acme)$ will be YES; the answer for the query $\sim buy\_stock(acme)$ will be NO; the answer for $buy\_stock(maybe)$ will be UNDECIDED; the answer for $\sim risky\_co(acme)$ will be YES; the answer for $risky\_co(acme)$ will be NO; and the answer for $\sim risky\_co(steel)$ will be YES.

5 DeLP Servers

In multi-agent systems (MAS), reasoning can be thought as a service that can be offered as part of a knowledge-based infrastructure providing a mechanism to exploit it. In this section we describe Defeasible Logic Programming Servers (or DeLP-Servers for short), that are the support of argumentative reasoning services conceived for a MAS setting. In such an environment, DeLP-Servers, introduced in [García et al., 2007b], provide client agents that can be distributed in remote hosts the ability to consult different reasoning services implemented as DeLP-Servers that can be distributed themselves. Figure 3 below schematically depicts a situation with five servers and three client agents spread over different hosts. Common (or public) knowledge can be stored in

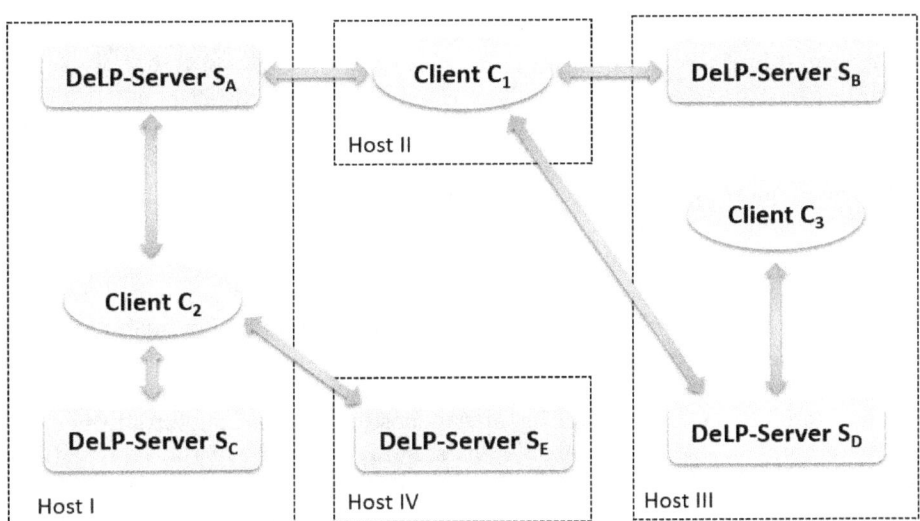

Figure 3. DeLP-Servers and clients spread over different hosts.

a server and represented as a DeLP-program; then, client agents may send queries to a DeLP-server and receive the corresponding answer. To respond to a query, a DeLP-server uses the knowledge stored in it; but, the server offers the possibility of integrating the knowledge stored in it with "private" knowledge the client making the query has the possibility to send as part of the query.

The following example was introduced in [García and Simari, 2014]. Consider a DeLP-server developed to handle knowledge of a real estate application domain. A program can be stored in that server with some facts that represent the renting availability of properties, and the necessary defeasible rules to represent the modeling of the useful common knowledge regarding how to select a suitable property. Client agents are able to consult the server for a property by submitting a query together with the private information that represents its current situation and/or requirements (*e.g.*, *has children, works at home, has no car, cannot afford a rent of* $3000 *or more*), structuring a composite query; we will refer to the private knowledge as the *context* of the composite query. When issuing a composite query, *i.e.*, the proper query plus the context, the latter part is temporarily added to the knowledge stored in the server, thus enriching knowledge stored in the server with the specific knowledge the agent considers necessary to answer the query properly; after the answer is returned, the context will be purged from the server not affecting future queries; thus, a client agent will not make permanent changes to the public DeLP-program stored in a server. Continuing with our running example, $Client_1$ can consult if a real estate property p is suitable for her, providing the contextual information that she is *single* and *has a car*. Another client, $Client_2$, can consult if p is suitable for him, providing a different context: he *has two children* and *works downtown*. Since queries do not effect permanent changes in the program stored in a server, both contextual queries can be received and processed simultaneously, and one will not affect the answer of the other.

The concept of *contextual query* will be extended and formalized below. As we have mentioned, a contextual query provides the server answering a particular DeLP-query, with contextual information in the form of DeLP-program, but there is also the matter of how the contextual information should be incorporated with the public program stored in the server.

Several agents can consult the same DeLP-server, and a particular agent can consult several DeLP-Servers, as shown in Figure 3. For instance, in a particular medical application, dedicated DeLP-Servers can maintain different programs, supporting particular shared knowledge of specific specialties of the medical profession. Hence, an agent can pose the same contextual query to different servers, and obtain a different answer from each of the servers. Our approach does not impose any restriction over the type, architecture, or implementation language of the client agents. We will now formalize the above mentioned concepts beginning with the notion of DeLP-Server.

Definition 5.1 *A DeLP-Server is a triple* $\ll \mathbf{I}, \mathbf{O}, \mathcal{P} \gg$ *where* $\mathbf{I}$ *is a DeLP-interpreter,* $\mathbf{O}$ *is a set of DeLP-program operators and* $\mathcal{P}$ *is a DeLP-program.*

Thus, a DeLP-Server $\ll \mathbf{I}, \mathbf{O}, \mathcal{P} \gg$ provides a DeLP-interpreter, the possibility of representing public knowledge in the form of a DeLP-program, and a set of operators that will handle the integration of the contextual information with the program stored $\mathcal{P}$.

Before providing more details on the first two elements in $\ll \mathbf{I}, \mathbf{O}, \mathcal{P} \gg$, let us consider our example for the real state application which will be expanded next. The DeLP-server will have stored public knowledge represented by the DeLP-program $\mathcal{P}_{RS}= (\Pi_{RS}, \Delta_{RS})$ which represents the third element in the triple. This program is shown below, where we are using the following shorthand notation: *av* (available), *br* (bedroom), *conv_br* (convenient number of bedrooms). The set Π_{RS} contains facts representing information about the name, number of bedrooms, location, and price of the properties available for renting; for instance, $av(p1, br(1), downtown, 2000)$ represents the fact that the property $p1$ is available, it has one bedroom, is located downtown, and its price is 2000.

$$\Pi_{RS} = \left\{ \begin{array}{ll} av(p1, br(1), downtown, 2000) & av(p2, br(2), suburbs, 4000) \\ av(p3, br(3), suburbs, 5000) & av(p4, br(3), downtown, 15000) \\ av(p5, br(2), suburbs, 10000) & av(p6, br(1), downtown, 7000) \end{array} \right\}$$

$$\Delta_{RS} = \left\{ \begin{array}{l} suitable(Property) \prec av(Property, br(B), \_, \_), conv\_br(B) \\ \sim suitable(Property) \prec expensive(Property) \\ expensive(Property) \prec av(Property, \_, \_, Price), afford(A), Price > A \\ conv\_br(X) \prec has\_children, X \geq 2 \\ conv\_br(X) \prec couple, X \geq 1 \\ conv\_br(X) \prec single, X \geq 1 \\ has\_children \prec grown\_children \\ \sim conv\_br(2) \prec grown\_children \\ \sim conv\_br(1) \prec work\_at\_home \end{array} \right\}$$

These defeasible rules should be interpreted as follows. A convenient number of bedrooms in any available property will provide reasons for considering it as a suitable property; observe that the reasons for and against a convenient number of bedrooms depends on the private information a client might send as a context. Properties are not considered suitable if they are expensive; to determine if a property is expensive for a client, the amount the client can afford to expend has to be provided in the contextual query.

The first element in the triple $\ll \mathbf{I}, \mathbf{O}, \mathcal{P} \gg$ is the DeLP-interpreter $\mathbf{I}$, which takes a DeLP-program, a DeLP-query (a literal), and returns one of the four DeLP possible answers: YES, NO, UNDECIDED, or UNKNOWN. Formally, let $\mathbb{D}$ be the domain of all possible DeLP-programs, and $\mathbb{L}$ the domain of potential DeLP-queries, then a DeLP-interpreter can be seen as a function:

$$\mathbf{I} : \mathbb{D} \times \mathbb{L} \longrightarrow \{\text{YES}, \text{NO}, \text{UNDECIDED}, \text{UNKNOWN}\}$$

The second element in the triple $\ll \mathbf{I}, \mathbf{O}, \mathcal{P} \gg$, the set $\mathbf{O}$, contains binary operators that take two DeLP-programs $\mathcal{P}_1$ and $\mathcal{P}_2$, returning a new DeLP-program; the resulting DeLP-program is the integration of $\mathcal{P}_1$ and $\mathcal{P}_2$ according to the definition of the operator used. Therefore, formally a DeLP-program operator $o \in \mathbf{O}$ can be seen as a function:

$$o : \mathbb{D} \times \mathbb{D} \longrightarrow \mathbb{D}$$

A contextual query to a DeLP-Server involves two elements: a particular DeLP-query Q to be answered by the DeLP-Server, and contextual information Co to be used together with the program stored in the to answer Q. This contextual information consists of both private knowledge in the form of a DeLP-program to be consider by the interpreter and appropriate operators that state how to integrate the private information with the public program stored at the server. Formally:

Definition 5.2 *A contextual query for a DeLP-Server $\ll \mathbf{I}, \mathbf{O}, \mathcal{P} \gg$ is a pair (Co, Q) where Q is a DeLP-query, and the context Co is represented as a sequence $[(\mathcal{P}_1, o_1), (\mathcal{P}_2, o_2), \ldots, (\mathcal{P}_n, o_n)]$, where each $\mathcal{P}_i$ is a DeLP-program and $o_i \in \mathbf{O}, 1 \leq i \leq n$, where the application of $[(\mathcal{P}_1, o_1), (\mathcal{P}_2, o_2), \ldots, (\mathcal{P}_n, o_n)]$ to $\mathcal{P}$ is recursively defined as:*

- $[\,](\mathcal{P}) = \mathcal{P}$,

- $[(\mathcal{P}_1, o_1), [(\mathcal{P}_2, o_2), \ldots, (\mathcal{P}_n, o_n)]](\mathcal{P}) = [(\mathcal{P}_2, o_2), \ldots, (\mathcal{P}_n, o_n)](\mathcal{P}\, o_1\, \mathcal{P}_1)$

The application of the context Co to the DLP-program $\mathcal{P}$ will be denoted $Co \diamond \mathcal{P}$.

In [García et al., 2007b] and [Tucat et al., 2009] several concrete examples of program operators were introduced. These operators are modular in the sense that each particular server can have a different set of operators designed specifically for the application domain in use. Recall that in DeLP the Π is non-contradictory; therefore, DeLP-program operators should return a program where Π is non-contradictory.

As an illustration of how the operators might be defined, below we include three simple and basic operators denoted $\oplus$, $\otimes$, and $\ominus$, that will be used through the examples below. To simplify the presentation, these three operators will be defined for a restricted DeLP-program (Π, Δ) where Π contains facts but no strict rules. We introduce the following auxiliary notation to use in the operators definition. Let x be a fact, we denote with $\overline{x}$ the complement of x with respect to strong negation; that is $\overline{a} = \sim a$ and $\overline{\sim a} = a$; and, if X is a set of facts, the complement of X is defined as $\overline{X} = \{\overline{x} \mid x \in X\}$. Previously, we have mentioned that a DeLP-program operator can be seen as a function $o : \mathbb{D} \times \mathbb{D} \longrightarrow \mathbb{D}$; with that in mind we define the operators $\oplus$, $\otimes$, and $\ominus$ as follows:

$$[((\Pi', \Delta'), \oplus)]((\Pi, \Delta)) = (\Pi, \Delta) \oplus (\Pi', \Delta') = ((\Pi \setminus \overline{\Pi'}) \cup \Pi', \Delta \cup \Delta')$$

$$[((\Pi', \Delta'), \otimes)]((\Pi, \Delta)) = (\Pi, \Delta) \otimes (\Pi', \Delta') = (\Pi \cup (\Pi' \setminus \overline{\Pi}), \Delta \cup \Delta')$$

$$[((\Pi', \Delta'), \ominus)]((\Pi, \Delta)) = (\Pi, \Delta) \ominus (\Pi', \Delta') = (\Pi \setminus \Pi', \Delta \setminus \Delta')$$

and one showing a possible combination of two of them:

$$[((\Pi',\Delta'),\ominus),((\Pi'',\Delta''),\oplus)]((\Pi,\Delta)) =$$
$$[((\Pi'',\Delta''),\oplus)]((\Pi,\Delta)\ominus(\Pi',\Delta')) =$$
$$((\Pi,\Delta)\ominus(\Pi',\Delta'))\oplus(\Pi'',\Delta'') =$$
$$((\Pi\setminus\Pi'),(\Delta\setminus\Delta'))\oplus(\Pi'',\Delta'') =$$
$$(((\Pi\setminus\Pi')\setminus\overline{\Pi''})\cup\Pi'',(\Delta\setminus\Delta')\cup\Delta'')$$

Notice that the algebraic properties of these operators will depend on the particular definition of each of them.

The first two operators, $\oplus$ and $\otimes$, integrate the defeasible part of the program representing the private knowledge and the defeasible part of the common knowledge stored in the server. The last one, $\ominus$, will eliminate the defeasible rules stored in the server that are in common with the defeasible part of the private knowledge present in the contextual query; the use of $\ominus$ in combination with the other two will allow to change the set of defeasible rules to be used in a particular situation where some defeasible rules become unnecessary or unrelated to the context of the query.

The first operator, denoted $\oplus$, will integrate the private knowledge of an agent received as a DeLP-program (Π',Δ') with the program stored in a server that contains (Π,Δ), giving priority to the information in Π'. For instance, if the public knowledge in the server is $(\Pi,\Delta) = (\{a,b\},\Delta)$ and the private information received in a contextual query is $(\Pi',\Delta') = (\{\sim a,c\},\Delta')$, then

$$(\{a,b\},\Delta)\oplus(\{\sim a,c\},\Delta') = ((\{a,b\}\setminus\overline{\{\sim a,c\}})\cup\{\sim a,c\},\Delta\cup\Delta') =$$
$$= ((\{a,b\}\setminus\{a,\sim c\})\cup\{\sim a,c\},\Delta\cup\Delta') = (\{b,\sim a,c\},\Delta\cup\Delta')$$

Note that a contradiction would arise if the private information would have been added directly since Π would have contained a and $\sim a$; by using $\oplus$, which gives priority to the received information, the fact $\sim a$ will remain, whereas a will be temporary removed, becoming ignored in answering the query.

In contrast, the operator denoted $\otimes$ gives priority to the facts that are part of the DeLP-program (Π,Δ) stored in the server; for instance, if the public knowledge in the server is as before $(\Pi,\Delta) = (\{a,b\},\Delta)$ and the received information is $(\Pi',\Delta') = (\{\sim a,c\},\Delta')$, then

$$(\{a,b\},\Delta)\otimes(\{\sim a,c\},\Delta') = (\{a,b\}\cup(\{\sim a,c\}\setminus\overline{\{a,b\}}),\Delta\cup\Delta') =$$
$$= (\{a,b\}\cup(\{\sim a,c\}\setminus\{\sim a,\sim b\}),\Delta\cup\Delta') = (\{a,b,c\}\Delta\cup\Delta')$$

That is, since the stored information has priority, the fact $\sim a$ will remain, but a will not be considered for answering the query.

Finally, the operator denoted $\ominus$ was developed with the purpose of not considering for a particular query part of the public information maintained in the server. For instance, if the public knowledge in the server is $(\Pi,\Delta) = (\{a,b,p\},\Delta)$ and the received information is $(\Pi',\Delta') = (\{a,b,\sim p\},\Delta')$, then

$$(\{a,b,p\},\Delta)\ominus(\{a,b,\sim p\},\Delta') = (\{a,b,p\}\setminus\{a,b,\sim p\},\Delta\setminus\Delta') = (\{p\},\Delta\setminus\Delta')$$

To provide an answer for a particular contextual query (Co, Q), first a DeLP-Server $\ll \mathbf{I}, \mathbf{O}, \mathcal{P} \gg$ has to integrate the context Co to the stored program $\mathcal{P}$ as we have described previously, denoted $Co \diamond \mathcal{P}$. Recall that DeLP-interpreter can be seen as a function $\mathbf{I}:(\mathbb{D} \times \mathbb{L}) \longrightarrow \{\text{YES}, \text{NO}, \text{UNDECIDED}, \text{UNKNOWN}\}$.

Definition 5.3 *Let* $\ll \mathbf{I}, \mathbf{O}, \mathcal{P} \gg$ *be a a DeLP-server and* (Co, Q) *a contextual query. Then, an answer for* (Co, Q) *is* $\mathbf{I}((Co \diamond \mathcal{P}), Q)$.

It is important to note that a context is a sequence and it will be processed by the DeLP-server in the order it appears in that sequence. Hence, different answers could be obtained depending on the order of the pairs in the context. Consider a DeLP-server $\ll \mathbf{I}, \{\oplus, \otimes, \ominus\}, (\Pi_{RS}, \Delta_{RS}) \gg$ for the real state application already introduced. We include some contextual queries that clients could send to the server and show the answers that will be returned:

$([[(single, afford(3000), \oplus)], suitable(p1))$
    answer: YES

$([[(single, afford(3000), \oplus)], suitable(p2))$
    answer: NO

$([[(single, afford(4000), \oplus)], suitable(p2))$
    answer: YES

$([[(grown\_children, afford(4000), \oplus)], suitable(p2))$
    answer: NO

$([[(grown\_children, afford(9000), \oplus)], suitable(p3))$
    answer: YES

$([[(grown\_children, afford(4000), \oplus)], suitable(p3))$
    answer: NO

$([[(grown\_children, afford(4000), \oplus),$
$(\sim conv\_br(2) \prec grown\_children, \ominus)], suitable(p2))$
    answer: YES

6 DeLP Extensions and Applications

Several extensions of DeLP have been proposed, and in this section we will briefly sketch a few these developments. The following are some of these research lines that are more related to applications.

The use of default negation [Alferes et al., 1996] was included in the original presentation of DeLP in [García and Simari, 2004]. When DeLP is extended to consider default negation, some characteristics of the formalism just described are affected and for a correct treatment of default negation in DeLP, further considerations will be required. Default negation will be allowed only preceding literals in the body of defeasible rules; for instance[3]

$$\sim cross\_railway\_tracks \prec not \sim train\_is\_coming.$$

The reason for not allowing default negation in strict rules is twofold: on one hand, a strict rule '$p \leftarrow not\,q$' is not strict in a proper sense, because the head 'p' will be derived assuming '$not\,q$' which is rather different from what happens in a strict rule; on the other hand, the set Π of strict rules and facts could become a contradictory set by assuming different literals. Therefore, a defeasible logic program will be defined by a set of facts, strict rules, and defeasible rules which possibly could contain literals in their bodies affected by the default negation '*not*'. This addition allows to handle assumptions in a defeasible logic program, and the corresponding definitions were adjusted accordingly; for instance, in this version of DeLP there are three types of defeaters: proper, blocking, and a defeater for an assumption. Given an assumption '*not L*', a warranted argument for literal L will be a defeater for the assumption. The reader interested in rest of the technical issues produced by the addition of '*not*' and how they are resolved is referred to the original paper [García and Simari, 2004].

In [Cohen *et al.*, 2016] an extension of DeLP, called *Extended Defeasible Logic Programming* (E-DeLP) was introduced. E-DeLP is a structured argumentation system enabling the possibility of representing reasons for and against the use of a defeasible rule; to achieve that, the formalism DeLP is extended incorporating two new types of rules: backing and undercutting rules. In that way, E-DeLP accounts for Toulmin's notion of backing and Pollock's notion of undercut, two important contributions within the field of argumentation; thus, E-DeLP is a structured argumentation system which provides an integrated setting for modeling Toulmin's and Pollock's ideas simultaneously.

Possibilistic Defeasible Logic Programming (P-DeLP) [Chesñevar *et al.*, 2004; Alsinet *et al.*, 2008], is an extension of DeLP that allows the treatment of possibilistic uncertainty and fuzzy knowledge at object-language level in a logic programming language. In P-DeLP, knowledge representation features are formalized based on a possibilistic logic based on a Horn-rule fragment of Gödel fuzzy logic (PGL). Formulas in PGL are built over fuzzy propositional variables and the certainty degree of formulas is expressed with a necessity measure. The proof method for PGL, in a logic programming setting, is based on a complete calculus for determining the maximum degree of possibilistic entailment of a fuzzy goal. In a multiagent context, it is possible to use P-DeLP to encode the agents' knowledge about the world, applying the argument and warrant computing procedure to obtain their inferences. A number of argument-based consequence operators which allow to model different aspects of the reasoning abilities in an intelligent agent have been formalized and studied, and the computational problem related to how answers to P-DeLP queries can be speeded

---

[3]This example was adapted from one attributed to John MacCarthy in [Gelfond and Lifschitz, 1990].

up by pruning the associated search space have been considered.

The consideration of time in argumentation is an important addition to the already reach area. Temporal Argumentation Frameworks (TAF), introduced in [Cobo et al., 2010; Cobo et al., 2011] extends Dung's abstract argumentation frameworks taking into account the temporal availability of arguments. In a TAF, arguments are valid during specific time intervals, called availability intervals, while the attack relation of the framework remains static and permanent in time; thus, which arguments are deemed acceptable will vary in time. Extended Temporal Argumentation Framework (E-TAF) [Budán et al., 2012] extends TAF introducing the capability of modeling the temporal availability of attacks among arguments. It interesting to observe that even when a particular argument is present in a time interval, it is possible that an attack which involves that argument it is only available in part of that time interval. E-TAF was enriched [Budán et al., 2015] by considering Structured Abstract Argumentation, using Dynamic Argumentation Frameworks [Moguillansky et al., 2013]. The resulting framework, E-TAF*, provides a suitable model for different time-dependent issues satisfying properties and equivalence results that permit to contrast the expressivity of E-TAF and E-TAF* with argumentation based on abstract frameworks. The representation of strength and time provide useful elements in many argumentation applications. The Strength and Time DeLP, ST-DeLP, is an instantiation of the abstract framework E-TAF based on DeLP that incorporates the representation of temporal availability and strength factors varying over time associated with the elements of the language of DeLP. It also specifies how arguments are built, and how the availability and strength of arguments are obtained from the corresponding information attached to the elements from which are built. After determining the availability of attacks by comparing strength of conflicting arguments over time, E-TAF definitions are applied to establish temporal acceptability of arguments. Thus, the main contribution of this development is the integration of time and strength in the context of argumentation systems. Other line of work related to this can be found in [Pardo and Godo, 2011].

In [García et al., 2008] an argumentation-based planning formalism has been introduced. This system can be used by an agent to construct plans using partial order planning technics, in such a manner that the traditional POP algorithm is extended to consider arguments as planning steps. When actions and arguments are combined to construct plans, new types of interferences appear and it is necessary to extend the standard notion of threat to consider: action-action, action-argument, and argument-argument threats. The resulting algorithm is called APOP, and extends the traditional POP algorithm to consider actions and arguments as planning steps and resolve the new types of threats using new methods.

A framework for ontology integration of DL ontologies based on DeLP was introduced in [Gómez et al., 2013]. In this work, the limitations for reasoning in the presence of inconsistent ontologies that Description Logic (DL) ontologies suffer are addressed using defeasible argumentation to model different DL reasoning capabilities when handling inconsistent ontologies, resulting in so-called d-ontologies. ONTOarg, provides a decision support framework for performing local-as-view integration of possibly inconsistent and incomplete ontologies in terms of DeLP. This ontology integration system is able to handle inconsistent DL-based ontologies by performing a dialectical analysis in order to determine the membership of individuals to concepts.

Another recent line of research deals with the problem of massively build arguments from premises obtained from relational databases and compute warrant [Deagustini et al., 2013]. In this setting, different databases can be updated by external, independent applications, leading to changes in the spectrum of available arguments. Algorithms for integrating a database management system with an argument-based inference engine have been designed and tested. The empirical results and running-time analysis associated with the approach show how, taking advantage of modern DBMS technologies, it is possible to efficiently implement process that requiere massive argumentation. This provides an interesting possibility for developing new architectures for knowledge-based applications, such as Decision Support Systems and Recommender Systems, that could efficiently use argumentation as the underlying inference model.

Acknowledgments

This work was partially supported by the Departement of Computer Science and Engineering, Universidad Nacional del Sur-Argentina (UNS) by PGI-UNS (grants 24/ZN32, 24/N040, 24/N035), by the Consejo Nacional de Ciencia y Técnica-Argentina (CONICET), and by EU H2020 Research and Innovation Programme under the Marie Sklodowska-Curie grant agreement No. 690974 for the project MIREL: MIning and REasoning with Legal texts.

BIBLIOGRAPHY

[Alferes et al., 1996] José J. Alferes, Luis Moniz Pereira, and Teodor C. Przymusinski. Strong and explicit negation in nonmonotonic reasoning and logic programming. *Lecture Notes in Computer Science*, 1126:143–163, 1996.

[Alsinet et al., 2008] Teresa Alsinet, Carlos I. Chesñevar, Lluis Godo, Sandra Sandri, and Guillermo R. Simari. Formalizing argumentative reasoning in a possibilistic logic programming setting with fuzzy unification. *Int. J. Approx. Reasoning*, 48(3):711–729, 2008.

[Baral and Gelfond, 1994] Chitta Baral and Michael Gelfond. Logic programming and knowledge representation. *J. Log. Program.*, 19/20:73–148, 1994.

[Baral, 2003] Chitta Baral. *Knowledge Representation, Reasoning and Declarative Problem Solving.* Cambridge University Press, Cambridge, UK, 2003.

[Bench-Capon and Dunne, 2007] Trevor J. M. Bench-Capon and Paul E. Dunne. Argumentation in artificial intelligence. *Artificial Intelligence*, 171(10-15):619–641, 2007.

[Besnard and Hunter, 2001] Philippe Besnard and Anthony Hunter. A Logic-Based Theory of Deductive Arguments. *Artif. Intell.*, 128(1-2):203–235, 2001.

[Besnard and Hunter, 2008] Philippe Besnard and Anthony Hunter. *Elements of Argumentation.* MIT Press, 2008.

[Bondarenko et al., 1993] Andrei Bondarenko, Francesca Toni, and Robert A. Kowalski. An assumption-based framework for non-monotonic reasoning. In *LPNMR*, pages 171–189, 1993.

[Bondarenko et al., 1997] Andrei Bondarenko, Phan Minh Dung, Robert A. Kowalski, and Francesca Toni. An abstract, argumentation-theoretic approach to default reasoning. *Artif. Intell.*, 93:63–101, 1997.

[Brewka and Eiter, 2000] Gerhard Brewka and Thomas Eiter. Prioritizing default logic. In Steffen Hölldobler, editor, *Intellectics and Computational Logic: Papers in Honor of Wolfgang Bibel*, pages 27–46. Kluwer Academic Publishers, Dordrecht, Boston, London, 2000.

[Bryant and Krause, 2008] Daniel Bryant and Paul J. Krause. A review of current defeasible reasoning implementations. *The Knowledge Engineering Review*, 23:1–34, 2008.

[Budán et al., 2012] Maximiliano C. Budán, Mauro J. Gómez Lucero, Carlos I. Chesñevar, and Guillermo R. Simari. Modelling time and reliability in structured argumentation frameworks. In Gerhard Brewka, Thomas Eiter, and Sheila A. McIlraith, editors, *KR*. AAAI Press, 2012.

[Budán et al., 2015] Maximiliano C. Budán, Mauro J. Gómez Lucero, Carlos I. Chesñevar, and Guillermo R. Simari. Modeling time and valuation in structured argumentation frameworks. *Inf. Sci.*, 290:22–44, 2015.

[Caminada and Amgoud, 2007] Martin Caminada and Leila Amgoud. On the evaluation of argumentation formalisms. *Artificial Intelligence*, 171(5-6):286–310, 2007.

[Chesñevar and Simari, 2007] Carlos I. Chesñevar and Guillermo R. Simari. Modelling inference in argumentation through labelled deduction: Formalization and logical properties. *Logica Universalis*, 1(1):93–124, 2007.

[Chesñevar et al., 2000] Carlos I. Chesñevar, Ana G. Maguitman, and Ronald P. Loui. Logical models of argument. *ACM Computing Surveys*, 32(4):337–383, 2000.

[Chesñevar et al., 2004] Carlos I. Chesñevar, Guillermo R. Simari, Teresa Alsinet, and Lluis Godo. A logic programming framework for possibilistic argumentation with vague knowledge. In David M. Chickering and Joseph Y. Halpern, editors, *UAI*, pages 76–84. AUAI Press, 2004.

[Cobo et al., 2010] Maria Laura Cobo, Diego C. Martínez, and Guillermo R. Simari. On admissibility in timed abstract argumentation frameworks. In Helder Coelho, Rudi Studer, and Michael Wooldridge, editors, *ECAI 2010 - 19th European Conference on Artificial Intelligence, Lisbon, Portugal, August 16-20, 2010, Proceedings*, volume 215 of *Frontiers in Artificial Intelligence and Applications*, pages 1007–1008. IOS Press, 2010.

[Cobo et al., 2011] Maria Laura Cobo, Diego C. Martínez, and Guillermo R. Simari. Acceptability in timed frameworks with intermittent arguments. In Lazaros S. Iliadis, Ilias Maglogiannis, and Harris Papadopoulos, editors, *Artificial Intelligence Applications and Innovations - 12th INNS EANN-SIG International Conference, EANN 2011 and 7th IFIP WG 12.5 International Conference, AIAI 2011*, volume 364 of *IFIP Advances in Information and Communication Technology*, pages 202–211. Springer, 2011.

[Cohen et al., 2016] Andrea Cohen, Alejandro J. García, and Guillermo R. Simari. A structured argumentation system with backing and undercutting. *Eng. Appl. of AI*, 49:149–166, 2016.

[Deagustini et al., 2013] Cristhian A. D. Deagustini, Santiago E. Fulladoza Dalibón, Sebastian Gottifredi, Marcelo A. Falappa, Carlos I. Chesñevar, and Guillermo R. Simari. Relational databases as a massive information source for defeasible argumentation. *Knowledge-Based Systems, (online)*, 2013.

[Dung, 1993] Phan Minh Dung. On the acceptability of arguments and its fundamental role in nonmonotonic reasoning and logic programming. In Ruzena Bajcsy, editor, *IJCAI*, pages 852–859, 1993.

[Dung, 1995] Phan Minh Dung. On the acceptability of arguments and its fundamental role in nonmonotonic reasoning, logic programming and n-person games. *Artificial Intelligence*, 77:321–357, 1995.

[Ferretti et al., 2008] Edgardo Ferretti, Marcelo Errecalde, Alejandro J. García, and Guillermo R. Simari. Decision rules and arguments in defeasible decision making. In Ph. Besnard, S. Doutre, and A. Hunter, editors, *Proc. 2nd Int. Conference on Computational Models of Arguments (COMMA)*, volume 172 of *Frontiers in Artificial Intelligence and Applications*, pages 171–182. IOS Press, 2008.

[Gabbay, 1985] Dov Gabbay. Theoretical foundations for non-monotonic reasoning in expert systems. In K. Apt, editor, *Logics and Models of Concurrent Systems*, pages 439–459. Springer-Verlag, 1985.

[García and Simari, 2004] Alejandro J. García and Guillermo R. Simari. Defeasible logic programming: An argumentative approach. *TPLP*, 4(1-2):95–138, 2004.

[García and Simari, 2014] Alejandro J. García and Guillermo R. Simari. Defeasible logic programming: DeLP-Servers, contextual queries, and explanations for answers. *Argument & Computation*, 5(1):63–88, 2014.

[García et al., 2007a] Alejandro J. García, Nicolás D. Rotstein, and Guillermo R. Simari. Dialectical explanations in defeasible argumentation. In Khaled Mellouli, editor, *ECSQARU*, volume 4724 of *Lecture Notes in Computer Science*, pages 295–307. Springer, 2007.

[García et al., 2007b] Alejandro J. García, Nicolás D. Rotstein, Mariano Tucat, and Guillermo R. Simari. An argumentative reasoning service for deliberative agents. In Zili Zhang and Jörg H. Siekmann, editors, *KSEM*, volume 4798 of *Lecture Notes in Computer Science*, pages 128–139. Springer, 2007.

[García et al., 2008] Diego R. García, Alejandro J. García, and Guillermo R. Simari. Defeasible reasoning and partial order planning. In S. Hartmann and G. Kern-Isberner, editors, *FoIKS*, volume 4932 of *Lecture Notes in Computer Science*, pages 311–328. Springer, 2008.

[García, 2000] Alejandro J. García. *Defeasible Logic Programming: Definition, Operational Semantics and Parallelism. (Ph.D. Thesis)*. PhD thesis, Computer Science and Engineering Department, Universidad Nacional del Sur, Bahía Blanca, Argentina, December 2000.

[Gelfond and Lifschitz, 1988] Michael Gelfond and Vladimir Lifschitz. The stable model semantics for logic programming. In *Logic Programming, Proceedings of the Fifth International Conference and Symposium, Seattle, Washington, 1988*, pages 1070–1080. MIT Press, 1988.

[Gelfond and Lifschitz, 1990] Michael Gelfond and Vladimir Lifschitz. Logic programs with classical negation. In David H. D. Warren and Péter Szeredi, editors, *Logic Programming, Proceedings of the Seventh International Conference*, pages 579–597. MIT Press, 1990.

[Gelfond, 2008] Michael Gelfond. Answer sets. In Frank van Harmelen, Vladimir Lifschitz, and Bruce Porter, editors, *Handbook of Knowledge Representation*, Foundations of Artificial Intelligence, chapter 7, pages 285–316. Elsevier, Amsterdam, The Netherlands, 2008.

[Gómez et al., 2013] Sergio A. Gómez, Carlos I. Chesñevar, and Guillermo R. Simari. Ontoarg: A decision support framework for ontology integration based on argumentation. *Expert Syst. Appl.*, 40(5):1858–1870, 2013.

[Kowalski, 2014] Robert Kowalski. History of logic programming. In Dov M. Gabbay, Jörg Siekmann, and John Woods, editors, *Computational Logic*, volume 9 of *Handbook of the History of Logic*, chapter 13, pages 523–569. North Holland, Amsterdam, The Netherlands, 2014.

[Kraus et al., 1990] Sarit Kraus, Daniel J. Lehmann, and Menachen Magidor. Nonmonotonic reasoning, preferential models and cumulative logics. *Artificial Intelligence*, 44:167–207, 1990.

[Leone et al., 2006] Nicola Leone, Gerald Pfeifer, Wolfgang Faber, Thomas Eiter, Georg Gottlob, Simona Perri, and Francesco Scarcello. The dlv system for knowledge representation and reasoning. *ACM Trans. Comput. Logic*, 7(3):499–562, July 2006.

[Lifschitz, 1996a] Vladimir Lifschitz. Foundations of logic programs. In G. Brewka, editor, *Principles of Knowledge Representation*, pages 69–128. CSLI Pub., 1996.

[Lifschitz, 1996b] Vladimir Lifschitz. Foundations of logic programs. In G. Brewka, editor, *Principles of Knowledge Representation*, pages 69–128. CSLI Pub., 1996.

[Lin and Shoham, 1989] Fangzhen Lin and Yoav Shoham. Argument systems: A uniform basis for nonmonotonic reasoning. In *Proceedings of the 1st International Conference on Principles of Knowledge Representation and Reasoning (KR'89). Toronto 1989.*, pages 245–255. Morgan Kaufmann, 1989.

[Loui, 1987] Ronald Prescott Loui. Defeat among arguments: a system of defeasible inference. *Computational Intelligence*, 3:100–106, 1987.

[Makinson, 1994] David Makinson. General patterns in nonmonotonic reasoning. In D. Gabbay, editor, *Handbook of Logic in Artificial Intelligence and Logic Programming (vol 3): Nonmonotonic and Uncertain Reasoning*, pages 35–110. Oxford University Press, 1994.

[McCarthy, 1980] John McCarthy. Circumscription - a form of non-monotonic reasoning. *Artif. Intell.*, 13(1-2):27–39, 1980.

[McDermott and Doyle, 1980] Drew V. McDermott and Jon Doyle. Non-monotonic logic i. *Artif. Intell.*, 13(1-2):41–72, 1980.

[Minker, 1994] Jack Minker. Overview of disjunctive logic programming. *Ann. Math. Artif. Intell.*, 12(1-2):1–24, 1994.

[Modgil and Prakken, 2013] Sanjay Modgil and Henry Prakken. A general account of argumentation with preferences. *Artif. Intell.*, 195:361–397, 2013.

[Moguillansky *et al.*, 2013] Martín O. Moguillansky, Nicolás D. Rotstein, Marcelo A. Falappa, Alejandro J. García, and Guillermo R. Simari. Dynamics of knowledge in *DeLP* through argument theory change. *TPLP*, 13(6):893–957, 2013.

[Moore, 1984] Robert C. Moore. Possible-world semantics for autoepistemic logic. In *NMR*, pages 344–354, 1984.

[Nute, 1987] Donald Nute. Defeasible reasoning. In *Proceedings of the 20th Hawaii International Conference on System Science*, pages 470–477, Hawaii, USA, 1987.

[Nute, 1988] Donald Nute. Defeasible reasoning: a philosophical analysis in PROLOG. In J. H. Fetzer, editor, *Aspects of Artificial Intelligence*, pages 251–288. Kluwer Academic Pub., 1988.

[Nute, 1994] Donald Nute. Defeasible logic. In D.M. Gabbay, C.J. Hogger, and J.A.Robinson, editors, *Handbook of Logic in Artificial Intelligence and Logic Programming, Vol 3*, pages 355–395. Oxford University Press, 1994.

[Pardo and Godo, 2011] Pere Pardo and Lluis Godo. t-delp: A temporal extension of the defeasible logic programming argumentative framework. In S. Benferhat and J. Grant, editors, *SUM*, volume 6929 of *Lecture Notes in Computer Science*, pages 489–503. Springer, 2011.

[Pollock, 1987] John L. Pollock. Defeasible reasoning. *Cognitive Science*, 11(4):481–518, 1987.

[Pollock, 1995] John L. Pollock. *Cognitive Carpentry: A Blueprint for How to Build a Person*. MIT Press, 1995.

[Pollock, 1996] John L. Pollock. A general-purpose defeasible reasoner. *Journal of Applied Non-Classical Logics*, 6(1):89–113, 1996.

[Prakken and Sartor, 1997] Henry Prakken and Giovanni Sartor. Argument-based logic programming with defeasible priorities. *J. of Applied Non-classical Logics*, 7(25-75), 1997.

[Prakken and Vreeswijk, 2002] Henry Prakken and Gerard Vreeswijk. Logics for defeasible argumentation. In D. Gabbay and F. Guenthner, editors, *Handbook of Philosophical Logic*, pages 218–319. Kluwer Academic Pub., 2002.

[Prakken, 2010] Henry Prakken. An abstract framework for argumentation with structured arguments. *Argument & Computation*, 1(2):93–124, 2010.

[Rahwan and Simari, 2009] Iyad Rahwan and Guillermo R. Simari. *Argumentation in Artificial Intelligence*. Springer, 2009.

[Reiter, 1980] Raymond Reiter. A logic for default reasoning. *Artif. Intell.*, 13(1-2):81–132, 1980.

[Schweimeier and Schroeder, 2005] Ralf Schweimeier and Michael Schroeder. A parameterised hierarchy of argumentation semantics for extended logic programming and its application to the well-founded semantics. *TPLP*, 5(1-2):207–242, 2005.

[Simari and Loui, 1992] Guillermo R. Simari and Ronald P. Loui. A Mathematical Treatment of Defeasible Reasoning and its Implementation. *Artificial Intelligence*, 53:125–157, 1992.

[Simari *et al.*, 1994] Guillermo R. Simari, Carlos I. Chesñevar, and Alejandro J. García. The role of dialectics in defeasible argumentation. In *Proc. of the XIV Int. Conf. of the Chilenean Computer Science Society*, pages 335–344, 1994.

[Simari, 1989] Guillermo R. Simari. *A Mathematical Treatment of Defeasible Reasoning and its Implementation*. PhD thesis, Washington University, Department of Computer Science (Saint Louis, Missouri, EE.UU.), 1989.

[Stolzenburg *et al.*, 2003] Frieder Stolzenburg, Alejandro J. García, Carlos I. Chesñevar, and Guillermo R. Simari. Computing generalized specificity. *Journal of Applied Non-Classical Logics*, 13(1):87–113, 2003.

[Teze et al., 2014] Juan Carlos Teze, Sebastian Gottifredi, Alejandro J. García, and Guillermo R. Simari. An approach to argumentative reasoning servers with multiple preference criteria. *Inteligencia Artificial, Revista Iberoamericana de Inteligencia Artificial*, 17(53):68–78, 2014.

[Teze et al., 2015] Juan Carlos Teze, Sebastian Gottifredi, Alejandro J. García, and Guillermo R. Simari. Improving argumentation-based recommender systems through context-adaptable selection criteria. *Expert Syst. Appl.*, 42(21):8243–8258, 2015.

[Tucat et al., 2009] Mariano Tucat, Alejandro J. Garcia, and Guillermo R. Simari. Using Defeasible Logic Programming with Contextual Queries for Developing Recommender Servers. In *AAAI Fall Symposium Series*, 2009.

[Turing, 1950] Alan M. Turing. I.–Computing Machinery and Intelligence. *Mind*, LIX(236):433, 1950.

Alejandro Javier García
Institute for Computer Science and Engineering (UNS-CONICET),
Department of Computer Science and Engineering,
Universidad Nacional del Sur,
Bahía Blanca, Argentina.
email: ajg@cs.uns.edu.ar

Guillermo Ricardo Simari
Institute for Computer Science and Engineering (UNS-CONICET),
Department of Computer Science and Engineering,
Universidad Nacional del Sur,
Bahía Blanca, Argentina.
email: grs@cs.uns.edu.ar

9
A Review of Argumentation Based on Deductive Arguments

PHILIPPE BESNARD, ANTHONY HUNTER

ABSTRACT. A deductive argument is a pair where the first item is a set of premises, the second item is a claim, and the premises entail the claim. This can be formalized by assuming a logical language for the premises and the claim, and logical entailment (or consequence relation) for showing that the claim follows from the premises. Examples of logics that can be used include classical logic, modal logic, description logic, temporal logic, and conditional logic. A counterargument for an argument A is an argument B where the claim of B contradicts the premises of A. Different choices of logic, and different choices for the precise definitions of argument and counterargument, give us a range of possibilities for formalizing deductive argumentation. Further options are available to us for choosing the arguments and counterarguments we put into an argument graph. If we are to construct an argument graph based on the arguments that can be constructed from a knowledgebase, then we can be exhaustive in including all arguments and counterarguments that can be constructed from the knowledgebase. But there are other options available to us. These include being selective in the arguments and counterargument we present according to a specified criterion. We consider some of the possibilities in this review and introduce properties and postulates for comparing proposals for deductive argumentation.

1 Introduction

In deductive reasoning, we start with some premises, and we derive a conclusion using one or more inference steps. Each inference step is infallible in the sense that it does not introduce uncertainty. In other words, if we accept the premises are valid, then we should accept that the intermediate conclusion of each inference step is valid, and therefore we should accept that the conclusion is valid. For example, if we accept that Philippe and Tony are having tea together in London is valid, then we should accept that Philippe is not in Toulouse (assuming the background knowledge that London and Toulouse are different places, and that nobody can be in different places at the same time). As another example, if we accept that Philippe and Tony are having an ice cream together in Toulouse is valid, then we should accept that Tony is not in London. Note, however, we do not need to believe or know that the premises are valid to apply deductive reasoning. Rather, deductive reasoning allows us to obtain conclusions that we can accept contingent on the validity of their premises. So for the first example above, the reader might not know whether or not Philippe and Tony are having tea together in London. However, the reader can accept that Philippe is not in Toulouse, contingent on the validity of these premises. Important alternatives to deductive reasoning in argumentation, include inductive reasoning, abductive reasoning, and analogical reasoning.

In this review, we assume that deductive reasoning is formalized by a monotonic logic. Each deductive argument is a pair where the first item is a set of premises that logically entails the second item according to the choice of monotonic logic. So we have a logical language to express the set of premises, and the claim, and we have a logical consequence relation to relate the premises to the claim.

Key benefits of deductive arguments include: (1) Explicit representation of the information used to support the claim of the argument; (2) Explicit representation of the claim of the argument; and (3) A simple and precise connection between the support and claim of the argument via the consequence relation. What a deductive argument does not provide is a specific proof of the claim from the premises. There may be more than one way of proving the claim from the premises, but the argument does not specify which is used. It is therefore indifferent to the proof used.

Deductive argumentation is formalized in terms of deductive arguments and counterarguments, and there are various choices for defining this [Besnard and Hunter, 2008]. Deductive argumentation offers a simple route to instantiating abstract argumentation which we will consider in this review chapter. Perhaps the first paper to consider this is by Cayrol who instantiated Dung's proposal with deductive arguments based on classical logic [Cayrol, 1995].

In the rest of this review, we will investigate some of the choices we have for defining arguments and counterarguments, and for how they can be used in modelling argumentation. We will focus on three choices for base logic. These are (1) simple logic (which has a language of literals and rules of the form $\alpha_1 \wedge \ldots \wedge \alpha_n \rightarrow \beta$ where $\alpha_1, \ldots, \alpha_n, \beta$ are literals, and modus ponens is the only proof rule), (2) classical logic (propositional and first-order classical logic), and (3) A conditional logic. Then for instantiating argument graphs (i.e. for specifying what the arguments and attacks are in an argument graph), we will consider descriptive graphs and generated graphs defined informally as follows.

- **Descriptive graphs** Here we assume that the structure of the argument graph is given, and the task is to identify the premises and claim of each argument. Therefore the input is an abstract argument graph, and the output is an instantiated argument graph. This kind of task arises in many situations: For example, if we are listening to a debate, we hear the arguments exchanged, and we can construct the instantiated argument graph to reflect the debate[1].

---

[1] When we listen to a debate (or similarly, when we read an article that discusses some topic), we use natural language processing to identify arguments and counterarguments in spoken (or written) language. At a high level, this involves determining the words used (i.e.

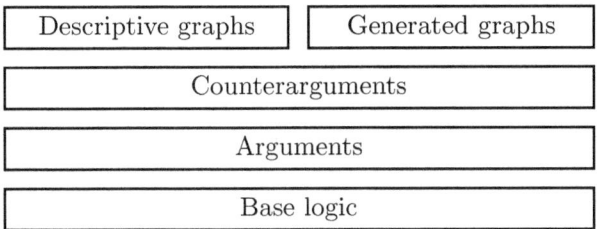

Figure 1. Framework for constructing argument graphs with deductive arguments: For defining a specific argumentation system, there are four levels for the specification: (1) A base logic is required for defining the logical language and the consequence or entailment relation (i.e. what inferences follow from a set of formlulae); (2) A definition of an argument $\langle \Phi, \alpha \rangle$ specified using the base logic (e.g. Φ is consistent, and Φ entails α); (3) A definition of counterargument specified using the base logic (i.e. a definition for when one argument attacks another); and (4) A definition of how the arguments and counterarguments are composed into an argument graph (which is either a descriptive graph or some form of generated graph).

- **Generated graphs** Here we assume that we start with a knowledgebase (i.e. a set of logical formulae), and the task is to generate the arguments and counterarguments (and hence the attacks between arguments). Therefore, the input is a knowledgebase, and the output is an instantiated argument graph. This kind of task also arises in many situations: For example, if we are making a decision based on conflicting information. We have various items of information that we represent by formulae in the knowledgebase, and we construct an instantiated argument graph to reflect the arguments and counterarguments that follow from that information.

For constructing both descriptive graphs and generated graphs, there may be a dynamic aspect to the process. For instance, when constructing descriptive graphs, we may be unsure of the exact structure of the argument graph, and it is only by instantiating individual arguments that we are able to say whether it is attacked or attacks another argument. As another example, when constructing generated graphs, we may be involved in a dialogue, and so through the dialogue, we may obtain further information which allows us to generate further arguments that can be added to the argument graph.

So in order to construct argument graphs with deductive arguments, we need to specify the choice of logic (which we call the base logic) that we use

---

the verbal description) for each argument and counterargument. This can be thought of as a form of argument mining. Then once the arguments and counterarguments have been identified, the actual logical structure of each argument can be determined from the verbal description.

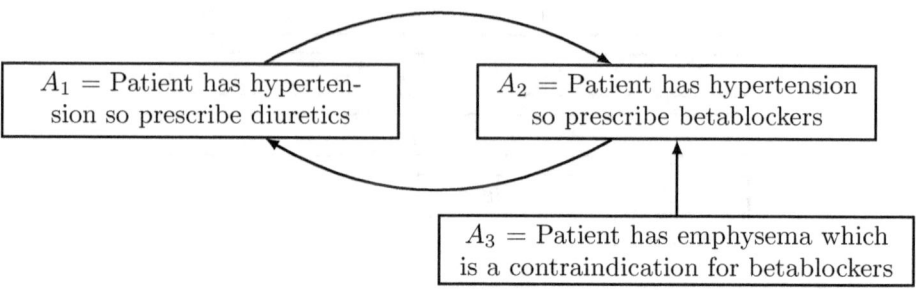

Figure 2. Example of an abstract argument graph which captures a decision making scenario where there are two alternatives for treating a patient, diuretics or betablockers. Since only one treatment should be given for the disorder, each argument attacks the other. There is also a reason to not give betablockers, as the patient has emphysema which is a contraindication for this treatment.

to define arguments and counterarguments, the definition for arguments, the definition for counterarguments, and the definition for instantiating argument graphs. For the latter, we can either produce a descriptive graph or a generated graph. We will explore various options for generated graphs. We summarize the framework for constructing argument graphs with deductive arguments in Figure 1.

We proceed as follows: (Section 2) We briefly review the definitions for abstract argumentation; (Section 3) We review the nature of base logics in argumentation; (Section 4) We consider options for arguments in deductive argumentation; (Section 5) We consider options for counterarguments in deductive argumentation; (Section 6) We consider options for constructing argument graphs instantiated with deductive arguments; (Section 7) We review some properties and postulates for argumentation based on deductive arguments; and (Section 8) We discuss the approach of deductive argumentation and provide suggestions for further reading.

2 Abstract argumentation

Abstract argumentation, as proposed by [Dung, 1995], provides a good starting point for formalizing argumentation. Dung proposed that a set of arguments and counterarguments could be represented by a directed graph. Each node in the graph denotes an argument and each arc denotes one argument attacking another. So if there is an arc from node A to node B, then A attacks B, or equivalently A is a counterargument to B. See Figure 2 for an example of an abstract argument graph.

An **abstract argument graph** is a pair $(\mathcal{A}, \mathcal{R})$ where $\mathcal{A}$ is a set and $\mathcal{R} \subseteq \mathcal{A} \times \mathcal{A}$. Each element $A \in \mathcal{A}$ is called an **argument** and $(A, B) \in \mathcal{R}$ means that A **attacks** B (accordingly, A is said to be an **attacker** of B) and so A is a **counterargument** for B. A set of arguments $S \subseteq \mathcal{A}$ **attacks** $A_j \in \mathcal{A}$ iff there is an argument $A_i \in S$ such that A_i attacks A_j. Also, S **defends** $A_i \in \mathcal{A}$ iff for each argument $A_j \in \mathcal{A}$, if A_j attacks A_i then S attacks A_j. A set $S \subseteq \mathcal{A}$ of arguments is **conflict-free** iff there are no arguments A_i and A_j in S such that A_i attacks A_j. Let Γ be a conflict-free set of arguments, and let Defended : $\wp(\mathcal{A}) \to \wp(\mathcal{A})$ be a function such that Defended(Γ) = $\{A \mid \Gamma$ defends $A\}$. We consider the following extensions: (1) Γ is a **complete extension** iff Γ = Defended(Γ); (2) Γ is a **grounded extension** iff it is the minimal (w.r.t. set inclusion) complete extension; (3) Γ is a **preferred extension** iff it is a maximal (w.r.t. set inclusion) complete extension; and (4) Γ is a **stable extension** iff it is a preferred extension that attacks every argument that is not in the extension.

Some argument graphs can be large, and yet we might only be interested in whether some subset of the arguments is in an extension according to some semantics. For this, we introduce the following definitions that lead to the notion of a focal graph.

Definition 2.1 *Let $G = (\mathcal{A}, \mathcal{R})$ be an argument graph. An argument graph $(\mathcal{A}', \mathcal{R}')$ is **faithful** with respect to $(\mathcal{A}, \mathcal{R})$ iff $(\mathcal{A}', \mathcal{R}')$ is a subgraph of $(\mathcal{A}, \mathcal{R})$ and for all arguments $A_i, A_j \in \mathcal{A}$, if $A_j \in \mathcal{A}'$ and $(A_i, A_j) \in \mathcal{R}$, then $A_i \in \mathcal{A}'$, and $\mathcal{R}' = \{(A_i, A_j) \mid \mathcal{R} \mid A_i, A_j \in \mathcal{A}'\}$.*

Example 2.2 *Consider the following graph G*

There are three subgraphs that are faithful with respect to G: (1) The graph G ; (2) The subgraph containing just the argument A_1; and (3) The following subgraph. All other subgraphs of G are not faithful.

A faithful subgraph has the same extensions as the graph modulo the arguments in the subgraph. So for every argument A in the subgraph, if A is in the grounded extension in the subgraph, then A is in the grounded extension of the graph, and vice versa. Similarly, for every argument A in the subgraph, if A is in a preferred extension of the subgraph, then A is in a preferred extension of the graph, and vice versa. This follows directly from the directionality criterion of [Baroni and Giacomin, 2007] that says that for a subgraph, arguments in the graph that do not attack any arguments in the subgraph have no affect on the extensions of the subgraph. Therefore, we can ignore the arguments that are not in a faithful subgraph.

Definition 2.3 *Let $\Pi \subseteq \mathcal{A}$ be a set of arguments of interest called the **focus**. A graph $(\mathcal{A}', \mathcal{R}')$ is the **focal graph** of graph $(\mathcal{A}, \mathcal{R})$ with respect to focus Π iff $(\mathcal{A}', \mathcal{R}')$ is the smallest subgraph of $(\mathcal{A}, \mathcal{R})$ such that $\Pi \subseteq \mathcal{A}'$ and $(\mathcal{A}', \mathcal{R}')$ is faithful with respect to $(\mathcal{A}, \mathcal{R})$.*

Example 2.4 *Continuing Example 2.2, if we let $\Pi = \{A_1, A_2\}$ be the focus, then the third subgraph (i.e. the faithful graph containing A_1, A_2, and A_3) is the focal graph.*

The motivation for finding the focal graph is that given a set of arguments Π as the focus, we want to just have those arguments and any arguments that may affect whether or not any of the arguments in Π are in an extension. By taking the directionality of the arcs into account (i.e. the directionality criteria [Baroni and Giacomin, 2007; Liao et al., 2011]), we can ignore the other arguments.

Even though abstract argumentation provides a clear and precise approach to formalizing aspects of argumentation, the arguments are treated as atomic. There is no formalized content to an argument, and so all arguments are treated as equal. Therefore if we want to understand individual arguments, we need to provide content for them. This leads to the idea of "instantiating" abstract argumentation with deductive arguments. Each deductive argument has some premises from which a claim is derived by deductive reasoning.

3 Base logics

Proposals for logic-based argumentation rely on an underlying logic, which we call a *base logic*, for generating logical arguments and for defining the counterargument relationships (using inference of conflict or existence of inconsistency).

The choice of base logic is an important design decision for a logic-based argumentation system. This then raises the questions of what are the minimal requirements for a base logic and what are the factors that need to be considered for a base logic?

In this chapter, we focus on three options for the base logic, namely simple logic, classical logic, and a conditional logic, but other options include modal logic, temporal logic, paraconsistent logic, description logics, and logic programming languages.

Let $\mathcal{L}$ be a language for a logic, and let $\vdash_i$ be the consequence relation for that logic. Therefore, $\vdash_i \subseteq \wp(\mathcal{L}) \times \mathcal{L}$. If α is an atom in $\mathcal{L}$, then α is a **positive literal** in $\mathcal{L}$ and, assuming a negation symbol $\neg$ is available in the language, $\neg \alpha$ is a **negative literal** in $\mathcal{L}$. For a literal β, the **complement** of β is defined as follows: If β is a positive literal, i.e. it is of the form α, then the complement of β is the negative literal $\neg \alpha$, and if β is a negative literal, i.e. it is of the form $\neg \alpha$, then the complement of β is the positive literal α.

The list of properties of a consequence relation given in Table 1 provides a good starting point for considering this question. These properties have been proposed as desirable conditions of a consequence relation. Furthermore, according to Gabbay [Gabbay, 1985] and Makinson [Makinson, 1994], the minimal properties of a consequence relation are reflexivity, monotonicity (or a variant of it) and cut, and the need for each of them can be justified as follows:

- Reflexivity captures the idea of "transparency"; If a formula α is assumed (i.e. $\alpha \in \Delta$), then α can be inferred (i.e $\Delta \vdash_x \alpha$).

- Monotonicity captures the idea of "irreversibility"; Once a formula α is inferred (i.e $\Delta \vdash_x \alpha$), then there is no assumption that can cause α to be withdrawn (i.e. there is no β such that $\Delta \cup \{\beta\} \nvdash_x \alpha$).

- Cut captures the idea of "equitability" of assumptions and inferences. Once a formula α is inferred (i.e $\Delta \vdash_x \alpha$), it can be used for further reasoning.

These three properties can be seen equivalently in terms of the following three properties based on the consequence closure C_x of a logic x [Makinson, 1994], where $C_x(\Delta) = \{\alpha \mid \Delta \vdash \alpha\}$: (inclusion) $\Delta \subseteq C_x(\Delta)$; (idempotence) $C_x(\Delta) = C_x(C_x(\Delta))$; and (monotony) $C_x(\Delta') \subseteq C_x(\Delta)$ whenever $\Delta' \subseteq \Delta$.

Classes of base logics can be identified using properties of the consequence relation, and then argument systems can be developed in terms of them. For instance, to instantiate abstract argumentation, in [Amgoud and Besnard, 2009], the class of Tarskian logics has been used. This is the class defined by inclusion, idempotence, finiteness (i.e. $C_x(\Delta)$ is the union of $C_x(\Gamma)$ for all finite subsets Γ of Δ), absurdity (i.e. $C_x(\{\phi\}) = \mathcal{L}$ for some ϕ in the language $\mathcal{L}$), and coherence (i.e. $C_x(\emptyset) \neq \mathcal{L}$). Classical logic is an example of a Tarskian logic.

4 Arguments

A **deductive argument** is an ordered pair $\langle \Phi, \alpha \rangle$ where $\Phi \vdash_i \alpha$. Φ is the support, or premises, or assumptions of the argument, and α is the claim, or conclusion, of the argument. The definition for a deductive argument only assumes that the premises entail the claim (i.e. $\Phi \vdash_i \alpha$). For an argument $A = \langle \Phi, \alpha \rangle$, the function $\mathsf{Support}(A)$ returns Φ and the function $\mathsf{Claim}(A)$ returns α.

Many proposals have further constraints for an ordered pair $\langle \Phi, \alpha \rangle$ to be an argument. The most commonly assumed constraint is the **consistency constraint**: An argument $\langle \Phi, \alpha \rangle$ satisfies this constraint when Φ is consistent (assuming that the base logic has a notion of consistency). For richer logics, such as classical logic, consistency is often regarded as a desirable property of a deductive argument because claims that are obtained with logics such as classical logic from inconsistent premises are normally useless as illustrated in the next example.

Example 4.1 *If we assume the consistency constraint, then the following are*

Name	Property
Reflexivity	$\Delta \cup \{\alpha\} \vdash_x \alpha$
Literal reflexivity	$\Delta \cup \{\alpha\} \vdash_x \alpha$ if α is a literal
Left logical equiv.	$\Delta \cup \{\beta\} \vdash_x \gamma$ if $\Delta \cup \{\alpha\} \vdash_x \gamma$ and $\vdash \alpha \leftrightarrow \beta$
Right weakening	$\Delta \vdash_x \alpha$ if $\Delta \vdash_x \beta$ and $\vdash \beta \rightarrow \alpha$
And	$\Delta \vdash_x \alpha \wedge \beta$ if $\Delta \vdash_x \alpha$ and $\Delta \vdash_x \beta$
Monotonicity	$\Delta \cup \{\alpha\} \vdash_x \beta$ if $\Delta \vdash_x \beta$
Cut	$\Delta \vdash_x \beta$ if $\Delta \vdash_x \alpha$ and $\Delta \cup \{\alpha\} \vdash_x \beta$
Conditionalization	$\Delta \vdash_x \alpha \rightarrow \beta$ if $\Delta \cup \{\alpha\} \vdash_x \beta$
Deduction	$\Delta \cup \{\alpha\} \vdash_x \beta$ if $\Delta \vdash_x \alpha \rightarrow \beta$
Contraposition	$\Delta \cup \{\alpha\} \vdash_x \beta$ if $\Delta \cup \{\neg\beta\} \vdash_x \neg\alpha$
Or	$\Delta \cup \{\alpha \vee \beta\} \vdash_x \gamma$ if $\Delta \cup \{\alpha\} \vdash_x \gamma$ and $\Delta \cup \{\beta\} \vdash_x \gamma$

Table 1. Some properties of a consequence relation $\vdash_x$ adapted from D. Makinson. General patterns in nonmonotonic reasoning. In D. Gabbay, C. Hogger, and J. Robinson, editors, *Handbook of Logic in Artificial Intelligence and Logic Programming*, volume 3, pages 35–110. Oxford University Press, 1994.

not arguments.

$\langle \{\text{study(Sid, logic)}, \neg\text{study(Sid, logic)}\},$
$\qquad\qquad \text{study(Sid, logic)} \leftrightarrow \neg\text{study(Sid, logic)}\rangle$

$\langle \{\text{study(Sid, logic)}, \neg\text{study(Sid, logic)}\}, \text{KingOfFrance(Sid)}\rangle$

In contrast, for weaker logics (such as paraconsistent logics), it may be desirable to not impose the consistency constraint. With such logics, a credulous approach could be taken so that pros and cons could be obtained from inconsistent premises (as illustrated by the following example).

Example 4.2 *If we assume the base logic is a paraconsistent logic (such as Belnap's four valued logic), and we do not impose the consistent constraint, then the following are arguments.*

$\langle \{\text{study(Sid, logic)} \wedge \neg\text{study(Sid, logic)}\}, \text{study(Sid, logic)}\rangle$

$\langle \{\text{study(Sid, logic)} \wedge \neg\text{study(Sid, logic)}\}, \neg\text{study(Sid, logic)}\rangle$

Another commonly assumed constraint is the **minimality constraint**: An argument $\langle \Phi, \alpha \rangle$ satisfies this constraint when there is no $\Psi \subset \Phi$ such that $\Psi \vdash \alpha$. Minimality is often regarded as a desirable property of a deductive argument because it eliminates irrelevant premises (as in the following example).

Example 4.3 *If we assume the minimality constraint, then the following is not an argument.*

$\langle \{\texttt{report(rain)}, \texttt{report(rain)} \to \texttt{carry(umbrella)}, \texttt{happy(Sid)}\},$
$\texttt{carry(umbrella)}\rangle$

When we construct a knowledgebase, with simple logic, classical logic, or other base logics, it is possible that some or all of the formulae could be incorrect. For instance, individual formulae may come from different and conflicting sources, they may reflect options that disagree, they may represent uncertain information. A knowledgebase may be inconsistent, and individual formulae may be contradictory. After all, if the knowledge is not inconsistent (i.e. it is consistent), then we will not have counterarguments. We may also include formulae that we know are not always correct. For instance, we may include a formula such as the following that says that a water sample taken from the Mediterranean sea in the summer will be above 15 degrees Celcius. While this may be a useful general rule, it is not always true. For instance, the sample could be taken when there is a period of bad weather, or the sample is taken from a depth of over 500 metres.

$\forall \texttt{X}, \texttt{Y}.\texttt{watersample}(\texttt{X}) \wedge \texttt{location}(\texttt{X}, \texttt{Mediterranean})$
$\wedge \texttt{season}(\texttt{X}, \texttt{summer}) \wedge \texttt{termperature}(\texttt{X}, \texttt{Y}) \to \texttt{Y} > 15$

In the following subsections, we define arguments based on simple logic and on classical logic as the base logic. Alternative base logics include description logic, paraconsistent logic, temporal logic, and conditional logic.

4.1 Arguments based on simple logic

Simple logic is based on a language of literals and simple rules where each **simple rule** is of the form $\alpha_1 \wedge \ldots \wedge \alpha_k \to \beta$ where α_1 to α_k and β are literals. A **simple logic knowledgebase** is a set of literals and simple rules. The consequence relation is modus ponens (i.e. implication elimination) as defined next.

Definition 4.4 *The* **simple consequence relation**, *denoted* $\vdash_s$, *which is the smallest relation satisfying the following condition, and where* Δ *is a simple logic knowledgebase:* $\Delta \vdash_s \beta$ *iff there is an* $\alpha_1 \wedge \cdots \wedge \alpha_n \to \beta \in \Delta$, *and for each* $\alpha_i \in \{\alpha_1, \ldots, \alpha_n\}$, *either* $\alpha_i \in \Delta$ *or* $\Delta \vdash_s \alpha_i$.

Note, the simple consequence relation does not satisfy reflexivity. We could slightly amend the definition so that it does satisfy reflexivity but the above definition will be useful to us later when we consider properties of the argumentation based on this base logic.

Example 4.5 *Let* $\Delta = \{a, b, a \wedge b \to c, c \to \neg d\}$. *Hence,* $\Delta \vdash_s c$ *and* $\Delta \vdash_s \neg d$. *However,* $\Delta \nvdash_s a$ *and* $\Delta \nvdash_s b$.

Definition 4.6 *Let Δ be a simple logic knowledgebase. For $\Phi \subseteq \Delta$, and a literal α, $\langle \Phi, \alpha \rangle$ is a* **simple argument** *iff $\Phi \vdash_s \alpha$ and there is no proper subset Φ' of Φ such that $\Phi' \vdash_s \alpha$.*

So each simple argument is minimal but not necessarily consistent (where consistency for a simple logic knowledgebase Δ means that for no atom α does $\Delta \vdash_s \alpha$ and $\Delta \vdash_s \neg \alpha$ hold). We do not impose the consistency constraint in the definition for simple arguments as simple logic is paraconsistent, and therefore can support a credulous view on the arguments that can be generated.

Example 4.7 *Let p_1, p_2, and p_3 be the following formulae. Then the following is a simple argument.*

$$\langle \{p_1, p_2, p_3\}, \text{goodInvestment(BP)} \rangle$$

Note, we use p_1, p_2, and p_3 as labels in order to make the presentation of the premises more concise.

$p_1 = \text{oilCompany(BP)}$
$p_2 = \text{goodPerformer(BP)}$
$p_3 = \text{oilCompany(BP)} \land \text{goodPerformer(BP)}) \rightarrow \text{goodInvestment(BP)}$

Simple logic is a practical choice as a base logic for argumentation. Having a logic with simple rules and modus ponens is useful for applications because the behaviour is quite predictable in the sense that given a knowledgebase it is relatively easy to anticipate the inferences that come from the knowlegebase. Furthermore, it is relatively easy to implement an algorithm for generating the arguments and counterarguments from a knowledgebase. The downside of simple logic as a base logic is that the proof theory is weak. It only incorporates modus ponens (i.e. implication elimination) and so many useful kinds of reasoning (e.g. contrapositive reasoning) are not supported.

4.2 Arguments based on classical logic

Classical logic is appealing as the choice of base logic as it better reflects the richer deductive reasoning often seen in arguments arising in discussions and debates.

We assume the usual propositional and predicate (first-order) languages for classical logic, and the usual the **classical consequence relation**, denoted $\vdash$. A **classical knowledgebase** is a set of classical propositional or predicate formulae.

Definition 4.8 *For a classical knowledgebase Φ, and a classical formula α, $\langle \Phi, \alpha \rangle$ is a* **classical argument** *iff $\Phi \vdash \alpha$ and $\Phi \not\vdash \bot$ and there is no proper subset Φ' of Φ such that $\Phi' \vdash \alpha$.*

So a classical argument satisfies both minimality and consistency. We impose the consistency constraint because we want to avoid the useless inferences that come with inconsistency in classical logic (such as via ex falso quodlibet).

Example 4.9 *The following classical argument uses a universally quantified formula in contrapositive reasoning to obtain the claim about number 77.*

$\langle \{\forall \mathtt{X}.\mathtt{multipleOfTen}(\mathtt{X}) \to \mathtt{even}(\mathtt{X}), \neg\mathtt{even}(77)\}, \neg\mathtt{multipleOfTen}(77) \rangle$

Given the central role classical logic has played in philosophy, linguistics, and computer science (software engineering, formal methods, data and knowledge engineering, artificial intelligence, computational linguistics, etc.), we should consider how it can be used in argumentation. Classical propositional logic and classical predicate logic are expressive formalisms which capture more detailed aspects of the world than is possible with restricted formalisms such as simple logic.

4.3 Arguments based on conditional logic

Conditional logics are a valuable alternative to classical logic for knowledge representation and reasoning. They can be used to capture hypothetical statements of the form "If α were true, then β would be true". This done by introducing an extra connective $\Rightarrow$ to extend a classical logic language. Informally, $\alpha \Rightarrow \beta$ is valid when β is true in the possible worlds where α is true. Representing and reasoning with such knowledge in argumentation is valuable because useful arguments exist that refer to fictitious and hypothetical situations (see [Besnard et al., 2013] for some examples).

In this review, we consider the well-known conditional logic MP which can be extended to give many other well-known conditional logics, and we follow the presentation for argumentation given by [Besnard et al., 2013]. The language of conditional logic is that of classical logic extended with the formulae of the form $\alpha \Rightarrow \beta$ where α and β are formulae in the language of conditional logic. The proof theory for the consequence relation $\vdash_c$ of MP is given by classical logic proposition extended by the following axiom schemas and inference rules.

$$RCEA \quad \dfrac{\vdash_c \alpha \leftrightarrow \beta}{\vdash_c (\alpha \Rightarrow \gamma) \leftrightarrow (\beta \Rightarrow \gamma)}$$

$$RCEC \quad \dfrac{\vdash_c \alpha \leftrightarrow \beta}{\vdash_c (\gamma \Rightarrow \alpha) \leftrightarrow (\gamma \Rightarrow \beta)}$$

$$CC \quad \vdash_c ((\alpha \Rightarrow \beta) \land (\alpha \Rightarrow \gamma)) \to (\alpha \Rightarrow (\beta \land \gamma))$$

$$CM \quad \vdash_c (\alpha \Rightarrow (\beta \land \gamma)) \to ((\alpha \Rightarrow \beta) \land (\alpha \Rightarrow \gamma))$$

$$CN \quad \vdash_c (\alpha \to \top)$$

$$MP \quad \vdash_c (\alpha \to \beta) \to (\alpha \to \beta)$$

Using the $\vdash_c$ consequence relation, we can define the notion of an argument with the same constraints as for classical logic arguments. For this, a conditional knowledgebase is a set of formulae of the language of conditional logic.

Definition 4.10 *For a conditional knowledgebase Delta, and a formula of conditional logic α, $\langle \Delta, \alpha \rangle$ is a* **conditional argument** *iff $\Delta \vdash_c \alpha$ and $\Delta \not\vdash_c \bot$ and there is no proper subset Δ' of Δ such that $\Delta' \vdash \alpha$.*

Example 4.11 *Let $\Delta = \{\texttt{matchIsStruck} \Rightarrow \texttt{matchLights}, \texttt{matchIsStruck}\}$. From this knowledgebase, we get the following argument.*

$$\langle \{\texttt{matchIsStruck} \Rightarrow \texttt{matchLights}, \texttt{matchIsStruck}\}, \texttt{matchLights} \rangle$$

Note, from Δ, we cannot get the following argument.

$$\langle \{\texttt{matchIsStruck} \land \texttt{matchIsWet} \Rightarrow \texttt{matchLights}, \texttt{matchIsStruck}\}, \texttt{matchLights} \rangle$$

Whereas if we consider $\Delta' = \{\texttt{matchIsStruck} \to \texttt{matchLights}, \texttt{matchIsStruck}\}$ where we use the classical implication for the formula, we get the classical argument from Δ'.

$$\langle \{\texttt{matchIsStruck} \land \texttt{matchIsWet} \to \texttt{matchLights}, \texttt{matchIsStruck}\}, \texttt{matchLights} \rangle$$

This is because from $\texttt{matchIsStruck} \to \texttt{matchLights}$, we can infer the following using classical logic.

$$\texttt{matchIsStruck} \land \texttt{matchIsWet} \to \texttt{matchLights}$$

So the above example illustrates how the proof theory (and the semantics) for conditional logic is more restricted for the $\Rightarrow$ connective than for the $\to$. This makes it useful for representing and reasoning with knowledge about fictitious and hypothetical situations [Cross and Nute, 2001; Girard, 2006].

5 Counterarguments

A counterargument is an argument that attacks another argument. In deductive argumentation, we define the notion of counterargument in terms of logical contradiction between the claim of the counterargument and the premises of claim of the attacked argument. We explore some of the kinds of counterargument that can be specified for simple logic, classical logic and classical logic.

5.1 Counteraguments based on simple logic

For simple logic, we consider two forms of counterargument. For this, recall that literal α is the complement of literal β if and only if α is an atom and β is $\neg \alpha$ or if β is an atom and α is $\neg \beta$.

Definition 5.1 *For simple arguments A and B, we consider the following type of* **simple attack***:*

- *A is a* **simple undercut** *of B if there is a simple rule $\alpha_1 \land \cdots \land \alpha_n \to \beta$ in $\mathsf{Support}(B)$ and there is an $\alpha_i \in \{\alpha_1, \ldots, \alpha_n\}$ such that $\mathsf{Claim}(A)$ is the complement of α_i*

- *A is a **simple rebut** of B if* Claim(*A*) *is the complement of* Claim(*B*)

Example 5.2 *The first argument A_1 captures the reasoning that the metro is an efficient form of transport, so one can use it. The second argument A_2 captures the reasoning that there is a strike on the metro, and so the metro is not an efficient form of transport (at least on the day of the strike). A_2 is a simple undercut of A_1.*

$A_1 = \langle \{\texttt{efficientMetro}, \texttt{efficientMetro} \to \texttt{useMetro}\}, \texttt{useMetro}\rangle$
$A_2 = \langle \{\texttt{strikeMetro}, \texttt{strikeMetro} \to \neg\texttt{efficientMetro}\}, \neg\texttt{efficientMetro}\rangle$

Example 5.3 *The first argument A_1 captures the reasoning that the government has a budget deficit, and so the government should cut spending. The second argument A_2 captures the reasoning that the economy is weak, and so the government should not cut spending. The arguments are simple rebuts of each other.*

$A_1 = \langle \{\texttt{govDeficit}, \texttt{govDeficit} \to \texttt{cutGovSpend}\}, \texttt{cutGovSpend}\rangle$
$A_2 = \langle \{\texttt{weakEconomy}, \texttt{weakEconomy} \to \neg\texttt{cutGovSpend}\}, \neg\texttt{cutGovSpend}\rangle$

So in simple logic, a rebut attacks the claim of an argument, and an undercut attacks the premises of the argument (by attacking one of the consequents of one of the rules in the premises).

Simple arguments and counterarguments can be used to model defeasible reasoning. For this, we use simple rules that are normally correct but sometimes are incorrect. For instance, if Sid has the goal of going to work, Sid takes the metro. This is generally true, but sometimes Sid works at home, and so it is no longer true that Sid takes the metro, as we see in the next example.

Example 5.4 *The first argument A_1 captures the general rule that if* workDay *holds, then* metro(Sid) *holds (denoting that Sid takes the metro). The use of the simple rule in A_1 requires that the assumption* normal *holds. This is given as an assumption. The second argument A_2 undercuts the first argument by contradicting the assumption that* normal *holds*

$A_1 = \langle \{\texttt{workDay}, \texttt{normal}, \texttt{workDay} \wedge \texttt{normal} \to \texttt{metro(Sid)}\}, \texttt{metro(Sid)}\rangle$
$A_2 = \langle \{\texttt{workAtHome(Sid)}, \texttt{workAtHome(Sid)} \to \neg\texttt{normal}\}, \neg\texttt{normal}\rangle$

Informally, if we start with just argument A_1, then A_1 is undefeated, and so metro(Sid) *is an acceptable claim. However, if we then add A_2, then A_1 is a defeated argument and A_2 is an undefeated argument. Hence, if we have A_2, we have to withdraw* metro(Sid) *as an acceptable claim.*

So by having appropriate conditions in the antecedent of a simple rule we can disable the rule by generating a counterargument that attacks it. This in effect stops the usage of the simple rule. This means that we have a convention to attack an argument based on the inferences obtained by the simple logic (e.g. as in Example 5.2 and Example 5.3), or on the rules used (e.g. Example 5.4).

This way to disable rules by adding appropriate conditions (as in Example 5.4) is analogous to the use abnormality predicates used in formalisms such as circumscription (see for example [McCarthy, 1980]). We can use the same approach to capture defeasible reasoning in other logics such as classical logic. Note, this does not mean that we turn the base logic into a nonmonotonic logic. Both simple logic and classical logic are monotonic logics. Hence, for a simple logic knowledgebase Δ (and similarly for a classical logic knowledgebase Δ), the set of simple arguments (respectively classical arguments) obtained from Δ is a subset of the set of simple arguments (respectively classical arguments) obtained from $\Delta \cup \{\alpha\}$ where α is a formula not in Δ. But at the level of evaluating arguments and counterarguments, we have non-monotonic defeasible behaviour. For instance in Example 5.2, with just A_1 we have the acceptable claim that useMetro, but then when we have also A_2, we have to withdraw this claim. In other words, if the set of simple arguments is $\{A_1\}$, then we can construct an argument graph with just A_1, and by applying Dung's dialectical semantics, there is one extension containing A_1. However, if the set of simple arguments is $\{A_1, A_2\}$, then we can construct an argument graph with A_1 attacked by A_2, and by applying Dung's dialectical semantics, there is one extension containing A_2. This illustrates the fact that the argumentation process is nonmonotonic.

5.2 Counterarguments based on classical logic

Given the expressivity of classical logic (in terms of language and inferences), there are a number of natural ways to define counterarguments.

Definition 5.5 *Let A and B be two classical arguments. We define the following types of* **classical attack**.

- A *is a* **classical defeater** *of B if* $\mathsf{Claim}(A) \vdash \neg \bigwedge \mathsf{Support}(B)$.

- A *is a* **classical direct defeater** *of B if* $\exists \phi \in \mathsf{Support}(B)$ *s.t.* $\mathsf{Claim}(A) \vdash \neg \phi$.

- A *is a* **classical undercut** *of B if* $\exists \Psi \subseteq \mathsf{Support}(B)$ *s.t.* $\mathsf{Claim}(A) \equiv \neg \bigwedge \Psi$.

- A *is a* **classical direct undercut** *of B if* $\exists \phi \in \mathsf{Support}(B)$ *s.t.* $\mathsf{Claim}(A) \equiv \neg \phi$.

- A *is a* **classical canonical undercut** *of B if* $\mathsf{Claim}(A) \equiv \neg \bigwedge \mathsf{Support}(B)$.

- A *is a* **classical rebuttal** *of B if* $\mathsf{Claim}(A) \equiv \neg \mathsf{Claim}(B)$.

- A *is a* **classical defeating rebuttal** *of B if* $\mathsf{Claim}(A) \vdash \neg \mathsf{Claim}(B)$.

A Review of Argumentation Based on Deductive Arguments

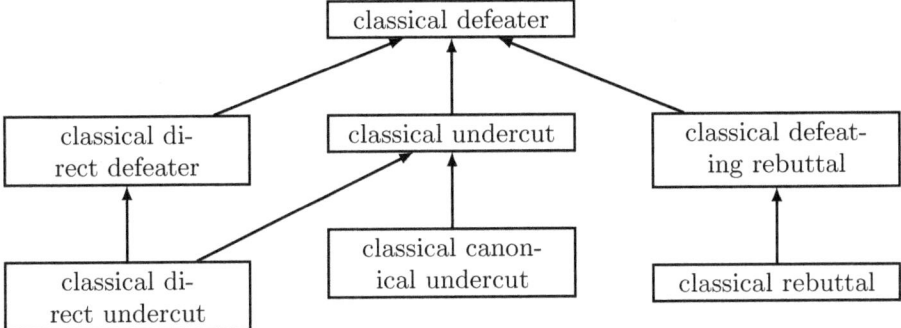

Figure 3. We can represent the containment between the classical attack relations as above where an arrow from R_1 to R_2 indicates that $R_1 \subseteq R_2$.

Note, in the rest of this section, we will drop the term "classical" when we discuss these types of attack.

To illustrate these different notions of classical counterargument, we consider the following examples, and we relate these definitions in Figure 3 where we show that classical defeaters are the most general of these definitions.

Example 5.6 Let $\Delta = \{a \vee b, a \leftrightarrow b, c \to a, \neg a \wedge \neg b, a, b, c, a \to b, \neg a, \neg b, \neg c\}$

$\langle \{a \vee b, c\}, (a \vee b) \wedge c \rangle$ is a defeater of $\langle \{\neg a, \neg b\}, \neg a \wedge \neg b \rangle$
$\langle \{a \vee b, c\}, (a \vee b) \wedge c \rangle$ is a direct defeater of $\langle \{\neg a \wedge \neg b\}, \neg a \wedge \neg b \rangle$
$\langle \{\neg a \wedge \neg b\}, \neg(a \wedge b) \rangle$ is a undercut of $\langle \{a, b, c\}, a \wedge b \wedge c \rangle$
$\langle \{\neg a \wedge \neg b\}, \neg a \rangle$ is a direct undercut of $\langle \{a, b, c\}, a \wedge b \wedge c \rangle$
$\langle \{\neg a \wedge \neg b\}, \neg(a \wedge b \wedge c) \rangle$ is a canonical undercut of $\langle \{a, b, c\}, a \wedge b \wedge c \rangle$
$\langle \{a, a \to b\}, b \vee c \rangle$ is a rebuttal of $\langle \{\neg a \wedge \neg b, \neg c\}, \neg(b \vee c) \rangle$
$\langle \{a, a \to b\}, b \rangle$ is a defeating rebuttal of $\langle \{\neg a \wedge \neg b, \neg c\}, \neg(b \vee c) \rangle$

Using simple logic, the definitions for counterarguments against the support of another argument are limited to attacking just one of the items in the support. In contrast, using classical logic, a counterargument can be against more than one item in the support. For example, in Example 5.7, the undercut is not attacking an individual premise but rather saying that two of the premises are incompatible (in this case that the premises lowCostFly and luxuryFly are incompatible).

Example 5.7 Consider the following arguments. A_1 is attacked by A_2 as A_2 is an undercut of A_1 though it is neither a direct undercut nor a canonical undercut. Essentially, the attack says that the flight cannot be both a low cost flight and a luxury flight.

$A_1 = \langle \{\text{lowCostFly}, \text{luxFly}, \text{lowCostFly} \land \text{luxFly} \to \text{goodFly}\}, \text{goodFly} \rangle$
$A_2 = \langle \{\neg\text{lowCostFly} \lor \neg\text{luxFly}\}, \neg\text{lowCostFly} \lor \neg\text{luxFly} \rangle$

Trivially, undercuts are defeaters but it is also quite simple to establish that rebuttals are defeaters. Furthermore, if an argument has defeaters then it has undercuts. It may happen that an argument has defeaters but no rebuttals as illustrated next.

Example 5.8 Let $\Delta = \{\neg\text{containsGarlic} \land \text{goodDish}, \neg\text{goodDish}\}$. Then the following argument has at least one defeater but no rebuttal.

$$\langle \{\neg\text{containsGarlic} \land \text{goodDish}\}, \neg\text{containsGarlic} \rangle$$

There are some important differences between rebuttals and undercuts that can be seen in the following examples.

Example 5.9 Consider the following arguments. The first argument A_1 is a direct undercut to the second argument A_2, but neither rebuts each other. Furthermore, A_1 "agrees" with the claim of A_2 since the premises of A_1 could be used for an alternative argument with the same claim as A_2.

$A_1 = \langle \{\neg\text{containsGarlic} \land \neg\text{goodDish}\}, \neg\text{containsGarlic} \rangle$
$A_2 = \langle \{\text{containsGarlic}, \text{containsGarlic} \to \neg\text{goodDish}\}, \neg\text{goodDish} \rangle$

Example 5.10 Consider the following arguments. The first argument is a rebuttal of the second argument, but it is not an undercut because the claim of the first argument is not equivalent to the negation of some subset of the premises of the second argument.

$A_1 = \langle \{\text{goodDish}\}, \text{goodDish} \rangle$
$A_2 = \langle \{\text{containsGarlic}, \text{containsGarlic} \to \neg\text{goodDish}\}, \neg\text{goodDish} \rangle$

So an undercut for an argument need not be a rebuttal for that argument, and a rebuttal for an argument need not be an undercut for that argument.

Arguments are not necessarily independent. In a sense, some encompass others (possibly up to some form of equivalence), which is the topic we now turn to.

Definition 5.11 An argument $\langle \Phi, \alpha \rangle$ is **more conservative** than an argument $\langle \Psi, \beta \rangle$ iff $\Phi \subseteq \Psi$ and $\beta \vdash \alpha$.

Example 5.12 $\langle \{a\}, a \lor b \rangle$ is more conservative than $\langle \{a, a \to b\}, b \rangle$.

Roughly speaking, a more conservative argument is more general: It is, so to speak, less demanding on the support and less specific about the claim.

Example 5.13 *Consider the following formulae.*

$$p_1 = \texttt{divisibleByTen(50)}$$
$$p_2 = \forall \texttt{X.divisibleByTen(X)} \rightarrow \texttt{divisibleByTwo(X)}$$
$$p_3 = \forall \texttt{X.divisibleByTwo(X)} \rightarrow \texttt{even(X)}$$

Hence, A_1 is an argument with the claim "The number 50 is divisible by 2", and A_2 is an argument with the claim "The number 50 is divisible by 2 and the number 50 is an even number". However, A_1 is more conservative than A_2.

$$A_1 = \langle \{p_1, p_2\}, \texttt{divisibleByTwo(50)} \rangle$$
$$A_2 = \langle \{p_1, p_2, p_3, \}, \texttt{even(50)} \wedge \texttt{divisibleByTwo(50)} \rangle$$

We can use the notion of "more conservative" to help us identify the most useful counterarguments amongst the potentially large number of counterarguments.

Example 5.14 *Let $\{a, b, c, \neg a \vee \neg b \vee \neg c\}$ be our knowledgebase. Suppose we start with the argument $\langle \{a, b, c\}, a \wedge b \wedge c \rangle$. Now we have numerous undercuts to this argument including the following.*

$$\langle \{b, c, \neg a \vee \neg b \vee \neg c\}, \neg a \rangle$$
$$\langle \{a, c, \neg a \vee \neg b \vee \neg c\}, \neg b \rangle$$
$$\langle \{a, b, \neg a \vee \neg b \vee \neg c\}, \neg c \rangle$$
$$\langle \{c, \neg a \vee \neg b \vee \neg c\}, \neg a \vee \neg b \rangle$$
$$\langle \{b, \neg a \vee \neg b \vee \neg c\}, \neg a \vee \neg c \rangle$$
$$\langle \{a, \neg a \vee \neg b \vee \neg c\}, \neg b \vee \neg c \rangle$$
$$\langle \{\neg a \vee \neg b \vee \neg c\}, \neg a \vee \neg b \vee \neg c \rangle$$

All these undercuts say the same thing which is that the set $\{a, b, c\}$ is inconsistent together with the formula $\neg a \vee \neg b \vee \neg c$. As a result, this can be captured by the last undercut listed above. Note this is the maximally conservative undercut amongst the undercuts listed, and moreover it is a canonical undercut. This example therefore illustrates how the canonical undercuts are the undercuts that (in a sense) represent all the other undercuts.

So choosing classical logic as the base logic gives us a wider range of choices for defining attacks. This has advantages if we want to better capture argumentation as arising in natural language, or to more precisely capture counterarguments generated from certain kinds of knowledge. However, it does also mean that we need to be aware of the consequences of our choice of definition for attacks when using a generated approach to instantiating argument graphs (as we will discuss in the next section).

5.3 Counterarguments based on conditional logic

In order to define counterarguments, we follow the proposal in [Besnard et al., 2013] for *conditional contrariety*. For this, we require the notion of an extended literal which is either of the form ϕ or $\neg\phi$ where ϕ is either an atom or a formula of the form $\alpha \Rightarrow \beta$. Then we exploit the fact that any formula of conditional logic can be rewritten using the proof rules into an equivalent set of disjunctive formulae of the form $\alpha_1 \vee \ldots \vee \alpha_n$ where each α_i is an extended literal. In particular, for the next definition, we are interested in the specific situation when a formula can be equivalently represented by a single disjunctive formula.

Definition 5.15 *Let α and β be two formulae of conditional logic such that the disjunctive form of α is $\alpha_1 \vee \ldots \vee \alpha_m$ and the disjunctive form of β is $\beta_1 \vee \ldots \vee \beta_n$. α is the **contrary** of β, denoted $\alpha \triangleright \beta$ iff for all $\alpha_i \in \{\alpha_1, \ldots, \alpha_m\}$, and for all $\beta_j \in \{\beta_1, \ldots, \beta_n\}$,*

1. *$\{\alpha_i, \beta_j\} \vdash_c \bot$*

2. *there exists $\gamma_1, \gamma_2, \delta_1, \delta_2$ in the language such that the following conditions are satisfied*

   (a) $\gamma_1 \vdash_c \delta_1$

   (b) $\delta_1 \not\vdash_c \gamma_1$

   (c) $\gamma_1 \not\vdash_c \gamma_2$

   (d) $\delta_2 \vdash_c \gamma_2$

   (e) $\gamma_2 \vdash_c \delta_2$

   where either $\{\alpha_i, \beta_j\} \vdash_c (\gamma_1 \Rightarrow \gamma_2) \wedge (\delta_1 \Rightarrow \delta_2)$ such that $\alpha \vdash_c \gamma_1 \Rightarrow \gamma_2$, or $\beta_j \vdash_c (\gamma_1 \Rightarrow \gamma_2) \wedge (\delta_1 \rightarrow \delta_2)$.

In the above definition, the first condition subsumes "classical contradiction", and the second condition captures situations where two rules conflict as illustated by the following examples.

- $a \wedge (a \wedge c \Rightarrow f)$ is a contrary of $a \rightarrow (a \Rightarrow f)$

- $a \wedge (a \wedge c \Rightarrow f)$ is a contrary of $\neg a \vee c \Rightarrow f$.

- $\neg a \wedge b$ and $a \wedge (a \wedge c \rightarrow b \vee \neg d)$ are the contrary of each other.

As another illustration of rules for which the second condition of the above definition applies is the following where the second rule is the contrary of the first. The intuition of the example is that the second rule "corrects" the circumstances under which John will go to watch the match at the stadium.

- matchTonight $\Rightarrow$ JohnGoesToTheStadium

- matchTonight $\wedge$ JohnHasEnoughMoney $\Rightarrow$ JohnGoesToTheStadium

Example 5.16 Let $\Delta = \{a \Rightarrow b, a \vee d \Rightarrow b \wedge c, a \Rightarrow c\}$. Let α be $a \Rightarrow b \wedge c$. Note that $\Delta \vdash_c \alpha$. However, $\alpha \triangleright \Delta$ because $\alpha \triangleright a \vee d \Rightarrow b \wedge c$.

The notion of contrary is extended to sets of conditional formulae Φ so that $\alpha \triangleright \Phi$ holds iff there exists a $\beta \in \Phi$ such that $\Phi \vdash_c \beta$ and $\alpha \triangleright \beta$ holds.

Example 5.17 Let $\Delta = \{a \Rightarrow b, a \vee d \Rightarrow b \wedge c, a \Rightarrow c\}$. Let α be $a \Rightarrow b \wedge c$. Note that $\Delta \vdash_c \alpha$. However, $\alpha \triangleright \Delta$ because $\alpha \triangleright a \vee d \Rightarrow b \wedge c$.

Now we can extend the definitions of counterargument given for classical logic. We just give two options to illustrate the space of possibilities.

Definition 5.18 Let $\langle \Phi, \alpha \rangle$ and $\langle \Psi, \beta \rangle$ be conditional logic arguments.

- $\langle \Psi, \beta \rangle$ is a **conditional rebuttal** for $\langle \Phi, \alpha \rangle$ iff $\beta \triangleright \alpha$.

- $\langle \Psi, \beta \rangle$ is a **conditional defeater** for $\langle \Phi, \alpha \rangle$ iff $\beta \triangleright \Phi$.

Example 5.19 Below, the second argument is a conditional rebuttal for the first argument.

$\langle \{\text{matchTonight} \Rightarrow \text{JohnGoesToTheStadium}\},$
$\quad \text{matchTonight} \Rightarrow \text{JohnGoesToTheStadium} \rangle$
$\langle \{\text{matchTonight} \wedge \text{JohnHasEnoughMoney} \Rightarrow \text{JohnGoesToTheStadium}\},$
$\quad \text{matchTonight} \wedge \text{JohnHasEnoughMoney} \Rightarrow \text{JohnGoesToTheStadium} \rangle$

Example 5.20 Some conditional defeaters for $\langle \{a \vee \neg d \Rightarrow b \wedge c, f \vee \neg b, b\}, f \wedge (a \vee \neg d \Rightarrow b \wedge c) \rangle$ are listed below.

$\langle \{\neg b\}, \neg b \rangle$
$\langle \{\neg b\}, \neg(\neg b \rightarrow b) \rangle$
$\langle \{\neg b, \neg a \rightarrow b\}, \neg b \wedge a \rangle$
$\langle \{e \wedge \neg d \Rightarrow b \wedge c\}, e \wedge \neg d \Rightarrow b \wedge c \rangle$
$\langle \{a \Rightarrow b, a \Rightarrow c\}, a \Rightarrow b \wedge c \rangle$
$\langle \{a \Rightarrow b, a \Rightarrow c\}, \neg\neg(a \Rightarrow b \wedge c) \rangle$
$\langle \{a \Rightarrow b, a \Rightarrow c\}, (a \Rightarrow b) \wedge (a \Rightarrow c) \rangle$

Example 5.21 Below, the second argument is a conditional defeater for the first argument.

$\langle \{\text{matchTonight}, \text{matchTonight} \Rightarrow \text{JohnGoesToTheStadium}\},$
$\quad \text{JohnGoesToTheStadium} \rangle$
$\langle \{\text{matchTonight} \wedge \text{JohnHasEnoughMoney} \Rightarrow \text{JohnGoesToTheStadium}\},$
$\quad \text{matchTonight} \wedge \text{JohnHasEnoughMoney} \Rightarrow \text{JohnGoesToTheStadium} \rangle$

By using conditional logic as a base logic, we have a range of options for more effective modelling complex real-world scenarios. Whilst many conditional logics extend classical logic, the implication introduced is normally more restricted than the strict implication used in classical logic. This means that many knowledge modelling situations, such as for non-monotonic reasoning, can be better captured by conditional logics (such as [Delgrande, 1987; Kraus et al., 1990; Arló-Costa and Shapiro, 1992]).

6 Argument graphs

We now investigate options for instantiating argument graphs. We start with descriptive argument graphs, and then turn to generated argument graphs, using simple logic, classical logic, and conditional logic.

6.1 Descriptive graphs

For the descriptive approach to argument graphs, we assume that we have some abstract argument graph as the input, together with some informal description of each argument. For instance, when we listen to a debate on the radio, we may identify a number of arguments and counterarguments, and for each of these we may be able to write a brief text summary. So if we then want to understand this argumentation in more detail, we may choose to instantiate each argument with a deductive argument. So for this task we choose the appropriate logical formulae for the premises and claim for each argument (compatible with the choice of base logic). Examples of descriptive graphs are given in Figure 4 using simple logic, and in Example 6.1 and Figure 5 using classical logic.

Example 6.1 *Consider the following argument graph where A_1 is "The flight is low cost and luxury, therefore it is a good flight", and A_2 is "A flight cannot be both low cost and luxury".*

$$\boxed{A_1} \longleftarrow \boxed{A_2}$$

For this, we instantiate the arguments in the above abstract argument graph to give the following descriptive graph. So in the descriptive graph below, A_2 is a classical undercut to A_1.

$$\boxed{A_1 = \langle \{\texttt{lowCostFly}, \texttt{luxFly}, \texttt{lowCostFly} \land \texttt{luxFly} \to \texttt{goodFly}\}, \texttt{goodFly} \rangle}$$

$$\uparrow$$

$$\boxed{A_2 = \langle \{\neg(\texttt{lowCostFly} \land \texttt{luxFly})\}, \neg\texttt{lowCostFly} \lor \neg\texttt{luxFly} \rangle}$$

So for the approach of descriptive graphs, we do not assume that there is an automated process that constructs the graphs. Rather the emphasis is on having a formalization that is a good representation of the argumentation. This is so that we can formally analyze the descriptive graph, perhaps as part of some sense-making or decision-making process. Nonetheless, we can envisage that in the medium term natural language processing technology will be able to parse the text or speech (for instance in a discussion paper or in a debate) in order to automatically identify the premises and claim of each argument and counterargument.

Since we are primarily interested in representational and analytical issues when we use descriptive graphs, a richer logic such as classical logic is a more appealing formalism than a weaker base logic such as simple logic. Given a set of real-world arguments, it is often easier to model them using deductive arguments with classical logic as the base logic than a "rule-based logic" like simple logic as the base logic. For instance, in Example 6.1, the undercut does not claim that the flight is not low cost, and it does not claim that it is not luxury. It only claims that the flight cannot be both low cost *and* luxury. It is natural to represent this exclusion using disjunction.

As another example of the utility of classical logic as base logic, consider the importance of quantifiers in knowledge which require a richer language such as classical logic for reasoning with them. Moreover, if we consider that many arguments are presented in natural language (spoken or written), and that formalizing natural language often calls for a richer formalism such as classical logic (or even richer), it is valuable to harness classical logic in formalizations of deductive argumentation.

Conditional logics are also important formalisms for capturing some of the subtleties of natural language as they can reflect hypothetical statements, and can often provide a better representation of non-monotonic statements. We give an example of descriptive graph using conditional logic in Figure 6.

6.2 Generated graphs based on simple logic

Given a knowledgebase Δ, we can generate an argument graph $G = (\mathcal{A}, \mathcal{R})$ where $\mathcal{A}$ is the set of simple arguments obtained from Δ as follows and $\mathcal{R} \subseteq \mathcal{A} \times \mathcal{A}$ is simple undercut.

Definition 6.2 *Let Δ be a simple logic knowledgebase. A* **simple exhaustive graph** *for Δ is an argument graph $G = (\mathcal{A}, \mathcal{R})$ where $\mathcal{A}$ is $\mathsf{Arguments}_s(\Delta)$ and $\mathcal{R}$ is $\mathsf{Attacks}_s(\Delta)$ defined as follows*

$\mathsf{Args}_s(\Delta) = \{\langle \Phi, \alpha \rangle \mid \Phi \subseteq \Delta \text{ and } \langle \Phi, \alpha \rangle \text{ is a simple argument }\}$
$\mathsf{Attacks}_s(\Delta) = \{(A, B) \mid A, B \in \mathsf{Args}_s(\Delta) \text{ and } A \text{ is a simple undercut of } B\}$

This is an exhaustive approach to constructing an argument graph from a knowledgebase since all the simple arguments and all the simple undercuts are in the argument graph. We give an example of such an argument graph in Figure 7.

Simple exhaustive graphs provide a direct and useful way to instantiate argument graphs. There are various ways the definitions can be adapted, such as defining the attacks to be the union of the simple undercuts and the simple rebuts.

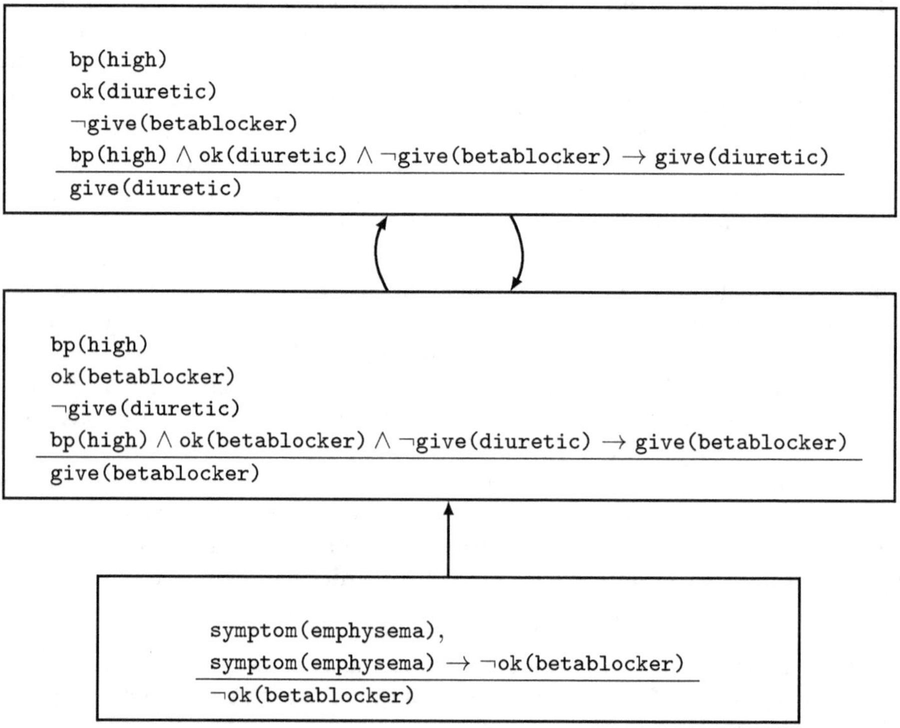

```
bp(high)
ok(diuretic)
¬give(betablocker)
bp(high) ∧ ok(diuretic) ∧ ¬give(betablocker) → give(diuretic)
─────────────────────────────────────────────────────────────
give(diuretic)
```

```
bp(high)
ok(betablocker)
¬give(diuretic)
bp(high) ∧ ok(betablocker) ∧ ¬give(diuretic) → give(betablocker)
─────────────────────────────────────────────────────────────
give(betablocker)
```

```
symptom(emphysema),
symptom(emphysema) → ¬ok(betablocker)
─────────────────────────────────────
¬ok(betablocker)
```

Figure 4. A descriptive graph representation of the abstract argument graph in Figure 2 using simple logic. The atom bp(high) denotes that the patient has high blood pressure. Each attack is a simple undercut by one argument on another. For the first argument, the premises include the assumptions ok(diuretic) and ¬give(betablocker) in order to apply its simple rule. Similarly, for the second argument, the premises include the assumptions ok(betablocker) and ¬give(diuretic) in order to apply its simple rule.

A Review of Argumentation Based on Deductive Arguments

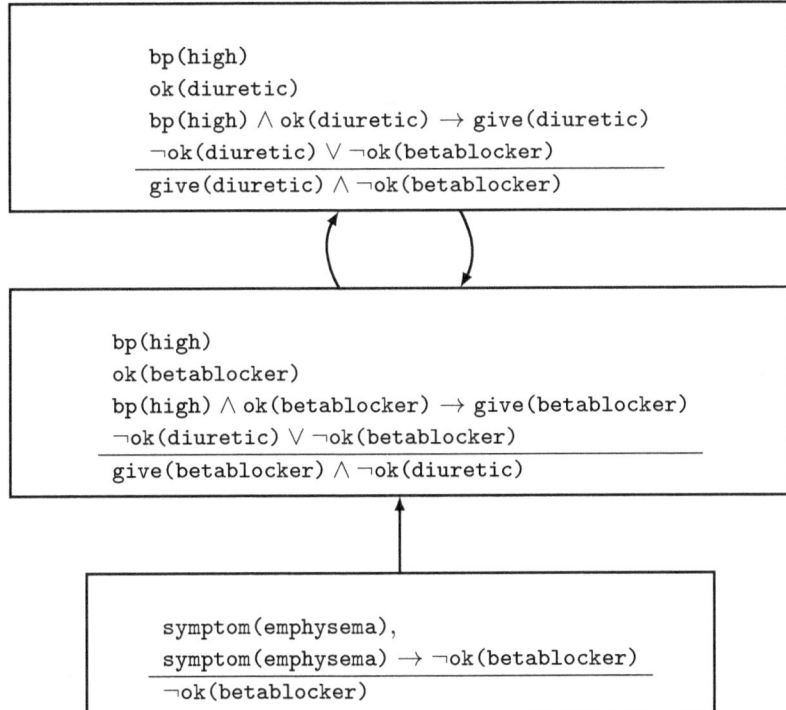

Figure 5. A descriptive graph representation of the abstract argument graph in Figure 2 using classical logic. The atom bp(high) denotes that the patient has high blood pressure. The top two arguments rebut each other (i.e. the attack is classical defeating rebut). For this, each argument has an integrity constraint in the premises that says that it is not ok to give both betablocker and diuretic. So the top argument is attacked on the premise ok(diuretic) and the middle argument is attacked on the premise ok(betablocker). So we are using the ok predicate as a normality condition for the rule to be applied (as suggested in Section 5.1).

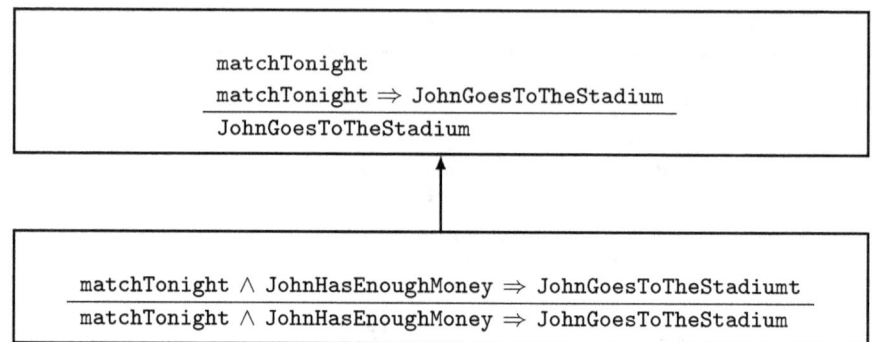

Figure 6. A descriptive graph that captures two arguments. The first argument says John will go to the stadium because there is a match tonight. The second corrects the first argument by correcting the circumstances under which John will go to watch the match at the stadium.

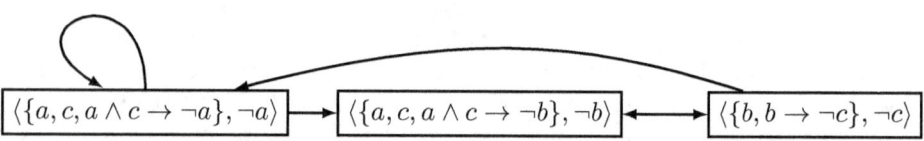

Figure 7. An exhaustive simple logic argument graph where $\Delta = \{a, b, c, a \wedge c \to \neg a, b \to \neg c, a \wedge c \to \neg b\}$. Note, that this exhaustive graph contains a self cycle, and an odd length cycle.

6.3 Generated graphs based on classical logic

In this section, we consider generated graphs for classical logic. We start with the classical exhaustive graphs which are the same as the simple exhaustive graphs except we use classical arguments and attacks. We show that whilst this provides a comprehensive presentation of the information, its utility is limited for various reasons. We then show that by introducing further information, we can address these shortcomings. To illustrate this, we consider a version of classical exhaustive graphs augmented with preference information. This is just one possibility for introducing extra information into the construction process.

6.3.1 Classical exhaustive graphs

Given a knowledgebase Δ, we can generate an argument graph $G = (\mathcal{A}, \mathcal{R})$ where $\mathcal{A}$ is the set of arguments obtained from Δ as follows and $\mathcal{R} \subseteq \mathcal{A} \times \mathcal{A}$ is one of the definitions for classical attack.

Definition 6.3 *Let Δ be a classical logic knowledgebase. A* **classical exhaustive graph** *is an argument graph $G = (\mathcal{A}, \mathcal{R})$ where $\mathcal{A}$ is* $\mathsf{Arguments}_c(\Delta)$ *and $\mathcal{R}$ is* $\mathsf{Attacks}_c^X(\Delta))$ *defined as follows where X is one of the attacks given in Definition 5.5 such as defeater, direct undercut, or rebuttal.*

$\mathsf{Arguments}_c(\Delta) = \{\langle \Phi, \alpha \rangle \mid \Phi \subseteq \Delta \text{ and } \langle \Phi, \alpha \rangle \text{ is a classical argument }\}$
$\mathsf{Attacks}_c^X(\Delta) = \{(A, B) \in \mathsf{Arguments}_c(\Delta) \times \mathsf{Arguments}_c(\Delta) \mid A \text{ is } X \text{ of } B\}$

This is a straightforward approach to constructing an argument graph from a knowledgebase since all the classical arguments and all the attacks (according to the chosen definition of attack) are in the argument graph as illustrated in Figure 8. However, if we use this exhaustive definition, we obtain infinite graphs, even if we use a knowledgebase with few formulae. For instance, if we have an argument $\langle \{a\}, a \rangle$, we also have arguments such as $\langle \{a\}, a \vee a \rangle$, $\langle \{a\}, a \vee a \vee a \rangle$, etc., as well as $\langle \{a\}, \neg \neg a \rangle$, etc.

Even though the graph is infinite, we can present a finite representation of it, by just presenting a representative of each class of structurally equivalent arguments (as considered in [Amgoud et al., 2011]), where we say that two arguments A_i and A_j are **structurally equivalent** in $G = (\mathcal{A}, \mathcal{R})$ when the following conditions are satisfied: (1) if A_k attacks A_i, then A_k attacks A_j; (2) if A_k attacks A_j, then A_k attacks A_i; (3) if A_i attacks A_k, then A_j attacks A_k; and (4) if A_i attacks A_k, then A_j attacks A_k. For example, in Figure 8, the argument A_4 is a representative for $\langle \{b\}, b \rangle$, $\langle \{b\}, a \vee b \rangle$, $\langle \{b\}, \neg a \vee b \rangle$, etc.

We can also ameliorate the complexity of classical exhaustive graphs by presenting a focal graph (as discussed in Section 2). We illustrate this in Figure 9.

To conclude our discussion of classical exhaustive graphs, the definitions ensure that all the ways that the knowledge can be used to generate classical arguments and classical counterarguments (modulo the choice of attack relation) are laid out. This may involve many arguments being presented. This

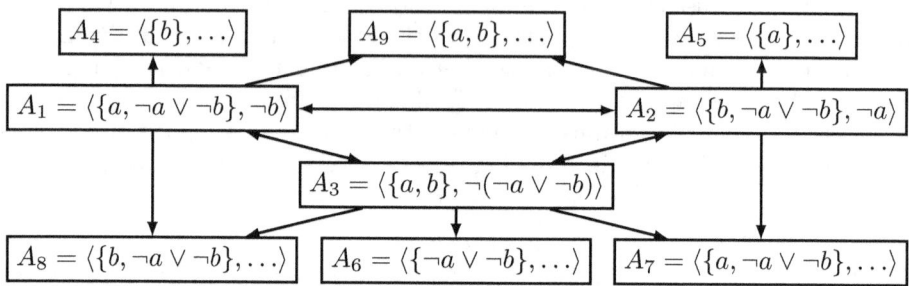

Figure 8. An exhaustive classical logic argument graph where $\Delta = \{a, b, \neg a \vee \neg b\}$ and the attack is direct undercut. Note, argument A_4 represents all arguments with a claim that is implied by b, argument A_5 represents all arguments with a claim that is implied by a, argument A_6 represents all arguments with a claim that is implied by $\neg a \vee \neg b$, argument A_7 represents all arguments with a claim that is implied by $a \wedge \neg b$ except $\neg b$ or any claim implied by a or any claim implied by $\neg a \vee \neg b$, argument A_8 represents all arguments with a claim that is implied by $\neg a \wedge b$ except $\neg a$ or any claim implied by b or any claim implied by $\neg a \vee \neg b$, and argument A_9 represents all arguments with a claim that is implied by $a \wedge b$ except $\neg(\neg a \vee \neg b)$ or any claim implied by a or any claim implied by b.

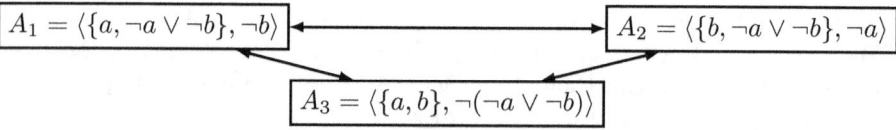

Figure 9. Focal graph formed from Figure 8 where focus is $\{A_1, A_2\}$.

can be addressed by the generation process discriminating between the arguments (and/or the attacks) based on extra information about the arguments and/or information about the audience. There are many ways that this can be done. In the next section, we consider a simple proposal for augmenting the generation process with preferences over arguments.

6.3.2 Preferential exhaustive graphs

One of the first proposals for capturing the idea of preferences in constructing argument graphs was preference-based argumentation frameworks (PAF) by [Amgoud and Cayrol, 2002]. This generalizes Dung's definition for an argument graph by introducing a preference relation over arguments that in effect causes an attack to be ignored when the attacked argument is preferred over the attacker. So in PAF, we assume a preference relation over arguments, denoted $\preceq$, as well as a set of arguments $\mathcal{A}$ and an attack relation $\mathcal{R}$. From this, we need to define a defeat relation $\mathcal{D}$ as follows, and then $(\mathcal{A}, \mathcal{D})$ is used as the argument graph, instead of $(\mathcal{A}, \mathcal{R})$, with Dung's usual definitions for extensions.

$$\mathcal{D} = \{(A_i, A_j) \in \mathcal{R} \mid (A_j, A_i) \notin \preceq\}$$

So with this definition for defeat, extensions for a preference-based argument graph $(\mathcal{A}, \mathcal{R}, \preceq)$ can be obtained as follows: For S denoting complete, preferred, stable or grounded semantics, $\Gamma \subseteq \mathcal{A}$, Γ is an extension of $(\mathcal{A}, \mathcal{R}, \preceq)$ w.r.t. semantics S iff Γ is an extension of $(\mathcal{A}, \mathcal{D})$ w.r.t. semantics S.

We now revise the definition for classical exhaustive graphs to give the following definition for preferential exhaustive graphs.

Definition 6.4 *Let Δ be a classical logic knowledgebase. A* **preferential exhaustive graph** *is an argument graph* $(\text{Arguments}_c(\Delta), \text{Attacks}^X_{c,\preceq}(\Delta))$ *defined as follows where X is one of the attacks given in Definition 5.5 such as defeater, direct undercut, or rebuttal.*

$\text{Arguments}_c(\Delta) = \{\langle \Phi, \alpha \rangle \mid \Phi \subseteq \Delta \ \& \ \langle \Phi, \alpha \rangle \text{ is a classical argument } \}$
$\text{Attacks}^X_{c,\preceq}(\Delta) = \{(A, B) \mid A, B \in \text{Arguments}_c(\Delta) \ \& \ A \text{ is } X \text{ of } B \ \& \ (B, A) \notin \preceq\}$

We give an illustration of a preferential exhaustive graph in Figure 10, and we give an illustration of a focal graph obtained from a preferential exhaustive graph in Example 6.5.

Example 6.5 *This example concerns two possible treatments for glaucoma caused by raised pressure in the eye. The first is a prostoglandin analogue (PGA) and the second is a betablocker (BB). Let Δ contain the following six formulae. The atom p_1 is the fact that the patient has glaucoma, the atom p_2 is the assumption that it is ok to give PGA, and the atom p_3 is the assumption that it is ok to give BB. Each implicational formula (i.e. p_4 and p_5) captures the knowledge that if a patient has glaucoma, and it is ok to give a particular*

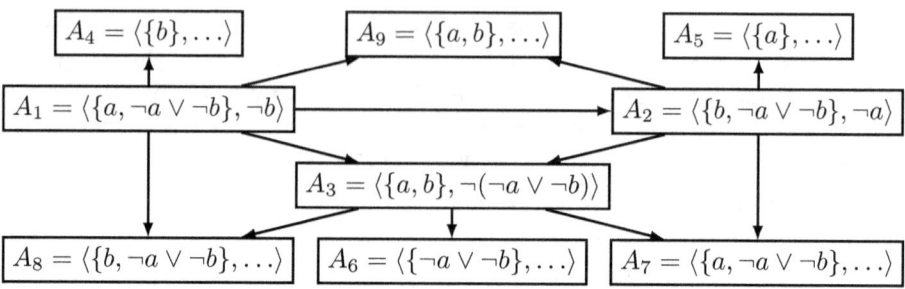

Figure 10. The preferential exhaustive graph where the knowledgesbase is $\Delta = \{a, b, \neg a \vee \neg b\}$ and the attack is direct undercut. This is the same knowledgebase and attack relation as in Figure 8. For the preference relation, A_1 is preferred to all other arguments, A_2 is preferred to all other arguments apart from A_1, and the remaining arguments are equally preferred. So for all i such that $i \neq 1$, $A_1 \prec A_i$, and for all i such that $i \neq 1$ and $i \neq 2$, $A_2 \prec A_i$. This results in three attacks in Figure 8 not appearing in this graph. The dropped attacks are A_2 on A_1, A_3 on A_1, and A_3 on A_1.

drug, then give that drug. Formula p_6 is an integrity constraint that ensures that only one treatment is given for the condition.

$$\begin{aligned}
&\mathsf{p}_1 = \mathtt{glaucoma} &&\mathsf{p}_4 = \mathtt{glaucoma} \wedge \mathtt{ok(PGA)} \rightarrow \mathtt{give(PGA)} \\
&\mathsf{p}_2 = \mathtt{ok(PGA)} &&\mathsf{p}_5 = \mathtt{glaucoma} \wedge \mathtt{ok(BB)} \rightarrow \mathtt{give(BB)} \\
&\mathsf{p}_3 = \mathtt{ok(BB)} &&\mathsf{p}_6 = \neg\mathtt{ok(PGA)} \vee \neg\mathtt{ok(BB)}
\end{aligned}$$

There are numerous arguments that can be constructed from this set of formulae such as the following.

$$\begin{aligned}
A_1 &= \langle \{\mathsf{p}_1, \mathsf{p}_2, \mathsf{p}_4, \mathsf{p}_6\}, \mathtt{give(PGA)} \wedge \neg\mathtt{ok(BB)} \rangle & A_5 &= \langle \{\mathsf{p}_1, \mathsf{p}_2, \mathsf{p}_4\}, \mathtt{give(PGA)} \rangle \\
A_2 &= \langle \{\mathsf{p}_1, \mathsf{p}_3, \mathsf{p}_5, \mathsf{p}_6\}, \mathtt{give(BB)} \wedge \neg\mathtt{ok(PGA)} \rangle & A_6 &= \langle \{\mathsf{p}_1, \mathsf{p}_3, \mathsf{p}_5\}, \mathtt{give(BB)} \rangle \\
A_3 &= \langle \{\mathsf{p}_2, \mathsf{p}_3\}, \mathtt{ok(PGA)} \wedge \mathtt{ok(BB)} \rangle & A_7 &= \langle \{\mathsf{p}_2, \mathsf{p}_6\}, \neg\mathtt{ok(BB)} \rangle \\
A_4 &= \langle \{\mathsf{p}_6\}, \neg\mathtt{ok(PGA)} \vee \neg\mathtt{ok(BB)} \rangle & A_8 &= \langle \{\mathsf{p}_3, \mathsf{p}_6\}, \neg\mathtt{ok(PGA)} \rangle
\end{aligned}$$

Let $\mathsf{Arguments}_c(\Delta)$ be the set of all classical arguments that can be constructed from Δ, and let the preference relation $\preceq$ be such that $A_i \preceq A_1$ and $A_i \preceq A_2$ for all i such that $i \neq 1$ and $i \neq 2$. Furthermore, let $\Pi = \{A_1, A_2\}$ be the focus (i.e. the arguments of interest). In other words, we know that each of these two arguments in the focus contains all the information we are interested in (i.e. we want to determine the options for treatment taking into account the integrity constraint). This would give us the following focal graph using the classical direct defeater definition for attack.

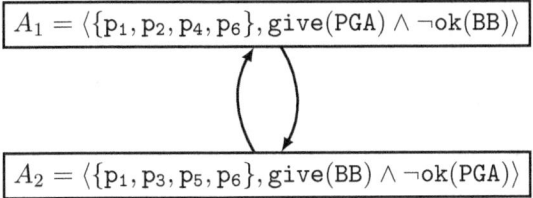

By taking this focal graph, we have ignored arguments such as A_3 to A_8 which do not affect the dialectical status of A_1 or A_2 given this preference relation.

Using preferences is a general approach. There is no restriction on what preference relation we use over arguments, and there are various natural interpretations for this ranking such as capturing belief for arguments (where the belief in the argument can be based on the belief for the premises and/or claim), and capturing the relative number of votes for arguments (where a group of voters will vote for or against each argument), etc.

To conclude, by introducing preferences over arguments, we can reduce the number of attacks that occur. Using preferences over arguments is a form of meta-information, and with the definition for preference-based argumentation (as defined by [Amgoud and Cayrol, 2002]), it supports selectivity in generating argument graphs that discriminates between arguments and thereby between attacks. With this definition more practical argument graphs can be constructed than with the definition for classical exhaustive graphs.

7 Properties of deductive arguments

In this section, we consider some properties of argumentation based on deductive arguments. Out focus is predominantly on classical logic. We consider postulates for counterarguments (i.e. for the attack relation), postulates for extensions, and properties of the structure of generated graphs.

7.1 Counterargument properties

We consider some desirable properties of attack relations in the form of postulates and classify several well-known attack relations from the literature with regards to the satisfaction of these postulates. We define these postulates in terms of function D where $D(A, B) = \top$ holds iff A attacks B. Different definitions of counterargument, give is different definitions of attack. So for example, if A is a defeater of B, then this denoted by $D_D(A, B) = \top$.

In Table 2, we review postulates relevant to attack functions. From now on, A, B, C and their primed versions will stand for arguments. We explain them as follows:

- (D0) This is a classic syntax-independence requirement: the syntax of the components of two arguments should not play a role in deciding whether there is an attack between those arguments;

- (D1) This mandates that if an argument attacks another, then it must be that the claim of the former is inconsistent with the support of the

Name	Property
D0	if $A \equiv A', B \equiv B'$ then $D(A,B) = D(A',B')$
D1	if $D(A,B) = \top$ then $\{\mathsf{Claim}(A)\} \cup \mathsf{Support}(B) \vdash \bot$
D2	if $D(A,B) = \top$ and $\mathsf{Claim}(C) \equiv \mathsf{Claim}(A)$ then $D(C,B) = \top$
D2i	if $D(A,B) = \top$ and $\mathsf{Claim}(C) \vdash \mathsf{Claim}(A)$ then $D(C,B) = \top$
D3	if $D(A,B) = \top$ and $\mathsf{support}(B) = \mathsf{support}(C)$ then $D(A,C) = \top$
D3i	if $D(A,B) = \top$ and $\mathsf{support}(B) \subseteq \mathsf{support}(C)$ then $D(A,C) = \top$
D4	if $\mathsf{Arcs}(G) = \emptyset$ then $\mathsf{MinIncon}(\Delta) = \emptyset$

Table 2. Postulates for an attack relation D. We denote an attack relation from A to B as holding when $D(A,B) = \top$.

latter. This requirement reflects a fundamental assumption in logical argumentation, namely that if two arguments are logically consistent there cannot be any attack between them.

- (D2) This imposes a certain fairness restrictions on existing attacks by requiring that all arguments that have equivalent claims with that of A should attack B.

- (D2i) is a strengthening of D2. It requires that any argument with a stronger claim than A, i.e., one that logically entails that of A, should also attack anything A attacks

- (D3) This requires that if A attacks B then all arguments with the same support with that of B should also be attacked by A.

- (D3i) is a strengthening of D3 proposed by Amgoud and Besnard [Amgoud and Besnard, 2009]. It mandates that any argument whose support is a superset of that of B, and thus is stronger than that of B, should also be attacked by A.

- (D4) This postulate can be read as follows: if we restrict D on the arguments that can be generated from Δ and find that no two such arguments attack each other, then it must be that Δ itself is consistent (hence it has no minimal inconsistent subsets).

Given the postulates in Table 2, we can classify the notions of counterargument given for classical logic. This classification is given in Table 3,

If we then impose further constraints as listed in Table 4, such as constraints on the claim of the attacker D1i or D1ii, and D5 to D5iii, and constraints forcing the existence of attacks D6 to D6iii, then we obtain the following proposition provides characterization results for the classical attack relations. This means we have alternative definitions for our attack relation that are specified entirely in terms of a set of properties. For the proofs, see [Gorogiannis and Hunter, 2011].

	D_D	D_{DD}	D_U	D_{DU}	D_{CU}	D_R	D_{DR}
D0	Yes	Yes	Yes	Yes	Yes	Yes	Yes
D1	Yes	Yes	Yes	Yes	Yes	Yes	Yes
D2	Yes	Yes	Yes	Yes	Yes	Yes	Yes
D2i	Yes	Yes	No	No	No	No	Yes
D3	Yes	Yes	Yes	Yes	Yes	No	No
D3i	Yes	Yes	Yes	Yes	No	No	No
D4	Yes	Yes	Yes	Yes	Yes	Yes	Yes

Table 3. Postulates satsified by attack functions

Name	Property
D1i	if $D(A,B) = \top$ then $\exists \phi \in \text{Support}(B)$ s.t. $\text{Claim}(A) \vdash \neg \phi$
D1ii	if $D(A,B) = \top$ then $\text{Claim}(A) \vdash \neg\text{Claim}(B)$
D5	if $D(A,B) = \top$ then $\neg\text{Claim}(A) \vdash \bigwedge \text{Support}(B)$
D5i	if $D(A,B) = \top$ then $\exists \phi \in \text{Support}(B)$ s.t. $\neg\text{Claim}(A) \vdash \phi$
D5ii	if $D(A,B) = \top$ then $\neg\text{Claim}(A) \vdash \text{Claim}(B)$
D5iii	if $D(A,B) = \top$ then $\exists X \subseteq \text{Support}(B)$ s.t. $\neg\text{Claim}(A) \equiv \bigwedge X$
D6	if $\{\text{Claim}(A)\} \cup \text{Support}(B) \vdash \bot$ then there exists C s.t. $\text{Claim}(A) \vdash \text{Claim}(C)$ and $D(C,B) = \top$
D6i	if $\exists \phi \in \text{Support}(B)$ s.t. $\text{Claim}(A) \vdash \neg \phi$ then there exists C s.t. $\text{Claim}(A) \vdash \text{Claim}(C)$ and $D(C,B) = \top$
D6ii	if $\text{Claim}(A) \vdash \neg\text{Claim}(B)$ then there exists C s.t. $\text{Claim}(A) \vdash \text{Claim}(C)$ and $D(C,B) = \top$
D6iii	if $\exists X \subseteq \text{Support}(B)$ s.t. $\text{Claim}(A) \equiv \neg \bigwedge X$ then $D(A,B) = \top$

Table 4. Further constraints on the attack relation.

Proposition 7.1 *Let D be an attack relation.*

- $D = D_D$ *is a defeater relation iff D satisfies D1, D2i and D6*
- $D = D_{DD}$ *is a direct defeater relation iff D satisfies D1i, D2i and D6i*
- $D = D_{DR}$ *is a defeating rebuttal relation iff D satisfies D1ii, D2i and D6ii*
- $D = D_{CU}$ *is a canonical undercut relation iff D satisfies D1, D2, D5 and D6*
- $D = D_{DU}$ *is a direct undercut relation iff D satisfies D1i, D2, D5i and D6i*
- $D = D_R$ *is a rebuttal relation iff D satisfies D1ii, D2, D5ii and D6ii*
- $D = D_U$ *is an undercut relation iff D satisfies D5iii and D6iii*

Since classical logic offers a variety of different options for defining a counterargument (i.e. an attack relation), it is helpful to characterize the options in terms of postulates. Furthermore, these postulates can be used or adapted for classifying and characterizing attack relations for a variety of base logics.

7.2 Extension properties

Various postulates have been proposed for classical exhaustive graphs (e.g. [Gorogiannis and Hunter, 2011]). Some of these are concerned with consistency of the set of premises (or set of claims) obtained from the arguments in an extension according to one of Dung's dialectical semantics. In the rest of this subsection, we restrict consideration to classical logic arguments, though the postulates can be adapted for other base logics.

To consider extension properties, we will introduce some subsidiary definitions. We start with sceptical acceptance of arguments and credulous acceptance of arguments defined as follows where G is an argument graph and $Y \in \{\mathsf{pr}, \mathsf{gr}, \mathsf{st}\}$ is a dialectical semantics (where pr denotes preferred, gr denotes grounded, and st denotes stable). and $\mathsf{Extensions}_Y(G)$ is the set of extensions obtained according to Y. If $\mathsf{Extensions}_Y(G) = \emptyset$, then $\mathsf{Sceptical}_Y(G) = \mathsf{Credulous}_Y(G) = \emptyset$, otherwise

$$\mathsf{Sceptical}_Y(G) = \bigcap_{S \in \mathsf{Extension}_Y(G)} S$$

$$\mathsf{Credulous}_Y(G) = \bigcup_{S \in \mathsf{Extension}_Y(G)} S$$

So, for example, we will say that an argument A in $\mathsf{Nodes}(G)$ is sceptically accepted in the preferred semantics if $A \in \mathsf{Sceptical}_{\mathsf{pr}}(G)$. Clearly, we have the following observations where $Y \in \{\mathsf{pr}, \mathsf{gr}, \mathsf{st}\}$

- Sceptical$_Y(G) \subseteq$ Credulous$_Y(G)$
- Credulous$_{gr}(G) =$ Sceptical$_{gr}(G)$

The definition of an exhaustive argument graph takes an attack function D and a knowledgebase Δ in order to produce the argument graph. Such an argument graph can be evaluated with choices for dialectical semantics (preferred, grounded, etc) and acceptability criteria (sceptical or credulous).

Since the arguments are logical, we can evaluate the logical properties of the extensions. We will review the free postulate, non-free postulate, and consistency postulates, in the rest of this section. For this, we require the following definition for the free formulae which is the set of formulae not in any minimal inconsistent subset of Δ.

$$\text{Free}(\Delta) = \{\alpha \in \Delta \mid \alpha \notin \bigcup_{\Gamma \in \text{MinIncon}(\Delta)} \Gamma\}$$

where

$$\text{MinIncon}(\Delta) = \{\Gamma \subseteq \Delta \mid \Gamma \vdash \bot \text{ and for all } \Gamma' \subset \Gamma, \Gamma' \not\vdash \bot\}$$

We identify the free arguments in a graph (i.e. the arguments with no premises involved in a minimal inconsistent subset of the knowledgebase) and the non-free arguments in a graph (i.e. the arguments with one or more premise involved in a minimal inconsistent subset of the knowledgebase) as follows.

$$\text{FreeArguments}(G) = \{A \in \text{Nodes}(G) \mid \text{Support}(A) \subseteq \text{Free}(\Delta)\}$$

$$\text{NonFreeArguments}(G) = \{A \in \text{Nodes}(G) \mid \text{Support}(A) \not\subseteq \text{Free}(\Delta)\}$$

Our first extension-based postulate is the free postulate (defined next) states that the free arguments are sceptical arguments (i.e. in all extensions of the graph). This encodes our expectation that since free arguments are uncontroversial, they should be in every extension.

Definition 7.2 *For a descriptive or generated argument graph G based on classical logic arguments, the* **free postulate** *is defined as follows, where $Y \in \{\text{pr}, \text{gr}, \text{st}\}$.*

$$\text{FreeArguments}(G) \subseteq \text{Sceptical}_Y(G)$$

For the proof of the following proposition, see [Gorogiannis and Hunter, 2011].

Proposition 7.3 *If the attack relation D satisfies D1, then D satisfies the free postulate.*

All the attack functions considered for classical logic in this chapter satisfy D1, and therefore satisfy the free postulate. Therefore, for all semantics considered, all extensions of G contain all free arguments.

Next, we define the non-free postulate. This states that there exists a knowledgebase that is inconsistent and for which some arguments are *not* credulously accepted.

Definition 7.4 *For a descriptive or generated argument graph G based on classical logic arguments, the* **non-free postulate** *is defined as follows, where $Y \in \{\mathsf{pr}, \mathsf{gr}, \mathsf{st}\}$.*

$$\exists \Delta \text{ s.t. } \bigcup_{A \in \mathsf{Nodes}(G)} \mathsf{Support}(A) \subseteq \Delta$$
$$\text{and } \mathsf{Nonfree}(G) \neq \emptyset \text{ and } \mathsf{Credulous}_X(G) \neq \mathsf{Nodes}(G)$$

Failure means that if $\mathsf{Support}(A) \subseteq \Delta$, then A is credulously accepted. So for any argument that can be formed from a knowledgebase, there is a preferred extension that contains that argument. So if it does fail for an attack function D and a dialectical semantics Y, then this indicates that the combination of D and Y is, in a sense, very credulous. For the proof of the following proposition, see [Gorogiannis and Hunter, 2011].

Proposition 7.5 *We consider the non-free postulate with respect to the semantics where the attack relation is undercut, direct undercut, canonical undercut, or rebuttal.*

- *For stable, preferred, and complete extensions, the non-free postulate is not satisfied.*

- *For grounded extensions, the non-free postulate is satisfied.*

Finally, we consider the consistency postulates. These postulates are variations of the requirement that certain arguments' supports or claims must be consistent together. The expectation is that once an extension is obtained, then the arguments contained in it present a somehow consistent set of assumptions. Applying this restriction to the supports of the arguments or to their claims, and to the sceptically accepted set of arguments or to all extensions individually, yields the versions of this principle listed below.

Definition 7.6 *For a descriptive or generated argument graph G based on classical logic arguments, the* **consistency postulates** *are defined as follows,*

Attack	CN1	CN1'	CN2	CN2'
Direct undercut	Yes	Yes	Yes	Yes
Direct defeat	Yes	Yes	Yes	Yes
Canonical undercut	Yes	Yes	Yes	Yes
Rebut	No	No	No	No

Table 5. Satisfaction of consistency postulates for grounded semantics

Attack	CN1	CN1'	CN2	CN2'
Direct undercut	Yes	Yes	Yes	Yes
Direct defeat	Yes	Yes	Yes	Yes
Canonical undercut	Yes	Yes	No	No
Rebut	No	No	No	No

Table 6. Satisfaction of consistency postulates for preferred and complete semantics

where $Y \in \{\mathsf{pr}, \mathsf{gr}, \mathsf{st}\}$.

$$(CN1) \bigcup_{A \in \mathsf{sceptical}_Y(G)} \mathsf{Support}(A) \not\vdash \bot$$

$$(CN2) \bigcup_{A \in S} \mathsf{Support}(A) \not\vdash \bot, \text{ for all } S \in \mathsf{Extension}_Y(G)$$

$$(CN1') \bigcup_{A \in \mathsf{Sceptical}_Y(G)} \mathsf{Claim}(A) \not\vdash \bot$$

$$(CN2') \bigcup_{A \in S} \mathsf{Claim}(A) \not\vdash \bot, \text{ for all } S \in \mathsf{Extension}_Y(G)$$

The reason we provide all four versions of the consistency postulates is that it is not yet clear whether one form of the postulate is more appropriate than others. For example, consistency postulates similar to CN1' and CN2' have been proposed in [Caminada and Amgoud, 2005] in the context of rule-based argumentation systems and versions of CN1 and CN2 have been proposed in [Amgoud and Besnard, 2009] for classical logics. It should be clear that CN2 entails CN1, CN2' entails CN1', CN1 entails CN1' and CN2 entails CN2'.

We summarize which attack relations satisfy which of the four consistency postulates in Table 7.2 for grounded semantics and in Table 7.2 for preferred, stable, and complete semantics. These results show that for some attack relation (e.g. rebuttal), the consistent extension property is not guaranteed (as in Example 7.7) whereas for other choices of attack relation (e.g. direct undercut), the consistent extension property is guaranteed. We illustrate a consistent set of premises obtained from arguments in a preferred extension in Example 7.8.

Example 7.7 *Let $\Delta = \{a \wedge b, \neg a \wedge c\}$. For the reviewed semantics for rebut, the following are arguments in any extension: $A_1 = \langle \{a \wedge b\}, b \rangle$ and $A_2 = \langle \{\neg a \wedge c\}, c \rangle$. Clearly $\{a \wedge b, \neg a \wedge c\} \vdash \bot$. Hence, the consistent extensions property fails for rebut.*

Example 7.8 *Consider the argument graph given in Figure 8. There are three preferred extensions $\{A_1, A_5, A_6, A_7\}$, $\{A_2, A_4, A_6, A_8\}$, and $\{A_3, A_4, A_5, A_9\}$. In each case, the union of the premises is consistent. For instance, for the first extension,*

$$\mathsf{Support}(A_1) \cup \mathsf{Support}(A_5) \cup \mathsf{Support}(A_6) \cup \mathsf{Support}(A_7) \not\vdash \bot$$

Example 7.9 *Consider the argument graph given in Figure 9. There are three preferred extensions $\{A_1\}$, $\{A_2\}$, and $\{A_3\}$. In each case, the union of the premises is consistent.*

The failure of the consistency postulates with some attack relations is an issue that may be interpreted as a weakness of the attack relation or of the specific semantics, and perhaps raises the need for alternatives to be identified. Another response is that it is not the attack relation and dialectical semantics that should be responsible for ensuring that all the premises used in the winning arguments are consistent together. Rather, it could be argued that checking that the premises used are consistent together should be the responsibility of something external to the defeat relation and dialectical semantics, and so knowing whether the consistent extensions property holds or not influences what external mechanisms are required for checking. Furthermore, checking consistency of premises of sets of arguments may be part of the graph construction process. For instance, in Garcia and Simari's proposal for dialectical trees [García and Simari, 2004], there are constraints on what arguments can be added to the tree based on consistency with the premises of other arguments in the tree.

7.3 Structural properties

Simple logic has the property that for any argument graph, there is a knowledgebase that can be used to generate it: Let (N, E) be a directed graph (i.e. N is a set of nodes, and E is a set of edges between nodes), then there is a simple logic knowledgebase Δ such that the generated argument graph $(\mathsf{Arguments}_s(\Delta), \mathsf{Attacks}_s(\Delta))$ is isomorphic to (N, E). So simple exhaustive graphs are said to be **constructively complete** for graphs.

To show that simple exhaustive graphs are constructively complete for graphs, we can use a coding scheme for the premises so that each argument is based on a single simple rule where the antecedent is a conjunction of one or more positive literals, and each consequent is a negative literal unique to that simple rule (i.e. it is an identifier for that rule and therefore for that argument). If we want one argument to attack another, and the attacking argument has the consequent $\neg \alpha$, then the attacked argument needs to have the positive literal α in the antecedent of its simple rule. The restriction of each rule to only have positive literals as conditions in the antecedent, and a negative literal as its consequent, means that the rules cannot be chained. This ensures that the premises of each argument has only one simple rule. We illustrate this in the following example.

Example 7.10 *Consider the following directed graph (N, E). Note, that it includes a self-attack, bidirectional attacks and uni-directional attacks.*

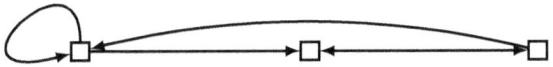

Let $\Delta = \{a, b, c, a \wedge c \to \neg a, a \wedge c \to \neg b, b \to \neg c\}$. *From this we can construct the following exhaustive argument graph which is isomorphic to the above directed graph. Note that each argument is identified by a single simple rule.*

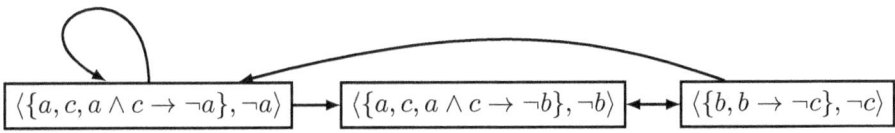

In contrast to simple logic, the definition for classical exhaustive graphs (i.e. classical logic, with any of the definitions for counterarguments), is not constructively complete for graphs. Since the premises of a classical argument are consistent, by definition, it is not possible for a classical argument to attack itself using the definitions for attack given earlier. But, there are many other graphs for which there is no classical logic knowledgebase that can be used to generate a classical exhaustive graph that is isomorphic to it. To illustrate this failure, we consider in Example 7.11 the problem of constructing a component with two arguments attacking each other. Note, this is not a pathological example as there are many graphs that contain a small number of nodes and that cannot be generated as a classical exhaustive graph.

Example 7.11 *Let $\Delta = \{a, \neg a\}$ be a classical logic knowledgebase. Hence, there are two classical arguments A_1 and A_2 that are direct undercuts of each other. Plus, there is the representative A_4 for arguments with a claim that is strictly weaker than a (i.e. the claim b is such that $\{a\} \vdash b$ and $\{b\} \not\vdash \{a\}$),* and there is the representative A_3 for arguments with a claim that is strictly weaker than $\neg a$ (i.e. the claim b is such that $\{\neg a\} \vdash b$ and $\{b\} \not\vdash \{\neg a\}$).

$$A_3 = \langle\{\neg a\}, \ldots\rangle \leftarrow A_1 = \langle\{a\}, a\rangle \rightleftarrows A_2 = \langle\{\neg a\}, \neg a\rangle \rightarrow A_4 = \langle\{a\}, \ldots\rangle$$

Given a set of directed graphs Φ we can define further properties. The set of directed graphs can be based on well-known definitions such as the set of bipartite graphs, the set of acyclic graphs, or the set of trees, or it can be defined in a domain specific way. A deductive argumentation system (i.e. a base logic, a definition for arguments, a definition for counterarguments, and a definition for a generated graphs) can then be evaluated with respect to Φ. We give these properties where we consider systems defined in terms of definitions for Arguments(Δ) (i.e. the definition for arguments that can be obtained from a knowledgebase Δ) and Attacks(Δ) (i.e. the definition for attacks that can be obtained from a knowledgebase Δ) with exhaustive graphs.

- A system (Arguments, Attacks) **constructively covers** Φ iff for all $G \in X$, there is a Δ and there is an $A \in$ Arguments(Δ), such that (Arguments(Δ), Attacks(Δ)) $= G$.

- A system (Arguments, Attacks) is **constructively covered** by Φ iff for all Δ and for all $A \in$ Arguments(Δ), if (Arguments(Δ), Attacks(Δ)) $= G$ then $G \in \Phi$.

- A system (Arguments, Attacks) is **constructively complete** for Φ iff (Arguments, Attacks) constructively covers Φ and (Arguments, Attacks) is constructively covered by Φ

The more general the class of graphs that a logical argument system can cover, the wider the range of argumentation situations the logical argument systems can capture. If one of these properties holds for a class of graphs, then it can be described as a kind of structural property of the system. If it fails then, it means that there are situations that cannot be captured by the system. This is, however, not necessarily bad news. In fact, it is known that the computational complexity of evaluating argumentation frameworks can be decreased if the class of graphs is restricted, for instance to acyclic, bipartite or symmetric graphs or to graphs which have certain parameters (like treewidth) fixed (see for example [Coste-Marquis et al., 2005; Dunne, 2007; Dvorák et al., 2012b; Dvorák et al., 2012a]).

7.4 Discussion of properties

In this section, we have considered a range of properties of argumentation based on deductive arguments. For some choices of attack relation, there is a question of consistency (which may be an issue if no further consistency checking is undertaken). Also, the definition for classical exhaustive graphs is not constructively complete for graphs (which means that many argument graphs cannot be generated as classical exhaustive graphs). Perhaps more problematical is that even for small knowledgebases, the classical exhaustive graphs that are generated are complex. Because of the richness of classical logic, the knowledge can be in different combinations to create many arguments. Whilst, we can ameliorate this problem by presenting argument graphs using a representative for a class of structurally equivalent arguments, and by using focal graphs, the graphs can still be large. What is evident from this is that there needs to be more selectivity in the process of generating argument graphs. The generation process needs to discriminate between the arguments (and/or the attacks) based on extra information about the arguments and/or information about the audience. There are many ways that this can be done.

8 Further reading

We provide further reading on formalization of deductive arguments and counterarguments, properties of exhaustive graphs, the importance of selectivity in generating argument graphs, and on automated reasoning.

8.1 Deductive arguments and counterarguments

There have been a number of proposals for deductive arguments using classical propositional logic [Cayrol, 1995; Besnard and Hunter, 2001; Amgoud and Cayrol, 2002; Gorogiannis and Hunter, 2011], classical predicate logic [Besnard and Hunter, 2005], description logic [Black et al., 2009; Moguillansky et al., 2010; Zhang et al., 2010; Zhang and Lin, 2013], temporal logic [Mann and Hunter, 2008], simple (defeasible) logic [Governatori et al., 2004; Hunter, 2010], conditional logic [Besnard et al., 2013], and probabilistic logic [Haenni, 1998; Haenni, 2001; Hunter, 2013].

There has also been progress in understanding the nature of classical logic in computational models of argument. Various types of counterarguments have been proposed including rebuttals [Pollock, 1987; Pollock, 1992], direct undercuts [Elvang-Gøransson et al., 1993; Elvang-Gøransson and Hunter, 1995; Cayrol, 1995], and undercuts and canonical undercuts [Besnard and Hunter, 2001]. In most proposals for deductive argumentation, an argument A is a counterargument to an argument B when the claim of A is inconsistent with the support of B. It is possible to generalize this with alternative notions of counterargument. For instance, with some common description logics, there is not an explicit negation symbol. In the proposal for argumentation with description logics, [Black et al., 2009] used the description logic notion of *incoherence* to define the notion of counterargument: A set of formulae in a description logic is incoherent when there is no set of assertions (i.e. ground literals) that would be consistent with the formulae. Using this, an argument A is a counterargument to an argument B when the claim of A together with the support of B is incoherent.

Meta-arguments for deductive argumentation was proposed by [Wooldridge et al., 2005], and the investigation of the representation of argument schemes in deductive argumentation was first proposed by [Hunter, 2008].

8.2 Properties of exhaustive argument graphs

In order to investigate how Dung's notion of abstract argumentation can be instantiated with classical logic, [Cayrol, 1995] presents results concerning stable extensions of argument graphs where the nodes are classical logic arguments, and the attacks are direct undercuts. As well as being the first paper to propose instantiating abstract argument graphs with classical arguments, it also showed how the premises in the arguments in the stable extension correspond to maximal consistent subsets of the knowledgebase, when the attack relation is direct undercut.

Insights into the options for instantiating abstract argumentation with classical logic can be based on postulates. [Amgoud and Besnard, 2009] have proposed a consistency condition and they examine special cases of knowledge bases and symmetric attack relations and whether consistency is satisfied in this context. Then [Amgoud and Besnard, 2010] extend this analysis by showing correspondences between the maximal consistent subsets of a knowledgebase and the maximal conflict-free sets of arguments.

Given the wide range of options for attack in classical logic, [Gorogiannis and Hunter, 2011] propose a series of desirable properties of attack relations to classify and characterize attack relations for classical logic. Furthermore, they present postulates regarding the logical content of extensions of argument graphs that may be constructed with classical logic, and a systematic study is presented of the status of these postulates in the context of the various combinations of attack relations and extension semantics.

Use of the notion of generated graphs then raises the question of whether for a specific logical argument system S, and for any graph G, there is a knowledgebase such that S generates G. If it holds, then it can be described as a kind of "structural" property of the system [Hunter and Woltran, 2013]. If it fails then, it means that there are situations that cannot be captured by the system. The approach of simple exhaustive graphs is constructively complete for graphs, whereas the approach of classical exhaustive graphs is not.

Preferences have been introduced into classical logic argumentation, and used to instantiate abstract argumentation with preferences by [Amgoud and Cayrol, 2002]. Amgoud and Vesic [Amgoud and Vesic, 2010] have shown how preferences can be defined so as to equate inconsistency handling in argumentation with inconsistency handling using Brewka's preferred sub-theories [Brewka, 1989].

8.3 Importance of selectivity in deductive argumentation

Some of the issues raised with classical exhaustive graphs (i.e. the lack of structural completeness, the failure of consistent extensions property for some choices of attack relation, and the correspondences with maximally consistent subsets of the knowledgebase) suggest that often we need a more sophisticated way of constructing argument graphs. In other words, to reflect any abstract argument graph in a logical argument system based on a richer logic, we need to be selective in the choice of arguments and counterarguments from those that can be generated from the knowledgebase. Furthermore, this is not just for theoretical interest. Practical argumentation often seems to use richer logics such as classical logic, and often the arguments and counterarguments considered are not exhaustive. Therefore, we need to better understand how the arguments are selected. For example, suppose agent 1 posits $A_1 = \langle \{b, b \to a\}, a \rangle$, and agent 2 then posits $A_2 = \langle \{c, c \to \neg b\}, \neg b \rangle$. It would be reasonable for this dialogue to stop at this point (since further arguments are only re-expressing the same conflict, and so, in a sense, they would be redundant) even though there are further arguments that can be constructed from the public knowledge such as $A_3 = \langle \{b, c \to \neg b\}, \neg c \rangle$. So in terms of constructing the constellation of arguments and counterarguments from the knowledge, we need to know what are the underlying principles for selecting arguments.

Selectivity in argumentation is an important and as yet under-developed topic [Besnard and Hunter, 2008]. Two key dimensions are selectivity based on object-level information and selectivity based on meta-level information.

- **Selectivity based on object-level information** In argumentation, object-level information is the information in the premises and claims of the arguments. So if these are generated by deductive reasoning from a knowledgebase, then the object-level information is the information in the knowledgebase. Selectivity based on object-level information is concerned with having a more concise presentation of arguments and counterarguments in an argument graph without changing the outcome of the argumentation. For instance, a more concise presentation can be obtained by removing structurally equivalent arguments or by using focal graphs (as discussed in Section 6.3.1).

- **Selectivity based on meta-level information** In argumentation, meta-level information is the information about the arguments and counterarguments (e.g. certainty and sources of the premises in arguments) and information about the participants or audience of the argumentation (e.g. the goals, beliefs, or biases of the audience). Selectivity based on meta-level information is concerned with generating an argument graph using the meta-level information according to sound principles. By using this extra information, a different argument graph may be obtained than would be obtained without the extra information. For instance, with a preference relation over arguments which is a form of meta-level information, preference-based argumentation offers a principled way of generating an argument graph that has potentially fewer attacks between arguments than obtained with the classical exhaustive argument graph (as discussed in Section 6.3.2).

Various kinds of meta-level information can be considered for argumentation including preferences over arguments, weights on arguments, weights on attacks, a probability distribution over models of the language of the deductive argumentation, etc. The need for meta-level information also calls for better modeling of the audience, of what they believe, of what they regard as important for their own goals, etc, are important features of selectivity (see for example [Hunter, 2004b; Hunter, 2004a]). Consider a journalist writing a magazine article on current affairs. There are many arguments and counterarguments that could be included, but the writer is selective. Selectivity may be based on what the likely reader already believes and what s/he may find interesting. Or, consider a lawyer in court, again there may be many arguments and counterarguments, that could be used, but only some will be used. Selection will in part be based on what could be believed by the jury, and convince them to take the side of that lawyer. Or, consider a politician giving a speech to an audience of potential voters. Here, the politician will select arguments based on what will be of more interest to the audience. For instance, if the audience is composed of older citizens, there may be more arguments concerning healthcare, whereas if the audience is composed of younger citizens, there may be more arguments concerning job opportunities. So whilst selectivity is clearly important in real-world argumentation, we need principled ways of bringing selectivity into structured argumentation such as that based on deductive argumentation.

8.4 Automated reasoning for deductive argumentation

For argumentation, it is computationally challenging to generate arguments from a knowledgebase with the minimality constraints using classical logic. If we consider the problem as an abduction problem, where we seek the existence of a minimal subset of a set of formulae that implies the consequent, then the problem is in the second level of the polynomial hierarchy [Eiter and Gottlob, 1995]. The difficult nature of argumentation has been underlined by studies concerning the complexity of finding individual arguments [Parsons *et al.*, 2003], the complexity of some decision problems concerning the instantiation of argument graphs with classical logic arguments and the direct undercut attack relation [Wooldridge *et al.*, 2006], and the complexity of finding argument trees [Hirsch and Gorogiannis, 2009]. Encoding of these tasks as quantified Boolean formulae also indicate that development of algorithms is a difficult challenge [Besnard *et al.*, 2009], and Post's framework, has been used to give a breakdown of where complexity lies in logic-based argumentation [Creignou *et al.*, 2011].

Despite the computational complexity results, there has been progress in developing algorithms for constructing arguments and counterarguments. One approach has been to adapt the idea of connection graphs to enable us to find arguments. A connection graph [Kowalski., 1975; Kowalski., 1979] is a graph where a clause is represented by a node and an arc (ϕ, ψ) denotes that there is a disjunct in ϕ with its complement being a disjunct in ψ. Essentially this graph is manipulated to obtain a proof by contradiction. Furthermore, finding this set of formulae can substantially reduce the number of formulae that need to be considered for finding proofs for a claim, and therefore for finding arguments and canonical undercuts. Versions for full propositional logic, and for a subset of first-order logic, have been developed and implemented [Efstathiou and Hunter, 2011].

Another approach for algorithms for generating arguments and counterarguments (canonical undercuts) has been given in a proposal that is based on a SAT solver [Besnard et al., 2010]. This approach is based on standard SAT technology and it is also based on finding proofs by contradiction.

8.5 Handling enthymemes

Real arguments (i.e. those presented by people in general) are normally enthymemes. We can consider two types which we will refer to as *implicit support enthymemes* and *implicit claim enthymemes*. An implicit support enthymeme only explicitly represents some of the premises for entailing its claim. An implicit claim enthymeme not only misses some of the premises for entailing its claim, but also does not explicitly represent its claim. So if Γ is the set of premises explicitly given for an implicit support enthymeme, and α is the claim, then Γ does not entail α, but there are some implicitly assumable premises Γ' such that $\Gamma \cup \Gamma'$ is a minimal consistent set of formulae that entails α. For example, for a claim that *you need an umbrella today*, a husband may give his wife the premise *the weather report predicts rain*. Clearly, the premise does not entail the claim, but it is easy for the wife to identify the common knowledge used by the husband (i.e. *if the weather report predicts rain, then you need an umbrella today*) in order to reconstruct the intended argument correctly.

If we want to build agents that can understand real arguments coming from humans, they need to identify the missing premises and missing claims with some reliability. And if we want to build agents that can generate real arguments for humans, they need to identify the premises and claims that can be missed without causing undue confusion. Clearly, deciding how to construct or deconstruct enthymemes is difficult, and proposals for logic-based formalisations of the process remain underdeveloped.

In [Hunter, 2007; Black and Hunter, 2008], we introduced a way for each agent in a dialogue to have information about what it can use as shared knowledge, and then a proponent can use this information to remove redundant premises from an intended argument (creating an implicit support enthymeme), and a recipient can use this information to identify the necessary premises in order to recover the intended argument. Then in [Black and Hunter, 2012], this proposal was extended by allowing each agent to also have a representation of information requirements. These are formulae that the agent would like to receive arguments about. So for example if an agent asks a question, it is making an explicit declaration of an information requirement. By introducing the notion of information requirements, we can formalise a key idea from relevance theory that the relevance of an utterance depends on maximising cognitive effect and minimising cognitive effort. This allows proponents to construct both implicit support and implicit claim enthymemes that are relevant for the intended recipient, and the recipient can deconstruct such enthymemes by using relevance criteria to overcome some of the ambiguities that normally arise when trying to understand enthymemes. This has been further developed for persuasion dialogues [Dupin de Saint-Cyr, 2011].

9 Discussion

Deductive argumentation is an appealing approach to instantiation of abstract argumentation. A deductive argument has all the premises explicitly in the support of the argument, and the claim is derived by the consequence relation of the base logic. Since established and well-understood logics can be used as a base logic, the semantics and proof theory for the individual arguments is inherited from the base logic. This is important if we want to harness developments in knowledge representation and in computational linguistics for specialized logics. So for example, if we want to represent a natural language argument as a deductive argument, we can use an appropriate logic from computational linguistics to represent the information.

The approach is also flexible since different constraints can put on an argument (e.g. consistency, minimality, etc) and on the definition for counterargument (e.g. defeater, undercut, direct undercut, canonical undercut, rebuttal, etc). Furthermore, we can use the approach for descriptive graphs and for generated graphs. Over the past few years, there has been substantial interest in bipolar argumentation (see for example [Cayrol and Lagasquie-Schiex, 2005; Oren and Norman, 2008; Nouioua and Risch, 2011]). In future work, we would like to generalize the use of deductive arguments for bipolar argumentation.

Acknowledgements

The authors are very grateful to Henry Prakken for some very interesting and valuable discussions on the nature of structural argumentation. The authors are also very grategul to the anonymous reviewers for their comprehensive and insightful feedback. This research was partly supported by EPSRC grant EP/N008294/1.

BIBLIOGRAPHY

[Amgoud and Besnard, 2009] L. Amgoud and Ph. Besnard. Bridging the gap between abstract argumentation systems and logic. In *Proceedings of the 3rd International Conference on Scalable Uncertainty Management (SUM'09)*, volume 5785 of *Lecture Notes in Computer Science*, pages 12–27. Springer, 2009.

[Amgoud and Besnard, 2010] L. Amgoud and Ph. Besnard. A formal analysis of logic-based argumentation systems. In *Proceedings of the 4th International Conference on Scalable Uncertainty Management (SUM'10)*, volume 6379 of *Lecture Notes in Computer Science*, pages 42–55. Springer, 2010.

[Amgoud and Cayrol, 2002] L. Amgoud and C. Cayrol. A reasoning model based on the production of acceptable arguments. *Annals of Mathematics and Artificial Intelligence*, 34:197–215, 2002.

[Amgoud and Vesic, 2010] L. Amgoud and S. Vesic. Handling inconsistency with preference-based argumentation. In *Proceedings of the 4th International Conference on Scalable*

Uncertainty Management (SUM'10), volume 6379 of *Lecture Notes in Computer Science*, pages 56–69. Springer, 2010.
[Amgoud et al., 2011] L. Amgoud, Ph. Besnard, and S. Vesic. Identifying the core of logic-based argumentation systems. In *Proceedings of the IEEE International Conference on Tools with Artificial Intelligence, (ICTAI'11)*, pages 633–636. IEEE Press, 2011.
[Arló-Costa and Shapiro, 1992] H. Arló-Costa and S. Shapiro. Maps between conditional logic and non-monotonic logic. In *Proceedings of the 3rd International Conference on Principles of Knowledge Representation and Reasoning (KR'92)*, page 553565. Morgan Kaufmann, 1992.
[Baroni and Giacomin, 2007] P. Baroni and M. Giacomin. On principle-based evaluation of extension-based argumentation semantics. *Artificial Intelligence*, 171:675–700, 2007.
[Besnard and Hunter, 2001] Ph. Besnard and A. Hunter. A logic-based theory of deductive arguments. *Artificial Intelligence*, 128:203–235, 2001.
[Besnard and Hunter, 2005] Ph Besnard and A Hunter. Practical first-order argumentation. In *Proceedings of the 20th American National Conference on Artificial Intelligence (AAAI'05)*, pages 590–595. MIT Press, 2005.
[Besnard and Hunter, 2008] Ph. Besnard and A Hunter. *Elements of Argumentation*. MIT Press, 2008.
[Besnard et al., 2009] Ph. Besnard, A. Hunter, and S. Woltran. Encoding deductive argumentation in quantified boolean formulae. *Artificial Intelligence*, 173(15):1406–1423, 2009.
[Besnard et al., 2010] Ph. Besnard, E. Gregoire, C. Piette, and B. Raddaoui. Mus-based generation of arguments and counter-arguments. In *Proceedings of the 11th IEEE International Conference on Information Reuse and Integration (IRI'10)*, pages 239–244. IEEE Press, 2010.
[Besnard et al., 2013] Ph. Besnard, E. Gregoire, and B. Raddaoui. A conditional logic-based argumentation framework. In *Proceedings of the 7th International Conference on Scalable Uncertainty Management (SUM'13)*, volume 7958 of *Lecture Notes in Computer Science*, pages 44–56. Springer, 2013.
[Black and Hunter, 2008] E. Black and A. Hunter. Using enthymemes in an inquiry dialogue system. In *Proceedings of the 7th International Conference on Autonomous Agents and Multiagent Systems (AAMAS'08)*, pages 437–444. IFAAMAS, 2008.
[Black and Hunter, 2012] E Black and A Hunter. A relevance-theoretic framework for constructing and deconstructing enthymemes. *Journal of Logic and Computation.*, 22(1):55–78, 2012.
[Black et al., 2009] E. Black, A Hunter, and J Pan. An argument-based approach to using multiple ontologies. In *Proceedings of the 3rd International Conference on Scalable Uncertainty Management (SUM'09)*, volume 5785 of *Lecture Notes in Computer Science*, pages 68–79. Springer, 2009.
[Brewka, 1989] G. Brewka. Preferred subtheories: An extended logical framework for default reasoning. In *Proceedings of the 11th International Joint Conference on Artificial Intelligence (IJCAI'89)*, pages 1043–1048. Morgan Kaufmann, 1989.
[Caminada and Amgoud, 2005] M. Caminada and L. Amgoud. An axiomatic account of formal argumentation. In *Proceedings of the 20th National Conference on Artificial Intelligence (AAAI'05)*, pages 608–613. AAAI Press, 2005.
[Cayrol and Lagasquie-Schiex, 2005] C. Cayrol and M-C Lagasquie-Schiex. On the acceptability of arguments in bipolar argumentation frameworks. In *Proceedings of the 8th Symbolic and Quantitative Approaches to Reasoning and Uncertainty (ECSQARU'05)*, volume 3571 of *LNCS*, pages 378–389. Springer, 2005.
[Cayrol, 1995] C. Cayrol. On the relation between argumentation and non-monotonic coherence-based entailment. In *Proceedings of the 14th International Joint Conference on Artificial Intelligence (IJCAI'95)*, pages 1443–1448, 1995.
[Coste-Marquis et al., 2005] S Coste-Marquis, C Devred, and P Marquis. Symmetric argumentation frameworks. In *Proceedings of the 8th European Conference on Symbolic and Quantitative Approaches to Reasoning with Uncertainty (ECSQARU'05)*, volume 3571 of *Lecture Notes in Computer Science*, pages 317–328. Springer, 2005.
[Creignou et al., 2011] N. Creignou, J. Schmidt, M. Thomas, and S. Woltran. Complexity of logic-based argumentation in Post's framework. *Argument & Computation*, 2(2-3):107–129, 2011.

[Cross and Nute, 2001] C. Cross and D. Nute. Conditional logic. In D. Gabbay, editor, *Handbook of Philosophical Logic*, volume IV. D. Reidel, 2001.

[Delgrande, 1987] J. Delgrande. A first-order logic for prototypical properties. *Artificial Intelligence*, 33:105–130, 1987.

[Dung, 1995] P. Dung. On the acceptability of arguments and its fundamental role in nonmonotonic reasoning, logic programming, and n-person games. *Artificial Intelligence*, 77:321–357, 1995.

[Dunne, 2007] P Dunne. Computational properties of argument systems satisfying graph-theoretic constraints. *Artificial Intelligence*, 171(10-15):701–729, 2007.

[Dupin de Saint-Cyr, 2011] F. Dupin de Saint-Cyr. Handling enthymemes in time-limited persuasion dialogs. In *Proceedings of the 5th International Conference on Scalable Uncertainty Management (SUM'11)*, volume 6929 of *Lecture Notes in Computer Science*, pages 149–162. Springer, 2011.

[Dvorák et al., 2012a] W. Dvorák, R. Pichler, and S. Woltran. Towards fixed-parameter tractable algorithms for abstract argumentation. *Artificial Intelligence*, 186:1–37, 2012.

[Dvorák et al., 2012b] W. Dvorák, S. Szeider, and S. Woltran. Abstract argumentation via monadic second order logic. In *Proceedings of the 6th International Conference on Scalable Uncertainty Management (SUM'12)*, volume 7520 of *LNCS*, pages 85–98. Springer, 2012.

[Efstathiou and Hunter, 2011] V. Efstathiou and A. Hunter. Algorithms for generating arguments and counterarguments in propositional logic. *International Journal of Approximate Reasoning*, 52:672–704., 2011.

[Eiter and Gottlob, 1995] T. Eiter and G. Gottlob. The complexity of logic-based abduction. *Journal of the ACM*, 42(1):3–42, 1995.

[Elvang-Gøransson and Hunter, 1995] M. Elvang-Gøransson and A. Hunter. Argumentative logics: Reasoning with classically inconsistent information. *Data & Knowledge Engineering*, 16(2):125–145, 1995.

[Elvang-Gøransson et al., 1993] M. Elvang-Gøransson, P. Krause, and J. Fox. Acceptability of arguments as 'logical uncertainty'. In *Proceedings of the 2nd European Conference on Symbolic and Quantitative Approaches to Reasoning and Uncertainty (ECSQARU'93)*, volume 747 of *Lecture Notes in Computer Science*, pages 85–90. Springer, 1993.

[Gabbay, 1985] D. Gabbay. Theoretical foundations for nonmonotonic reasoning in expert systems. In K. Apt, editor, *Logic and Models of Concurrent Systems*. Springer, 1985.

[García and Simari, 2004] A. García and G. Simari. Defeasible logic programming: An argumentative approach. *Theory and Practice of Logic Programming*, 4:95–138, 2004.

[Girard, 2006] P. Girard. From onions to broccoli: Generalizing lewis' counterfactual logic. *Journal of Applied Non-Classical Logic*, 17(2):213 – 229, 2006.

[Gorogiannis and Hunter, 2011] N. Gorogiannis and A. Hunter. Instantiating abstract argumentation with classical logic arguments: Postulates and properties. *Artificial Intelligence*, 175(9-10):1479–1497, 2011.

[Governatori et al., 2004] G. Governatori, M. Maher, G. Antoniou, and D. Billington. Argumentation semantics for defeasible logic. *Journal of Logic and Computation*, 14(5):675–702, 2004.

[Haenni, 1998] R. Haenni. Modelling uncertainty with propositional assumptions-based systems. In *Applications of Uncertainty Formalisms*, volume 1455 of *Lecture Notes in Computer Science*, pages 446–470. Springer, 1998.

[Haenni, 2001] R. Haenni. Cost-bounded argumentation. *International Journal of Approximate Reasoning*, 26(2):101–127, 2001.

[Hirsch and Gorogiannis, 2009] R. Hirsch and N. Gorogiannis. The complexity of the warranted formula problem in propositional argumentation. *Journal of Logic and Computation*, 20(2):481–499, 2009.

[Hunter and Woltran, 2013] A Hunter and S Woltran. Structural properties for deductive argument systems. In *Proceedings of the 12th European Conference on Symbolic and Quantitative Approaches to Reasoning and Uncertainty (ECSQARU'13)*, volume 7958 of *Lecture Notes in Computer Science*, pages 278–289. Springer, 2013.

[Hunter, 2004a] A. Hunter. Making argumentation more believable. In *Proceedings of the 19th National Conference on Artificial Intelligence (AAAI'04)*, pages 269–274. MIT Press, 2004.

[Hunter, 2004b] A. Hunter. Towards higher impact argumentation. In *Proceedings of the 19th National Conference on Artificial Intelligence (AAAI'04)*, pages 275–280. MIT Press, 2004.

[Hunter, 2007] A. Hunter. Real arguments are approximate arguments. In *Proceedings of the 22nd AAAI Conference on Artificial Intelligence (AAAI'07)*, pages 66–71. MIT Press, 2007.

[Hunter, 2008] A Hunter. Reasoning about the appropriateness of proponents for arguments. In *Proceedings of the 23rd AAAI Conference on Artificial Intelligence (AAAI'08)*, pages 89–94. MIT Press, 2008.

[Hunter, 2010] A. Hunter. Base logics in argumentation. In *Proceedings of the 3rd International Conference on Computational Models of Argument (COMMA'10)*, volume 216 of *Frontiers in Artificial Intelligence and Applications*, pages 275–286. IOS Press, 2010.

[Hunter, 2013] A Hunter. A probabilistic approach to modelling uncertain logical arguments. *International Journal of Approximate Reasoning*, 54(1):47–81, 2013.

[Kowalski., 1975] R. Kowalski. A proof procedure using connection graphs. *Journal of the ACM*, 22:572–595, 1975.

[Kowalski., 1979] R. Kowalski. *Logic for Problem Solving*. North-Holland Publishing, 1979.

[Kraus et al., 1990] S. Kraus, D. Lehmann, and M. Magidor. Non-monotonic reasoning, preferential models and cumulative logics. *Artificial Intelligence*, 44:167–207, 1990.

[Liao et al., 2011] B. Liao, L. Jin, and RC. Koons. Dynamics of argumentation systems: A division-based method. *Artificial Intelligence*, 175(11):1790–1814, 2011.

[Makinson, 1994] D. Makinson. General patterns in nonmonotonic reasoning. In D. Gabbay, C. Hogger, and J. Robinson, editors, *Handbook of Logic in Artificial Intelligence and Logic Programming*, volume 3, pages 35–110. Oxford University Press, 1994.

[Mann and Hunter, 2008] N. Mann and A. Hunter. Argumentation using temporal knowledge. In *Proceedings of the 2nd Conference on Computational Models of Argument (COMMA'08)*, volume 172 of *Frontiers in Artificial Intelligence and Applications*, pages 204–215. IOS Press, 2008.

[McCarthy, 1980] J. McCarthy. Circumscription: A form of non-monotonic reasoning. *Artificial Intelligence*, 13(1-2):23–79, 1980.

[Moguillansky et al., 2010] M. Moguillansky, R. Wassermann, and M. Falappa. An argumentation machinery to reason over inconsistent ontologies. In *Advances in Artificial Intelligence (IBERAMIA 2010)*, volume 6433 of *LNCS*, pages 100–109. Springer, 2010.

[Nouioua and Risch, 2011] F. Nouioua and V. Risch. Argumentation frameworks with necessities. In *Proceedings of the 5th International Conference on Scalable Uncertainty Mangement (SUM'11)*, volume Lecture Notes in Computer Science, pages 163–176. Springer, 2011.

[Oren and Norman, 2008] N. Oren and T. Norman. Semantics for evidence-based argumentation. In *Proceedings of the 2nd International Conference Computational Models of Argument (COMMA'08)*, pages 276–284. IOS Press, 2008.

[Parsons et al., 2003] S. Parsons, M. Wooldridge, and L. Amgoud. Properties and complexity of some formal inter-agent dialogues. *Journal of Logic and Computation*, 13(3):347–376, 2003.

[Pollock, 1987] J.L. Pollock. Defeasible reasoning. *Cognitive Science*, 11(4):481–518, 1987.

[Pollock, 1992] J.L. Pollock. How to reason defeasibly. *Artificial Intelligence*, 57(1):1–42, 1992.

[Wooldridge et al., 2005] M. Wooldridge, P. McBurney, and S. Parsons. On the meta-logic of arguments. In *Argumentation in Multi-agent Systems*, volume 4049 of *Lecture Notes in Computer Science*, pages 42–56. Springer, 2005.

[Wooldridge et al., 2006] M. Wooldridge, P. Dunne, and S. Parsons. On the complexity of linking deductive and abstract argument systems. In *Proceedings of the 21st National Conference on Artificial Intelligence (AAAI'06)*, pages 299–304. AAAI Press, 2006.

[Zhang and Lin, 2013] X. Zhang and Z. Lin. An argumentation framework for description logic ontology reasoning and management. *Journal of Intelligent Information Systems*, 40(3):375–403, 2013.

[Zhang et al., 2010] X. Zhang, Z. Zhang, D. Xu, and Z. Lin. Argumentation-based reasoning with inconsistent knowledge bases. In *Advances in Artificial Intelligence*, volume 6085 of *Lecture Notes in Computer Science*, pages 87–99. Springer, 2010.

Philippe Besnard
IRIT
Université Paul Sabatier
F-31062, Toulouse, France
Email: philippe.besnard@irit.fr

Anthony Hunter
Department of Computer Science
University College London
London, WC1E 6BT, UK
Email: anthony.hunter@ucl.ac.uk

PART C

ARGUMENTATION AND DIALOGUES

10
Argumentation Semantics as Formal Discussion

MARTIN CAMINADA

ABSTRACT. In the current chapter, we interpret a number of mainstream argumentation semantics by means of structured discussion. The idea is that an argument is justified according to a particular argumentation semantics iff it is possible to win a discussion of a particular type. Hence, different argumentation semantics correspond to different types of discussion. Our aim is to provide an overview of what these discussions look like, and their formal correspondence to argumentation semantics.

1 Introduction

The term "argumentation", when used in an informal way, calls upon intuitions of arguments being exchanged in some kind of interactive discussion. Yet, the notion of discussion plays a relatively limited role in abstract argumentation theory, which mainly focuses on various principles (called "argumentation semantics") for selecting nodes from a graph. As such, there seems to be quite a gap between (abstract) argumentation theory as described in much of the literature,[1] as it occurs in everyday life.[2]

In order to address this gap, attempts have been made to express argumentation semantics in terms of structured discussion. More precisely, the idea is that an argument is accepted w.r.t. a particular argumentation semantics iff it is possible to successfully defend the argument using a particular kind of discussion. In the current chapter we provide an overview of what the different kinds of discussion are, and how they formally relate to their associated argumentation semantics.

Although the discussion protocols (which we will often refer to as "discussion games") can serve as proof procedures of their associated argumentation semantics, their potential application is much wider than that. One could for instance use the discussion games for the purpose of human computer interaction. Suppose a knowledge-based system has determined that a particular argument (say, about how to treat a patient) should be accepted, and communicates this to its user (say, a doctor). When the user asks why this is the

---

[1] As for instance described in chapters 4 ("Abstract argumentation frameworks and their semantics"), 5 ("Abstract dialectical frameworks"), 6 ("Abstract rule-based argumentation") 9 ("A review of argumentation based on deductive arguments") and 19 ("Argumentation, nonmonotonic reasoning and argumentation and logic") of this volume.

[2] As for instance described in Chapter 12 ("Processing natural language argumentation") of this volume.

case, what should probably be avoided is a highly technical answer of the form "because the argument is in the minimal fixpoint of monotonic function F".[3] Instead, one would like the user to critically question the answer,[4] and be able to utter counter arguments to see whether these are properly addressed (by the system providing counter counter arguments). As an example of such a human-computer discussion, consider the following dialogue:

System: The patient is best off with medicine X, because this is the most effective.

User: But the patient is diabetic, for which medicine X could have side effects.

System: Recent studies have shown that these side effects are relatively minor.

So instead of the system immediately providing the full justification for its answer (say, by providing the entire grounded extension) in engages in a discussion with its user. Ideally, such a discussion should be "natural" in the sense that the human-computer interaction looks as much as possible as human-human interaction (say, if the doctor were to discuss the case with a more senior colleague).

Apart from being natural, the discussion should also be sound and complete. That is, the ability to win the discussion for a particular argument (that is, to have a winning strategy for the argument in the discussion game) should coincide with the argument being justified according to a pre-defined argumentation semantics. Soundness and completeness imply that if the system provides an answer ("argument A is (or is not) justified according to a particular argumentation semantics") the system can successfully defend itself in the discussion with the user. When this discussion is also perceived as natural by the user, this will hopefully increase the user's confidence in the system's answer.

Soundness and completeness also imply that what we are looking for are essentially proof procedures for particular argumentation semantics. Several of these have been stated in the literature. Inclusion in the current book chapter is done based on two criteria:

(1) does the discussion game have any link with natural discussion concepts, like described in philosophy or linguistics?

(2) is the discussion game such that it guarantees the absence of any exponential blowups, in either time or space?

Criterion (1) is the reason why for instance we have not included any discussion games for sceptical preferred semantics (like those of Doutre and Mengin [2004] and Dung and Thang [2007]). Criterion (2) is the reason why we did not include a detailed treatment of tree-based discussion games (like those of Prakken and Sartor [1997], Caminada [2004], Modgil and Caminada [2009] and Dung *et al.*

---

[3]which basically says the argument is in the grounded extension.
[4]See for instance Chapter 11 ("Argumentation schemes") of this volume.

[2007]).[5]

The remaining part of this chapter is structured as follows. First, in Section 2 we briefly recall some basic definitions and results from abstract argumentation theory. Then, in Section 3 we describe a discussion game for (credulous) preferred semantics [Caminada et al., 2014], and explain that it contains aspects of Socratic discussion. Then, in Section 4 we briefly state how this discussion game can be reapplied in the context of ideal semantics [Caminada et al., 2014]. In Section 5 we subsequently describe a discussion game for stable semantics [Caminada and Wu, 2009], basically by making minor modifications to the earlier described discussion game for (credulous) preferred semantics. In Section 6 we then describe a different discussion game in the context of grounded semantics [Caminada, 2015a] and explain its relationship with persuasion dialogue. Then, in Section 7 we briefly examine tree-based discussion games and explain one of their main disadvantages: the possibility of an exponential blowup in time or space. We round off with a discussion in Section 8.

2 Formal Preliminaries

In the current section, we briefly recall some basic definitions from abstract argumentation theory.[6] For current purposes, we restrict ourselves to finite argumentation frameworks.

Definition 1 (argumentation framework) *An* argumentation framework *is a pair* (Ar, att) *where* Ar *is a finite set of entities called* arguments *and* att *is a binary relation on* Ar.

Given an argumentation framework (Ar, att), $A, A' \in Ar$ and $Args, Args' \subseteq Ar$, we say that (1) A attacks A' iff $(A, A') \in att$, (2) A attacks $Args$ iff A attacks some argument in $Args$, (3) $Args$ attacks A iff some argument in $Args$ attacks A, and (4) $Args$ attacks $Args'$ iff some argument in $Args$ attacks some argument in $Args'$.

Definition 2 (preliminaries, extension-based) *Let* (Ar, att) *be an argumentation framework. A set* $Args \subseteq Ar$ *is* conflict-free *iff* $Args$ *does not attack itself. A set* $Args \subseteq Ar$ *defends* $A \in Ar$ *iff for each* $B \in Ar$ *that attacks* A, *it holds that* $Args$ *attacks* B.

Definition 3 (admissibility, extension-based) *Let* (Ar, att) *be an argumentation framework. A set* $Args \subseteq Ar$ *is* admissible *iff* $Args$ *is conflict-free and each* $A \in Args$ *is defended by* $Args$.

Definition 4 (strong admissibility, extension-based) *Let* (Ar, att) *be an argumentation framework. A set* $Args \subseteq Ar$ *is* strongly admissible *iff each*

---

[5] How tree-based discussion games can lead to an exponential blowup is explained in Section 7.

[6] We refer Chapter 4 ("Abstract argumentation frameworks and their semantics") of this volume for a more thorough discussion.

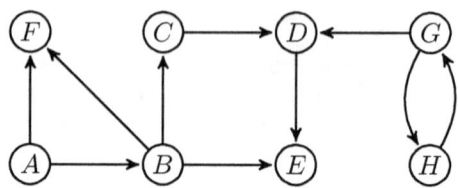

Figure 1. An argumentation framework to illustrate strong admissibility.

$A \in \mathcal{A}rgs$ is defended by some $\mathcal{A}rgs' \subseteq \mathcal{A}rgs \setminus \{A\}$ which in its turn is again strongly admissible.

It has been proved that each strongly admissible set is conflict-free as well as admissible [Baroni and Giacomin, 2007; Caminada, 2014].

As an example, consider the argumentation framework of Figure 1. Here, the set $\{A, C\}$ is strongly admissible as A is defended by $\emptyset \subseteq \{A, C\} \setminus \{A\}$ which is trivially strongly admissible, and C is defended by $\{A\} \subseteq \{A, C\} \setminus \{C\}$ which is strongly admissible (as A is defended by $\emptyset \subseteq \{A\} \setminus \{A\}$). The set $\{G\}$, however, is admissible but not strongly admissible as G is not defended by any subset of $\{G\} \setminus \{G\}$.

Definition 5 (completeness, extension-based) *Let (Ar, att) be an argumentation framework. A set $\mathcal{A}rgs \subseteq Ar$ is a complete extension iff $\mathcal{A}rgs$ is conflict-free and the set of arguments defended by $\mathcal{A}rgs$ is equal to $\mathcal{A}rgs$.*

Definition 6 (semantics, extension-based) *Let (Ar, att) be an argumentation framework. A set $\mathcal{A}rgs \subseteq Ar$ is called*

1. *a grounded extension iff $\mathcal{A}rgs$ is the minimal (w.r.t. $\subseteq$) complete extension*

2. *a preferred extension iff $\mathcal{A}rgs$ is a maximal (w.r.t. $\subseteq$) complete extension*

3. *a stable extension iff $\mathcal{A}rgs$ is a complete extension that attacks each argument in $Ar \setminus \mathcal{A}rgs$*

4. *an ideal extension iff $\mathcal{A}rgs$ is the maximal (w.r.t. $\subseteq$) complete extension that is not attacked by any complete extension*

We recall that each argumentation framework has precisely one grounded extension, precisely one ideal extension, one or more preferred extensions and zero or more stable extensions.

The above definition describes grounded, preferred, stable and ideal semantics uniformly in terms of complete semantics. However, for our purposes it is sometimes useful to describe these semantics in terms of (strong) admissibility.

Theorem 1 (semantics, extension-based) *Let (Ar, att) be an argumentation framework. A set $Args \subseteq Ar$ is*

1. *a preferred extension iff $Args$ is a maximal (w.r.t. $\subseteq$) admissible set*
2. *a grounded extension iff $Args$ is the maximal (w.r.t. $\subseteq$) strongly admissible set*
3. *a stable extension iff $Args$ is an admissible set that attacks each argument in $Ar \setminus Args$*
4. *an ideal extension iff $Args$ is the maximal (w.r.t. $\subseteq$) admissible set that is not attacked by any admissible set*

Apart from the extension-based view on argumentation semantics, there is also the labelling-based view [Caminada, 2006; Caminada and Gabbay, 2009; Caminada, 2011; Baroni et al., 2011] of which we now provide a brief overview.

Definition 7 (preliminaries, labelling-based) *Let (Ar, att) be an argumentation framework. An argument labelling is a function $\mathcal{L}ab : Ar \to \{\text{in}, \text{out}, \text{undec}\}$. We define $\text{in}(\mathcal{L}ab)$ as $\{A \in Ar \mid \mathcal{L}ab(A) = \text{in}\}$, $\text{out}(\mathcal{L}ab)$ as $\{A \in Ar \mid \mathcal{L}ab(A) = \text{out}\}$ and $\text{undec}(\mathcal{L}ab)$ as $\{A \in Ar \mid \mathcal{L}ab(A) = \text{undec}\}$. We sometimes write a labelling as a triple $(\text{in}(\mathcal{L}ab), \text{out}(\mathcal{L}ab), \text{undec}(\mathcal{L}ab))$. If $\mathcal{L}ab_1$ and $\mathcal{L}ab_2$ are labellings, we write $\mathcal{L}ab_1 \sqsubseteq \mathcal{L}ab_2$ when $\text{in}(\mathcal{L}ab_1) \subseteq \text{in}(\mathcal{L}ab_2)$ and $\text{out}(\mathcal{L}ab_1) \subseteq \text{out}(\mathcal{L}ab_2)$. Moreover, we write $\mathcal{L}ab_1 \approx \mathcal{L}ab_2$ when $\text{in}(\mathcal{L}ab_1) \cap \text{out}(\mathcal{L}ab_2) = \emptyset$ and $\text{out}(\mathcal{L}ab_1) \cap \text{in}(\mathcal{L}ab_2) = \emptyset$.*

Definition 8 (admissibility, labelling-based) *Let $\mathcal{L}ab$ be a labelling of argumentation framework (Ar, att). $\mathcal{L}ab$ is called an admissible labelling iff for each $A \in Ar$ it holds that*

1. *if $\mathcal{L}ab(A) = \text{in}$ then for each $B \in Ar$ that attacks A it holds that $\mathcal{L}ab(B) = \text{out}$*
2. *if $\mathcal{L}ab(A) = \text{out}$ then there exists a $B \in Ar$ that attacks A such that $\mathcal{L}ab(B) = \text{in}$*

In order to define strong admissibility in the context of argument labellings, we first need to introduce the concept of a min-max numbering.

Definition 9 (min-max numbering) *Given an admissible labelling $\mathcal{L}ab$ of argumentation framework (Ar, att), a min-max numbering is a function $\mathcal{MM}_{\mathcal{L}ab} : \text{in}(\mathcal{L}ab) \cup \text{out}(\mathcal{L}ab) \to \mathbb{N} \cup \{\infty\}$ such that for each $A \in \text{in}(\mathcal{L}ab) \cup \text{out}(\mathcal{L}ab)$ it holds that*

- *if $\mathcal{L}ab(A) = \text{in}$ then $\mathcal{MM}_{\mathcal{L}ab}(A) = \max(\{\mathcal{MM}_{\mathcal{L}ab}(B) \mid B \text{ attacks } A \text{ and } \mathcal{L}ab(B) = \text{out}\}) + 1$ (with $\max(\emptyset)$ defined as 0)*

- if $\mathcal{L}ab(A) = \mathtt{out}$ then $\mathcal{MM}_{\mathcal{L}ab}(A) = \min(\{\mathcal{MM}_{\mathcal{L}ab}(B) \mid B$ attacks A and $\mathcal{L}ab(B) = \mathtt{in}\}) + 1$ (with $\min(\emptyset)$ defined as ∞)

It can be proved that each admissible labelling has a unique min-max numbering [Caminada, 2014].[7]

Definition 10 (strong admissibility, labelling-based) *Let $\mathcal{L}ab$ be a labelling of argumentation framework (Ar, att). $\mathcal{L}ab$ is called a strongly admissible labelling iff it is an admissible labelling whose associated min-max numbering yields natural numbers only (so no argument is numbered ∞).*

From Definition 10 it trivially follows that each strongly admissible labelling is also an admissible labelling.

As an example, consider the argumentation framework shown in Figure 1. Here $\mathcal{L}ab_1 = (\{A, C, E, G\}, \{B, D, H\}, \{F\})$ is an admissible labelling with associated min-max numbering $\mathcal{MM}_{\mathcal{L}ab_1} = \{(A\!:\!1), (B\!:\!2), (C\!:\!3), (D\!:\!4), (E\!:\!5), (G\!:\!\infty), (H\!:\!\infty)\}$, which implies that $\mathcal{L}ab_1$ is not strongly admissible. Furthermore, $\mathcal{L}ab_2 = (\{A, C, E\}, \{B, D, F\}, \{G, H\})$ is an admissible labelling with associated min-max numbering $\mathcal{MM}_{\mathcal{L}ab_2} = \{(A\!:\!1), (B\!:\!2), (C\!:\!3), (D\!:\!4), (E\!:\!5), (F\!:\!2)\}$, which implies that $\mathcal{L}ab_2$ is indeed a strongly admissible labelling.

Definition 11 (completeness, labelling-based) *Let $\mathcal{L}ab$ be a labelling of argumentation framework (Ar, att). $\mathcal{L}ab$ is called a complete labelling iff for each $A \in Ar$ it holds that*

1. *if $\mathcal{L}ab(A) = \mathtt{in}$ then for each $B \in Ar$ that attacks A it holds that $\mathcal{L}ab(B) = \mathtt{out}$*

2. *if $\mathcal{L}ab(A) = \mathtt{out}$ then there exists a $B \in Ar$ that attacks A such that $\mathcal{L}ab(B) = \mathtt{in}$*

3. *if $\mathcal{L}ab(A) = \mathtt{undec}$ then not for each $B \in Ar$ that attacks A it holds that $\mathcal{L}ab(B) = \mathtt{out}$ and there does not exist a $B \in Ar$ that attacks A such that $\mathcal{L}ab(B) = \mathtt{in}$*

Definition 12 (semantics, labelling-based) *Let (Ar, att) be an argumentation framework. A labelling $\mathcal{L}ab$ is called*

1. *a grounded labelling iff it is the minimal (w.r.t. $\sqsubseteq$) complete labelling*

2. *a preferred labelling iff it is a maximal (w.r.t. $\sqsubseteq$) complete labelling*

---

[7]The min-max numbering can be constructed in an iterative way, starting from the unnumbered in-labelled arguments without attackers (these are numbered 1), then the unnumbered out-labelled arguments that are attacked by these (these are numbered 2), etc. When a particular iteration provides no new argument numbers, the remaining unnumbered in and out-labelled arguments are numbered ∞. See the work of Caminada [2014] for details.

3. a stable labelling *iff it is a complete labelling with* undec($\mathcal{L}ab$) = ∅

4. an ideal labelling *iff it is the maximal (w.r.t.* $\sqsubseteq$*) complete labelling that is compatible (*≈*) with every complete labelling*

We recall that each argumentation framework has precisely one grounded labelling, precisely one ideal labelling, one or more preferred labellings and zero or more stable labellings.

The above definition describes grounded, preferred, stable and ideal semantics in terms of complete labellings. However, it is sometimes useful to be able to describe these semantics in terms of (strong) admissibility, similar to what was done earlier for the extension-based semantics.

Theorem 2 (semantics, labelling-based) *Let* (Ar, att) *be an argumentation framework. A labelling $\mathcal{L}ab$ is called*

1. a preferred labelling *iff it is a maximal (w.r.t.* $\sqsubseteq$*) admissible labelling*

2. a grounded labelling *iff it is the maximal (w.r.t.* $\sqsubseteq$*) strongly admissible labelling*

3. a stable labelling *iff it is an admissible labelling with* undec($\mathcal{L}ab$) = ∅

4. an ideal labelling *iff it is the maximal (w.r.t.* $\sqsubseteq$*) admissible labelling that is compatible (*≈*) with every admissible labelling*

To be able to easily switch between the labelling-based approach and the extension-based approach, we introduce two functions Lab2Ext and Ext2Lab, such that for an admissible labelling $\mathcal{L}ab$, Lab2Ext($\mathcal{L}ab$) is defined as in($\mathcal{L}ab$), and for an admissible set $\mathcal{A}rgs$, Ext2Lab($\mathcal{A}rgs$) is defined as ($\mathcal{A}rgs$, {$A \in Ar \mid A$ is attacked by $\mathcal{A}rgs$}, {$A \in Ar \mid A \notin \mathcal{A}rgs$ and A is not attacked by $\mathcal{A}rgs$}) where Ar is the set of all arguments in the argumentation framework. It holds that if $\mathcal{L}ab$ is a (strongly) admissible labelling (resp. a complete, grounded, preferred, stable or ideal labelling) then Lab2Ext($\mathcal{L}ab$) is a (strongly) admissible set (resp. a complete, grounded, preferred, stable or ideal extension). It also holds that if $\mathcal{A}rgs$ is a (strongly) admissible set (resp. a complete, grounded, preferred, stable or ideal extension) then Ext2Lab($\mathcal{A}rgs$) is a (strongly) admissible labelling (resp. complete, grounded, preferred, stable or ideal labelling). Moreover, when restricted to complete (or resp. grounded, preferred, stable or ideal) extensions and labellings, the functions Lab2Ext and Ext2Lab become bijections that are each other's inverses [Caminada, 2006; Caminada and Gabbay, 2009].

The above results imply that:

- in order to determine whether an argument is in a preferred extension, it suffices to determine whether the argument is labelled **in** by an admissible labelling

- in order to determine whether an argument is in the grounded extension, it suffices to determine whether the argument is labelled **in** by a strongly admissible labelling

- in order to determine whether an argument is in a stable extension, it suffices to determine whether the argument is labelled **in** by an admissible labelling without **undec**

- in order to determine whether an argument is in the ideal extension, it suffices to determine whether the argument is labelled **in** by an admissible labelling that is compatible with every admissible labelling

In the sections that follow, we will apply the above observations to provide discussion games for preferred, grounded, stable and ideal semantics.

3 Preferred Semantics

In the current section, we describe the discussion game for preferred semantics as stated by Caminada *et al.* [2014].[8] The idea of the preferred discussion game is to show membership of a preferred extension by constructing an admissible labelling where the argument in question is labelled **in**.

The preferred discussion game has two players which we will refer to as M and S. Player M starts; his task is to defend the fact that he has a reasonable position (admissible labelling) in which a particular argument is accepted (labelled **in**). Player S then tries to confront M with the consequences of M's own position, and asks for these consequences to be resolved. Player M is successful if he is able to address all the issues pointed out by player S, without being led to a contradiction.

As an example of how such a discussion can take place, consider the argumentation framework of Figure 2.

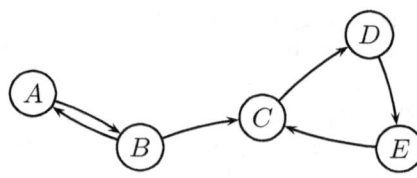

Figure 2. An argumentation framework

Here, the player M can win the discussion game for argument D in the following way.

---

[8]The discussion game of Caminada *et al.* [2014] consists of a labelling-based reinterpretation of the work of Vreeswijk and Prakken [2000].

Example 1
M: in(D)
"I have an admissible labelling in which D is labelled in."
S: out(C)
"But then in your labelling it must also be the case that D's attacker C is labelled out. Based on which grounds?"
M: in(B)
"C is labelled out because B is labelled in."
S: out(A)
"But then in your labelling it must also be the case that B's attacker A is labelled out. Based on which grounds?"
M: in(B)
"A is labelled out because B is labelled in."

As is shown in the above example, the discussion moves of player M are statements that particular arguments are labelled in in M's labelling. The moves of player S, on the other hand, are meant to confront M with the consequences of his own position: "if you think that argument X is labelled in then you must also hold that X's attacker Y is labelled out in your labelling." That is, by uttering out(Y), player S points out that player M is implicitly committed to the fact that Y should be rejected. This means that player M has to explain *why* Y should be rejected. That is, the moves of player S can be seen as *questions* about why a particular argument Y should be labelled out. The moves of player M (except his first move) can then be interpreted as the *answers* to the questions of player S. Each answer follows directly to the question raised by player S. That is:

Each move of M (except the first) contains an attacker of the argument in the directly preceding move of S. (1)

Every time player M claims that an argument is labelled in, player S should be given the opportunity to state that as a consequence of this, player M is implicitly committed that *all* attackers of the argument are labelled out. The problem, however, is that each move of player S is a statement about just *one* argument. In order to deal with this problem, player S should be given the opportunity to react on the same in-labelled argument several times, each time confronting player M with a different out-labelled argument. This means that player S should be allowed to react not just on the immediately preceding move of player M, but on *any* previous move of player M.

Each move of player S contains an attacker of an argument contained in some (not necessarily the directly preceding) move of player M. (2)

Another issue is whether player S should be allowed to repeat his own moves. Recall that each move essentially contains a question ("Based on which grounds is argument Y labelled out?"). At the moment player S repeats one of his moves, this question has already been answered by player M, so there is no good reason to ask again. In order to avoid the discussion from going round in circles, it does not make sense to allow player S to repeat his moves.

Player S is not allowed to repeat his moves. (3)

On the other hand, Example 1 does illustrate the need for player M to be able to repeat his moves (like in(B)). This is because some of the questions of S (like "why is argument C out" and "why is argument A out") can have the same answer ("because argument B is in").

Player M is allowed to repeat his moves. (4)

The argumentation framework of Figure 2 can also be used for an example of a game won by player S:

Example 2
    M: in(E)
           "I have an admissible labelling in which E is labelled in."
    S: out(D)
           "But then in your labelling it must be the case that E's attacker D is labelled out. Based on which grounds?"
    M: in(C)
           "D is labelled out because C is labelled in."
    S: out(E)
           "But then in your labelling it must be the case that C's attacker E is labelled out. This contradicts with your earlier claim that E is labelled in."

The above example illustrates that when player S manages to use an argument uttered previously by player M, player S has won the game. After all, if player M claims an argument to be in and player S subsequently manages to confront player M with the fact that in M's own position, the same argument should be labelled out, then player S has successfully pointed out a contradiction in M's position.

If player S uses an argument previously used by player M, then player S wins the discussion game. (5)

One can ask a similar question regarding what happens when player M uses one of the arguments previously used by player S. The fact that player S performed an out move means that the argument must be labelled out in the labelling of player M. If player M then subsequently claims that the same argument is labelled in, then he has directly contradicted himself.

If player M uses an argument previously used by player S, then player S wins the discussion game. (6)

There also exists a third condition under which player S wins the game. This is when player M is unable to answer one of the questions of S. This can be the case when there exists no attacker against an argument uttered by player S. Hence, player S asks why a particular argument is labelled out but player M is unable to come up with any attacker to be labelled in. In that case, player M has lost the game, for not being able to answer the critical questions of player S.

If player M cannot make a move any more, player S wins the discussion game. (7)

Similarly, one might examine what happens when it is player S who cannot make a move any more. This essentially means that player S has run out of questions. All possible relevant questions have already been asked; all relevant issues have already been raised. Moreover, player M has managed to answer all questions in a satisfactory way. Therefore, player M has survived the process of critical questioning, hence winning the discussion.

If player S cannot make a move any more, player M wins the discussion game. (8)

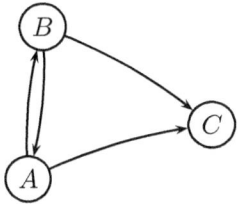

Figure 3. An argumentation framework with floating attack

As a last illustration of how the discussion game functions, consider the argumentation framework of Figure 3. Argument C is not in any admissible set. It is illustrative to see what happens if player M tries to defend C.

Example 3
M: in(C)
     "I have an admissible labelling in which C is labelled in."
S: out(A)
     "But then in your labelling C's attacker A must be labelled out. Based on which grounds?"
M: in(B)
     "A is labelled out because B is labelled in."
S: out(B)
     "But from the fact that you hold C to be in, it follows that C's attacker B must be labelled out. This contradicts with your earlier claim that B is labelled in."

The above example illustrates the need for player S to be able to respond not only to the immediately preceding move, but to any past move of player M; in the example, out(B) is a response to in(C). This is because, as we have mentioned before, for an argument to be labelled in, *all* its attackers have to be out, so player S may need to respond to a move of player M with more than one countermove.

When putting observations (1) to (8) together, we obtain the following description of the discussion game

Definition 13 *Let (Ar, att) be an argumentation framework. A preferred discussion is a sequence of moves $[\Delta_1, \Delta_2, \ldots, \Delta_n]$ ($n \geq 0$) such that:*

- *each move Δ_i ($1 \leq i \leq n$) where i is odd is called an M-move and is of the form in(A), where $A \in Ar$*

- *each move Δ_i ($1 \leq i \leq n$) where i is even is called an S-move and is of the form out(A), where $A \in Ar$*

- *for each S-move $\Delta_i = $ out(A) ($2 \leq i \leq n$) there exists an M-move $\Delta_j = $ in(B) ($j < i$) such that A attacks B*

- *for each M-move $\Delta_i = $ in(A) ($3 \leq i \leq n$) it holds that Δ_{i-1} is of the form out(B), where A attacks B*

- *there exist no two S-moves Δ_i and Δ_j with $i \neq j$ and $\Delta_i = \Delta_j$*

A preferred discussion $[\Delta_1, \Delta_2, \ldots, \Delta_n]$ is said to be finished iff (1) there exists no Δ_{n+1} such that $[\Delta_1, \Delta_2, \ldots, \Delta_n, \Delta_{n+1}]$ is a preferred discussion, or there exists an M-move and an S-move containing the same argument, and (2) no subsequence $[\Delta_1, \ldots, \Delta_m]$ ($m < n$) is finished. A finished preferred discussion is won by player S if there exist an M-move and an S-move containing the same argument. Otherwise, it is won by the player making the last move (Δ_n).

The soundness and completeness of the game described above is stated in the following theorem.

Theorem 3 ([Caminada and Wu, 2009; Caminada et al., 2014])
Let (Ar, att) be an argumentation framework and $A \in Ar$.

1. *If there exists a preferred discussion for A that is won by player M, then there exists a preferred extension that contains A.*

2. *If there exists a preferred extension that contains A then player M has a winning strategy[9] for the preferred discussion game.*

The correctness of Theorem 3 can be seen as follows. As for point 1, it has to be observed that what the game essentially does is to build an admissible labelling of which the in-labelled arguments coincide with the M-moves and the out-labelled arguments coincide with the S-moves (all the other arguments are labelled undec). The resulting labelling is well-defined in the sense that no argument is labelled both in and out (otherwise there would be an argument that is subject to both an M-move and an S-move, in which case player S would have won the discussion). Moreover, the fact that player M wins the discussion also means that he made the last move, which implies that (i) each out-labelled argument has an in-labelled attacker. Also, the fact that player S cannot move anymore implies that (ii) each in-labelled argument has all its attackers labelled out. From (i) and (ii) it follows that the labelling yielded by the game is indeed an admissible one, satisfying the conditions of Definition 8. In this admissible labelling, argument A is labelled in (since A was the subject of the first M-move). This implies that A is element of an admissible set, and therefore also element of a preferred extension.

As for point 2, it should be mentioned that the fact that A is in a preferred extension by definition implies that A is in an admissible set ($Args$), which then implies that A is labelled in by an admissible labelling $Lab = \text{Ext2Lab}(Args)$. This makes it possible for player M to win the game simply by staying within the borders of admissible labelling Lab. That is, as long as player M only plays arguments that are labelled in by Lab, each move of player S has to be an argument that is labelled out by Lab, which then implies that player M can always react with an argument that is labelled in by Lab, etc. If player M follows such a strategy, there will never be an M-move and an S-move for the same argument (this is because Lab is a well-defined labelling, meaning that no argument is labelled both in and out). Moreover, the fact that player S cannot repeat himself means that the game has to finish in a finite number of moves. As player M can *always* react on a move of player S, this means that the last move has to be an M-move. Hence, player M wins the game.

From points 1 and 2 together, it follows that if there is at least one preferred discussion that is won by player M, then M has a winning strategy for the preferred discussion game. This is not the case in alternative discussion games

---

[9]Winning strategy in the sense of [Caminada et al., 2014, Definition 5.6]. Informally this means that player M has a way of winning the discussion, regardless of what moves player S decides to play.

for preferred semantics, like the one described by Modgil and Caminada [2009]. In their approach, a single discussion game does not prove membership (for this, the presence of a winning strategy is really necessary). From informal perspective, this is rather odd, as in everyday life the aim of a (persuasion) discussion is to convince the other party in a single discussion. This means that at the end of the discussion, the other party has to have heard sufficient evidence to accept the main claim. This is the case in the above described preferred discussion game, but not in the alternative discussion game of Modgil and Caminada [2009].

As we have observed, an admissible labelling can serve as a "roadmap" for winning the preferred discussion game.[10] However, an argument can be labelled in by more than one admissible labelling, which raises the question of which admissible labelling to choose as a basis to play the game. It can be verified that given an admissible labelling $\mathcal{L}ab$ (with $\mathcal{L}ab(A) = \text{in}$ and $\text{out}(\mathcal{L}ab)$ being minimal w.r.t. set inclusion) the number of moves required in the game for main argument A is $2 \cdot |\text{out}(\mathcal{L}ab)| + 1$ (see [Caminada et al., 2014] for details). Hence, in order to be able to finish the game in as few moves as possible (which could be desirable from the perspective of human-computer interaction if the aim of the game is to convince a human user) one should try to find an admissible labelling $\mathcal{L}ab$ where $|\text{out}(\mathcal{L}ab)|$ is minimal. This is a computationally hard problem, as even verifying whether a particular admissible labelling has this property is coNP complete [Caminada et al., 2014].

The essential nature of the preferred discussion game is that of critically questioning a particular position, and to see whether the proponent of this position (player M) can avoid being led to a contradiction (by player S). As such, the preferred discussion game bears a close resemblance to the concept of Socratic discussion, as well as to its modern variants like critical interviews or cross-examinations in court.[11] The general idea is to have somebody take a position and then iteratively confront him (through questioning) with what appears to be the consequences of this position, in the hope of ultimately leading him to a contradiction. We refer to the work of Caminada et al. [2014] for a details.

4 Ideal Semantics

An ideal set of arguments, as was originally defined by Dung et al. [2007], is an admissible set that is a subset of each preferred extension. It can be proved that the maximal ideal set (commonly known as the *ideal extension*) is unique and is a complete extension as well.

An alternative but equivalent way of characterising the ideal extension is as the maximal admissible set that is not attacked by any admissible set (like is done in Theorem 1) or as the maximal complete extension that is not attacked

---

[10] For details, we refer to the work of Caminada et al. [2014].

[11] In fact, in the work of Caminada et al. [2014] player S stands for Socrates and player M stands for Menexenus, which is one of Socrates's historic discussion partners.

by any complete extension (like is done in Definition 6). It can be proved that for each admissible sets $Args_1$ and $Args_2$ it holds that $Args_1$ attacks $Args_2$ iff $Args_2$ attacks $Args_1$. This gives rise to the labelling-based descriptions of ideal semantics of Theorem 2 and Definition 12.[12]

For current purposes, our characterisation of the ideal extension is as the maximal admissible set that is not attacked by any admissible set. To determine membership of the ideal extension, one then needs to find an admissible set (although not necessarily the maximal one) that contains the argument in question and is not attacked by any admissible set. This makes it possible to express ideal semantics using the preferred discussion game. Basically, the discussion whether an argument is in an ideal extension consists of two phases. In the first phase, one runs the preferred discussion game, as is described in the previous section. This is to determine whether the argument is in an admissible set. Then, in the second phase of the discussion, one needs to determine whether this set is attacked by another admissible set. This is done by again running the preferred discussion game for each of the arguments that were rejected (labelled out) during the first phase of the discussion, this time trying to defend (label in) the argument.

As an example, consider again the argumentation framework of Figure 2. Now consider the question of whether argument D is in an ideal set. The first phase of the discussion would be like Example 1 (page 495). Then, in the second phase of the discussion, one has to try to find an argument that was labelled out during the first phase[13] (say A) and can be defended in a new preferred discussion game. Such a game would be as follows.

M: in(A)
"I have a reasonable position (admissible labelling) in which A is accepted (labelled in)."

S: out(B)
"Then in your position, argument B must be rejected (labelled out). Based on which grounds?"

M: in(A)
"B is rejected (labelled out) because A is accepted (labelled in)."

Hence, we have an admissible set $\{A\}$ that attacks the admissible set $\{B, D\}$ found during the first phase, so the admissible set $\{B, D\}$ of the first phase is not an ideal set.[14]

The overall procedure for ideal semantics puts an extra burden on the proponent of the argument. Not only does he have to win the preferred discussion game in the first phase, but he has to win it in such a way[15] that the resulting

---

[12]Recall that each complete extension (labelling) is also an admissible set (labelling).

[13]Recall that the preferred game is such that the out-labelled arguments are the attackers of the in-labelled arguments (which is not necessarily the case for admissible labellings in general).

[14]In fact, for the argumentation framework of Figure 2, the only ideal set is the empty set.

[15]Since an argument can be element of more than one admissible set, there can be different

position (labelling) cannot be argued against in the second phase.

5 Stable Semantics

In the current section, we describe a discussion game for credulous stable semantics based on the work of Caminada and Wu [2009]. Before doing so, it may be illustrative to see why the preferred discussion game does not work for stable semantics. Consider again the argumentation framework of Figure 2. Even though A is in an admissible set and in a preferred extension ($\{A\}$), A is not in a stable extension. To see why A is in an admissible set, consider the following discussion:
M: in(A) "I have an admissible labelling where A is labelled in."
S: out(B) "Then in your labelling, argument B must be labelled out. Based on which grounds?"
M: in(A) "B is labelled out because A is labelled in."
The point is, however, that once it has been decided that A is labelled in and B is labelled out, it is not possible anymore to label the remaining arguments such that final result will be a stable labelling. This can be seen as follows. Suppose C is labelled in. Then E must be labelled out, so D should be labelled in, which means that C would be labelled out. Contradiction. Similarly, suppose that C is labelled out. Then E must be labelled in, so D should be labelled out, so C should be labelled in. Again, contradiction.

In general, there are many ways to characterize a stable extension [Caminada and Gabbay, 2009]. For our purposes, the most useful characterization is that of an admissible set which attacks every argument that is not in it (Theorem 1). When one translates this to labellings, one obtains an admissible labelling where each argument is labelled either in or out (that is, no argument is labelled undec, Theorem 2).

It appears that a discussion game for stable semantics requires an additional type of move: question. To illustrate the role of this new move, imagine a politician being interviewed for TV. At first the discussion may be about financial matters (say, whether the banking system should be nationalized). Then, the discussion may be about the consequences of the politician's opinion ("If you accept to nationalize the banks, then you must reject the possibility to improve healthcare, because there will not be enough money left to do so."). However, at some moment, the interviewer could choose to totally change topic ("By the way, what are your opinions about abortion?"). It is this change of topic that is enabled by the question move.[16]

For the discussion game for stable semantics, we use the question move to

---

ways to win the preferred discussion game.

[16]One of the reasons the question move is needed is because stable semantics does not satisfy the property of *directionality* [Baroni and Giacomin, 2007]. This means that for determining the status of an argument, not just the "ancestors" (the attackers, the attackers of these attackers, etc) are relevant but also the "offspring" (the attacked, the attacked of the attacked, etc) as well as arguments from unconnected parts of the graph. See also Chapter 16 ("The principle-based approach to abstract argumentation semantics") of this volume.

involve those arguments that have never been uttered before so that we are able to label all the arguments in Ar. By questioning an argument (question(A)), player S (the opponent) asks player M (the proponent) to give an explicit opinion on whether A should be labelled in or out. If player M thinks that A should be labelled in then he should respond with in(A). If, on the other hand, player M thinks that A should be labelled out then he should respond with in(B) where B is a attacker of A. The discussion game for stable semantics can thus be described as follows:

- Player M (the proponent) and player S (the opponent) take turns. Player M starts.

- Each move of player S is either of the form out(A), where A is a attacker of some (not necessarily the directly preceding) move of player M, or of the form question(A), where A is an argument that has not been uttered in the discussion before (by either player M or player S).

- The first move of player M is of the form in(A), where A is the main argument of the discussion. The following moves of player M are also of the form in(A) although A no longer needs to be the main claim. If the directly preceding move of player S is of the form out(B) then A is a attacker of B. If the directly preceding move of player S is of the form question(B) then A is either equal to B or a attacker of B.

- Player S is not allowed to repeat any of his out moves.

- Player M is allowed to repeat his own in moves.

Player S wins if there is an argument A that has been subject to both an in move (by player M) and an out move (by player S). Otherwise, the discussion continues until one of the players cannot move anymore, in which case the discussion is won by the player making the last move.

To illustrate the use of the discussion game, consider the argumentation framework depicted in Figure 4.

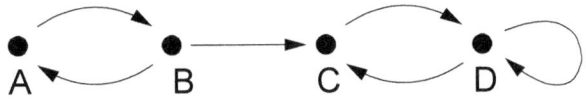

Figure 4. Another argumentation framework

Suppose player M would like to start a discussion about A.
M: in(A) "I have a stable labelling in which A is labelled in."
S: out(B) "Then in your labelling, A's attacker B must be labelled out. Based on which grounds?"
M: in(A) "B is labelled out because A is labelled in."

S: question(C) "What about C?"
M: in(C) "C is labelled in."
S: out(D) "Then C's attacker D must be labelled out. Based on which grounds?"
M: in(C) "D is labelled out because C is labelled in."
Player M wins the discussion, since player S cannot move anymore.

The above example also shows that the outcome of a discussion may depend on player M's response to a question move. For instance, if player M would have replied to question(C) with in(D), then he would have lost the discussion, since player S would then move out(D).

As an example of a discussion that cannot be won by player M, consider the discussion for argument B. This discussion has to be lost by player M since the argumentation framework of Figure 4 has only one stable extension: $\{A, C\}$, which does not include B.

M: in(B) "I have a stable labelling in which B is labelled in."
S: out(A) "Then in your labelling, B's attacker A must be labelled out. Based on which grounds?"
M: in(B) "A is labelled out because B is labelled in."
S: question(C) "What about C?"
M: in(D) "C is labelled out because its attacker D is labelled in."
S: out(D) "Then D's attacker D (itself) must be labelled out. Contradiction."
Player M would still have lost the discussion if he had responded to question(C) with in(C) instead of with in(D). This is because then player S would have reacted with out(B) and would therefore still have won the discussion.

Formally, the stable discussion game can be described as follows.

Definition 14 *Let (Ar, att) be an argumentation framework. A stable discussion is a sequence of moves $[\Delta_1, \Delta_2, \ldots, \Delta_n]$ ($n \geq 0$) such that:*

- *each Δ_i ($1 \leq i \leq n$) where i is odd (which is called an M-move) is of the form in(A), where $A \in Ar$.*

- *each Δ_i ($1 \leq i \leq n$) where i is even (which is called an S-move) is of the form out(A) where $A \in Ar$, or of the form question(A) where $A \in Ar$.*

- *For each S-move $\Delta_i = $ out(A) ($2 \leq i \leq n$) there exists an M-move $\Delta_j = $ in(B) ($j < i$) where A attacks B.*

- *For each M-move $\Delta_i = $ in(A) ($3 \leq i \leq n$) it either holds that (1) $\Delta_{i-1} = $ out(B) where A attacks B, or (2) $\Delta_{i-1} = $ question(B) where either $A = B$ or A attacks B.*

- *For each S-move $\Delta_i = $ out(A) ($1 \leq i \leq n$) there does not exist an S-move $\Delta_j = $ out(A) with $j < i$.*

- *For each S-move $\Delta_i = $ question(A) ($1 \leq i \leq n$) there does not exist any move Δ_j ($j < i$) of the form in(A), out(A) or question(A).*

- For each M-move $\Delta_i = \text{in}(A)$ $(1 \leq i \leq n)$ there does not exist an S-move $\Delta_j = \text{out}(A)$ with $j < i$.

A stable discussion $[\Delta_1, \Delta_2, \ldots, \Delta_n]$ is said to be finished iff (1) there exists no Δ_{n+1} such that $[\Delta_1, \Delta_2, \ldots, \Delta_n, M_{n+1}]$ is a stable discussion, or there exists an M-move $\text{in}(A)$ and an S-move $\text{out}(A)$ for the same argument A, and (2) no subsequence $[\Delta_1, \ldots, \Delta_m]$ $(m < n)$ is finished. A finished stable discussion is won by player S if there exists an M-move $\text{in}(A)$ and an S-move $\text{out}(A)$ for the same argument A. Otherwise it is won by the payer making the last move Δ_n.

It turns out that an argument is in at least one stable extension iff the proponent can win the stable discussion game for it.

Theorem 4 *Let (Ar, att) be an argumentation framework and $A \in Ar$.*

1. *If there exists a stable discussion for A that is won by player M, then A is in a stable extension.*

2. *If A is in a stable extension, then player M has a winning strategy for the stable discussion game.*

As for point 1, it can be observed that what the discussion game essentially does is to build a stable labelling $\mathcal{L}ab$ with $\text{in}(\mathcal{L}ab) = \{A \mid \text{there exists an M-move } \text{in}(A)\}$ and $\text{out}(\mathcal{L}ab) = \{A \mid \text{there exists an S-move } \text{out}(A)\} \cup \{A \mid \text{there exists an S-move question}(A) \text{ that was responded to with } \text{in}(B) \text{ where } B \text{ attacks } A\}$. It can be verified that $\mathcal{L}ab$ is an admissible labelling without any argument being labelled undec. Hence, $\mathcal{L}ab$ is a stable labelling in the sense of Theorem 2. As A is labelled in by $\mathcal{L}ab$ (since A is the subject of the first M-move) it holds that A is in $\text{Lab2Ext}(\mathcal{L}ab)$. Hence, A is in a stable extension.

As for point 2, it should be mentioned that player M can win the game simply by staying within the borders of the stable labelling $\mathcal{L}ab = \text{Ext2Lab}(\mathcal{A}rgs)$ (with $\mathcal{A}rgs$ being the stable extension that contains A, the argument that the discussion will start with). That is, as long as player M only plays arguments that are labelled in by $\mathcal{L}ab$, each out move of player S will be labelled out by $\mathcal{L}ab$, which then implies that player M can always react with an argument that is labelled in by $\mathcal{L}ab$, etc. Moreover, when player S does a question(A) move, either A itself or an attacker of A is labelled in by $\mathcal{L}ab$, which again means that player M can always respond with an argument that is labelled in by $\mathcal{L}ab$. As the argumentation framework is finite and player S cannot repeat himself, it follows that the game will finish in a finite number of moves. As player M can *always* react to the moves of player S, this means that the last move has to be an M-move. Hence, player M wins the game.[17]

Definition 14 describes the discussion game for credulous stable semantics (that is, it can be used to determine whether an argument is in at least one stable

---

[17] A more elaborate proof can be found in [Caminada and Wu, 2009].

extension). It is, however, relatively straightforward to re-apply this game in the context of sceptical stable semantics (that is, to determine whether an argument is in every stable extension). The idea is that an argument A is in each stable extension iff no attacker of A is in any stable extension. So in order to determine whether A is in every stable extension, one could try to play the stable discussion game for each attacker of A. If for none of these attackers the discussion game can be won, argument A is in each stable extension.

6 Grounded Semantics

So far, we have mainly focussed on the preferred discussion game and its slightly modified variants for ideal and stable semantics. In the current section we will focus on a fundamentally different type of discussion game, in the context of grounded semantics.

One of the main differences between the preferred discussion game and the grounded discussion game to be introduced in the current section is a conceptual one. To properly understand this difference, it is useful to take the perspective of complete labellings. We recall that a complete labelling (Definition 11) is a labelling where one has reasons for each argument one accepts (because all its attackers are rejected), reasons for each argument one rejects (because it has an attacker that is accepted), and reasons for each argument one abstains from having an explicit opinion about (because there are insufficient grounds to accept it and insufficient grounds to reject it). As such, a complete labelling can be seen as a reasonable position on how to evaluate the conflicting information represented in the argumentation framework. The preferred discussion game determines whether an argument is accepted (labelled in) by *at least one* such reasonable position.[18] The grounded discussion game, to be introduced in the current section, determines whether an argument is accepted (labelled in) by *every* such reasonable position.[19] That is, from the perspective of complete labellings, the preferred discussion game is about whether an argument *can be* accepted, whereas the grounded discussion game is about whether an argument *has to be* accepted.

The difference between determining whether an argument can be accepted and whether an argument has to be accepted is reflected in the nature of the associated discussion game. If the discussion is merely about whether an argument can be accepted (that is, about whether there exists a reasonable position in which the argument is accepted) then arguing against this means pointing out that any position in which the argument is accepted is somehow not reasonable. That is, the opponent tries to lead the proponent of such a position towards a contradiction.[20] Hence, the admissible discus-

---

[18]This is because an argument is labelled in by some admissible labelling iff it is labelled in by some complete labelling.

[19]This is because an argument is labelled in by the grounded labelling iff it is labelled in by every complete labelling.

[20]like saying, "if you think that argument X is labelled in, then it follows that X's attacker Y should be labelled out, but previously you claimed that Y should be labelled in."

sion game has at least some properties of Socratic discussion [Caminada, 2008; Caminada *et al.*, 2014]. If, on the other hand, the discussion is about whether an argument has to be accepted (that is, about whether the argument is accepted in each reasonable position) then the discussion gets a totally different nature. If an argument is accepted in each reasonable position, then in particular one's discussion partner, by being reasonable, should accept the argument. So the discussion becomes one of trying to *convince* the discussion partner that he has to accept a particular argument. That is, the discussion partner should be shown that by being reasonable, he cannot avoid having to accept the argument in question. As such, the nature of the discussion becomes that of persuasion dialogue [Walton and Krabbe, 1995].

Now that the conceptual difference between the preferred discussion game and the grounded discussion game has been explained, we will take a closer look at the technical differences. Although the preferred discussion game is used to determine membership of a preferred extension, it does so by determining membership of an admissible set (labelling).[21] This has the advantage of not having to construct the entire preferred extension (labelling), as constructing an admissible set (labelling) will be sufficient. Similarly, although the grounded discussion game is used to determine membership of the grounded extension, it does so by determining membership of a strongly admissible set (labelling) [Baroni and Giacomin, 2007; Caminada, 2014].[22] This has the advantage of not having to construct the entire grounded extension (labelling) as constructing a strongly admissible set (labelling) will be sufficient.

The grounded discussion game [Caminada, 2015a; Caminada, 2015b] that we will described in the current section has two players (proponent and opponent) and is based on four different moves, each of which has an argument as a parameter.

$HTB(A)$ ("*A* has to be the case")
  With this move, the proponent claims that A has to be labelled **in** by every complete labelling, and hence also has to be labelled **in** by the grounded labelling.

$CB(B)$ ("*B* can be the case, or at least cannot be ruled out")
  With this move, the opponent claims that B does not have to be labelled **out** by every complete labelling. That is, the opponent claims there exists a complete labelling where B is labelled **in** or **undec**, and that B is therefore not labelled **out** by the grounded labelling.

$CONCEDE(A)$ ("I agree that *A* has to be the case")
  With this move, the opponent indicates that he now agrees with the

---

[21] Recall that an admissible set (labelling) can always be extended to a preferred extension (labelling), as a preferred extension (labelling) is a maximal admissible set (labelling).

[22] Recall that a strongly admissible set (labelling) can always be extended to the grounded extension (labelling), as the grounded extension (labelling) is the maximal strongly admissible set (labelling) (see Theorem 2 and the work of Baroni and Giacomin [2007] and Caminada [2014]).

proponent (who previously did an $HTB(A)$ move) that A has to be the case (labelled **in** by every complete labelling, including the grounded).

$RETRACT(B)$ ("I give up that B can be the case")
With this move, the opponent indicates that he no longer believes that B can be **in** or **undec**. That is, the opponent acknowledges that B has to be labelled **out** by every complete labelling, including the grounded.

One of the key ideas of the discussion game is that the proponent has burden of proof. He has to establish the acceptance of the main argument and make sure the discussion does not go around in circles. The opponent merely has to cast sufficient doubts.

The game starts with the proponent uttering an HTB statement. After each HTB statement (either the first one or a subsequent one) the opponent utters a sequence of one or more CB, $CONCEDE$ and $RETRACT$ statements, after which the proponent again utters an HTB statement, etc. In the argumentation framework of Figure 1 the discussion could go as follows.

(1) P: $HTB(C)$ (4) O: $CONCEDE(A)$
(2) O: $CB(B)$ (5) O: $RETRACT(B)$
(3) P: $HTB(A)$ (6) O: $CONCEDE(C)$

In the above discussion, C is called *the main argument* (the argument the discussion starts with). The discussion above ends with the main argument being conceded by the opponent, so we say that the proponent wins the discussion.

As an example of a discussion that is lost by the proponent, it can be illustrative to examine what happens if the proponent claims that B has to be the case.

(1) P: $HTB(B)$ (2) O: $CB(A)$

After the second move, the discussion is terminated, as the proponent cannot make any further move, since A does not have any attackers. This brings us to the precise preconditions of the discussion moves.

$HTB(A)$ Either this is the first move, or the previous move was $CB(B)$, where A attacks B, and no $CONCEDE$ or $RETRACT$ move is applicable.

$CB(A)$ A is an attacker of the last $HTB(B)$ statement that is not yet conceded, the directly preceding move was not a CB statement, argument A has not yet been retracted, and no $CONCEDE$ or $RETRACT$ move is applicable.

$CONCEDE(A)$ There has been an $HTB(A)$ statement in the past, of which every attacker has been retracted, and $CONCEDE(A)$ has not yet been moved.

$RETRACT(A)$ There has been a $CB(A)$ statement in the past, of which there exists an attacker that has been conceded, and $RETRACT(A)$ has not yet been moved.

Apart from the preconditions mentioned above, all four statements also have the additional precondition that no HTB-CB repeats have occurred. That is, there should be no argument for which HTB has been uttered more than once, CB has been uttered more than once, or both HTB and CB have been uttered. In the first and second case, the discussion is going around in circles, which the proponent has to prevent as he has burden of proof. In the third case, the proponent has been contradicting himself, as his statements are not conflict-free. In each of these three cases, the discussion comes to an end with no move being applicable anymore. The above conditions are made formal as follows.

Definition 15 *Let $AF = (Ar, att)$ be an argumentation framework. A grounded discussion is a sequence of discussion moves constructed by applying the following principles.*

BASIS (HTB) *If $A \in Ar$ then $[HTB(A)]$ is a grounded discussion.*

STEP (HTB) *If $[M_1, \ldots, M_n]$ $(n \geq 1)$ is a grounded discussion without HTB-CB repeats,[23] and no $CONCEDE$ or $RETRACT$ move is applicable,[24] and $M_n = CB(A)$ and B is an attacker of A then $[M_1, \ldots, M_n, HTB(B)]$ is also a grounded discussion.*

STEP (CB) *If $[M_1, \ldots, M_n]$ $(n \geq 1)$ is a grounded discussion without HTB-CB repeats, and no $CONCEDE$ or $RETRACT$ move is applicable, and M_n is not a CB move, and there is a move $M_i = HTB(A)$ $(i \in \{1 \ldots n\})$ such that the discussion does not contain $CONCEDE(A)$, and for each move $M_j = HTB(A')$ $(j > i)$ the discussion contains a move $CONCEDE(A')$, and B is an attacker of A such that the discussion does not contain a move $RETRACT(B)$, then $[M_1, \ldots, M_n, CB(B)]$ is a grounded discussion.*

STEP $(CONCEDE)$ *If $[M_1, \ldots, M_n]$ $(n \geq 1)$ is a grounded discussion without HTB-CB repeats, and $CONCEDE(B)$ is applicable then $[M_1, \ldots, M_n, CONCEDE(B)]$ is a grounded discussion.*

STEP $(RETRACT)$ *If $[M_1, \ldots, M_n]$ $(n \geq 1)$ is a grounded discussion without HTB-CB repeats, and $RETRACT(B)$ is applicable then $[M_1, \ldots, M_n, RETRACT(B)]$ is a grounded discussion.*

It can be observed that the preconditions of the moves are such that a proponent move (HTB) can never be applicable at the same moment as an opponent move

---

[23] We say that there is a HTB-CB repeat iff $\exists i, j \in \{1 \ldots n\} \exists A \in Ar : (M_i = HTB(A) \lor M_i = CB(A)) \land (M_j = HTB(A) \lor M_j = CB(A)) \land i \neq j$.

[24] A move $CONCEDE(B)$ is applicable iff the discussion contains a move $HTB(A)$ and for every attacker A of B the discussion contains a move $RETRACT(B)$, and the discussion does not already contain a move $CONCEDE(B)$. A move $RETRACT(B)$ is applicable iff the discussion contains a move $CB(B)$ and there is an attacker A of B such that the discussion contains a move $CONCEDE(A)$, and the discussion does not already contain a move $RETRACT(B)$.

(CB, $CONCEDE$ or $RETRACT$). That is, proponent and opponent essentially take turns in which each proponent turn consists of a single HTB statement, and every opponent turn consists of a sequence of $CONCEDE$, $RETRACT$ and CB moves.

Definition 16 *A grounded discussion* $[M_1, \ldots, M_n]$ *is called* terminated *iff there exists no move* M_{n+1} *such that* $[M_1, \ldots, M_n, M_{n+1}]$ *is a grounded discussion. A terminated grounded discussion (with A being the main argument) is won by the proponent iff the discussion contains* $CONCEDE(A)$, *otherwise it is won by the opponent.*

To illustrate why the discussion has to be terminated after the occurrence of an HTB-CB repeat, consider the following discussion in the argumentation framework of Figure 1.

(1) P: $HTB(G)$ (3) P: $HTB(G)$
(2) O: $CB(H)$

At the third move, an HTB-CB repeat occurs and the discussion is terminated (opponent wins). Hence, termination after an HTB-CB repeat is necessary to prevent the discussion from going on perpetually.

Theorem 5 *Every discussion will terminate after a finite number of steps.*

From the fact that a discussion terminates after an HTB-CB repeat, the following result follows.

Lemma 1 *No discussion can contain a CONCEDE and RETRACT move for the same argument.*

The soundness and completeness of the game described above is stated in the following theorem.

Theorem 6 ([Caminada, 2015a]) *Let* (Ar, att) *be an argumentation framework and let* $A \in Ar$.

1. *If there exists a grounded discussion for A that is won by player P, then A is labelled* in *by the grounded labelling.*

2. *If A is labelled* in *by the grounded labelling, then player P has a winning strategy for A in the grounded discussion game.*

The correctness of Theorem 6 can be seen as follows. As for point 1, it can be observed that what the discussion game actually does is to construct a strongly admissible labelling of which the in-labelled arguments coincide with the $CONCEDE$ moves, and the out-labelled arguments coincide with the $RETRACT$ moves. In fact, it can be proved by induction that at each state of the discussion, the labelling where each $CONCEDE$ move is labelled in and each retract move is labelled out is strongly admissible [Caminada, 2015b]. The fact that the discussion is won by player P implies that the main argument (A) has been conceded. So at the end of the discussion, we have a strongly admissible labelling where argument A is labelled in. Hence, by Theorem 2, A is labelled in by the grounded labelling.

As for point 2, it should be mentioned that a strongly admissible labelling (for instance the grounded labelling) with its associated min-max numbering can serve as a roadmap for winning the discussion. The proponent will be able to win if, whenever he has to do an HTB move, he prefers to use an in argument with the lowest min-max number that attacks the directly preceding CB move. We refer to this as a *lowest number strategy*.[25]

It turns out that when applying such a strategy, the game stays within the boundaries of the strongly admissible labelling (that is, within its in and out labelled part). As long as each HTB move of the proponent is related to an in-labelled argument, it follows that all the attackers are labelled out (Definition 8, first bullet) so each CB move the opponent utters in response will be related to an out-labelled argument. This out-labelled argument will then have at least one in-labelled attacker (Definition 8, second bullet) as a candidate for the proponent's subsequent HTB move.

The next thing to be observed is that when the proponent applies a lowest number strategy, the game will not terminate due to any HTB-CB repeats. This is due to the facts that (1) after a move $HTB(A)$ is played (for some argument A) all subsequent CB and HTB moves will be related to arguments with lower min-max numbers than A until a move $CONCEDE(A)$ is played, and (2) after a move $CB(A)$ is played (for some argument A), all subsequent HTB and CB moves will be related to arguments with lower min-max numbers than A, until a move $RETRACT(A)$ is played. We refer to [Caminada, 2015b] for details.

7 Tree-Based Discussion Games

The discussion games that were described in the previous sections are not the only ones that have been stated for preferred, stable, ideal and grounded semantics. In fact, various alternative dialectical proof procedures can be found in the literature, many of them are based on the concept of dialectical trees [Dung et al., 2007; Modgil and Caminada, 2009; Thang et al., 2009]. In the current section, we aim to provide an impression of these tree-based discussion games, and explain some of their disadvantages compared to the discussion games described in the previous sections. Rather than giving an overview of all tree-based discussion games that have been stated in the literature, we will focus our attention on one of them: the Standard Grounded Game [Prakken and Sartor, 1997; Caminada, 2004; Modgil and Caminada, 2009].

---

[25] We write "*a* lowest number strategy" instead of "*the* lowest number strategy" as a lowest number strategy might not be unique due to different lowest numbered in-labelled arguments being applicable at a specific point. In that case it is sufficient to pick an arbitrary one.

The Standard Grounded Game (SGG) [Prakken and Sartor, 1997; Caminada, 2004; Modgil and Caminada, 2009] is one of the earliest dialectical proof procedures for grounded semantics. Each game[26] consists of a sequence $[A_1, \ldots, A_n]$ $(n \geq 1)$ of arguments, moved by the proponent and opponent taking turns, with the proponent starting. That is, a move A_i $(i \in \{1 \ldots n\})$ is a proponent move iff i is odd, and an opponent move iff i is even. Each move, except the first one, is an attacker of the previous move. In order to ensure termination even in the presence of cycles, the proponent is not allowed to repeat any of his moves. A game is terminated iff no next move is possible; the player making the last move wins. Formally, the Standard Grounded Game can be defined as follows.

Definition 17 *A* discussion *in the Standard Grounded Game is a finite sequence* $[A_1, \ldots, A_n]$ $(n \geq 1)$ *of arguments (sometimes called* moves*), of which the odd moves are called P-moves (Proponent moves) and the even moves are called O-moves (Opponent moves), such that:*

1. *every O-move is an attacker of the preceding P-move (that is, every A_i where i is even and $2 \leq i \leq n$ attacks A_{i-1})*

2. *every P-move except the first one is an attacker of the preceding O-move (that is, every A_i where i is odd and $3 \leq i \leq n$ attacks A_{i-1})*

3. *P-moves are not repeated (that is, for every odd $i, j \in \{1, \ldots, n\}$ it holds that if $i \neq j$ then $A_i \neq A_j$)*

A discussion is called terminated *iff there is no A_{n+1} such that $[A_1, \ldots, A_n, A_{n+1}]$ is a discussion. A terminated discussion is said to be* won *by the player making the last move.*

As an example, in the argumentation framework of Figure 1 $[C, B, A]$ is terminated and won by the proponent (as A has no attackers, the opponent cannot move anymore) whereas $[G, H]$ is terminated and won by the opponent (as the only attacker of H is G, which the proponent is not allowed to repeat). It is sometimes possible for the proponent to win a game even if the main argument is not in the grounded extension. An example would be $[F, B, A]$. This illustrates that in order to show that an argument is in the grounded extension, a single game won by the proponent is not sufficient. Instead, what is needed is a *winning strategy*. This is essentially a tree in which each node is associated with an argument such that (1) each path from the root to a leaf constitutes a terminated discussion won by the proponent, (2) the children of each proponent node (a node corresponding with a proponent move) coincide with all attackers of the associated argument, and (3) each opponent node (a node corresponding with an opponent move) has precisely one child, whose argument attacks the argument of the opponent node.

---

[26] What we call an SGG game is called a "line of dispute" in [Modgil and Caminada, 2009].

Formally, *argument tree* is a tree of which each node (**n**) is labelled with an argument ($Arg(\mathbf{n})$). The *level* of a node is the number of nodes in the path to the root. This leads to the following formal definition of a winning strategy in the context of the Standard Grounded Game.

Definition 18 *A winning strategy of the Standard Grounded Game for argument A is an argument tree, where the root is labelled with A, such that*

1. *for each path from the root ($\mathbf{n}_{root}$) to a leaf node ($\mathbf{n}_{leaf}$) it holds that the arguments on this path form a terminated discussion won by P*

2. *for each node at odd level $\mathbf{n}_P$ it holds that $\{Arg(\mathbf{n}_{child}) \mid \mathbf{n}_{child}$ is a child of $\mathbf{n}_P\} = \{B \mid B$ attacks $Arg(\mathbf{n}_P)\}$ and the number of children of $\mathbf{n}_P$ is equal to the number of attackers of $Arg(\mathbf{n}_P)$*

3. *each node of even level $\mathbf{n}_O$ has precisely one child $\mathbf{n}_{child}$, and $Arg(\mathbf{n}_{child})$ attacks $Arg(\mathbf{n}_O)$*

It has been proved that an argument is in the grounded extension iff the proponent has a winning strategy for it in the SGG [Prakken and Sartor, 1997; Caminada, 2004]. Moreover, it has also been shown that an SGG winning strategy defines a strongly admissible labelling, when each argument of a proponent node is labelled **in**, each argument of an opponent node is labelled **out** and all remaining arguments are labelled **undec** [Caminada, 2014].

As an example, in the argumentation framework of Figure 1 the winning strategy for argument E would be the tree consisting of the two branches $E-B-A$ and $E-D-C-B-A$, thus proving its membership of the grounded extension by yielding the strongly admissible labelling ($\{A, C, E\}, \{B, D\}, \{F, G, H\}$).

As can be observed from this example, a winning strategy of the SGG can contain some redundancy when it comes to multiple occurrences of the same arguments in different branches. In the current example, the redundancy is relatively mild (consisting of just the two arguments A and B) but other cases exist where the SGG requires a number of moves in the winning strategy that is *exponential* w.r.t. the size of the strongly admissible labelling the winning strategy is defining. As an example, consider the argumentation framework of Figure 5 (top left). The winning strategy of the SGG is in the same figure (top right). Now consider what would happen if one would start to extend the argumentation framework by duplicating the middle part. That is, suppose we have arguments $B_1, \ldots, B_n$ and $C_1, \ldots, C_n$ (with n being an odd number), as well as arguments A and D. Suppose that for every $i \in \{1, \ldots, n-1\}$ B_{i+1} attacks B_i, and C_{i+1} attacks C_i, and that for each even $i \in \{2, \ldots n-1\}$ B_{i+1} attacks C_i, and C_{i+1} attacks B_i, and that B_1 and C_1 attack A, and that D attacks B_n and C_n. In that case, the branches in the SGG winning strategy would split at every O-move. So for $n = 3$ (as is the case in Figure 5) the number of branches is four, for $n = 5$ it is eight, etc. In general, the number of

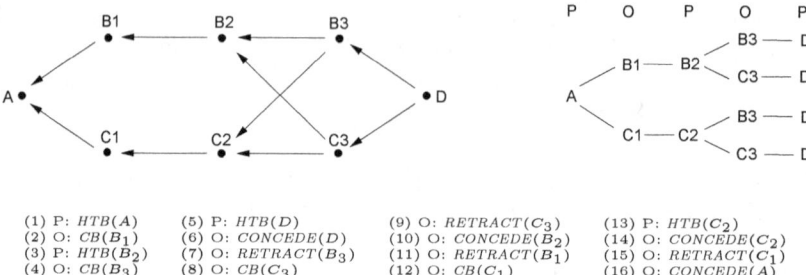

Figure 5. The Standard Grounded Game (SGG) versus the Grounded Discussion Game (GDG).

branches in the SGG winning strategy is $2^{(n+1)/2}$, with the number of nodes in the SGG winning strategy being $1 + 2\Sigma_{i=1}^{(n+1)/2} 2^i$. Hence, the number of steps needed in a winning strategy of the SGG can be *exponential* in relation to the size (number of in and out labelled arguments) of the strongly admissible labelling that the SGG winning strategy is constructing.[27]

As for the Grounded Discussion Game (GDG) as described in Section 6, the situation is different. As was mentioned in Section 6, what the GDG essentially does is to construct a strongly admissible labelling of which the in labelled arguments coincide with the *CONCEDE* moves and the out labelled arguments coincide with the *RETRACT* moves. It can be observed that no argument occurs in both a *CONCEDE* and *RETRACT* move (otherwise the argument would also have occurred in both an *HTB* and *CB* move, and the discussion would have terminated before reaching the *CONCEDE* and *RETRACT* moves) and that for each argument there exists at most one *CONCEDE* move and at most one *RETRACT* move. As we assume the game is won by the proponent, who is playing a lowest number strategy, there will be no *HTB-CB* repeats. This implies that for each *CONCEDE* move, there exists precisely one *HTB* move, and for each *RETRACT* move, there exists precisely one *CB* move. This means that the total number of moves (in a game won by the proponent, who is applying a lowest number strategy) is two times the number of in labelled arguments (which accounts for the *HTB* and *CONCEDE* moves) plus two times the number of out labelled arguments (which accounts for the *CB* and *RETRACT* moves). Hence, the number of moves in the game is *linear* in relation to the size (number of in and out labelled arguments) of the strongly admissible labelling the GDG is constructing.[28]

Hence, whereas for the Grounded Discussion Game, constructing a strongly admissible labelling (which is needed to show membership of the grounded extension) requires a linear number of moves, for the Standard Grounded Game this requires a potentially exponential number of moves. This makes the GDG

---

[27]We thank Mikołaj Podlaszewski for this example.
[28]See [Caminada, 2015a] for details.

a better choice for purposes of human-computer interaction, assuming that the human user's time is precious.

It should be mentioned that the possibility of an exponential blowup in the number of moves is not restricted to the SGG, but is a feature of tree-based discussion games in general. For instance, the above sketched example also leads to an exponential number of moves in the preferred semantics game of Modgil and Caminada [2009] and in the ideal semantics game of Dung et al. [2007]. The key feature of these approaches is that they require a winning strategy to show membership of a (grounded, preferred or ideal) extension. It is this winning strategy that is responsible for the exponential blowup. In the discussion games described in sections 3, 5 and 6, however, no winning strategy is required, as just a single game won by the proponent is sufficient to prove membership of a (preferred, stable or grounded) extension.[29]

8 Discussion

What the above described discussion games for preferred semantics (Section 3), stable semantics (Section 5) and grounded semantics (Section 6) have in common is that (1) a single game won by the proponent is sufficient to prove membership of a (preferred, stable or grounded) extension, and (2) if an argument is member of a (preferred, stable or grounded) extension then the proponent has a winning strategy for it. This is evidenced by theorems 3, 4 and 6. In tree-based discussion games, like those of Dung et al. [2007], Modgil and Caminada [2009] and Thang et al. [2009] point (1) is altered such that a single game won by the proponent is *not* sufficient to prove membership of an extension; for this a winning strategy is needed. Having to provide such a winning strategy in a dialectical way can be troublesome for two reasons. First of all, the tree of the winning strategy would need to be "linearized" as discussions take place not in branching time but in linear time. But even if linearization takes place, one still has to deal with the fact that the original (tree-based) winning strategy could have a size that is exponentially related to the (strongly) admissible labelling it is based on. The discussion games presented in sections 3, 5 and 6 have the advantage that they are not tree-based and hence do not have these problems.

One can ask the question whether it is always possible (for any argumentation semantics) to define a discussion game that satisfies the points (1) and (2) mentioned above. For instance, the procedure sketched in Section 4 (ideal semantics) does not satisfy point (1). This is because in the second phase of the discussion, when trying to find an admissible set that attacks the admissible set obtained in the first phase of the discussion, not finding such a set could be due to the proponent making the "wrong" choices during the second phase, rather than due to the actual absence of such a set. It would be a challenge

---

[29]It can be proved that the preferred discussion game (Section 3) is linear in the number of moves required. See [Caminada et al., 2014] for details. Using similar techniques one can also prove that the stable discussion game (Section 5) requires only a linear number of moves.

to change the discussion procedure for ideal semantics such that both points (1) and (2) are satisfied. An even greater challenge would be to formulate discussion games (still satisfying points (1) and (2)) for semi-stable, stage or even CF2 semantics.

As the tree-based discussion games of Dung et al. [2007], Modgil and Caminada [2009] and Thang et al. [2009] violate point (1) but satisfies point (2), one can ask the question of whether there also exists a discussion game that satisfies point (1) but violates point (2). The answer is affirmative, as is evidenced by the work of Caminada and Podlaszewski [2012a; 2012b]. Here, the ability to win the discussion game might depend on cooperation of the opponent. So even though an argument being in the grounded extension implies the existence of a discussion for it that is won by the proponent, it does not imply that the proponent also has a winning strategy.[30] For the purpose of human-computer interaction, this property is undesirable, as the computer should be able to win the discussion (for an argument that is actually in the grounded extension) regardless of how the human user choses to utter the possible counterarguments.

The discussion games presented in the current chapter have been stated in the context of *abstract* argumentation theory. This raises the question of whether these discussion games are also suitable in the context of *instantiated* argumentation, like ASPIC+ [Modgil and Prakken, 2014][31], ABA [Toni, 2014][32] or logic-based argumentation [Gorogiannis and Hunter, 2011][33]. Technically, this should not be a problem, as each of these formalisms provides an instantiation of Dung's abstract argumentation theory. That is, each of these formalisms specifies what arguments can be constructed and how these attack each other, starting from a particular knowledge base. Although applying the discussion games in the context of instantiated argumentation is technically straightforward, there is a catch. The question is whether the notion of attack of the instantiated argumentation formalism is defined in such a way that it allows for moves that can be considered as intuitive during the course of the discussion. For instance, in ASPIC+ it can be the case that a discussion partner utters an argument with conclusion c, which cannot be replied to with an argument for conclusion $\neg c$ (even though such an argument is well-formed and perhaps even justified) because the definition of attack is such that it does not attack the argument with conclusion c. This is like having your discussion partner uttering an argument for a claim (c) which you know is not the case, but you're not allowed to reply with an argument that directly rebuts this claim. We refer to Chapter 15 ("Rationality postulates: applying argumentation theory for non-monotonic reasoning") of this volume for details.

As mentioned in the introduction, one of the possible applications of the

---

[30]We refer to [Caminada, 2015a] for a specific example.

[31]See also Chapter 6 ("Abstract rule-based argumentation") of this volume.

[32]See also Chapter 7 ("assumption-based argumentation: disputes, explanations, preferences") of this volume.

[33]See also Chapter 9 ("A review of argumentation based on deductive arguments") of this volume.

discussion games is for the purpose of human-computer interaction. The context here is that of a shared knowledge base[34] (say, of medical research and clinical evidence) that allows for the construction of arguments (say, regarding to how to treat a particular patient). As the knowledge base can be complex and huge, it is not always directly obvious what the justified arguments are. Although a software implementation of (instantiated) argumentation theory can help to provide an answer, the correctness of this answer might need to be explained to a human user. Our hypothesis is that human-computer discussion can contribute to acceptance of argument-based entailment. In order to test this hypothesis, one would need to perform experiments in which the user's confidence in the argument-based entailment is tested, before and after performing the discussion game. Experiments like these is what we would like to perform in the near future.

BIBLIOGRAPHY

[Baroni and Giacomin, 2007] P. Baroni and M. Giacomin. On principle-based evaluation of extension-based argumentation semantics. *Artificial Intelligence*, 171(10-15):675–700, 2007.

[Baroni et al., 2011] P. Baroni, M.W.A. Caminada, and M. Giacomin. An introduction to argumentation semantics. *Knowledge Engineering Review*, 26(4):365–410, 2011.

[Caminada and Gabbay, 2009] M.W.A. Caminada and D.M. Gabbay. A logical account of formal argumentation. *Studia Logica*, 93(2-3):109–145, 2009. Special issue: new ideas in argumentation theory.

[Caminada and Podlaszewski, 2012a] M.W.A. Caminada and M. Podlaszewski. Grounded semantics as persuasion dialogue. In Bart Verheij, Stefan Szeider, and Stefan Woltran, editors, *Computational Models of Argument - Proceedings of COMMA 2012*, pages 478–485, 2012.

[Caminada and Podlaszewski, 2012b] M.W.A. Caminada and M. Podlaszewski. User-computer persuasion dialogue for grounded semantics. In Jos W.H.M. Uiterwijk, Nico Roos, and Mark H.M. Winands, editors, *Proceedings of BNAIC 2012; The 24th Benelux Conference on Artificial Intelligence*, pages 343–344, 2012.

[Caminada and Sakama, 2015] M.W.A. Caminada and Ch. Sakama. On the issue of argumentation and informedness. In *2nd International Workshop on Argument for Agreement and Assurance*, 2015.

[Caminada and Wu, 2009] M.W.A. Caminada and Y. Wu. An argument game of stable semantics. *Logic Journal of IGPL*, 17(1):77–90, 2009.

[Caminada et al., 2014] M.W.A. Caminada, W. Dvořák, and S. Vesic. Preferred semantics as socratic discussion. *Journal of Logic and Computation*, 2014. (in print).

[Caminada, 2004] M.W.A. Caminada. *For the sake of the Argument. Explorations into argument-based reasoning*. PhD thesis, Vrije Universiteit Amsterdam, 2004.

[Caminada, 2006] M.W.A. Caminada. On the issue of reinstatement in argumentation. In M. Fischer, W. van der Hoek, B. Konev, and A. Lisitsa, editors, *Logics in Artificial Intelligence; 10th European Conference, JELIA 2006*, pages 111–123. Springer, 2006. LNAI 4160.

[Caminada, 2008] M.W.A. Caminada. A formal account of socratic-style argumentation. *Journal of Applied Logic*, 6(1):109–132, 2008.

[Caminada, 2011] M.W.A. Caminada. A labelling approach for ideal and stage semantics. *Argument & Computation*, 2:1–21, 2011.

---

[34] A particularly interesting situation is where such a shared knowledge base is absent, that is, where proponent and opponent each have their own private knowledge base and associated argumentation framework. In that case, both proponent and opponent learn new information from each other during the course of the discussion. This puts additional constraints on the discussion protocol. We refer to [Caminada and Sakama, 2015] for details.

[Caminada, 2014] M.W.A. Caminada. Strong admissibility revisited. In Simon Parsons, Nir Oren, Chris Reed, and Frederico Cerutti, editors, *Computational Models of Argument; Proceedings of COMMA 2014*, pages 197–208. IOS Press, 2014.

[Caminada, 2015a] M.W.A. Caminada. A discussion game for grounded semantics. In Elizabeth Black, Sanjay Modgil, and Nir Oren, editors, *Theory and Applications of Formal Argumentation (proceedings TAFA 2015)*, pages 59–73. Springer, 2015.

[Caminada, 2015b] M.W.A. Caminada. A discussion protocol for grounded semantics (proofs). Technical report, University of Aberdeen, 2015.

[Doutre and Mengin, 2004] S. Doutre and J. Mengin. On sceptical versus credulous acceptance for abstract argument systems. In *Proceedings of the 9th European Conference on Logics in Artificial Intelligence (JELIA-2004)*, pages 462–473, 2004.

[Dung and Thang, 2007] P.M. Dung and P.M. Thang. A sound and complete dialectical proof procedure for sceptical preferred argumentation. In *Proc. of the LPNMR Workshop on Argumentation and Nonmonotonic Reasoning (ArgNMR07)*, pages 49–63, 2007.

[Dung et al., 2007] P. M. Dung, P. Mancarella, and F. Toni. Computing ideal sceptical argumentation. *Artificial Intelligence*, 171(10-15):642–674, 2007.

[Gorogiannis and Hunter, 2011] N. Gorogiannis and A. Hunter. Instantiating abstract argumentation with classical logic arguments: Postulates and properties. *Artificial Intelligence*, 175(9-10):1479–1497, 2011.

[Modgil and Caminada, 2009] S. Modgil and M.W.A. Caminada. Proof theories and algorithms for abstract argumentation frameworks. In I. Rahwan and G.R. Simari, editors, *Argumentation in Artificial Intelligence*, pages 105–129. Springer, 2009.

[Modgil and Prakken, 2014] S. Modgil and H. Prakken. The ASPIC+ framework for structured argumentation: a tutorial. *Argument & Computation*, 5:31–62, 2014. Special Issue: Tutorials on Structured Argumentation.

[Prakken and Sartor, 1997] H. Prakken and G. Sartor. Argument-based extended logic programming with defeasible priorities. *Journal of Applied Non-Classical Logics*, 7:25–75, 1997.

[Thang et al., 2009] Phan Minh Thang, Phan Minh Dung, and Nguyen Duy Hung. Towards a common framework for dialectical proof procedures in abstract argumentation. *Journal of Logic and Computation*, 19(6):1071–1109, 2009.

[Toni, 2014] F. Toni. A tutorial on assumption-based argumentation. *Argument & Computation*, 5:89–117, 2014. Special Issue: Tutorials on Structured Argumentation.

[Vreeswijk and Prakken, 2000] G.A.W. Vreeswijk and H. Prakken. Credulous and sceptical argument games for preferred semantics. In *Proceedings of the 7th European Workshop on Logic for Artificial Intelligence (JELIA-00)*, number 1919 in Springer Lecture Notes in AI, pages 239–253, Berlin, 2000. Springer Verlag.

[Walton and Krabbe, 1995] D. N. Walton and E. C. W. Krabbe. *Commitment in Dialogue: Basic Concepts of Interpersonal Reasoning*. SUNY Series in Logic and Language. State University of New York Press, Albany, NY, USA, 1995.

Martin Caminada
Cardiff University
Email: CaminadaM@cardiff.ac.uk

11
Argumentation Schemes
FABRIZIO MACAGNO, DOUGLAS WALTON, CHRIS REED

ABSTRACT. The purpose of this chapter is threefold: 1) to describe the schemes, showing how they evolved and how they have been classified in the traditional and the modern theories; 2) to propose a method for classifying them based on ancient and modern developments; and 3) to outline and show how schemes can be used to describe and analyze or produce real arguments. To this purpose, we will build on traditional distinctions for building dichotomic classifications of schemes, and we will advance a modular approach to argument analysis, in which different argumentation schemes are combined together in order to represent each step of reasoning on which a complex argument relies. Finally, we will show how schemes are applied to formal systems, focusing on their applications to Artificial Intelligence, AI & Law, argument mining, and formal ontologies.

1 Introduction

The purpose of this chapter is threefold: 1) to describe the schemes, showing how they evolved and how they have been classified in the traditional and the modern theories; 2) to propose a method for classifying them based on ancient and modern developments; and 3) to outline and show how schemes are interrelated and can be organized in a modular way to describe natural arguments or produce complex arguments. Historically, the schemes evolved from the Aristotelian topics, the so-called places to find arguments. But looking over the descriptions Aristotle presented of them in the *Topics*, for the most part they do not appear to very much resemble the argumentation schemes in the contemporary list of Walton, Reed and Macagno [Walton et al., 2008]. Of course there are exceptions, such as the topic for argument from analogy described in Aristotle, which is recognizable as standing for the same kind of argument as the current scheme for argument from analogy, even though the detailed description of it is quite different.

Argumentation schemes are instruments for argumentation, involving the activity of critically evaluating a viewpoint and the reasons given in its support. For this reason, every scheme has a corresponding set of critical questions, representing its defeasibility conditions and the possible weak points that the interlocutor can use to question the argument and evaluate its strength. A critic who has no counterarguments ready to hand can search through the list of critical questions matching the argument he is confronted with in order to look for clues on how the argument can be attacked that might suggest sources of evidence that could be used to build up a whole line of argumentation that furnishes a way of refuting the argument.

The fundamental challenge that a theory of argumentation schemes needs to face is the problem of finding a useful and sound classification system. The schemes need to be usable, easily identifiable, and at the same time they need to allow the user to detect the most specific pattern of argument that can fit the text or that can be employed for producing an argument suitable to the circumstances and the purpose. In any classification system, entities can be classified in many different ways, depending on the purpose of the classification. The purpose of the classification system will determine the criteria for classification that are adopted in that system. For example, a much more detailed classification of animals may be useful in biology than the kind of classification that might be useful for law, or for classifying animals as they are spoken and written about in everyday conversational English. We need to begin by specifying the purpose of the classification, so that some guidance can be given on how to identify the criteria used in the classification system. From this perspective it is useful to examine how the study of argumentation schemes evolved.

2 Introducing argumentation schemes

Argumentation schemes represent forms of argument that are widely used in everyday conversational argumentation, and in other contexts such as legal and scientific argumentation. But for the most part these arguments are not adequately modeled by deductive forms of reasoning of the kind familiar in classical logic or as statistical inferences based on the standard Bayesian account of probability. They represent the premise-conclusion structure of an argument, and they are defeasible. Their defeasibility conditions are shown as a set of critical questions, dialectical instruments to help begin the procedure of testing the strength and acceptability of an argument by weighing the pro and con arguments.

2.1 Nature of the schemes

Argumentation schemes are stereotypical patterns of inference, combining semantic-ontological relations with types of reasoning and logical axioms and representing the abstract structure of the most common types of natural arguments [Macagno and Walton, 2015]. The argumentation schemes provided in [Walton et al., 2008] describe the patterns of the most typical arguments, without drawing distinctions between material relations (namely relations between concepts expressed by the warrant of an argument), types of reasoning (such as induction, deduction, abduction), and logical rules of inference characterizing the various types of reasoning (such as *modus ponens*, *modus tollens*, etc.). For this reason, argumentation schemes fall into distinct patterns of reasoning such as abductive, analogical, or inductive ones, and ones from classification or cause to effect.

In order to design a system for classifying the schemes, it is useful to understand their limits, and investigate how the dimensions of an argument (material relation and logical form) are merged. For example, consider argument from cause to effect [Walton et al., 2008, p.328]:

Table 1. Argument from cause to effect

Major premise	Generally, if A occurs, then B will (might) occur.
Minor premise	In this case, A occurs (might occur).
Conclusion	Therefore, in this case B will (might) occur.

This argumentation scheme is based on a defeasible *modus ponens* scheme [Verheij, 2003a] which is combined with a semantic causal relation between two events. The material (semantic) relation is merged with the logical one. However, this combination represents only one of the possible types of inferences that can be drawn from the same semantic-ontological connection. The actual relationship between the material and the logical relation is much more complex. For example, we consider the classic Aristotelian causal link between "having fever" and "breathing fast," and see how this cause-effect relation can be used to draw a conclusion based on different logical rules ([Macagno and Walton, 2015]; [Macagno, 2015]):

1. He had fever. (Fever causes breathing fast). Therefore, he (must have) breathed fast.

2. He did not breathe fast. (Fever causes breathing fast). Therefore, he had no fever.

3. He is breathing fast. (Fever causes breathing fast). Therefore, he might have fever.

4. He has no fever. (Fever causes breathing fast). Therefore, he may be not breathing fast.

5. You may have fever. When I had fever, I was breathing fast, and you are breathing fast.

Cases (1) and (2) proceed logically from defeasible deductive axioms, i.e. the defeasible *modus ponens* (in 1), and the defeasible *modus tollens* (in 2). Cases 3 and 4 proceed from abductive reasoning. In (3) the conclusion is drawn by affirming the consequent, while in (4) the denial of the antecedent can be rephrased by contraposition as "not breathing fast is caused by having no fever," leading to a conclusion drawn abductively [Walton et al., 2008, pp.169–173]. In (5) the conclusion is based on an inductive generalization from one single case.

Schemes represent only the prototypical matching between semantic relations and logical rules (types of reasoning and axioms). This matching is, however, only the most common one. The material and the logical relations can combine in several different ways. Hence this distinction needs to be taken into account order to classify the schemes.

2.2 Why schemes are important

Critics often ask how these schemes can be justified, given that they resisted analysis as deductive or inductive forms of argument of the kind recognized as valid in the dominant 20th-century logic tradition [Walton and Sartor, 2013].

Schemes are becoming extremely important for practical reasons. First, argumentation schemes are instruments for analyzing and recognizing natural arguments occurring in ordinary and specialized discourse. For example, arguments from political discourse have been analyzed using the schemes, and the argumentative profiles of the candidates have been brought to light considering their preferences of the types of arguments used (Hansen & Walton, 2013). Thousands of real examples of these forms of argument have been analyzed in the argumentation literature, such as the considerable literature on fallacies, with the aid of tools like argument mapping ([Reed et al., 2007][Rowe et al., 2006]). On this basis, the structure, use, and importance of schemes for argumentation studies have been justified inductively. This method consists in the following steps:

1. The structure of a scheme is outlined considering the literature on the topic.

2. A significant mass of examples of arguments is analyzed using the scheme, adapting and modifying the scheme so that it can best describe the specific natural arguments.

3. It is shown that the form of argument represented by the scheme under analysis is significantly important for the study of argumentation as it occurs in natural language discourse (and other specialized contexts such as legal discourse)

4. Empirical justification is given that this form of argument needs to be recognized as a basic scheme for argumentation.

Second, schemes are instruments that can be used for the purpose of teaching critical thinking. Informal logic is a field is known for having grown from its origins in textbooks that departed from formal logic and instead proceeded on the basis of analyzing numerous examples of arguments from ordinary discourse, such as those taken from magazines and newspapers. There is an abundance of such textbooks full of examples of everyday arguments related to topics such as the informal fallacy of appeal to authority, false cause, and so forth. During its growth stage and subsequent theoretical flowering, the field followed this trend by stressing the importance of analyzing real arguments "on the hoof". For example, the handbook *Informal Logic* [Walton, 1989] was based on 150 key examples, many of them illustrating forms of argument now identified with argumentation schemes, including personal attack, uses and abuses of expert opinion, arguments from analogy, arguments from correlation to cause, and so forth. These textbooks and continued academic writings on informal logic contained a very large number of such examples, often analyzed in minute detail. Argumentation schemes, such as argument from expert opinion, are tested against the real examples, to discuss the respects in which the abstract scheme fits or does not fit the vagaries of the real-life example. This body of data confirms that certain types of arguments, mainly the ones subsequently identified as argumentation schemes, are not only extremely common, but are also highly influential in daily practices of argumentation.

Third, schemes can be used in education both for teaching students how to argue and for learning through argumentation ([Erduran and Jimenez-Aleixandre, 2007]; [Erduran and Jiménez Aleixandre, 2012]; [Rapanta and Walton, 2016]). The interest in argumentation and the patterns for representing natural arguments is growing [Rapanta et al., 2013]. The argumentation schemes illustrated in ([Walton, 1995]; [Walton et al., 2008]) have been applied to science education in order to represent students' arguments and improve the quality thereof ([Rapanta and Macagno, 2016]), retrieve the implicit premises, and assess and rebut their reasoning in a systematic fashion ([Macagno and Konstantinidou, 2013]), or to assess the quality of argumentation ([Duschl et al., 1999]; [Ozdem et al., 2013]). However, a crucial problem arising out of the use of schemes in education is their differentiation ([Kim et al., 2010]; [Nussbaum and Edwards, 2011]). Students often fail to understand the differences between various types of arguments, and the recent developments in education tend to conflate the schemes instead of providing criteria for classifying or distinguishing between them.

Fourth, schemes have now been recognized as important for argument mining, and it has also been recognized that there are too many schemes for handy use [Mochales Palau and Moens, 2009]; [Mochales Palau and Moens, 2011]). Configuring the relationships between clusters of them, and the internal structure of each cluster, would help in the research efforts to apply the schemes as working tools to a broader range of problems as the field of computational linguistics has moved forward.

From a theoretical point of view, schemes fit into current formal argumentation models such as ASPIC+ ([Prakken et al., 2015]), DefLog ([Verheij, 2003a]) and the Carneades Argumentation System ([Walton and Gordon, 2012]). Among the basic schemes presented in the list of 60+ schemes in chapter 9 of ([Walton et al., 2008]) are argument from expert opinion, argument from sign, argument from example, argument from commitment, argument from position to know, argument from lack of knowledge, practical reasoning (argument from goal to action), argument from cause to effect, the sunk costs argument, argument from analogy, *ad hominem* argument, and the slippery slope argument. These schemes are at this point well enough recognized in the argumentation literature that no detailed account of them needs to be given in this chapter, except for the ones that we will focus on to illustrate general characteristics of schemes discussed in detail in the chapter.

Moreover, Walton and Sartor ([Walton and Sartor, 2013]) have shown that the basic defeasible schemes can be justified within a teleological argumentation framework. According to this reasoning, the use of a specific scheme is warranted by the fact that it can serve an agent's goals better than using nothing, and better than other alternative schemata the agent has at its disposal. This kind of justification of basic schemes is essentially a practical one saying that these schemes, even at their current state of development, are proving to be useful in such areas as artificial intelligence and multiagent computing. Defeasible schemes allow agents to arrive at a presumptive conclusion on how to proceed in a situation where continuing to collect evidence may cause delay, taking time and costing money.

This form of justification of schemes applies both to goals of epistemic cognition (getting to the truth of a matter) and goals of practical cognition (making the best choice in given circumstances). The importance of the schemes has also been acknowledged in the history of dialectics. The forms of argument, their critical and defeasible dimension, and their structure were long ago acknowledged in the earlier concerns of the Sophists, who pointed out forms of argument useful for persuasion and deliberation ([Schiappa, 1999]; [Tindale, 2010]). In the *Topics* [Aristotle, 1991b] and in the *Rhetoric* [Aristotle, 1991a], Aristotle set out a list of topics that, providing the abstract and general hypothetical premises of dialectical syllogisms, can be considered to be the predecessors of the argument patterns developed in modern times ([Macagno *et al.*, 2014]; [Rubinelli, 2009]; [Macagno *et al.*, 2014]).

The tradition of the topics was continued through the Middle Ages, with various theories aimed at providing a classification and an analysis of the nature of the schemes [Bird, 1962]; [Gabbay and Woods, 2008]; [Green-Pedersen, 1984]; [Green-Pedersen, 1987]; [Stump, 1982]; [Stump, 1989]). Study of the kinds of schemes that are the focus of this chapter was eclipsed during the Enlightenment, as the dominant view became firmly entrenched that the only forms of reasoning that can be identified with rational thinking are those of deductive logic, and inductive reasoning of the kind used in games of chance. But the study of these schemes made a comeback in the 20th century at the beginning of, and after the rise of argumentation studies as a respectable discipline, once the basic schemes were identified by Hastings ([Hastings, 1963]), Perelman and Olbrechts-Tyteca ([Perelman and Olbrechts-Tyteca, 1969]), Kienpointner ([Kienpointner, 1992]), Walton ([Walton, 1995]), Grennan ([Grennan, 1997]), and Walton, Reed and Macagno ([Walton *et al.*, 2008]). From that point onwards, the study of schemes has been recognized as important for building computational models of argumentation, and especially for applying these models to argumentation in natural language discourse.

2.3 Classification of the schemes: how to proceed

In this chapter, it is shown how the complex project of classifying schemes needs to proceed by matching a top-down approach with a bottom-up approach, and in particular that this bottom-up approach needs to begin by studying relationships between clusters of nested schemes. From a top-down approach, dichotomic criteria of classification need to be found, allowing the user to decide the scheme needed, both by direct identification and by exclusion. For this purpose, an overview of the existing classification systems developed in the tradition and in the recent theories can provide useful criteria. From a bottom-up approach, relationships within groups of schemes need to be studied, and then how one group fits with another can be studied. Walton [Walton, 2012] took a bottom-up approach that began with some examples at the ground level of cases where two schemes seem to apply to the same real example of an argument found in a text, leading to a difficulty of determining which scheme fits the argument. Working from there, we identify clusters of schemes that fit together, and then at the next step, we examine how these clusters can be fitted together. Once clusters of schemes are fitted together into larger groups, we can gradually learn how they fit into an overarching system.

3 The topics in the dialectical and rhetorical tradition

Argumentation schemes describe patterns from which specific arguments can be drawn. In this sense, they can be seen as the modern development of the traditional concept of *topos*, the conditional expressing a generic principle from which some of the specific premises warranting the conclusion in an argument can be drawn. The purpose of this section is to show how the ancient account of *topoi* and *loci* can be considered as the ground and the predecessor of the modern theory of schemes.

3.1 Aristotle

The idea of providing general principles of inference from which various arguments can be drawn was the aim of Aristotle's *Topics* and *Rhetoric*. The Aristotelian *topoi* can be conceived as principles [De Pater, 1965, pp.150–159] having often the form of "If P, then Q." The various semantic (material) relations between P and Q, or the "nature of the things which the terms of the argument represent or stand for" [Green-Pedersen, 1987, p.413], constitute the differences between the various *topoi*. For example, P and Q can be related by a relation of genus-species, *definiens-definiendum*, contraries, similarity, etc. The function of the *topoi* in the mechanism of argument production can be explained as follows [Slomkowski, 1997, p.45]:

> The enthymemes seem to be instances of *topoi*; or, expressed differently, enthymemes are arguments which are warranted by the principle expressed in the *topos*. Thus hypothetical syllogism would fall under a *topos* insofar as it falls under its major premiss in which the essence of the hypothetical syllogism is expressed.

Topoi can be considered as the external general rules of reasoning of an enthymeme, or the genera of the major premises of dialectical and rhetorical syllogisms. *Topoi* can work as rules, namely as the principle of inference guaranteeing the passage from an enthymematic premise to the conclusion. For example, we consider the following enthymeme [Slomkowski, 1997, p.51]:

Doing greater injustice is a greater evil.

From "what is more A is more B," you may infer: "A is B."

Doing injustice is an evil.

The *topos* can be also used as a general principle from which it is possible to draw the specific premises of a hypothetical syllogism ([Bird, 1960]; [Bird, 1962]; [Macagno et al., 2014]). For example, the same argument mentioned above can be completed by adding the major premise that is an instantiation (an axiom-instance) of the *topos* from the more [Slomkowski, 1997, p.53] (Table 2):

Table 2. *Topoi* as general principles of inference

General principle	If being more A is more B, then A is B.
Specific instantiation of the *topos* as a premise	If doing greater injustice (A) is a greater evil (B), then doing injustice (A) is an evil (B).
Minor premise	Doing greater injustice (A) is a greater evil (B).
Conclusion	Doing injustice (A) is an evil (B)

The aforementioned mechanism of specification (or instantiation) of the *topoi* brings to light a fundamental distinction that Aristotle draws between generic *topoi* and the *idia* (the specific topics) [Rubinelli, 2009, pp.5970]. While generic *topoi* are abstract and commonly shared conditionals under which specific premises can be found, the specific *topoi* represent premises warranting the conclusion ([De Pater, 1965, p.134]; [Stump, 1989, p.29]) that are accepted within specific disciplines, such as ethics, law, or medicine. For example, consider the following specific topic [Lawson, 1885, p.262]:

> Where a person does an act, he is presumed in so doing to have intended that the natural and legal consequences of his act shall result.

In specific domains of knowledge, specific *topoi* can be listed as instruments of invention, premises that can be used to construct arguments in support of typical conclusions.

Generic topics can be considered as abstractions from the specific ones, or more correctly, an abstraction from a large number of specific topics. They provide classes of both necessary and defeasible inferences ([Bird, 1960];[Bird, 1962];[Christensen, 1988];[Drehe, 2011];[Stump, 2004]). In the first class fall some maxims setting out definitional properties of meta-semantic concepts, i.e. concepts representing semantic relations between concepts, such as definition, genus, and property. For example the *locus* from definition, which establishes the convertibility between definition and *definiendum*, represents also the essential logical characteristic that a predicate needs to have in order be considered as a "discourse signifying what a thing is." Other *loci*, such as the ones based on analogy or the more and the less, are only defeasible, as they represent only commonly accepted relationships. In the *Topics* [Aristotle, 1991b], Aristotle focuses most of his analysis on the topics governing the meta-semantic relations between concepts, i.e. genus, property, definition, and accident. The Aristotelian account was developed in the Latin and medieval dialectical tradition, which developed classifications of the topics (called *loci*) based on the type of material relation they represent.

3.2 Cicero

Cicero [Cicero, 2003] reduced the Aristotelian list of *topoi* to 20 *loci* or maxims, grouping them in generic categories (differences) and dividing them in two broad classes, the intrinsic and the extrinsic topics [Stump, 1989]. While the first ones proceed directly from the subject matter at issue (for instance, its semantic properties), the external topics (the Aristotelian arguments from authority) support the conclusion through contextual elements (for instance, the source of the speech act expressing the claim) (Cicero, *Topica*, 8, 34). In between there are the topics that concern the relationship between a predicate and the other predicates of a linguistic system (for instance, its relations with its contraries or alternatives). We represent the topics of Cicero in Figure 1 below.

Intrinsic		Extrinsic
Directly from the subject matter	*From things somehow related to the subject matter*	
1. *definitio* • By material parts (whole-part definition) • By essential parts (genus-species definition) 2. *notatio* (etymological relation)	1. *Coniugata* (inflectional relations) 2. *Genus* (genus-species relation) 3. *Forma* (species-genus relation) 4. *Similitudo* (similarity relation) 5. *Differentia* (difference relation) 6. *Contraria* (4 types of opposite relation) 7. *Adiuncta* (relation of concomitance) 8. *Antecedentia* 9. *Consequentia* 10. *Repugnantia* (incompatibles) 11. *Efficentia* (cause-effect relation) 12. *Effecta* (effect-cause relation) 13. *Ex comparatione maiorum, minorum, parium* (comparison)	Authority

Figure 1. Cicero - Classification of generic topics

Cicero pointed out some *loci* that, on his view, are principally used by dialecticians. Such topics, named *loci* from antecedents, consequents, and incompatibles (no. 8, 9, and 10 in Figure 1), represent patterns of reasoning based only on the meaning of the connector of the hypothetical premise (if...then).

For instance, if such a premise holds, and the antecedent is affirmed, the consequent follows necessarily (topic from antecedents) (Cicero, *Topica*, 53, 1-25). These *loci* seem to be aimed at establishing commitments based on previous commitments. In other words, instead of increasing the acceptability of a viewpoint based on the acceptability of the content of the premises on which it is grounded, such topics lead the interlocutor to the acceptance of a conclusion because of his previous acceptance of other propositions [Green-Pedersen, 1984, p.256].

Cicero connected the theory of topics to the division of discourse according to the Hermagoras stasis, the issue of the discussion, formulating the proposition to be proved or disputed [Kennedy, 1963, p.303]. He provided a classification of the topics according to their function for addressing a specific type of issue, namely conjecture, definition, and qualification (Cicero, *Topica*, 87) (Table 3).

Table 3. Cicero - Division of topics by issue

Conjecture	Definition	Qualification
Cause, effect, circumstances	Definition, description, notation, division, partition, consequent, antecedent, inconsistencies, cause and effect, *adiuncta*.	Comparison

Cicero's classification of topics became the ground for Boethius' works, which are the basis of the medieval dialectical tradition ([Stump, 1982]; [Stump, 1989]; [Stump, 2004]).

3.3 Boethius

Boethius commented on and organized Cicero's *loci* in his *In Ciceronis Topica* and *De Topicis Differentiis*, distinguishing between necessary and plausible connections and between dialectical and rhetorical *loci*. The treatise on *De Topicis Differentiis* includes *loci* that in Cicero and previously in Aristotle were distinguished as dialectical and rhetorical topics.

Boethius underscored how while dialectical *loci* stem from the rules of prediction and the logic-semantic properties of the predicates, rhetorical *topoi* represent the possible connections between things having different qualities (*De Topicis Differentiis*,1215C).[1] Some dialectical topics, such as topics from definition or genus and species, are necessary [Macagno and Walton, 2014, Ch.3], while others (for instance, from *adiuncta*) represent only frequent connections. This relation between probable and necessary consequence was studied in the Middle Ages. Garlandus Compotista classified topics according to their logical

---

[1] Rhetorical *loci* are similar in form to the dialectical ones, but they proceed from frequent connections between things, from stereotypes and not from semantic properties of concepts (for instance, usually people addicted to alcohol are dissolute, this person is alcoholic, therefore he is dissolute. See Boethius *De Topicis Differentiis*1215b).

(demonstrative) role. Topics from whole (which includes definition and genus), along with part and equal became the foundations of categorical syllogism [Stump, 1982, p.277].

In Boethius the Aristotelian *topoi* are interpreted as *maximae propositiones* falling under *differentiae*, genera of these maxims. *Maximae propositiones* are general principles, also called axioms. They are general (indefinite in respect to particulars) and generic propositions that several arguments can instantiate, and they have warranting the conclusion in an argument as a primary role. The relationship between the terms of the premises and the conclusion, namely the respect under which they are regarded, is called *differentia*, representing the criterion of appropriateness or the genus of maxims. The maxim is found from the genus of the *maximae propositiones* and the relationship between the terms of the first premise [Stump, 1989, p.6]. The structure of a topic is illustrated in Table 4.

Table 4. Argument and maxim in Boethius

First term:	Every virtue is advantageous.
Middle term:	Justice is a virtue.
Second term:	Therefore justice is advantageous.
Maxim:	What belongs to the genus, belongs to the species.
Differentia:	From the whole, i.e. the genus

Topoi are divided into three main categories: intrinsic, extrinsic and intermediate. While the first two categories are similar to Cicero's organization, the third is based on different principles. *Loci medii* represent semantic connections of grammatical relations, such as from words stemming from the same root, or semantic relations of division underlying the definition of the word (Figure 2).

Boethius distinguishes the dialectical *loci* from the rhetorical ones. Rhetorical topics are drawn from not from the concepts (representing the abstract relations between concepts), but from the things and how things usually are. For example, while the dialectical topic from genus proceeds from the definition of a concept (if a person is drunk, he is also intoxicated), the rhetorical one concerns how a more generic concept is usually related to a more specific one (usually if someone is not dissipated, he does not get drunk). Boethius takes from Cicero the rhetorical topics, not dealing with the abstract principles of inference concerning concepts, but with the circumstances concerning the specific cases[2]. For instance, reasoning from place, name, time depends on

---

[2]They are different from the preceding topics, because the preceding topics either contained deeds or adhered to deeds in such a way that they could not be separated, as place, time, and the rest, which do not desert the action performed. But those things that are associated with the action do not adhere to the action itself but are accidents of the circumstances, and they provide an argument only when they enter into comparison. The arguments, however, are taken not from contrariety but from a contrary, and not from similarity but from a similar, so that the argument seems to be taken not from a relationship [such as contrariety] but from things associated with the action [such as contraries]. Those things are associated

Intrinsic Loci		
From substance	From things accompanying the substance	
•From the definition •From the description •From the explanation of the name	•From the whole (genus) •From the integral whole •From a part (species) •From the parts of an integral whole •From efficient cause •From the matter	•From the end •From the form •From the generation (effects) •From the corruption •From uses •From associated accidents
Intermediate Loci	Extrinsic Loci	
•From inflections •From coordinates •From division	•From estimation about a thing •From similar •From what is more •From things that are less •From proportion	•From contraries •From opposites with reference to privation and possession •From relative opposites •From opposites with reference to affirmation and negation •From transumption

Figure 2. Boethius - Division of the dialectical *loci*

the fact, stem from the factors of the event and not from the logic-semantic relations between concepts. The rhetorical topics are organized into the four classes pointed out by Cicero (*De Topicis Differentiis*,1212A-1214A) (Figure 3).

3.4 Abaelardus

During the Middle Ages, the focal point of the study of argument was the connection between dialectics and demonstration. Beginning with the XI century, Garlandus Compotista analysed the categorical syllogisms as proceeding from topics from whole, part, and equal. On the other hand, he conceived all the topics under the logical forms of topics from antecedent and consequent, whose *differentiae* (the *genera* of *maximae propositiones*) are the syllogistic rules [Stump, 1982, p.277]. In the XII century, Abelard in his *Dialectica* examined the structure of dialectical consequence in its components for the first time [Kienpointner, 1987, p.283]).

Abelard described topics as imperfect inferences, different from valid categorical syllogisms. In this work, the *maxima propositio*, expressing a principle of inference, is related to the function of invention. The *maxima* is the gen-

---

with the action which are related to the very action at issue (*De Topicis Differentiis*,1214B 6-1214C 19).

Intrinsic Loci			
Person		Action	
• Name (Verres) • Natura (Barbar) • Mode of life (Friend of nobles) • Fortune (Rich) • Studies (Architect)	• Luck (Exiled) • Feelings (Lover) • Disposition (Wise) • Purpose • Deeds • Words	• Gist of the deed (Murder of a relative) • Before the deed (He stole a sword) • While the deed occurs (He struck violently) • After the deed (He hid him in a secret place)	• When: Time (night) and opportunity (people were sleeping) • Where: Place (bedroom) • How: Method (secretly) • With the aid: Means (with many men)
Comparing circumstances		Extrinsic Loci	
• Species • Genus • Contrary • Result • Greater • Lesser • Equal			• By what name to call what has been done? • Who are the doers of the deed? • Who approve of its having been thought up? • What is the law, custom, agreement, judgment, opinion, and theory for the thing? • Whether the thing is contrary to custom. • Whether men generally agree to these things.

Figure 3. Boethius - Division of the rhetorical *loci*

eral principle that is useful for finding the propositions accepted by everybody or the by the wise (the *endoxa*) relative to the subject dealt with in the argument. From this perspective, the structure of an argument is similar to that of a syllogism. The main difference lies in the nature of the assumptions, the propositions connecting the general principles to the subject of the reasoning. While dialectical inferences depend on the content of the propositions (or, rather, on the terms and their connections), syllogisms depend only on the form. The difference between form and content can be explained with the following cases. A syllogism such as:

> Every man is an animal
> But every animal is animate
> Therefore, every man is animate

depends on a rule of inference, that is [Abaelardus, 1970, p.262]:

> *posito antecedenti ponitur consequens* (if the antecedent is affirmed, the consequent is affirmed as well)

The connection between the terms of the inference depends only on their position in the propositions. On the other hand, dialectical inferences cannot be resolved only by considering the positions of the terms. These inferences are imperfect, since assumptions are needed for the conclusion to follow from the premises. For instance, the consequence

> If he is a man, he is an animate being

is necessarily valid since it is known that "animate being" is the genus of man and "whatever is predicated of the species is predicated of the genus as well." The inference depends on the local connection between the terms, on the *habitudo*. The *habitudo* is the topical relation, the semantic-ontological respect under which the terms are connected to each other in a (dialectical) syllogism ([Green-Pedersen, 1984, p.185]; [Green-Pedersen, 1987, p.415]), and on which the strength of the inference depends [Abaelardus, 1970, pp.254-257]. The mechanism of an argument scheme can be shown by the ancient model of Abelard [Abaelardus, 1970, p.315], in which the assumptions were connected to the axioms, to the maxims the *locus* proceeded from [Stump, 1989, p.36] (Figure 4).

Consequence	If Socrates is a man, he is an animate being.
Maxim	What the species is said of, the genus is said of as well.
Assumption	But "man," which is the species of "animate being" is said of Socrates; also therefore "animate being," which is clearly its genus.
Assumption 1	"Man" is a species of "animate being."
Syllogism 1	• What the species is said of, the genus is said of as well. • Man is species of "animate being". • Therefore, if man is said of anything, "animate being" is said of it as well.
Syllogism 2	• If "man" is said of anything, "animate being" is said of it as well. • Socrates is a man. • Therefore Socrates is an animate being.

Figure 4. Rules of inference and the material structure of arguments in Abelard

In the example above, the passage from the predicate "to be a man" attributed to the subject to the different predicate "to be an animate being" is grounded on a relation of semantic inclusion between these two predicates, i.e. a genus-species relation [Bird, 1962]. This relationship guarantees the inference based on a rule (the maxim) that expresses a necessary consequence of the concept of genus itself. The genus expresses the generic fundamental features of a concept, answering to the question "what is it?" and is attributed to all the concepts different in kind (Aristotle, *Topics* 102a 31-32). For this reason, it is predicated of what the species is predicated of.

After Abelard, in the 12th century, the notion of form of inference was developed into a reduction of all topical inferences to syllogisms. Later on, in the 13th century analytical consequences were analysed as following from topics "*dici de omni*" and "*dici de nullo*" (Every A is B, Every B is C, therefore every A is C). Demonstration is for this reason based on a topical relation (from the whole)[Green-Pedersen, 1984, p.256].

4 Modern Theories of Schemes

In the modern and contemporary theories on argumentation (or argument) schemes, several types of classification have been advanced [Walton et al., 2008]. In this section, the most relevant theories on schemes and the classification thereof will be summarized.

4.1 Perelman and the New Rhetoric

Perelman and Olbrechts-Tyteca divided their system of *topoi* into two broad categories, defined based on the two purposes that they considered to be the basic ones, finding associations and dissociations between concepts [Perelman and Olbrechts-Tyteca, 1969, p.190]. According to the *New Rhetoric*, arguments from association are divided into three main classes: Quasi-logical Arguments, Relations Establishing the Structure of Reality and Arguments based on the Structure of Reality, while dissociation constitutes a distinct class. This classification can be represented in Figure 5.

Quasi-Logical Arguments	*The Relations Establishing the Structure of Reality*	
• Contradiction and Incompatibility • Identity and Definition • Analyticity, Analysis and Tautology • The Rule of Justice • Arguments of Reciprocity • Arguments of Transitivity • Inclusion of the Part in the Whole • Division of the Whole into its Parts • Arguments by Comparison • Argumentation by Sacrifice	*Establishment through Particular Case* • Example • Illustration • Model and Anti-model	*Reasoning by Analogy* • Analogy • Metaphor
Arguments based on the Structure of Reality		
Sequential Relations	*The Relations of Coexistence*	*Double Hierarchy Argument*	*Differences of Degree and Order*
• Causal Link • Pragmatic Argument • Ends and Means • Argument of Waste • Argument of Direction • Unlimited Development	• Analogy • The person and His Acts • Argument from Authority • The Speech as an Act of the Speaker • The Group and its Members • Act and Essence • Symbolic Relation		

Figure 5. Classification of the arguments in the New Rhetoric

This classification is based several criteria, namely on the conceptual/ontological structure (association-dissociation; the reference to the structure of reality), the logical structure (quasi-logical vs. non-logical arguments), and the type of relations between concepts (sequential vs. coexistence). However, the interrelation between all these criteria is not specified, and there is not a unique rationale linking all such different arguments.

4.2 Toulmin

A different approach is provided by Toulmin, Rieke and Janik (1984), in which they classified arguments based on the basic functions of the warrants on which the arguments are grounded. Nine general classes of arguments were distinguished, subdivided into subclasses [Toulmin et al., 1984], shown in Figure 6.

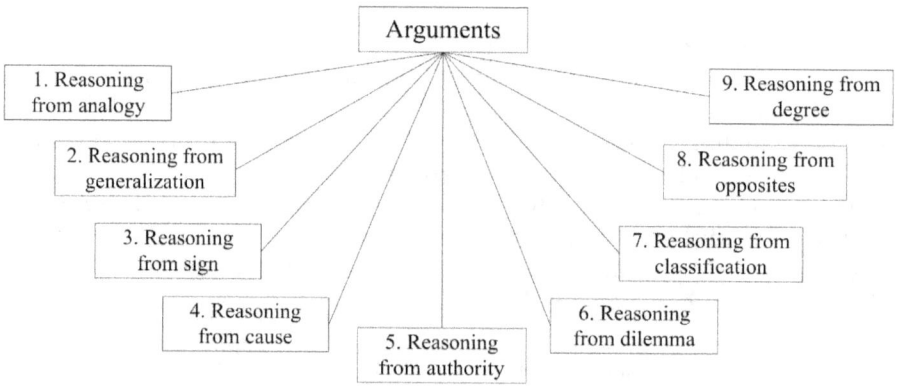

Figure 6. Classification of the arguments in Toulmin

Also in this case, different criteria are used in the classification. Some schemes represent types of reasoning (such as generalization, sign, or analogy); others are characterized by logical rules of inference (dilemma, opposites); others refer to the content of the argument (authority, classification, cause, degree). The relationship between the various criteria is not given.

4.3 Kienpointner

Kienpointner in *Alltagslogik* provides a complex and fine-grained classification, based on four criteria: 1) the type of inference; 2) the epistemic nature of the premises; 3) the dialectical function of the conclusion; and 4) the pragmatic function of the conclusion. On his view, every scheme 1) can proceed from different logical rules; 2) must be real (namely based upon the truth or likeliness of the premises), or fictive (grounded upon the mere possibility) (epistemic nature of the premises); 3) it must be pro or contra a certain thesis (dialectical function); and 4) it must have either a descriptive or a normative conclusion (pragmatic function) [Kienpointner, 1992, p.241]. In this sense, all the schemes can have descriptive or normative, pro or contra, real or fictive variants. The classification provided in *Alltagslogik* groups 21 schemes in three abstract classes characterized by the typology of the inferential rule: argument schemes using a rule; argument schemes establishing a rule by means of induction; argument schemes both using and establishing a rule (Figure 7).

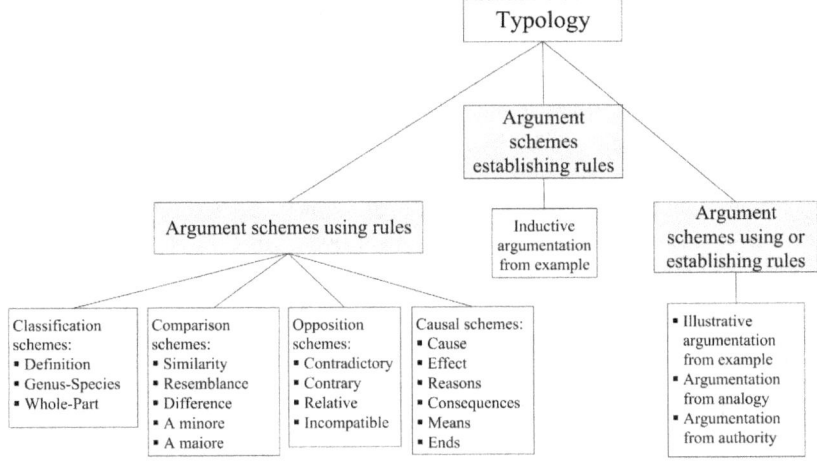

Figure 7. Classification of the arguments in Kienpointner

The first class, as shown in Figure 7, is subdivided in its turn in four content-based categories: classification, comparison, opposition, and causal schemes [Kienpointner, 1992, p.246]. Based on the aforementioned criteria, all the argument schemes may in turn have descriptive or normative variants, different logical forms (*Modus Ponens, Modus Tollens*, Disjunctive Syllogism, etc.), different dialectical purposes (establishing or countering a viewpoint), and different word-world relation (fictive   real).

This system of classification is aimed at distinguishing first the type of reasoning (induction, deduction), and then differentiating between the various material relations. The possible limitation of this system is that while the material relation of many deductive schemes is specified and distinguished, the content dimension of the inductive schemes is not pointed out.

4.4 Pragma-Dialectics

The pragma-dialectical system of classification of schemes consists of three basic schemes [Van Eemeren and Grootendorst, 1992]: 1) symptomatic argumentation; 2) argumentation based on similarities; and 3) the instrumental argumentation. The first scheme represents type of argumentation in which the speaker tries to convince his interlocutor "by pointing out that something is symptomatic of something else." In this type of pattern, what is stated in the argument premise is a sign or symptom of what is stated in the conclusion. The second scheme is grounded on a relation of analogy between what is stated in the argument premise and what is stated in the conclusion. In the third type of scheme the argument and the conclusion are linked by a very broad relation of causality. Other arguments are classified under these categories [Van Eemeren and Grootendorst, 1992]. For instance, arguments based on inherent qualities or a characteristic part of an entity or from authority are regarded as belonging to the symptomatic argumentation; arguments pointing out the consequences of an action or based on the means-end relationship are considered as subclasses of causal arguments [Garssen, 2001].

This system of classification is grounded on a twofold criterion. While causal argumentation is characterized by a material relation, analogical argumentation represents a type of reasoning independent from the specific content of the premises and conclusion. Symptomatic argumentation is a combination of these two criteria, as a sign or a symptom presupposes an abductive pattern and a material causal relation.

4.5 Grennan

In Grennan's [Grennan, 1997, pp.163-165] typology, the structurally valid inductive[3] inference patterns are classified according to 9 warrant types, derived from the Ehninger and Brockreide's typology [Brockriede and Ehninger, 1963]. The warrant types include possible reasons for inferring conclusions from premises, all belonging to the "logical mode" (and not to other types of motivations, such as emotions). The argument patterns can be summarized as follows:

1. **Cause to Effect**: The phenomenon mentioned in P produces the one in C.
2. **Effect to Cause**: The phenomenon mentioned in P is best explained by C.
3. **Sign**: The phenomenon mentioned in P is symptomatic (naturally or conventionally) of one reported in C.
4. **Sample to Population**: What is true of sample of X is also true of other X's.
5. **Parallel Case**: What is true of the referent of P is also true of other X's.
6. **Analogy**: $B1$ is to $B2$ in C as $A1$ is to $A2$ in P.
7. **Population to Sample**: What is true of Known X's is also true of this X.
8. **Authority**: S (the assertor of C) is a reliable source.
9. **Ends-Means**: The action mentioned in C generally achieves the end mentioned in P.

The patterns mentioned above are individuated on the basis of the warrant type. Together with this criterion of argument classification, Grennan presents a typology of claims. Each argument can be analysed relative to the type of warrant and to the kind of conclusion to be supported. The types of claim identified by Grennan [Grennan, 1997, p.162] can be represented in Table 5.

---

[3] Inferences, in an informal logic perspective, are considered inductive, since argumentation does not deal with deductive validity. The criterion for discriminating between acceptable and unacceptable patterns is provided by a logical intuition.

Table 5. Grennan: Classification of schemes

Type of Claim	Example
1. *Obligation Claims:* X must do A.	"Sam must apologize"
2. *Supererogatory Actuative Claims:* X ought to do A (they express a judgment that is in the interests of someone other than X for X to do A)	"I ought to help the needy in this area"
3. *Prudential Actuative Claims:* X ought to do A.	"Canadians ought to avoid heart diseases"
4. *Evaluative Claims,* of which there are three kinds: grading, rating, and comparison.	"This is a good cantaloupe"; "Steffi Graf is the best female tennis player at this time"; "Gretzky is a better hockey player than Howe was"
5. *Physical-World Claims,* which include both physical brute facts and institutional facts. "	The sun is setting" "The Dodgers beat the Giants three to two in eleven innings"
6. *Mental-World Claims,* which ascribe mental phenomena.	"He is upset"
7. *Constitutive-Rule Claims,* which are based on definitions and other necessary truths and falsehoods.	"In this election, majority should be defined as a majority of members present and voting" "Solid iron does not float in water"
8. *Regulative-Rule Claims,* which express obligations and prohibitions.	"Driving on the right is obligatory"

The types of warrant and the types of claim are the two criteria underlying Grennan's typology of argument patterns, each characterized by a premise, a warrant, and a conclusion. In the diagram below are represented the valid and useful patterns of arguments for obligation claims resulting from this classification [Grennan, 1997, p.162] (Figure 8).

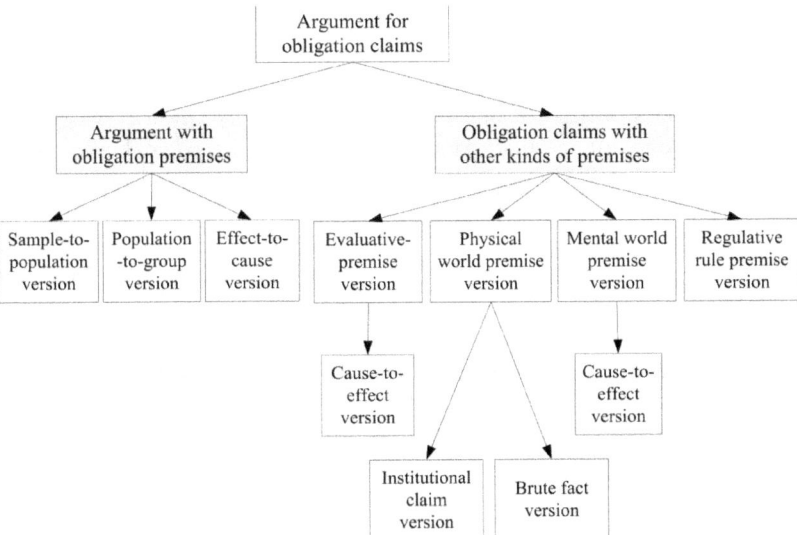

Figure 8. Classification of the arguments for obligation claims in Grennan

Grennan's typology develops the distinction between the warrant type and the kind of conclusion. The typology is extremely deep as regards the relation between speech acts and argument, but is limited to eight warrant types.

4.6 Katzav and Reed

Rooted in the schemes presented by Walton [Walton, 1995], the classification system of Katzav and Reed [Katzav and Reed, 2004b] aims to classify an argument by virtue of the 'relation of conveyance' that the complex proposition constituting the argument represents. These relations of conveyance describe how it is that one fact necessitates another, such as in the following example [Katzav and Reed, 2004a, p.2]:

> Consider, by way of illustration, a case in which the causal relation is operative: in the circumstances, the fact that the US military attacked Iraq caused the fall of Saddam's regime. Thus, in the circumstances, and via or in virtue of the obtaining of a causal relation, the fact that the US military attacked Iraq necessitated, or made it liable that, Saddam's regime fell. Using the causal relation and the above statements about Saddam's regime, we can construct the following simple argument:
>
> (1) Saddam's regime fell, because the US military attacked Iraq and if the US military were to attack Iraq, Saddam's regime would fall.
>
> In (1), the fact that the US military attacked Iraq is represented as conveying, via the causal relation, the fact that Saddam's regime fell. That the relation of conveyance represented is the causal relation is implicit in the subjunctive conditional 'if the US military were to attack Iraq, Saddam's regime would fall.'

In [Katzav and Reed, 2004b] the nature of such relations of conveyance is unpacked and connected to the concepts of warrant and scheme and to the work of Kienpointner [Kienpointner, 1992] and Walton [Walton, 1995] in particular. In [Katzav and Reed, 2004a], they sketch a high-level classification of relations of conveyance. At the topmost level, they distinguish between 'internal' and 'external' relations, whereby the former depend solely upon intrinsic features (and therefore encompass definitional, cladistic, mereological and normative relations, amongst others), whilst external relations depend upon extrinsic features (thereby covering such as spatiotemporal and casual relations, amongst others). Beneath this, the classification is further broken down into groups of schemes: those of specification, constitution, analyticity and identity under intrinsic relations and causal and non-causal under extrinsic (due largely to the fact that so many schemes rely upon causal relations). The full top-level classification tree (which identifies the main branches but does not give an exhaustive specification) is given in the scheme below:

Internal relation of conveyance
Relation of specification

- Relation of species to genus
- Relation of species to genus

- Relation of genus to species
- Determinable-determinate
- Etc.

Relation of constitution

- Abstract fact constitution
- Constitution of normative facts
- Constitution of positive normative facts
- Constitution of negative normative facts
- Constitution of non-normative abstract facts
- Constitution of necessary conditions
- Constitution of causal law
- Constitution of singular causal conditionals
- Constitution of constitution facts
- Constitution of Possibility
- Constitution of Impossibility
- Etc.

Concrete fact constitution

- Species/kind instance constitution
- Property instance constitution
- Property constitution by properties
- Property constitution by particulars
- Etc.
- Constitution of singular causal facts
- Relation of a part to a whole
- Relation of whole to one of its parts
- Etc.

Relation of analyticity

- Relation of sameness of meaning
- Relation of stipulative definition
- Relation of implication

Relation of identity

- Relation of qualitative identity
- Relation of numerical identity
- Etc.

External relation of conveyance
Non-causal dependence

- Non-causal law
- Conservation
- Conserved quantity
- Conserved quality
- Etc.
- Symmetry
- Spatial symmetry
- Etc.
- Nomological incompatibility
- Thing location incompatibility
- Thing type incompatibility
- Etc.
- Topological structure conveyance

Causal dependence

- Efficient cause conveyance
- Causal law
- Singular cause to effect
- Singular effect to cause
- Common cause
- Final cause conveyance

Though the mapping from individual relations of conveyance in this classification to the argumentation schemes in [Walton, 1995] and particularly [Walton et al., 2008] is not a trivial 1-to-1 correspondence, those schemes have been slotted in successfully in later work with a computational focus such as [Bex and Reed, 2011].

4.7 Lumer and Dove

The last system of classification that we consider was provided by Lumer and Dove (Lumer & Dove, 2011), using three general classes, each including subclasses:

1. Deductive argument schemes
   - Elementary deductive argument schemes;
   - Analytical arguments:
   - Definitoric arguments
   - Subsuming legal arguments:

2. Probabilistic argument schemes

   - Pure probabilistic argument schemes (statistics, signs);
   - Impure probabilistic argument schemes (best explanation);

3. Practical argument schemes

   - Pure practical argument for pure evaluations;
   - Impure practical argument schemes (for justification of actions; justification of instruments);
   - Arguments for evaluations based on adequacy conditions;
   - Arguments for welfare-ethical value judgements;
   - Practical arguments for theoretical theses.

This system consists of a mix of two distinct criteria, logical and pragmatic. While the first two classes are characterized by the type of reasoning, the last one is a type of argument with a specific pragmatic purpose, recommending a course of action. Moreover, the subclasses are defined based on both logic-based and content-based criteria, where together with distinctions based on the logical form (analytic schemes; probabilistic schemes) there are subclasses based on the nature of the premises (definitoric; subsuming).

All these types of classification show how a sole criterion is not sufficient for providing a clear and comprehensive classification of schemes. In order to understand what criteria can be used and in what abstract categories can be considered as the most basic ones, it is necessary to analyze the structure of the schemes. Once the common components of these heterogeneous combinations of premises and conclusions are brought to light, it is possible to find criteria for organizing them for specific purposes.

5 Using the schemes: A classification system

Argumentation schemes can be conceived as the prototypical combination of semantic (or topical) relations with logical rules of inference ([Macagno and Walton, 2015]; [Macagno et al., 2016]; [Walton and Macagno, 2015]). A classification based on the semantic link can provide an instrument for bringing to light the material relation between premises and conclusion, but the same semantic relation can be combined with types and rules of reasoning, and lead to various types of conclusion. For instance, causal relations are the ground of the argument from cause to effect, but also of arguments from sign and practical reasoning. Argumentation schemes merge the most common combinations between types of reasoning and material relations. For this reason, we need first to distinguish between these two levels, distinguishing between the various types of reasoning in Figure 9.

Semantic relations \ Types of reasoning	Deductive axioms	Induction/ Analogy	Abduction
Definitional	Argument from definition, genus...	Argument from example	Argument from (improper) signs
Causal	Argument from cause to effect	...	Argument from best explanation
Practical	Argument from negative consequences	...	Practical reasoning
Authority	Argument from expert opinion	...	...

Figure 9. Types of argument and types of reasoning

A multi-logical perspective needs to be taken into account as a classification criterion, in which the logical form can be described using distinct types of reasoning, which in turn can include various logical rules of inference (MP, MT). However, in the Latin and Medieval tradition, the formal rules of inference are treated as maxims and not as distinct levels of abstraction. For this reason, the two levels of the general, semantic topics and of the logical rules are not distinguished, and the possible interconnections between them are not taken into account. The modern theories of argumentation schemes propose classifications essentially mirroring the ancient approach. The logical rules are treated at the same level as the semantic-ontological topics, and not as distinct levels of abstraction. A possible solution is to acknowledge the discrepancy between logical form and semantic content as a divergence in kind, and try to show how these two levels can be interconnected.

A possible overarching principle can be found in the pragmatic function of the schemes, namely what they have been intended for. Argumentation schemes can be thought of as instruments for reconstructing and building arguments (intended as discourse moves), i.e. analytical or invention tools. For this reason, in order to provide a classificatory system to retrieve and detect the needed scheme it can be useful to start from the intended purpose of an argumentation scheme. From an analytical point of view, the analysis of an argument in a discourse, a text, or dialogue presupposes a previous understanding of the communicative goal (and, therefore, the "pragmatic" meaning) of the argument and the components thereof. For example, an argument can be aimed at classifying a state of affairs, supporting the existence of a state of affairs, or influencing a decision-making process.

This teleological classification needs to be combined with a practical one. The generic purposes of a move need to be achieved by means of an inferential passage. In this sense, the classificatory system needs to account for the possible (argumentative) means to achieve the pragmatic purpose of an argument.

Not all the semantic relations underlying the schemes can support all the possible conclusions or purposes of an argument. Definitional schemes are aimed at supporting the classification of a state of affairs; they cannot lead to the prediction or retrodiction of an event. Similarly, a pattern of reasoning based on the evaluation of the consequences of an action or an event can be used to establish the desirability of a course of action brining it about. However, it cannot be reasonably used to establish the truth or falsity (or acceptability) of a proposition. For this reason, the analysis of the pragmatic meaning (i.e. the purpose) of an argument provides a criterion for restricting the paradigm of the possible means to achieve it. The crucial problem is to find categories of argument purposes that can establish criteria for distinguishing among classes of semantic relations, which in turn can be specified further according to the means to achieve such goals.

The first distinction to be made is based on the nature of the subject matter, which can be 1) a course of action or 2) a state of affairs. In the first case, the goal is to support the desirability or non-desirability of an action; in the second case, the schemes are aimed at providing grounds for the acceptability of a judgment on a state of affairs. The ancient dialectical accounts (Cicero, *Topica*; Boethius, *De Topicis Differentiis*) distinguished between two types of argumentative "means" to support a conclusion, namely the "internal" and the "external" arguments. The first ones are based on the characteristics of the subject matter (such as arguments from definition or cause). The latter derive their force from the source of the statement, namely from the authority of who advances the judgment or the proposal (arguments from authority). This first distinction can be represented as shown in Figure 10.

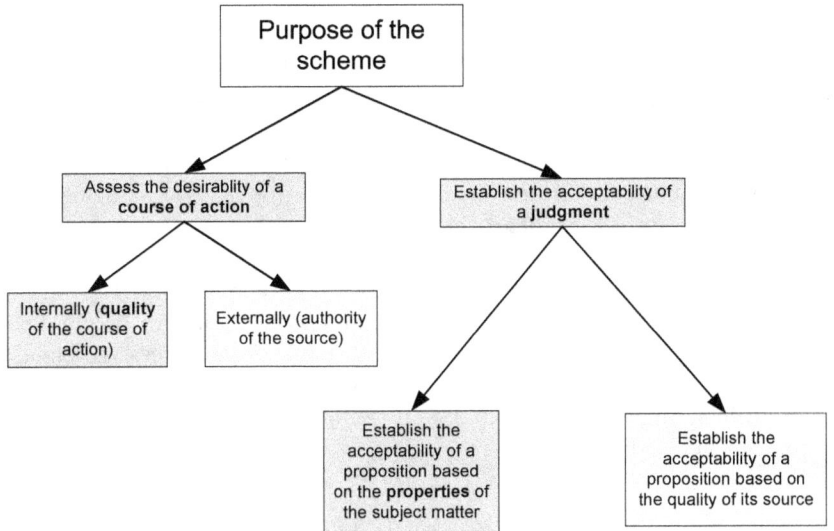

Figure 10. Purposes of an argument

The acceptability of a conclusion can be supported externally in two ways. If the argument is aimed at establishing the desirability of a course of action, the authority can correspond to the role of the source ("You should do it because he told you that!"). Otherwise, the popular practice can be a reason for pursuing a course of action ("We should buy a bigger car. Everyone drives big cars here!"). External arguments can be represented in Figure 11.

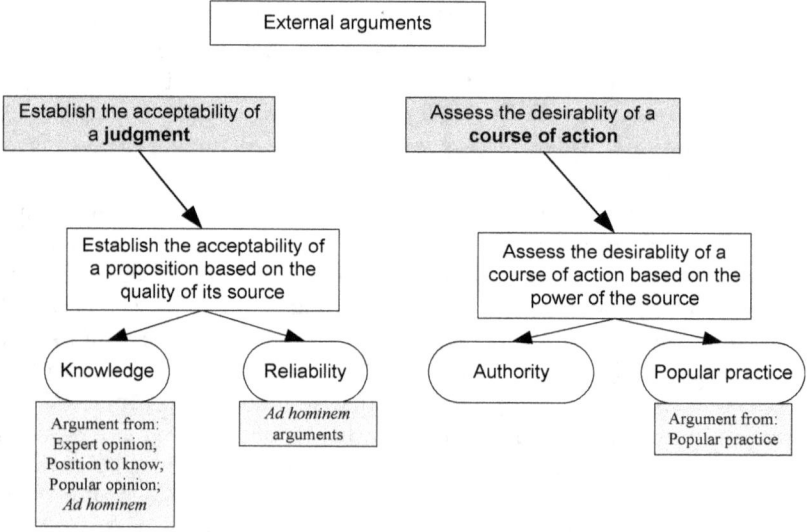

Figure 11. External arguments

When external arguments are used to support also a judgment on a state of affairs, the relevant quality of the source is not the speaker's authority (connected with the consequences of not complying with the orders/conforming to common behavior) but rather with the source's superior knowledge. The quality of the source can be also used negatively to show that a source is not reliable (it is not a good source), and that consequently the conclusion itself should be considered as doubtful (*ad hominem* arguments).

Internal arguments can be divided into the two categories of arguments aimed at assessing the desirability of a course of action, and the ones supporting the acceptability of a judgment. Courses of action can be classified as desirable or not depending on the quality of their consequences (the course of action is a condition of a resulting positive or negative state of affairs) or their function in bringing about a desired goal (an action is productive of a pursued state of affairs) (Figure 12).

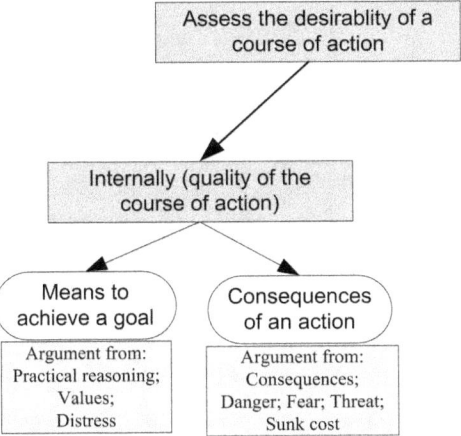

Figure 12. Internal practical arguments

The arguments used to provide grounds for a judgment on a state of affairs can be divided according to the nature of the predicate that is to be attributed. The most basic differentiation can be traced between the predicates that attribute the existence of a state of affairs (the occurrence of an event or the existence of an entity in the present, the past, or the future), and the ones representing factual or evaluative properties.

The arguments supporting a prediction or a retrodiction are aimed at establishing whether or not an event has occurred or will occur, or whether an entity was or will be present (existent). The arguments proceeding from casual relations (in particular from material and efficient causes) bear out this type of conclusion. The other type of predicates can be divided in two categories: factual judgments and value judgments. The first type of predicates can be attributed by means of reasoning from classification, grounded on descriptive (definitional) features and supporting the attribution of a categorization to an entity or an event (Bob is a man; Tom is a cat). Value judgments are classifications that are not based on definitions of categorical concepts (to be a cat) but rather on values, or rather hierarchies of values. Such judgments proceed from criteria (or more specifically, criteria of importance to the audience to whom the argument is presented) for classifying what is commonly considered to be "good" or "bad." Also the reasoning underlying the attribution of evaluative predicates, such as "to be a criminal," can be considered as belonging to this group of arguments. These latter patterns are grounded on signs of an internal disposition of character, which in its turn is evaluated. The distinctions discussed above are summarized in Figure 13 below.

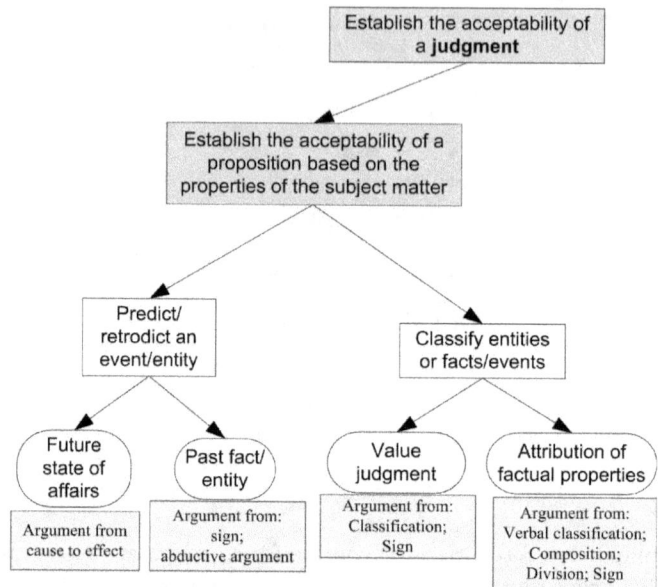

Figure 13. Establishing the acceptability of a judgment (SoA)

This system of classification of argumentation schemes is based on the interaction between two criteria, the (pragmatic) purpose of an argument and the means to achieve it. This tree model can be used both for analytical and production purposes. In the first case, the speaker's intention is reconstructed by examining the generic purpose of his move, and then the possible choices that he made to support it, based on the linguistic elements of the text. Depending on the desired level of precision, the analysis can be narrowed down until detecting the specific scheme, namely the precise combination of the semantic principle and the logical rule supporting the conclusion. In this fashion, the analyst can decide where to stop his reconstruction. This analytical model can be of help also for educational purposes, as it can be adapted to various teaching needs and levels. For production purposes, the nature of the viewpoint to be supported can be analyzed using the most generic criteria set out above (What is under discussion, a decision or a fact? The occurrence of an event or its classification? The naming of a state of affairs or its qualification?). Such questions closely resemble the ones that were at the basis of the rhetorical theory of *stasis*, namely the issues that can be discussed [Heath, 1994]. These distinctions are then combined with the specific alternative strategies to support the defended viewpoint.

The aforementioned system of classification can also account for the interrelation between the semantic relation and the different types of reasoning, namely logical forms. For example, the desirability of a course of action can be assessed internally by taking into consideration the means to achieve a goal.

This pattern of reasoning can be stronger or weaker depending on whether there is only one or several alternatives. The paradigm of the possible means will determine whether the reasoning is abductive or deductive, resulting in a more or less defeasible conclusion. The same principle applies to the other semantic relations, such as the ones proceeding from cause or classification, which can be shaped logically according to inductive, analogical, deductive, or abductive types of reasoning.

6 A bottom-up approach to classification: Clusters of decision-making schemes

Argumentation schemes are characterized by both "family" resemblances and actual interconnections [Walton and Macagno, 2015]. Practical reasoning, value-based reasoning, value-based practical reasoning, argument from positive consequences, argument from negative consequences, and the slippery slope argument are related by the same similar structure based on value judgments and practical outcome. Such schemes are often also interconnected when we analyze the structure of actual arguments. However, in order to understand and choose between similar and interrelated schemes, it is necessary to examine their relations and their differences. The simplest and most intuitive version of the scheme for practical reasoning (Table 6) uses the first-person pronoun "I" to represent a rational agent, an entity that has goals, some knowledge of its circumstances, and the capability of taking action to change those circumstances. It also has sensors to perceive its circumstances, and to perceive at least some of the consequences of its actions when it acts to change its circumstances. Such a rational agent also therefore has the capability for feedback. When it perceives changes in its circumstances due to its own actions, it can modify its actions or goals accordingly, depending on whether the consequences of its actions are deemed to contribute to its goals or not. This simplest form of practical reasoning [Walton et al., 2008, p.95] can be described as a fast and frugal heuristic for jumping to a quick conclusion that may later need to be retracted in the light of further considerations (Table 6).

Table 6. Argument from Practical reasoning

Major Premise:	I have a goal G.
Minor Premise:	Carrying out this action A is a means to realize G.
Conclusion:	Therefore, I ought (practically speaking) to carry out this action A.

The defeasible nature of this simple form of practical reasoning is brought out by the observation that it typically provides a starting point for action that needs to be challenged by the asking of critical questions as the agent moves ahead. Below is the standard set of critical questions matching this scheme.

CQ1 What other goals do I have that should be considered that might conflict with G?

CQ2 What alternative actions to my bringing about A that would also bring about G should be considered?

CQ3 Among bringing about A and these alternative actions, which is arguably the most efficient?

CQ4 What grounds are there for arguing that it is practically possible for me to bring about A?

CQ5 What consequences of my bringing about A should also be taken into account?

The last critical question, CQ5, often called the side effects question, concerns assessment of the potential negative consequences of carrying out the action described in the conclusion of the scheme. If negative consequences of this course of action are identified, that is a reason for withdrawing the conclusion and considering an alternative course of action that might avoid the negative consequences. Use of the term 'negative' implies that values are involved, and that a rational agent is assumed to have values as well as goals that it bases its practical reasoning on.

A complication is that there is another closely related argumentation scheme associated with this critical question, Argument from negative consequences. This scheme, widely recognized in the literature, cites known or estimated consequences of a proposed course of action as presenting a reason, or set of reasons, against taking the course of action initially indicated by the practical reasoning scheme. Argument from negative consequences also has a positive form. According to the scheme for argument from positive consequences, known or estimated consequences that have a positive value for the agent are cited as a reason, or set of reasons, supporting the carrying out of the action initially considered. Below the versions of the two basic argumentation schemes for arguments from consequences are formulated as they were in [Walton et al., 2008, p.101]. The first one is called argument from positive consequences (Table 7).

Table 7. Argument from positive consequences

Premise:	If A is brought about, good consequences will plausibly occur.
Conclusion:	Therefore A should be brought about.

The second one is called argument from negative consequences (Table 8).

Table 8. Argument from negative consequences

Premise:	If A is brought about, bad consequences will plausibly occur.
Conclusion:	Therefore A should not be brought about.

In both instances, an implicit premise could be made explicit in the scheme stating that if good (bad) consequences will plausibly occur, A should (not) be brought about. As with the basic form of practical reasoning, arguments from positive or negative consequences are defeasible. The premise offers a reason to accept a proposal for action tentatively, subject to exceptions as new circumstances come to be known by the agent. In these formulations, the expression "good consequences" refers to consequences taken by the agent to have positive value, and the expression "bad consequences" refers to actions taken to have negative value. These observations bring us to another pair of schemes closely related to the ones for argument from positive consequences and argument from negative consequences.

The relationship between a state of affairs, its classification according to a value, and the commitment to an action is represented in terms of value. Values (differently from [Atkinson et al., 2005];[Bench-Capon, 2003]) are regarded as grounds for a type of reasoning independent from and related to (or rather, presupposed by) practical reasoning. This reasoning guarantees the so-called "practical classification" [Westberg, 2002, p.163] of a state of affairs and the commitment thereto. The scheme for argument from positive value is formulated in Table 9 as in [Walton et al., 2008, p.321]:

Table 9. Argument from positive value

Premise 1:	Value V is positive as judged by agent A.
Premise 2:	If V is positive, it is a reason for A to commit to goal G.
Conclusion:	V is a reason for A to commit to goal G.

The corresponding scheme representing argument for argument from negative value is formulated in Table 10.

Table 10. Argument from negative value

Premise 1:	Value V is negative as judged by agent A.
Premise 2:	If V is negative, it is a reason for retracting commitment to goal G.
Conclusion:	V is a reason for retracting commitment to goal G.

Argument from positive consequences typically supports an argument taking the form of basic practical reasoning by giving justification for going ahead with the contemplated action. Argument from negative consequences presents a reason against taking the action being considered by citing consequences of it that would contravene the values of the agent.

Another more complex argumentation scheme has also been recognized in the literature [Bench-Capon, 2003] that combines all the schemes mentioned above. This scheme describes a form of argument called goal-based practical reasoning that combines basic practical reasoning with value-based reasoning. The version of this scheme (Table 11) is from [Walton et al., 2008, p.324].

Table 11. Argument from goal-based practical reasoning

Premise 1:	I have a goal G.
Premise 2:	G is supported by my set of values, V.
Premise 3:	Bringing about A is necessary (or sufficient) for me to bring about G.
Conclusion:	Therefore, I should (practically ought to) bring about A.

The scheme for value-based practical reasoning can also be formulated in a more explicit way that brings out an important aspect of practical reasoning, namely the circumstances of the case that can be observed by the agent and used by as a basis for reaching a decision on what to do. According to the version of the scheme formulated in [Atkinson et al., 2005], any action the agent takes can be seen as a transition from the current set of circumstances to a new set of circumstances, as the agent moves forward to attempt to realize its goal.

The last decision-making argument is the slippery slope argument, sometimes also called the wedge argument. Different varieties of slippery slope argument have been recognized, such as the causal slippery slope argument, the precedent slippery slope argument, the linguistic slippery slope argument, which depends on the vagueness of terms or concepts, and a more complex (all-in) form of slippery slope argument that combines the simpler variants. A good place to start is a simple version of the slippery slope type of argument formulated as the basic scheme in [Walton et al., 2008, p.340] (Table 12).

According to [Walton et al., 2008, p.340], the following three critical questions match this basic scheme.

CQ1 What intervening propositions in the sequence linking up A_0 with A_n are actually given?

CQ2 What other steps are required to fill in the sequence of events, to make it plausible?

CQ3 What are the weakest links in the sequence, where specific critical questions should be asked on whether one event will really lead to another?

Table 12. Argument from goal-based practical reasoning

First Step Premise:	A_0 is up for consideration as a proposal that seems initially like something that should be brought about.
Recursive Premise:	Bringing up A_0 would plausibly lead (in the given circumstances) to A_1, which would in turn plausibly lead to A_2, and so forth, through the sequence $A_2 \ldots A_n$.
Bad Outcome Premise:	A_n is a horrible (disastrous, bad) outcome.
Conclusion:	A_0 should not be brought about.

So here we have a cluster of schemes all closely related to each other. The argument from negative consequences is one of the critical questions matching the basic scheme, but the scheme for argument from negative consequences is itself based on the closely related scheme for argument from negative values.

Clarifying the relationships among this cluster of schemes enables us to draw an important distinction widely discussed in the philosophical literature on practical reasoning between two distinct types of practical reasoning: instrumental practical reasoning and value-based practical reasoning. When it comes to classifying the arguments within this cluster of schemes, it would seem reasonable to venture as a hypothesis that the basic scheme for practical reasoning is the simplest form of it, while the scheme for value-based practical reasoning is a more complex variant of the scheme. It combines the basic scheme with the schemes for argument from values. On this approach to drawing distinctions within the cluster, arguments from positive consequences can be taken as species of arguments from positive value, and arguments from negative consequences can be taken as species of arguments from negative value. Practical experience in using assistants to use argumentation schemes to identify types of arguments in natural language text suggests that the assistants sometimes find it difficult to classify a particular argument identified in a text as fitting one or more of these schemes. It can be helpful for this purpose to give the assistants identification conditions that attempt to formulate key essential requirements of the type of argument represented by a particular scheme.

The following is a set of three identification conditions for the type of argument matching the scheme for instrumental practical reasoning: (1) An agent (or group of agents in the case of multiagent reasoning) is attempting to arrive at a reasoned decision on what course of action to take in a given set of circumstances requiring some action, (2) the circumstances provide evidence on which to build pro and con arguments, arguments for and against the course of action being considered, (3) the agent is basing its decision on its goals, as well as its perception of the circumstances of the case, (4) arguments need to be weighed against each other as stronger or weaker reasons for taking this action or not, and (5) the agent purports to be using this evaluation of the stronger or weaker reasons as its basis for taking the action or not. Here the four conditions describe an agent deciding whether to take a particular course of action or not. But it needs to be recognized that in some situations there may be several alternative courses of action to be considered, and the agent is trying to decide which of them would be the best course of action, based on the reasons provided by its goals and the circumstances of the case.

The identification conditions for the value-based species of practical reasoning are the same as the five identification conditions for instrumental practical reasoning, except that another condition needs to be added: (6) the agent is justifying its decision based on its values, as well as on its goals and its perception of the circumstances of the case. The aforementioned cluster of arguments is characterized by several types of relations, which can be of help in distinguishing them and detecting their possible nets. For example, argument from negative consequences is one of the questions matching the scheme for argument from practical reasoning. So this relationship could be described by saying that argument from negative consequences is a counterargument, a rebuttal or undercutter that can defeat an argument from practical reasoning in a given case, provided that the negative consequences can be specified, and provided that it can be shown that these consequences are indeed negative.

Already from these remarks one relationship emerges. Argument from negative consequences is based on argument from values, and is a species of argument from values. Another relationship already shown above, is that value-based practical reasoning is a more complex form of argument than instrumental practical reasoning. Value-based practical reasoning is a species of instrumental practical reasoning with argument from values added onto it.

Another relationship that emerges is that the slippery slope type of argument is clearly a subtype and special instance of argument from negative consequences. It is less evident that the slippery slope argument is also a species of value-based practical reasoning. However, it can be seen that it is. In the case of the slippery slope argument, the agent doing the decision-making must be assumed to have some goals and values in mind that the other party, the agent attacking its argument, can appeal to when mounting a slippery slope argument. Let's call the two parties the agent and critic. The slippery slope type of argument is inherently negative. The critic is using the argument to warn the agent that if it takes a first step, or continues a series of steps that it has already started, these steps will lead to a loss of control that cannot be anticipated in advance so that the sequence of actions will ultimately result in a catastrophic outcome. The critic has to assume that the agent has some values that both of them share, so that they can both agree that the outcome warned of by the critic is catastrophic, that is highly negative and worth avoiding. The critic has to assume that the agent has some goals and is acting in a rational manner so that it is trying to either achieve or at least be consistent with these goals as it carries out action supposedly designed to fulfill them. Otherwise the critic's argument is not going to have much force and will be unlikely to deter the agent from moving ahead.

What especially distinguishes the slippery slope as a distinctive type of argument are three premises, the recursive premise, the gray zone premise and the loss of control premise. Given these observations, we can see how the value-based practical reasoning argument is embedded into the basic slippery slope argument and is a part of it. As shown in Figure 14, the basic slippery slope type of argument, represented by the scheme formulated above, is at the center of a cluster of other related schemes.

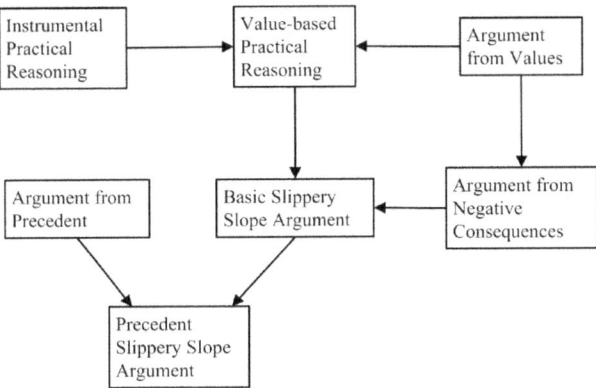

Figure 14. Cluster of Schemes

The basic slippery slope argument is derived from value-based practical reasoning as its core argument structure, where value-based practical reasoning is a combination of instrumental practical reasoning and argument from values. So here it is shown how these schemes are structured together into a cluster. It is also shown that the basic slippery slope argument is a species of argument from negative consequences, as scheme that is in turn built partly from the scheme for argument from values. So these five schemes form a cluster. But the basic slippery slope argument also has several subtypes, including the precedent slippery slope argument, the causal slippery slope argument, and the variety of slippery slope arguments deriving from vagueness of a verbal criterion. According to the analysis of the slippery slope argument given in [Walton, 1992] these four species of slippery slope argument are subtypes of a more general form of argument called the all-in slippery slope argument.

Here we put forward the hypothesis that there is a basic, minimal type of slippery slope argument from which these other more specialized variants are derived. To indicate the existence of such connections in Figure 14, we have inserted the name of the scheme for the precedent slippery slope argument underneath the schemes for argument from precedent and the basic slippery slope argument. This classification indicates another aspect of the cluster of schemes surrounding the category of slippery slope arguments.

7 Using argumentation schemes: Nets of Argumentation Schemes

Argumentation schemes are imperfect bridges between the logical (or quasi-logical) level and the conceptual one ([Macagno and Walton, 2015]; [Macagno, 2015]). From a conceptual (material) point of view, schemes usually represent an inferential step from a specific type of premise to a specific type of conclusion. However, there is a crucial gap between the complexity of natural argumentation, characterized by several conceptual passages leading to a conclusion, and the schemes. In order to reason from consequences, we need to classify a state of affairs, evaluate it positively or negatively, and then suggest a suitable course of action, which can lead to further reasoning steps, for example from commitment. A single argumentation scheme cannot capture the complexity of such real argumentation. For this reason, we need to conceive the relationship between arguments and schemes in a modular way, in terms of nets of schemes.

A real argument can be described through interconnected and interdependent argumentation schemes, each of them bringing to light a single argumentative step that can be explicit, presupposed, or simply implied. In order to explain the idea of nets of schemes, we consider the following example taken from the debates during the conflict between Russia and Ukraine in 2014. In this case, the British Foreign secretary William Hague commented on Russian intervention in Crimea and Ukraine as follows[4]:

Example 7.1 (The Hague Speech) *Be in no doubt, there will be consequences. The world cannot say it is OK to violate the sovereignty of other nations. This clearly is a violation of the sovereignty independence and territorial integrity of Ukraine. If Russia continues on this course we have to be clear this is not an acceptable way to conduct international relations.*

This example is apparently an easy case of argument from consequences, in which Russia's continuation of its military operations is depicted by the British Foreign Secretary as leading to undesirable consequences. However, this reasoning involves also a classification of Russia's behavior as a "violation of the sovereignty independence and territorial integrity of Ukraine," and a qualification of this behavior as unacceptable by the UK and the "world." By pointing out the shared values to which the world countries are committed (the sovereignty of other nations cannot be violated), the speaker makes explicit the commitment against Russia's behavior, which is represented by the vague notion of "consequences." We represent this structure in Figure 15.

---

[4]Ukraine crisis: William Hague warns Russia of economic fallout. *The Guardian*, 3 March 2014. Retrieved from: https://is.gd/Kw8Vax. (Accessed on 15 May 2017)

Figure 15. Net of arguments in the Hague Example

In Figure 15 the dotted boxes represent the tacit premises and the tacit ultimate conclusion, which are taken for granted by the speaker but are needed for reconstructing his reasoning. The classification, the reasoning from commitment, and the argument from consequences are deeply interconnected. The alleged world's commitment to consequences against Russia depends on the classification of the state of affairs ([Macagno and Walton, 2014]; [Walton and Macagno, 2009]), which fits into the value of "protecting nations' sovereignty." This commitment leads to an implicit threat, namely a consequence that is presupposed to be negatively evaluated by Russia.

This analysis can be applied to the structure of a slippery slope argument, such as the one advanced by the Russian defense analysts in reply to the help provided by the United States to Ukraine (which includes weapons and hardware)[5]:

Example 7.2 (The Global Escalation) *U.S. provision of military aid to Ukraine would be seen by Moscow as a declaration of war and spark a global escalation of Ukraine's separatist conflict, Russian defense analysts said.*

This argument stems from a classification (US provision of military help is a declaration of war), and leads to a chain of negative consequences (global escalation) that ultimately are going to affect the Western countries. Also in this case, the central argument (the slippery slope) is associated with other

---
[5]Russia Would See U.S. Moves to Arm Ukraine as Declaration of War. *The Moscow Times*, 9 February 2015. Retrieved from: https://is.gd/hxO6MW (Accessed on 15 May 2017)

arguments (argument from classification and from values), resulting in the net shown in the graph in Figure 16.

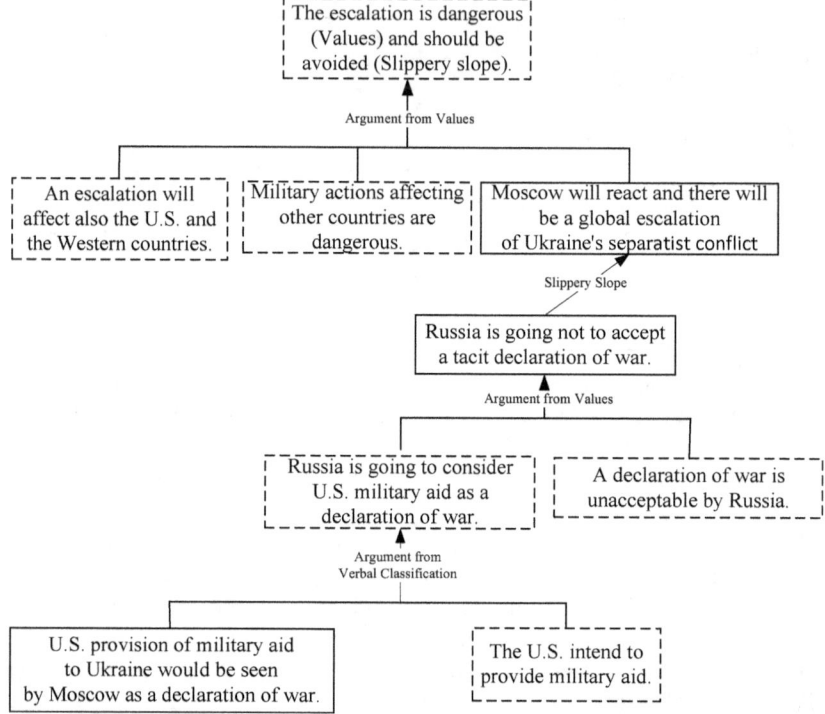

Figure 16. Net of arguments in the global escalation example

In this case, the classification justifies the slippery slope, whose force partially depends on the fact that the escalation is claimed to be global, affecting also other countries. The evaluation of this consequence therefore combines with the chain of events claimed by the analysts, and leads to the practical conclusion of avoiding the provision of military aid.

A special feature of this example is its compressed style of presentation. Slippery slope is a complex form of argument built around a connected sequence of actions and consequences starting from an initial action or policy and then proceeding through a sequence to an eventual outcome. However in many examples, the intervening sequence is left implicit, concealing a chain of intervening propositions that have to be filled in as implicit assumptions of the argument. These implicit assumptions are needed to make it fit the scheme for the slippery slope type of argument. The example really is a slippery slope argument, but in order to prove that it is, several implicit premises or conclusions have to be filled in that are essential. These intervening links are basically filled in by common knowledge concerning the normal way we expect military inventions to take place and to have consequences. By using an argument map that reveals the network of argumentation into which the given slippery slope argument fits, the puzzle of unraveling the network of argumentation using a cluster can be solved in any given case of argument interpretation.

On the perspective presented in this section, we notice that argumentation schemes appear in nets instead of in clear and independent occurrences. A scheme can capture only one passage of reasoning, while the nets can map a more complex argumentative strategy, involving distinct and interdependent steps.

8 Using Argumentation Schemes in AI and law

In the sections above we have shown how argumentation schemes have been developed theoretically, providing a system of classification and representation thereof. One of the most important areas of application of the schemes is computing, and in particular artificial intelligence. In this section, we will show very briefly how argumentation schemes have been used in AI and AI and Law, and in particular the principles guiding the formalization thereof. It is far from a complete survey, but merely attempts to show how schemes are currently being applied and modeled. It also tries to convey very briefly how schemes have evolved as they have been used for different purposes in different AI systems and areas. The discussion includes the problem of how to model critical questions matching each scheme, and how schemes are being used in AI and Law in argument mining, case-based reasoning and statutory interpretation.

The paper that introduced argumentation schemes to the AI and law community was [Verheij, 2003b]. This paper proposed the use of argumentation schemes, as a main tool for analysis in AI and law, stating [Verheij, 2003b, p.168] that the argumentation scheme is "a concept borrowed from the field of argumentation theory." Verheij investigated how argumentation schemes could be formalized for use in computational settings. He proposed [Verheij, 2003b, p.176] that any argumentation scheme can be expressed in the following format: Premise 1, Premise 2. . . . , Premise n, therefore Conclusion. Verheij visually represented the graph structure of an argumentation scheme by building an argument mapping software tool called ArguMed.

A formal analysis of argumentation schemes of Reed and Walton [Reed and Walton, 2005] defined a set of attributes, T, associated with propositions by a typing relation that associates every proposition to a set of attributes called a type. On this analysis a scheme is comprised of a set of tuples ¡SName, SConclusion, SPremises¿ where SName is some arbitrary token [Reed and Walton, 2005, p.179]. The gist of the analysis is that a particular scheme is given a unique name which is associated with a conclusion type and a set of premise types. An instantiation of a scheme of a type represented by a unique name must have a conclusion of the right type, and each premise must also be of the right type.

Prakken [Prakken, 2005, p.34] remarked that schemes act very much like the rules used in rule-based computer systems. The problem was that AI systems, as well as argument mapping tools of the kind used in argumentation theory, including the software systems developed in AI to assist with the building of argument diagrams, use a model of argument where the premises and conclusions are propositions. Along these lines, the structure is basically a graph with arcs joining the various points representing the propositions that can be identified as premises or conclusions. So far then it seemed that schemes were amenable to being fitted into AI systems without undue difficulty, but the central problem posed at that point was how to model the distinctive set of critical questions matching each scheme. One proposal, commented on below, is to model the critical questions as additional premises of an argument fitting a scheme.

But there was a big problem with this way of proceeding because different critical questions act in different ways in this regard. Sometimes merely asking a critical question is enough to defeat the target argument, whereas in other instances the asking of the question does not defeat the target argument unless some evidence is offered. The issue turned out to be one of burden of proof [Gordon et al., 2007]. In some instances, merely asking a critical question is enough to shift the burden of proof onto the proponent who put forward the argument. In other instances, the burden of proof does not shift unless the questioner can provide some backup evidence to support the question.

Verheij [Verheij, 2003b] noted that there were variations on how the critical questions work in this regard. He noted that critical questions that point to exceptions to a general rule only undercut an argument while others could be seen refuting the argument in one of two different ways. One way is to deny an implicit assumption on which the argument depends. Another is to point to counter-arguments that can be used to attack the given argument. Verheij [Verheij, 2003b, p.180] showed that critical questions can perform four distinctively different kinds of roles:

1. They can be used to question whether a premise of a scheme holds.

2. They can point to exceptional situations in which a scheme defaults.

3. They can frame conditions for the proper use of a scheme.

4. They can indicate other arguments that might be used to attack the scheme.

It is currently widely assumed in AI that there are three ways you can attack an argument. You can attack one or more of the premises (premise attack), you can attack the conclusion (conclusion attack), or you can attack the inferential link joining the premises to the conclusion (for example by arguing that an exception applies). The last mode of attack is called undercutting [Pollock, 1995]. The first role would be that of a premise attack. The second and third roles would be undercutting attacks. The fourth role might refer to an undercutter but could also perhaps be taken to refer to a conclusion attack. So here the problem is posed of how to model critical questions given that critical questions can perform more than one function.

ASPIC+ [Prakken, 2010] is a formal argumentation system that consists of a logical language L with a binary contrariness relation that operates like negation along with two kinds of inference rules, strict and defeasible, defined over L. ASPIC+ is based on the abstract argumentation framework (Dung, 1995) which can be defined as a pair (Args, R), where Args is a set of arguments and a binary relation R on Args is called attack relation. The underlying idea of the formalism is that each argument in a sequence of argumentation forming a directed graph structure can be defeated by other arguments so that a_2 defeats a_1, a_3 defeats a_2, . . ., and defeats a_{n-1}. Arguments in the graph can be labeled as *in* or *out*. An argument is rejected (*out*) if it is attacked by any other argument that is *in*. An argument is accepted if it is not attacked by any other argument that is *in*. Note that the notions of argument and argument attack are taken as primitive in an abstract argumentation system, so that such a system by itself provides no way of modeling the premises and the conclusion. In the system developed in [Prakken et al., 2015] for case-based reasoning, preferences among factors are established in the present case, and then these preferences can be applied to the current case. One of the argumentation schemes (CS1) of can be used to briefly explain how such schemes are meant to be used in legal arguments from precedent. In all these schemes, for purposes of presentation, it is assumed that the arguer is putting forward the current argument (curr) to support the side of the plaintiff.

$$\frac{\text{commonPfactors(curr; prec)} = p,}{\text{commonDfactors(curr; prec)} = d,}$$
$$\text{preferred}(p; d)$$

outcome(curr) = Plaintiff

According to this scheme, the current argument should be decided for the plaintiff because the common p factors were preferred to the common d factors in the precedent argument. [Prakken et al., 2015] uses a running example to illustrate how an argument fitting a scheme can be attacked by other arguments in the formal system representing the argumentation in a legal case.

Argument schemes are being used in AI and Law for argument mining. Moens, Mochales Palau, Boiy and Reed devised techniques for automatically classifying arguments in legal texts by using indicators of rhetorical structure expressed by conjunctions and adverbial groupings [Moens et al., 2007, p.226]. They identify words, pairs of successive words, sequences of three successive words, adverbs, verbs and modal auxiliary verbs. This work has been applied to legal argumentative texts ([Mochales Palau and Moens, 2009]; [Mochales Palau and Moens, 2011]). By classifying types of arguments using argumentation schemes they built a system for searching for arguments in legal cases [Mochales

Palau and Moens, 2008]. The project used human annotators supervised by legally trained personnel to identify arguments in texts of the European Court of Human Rights [Mochales Palau and Ieven, 2009]. Their results suggested that it would help to have additional criteria that can be applied to judge whether a given argument fits a particular scheme.

Rahwan et al. [Rahwan *et al.*, 2011] carried forward research on the automated identification of particular schemes by developing an OWL-based ontology of argumentation schemes in description logic that showed how description logic inference techniques can be used to reason about automatic argument classification. Their method of identifying schemes has been implemented in a web-based system called Avicenna (Rahwan et al., 2011, pp. 1113). A user can search arguments by using schemes along with other tools.

Gordon and Walton [Gordon and Walton, 2006] proposed a solution to the problem of how to model critical questions by using three kinds of premises (ordinary premises, assumptions and exceptions) in the Carneades Argumentation System. This solution used information about the dialectical status of statements (undisputed, at issue, accepted or rejected) to model critical questions in such a way as to allow the burden of proof to be allocated to the proponent of the argument or the critical questioner as appropriate for the case in point. On this way of proceeding, ordinary premises need to be supported by further arguments even if they have not been questioned. In the case of exceptions, however, the critical questioner is the one who has to offer evidential support to make his criticism defeat the argument.

Version 4 is the current implemented formal and computational system of Carneades, based on the formal model of argument [Gordon and Walton, 2016] called Carneades 2. Carneades 2 provides support for cumulative arguments, cyclic argument graphs, practical reasoning, and multi-criteria decision analysis. The source code of all four versions can be accessed on the Internet[6]. Carneades 4 is now online [7]. Carneades 2, as implemented in version 4 of Carneades, provides a formal model that uses argumentation schemes.

In the Carneades 2 model [Gordon and Walton, 2016] an argumentation scheme is defined as a tuple (e, v, g), where e is a function for weighing arguments which instantiate a scheme, v is a function for validating arguments, to test whether they properly instantiate an argumentation scheme, and g is a function for generating arguments by instantiating the scheme. The validation function tells us whether the argument instantiates a particular scheme, but then, once a set of schemes has been specified, the system can apply their validation functions to given argument to whether that scheme is instantiated, or not, by the given argument.

An argument is defined as a tuple (S, P, C, U), where S is the scheme instantiated by the argument; P, a finite subset of L, is the set of premises of the argument; C, a member of L, is the conclusion of the argument. U is

---

[6] Retrieved from: https://github.com/carneades
[7] Retrieved from: http://carneades.fokus.fraunhofer.de/carneades

an undercutter of the argument [Pollock, 1995]. In version 4 of Carneades an issue is defined as a tuple (O, F), where O represents the options (called the alternative positions) of the issue, and F is the proof standard of the issue. Argument graphs in Carneades version 4 are tripartite, rather than bipartite, as in the previous versions, with separate nodes for statements, arguments and issues. Argument diagrams in version 4 are extended with a new node type, diamonds, for representing issues. There can be any number of issues you like in a single diagram. Argument evaluation is carried out by labeling statements *in, out* or *undecided*. A statement is *in* if and only if it has been assumed to be acceptable to a rational audience, or has been derived from such assumptions via the application of the arguments, argument weighing functions and proof standards used in Carneades. A statement is *out* if and only if it is neither assumed nor supported by arguments and would therefore be rejected by a rational audience. A statement is *undecided* if it is neither *in* nor *out*.

Carneades 3 uses backwards-chaining, in a goal-directed way, whereas Carneades 4 uses forwards-reasoning to derive arguments from argumentation schemes and assumptions. Both strategies, forwards and backwards reasoning, have their advantages. Forwards reasoning allows Carneades to invent arguments using argumentation schemes, such as the scheme for argument from expert opinion, where the conclusion is a second-order variable ranging over propositions. Only Carneades 4 can construct arguments using formalizations of all of the twenty or so schemes currently built into the system.

Case-based reasoning (CBR) is vitally important for AI and Law and for understanding legal reasoning generally. CBR evaluates an argument in a given case by comparing and contrasting its features to those of prior cases that have already been evaluated [Aleven, 1997]. These prior cases are stored in a knowledge base which supplies similar precedent cases that can be pro or con the evaluation being considered in the given case. In some systems widely known in AI and law [Ashley, 1990], judgments of similarity between a pair of cases are decided by the factors that they share. Special argumentation schemes have been built to model arguments from precedent using factors in case-based reasoning ([Gordon and Walton, 2009]; [Wyner and Bench-Capon, 2007]; [Wyner et al., 2011]). Prakken et al. [Prakken et al., 2015] offered a formal version of these legal case-based argumentation schemes using ASPIC+.

Walton, Sartor and Macagno [Walton et al., 2016] showed how canons of interpretation can be translated into argumentation schemes. This project was carried out by by analyzing the most common types of statutory arguments found in legal examples and certain key forms of interpretive legal argumentation found in the work of Tarello [Tarello, 1980] and McCormick and Summers [MacCormick and Summers, 1991]. Steps were carries out to show how these legally recognizable forms of argument can be formulated as argumentation schemes. Among the schemes modeled are argument from ordinary meaning, argument from technical meaning, argument from precedent, argument from purpose, *a contrario* argument, historical argument and the non-redundancy argument. It was shown using classical examples of statutory interpretation in law how these schemes (and others) can be incorporated into computational argumentation systems such as Carneades and ASPIC+ and applied to displaying the pro-contra structure argumentation in legal cases using argument mapping tools.

In the following sections we will illustrate shortly two other computational applications of argumentation schemes, namely their role in argument mining and formal ontologies.

9 Using Schemes for Argument Mining

Argumentation schemes also have an important role to play in a major new area of computational research into argumentation: argument mining. Argument mining focuses on the development of algorithms and techniques for the automatic extraction of argument structure from natural language text. Though it has connections to areas such as sentiment analysis and opinion mining, it represents a substantially more demanding task. There are two features that make argument mining so difficult. The first concerns the availability of data and the second, the limits of statistical approaches to language understanding.

Many approaches to mining syntactic and semantic structure from unrestricted natural language have, since the late 1990s, been based heavily in statistical analysis: essentially, modelling the regularities in language by examining and comparing many, many different examples. The most robust syntactic parsers, for example, are based not on theoretical linguistic analysis which proved on the whole to be too limited and too brittle, but on statistical models based on corpora typically comprising millions of examples [Koehn et al., 2003]. Though the machine learning mechanisms upon which such techniques depend vary, one feature that they share is the need for such large datasets from which to draw regularities. If, therefore, argument mining is to be able to deploy the same techniques, it requires large datasets, and datasets not just of argumentation *per se*, by argumentation that has been analysed for its structure. As anyone involved in the teaching of critical thinking skills will attest, such analysis of argument structure is both demanding and extremely time consuming. Until very recently there were few datasets, and those that did exist were available in idiosyncratic representation languages, with little re-use between research teams and projects, so what effort was invested in data collection and analysis was regularly lost. Two approaches have started to change this.

First, there have been attempts to collect datasets specifically for community use. The first example is the Internet Argument Corpus, IAC [Walker et al., 2012], which collects 390,000 examples. The problem facing the IAC is that it is designed primarily from a text-processing viewpoint, with little argumentation theory sitting behind it. As a result, the conception of argument that it embodies is very thin and more or less unrecognisable to researchers from argumentation theory and computational models of argument, viz., quote-response pairs with associated polarity (additional features including sarcasm and nastiness are marked for subsets). A second example is more directly rooted in informed models of argumentation. The Potsdam Microtext Corpus [Peldszus and Stede, 2016] provides artificially constrained but completely human-generated, natural language arguments that are structured according to the work of Freeman [Freeman, 1991] with explicit distinction between, for example, linked and convergent arguments, undercutting and rebutting attacks and so on. Another unique advantage of the Microtext Corpus is that it has been professionally translated so that both English and German versions exist: to our knowledge this is the first parallel corpus of argumentation. On the other hand, the fact that every argument is required to contain a total of five components (premises and conclusions), whilst providing a vitally useful 'laboratory' for testing techniques, risks placing a severe limitation on the subsequent generalizability of those techniques to unrestricted arguments in the wild. The limited size of the corpus (just 130 examples) also places limitations on what can be accomplished using traditional statistical machine learning techniques.

The second approach has been to provide infrastructure specifically for collecting, publishing, sharing and re-using corpora. Whilst there are now several platforms for online analysis of argument (argunet [8], debategraph [9], AGORA-net [10], RationaleOnline [11], etc.) none provide open access to the data in machine processable ways, except, as far as we are aware, for the infrastructure offered by the Argument Web ([Rahwan et al., 2011]; [Bex et al., 2013]). The Argument Web is a vision for an interconnected web of arguments and debates, regardless of the software used to create them, analyse them or extract them, and regardless, too, of the uses academic, social or commercial to which they might be put. The vision supports, for example, the academic analysis of an argument presented in a political broadcast; the automated analysis of responses to it on social media; the deployment of automated dialogue games for online users to interact with both original and responses; the automated summary of the status of the debate to a government policy department; and the delivery of a corpus comprising the debate to researchers in argument mining. Argumentation schemes in the style of [Walton et al., 2008] form a cornerstone of the Argument Web, as a way of providing a rich ontology of reasoning forms. Further details of this ontology occur in the next section; here we focus on the tools and the ways in which they can be used to develop corpora.

Though the first publicly available corpus of argumentation was developed using Araucaria (viz. AraucariaDB, see [Reed and Walton, 2005]), the software itself is now very old and virtually obsolete. Though it remains the only software to handle large analyses, such as the ones developed by Wigmore for mapping cases, and the only to interchange between Wigmore, Toulmin and Freeman styles of analysis, it has been superseded in its core functionality by the Online Visualisation of Argument tool, OVA [Janier et al., 2014].

---

[8]Retrieved from: http://www.argunet.org/ (Accessed on 10 May 2016)
[9]Retrieved from: http://debategraph.org/Stream.aspx?nid=61932&vt=ngraph&dc=focus (Accessed on 10 May 2016)
[10]http://agora.gatech.edu/ (Accessed on 10 May 2016)
[11]https://www.rationaleonline.com/ (Accessed on 10 May 2016)

OVA provides a simple-to-use interface for analysing existing argumentation in both monologue and, in the extended OVA+, also dialogue. It supports enthymeme reconstruction; argumentation scheme analysis; critical question processing; serial, linked, convergent and divergent structures; undercutting, rebutting and undermining attacks; and in OVA+, locution analysis; dialogue game rule analysis; illocutionary force identification; the role of *ethos* [Duthie et al., 2016] and personal attacks; and ultimately, full Inference Anchoring Theory analysis [Budzynska and Reed, 2011]. Analyses from OVA can be stored in AIFdb, a database infrastructure fabric for storing and accessing argument data [Lawrence et al., 2012].

One side effect of using AIFdb is that the data is easily transportable to other forms, both representational (in being able to convert to formats required for Carneades [Walton and Gordon, 2012] and Rationale [van Gelder, 2007], for example), and processable, in being able to convert via ASPIC+ [Modgil and Prakken, 2013] to abstract frameworks [Dung, 1995] via formal equivalences established in [Bex et al., 2013]. More importantly for our current purposes, sets of analyses in AIFdb can be configured to constitute a corpus using the AIFdb corpus management tools [Lawrence and Reed, 2015] available online at corpora.aifdb.org, and AIFdb current constitutes the largest publicly available dataset of analysed argumentation. These tools enable research teams to define corpora comprising both analysed argumentation and raw text; both argumentative and non-argumentative source material; both raw data and metadata. Corpora themselves are aggregable providing flexible structuring options to manage dependencies between teams, projects, and objectives. The original AraucariaDB corpus is available on this infrastructure, but so too are smaller datasets focusing specifically on argumentation schemes, such as the Argument Schemes in the Moral Maze, comprising excerpts from the BBC Moral Maze radio programme that involve 35 instances of argumentation schemes and the ExpertOpinion-PositiveConsequences corpus comprising 71 examples of just these two schemes.

With the availability of appropriate datasets becoming less of an impediment, various approaches to automatically recognising argument structure have been developed. The majority have been focused specifically on statistical models, which brings us to the second major challenge facing argument mining: the limits of such models. Whilst it is certainly the case that statistical approaches are starting to deliver results for argument mining, and will undoubtedly continue to do so, it is also the case that the more sophisticated conceptions of argument developed in argumentation theory remain extraordinarily demanding. The reason for this lies precisely in their sophistication. With so many patterns of argumentation, so many structures, so many ways in which components can be left implicit, so many types of reasoning, the amount of data required to train statistical models becomes not just unwieldy but unreasonable and, quite probably, unattainable.

Consider a comparison with syntactic analysis, where statistical models have been so successful. The number of rules governing how different parts of speech can be legally combined run in theoretical linguistics to tens of examples. In statistical models, it is hundreds (which is why they are so successful). The number of rules governing how argument components can be assembled (and left implicit) runs, by combination across argumentation schemes, to thousands or more. So whilst we might expect statistically oriented techniques to deliver us good results on simple and strongly generalizable aspects of argument recognition, for the type of analysis that is typically taught to students of critical thinking classes, more is required. It is looking increasingly likely that having strong, well defined conceptions of argument, dialogue and argument schemes provide exactly the sort of additional information required to guide machine learning processes by acting, in essence, as priors to that process: defining expectations about what is likely to be seen. This combination of statistical and structural approaches is looking very promising. In particular, we provide examples here that tap in specifically to structure provided by argument schemes.

Feng and Hirst [Wei Feng and Hirst, 2011] aimed to classify arguments into the type of scheme employed. Like some of the earliest work in argument mining, such as [Moens and Mochales Palau, 2007], they also used the AraucariaDB corpus as a starting point, because it was the only dataset at that time with annotated examples of argumentation schemes. They used the 65 argumentation schemes from [Walton et al., 2008], but emphasized the importance of the five schemes they found to be the most commonly used ones in their corpus: argument from example, argument from cause to effect, practical reasoning, argument from consequences and argument from verbal classification [Wei Feng and Hirst, 2011]. The number of occurrences of these most common five schemes constituted 61% of the kinds of arguments identified in their database [Wei Feng and Hirst, 2011, p.998]. They used a variety of features with which to train the machine learning classifiers including key words and phrases as textual indicators of argumentation schemes. They identified, for example, twenty-eight keywords and phrases associated with the scheme for practical reasoning, including 'want', 'aim', 'objective', and modal verbs like 'should', 'must' and 'need' [Wei Feng and Hirst, 2011, p.991]. Their results were extremely promising, providing classification accuracies ranging from 0.64 to 0.98.

Building on this approach, Lawrence and Reed ([Lawrence and Reed, 2015] extended the model to use argumentation schemes not just as a target for machine learning but to aid the very process of identifying argumentative structure (rather than presupposing it as input, as in Feng and Hirst). The intuition is that argumentation schemes do not connect propositions that are all alike, but rather are associated with particular types of propositions. In this way, arguments from positive consequence will typically conclude with a normative statement in the subjunctive mood; arguments from expert opinion will typically have a premise which reports, either directly or indirectly, the speech of another; arguments from analogy will include a premise which attributes some property to some individual; and so on. If it is possible to identify instances of some of these types, it will constrain the potential argument structures that can be reassembled. If, for example, an automatic algorithm can spot the lexeme said, there is a reasonable chance that we have reported speech, which in turn increases the chance that it is part of an expert opinion argument. If we can find the lexeme expert in a sentence close by, we can be even more sure we have argument from expert opinion and can start looking nearby for a conclusion and that conclusion is likely to be a sentence which has strong semantic similarity with the clause that follows "said". In this way, knowing *a priori* about argumentation scheme structure helps to constrain the problem of automatically recognising the argument structure. It turns out that this hypothesis is borne out by results.

Lawrence and Reed report (ibid.) results ranging from an F1 performance of 0.59 to 0.91 for detecting scheme components and of 0.62 to 0.88 for identifying scheme instances. Operationalising argumentation scheme structure in this way depends, however, upon 'knowledge engineering', or, more specifically, 'ontology engineering' the construction of explicit computational models that capture scheme structure and the commonalities, similarities and classificatory relationships between schemes. It is to this question that we turn next.

10 Schemes in Formal Ontologies

The Argument Interchange Format, AIF, is not just a representation language for argument structure; it also has a formal definition rooted in description logic; that is to say, it provides a core ontology for describing argument (though that core is rather compact, it admits of extension using 'adjunct ontologies' that extend it to handle features such as dialogical interaction, user- and social-oriented features, and so on). The AIF was laid out initially in [Chesñevar *et al.*, 2006] and extended in its description logic specification in [Rahwan *et al.*, 2007]. Given this basis, it is then rather straightforward to extend it to further specify not just that two propositions might be linked by an application of a rule of inference (or 'RA'), but to also specify the different types of such rules of inference, that is to define an ontology of argumentation schemes. This ontology not only describes the structure of argumentation schemes in machine-processable form, but also defines relationships between schemes (such as generalisation-specification relationships) and relationships between scheme components (such as that knowledge assertions occur as premises in several different schemes). By way of example, snippets of the ontology concerned with the argumentation scheme from expert opinion are shown in Figure 17 below.

```xml
<Class IRI="#ExpertOpinion_Inference"/>
    <ObjectIntersectionOf>
        <Class IRI="#Presumptive_Inference"/>
        <ObjectSomeValuesFrom>
          <ObjectProperty IRI="#hasConclusion"/>
          <Class IRI="#KnowledgePosition_Statement"/>
        </ObjectSomeValuesFrom>
        <ObjectSomeValuesFrom>
          <ObjectProperty IRI="#hasFieldExpertise_Premise"/>
          <Class IRI="#FieldExpertise_Statement"/>
        </ObjectSomeValuesFrom>
        <ObjectSomeValuesFrom>
          <ObjectProperty IRI="#hasKnowledgeAssertion_Premise"/>
          <Class IRI="#KnowledgeAssertion_Statement"/>
        </ObjectSomeValuesFrom>
    </ObjectIntersectionOf>

<Class IRI="#ExpertOpinion_Inference"/>
    <ObjectIntersectionOf>
        <ObjectSomeValuesFrom>
          <ObjectProperty IRI="#hasCredibilityOfSource_Presumption"/>
          <Class IRI="#CredibilityOfSource_Statement"/>
        </ObjectSomeValuesFrom>
        <ObjectSomeValuesFrom>
          <ObjectProperty IRI="#hasExpertiseBackUpEvidence_Presumption"/>
          <Class IRI="#ExpertiseBackUpEvidence_Statement"/>
        </ObjectSomeValuesFrom>
        <ObjectSomeValuesFrom>
          <ObjectProperty IRI="#hasExpertiseInconsistency_Exception"/>
          <Class IRI="#ExpertiseInconsistency_Statement"/>
        </ObjectSomeValuesFrom>
        <ObjectSomeValuesFrom>
          <ObjectProperty IRI="#hasLackOfReliability_Exception"/>
          <Class IRI="#LackOfReliability_Statement"/>
        </ObjectSomeValuesFrom>
    </ObjectIntersectionOf>

<Class IRI="#ExpertiseInconsistency_Conflict"/>
    <ObjectIntersectionOf>
        <Class IRI="#Exception_Conflict"/>
        <ObjectExactCardinality cardinality="1">
          <ObjectProperty IRI="#hasConflictedElement"/>
          <Class IRI="#ExpertOpinion_Inference"/>
        </ObjectExactCardinality>
        <ObjectExactCardinality cardinality="1">
          <ObjectProperty IRI="#hasExpertiseInconsistency_Exception"/>
          <Class IRI="#ExpertiseInconsistency_Statement"/>
        </ObjectExactCardinality>
    </ObjectIntersectionOf>
```

Figure 17. Snippet of Argumentation Scheme Ontology

In the first stanza, the conclusion and premises concerning the knowledge assertion (that the expert said something) and the field expertise (that the speaker is indeed an expert) are set up. In the second stanza, the remaining premises (those captured as presumptions and exceptions) are added in, covering credibility, backup evidence, consistency between experts and expert reliability. The third stanza shows how one of these, consistency between experts, can be used to drive a stereotypical way of attacking this inference i.e. the posing of a critical question (see [Reed and Walton, 2005] for the mechanics of operationalizing critical questions in this way). The aim here is just to give a flavour of how all the important components of argumentation schemes structure, description and critical questions can be captured in a formal ontology. The full ontology is available online at http://arg.tech/aif.owl, and is used by many of the Argument Web online services.

Two benefits of this approach are demonstrated in [Rahwan et al., 2011]. The first is an economy in specification, that allows more specific schemes to be defined in terms of minor additions to more general ones. The second, much more importantly, is that these structures support automated reasoning, in three distinct ways. First, it becomes possible to reason across argument structures, identifying, for example, transitivity of inferences, so that if X is used to infer Y and Y to infer Z, the dependence of X on Z can be inferred automatically. Of course such reasoning is not at all unique to ontologically based systems, but is a convenient side benefit. An ontologically more interesting way of performing automated reasoning is to perform automatic classification. This is where formal ontologies, and the reasoning systems constructed on top of them, excel. Rahwan et al., exemplify this technique by showing how fear appeal arguments are naturally classifiable as a subset of negative consequence arguments. The third and final way of performing automated reasoning is also of use in designing and implementing dialogue systems. By virtue of hierarchical relationships between schemes that are represented in, or inferable from, the ontology, it also becomes possible to infer appropriate critical questions that might be asked of a given argument. Thus, for example, all of the critical questions of a superclass can be asked of an instance of a sub-class of argumentation schemes. In these ways, formal representation of argumentation schemes in an explicit ontology can contribute to the computational techniques for analysing, processing and interacting with arguments.

11 Conclusions

Argumentation schemes represent the abstract structures of the most common and stereotypical arguments used in everyday conversation and specific fields, such as law, science and politics. They appear as a set of premises having an abstract form with variables and constants, leading to an abstract conclusion. They are abstract in the sense that they provide a form for structuring inferential relation between the premises and the conclusion. Some schemes are based on the most abstract relations (classification, cause, authority), while others specify the most abstract premises including some further detail (negative consequences; expert opinion *ad populum* argument) [Walton et al., 2008].

The abstract nature of the schemes allows the analyst to detect the structure of natural arguments, and recognize patterns occurring in everyday reasoning. This chapter has shown how they can be applied to real arguments in natural language discourse, and in technical discourse as well, for that matter. In this chapter it has been shown how these schemes, at their current state of development can be used as tools to identify kinds of arguments in a text, and beyond that how they can be an important part of argument evaluation.

Throughout the history of logic and rhetoric there has always been some uncertainty about the role of the topics [Bird, 1962]. Some have seen them as forms of logical inference that can be used to show that arguments are valid, where the term 'valid' is used in a wider sense that can include not only deductively valid arguments but also defeasible arguments that have an identifiable structure as fitting a particular topic. Others have seen the topics of as having a search function that can be used to find arguments to prove a designated conclusion. The search function is supposed to help an arguer select arguments that have premises accepted by the audience to whom the argument is directed [Kienpointner and Kindt, 1997].

Schemes can also be used for argument construction. As we saw in this chapter, an argumentation scheme is taken to have a warranting function that enables an inference to be drawn from a set of premises to a conclusion. This practical way of justifying schemes indicates their usefulness not only for argument evaluation, but also for argument construction, also called argument invention in the long history of the subject tracing back to the Sophists and Aristotle. An argument invention device would enable an arguer to search for an argument that could be used to support a claim s/he wants to prove [Kienpointner, 1987]. When viewed in this way, topics can be seen to have a use as components of an argument construction function, for use in a system for finding arguments. The schemes can be used as instruments for producing arguments, allowing the user to decide the type of argument he considers the most applicable to his purpose, and then develop a specific line of reasoning from the premises or evidential facts he has to the conclusion he needs to prove. In this guise, the schemes are dialectical instruments for use in the task of argument construction.

The advent of IBM's new Watson Debater tool [Aharoni et al., 2014] is a leap forward for argument invention because it enables a user to quickly search through a database such as Wikpedia and find useful pro and con arguments supporting or attacking a designated claim. Once this tool comes onto the market, it will greatly stimulate research on argument invention in argumentation studies. The Debater tool does not (so far) use argumentation schemes, but there is a formal and computational argumentation system, the Carneades Argumentation System [12]. By inputting information into the Carneades find arguments assistant, a user who has a database containing propositions recording the commitments of the audience, the automated assistant constructs a chain of argumentation where the conclusion of the chain is the proposition is the goal proposition that the speaker wants to persuade the audience to accept, called the arguer's ultimate claim, or ultimate *probandum*, the proposition to be proved, in the language of the ancient *stasis* theory [Walton and Gordon, 2012]. The argument assistant searches through the commitments of the audience and uses a repository of argumentation schemes in its knowledge base to collect a set of arguments moving from these premises to the ultimate claim.

---

[12] https://carneades.github.io/ (Accessed on 10 May 2016)

If there are such arguments available the assistant gives that information, but may also suggest a partial way forward.

Argumentation schemes are instruments that can be applies in different ways in many disciplines addressing the analysis of discourse in general, and reasoned discourse in particular. The current research on schemes can improve noticeably the field of application and make this tool crucial for a deeper analysis of argumentative exchanges. To this purpose, argumentation schemes need first to be integrated within a theory of discourse interpretation. Schemes can be powerful instruments for representing arguments and relations between sentences. However, at present they presuppose an interpretation of discourse. This line of research could show how schemes can represent interpretation, and how they can be used to assess what interpretation is the best one [Macagno, 2012]. A second challenge in this area is to link the theory of dialogue types and discourse moves (utterances) to argumentation schemes [Macagno and Bigi, 2017]. By showing how certain schemes are the most adequate to pursue specific dialogical ends, it is possible to map not only a set of useful tools for argument production, but also a set of presumptions for interpreting and classifying arguments based on the type of dialogue.

Acknowledgments

Fabrizio Macagno would like to thank the Fundação para a Ciência e a Tecnologia for the research grants no. IF/00945/2013, PTDC/IVC-HFC/1817/2014, and PTDC/MHC-FIL/0521/2014.

Douglas Walton would like to thank the Social Sciences and Humanities Research Council of Canada for Insight Grant 435-2012-0104.

BIBLIOGRAPHY

[Abaelardus, 1970] Petrus Abaelardus. *Dialectica*. Van Gorcum, Assen, 1970.

[Aharoni et al., 2014] Ehud Aharoni, Carlos Alzate, Roy Bar-Haim, Yonatan Bilu, Lena Dankin, Iris Eiron, Daniel Hershcovich, and Shay Hummel. Claims on demand–an initial demonstration of a system for automatic detection and polarity identification of context dependent claims in massive corpora. In *COLING 2014, the 25th International Conference on Computational Linguistics: System Demonstrations*, pages 6–9, 2014.

[Aleven, 1997] Vincent Aleven. *Teaching case-based argumentation through a model and examples*. PhD thesis, University of Pittsburgh, 1997.

[Aristotle, 1991a] Aristotle. Rhetoric. In Jonathan Barnes, editor, *The complete works of Aristotle, vol. I*. Princeton University Press, Princeton, 1991.

[Aristotle, 1991b] Aristotle. Topics. In Jonathan Barnes, editor, *The complete works of Aristotle, vol. I*. Princeton University Press, Princeton, 1991.

[Ashley, 1990] Kevin Ashley. *Modeling Legal Argument: Reasoning with Cases and Hypotheticals*. MIT Press, Cambridge, Mass., 1990.

[Atkinson et al., 2005] Katie Atkinson, Trevor Bench-Capon, and Peter McBurney. Arguing about cases as practical reasoning. In Giovanni Sartor, editor, *Proceedings of the 10th international conference on Artificial intelligence and law*, pages 35–44, New York, 2005. ACM Press.

[Bench-Capon, 2003] Trevor Bench-Capon. Persuasion in practical argument using value-based argumentation frameworks. *Journal of Logic and Computation*, 13(3):429–448, 2003.

[Bex and Reed, 2011] Floris Bex and Christopher Reed. Schemes of inference, conflict, and preference in a computational model of argument. *Studies in Logic, Grammar and Rhetoric*, 23(36):39–58, 2011.
[Bex et al., 2013] Floris Bex, John Lawrence, Mark Snaith, and Chris Reed. Implementing the argument web. *Communications of the ACM*, 56(10):66–73, oct 2013.
[Bird, 1960] Otto Bird. The formalizing of the topics in mediaeval logic. *Notre Dame Journal of Formal Logic*, 1(4):138–149, 1960.
[Bird, 1962] Otto Bird. The tradition of the logical Topics: Aristotle to Ockham. *Journal of the History of Ideas*, 23(3):307–323., 1962.
[Brockriede and Ehninger, 1963] Wayne Brockriede and Douglas Ehninger. *Decision by debate*. Dodd, Mead & Co,, New York, 1963.
[Budzynska and Reed, 2011] Katarzyna Budzynska and Chris Reed. Speech acts of argumentation: Inference anchors and peripheral cues in dialogue. In *Computational models of natural argument: Papers from the 2011 AAAI workshop*, pages 3–10, Menlo Park, 2011. AAAI Press.
[Chesñevar et al., 2006] Carlos Chesñevar, Sanjay Modgil, Iyad Rahwan, Chris Reed, Guillermo Simari, Matthew South, Gerard Vreeswijk, and Steven Willmott. Towards an argument interchange format. *The Knowledge Engineering Review*, 21(4):293–316, dec 2006.
[Christensen, 1988] Johnny Christensen. The formal character of *koinoi topoi* in Aristotle's rhetoric and dialectic. Illustrated by the list in Rhetorica II, 23. *Cahiers de l'institut du moyen âge grec et latin*, (57):3–10, 1988.
[Cicero, 2003] Marcus Tullius Cicero. *Topica*. Oxford University Press, Oxford, 2003.
[De Pater, 1965] Wilhelm A De Pater. *Les Topiques d'Aristote et la dialectique platonicienne*. Éditions de St. Paul, Fribourg, 1965.
[Drehe, 2011] Iovan Drehe. The Aristotelian dialectical topos. *Argumentum*, 9(2):129–139, 2011.
[Dung, 1995] Phan Minh Dung. On the acceptability of arguments and its fundamental role in nonmonotonic reasoning, logic programming and n-person games. *Artificial Intelligence*, 77(2):321–357, 1995.
[Duschl et al., 1999] Richard Duschl, Kirsten Ellenbogen, and Sibel Erduran. Promoting argumentation in middle school science classrooms: A project SEPIA evaluation. In *Annual Meeting of the National Association for Research in Science Teaching*, pages 1–19, Boston, 1999. ERIC Document Reproduction Service No.ED453050.
[Duthie et al., 2016] Rory Duthie, Katarzyna Budzynska, and Chris Reed. Mining *ethos* in political debate. In *Proceedings of the Sixth International Conference on Computational Models of Argument*, pages 299–310, Amsterdam, 2016. IOS Press.
[Erduran and Jimenez-Aleixandre, 2007] Sibel Erduran and M Pilar Jimenez-Aleixandre, editors. *Argumentation in science education: perspectives from classroom-based research*. Springer, Dordrecht, 2007.
[Erduran and Jiménez Aleixandre, 2012] Sibel Erduran and Maria Pilar Jiménez Aleixandre. Argumentation in science education research. In Doris Jorde and Justin Dillon, editors, *Science Education Research and Practice in Europe: Retrospective and Prospective*, pages 253–289. SensePublishers, Rotterdam, 2012.
[Freeman, 1991] James Freeman. *Dialectics and the Macrostructure of Arguments. A Theory of Argument Structure*. Foris, Berlin/New York, 1991.
[Gabbay and Woods, 2008] Dov Gabbay and John Woods, editors. *Handbook of the history of logic. Volume 2. Mediaeval and Renaissance Logic*. Elsevier, Amsterdam, 2008.
[Garssen, 2001] Bart Garssen. Argumentation schemes. In Frans Van Eemeren, editor, *Crucial concepts in argumentation theory*, pages 81–99. Amsterdam University Press, Amsterdam, 2001.
[Gordon and Walton, 2006] Thomas Gordon and Douglas Walton. The Carneades argumentation framework: Using presumptions and exceptions to model critical questions. In *Proceedings of the 2006 conference on Computational Models of Argument: Proceedings of COMMA 2006*, pages 195–207, Amsterdam, 2006. IOS Press.
[Gordon and Walton, 2009] Thomas Gordon and Douglas Walton. Legal reasoning with argumentation schemes. In Carole D. Hafner, editor, *Proceedings of the 12th International Conference on Artificial Intelligence and Law*, pages 137–146, New York, 2009. ACM.

[Gordon and Walton, 2016] Thomas Gordon and Douglas Walton. Formalizing Balancing Arguments. In *Proceedings of the 2016 Conference on Computational Models of Argument (COMMA 2016)*, pages 327–338, Amsterdam, 2016. IOS Press.

[Gordon et al., 2007] Thomas Gordon, Henry Prakken, and Douglas Walton. The Carneades model of argument and burden of proof. *Artificial Intelligence*, 171(10-15):875–896, jul 2007.

[Green-Pedersen, 1984] Niels Jørgen Green-Pedersen. *The tradition of the Topics in the Middle Ages: The commentaries on Aristotle's and Boethius' Topics*. Philosophia, Munich, 1984.

[Green-Pedersen, 1987] Niels Jørgen Green-Pedersen. The topics in medieval logic. *Argumentation*, 1(4):407–417, 1987.

[Grennan, 1997] Wayne Grennan. *Informal logic*. McGill-Queen's University Press, Montreal, Quebec, 1997.

[Hastings, 1963] Arthur Hastings. *A reformulation of the modes of reasoning in argumentation*. Ph.D. Dissertation, Northwestern University, Evanston, Illinois, 1963.

[Heath, 1994] Malcolm Heath. The Substructure of stasis-theory from Hermagoras to Hermogenes. *The Classical Quarterly*, 44(01):114, may 1994.

[Janier et al., 2014] Mathilde Janier, John Lawrence, and Chris Reed. OVA+: An argument analysis interface. In *Proceedings of the Fifth International Conference on Computational Models of Argument (COMMA 2014)*, pages 463–464, Amsterdam, 2014. IOS Press.

[Katzav and Reed, 2004a] Joel Katzav and Chris Reed. A Classification System for Arguments. Technical report, University of Dundee, Dundee, 2004.

[Katzav and Reed, 2004b] Joel Katzav and Chris Reed. On argumentation schemes and the natural classification of arguments. *Argumentation*, 18(2):239–259, 2004.

[Kennedy, 1963] George Kennedy. *The art of persuasion in ancient Greece*. Princeton University Press, Princeton, 1963.

[Kienpointner and Kindt, 1997] Manfred Kienpointner and Walther Kindt. On the problem of bias in political argumentation: An investigation into discussions about political asylum in Germany and Austria. *Journal of Pragmatics*, 27(5):555–585, 1997.

[Kienpointner, 1987] Manfred Kienpointner. Towards a typology of argumentative schemes. In Frans van Eemeren, Rob Grootendorst, Anthony Blair, and Charles Willard, editors, *Argumentation: Across the lines of discipline*, pages 275–287. Foris, Dordrech, 1987.

[Kienpointner, 1992] Manfred Kienpointner. *Alltagslogik: Struktur und Funktion von Argumentationsmustern*. Fromman-Holzboog, Stuttgart, Germany, 1992.

[Kim et al., 2010] Mijung Kim, Robert Anthony, and David Blades. Argumentation as a Tool to Understand Complexity of Knowledge Integration. In *Proceedings of the 2nd International STEM in Education Conference Beijing, China 24-27 November 2012*, pages 154–160, Beijing, 2010. Beijing Normal University.

[Koehn et al., 2003] Philipp Koehn, Franz Josef Och, and Daniel Marcu. Statistical phrase-based translation. In *Proceedings of the 2003 Conference of the North American Chapter of the Association for Computational Linguistics on Human Language Technology - NAACL '03*, volume 1, pages 48–54, Morristown, NJ, USA, 2003. Association for Computational Linguistics.

[Lawrence and Reed, 2015] John Lawrence and Chris Reed. Combining Argument Mining Techniques. In *Proceedings of the 2nd Workshop on Argumentation Mining*, pages 127–136, Stroudsburg, PA, USA, 2015. Association for Computational Linguistics.

[Lawrence et al., 2012] John Lawrence, Floris Bex, Chris Reed, and Mark Snaith. AIFdb: Infrastructure for the argument web. In *Proceedings of the Fourth International Conference on Computational Models of Argument (COMMA 2012)*, pages 515–516, Amsterdam, 2012. IOS Press.

[Lawson, 1885] John Lawson. *The law of presumptive evidence*. Bancroft & Co., San Francisco, 1885.

[Macagno and Bigi, 2017] Fabrizio Macagno and Sarah Bigi. Analyzing the pragmatic structure of dialogues. *Discourse Studies*, 19(2):148–168, apr 2017.

[Macagno and Konstantinidou, 2013] Fabrizio Macagno and Aikaterini Konstantinidou. What students' arguments can tell us: Using argumentation schemes in science education. *Argumentation*, 27(3):225–243, 2013.

[Macagno and Walton, 2014] Fabrizio Macagno and Douglas Walton. *Emotive Language in Argumentation*. Cambridge University Press, Cambridge, 2014.

[Macagno and Walton, 2015] Fabrizio Macagno and Douglas Walton. Classifying the patterns of natural arguments. *Philosophy and Rhetoric*, 48(1):26–53, 2015.
[Macagno et al., 2014] Fabrizio Macagno, Douglas Walton, and Christopher Tindale. Analogical reasoning and semantic rules of inference. *Revue internationale de philosophie*, 270(4):419–432, 2014.
[Macagno et al., 2016] Fabrizio Macagno, Douglas Walton, and Christopher Tindale. Analogical Arguments: Inferential Structures and Defeasibility Conditions. *Argumentation*, pages 1–23, 2016.
[Macagno, 2012] Fabrizio Macagno. Presumptive reasoning in interpretation. Implicatures and conflicts of presumptions. *Argumentation*, 26(2):233–265, 2012.
[Macagno, 2015] Fabrizio Macagno. A means-end classification of argumentation schemes. In Frans van Eemeren and Bart Garssen, editors, *Reflections on theoretical issues in argumentation theory*, pages 183–201. Springer, Cham, 2015.
[MacCormick and Summers, 1991] Neil MacCormick and Robert Summers. *Interpreting Statutes: A Comparative Study*. Dartmouth, Aldershot, 1991.
[Mochales Palau and Ieven, 2009] Raquel Mochales Palau and Aagje Ieven. Creating an argumentation corpus: do theories apply to real arguments?: a case study on the legal argumentation of the ECHR. In *Proceedings of the 12th International Conference on Artificial Intelligence and Law*, pages 21–30, New York, 2009. ACM Press.
[Mochales Palau and Moens, 2008] Raquel Mochales Palau and Marie-Francine Moens. Study on the structure of argumentation in case law. In Enrico Francesconi, Giovanni Sartor, and Daniela Tiscornia, editors, *Proceedings of the 2008 Conference on Legal Knowledge and Information Systems*, pages 11–20, Amsterdam, 2008. IOS Press.
[Mochales Palau and Moens, 2009] Raquel Mochales Palau and Marie Francine Moens. Argumentation Mining: the detection, classification and structuring of arguments in text. In *Belgian/Netherlands Artificial Intelligence Conference*, pages 351–352, New York, 2009. ACM.
[Mochales Palau and Moens, 2011] Raquel Mochales Palau and Marie Francine Moens. Argumentation mining. *Artificial Intelligence and Law*, 19(1):1–22, 2011.
[Modgil and Prakken, 2013] Sanjay Modgil and Henry Prakken. A general account of argumentation with preferences. *Artificial Intelligence*, 195:361–397, 2013.
[Moens and Mochales Palau, 2007] Marie-Francine Moens and Raquel Mochales Palau. Study on Sentence Relations in the Automatic Detection of Argumentation in Legal Cases. In Arno Lodder and Laurens Mommers, editors, *Legal Knowledge and Information Systems: JURIX 2007*, pages 89–98, Amsterdam, 2007. IOS Press.
[Moens et al., 2007] Marie-Francine Moens, Erik Boiy, Raquel Mochales Palau, and Chris Reed. Automatic detection of arguments in legal texts. In *Proceedings of the 11th international conference on Artificial intelligence and law*, pages 225–230, New York, 2007. ACM Press.
[Nussbaum and Edwards, 2011] Michael Nussbaum and Ordene V. Edwards. Critical questions and argument stratagems: A framework for enhancing and analyzing students' reasoning practices. *Journal of the Learning Sciences*, 20(3):443–488, 2011.
[Ozdem et al., 2013] Yasemin Ozdem, Hamide Ertepinar, Jale Cakiroglu, and Sibel Erduran. The nature of pre-service science teachers' argumentation in inquiry-oriented laboratory contex. *International Journal of Science Education*, 35(15):2559–2586, 2013.
[Peldszus and Stede, 2016] Andreas Peldszus and Manfred Stede. An annotated corpus of argumentative microtexts. In *Argumentation and Reasoned Action: Proceedings of the 1st European Conference on Argumentation*, pages 801–815, London, 2016. College Publications.
[Perelman and Olbrechts-Tyteca, 1969] Chaim Perelman and Lucie Olbrechts-Tyteca. *The New Rhetoric: A treatise on argumentation*. University of Notre Dame Press, Notre Dame, Ind., 1969.
[Pollock, 1995] John Pollock. *Cognitive Carpentry*. MIT Press, Cambridge, Mass., 1995.
[Prakken et al., 2015] Henry Prakken, Adam Wyner, Trevor Bench-Capon, and Katie Atkinson. A formalization of argumentation schemes for legal case-based reasoning in ASPIC+. *Journal of Logic and Computation*, 25(5):1141–1166, oct 2015.
[Prakken, 2005] Henry Prakken. AI & Law, logic and argument schemes. *Argumentation*, 19(3):303–320, 2005.

[Prakken, 2010] Henry Prakken. An abstract framework for argumentation with structured arguments. *Argument & Computation*, 1(2):93–124, 2010.

[Rahwan et al., 2007] Iyad Rahwan, Fouad Zablith, and Chris Reed. Laying the foundations for a World Wide Argument Web. *Artificial Intelligence*, 171(10-15):897–921, jul 2007.

[Rahwan et al., 2011] Iyad Rahwan, Bita Banihashemi, Christopher Reed, Douglas Walton, and Sherief Abdallah. Representing and classifying arguments on the semantic web. *The Knowledge Engineering Review*, 26(04):487–511, 2011.

[Rapanta and Macagno, 2016] Chrysi Rapanta and Fabrizio Macagno. Argumentation methods in educational contexts: Introduction to the special issue. *International Journal of Educational Research*, 79:online first, 2016.

[Rapanta and Walton, 2016] Chrysi Rapanta and Douglas Walton. The use of argument maps as an assessment tool in higher education. *International Journal of Educational Research*, 79:211–221, 2016.

[Rapanta et al., 2013] Chrysi Rapanta, Merce Garcia-Mila, and Sandra Gilabert. What is meant by argumentative competence? An integrative review of methods of analysis and assessment in education. *Review of Educational Research*, 83(4):483–520, 2013.

[Reed and Walton, 2005] Chris Reed and Douglas Walton. Towards a Formal and Implemented Model of Argumentation Schemes. In Iyad Rahwan, Pavlos Moraitis, and Chris Reed, editors, *Agent Communication, Argumentation in Multi-Agent Systems*, pages 19–30. Springer-Verlag, Berlin/Heidelberg, 2005.

[Reed et al., 2007] Christopher Reed, Douglas Walton, and Fabrizio Macagno. Argument diagramming in logic, law and artificial intelligence. *Knowledge Engineering Review*, 22(1):87–109, 2007.

[Rowe et al., 2006] Glenn Rowe, Fabrizio Macagno, Christopher Reed, and Douglas Walton. Araucaria as a tool for diagramming arguments in teaching and studying philosophy. *Teaching Philosophy*, 29(2):111–124, 2006.

[Rubinelli, 2009] Sara Rubinelli. *Ars Topica: The classical technique of constructing arguments from Aristotle to Cicero*, volume 15. Springer, Amsterdam, 2009.

[Schiappa, 1999] Edward Schiappa. *The Beginnings of Rhetorical Theory in Classical Greece*. Yale University Press, New Haven & London, 1999.

[Slomkowski, 1997] Paul Slomkowski. *Aristotele's Topics*. Brill, Leiden, 1997.

[Stump, 1982] Eleonore Stump. Topics: their development and absorption into consequences. In Norman Kretzmann, Anthony Kenny, and Jan Pinborg, editors, *Cambridge History of Later Medieval Philosophy*, pages 273–299. Cambridge University Press, Cambridge, 1982.

[Stump, 1989] Eleonore Stump. *Dialectic and its place in the development of Medieval logic*. Cornell University Press, Ithaca and London, 1989.

[Stump, 2004] Eleonore Stump. *Boethius's "De topicis differentiis"*. Cornell University Press, Ithaca and London, 2004.

[Tarello, 1980] Giovanni Tarello. *L'interpretazione della legge*. Giuffrè, Milano, 1980.

[Tindale, 2010] Christopher Tindale. *Reason's Dark Champions: Constructive Strategies of Sophistical Argument*. University of South Carolina Press, Columbia, 2010.

[Toulmin et al., 1984] Stephen Toulmin, Richard Rieke, and Allan Janik. *An introduction to reasoning*. Number Sirsi) i9780024211606. Macmillan Publishing Company, New York, 1984.

[Van Eemeren and Grootendorst, 1992] Frans Van Eemeren and Rob Grootendorst. *Argumentation, communication and fallacies*. Erlbaum, Hillsdale, NJ, 1992.

[van Gelder, 2007] Tim van Gelder. The rationale for RationaleTM. *Law, Probability and Risk*, 6(1-4):23–42, oct 2007.

[Verheij, 2003a] Bart Verheij. Deflog: on the logical interpretation of prima facie justified assumptions. *Journal of Logic and Computation*, 13(3):319–346, 2003.

[Verheij, 2003b] Bart Verheij. Dialectical argumentation with argumentation schemes. *Artificial Intelligence and Law*, 11(2):167–195, 2003.

[Walker et al., 2012] Marilyn Walker, Jean Fox Tree, Pranav Anand, Rob Abbott, and Joseph King. A Corpus for Research on Deliberation and Debate. In Nicoletta Calzolari, Khalid Choukri, Thierry Declerck, Mehmet Uur Doan, Bente Maegaard, Joseph Mariani, Asuncion Moreno Piperidis, Jan Odijk, and Stelios Piperidis, editors, *Proceedings of the 8th International Conference on Language Resources and Evaluation (LREC)*,, pages 812–817. European Language Resources Association, 2012.

[Walton and Gordon, 2012] Douglas Walton and Thomas Gordon. The Carneades model of argument invention. *Pragmatics & Cognition*, 20(1):1–26, 2012.

[Walton and Macagno, 2009] Douglas Walton and Fabrizio Macagno. Reasoning from classifications and definitions. *Argumentation*, 23(1):81–107, 2009.

[Walton and Macagno, 2015] Douglas Walton and Fabrizio Macagno. A classification system for argumentation schemes. *Argument and Computation*, 6(3):219–245, 2015.

[Walton and Sartor, 2013] Douglas Walton and Giovanni Sartor. Teleological Justification of Argumentation Schemes. *Argumentation*, 27(2):111–142, 2013.

[Walton et al., 2008] Douglas Walton, Christopher Reed, and Fabrizio Macagno. *Argumentation Schemes*. Cambridge University Press, New York, 2008.

[Walton et al., 2016] Douglas Walton, Giovanni Sartor, and Fabrizio Macagno. An argumentation framework for contested cases of statutory interpretation. *Artificial Intelligence and Law*, 24(1):51–91, 2016.

[Walton, 1989] Douglas Walton. *Informal logic*. Cambridge University Press, New York, 1989.

[Walton, 1992] Douglas Walton. *Slippery Slope Arguments*. Oxford University Press, Oxford, 1992.

[Walton, 1995] Douglas Walton. *Argumentation Schemes for Presumptive Reasoning*. Routledge, Mahwah, dec 1995.

[Walton, 2012] Douglas Walton. Using argumentation schemes for argument extraction: a bottom-up method. *International Journal of Cognitive Informatics and Natural Intelligence*, 6(3):33–61, 2012.

[Wei Feng and Hirst, 2011] Vanessa Wei Feng and Graeme Hirst. Classifying arguments by scheme. In *Proceedings of the 49th Annual Meeting of the Association for Computational Linguistics: Human Language Technologies-Volume 1*, pages 987–996, Stroudsburg, 2011. Association for Computational Linguistics.

[Westberg, 2002] Daniel Westberg. *Right practical reason: Aristotle, action, and prudence in Aquinas*. Clarendon Press, Oxford, 2002.

[Wyner and Bench-Capon, 2007] Adam Wyner and Trevor Bench-Capon. Argument Schemes for Legal Case-based Reasoning. In Arno Lodder and Laurens Mommers, editors, *Proceedings of the 2007 Conference on Legal Knowledge and Information Systems: JURIX 2007: The Twentieth Annual Conference*, pages 139–149, Amsterdam, 2007. IOS Press.

[Wyner et al., 2011] Adam Wyner, Trevor Bench-Capon, and Katie Atkinson. Towards formalising argumentation about legal cases. In *ICAIL '11 Proceedings of the 13th International Conference on Artificial Intelligence and Law*, pages 1–10, New York, 2011. ACM Press.

Fabrizio Macagno
ArgLab, Instituto de Filosofia da Nova (IFILNOVA)
Facultade de Cincias Sociais e Humanas,
Universidade Nova de Lisboa
Av. de Berna 26C
1069-061 Lisboa
Portugal
Email:fabriziomacagno@hotmail.com

Douglas Walton
CRRAR, Philosophy Dept.
401 Sunset Avenue
University of Windsor
Windsor, Ontario
N9B 3P4 Canada
Email: dwalton@uwindsor.ca

Chris Reed
Computing - Queen Mother Building
University of Dundee
Dundee DD1 4HN
Scotland, UK
Email: chris@computing.dundee.ac.uk

12
Processing Natural Language Argumentation

Katarzyna Budzynska, Serena Villata

ABSTRACT. Although natural language argumentation has attracted the attention of philosophers and rhetoricians since Greek antiquity, it is only very recently that the methods and techniques of computational linguistics and machine learning have become sufficiently mature to tackle this extremely challenging topic. *Argument mining*, the new and rapidly growing area of natural language processing and computational models of argument, aims at automatic recognition of argument structures in large resources of natural language texts. The goal of this chapter is to familiarise the reader focused on formal aspects of argumentation with this approach, and to show how argument structures, e.g. those studied in abstract argumentation frameworks, can be extracted, providing a bridge between mathematical models and natural language. To this end, we first describe the typical argument mining pipeline and related tasks, and then present in more detail a specific example of work in this area.

1 Introduction

Argument mining[1] (see e.g. [Moens, 2013; Peldszus and Stede, 2013; Lippi and Torroni, 2015a] for an overview) aims to develop methods and techniques which allow for automation of the process of identification and extraction of argument data from large resources of natural language texts. This task is split into two parts: linguistic and computational. The linguistic part consists of the analysis (called annotation) of natural language texts of a selected type of discourse (called a genre) and the association of tags with segments of the texts (elementary discourse units, EDUs). For example, if we want to automatically recognise the categories of premise and conclusion, then the annotator analysing the text should assign the tags "premise" and "conclusion" to identified EDUs. The set of texts with tags, i.e. the corpus, constitutes an input for the computational part of the process. The tagging is used to teach algorithms to automatically identify categories or aspects of argumentation such as "premise" and "conclusion". The techniques are based on either a structural or a statistical approach. In the structural approach, a linguist analyses part of the corpus and identifies patterns between the tagged categories and various cues present, e.g. at the linguistic surface of the text. For instance, the linguist may observe that whenever the word "since" occurs, the following segment of

---
[1] Also referred to or associated with argumentation mining, computational argumentation or debating technologies.

text is a premise. The resulting set of such patterns is called a grammar. In the statistical approach, the task of recognising patterns is automatised, i.e. a machine learns these patterns using statistical correlations. The developed or trained algorithms can then be used to mine arguments in large resources of natural language texts such as Wikipedia.

The area of argument mining continues and extends two other areas of natural language processing – sentiment analysis and opinion mining[2], which have proved to be very successful and important both academically and commercially. *Sentiment analysis* develops techniques for fast, automatic processing of texts, such as posts on social media, in order to detect positive and negative sentiments expressed on the Internet towards various products, companies, or people. These techniques are widely used in predicting stock market trends. *Opinion mining* moves from searching for general attitudes ("I like Mercedes") to recognising specific beliefs and opinions ("Mercedes cars are reliable"). Its techniques have found an important application in media analysis, where automation significantly accelerates analysis of natural language texts as compared to manual analysis and consequently enables more accurate summaries of people's opinions, e.g. what they think about a candidate for the presidency. Argument mining expands these research tasks further, allowing us to extract valuable information not only about what attitudes and opinions people hold, but also about the arguments they give in favour of (supporting arguments) and against (conflicting arguments) these attitudes and opinions.

The chapter consists of two parts. First, after a brief introduction to the history (Section 1.1) and background (Section 1.2) of argument mining and the types of arguments that are mined (Section 1.3), in Section 2 we present the typical pipeline of Natural Language Processing (NLP) methods and techniques employed for automatic recognition and extraction of argumentation. In Section 3, we then describe in detail an example of work in argument mining [Cabrio and Villata, 2012b; Cabrio and Villata, 2013] in order to provide a better understanding of how this pipeline is actually used to process argumentation in natural language texts. This chapter does not aim to present the area in a detailed and exhaustive way – we recommend interested readers to have a look at already existing, excellent surveys such as [Moens, 2013; Peldszus and Stede, 2013; Lippi and Torroni, 2015a]. Our goal here is to provide a roadmap of argument mining which shows the most typical approaches, makes links to the work presented in other chapters of this handbook, and offers pointers to a variety of papers for further relevant reading.

---

[2]In the literature, the terms "sentiment analysis" and "opinion mining" either are treated as names referring to two different areas of study or are used interchangeably. Here we follow the first convention, in order to highlight the nuanced distinction between people's general attitude (sentiment) and specific attitude (opinion) to products, institutions, persons, etc.

1.1 A short overview of studies on natural language argumentation

The communicative phenomenon of natural language argumentation has attracted attention since the very beginning of science. Aristotle popularised argument studies in his *Rhetoric* [Aristotle, 2004], distinguishing three means of persuasion: *logos*, i.e. the argumentation or propositional content of the message; *ethos*, i.e. the speaker's character or credibility; and *pathos*, the emotions of the audience. Since that time rhetoric has continued to expand and evolve over many centuries.

At the end of the nineteenth century, the success of mathematical logic drew attention away from natural reasoning to formal reasoning. The dominance of this approach, however, did not last long – in the 1950s it was criticised by a number of researchers, such as Perelman & Olbrechts-Tyteca [1958] and Toulmin [1958], who pointed out the limitations of deduction in modelling real-life communication. Since this turning point, the theory of natural language argumentation has become an important area of research in the disciplines of philosophy, linguistics, discourse analysis, communicative studies, and rhetoric, focusing on topics such as argument structure and schemes, fallacies, enthymemes, dialogue games, counter-argumentation of rebuttals, and undercutters (see e.g. [Groarke, 1996; van Eemeren *et al.*, 2014] for a detailed overview). To facilitate and foster the exchange of ideas amongst members of this interdisciplinary community, two international associations were founded: the International Society for the Study of Argumentation (ISSA), which has organised conferences every four years since 1986[3]; and the Ontario Society for the Study of Argumentation (OSSA), which has held conferences every two years since 1995[4]. Since 2015, another large biennial conference, the European Conference on Argumentation (ECA), has been organised[5]. These conferences gather hundreds of academics from all around the world and various disciplines.

Natural language argumentation has recently begun to attract increasing attention from researchers with a computationally oriented perspective, in particular from the communities of Computational Models of Argument (COMMA)[6] and Workshops on Argument Mining (co-organised with NLP conferences such as the Association for Computational Linguistics, ACL)[7]. The year 2014 was an important milestone for argument mining, resulting in many publications and events, e.g. (1) "From Real Data to Argument Mining" at the 12th ArgDiaP workshop, Polish Academy of Sciences, Warsaw, Poland, 23-24 May 2014 (22 papers); (2) the First Workshop on Argumentation Mining at the 52nd Annual Meeting of the Association for Computational Linguistics, Baltimore, Maryland, USA, June 26, 2014 (18 papers); (3) the SICSA Workshop on Argument Mining: Perspectives from Information Extraction, Information Retrieval

---

[3] http://cf.hum.uva.nl/issa/
[4] http://scholar.uwindsor.ca/ossaarchive/
[5] http://ecargument.org/
[6] http://www.comma-conf.org/
[7] http://argmining2016.arg.tech/

and Computational Linguistics, Centre for Argument Technology ARG-tech, Dundee, Scotland, 9-10 July 2014 (about 25 participants); (4) "Frontiers and Connections between Argumentation Theory and Natural Language Processing", Bertinoro (Forli-Cesena), Italy, 21-25 July 2014 (about 30 participants); (5) the Second Workshop on Argumentation Mining, 53rd Annual Meeting of the Association for Computational Linguistics, Denver, Colorado, USA, June 04, 2015 (16 papers); (6) "Debating Technologies", Dagstuhl Seminar, December 14-18, 2015 (about 25 participants); (7) "Arguments in Natural Language: The Long Way to Analyze the Reasons to Believe and the Reasons to Act" at the First European Conference on Argumentation, Lisbon, 9-12 July 2015; (8) "Natural Language Argumentation: Mining, Processing, and Reasoning over Textual Arguments", Dagstuhl Seminar, April 17-22, 2016 (about 40 participants); and (9) the Third Workshop on Argumentation Mining at the 54th Annual Meeting of the Association for Computational Linguistics, Berlin, August 2016 (20 papers and about 70 participants). In 2016, two tutorials on argument mining were accepted for presentation at the top conferences in artificial intelligence and computational linguistics: the 25th International Joint Conference on Artificial Intelligence (IJCAI-16), New York, NY, USA, 11 July 2016[8] and the annual meeting of the Association for Computational Linguistics (ACL 2016), Berlin 7 August 2016[9]. In 2017 three further tutorials have been delivered at the main graduate school dedicated to language and communication: the 29th European Summer School in Logic, Language, and Information (ESSLLI 2017), Toulouse (France), 17-28 July, 2017. Finally, projects on argument mining have been recognised as important by several funding institutions, e.g. the project "Robust Argumentation Machines" has been successfully selected as one of the seventeen German DFG priority programs which start in 2017[10]; "Argument Mining" is a British EPSRC-funded project implemented by the Centre for Argument Technologies in collaboration with IBM[11]; and "MIning and REasoning with legal text" is the EU H2020 Research and Innovation Staff Exchange project MIREL, in which mining legal arguments is one of the main tasks[12].

1.2 Explosion in the amount of data

Argument mining is an area of text mining which aims to develop techniques and methods for automated extraction of a variety of valuable information from large datasets of texts in natural language. The increase in the importance of text mining is the result of the challenge posed by the explosion of data available on the Internet. While having a vast amount of information is unquestionably of value, such resources become less useful, or even useless, if we cannot process the data efficiently and quickly enough. For example, customer feedback can be

---

[8] http://www.i3s.unice.fr/~villata/tutorialIJCAI2016.html
[9] http://acl2016tutorial.arg.tech/
[10] See e.g. http://www.eurekalert.org/pub_releases/2016-03/df-de1032416.php
[11] http://arg.tech/argument-mining
[12] http://mirelproject.eu/

an important source of information for companies such as Amazon or eBay, but manual analysis of this feedback will become ineffective if it is posted on the company's website faster than it can be processed by human analysts. This is often referred to as the Big Data problem. Text mining aims to develop robust tools and techniques to address this challenge and speed up the process of processing, interpreting and making sense of the large amount of natural language data.

Text mining focuses on several valuable types of information to be mined from natural language texts. Sentiment analysis aims to extract people's attitudes (positive, neutral or negative) towards persons, institutions, events and products, e.g. U.S. presidential candidates. It is applied in stock markets to quickly process vast resources, such as news and social media, to extract information about market trends and predict changes in stock prices. Opinion mining aims to mine opinions about persons, institutions, events and products, e.g. the opinion that the UK economy will be stronger without contributing money to the EU budget, or the contrasting opinion that the UK economy will be weakened without access to the common EU market. Its main commercial application is media analysis, which monitors various media to identify people's reactions to new products or political candidates. Finally, argument mining can be used to recognise not only *what* attitudes and opinions people hold, but also *why* they hold them. For example, people may think that the UK economy will be weakened without access to the common market because the EU will introduce high taxes for goods coming from the UK. The interest of various industries in this technology is manifested in the involvement of companies in a number of academic projects, as well as the development of techniques for argument mining in industry, such as IBM's Watson Debater, which searches for supporting and conflicting arguments on a given issue in Wikipedia articles[13]. One of the reasons for the huge upswing in interest in this area lies in the maturing technology and available returns. Opinion mining has transformed the way market research and PR are carried out, whilst sentiment analysis has had an even greater impact in predicting financial markets. Argument mining is the natural evolution of these technologies, providing a step change in the level of detail available – and this is why major organisations such as IBM are so interested in the technology.

1.3 Objects to be extracted in argument mining

Although argument mining always aims to extract arguments, the concept of "argument" itself may vary from one approach to another. This is the result of different applications, domains and theoretical foundations selected by different researchers. Two basic types of arguments are studied in the literature: supporting arguments and conflicting arguments.

In this section, we present an example of specification of these two notions in order to give the reader an intuitive understanding of the approach to argu-

---

[13]See e.g. http://arg.tech/ibmdebater

mentation in argument mining. Nevertheless, it is important to bear in mind that they can be defined in different ways in different argument mining work (see also subsection 2.2 for a few other examples of how these notions can be defined). This variety is important, as it allows the research to be adapted to different research tasks and applications. For example, in some cases it might be sufficient to identify only the two basic categories of "premise" and "conclusion", while for other research goals it might be necessary to recognise finer structures of supporting arguments, such as linked arguments (all the premises jointly support the conclusion), convergent arguments (the premises independently support the conclusion) and serial arguments (a chain of arguments). This variety, however, does have a drawback, as an analysis of argument structures (either manual or automatic) using one conceptualisation of argumentation might not be comparable to or reusable by another approach. To address this problem, the Argument Interchange Format (AIF) (see e.g. [Rahwan et al., 2007; Chesnevar et al., 2006; Bex et al., 2012; Bex et al., 2013; Budon et al., 2014]) provides a standard for argument representation by capturing common argument structures shared by different approaches. An argument is here represented as a directed graph with two basic types of nodes: information – I-Nodes – and instances of schemes – S-Nodes, with sub-types of S-Nodes representing the application of rules of inference – RA-Nodes – and rules of conflict – CA-Nodes.

A *supporting argument* (also called an inference, pro-argument or argumentation) can be defined as a relation between propositions where p is used to support q[14], when p provides a reason to accept q. In other words, p can be used as a reply to the question "Why q?" ("Because p"). The concept of supporting argument is related to the traditional notion of (non-deductive) reasoning (see e.g. [Groarke, 1996; van Eemeren et al., 2014]). In example (1), taken from the Regulation Room Divisiveness corpus[15], the annotator has interpreted the post by user sbarb95(272) as an argument in which the claim "I wonder if delays of 30 minutes would actually affect passenger behavior" is supported by the user's personal experience "In my experience it usually takes about 30 minutes to get to major airports".

(1)   sbarb95(272): *(...) I wonder if delays of 30 minutes would actually affect passenger behavior. In my experience it usually takes about 30 minutes to get to major airports.*

In the Argument Interchange Format, the supporting argument is modelled as a graph consisting of two information nodes (in this example, a node repre-

---

[14]Note that there can be more than one support for q. In this case, we would have p_1, p_1, ..., p_n.

[15]In the US, online deliberative democracy or e-rulemaking, e.g. RegulationRoom (http://RegulationRoom.org), has been introduced as a multi-step process of social media outreach that federal agencies use to consult with citizens on new regulations on health and safety, finance, and other complex topics (see [Lawrence et al., 2017] for the description of the corpus available at http://arg.tech/rrd).

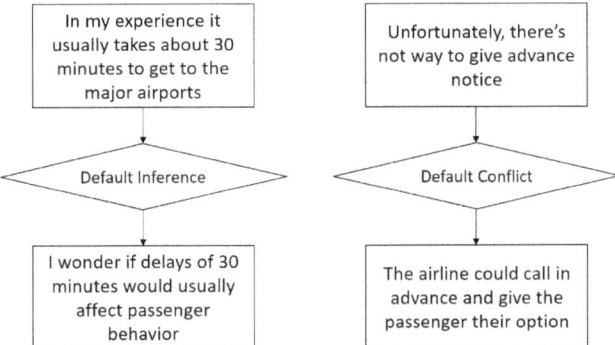

Figure 1. Supporting and conflicting arguments visualised according to the Argument Interchange Format (AIF) stored in the AIFdb database in the Regulation Room Divisiveness corpus, fragment of maps #4900 and #4891

senting the proposition "I wonder if delays of 30 minutes would actually affect passenger behavior" and a node representing "In my experience it usually takes about 30 minutes to get to major airports"); one node (RA-Node) representing the relation of support, called *Default Inference*; and an arrow representing the direction of reasoning – from premise(s) to conclusion (see the argument structure on the left side of Figure 1).

Supporting arguments are sometimes treated as a general, natural-language expression of support, even if actual reasons to accept a proposition are not provided, e.g. "Peanuts should be served on the plane" or "Peanuts are served on board on AirAlbatross flights". This broad notion is very close to the concept of support relation used in bipolar abstract argumentation [Cayrol and Lagasquie-Schiex, 2010].

A *conflicting argument* (also called an attack, con-argument or counter-argumentation) can be characterised as a relation between two propositions in which one proposition is used in order to provide an incompatible alternative to another proposition. Speakers use conflicting propositions to attack another speaker's claims with counter-claims. The concept of conflicting argument is close to the concept of attack relation used in abstract argumentation frameworks [Dung, 1995]. In a fragment of the discussion on the RegulationRoom platform (2), the analyst identified a conflicting argument between two forum users, AK traveler(287) and SofieM(947), in which "Unfortunately, there's no way to give advance notice" is used to attack the claim "The airline could call in advance and give the passenger their options".

(2) a. AK traveler(287): *(...) The airline could call in advance and give the passenger their options (...)*
     b. SofieM(947): *(...) Unfortunately, there's no way to give advance notice.*

In AIF graphs, the conflicting argument is represented by a CA-node called *Default Conflict* (see the argument structure on the right side of Figure 1).

The conflict does not have to be restricted to the logical contradiction of *p* and not-*p*, but can be treated as a more general, natural-language expression of opposition, e.g. "Peanuts should be served on the plane" and "Peanuts cause allergic reactions". These two more general relations are used, for example, to identify conflicting and supporting viewpoints in online debates [Cabrio and Villata, 2012a].

Some approaches to formal argumentation focus only on one type of argument structure, such as the original abstract argumentation framework [Dung, 1995], while some work uses both supporting and conflicting arguments, such as ASPIC+ [Prakken, 2010].

2 Pipeline of natural language processing techniques applied to argument mining

In this section, we describe a pipeline which is typically used for automatic mining of arguments. The pipeline comprises two parts, linguistic and computational, which can be viewed as analogues (see Figure 2). The linguistic part aims to develop large corpora, which are datasets of manually annotated (analysed) argument data, evaluated by measuring the level of inter-annotator agreement. The computational part of the argument mining pipeline aims to develop grammars (structural approach) and classifiers (statistical approach) to create automatically annotated corpora of arguments, and the performance of the system is then evaluated by measures such as accuracy or F_1- score. The ultimate goal of the pipeline is to process real arguments in natural language texts (such as arguments formulated on Wikipedia) so that the output is only the information that is valuable to us, i.e. structured argument data. In subsequent subsections we briefly describe how each step of this process is typically developed and give a few examples of specific work and approaches, to help the reader to better understand what types of methods, techniques and results are likely to be found in papers in this area.

2.1 Databases of texts in natural language

The argument mining pipeline starts with the task of collecting large resources of natural language texts (see "large resources of NL text" box in Figure 2), which then can be used for training and testing of argument mining algorithms. Texts sometimes need to be pre-processed in order to obtain a common and processable format.

The size of the datasets varies from project to project, but in the case of approaches which apply machine learning techniques (see Section 2.5), the dataset should be relatively large, because the algorithms learn rules for argument mining using statistical correlations. In the case of structural approaches, such rules (a grammar) are developed by a linguist and thus the dataset can be smaller, as its size is complemented by the linguist's expertise and knowledge about

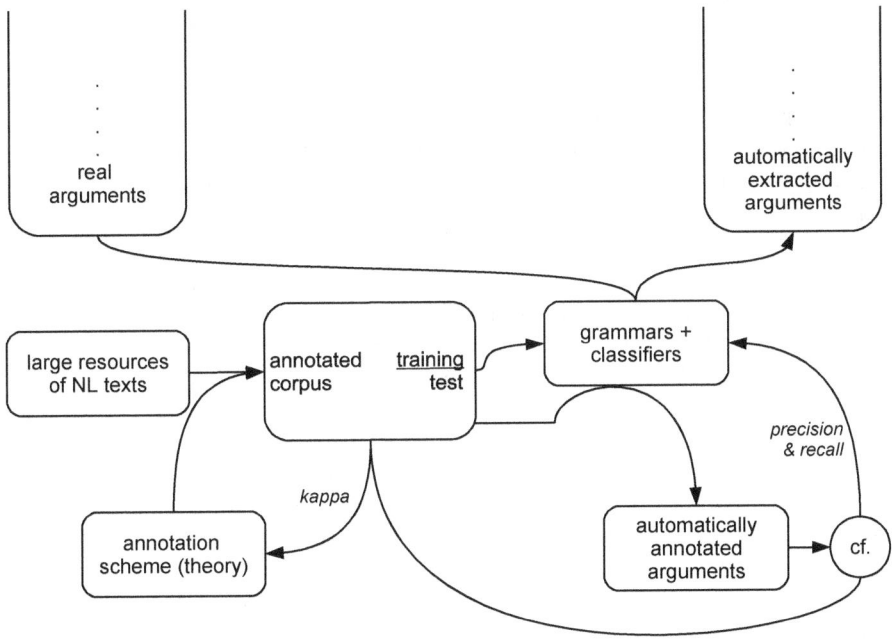

Figure 2. A pipeline of natural language processing techniques applied to argument mining (all steps in the pipeline will be explained in subsequent subsections of this section)

language.

For example, in the case of the statistical approach, Palau and Moens used a dataset consisting of 92,190 words (2,571 sentences) in 47 documents from the European Court of Human Rights [Palau and Moens, 2009], and Habernal and Gurevych collected a database comprising 90,000 words in 340 documents of user-generated web discourse [Habernal and Gurevych, 2016a]. In the case of the structural approach, Garcia-Villalba and Saint-Dizier used 21,500 words in 50 texts as a test corpus [Villalba and Saint-Dizier, 2012], and Budzynska, Janier, Reed and Saint-Dizier collected a dataset consisting of 24,000 words [Budzynska et al., 2016].

Typically, the task of argument mining is narrowed down to a specific type of discourse (genre), since algorithms use the linguistic surface for argument recognition with little or no knowledge about the world, the discourse context or the deeper pragmatic level of a text. In other words, the way people express argumentation in language depends on the type of discourse they are engaged in; for example, we can expect a lawyer in court to argue in a different linguistic style than a politician during a presidential debate or a scientist at a conference.

Genres that have been studied include legal texts [Moens et al., 2007; Reed et al., 2008; Palau and Moens, 2009; Ashley and Walker, 2013; Wyner et al., 2016]; mediation [Janier et al., 2015; Janier and Reed, 2015]; scientific papers [Teufel et al., 1999; Teufel et al., 2009; Kirschner et al., 2015]; student essays [Nguyen and Litman, 2015; Nguyen and Litman, 2016]; online comments [Villalba and Saint-Dizier, 2012; Park and Cardie, 2014b; Park and Cardie, 2014a; Habernal et al., 2014; Wachsmuth et al., 2015; Konat et al., 2016]; political debates [Hirst et al., 2014; Naderi and Hirst, 2015; Duthie et al., 2016b]; philosophical texts [Lawrence et al., 2014]; technical texts [Marco et al., 2006; Saint-Dizier, 2014]; moral debate on the radio [Budzynska et al., 2014b]; online debates [Swanson et al., 2015; Walker et al., 2012; Cabrio and Villata, 2014; Sridhar et al., 2015; Boltuzic and Snajder, 2016; Habernal and Gurevych, 2016c]; persuasive essays [Stab and Gurevych, 2014; Ghosh et al., 2016]; and Wikipedia articles [Aharoni et al., 2014; Levy et al., 2014; Lippi and Torroni, 2015b; Roitman et al., 2016]. The choice of genre is the researcher's subjective decision, but is typically motivated by the presumed likelihood of finding a large amount of non-sparse (not rare) argument data in a given type of discourse (in other words, how argumentative a given genre seems to be). Sometimes, if a project is run in cooperation with an industrial partner, the choice also depends on commercial applications.

2.2 Theories & annotation schemes

The next step in the argument mining pipeline consists of choosing a model of argumentation, which is then used to develop an annotation scheme for analysing argumentation in natural language texts. An annotation scheme is a set of labels (tags) which defines argumentation and its aspects, as well as which annotators (analysts) will be used to structure the dataset.

This task is important but also challenging, as a faulty conceptualisation of the phenomenon of argumentation results in erroneous and inconsistent annotations (analysts tend to make more errors if they do not understand the annotation scheme, and different analysts will make different decisions; see Section 2.4), which will then have negative consequences for the performance of the automation process (see Section 2.5). In other words, if the human analyst struggles with understanding a given annotation scheme, then we should not expect the algorithm to be able to learn it efficiently. As a result, the extraction of arguments will not be accurate (see Section 2.7), as the algorithm is unable to detect the correct patterns between the annotated tags and the linguistic surface of the annotated texts.

The literature contains a variety of annotation schemes which aim to achieve a balance between efficiency (simpler schemes will be quicker and easier to annotate) and adequacy (more specific sets of labels will be better tailored to describing given aspects of argumentation or a given genre). In one of the first works in argument mining, [Moens et al., 2007; Palau and Moens, 2009], Palau and Moens choose a basic, intuitive conceptualisation of argument structure:

- Premise: statement which provides support

- Conclusion: statement which is supported

- Argument: a full structure comprising the premises and conclusion

Such a set of labels is unquestionably the most natural and obvious choice to start with for the purpose of mining argumentation. It is also a general, high-level description and can be reused for any other genre. The only limitation that must be accounted for is that the property of being a premise and the property of being a conclusion are in fact relations, i.e. a premise cannot merely provide support – it always provides support for something. For instance, the sentence "home ownership may not be a basic human right" uttered in isolation is just a statement (an assertion) of someone's belief; it becomes a premise only if it is a response to or a continuation of another utterance, such as "more successful countries, like Switzerland and Germany, prefer to rent". Thus, we need another way to account for this relational context, so that the algorithm has enough information to learn argument structure properly.

In her work on Argumentative Zoning [Teufel et al., 1999; Teufel et al., 2009], Teufel uses a more complex set of labels specifically tailored for mining argumentation in scientific texts:

- Background: general scientific background

- Other: neutral descriptions of other people's work

- Own: neutral descriptions of our own, new work

- Aim: statements of the particular aim of the current paper

- Textual: statements characterising the textual organisation of the current paper (e.g. "In chapter 1, we introduce...")

- Contrast: contrastive or comparative statements about other work; explicit mention of weaknesses of other work

- Basis: statements that our own work is based on other work

The advantage of an annotation scheme tailored to a specific genre is that specific categories are able to capture more detailed information about typical language constructions in a given type of discourse, and thus to enhance the development of rules or the learning process. This advantage, however, must be balanced against the limited reusablity of such an annotation scheme for other genres.

Peldszus and Stede [Peldszus and Stede, 2013] introduce an annotation scheme drawing on different ideas from the literature and their own practical experience analysing texts in the Potsdam Commentary Corpus [Stede,

2004]. The schema follows Freeman's idea of using the moves of the proponent and challenger in a basic dialectical situation as a model of argument structure [Freeman, 1991; Freeman, 2011]. The authors define an argument as a non-empty set of premises supporting some conclusion, using the term "argument" not for premises, but for the complex of one or more premises put forward in favour of a claim. Premises and conclusions are propositions expressed in text segments. If an argument involves multiple premises that jointly support the conclusion, a linked structure is identified (i.e. none of the linked premises would be able to support the conclusion on its own). In the basic dialectical situation, a linked structure is induced by the challenger's question as to why a premise is relevant to the claim. The proponent then answers by presenting another premise explicating the connection. Building a linked structure is conceived as completing an argument. In this scheme, the label "argumentation" refers to the structure that emerges when multiple arguments are related to each other and form larger complexes. The manner in which arguments combine into larger complexes can be generally described as either supporting, attacking or counter-attacking. More precisely, the scheme considers five kinds of support:

- basic argument
- linked support
- multiple support
- serial support
- example support

four kinds of attacks on the proponent's argument by the challenger:

- rebut a conclusion
- rebut a premise
- undercut an argument
- support a rebutter

and four counter-attacks on the challenger's attack by the proponent:

- rebut a rebutter
- rebut an undercutter
- undercut a rebutter
- undercut an undercutter

Stab and Gurevych [Stab and Gurevych, 2014] propose an annotation scheme whose goal is to model argument components as well as the argumentative relations that constitute the argumentative discourse structure in persuasive essays[16]. Essays of this kind exhibit a common structure: the introduction typically includes a major claim that expresses the author's stance on the topic, and the major claim is supported or attacked by arguments (composed of premises and a claim) covering certain aspects of the stance in subsequent paragraphs. The major claim is the central component of an argument. It is a controversial statement that is either true or false and should not be accepted by the reader without additional support. The premise underpins the validity of the claim and is a reason given by the author to persuade the reader. Argumentative relations model the discourse structure of arguments and indicate which premises belong to a given claim. In this scheme, two directed relations between argument components are annotated: the support relation and the attack relation. Both relations can hold between a premise and another premise, a premise and a claim, or a claim and a major claim. An argumentative relation between two components indicates that the source component is a reason for or refutation of the target component. The annotation guidelines proposed in [Stab and Gurevych, 2014] consist of the following steps:

- Topic and stance identification: before starting the annotation process, annotators identify the topic and the author's stance by reading the entire essay.

- Annotation of argument components: the major claim is identified in either the introduction or the conclusion of the essay. Subsequently, annotators identify the claims and premises in each paragraph.

- Annotation of argumentative relations: finally, the claims and premises are linked within each paragraph, and the claims are linked to the major claim with either a support relation or an attack relation.

Finally, an annotation scheme which considers the broad dialogical context of argumentation can be found in [Budzynska et al., 2014b]. Building upon Inference Anchoring Theory (IAT) [Budzynska and Reed, 2011] as a theoretical foundation, they propose to extend the set of tags for supporting and conflicting arguments with dialogical and illocutionary structures (the latter are modelled as illocutionary forces introduced in [Austin, 1962; Searle, 1969; Searle and Vanderveken, 1985]). The illocutionary structures are described by two groups of tags:

- Tags associated with a speaker's individual moves in the dialogue:
    - Asserting: The speaker S *asserts* p to communicate his opinion on p.

---

[16]The corpus and annotated guidelines are available at https://www.ukp.tu-darmstadt.de/data/argumentation-mining.

- Questioning: S *questions whether p* when S formulates p as an interrogative sentence of the form "Is/Isn't p the case?" Three categories of questioning are distinguished: Pure Questioning (PQ), Assertive Questioning (AQ), and Rhetorical Questioning (RQ). In the case of PQ, S is asking for the hearer H's opinion on p: whether H believes, disbelieves, or has no opinion on p. AQ and RQ, in contrast, carry some degree of assertive force and thus can be a part of the argumentation itself. For AQ, S not only seeks H's opinion on p, but also indirectly declares publicly his own opinion on p. For RQ, S is grammatically stating a question, but in fact he conveys that he does (or does not) believe p.

- Challenging: When S *challenges p*, S declares that he is seeking (asking about) the grounds for H's opinion on p. Challenges are a dialogical mechanism for triggering argumentation. Like questions, challenges form a continuum from Pure Challenging (PCh) to Assertive Challenging (ACh) to Rhetorical Challenging (RCh).

- Popular Conceding: Through *popular conceding*, S communicates some sort of general knowledge which is assumed to be obvious and as such does not need to be defended (does not place a burden of proof on S).

- Tags associated with the interactions between speaker(s)' moves in the dialogue:

  - Agreeing: Agreeing is used to express a positive reaction, i.e. when the speaker S declares that he shares the opinion of the hearer. This can take the form of signalling such as "Yes", "Indeed", "Most definitely" or "Sure", but may also be a complete sentence.

  - Disagreeing: Disagreeing is used to express a negative reaction, i.e. when S declares that he does not share H's opinion. This can take the form of utterances which have similar meaning to "No" (e.g. "I'm not saying that", "Actually, that's not correct", "Definitely not" or "No it's not"), or it can be an utterance with complete propositional content.

  - Arguing: S is *arguing* when he defends a standpoint. This illocution is sometimes signalled by linguistic cues such as "therefore", "since" or "because", but these indicators rarely occur in natural spoken language.

The complexity of the annotation scheme allows more aspects of natural communication to be captured (in the case of this project, both argument and dialogical structures, as well as the relations between them), but on the other hand, the annotation process becomes a more time-consuming task and requires well trained and experienced annotators.

2.3 Manual annotation & corpora

Once we collect a database of texts for our genre and select or develop an annotation scheme, we need to combine these two together, i.e. an annotator manually assigns a label from the scheme to a segment of the text. The resulting database of tagged texts is called a corpus.

More specifically, the annotation process starts with segmentation (splitting) of the text into elementary discourse units (EDUs; see e.g. [Lawrence et al., 2014]) or, more precisely, argumentative discourse units (ADUs; see e.g. [Peldszus and Stede, 2013; Ghosh et al., 2014]). ADUs present an additional challenge for the segmentation, since they require the analyst not only to find spans of text which are minimal meaningful building blocks of a discourse (EDU)[17], but also to recognise whether or not they carry an argumentative function (whether or not an EDU is an ADU). Next, the annotator assigns a label from the scheme to the ADU following the guidelines, i.e. the definitions of the labels.

Annotators typically use software tools, such as the **arggraph** DTD[18], the RSTTool[19] and the Glozz annotation tool[20], to help them assign labels from the annotation scheme to ADUs directly in a code. The method often uses the XML format of tagging text. Take this example of an ADU:

(3) Ruth Levitas: *No parent in the family is in work since we have a huge problem with unemployment.*

This ADU can be represented by the following string of XML code:

```
<argumentation><conclusion>No parent in the family is in
work</conclusion> since <premise>we have a huge problem with
unemployment</premise></argumentation>
```

Another approach would be to exploit a visualisation software tool such as GraPAT[21] [Sonntag and Stede, 2014], which is a graph-based, web-based tool suited for annotation of sentiment and argumentation structure. The tool provides a graph structure annotation for the selected text. Certain parts of the text can be connected to nodes in the graph, and those nodes can be connected via the relations of the selected annotation scheme. Edges also have attributes to describe them. Figure 3 shows an example of annotation of argumentation structure using the GraPAT tool.

---

[17] There is no general agreement on one standard definition of EDU.
[18] https://github.com/peldszus/arg-microtexts/blob/master/corpus/arggraph.dtd
[19] http://www.wagsoft.com/RSTTool/
[20] http://www.glozz.org
[21] Available at: http://angcl.ling.uni-potsdam.de/resources/grapat.html

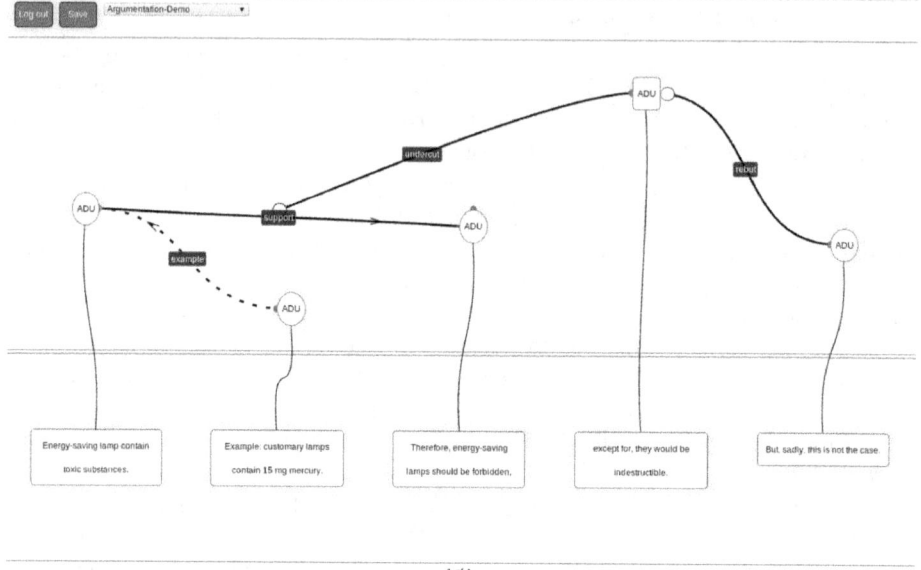

Figure 3. Annotation of an argumentation structure in GraPAT

To capture more detailed information about the dialogical context of argumentation, the Online Visualisation of Arguments tool (OVA+)[22] [Janier et al., 2014] can be used (see Figure 4). In [Budzynska et al., 2016], this tool was used to analyse argument structures in broadcast debates from the BBC Radio 4 programme *Moral Maze*. OVA+ allows the annotator to represent the communicative intention of arguing in Example (3) as a relation between the *inference relation* (Default Inference in Figure 5) and the *dialogical relation* (Default Transition in Figure 5)[23]. In other words, we can say that the inference is the content of arguing, while the transition between moves in the dialogue anchors this intention. In a similar manner, we can annotate an illocution of disagreeing in order to link a conflict (a conflicting argument) between propositions with the dialogical relation of transition (see Figure 6).

Finally, the annotated data must be stored as a corpus ([Reed, 2006] is one of the first corpora of analysed arguments). For example, the IBM Debating

---

[22] Available at: http://ova.arg-tech.org
[23] More precisely, the graph illustrates that the conclusion (top left node in Figure 5) is introduced by Levitas' first utterance (top right node) via the communicative intention of asserting (middle top node); the premise (bottom left node) is introduced by her second utterance (bottom right node), also via the communicative intention of asserting (middle bottom node), while the relation of inference (*Default Inference* node in Figure 5) is triggered by the interaction between these two moves (*Default Transition* node) via the communicative intention of arguing.

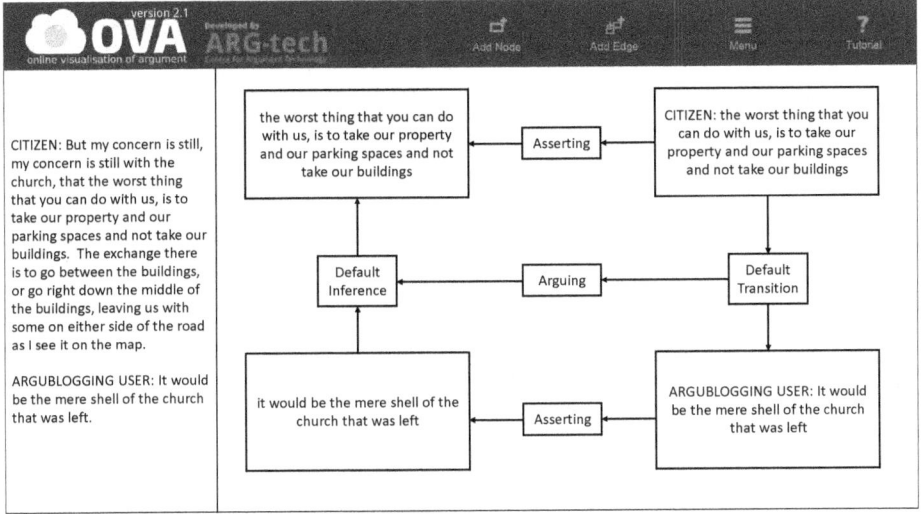

Figure 4. Annotation of an argumentation structure in OVA+. The panel on the left side shows the raw text to be annotated, and the panel on the right shows the visualisation of the annotation, i.e. argument and dialogue structures represented as a directed graph.

Technologies corpus[24] contains three different datasets: a dataset for automatic detection of claims and evidence in the context of controversial topics (1,392 labelled claims for 33 different topics) [Aharoni et al., 2014]; its extended version (2,294 labelled claims and 4,690 labelled evidence for 58 different topics) [Rinott et al., 2015]; and a dataset of multi-word term relatedness (term-relatedness values for 9,856 pairs of terms) [Levy et al., 2015].

Another important resource is the Internet Argument Corpus (IAC), which provides analyses of political debate on Internet forums. It consists of 11,000 discussions and 390,000 posts annotated for characteristics such as topic, stance, degree of agreement, sarcasm, and nastiness, among others [Walker et al., 2012]. It does not include argument components but includes several properties of arguments, and it can be used for argument attribution tasks. The UKPConvArg1[25] corpus is another recently released dataset, composed of 16,000 pairs of arguments on 32 topics annotated with the relation "A is more convincing than B" [Habernal and Gurevych, 2016c].

The application of labels and Inference Anchoring Theory to the dataset of transcripts of the BBC Radio 4 programme *Moral Maze* (the labels were described in Section 2.2) resulted in the MM2012 corpus [Budzynska et al., 2014b] (see Figure 8 for an example[26]). Figures 5 and 6 depict fragments of

---
[24] http://researchweb.watson.ibm.com/haifa/dept/vst/mlta_data.shtml
[25] https://github.com/UKPLab/acl2016-convincing-arguments
[26] Available at: http://corpora.aifdb.org/mm2012

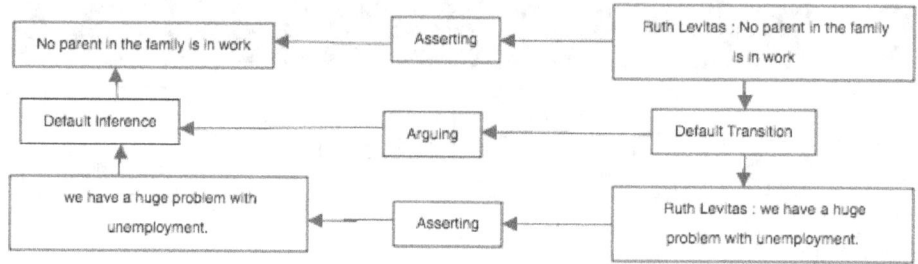

Figure 5. Annotation of *Arguing* and *Inference* using the IAT annotation scheme and the OVA+ software tool, stored in the AIFdb database in the Moral Maze 2012 corpus (fragment of map #6306)

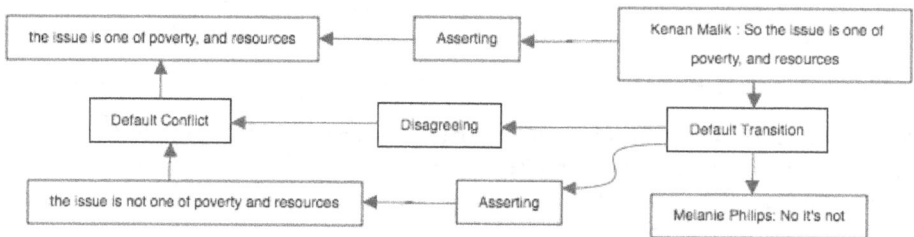

Figure 6. Annotation of *Disagreeing* and *Conflict* using the IAT annotation scheme and the OVA+ software tool, stored in the AIFdb database in the Moral Maze 2012 corpus (fragment of map #6331)

maps representing structure of argument data[27]. Table 1 shows the frequency of tags associated with the interactions between moves of speaker(s) in the dialogue: 1,191 such tags were used to annotate the MM2012 corpus, and arguing was the most frequently occurring tag (72% of cases).

As manual annotation is a highly time-consuming task, sharing and reusing annotated data is of real value[28]. This is the aim of the freely accessible

---

[27] Available at: http://corpora.aifdb.org/mm2012
[28] See for example the "Unshared task" session at the 3rd Argument Mining workshop or " Unshared untask" sessions at Dagstuhl's *Natural Language Argumentation* Seminar.

IC type	Occurrences	Kappa κ
Agreeing	119 (10%)	
Disagreeing	219 (18%)	
Arguing	853 (72%)	
TOTAL	1,191 (100%)	.76

Table 1. The distribution of illocutionary connections anchored in transitions in the MM2012 corpus

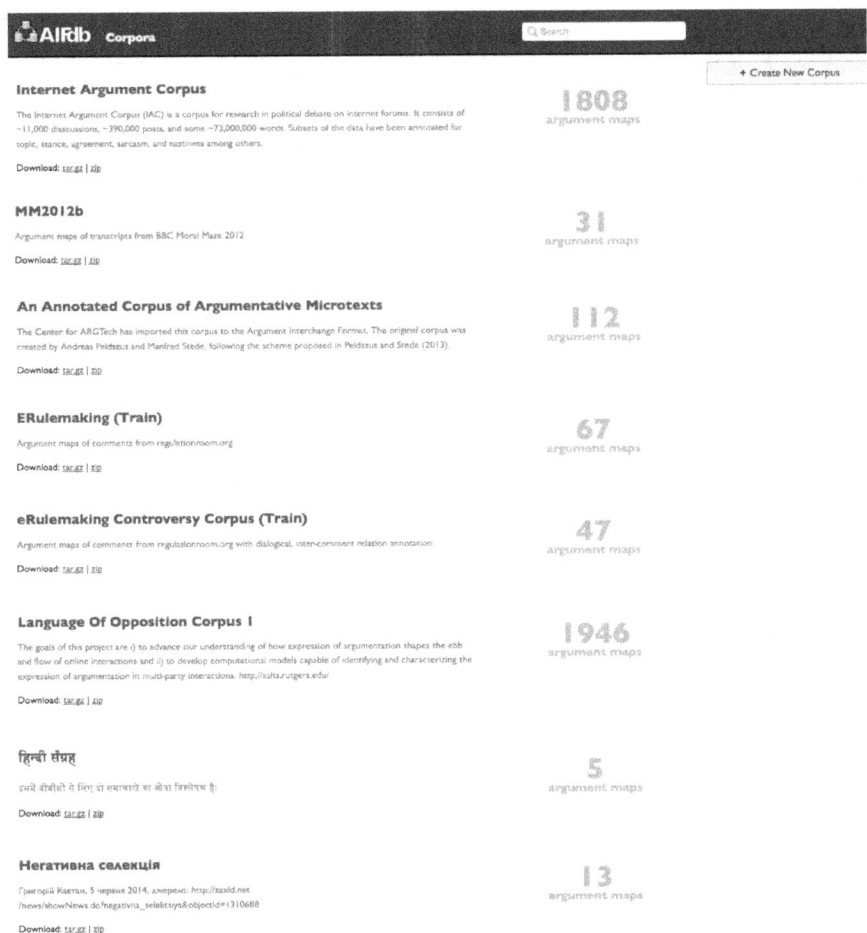

Figure 7. Freely available AIFdb corpora

database AIFdb [Lawrence et al., 2012][29], which hosts multiple corpora [Lawrence et al., 2015][30]. The key advantage of AIFdb is that it uses a standard for argument representation – the Argument Interchange Format (AIF) [Rahwan et al., 2007; Chesnevar et al., 2006]. The database stores corpora which were either originally annotated according to this format, such as the MM2012 corpus described above, or imported to the AIFdb, such as the Internet Argument Corpus developed in Santa Cruz [Walker et al., 2012], the Microtext corpus developed in Potsdam [Peldszus and Stede, 2015], the Erulemaking corpus from Cornell [Park and Cardie, 2014b], and the Language of Opposition corpus de-

---

[29] http://aifdb.org
[30] http://corpora.aifdb.org/

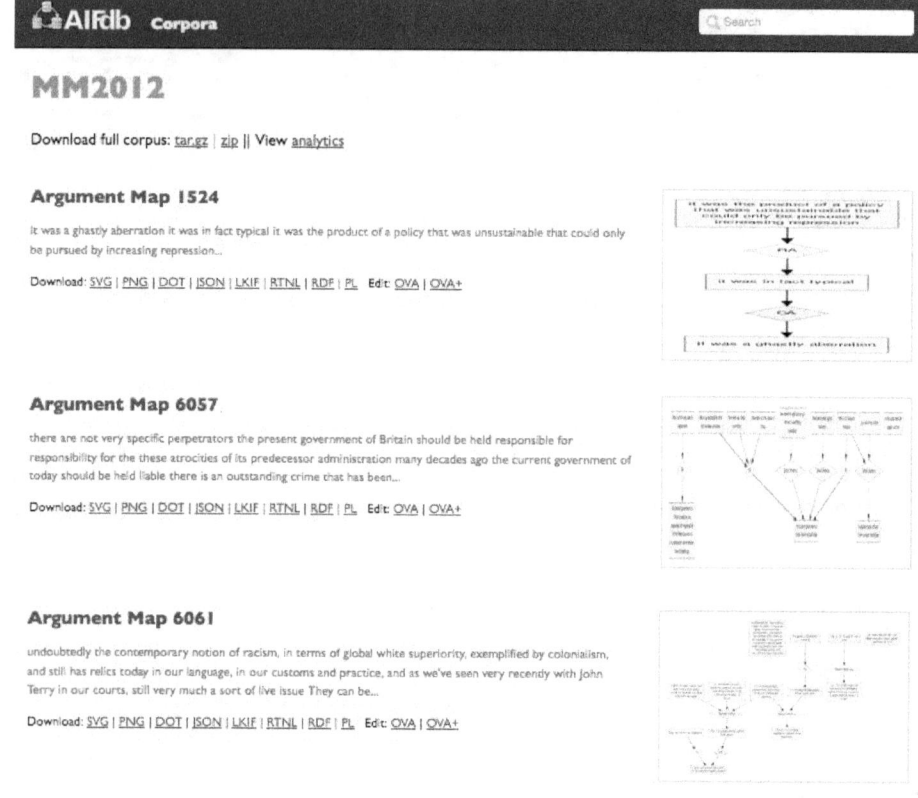

Figure 8. The Moral Maze 2012 corpus created, stored and managed on the AIFdb corpora platform

veloped at Rutgers and Columbia [Ghosh et al., 2014; Wacholder et al., 2014][31] (see Figure 7). The AIF standard allows for the expression of a variety of argument aspects shared by many approaches to argumentation, and thus for the reusability of the resources analysed according to this format. The AIFdb corpora provide the service of creating (see the button in the top right hand corner in Figure 7), storing and managing the annotator's own corpus and allow the stored data to be reused by other researchers and projects. Currently the AIFdb database has 300-500 unique users per month and stores 1,500,000 words and 50,000 annotated arguments in 15 languages (statistics obtained in July 2016).

2.4 Evaluation of the manual annotation step

The last step of the linguistic part of the argument mining pipeline is the evaluation of the corpus. This task aims to reduce the propagation of errors

---

[31]The data and coding are available at: http://salts.rutgers.edu/.

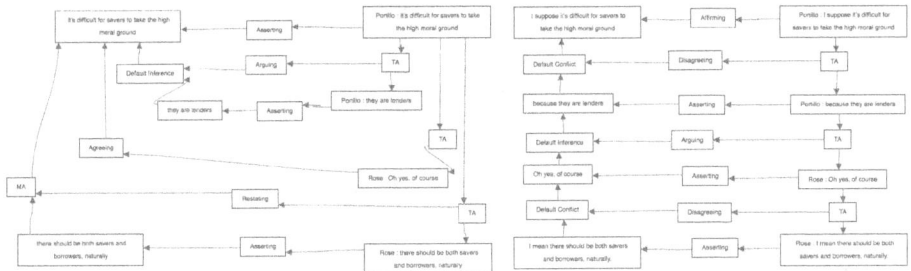

Figure 9. An evaluation of the quality of manual annotation using inter-annotator agreement: the argument map on the left analysed by one annotator is compared with the argument map on the right analysed by another annotator.

from the manual annotation to the automatic annotation in order to improve the performance of the system.

The quality of annotation is evaluated using a measure of inter-annotator agreement, i.e. by comparing analyses of the same texts done by different annotators. Most typically, the comparison is calculated for a subset of the corpus – one annotator analyses the full database, while the other analyses just a subset of this database. For example, we can look for matches between the map on the left hand side and the map on the right in Figure 9, counting how many nodes and relations were analysed by both annotators in the same way.

The two comparison measures used most often are agreement and kappa κ. Agreement is a simple proportion (percentage) of matches. This measure, however, does not take into account the possibility of random matches, as if the annotators were tossing a coin and then assigning labels according to the result. For this reason the more informative kappa measure was introduced. Its most popular version is Cohen's kappa [Cohen, 1960], which shows the agreement between two annotators who each classify N items (e.g. ADUs) into C mutually exclusive categories (tags). The equation for κ is as follows:

$$\kappa = \frac{\Pr(a) - \Pr(e)}{1 - \Pr(e)}$$

where $Pr(a)$ is the relative observed agreement among raters and $Pr(e)$ is the hypothetical probability of chance agreement. The work [Landis and Koch, 1977] proposes a scale for κ results to help to interpret the level of agreement. According to these authors, 0.41-0.60 represents moderate agreement, 0.61-0.80 is substantial agreement, and 0.81-1 is almost perfect agreement. It should be noted, however, that this scale is not widely accepted.

Most recently, Duthie, Lawrence, Budzynska and Reed proposed a new met-

Method	Overall Score
Cohen's κ	0.44
CASS-κ	0.59
$F1$	0.66
CASS-$F1$	0.74

Table 2. Scores are provided for Cohen's kappa and $F1$ score, for both segmentation and structure, and their CASS equivalence.

ric [Duthie et al., 2016a]. The Combined Argument Similarity Score (CASS) technique aims to account for the problem of over-penalising or double-penalising small differences between two manual annotations or between manual and automatic annotation. This technique looks to break down the argument structure into three components: segmentation, argumentative relations and dialogical relations. Segmentation calculations are particularly important when performing argument mining on unaltered natural language text. Arguments do not always span full sentences and automatic solutions may miss some tokens, which can then have a knock-on effect on the argumentative or dialogical structure, with the same ADU segmented in a slightly different way by two different analysts and κ penalising for this twice. CASS offers the Segmentation Similarity metric [Fournier and Inkpen, 2012] to give a score purely for segmentation, which does not penalise as heavily for near-misses. Further, it uses the Levenshtein distance [Levenshtein, 1966] together with word positions to match text spans and compare the relations between them in order to work out scores for both argumentative structure and dialogical structure. CASS then takes the mean of the argumentative and dialogical scores and the harmonic mean to calculate the CASS score. The resulting differences for standard metrics and the new CASS metrics can be seen in Table 2.

CASS is implemented and made available as part of Argument Analytics [Lawrence et al., 2016], a suite of techniques which provides an interpretation of and insight into large-scale argument data stored in AIFdb corpora for both specialist and general audiences. Currently, it has modules available for viewing the following: simple statistical data, which give both an overview of the argument structure and frequencies of patterns such as argumentation schemes; comparative data, which can be quantified by a range of measures (including the CASS metric) describing the similarity of two annotations; dialogical data showing the behaviour of participants of the dialogue; and real-time data allowing for the graphical representation of an argument structure developing over time.

In the eRulemaking corpus [Park and Cardie, 2014a], inter-annotator agreement was measured on 30% of the data for two annotators, resulting in a Cohen's κ of 0.73; in the Moral Maze corpus [Budzynska et al., 2014b], Cohen's kappa for two annotators and three types of illoctionary connections (arguing, agreeing and disagreeing) was $\kappa = .76$ (see also Table 1); in the Persuasive

Essays corpus [Stab and Gurevych, 2014], inter-annotator agreement was measured on 90 persuasive essays for three annotators, resulting in a Krippendorff's inter-rater agreement of $\alpha = 0.81$[32]; and in the argumentative microtexts corpus [Peldszus and Stede, 2015], three annotators achieved an agreement of Fleiss' $\kappa = 0.83$[33] for the full task.

2.5 NLP techniques

The next step initiates the computational part of the argument mining pipeline (see the "grammars + classifiers" box in Figure 2). In principle, there are two basic styles of automation (in practice, they in fact share some techniques, to the extent that they sometimes form a hybrid approach): the structural approach, i.e. grammars (hand-coded sets of rules) and the statistical approach, i.e. machine learning (general learning algorithms).

In the first case, the linguist looks through a selected fragment of a corpus and aims to find patterns in the expression of argumentation in natural language. For instance, in a given corpus it might be observed that arguments are linguistically signalled by words such as "because", "since" or "therefore", or that assertive questions typically start with "Isn't it..." or "Can we agree that...", etc. In such a case the linguist will formulate rules describing these patterns and add them to the grammar. Then, the grammar will be used by a system for automatic recognition of categories of argumentation and assertive questioning.

The statistical approach replaces a linguist with an algorithm. Like a human being, a system will also look for patterns, but this time on a larger sample of the training corpus. The programmer chooses what features the machine should process to check whether they coexist with the categories from the annotation scheme. Commonly chosen features include the following:

- unigrams: each word in the sentence

- bigrams: each pair of successive words

- text statistics: sentence length, average word length, number of punctuation marks

- punctuation: the sequence of punctuation marks present in the sentence

- keywords: words which are typically associated with a category, e.g. "but" and "however" for conflicting argument, or "because", "since" and "therefore" for supporting argument.

Both methods have their advantages. The structural approach exploits the expert's knowledge and experience with language analysis, which is helpful in

---

[32] Krippendorff's α is a statistical measure of the agreement achieved when coding a set of units of analysis in terms of the values of a variable.
[33] Fleiss' κ assesses the reliability of agreement between a fixed number of raters when assigning categorical ratings to a number of items.

formulating correct patterns and, at the same time, to ignore random statistical correlations. On the other hand, a machine can easily spot new and infrequent patterns which are difficult for a linguist to spot due to the smaller sample size that can be analysed by a human being or to bias arising from prior knowledge and expectations. Algorithms are also able to quickly and easily update the learning process, while extending a grammar requires more effort. A lot of work in argument mining applies typical, "off-the-shelf" NLP methods and techniques, which in order to improve the performance of argument mining systems are adapted to the domain of argumentation by further enriching them with discourse indicators [Knott, 1996; Henkemans et al., 2007], argumentation schemes [Feng and Hirst, 2011; Lawrence and Reed, 2016], frames (aspects of an issue) [Naderi and Hirst, 2015], dialogical context [Budzynska et al., 2014a; Budzynska et al., 2015], semantic context [Cabrio and Villata, 2012a], and combinations of various cues and techniques [Lawrence and Reed, 2015].

An example of the structural approach is the framework proposed by Garcia-Villalba and Saint-Dizier [2012], which aims to extract arguments in opinion texts. The authors discuss how automatic recognition of arguments can be implemented in the Dislog programming language on the <TextCoop> platform. The main idea of this work is that argument mining techniques make it possible to capture the underlying motivations consumers express in reviews, which provide more information than a basic attitude of the type "I do/don't like product A". Extraction of arguments associated with evaluative expressions provides a deeper understanding of consumer motivations and preferences. Evaluative expressions, more precisely, are accompanied by additional elements such as comments, elaborations, comparisons, illustrations, etc., which are forms of explanation. According to the authors, explanation is not a basic rhetorical relation, but a very generic construction which covers a large number of communication situations and which is realised in language by means of several types of rhetorical relations (e.g. elaboration, arguments or illustration). Garcia-Villalba and Saint-Dizier argue that rhetorical elements related to explanation behave as argument supports and make explicit the semantic and pragmatic function of the support: they justify, illustrate and develop the evaluative expression. The main hypothesis is that while elaborations, illustrations and other rhetorical relations related to explanation have no argumentative orientation, they acquire such orientation when combined with an evaluative expression. In the example given by the authors, a justification gives a reason for the evaluation expressed in the review: "The hotel is 2 stars [JUSTIFICATION due to the lack of bar and restaurant facilities]" can be classified as a justification whose general abstract schema is "X is Eval because of Fact*", where Eval denotes the evaluative expression and Fact* is a set of facts acting as justifications. The contrast relation introduces a statement which is somewhat symmetric but in opposition with the evaluation, e.g. "It was clean and comfortable, [CONTRAST but the carpet was in need of replacing]" whose schema is "X is Eval but B*", for which examples of linguistic markers include "whereas", "but" and "while".

In order to process these relations, the authors use <TextCoop>, a platform designed for discourse analysis with a logic and linguistics perspective. Argument extraction rules correspond to the types of discourse structures that appear in consumer opinion texts. These rules have been adapted from a rule repository dedicated to explanation structures by developing additional lexical resources (terms expressing polarity, adverbs of intensity or domain verbs) and revising the structure of rules. For example, for the contrast relation, here are the six rules that have been defined:

Contrast ⇒ conn(opposition whe), gap(G), ponct(comma). / conn(opposition_whe), gap(G), end. / conn(opposition_how), gap(G), end.

where conn(opposition_whe): whereas, but whereas, but while, and conn(opposition_how): however.

The majority of the work in argument mining employs the statistical approach. For instance, binary supervised classifiers are trained to distinguish argumentative sentences from non-argumentative ones. Several standard machine learning algorithms have been exploited for argument mining tasks, such as Support Vector Machines (SVM), Logistic Regression, Naive Bayes classifiers, Decision Trees, and Random Forests. They are trained in a supervised setting, starting from a set of annotated examples (the training set of the corpus), and are then tested on unseen examples (the test set of the corpus). The annotated instances are provided to the system in the form of a feature vector, and each example is associated with the appropriate class. Concerning the choice of features, many of the existing approaches in the literature share almost the same set of features, including classical features for text representation (such as those mentioned above), sentiment-based features, and Bag-of-Words (BoW) representations, where the sentence is encoded as a vector of binary values over huge dictionaries. BoWs for bigrams and trigrams are also possible. Recent approaches to argument mining are also beginning to exploit deep neural architectures, such as Recurrent Neural Networks (RNN), a popular model in NLP tasks. In a nutshell, the idea behind RNN is to make use of sequential information, i.e. to perform the same task for every element of a sequence, with the output depending on the previous computations. The most commonly used type of RNNs are Long Short-Term Memory neural networks (LSTMs), which are better at capturing long-term dependencies than basic RNNs.

Among these, Lippi and Torroni [Lippi and Torroni, 2015b] present a framework to detect claims in unstructured corpora without necessarily resorting to contextual information. Their methodology is driven by the observation that argumentative sentences are often characterised by common rhetorical structures. As the structure of a sentence can be highly informative for argument detection, and in particular for identification of a claim, the authors choose constituency parse trees to represent such information. They therefore build a claim detection system based on an SVM classifier which aims at capturing similarities between parse trees through Tree Kernels, a method used to measure the similarity between two trees by evaluating the number of their common substructures.

Another recent statistical approach to argument mining has been presented by Habernal and Gurevych [Habernal and Gurevych, 2016b], who tackle a new task in computational argumentation, aiming to assess the qualitative properties of arguments in order to explain why one argument is more convincing than another. Based on a corpus of 26,000 annotated explanations written in natural language, two tasks are proposed on this dataset: prediction of the full label distribution and classification of the types of flaws in less convincing arguments. They define a framework composed of feature-rich SVM learners and Bidirectional LSTM neural networks with convolution.

In Section 3, we present an example of the statistical approach in more detail.

2.6 Automatically annotated data

A system developed in the NLP stage is then used to process raw, unannotated text in order to automatically extract arguments. The text is taken from the test corpus with all the tags removed. This step can be viewed as an automated equivalent to the manual annotation and corpus development described in Section 2.3. In other words, the NLP system creates an automatically annotated corpus.

In the work by Lippi and Torroni [Lippi and Torroni, 2015b] described in the previous section, automatically annotated data are visualised using colour highlighting (see Figure 10): the claims detected by their system are highlighted in red; false positives (detected by the system but not labelled as context-dependent claims for a given topic in the manually annotated IBM corpus) are blue; and false negatives (not detected by the system but labelled as positives in the manually annotated IBM corpus) are green.

Figure 11 shows another example of the output of a software tool. The <TextCoop> platform [Saint-Dizier, 2012], described above, produces automatic segmentation and annotation. The text is split into argumentative discourse units (ADUs) which contain minimal meaningful building blocks of a discourse with argumentative function. These propositional contents are presented as text in purple. Then, the system assigns illocutionary connections (text in green) to ADUs which are assertions, rhetorical questions (RQ), and so on, as well as polymorphic types to represent ambiguity (or underspecification), such as RQ-AQ, which means that an ADU can be anchored in the discourse via either rhetorical questioning or assertive questioning (AQ).

Kingdom, its shared nature, or both, cannot be representative of the Canadian nation. Their position is that because of its hereditary aspects, the sovereign's role as Supreme Governor of the Church of England (in England only), and the provisions of the Act of Settlement, 1701, that currently bar Roman Catholics from the line of succession, the monarchy is inherently contrary to egalitarianism and multiculturalism. Further, though it diverges from both the official position of the Canadian government and the opinions of some judges, legal scholars, and members of the Royal Family themselves, republicans deem the King or Queen of Canada to be either a solely British or English individual representing a British institution foreign to Canada. Founded on this perception is the republican assertion that national pride is diminished by the monarchy, its presence negating the country's full independence achieved in 1982, and makes Canada appear colonial and subservient to the United Kingdom, under which they feel Canadians suffered " military, economic, and cultural subjugation. "Instead, equating anti-monarchism with patriotism, they desire a Canadian citizen to act as head of state, and promote the national flag and/or the " country " as a more fitting locus of allegiance. This questioning of the monarchy's role in Canadian identity arose as a part of wider cultural changes that followed the evolution of the British Empire into the Commonwealth of Nations, the rise of anti-establishmentism, the creation of multiculturalism as an official policy in Canada, and the blossoming of Quebec separatism; the latter becoming the major impetus of political controversy around the Crown. Quebec nationalists agitated for an independent Quebec republicsuch as the Marxist form desired by the Front de liberation du Quebec and the monarchy was targeted as a symbol of anti-Anglophone demonstration, notably when assassination threats were in 1964 made against Queen Elizabeth II and Quebecers turned their backs on her procession when she toured Quebec City that year. In a 1970 speech to the Empire Club of Canada, Former Governor General Roland Michener summed up the contemporary arguments against the Crown: From its opponents, he said, came the claims that monarchies are unfashionable, republicsother than those with oppressive regimesoffer more freedom, people are given greater dignity from choosing their head of state, the monarchy is foreign and incompatible with Canada's multicultural society, and that there should be change for the sake of change alone. However, though it was later thought the Quiet Revolution and the period

Figure 10. Example of data automatically annotated using the framework of Lippi and Torroni on the Wikipedia page "Republicanism_in_Canada".

2.7 Evaluation of the automatic annotation step

The last step in the argument mining pipeline is evaluation of the automatic annotation. This is an analogue of the step of evaluating manual annotation: in the previous, linguistic part of the process we compared analyses of the same set of texts tagged by different annotators, and here – in a similar way – we compare an analysis created by a human annotator and stored in our test corpus with the automatic annotation created by the system. A key difference between these two types of comparison (manual vs. manual and manual vs. automatic) is that it is assumed that neither of the human annotators is better than the other (i.e. if they create two different argument maps, we do not assume that one of them is correct), whereas in the second case we treat the manual analysis as a gold standard; i.e. whenever there is a difference between human annotation and machine annotation, we assume that the manual analysis is the correct one.

A simple measure often used for this task is accuracy, i.e. the proportion (percentage) of matches between manual and machine assignment of labels. If we want to capture more detailed information about how well the system has performed in mining arguments, a group of metrics – recall, precision and F_1 score – can be used. These are defined using the following scores: true positives, tp, when the machine has assigned a label to the same text span as the human analyst; true negatives, tn, when neither the machine nor the human analyst has assigned a label to a given segment; false positives, fp, when the machine has assigned a label to a given text span while the human being did not; and false negatives, fn, when the machine did not assign a label to a segment to

```
< utterance speaker = "lj" illoc = " standard_assertion " > < textunit nb = " 215 "
> it was a ghastly aberration < /textunit> < /utterance>

<utterance speaker = "cl" illoc = "RQ"> <textunit nb= "216"> or was it in fact
typical ? < /textunit> </utterance> .

<utterance speaker = "cl" illoc = "RQ-AQ"> <textunit nb = "217">
was it the product of a policy that was unsustainable that could
only be pursued by increasing repression? < /textunit> < /utterance>.
```

Figure 11. Example of data automatically annotated using the <TextCoop> platform for discourse processing of dialogical arguments in natural-language transcribed texts of the BBC Radio 4 programme *Moral Maze*

which the human being made an assignment. Then:
- **recall** measures how many times the system failed to recognise ("missed out") arguments:

$$R = \frac{tp}{tp + fn}$$

- **precision** shows how many times the program incorrectly identified a text span as an argument:

$$P = \frac{tp}{tp + fp}$$

- **F_1 score** (F-score, F-measure) provides the harmonic mean of precision and recall:

$$F_1 = 2 \cdot \frac{P \cdot R}{P + R}$$

If the matrices are computed and the performance of the system proves to be low, then we need to repeat the computational part of the process of argument mining, attempting to improve the NLP techniques and methods we are using.

For example, in their work [Moens et al., 2007; Palau and Moens, 2009] Palau and Moens obtain the following F_1-scores: 0.68 for the classification of premises; 0.74 for the classification of conclusions; and 0.60 for the determination of argument structures. For the eRulemaking corpus [Park and Cardie, 2014a], the system developed by Park and Cardie has the performance $P = 86.86$, $R = 83.05$, $F_1 = 84.91$ for the label Unverifiable Proposition (UnVerif) vs All; and $P = 49.88$, $R = 55.14$ and $F_1 = 52.37$ for the label Verifiable Proposition Non-experiential (VERIFnon) vs All. Finally, in the work [Lawrence and Reed, 2016], Lawrence and Reed aim to use argumentation schemes and combine different techniques in order to improve the success of recognising argument structure, achieving the following results: for the technique of Discourse Indicators the system delivers precision of 1.00, recall of 0.08, and an F_1-score of 0.15; for the technique of Topic Similarity the system has precision of 0.70, recall of 0.54 and an F_1-score of 0.61; for the technique of Schematic Structure the system delivers precision of 0.82, recall of 0.69, and an F_1-score of 0.75; and finally, for the combination of these techniques the system improves performance and delivers precision of 0.91, recall of 0.77, and an F_1-score of 0.83.

3 An example: predicting argument relations

As we discussed in the introduction of this chapter, there are many approaches to argument mining, from the perspectives of computational models of arguments or of computational linguistics. In this section, we highlight the main features of an approach to argument mining where the selected argumentation model is the abstract argumentation framework [Dung, 1995; Cayrol and Lagasquie-Schiex, 2013] and the NLP method applied is Textual Entailment.

More precisely, one of the main goals in the argument mining pipeline is to predict the relation holding between pairs of arguments. In this section, we describe in detail one of the few approaches proposed in the argument mining community to address this task [Cabrio and Villata, 2012b; Cabrio and Villata, 2013], while also showing the execution and implementation of the argument pipeline presented in Section 2. The idea is to predict positive and negative relations between arguments, i.e. *support* and *attack* relations, using textual entailment. Alongside formal approaches to semantic inference that rely on logical representation of meaning, the notion of Textual Entailment (TE) has been proposed as an applied framework to represent major semantic inferences across applications in the field of computational linguistics [Dagan *et al.*, 2009]. The development of the Web has led to a paradigm shift, due to the need to process a huge amount of available (but often noisy) data. TE is a generic framework for applied semantics, where linguistic objects are mapped by means of semantic inferences at a textual level.

3.1 The NoDE dataset

The NoDE dataset (Natural language arguments in online DEbates) is a benchmark of natural language arguments extracted from different kinds of textual sources[34]. It is composed of three datasets of natural language arguments, released in two machine-readable formats: the standard XML format and the XML/RDF format adopting the SIOC-Argumentation ontology, which has been extended in [Cabrio *et al.*, 2013] to deal with bipolar abstract argumentation.

We have identified three different scenarios for data extraction: *(i)* the online debate platforms Debatepedia[35] and ProCon[36] present a set of topics to be

---
[34]Available at: www.inria.fr/NoDE
[35]http://idebate.org/debatabase
[36]http://www.procon.org/

discussed, and participants argue about issues related to such topics that are raised on the platform, pointing out whether their arguments are in favour of or against a particular position on a given issue; *(ii)* in the script of the famous play *Twelve Angry Men*, the jurors of a trial discuss whether or not a boy is guilty, and at the end of each act they vote to determine whether they all agree; and *(iii)* the Wikipedia revision history of an article shows its evolution over time (in our case, a four-year period); we focused on the five most revised articles. These three scenarios lead to three different resources: the online debates resource collects arguments about a debated issue or other arguments into small bipolar argumentation graphs. The same happens for the Wikipedia dataset, with the revisions of the articles used to build small bipolar argumentation graphs. The "Twelve Angry Men" resource also collects pro and con arguments, but the resulting three bipolar argumentation graphs present a higher complexity than the debate graphs.

Methodology. Given a set of linked arguments (e.g. in a debate), we proceed as follows:

1. We couple each argument with the argument it is related to (which it attacks or supports). The first layer of the dataset is therefore composed of pairs of arguments (each one labelled with a unique ID) annotated by the semantic relations linking them (i.e. *attack* or *support*);

2. starting from the pairs of arguments in the first layer, we then build a bipolar argumentation graph for each topic in the dataset. In the second layer of the dataset, we therefore find argument graphs.

To create the dataset of argument pairs, we followed the criteria defined and used by the organisers of the Recognizing Textual Entailment challenge [Dagan *et al.*, 2009][37]. To test the progress of Textual Entailment (TE) systems in a comparable setting, RTE participants are provided with datasets composed of pairs of textual snippets (the Text T and the Hypothesis H) involving various levels of entailment reasoning (e.g. lexical or syntactic). The TE systems are required to produce a correct judgement on the given pairs (i.e. to say if the meaning of H can be inferred from the meaning of T). Two kinds of judgements are allowed: two-way (yes-or-no entailment) or three-way judgement (entailment, contradiction or unknown). To perform the latter, if there is no entailment between T and H, systems must be able to distinguish whether the truth of H is contradicted by T or remains unknown on the basis of the information contained in T. To correctly judge each single pair inside the RTE datasets, systems are expected to cope with both the different linguistic phenomena involved in TE and the complex ways in which they interact.

---

[37]Since 2004, RTE Challenges have promoted research in RTE http://www.nist.gov/tac/2010/RTE/.

Data format. As regards the choice of format, the NoDE dataset also uses the one proposed by the RTE challenges to annotate the first layer of the data in NoDE. It is an XML format, where each pair is identified by a unique ID and by the task (in this case, *argumentation*). The element *entailment* contains the annotated relation of entailment/non-entailment between the two arguments in the pair. Unlike the RTE dataset, in NoDE the element *topic* is added to identify the graph name to which the pair belongs, as well as an ID to keep track of each text snippet (i.e. each argument). The argument IDs are unique within each graph. An example pair from the Debatepedia dataset is as follows:

```
<entailment-corpus>
  <pair task="ARG" id="1" topic="Violentgames" entailment=
  "NONENTAILMENT">
      <t id="2">Violent video games do not increase aggression.
      </t>
      <h id="1">Violent games make youth more aggressive/violent.
      </h>
  </pair>
   ...
</entailment-corpus>
```

For the second layer, the XML/RDF format adopted in NoDE relies on the SIOC-Argumentation extended vocabulary[38].

Each argument is a sioc_arg:Argument, and the two relations of *support* and *attack* are respectively characterised by the properties sioc_arg:supportsArg and sioc_arg:challengesArg (mapped on the *entailment* and *nonentailment* relations, respectively). An example from the Twelve Angry Men XML/RDF dataset is as follows:

```
<http://example.org/12AngryMen/pair1t> rdf:type sioc_arg:Argument;
   sioc:content "Ever since he was five years old his father beat
                him up regularly. He used his fists." ;
   sioc_arg:challengesArg <http://example.org/12AngryMen/pair1h> .

<http://example.org/12AngryMen/pair1h> rdf:type sioc_arg:Argument;
   sioc:content "Look at the kid's record. At fifteen he was
   in reform  school. He stole a car. He was picked up for
   knife-fighting. I think they said he stabbed somebody in the
   arm. This is a very fine boy." .
```

All the abstract bipolar argumentation graphs resulting from the datasets of the benchmark are also available for visualisation as **png** images.

---

[38] http://bit.ly/SIOC_Argumentation

Debatepedia/ProCon dataset. To build the first benchmark of natural language arguments, the Debatepedia and ProCon encyclopaedia of pro and con arguments on critical issues were selected as data sources. To fill in the first layer of the dataset, a set of topics (Table 5 column *Topics*) of Debatepedia/ProCon debates was selected, and for each topic, the following procedure was applied:

1. The main issue (i.e. the title of the debate expressed as an affirmative statement) is considered as the starting argument.

2. Each user opinion is extracted and considered an argument.

3. Since *attack* and *support* are binary relations, the arguments are coupled with either

    (a) the starting argument, or

    (b) other arguments in the same discussion to which the most recent argument refers (i.e., when a user's opinion supports or attacks an argument previously expressed by another user, the former is coupled with the latter), in chronological order to maintain the dialogue structure.

4. The resulting pairs of arguments are then tagged with the appropriate relation, i.e. *attack* or *support*.

The use of Debatepedia/ProCon as a case study provides us with arguments that have already been annotated (*pro* ⇒ *entailment*, and *con* ⇒ *contradiction*) and casts our task as a yes/no entailment task. To show a step-by-step application of the procedure, let us consider the debated issue *Can coca be classified as a narcotic?* In step 1, we transform its title into an affirmative form and consider it to be the starting argument (a). Then, in step 2, we extract all the users' opinions concerning this issue (both pro and con), e.g. (b), (c) and (d):

(4)   a.   *Coca can be classified as a narcotic.*
          b.   *In 1992 the World Health Organization's Expert Committee on Drug Dependence (ECDD) undertook a "prereview" of coca leaf. The ECDD report concluded that, "the coca leaf is appropriately scheduled as a narcotic under the Single Convention on Narcotic Drugs, 1961, since cocaine is readily extractable from the leaf." This ease of extraction makes coca and cocaine inextricably linked. Therefore, because cocaine is defined as a narcotic, coca must also be defined in this way.*
          c.   *Coca in its natural state is not a narcotic. What is absurd about the 1961 convention is that it considers the coca leaf in its natural, unaltered state to be a narcotic. The paste or the concentrate that is*

extracted from the coca leaf, commonly known as cocaine, is indeed a narcotic, but the plant itself is not.

d. *Coca is not cocaine. Coca is distinct from cocaine. Coca is a natural leaf with very mild effects when chewed. Cocaine is a highly processed and concentrated drug using derivatives from coca, and therefore should not be considered as a narcotic.*

In step 3a we couple arguments (b) and (d) with the starting issue, since they are directly linked to it, and in step 3b we couple argument (c) with argument (b) and argument (d) with argument (c), since they follow one another in the discussion (i.e. the user expressing argument (c) answers the user expressing argument (b), so that the arguments are concatenated; the same applies to arguments (d) and (c)).

In step 4, the resulting pairs of arguments are then tagged with the appropriate relation: (**b**) *supports* (**a**), (**d**) *attacks* (**a**), (**c**) *attacks* (**b**) and (**d**) *supports* (**c**).

We collected 260 T-H pairs (Tables 1, 4). The training set is composed of 85 entailment and 75 contradiction pairs and the test set consists of 55 entailment and 45 contradiction pairs. Pairs in the test set are extracted from topics different from those of the training set.

Training set					Test set				
Topic	#arg	#pairs			Topic	#arg	#pairs		
		tot	yes	no			tot	yes	no
Violent games	16	15	8	7	Ground zero mosque	9	8	3	5
China 1-child policy	11	10	6	4	Military service	11	10	3	7
Coca as a narcotic	15	14	7	7	Libya no fly zone	11	10	6	4
Child beauty contests	12	11	7	4	Airport security prof.	9	8	4	4
Arming Libyan rebels	10	9	4	5	Solar energy	16	15	11	4
Alcohol breath tests	8	7	4	3	Natural gas vehicles	12	11	5	6
Osama death photo	11	10	5	5	Cell phones/driving	11	10	5	5
Private social security	11	10	5	5	Legalize marijuana	17	16	10	6
Internet as a right	15	14	9	5	Gay marriages	7	6	4	2
Tablets vs. Textbooks	22	21	11	10	Vegetarianism	7	6	4	2
Obesity	16	15	7	8					
Abortion	25	24	12	12					
TOTAL	172	160	85	75	TOTAL	110	100	55	45

Table 3. The Debatepedia dataset

Based on the TE definition [Dagan et al., 2009], an annotator with skills in linguistics carried out the first phase of annotation of the Debatepedia/ProCon dataset. Then, to assess the validity and reliability of the annotation, the same annotation task was independently carried out by a second annotator, so that inter-annotator agreement could be computed. As discussed in Section 2.4, Cohen's kappa was used to calculate inter-rater agreement on the NoDE dataset. Inter-rater agreement results in $\kappa = 0.7$ (see Table 4). As a rule of thumb, this is a satisfactory level of agreement, so we consider these annotated datasets as the *gold standard*, i.e. a reference dataset to which the performances of automated systems can be compared.

To fill the second layer of the Debatepedia/ProCon dataset, the pairs annotated in the first layer are combined to build a bipolar argumentation graph for each topic (12 topics in the training set and 10 topics in the test set, listed in Table 5).

Twelve Angry Men dataset. As a second scenario for extraction of natural language arguments, the NoDE dataset selected the script of the play *Twelve Angry Men*. The play concerns the deliberations of a jury in a homicide trial. The story begins after closing arguments have been presented in the homicide case. At first, the jurors have a nearly unanimous decision that the defendant is guilty, with a single dissenter voting "not guilty" and sowing a seed of reasonable doubt throughout the play.

The play is divided into three acts, and at the end of each act the jury votes, until they reach unanimity. For each act, we manually identified the arguments (excluding sentences which cannot be considered self-contained arguments) and coupled each argument with the argument it supports or attacks in the dialogue flow (Examples (5) and (6), respectively). In the discussions, one character's argument comes after another's (entailing or contradicting one of the arguments previously expressed by another character); therefore, pairs are created in the graph connecting the former to the latter (more recent arguments are considered to be T and the argument with respect to which we want to detect the relation is considered to be H). In Example (7), juror 1 claims argument (7-c) and is attacked by juror 2, claiming argument (7-b). Juror 3 then claims argument (7-a) to support juror 2's opinion. In the dataset, the following couples were annotated: (7-c) is contradicted by (7-b) and (7-b) is entailed by (7-a). In Example (8), juror 1 claims argument (8-c) supported by juror 2 (argument (8-b)) and juror 3 attacks juror's 2 opinion with argument (8-a). More specifically, (8-c) is entailed by (8-b) and (8-b) is contradicted by (8-a).

(5) a. *Maybe the old man didn't hear the boy yelling "I'm going to kill you". I mean with the el noise.*
     b. *I don't think the old man could have heard the boy yelling.*

(6) a. *I never saw a guiltier man in my life. You sat right in court and heard the same thing I did. The man's a dangerous killer.*
     b. *I don't know if he is guilty.*

(7) a. *Maybe the old man didn't hear the boy yelling "I'm going to kill you". I mean with the el noise.*
     b. *I don't think the old man could have heard the boy yelling.*
     c. *The old man said the boy yelled "I'm going to kill you" out. That's enough for me.*

(8) a. *The old man cannot be a liar, he must have heard the boy yelling.*
     b. *Maybe the old man didn't hear the boy yelling "I'm going to kill you". I mean with the el noise.*
     c. *I don't think the old man could have heard the boy yelling.*

Given the complexity of the play and the fact that in human linguistic interactions much is left implicit, we simplified the arguments as follows: *i)* by adding the required context in T to make the pairs self-contained (in the TE framework entailment is detected based on the evidence provided in T); and *ii)* by solving intra-document co-references, as in *Nobody has to prove that!*, transformed into *Nobody has to prove [that he is not guilty]*.

A total of 80 T-H pairs were collected: 25 entailment pairs, 41 contradiction pairs and 14 unknown pairs (contradiction and unknown pairs can be collapsed in the judgement *non-entailment* for the two-way classification task). Table 4 shows the inter-annotator agreement.

To fill the second layer of the Twelve Angry Men dataset, again the pairs annotated in the first layer are combined to build a bipolar argumentation graph for each topic in the dataset (the three acts of the play). The complexity of the graphs obtained for this scenario is higher than in the debate graphs (on average, 27 links per graph as compared to 9 links per graph in the Debatepedia dataset).

Wikipedia revision history dataset. We selected four dumps of English Wikipedia (*Wiki 09* dated 6.03.2009, *Wiki 10* dated 12.03.2010, *Wiki 11* dated 9.07.2011, and *Wiki 12* dated 6.12.2012), and NoDE focuses on the five most revised pages[39] at that time (George W. Bush, United States, Michael Jackson, Britney Spears, and World War II).

After extracting plain text from these pages, each document was sentence-split for both *Wiki 09* and *Wiki 10*, and the sentences of the two versions were automatically aligned to create pairs. Then, following [Cabrio et al., 2012], the *Position Independent Word Error Rate (PER)* – a metric based on calculation of the number of words which differ between a pair of sentences – was adopted in order to measure the similarity between the sentences in each pair. Only pairs composed of sentences where major editing was carried out ($0.2 < PER < 0.6$), but still describing the same event, were selected. For each pair of extracted sentences, TE pairs were created by setting the revised sentence (from *Wiki 10*) as T and the original sentence (from *Wiki 09*) as H. Starting from such pairs composed of the same revised argument, we checked the more recent Wikipedia versions (i.e. *Wiki 11* and *Wiki 12*) to see whether the arguments had been further modified. If so, another T-H pair based on the same assumptions as before was created by setting the revised sentence as T and the older sentence as H (see Example (9)).

(9) a. Wiki12: *The land area of the contiguous United States is 2,959,064 square miles (7,663,941 km2).*
 b. Wiki11: *The land area of the contiguous United States is approximately 1,900 million acres (7,700,000 km2).*
 c. Wiki10: *The land area of the contiguous United States is approximately 1.9 billion acres (770 million hectares).*
 d. Wiki09: *The total land area of the contiguous United States is ap-*

---

[39] http://bit.ly/WikipediaMostRevisedPages

proximately 1.9 billion acres.

Such pairs were then annotated with respect to the TE relation, following the criteria defined for the two-way judgement TE task. As a result of the first step (extraction of the revised arguments in (9-d) and (9-c)), 280 T-H pairs were collected. After applying the procedure to the same arguments in (9-b) and (9-a), the total number of collected pairs was 452 (training set composed of 114 entailment and 114 non-entailment pairs and test set of 101 entailment and 123 non-entailment pairs); see Table 4. To enable correct training of automatic systems, the dataset was balanced with respect to the percentage of yes/no judgements. In Wikipedia, the actual distribution of attacks and supports among revisions of the same sentence is slightly unbalanced, as users generally edit a sentence to add different information or correct it. Inter-annotator agreement is reported in Table 4.

Dataset	pairs	supp	att	graphs	Inter-annotator agreement		
					annotators	pairs	κ
Debatepedia	260	140	120	22	2	100	0.7
12 Angry Men	80	25	55	3	2	40	0.74
Wiki revisions	452	215	237	416	2	140	0.7

Table 4. Summary of the datasets composing NoDE, where *supp* means support and *att* means attack

3.2 Textual Entailment for relation prediction in abstract argumentation

In the reminder of this section, we will provide some basic information about Textual Entailment and abstract bipolar argumentation theory, and then we will detail how relation prediction in the argument mining pipeline can be addressed using TE systems. An evaluation of the results obtained by such a method are reported as well.

Textual Entailment: background and main insights. Classical approaches to semantic inference rely on logical representations of meaning that are external to the language itself and are typically independent of the structure of any particular natural language. Texts are first translated or interpreted into some logical form and then new propositions are inferred from interpreted texts by a logic theorem prover. But, especially since the development of the Web, we have witnessed a paradigm shift, due to the need to process a huge amount of available (but often noisy) data. Addressing the inference task by means of logic theorem provers in automated applications aimed at natural language understanding has shown several intrinsic limitations [Blackburn et al., 2001].

Especially in data-driven approaches, such as the one presented in this work, where patterns are learnt from large-scale naturally-occurring data, we can accept approximate answers provided by efficient and robust systems, even at the price of logical unsoundness or incompleteness. Starting from these considerations, Monz and de Rijke [Monz and de Rijke, 2001] propose addressing the inference task directly at the textual level instead, exploiting currently available NLP techniques. While methods for automated deduction assume that the arguments in the input are already expressed in some formal representation of meaning (e.g. first-order logic), addressing the inference task at a textual level opens new and different challenges from those encountered in formal deduction. More emphasis is placed on informal reasoning, lexical semantic knowledge, and the variability of linguistic expressions (see Section 2).

The notion of Textual Entailment has been proposed as an applied framework to detect major semantic inferences across applications in NLP [Dagan et al., 2009]. It is defined as a relation between a coherent textual fragment (the Text T) and a language expression, which is considered the Hypothesis (H). Entailment holds (i.e. $T \Rightarrow H$) if the meaning of H can be inferred from the meaning of T, as interpreted by a typical language user. The TE relationship is directional, since the meaning of one expression usually entails the other, while the opposite is much less certain. Consider the pairs in Examples (10) and (11).

(10) a. **T1** *Internet access is essential now; must be a right. The internet is only that wire that delivers freedom of speech, freedom of assembly, and freedom of the press in a single connection.*
 b. **H** *Making Internet a right only benefits society.*

(11) a. **T2** *Internet not as important as real rights. We may think of such trivial things as a fundamental right, but consider the truly impoverished and what is most important to them. The right to vote, the right to liberty and freedom from slavery or the right to elementary education.*
 b. **H** *Making Internet a right only benefits society.*

A system aimed at recognising TE should detect an inference relation between T1 and H (i.e. the meaning of H can be derived from the meaning of T) in Example (10), while it should not detect an entailment between T2 and H in Example (11). The definition of TE is based on (and assumes) common human understanding of language, as well as common background knowledge. However, the entailment relation is said to hold only if the statement in the text licenses the statement in the hypothesis, meaning that the content of T and common knowledge together should entail H, and not background knowledge alone. In this applied framework, inferences are performed directly over lexical-syntactic representations of the texts.

For pairs where the entailment relation does not hold between T and H, systems are required to make a further distinction between pairs where the entailment does not hold because the content of H is contradicted by the content of T (*contradiction*, see Example (11)), and pairs where the entailment cannot be determined because the truth of H cannot be verified on the basis of the content of T (*unknown*, see Example (12)). De Marneffe and colleagues [de Marneffe et al., 2008] provide a definition of contradiction for the TE task, claiming that it occurs when two sentences *i)* are extremely unlikely to be true simultaneously and *ii)* involve the same event. This three-way judgment task (*entailment* vs *contradiction* vs *unknown*) was introduced for the RTE-4 challenge. Before RTE-4, TE was considered a two-way decision task (*entailment* vs *no entailment*). However, the classic two-way task is also offered as an alternative in recent editions of the evaluation campaign (*contradiction* and *unknown* judgments are collapsed into the judgment *no entailment*).

(12)   a.   **T3** *Internet "right" means denying parents' ability to set limits. Do you want to make a world when a mother tells her child: "you cannot stay on the internet anymore" that she has taken a right from him? Compare taking the right for a home or for education with taking the "right" to access the internet.*
       b.   **H** *Internet access is essential now; must be a right. The internet is only that wire that delivers freedom of speech, freedom of assembly, and freedom of the press in a single connection.*

Abstract bipolar argumentation. Abstract argumentation frameworks consider arguments as abstract entities, deprived of any structural property and of any relations but attack [Dung, 1995]. However, in these frameworks, the assessment of argument acceptability depends only on the attack relation in abstract terms, while in other application domains other relations may be required. In particular, abstract bipolar argumentation frameworks, first proposed by [Cayrol and Lagasquie-Schiex, 2005], extend Dung's abstract framework by taking into account both the attack relation and the support relation. An abstract bipolar argumentation framework is a labelled directed graph, with two labels, indicating either attack or support. We represent the attack relation by $a \rightarrow b$ and the support relation by $a \dashrightarrow b$.

Casting bipolar argumentation as a TE problem. The issue of predicting the relation holding between two arguments can be instantiated in the task of predicting whether the relation between two arguments is a support or attack relation. In the remainder of this section, we present and evaluate this approach on the data of the NoDE dataset. For more details about this task and its main insights, we refer the reader to [Cabrio and Villata, 2012a; Cabrio and Villata, 2013; Cabrio et al., 2013; Cabrio and Villata, 2014].

The general goal of our work is to propose an approach to help the participants in forums or debates (e.g. on Debatepedia or Twitter) detect which arguments about a certain topic are accepted. As a first step, we need to *(i)* automatically generate the arguments (i.e., recognise a participant's opinion on a certain topic as an argument) and *(ii)* detect their relation to the other arguments. We cast the problem as a TE problem, where the T-H pair is a pair of arguments expressed by two different participants in a debate on a certain topic. For instance, given the argument "Making Internet a right only benefits society" (which we regard as H), participants can be in favour of it (expressing arguments from which H can be inferred, as in Example (10)), or can contradict it (expressing an opinion against it, as in Example (11)). Since in debates one participant's argument comes after another's, we can extract such arguments and compare them both w.r.t. the main issue and w.r.t. the other participants' arguments (when the new argument entails or contradicts one of the arguments previously expressed by another participant). For instance, given the same debate as before, a new argument T3 may be expressed by a third participant to contradict T2 (which becomes the new H (H1) in the pair), as shown in Example (13).

(13)   a.   **T3** *I've seen the growing awareness within the developing world that computers and connectivity matter and can be useful. It's not that computers matter more than water, food, shelter and healthcare, but that the network and PCs can be used to ensure that those other things are available. Satellite imagery sent to a local computer can help villages find fresh water, mobile phones can tell farmers the prices at market so they know when to harvest.*

       b.   **T2 ≡ H1** *Internet not as important as real rights. We may think of such trivial things as a fundamental right, but consider the truly impoverished and what is most important to them. The right to vote, the right to liberty and freedom from slavery or the right to elementary education.*

TE provides us with techniques to identify the arguments in a debate and to detect which kind of relation underlies each pair of arguments. A TE system returns a judgement (entailment or contradiction) on the argument pairs related to a certain topic, which are used as input to build the abstract argumentation framework. Example (14) presents how TE is combined with bipolar argumentation to compute the set of accepted arguments at the end of the argument mining pipeline.

(14)   *The textual entailment phase returns the following pairs for the natural language opinions detailed in Examples (10), (11), and (13):*

- *T1 entails H*

- *T2 attacks H*

- *T3 attacks H1 (i.e. T2)*

Given this result, the argumentation module of our framework maps each element to its corresponding argument: $H \equiv A_1$, $T1 \equiv A_2$, $T2 \equiv A_3$, and $T3 \equiv A_4$. The accepted arguments (using admissibility-based semantics) are $\{A_1, A_2, A_4\}$. This means that the issue "Making Internet a right only benefits society"' A_1 is considered to be accepted.

Experimental evaluation: the Debatepedia/ProCon dataset of NoDE.
Following the methodology described in Section 2.7, in the first set of experiments the 200 T-H pairs of the Debatepedia/ProCon dataset have been divided into 100 for training and 100 for testing of the TE system (each dataset is composed of 55 entailment and 45 contradiction pairs). The pairs considered for the test set concern completely new topics, never seen by the system. Table 5 shows the topics used to train the system and those used to test it.

Training set				
Topic	#argum	#pairs		
		TOT.	yes	no
Violent games boost aggressiveness	16	15	8	7
China one-child policy	11	10	6	4
Consider coca as a narcotic	15	14	7	7
Child beauty contests	12	11	7	4
Arming Libyan rebels	10	9	4	5
Random alcohol breath tests	8	7	4	3
Osama death photo	11	10	5	5
Privatizing social security	11	10	5	5
Internet access as a right	15	14	9	5
TOTAL	109	**100**	**55**	**45**
Test set				
Topic	#argum	#pairs		
		TOT.	yes	no
Ground zero mosque	9	8	3	5
Mandatory military service	11	10	3	7
No fly zone over Libya	11	10	6	4
Airport security profiling	9	8	4	4
Solar energy	16	15	11	4
Natural gas vehicles	12	11	5	6
Use of cell phones while driving	11	10	5	5
Marijuana legalization	17	16	10	6
Gay marriage as a right	7	6	4	2
Vegetarianism	7	6	4	2
TOTAL	110	**100**	**55**	**45**

Table 5. The Debatepedia dataset used in our experiments

To detect which kind of relation underlies each pair of arguments, we adopted the modular architecture of the EDITS system (Edit Distance Textual Entailment Suite), version 3.0. EDITS is an open-source software package for recognising TE[40] [Kouylekov and Negri, 2010] which implements a distance-based

---
[40] http://edits.fbk.eu/

		Train			Test		
	rel	Pr.	Rec.	Acc.	Pr.	Rec.	Acc.
EDITS	yes	0.71	0.73	0.69	0.69	0.72	**0.67**
	no	0.66	0.64	**0.69**	0.64	0.6	
WordOverl.	yes	0.64	0.65	0.61	0.64	0.67	0.62
	no	0.56	0.55		0.58	0.55	

Table 6. System performance on the Debatepedia/ProCon dataset of NoDE (precision, recall and accuracy)

framework assuming that the probability of an entailment relation between a given T-H pair is inversely proportional to the distance between T and H (i.e., the higher the distance, the lower the probability of entailment)[41].

A two-step evaluation is then carried out. First, we need to assess the performance of the TE system in assigning entailment and contradiction relations to the pairs of arguments in the Debatepedia/ProCon dataset. Then, we evaluate to what degree its performance affects the application of the argumentation theory module, i.e. to what degree an incorrect assignment of a relation to a pair of arguments is propagated in the argumentation framework.

For the first evaluation, we run EDITS on the Debatepedia/ProCon training set so it can learn the model, and then test it on the test set. EDITS was tuned in the following configuration: *i)* cosine similarity as the core distance algorithm, *ii)* distance calculated on lemmas, and *iii)* a stopword list defined to set no distance between stopwords. We used the system off-the-shelf, applying one of its basic configurations. Table 6 reports on the results obtained both using EDITS and using a baseline that applies a Word Overlap algorithm on tokenised text. Even where a basic configuration of EDITS and a small data set (100 pairs for training) were used, performance on the Debatepedia/ProCon test set is promising and in line with those of TE systems on RTE data sets (usually containing about 1,000 pairs for training and 1,000 for testing).

As a second step in the evaluation phase, we need to consider the impact of EDITS performances on the acceptability of the arguments, i.e. to what extent incorrect assignment of a relation to a pair of arguments affects the acceptability of the arguments in the argumentation framework. Admissibility-based semantics are then used to identify the accepted arguments for the argumentation frameworks of the gold standard and those resulting from the TE system. The precision of the combined approach to identification of accepted arguments (i.e. arguments accepted by the combined system and by the gold standard with regard to a given Debatepedia/ProCon topic) is on average 0.74, and the

---

[41] In previous RTE challenges, EDITS has always ranked among the 5 best participating systems out of an average of 25 systems, and is one of the few RTE systems available as open source http://aclweb.org/aclwiki/index.php?title=Textual_Entailment_Resource_Pool.

Table 7. System performances on the Wikipedia revision history dataset of NoDE

EDITS conf.	rel	Train			Test		
		Pr.	Rec.	Acc.	Pr.	Rec.	Acc.
WordOverlap	yes	0.83	0.82	**0.83**	0.83	0.82	**0.78**
	no	0.76	0.73		0.79	0.82	
CosineSimilarity	yes	0.58	0.89	0.63	0.52	0.87	0.58
	no	0.77	0.37		0.76	0.34	

recall (arguments accepted in the gold standard and retrieved as accepted by the combined system) is 0.76. Its accuracy (the ability of the combined system to accept some arguments and discard others) is 0.75, meaning that the TE system's mistakes in assigning relations propagate in the argumentation framework, but the results are still satisfactory.

Experimental evaluation: the Wikipedia revision history dataset of NoDE. To evaluate the results of relation prediction by means of Textual Entailment with respect to the Wikipedia revision history dataset of NoDE, we again need to carry out a two-step evaluation: first, we must assess the performance of EDITS in assigning the *entailment* and the *no entailment* relations to pairs of arguments with respect to the Wikipedia dataset, and then we can evaluate the extent to which the performance affects the application of the argumentation theory module, i.e. the extent to which incorrect assignment of a relation to a pair of arguments is propagated in the argumentation framework. For the first evaluation, EDITS was run on the Wikipedia training set to learn the model, and then tested on the test set. In the configuration of EDITS, the distance entailment engine applies *cosine similarity* and *word overlap* as the core distance algorithms. In both cases, distance is calculated on lemmas and a stopword list is defined to have no distance value between stopwords.

The results are reported in Table 7. Due to the specificity of this dataset (i.e., it is composed of revisions of arguments), the *word overlap* algorithm outperforms *cosine similarity*, as there is high similarity between the revised and original arguments (in most of the positive examples the two sentences are very close or there is an almost perfect inclusion of H in T). For the same reason, the results are higher than, for example, in Cabrio and Villata [Cabrio and Villata, 2012a], and higher than the results obtained on average in RTE challenges. For these runs, we used the system off-the-shelf, applying its basic configuration. As a second step in the evaluation phase, we consider the impact of EDITS performances (obtained using word overlap, since it provided the best results) on the acceptability of the arguments, i.e. the extent to which incorrect assignment of a relation to a pair of arguments affects the acceptability of the arguments in the argumentation framework. We again use admissibility-based semantics [Dung, 1995] to identify the accepted arguments both with respect to the correct argumentation frameworks of each Wikipedia revised argument (where entailment/contradiction relations are correctly assigned, i.e. the gold standard) and on the frameworks generated by assigning the relations resulting from the TE system judgements. The precision of the combined approach in identifying accepted arguments (i.e. arguments accepted by the combined system and by the gold standard with respect to a specific Wikipedia revised argument) is on average 0.90 and the recall (arguments accepted in the gold standard and retrieved as accepted by the combined system) is 0.92. The F-measure is 0.91, meaning that the TE system's mistakes in relation assignment propagate in the argumentation framework, but results are still satisfactory.

4 Conclusions

In this chapter, we have presented the entire argument mining pipeline with an in-depth look at two approaches proposed in the literature, applied to structured and abstract argumentation (supporting arguments and conflicting arguments). A case study of automatic extraction of argument relations is then discussed in Section 3.

This chapter has highlighted rising trends in the very new field of argument mining research. We can summarise the main steps of the argument mining pipeline as follows: *(i)* We must select a precise argument mining task and define precise guidelines for annotation of the data to be used by our system. If the guidelines are not sufficiently clear, the results of the annotation phase will be poor, leading to unsatisfactory inter-annotator agreement. Defining precise annotation guidelines is a time-consuming task, but it ensures the reliability of the resources produced when the guidelines are followed. *(ii)* When the guidelines have been established, the corpus of natural language arguments must be annotated according to these guidelines. After the annotation process, inter-annotator agreement must be evaluated to ensure the reliability of the resources. *(iii)* Now that the data is ready, we must choose or define the best solution for the task we are addressing, e.g. argument classification or relation prediction. *(iv)* Finally, the results must be evaluated to ensure that the proposed solution is correct and scalable.

Some conclusions can be drawn from this overview of the field of argument mining. First of all, it is important to distinguish between the well-known NLP research field of *opinion mining* (or *sentiment analysis*) and argument mining. Apart from some minor differences, the main point here is that the goal of opinion mining is to understand *what* people think about something, while the goal of argument mining is to understand *why* people think something about a given topic [Habernal *et al.*, 2014]. This key difference between these two research areas characterises the main feature of argumentation theory in general, which is the ability to explain and to justify different viewpoints. Second, argumentation theory is traditionally discussed in correlation with what is called *critical thinking*, i.e. the intellectual process of objective analysis and evaluation of an issue in order to form a judgement. However, argument mining approaches can support formal argumentation approaches in order to define formal models that are closer to human reasoning, in which the fuzziness and ambiguity of natural language play an important role and the intellectual process is not always completely rational and objective. Argument mining can provide greater insight into the answers to such questions as "What are the best arguments to influence a real audience?" or "What is the role of emotions in the argumentation process?"

As discussed in recently published surveys of argument mining [Peldszus and Stede, 2013; Lippi and Torroni, 2015a], argument mining approaches currently face two main issues: big data and deep learning. Concerning the first point, a huge amount of data is now available on the Web, such as social network posts, forums, blogs, product reviews, or user comments to newspapers articles, and it must be analysed automatically as it far exceeds human capabilities to parse and understand it without an automatic support tool. Argument mining can make the difference here, and can exploit the Web to perform crowd-sourced annotations for very large corpora. A first step in this direction is described in the use case in this chapter, based on the NoDE dataset, whose texts are extracted from various online sources. As shown in this example, classical semantic approaches to NLP cannot cope with the variability and noise present in such texts and the huge quantity of data that must be processed. Hence the need to apply new methods such as Textual Entailment. Concerning the second point, deep learning methods, i.e. fast and efficient machine learning algorithms such as word embeddings[42], can be exploited in the argument mining pipeline to deal with large corpora and unsupervised learning. An issue associated with data on the Web is multilingualism, which poses a challenge to argument mining approaches. As highlighted in this chapter, the vast majority of research proposed in the field of argument mining deals with English data only. This is because far better algorithms are available for NLP tools for English texts than for other languages. However, the study of other languages may lead to the identification of language-specific argumentation patterns that would be useful for further analysis of text snippets. Other issues to be considered are how to extract an uncertainty measure from natural language arguments, e.g. those exchanged in a debate, such that this measure can be used to weight the acceptability of an argument, and determination of the most suitable visual representation of the results of the argument mining pipeline, e.g. identified argument components, argument boundaries, and relations between arguments, in order to support human decision-making processes.

Although natural language argumentation has been investigated for many centuries in philosophy and rhetoric, and has recently been studied extensively in natural language processing and computational models of argument as well, argument mining has in fact just begun to tackle this important yet challenging

---

[42]Automatically learned feature spaces encoding high-level, rich linguistic similarity between terms.

topic, leaving us with many interesting research questions and tasks for the future work.

Acknowledgements

Some research reported in this chapter was supported in part by the EPSRC in the UK under grant EP/M506497/1 and in part by the European Union's Horizon 2020 research and innovation programme under Marie Sklodowska-Curie grant agreement No 690974 for the project "MIREL: MIning and REasoning with Legal texts". We would also like to thank Rory Duthie and Elena Cabrio for their useful comments and consultation.

BIBLIOGRAPHY

[Aharoni et al., 2014] Ehud Aharoni, Anatoly Polnarov, Tamar Lavee, Daniel Hershcovich, Ran Levy, Ruty Rinott, Dan Gutfreund, and Noam Slonim. A benchmark dataset for automatic detection of claims and evidence in the context of controversial topics. In *Proceedings of the First Workshop on Argumentation Mining*, pages 64–68, Baltimore, Maryland, June 2014. Association for Computational Linguistics.

[Aristotle, 2004] Aristotle. *Rhetoric*. Dover Publications, 2004.

[Ashley and Walker, 2013] Kevin D. Ashley and Vern R. Walker. Toward constructing evidence-based legal arguments using legal decision documents and machine learning. In Enrico Francesconi and Bart Verheij, editors, *International Conference on Artificial Intelligence and Law, ICAIL '13, Rome, Italy, June 10-14, 2013*, pages 176–180. ACM, 2013.

[Austin, 1962] J. L. Austin. *How to Do Things with Words*. Oxford: Clarendon, 1962.

[Bex et al., 2012] Floris Bex, Thomas Gordon, John Lawrence, and Chris Reed. Interchanging arguments between Carneades and AIF. In B. Verheij, S. Szeider, and S. Woltran, editors, *Computational Models of Argument (COMMA)*, volume 245, pages 390–397. IOS Press, 2012.

[Bex et al., 2013] F. Bex, S. Modgil, H. Prakken, and Chris Reed. On logical specifications of the Argument Interchange Format. *Journal of Logic and Computation*, 23(5):951–989, 2013.

[Blackburn et al., 2001] P. Blackburn, J. Bos, M. Kohlhase, and H. de Nivelle. Inference and computational semantics. *Studies in Linguistics and Philosophy, Computing Meaning*, 77(2):1128, 2001.

[Boltuzic and Snajder, 2016] Filip Boltuzic and Jan Snajder. Fill the gap! analyzing implicit premises between claims from online debates. In *Proceedings of the 3rd Workshop on Argument Mining*, 2016.

[Budon et al., 2014] M. Budon, M. Lucero, and G. Simari. An aif-based labeled argumentation framework. In *Proc. of Foundations of Information and Knowledge Systems (FoIKS 2014)*, volume 8367, pages 117–135. LNCS Springer, 2014.

[Budzynska and Reed, 2011] Katarzyna Budzynska and Chris Reed. Whence inference. Technical report, University of Dundee, 2011.

[Budzynska et al., 2014a] Katarzyna Budzynska, Mathilde Janier, Juyeon Kang, Chris Reed, Patrick Saint Dizier, Manfred Stede, and Olena Yaskorska. Towards Argument Mining from Dialogue. In *Frontiers in Artificial Intelligence and Applications*, volume 266, pages 185–196. Computational Models of Argument (COMMA14), IOS Press, September 2014.

[Budzynska et al., 2014b] Katarzyna Budzynska, Mathilde Janier, Chris Reed, Patrick Saint-Dizier, Manfred Stede, and Olena Yaskorska. A model for processing illocutionary structures and argumentation in debates. In *Proceedings of the 9th edition of the Language Resources and Evaluation Conference (LREC)*, pages 917–924, 2014.

[Budzynska et al., 2015] Katarzyna Budzynska, Mathilde Janier, Juyeon Kang, Chris Reed, Patrick Saint-Dizier, Manfred Stede, Olena Yaskorska, and Barbara Konat. Automatically identifying transitions between locutions in dialogue. In *European Conference on Argumentation (ECA)*, 2015.

[Budzynska et al., 2016] Katarzyna Budzynska, Mathilde Janier, Chris Reed, and Patrick Saint-Dizier. Theoretical foundations for illocutionary structure parsing. *Argument and Computation*, 7(1):91–108, 2016.

[Cabrio and Villata, 2012a] E. Cabrio and S. Villata. Natural language arguments: A combined approach. In *Procs of ECAI, Frontiers in Artificial Intelligence and Applications 242*, pages 205–210, 2012.

[Cabrio and Villata, 2012b] Elena Cabrio and Serena Villata. Generating abstract arguments: a natural language approach. In *Proceedings of the Fourth International Conference on Computational Models of Argument (COMMA 2012)*, pages 454–461. IOS Press, 2012.

[Cabrio and Villata, 2013] E. Cabrio and S. Villata. A natural language bipolar argumentation approach to support users in online debate interactions;. *Argument & Computation*, 4(3):209–230, 2013.

[Cabrio and Villata, 2014] Elena Cabrio and Serena Villata. Node: A benchmark of natural language arguments. In Simon Parsons, Nir Oren, Chris Reed, and Federico Cerutti, editors, *Computational Models of Argument - Proceedings of COMMA 2014, Atholl Palace Hotel, Scottish Highlands, UK, September 9-12, 2014*, volume 266 of *Frontiers in Artificial Intelligence and Applications*, pages 449–450. IOS Press, 2014.

[Cabrio et al., 2012] E. Cabrio, B. Magnini, and A. Ivanova. Extracting context-rich entailment rules from wikipedia revision history. In *The People's Web Meets NLP Workshop*, 2012.

[Cabrio et al., 2013] E. Cabrio, S. Villata, and F. Gandon. A support framework for argumentative discussions management in the web. In *Procs of The Semantic Web: Semantics and Big Data, ESWC 2013*, pages 412–426, 2013.

[Cayrol and Lagasquie-Schiex, 2005] C. Cayrol and M.C. Lagasquie-Schiex. On the acceptability of arguments in bipolar argumentation frameworks. In *Procs of ECSQARU, LNCS 3571*, pages 378–389, 2005.

[Cayrol and Lagasquie-Schiex, 2010] C. Cayrol and M.C. Lagasquie-Schiex. Coalitions of arguments: A tool for handling bipolar argumentation frameworks. *Int. J. Intell. Syst.*, 25(1):83–109, 2010.

[Cayrol and Lagasquie-Schiex, 2013] C. Cayrol and M.-C. Lagasquie-Schiex. Bipolarity in argumentation graphs: Towards a better understanding. *Int. J. Approx. Reasoning*, 54(7):876–899, 2013.

[Chesnevar et al., 2006] C. Chesnevar, J. McGinnis, S. Modgil, I. Rahwan, C. Reed, G. Simari, M. South, G. Vreeswijk, and S. Willmott. Towards an argument interchange format. *The Knowledge Engineering Review*, 21(4):293–316, 2006.

[Cohen, 1960] Jacob Cohen. A coefficient of agreement for nominal scales. *Educational and Psychological Measurement*, 20:3746, 1960.

[Dagan et al., 2009] I. Dagan, B. Dolan, B. Magnini, and D. Roth. Recognizing textual entailment: Rational, evaluation and approaches. *Natural Language Engineering (JNLE)*, 15(04):i–xvii, 2009.

[de Marneffe et al., 2008] Marie-Catherine de Marneffe, Anna N. Rafferty, and Christopher D. Manning. Finding contradictions in text. In *Procs of ACL*, 2008.

[Dung, 1995] P. M. Dung. On the acceptability of arguments and its fundamental role in nonmonotonic reasoning, logic programming and n-person games. *Artificial Intelligence*, 77(2):321–357, 1995.

[Duthie et al., 2016a] R. Duthie, J. Lawrence, K Budzynska, and C. Reed. The CASS technique for evaluating the performance of argument mining. In *Proceedings of the Third Workshop on Argumentation Mining*, Berlin, 2016. Association for Computational Linguistics.

[Duthie et al., 2016b] Rory Duthie, Katarzyna Budzynska, and Chris Reed. Mining ethos in political debate. In *Proceedings of 6th International Conference on Computational Models of Argument (COMMA 2016)*. IOS Press, Frontiers in Artificial Intelligence and Applications, 2016.

[Feng and Hirst, 2011] Vanessa Wei Feng and Graeme Hirst. Classifying arguments by scheme. In *Proceedings of the 49th Annual Meeting of the Association for Computational Linguistics: Human Language Technologies (ACL-2011)*, pages 987–996, 2011.

[Fournier and Inkpen, 2012] Chris Fournier and Diana Inkpen. Segmentation similarity and agreement. In *Proceedings of the 2012 Conference of the North American Chapter of the Association for Computational Linguistics: Human Language Technologies*, page 152161. Association for Computational Linguistics, 2012.

[Freeman, 1991] James B Freeman. *Dialectics and the macrostructure of arguments: A theory of argument structure*, volume 10. Walter de Gruyter, 1991.

[Freeman, 2011] James B Freeman. *Argument Structure: Representation and Theory*. Springer, 2011.

[Ghosh et al., 2014] Debanjan Ghosh, Smaranda Muresan, Nina Wacholder, Mark Aakhus, and Matthew Mitsui. Analyzing argumentative discourse units in online interactions. In *Proceedings of the First Workshop on Argumentation Mining*, page 3948. Association for Computational Linguistic, 2014.

[Ghosh et al., 2016] Debanjan Ghosh, Aquila Khanam, and Smaranda Muresan. Coarse-grained argumentation features for scoring persuasive essays. In *Proceedings of ACL 2016*, 2016.

[Groarke, 1996] L. Groarke. Informal logic. $http://plato.stanford.edu/entries/logic-informal/$, 1996.

[Habernal and Gurevych, 2016a] Ivan Habernal and Iryna Gurevych. Argumentation mining in user-generated web discourse. *Computational Linguistics*, page (in press), 2016. Submission received: 2 April 2015; revised version received: 20 April 2016; accepted for publication: 14 June 2016. Pre-print available at http://arxiv.org/abs/1601.02403.

[Habernal and Gurevych, 2016b] Ivan Habernal and Iryna Gurevych. What makes a convincing argument? empirical analysis and detecting attributes of convincingness in web argumentation. In *Proceedings of the 2016 Conference on Empirical Methods in Natural Language Processing (EMNLP)*, page (to appear). Association for Computational Linguistics, November 2016.

[Habernal and Gurevych, 2016c] Ivan Habernal and Iryna Gurevych. Which argument is more convincing? Analyzing and predicting convincingness of Web arguments using bidirectional LSTM. In *Proceedings of the 54th Annual Meeting of the Association for Computational Linguistics (Volume 1: Long Papers)*, pages 1589–1599, Berlin, Germany, 2016. Association for Computational Linguistics.

[Habernal et al., 2014] Ivan Habernal, Judith Eckle-Kohler, and Iryna Gurevych. Argumentation mining on the web from information seeking perspective. In *Proceedings of the Workshop on Frontiers and Connections between Argumentation Theory and Natural Language Processing, Forlì-Cesena, Italy, July 21-25, 2014.*, 2014.

[Henkemans et al., 2007] Francisca Snoeck Henkemans, Frans van Eemeren, and Peter Houtlosser. *Argumentative Indicators in Discourse. A Pragma-Dialectical Study*. Dordrecht: Springer, 2007.

[Hirst et al., 2014] Graeme Hirst, Vanessa Wei Feng, Christopher Cochrane, and Nona Naderi. Argumentation, ideology, and issue framing in parliamentary discourse. In *Proceedings of the Workshop on Frontiers and Connections between Argumentation Theory and Natural Language Processing, Forlì-Cesena, Italy, July 21-25, 2014.*, 2014.

[Janier and Reed, 2015] Mathilde Janier and Chris Reed. Towards a theory of close analysis for dispute mediation discourse. *Argumentation*, 0.1007/s10503-015-9386-y, 2015.

[Janier et al., 2014] Mathilde Janier, John Lawrence, and Chris Reed. Ova+: An argument analysis interface. In *Computational Models of Argument (COMMA)*, 2014.

[Janier et al., 2015] Mathilde Janier, Mark Aakhus, Katarzyna Budzynska, and Chris Reed. Modeling argumentative activity in mediation with Inference Anchoring Theory: The case of impasse. In *European Conference on Argumentation (ECA)*, 2015.

[Kirschner et al., 2015] Christian Kirschner, Judith Eckle-Kohler, and Iryna Gurevych. Linking the thoughts: Analysis of argumentation structures in scientific publications. In *Proceedings of the 2nd Workshop on Argumentation Mining*, pages 1–11, Denver, CO, June 2015. Association for Computational Linguistics.

[Knott, 1996] Alister Knott. *A Data-Driven Methodology for Motivating a Set of Coherence Relations*. PhD thesis, Department of Artificial Intelligence, University of Edinburgh, 1996.

[Konat et al., 2016] Barbara Konat, John Lawrence, Joonsuk Park, Katarzyna Budzynska, and Chris Reed. A corpus of argument networks: Using graph properties to analyse divisive issues. In *Proc. of the 10th edition of the Language Resources and Evaluation Conference (LREC 2016)*, 2016.

[Kouylekov and Negri, 2010] M. Kouylekov and M. Negri. An open-source package for recognizing textual entailment. In *Procs of ACL 2010 System Demonstrations*, pages 42–47, 2010.

[Landis and Koch, 1977] J.R. Landis and G.G. Koch. The measurement of observer agreement for categorical data. *Biometrics*, 33:159174, 1977.

[Lawrence and Reed, 2015] John Lawrence and Chris Reed. Combining argument mining techniques. In *Proceedings of the 2nd Workshop on Argumentation Mining*, pages 127–136, Denver, CO, June 2015. Association for Computational Linguistics.

[Lawrence and Reed, 2016] J. Lawrence and C.A. Reed. Argument mining using argumentation scheme structures. In P. Baroni, M. Stede, and T. Gordon, editors, *Proceedings of the Sixth International Conference on Computational Models of Argument (COMMA 2016)*, Berlin, 2016. IOS Press.

[Lawrence et al., 2012] John Lawrence, Floris Bex, Chris Reed, and Mark Snaith. AIFdb: Infrastructure for the argument web. In *Proceedings of the Fourth International Conference on Computational Models of Argument (COMMA 2012)*, pages 515–516, 2012.

[Lawrence et al., 2014] John Lawrence, Chris Reed, Colin Allen, Simon McAlister, and Andrew Ravenscroft. Mining arguments from 19th century philosophical texts using topic based modelling. In *Proceedings of the First Workshop on Argumentation Mining*, pages 79–87, Baltimore, Maryland, June 2014. Association for Computational Linguistics.

[Lawrence et al., 2015] J. Lawrence, M. Janier, and C. Reed. Working with open argument corpora. In *Proceedings of the 1st European Conference on Argumentation (ECA 2015)*, Lisbon, 2015. College Publications.

[Lawrence et al., 2016] John Lawrence, Rory Duthie, Katarzyna Budzynska, and Chris Reed. Argument analytics. In *Computational Models of Argument (COMMA)*. IOS Press, 2016.

[Lawrence et al., 2017] J. Lawrence, J. Park, B. Konat, K. Budzynska, C. Cardie, and C. Reed. Mining arguments for mapping controversies in online deliberative democracy and erulemaking. *ACM Transactions on Internet Technology*, 17(3), 2017.

[Levenshtein, 1966] V. I. Levenshtein. *Binary Codes Capable of Correcting Deletions, Insertions and Reversals*, volume 10:707. Soviet Physics Doklady, 1966.

[Levy et al., 2014] Ran Levy, Yonatan Bilu, Daniel Hershcovich, Ehud Aharoni, and Noam Slonim. Context dependent claim detection. In *COLING 2014, 25th International Conference on Computational Linguistics, Proceedings of the Conference: Technical Papers, August 23-29, 2014, Dublin, Ireland*, pages 1489–1500, 2014.

[Levy et al., 2015] Ran Levy, Liat Ein-Dor, Shay Hummel, Ruty Rinott, and Noam Slonim. TR9856: A multi-word term relatedness benchmark. In *Proceedings of the 53rd Annual Meeting of the Association for Computational Linguistics and the 7th International Joint Conference on Natural Language Processing of the Asian Federation of Natural Language Processing, ACL 2015, July 26-31, 2015, Beijing, China, Volume 2: Short Papers*, pages 419–424, 2015.

[Lippi and Torroni, 2015a] Marco Lippi and Paolo Torroni. Argumentation mining: State of the art and emerging trends. *ACM Transactions on Internet Technology*, 16(2), 2015.

[Lippi and Torroni, 2015b] Marco Lippi and Paolo Torroni. Context-independent claim detection for argument mining. In Qiang Yang and Michael Wooldridge, editors, *Proceedings of the Twenty-Fourth International Joint Conference on Artificial Intelligence, IJCAI 2015, Buenos Aires, Argentina, July 25-31, 2015*, pages 185–191. AAAI Press, 2015.

[Marco et al., 2006] Chrysanne Di Marco, Donald D. Cowan, Peter Bray, H. Dominic Covvey, Vic Di Ciccio, Eduard H. Hovy, Joan Lipa, and Douglas W. Mulholland. A physician's authoring tool for generation of personalized health education in reconstructive surgery. In *Argumentation for Consumers of Healthcare, Papers from the 2006 AAAI Spring Symposium, Technical Report SS-06-01, Stanford, California, USA, March 27-29, 2006*, pages 39–46. AAAI, 2006.

[Moens et al., 2007] Marie-Francine Moens, Erik Boiy, Raquel M. Palau, and Chris Reed. Automatic detection of arguments in legal texts. In *Proceedings of the 11th international conference on Artificial intelligence and law*, pages 225–230. ACM, 2007.

[Moens, 2013] Marie-Francine Moens. Argumentation mining: Where are we now, where do we want to be and how do we get there? In *FIRE '13 Proceedings of the 5th 2013 Forum on Information Retrieval Evaluation*, 2013.

[Monz and de Rijke, 2001] C. Monz and M. de Rijke. Light-weight entailment checking for computational semantics. In *Procs of Inference in Computational Semantics (ICoS-3)*, pages 59–72, 2001.

[Naderi and Hirst, 2015] Nona Naderi and Graeme Hirst. Argumentation mining in parliamentary discourse. In *Proceedings of the 15th Workshop on Computational Models of Natural Arguments (CMNA 2015)*, 2015.

[Nguyen and Litman, 2015] Huy Nguyen and Diane Litman. Extracting argument and domain words for identifying argument components in texts. In *Proceedings of the 2nd Workshop on Argumentation Mining*, pages 22–28, Denver, CO, June 2015. Association for Computational Linguistics.

[Nguyen and Litman, 2016] Huy Nguyen and Diane J. Litman. Improving argument mining in student essays by learning and exploiting argument indicators versus essay topics. In Zdravko Markov and Ingrid Russell, editors, *Proceedings of the Twenty-Ninth International Florida Artificial Intelligence Research Society Conference, FLAIRS 2016, Key Largo, Florida, May 16-18, 2016.*, pages 485–490. AAAI Press, 2016.

[Palau and Moens, 2009] Raquel M. Palau and Marie-Francine Moens. Argumentation mining: the detection, classification and structure of arguments in text. In *Proceedings of the 12th international conference on artificial intelligence and law*, pages 98–107. ACM, 2009.

[Park and Cardie, 2014a] Joonsuk Park and Claire Cardie. Assess: A tool for assessing the support structures of arguments in user comments. In *Proc. of the Fifth International Conference on Computational Models of Argument*, pages 473–474, 2014.

[Park and Cardie, 2014b] Joonsuk Park and Claire Cardie. Identifying appropriate support for propositions in online user comments. In *Proceedings of the First Workshop on Argumentation Mining*, pages 29–38, Baltimore, Maryland, June 2014. Association for Computational Linguistics.

[Peldszus and Stede, 2013] Andreas Peldszus and Manfred Stede. From argument diagrams to argumentation mining in texts: a survey. *International Journal of Cognitive Informatics and Natural Intelligence (IJCINI)*, 7(1):1–31, 2013.

[Peldszus and Stede, 2015] Andreas Peldszus and Manfred Stede. An annotated corpus of argumentative microtexts. In *First European Conference on Argumentation: Argumentation and Reasoned Action*, 2015.

[Perelman and Olbrechts-Tyteca, 1958] Ch. Perelman and L. Olbrechts-Tyteca. *Traite de l'argumentation: La nouvelle rhetorique*. Paris: Presses Universitaires de France, 1958.

[Prakken, 2010] H. Prakken. An abstract framework for argumentation with structured arguments. *Argument and Computation*, 1:93–124, 2010.

[Rahwan et al., 2007] I. Rahwan, F. Zablith, and C. Reed. Laying the foundations for a World Wide Argument Web. *Artificial Intelligence*, 171(10-15):897–921, 2007.

[Reed et al., 2008] Chris Reed, Raquel Mochales Palau, Glenn Rowe, and Marie-Francine Moens. Language resources for studying argument. In *Proceedings of the 6th Language Resources and Evaluation Conference (LREC-2008)*, pages 91–100, Marrakech, 2008.

[Reed, 2006] Chris Reed. Preliminary results from an argument corpus. In Elona Miyares Bermdez and Leonel Ruiz Miyares, editors, *Linguistics in the twenty-first century*, pages 185–196. Cambridge Scholars Press, 2006.

[Rinott et al., 2015] Ruty Rinott, Lena Dankin, Carlos Alzate Perez, Mitesh M. Khapra, Ehud Aharoni, and Noam Slonim. Show me your evidence - an automatic method for context dependent evidence detection. In *Proceedings of the 2015 Conference on Empirical Methods in Natural Language Processing, EMNLP 2015, Lisbon, Portugal, September 17-21, 2015*, pages 440–450, 2015.

[Roitman et al., 2016] Haggai Roitman, Shay Hummel, Ella Rabinovich, Benjamin Sznajder, Noam Slonim, and Ehud Aharoni. On the retrieval of wikipedia articles containing claims on controversial topics. In Jacqueline Bourdeau, Jim Hendler, Roger Nkambou, Ian Horrocks, and Ben Y. Zhao, editors, *Proceedings of the 25th International Conference on World Wide Web, WWW 2016, Montreal, Canada, April 11-15, 2016, Companion Volume*, pages 991–996. ACM, 2016.

[Saint-Dizier, 2012] Patrick Saint-Dizier. Processing natural language arguments with the <textcoop> platform. *Journal of Argument and Computation*, 3(1):49–82, 2012.
[Saint-Dizier, 2014] Patrick Saint-Dizier. *Challenges of Discourse processing: the case of technical documents*. Cambridge Scholars Publishing, 2014.
[Searle and Vanderveken, 1985] J. Searle and D. Vanderveken. *Foundations of Illocutionary Logic*. Cambridge University Press, 1985.
[Searle, 1969] J. Searle. *Speech Acts: An Essay in the Philosophy of Language*. Cambridge University Press, New York, 1969.
[Sonntag and Stede, 2014] Jonathan Sonntag and Manfred Stede. GraPAT: a tool for graph annotations. In Nicoletta Calzolari (Conference Chair), Khalid Choukri, Thierry Declerck, Hrafn Loftsson, Bente Maegaard, Joseph Mariani, Asuncion Moreno, Jan Odijk, and Stelios Piperidis, editors, *Proceedings of the Ninth International Conference on Language Resources and Evaluation (LREC'14)*, Reykjavik, Iceland, may 2014. European Language Resources Association (ELRA).
[Sridhar et al., 2015] Dhanya Sridhar, James R. Foulds, Bert Huang, Lise Getoor, and Marilyn A. Walker. Joint models of disagreement and stance in online debate. In *Proceedings of the 53rd Annual Meeting of the Association for Computational Linguistics and the 7th International Joint Conference on Natural Language Processing of the Asian Federation of Natural Language Processing, ACL 2015, July 26-31, 2015, Beijing, China, Volume 1: Long Papers*, pages 116–125. The Association for Computer Linguistics, 2015.
[Stab and Gurevych, 2014] Christian Stab and Iryna Gurevych. Annotating argument components and relations in persuasive essays. In *COLING 2014, 25th International Conference on Computational Linguistics, Proceedings of the Conference: Technical Papers, August 23-29, 2014, Dublin, Ireland*, pages 1501–1510, 2014.
[Stede, 2004] Manfred Stede. The Potsdam Commentary Corpus. In *Proceedings of the ACL Workshop on Discourse Annotation*, pages 96–102, 2004.
[Swanson et al., 2015] Reid Swanson, Brian Ecker, and Marilyn A. Walker. Argument mining: Extracting arguments from online dialogue. In *Proc. of 16th Annual Meeting of the Special Interest Group on Discourse and Dialogue*, pages 217–225, 2015.
[Teufel et al., 1999] Simone Teufel, Jean Carletta, and Marie-Francine Moens. An annotation scheme for discourse-level argumentation in research articles. In *Proceedings of the ninth conference on European chapter of the Association for Computational Linguistics*, pages 110–117. Association for Computational Linguistics, 1999.
[Teufel et al., 2009] Simone Teufel, Advaith Siddharthan, and Colin Batchelor. Towards discipline-independent argumentative zoning: Evidence from chemistry and computational linguistics. In *Proceedings of the 2009 Conference on Empirical Methods in Natural Language Processing*, pages 1493–1502. Association for Computational Linguistics, 2009.
[Toulmin, 1958] Stephen Toulmin. *The Uses of Argument*. Cambridge University Press, 1958.
[van Eemeren et al., 2014] Frans H. van Eemeren, Bart Garssen, Eric C.W. Krabbe, A.Francisca Snoeck Henkemans, Bart Verheij, and Jean H.M. Wagemans. *Handbook of Argumentation Theory*. Springer, 2014.
[Villalba and Saint-Dizier, 2012] Maria Paz G. Villalba and Patrick Saint-Dizier. Some facets of argument mining for opinion analysis. In *Proceedings of the Fourth International Conference on Computational Models of Argument (COMMA 2012)*, pages 23–34, 2012.
[Wacholder et al., 2014] Nina Wacholder, Smaranda Muresan, Debanjan Ghosh, and Mark Aakhus. Annotating multiparty discourse: Challenges for agreement metrics. In *Proceedings of the 8th Linguistic Annotation Workshop, COLING*, pages 120–128, 2014.
[Wachsmuth et al., 2015] Henning Wachsmuth, Johannes Kiesel, and Benno Stein. Sentiment flow - A general model of web review argumentation. In *Proceedings of the 2015 Conference on Empirical Methods in Natural Language Processing, EMNLP 2015, Lisbon, Portugal, September 17-21, 2015*, pages 601–611, 2015.
[Walker et al., 2012] Marilyn A. Walker, Jean E. Fox Tree, Pranav Anand, Rob Abbott, and Joseph King. A corpus for research on deliberation and debate. In *LREC*, pages 812–817, 2012.
[Wyner et al., 2016] Adam Wyner, Tom van Engers, and Anthony Hunter. Working on the argument pipeline: Through flow issues between natural language argument, instantiated arguments, and argumentation frameworks. *Argument & Computation*, 7:6989, 2016.

Katarzyna Budzynska
Centre for Argument Technology
Polish Academy of Sciences, Poland
& University of Dundee, UK
Email: budzynska.argdiap@gmail.com

Serena Villata
Université Cte dAzur, CNRS, Inria, I3S
France
Email: villata@i3s.unice.fr

PART D

ALGORITHMS AND IMPLEMENTATIONS

13
Computational Problems in Formal Argumentation and their Complexity
WOLFGANG DVOŘÁK, PAUL E. DUNNE

ABSTRACT. In this chapter we give an overview of the core computational problems arising in formal argumentation together with a complexity analysis highlighting different sources of computational complexity. To this end we consider three of the previously discussed formalisms, that are Dung's abstract argumentation frameworks, assumption-based argumentation, and abstract dialectical frameworks, each of which allows to highlight different sources of computational complexity in formal argumentation. As most of these problems turn out to be of high complexity we also consider properties of instances, like being in a specific graph class, that reduce the complexity and thus allow for more efficient algorithms. Finally, we also show how to apply techniques from parametrized complexity that allow for a more fine-grained complexity classification.

1 Introduction

In the previous chapters of this handbook several models for formal argumentation were discussed. They propose different ways to construct arguments, draw conclusions and each of these models comes with several proposals for semantics as to how coherent sets of arguments or statements should be selected. In this chapter we address the computational issues appearing in argumentation formalisms and that have to be tackled when implementing argumentation systems. That is, we will identify core problems of abstract argumentation and present basic procedures to solve them together with hardness results, based on computational complexity theory, that show some problems to have inherent complexity that cannot be circumvented by any algorithm.

The computational problems of formal argumentation occur at several places in the argumentation process. (A1) First, when instantiating argumentation frameworks from knowledge bases one has to deal with the task of constructing arguments and identifying conflicts (or even more complex) relations between arguments. (A2) Second, given the arguments and conflicts between them one has to find coherent sets of arguments that can be simultaneously accepted (w.r.t. a selected semantics). (A3) Finally, given the coherent sets of arguments one has to draw conclusions based on this selection. The computations in the first and third item often correspond to problems that are purely located in the underlying logic, e.g., to evaluating an inference operator. The second item is at the core of formal argumentation. That is, we are given arguments and relations (e.g. attacks) between them, and have to evaluate them w.r.t. to an argumentation semantics. This may require computing all extensions of a given semantics, the acceptance status of some argument w.r.t. some semantics, or finding some witness or counter example for a claim.

In this chapter we consider computational problems in three argumentation formalisms introduced in earlier chapters of this handbook, i.e. Dung's abstract argument frameworks, assumption-based argumentation and abstract dialectical frameworks. Dung's abstract argument frameworks are a model for (A2) which only consists of abstract entities called arguments and a binary attack relation between them. There is no instantiation process or computation of conclusions. Thus, it is perfectly suited for studying the computational issues involved in (A2). Assumption-based argumentation models the whole process (A1), (A2) and (A3), starting from a knowledge base, constructing arguments and conflicts, and finally returning conclusions. By comparing assumption-based argumentation with the results for Dung's abstract argumentation we are able to highlight the computational costs for (A1) and (A3). Finally, we consider abstract dialectical frameworks (ADFs) which are a richer model for (A2). As for Dung's abstract argument frameworks, here only abstract entities are considered but instead of just a binary attack relation ADFs allow for more complex relations between these entities. On the basis of ADFs we will highlight the impact of the allowed relations between arguments on the computational complexity of the reasoning tasks.

Notice by the term *"computational problem"* we mean the task of when presented with a *description* of some *input*, e.g., the vertices and edges in a graph, a collection of numeric values, producing an output related in a specified way to this input. For example: reporting the set of vertices forming the end-points of at least two edges, returning the collection of numerical values *sorted* in increasing order. One special type of computational problem is of particular interest: the class of so-called *decision* problems. These concern determining whether the given input structure has a particular property of interest, e.g. given a graph as before does it contain a cycle?, given a list of numbers, does the largest exceed 100?

The formal study of computational problems has two principal foci:

A. The construction of ("efficient") *algorithms* to *solve* the problem. That is methods which when presented with an input instance *always* report the *correct* output.

B. To categorise collections of computational problems that are "similar" in terms of their "best" algorithms, and thence provide a formal proof that *every* algorithmic approach must take some number of steps.

Thus (A) is concerned with positive constructive demonstration of an *upper* bound on a problem's *computational complexity* while (B) is a (more negative) statement prescribing *lower* bounds on computational complexity.

Why are these focal points of importance? To gain some insight to this consider the well-known computational problem of *sorting*: given a collection of N numbers $< a_1, a_2, \ldots, a_n >$ return this collection in increasing order of its members. Here are three informally presented "sorting algorithms":

S1 Generate each possible ordering, π, of $< a_1, a_2, \ldots, a_N >$ in turn: return the first ordering found that is correct.

S2 Form a new ordering by comparing for each $i > 1$ the (current) a_{i-1} and a_i: if $a_i > a_{i-1}$ exchange the pair. Repeat with the new ordering produced until the collection is sorted.

S3 If $N = 1$ the list is already sorted. Otherwise (recursively) sort the two list $< a_1, \ldots, a_{N/2} >$ and $< a_{N/2+1}, \ldots, a_N >$ and "merge" the two sorted lists to give the final output.

On the surface, in the sense that all three methods are correct there appears to be little to choose between these three methods. If, however, we examine their performance a very different picture emerges.

Method (S1) in the worst case (no matter how the successive ordering are produced) requires $N!$ steps: if $N = 100$ this is roughly 10^{200}

Method (S2) in the worst case needs N^2 steps: for $N = 100$ this is 10^4.

Method (S3) takes of the order of $N \log_2 N$ steps: with $N = 100$ this is about $10^{2.5}$.

Now (S1) is unusable as a *realistic* algorithm: even with a high-performance computer implementation capable of executing 10^{12} operations per second, in the worst case (S1) will require 10^{150} *years*. On a much slower machine (say 100 operations per second) even a "naive" implementation of (S2) will have finished in about 2 *minutes* and (S3) in just over 1 second.

Although a quite extreme case is being considered, this overview of one particular range of algorithmic methods for a computational problem does highlight two significant issues:

H1. The efficiency of an algorithm is a crucial factor in determining its practical usability: if (S1) were the *only known* sorting method, tasks such as organising records in a database would not be possible.

H2. Developments in technology – the platforms on which algorithms are realised – have minimal impact: a reasonable algorithm (S2 or S3) even running on an antiquated very slow machine (100 ops/sec) will easily outperform a very inefficient approach (such as S1) even if this is run on a machine with significant computational power (10^{12} ops/sec)

The study of algorithms for computational problems in argumentation has made notable advances over the last twenty years. There is, however, a significant issue that besets many of its computational concerns: that within the technical classifications of problem difficulty presented in the field of *computational complexity theory* there is powerful evidence that the prospects for identifying efficient solution methods are extremely limited: that is to say, in terms of the sorting method example given, the status of best known worst-case methods for important computational problems in argumentation is more likely to be characterised by (S1) than (S2) or (S3).

Our intention in this chapter is to present a survey of computational complexity results that have been obtained within formal argumentation.

Prior to embarking on this overview, in order to provide some necessarily technical background, we give an very informal basic introduction to the ideas and techniques used in this study.

From a practical point, complexity classification is in particular crucial when one considers implementing argumentation reasoning tasks by a *reduction approach*. That is, instead of designing and implementing complex algorithms and systems from scratch, one might reduce the new reasoning tasks to related formalisms where sophisticated solvers already exist. For instance, for a broad range of argumentation semantics one can reduce the task of computing a set of coherent arguments of an argumentation framework to computing a model of a propositional formula that can be efficiently constructed from the argumentation framework [Besnard and Doutre, 2004]. Now one can exploit the sophisticated systems to deal with propositional formulae to get an efficient system for the encoded argumentation semantics with relatively small effort. In the reduction approach the complexity of the actual problem and the corresponding problem in the target formalism are crucial for the following reasons: given that an actual problem has higher complexity than the designated target problem we know that there is no efficient encoding of our problem and we might consider a different target formalism. On the other hand if the target problem is of higher complexity we may end up with unnecessarily high computational costs. In such a case it might be a good idea to encode the problem within a restriction of the target formalism, providing lower complexity.

The remainder of the chapter is organised as follows. In Section 2 we give a brief introduction to computational complexity. That is we introduce the techniques and complexity classes we will use in the later parts of the chapter. In Section 3 we consider Dung's abstract argumentation frameworks and the main computational problems thereof. In Section 4 we consider computational problems in assumption-based argumentation. In Section 5 we consider computational problems in abstract dialectical frameworks. Finally, in Section 6 we summarise and discuss the presented results as well as related results not covered by this chapter.

2 A brief Introduction to Computational Complexity Theory

In very informal terms, computational complexity theory is the field of computer science concerned with grouping computational problems (in the sense we introduced above) into so-called "complexity classes". Such classes are captured by different resource requirements, typically measured by quantities such as Time (number of steps taken by an algorithm) or Space (amount of "memory" needed). Thus a *complexity class* $\mathcal{C}$ is a *set* of computational problems, and when we say that "problem P is in the complexity class $\mathcal{C}$" (or P has complexity $\mathcal{C}$) this indicates that there *exists* an algorithm that solves P and meets the resource criteria prescribed by $\mathcal{C}$. For example, as illustrated by the methods discussed in the introduction, the computational problem of "sorting n numbers" is in the (function) complexity class of problems solvable in time $n \log n$ (evidenced by method S3).

Now already this basic description raises many issues, among which we have:

a. How do we avoid proliferating "complexity classes" because of different technological capabilities, i.e., having to formulate a "complexity theory for Apple Mac machines", another for IBM hardware, and yet another for Windows O/S, etc. etc.?

b. How do we formalise notions of "input size" and relate such to the computational complexity of a problem?

c. How do we, in a precise sense, group distinct computational problems into collections of similar behaviour?

Before developing these questions further, we observe that it is convenient to focus on *decision problems*. That is to say, problems that separate input *instances* into two disjoint sets:

- **Positive** instances x of problem P: those on which P reports the answer **true** (equivalently, 1 or **yes**).

- **Negative** instances x of problem P: those on which P reports the answer **false** (equivalently, 0 or **no**).

In order to abstract away from the trivialities of platform specifics, algorithms are considered as realised on some standard *"model of computation"*. While a huge number of such models have appeared in the technical literature[1] those adopted in computational complexity, ultimately, derive from *Turing machine (*TM*) programs*. The exact specification of these is unimportant for the purposes of this overview. The interested reader is referred to any standard textbook for further details (e.g., [Papadimitriou, 1994; Arora and Barak, 2009]).

By fixing a standard basis for specifying algorithms (that is, TM programs) we obtain methods for addressing questions (b) and (c). At the most rarefied abstract level of TM operation, "input size" is simply the total number of *characters* (symbols) appearing in the input data. Usually (although not invariably) this will take the form of a sequence of *binary* "digits". The important feature is that the input sequence uses only characters from a *fixed finite*

---

[1] In one form or another the abstraction "model of computation" can be traced back almost a hundred years: its first appearance being with respect to capturing the notion of "computational problems that *can* be solved".

set or *alphabet* no matter whether these characters are digits, letters, or any other type of characters. [2]

While "length of the input string" offers a common basis for comparison, it can be somewhat cumbersome for practical analysis. Fortunately (and certainly in the case of abstract argumentation problems which are our principal interest) there is, usually, some supporting structure to a problem instance which can serve as a size parameter. For example, returning to the example of "sorting", instead of considering the total number of *bits* to represent instances (which could be $n \log_2 k$ when non-negative integers of value $< 2^k$ are involved), since most sorting methods work at the level of numeric comparisons (as opposed to individual bit-level manipulation), the size of an instance can reasonably be viewed as the number of values (N) to be sorted. In the consideration of decision problems arising in Dung's formalism a typical instance will specify an argumentation framework, that is to say a *directed graph*, (A, R) and a subset S of arguments: hence the "obvious" input size parameter is simply "the number of arguments in A". Notice that, total input size is bounded polynomially in n, as the size of each part of input, i.e., of A, R and S, is bounded polynomially in n.

We can now deal with the second part of (b): relating such notions of "size" to problem complexity, in particular precise interpretations of "problem P has lower (time) complexity than problem Q". Notice that such statements combine *two separate* claims:

C1. That there *exists* an algorithm A_P solving P that runs in time $T_P(n)$ on instances of size n.

C2. That *every* algorithm A_Q solving Q takes times at least $T_Q(n)$ on instances of size n and $T_Q(n)$ is "larger" than $T_P(n)$.[3]

Let us focus now on problems concerning AFs in which the dominant input component is a directed graph, (A, R). Considering the character of the algorithm, A_P, associated with this there is an infinite sequence,

$$\{<A_P, R_P>_{(1)}, <A_P, R_P>_{(2)}, \ldots, <A_P, R_P>_{(k)}, \ldots\}$$

for which $<A_P, R_P>_k$ is an AF having exactly k arguments. In addition, the number of steps (run-time) of A_P on the instance $<A_P, R_P>_k$ is not exceeded by any other instance (A, R) in which A has exactly k arguments. Such an instance, $<A_P, R_P>_k$ is called a "*worst-case* input for A_P". In this way the run-time function, T_P, is just the

$T_P(n) =_{\text{def}}$ The number of steps A_P takes when given the input $<A_P, R_P>_{(n)}$

---

[2] In complexity matters, if such a set contains at least 2 distinct symbols, it makes little difference whether the alphabet has 2 or 1000 or more symbols. In contrast, however, *unary* (single symbol) encodings may lead to notably different algorithmic behaviour.

[3] Typically, one is interested in the asymptotic behaviour with growing input size n, i.e., whether there is an there exists n_0 such that $T_Q(n) >= T_P(n)$ for every $n >= n_0$.

Now, unless we are dealing with a highly artificial and contrived problem, P, one will typically have $T_P(n+1) > T_P(n)$.[4] This clarifies the precise meaning of (C1), and by a similar analysis we can associate run-time functions, T_Q with every algorithm solving Q. The statement "P has smaller complexity than Q" is thus a *positive* (upper bound) claim about algorithms for problem P and a *negative* lower bound claim about all algorithms for Q: as the number of arguments in A increases we will see a growing disparity between the worst-case time that P requires to deliver an answer (using A_P) compared to the worst-case time that *any algorithm*, A_Q takes to deliver its answer.

Much of the focus of computational complexity theory is in grouping problems into classes where this disparity is at its most extreme: these extremes and the techniques for placing problems at either end of the spectrum of difficulty are the subject of the next subsection.

2.1 Basic Complexity Classes

Here we briefly review the *complexity classes* used in this work and their relations. As discussed above the high-level idea of complexity theory is to group problems with similar resource requirements in complexity classes and also put these classes into an order so that we can distinguish between "easier" and "harder" problems.

2.1.1 Polynomial-Time

By convention, a problem is viewed as having an efficient algorithmic solution if it can be placed into the class P (*polynomial-time*) of all problems that have a polynomial-time algorithm, i.e., an algorithm that for each instance x (of size $|x|$) produces its answer after at most $|x|^k$ steps, for a fixed constant k. It is noted that this a rather coarse-grained classification: problems whose fastest algorithm runs in time n^{100} are considered to be "efficiently solvable". This may seem rather arbitrary, however, there is a very noticeable performance difference between methods whose run-time is bounded by n^k and those that cannot be so bounded.

An important subclass we will consider is L (*logarithmic space*), which consists of the problems that can be solved in logarithmic space (not counting input and output) and polynomial-time. Just as P is seen as the class of computational problems with efficient "sequential" algorithms, so L is the class having efficient "parallel" algorithms (see, e.g., [Greenlaw et al., 1995]).

We consider problems in the classes L, P to be computationally tractable, while we will consider problems in all the other classes in this chapter to be intractable or computationally hard.

---

[4] We are, of course, ignoring minor issues whereby P requires (A, R) to have a particular structure rendering frameworks with some numbers of arguments unsuitable, e.g., problems in which A must have an even number of arguments and are ill-defined when the size of A is an odd number.

2.1.2 The classes NP, coNP and DP

Often a decision question can be solved by finding a witness for the instance satisfying the questioned property. For instance if we ask whether an AF has a stable extension, a way to answer that positively would be to actually compute a stable extension as witness.

Taking this view, we may associate with any instance x of a decision problem Q, a set $W(x)$ of potential *witnesses* that x has the property of interest. For example, for admissibility semantics if we are interested in whether a specified argument p is credulously accepted with respect to admissibility the instances have the form $((A, R), p)$ (with $p \in A$) and potential witnesses are all subsets of $A \smallsetminus \{p\}$. A witness S in this set is valid for the instance if and only if the set $S \cup \{p\}$ is admissible.

The class NP. The complexity class NP (*non-deterministic polynomial-time*) can be characterised by such witnesses. A decision problem is in the class NP if (i) for each instance x there is a set $W(x)$ of potential witnesses, which are of polynomial size in $|x|$, such that (ii) one can verify that a $y \in W(x)$ is actually a witness for x in polynomial time and (iii) x is a "yes" instance if and only if at least one $y \in W(x)$ is a witness for x.

In the above example for an AF F the potential witnesses $W(F)$ would be all the subsets of arguments. Verifying whether a set is admissible is in polynomial time and F is a positive instance iff [5] at least one of these sets is a stable extension.

Formally the specification of a decision problem in terms of witness sets can be seen in the following way. Let x be an instance of a (decision) problem Q we write, $Q(x) = 1$ if x is a "yes" instance of Q, and $Q(x) = 2$ if x is a "no" instance of Q. We have a *binary* relation $W_Q(x, y)$ for which $<x, y> \in W_Q$ iff y is a valid witness that x is a positive instance of the (decision) problem Q. This yields,

$$Q(x) = 1 \Leftrightarrow \exists\, y \in W(x): \ <x, y> \in W_Q$$

Thus the class NP can be interpreted as those decision problems, Q, for which the membership problem $<x, y> \in W_Q$ can be decided in time polynomial in the size of x. Notice that this constraint immediately forces y (a valid witness) also to have size polynomial in $|x|$.

The class coNP. The quantifier in our formalisation of NP is an existential one. If we modify this to

$$\forall\, y \in W(x) \ <x, y> \notin W_Q$$

then we obtain the important class coNP capturing instances that *do not* have the property of interest. For example if we wish to demonstrate that an argument, x is inadmissible then it suffices to show "for every subset S of A the set $S \cup \{x\}$ either is not conflict-free or has an undefended argument".

---

[5]We will frequently use "iff" as short form for "if and only if".

We, now, briefly summarise some developments of this view of "decision problems as witness testing".

The first of these is the concept of "oracles": in an *oracle* computation we are provided with a "black-box" for witness testing which given a problem instance x provides the answer for "$Q(x) = 1$?" in a single computational step. Now such oracle machines may be considered with respect to arbitrary complexity classes, so P^A describes the class of "decision problems that have a polynomial-time algorithm *that makes use of an oracle for a decision problem in the complexity class A*".

For example, for an NP oracle we might use "existence of a stable extension". In exploiting such an oracle to solve another decision problem B "in polynomial-time" we might use an algorithm which, given an instances p of B, constructs one or more (but at most polynomial in $|p|$) frameworks F_1^p, F_2^p, etc., using the answer to "does F_k^p have a stable extension?" to determine if p should be accepted as an instance of B.

The class DP. A number of important classes have been found to occur in complexity analysis of argumentation via such oracles. Among them we have DP, the so-called "difference class" of decision problems whose members are captured by the intersection of instances x accepted by a problem L_1 (with $L_1 \in$ NP) and x accepted by a problem L_2 (with $L_2 \in$ coNP). For example the set of pairs of propositional formulae $<\varphi_1, \varphi_2>$ in which φ_1 is satisfiable and φ_2 is not so (the SAT-UNSAT problem) is in DP since its positive instances are the intersection of

$$L_1 = \{ <\varphi, \psi> \ : \ \varphi \text{ is satisfiable} \}$$
$$L_2 = \{ <\varphi, \psi> \ : \ \psi \text{ is unsatisfiable} \}$$

2.1.3 The Polynomial-time Hierarchy

The notion of "oracle" can also be used in defining the important "Polynomial-time Hierarchy" (PH). Consider the quantifier formulation of NP and coNP

$$\exists y \ <x, y> \in W_Q$$
$$\forall y \ <x, y> \notin W_Q$$

This uses a (polynomial) time decidable *binary* relation and a *single* quantifier. We could, however, extend this further, e.g.

$$\exists y_1 \ \forall y_2 \ W_Q^2(x, y_1, y_2)$$
$$\forall y_1 \ \exists y_2 \ \neg W_Q^2(x, y_1, y_2)$$

or even

$$\exists y_1 \ \forall y_2 \ \exists y_3 \ W_Q^3(x, y_1, y_2, y_3)$$
$$\forall y_1 \ \exists y_2 \ \forall y_3 \ \neg W_Q^3(x, y_1, y_2, y_3)$$

and, generally

$$Q_1 \ y_1 \ Q_2 \ y_2 \ \ldots \ Q_k \ y_k \ W_Q^k(x, y_1, y_2, \ldots, y_k)$$

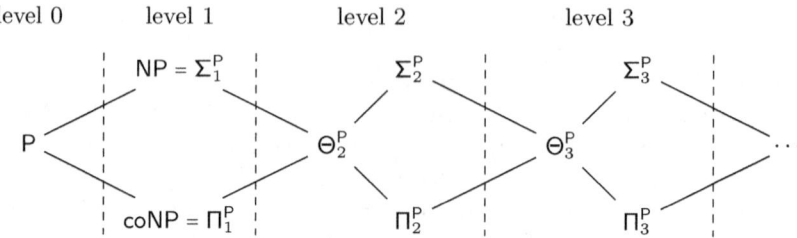

Figure 1. Levels of the polynomial-hierarchy. An edge denotes that all problems in the class on the left side are also contained in the class on the right side. Notice that only classes relevant for this chapter are shown.

In the last case we have k *alternating* quantifiers (that is $\exists$ is followed by $\forall$ and vice-versa) and the predicate $W_Q^k(x, y_1, y_2, \ldots, y_k)$ is decidable in time polynomial in $|x|$. When the opening (Q_1) quantifier is $\exists$ this defines the class of languages Σ_k^P; when this quantifier is $\forall$ we have Π_k^P. The *polynomial-hierarchy* (PH) is

$$\text{PH} = \bigcup_{k=0}^{\infty} \Sigma_k^P = \bigcup_{k=0}^{\infty} \Pi_k^P$$

We sometimes refer to *levels of the polynomial-hierarchy*, where the k-th level is formed by the classes Σ_k^P and Π_k^P. For instance on the first level there are the classes NP and coNP while on the second level there are the classes Σ_2^P and Π_2^P. Moreover, we will later introduce a further family of complexity classes Θ_k^P, and will consider the class Θ_k^P to be in the k-th level of the polynomial hierarchy (cf. Figure 1).

How does this relate to the concept of "oracle machines"? The answer to this is given by examining the quantifier structure in more depth. We have required the inner most $(k+1)$-ary predicate W_Q^k to be (deterministic) polynomial-time computable. If we have an oracle for the decision problem implied by removing the *first* quantifier then this class of languages (when $Q_1 = \exists$) is formed by languages which belong to NP given access to a Σ_{k-1}^P oracle, conventionally denoted $\text{NP}^{\Sigma_{k-1}^P}$, while Π_k^P (first quantifier is $\forall$) are those problems computable in coNP with a Σ_{k-1}^P oracle.

For example, consider the "quantified SAT problem" one version of which involves two disjoint sets of propositional variables, X and Y, and asks of a given formula $\varphi(X, Y)$ whether $\exists \alpha_X \forall \beta_Y \varphi(\alpha_X, \beta_Y)$, that is, "can we find an assignment of values to the X variables (α_X) which renders the formula $\varphi(\alpha_X, Y)$ a tautology?". Given an oracle for satisfiability we can test $\varphi(\alpha_X, Y) \equiv \top$ to be a "single step", by testing the negated formula for satisfiability. The implied NP question ("can we find …") is handled by a "polynomial" algorithm with access to this oracle so that $\Sigma_2^P = \text{NP}^\text{NP}$. Notice that, in our example we can also directly use an oracle for the coNP problem of tautology which gives us $\text{NP}^\text{NP} = \text{NP}^\text{coNP}$. That is, for an oracle machine it does not matter whether it has access to a NP or coNP oracle (or more generally to a Σ_{k-1}^P or Π_{k-1}^P oracle) as it can easily switch "yes" and "no" answers after an oracle call.

In total we can treat PH as groups of problems described via alternation of a fixed number (k) of quantifiers or in terms of polynomial-time oracle machines exploiting oracles to the immediately lower level, i.e. both Σ_k^P and Π_k^P use access to a Σ_{k-1}^P oracle.

Moreover, we consider related oracle complexity classes that have only restricted access to their oracle. Concretely, the class $\Theta_k^P = P^{\Sigma_{k-1}^P[\log(|x|)]}$ contains problems decidable by a deterministic polynomial-time algorithm that is allowed to make a logarithmic number (w.r.t. input size) of Σ_{k-1}^P-oracle calls. An alternative characterisation for Θ_k^P is that the deterministic algorithm is allowed to make linearly (in the input size) so called non-adaptive calls to the Σ_{k-1}^P-oracle, that is all oracle calls are evaluated in parallel. When using this alternative characterisation the class Θ_k^P is sometimes also denoted as $P_\|^{\Sigma_{k-1}^P}$.

Notice that all complexity classes we consider can be solved by in worst-case exponential time algorithms that only require polynomial space. However, problems on different levels of the polynomial hierarchy behave quite differently, and methods that work reasonable for problems at the NP, coNP level might not work as well for Σ_2^P or Π_2^P-hard problems.[6]

2.2 Reductions, Hardness and Completeness

At the conclusion of the preceding sub-section we referred to particular problems as "among the hardest Π_2^P problems". This (at the time of writing) does *not* mean we can formally demonstrate that *every* problem that can be classified as belonging to Π_2^P may be solved by a (deterministic) algorithm whose run-time is no worse than that of the best algorithm for, e.g., semi-stable skeptical reasoning. It does, however, mean the following: *if* we can find an NP (or even P) algorithm for skeptical semi-stable reasoning *then* we can construct NP (resp. P) algorithms *for every problem in the class* Π_2^P, i.e. it would follow that the classes Π_2^P and NP (resp. P) contained *exactly* the same decision problems. Despite this, throughout this work we will follow the standard assumptions in computational complexity theory and consider problems in higher levels of the polynomial hierarchy to be harder than problems in the lower levels of the polynomial-hierarchy.[7]

2.2.1 Polynomial Reducibility

The key idea used to support this claim is that of *polynomial reducibility*. Suppose we have two decision problems – F and G say. These have sets of instances I_F and and I_G. Now, while we may not be able to formally prove

---

[6] In the context of formal argumentation such a behaviour can be observed at the results of the First International Competition on Computational Models of Argumentation [Thimm and Villata, 2015; Thimm et al., 2016].

[7] This relates to two famous open problems in complexity theory, namely to show that P ≠ NP and to show that the polynomial hierarchy is an infinite hierarchy and does not collapse at a certain level, i.e $\Sigma_k^P \neq \Sigma_{k+1}^P$ for all $k > 0$. Both statement are widely believed but (at the time of writing) there are no formal proofs.

that either problem is intractable we can argue, using the following approach, that if G is decidable in polynomial time then F is also.

Build an *efficient* procedure, τ, transforming any instance of F into an instance of G, i.e., $\tau : I_F \to I_G$ and with the property that $x \in I_F$ is a positive instance of F iff $\tau(x) \in I_G$ is a positive instance G.

With such a transformation procedure any algorithm for G can be used as a sub-routine to give an algorithm for F. So were it the case that $G \in \mathsf{P}$, as τ is efficient, it follows that $F \in \mathsf{P}$ also. By contraposition, it can be shown that if $F \notin \mathsf{P}$ it must be the case that $G \notin \mathsf{P}$. When such a transformation can be found between decision problems F and G as above, we say that "F is *polynomially-reducible* to G" using the notation $F \leq_p G$ to describe this relationship.

Notice that the form of instances for F and G do not have to be identical: G could, for example, be a decision problem concerning propositional formulae and F one whose instances are AFs: a transformation between the two would define how a formula is constructed from a given AF.

2.2.2 Hardness and Completeness

The concept of reducibility offers a means to argue that the class NP differs from the class P and formalise the notion of "hardest" problem of a complexity class. Intuitively we consider a problem to be among the "hardest" problems of a complexity class if an efficient method for the problem would yield efficient methods for *all* problems in the class. That is, an efficient method for just *one* of the "hardest" problems would yield efficient methods for *all* problems in the class. Formally, for any complexity class, $\mathcal{C}$, a decision problem G is said to be $\mathcal{C}$-*hard* if

$$\forall \; F \in \mathcal{C} \quad F \leq_p G$$

If, in addition $G \in \mathcal{C}$ then G is said to be $\mathcal{C}$-*complete*.

So the class of NP-complete problems are those problems in NP to which any other problem in NP can be polynomially reduced. The class of known NP-complete problems includes many well-studied combinatorial, logic, and graph problems for which no efficient algorithm has been discovered, in some cases after several centuries of study. Among these are: deciding if a propositional formula has a model (SAT); deciding if a graph has a path that contains every vertex exactly once (a variant of the so-called Travelling Salesperson Problem), deciding if a given argument is acceptable w.r.t. Dung's stable semantics.

It is considered highly unlikely that every single one of these problems can be solved efficiently. In order to prove that no NP-complete problem can be solved in polynomial time it would suffice to show that just *one* could not be.

Thus, a proof that a problem G is NP-complete is seen as very strong evidence that F is intractable. Given the transitivity of $\leq_p$ all that is required to proof NP-hardness is a known NP–hard problem (F say) and a transformation, τ, to witness $F \leq_p G$. In order to obtain NP-completeness one has to additionally give a procedure that decides G and fits the definition of NP, we sometimes call such a procedure a NP-algorithm (more generally $\mathcal{C}$-algorithm for complexity class $\mathcal{C}$).

Next let us briefly reconsider our restrictions on reductions. All the complexity classes $\mathcal{C}$ considered in this handbook chapter, except L, are *closed under polynomial reductions*, that is whenever a problem A can be polynomial-time reduced to a problem $B \in \mathcal{C}$ then also A belongs to $\mathcal{C}$. Notice that any problem in the class P and in particular those in the class L would be complete for P with respect to polynomially-reducibility. Thus when differentiating between problems in L and P one uses the concept of *logspace-reducibility* where the transforming procedure is required to work in logarithmic space. In particular, P-completeness results are stated w.r.t. logspace-reducibility.

2.2.3 Complete Problems for the Polynomial Hierarchy.

To show that a problem A is hard for a specific complexity class $\mathcal{C}$ one typically starts from a problem B that is complete for the class $\mathcal{C}$ and provides a reduction from B to A. In the following we briefly introduce some canonical complete problems for the complexity classes in the polynomial-hierarchy.

As already mentioned a famous NP-complete problem is deciding if a propositional formula has a model (SAT). On the other side standard coNP-complete problems are verifying that a propositional formula is a tautology (TAUT) or that a propositional formula has no model (UNSAT). The canonical DP-complete problem is the earlier mentioned SAT–UNSAT problem.

The complete problems for classes Σ_k^P and Π_k^P are given by quantified SAT problems (cf. Section 2.1.3). That is, one is given a propositional formula φ whose variables are split up in k disjoint sets $X_1, \ldots X_k$ and the possible assignments for these sets X_1 are quantified with alternating existential and universal quantifiers. A quantified boolean formula (QBF) is then of the form

$$Q_1 X_1 Q_2 X_2 \ldots Q_k X_k \, \varphi(X_1, \ldots X_k)$$

with Q_i being alternating $\exists, \forall$ quantifiers (i.e., $\exists$ is followed by $\forall$ and vice versa). Deciding whether a QBF with k quantifiers and $Q_1 = \exists$ is valid is the canonical Σ_k^P-complete problem while deciding whether a QBF with k quantifiers and $Q_1 = \forall$ is valid is the canonical Π_k^P-complete problem.

As the second level of the polynomial-hierarchy is of special interest in the setting of formal argumentation we next introduce minimal model satisfiability (MINSAT) as another problem that is Σ_2^P-complete [Eiter and Gottlob, 1993]. In the MINSAT problem one is given a propositional formula φ over variables X and a variable x thereof and has to decide whether the variable is true in some minimal model of φ.

2.3 Parametrized Complexity

Classical complexity theory deals with the complexity of problems w.r.t. the size of the instance. However, often the complexity of a problem does not mainly depend on the size of an instance but on some (structural) properties of the instance. That is, we can solve huge instances efficiently as long as some property is satisfied or the obstacles in the structure are bounded independent of the size. The field of parametrized complexity theory[8] deals with this observation. The idea is to consider parametrized problems, i.e., the problem description contains a designated parameter (typically an integer) which is instantiated by each problem instance. An example for a parametrized problem is given a graph G and an integer parameter k deciding whether G has a clique of size k.

Definition 2.1 *A parametrized (decision) problem is called* fixed-parameter tractable *(or in* FPT*) if it can be determined in time* $f(k) \cdot |x|^{O(1)}$ *for a computable function* f.

Now given that a problem is in FPT and just consider those instances where the parameter is bounded by some constant then we can decide an instance with a polynomial-time algorithm. Only the constants in the polynomial-time bound are affected by the parameter, but not the order of the polynomial.

Beside FPT there is also a weaker form of tractability w.r.t. a parameter allowing the order of the polynomial to depend on the parameter.

Definition 2.2 *A parametrized (decision) problem is* slice-wise polynomial *(or in* XP*) if it can be determined in time* $f(k) \cdot |x|^{g(k)}$ *for computable functions* f, g.

A problem in XP can be solved in polynomial time if we bound the parameter, but distinguishing it from FPT the order of the polynomial may highly depend on the bound of the parameter.

Let us briefly present the relations between the classes FPT, XP and P:

$$P \subseteq FPT \subseteq XP$$

When considering unparametrized problems and talking about FPT we have to mention the used parameter explicitly. Thus we say a problem P is fixed-parameter tractable w.r.t. the parameter k iff the corresponding parametrized problem (P, k) is fixed-parameter tractable.

3 Complexity of Dung's Abstract Argumentation

We start our analysis with Dung's Abstract Argumentation Frameworks. These frameworks consist of a set of abstract arguments and a relation representing directed conflicts or attacks between these arguments. Then rules, so called semantics, are defined to select coherent sets of arguments that can be accepted simultaneously. That is, abstract argument frameworks focus on the core issue of argumentation, i.e., resolving conflicts between arguments.

This part of the chapter is organised as follows: In Section 3.1 we recall the basic definitions of Dung's Abstract Argumentation Frameworks and the

---

[8]We just briefly introduce the concepts relevant for this chapter; for comprehensive introductions to parametrized complexity the reader is referred to [Flum and Grohe, 2006; Niedermeier, 2006; Cygan et al., 2015].

most popular semantics for it. That is, beside the semantics introduced by Dung [Dung, 1995], we consider ideal [Dung et al., 2007], semi-stable [Verheij, 1996; Caminada et al., 2012], stage [Verheij, 1996] and cf2 [Baroni et al., 2005] semantics. Then in Section 3.2 we discuss the core computational Problems of Abstract Argumentation and define formal variants that serve as basis for the complexity analysis in Section 3.3. In Section 3.4 we consider potential computational advantages when the argumentation frameworks fall into some specific graph class. The potential of techniques from parametrized complexity theory is discussed in Section 3.5. In Section 3.6 we discuss some computational issues specific to labelling-based argumentation semantics. Finally, in Section 3.7 we summarise and discuss the presented results and give additional pointers to literature.

3.1 Dung's Abstract Argumentation Frameworks

In this section we introduce (abstract) argumentation frameworks [Dung, 1995] and recall the semantics we study (for a comprehensive introduction the reader is referred to [Baroni et al., 2011a] or chapter 4 of this handbook dedicated to Dung's Abstract Argumentation Frameworks).

Definition 3.1 *An argumentation framework (AF) is a pair $F = (A, R)$ where A is a (finite) set of arguments and $R \subseteq A \times A$ is the attack relation. The pair $(a, b) \in R$ means that a attacks b. We say that an argument $a \in A$ is defended (in F) by a set $S \subseteq A$ if, for each $b \in A$ such that $(b, a) \in R$, there exists $c \in S$ such that $(c, b) \in R$.*

Indeed when studying computational complexity we are only interested in AFs where the set A is finite.

Semantics for argumentation frameworks are defined as functions σ which assign to each AF $F = (A, R)$ a set $\sigma(F) \subseteq 2^A$ of extensions. We consider for σ the functions na, gr, st, ad, co, $cf2$, id, pr, sst and stg which stand for naive, grounded, stable, admissible, complete, cf2, ideal, preferred, semi-stable and stage semantics, respectively. Towards the definition of these semantics we have to introduce a few more formal concepts.

Definition 3.2 *Given an AF $F = (A, R)$, the characteristic function $F_F : 2^A \to 2^A$ of F is defined as $F_F(S) = \{x \in A \mid x \text{ is defended by } S\}$.*

Definition 3.3 *For a set $S \subseteq A$ and an argument $a \in A$, we say S attacks a (resp. a attacks S) in case there is an argument $b \in S$, such that $(b, a) \in R$ (resp. $(a, b) \in R$). Moreover, for a set $S \subseteq A$, we denote the set of arguments attacked by (resp. attacking) S as $S_R^+ = \{x \mid S \text{ attacks } x\}$ (resp. $S_R^- = \{x \mid x \text{ attacks } S\}$), and define the range of S as $S_R^\oplus = S \cup S_R^+$.*

We are now prepared to give the formal definitions of the abstract argumentation semantics we will consider. Notice that we restrict ourselves to extension-based semantics, but some aspects of labelling-based semantics are discussed in Section 3.6).

Definition 3.4 Let $F = (A, R)$ be an AF. A set $S \subseteq A$ is conflict-free (in F), if there are no $a, b \in S$, such that $(a, b) \in R$. $cf(F)$ denotes the collection of conflict-free sets of F. For a conflict-free set $S \in cf(F)$, it holds that

- $S \in na(F)$, if there is no $T \in cf(F)$ with $T \supset S$;
- $S \in st(F)$, if $S_R^+ = A \smallsetminus S$;
- $S \in ad(F)$, if $S \subseteq F_F(S)$;
- $S \in co(F)$, if $S = F_F(S)$;
- $S \in gr(F)$, if $S \in co(F)$ and there is no $T \in co(F)$ with $T \subset S$;
- $S \in pr(F)$, if $S \in ad(F)$ and there is no $T \in ad(F)$ with $S \subset T$;
- $S \in id(F)$ if S is $\subseteq$-maximal among $\{S' \mid S' \in ad(F)$ and $S' \subseteq E$ for each $E \in pr(F)\}$.
- $S \in sst(F)$, if $S \in ad(F)$ and there is no $T \in ad(F)$ with $S_R^\oplus \subset T_R^\oplus$;
- $S \in stg(F)$, if there is no $T \in cf(F)$, with $S_R^\oplus \subset T_R^\oplus$.

We recall that for each AF F, the grounded semantics yields a unique extension, the grounded extension, which is the least fixed-point of the characteristic function F_F.

Finally, we give the recursive definition of cf2 semantics (see [Baroni et al., 2005; Gaggl and Woltran, 2013] for further reference).

Definition 3.5 Given an argumentation framework $F = (A, R)$, then $E \in cf2(F)$, if

- $E \in na(F)$ if $|SCCs_F| = 1$, and
- $\forall S \in SCCs_F$ $(E \cap S) \in cf2(F\!\downarrow_{UP_F(S,E)})$ otherwise.

Here $SCCs_F$ denotes the set of strongly connected components of F, and for any $E, S \subseteq A$, $UP_F(S, E) = \{a \in S \mid \nexists b \in E \smallsetminus S : (b, a) \in R\}$. Moreover, for $S \subseteq A$ we use $F\!\downarrow_S$ to denote the AF $(A \cap S, R \cap S \times S)$, i.e., the AF that one obtains when restricting F to the arguments in S.

We recall some basic properties of these semantics. For each AF F we have the following subset relations:

$$st(F) \subseteq stg(F) \subseteq na(F) \subseteq cf(F),$$

$$st(F) \subseteq sst(F) \subseteq pr(F) \subseteq co(F) \subseteq ad(F) \subseteq cf(F),$$

and $st(F) \subseteq cf2(F) \subseteq na(F)$. Furthermore, for any of the considered semantics σ except stable semantics we have that $\sigma(F) \neq \emptyset$ holds, i.e., these semantics always propose at least one extension. Grounded and ideal semantics always yield exactly one extension, thus we also say that they are unique status semantics, and the ideal extension is always a complete extension. With slight abuse of notation we sometimes use $gr(F)$, resp. $id(F)$, to refer to the unique grounded, resp. ideal, extension of F. Moreover, stable, semi-stable, and stage semantics coincide for AFs with at least one stable extension.

3.2 Computational Problems

In general an argumentation semantics assigns several extensions to a single framework, but at the end of the day we want to make a conclusion about arguments. There are different ways to aggregate the acceptance status of an argument from the set of extensions, which mirrors different levels of scepticism. First it is quite clear that an argument which is in no extension at all should not be accepted, but in certain situations it might be fine to accept an argument that appears in just one extension, this is what we will call *credulous reasoning*. On the other hand in situations where one has to be cautious one might demand that an argument is in all extensions, we refer to this as *skeptical reasoning*.

These reasoning modes give rise to the following computational problems for argumentation semantics σ.

- *Credulous Acceptance* $Cred_\sigma$: Given AF $F = (A, R)$ and an argument $a \in A$. Is a contained in some $S \in \sigma(F)$?

- *Skeptical Acceptance* $Skept_\sigma$: Given AF $F = (A, R)$ and an argument $a \in A$. Is a contained in each $S \in \sigma(F)$?

If an AF has no stable extensions, according to our definition of skeptical acceptance, all arguments are skeptically accepted. This may be unwanted and hence one might consider a variation of the skeptical acceptance problem asking whether an argument is contained in all extensions and there exists at least one extension [Dunne and Wooldridge, 2009].

In practice, one often is interested in computing all extensions or a certain number of extensions. However, complexity theory provides much better tools for decision problems than for function problems and thus one usually sticks to decision problems when analysing the computational problems, in our case credulous and skeptical acceptance. Nevertheless, the complexities of credulous and skeptical acceptance together give a good impression of the complexity to actually compute the extensions.

Beside these reasoning problems there are also several other computational problems in the field of abstract argumentation. In this work we consider the most prominent ones of them. First of all one might be interested in verifying given extensions, which may come from another agent or potentially corrupted file, or simply as part of a reasoning algorithm.

- *Verification of an extension* Ver_σ: Given AF $F = (A, R)$ and a set of arguments $S \subseteq A$. Is $S \in \sigma(F)$?

Another task is deciding whether an AF provides any coherent conclusion. That can be deciding whether it has at least one extension, in the case of stable semantics, or whether it has an extension different from the empty set, for all the other semantics under our consideration.

- *Existence of an extension* $Exists_\sigma$: Given AF $F = (A, R)$. Is $\sigma(F) \neq \emptyset$?

- *Existence of a non-empty extension* $Exists_\sigma^{\neg\emptyset}$: Given AF $F = (A, R)$. Does there exist a set $S \neq \emptyset$ such that $S \in \sigma(F)$?

Finally, we will also consider the problem of deciding whether a semantics yields a unique extension for a given an AF (cf. Chapter 17 of this handbook).

- *Uniqueness of the solution* $Unique_\sigma$: Given AF $F = (A, R)$. Is there a unique set $S \in \sigma(F)$, i.e., is $\sigma(F) = \{S\}$?

3.3 Computational Complexity

A typical complexity analysis of a problem consists of two parts. First, we have to give an upper bound for the complexity of the problem. That is, we have to either give an algorithm showing the problem can be solved within a class $\mathcal{C}$ or we reduce the problem to another problem already shown to be in the class $\mathcal{C}$. Second, we want to prove lower bounds for the complexity of the problem. That is, we consider a problem that was shown to be hard for some complexity class $\mathcal{C}'$ and reduce it to the current problem. That is, we show the problem to be $\mathcal{C}'$-hard. In case that the classes $\mathcal{C}$ and $\mathcal{C}'$ coincide we obtain that the studied problem is $\mathcal{C}$-complete, and have an exact classification of the complexity of the problem.

The complexity landscape of abstract argumentation semantics is given in Table 1 and discussed below. For Dung's semantics the "in P" and "trivial" are immediately by properties of the corresponding semantics [Dung, 1995]; results for naive semantics are due to Coste-Marquis et al. [2005]; results for stable, admissible and preferred semantics follow from results on logic programs by Dimopoulos and Torres [1996], except for the Π_2^P-completeness of $Skept_{pr}$ which is due to Dunne and Bench-Capon [2002]; the complexity of ideal semantics is due to Dunne [2009]; results for complete semantics are due to Coste-Marquis et al. [2005]; results for semi-stable and stage semantics are due to Caminada et al. [2012] and Dvořák and Woltran [2010]; the results for cf2 semantics are due to Gaggl and Woltran [2013] and the analysis of polynomial-time problems that distinguishes problems that can be solved in L from problems that are P-complete is due to Dvořák and Woltran [2011] [Dvořák, 2012a].

In accordance with the above we will first consider upper bounds for the introduced reasoning problems and then discuss hardness results for them.

3.3.1 Upper Bounds for the Computational Complexity

Most of the problems we introduced in the previous section will fall into one of the complexity classes based on non-deterministic algorithms, e.g. NP and

Table 1. Complexity of Dung's abstract argumentation ($\mathcal{C}$-c denotes completeness for class $\mathcal{C}$).

σ	$Cred_\sigma$	$Skept_\sigma$	Ver_σ	$Exists_\sigma$	$Exists_\sigma^{\neg\emptyset}$	$Unique_\sigma$
cf	in L	trivial	in L	trivial	in L	in L
na	in L	in L	in L	trivial	in L	in L
gr	P-c	P-c	P-c	trivial	in L	trivial
st	NP-c	coNP-c	in L	NP-c	NP-c	DP-c
ad	NP-c	trivial	in L	trivial	NP-c	coNP-c
co	NP-c	P-c	in L	trivial	NP-c	coNP-c
cf2	NP-c	coNP-c	in P	trivial	in L	in P
id	Θ_2^P-c	Θ_2^P-c	Θ_2^P-c	trivial	Θ_2^P-c	trivial
pr	NP-c	Π_2^P-c	coNP-c	trivial	NP-c	coNP-c
sst	Σ_2^P-c	Π_2^P-c	coNP-c	trivial	NP-c	in Θ_2^P
stg	Σ_2^P-c	Π_2^P-c	coNP-c	trivial	in L	in Θ_2^P

coNP, and thus most of the upper bound are by guess and check algorithms that first non-deterministically guess a potential extension and then verify that it is indeed an extension and satisfies the desired properties.

Standard Reasoning Procedures. The standard algorithm for credulous acceptance first non-deterministically guesses a set of arguments, and then verifies that the set is an extension for the considered semantics and contains the argument under question. The answer to the credulous acceptance query is yes if at least one of the possible guesses evaluates to true. Now let V be the complexity for verifying an extension then the above gives use a NP^V algorithm for credulous acceptance. We next consider skeptical acceptance and show that it has a $coNP^V$ algorithm. To show that a problem falls into a $coNP^\mathcal{C}$ class one could follow the definition of coNP and give an algorithm that first guesses a potential witness and then tests whether the potential witness satisfies certain conditions that can be tested in P^V. This conditions have to be such that an instance is positive iff all possible guesses evaluate to true. However, often it is more convenient to consider the complementary problem and provide a $NP^\mathcal{C}$ algorithm for that problem. That is, instead of skeptical acceptance we consider the problem of showing that an argument is not skeptically accepted. The standard algorithm non-deterministically guesses a set of arguments, and then verifies that it is an extension and does not contain the argument under question. The answer to the skeptical acceptance query is yes only if each possible guess evaluates to false. Let V be again the complexity for verifying an extension then the above gives use a $coNP^V$ algorithm for skeptical acceptance. Towards upper bounds for *Cred* and *Skept* we next consider upper bounds for the verification problems.

Verifying Extensions. For conflict-free, naive, stable, admissible and complete semantics we only have to check whether for the given set certain attacks exist, respectively do not exist. For instance to verify a stable extension we have to verify that (a) between arguments in the extension there is no attack and (b) that all arguments not in the extension are attacked by at least one argument in the extension. This can clearly be done in polynomial time and as it only needs two pointers to arguments also in logarithmic space.[9] Since polynomial time oracles do not add any computational power they can be neglected, i.e., $NP^P = NP$ and $coNP^P = coNP$. This gives NP, resp. coNP, upper bounds for credulous and skeptical acceptance under these semantics.

Next consider semantics that require maximisation, that are pr, sst, stg (we will see later that for ideal semantics no maximisation is required). Again the basic criterion of being admissible or conflict-free can be easily checked in polynomial time but the maximality criterion adds some complexity. To show that checking whether a set S is an extension is in coNP we again give a non-deterministic algorithm for the complimentary problem, of falsifying the set S to be an extension. This is done by first testing whether S is not admissible (for pr, sst) or not conflict-free (for stg) and then guessing a set $T \supset S$ and testing whether it is admissible (for pr, sst) or conflict-free (for stg). The algorithm successfully falsifies the set S to be an extension iff the first test succeeds or the second test succeeds for at least one guess. In other words, the set S is an extension only if the first and the second tests fails for all possible guessed sets T. Combined with the NP^V, $coNP^V$ resp., algorithm for credulous, skeptical resp., acceptance the above coNP-algorithms for verification give Σ_2^P, resp. Π_2^P algorithms, for the credulous, resp. skeptical, acceptance problems.

Improved Procedures. For many semantics the above upper bounds are already optimal, but there are some cases where we can improve over them.

First, consider *conflict-free* and naive sets. If an argument is not self-attacking then it certainly will appear in a conflict-free set and thus also in a naive extension. Thus credulous acceptance can be decided by just testing whether the argument under question is self-attacking. Considering skeptical acceptance we have that the empty set is always conflict-free and admissible. Thus for conflict-free and admissible semantics we can reply "no" to each skeptical acceptance query without looking at the actual framework.

For *naive sets* we know that an argument is in a naive set iff it has no self-attack and none of its neighbours is in the set. Thus for skeptical acceptance we just have to test whether the argument is not self-attacking and all of its neighbours are not credulously accepted, i.e., they are self-attacking.

The *grounded* semantics can be computed by iterating the characteristic function until the least fixed-point is reached [Dung, 1995]. The characteristic function can be computed in polynomial time and, as the least fixed-point is reached after at most linearly many iterations. That is, the grounded extension

---

[9] For the corresponding result for cf2 semantics see [Nieves *et al.*, 2009; Gaggl and Woltran, 2013].

can be computed in polynomial time and the decision problems can then be easily answered. Moreover, as the grounded extension is the unique minimal complete extension skeptical acceptance for complete semantics is exactly the problem of testing whether an argument is contained in the grounded extension and thus in polynomial time.

As each admissible set can be extended to a preferred extension and each preferred extension is admissible we have $Cred_{ad} = Cred_{pr}$. That is, for *credulous acceptance* under *preferred* semantics it suffices to consider admissible sets and thus an NP algorithm suffices.

Finally, for *ideal* semantics there is an alternative characterisation that allows for a Θ_2^P algorithm [Dunne, 2009]. That is, the ideal extension is the maximal admissible set that is not attacked by any other admissible set. The algorithm first computes the credulously accepted arguments (w.r.t. preferred semantics) via an NP-oracle and then considers the set of arguments that are credulously accepted but not attacked by any credulous accepted argument. Within this set one then computes the ideal extension by a polynomial-time algorithm that iteratively removes arguments which are not defended.

3.3.2 Hardness results

Given the complexity upper bounds from above we are now going for hardness results that show that these upper bounds are optimal. We start with what we call the *standard translation* from propositional formulae to argumentation frameworks and then discuss some prototypical hardness results that extend the standard translation.

Standard Translation. On the one hand the standard translation will give us our first hardness results and on the other hand it is part of almost all reductions in abstract argumentation. To show hardness one typically starts from the standard translation and adds modifications to match the actual problem and semantics.

Reduction 3.6 *Given a propositional formula φ in CNF given by a set of clauses C over the atoms Y, we define the standard translation from φ as $F_\varphi = (A, R)$, where*

$$A = \{\varphi\} \cup C \cup Y \cup \bar{Y}$$
$$R = \{(c, \varphi) \mid c \in C\} \cup$$
$$\{(x, c) \mid x \in c, c \in C\} \cup \{(\bar{x}, c) \mid \bar{x} \in c, c \in C\} \cup$$
$$\{(x, \bar{x}), (\bar{x}, x) \mid x \in Y\}$$

The AF F_φ from Reduction 3.6 is illustrated in Figure 2. The intuition behind the construction is as follows. Assume we are arguing whether the formula φ is true. For an atom y_i we have two arguments, y_i claiming the atom is true, $\bar{y}_i$ claiming the atom is false and thus $\neg y_i$ is true. As exactly one of y_i and $\bar{y}_i$ is true they are mutually attacking. Now consider the argument φ,

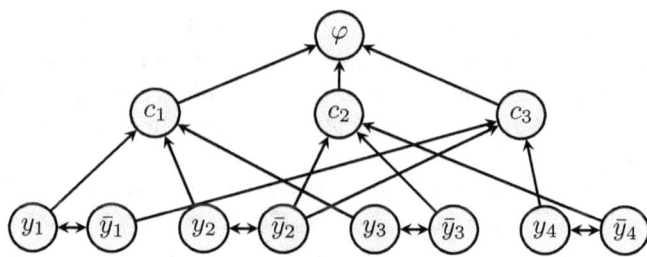

Figure 2. Illustration of the standard translation F_φ, for the propositional formula φ with clauses $\{\{y_1, y_2, y_3\}, \{\bar{y}_2, \bar{y}_3, \bar{y}_4\}\}, \{\bar{y}_1, \bar{y}_2, y_4\}\}$.

which can be interpreted as "the formula φ is true". This argument is attacked by the arguments c_i which can be read as "clause c_i is not satisfied". Clearly if one clause is false the whole formula is not satisfied. Finally, if one of the literals in a clause is true the whole clause is true and thus an argument c_i is attacked by all arguments corresponding to literals in c_i.

Credulous Acceptance. For the NP-hardness of credulous acceptance consider the AF F_φ constructed by the above reduction. It is not to hard to show that each model I_Y[10] of φ corresponds to a stable extension of F_φ that consists of the argument φ, the arguments y_i for $y_i \in I_Y$, and the arguments $\bar{y}_i$ for $y_i \in Y \setminus I_Y$. Moreover also the converse holds, i.e., each stable extension of F_φ containing the argument φ corresponds to a model of φ. Thus, F_φ has a stable extension containing φ iff φ has a model. The same holds for admissible sets, complete, preferred and cf2 extensions. Thus Reduction 3.6 is a reduction from SAT to credulous reasoning under these semantics and as it can be clearly performed in polynomial time we obtain that credulous reasoning under these semantics is NP-hard.

Skeptical Acceptance. To show coNP-hardness of skeptical acceptance for stable, preferred and cf2 semantics we extend the standard translation F_φ by an additional argument $\bar{\varphi}$ that is attacked by φ. In the resulting AF G_φ the argument $\bar{\varphi}$ is skeptically accepted w.r.t. the mentioned semantics iff φ is not credulously accepted iff φ is unsatisfiable [Dimopoulos and Torres, 1996; Gaggl and Woltran, 2013]. Thus we have a reduction from UNSAT to skeptical acceptance showing coNP-hardness. Notice that this reduction does not work for admissible and complete semantics as for both the empty set is an extension neither containing φ nor $\bar{\varphi}$.

Skeptical Acceptance with Preferred Semantics. Skeptical acceptance with preferred semantics is a prototypical problem for the second level of the polynomial-hierarchy. The hardness proof is reported in [Dunne and Bench-Capon, 2002] and we next discuss a slight variation of the reduction presented

---

[10] A model I_Y is a subset of the variables Y such that if we set all variables in I_Y to true and all arguments in $Y \setminus I_Y$ to false the formula φ evaluates to true.

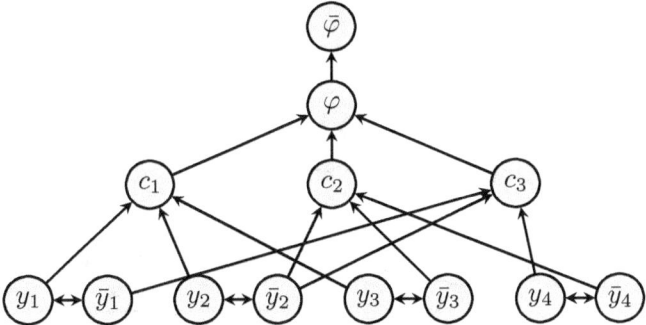

Figure 3. Illustration of the reduction G_φ, for the propositional formula φ with clauses $\{\{y_1, y_2, y_3\}, \{\bar{y}_2, \bar{y}_3, \bar{y}_4\}\}, \{\bar{y}_1, \bar{y}_2, y_4\}\}$.

there. That is, we give a reduction from the Π_2^P-complete problem $QSAT_\forall^2$ of deciding whether a $QBF_\forall^2$ formula is valid to skeptical acceptance with preferred semantics. That is, given a $QBF_\forall^2$ formula $\forall Y \exists Z\, \varphi(Y, Z)$ with φ being a CNF formula we construct an AF as follows. We first apply the standard reduction from propositional CNF formulae to AFs and then add an additional argument $\bar{\varphi}$ which is attacked by φ, attacks itself and attacks all arguments z, $\bar{z}$ for $z \in Z$ (but not the arguments y, $\bar{y}$ for $y \in Y$). The full reduction is given below and illustrated in Figure 4.

Reduction 3.7 *Given a $QBF_\forall^2$ formula $\Phi = \forall Y \exists Z\, \varphi(Y, Z)$ with φ being a CNF formula given by a set of clauses C over atoms $X = Y \cup Z$, we define the following translation from Φ to $H_\Phi = (A, R)$, where*

$$A = \{\varphi, \bar{\varphi}\} \cup C \cup X \cup \bar{X}$$
$$\begin{aligned}R = &\{(c, \varphi) \mid c \in C\} \cup \{(x, \bar{x}), (\bar{x}, x) \mid x \in X\} \cup \\ &\{(x, c) \mid x \in c, c \in C\} \cup \{(\bar{x}, c) \mid \bar{x} \in c, c \in C\} \cup \\ &\{(\varphi, \bar{\varphi}), (\bar{\varphi}, \bar{\varphi})\} \cup \{(\bar{\varphi}, z), (\bar{\varphi}, \bar{z}) \mid z \in Z\}\end{aligned}$$

In the Reduction 3.7 we have that each interpretation $I_Y \subseteq Y$ corresponds to an admissible set $\{y \mid y \in I_Y\} \cup \{\bar{y} \mid y \in Y \setminus I_Y\}$ in H_Φ while arguments z and $\bar{z}$ are attacked by φ and thus can only be in an admissible set if also φ is in that set. To make φ admissible we have to find $I_Y \subseteq Y$ and $I_Z \subseteq Y$ that together satisfy φ. Moreover, as each $c \in C$ is in conflict with φ and attacked by either z and $\bar{z}$ for some $z \in Z$ none of them can be in an admissible set. We then have that a set $\{y \mid y \in I_Y\} \cup \{\bar{y} \mid y \in Y \setminus I_Y\}$ is a preferred extension, i.e., a subset maximal admissible set, iff there is no I_Z such $I_Y \cup I_Z$ satisfies φ. That is, there is a preferred extension in H_Φ not containing the argument φ iff the $QBF_\forall^2$ formula $\forall Y \exists Z\, \varphi(Y, Z)$ is false. Thus, we have a polynomial reduction from the Π_2^P-complete problem of $QSAT_\forall^2$ to skeptical acceptance with preferred semantics which proves the Π_2^P-hardness of the latter.

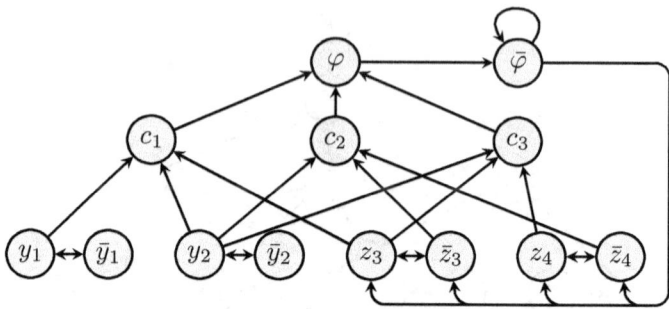

Figure 4. Illustration of the reduction for Π_2^P-hardness of $Skept_{pr}$. The AF H_Φ, for $\Phi = \forall y_1 y_2 \exists z_3 z_4 \left((y_1 \vee y_2 \vee z_3) \wedge (y_2 \vee \neg z_3 \vee \neg z_4) \wedge (y_2 \vee z_3 \vee z_4) \right)$.

Acceptance with Grounded Semantics. Here we consider the problem of deciding whether an argument is in the grounded extension and show that it is P-hard. To this end we first have to introduce the P-complete problem HORNSAT. A *definite Horn-clause* c is a disjunction over literals from a countable domain U such that c contains exactly one positive literal. A definite Horn-formula is the conjunction over definite Horn-clauses. For example consider the definite Horn-formula $\varphi = x \wedge (\neg x \vee \neg y \vee z) \wedge (\neg y \vee \neg z \vee x)$. A more convincing way to denote definite Horn-formulae is as set of clauses and moreover denoting clauses as (logically equivalent) rules. Thus, our example formula φ can be denoted as $\varphi = \{\rightarrow x, x \wedge y \rightarrow z, y \wedge z \rightarrow x\}$. It is well-known that a definite Horn-formula has a unique minimal model which can be computed in polynomial time. Moreover, the problem HORNSAT of deciding whether an atom is in the minimal model of a definite Horn formula is P-complete [Kasif, 1986].

Next, in order to show P-hardness of $Cred_{gr}$, we give a logspace-reduction from the P-complete problem HORNSAT, to $Cred_{gr}$ (see Figure 5). Starting from a definite Horn formula one constructs an AF with one argument for each Horn clause; one argument $\bar{x}$ for each variable x; and an additional argument z for the variable that we are asking for being in the minimal model. All the arguments $\bar{x}$ are self attacking. Each argument corresponding to a rule is attacked by all arguments $\bar{x}$ for which the variable x is in the body of the rule and the argument attacks the argument $\bar{h}$ where h is the head of the rule. Finally, the argument z is only attacked by $\bar{z}$ but attacks all $\bar{x}$ arguments. The reduction is formally stated below.

Reduction 3.8 *Let* $\varphi = \{r_l : b_{l,1} \wedge \cdots \wedge b_{l,i_l} \rightarrow h_l \mid 1 \leq l \leq n\}$ *be a definite Horn*

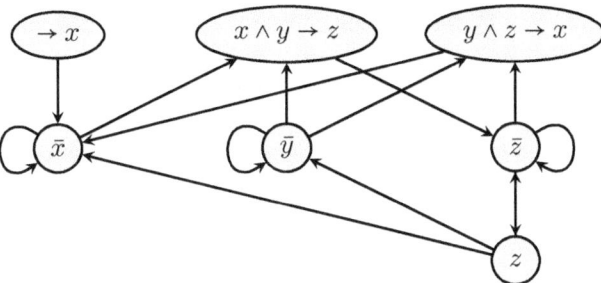

Figure 5. Illustration of the reduction for P-hardness of $Cred_{gr}$, that is $F_{\varphi,z}$ for $\varphi = \{\to x, x \wedge y \to z, y \wedge z \to x\}$.

theory over atoms X. We construct the AF $F_{\varphi,z} = (A, R)$ as follows:

$$A = \varphi \cup \bar{X} \cup \{z\}$$
$$R = \{(\bar{x}, \bar{x}), (z, \bar{x}) \mid x \in X\} \cup \{(\bar{z}, z)\} \cup$$
$$\{(r_l, h_l), (b_{l,j}, r_l) \mid r_l \in \varphi, 1 \le j \le i_l)\}$$

The intuition behind the above reduction is that an argument corresponding to a rule is in the grounded extension only if all atoms in the rule body are in the minimal model of φ and an argument $\bar{x}$ is attacked by the grounded extension only if x is in the minimal model. That is, when computing the grounded extension via iteratively applying the characteristic function we simulate the following algorithm for deciding whether z is in the minimal model of φ. The algorithm starts with the rules with empty body and adds their rule heads to the minimal model. Then it iteratively considers all rules with the body already being part of the minimal model and adds their heads to the minimal model until either z is added or a fixed-point is reached. Notice that as soon as z is added to the grounded extension all arguments corresponding to rules are defended and thus also added to the grounded extension. We then have that z is in the minimal model of the Horn-formula φ iff z is in the grounded extension of $F_{\varphi,z}$ iff $\{r_l \mid 1 \le l \le n\} \cup \{z\}$ is the grounded extension of $F_{\varphi,z}$. This shows the P-hardness of credulous acceptance as well as of verifying the grounded extension.

3.3.3 Existence and Uniqueness of Extensions

Next, let us consider the results for the existence of (non-trivial) extensions and the uniqueness of solutions in Table 1.

Existence problems. First notice that the $Exists_\sigma$ problem is only relevant for stable semantics as all the other semantics always lead at least one extensions. Moreover, for AFs with at least one argument the problems $Exists_{st}$ and $Exists_{st}^{\neg\emptyset}$ coincide. The standard (non-deterministic) algorithm for $Exists_\sigma^{\neg\emptyset}$ first guesses a non-empty set and then verifies that it is an extension. Now let be V be the complexity for verifying an extension then the above gives use a NP^{V} algorithm for $\mathit{Exists}_{\sigma}^{\neg \emptyset}$. However, the algorithm is only adequate for st, ad, co, and id semantics, for the other semantics the problem can be solved more efficiently: for cf and na semantics it suffices to find one argument that is no self-attacking; for gr semantics one tests whether there is an argument that is not attacked by other arguments; for $cf2$ and stg semantics the problem reduces to test whether there is a non-empty conflict-free set; and finally for pr and sst semantics the problem reduces to check whether there is a non-empty admissible set.

Uniqueness. When testing for the uniqueness of extensions again stable semantics has a special behaviour. While for all the other semantics we are guaranteed that there is at least one extension for stable semantics we have to perform an additional check that there exists an extension. To check that there are not two (or more) extensions we use the following NP^{V} procedure that shows that an AF has at least two extensions. It first non-deterministically guesses two sets and then verifies that they are different from each other and both are extensions (for the latter the V oracle is used). That is we have an $\mathsf{coNP}^{\mathsf{V}}$ for testing that an AF has at most one extension, which for all semantics, except stable, is equivalent to Unique_{σ}.

Let us now briefly discuss the results for the specific semantics listed in Table 1 (cf. [Dvořák, 2017]). First, cf semantics yields a unique extension iff all arguments in the AF are self-attacking, and naive semantics yield a unique extensions if there is no conflict between non-self-attacking arguments. Both criteria can be easily tested in L. Second, gr and id always yield a unique extension and thus an algorithm can answer "yes" without any computation. For st, ad, and co we can use the $\mathsf{coNP}^{\mathsf{V}}$ algorithm. However, for stable we have to use an additional NP-algorithm to test whether there exists an extension, resulting in a DP algorithm for Unique_{st}. The situation of $cf2$ is different. As shown in [Kröll et al., 2017] one can enumerate $cf2$ extensions with polynomial delay and thus can also test uniqueness in polynomial time (by computing the first two $cf2$ extensions). For pr semantics the $\mathsf{coNP}^{\mathsf{V}}$ algorithm can be improved by the observations that an AF has two (or more) preferred extensions iff it has two admissible sets that are in conflict with each other. Thus it suffices to guess two sets, and verify that both sets are admissible and there is a conflict between the two sets. For sst, and stg the exact complexity is still open but one can also do better than the standard algorithm. That is, the standard algorithm would give a Π_2^P-algorithm but one can actually decide uniqueness with a Θ_2^P-algorithm [Dvořák, 2017].

3.4 Computational Advantages of Specific Graph-Classes

As most of the reasoning tasks are hard for most of the semantics, one is interested in criteria that make concrete instances tractable. Here we consider special graph classes such that abstract argumentation frameworks within this graph class can be evaluated efficiently. However, these tractability results often only hold for specific semantics and not for the others. This section is based on [Dunne, 2007] and follow up work. In the following we omit semantics where the reasoning tasks are already in L in the general case.

3.4.1 Acyclic AFs

For acyclic AFs we have that each argument is either contained in the grounded extension or attacked by an argument in the grounded extension. Thus the grounded extension is the only stable extension and all the semantics under our consideration coincide. Thus, for all semantics reasoning reduces to computing the grounded extension, but which itself remains P-hard even for acyclic bipartite AFs [Dvořák, 2012a]. The results are summarised in Table 2. Notice that $Skept_{ad}$ is trivially false even in the general case.

Table 2. Complexity for acyclic AFs.

σ	gr	st	ad	co	pr	sst	stg	cf2	id
$Cred_\sigma$	P-c	P-c	P-c	P-c	P-c	P-c	P-c	P-c	P-c
$Skept_\sigma$	P-c	P-c	trivial	P-c	P-c	P-c	P-c	P-c	P-c

For admissibility based semantics there is a conceptual difference how they deal with even (length) and odd (length) cycles. In an even-cycle there are three admissible sets, the empty set, the set of odd numbered arguments and the set of even numbered arguments, while for an odd-cycle the only admissible set is the empty set. Due to different treatments even and odd-cycles have a quite different impact on the computational complexity.

Even-cycle free AFs. Let us first consider the impact of even-cycles for admissibility based semantics. By an observation in [Dunne and Bench-Capon, 2001] each AF with at least two preferred extensions has an even-cycle. This even holds for complete extensions, i.e., each AF with two complete extensions has an even-cycle [Dvořák, 2012a]. The number of even-cycles in an AF bounds the number of complete and thus also preferred extensions. Thus if an AF has no even-cycles the grounded extension is as well the unique preferred extension and therefore the only candidate for being a stable extension. Again the reasoning tasks for the admissibility based semantics reduce to computing the grounded extension.

Table 3. Complexity results for even-cycle free AFs.

σ	gr	st	ad	co	pr	sst	stg	cf2	id
$Cred_\sigma$	P-c	P-c	P-c	P-c	P-c	P-c	Σ_2^P-c	NP-c	P-c
$Skept_\sigma$	P-c	P-c	trivial	P-c	P-c	P-c	Π_2^P-c	coNP-c	P-c

The picture is different for stage and cf2 semantics which are not based on admissibility and handle odd and even-cycles in a similar way. Both maintain their full complexity for even-cycle free AFs [Dvořák and Gaggl, 2016]. The results for even-cycle free AFs are summarised in Table 3.

Odd-cycle free AFs. Odd-cycles are of interest as they distinguish stable from preferred semantics. By a result from [Dung, 1995] in the absence of odd-cycles stable and preferred semantics coincide, i.e., the AF is coherent. As this implies that there is at least one stable extension also semi-stable and stage semantics coincide with stable and preferred semantics in odd-cycle free AFs. But then the complexity of preferred, semi-stable and stage drops down to the complexity of stable, which however stays the same as in the general case. Also admissible, complete and cf2 are not profiting from the absence of odd-cycles, which is proven by the fact that both the standard translation and the modification for skeptical acceptance do not make use of odd-cycles [Dimopoulos and Torres, 1996; Gaggl and Woltran, 2013]. An overview is given in Table 4.[11]

Table 4. Complexity results for odd-cycle free AFs.

σ	gr	st	ad	co	pr	sst	stg	cf2	id
$Cred_\sigma$	P-c	NP-c	NP-c	NP-c	NP-c	NP-c	NP-c	NP-c	coNP-c
$Skept_\sigma$	P-c	NP-c	trivial	coNP-c	coNP-c	coNP-c	coNP-c	coNP-c	coNP-c

3.4.2 Bipartite AFs

Bipartite AFs, AFs where the arguments can be partitioned on two conflict-free sets, are a special case of odd-cycle free AFs and thus again stable, preferred, semi-stable, and stage semantics coincide. There is a polynomial time algorithm for computing the credulously accepted arguments [Dunne, 2007], which is based on the following observation. Let the arguments be partitioned in two conflict-free sets A, B. Arguments in A are only attacked by arguments in B and can only be defended by arguments in A. Now the algorithms starts with the set A and tests if it is admissible. If yes then all arguments in the set A are credulously accepted, otherwise all arguments in A which are not defended can not be in any admissible set, i.e., they are not credulously accepted. In the latter case the algorithm removes the undefended arguments and tests the new set for being admissible. It proceeds until it reaches an admissible set (which might be empty). At the end we have that all arguments which are in the computed admissible set are credulously accepted and the remaining arguments in A are not. We then apply the same algorithm to the set B to compute the remaining credulously accepted arguments.

To decide skeptical acceptance we can use that for stable semantics an argument is skeptically accepted iff none of its attackers is credulously accepted.[12] Hence given that we can compute all credulously accepted arguments in polynomial time we can also decide skeptical acceptance in P.

---

[11]The result for ideal has not been stated before, but is immediate by a generic result in [Dunne et al., 2013] stating that $Cred_{id}$ belongs to coNP^V where V is the complexity of Ver_{pr} and the fact that the reduction for coNP hardness in [Dunne, 2008] constructs an odd-cycle free AF.

[12]Each stable extension has to either contain the argument or one of its attackers.

In [Dvořák and Gaggl, 2016] it was shown that the above algorithm also works for cf2 semantics. The result for ideal semantics follows from the result for preferred semantics and the polynomial-time algorithm in [Dunne et al., 2013] that computes the ideal extension given the skeptically accepted arguments w.r.t. preferred semantics.

The results are summarised in Table 5. Notice that while in bipartite AFs we can efficiently compute credulous and skeptical acceptance, in contrast to previous tractable fragments, we cannot compute all extensions nor have a good handle on them. This is mirrored by the fact that deciding whether two arguments appear together in one extension is NP-hard [Dunne, 2007]. One can also imagine to consider generalisations of bipartite graphs, so called k-partite graphs, where the arguments can be divided into k conflict-free sets. However, for $k \geq 3$ there are no computational advances from k-partite graphs [Dunne, 2007].

Table 5. Complexity results for bipartite AFs.

σ	gr	st	ad	co	pr	sst	stg	cf2	id
$Cred_\sigma$	P-c	P-c	P-c	P-c	P-c	P-c	P-c	P-c	P-c
$Skept_\sigma$	P-c	P-c	trivial	P-c	P-c	P-c	P-c	P-c	P-c

3.4.3 Symmetric AFs

Here we consider AFs where each attack is symmetric. As each attacker is immediately defended by the symmetric attack the notion of admissibility reduces to conflict-freeness. Considering grounded semantics, in each non-trivial connected component of arguments all arguments are attacked by at least one other argument and thus none of them can be in the grounded extension. Thus computing the grounded extension reduces to find all isolated arguments which can be done in L. Hence all reasoning tasks for grounded, admissible, complete, and preferred semantics are in L. Moreover, also cf2 coincides with naive semantics [Dvořák and Gaggl, 2016]. Finally for symmetric AFs the set of skeptically accepted arguments is always admissible and thus the skeptically accepted arguments coincide with the ideal extension.

Table 6. Complexity results for symmetric AFs.

σ	gr	st	ad	co	pr	sst	stg	cf2	id
$Cred_\sigma$	L	L/NP-c	L	L	L	L/Σ_2^P-c	L/Σ_2^P-c	L	L
$Skept_\sigma$	L	L/coNP-c	trivial	L	L	L/Π_2^P-c	L/Π_2^P-c	L	L

For symmetric AFs one often also requires [Coste-Marquis et al., 2005] that the AF is irreflexive, i.e., it has no self attacks. In that case each naive extension is also a stable extension and thus also semi-stable, and stage coincides with

Table 7. Complexity of acceptance problems, parametrized by the distance from graph classes that allows for efficient algorithms.

graph class	ad	co	pr	sst	st	stg	cf2	id
acyclic	FPT	FPT	FPT	FPT	FPT	hard	hard	FPT
noeven	XP	XP	XP	XP	XP	hard	hard	XP
bipartite	hard	hard	hard	hard	hard	hard	hard	hard
symmetric	hard	hard	hard	hard	hard	hard	hard	hard

naive. That is, the corresponding reasoning tasks become tractable. However, if we allow self-attacks in the framework then these three semantics maintain their full complexity [Dvořák, 2012a].

3.5 Fixed-Parameter Tractable Fragments

In the previous section we considered properties of the graph structure that make argumentation problems tractable. In this section we study parameters that are quantitative measures for some kind of structure in the graph, with the goal to find parameters such that the complexity of the problem rather scales with the parameter than with input size. That is, we are looking for parameters that allow for fixed-parameter tractable algorithms.

Backdoors for abstract argumentation. One approach to fixed-parameter tractable algorithms is the so called backdoor approach [Dvořák et al., 2012a]. The idea is to start from a tractable fragment, define some kind of distance to the tractable fragment, and then use the distance as a parameter for the reasoning problem. The hope behind this approach is that the running time will scale with the distance to the tractable fragment instead of jumping instantly to the full problem complexity when leaving the fragment. In argumentation one can use the graph classes discussed above as tractable fragments and as distance one considers the number of arguments that have to be deleted from an AF to fall into the graph class.

Definition 3.9 *Let $\mathcal{G}$ be a graph class and $F = (A, R)$ an AF. We define $dist_\mathcal{G}(F)$ as the minimal number k such that there exists a set $S \subseteq A$ (the backdoor set) with $|S| = k$ and $(A \smallsetminus S, R \cap ((A \smallsetminus S) \times (A \smallsetminus S)) \in \mathcal{G}$. If there is no such set S we define $dist_\mathcal{G}(F) = \infty$.*

We will see that this parametrization only works for certain fragments and semantics, while for other fragments and semantics we have the full complexity even for AFs with constant distance to the fragment. Table 7 summarises the results for the semantics under our considerations (again we omit semantics already tractable in the general case). All results are due to [Ordyniak and Szeider, 2011; Dvořák et al., 2012a] [13], except the results for cf2 semantics

---
[13] Notice, that ideal semantics is not explicitly mentioned in [Dvořák et al., 2012a], but the

which are due to [Dvořák and Gaggl, 2016]. The entries in Table 7 are to be read as follows. FPT: all reasoning tasks are in FPT; XP: all reasoning tasks are in XP; hard: all problems are as hard as for general graphs even for instances with a fixed distance to the fragment (and are at least NP/coNP-hard for distance 1).

In the remainder of the section we will first present the algorithm that underlies the FPT and XP results and then exemplify some hardness results.

FPT backdoor-algorithms. The positive results are all based on the fact that the number of complete extensions is small and we can compute them efficiently. As soon as we have the complete extensions all the reasoning tasks for admissibility based semantics can be answered efficiently. The algorithms for complete semantics consist of two parts: first one has to compute the backdoor set; second given the backdoor set one has to compute the complete extensions.

Let us first consider computing a backdoor. The detection of *acyc*-backdoors for AFs is equivalent to the so-called *directed feedback vertex set* problem in graph theory, which is known to be fixed-parameter tractable [Chen *et al.*, 2008]. For detecting *noeven*-backdoors in AFs the following algorithm is known, which only shows the problem to be in XP. By a result of Robertson *et al.* [1999] one can test in polynomial time whether a graph is in *noeven* or not. Now, to find a backdoor of size k one can simply iterate over all sets of size k and test whether removing these arguments break all even-cycles. As there are $\Theta(n^k)$ many such sets the algorithm is not fixed-parameter tractable and thus only shows the problem to be in XP.

Now let us assume we already have a backdoor set. We consider labels for arguments that correspond to their status in the extension. An argument is labelled **in** if it is in the extension, **out** if it is not in the extension but attacked by an argument in the extension, and **undec** otherwise. The algorithm tests all possible assignments of labels to the arguments in the backdoor set (these are 3^k many) and for each of them propagates the labels to the remaining AF (which is acyclic or noeven) according to the characteristic function. That is, a node gets label **out** as soon as one attacker is labelled **in**, label **in** if all attackers are labelled **out** and label **undec** if all attackers are labelled and none of the above applies. Finally, one considers the set of arguments labelled **in** and keeps the set if it is a complete extension or withdraws them otherwise. As we do this for each possible labelling of the backdoor set we finally get the set of all complete extensions, which is of size at most 3^k. As one can propagate the labels in polynomial time the, total running time of the algorithm is 3^k multiplied by some polynomial and thus in FPT.

Finally combining the results for computing a backdoor set and for evaluating an AF given a backdoor we have an FPT algorithm for acyclic AFs and an XP algorithm for noeven AFs. Notice that the XP complexity for noeven AFs comes solely from the algorithm for computing the backdoor, the evaluation itself is in FPT.

---

results follow immediately from the results presented for preferred semantics.

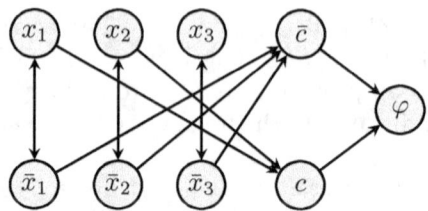

Figure 6. Hardness reduction for $Cred_{ad}$ and backdoors to bipartite graphs, illustrated for the propositional formula φ, with clauses $c = \{x_1, x_2\}$, and $\bar{c} = \{\bar{x}_1, \bar{x}_2, \bar{x}_3\}$.

Hardness Results. The hardness proofs work very much like for the general case, one has to give a reduction from a hard problem but additionally take into account the graph structure [Dvořák et al., 2012a; Dvořák, 2012c; Dvořák et al., 2014a; Dvořák and Gaggl, 2016]. We exemplify such a reduction for credulous reasoning under admissible, complete, preferred and cf2 semantics and backdoors for bipartite graphs. To this end consider the standard translation from proposition logic and the NP-hard problem monotone SAT, of deciding whether a formula in CNF where each clause either contains solely positive or solely negative literals is satisfiable. As each clause either contains solely positive literals or solely negative literals the graph constructed by the standard translation is almost bipartite (cf. Figure 6). That is, there are no edges between the arguments corresponding to positive literals and negative clauses and no edges between the arguments corresponding to negative literals and positive clauses. Thus, when deleting φ from the graph the graph becomes bipartite with two independent sets, one containing the positive literals and the negative clauses, and one containing the negative literals and the positive clauses. We obtain that credulous reasoning is NP-hard even for graphs with distance 1 to bipartite graphs.

Further FPT Results. Besides backdoors to tractable fragments several other approaches for parametrizations can be found in the literature. One approach is to consider graph parameters that measure structural properties, most prominently tree-width, a parameter that, roughly speaking, measures how tree-like a graph is. Results for tree-width (and the related parameter clique-width) can be either obtained by dynamic programming algorithms that exploit the structural properties or by powerful meta-theorems. These meta-theorems basically say that every property which can be characterised by a formula from monadic-second order logic (MSO) over a graph structure can be tested in FPT w.r.t. tree-width and clique-width. Results via the MSO meta-theorems are given in [Dunne, 2007; Dvořák et al., 2012c] concrete dynamic programming algorithms are given in [Dvořák et al., 2012b; Charwat, 2012] for tree-width and in [Dvořák et al., 2010] for clique-width. Moreover, in [Dvořák et al., 2012b] a lot of parameters specific to directed graphs, e.g. directed tree-width, are shown to be not applicable for FPT algorithms in abstract argumentation.

Finally for semantics harder than NP one can also think about backdoors to graph classes that allow to solve problems in NP or coNP [Dvořák et al., 2014a]. While this does not give FPT results it still reduces complexity, with notable effects on the practical resolvability.

3.6 Computational Problems related to Labelling-Based Semantics

So far our complexity analysis was in terms of extension-based semantics (which is in accordance with the literature), in this section we discuss some computational aspects related to labelling-based semantics.

Labelling-based semantics. Beside the so-called extension-based semantics we have considered so far, there are several approaches defining argumentation semantics via certain kinds of argument labellings. As an example we consider the popular approach of 3-valued labellings by Caminada and Gabbay [2009] and in particular their complete labellings. Basically, such a labelling is a three-valued function $\mathcal{L}ab$ that assigns one of the labels in, out and undec to each argument, with the intuition behind these labels being the following. An argument is labelled with: in if it is accepted, i.e., it is defended by the in labelled arguments; out if there are strong reasons to reject it, i.e., it is attacked by an accepted argument; undec if the argument is undecided, i.e., neither accepted nor attacked by accepted arguments. Complete labellings can be one-to-one mapped to complete extensions by considering the set of in labelled arguments and vice versa, by labelling all arguments in the extension with in all arguments attacked by the extension with out and the remaining arguments with undec [Caminada and Gabbay, 2009]. Notice that this is not only a property of complete semantics but this one-to-one correspondence holds for most argumentation semantics.

Computational problems. Given the above correspondence between labellings and extensions, the tasks of computing all labellings and all extensions are, from a computational point of view, equivalent and the same holds for credulous and skeptical reasoning. However, three-valued labellings allow for more fine-grained acceptance statuses of arguments. Wu and Caminada 2010 introduced the notion of justification status of an argument w.r.t. a semantics which is given by the set of labels that are assigned by at least one labelling of the semantics. That is, given an AF F and a labelling-based semantics $\sigma_{\mathcal{L}ab}$ the *justification status* $\mathcal{JS}_{\sigma_{\mathcal{L}ab}}(F,a)$ of an argument a in F is given by $\mathcal{JS}_{\sigma_{\mathcal{L}ab}}(F,a) = \{\mathcal{L}ab(a) \mid \mathcal{L}ab \in \sigma_{\mathcal{L}ab}(F)\}$. The above definition gives rise to eight different justification statuses, most prominently the set {in} called strong accept, which corresponds to skeptical acceptance, and the set {in, undec} called weak accept.[14] We are now faced with the computational problem of *verifying the justification status* of an argument, which was studied in [Dvořák, 2012b].

---

[14] Notice that credulous acceptance of argument a corresponds to the query in $\in \mathcal{JS}_\sigma(F,a)$.

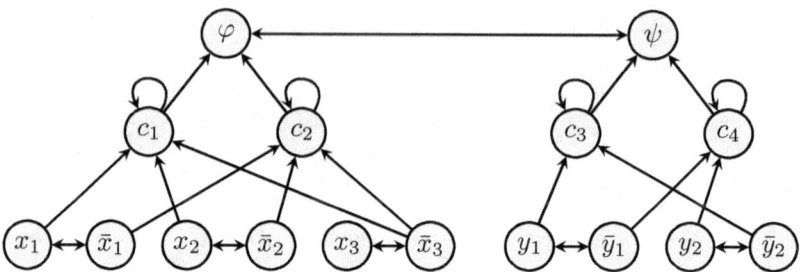

Figure 7. Reduction for showing the DP-hardness of weak acceptance. AF $F_{\varphi,\psi}$ for the propositional formulae φ, with clauses $\{x_1, x_2, \bar{x}_3\}, \{\bar{x}_1, \bar{x}_2, \bar{x}_3\}$, and ψ, with clauses $\{y_1, \bar{y}_2\}, \{\bar{y}_1, y_2\}$.

Algorithms. Compared with credulous and skeptical acceptance, where we either search for an extension containing a specific argument or for an extension not containing a specific argument, the problem of testing whether an argument has a specific justification status, e.g., whether it is a weakly accepted, has two sources of complexity. First, we have to search for labellings that assign the labels appearing in the justification status, e.g., in and undec for weak acceptance, and second we have to make sure that no labelling assigns one of the labels not in the in the justification status, e.g., out for weak acceptance, to a. For complete semantics this means we have to perform both an NP search for the good labels and a coNP search for the bad labels, which together gives a DP algorithm.

Hardness. To prove DP-hardness of weak acceptance w.r.t. complete semantics one starts from an instance (φ, ψ) of the DP-complete SAT–UNSAT problem and constructs the AF $F_{\varphi,\psi}$ (see Figure 7) as follows. First one applies the standard translation to each of the two formulae and then makes the arguments c corresponding to clauses unacceptable, by adding self-attacks. Finally the arguments φ and ψ are connected by a mutual attack. As in the standard translation we have that φ, respectively ψ, is credulously accepted iff φ, respectively ψ, is satisfiable. Moreover, (a) φ is labelled out iff ψ is labelled in by some labelling, i.e., if ψ is credulously accepted, and (b) the grounded labelling maps all arguments to undec and thus undec $\in \mathcal{JS}_{\sigma_{Lab}}(F_{\varphi,\psi}, \varphi)$. We then have that the argument φ is weakly accepted iff φ is satisfiable and ψ is unsatisfiable, that is iff (φ, ψ) is a "yes" instance of SAT–UNSAT. Thus we have a reduction from SAT–UNSAT to weak acceptance and can conclude that also the latter is DP-hard.

3.7 Discussion

As illustrated in Table 1 there is a significant difference in the computational complexity between the different semantics. Let us first consider the polynomial-time computable semantics. Grounded semantics distinguishes itself from the remaining semantics by the fact that it has a unique extension which can be efficiently computed in an iterative fashion by applying the characteristic function. For conflict-free and naive sets the good complexity comes from the fact that we can decide the reasoning problems without computing the actual conflict-free, respectively naive, sets. However, there are AFs with exponentially many conflict-free, respectively naive, sets and there are non-standard problems the are computationally hard, for instance counting the number of conflict-free, respectively naive, sets [Baroni et al., 2010].

On the NP, coNP layer of the polynomial-hierarchy we have semantics with potentially exponentially many extensions but where each set itself can be easily tested to be an extension. That is, the source of the computational hardness is the fact that one, in the worst case, has to check many sets to find a witness for credulous acceptance, respectively to find a counter-example for skeptical acceptance. However, these problems can be efficiently encoded in formalisms where the corresponding problems are NP- and coNP-hard, like propositional logic, and then can be evaluated with corresponding systems for these formalisms [Besnard and Doutre, 2004].

Finally, we have semantics that require some sort of subset maximisation which adds an additional source of complexity. Thus, these semantics are harder than NP and located at the second level of the polynomial-hierarchy. For reduction-based approaches this implies that one cannot efficiently translate them to a single instance of propositional logic but has either to consider richer formalisms like QBFs [Egly and Woltran, 2006; Arieli and Caminada, 2012] or ASP [Egly et al., 2010] or consider iterative approaches [Cerutti et al., 2014; Dvořák et al., 2014a] that make several calls to a SAT-Solver. The different levels of hardness of different semantics are also mirrored by the results of the First International Competition on Computational Models of Argumentation [Thimm and Villata, 2015; Thimm et al., 2016], where the computational tasks for preferred semantics appear significantly harder than the corresponding tasks for stable or complete semantics.

Notice that there are several established semantics which are beyond the scope of this chapter. First there is the scheme of resolution-based semantics [Baroni et al., 2011c], with resolution-based grounded semantics being the most prominent instantiation. A comprehensive complexity analysis for resolution-based grounded semantics can be found in [Baroni et al., 2011c], which is complemented by results in [Dvořák et al., 2012c; Dvořák et al., 2014b]. Another semantics we neglected is eager semantics [Caminada, 2007], whose complexity was studied in the generalised setting of parametrized ideal semantics [Dunne et al., 2013].

4 Complexity of Assumption-based Argumentation

With Dung's abstract argumentation frameworks we focused on the issue of finding coherent sets of simultaneously acceptable arguments, but neglected the effort for constructing these frameworks and for drawing conclusions from the accepted arguments. With Assumption-based Argumentation [Bondarenko et al., 1997] we now switch to a formalism that covers the whole argumentation process. That is, arguments and conflicts are constructed from a knowledge base, then acceptable sets, i.e., extensions, are identified, and finally one draws conclusions from the extensions. We are in particular interested in how these additional steps affect the overall computational complexity.

In this section will discuss complexity results for Assumption-based Argumentation which are due to the work of Dimopoulos et al. [1999; 2000; 2002] and the later work on ideal semantics [Dunne, 2009].[15] We first briefly introduce assumption-based frameworks and the different semantics thereof and define the core reasoning problems in assumption-based argumentation. We then discuss procedures to solve the reasoning problems, which give us upper bounds for the computational complexity. As most of these procedures are of high complexity we also discuss the special case of flat ABFs which allows for a milder complexity. Finally, we discuss some hardness results showing that the presented procedures are essentially optimal.

4.1 Assumption-based Argumentation

We first briefly recall the definitions of assumption-based argumentation, for a comprehensive introduction the reader is referred to [Toni, 2014] or chapter 7 on Assumption-based Argumentation in this handbook.

For an assumption-based framework we assume a *deductive system* $(\mathcal{L}, \mathcal{R})$, where $\mathcal{L}$ is a formal language and $\mathcal{R}$ a set of inference rules that induces a derivability relation $\vdash$. Given a theory $T \subseteq \mathcal{L}$ the *deductive closure* $Th(T)$ of T is defined as $Th(T) = \{\alpha \in \mathcal{L} \mid T \vdash \alpha\}$.

Definition 4.1 *An* abstract *assumption-based framework (ABF) is a tuple* $\langle \mathcal{L}, \mathcal{R}, A, ^- \rangle$ *with* $(\mathcal{L}, \mathcal{R})$ *a deductive system,* $A \subseteq \mathcal{L}$ *is a (non-empty finite) set, with elements referred to as assumptions; and the* contrary function $^-$, *a total mapping from A into $\mathcal{L}$.*

An extension of an ABF is a set of assumptions $\Delta \subseteq A$ meeting some requirements.

Definition 4.2 *Given an ABF and an assumption set $\Delta \subseteq A$ we say that Δ* attacks *an assumption $\alpha \in A$ if $\bar{\alpha} \in Th(\Delta)$. Further we say that an assumption set Δ attacks an assumption set Δ' if Δ attacks at least one $\alpha \in \Delta'$*

We will further require that assumptions sets are closed, i.e., we can not derive additional assumptions.

---

[15]The complexity of Assumption-based Argumentation was also briefly discussed in the earlier survey on the complexity of argumentation [Dunne and Wooldridge, 2009].

Definition 4.3 *We call an assumption set Δ closed if $Th(\Delta) \cap A = \Delta$.*

It is often the case that the derivability relation is such that all assumption sets are closed, in that case we call the ABF *flat*.

We are now prepared to define the standard semantics for ABFs.

Definition 4.4 *Given an ABF F and an assumption set $\Delta \subseteq A$. Δ is called*

- *stable extension ($\Delta \in st(F)$), if Δ is closed, Δ does not attack itself, and Δ attacks each assumption $\alpha \in A \smallsetminus \Delta$.*

- *admissible set ($\Delta \in ad(F)$), if Δ is closed, Δ does not attack itself, and for all closed assumption sets $\Delta' \subseteq A$, if Δ' attacks Δ then also Δ attacks Δ'.*

- *preferred extension ($\Delta \in pr(F)$), if Δ is a subset-maximal admissible assumption set.*

Moreover, for flat frameworks also *ideal semantics* can be defined [Dung et al., 2006; Dung et al., 2007]. The unique ideal extensions $id(F)$ is the maximal admissible set Δ that is contained in all preferred extensions.[16]

We have that every stable assumption set is also a preferred assumption set, and every preferred assumption set is an admissible assumptions set, but not vice versa. However, each admissible assumption set is a subset of some preferred assumption set. Moreover, if the ABF is flat the empty assumption set is always admissible.

4.2 Reasoning Problems

As for abstract argumentation we are mainly interested in computing acceptance statuses of statements instead of extensions. However, the reasoning tasks we consider will give us a good impression of the complexity of computing extensions. That is, we again consider credulous and skeptical acceptance but now of a sentence $\varphi \in \mathcal{L}$ instead of an argument. More concretely we either want to decide whether there is at least one extension that entails φ (credulous reasoning) or whether φ is entailed by each extension. This gives rise to the following computational problems for an assumption-based argumentation semantics σ.

- *Credulous Acceptance $Cred_\sigma$*: Given ABF F and a sentence $\varphi \in \mathcal{L}$. Is $\varphi \in Th(\Delta)$ for some assumption set $\Delta \in \sigma(F)$?

- *Skeptical Acceptance $Skept_\sigma$*: Given ABF F and a sentence $\varphi \in \mathcal{L}$. Is $\varphi \in Th(\Delta)$ for all assumption sets $\Delta \in \sigma(F)$?

Beside the above reasoning problems we again consider the task of verifying extensions, i.e., one is given an assumption set and has to verify that it is an extension of a given semantics σ.

---

[16] Notice that uniqueness and other properties of the ideal extension are only guaranteed for flat ABFs.

Table 8. Complexity upper bounds for different types of ABFs. $\mathcal{C}$ denotes the complexity of deciding the $\vdash$ relation.

	General ABFs			Flat ABFs		
σ	$Cred_\sigma$	$Skept_\sigma$	Ver_σ	$Cred_\sigma$	$Skept_\sigma$	Ver_σ
st	$\mathsf{NP}^\mathcal{C}$	$\mathsf{coNP}^\mathcal{C}$	$\mathsf{P}^\mathcal{C}$	$\mathsf{NP}^\mathcal{C}$	$\mathsf{coNP}^\mathcal{C}$	$\mathsf{P}^\mathcal{C}$
ad	$\mathsf{NP}^{\mathsf{NP}^\mathcal{C}}$	$\mathsf{coNP}^{\mathsf{NP}^\mathcal{C}}$	$\mathsf{coNP}^\mathcal{C}$	$\mathsf{NP}^\mathcal{C}$	$\mathcal{C}$	$\mathsf{P}^\mathcal{C}$
id	–	–	–	$\mathsf{P}_{\|}^{\mathsf{NP}^\mathcal{C}}$	$\mathsf{P}_{\|}^{\mathsf{NP}^\mathcal{C}}$	$\mathsf{P}_{\|}^{\mathsf{NP}^\mathcal{C}}$
pr	$\mathsf{NP}^{\mathsf{NP}^\mathcal{C}}$	$\mathsf{coNP}^{\mathsf{NP}^{\mathsf{NP}^\mathcal{C}}}$	$\mathsf{NP}^{\mathsf{NP}^\mathcal{C}}$	$\mathsf{NP}^\mathcal{C}$	$\mathsf{coNP}^{\mathsf{NP}^\mathcal{C}}$	$\mathsf{NP}^\mathcal{C}$

- *Verification of an Extension* Ver_σ: Given ABF $F = \langle T, A, ^- \rangle$ and an assumption set $\Delta \subseteq A$. Is $\Delta \in \sigma(F)$?

4.3 Procedures to solve ABA Reasoning Problems

In ABA, new computational challenges come up when compared with Dung's abstract argumentation. While in Dung's abstract argumentation arguments and attacks are given explicitly, they are only given implicitly in ABFs and depend on the set of assumptions and the derivability relation $\vdash$. That is, we get two additional sources of complexity: (1) the construction of arguments, and (2) the identification of conflicts between them. Both highly depend on the complexity of deciding the derivability relation $\vdash$. Thus, upper bounds for the complexity in assumption-based argumentation usually assume that the derivability relation $\vdash$ can be decided in some complexity class $\mathcal{C}$ and the actual complexity results are then given in terms of some $\mathcal{C}$-oracle complexity classes.

Verifying an Assumption Set. First, we consider the problem of verifying an assumption set Δ as an extension and start with stable semantics. We have to check that (i) Δ is closed, (ii) Δ is conflict-free, and (iii) Δ attacks every assumption $\alpha \in A \setminus \Delta$. Each of these checks can be done in $\mathsf{P}^\mathcal{C}$ as follows: For (i) one has to check whether $\Delta \vdash \alpha$ for $\alpha \in A \setminus \Delta$ which just requires a linear number of $\vdash$ computations. For (ii) one has to check whether $\Delta \not\vdash \bar{\alpha}$ for $\alpha \in \Delta$ which again just requires a linear number of $\vdash$ computations. Finally, (iii) can also be checked by a linear number of $\vdash$ computations and thus a stable set can be verified in $\mathsf{P}^\mathcal{C}$. For admissible semantics verification is a bit harder. Here instead of condition (iii) we have to verify that for all closed assumption sets $\Delta' \subseteq A$, if Δ' attacks Δ then also Δ attacks Δ'. This can be done with a $\mathsf{coNP}^\mathcal{C}$-algorithm that guesses a counter-example Δ' and then verifies via the $\mathcal{C}$ oracle that Δ' is closed, Δ' attacks Δ, and Δ' is not attacked by Δ. In total we have that verifying an admissible extension is in $\mathsf{coNP}^\mathcal{C}$. For preferred semantics we additionally have to take into account the maximality check which leads to a $\mathsf{coNP}^{\mathsf{NP}^\mathcal{C}}$-algorithm.

Reasoning. The complexity upper bounds for skeptical and credulous reasoning are immediate by the algorithms for verifying extensions. We can decide the acceptance of a sentence by first guessing an assumption set, second verifying that the guessed set is an extension and finally deciding via a $\mathcal{C}$ oracle whether the extension entails the queried sentence. The corresponding complexity results are given in the left part of Table 8 (recall that ideal semantics was only introduced for flat ABFs). As for Dung's AFs, credulous reasoning with preferred semantics reduces to credulous reasoning with admissible semantics and thus has a lower complexity than skeptical reasoning. Moreover, in contrast to Dung's AFs, we have a complexity gap between stable and admissible semantics, which is due to the fact that for admissible extensions for each attacking assumption set we have to test whether it is closed or not.

Flat ABFs. Flat ABFs as a special class of ABFs that provide milder complexity. Recall that in flat ABFs each assumption set is already closed and we thus do not have to check this in the algorithms. Let us now reconsider the problem of verifying an admissible extension Δ. As Δ is closed we only have to check whether (i) Δ is conflict-free, and (ii) for all assumption sets $\Delta' \subseteq A$, if Δ' attacks Δ then also Δ attacks Δ'. The latter simplifies to checking whether $\{\alpha \in A \mid \Delta \not\vdash \bar{\alpha}\}$ does not attack Δ, which can be decided in $\mathsf{P}^\mathcal{C}$. Thus, verifying admissible extensions in flat ABFs is in $\mathsf{P}^\mathcal{C}$ and hence also verifying preferred extensions is in $\mathsf{coNP}^\mathcal{C}$. This gives improved complexity bounds for credulous and skeptical acceptance listed in in Table 8 in the column *flat*. Finally, notice that in flat ABFs the empty set is always admissible and thus only the assumptions contained in $Th(\emptyset)$ are skeptically accepted. That is, skeptical reasoning reduces to testing whether $\varphi \in Th(\emptyset)$, which is in $\mathcal{C}$.

The *ideal extension* can be computed by the same algorithm as for Dung AFs [Dunne, 2009]. That is, one first determines the credulously accepted assumptions w.r.t. admissible semantics that are not attacked by other credulously accepted assumptions. Given those arguments one iteratively removes assumptions that can not be defended until an admissible set, the ideal extension, is reached. Overall, this gives an $\mathsf{P}_{\parallel}^{\mathsf{NP}^\mathcal{C}}$ algorithm for credulous and skeptical reasoning as well as for the verification problem.

4.4 Complexity lower bounds

While the upper bounds can be given in a generic fashion, which immediately gives upper bounds/algorithms for each instantiation, hardness results only exist for concrete formalisms. However, the complexity results for the concrete instantiations [Dimopoulos *et al.*, 2002] show that the generic upper bounds are tight in the sense that there are formalisms where the lower bounds match the generic upper bounds. In Table 9 we list the complexity results for Autoepistemic Logic (AEL) [Moore, 1985], Logic Programming (LP) [Gelfond and Lifschitz, 1988], and Default Logic (DL) [Reiter, 1980] all the results are due to Dimopoulos *et al.* [2002] and Dunne [2009]. For Autoepistemic Logic we have that the ABF is not flat and deciding the $\vdash$ relation is coNP-complete. Thus the complexity results in Table 9 exactly match the generic upper bounds for general ABFs. In contrast, Logic Programming and Default Logic result flat ABFs and for the former the ⊢ relation is in P and for the latter the ⊢ relation is coNP-complete. In both cases the complexity results in Table 9 exactly match the generic upper bounds for flat ABFs.

Table 9. Completeness results for instantiations of ABA.

	type	stability cred.	stability skept.	Admissibility cred.	Admissibility skept.	Preferability cred.	Preferability skept.	Ideal cred.
AEL	general	Σ_2^P-c	Π_2^P-c	Σ_3^P-c	Π_3^P-c	Σ_3^P-c	Π_4^P-c	–
LP	flat	NP-c	coNP-c	NP-c	P-c	NP-c	Π_2^P-c	Θ_2^P-c
DL	flat	Σ_2^P-c	Π_2^P-c	Σ_2^P-c	coNP-c	Σ_2^P-c	Π_3^P-c	Θ_3^P-c

For a hardness proof in ABA one has to construct a certain knowledge base in the considered formalism instead of arguments interlinked with conflicts. Thus hardness proofs in the context of ABA are of a different nature than in Dung's abstract argumentation. To exemplify such an hardness proof we next present the hardness result for credulous admissible reasoning in Default Logic which is Σ_2^P-complete [Dimopoulos et al., 2002].

Default logic as ABA. A Default theory (W, D) that consists of a set W of propositional formulae[17], called background theory, and a set D of default rules of the form $\frac{\alpha : M\beta_1 \ldots M\beta_n}{\gamma}$, where α, β_i, γ are sentences in propositional logic, can be interpreted as assumption-based framework $\langle \mathcal{L}, \mathcal{R}, A, ^{-} \rangle$ [Bondarenko et al., 1997]. As deductive system one uses the deductive system of propositional logic extended by the set D of default rules, where the intuitive meaning of a default rule is that if we know α is the case and have no basis on which to suppose any $\neg \beta_i$ holds it is reasonable to assume γ. The ABF is now built as follows: the set of assumptions A consists of the expressions of the form $M\beta$, the contrary $\overline{M\beta}$ of an assumption $M\beta$ is $\neg\beta$ and the derivability relation ⊢ is given by $\Delta \vdash \phi$ iff $\phi \in Th_{DL}(W \cup \Delta)$ where Th_{DL} is the deductive closure of the deductive system described above.

Hardness of credulous admissible reasoning in DL. To show hardness for credulous admissible reasoning in Default Logic, we give a reduction from the Σ_2^P-hard problem $QSAT_\exists^2$ of deciding whether a $QBF_\exists^2$ is valid. That is, we start with a QBF $\exists Y \forall Z \, \varphi(Y, Z)$ and construct a DL theory $(\emptyset, D)$ and thus the corresponding ABF F as follows. To construct the set of default rules D, we add the two default rules (i) $\frac{My}{y}$ and (ii) $\frac{M\neg y}{\neg y}$ for each variable $y \in Y$. This corresponds to the ABF F with $A = \{My, M\neg y \mid y \in Y\}$ and $\overline{My} = \neg y$, $\overline{M\neg y} = y$ for all $y \in Y$. By that we have that an admissible set can only contain either My or $M\neg y$ but not both, and thus that the admissible sets correspond to the

---

[17] Notice that instead of using propositional logic one could also define Default logic on top of first-order logic or any other formal logic.

partial truth assignments of Y. Moreover for an admissible set E we have that $E \vdash \varphi$ iff $\varphi(Y, Z)$ is true for all assignments Z under the partial assignment given for Y. That is, φ is credulously accepted iff there is a partial assignment of Y such that for each assignment to Z the formula $\varphi(Y, Z)$ evaluates to true, that is iff $\exists Y \forall Z \varphi(Y, Z)$ is valid.

Example 4.5 *Consider the QBF $\Phi = \exists y_1, y_2 \forall z_1, z_2 \, (y_1 \vee z_2 \vee \neg z_3) \wedge (\neg y_2 \vee z_3)$. The above reduction would construct*

- *the default rules $\frac{My_1}{y_1}$, $\frac{My_2}{y_2}$, $\frac{M \neg y_1}{\neg y_1}$ and $\frac{M \neg y_1}{\neg y_1}$;*

- *the assumption set $A = \{My_1, My_2, M \neg y_1, M \neg y_2\}$; and*

- *the contrary function $\overline{}$ with $\overline{My_1} = \neg y_1$, $\overline{My_2} = \neg y_2$, $\overline{M \neg y_1} = y_1$, and $\overline{M \neg y_2} = \neg y_2$.*

Now, by the above, it must be that Φ is valid if and only if there is an admissible set $E \subseteq A$ such that $E \vdash (y_1 \vee z_2 \vee \neg z_3) \wedge (\neg y_2 \vee z_3)$. The formula Φ is valid as setting y_1 to true and y_2 to false makes both clauses true no matter which truth value is assigned to the variables z_1 and z_2. On the other hand also the set $E = \{My_1, M \neg y_2\}$ is admissible and, by our default rules, we have $E \vdash (y_1 \vee z_2 \vee \neg z_3) \wedge (\neg y_2 \vee z_3)$. ◊

4.5 Discussion

The upper bounds for the complexity of the reasoning problems in Table 8 indicate that assumption-based argumentation indeed has a higher complexity than just Dung style argumentation. However, by the discussed results for flat argumentation in Table 8 and the concrete instantiations of ABA in Table 9 one can see that the actual complexity heavily depends on the complexity of derivability relation $\vdash$ and the type of the assumption-based framework. For instance for LP we have a flat assumption-based framework and a tractable derivability relation and end up with the same complexity bounds as for Dung's abstract argumentation frameworks. The complexity of deciding the derivability relation directly corresponds to the costs of constructing an argument, or drawing some conclusion when already given an extension. Thus the parameter $\mathcal{C}$ in Table 8 can be interpreted as the costs of these two steps, i.e., constructing arguments and drawing conclusions, in the argumentation process.

For general ABFs the complexity of assumption-based argumentation is quite high and thus it is promising to consider some restrictions of the formalism to get better algorithms. In this chapter we considered flat ABFs which reduced the complexity significantly. In [Dimopoulos et al., 2002] also two other classes, namely so called simple and normal ABFs, are studied and shown to have computational advantages for certain problems.

5 Computational Problems in Abstract Dialectical Frameworks

In this section we consider abstract dialectical frameworks, a generalisation of Dung style abstract argumentation frameworks. While arguments are still abstract entities abstract dialectical frameworks allow for more complex relations between the arguments. That is, each abstract dialectical framework has a link relation between the arguments, which is not necessarily an attack relation. The semantics of the links is given by acceptance conditions for each argument that define the acceptance status of an argument in dependence on the acceptance status of the predecessor arguments. This allows for classical binary attacks between arguments but also for joint attacks, support and more complex dependencies.

This section is based on the works of Brewka et al. [2013], Wallner [2014, Chapter 4], and Strass and Wallner [2015] and organised as follows. We first define abstract dialectical frameworks and semantics thereof. We then discuss and formally define the core computational problems and consider the general computational complexity of abstract dialectical frameworks. Moreover, we discuss a restricted class of ADFs, so called bipolar ADFs, that only allow for links that are attacking or supporting (but might be both), and their computational advantages.

5.1 Abstract Dialectical Frameworks (ADFs)

Here we give a very brief discussion of Abstract Dialectical Frameworks.[18] Notice that in the literature there are several proposals how to define semantics for ADFs, here we will follow the lines of Brewka et al. [Brewka et al., 2013].

Definition 5.1 ([Brewka et al., 2013]) *An abstract dialectical framework is a tuple $D = (S, L, C)$ where*

- *S is a (finite) set of abstract arguments / statements,*
- *$L \subseteq S \times S$ is a set of links,*
- *$C = \{C_s\}_{s \in S}$ is a set of total functions $C_s : 2^{par(s)} \to \{\mathbf{t}, \mathbf{f}\}$, one for each statement $s \in S$. C_s is called acceptance condition of s.*

Here, we will assume that each acceptance condition C_s is given by a propositional formula φ_s over the predecessors of s. An example is provided in Figure 8.

As first semantics we define (two-valued) *models* of ADFs. To this end we consider two-valued interpretations I that to each $s \in S$ assign either $\mathbf{t}$ or $\mathbf{f}$. Given an interpretation I we will use $I^{\mathbf{t}}$ to denote the set $\{s \in S \mid I(s) = \mathbf{t}\}$ and $I^{\mathbf{f}}$ to denote the set $\{s \in S \mid I(s) = \mathbf{f}\}$.

---

[18]For a more detailed discussion the reader is referred to chapter 5 of this handbook dedicated to ADFs or [Brewka et al., 2013].

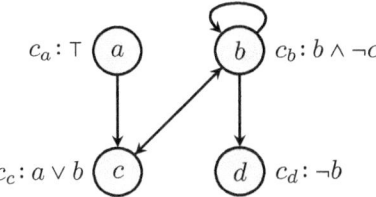

Figure 8. Illustration of an ADF $D = (S, L, C)$ with $S = \{a, b, c, d\}$, $L = \{(a,c),(b,b),(b,c),(c,b),(b,d)\}$, and $C = \{c_a : \top, c_b : b \wedge \neg c, c_c : a \vee b, c_d : \neg b\}$).

Definition 5.2 *Let $D = (S, L, C)$ be an ADF, a two-valued interpretation I defined over S is a two-valued model of D if $I \models \varphi_s$ for each $s \in I^t$ and $I \not\models \varphi_s$ for each $s \in I^f$.*

Most of the ADF semantics are based on *3-valued interpretations* [Kleene, 1952] that map each argument in S to one of the values **t**, **f** and **u**. The three values **t**, **f**, **u** are ordered, by $<_i$, such that $\mathbf{u} <_i \mathbf{t}$, $\mathbf{u} <_i \mathbf{f}$, and **t**,**f** are incomparable. This ordering is then extended to interpretations such that for 3-valued interpretations I, J we have $I \leq_i J$ iff $I(s) \leq_i J(s)$ for all $s \in S$. We say that a two-valued interpretation I extends a 3-valued interpretation J iff $I \leq_i J$. That is, all arguments mapped to **f** or **t** by J are mapped to the same by I and all arguments that are mapped to **u** by J are mapped to either **t** or **f** by I. Given a 3-valued interpretation J, by $[J]_2$ we denote the set of all two-valued interpretations that extend J.

In Dung's abstract argumentation frameworks the characteristic function and its fixed-points are central in the definition of the semantics. We next define the operator Γ_D that will be central in our definitions of ADF semantics. Γ_D generalises the characteristic function in two directions: (i) it gives a three valued assignment on arguments, i.e., beside marking arguments as accepted it also explicitly marks arguments as rejected; and (ii) it allows for the more general acceptance conditions of ADFs.[19] Given an interpretation I, the operator Γ_D computes the arguments that should be set to **t** or **f** under the current interpretation I.

Definition 5.3 *For an ADF D and a three-valued interpretation I, the interpretation $\Gamma_D(I)$ is given by*

$$\Gamma_D(I)(s) = \bigsqcap \{w(\varphi_s) \mid w \in [I]_2\}$$

*where $\sqcap$ is the consensus operation that assigns $\mathbf{t} \sqcap \mathbf{t} = \mathbf{t}$, $\mathbf{f} \sqcap \mathbf{f} = \mathbf{f}$, and assigns **u** otherwise.*

We are now prepared to define admissibility based semantics.

---

[19] For a in depth discussion of the relation between the characteristic function in AFs and the Γ_D operator in ADFs in the context of approximation fixed-point theory the reader is referred to chapter 5 "Abstract Dialectical Frameworks" in this handbook.

Definition 5.4 ([Brewka et al., 2013]) *A three-valued interpretation I for an ADF D is*

- *the grounded interpretation iff it is the least fixed point of Γ_D.*
- *admissible iff $I \leq_i \Gamma_D(I)$;*
- *complete iff $I = \Gamma_D(I)$.*
- *preferred iff it is $\leq_i$-maximal admissible.*

Finally, one can define stable semantics which, in order to avoid cyclic support, makes use of a reduced ADF in the definition.

Definition 5.5 ([Brewka et al., 2013]) *Let $D = (S, L, C)$ be an ADF with $C = \{\varphi_s\}_{s \in S}$. A two-valued model I of D is a stable model of D iff $E_I = \{s \in S : I(s) = \mathbf{t}\}$ equals the set of statements that are $\mathbf{t}$ in the grounded interpretation of the reduced ADF $D^I = (E_I, L^I, C^I)$, where $L^I = L \cap (E_I \times E_I)$ and for $s \in E_I$ we set $\varphi_s^I = \varphi_s[b/\mathbf{f} : I(b) = \mathbf{f}]$.*

5.2 Computational Problems

As the nature of abstract dialectical frameworks is quite similar to the nature of Dung's abstract argumentation frameworks also the core computational problems coincide. That is, we first have *credulous reasoning*, i.e., an argument is accepted if it is mapped to $\mathbf{t}$ by at least one interpretation, and *skeptical reasoning*, i.e., an argument is accepted only if it is mapped to $\mathbf{t}$ by all interpretations. These two reasoning modes again give rise to the following computational problems for argumentation semantics σ.

- *Credulous Acceptance Cred_σ:* Given ADF $D = (S, L, C)$ and an argument $a \in S$. Is there an interpretation $I \in \sigma(D)$ with $I(a) = \mathbf{t}$?

- *Skeptical Acceptance Skept_σ:* Given ADF $D = (S, L, C)$ and an argument $a \in S$. Is $I(a) = \mathbf{t}$ for each interpretation $I \in \sigma(D)$?

Beside these reasoning problems we also consider the problem of *verifying* a given interpretation, and deciding whether an ADF provides any *coherent conclusion*. Depending on the actual semantics the latter can corresponds to deciding whether the ADF has at least one interpretation, or whether the ADF has an interpretation that maps at least one statement to either $\mathbf{t}$ or $\mathbf{f}$.

- *Verification of an interpretation Ver_σ:* Given an ADF $D = (S, L, C)$ and an interpretation I. Is $I \in \sigma(F)$?

- *Existence of an interpretation Exists_σ:* Given an ADF $D = (S, L, C)$. Is $\sigma(F) \neq \emptyset$?

- *Existence of a non-trivial interpretation $\mathrm{Exists}_\sigma^{\neg \emptyset}$:* Given an ADF $D = (S, L, C)$. Does there exist an interpretation I with $I(a) \in \{\mathbf{t}, \mathbf{f}\}$ for some argument $a \in S$.

Table 10. Complexity of ADFs ($\mathcal{C}$-c denotes completeness for class $\mathcal{C}$).

σ	$Cred_\sigma$	$Skept_\sigma$	Ver_σ	$Exists_\sigma$	$Exists_\sigma^{\neg\emptyset}$
gr	coNP-c	coNP-c	DP-c	trivial	coNP-c
model	NP-c	coNP-c	in P	NP-c	NP-c
st	Σ_2^P-c	Π_2^P-c	coNP-c	Σ_2^P-c	Σ_2^P-c
ad	Σ_2^P-c	trivial	coNP-c	trivial	Σ_2^P-c
co	Σ_2^P-c	coNP-c	DP-c	trivial	Σ_2^P-c
pr	Σ_2^P-c	Π_3^P-c	Π_2^P-c	trivial	Σ_2^P-c

5.3 Complexity Results for ADFs

The ability of ADFs to express more complex relations between arguments resulted in more evolved definitions of the semantics and as we will discuss next also increases the complexity of the core reasoning tasks. As the complexity results for ADFs in Table 10 show, all the non-trivial reasoning tasks are one level higher in the polynomial-hierarchy than for Dung's AFs. The main reason for that is the complexity of the Γ_D operator which replaces the characteristic function. While the characteristic function can be evaluated in polynomial time (and even logarithmic space) deciding problems associated with the Γ_D operator are in general NP/coNP-hard.[20]

In this section we discuss the complexity of admissible and grounded semantics in more detail.

Complexity of admissible semantics. Again the most fundamental problem is to verify that an interpretation is admissible. To show that the problem is in coNP we give an NP algorithm [Wallner, 2014] to falsify the admissibility of an interpretation I. Such an algorithm would guess an argument s, such that (i) $I(s) = \mathbf{t}$ or (ii) $I(s) = \mathbf{f}$, and a 2-valued interpretation $J \in [I]_2$ extending I such that either, in case (i), $J(\varphi_s) = \mathbf{f}$ or, in case (ii), $J(\varphi_s) = \mathbf{t}$. As I is admissible iff no such pair s, J exists this is a NP algorithm that falsifies I being admissible and thus the complementary problem of verifying an admissible interpretation is in coNP.

The coNP-hardness is by reduction from the coNP-complete UNSAT problem [Wallner, 2014] of testing whether a propositional formula is unsatisfiable. To this end consider a propositional formula φ over variables X and construct an ADF $D = (S, L, C)$ as follows: The set of arguments S consists of X and an additional argument a, each $x \in X$ is linked towards a, and the acceptance conditions are given by $c_x = x$ for $x \in X$ and $c_a = \varphi$. Now we consider the interpretation I mapping all $x \in X$ to $\mathbf{u}$ and a to $\mathbf{f}$. We have that the two-valued interpretations $J \in [I]_2$ correspond to the two-valued interpretations of φ and thus $\Gamma_D(I)(a) = \mathbf{f}$ iff φ has no model. That is, I is admissible iff

---
[20] For instance testing whether an argument is mapped to $\mathbf{t}$ is basically the validity problem of propositional logic.

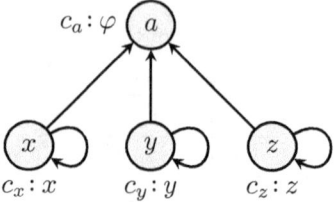

Figure 9. Illustration of the ADF constructed in the reduction from UNSAT to the problem of verifying an admissible interpretation in an ADF, for a propositional formula φ over atoms x, y, z.

φ is unsatisfiable and, as the ADF D can be constructed in polynomial time, coNP-hardness follows.

Combining the coNP verification algorithm for admissible semantics with the standard guess and check algorithms gives a Σ_2^P upper bound for credulous reasoning with admissible, complete and preferred semantics, and a Π_2^P upper bound for verifying a preferred interpretation. The latter then gives a Π_3^P algorithm for skeptical reasoning with preferred semantics.

Complexity of grounded semantics. The computational properties of grounded semantics in ADFs are quite in contrast to the computational properties of grounded semantics in AFs. When considering grounded semantics in ADFs, a straight forward algorithm is, starting from the three-valued model mapping all arguments to $\mathbf{u}$, and then iteratively apply the operator Γ_D until a fixed-point is reached. The straight forward algorithm is only a P^{NP}-algorithm, because of the costly evaluation of Γ_D. However, due to a sophisticated characterisation of grounded semantics [Wallner, 2014] there is a more efficient way to test whether an argument is mapped to $\mathbf{t}$ in the grounded interpretation of an ADF. Also notice that verifying the grounded interpretation is in DP as we have to do verify both the $\mathbf{t}, \mathbf{f}$ assignments and the $\mathbf{u}$ assignments.

The DP-hardness of verifying the grounded interpretation is by a reduction from the DP-complete SAT–UNSAT problem [Brewka and Woltran, 2010; Wallner, 2014]. To this end consider an instance (φ, ψ) of SAT–UNSAT where φ is a propositional formula over atoms X and ψ is a propositional formula over different atoms Y. In polynomial time we construct the ADF $D = (S, L, C)$ with $S = X \cup Y \cup \{d, s, v\}$, $L = \{(x, s) \mid x \in X\} \cup \{(y, v) \mid y \in Y\} \cup \{(d, s)\}$ and the acceptance conditions $c_x = x$ for $x \in X \cup Y$, $c_d = d$, $c_s = \varphi \wedge d$ and $c_v = \psi$. Now we consider the interpretation I with $I(v) = \mathbf{f}$ and $I(a) = \mathbf{u}$ for all the other arguments $a \in S \setminus \{v\}$. We next argue that I is the grounded model iff (φ, ψ) is a "yes" instance of SAT–UNSAT. Let G be the grounded model. First notice that the arguments $a \in X \cup Y \cup \{d\}$ do not have incoming edges from other arguments. Whenever $J(a) = \mathbf{u}$ then there are both an $I_1 \in [J]_2$ with $I_1(a) = \mathbf{t}$ and an $I_2 \in [J]_2$ with $I_2(a) = \mathbf{f}$, and thus also $\Gamma_D(J)(a) = \mathbf{u}$. That is, the grounded model G maps all arguments in $X \cup Y \cup \{d\}$ to $\mathbf{u}$. Now consider the

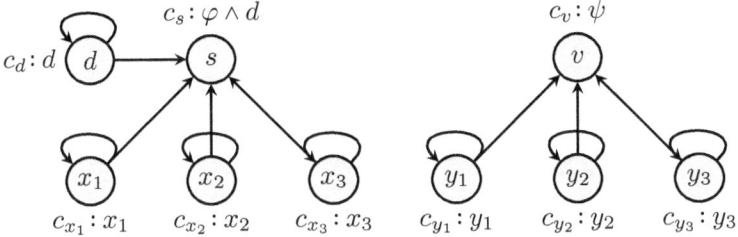

Figure 10. Illustration of the ADF constructed in the reduction from SAT-UNSAT to the problem of verifying the grounded model of an ADF, for an propositional formulae φ, ψ over atoms $X = \{x_1, x_2, x_3\}$ and $Y = \{y_1, y_2 y_3\}$ respectively.

argument s and $c_s = \varphi \wedge d$. The two-valued interpretations $J \in [G]_2$ correspond to the two-valued interpretations over $X \cup Y \cup \{d\}$. That is, either (a) φ has a model and we can satisfy $\varphi \wedge d$ by setting d to $\mathbf{t}$ as well as falsify $\varphi \wedge d$ by setting d to $\mathbf{f}$ and thus $\Gamma_D(G)(s) = G(s) = \mathbf{u}$, or (b) φ is unsatisfiable and thus $\Gamma_D(J)(s) = G(s) = \mathbf{f}$. One the other hand, for v and $c_v = \psi$, we have that either ψ is unsatisfiable and $\Gamma_D(G)(v) = G(v) = \mathbf{f}$, ψ is satisfiable but not valid and $\Gamma_D(G)(v) = G(v) = \mathbf{u}$, or ψ is valid and $\Gamma_D(G)(v) = G(v) = \mathbf{t}$. Hence, we have that $G = I$ iff φ is satisfiable and ψ is unsatisfiable.

5.4 Complexity of Bipolar ADFs with Known Link Types

Again there are certain instances of ADFs that do not have the worst-case complexity, but can be processed with milder complexity. Here we discuss so called *Bipolar ADFs* which put some restriction on the link structure, i.e., each link has to be supporting or attacking (but might be both). For a given set $X \subseteq S$ let I_X be the two-valued interpretation with $I^t = X$ and $I^f = S \setminus X$. A link (a, b) is called *supporting* if there is no $X \subseteq S$ such that $I_X \models \varphi_b$ and $I_{X \cup \{a\}} \not\models \varphi_b$; whereas it is called *attacking* if there is no $X \subseteq S$ such that $I_X \not\models \varphi_b$ and $I_{X \cup \{a\}} \models \varphi_b$. While in general testing the link type is itself coNP-complete [Brewka and Woltran, 2010; Ellmauthaler, 2012] there are certain applications of ADFs where the link type is known beforehand [Brewka and Gordon, 2010; Strass, 2013]. This motivates the research on *bipolar AFs with know link types* which we discuss in the remainder of this section.

The main observation that leads to the better complexity results for bipolar AFs (see Table 11) is that the operator Γ_D can be efficiently computed when all the links are attacking or supporting. The matching hardness results are then by the lower bounds for Dung's abstract argumentation (cf. Table 1) and the observation that AFs can be interpreted as bipolar ADFs with known link types [Brewka et al., 2013] as follows.[21] Given an AF (A, R) the equivalent

---

[21] Notice that both ADF semantics models and stable models are generalisations of Dung's stable semantics.

Table 11. Complexity of Bipolar ADFs with know link types ($\mathcal{C}$-c denotes completeness for class $\mathcal{C}$).

σ	$Cred_\sigma$	$Skept_\sigma$	Ver_σ	$Exists_\sigma$	$Exists_\sigma^{\neg\emptyset}$
gr	P-c	P-c	P-c	trivial	in P
$model$	NP-c	coNP-c	in P	NP-c	NP-c
st	NP-c	coNP-c	in P	NP-c	NP-c
ad	NP-c	trivial	in P	trivial	NP-c
co	NP-c	P-c	P-c	trivial	NP-c
pr	NP-c	Π_2^P-c	coNP-c	trivial	NP-c

ADF is given by (A, R, C) with $C = \{c_a : \bigwedge_{(b,a)\in R} \neg b \mid a \in A\}$. Notice that all the links are indeed attacking.

To compute $\Gamma_D(I)$ in general ADFs, we have to consider all 2-valued interpretations that extend I, which is coNP-hard, but given the link type of each link we only have to check two 2-valued interpretations for each argument as follows [Wallner, 2014; Strass and Wallner, 2014]. Let $Supp$ be the set of supporting links and Att the set of attacking links, but there might be links that are both supporting and attacking (such links are called redundant links). For $s \in S$ consider the the formula φ_s and the interpretations J_1, J_2 as follows.

1. $J_1(s) = \begin{cases} \mathbf{t} & \text{if } I(s) = \mathbf{t}, \text{ or } I(s) = \mathbf{u} \text{ and } (s, a) \in Supp \\ \mathbf{f} & \text{if } I(s) = \mathbf{f}, \text{ or } I(s) = \mathbf{u} \text{ and } (s, a) \in Att \setminus Supp \end{cases}$

2. $J_2(s) = \begin{cases} \mathbf{t} & \text{if } I(s) = \mathbf{t}, \text{ or } I(s) = \mathbf{u} \text{ and } (s, a) \in Att \\ \mathbf{f} & \text{if } I(s) = \mathbf{f}, \text{ or } I(s) = \mathbf{u} \text{ and } (s, a) \in Supp \setminus Att \end{cases}$

The interpretation J_1 sets all yet undecided supporters to true and all yet undecided (non-redundant) attackers to false. If $J_1(\varphi_s) = \mathbf{f}$ then no 2-valued interpretation extending I satisfies φ_s, and thus $\Gamma_D(I)(s) = \mathbf{f}$. Otherwise if $J_1(\varphi_s) = \mathbf{t}$ then clearly $\Gamma_D(I)(s) \neq \mathbf{f}$. The interpretation J_2 sets all yet undecided (non-redundant) supporters to false and all yet undecided attackers to true. Now, whenever $J_2(\varphi_s) = \mathbf{t}$ then all 2-valued interpretations extending I satisfy φ_s, and thus $\Gamma_D(I)(s) = \mathbf{t}$. Otherwise if $J_2(\varphi_s) = \mathbf{f}$ then clearly $\Gamma_D(I)(s) \neq \mathbf{t}$. Hence, we can compute $\Gamma_D(I)$ by just considering J_1 and J_2 and set $\Gamma_D(I)(s) = \mathbf{t}$ if $J_2(\varphi_s) = \mathbf{t}$; $\Gamma_D(I)(s) = \mathbf{f}$ if $J_1(\varphi_s) = \mathbf{f}$; and $\Gamma_D(I)(s) = \mathbf{u}$ otherwise.

Now, as $\Gamma_D(I)$ can be computed in polynomial time, we can also (i) efficiently compute the grounded model by iteratively applying $\Gamma_D(I)$. Moreover, (ii) verifying an admissible or complete interpretation just requires to apply the Γ_D operator once and thus can be done in polynomial time. The remaining results in Table 11 are by the combination of the polynomial-time verification algorithms with the standard guess and check algorithms as used for Dung's AFs.

6 Discussion

In this chapter we presented complexity results for the three argumentation formalisms of Dung's Abstract Argumentation Frameworks, Assumption-based Argumentation and Abstract Dialectical Frameworks. We have identified several sources of computational complexity: (i) the construction of arguments and the interlinking structure, e.g., the attack relation, (ii) the search for coherent sets of arguments, and (iii) the decision about certain conclusions. Points (i) and (iii) are present in the complexity results for Assumption-based Argumentation where the complexity of algorithms heavily depends on the complexity of the derivability relation, which is the essential ingredient to build arguments, identify conflicts, and draw conclusions. Point (ii) is present in all three formalisms. The discussed results show that the actual computational complexity in this step may highly depend on the chosen semantics and reasoning task. Moreover, faced with the typically high complexity we discussed approaches to identify instances with lower complexity and solve them more efficiently.

Implications for the Design of Systems and Algorithms. First given the upper bounds of the complexity analysis we have first guidelines how to implement argumentation semantics, and on the computational resources required for that. An efficient system should process any instance within the resources given by the upper bound but moreover also should perform better on easier instances. In particular an efficient system should also be able to process instances that fall into one of the tractable fragments with milder complexity more efficiently. A system for abstract argumentation that is explicitly built around this idea is CEGARTIX [Dvořák et al., 2014a], that is based on easier fragments for semantics on the second level of the polynomial-hierarchy.

The complexity classification of a semantics is also crucial for reduction-based implementations. To get an appropriate reduction the target formalism should have a similar complexity as the argumentation semantics, or one should only use a fragment of the target formalism with similar complexity. One example is the ASPARTIX [Egly et al., 2010] system that encodes abstract argumentation problems in logic-programming in a query-based fashion. That is, the system provides fixed encodings for the supported argumentation semantics (the queries) that are then evaluated on the encoding of considered AF (the input data).[22] The polynomial-time computable grounded semantics is encoded as stratified logic program, a fragment whose data-complexity is in polynomial time (even P-complete), the semantics at the NP, coNP level are encoded as programs without disjunction in the rule heads, the data-complexity of this kind of logic programs is on the NP / coNP level, and the full expressiveness of disjunctive logic programs is only used for the argumentation semantics whose

---

[22]The specific encoding of an AF as logic program has became popular beyond the logic programming setting as the so-called ASPARTIX-format for encoding AFs.

complexity is at the second level of the polynomial-hierarchy.[23]

Another example is the work on intertranslatability of abstract argumentation semantics where one aims to efficiently translate one argumentation semantics to another, by modifying the argumentation framework [Dvořák and Woltran, 2011; Dvořák and Spanring, 2016]. Here a gap in the complexities of the semantics immediately gives a negative result.

Function Complexity. In this chapter we restricted ourselves to what we consider to be the core computational problems and in particular to decision problems. In terms of computational complexity function problems, problems where one wants to compute a number, extensions or the set of extensions, are only rarely studied, notable exceptions are the research line on ideal semantics [Dunne, 2009; Dunne et al., 2013], the work on counting the number of extensions [Baroni et al., 2010], and the work on computing an admissible set that results in a minimal socratic discussion [Caminada et al., 2016]. Recently, Kröll et al. started the research on enumeration complexity in abstract argumentation [Kröll et al., 2017], where one is interested in the computational cost per extension.

Fine-Grained Lower Bounds. Lower bounds from classical computational complexity theory like NP-hardness indicate that there are no polynomial-time algorithms. However, they neither indicate lower bounds for the constants in the exponent of exponential running times nor rule out subexponential algorithms at all. That is, there is still some gap between the best known algorithms for the hard problems, they are exponential-time (see, e.g., [Nofal et al., 2014]), and the existing lower bounds. To overcome this gap, in the field of combinatorial algorithms, so called conditional lower bounds are studied (see, e.g., [Abboud and Williams, 2014]). That is, one uses conjectures about lower bounds for well studied algorithmic problems. To obtain a lower bound for a new problem one then reduces the problem from the conjecture to the problem under question such that a faster algorithm for the new problem would imply a faster algorithm for the original problem and thus would contradict the conjecture. By that one gets an algorithmic lower bound for the new problem conditioned on the original conjecture. Probably the most prominent such conjecture is the (strong) exponential-time hypothesis (S)ETH [Impagliazzo and Paturi, 1999], with ETH conjecturing that there is no subexponential algorithm for 3-SAT, and SETH conjecturing that there is no algorithm for CNF-SAT that runs in time $2^{(1-\epsilon)n} \cdot poly(n,m)$, for every constant $\epsilon > 0$ and polynomial $poly(n,m)$. As many of the existing reductions in formal argumentation are based on propositional logic (S)ETH is also a promising starting point for closing these complexity gaps in formal argumentation.

Complexity Analysis of further Argumentation Formalisms. In this chapter we only cover three argumentation formalisms, while there are many more around and many of them come with a complexity analysis. Below we give a brief overview and pointers to the relevant literature. First, there are

---

[23]For a survey on the complexity of logic-programs see [Dantsin et al., 2001].

formalisms, e.g. AFRAs [Baroni et al., 2011b], that extend Dung's Abstract argumentation frameworks and can be efficiently reduced to them. For such formalisms the complexity results for AFs directly extend to the new formalism. Second, there are extensions of AFs that can not be reduced in such a direct way and thus need their own complexity analysis. Most prominently: The complexity of Extended argumentation frameworks was studied in [Dunne et al., 2010] and later complemented by a result in [Dvořák et al., 2015]; Valued-based argumentation has been discussed in an earlier survey [Dunne and Wooldridge, 2009] on the complexity of abstract argumentation and more recent results can be found in, e.g., [Dunne, 2010; Kim et al., 2011]; weighted argumentation systems and their complexity have been studied in [Dunne et al., 2011]; and Constrained Argumentation Frameworks [Coste-Marquis et al., 2006]. Finally, there are complexity results for logic-based argumentation formalisms. Complexity aspects of Deductive Argumentation were for instance considered in [Besnard et al., 2009; Creignou et al., 2011], while the complexity of Defeasible Logic Programming (DeLP) was studied in [Cecchi et al., 2006].

Complexity Analysis in Formal Argumentation. At the current state of the field for most argumentation formalisms we already have a good understanding of the computational complexity of the fundamental problems and the important semantics. However, this by no means says that all research questions in that direction are solved. Indeed the field of formal argumentation is very active and with almost every new research topic there come associated computational problems that should be analysed w.r.t. their computational complexity. Let us exemplify three such occasion where a complexity analysis can deepen our understanding: (a) For a newly proposed semantics the complexity of the fundamental reasoning problems should be analysed in order to compare it with existing semantics and identified computational benefits/drawbacks. (b) When expanding existing argumentation formalisms with additional (syntactic) concepts one is interested in the (additional) computational costs of these concepts. That is by how much the complexity increases or whether one can add these concepts without any computational drawbacks. (c) When considering novel tasks for argumentation systems a complexity classification gives a first impression on the feasibility of the new task and guides the way to efficient implementations. For instance, recently the field of dynamics of argumentation received some attention [Diller et al., 2015; Snaith and Reed, 2016; Wallner et al., 2016; Kim et al., 2013] and raised a couple of computational problems, e.g., the so-called extension enforcement problem [Baumann and Brewka, 2010] where one aims to modify an AF such that a certain set of arguments becomes acceptable. The work of [Wallner et al., 2016] first gives a comprehensive complexity analysis of the enforcement problem and then turns these results into algorithms and the prototype system Pakota [24].

---

[24] https://www.cs.helsinki.fi/group/coreo/pakota/

Acknowledgments

The authors are grateful to Hannes Strass, Johannes P. Wallner and an anonymous reviewer for their thoughtful comments on earlier versions of this chapter which helped to improve the quality of presentation.

BIBLIOGRAPHY

[Abboud and Williams, 2014] Amir Abboud and Virginia Vassilevska Williams. Popular conjectures imply strong lower bounds for dynamic problems. In *55th IEEE Annual Symposium on Foundations of Computer Science, FOCS 2014, Philadelphia, PA, USA, October 18-21, 2014*, pages 434–443. IEEE Computer Society, 2014.

[Arieli and Caminada, 2012] Ofer Arieli and Martin W. A. Caminada. A general QBF-based formalization of abstract argumentation theory. In Bart Verheij, Stefan Szeider, and Stefan Woltran, editors, *Computational Models of Argument: Proceedings of COMMA 2012, Vienna, Austria, September 10-12, 2012*, volume 245 of *Frontiers in Artificial Intelligence and Applications*, pages 105–116. IOS Press, 2012.

[Arora and Barak, 2009] Sanjeev Arora and Boaz Barak. *Computational Complexity: A Modern Approach*. Cambridge University Press, New York, NY, USA, 1st edition, 2009.

[Baroni et al., 2005] Pietro Baroni, Massimiliano Giacomin, and Giovanni Guida. SCC-recursiveness: A general schema for argumentation semantics. *Artif. Intell.*, 168(1-2):162–210, 2005.

[Baroni et al., 2010] Pietro Baroni, Paul E. Dunne, and Massimiliano Giacomin. On extension counting problems in argumentation frameworks. In Pietro Baroni, Federico Cerutti, Massimiliano Giacomin, and Guillermo Ricardo Simari, editors, *Computational Models of Argument: Proceedings of COMMA 2010, Desenzano del Garda, Italy, September 8-10, 2010*, volume 216 of *Frontiers in Artificial Intelligence and Applications*, pages 63–74. IOS Press, 2010.

[Baroni et al., 2011a] Pietro Baroni, Martin Caminada, and Massimiliano Giacomin. An introduction to argumentation semantics. *Knowledge Eng. Review*, 26(4):365–410, 2011.

[Baroni et al., 2011b] Pietro Baroni, Federico Cerutti, Massimiliano Giacomin, and Giovanni Guida. AFRA: Argumentation framework with recursive attacks. *Int. J. Approx. Reasoning*, 52(1):19–37, 2011.

[Baroni et al., 2011c] Pietro Baroni, Paul E. Dunne, and Massimiliano Giacomin. On the resolution-based family of abstract argumentation semantics and its grounded instance. *Artif. Intell.*, 175(3-4):791–813, 2011.

[Baumann and Brewka, 2010] Ringo Baumann and Gerhard Brewka. Expanding argumentation frameworks: Enforcing and monotonicity results. In Pietro Baroni, Federico Cerutti, Massimiliano Giacomin, and Guillermo Ricardo Simari, editors, *Computational Models of Argument: Proceedings of COMMA 2010, Desenzano del Garda, Italy, September 8-10, 2010*, volume 216 of *Frontiers in Artificial Intelligence and Applications*, pages 75–86. IOS Press, 2010.

[Besnard and Doutre, 2004] Philippe Besnard and Sylvie Doutre. Checking the acceptability of a set of arguments. In James P. Delgrande and Torsten Schaub, editors, *Proceedings of the 10th International Workshop on Non-Monotonic Reasoning (NMR 2004)*, pages 59–64, 2004.

[Besnard et al., 2009] Philippe Besnard, Anthony Hunter, and Stefan Woltran. Encoding deductive argumentation in quantified boolean formulae. *Artificial Intelligence*, 173(15):1406–1423, 2009.

[Bondarenko et al., 1997] Andrei Bondarenko, Phan Minh Dung, Robert A. Kowalski, and Francesca Toni. An abstract, argumentation-theoretic approach to default reasoning. *Artif. Intell.*, 93:63–101, 1997.

[Brewka and Gordon, 2010] Gerhard Brewka and Thomas F. Gordon. Carneades and abstract dialectical frameworks: A reconstruction. In Pietro Baroni, Federico Cerutti, Massimiliano Giacomin, and Guillermo Ricardo Simari, editors, *Computational Models of Argument: Proceedings of COMMA 2010, Desenzano del Garda, Italy, September 8-10, 2010*, volume 216 of *Frontiers in Artificial Intelligence and Applications*, pages 3–12. IOS Press, 2010.

[Brewka and Woltran, 2010] Gerhard Brewka and Stefan Woltran. Abstract dialectical frameworks. In Fangzhen Lin, Ulrike Sattler, and Miroslaw Truszczynski, editors, *Principles of Knowledge Representation and Reasoning: Proceedings of the Twelfth International Conference, KR 2010, Toronto, Ontario, Canada, May 9-13, 2010*, pages 780–785. AAAI Press, 2010.

[Brewka et al., 2013] Gerhard Brewka, Hannes Strass, Stefan Ellmauthaler, Johannes Peter Wallner, and Stefan Woltran. Abstract dialectical frameworks revisited. In Francesca Rossi, editor, *IJCAI 2013, Proceedings of the 23rd International Joint Conference on Artificial Intelligence, Beijing, China, August 3-9, 2013*, pages 803–809. IJCAI/AAAI, 2013.

[Caminada and Gabbay, 2009] Martin Caminada and Dov M. Gabbay. A logical account of formal argumentation. *Studia Logica*, 93(2):109–145, 2009.

[Caminada et al., 2012] Martin Caminada, Walter A. Carnielli, and Paul E. Dunne. Semi-stable semantics. *J. Log. Comput.*, 22:1207–1254, 2012.

[Caminada et al., 2016] Martin Caminada, Wolfgang Dvořák, and Srdjan Vesic. Preferred semantics as socratic discussion. *J. Log. Comput.*, 26:1257–1292, 2016.

[Caminada, 2007] Martin Caminada. Comparing two unique extension semantics for formal argumentation: ideal and eager. In *Proceedings of the 19th Belgian-Dutch Conference on Artificial Intelligence (BNAIC 2007)*, pages 81–87, 2007.

[Cecchi et al., 2006] Laura A. Cecchi, Pablo R. Fillottrani, and Guillermo R. Simari. On the complexity of DeLP through game semantics. In *11th. Intl. Workshop on Nonmonotonic Reasoning*, pages 386–394, 2006.

[Cerutti et al., 2014] Federico Cerutti, Massimiliano Giacomin, Mauro Vallati, and Marina Zanella. An SCC recursive meta-algorithm for computing preferred labellings in abstract argumentation. In Chitta Baral, Giuseppe De Giacomo, and Thomas Eiter, editors, *Principles of Knowledge Representation and Reasoning: Proceedings of the Fourteenth International Conference, KR 2014, Vienna, Austria, July 20-24, 2014*, pages 42–51. AAAI Press, 2014.

[Charwat, 2012] Günther Charwat. Tree-decomposition based algorithms for abstract argumentation frameworks. Master's thesis, Vienna University of Technology, 2012. Stefan Woltran and Wolfgang Dvořák advisors.

[Chen et al., 2008] Jianer Chen, Yang Liu, Songjian Lu, Barry O'Sullivan, and Igor Razgon. A fixed-parameter algorithm for the directed feedback vertex set problem. *J. ACM*, 55(5):21:1–21:19, 2008.

[Coste-Marquis et al., 2005] Sylvie Coste-Marquis, Caroline Devred, and Pierre Marquis. Symmetric argumentation frameworks. In Lluis Godo, editor, *Proceedings of the 8th European Conference on Symbolic and Quantitative Approaches to Reasoning with Uncertainty (ECSQARU 2005)*, volume 3571 of *Lecture Notes in Computer Science*, pages 317–328. Springer, 2005.

[Coste-Marquis et al., 2006] Sylvie Coste-Marquis, Caroline Devred, and Pierre Marquis. Constrained argumentation frameworks. In Patrick Doherty, John Mylopoulos, and Christopher A. Welty, editors, *Proceedings, Tenth International Conference on Principles of Knowledge Representation and Reasoning, Lake District of the United Kingdom, June 2-5, 2006*, pages 112–122. AAAI Press, 2006.

[Creignou et al., 2011] Nadia Creignou, Johannes Schmidt, Michael Thomas, and Stefan Woltran. Complexity of logic-based argumentation in Post's framework. *Argument & Computation*, 2(2-3):107–129, 2011.

[Cygan et al., 2015] Marek Cygan, Fedor V. Fomin, Lukasz Kowalik, Daniel Lokshtanov, Dániel Marx, Marcin Pilipczuk, Michal Pilipczuk, and Saket Saurabh. *Parameterized Algorithms*. Springer, 2015.

[Dantsin et al., 2001] Evgeny Dantsin, Thomas Eiter, Georg Gottlob, and Andrei Voronkov. Complexity and expressive power of logic programming. *ACM Comput. Surv.*, 33(3):374–425, 2001.

[Diller et al., 2015] Martin Diller, Adrian Haret, Thomas Linsbichler, Stefan Rümmele, and Stefan Woltran. An extension-based approach to belief revision in abstract argumentation. In Qiang Yang and Michael Wooldridge, editors, *Proceedings of the Twenty-Fourth International Joint Conference on Artificial Intelligence, IJCAI 2015, Buenos Aires, Argentina, July 25-31, 2015*, pages 2926–2932. AAAI Press, 2015.

[Dimopoulos and Torres, 1996] Yannis Dimopoulos and Alberto Torres. Graph theoretical structures in logic programs and default theories. *Theor. Comput. Sci.*, 170(1-2):209–244, 1996.

[Dimopoulos et al., 1999] Yannis Dimopoulos, Bernhard Nebel, and Francesca Toni. Preferred arguments are harder to compute than stable extension. In Thomas Dean, editor, *Proceedings of the Sixteenth International Joint Conference on Artificial Intelligence, IJCAI 99, Stockholm, Sweden, July 31 - August 6, 1999. 2 Volumes, 1450 pages*, pages 36–43. Morgan Kaufmann, 1999.

[Dimopoulos et al., 2000] Yannis Dimopoulos, Bernhard Nebel, and Francesca Toni. Finding admissible and preferred arguments can be very hard. In Anthony G. Cohn, Fausto Giunchiglia, and Bart Selman, editors, *KR 2000, Principles of Knowledge Representation and Reasoning Proceedings of the Seventh International Conference, Breckenridge, Colorado, USA, April 11-15, 2000.*, pages 53–61. Morgan Kaufmann, 2000.

[Dimopoulos et al., 2002] Yannis Dimopoulos, Bernhard Nebel, and Francesca Toni. On the computational complexity of assumption-based argumentation for default reasoning. *Artif. Intell.*, 141(1/2):57–78, 2002.

[Dung et al., 2006] Phan Minh Dung, Paolo Mancarella, and Francesca Toni. A dialectic procedure for sceptical, assumption-based argumentation. In Paul E. Dunne and Trevor J. M. Bench-Capon, editors, *Computational Models of Argument: Proceedings of COMMA 2006, September 11-12, 2006, Liverpool, UK*, volume 144 of *Frontiers in Artificial Intelligence and Applications*, pages 145–156. IOS Press, 2006.

[Dung et al., 2007] Phan Minh Dung, Paolo Mancarella, and Francesca Toni. Computing ideal sceptical argumentation. *Artif. Intell.*, 171(10-15):642–674, 2007.

[Dung, 1995] Phan Minh Dung. On the acceptability of arguments and its fundamental role in nonmonotonic reasoning, logic programming and n-person games. *Artif. Intell.*, 77(2):321–358, 1995.

[Dunne and Bench-Capon, 2001] Paul E. Dunne and Trevor J. M. Bench-Capon. Complexity and combinatorial properties of argument systems. Technical report, Dept. of Computer Science, University of Liverpool, 2001.

[Dunne and Bench-Capon, 2002] Paul E. Dunne and Trevor J. M. Bench-Capon. Coherence in finite argument systems. *Artif. Intell.*, 141(1/2):187–203, 2002.

[Dunne and Wooldridge, 2009] Paul E. Dunne and Michael Wooldridge. Complexity of abstract argumentation. In Guillermo Simari and Iyad Rahwan, editors, *Argumentation in Artificial Intelligence*, pages 85–104. Springer US, 2009.

[Dunne et al., 2010] Paul E. Dunne, Sanjay Modgil, and Trevor J. M. Bench-Capon. Computation in extended argumentation frameworks. In *ECAI 2010 - 19th European Conference on Artificial Intelligence, Lisbon, Portugal, August 16-20, 2010, Proceedings*, volume 215 of *Frontiers in Artificial Intelligence and Applications*, pages 119–124. IOS Press, 2010.

[Dunne et al., 2011] Paul E. Dunne, Anthony Hunter, Peter McBurney, Simon Parsons, and Michael Wooldridge. Weighted argument systems: Basic definitions, algorithms, and complexity results. *Artif. Intell.*, 175(2):457–486, 2011.

[Dunne et al., 2013] Paul E. Dunne, Wolfgang Dvořák, and Stefan Woltran. Parametric properties of ideal semantics. *Artif. Intell.*, 202(0):1 – 28, 2013.

[Dunne, 2007] Paul E. Dunne. Computational properties of argument systems satisfying graph-theoretic constraints. *Artif. Intell.*, 171(10-15):701–729, 2007.

[Dunne, 2008] Paul E. Dunne. The computational complexity of ideal semantics I: Abstract argumentation frameworks. In Philippe Besnard, Sylvie Doutre, and Anthony Hunter, editors, *Computational Models of Argument: Proceedings of COMMA 2008, Toulouse, France, May 28-30, 2008*, volume 172 of *Frontiers in Artificial Intelligence and Applications*, pages 147–158. IOS Press, 2008.

[Dunne, 2009] Paul E. Dunne. The computational complexity of ideal semantics. *Artif. Intell.*, 173(18):1559–1591, 2009.

[Dunne, 2010] Paul E. Dunne. Tractability in value-based argumentation. In Pietro Baroni, Federico Cerutti, Massimiliano Giacomin, and Guillermo Ricardo Simari, editors, *Computational Models of Argument: Proceedings of COMMA 2010, Desenzano del Garda, Italy, September 8-10, 2010*, volume 216 of *Frontiers in Artificial Intelligence and Applications*, pages 195–206. IOS Press, 2010.

[Dvořák and Gaggl, 2016] Wolfgang Dvořák and Sarah Alice Gaggl. Stage semantics and the SCC-recursive schema for argumentation semantics. *J. Log. Comput.*, 26(4):1149–1202, 2016.
[Dvořák and Spanring, 2016] Wolfgang Dvořák and Christof Spanring. Comparing the expressiveness of argumentation semantics. *J. Log. Comput.*, in press (available online), 2016.
[Dvořák and Woltran, 2010] Wolfgang Dvořák and Stefan Woltran. Complexity of semi-stable and stage semantics in argumentation frameworks. *Inf. Process. Lett.*, 110(11):425–430, 2010.
[Dvořák and Woltran, 2011] Wolfgang Dvořák and Stefan Woltran. On the intertranslatability of argumentation semantics. *J. Artif. Intell. Res. (JAIR)*, 41:445–475, 2011.
[Dvořák et al., 2010] Wolfgang Dvořák, Stefan Szeider, and Stefan Woltran. Reasoning in argumentation frameworks of bounded clique-width. In Pietro Baroni, Federico Cerutti, Massimiliano Giacomin, and Guillermo Ricardo Simari, editors, *Computational Models of Argument: Proceedings of COMMA 2010, Desenzano del Garda, Italy, September 8-10, 2010*, volume 216 of *Frontiers in Artificial Intelligence and Applications*, pages 219–230. IOS Press, 2010.
[Dvořák et al., 2012a] Wolfgang Dvořák, Sebastian Ordyniak, and Stefan Szeider. Augmenting tractable fragments of abstract argumentation. *Artif. Intell.*, 186(0):157–173, 2012.
[Dvořák et al., 2012b] Wolfgang Dvořák, Reinhard Pichler, and Stefan Woltran. Towards fixed-parameter tractable algorithms for abstract argumentation. *Artif. Intell.*, 186(0):1 – 37, 2012.
[Dvořák et al., 2012c] Wolfgang Dvořák, Stefan Szeider, and Stefan Woltran. Abstract argumentation via monadic second order logic. In Eyke Hüllermeier, Sebastian Link, Thomas Fober, and Bernhard Seeger, editors, *Scalable Uncertainty Management - 6th International Conference, SUM 2012, Marburg, Germany, September 17-19, 2012. Proceedings*, volume 7520 of *Lecture Notes in Computer Science*, pages 85–98. Springer, 2012.
[Dvořák et al., 2014a] Wolfgang Dvořák, Matti Järvisalo, Johannes Peter Wallner, and Stefan Woltran. Complexity-sensitive decision procedures for abstract argumentation. *Artif. Intell.*, 206(0):53 – 78, 2014.
[Dvořák et al., 2014b] Wolfgang Dvořák, Thomas Linsbichler, Emilia Oikarinen, and Stefan Woltran. Resolution-based grounded semantics revisited. In Simon Parsons, Nir Oren, Chris Reed, and Federico Cerutti, editors, *Computational Models of Argument: Proceedings of COMMA 2014, Atholl Palace Hotel, Scottish Highlands, UK, September 9-12, 2014*, volume 266 of *Frontiers in Artificial Intelligence and Applications*, pages 269–280. IOS Press, 2014.
[Dvořák et al., 2015] Wolfgang Dvořák, Sarah Alice Gaggl, Thomas Linsbichler, and Johannes Peter Wallner. Reduction-based approaches to implement Modgil's extended argumentation frameworks. In Thomas Eiter, Hannes Strass, Miroslaw Truszczynski, and Stefan Woltran, editors, *Advances in Knowledge Representation, Logic Programming, and Abstract Argumentation - Essays Dedicated to Gerhard Brewka on the Occasion of His 60th Birthday*, volume 9060 of *Lecture Notes in Computer Science*, pages 249–264. Springer, 2015.
[Dvořák, 2012a] Wolfgang Dvořák. *Computational Aspects of Abstract Argumentation*. PhD thesis, Vienna University of Technology, Institute of Information Systems, 2012.
[Dvořák, 2012b] Wolfgang Dvořák. On the complexity of computing the justification status of an argument. In Sanjay Modgil, Nir Oren, and Francesca Toni, editors, *Theory and Applications of Formal Argumentation - First International Workshop, TAFA 2011. Barcelona, Spain, July 16-17, 2011, Revised Selected Papers*, volume 7132 of *Lecture Notes in Computer Science*, pages 32–49. Springer, 2012.
[Dvořák, 2012c] Wolfgang Dvořák. Technical note: Exploring Σ_2^P / Π_2^P -hardness for argumentation problems with fixed distance to tractable classes. *CoRR*, abs/1201.0478, 2012.
[Dvořák, 2017] Wolfgang Dvořák. Technical note: On the complexity of the uniqueness problem in abstract argumentation. Technical Report DBAI-TR-2008-xy, Technische Universität Wien, 2017.
[Egly and Woltran, 2006] Uwe Egly and Stefan Woltran. Reasoning in argumentation frameworks using quantified boolean formulas. In Paul E. Dunne and Trevor J. M. Bench-Capon, editors, *Computational Models of Argument: Proceedings of COMMA 2006, September*

11-12, 2006, Liverpool, UK, volume 144 of *Frontiers in Artificial Intelligence and Applications*, pages 133–144. IOS Press, 2006.

[Egly et al., 2010] Uwe Egly, Sarah Alice Gaggl, and Stefan Woltran. Answer-set programming encodings for argumentation frameworks. *Argument and Computation*, 1(2):147–177, 2010.

[Eiter and Gottlob, 1993] Thomas Eiter and Georg Gottlob. Propositional circumscription and extended closed-world reasoning are Π_2^P-complete. *Theor. Comput. Sci.*, 114(2):231–245, 1993.

[Ellmauthaler, 2012] Stefan Ellmauthaler. Abstract dialectical frameworks: Properties, complexity, and implementation. Master's thesis, Vienna University of Technology, 2012. Stefan Woltran and Johannes Peter Wallner advisors.

[Flum and Grohe, 2006] Jörg Flum and Martin Grohe. *Parameterized Complexity Theory*. Texts in Theoretical Computer Science. An EATCS Series. Springer, 2006.

[Gaggl and Woltran, 2013] Sarah Alice Gaggl and Stefan Woltran. The cf2 argumentation semantics revisited. *J. Log. Comput.*, 23(5):925–949, 2013.

[Gelfond and Lifschitz, 1988] Michael Gelfond and Vladimir Lifschitz. The stable model semantics for logic programming. In Robert A. Kowalski and Kenneth A. Bowen, editors, *Logic Programming, Proceedings of the Fifth International Conference and Symposium, Seattle, Washington, August 15-19, 1988 (2 Volumes)*, pages 1070–1080. MIT Press, 1988.

[Greenlaw et al., 1995] Raymond Greenlaw, H.James Hoover, and Walter L. Ruzzo. *Limits to parallel computation: P-completeness theory*. Oxford University Press, 1995.

[Impagliazzo and Paturi, 1999] Russell Impagliazzo and Ramamohan Paturi. Complexity of k-sat. In *Proceedings of the 14th Annual IEEE Conference on Computational Complexity, Atlanta, Georgia, USA, May 4-6, 1999*, pages 237–240. IEEE Computer Society, 1999.

[Kasif, 1986] Simon Kasif. On the parallel complexity of some constraint satisfaction problems. In Tom Kehler, editor, *Proceedings of the 5th National Conference on Artificial Intelligence. Philadelphia, PA, August 11-15, 1986. Volume 1: Science*, pages 349–353. Morgan Kaufmann, 1986.

[Kim et al., 2011] Eun Jung Kim, Sebastian Ordyniak, and Stefan Szeider. Algorithms and complexity results for persuasive argumentation. *Artif. Intell.*, 175(9-10):1722–1736, 2011.

[Kim et al., 2013] Eun Jung Kim, Sebastian Ordyniak, and Stefan Szeider. The complexity of repairing, adjusting, and aggregating of extensions in abstract argumentation. In Elizabeth Black, Sanjay Modgil, and Nir Oren, editors, *Theory and Applications of Formal Argumentation - Second International Workshop, TAFA 2013, Beijing, China, August 3-5, 2013, Revised Selected papers*, volume 8306 of *Lecture Notes in Computer Science*, pages 158–175. Springer, 2013.

[Kleene, 1952] S.C. Kleene. *Introduction to Metamathematics*. Bibliotheca Mathematica. North-Holland, 1952.

[Kröll et al., 2017] Markus Kröll, Reinhard Pichler, and Stefan Woltran. On the complexity of enumerating the extensions of abstract argumentation frameworks. In *IJCAI 2017 (to appear)*, 2017.

[Moore, 1985] Robert C. Moore. Semantical considerations on nonmonotonic logic. *Artif. Intell.*, 25(1):75–94, 1985.

[Niedermeier, 2006] Rolf Niedermeier. *Invitation to fixed-parameter algorithms*, volume 31. Oxford University Press, USA, 2006.

[Nieves et al., 2009] Juan Carlos Nieves, Mauricio Osorio, and Claudia Zepeda. Expressing extension-based semantics based on stratified minimal models. In Hiroakira Ono, Makoto Kanazawa, and Ruy J. G. B. de Queiroz, editors, *Logic, Language, Information and Computation, 16th International Workshop, WoLLIC 2009, Tokyo, Japan, June 21-24, 2009. Proceedings*, volume 5514 of *Lecture Notes in Computer Science*, pages 305–319. Springer, 2009.

[Nofal et al., 2014] Samer Nofal, Katie Atkinson, and Paul E. Dunne. Algorithms for decision problems in argument systems under preferred semantics. *Artif. Intell.*, 207:23–51, 2014.

[Ordyniak and Szeider, 2011] Sebastian Ordyniak and Stefan Szeider. Augmenting tractable fragments of abstract argumentation. In Toby Walsh, editor, *IJCAI 2011, Proceedings of the 22nd International Joint Conference on Artificial Intelligence, Barcelona, Catalonia, Spain, July 16-22, 2011*, pages 1033–1038. IJCAI/AAAI, 2011.

[Papadimitriou, 1994] Christos H. Papadimitriou. *Computational Complexity*. Addison-Wesley, Reading, Massachusetts, 1994.
[Reiter, 1980] Raymond Reiter. A logic for default reasoning. *Artif. Intell.*, 13(1-2):81–132, 1980.
[Robertson et al., 1999] Neil Robertson, P. D. Seymour, and Robin Thomas. Permanents, Pfaffian orientations, and even directed circuits. *Ann. of Math. (2)*, 150(3):929–975, 1999.
[Snaith and Reed, 2016] Mark Snaith and Chris Reed. Argument revision. *J. Log. Comput.*, in press (available online), 2016.
[Strass and Wallner, 2014] Hannes Strass and Johannes P. Wallner. Analyzing the Computational Complexity of Abstract Dialectical Frameworks via Approximation Fixpoint Theory. In Chitta Baral, Giuseppe De Giacomo, and Thomas Eiter, editors, *Proceedings of the 14th International Conference on Principles of Knowledge Representation and Reasoning, KR 2014*, pages 101–110. AAAI Press, 2014.
[Strass and Wallner, 2015] Hannes Strass and Johannes Peter Wallner. Analyzing the computational complexity of abstract dialectical frameworks via approximation fixpoint theory. *Artif. Intell.*, 226:34–74, 2015.
[Strass, 2013] Hannes Strass. Instantiating knowledge bases in abstract dialectical frameworks. In João Leite, Tran Cao Son, Paolo Torroni, Leon van der Torre, and Stefan Woltran, editors, *Computational Logic in Multi-Agent Systems - 14th International Workshop, CLIMA XIV, Corunna, Spain, September 16-18, 2013. Proceedings*, volume 8143 of *Lecture Notes in Computer Science*, pages 86–101. Springer, 2013.
[Thimm and Villata, 2015] Matthias Thimm and Serena Villata. System descriptions of the first international competition on computational models of argumentation (iccma'15). *CoRR*, abs/1510.05373, 2015.
[Thimm et al., 2016] Matthias Thimm, Serena Villata, Federico Cerutti, Nir Oren, Hannes Strass, and Mauro Vallati. Summary report of the first international competition on computational models of argumentation. *AI Magazine*, 37(1):102, 2016.
[Toni, 2014] Francesca Toni. A tutorial on assumption-based argumentation. *Argument & Computation*, 5(1):89–117, 2014.
[Verheij, 1996] Bart Verheij. Two approaches to dialectical argumentation: admissible sets and argumentation stages. In J. Meyer and L. van der Gaag, editors, *Proceedings of the 8th Dutch Conference on Artificial Intelligence (NAIC'96)*, pages 357–368, 1996.
[Wallner et al., 2016] Johannes Peter Wallner, Andreas Niskanen, and Matti Järvisalo. Complexity results and algorithms for extension enforcement in abstract argumentation. In Dale Schuurmans and Michael P. Wellman, editors, *Proceedings of the Thirtieth AAAI Conference on Artificial Intelligence, February 12-17, 2016, Phoenix, Arizona, USA.*, pages 1088–1094. AAAI Press, 2016.
[Wallner, 2014] Johannes P. Wallner. *Complexity Results and Algorithms for Argumentation - Dung's Frameworks and Beyond*. PhD thesis, Vienna University of Technology, Institute of Information Systems, 2014.
[Wu and Caminada, 2010] Yining Wu and Martin Caminada. A labelling-based justification status of arguments. *Studies in Logic*, 3(4):12–29, 2010.

Wolfgang Dvořák
Institute of Information Systems
TU Wien
Vienna, Austria
Email: dvorak@dbai.tuwien.ac.at

Paul E. Dunne
Department of Computer Science
University of Liverpool
Liverpool, United Kingdom
Email: ped@csc.liv.ac.uk

14
Foundations of Implementations for Formal Argumentation

FEDERICO CERUTTI, SARAH A. GAGGL, MATTHIAS THIMM, JOHANNES P. WALLNER

ABSTRACT. We survey the current state of the art of general techniques, as well as specific software systems for solving tasks in abstract argumentation frameworks, structured argumentation frameworks, and approaches for visualizing and analysing argumentation. Furthermore, we discuss challenges and promising techniques such as parallel processing and approximation approaches. Finally, we address the issue of evaluating software systems empirically with links to the International Competition on Computational Models of Argumentation.

1 Introduction

Compared to related areas such as argumentation theory [van Eemeren *et al.*, 2014], research conducted in the formal argumentation community seeks *formal* accounts of argumentation with explicit links to knowledge representation and reasoning, and artificial intelligence [Brachman and Levesque, 2004; Russell and Norvig, 2003]. An important feature for these accounts is *computability*, i.e., the possibility to provide algorithmic methods to solve problems.

In this chapter, we survey general computational techniques and concrete implementations for solving problems related to formal argumentation. We distinguish between: (1) Approaches to abstract argumentation frameworks, (2) Approaches to structured argumentation frameworks (such as ASPIC+ and DeLP), and (3) Other approaches, including semi-formal systems related to visualization of argumentation processes or exchange of arguments on the web.

Between them, the most active research direction within the formal argumentation community[1] is devoted to the first category—algorithms and systems for abstract argumentation frameworks—reviewed in Section 2. The relevant computational problems and their (high) computational complexity

---

[1] Approaches in the third category are also addressed by other research communities such as human-computer-interaction and web science.

have been already introduced in Chapter 13. Here, we focus on the algorithmic issues and techniques to handle the high computational complexity of some of those problems. The development of implementations has accelerated recently, also due to the foundation of the *International Competition on Computational Models of Argumentation* (ICCMA):[2] besides discussing general techniques we will also survey concrete systems.

We will also look at techniques and systems solving problems for structured approaches to formal argumentation. Due to the multitude of different approaches to structured argumentation, computational techniques and algorithms are usually tailored towards specific approaches. We will discuss them in Section 3 following the presentation in Chapters 6, 7, 8, and 9.

In order to complement our survey we will also have a brief look at other systems that incorporate some kind of (semi-)formal argumentation such as argument schemes and argumentation technologies (or *debating technologies*) which are popular in many other fields besides the formal argumentation community. In contrast to the perspective of artificial intelligence and knowledge representation usually taken by researchers in the formal argumentation community, the focus of the systems in this third category is on human-computer interaction and supporting critical thinking. We will discuss these systems in Section 4, concluding the survey part of this chapter.

In Section 5 we will look beyond the current state of the art of algorithms and systems and current challenges for the development of systems, such as parallelization and approximation algorithms, focusing on abstract and structured argumentation approaches. A recent effort to promote the development of systems for solving argumentation tasks is the ICCMA: the first instance of the competition took place in 2015 [Thimm et al., 2016]. We will discuss this competition and general methods for empirically evaluating systems in Section 6.

2 Abstract Argumentation Implementations

In this section we will give an overview of implementations for abstract Argumentation Frameworks (AFs) following the approach from Dung [Dung, 1995] (see Chapter 4) and give an overview of existing systems for Dung's framework as well as for some related formalisms.

One can divide the implementations for abstract AFs into two categories: the *reduction-based approach* and the *direct approach*. The former one reduces the problem at hand into another formalism to exploit existing solvers from the other formalism. We will discuss this method and the dedicated

---

[2]http://argumentationcompetition.org (on 27/04/2017).

implementations in the following subsection. The other possibility is to design algorithms to directly solve the problem. This implementation method will be presented in Subsection 2.2. For a more detailed discussion on implementation methods for AFs we refer to [Charwat et al., 2015].

Before we go into details on the different approaches we briefly introduce the background on abstract argumentation [Dung, 1995] and the notation we will use in this section. For comprehensive surveys on argumentation semantics the interested reader is referred to [Baroni et al., 2011a] (see also Chapter 4).

Definition 2.1 *An argumentation framework (AF) is a pair $AF = \langle Ar, att \rangle$, where Ar is a finite set of arguments and $att \subseteq Ar \times Ar$ is the attack relation. The pair $\langle a, b \rangle \in Ar$ means that a attacks b. A set $S \subseteq Ar$ of arguments attacks b (in AF), if there is an $a \in S$, such that $\langle a, b \rangle \in att$. An argument $a \in Ar$ is defended by $S \subseteq Ar$ (in AF) iff, for each $b \in Ar$, it holds that, if $\langle b, a \rangle \in att$, then S attacks b (in AF). Given a set $S \subseteq Ar$, $S^+ = \{a \in Ar \mid \langle b, a \rangle \in att, b \in S\}$, and $S^- = \{a \in Ar \mid \langle a, b \rangle \in att, b \in S\}$.*

The inherent conflicts between the arguments are solved by selecting subsets of arguments, where a semantics σ assigns a collection of sets of arguments to an argumentation framework AF. The basic requirement for all semantics is that none of the selected arguments attack each other[3].

Definition 2.2 *Let $AF = \langle Ar, att \rangle$ be an AF. A set $S \subseteq Ar$ is said to be conflict-free (in AF), if there are no $a, b \in S$, such that $\langle a, b \rangle \in att$. We denote the collection of sets which are conflict-free (in AF) by $cf(F)$.*

Definition 2.3 *Let $AF = \langle Ar, att \rangle$ be an AF, then $S \in cf(AF)$ is*

- *a stable extension, i.e. $S \in \mathcal{E}_{ST}(AF)$, if each $a \in Ar \setminus S$ is attacked by S in AF;*

- *an admissible extension, i.e. $S \in \mathcal{E}_{AD}(AF)$, if each $a \in S$ is defended by S;*

- *a preferred extension, i.e. $S \in \mathcal{E}_{PR}(AF)$, if $S \in \mathcal{E}_{AD}(AF)$ and for each $T \in \mathcal{E}_{AD}(AF)$, $S \not\subset T$;*

---

[3]We concentrate here on the basic Dung-style argumentation framework, and do not consider approaches like value-based argumentation frameworks (VAFs) [Bench-Capon, 2003] or inconsistency tolerant semantics [Dunne et al., 2009] (where this requirement does not hold), as our main focus is on implementation methods.

- a complete *extension*, i.e. $S \in \mathcal{E}_{\mathcal{CO}}(AF)$, if $S \in \mathcal{E}_{\mathcal{AD}}(AF)$ and for each $a \in Ar$ defended by S it holds that $a \in S$;

- the grounded *extension (of AF)*, i.e. the unique set $S = \mathcal{E}_{\mathcal{GR}}(AF)$, if $S \in \mathcal{E}_{\mathcal{CO}}(AF)$ and for each $T \in \mathcal{E}_{\mathcal{CO}}(AF)$, $T \not\subset S$.

The typical problems of interest in abstract argumentation are the following decision problems for given $AF = \langle Ar, att \rangle$, a semantics σ, $a \in Ar$ and $S \subseteq Ar$:

- Verification Ver_σ: is $S \in \mathcal{E}_\sigma(AF)$?

- Credulous acceptance $Cred_\sigma$: is a contained in at least one σ extension of AF?

- Skeptical acceptance $Skept_\sigma$: is a contained in every σ extension of AF?

- Non-emptiness $Exists_\sigma^{\neg\emptyset}$: is there any $S \in \mathcal{E}_\sigma(AF)$ for which $S \neq \emptyset$?

Computational complexity of decision problems on AFs is well-studied. For an overview see Chapter 13.

2.1 Reduction-based Implementations

Reduction-based implementations are a very common approach as one benefits from very sophisticated solvers developed and improved by several communities. The underlying idea is to exploit existing efficient software which has originally been developed for other purposes. To this end, one has to formalize the reasoning problems within other formalisms such as constraint-satisfaction problems (CSP) [Rossi et al., 2006], propositional logic [Biere et al., 2009] or answer-set programming (ASP) [Brewka et al., 2011]. The general methodology of the reduction-based approach is to reduce the problem at hand to the target formalism, run the solver (of the target formalism) and interpret the output as the solutions of the original problem, as depicted in Figure 1.

2.1.1 SAT-based Approach

Reductions to SAT have been first advocated in [Dunne and Bench-Capon, 2002] and [Dunne and Bench-Capon, 2003] and then further developed by Besnard and Doutre [Besnard and Doutre, 2004], and later extended by means of quantified propositional logic [Arieli and Caminada, 2013; Egly and Woltran, 2006]. Several prominent systems use reductions to SAT, such as **Cegartix** [Dvořák et al., 2014] and **{j}ArgSemSAT** [Cerutti et al., 2014c; Cerutti et al., 2016b; Cerutti et al., 2017] that both rely on

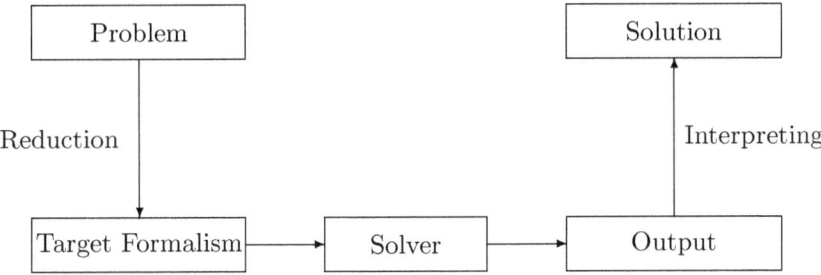

Figure 1. Reduction-based approach.

iterative calls to SAT solvers for argumentation semantics of high complexity (i.e. being located on the second level of the polynomial hierarchy). Further SAT-based systems include **prefMaxSAT** [Vallati et al., 2015; Faber et al., 2016], which uses the MaxSAT approach for the computation of preferred semantics; the **LabSATSolver** [Beierle et al., 2015], which uses propositional formulas based on labellings and, for the subset maximization task, the PrefSat Algorithm [Cerutti et al., 2014a] that then become **{j}ArgSemSAT**. The system **CoQuiAAS** [Lagniez et al., 2015], which also uses SAT encodings for some semantics, will be explained in Subsection 2.1.2, as the maximization task necessary for instance for preferred semantics is performed by means of constraint programming.

Background. Let us consider a set of propositional variables (or atoms) $\mathcal{P}$ and the connectives $\wedge, \vee, \rightarrow$ and $\neg$, denoting respectively the logical conjunction, disjunction, material implication and negation. The constants $\top$ and $\bot$ denote respectively *true* and *false*. In addition, we consider quantified Boolean formulae (QBF) with the universal quantifier $\forall$ and the existential quantifier $\exists$ (both over atoms), that is, given a formula ϕ, then $Qp\phi$ is a QBF, with $Q \in \{\forall, \exists\}$ and $p \in \mathcal{P}$. $Q\{p_1, \ldots, p_n\}\phi$ is a shorthand for $Qp_1 \cdots Qp_n\phi$. A propositional variable p in a QBF ϕ is free if it does not occur within the scope of a quantifier Qp and bound otherwise. If ϕ contains no free variable, then ϕ is said to be closed and otherwise open. We will write $\phi[p/\psi]$ to denote the result of uniformly substituting each free occurrence of p with ψ in formula ϕ.

An interpretation $I \subseteq \mathcal{P}$ defines for each propositional variable a truth assignment where $p \in I$ indicates that p evaluates to true while $p \notin I$ indicates that p evaluates to false. This generalizes to arbitrary formulae in the standard way: Given a formula ϕ and an interpretation I, then ϕ

evaluates to true under I (i.e., I satisfies ϕ) if one of the following holds (with $p \in \mathcal{P}$).

- $\phi = p$ and $p \in I$
- $\phi = \neg p$ and $p \notin I$
- $\phi = \psi_1 \wedge \psi_2$ and both ψ_1 and ψ_2 evaluate to true under I
- $\phi = \psi_1 \vee \psi_2$ and one of ψ_1 and ψ_2 evaluates to true under I
- $\phi = \psi_1 \rightarrow \psi_2$ and ψ_1 evaluates to false or ψ_2 evaluates to true under I
- $\phi = \exists p \psi$ and one of $\psi[p/\top]$ and $\psi[p/\bot]$ evaluates to true under I
- $\phi = \forall p \psi$ and both $\psi[p/\top]$ and $\psi[p/\bot]$ evaluate to true under I.

If an interpretation I satisfies a formula ϕ, denoted by $I \models \phi$, we say that I is a model of ϕ.

Reductions to propositional logic. The first reduction-based approach [Besnard and Doutre, 2004; Egly and Woltran, 2006] we consider here uses propositional logic formulae (without quantifiers) to encode the problem of finding admissible sets. Given an AF $AF = \langle Ar, att \rangle$, for each argument $a \in Ar$ a propositional variable v_a is used. Then, $S \subseteq Ar$ is an extension under semantics σ iff $\{v_a \mid a \in S\} \models \phi$, with ϕ being a propositional formula that evaluates AF AF under semantics σ (below we will present in detail how to translate AFs into formulae). Formally, the correspondence between sets of extensions and models of a propositional formula can be defined as follows.

Definition 2.4 *Let $\mathcal{T} \subseteq 2^{Ar}$ be a collection of sets of arguments and let $\mathcal{I} \subseteq 2^{\mathcal{P}}$ be a collection of interpretations. We say that $\mathcal{T}$ and $\mathcal{I}$ correspond to each other, in symbols $\mathcal{T} \cong \mathcal{I}$, if*

1. *for each $S \in \mathcal{T}$, there exists an $I \in \mathcal{I}$, such that $\{a \mid v_a \in I, a \in Ar\} = S$;*

2. *for each $I \in \mathcal{I}$, there exists an $S \in \mathcal{T}$, such that $\{a \mid v_a \in I, a \in Ar\} = S$; and*

3. *$|\mathcal{T}| = |\mathcal{I}|$.*

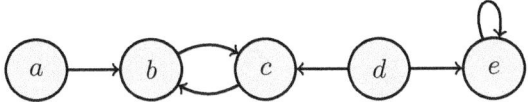

Figure 2. Example argumentation framework.

Given an AF $AF = \langle Ar, att \rangle$, the following formula can be used to solve the enumeration problem of admissible semantics.

(1)
$$adm_{Ar,att} := \bigwedge_{a \in Ar} \left((v_a \to \bigwedge_{\langle b,a \rangle \in att} \neg v_b) \land (v_a \to \bigwedge_{\langle b,a \rangle \in att} (\bigvee_{\langle c,b \rangle \in att} v_c)) \right)$$

Note that an empty conjunction is treated as $\top$, whereas the empty disjunction is treated as $\bot$.

The models of $adm_{Ar,att}$ now correspond to the admissible sets of AF, i.e., we have $\mathcal{E}_{\mathcal{AD}}(AF) \cong \{M \mid M \models adm_{Ar,att}\}$. The first conjunction in (1) ensures that the resulting set of arguments is conflict-free, that is, whenever we accept an argument a (i.e., v_a evaluates to true under a model), all its attackers cannot be accepted. The second conjunct expresses the defense of arguments by stating that, if we accept a, then for each attacker b, some defender c must be accepted as well.

Example 2.5 Let $AF = \langle Ar, att \rangle$ be an AF with $Ar = \{a, b, c, d, e\}$ and $att = \{\langle a,b \rangle, \langle b,c \rangle, \langle c,b \rangle, \langle d,c \rangle, \langle d,e \rangle, \langle e,e \rangle\}$ as depicted in Figure 2. The corresponding propositional formula $adm_{Ar,att}$ is as follows.

$$\begin{aligned}
adm_{Ar,att} \equiv & (v_a \to \top) \land \\
& (v_b \to (\neg v_a \land \neg v_c)) \land \\
& (v_c \to (\neg v_b \land \neg v_d)) \land \\
& (v_d \to \top) \land \\
& (v_e \to (\neg v_d \land \neg v_e)) \land \\
& (v_a \to \top) \land \\
& (v_b \to (\bot \land (v_b \lor v_d))) \land \\
& (v_c \to ((v_a \lor v_c) \land \bot)) \land \\
& (v_d \to \top) \land \\
& (v_e \to (\bot \land v_d))
\end{aligned}$$

It is easy to see that $\mathcal{I} = \{I_1, I_2, I_3, I_4\}$ represents the set of models of $adm_{Ar,att}$, where

$$I_1 = \{v_a \mapsto \bot, v_b \mapsto \bot, v_c \mapsto \bot, v_d \mapsto \bot, v_e \mapsto \bot\},$$
$$I_2 = \{v_a \mapsto \top, v_b \mapsto \bot, v_c \mapsto \bot, v_d \mapsto \bot, v_e \mapsto \bot\},$$
$$I_3 = \{v_a \mapsto \bot, v_b \mapsto \bot, v_c \mapsto \bot, v_d \mapsto \top, v_e \mapsto \bot\},$$
$$I_4 = \{v_a \mapsto \top, v_b \mapsto \bot, v_c \mapsto \bot, v_d \mapsto \top, v_e \mapsto \bot\}.$$

As $\mathcal{T} = \{S_1, S_2, S_3, S_4\}$, with $S_1 = \{\}$, $S_2 = \{a\}$, $S_3 = \{d\}$ and $S_4 = \{a, d\}$, is the set of all admissible sets of AF we clearly have the correspondence $\mathcal{I} \cong \mathcal{T}$ as desired.

Reductions to quantified Boolean formulas. For problems beyond NP we require a more expressive formalism than propositional logic. For this purpose we consider QBFs. In the following we will show how to reduce a given AF into a QBF such that the models of the QBF correspond to the preferred extensions of the AF [Egly and Woltran, 2006].

In order to realize the maximality check for preferred semantics we need to be able to compare two sets of atoms w.r.t. set inclusion. Consider the formula

$$Ar < Ar' := \bigwedge_{a \in Ar} (v_a \to v_{a'}) \wedge \neg \bigwedge_{a' \in Ar'} (v_{a'} \to v_a),$$

where $Ar' = \{a' \mid a \in Ar\}$. This formula ensures that any model $M \models (Ar < Ar')$ satisfies $\{a \in Ar \mid v_a \in M\} \subset \{a \in Ar \mid v_{a'} \in M\}$. Now we can state the QBF $prf_{Ar,att}$ for preferred extensions. Let the quantified variables be $Ar'_v = \{v_{a'} \mid a' \in Ar'\}$ and $att' = \{\langle a', b'\rangle \mid \langle a, b\rangle \in att\}$. Then

(2) $\quad prf_{Ar,att} := adm_{Ar,att} \wedge \neg \exists Ar'_v((Ar < Ar') \wedge adm_{Ar',att'})$.

Thus, for any AF $AF = \langle Ar, att\rangle$ an interpretation I is a model of $prf_{Ar,att}$ iff it satisfies the formula for admissible sets and there exists no "bigger" interpretation I' that also satisfies the the corresponding formula for admissible sets.

Example 2.5 (continued) There, I_4 is the only interpretation which satisfies the QBF $prf_{Ar,att}$ and the corresponding set S_4 is the only preferred extension of AF.

Similar approaches have been proposed by Arieli and Caminada in [Arieli and Caminada, 2013] and for Abstract Dialectical Frameworks by Diller et al. in **QADF** [Diller et al., 2015].

Iterative application of SAT solvers. The last approach we outline here is based on the idea of iteratively searching for models of propositional formulae and has been instantiated in the systems **{j}ArgSemSAT** [Cerutti et al., 2014a; Cerutti et al., 2014c; Cerutti et al., 2016b] and **Cegartix** [Dvořák et al., 2014; Dvořák et al., 2014]. The idea is to use an algorithm which iteratively constructs formulae and searches for models of these formulae. A new formula is generated based on the model of the previous one (or based on the fact that the previous formula is unsatisfiable). At some point the algorithm reaches a final decision and terminates.

The iterative approach is suitable when the problem to be solved cannot be decided in general—under standard complexity theoretic assumptions—by the satisfiability of a single propositional formula, constructible in polynomial time without quantifiers. This is, for instance, the case with skeptical acceptance under preferred semantics, where the corresponding decision problem is Π_2^P-complete. Instead of reducing the problem to a single QBF formula, the solving task is delegated to the iterative scheme of an algorithm querying a SAT solver multiple times.

The algorithms for preferred semantics work roughly as follows. To compute preferred extensions we traverse the search space of a computationally simpler semantics. For instance, we can iteratively search for admissible sets or complete extensions and iteratively extend them until we reach a maximal set, which is a preferred extension. By generating a new candidate for an admissible set or a complete extension, which is not contained in an already visited preferred extension, we can enumerate all preferred extensions in this manner. This allows answering both credulous and skeptical reasoning problems as well.

For deciding e.g. skeptical acceptance of an argument under preferred semantics one requires, in the worst case, an exponential number of calls to the SAT solver—under standard complexity-theoretic assumptions. However, the actual number of SAT calls in the iterative SAT scheme depends on the number of preferred extensions of the given AF, see [Dvořák et al., 2014].

In the following, we sketch the **Cegartix** approach from [Dvořák et al., 2014] for skeptical acceptance of an argument under preferred semantics. The algorithm returns YES if a is skeptically accepted, NO otherwise. To do so we try to construct a preferred extension which does not contain a. If this is possible we know that a is not skeptically accepted under preferred semantics, otherwise the algorithm returns YES.

1) Check if there is an interpretation I satisfying the formula ϕ (initially $\phi = adm_{Ar,att} \wedge \neg v_a$). If such an interpretation I exists, go to Step 2. Otherwise there is no admissible set which does not contain a, and

the algorithm returns YES.

2) Try to add new arguments to I by updating it (as long as possible) with interpretations satisfying the formula

$$adm_{Ar,att} \land \neg v_a \land (\bigwedge_{a \in Ar, v_a \in I} v_a) \land (\bigvee_{a \in Ar, v_a \notin I} v_a).$$

3) For the maximized interpretation I, check if it is possible to add the argument a to it by checking for models of the formula

$$\phi' = adm_{Ar,att} \land (\bigwedge_{a \in Ar, v_a \in I} v_a) \land (\bigvee_{a \in Ar, v_a \notin I} v_a).$$

If there is an interpretation I' satisfying ϕ', there is a preferred extension which contains a. Otherwise, there is a preferred extension, namely the one represented by the interpretation I, which does not contain the argument a. In this case the algorithm outputs NO and terminates.

4) The algorithm continues with the search for a different preferred extension which does not contain the arguments of I by modifying the formula ϕ as follows:

$$\phi' = \phi \land (\bigvee_{a \in Ar, v_a \notin I} v_a).$$

Go to Step 1.

Example 2.5 (continued) *Let us exemplify the algorithm of **Cegartix** on our AF from Example 2.5, where we want to decide skeptical acceptance of the argument d. We know that there are four interpretations satisfying the formula for admissible sets and only I_1 and I_2 satisfy the formula $\phi = adm_{Ar,att} \land \neg v_d$ of Step 1. Let us continue with $I = I_1$ which represents the admissible set $S_1 = \{\}$. In Step 2, we update I by setting v_a to $\top$. Remember, we cannot set v_d to $\top$ as ϕ contains the clause $\neg v_d$. In Step 3 we check if there is an I' satisfying the formula $\phi' = adm_{Ar,att} \land v_a \land (v_b \lor v_c \lor v_d \lor v_e)$. Indeed $I' = \{v_a \mapsto \top, v_b \mapsto \bot, v_c \mapsto \bot, v_d \mapsto \top, v_e \mapsto \bot\}$ is a model of ϕ', thus we constructed a preferred extensions, namely $S = \{a, d\}$ containing the argument a. In Step 4 we update our formula to $\phi = adm_{Ar,att} \land \neg v_d \land (v_b \lor v_c \lor v_d \lor v_e)$ and go to Step 1. In the next iteration, we check the new formula ϕ for models, but as ϕ is not satisfiable the algorithm outputs YES and terminates.*

One can use a modified version of the above algorithm to enumerate all preferred extensions. More concretely, one can add the obtained preferred extension from Step 2 to the output-set and then update the formula as in Step 4, while omitting Step 3. Further, the conjunct containing a negated variable for the queried argument must be removed. The PrefSat approach [Cerutti et al., 2014a] as implemented in the system **{j}ArgSemSAT** [Cerutti et al., 2014c; Cerutti et al., 2016b] uses this method to compute all preferred labellings.

2.1.2 Reductions to Constraint Satisfaction Problems

In the following we introduce reductions to another target formalism, namely Constraint Satisfaction Problems (CSPs) [Rossi et al., 2006], which allow to solve combinatorial search problems. Reductions to CSP have been addressed by Amgoud and Devred [Amgoud and Devred, 2011] and Bistarelli, Pirolandi, and Santini [Bistarelli et al., 2009; Bistarelli and Santini, 2010; Bistarelli and Santini, 2011; Bistarelli and Santini, 2012b; Bistarelli and Santini, 2012a]; the latter works led to the development of the **ConArg** system. Further systems based on CSP are **CoQuiAAS** [Lagniez et al., 2015] and **ASGL** [Sprotte, 2015]. The approach of CSP is inherently related to propositional logic reductions as introduced in Subsection 2.1.1, see also [Walsh, 2000] for a formal analysis of the relation between the two approaches.

A CSP can generally be described by a triple (X, D, C), where $X = \{x_1, \ldots, x_n\}$ is the set of variables, $D = \{D_1, \ldots, D_n\}$ is a set of finite domains for the variables and $C = \{c_1, \ldots, c_m\}$ a set of constraints. Each constraint c_i is a pair (h_i, H_i) where $h_i = (x_{i1}, \ldots, x_{ik})$ is a k-tuple of variables and H_i is a k-ary relation over D. In particular, H_i is a subset of all possible variable values representing the allowed combinations of simultaneous values for the variables in h_i. An assignment v is a mapping that assigns to every variable $x_i \in X$ an element $v(x_i) \in D_i$. An assignment v satisfies a constraint $((x_{i1}, \ldots, x_{ik}), H_i) \in C$ iff $(v(x_{i1}), \ldots, v(x_{ik})) \in H_i$. Finally, a solution is an assignment v to all variables such that all constraints are satisfied, denoted by $(v(x_1), \ldots, v(x_n))$.

Finding a valid assignment of a CSP is in general NP-complete. Nevertheless, several programming libraries support constraint programming, like ECLiPSe,[4] SWI Prolog,[5] Gecode,[6] JaCoP,[7] Choco,[8] Turtle[9] (just to men-

---

[4] http://eclipseclp.org/ (on 27/04/2017).
[5] http://www.swi-prolog.org/ (on 27/04/2017).
[6] http://www.gecode.org/ (on 27/04/2017).
[7] https://github.com/radsz/jacop (on 27/04/2017).
[8] http://www.choco-solver.org/ (on 27/04/2017).
[9] https://github.com/timfel/turtle (on 27/04/2017).

tion some of them) and allow for efficient implementations of CSPs. These constraint programming solvers make use of techniques like backtracking and local search.

Given an AF $AF = \langle Ar, att \rangle$, the associated CSP (X, D, C) is specified as $X = Ar$ and for each $a_i \in X$, $D_i = \{0, 1\}$. The constraints are formulated depending on the specific semantics σ. For example, solutions that correspond to conflict-free sets can be obtained by defining a constraint for each pair of arguments a and b with $\langle a, b \rangle \in att$, where the two variables may not be set to 1 at the same time. Here, the constraint is of the form $((a, b), ((0, 0), (0, 1), (1, 0)))$ which is equivalent to the cases when the propositional formula $(a \to \neg b)$ evaluates to true.

In the following, we will use the notation from [Amgoud and Devred, 2011], because it reflects the similarities between the CSP approach and the reductions to propositional logic as outlined above.

For admissible semantics we get the following constraints.

$$(3) \quad C_{AD} = \left\{ (a \to \bigwedge_{b:\langle b,a \rangle \in att} \neg b) \wedge (a \to \bigwedge_{b:\langle b,a \rangle \in att} (\bigvee_{c:\langle c,b \rangle \in att} c)) \,\Big|\, a \in Ar \right\}$$

The first part ensures conflict-free sets and the second part encodes the defense of arguments. Then, for an AF $AF = \langle Ar, att \rangle$ and its associated admissible CSP (X, D, C_{AD}), $(v(x_1), \ldots, v(x_n))$ is a solution of the CSP iff the set $\{x_j, \ldots, x_k\}$ s.t. $v(x_i) = 1$ is an admissible set in AF.

Example 2.5 (continued) *For our AF we obtain the following admissible CSP (X, D, C_{AD}). $X = A$, for each $a_i \in X$ we have $D_i = \{0, 1\}$ and*

$$\begin{aligned} C_{AD} = \{ &(a \to \top) \wedge (a \to \top), (b \to \neg a \wedge \neg c) \wedge (b \to \bot \wedge d), \\ &(c \to \neg b \wedge \neg d) \wedge (c \to (a \vee c) \wedge \bot), (d \to \top) \wedge (d \to \top), \\ &(e \to \neg d \wedge \neg e) \wedge (e \to \bot \vee d) \}. \end{aligned}$$

This CSP has the following solutions: $(0, 0, 0, 0, 0)$, $(1, 0, 0, 0, 0)$, $(0, 0, 0, 1, 0)$, $(1, 0, 0, 1, 0)$ which correspond to the admissible sets of AF, namely $\{\}, \{a\}, \{d\}$ and $\{a, d\}$.

Most CSP solvers do not support subset maximization. Thus, for preferred semantics, Bistarelli and Santini [2012a] propose an approach that iteratively computes admissible/complete extensions and adds constraints to exclude certain sets, such that one finally obtains the preferred extensions.

Reductions to Weighted Partial Max-SAT. This approach has been implemented in **CoQuiAAS** [Lagniez et al., 2015] and in **prefMaxSAT** [Vallati *et al.*, 2015; Faber *et al.*, 2016] and is particularly tailored to maximization problems as needed to compute preferred semantics. A *Weighted Partial Max-SAT* problem is a problem which maximizes the sum of weights associated to constraints, where the term *partial* means that some constraints have an infinite weight, which means they need to be satisfied. The system **CoQuiAAS** uses a SAT-Solver but the problem of Weighted Partial Max-SAT is more related to Constraint Programming, therefore we discuss this approach in this section, but of course it is also closely related to the previous section.

The computation of preferred extensions in [Lagniez *et al.*, 2015] is based on complete extensions which are obtained as follows. For an AF $AF = \langle Ar, att \rangle$ and for each $a \in Ar$ we use a boolean variable v_a.

$$comp_{Ar,att} := \bigwedge_{a \in Ar} \left(v_a \to \left(\bigwedge_{b \in Ar : \langle b,a \rangle \in att} \neg v_b \right) \land \left(v_a \leftrightarrow \left(\bigwedge_{b \in Ar : \langle b,a \rangle \in att} \bigvee_{c \in Ar : \langle c,b \rangle \in att} v_c \right) \right) \right)$$

The models of $comp_{Ar,att}$ correspond to the complete extensions of AF, i.e., we have $\mathcal{E}_{CO}(F) \cong \{M | M \models comp_{Ar,att}\}$. Then, the maximal models of $comp_{Ar,att}$ correspond to the preferred extensions of AF. To obtain these one uses the concept of a *maximal satisfiable subset* (MSS). For a set of formulas $\mathcal{F}$ the set of formulas $\mathcal{S} \subseteq \mathcal{F}$ is a MSS iff $\mathcal{S}$ is satisfiable and for each $c \in \mathcal{F} \setminus \mathcal{S}$, $\mathcal{S} \cup \{c\}$ is unsatisfiable.

Now, the computation of preferred extension reduces to the computation of MSSs of the sets of weighted formulas

$$prf_{Ar,att} = \{(comp_{Ar,att}, +\infty), (a_1, 1), \ldots, (a_n, 1)\}$$

where $a_1, \ldots, a_n \in Ar$.

2.1.3 Reductions to Answer Set Programming

The use of logic programming to solve abstract argumentation problems has been initiated by several authors (the survey article by Toni and Sergot [Toni and Sergot, 2011] provides a good overview), including the approach proposed by Nieves *et al.* [Nieves *et al.*, 2008], where the program is re-computed for every input instance; Wakaki and Nitta [Wakaki and Nitta, 2008], who use labelling-based semantics; and the approach by Egly *et al.* [Egly *et al.*, 2010a], which follows extension-based semantics. Here, we focus on the latter—the ASPARTIX approach— [Egly *et al.*, 2010a; Dvořák *et al.*, 2013a; Gaggl *et al.*, 2015], which relies on a query-based implementation where the argumentation framework to be evaluated is provided as an input database. From this point of view, the SAT or CSP methods can be seen as a compiler-like approach to abstract argumentation, while the ASP method acts like an interpreter.

A large collection of such ASP queries is provided by the **ASPARTIX-D** and **ASPARTIX-V** systems. Furthermore, the **DIAMOND** system [Ellmauthaler and Strass, 2014] for *Abstract Dialectical Frameworks* (ADFs), as well as the **GERD** system [Dvořák et al., 2015] for *extended argumentation frameworks* (EAFs) are based on ASP. In the following, we first give a brief introduction to ASP. We then present how the computation of admissible sets can be encoded in ASP. In order to obtain preferred extensions, it is necessary to check for subset-maximality of admissible sets. We will give pointers to the literature on several approaches for the subset-maximality check and refer to [Charwat et al., 2015] for a detailed discussion.

Background. Let us consider disjunctive logic program under the answer-set semantics [Gelfond and Lifschitz, 1991].[10] We fix a countable set $\mathcal{U}$ of *(domain) elements*, also called *constants*, and suppose a total order $<$ over the domain elements. An *atom* is an expression $p(t_1, \ldots, t_n)$, where p is a *predicate* of arity $n \geq 0$ and each t_i is either a variable or an element from $\mathcal{U}$. An atom is *ground* if it is free of variables. $B_\mathcal{U}$ denotes the set of all ground atoms over $\mathcal{U}$.

A *(disjunctive) rule* r with $n \geq 0$, $m \geq k \geq 0$, $n + m > 0$ is of the form

$$a_1 \vee \cdots \vee a_n \leftarrow b_1, \ldots, b_k, \mathit{not}\, b_{k+1}, \ldots, \mathit{not}\, b_m$$

where $a_1, \ldots, a_n, b_1, \ldots, b_m$ are atoms, and "*not*" stands for *default negation*. An atom a is a positive literal, while $\mathit{not}\, a$ is a default-negated literal. The *head* of r is the set $H(r) = \{a_1, \ldots, a_n\}$ and the *body* of r is $B(r) = B^+(r) \cup B^-(r)$ with $B^+(r) = \{b_1, \ldots, b_k\}$ and $B^-(r) = \{b_{k+1}, \ldots, b_m\}$. A rule r is *normal* if $n \leq 1$ and a *constraint* if $n = 0$. A rule r is *safe* if each variable in $H(r)$ occurs in $B^+(r)$. A rule r is *ground* if no variable occurs in r. A *fact* is a ground rule with a single literal in the head and with an empty body. An *(input) database* is a set of facts. A program is a finite set of safe disjunctive rules. For a program π and an input database D, we often write $\pi(D)$ instead of $D \cup \pi$. If each rule in a program is normal (resp. ground), we call the program normal (resp. ground).

For any program π, let U_π be the set of all constants appearing in π. $Gr(\pi)$ is the set of rules $r\tau$ obtained by applying, to each rule $r \in \pi$, all possible substitutions τ from the variables in r to elements of U_π. An *interpretation* $I \subseteq B_\mathcal{U}$ *satisfies* a ground rule r iff $H(r) \cap I \neq \emptyset$ whenever $B^+(r) \subseteq I$ and $B^-(r) \cap I = \emptyset$. I satisfies a ground program π, if each $r \in \pi$ is satisfied by I. A non-ground rule r (resp. a program π) is satisfied by an

---

[10]For further background, see [Eiter et al., 1997; Brewka et al., 2011].

interpretation I iff I satisfies all groundings of r (resp. $Gr(\pi)$). $I \subseteq B_{\mathcal{U}}$ is an *answer set* of π iff it is a subset-minimal set satisfying the *Gelfond-Lifschitz reduct* $\pi^I = \{H(r) \leftarrow B^+(r) \mid I \cap B^-(r) = \emptyset, r \in Gr(\pi)\}$. For a program π, we denote the set of its answer sets by $\mathcal{AS}(\pi)$.

Reduction to ASP. We now provide fixed queries for admissible sets in such a way that an argumentation framework AF is given as an input database $\widehat{F}$ and the answer sets of the program $\pi_e(\widehat{F})$ are in a certain one-to-one correspondence with the respective extensions, where $e \in \{\mathcal{AD}, \mathcal{PR}\}$. For an AF $AF = \langle Ar, att \rangle$, we define

$$\widehat{F} = \{\ \arg(a) \mid a \in Ar\} \cup \{\text{att}(a,b) \mid \langle a,b \rangle \in att\ \}.$$

We have to guess candidates for the selected type of extensions and then check whether a guessed candidate satisfies the corresponding conditions, where default negation is an appropriate concept to formulate such a guess within a query. In what follows, we use unary predicates in($\cdot$) and out($\cdot$) to perform a guess for a set $S \subseteq Ar$, where in(a) means $a \in S$.

Similar to Definition 2.4, we define the subsequent notion of correspondence which is relevant for our purposes.

Definition 2.6 *Let $\mathcal{T} \subseteq 2^{\mathcal{U}}$ be a collection of sets of domain elements and let $\mathcal{I} \subseteq 2^{B_{\mathcal{U}}}$ be a collection of sets of ground atoms. We say that $\mathcal{T}$ and $\mathcal{I}$ correspond to each other, in symbols $\mathcal{T} \cong \mathcal{I}$, iff*

1. *for each $S \in \mathcal{T}$, there exists an $I \in \mathcal{I}$, such that $\{a \mid \text{in}(a) \in I\} = S$;*

2. *for each $I \in \mathcal{I}$, there exists an $S \in \mathcal{T}$, such that $\{a \mid \text{in}(a) \in I\} = S$; and*

3. $|\mathcal{T}| = |\mathcal{I}|$.

Let $AF = \langle Ar, att \rangle$ be an argumentation framework. The following program fragment guesses, when augmented by $\widehat{F}$, any subset $S \subseteq A$ and then checks whether the guess is conflict-free in AF:

$$\begin{aligned}\pi_{cf} = \{\ &\text{in}(X) \leftarrow not\ \text{out}(X), \arg(X); \\ &\text{out}(X) \leftarrow not\ \text{in}(X), \arg(X); \\ &\leftarrow \text{in}(X), \text{in}(Y), \text{att}(X,Y)\ \}.\end{aligned}$$

The program module $\pi_{\mathcal{AD}}$ for the admissibility test is as follows:

$$\begin{aligned}\pi_{\mathcal{AD}} = \pi_{cf} \cup \{\ &\text{defeated}(X) \leftarrow \text{in}(Y), \text{att}(Y,X); \\ &\leftarrow \text{in}(X), \text{att}(Y,X), not\ \text{defeated}(Y)\ \}.\end{aligned}$$

For each conflict-free set one computes the arguments defeated by the set via the predicate defeated/1. The constraint then rules out those sets where an argument in the guessed set is attacked by an argument which is not defeated by the set, thus there is an argument in the conflict-free set which is not defended.

For any AF $AF = \langle Ar, att \rangle$, the admissible sets of AF correspond to the answer sets of $\pi_{\mathcal{AD}}$ augmented by $\widehat{F}$, i.e. $\mathcal{E}_{\mathcal{AD}}(AF) \cong \mathcal{AS}(\pi_{\mathcal{AD}}(\widehat{F}))$.

For semantics beyond NP we need to make use of *disjunction* in the logic program. There are several different ways how to encode these semantics. The first approach was to use the so called *saturation encodings* as pointed out in [Egly et al., 2010a] which are part of **ASPARTIX**. Other encodings also incorporated in **ASPARTIX** are the *metasp encodings* [Dvořák et al., 2013a], and the recently proposed encodings based on *conditional disjunction* which make use of a particular property of preferred semantics as shown in [Gaggl et al., 2015].

2.2 Direct Implementations

A direct implementation refers to a dedicated algorithm for a reasoning problem of a specific semantics. The advantage is that direct implementations directly incorporate some problem-specific shortcuts, which is often not possible—or it leads to limited improvement—in the case of reduction-based implementations.

2.2.1 Labelling-based Algorithms

Many direct implementations are based on an alternative characterization for semantics using certain labelling functions for arguments [Verheij, 1996b; Doutre and Mengin, 2001; Modgil and Caminada, 2009; Nofal et al., 2014b; Nofal et al., 2014a; Verheij, 2007] (see also Chapter 4). A labelling usually assigns each argument one of the following labels $\Lambda = \{\texttt{in}, \texttt{out}, \texttt{undec}\}$, which stand for accepted, rejected and undecided arguments. A labelling is a total function $\mathcal{L}ab : Ar \to \Lambda$. In the following we write $\texttt{x}(\mathcal{L}ab)$ for $\{a \in Ar \mid \mathcal{L}ab(a) = \texttt{x}\}$. For instance, $\texttt{in}(\mathcal{L}ab)$ is the set of all $\texttt{in}$-labeled arguments. Sometimes we will also represent a labelling $\mathcal{L}ab$ as the triple $\langle \texttt{in}(\mathcal{L}ab), \texttt{out}(\mathcal{L}ab), \texttt{undec}(\mathcal{L}ab) \rangle$.

One advantage of labellings is that the label of one argument has an immediate consequence to its neighbours. For example, if an argument a is labeled with $\texttt{in}$, all arguments attacked by a will be labeled with $\texttt{out}$. Such labelling-based algorithms have been materialized in several systems, see Table 1.

Enumeration. Several labelling-based algorithms to enumerate all extensions for various semantics have been proposed. For instance, the algorithm in [Nofal et al., 2014a] makes use of five labels, namely $\Lambda = \{\text{in},$ out, must_out, blank, undec$\}$, where the additional label blank denotes the not yet labeled arguments and must_out is assigned to arguments that attack in-labeled arguments. Initially all arguments are labeled with blank. Then, the algorithm selects an $a \in \text{blank}(\mathcal{L}ab)$ which is labeled with in in the left branch and undec in the right branch of the search tree. Every time an argument a is labeled with in all arguments attacked by it are labeled out and all remaining arguments which attack a are labeled with must_out. These steps are repeated until there are no arguments left to be labeled. The algorithm stores a preferred extension in one branch if each argument has one of the labels in, out and undec and the in-labeled arguments are not a subset of a previously stored preferred extension. Then, the algorithm backtracks to try to find all preferred extensions.

For the selection of the next argument to be labeled out from blank($\mathcal{L}ab$) the following heuristics are used.

- Don't pick an argument a to label it in iff there is a $b \in \{a\}^-$ such that $\mathcal{L}ab(b) \neq \text{out}$ and there is no $c \in \{b\}^-$ with $\mathcal{L}ab(c) = \text{blank}$.

- Don't pick an argument a to label it undec iff each $b \in \{a\}^-$ is either labeled with out or must_out.

- First select those blank-labeled argument to be labeled in which are not attacked at all or all its attacker are labeled with out or must_out.

- Otherwise, select a blank-labeled argument to be labeled in which attacks the most not out-labeled arguments.

Here we have only considered the case of preferred semantics, but for most of the semantics labelling-based algorithms have been proposed in the literature: algorithms for grounded and stable semantics are given in [Modgil and Caminada, 2009]; algorithms for semi-stable and stage semantics can be found in [Caminada, 2007; Caminada, 2010; Modgil and Caminada, 2009]. Recently [Nofal, 2013] studied improved algorithms for enumerating grounded, complete, stable, semi-stable, stage and ideal semantics. Labelling-based Algorithms are implemented in the **ArguLab** [Podlaszewski et al., 2011] system as well as in the **ArgTools** [Nofal et al., 2012].

Decision Procedures. In the following we will exemplify the use of labellings in an algorithm dedicated to credulous reasoning with preferred semantics, following the work of [Verheij, 2007], which is implemented in the **CompArg** system. In credulous reasoning one is only interested if a particular argument is accepted in at least one extension, thus we try to produce a witness (or counter-example) for this argument, instead of computing all extensions.

The algorithm starts with labelling the queried argument with in and all the other arguments with undec. Then, it iterates the following two steps. Firstly, it checks whether the set of in-labeled arguments is conflict-free and if so label all arguments attacking them with out. Otherwise terminate the branch of the algorithm. Secondly, for each argument a which is labeled out but not attacked by an argument labelled in, it picks an undec labeled attacker b of a and label it with in. In case there are several such arguments, it starts a new branch of the algorithm for each choice. If no such argument exists it terminates the branch. It stops a branch as soon as no more changes to labellings are made. In that case, it has reached an admissible labelling acting as proof for the credulous acceptance of the queried argument.

Consider the AF of Example 2.5 and the argument c. In the first step we obtain the following intermediate labelling

$$\mathcal{L}ab_1 = \langle\{c\}, \{\}, \{a, b, d, e\}\rangle.$$

As in($\mathcal{L}ab_1$) is conflict-free, we label all arguments attacking c with out:

$$\mathcal{L}ab_2 = \langle\{c\}, \{b, d\}, \{a, e\}\rangle.$$

Next we need to make arguments b and d *legally out* by labelling at least one of their attacker with in. In case of b this is already fulfilled as c is labeled with in. However, the argument d has no attacker, so the algorithm stops. We could not construct an admissible labelling for accepting the argument c, thus it is not credulously accepted under preferred semantics.

2.2.2 Dynamic Programming-based Approaches

We briefly mention the dynamic programming-based approach, which is defined on tree decompositions of argumentation frameworks. Many argumentation problems have been shown to be solvable in linear time for AFs of bounded tree-width [Dunne, 2007; Dvořák et al., 2012c; Courcelle, 1989].

First introduced in [Dvořák et al., 2012b], this approach especially aims at the development of efficient algorithms that turn complexity-theoretic results into practice. The algorithms from [Dvořák et al., 2012b] are capable of solving credulous and skeptical reasoning problems under admissible and preferred semantics. Later, this approach was extended to work with stable and complete semantics [Charwat, 2012]. Further fixed-parameter tractability results were obtained for AFs with bounded clique-width [Dvořák et al., 2010] and in the work on backdoor sets for argumentation [Dvořák et al., 2012a]. Negative results for other graph parameters like bounded cycle-rank, directed path-width, and Kelly-width can be found in [Dvořák et al., 2012b].

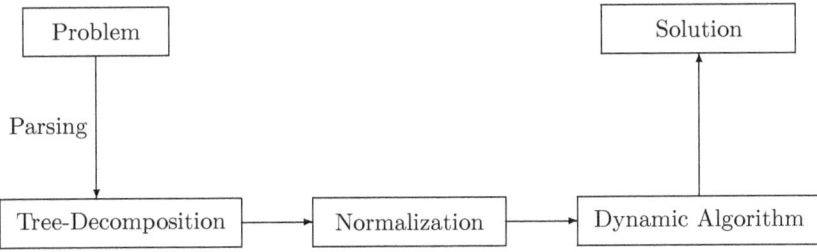

Figure 3. Dynamic-programming approach based on tree-decompositions.

Systems implemented towards this approach are **dynPARTIX** [Charwat, 2012; Dvořák et al., 2013b] as well as the **D-FLAT** system [Bliem, 2012; Bliem et al., 2012]. **D-FLAT** is a general-purpose system that is capable of solving problems from multiple domains. The methodology underlying both systems is to build a tree-decomposition of a framework and then run a dynamic programming algorithm on the tree-decomposition to obtain the extensions of the desired semantics, as depicted in Figure 3. For an extensive discussion of the approach we refer to [Charwat et al., 2015].

2.3 Summary

In this section we discussed the two main approaches to implement abstract argumentation frameworks, namely the reduction-based and the direct implementation approach. Systems which implement the reduction-based approach are very popular, as they benefit from highly sophisticated solvers. One can say that they delegate the difficult part of the design of an efficient algorithm to the solvers of the target formalism. This might be the reason why so many solvers make use of this approach (see Table 1). On the other side the direct implementations can incorporate shortcuts if specific properties for certain structures in AFs are known, and in particular when it comes to the reasoning problems of skeptical and credulous acceptance, these algorithms can benefit from them. Many direct implementation algorithms make use of labellings. Table 1 summarizes all systems.

3 Structured Argumentation Implementations

This section gives an overview of algorithmic approaches to structured argumentation [Besnard et al., 2014] and their respective systems. In contrast to abstract argumentation where arguments are interpreted as abstract entities and only logical relationships between arguments are taken into account, structured argumentation considers an argument's internal structure for sev-

	Direct	Reduction-Based	Type	Reference
{j}ArgSemSAT		Yes	SAT	[Cerutti et al., 2014c; Cerutti et al., 2016b; Cerutti et al., 2017]
ArgTools	Yes		Labellings	[Nofal et al., 2014b]
ArguLab	Yes		Labellings	[Podlaszewski et al., 2011]
ASGL		Yes	CSP	[Sprotte, 2015]
ASPARTIX-D		Yes	ASP, SAT	[Egly et al., 2010a; Gaggl and Manthey, 2015]
ASPARTIX-V		Yes	ASP	[Gaggl et al., 2015]
ASSA	Yes		Matrices	[Hadjisoteriou, 2015]
Carneades	Yes		Labellings	[Gordon et al., 2007]
Cegartix		Yes	SAT	[Dvořák et al., 2014]
CompArg	Yes		Labellings	[Verheij, 2007]
ConArg		Yes	CSP	[Bistarelli et al., 2015]
CoQuiAAS		Yes	SAT	[Lagniez et al., 2015]
DIAMOND		Yes	ASP	[Ellmauthaler and Strass, 2014]
Dungell	Yes		Haskell	[van Gijzel and Nilsson, 2013]
EqArgSolver		Yes	Equations, Labellings	[Rodrigues, 2016]
GERD		Yes	ASP	[Dvořák et al., 2015]
GRIS		Yes	Equations, Labellings	[Gabbay and Rodrigues, 2015]
LabSATSolver		Yes	SAT, Labellings	[Beierle et al., 2015]
LamatzSolver	Yes			[Lamatz, 2015]
prefMaxSAT		Yes	SAT	[Vallati et al., 2015; Faber et al., 2016]
ProGraph	Yes			[Groza and Groza, 2015]
QADF		Yes	QBF, Labellings	[Diller et al., 2015]
ZJU-ARG	Yes		Labellings	[Liao et al., 2013]

Table 1. Summary of abstract argumentation implementations.

eral aspects including evaluation. Within formal argumentation, formalisms for structured argumentation assume a formalized knowledge base, often in a logical or rule-based form, from which arguments and their relations are constructed. Conceptually, formalisms for structured argumentation often follow the steps of the so-called argumentation process or argumentation pipeline (see e.g. [Dung, 1995, Sections 4 and 5] and [Caminada and Amgoud, 2007, Section 2]):

1. argument construction;

2. determining conflicts among arguments;

3. evaluation of acceptability of arguments; and

4. drawing conclusions.

Argument construction typically refers to the task of building arguments composed of a claim and a derivation of that claim (e.g. a proof tree) from the given knowledge base. Moreover, conflicts need to be recorded, e.g., when claims of two arguments are contradictory, or when the derivation of an argument's claim contradicts with the claim of another argument. Evaluation of acceptability refers to formal means of finding acceptable arguments, and finally conclusions can be drawn from the acceptable arguments.

From a computational point of view, all of the steps of the process taken individually can be quite computationally expensive: for instance even construction of single arguments may be computationally complex (NP-hard in cases); a large number of arguments may be constructed; finding conflicts can be non-trivial; and evaluation of acceptability has in general a high complexity, as in the case of abstract argumentation.

Several algorithmic approaches have been proposed, which result in a quite heterogeneous and evolving field comprising of many different solutions. In the following we highlight properties that distinguish algorithms for structured argumentation from each other.

Reasoning on structural or abstract representation. The first aspect that distinguishes algorithms and systems for structured argumentation is that they may deviate from the conceptual argumentation process. In particular, the approaches can be roughly categorized whether they perform

- (query-based) structural reasoning; or

- reasoning on an abstract representation.

The latter classification encompasses algorithms that explicitly construct an abstract representation, e.g. an AF, and perform reasoning solely on

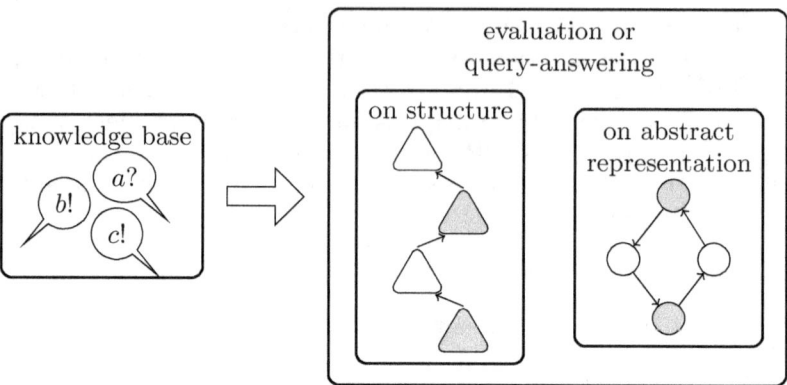

Figure 4. Argumentation process from a computational point of view

that representation. Algorithms following the other approach construct no such representation, but combine argument construction, conflict discovery, and argument evaluation in possibly interleaving steps and take structured information from the input knowledge base into consideration in possibly every step.

Algorithms that perform structural reasoning are typically query-based, i.e., decide acceptability of a certain claim, and construct arguments for and counterarguments against the queried claim from the knowledge base. A structural approach can restrict argument construction more easily than the abstract approach, in particular for query-based reasoning, since structural information can be used to determine which arguments have an effect on the query or the currently processed argument.

On the other hand, the abstract approach first "compiles" the structured knowledge base and subsequently all reasoning can be performed on the abstraction. In some cases "full" knowledge of all arguments occurring in the abstract representation is required to perform reasoning, e.g. for stable semantics. Conceptually, the abstract approach follows more closely the argumentation process. We illustrate structural and abstract approaches to algorithms for structured argumentation in Figure 4. In this figure triangles are arguments with internal structure and round vertices are abstract arguments.

Dedicated and reduction-based approaches. Similarly as for approaches to implement abstract argumentation, we can distinguish between direct or dedicated approaches and reduction-based approaches to implement structured argumentation. An approach is reduction-based if the input is translated to a problem of another target formalism with available solvers for that problem. Direct algorithms solve the problem at hand with a domain-specific dedicated algorithm. Direct algorithms have the benefit of incorporating domain-specific properties and optimizations more easily. On the other hand, reduction approaches can re-use off-the-shelf solvers. Reduction-based approaches for structured argumentation typically incorporate all involved tasks, i.e., argument construction, conflict evaluation, and deciding acceptability of arguments. When constructing an abstract representation, approaches to structured argumentation can also be hybrid systems, i.e., providing a direct or reduction-based approach for constructing the abstraction, and providing another for abstract reasoning. Usual target systems for reduction-based approaches are Prolog systems, solvers for Boolean satisfiability (SAT) and related formalisms, and solvers for answer-set programming (ASP) [Brewka et al., 2011]. We also call an algorithm or system reduction-based if it incorporates a translation of subproblems to a target language with available solvers.

Considered Approaches. In the following we overview concrete algorithmic approaches to structured argumentation, introducing them with examples and discussing the main computational problems, properties of interest from a computational point of view, and algorithms and systems proposed to solve the problem.[11] We focus on implemented algorithms for abstract rule-based argumentation (in particular concrete instantiations of the general ASPIC+ formalism) [Prakken, 2010; Modgil and Prakken, 2014], assumption-based argumentation (ABA) [Bondarenko et al., 1997; Toni, 2014], argumentation based on logic programming, in particular based on defeasible logic programs (DeLPs) [García and Simari, 2004; García and Simari, 2014], argumentation based on classical logic [Besnard and Hunter, 2008], and Carneades [Gordon et al., 2007]. Further information on the formalisms can also be found in Chapters 6, 7, 8, and 9. Complementing information can be found in a review of implementations for defeasible reasoning [Bryant and Krause, 2008], in particular sections 4.2.7, 4.3.1, 4.3.2, 4.3.3, and 4.3.4; in the review for argumentation for the social web [Schneider et al., 2013]; and in the overview on research in argumentation systems given by [Simari, 2011].

3.1 Abstract Rule-Based Argumentation

In this section we focus on systems for abstract rule-based argumentation, in particular concrete instantiations of the ASPIC+ [Prakken, 2010;

---

[11]Tools presented and referenced within the following subsections sometimes do not solve the same reasoning tasks proposed for a formalism. We refer the reader to the references for each algorithm and tool for the exact problem definitions that are solved.

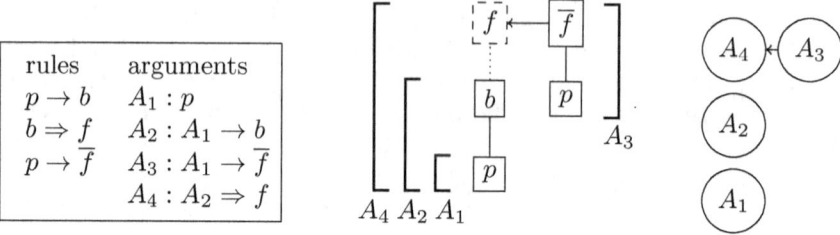

Figure 5. Tweety example knowledge base in ASPIC+ (left) with axiom p, structure of corresponding arguments (middle), and AF (right).

Modgil and Prakken, 2014] formalism (see also Chapter 6). We begin with a brief introduction to a concrete instantiation of ASPIC+ following notation of [Modgil and Prakken, 2014]. Input in this formalism is a knowledge base consisting of several components, central among them are (ordinary) premises and axioms, defeasible and strict rules, and preferential information. Semantics are specified via a translation to an abstract argumentation framework. Arguments are constructed by chaining premises or axioms with rules. Conflicts among arguments are defined via so-called undercuts, rebuts, and undermining among arguments, all respecting the preferential information.

We illustrate the concepts in a toy example knowledge base in Figure 5.

Example 3.1 *Figure 5 shows two strict rules (with a simple arrow $\rightarrow$) and one defeasible rule (using a double-lined arrow $\Rightarrow$), and assuming p (Tweety is a penguin) to be an axiom, one can infer the four arguments shown in the figure, namely by a strict rule that Tweety is a bird (b), that birds normally fly (via a defeasible rule inferring f), and that penguins do not fly (via a strict rule inferring $\overline{f}$; note that overlining indicates contrariness). The structure of the arguments is visualized in the middle of Figure 5 where we also see the only conflict in this example, namely that argument A_3 attacks A_4 via rebut (contradictory conclusions). On the right of Figure 5 the abstract AF is shown.*

Computational problems for abstract rule-based argumentation include argument construction, conflict discovery, and semantic evaluation. These problems may be tackled in an intertwined way, for instance interleaving construction and evaluation or following more closely the argumentation process step-by-step and thus firstly constructing the abstract argumentation framework and then proceeding by semantical evaluation.

As a rough and general outline for algorithms based on structural reasoning, given a potential conclusion (e.g. Tweety can fly in example Figure 5), arguments can be constructed via backward chaining using rules until premises or axioms are found. For instance, argument A_4 can be constructed from conclusion f and back-chaining of two rules until axiom p is reached. Counterarguments can be found in a similar manner by back-chaining from conclusions of arguments that would attack the arguments constructed so far. The so constructed arguments, i.e., arguments in favor of the queried claim and the counterarguments, corresponds to a game-theoretic approach to compute acceptability of the given query (and one of its argument in favor) under the specified semantics. For instance, one can conclude that A_3 is contained in an admissible set $\{A_3\}$.

We begin our survey of systems for abstract rule-based argumentation with the **TOAST** system[12] [Snaith and Reed, 2012]. **TOAST** directly follows the steps of the argumentation pipeline by constructing an abstract AF from given input knowledge base and delegates the reasoning tasks to a dedicated AF reasoner, namely the Dung-O-Matic web service [Snaith et al., 2010]. As an example, given the input in Figure 5 (left) the system would return a semantical evaluation of the AF shown on the right of that figure. The **TOAST** and Dung-O-Matic system together provide a system supporting axioms, premises, assumptions, and preferential information (last link and weakest link principles, see also [Modgil and Prakken, 2014]), rules, and a user-specified contrariness relation. The system further supports reasoning on the resulting AF under grounded, preferred, semi-stable, and stable semantics. **TOAST** is available as both a Java-based web service and web form.

Next we overview contributions to systems for abstract rule-based argumentation by Vreeswijk, which influenced subsequent successor systems. These systems follow query-based structural reasoning. Vreeswijk's works for argumentation systems are well summarized in the survey of [Bryant and Krause, 2008, Sections 4.3.1, 4.3.2, 4.3.3, and 4.3.4]. A system that resulted from Vreeswijk's PhD thesis [Vreeswijk, 1993], **IACAS** (InterActive Argumentation System), was written in LISP and is one of the earliest implementations of structured argumentation that is capable of handling input with strict and defeasible rules. This system allows for argument generation for or against a queried claim, and concluding its acceptability taking all the arguments into consideration. Vreeswijk's argumentation system (**AS**) is a Ruby-based implementation that handles strict and defeasible rules and tries to construct an admissible set containing an argument that concludes the queried claim. Two systems based on Vreeswijk's **AS** have been developed, namely the **ASPIC Inference Engine** and **Argue tuProlog** [Bryant et

---

[12]http://www.arg.dundee.ac.uk/toast/ (on 27/04/2017).

al., 2006].

The **ASPIC Inference Engine** is available from the ASPIC resources at the Cancer Research UK's Advanced Computation Laboratory.[13] It provides both a web-based front-end and a Java-based system that implement query-based structural reasoning under grounded and (credulous) admissible semantics. The Java-based implementation offers a graphical user-interface.

A reduction approach to the Prolog language is used in **Argue tuProlog** and the system is presented in [Bryant et al., 2006]. The reduction utilizes a game-theoretic approach for implementing ASPIC, similarly as the previous approaches. In contrast to reduction approaches for other formalisms, **Argue tuProlog** reduces the input to several Prolog queries, i.e., every query for an argument for each player is instantiated as a separate Prolog call and thus the dialogue can be terminated at any time.

We conclude this section with Wietske Visser's Epistemic and Practical Reasoner (**EPR**)[14] [Visser, 2008] which is a direct Java-based implementation that implements query-based reasoning under grounded semantics, (credulous) admissible semantics, and e-p semantics [Prakken, 2006]. The system provides a graphical user-interface, and is documented in detail in Wietske Visser's master's thesis [Visser, 2008].

3.2 Assumption based argumentation

In assumption-based argumentation (ABA) [Bondarenko et al., 1997; Toni, 2014] (see also Chapter 7), arguments and conflicts are drawn from three main components: a knowledge base, a set of assumptions, and a contrariness relation. We illustrate these concepts in Figure 6. On the left of Figure 6 we see an ABA framework, with four rules, the set of assumptions A containing a and e, and the contrariness relation relating the two assumptions to be contrary to f and d respectively (denoted via $\overline{a} = f$ and $\overline{e} = d$). Arguments (in squares) and conflicts (with solid arrows) that can be drawn from this framework are shown on the right of the figure. These arguments correspond to proof trees of claims. More concretely, the arguments' structure is based on the rules with the conclusion shown on the top of the squares and attacks take place based on assumptions and their contraries. For instance, the argument with f as the conclusion attacks the argument with conclusion b, since this argument requires the assumption a which is the contrary of f ($\overline{a} = f$). Arguments without assumptions are not attacked, e.g. argument with conclusion c.

Semantics of ABA can be defined via extensions as sets of arguments

---

[13]http://aspic.cossac.org (on 27/04/2017).
[14]http://www.wietskevisser.nl/research/epr/ (on 27/04/2017).

Foundations of Implementations for Formal Argumentation 715

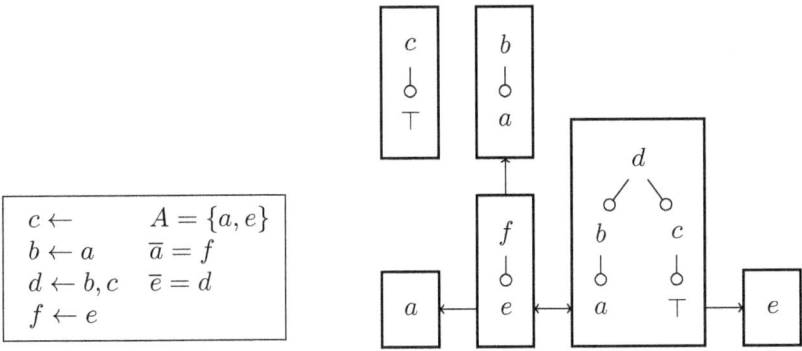

Figure 6. ABA framework (left) and its corresponding arguments and attacks (right)

or, equivalently, as sets of assumptions. For instance, in the example in Figure 6 the set of arguments with claims for c, f, and e (that in this instance uniquely determine the corresponding arguments) is an admissible extension of the ABA framework (no attacks between these arguments are present and all attackers from outside are counterattacked). The corresponding set of assumptions is $\{e\}$.

A typical reasoning task for ABA frameworks is to check whether an argument for a given claim is contained in an extension under a specified semantics. The computational complexity for reasoning with an abstract ABA formalism has been investigated in [Bondarenko et al., 1997]. In [Bondarenko et al., 1997] decision problems for credulous and skeptical acceptance are studied and the complexity ranges from polynomial-time decidable to completeness for Σ_4^P, a class on the fourth level of the polynomial hierarchy. Complexity of assumption-based argumentation is also discussed in this volume in Chapter 13.

Common to several algorithms for computing acceptability of a given claim under a specified semantics in a given ABA framework are so-called *dispute derivations* [Craven and Toni, 2016; Dung et al., 2006; Dung et al., 2007; Gaertner and Toni, 2007b; Gaertner and Toni, 2008; Toni, 2013]. Intuitively, dispute derivations can be seen as a game-theoretic constructive proof of acceptability of the given claim by constructing (part of) the argument in favor of the claim as well as constructing (parts of) its counterarguments and their counterarguments. Dispute derivations were proposed for grounded, admissible, and ideal semantics, called respectively GB, AB, and IB[15] dispute derivations [Dung et al., 2007], which are an advancement of the proof trees proposed in [Dung et al., 2006]. In [Gaertner and Toni, 2007b; Gaertner and Toni, 2008] *structured* dispute derivations were proposed that explicitly compute the dialectical structure hidden in dispute derivations, e. g., computing the attack structure explicitly. A parametrized version of dispute derivations was proposed in [Toni, 2013] that have a richer output incorporating both equivalent views of semantics of ABA, namely the view of extensions as sets of arguments and sets of assumptions.

In this chapter we illustrate concepts of dispute derivations by showing GB-dispute derivations [Dung et al., 2007]. In Figure 7 we see on the left a representation of a simple ABA framework with assumptions $A = \{b, c\}$ and a rule that infers a without assumptions. The grounded extension of this ABA framework contains the arguments for c and a, which are uniquely determined in this particular framework. A GB-dispute derivation is a sequence of quadruples (P_i, O_i, A_i, C_i) with integer i denoting the sequence or step. The ingredients for a step are the sentences or nodes for proponent (P_i) and opponent (O_i), the assumptions for defense of the queried claim (A_i) and assumptions for the opponent, so-called culprits (C_i). The component P_i is a set of sentences and both A_i and C_i are sets of assumptions. The second component of the quadruple, O_i, is a set of sets containing sentences. For querying acceptability for a claim α we initialize with $P_0 = \{\alpha\}$, $A_0 = \alpha \cap A$, and empty O_0 and C_0, where A is the set of assumptions in the ABA framework. We next illustrate the basics of GB-dispute derivations by recalling the corresponding sequences from [Dung et al., 2007], where we assume a selection function f that selects at each step either an element in P_i or in O_i and in the latter case an element of the set selected. For a given ABA framework and a selection function f, a GB-dispute derivation of a defense set D for sentence α is a finite sequence of quadruples

$$(P_0, O_0, A_0, C_0), \ldots, (P_i, O_i, A_i, C_i), \ldots, (P_n, O_n, A_n, C_n)$$

with $P_0 = \{\alpha\}$, $A_0 = \alpha \cap A$, and empty O_0 and C_0; $P_n = O_n = \emptyset$ and $A_n = D$; and for every $0 \leq i < n$ and $X = f(P_i, O_i, A_i, C_i)$ the selected element s. t.

1. if $X \in P_i$ then

    (a) if $X \in A$ then
    $$\begin{aligned} P_{i+1} &= P_i \setminus X, & A_{i+1} &= A_i, \\ C_{i+1} &= C_i, & O_{i+1} &= O_i \cup \{\{\overline{X}\}\} \end{aligned}$$

---

[15] Here, the "B" stands for belief.

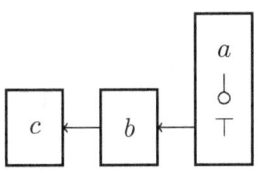

i	P_i	O_i	A_i	C_i	selected
0	$\{c\}$	$\emptyset$	$\{c\}$	$\emptyset$	–
1	$\emptyset$	$\{\{b\}\}$	$\{c\}$	$\emptyset$	c
2	$\{a\}$	$\emptyset$	$\{c\}$	$\{b\}$	$\{b\}, b$
3	$\{\top\}$	$\emptyset$	$\{c\}$	$\{b\}$	a

Figure 7. ABA with $A = \{b, c\}$, $\bar{b} = a$, $\bar{c} = b$, and rule $a \leftarrow$ (left); GB-dispute derivation for c (right)

(b) else (there exists a rule $X \leftarrow R$ with body R s.t. $C_i \cap R = \emptyset$)
$$P_{i+1} = (P_i \setminus X) \cup R, \quad A_{i+1} = A_i \cup (A \cap R),$$
$$C_{i+1} = C_i, \quad O_{i+1} = O_i$$

2. else ($T \in O_i$ is selected with $X \in T$)

    (a) if $X \in A$ then
    $$P_{i+1} = P_i \cup \{\overline{X}\}, \quad A_{i+1} = A_i \cup (\{\overline{X}\} \cap A),$$
    $$C_{i+1} = C_i \cup \{X\}, \quad O_{i+1} = O_i \setminus \{T\}$$

    (b) else
    $$P_{i+1} = P_i, \quad A_{i+1} = A_i,$$
    $$C_{i+1} = C_i, \quad O_{i+1} = (O_i \setminus \{T\}) \cup$$
    $$\{T \setminus \{X\} \cup R \mid X \leftarrow R \in \mathcal{R}\}$$

with $\mathcal{R}$ the set of rules of the given ABA framework. In Figure 7 we see on the right a sequence of a GB-dispute derivation. Briefly put, in each step in the sequence we select either an element of proponent or opponent, which in turn can either be assumptions or non-assumptions. Depending on the choice, different updates to the step have to be applied. For instance, if we choose an assumption of the proponent, then we remove that assumption from the sentence the proponent holds and add the contrary to the opponent who may construct an argument in favor of the contrary. We can note that each step in the sequence individually is straightforward to compute, however computation relies heavily on the selection function (also on selecting a rule in one case), which is discussed in more detail e.g. in [Gaertner and Toni, 2007b; Craven and Toni, 2016], which also highlights design choices for an algorithm based in dispute derivations.

Several systems have been developed implementing algorithms based on variants of dispute derivations. Current state of the art of dispute-derivation-based algorithms and systems for ABA are query-based and reason on the structural level and generally do not construct the full abstract representation to perform reasoning. Interestingly, most implementations, that build upon dispute derivations, rely on a reduction to Prolog with one exceptions **sxdd** [Craven et al., 2012], which is an implementation in C++.

The system **CaSAPI**,[16] which stands for "Credulous and Sceptical Argumentation: Prolog Implementation", is, as the name suggests, an implementation for ABA in Prolog. In version 2.0 [Gaertner and Toni, 2007a], **CaSAPI** implements GB, AB, and IB dispute derivations to perform query-based structural reasoning. Further, in versions 3.0 [Gaertner and Toni, 2007b] and 4.3 [Gaertner and Toni, 2008; Dung et al., 2007] structured dispute derivations are employed. Nowadays, **CaSAPI** acts as a precursor system for more recent systems.

Several tools with refined dispute derivations and reduction to Prolog have been proposed and implemented to perform query-based structural reasoning for ABA.[17] In the tool **proxdd** [Toni, 2013] the parametrized versions of dispute derivations are used. Graph-based versions of dispute derivations have been applied in the systems **grapharg** [Craven et al., 2013] and its follow-up system **abagraph** [Craven and Toni, 2016]. These tools include graphical visualization.

Recently, two systems for ABA were developed which are not based on dispute derivations: **ABAplus**[18] and the system from [Lehtonen et al., 2017], which we call here **ABAToAF**. Both of these systems compute semantics of ABA frameworks via an AF reasoner, ASPARTIX [Egly et al., 2010a], on an abstract representation of the ABA framework.

The system **ABAplus** implements ABA^+ [Cyras and Toni, 2016a], an extension of ABA with preferences. More concretely, this system provides computations for flat ABA^+ frameworks satisfying the axiom of weak contraposition [Cyras and Toni, 2016b] (this class subsumes flat ABA frameworks). The system **ABAplus** is capable of enumeration of extensions (as sets of assumptions together with their conclusions) under grounded, complete, preferred, stable, and ideal semantics. In contrast to systems described above, **ABAplus** constructs an abstract AF to reason on the ABA, with arguments being sets of assumptions, with the AF being solved via encodings of ASPARTIX. The system **ABAplus** generates arguments, using Python, based on (i) sets of assumptions that deduce contraries of assumptions and (ii) singleton sets of assumptions. Both the ABA^+ framework and the enumerated extensions are visualized in a web frontend.

The other system for ABA that relies on an AF reasoner, **ABAToAF**,

---

[16]http://www.doc.ic.ac.uk/~ft/CaSAPI/ (on 27/04/2017).
[17]Available at http://www.doc.ic.ac.uk/~rac101/proarg/ (on 27/04/2017).
[18]Web front end available at http://www-abaplus.doc.ic.ac.uk/ (on 27/04/2017) and stand-alone version at https://github.com/zb95/2016-ABAPlus/ (on 27/04/2017).

constructs arguments and attacks, similarly to **ABAplus**, based on sets of assumptions and derived sentences. Argument construction, implemented in Java 8, approximates here the restriction to generate arguments only for those sets of assumptions where at least one sentence can be derived from such a set, but not any proper subset. The system **ABAToAF** solves credulous (under admissible and stable semantics) and skeptical (under stable semantics) acceptance queries via calling an ASP solver on modified ASPARTIX encodings on the constructed AF.

Empirical evaluations of systems for ABA have been carried out for **sxdd** [Craven et al., 2012], **grapharg** [Craven et al., 2013], **abagraph** [Craven and Toni, 2016], and **ABAToAF** [Lehtonen et al., 2017].

The work of [Craven and Toni, 2016], based on preliminary research of [Craven et al., 2013], improves on several computational aspects of dispute derivations by altering the arguments' tree-structure to general graphs and introducing graphical dispute derivations (graph-DDs). In addition to tackle certain circularity questions for computation, in [Craven and Toni, 2016] an improvement for the problems of so-called flabbiness and bloatedness is provided. Briefly put, flabbiness refers to the potential shortcoming that the same sentence or claim is proved in several different ways, and bloatedness talks about deriving a claim in multiple ways in different arguments in an extension. That is, the former talks about computation of claims for individual arguments and the latter talks about computation of extension-based acceptability questions incorporating redundancy. In [Craven and Toni, 2016] graph-DDs are proposed for admissible and grounded semantics.

3.3 Argumentation based on logic programming

In this section we focus on algorithms and systems for argumentation based on logic programming, in particular defeasible logic programming [García and Simari, 2004; García and Simari, 2014]. More details on DeLP and semantics can also be found in Chapter 8. A defeasible logic program (DeLP) consists of strict ($\leftarrow$) and defeasible ($\prec$) rules as illustrated in Figure 8. Arguments in a DeLP are composed of a claim (a literal) and a set of defeasible rules. Acceptance of arguments is decided via a dialectical tree, see Figure 8 (right) for an example which includes an argument (A, a) that argues for literal a with set of rules A, arguments $(B_1, \sim b)$ and $(B_2, \sim b)$ that argue for (strongly) negated b, and argument $(E, \sim e)$ that argues for (strongly) negated e. Argument $(B_2, \sim b)$ defeats (A, a) because the former contradicts a subargument of the latter (arguing for b). Such a dialectical tree is then marked conceptually in a bottom-up manner with undefeated U and defeated D, i.e., leaves are undefeated and arguments are defeated if at least one child node is un-

A	B_1	B_2	E
$a \leftarrowtail b$	$\sim b \leftarrowtail d$	$\sim b \leftarrowtail e$	$\sim e \leftarrowtail g$
$b \leftarrowtail c$	$d \leftarrow$	$e \leftarrowtail f$	$g \leftarrow$
$c \leftarrow$		$f \leftarrow$	

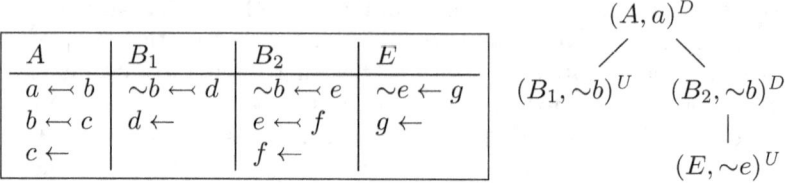

Figure 8. DeLP knowledge base (left) and dialectical tree (right)

defeated. Arguments are undefeated if all its children are defeated. Important for determining conflicts are preference relations which can either be given as input or derived via specificity, see [García and Simari, 2004; Stolzenburg et al., 2003] for details. In our example, the argument (A, a) is not warranted, simply because it is defeated by $(B_1, \sim b)$. If the rules used in argument $(B_1, \sim b)$ would be removed from the input DeLP, then argument (A, a) would be warranted.

Complexity of decision problems in DeLP has been studied in [Cecchi et al., 2006], showing complexity results for problems of deciding whether a given structure is an argument in a given DeLP (polynomial-time decidable), existence of arguments (a problem in NP), and further results regarding data complexity.

Algorithms for DeLP, which are based on dialectical trees, inherently solve query-based structural reasoning and check whether the queried claim is acceptable or warranted in a dialectical tree. Regarding enhancements for algorithms for computing acceptance of DeLPs, as stated in the survey of [Bryant and Krause, 2008], three concepts have been proposed to optimize efficiency for deciding acceptance in DeLPs: (i) pruning of dialectical trees [Chesñevar et al., 2000], (ii) using pre-compiled arguments in a dialectical database [Capobianco et al., 2004], and (iii) using parallelism [García and Simari, 2000]. We briefly illustrate these concepts and also refer the reader to the survey [Bryant and Krause, 2008] which includes a section on DeLP (Section 4.2.7).

For pruning of dialectical trees, as can be seen in the example dialectical tree of Figure 8, we do not need to consider all arguments in the tree to determine the dialectical status of the root argument. In particular, since argument $(B_1, \sim b)$ is undefeated, it is immediate that the top argument in this case is defeated. Therefore the right subtree is not relevant for concluding the overall result. Details on general pruning procedures for DeLP can be found in [Chesñevar et al., 2000], in particular how to "choose" the most promising argumentation line (path from root to a leaf in a dialectical tree) that determines an answer to the acceptability question as soon as possible.

In [Capobianco et al., 2004] for speeding up algorithms for ODeLP, a precompiled so-called dialectical database is suggested. Briefly put, potential arguments and defeats from the initial knowledge base are pre-compiled. In this way queries can incorporate first look-ups in the pre-compiled dialectical database.

For exploiting parallelism, in [García and Simari, 2000] it is suggested to parallelize computation for (i) finding several arguments for the same conclusion, (ii) discovering several defeaters for an argument, and (iii) finding several argumentation lines.

For concrete systems, DeLP reasoning has been implemented in Prolog accessible via the **DeLP client**,[19] and in the general-purpose libraries of **Tweety**[20] [Thimm, 2014]. In **Tweety** both the algorithm outlined in [García and Simari, 2004] for marking a dialectial tree and a translation to an AF have been implemented (the latter does not preserve the dialectical semantics of DeLP and only interprets the arguments and counterargument relationship within an abstract framework). **Tweety** also provides a web-interface for DeLP. Also, an abstract machine called JAM (justification abstract machine) [García, 1997] has been designed for DeLP. Furthermore, a reduction to ASP is given in [Thimm and Kern-Isberner, 2008].

Two further notable reduction-based approaches for extensions of DeLP have been proposed and implemented.[21] Possibilistic DeLP (P-DeLP) extends DeLP rules by attaching levels of strength. In [Alsinet et al., 2010] a recursive semantics for P-DeLP has been proposed, the corresponding framework is called RP-DeLP. An ASP-based approach to compute queries for RP-DeLP, i.e., to decide if a literal is warranted in the framework, is presented and experimentally evaluated in [Alsinet et al., 2012], which is based on results and complexity bounds of [Alsinet et al., 2011]. We call the corresponding system **ASP-RP-DeLP**. A reduction-based approach to SAT for multiple outputs of R-DeLP, we call the system **SAT-R-DeLP**, has been presented in [Alsinet et al., 2013] and also experimentally evaluated in that paper. The SAT approach is based on results of [Alsinet et al., 2011].

3.4 Argumentation based on classical logic

In argumentation based on classical logic, also called deductive argumentation, arguments and conflicts are generated from a (classical) logic knowledge base [Besnard and Hunter, 2008]. More information on argumentation

---

[19]Web interface available at http://lidia.cs.uns.edu.ar/delp_client/ (on 27/04/2017).
[20]http://tweetyproject.org (on 27/04/2017).
[21]Available via web-front-end at http://arinf.udl.cat/rp-delp (on 27/04/2017).

knowledge base $\{a, a \to b, \neg b\}$ $(\{a, a \to b\}, b)$
a
$a \to b$ | ↑
$\neg b$ $\{\neg b, \neg b \to \neg a, a\}$ $(\{\neg b\}, \neg(a \wedge (a \to b)))$
$\neg b \to \neg a$ ↑
                                                              $(\{\neg b \to \neg a, a\}, b)$

(a)   (b)   (c)

Figure 9. Knowledge base for deductive argumentation (a), inconsistent subsets of that knowledge base (b), and argument tree based on the inconsistent subsets as constructed by compilation-based approach (c)

based on classical logic can be found in Chapter 9. A knowledge base is here a set of formulas and arguments are pairs (S, C) of support S and claim C. The first component is a consistent, minimal (w.r.t. $\subseteq$) subset of the knowledge base that entails the claim, which in turn is a formula. Arguments can be compared w.r.t. conservativeness, i. e., (S, C) is more conservative than (S', C') iff $S \subseteq S'$ and $C' \models C$. Several notions of conflicts among arguments have been studied [Gorogiannis and Hunter, 2011]. We illustrate here the notion of (canonical) undercuts. Argument (S, C) undercuts (S', C') if $C = \neg(\phi_1 \wedge \cdots \wedge \phi_n)$ with $\{\phi_1, \ldots, \phi_n\} \subseteq S'$. Canonical undercuts incorporate notions of maximal conservativeness and canonical enumeration of formulas, i. e., the sequence of formulas ϕ_i in the conjunction C does not matter. In Figure 9 we see on the left (a) a knowledge base and on the right (c) three arguments where the middle one is a canonical undercut of the top one and the bottom one a canonical undercut of the middle one. Note that in contrast to other structured approaches to argumentation, the arrows in formulas in this section denote logical (material) implication, i. e., within formulas $a \to b$ is logically equivalent to $\neg a \leftarrow \neg b$ and $\neg a \vee b$. A further important notion is that of (complete) argument trees. A given argument is the root of an argument tree, for each node its children are its canonical undercuts, and the support of no node is a subset of the union of supports of all its ancestor nodes.

Computational complexity is in general very high for deductive argumentation [Parsons et al., 2003; Hirsch and Gorogiannis, 2010; Wooldridge et al., 2006; Creignou et al., 2011], as can be intuitively explained from the definitions which incorporate both minimality and entailment properties.[22]

---

[22] Another explanation for complexity of deductive argumentation is to consider its connection to (propositional) abduction, see [Besnard and Hunter, 2014, Section 7.4].

Complexity of finding individual arguments has been analyzed in [Parsons et al., 2003], decisions problems concerning instantiation of argument graphs with classical logic in [Wooldridge et al., 2006], and finding argument trees in [Hirsch and Gorogiannis, 2010]. Complexity for problems for deductive argumentation based on propositional logic can reach up to PSPACE.

Proposed algorithms and systems for deductive argumentation are based on minimal unsatisfiable subsets (MUSes) of formulas [Besnard and Hunter, 2006; Besnard et al., 2010], connection graphs [Efstathiou and Hunter, 2011; Efstathiou and Hunter, 2008], reductions to QBF [Besnard et al., 2009] and ASP [Charwat et al., 2012], so-called "contours" [Hunter, 2006b] and approximate arguments [Hunter, 2006a]. Algorithms that utilize contours, approximate arguments, and one MUS-based approach [Besnard and Hunter, 2006] are also discussed in detail in the book [Besnard and Hunter, 2008].

We begin with our algorithmic overview with two MUS-based approaches. The first one [Besnard and Hunter, 2006] falls into the general scheme of knowledge compilation [Darwiche and Marquis, 2002] where a given input is compiled into a structure to which one can pose queries that are computationally easier to compute on that structure compared to the original input. For deductive argumentation, the input knowledge base is compiled into a graph consisting of minimal inconsistent subsets of the knowledge base as the vertices and edges between non-disjoint subsets.

In Figure 9 we see in the middle (b) the compiled graph from knowledge base in the left (a). Given an argument, say ($\{a, a \to b\}, b$) (top right of Figure 9) one can construct an argument tree for this argument using the inconsistent subsets. Note that the support $\{a, a \to b\}$ of this argument is contained in a MUS. The remainder of that MUS ($\neg b$) then is the support for a canonical undercut of the argument, since both parts of the MUS, $\{a, a \to b\}$ and $\{\neg b\}$, each entail a negated conjoined subset of the other, e. g. $\{\neg b\}$ entails $\neg(a \wedge (a \to b))$. Using this line of reasoning recursively, one can construct all counterarguments and in turn the argument tree (shown on the right of Figure 9). For details on the algorithm see [Besnard and Hunter, 2006]. The compilation-based approach has been implemented in the **Tweety** libraries [Thimm, 2014] which can be configured to use different MUS solvers, for instance MARCO [Liffiton et al., 2016] or MIMUS [McAreavey et al., 2014].

Another approach using MUSes [Besnard et al., 2010] directly constructs arguments and counterarguments with a MUS solver, without an "offline" compilation beforehand.

The idea underlying argument construction of [Besnard et al., 2010] is

---

Complexity of propositional abduction is analyzed in [Eiter and Gottlob, 1995], with problems complete for Σ_2^P a class that is presumably more complex than the class NP.

that (S, C) is an argument iff $S \cup \{\neg C\}$ is a MUS of the knowledge base together with $\neg C$. Conditions of minimality and entailment for argument (S, C) follow from the fact that if $S \cup \{\neg C\}$ is a MUS, then S is consistent and entails C and S' with $S' \subset S$ does not entail C. The algorithms for argument construction and argument tree generation proposed in [Besnard et al., 2010], **BA** and **BT**, follow this line of reasoning and directly incorporate algorithmic issues like construction of formulas in conjunctive normal form. Algorithm **BA** has been implemented with the MUS solver HYCAM [Grégoire et al., 2009] and experimentally evaluated in [Besnard et al., 2010].

A different approach for generating argument trees for a given claim is proposed in [Efstathiou and Hunter, 2011], building on earlier work in [Efstathiou and Hunter, 2008] which utilizes connection graphs. Connection graphs consist of clauses as vertices and edges between clauses with complementary literals. Briefly put, for a given claim one can reduce the connection graph in such a way that, if non-empty, a support for the claim is contained in the reduced connection graph. In [Efstathiou and Hunter, 2011] this idea is used to construct argument trees. The approach has been implemented in **Java** in the tool **JArgue** and experimentally evaluated.

Reduction-based approaches are given in [Besnard et al., 2009; Charwat et al., 2012]. The former is a reduction to QBF and the latter to ASP. The latter has been implemented in the system called **vispartix**[23] within the tool ARVis [Ambroz et al., 2013] for visualizing relations between answersets of an ASP encoding. In **vispartix** an AF is generated from a given knowledge base and pre-specified set of claims, and conflicts are constructed as specified in [Gorogiannis and Hunter, 2011], thus partially deviating from other works in this section. The construction process is done via two ASP calls, the first constructing the arguments and the second constructing the attacks. In a final step the AF is visualized. Semantics can be computed via tools developed for AFs.

Algorithms following the concept of contours [Hunter, 2006b] are based on the idea of providing boundaries of what is provable in a knowledge base. Briefly put, an upper (lower) contour stores for a given formula which subsets of the knowledge base entail (do not entail) the formula. Finally, algorithms for approximate arguments [Hunter, 2006a] are based on the idea of relaxing one of the conditions for arguments (consistency, entailment, or minimality).

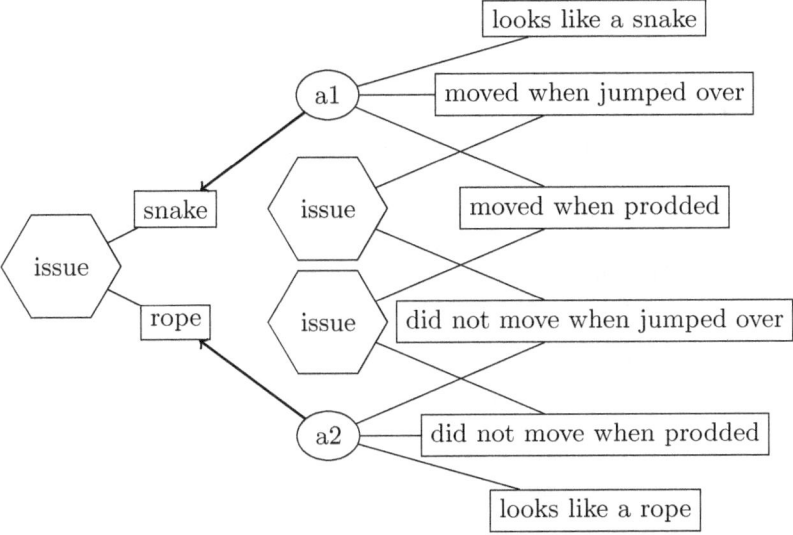

Figure 10. Example Carneades argument graph

3.5 Carneades

Carneades [Gordon and Walton, 2016; Gordon et al., 2007] is both a formal model of argument structure and evaluation, and a system[24] implementing the model. Evaluation of acceptance incorporates proof standards [Freeman and Farley, 1996], argument strength, and several ingredients available to a user. We illustrate briefly some of the capabilities of Carneades in a simple example[25] in Figure 10 and refer the reader for more details on the language and acceptability definitions to the literature [Gordon and Walton, 2016]. On the right part of Figure 10 there are six statements, i.e., that an object looks like a snake or a rope, and whether the object moved when jumped over or prodded. *Issue nodes* connect contradictory statements. Two arguments are formed (a1 and a2), which build on their premises (right of the figure) to conclude (left of the figure) that the seen object is indeed a snake or a rope. Let us assume that the object indeed looks like a snake and a rope (e.g. due to poor illumination), but neither did the object move when prodded with a stick nor when jumped over (e. g. by an adventurous person). In this case we conclude that the object is indeed a rope and not a snake (all premises

---

[23] http://www.dbai.tuwien.ac.at/proj/argumentation/vispartix/ (on 27/04/2017).
[24] https://carneades.github.io/ (on 27/04/2017).
[25] Example taken from http://carneades.github.io/ (on 27/04/2017). Variants of this example are discussed in [Walton et al., 2014].

of argument a2 are given but only one for a1).

The system Carneades (currently in version 4.2), features collaborative argument construction, argument visualization, and argument evaluation both for the structured arguments like we have seen in Figure 10 and also for Dung's AFs under grounded, complete, preferred, and stable semantics. Construction of structured arguments relies partially on internal calls to Prolog, and evaluation in the Carneades system can be classified as structural reasoning, since explicit abstract representation in the form of an AF is not utilized. Carneades is also available as a web-service and front-end [Gordon, 2012; Gordon, 2013], and includes a detailed manual.

3.6 Further implementations

Here we give pointers to related algorithms and implementations for structured argumentation that fall outside the previous sections.

Besides other approaches to structured argumentation, **Tweety** [Thimm, 2014] features an implementation to structured argumentation as proposed in [Thimm and García, 2010]. Further, Wyner et al's [Wyner et al., 2013] approach to instantiate rule-based knowledge bases with strict and defeasible rules as AFs has been encoded in ASP[26] [Strass, 2014].

A translational approach[27] to implement structured argumentation formalisms has been proposed in [van Gijzel and Nilsson, 2014] using Haskell as the programming language to capture definitions of these formalisms as directly as possible inside the programming language. For instance, in [van Gijzel and Nilsson, 2014] it is shown how to utilize this approach to translate Carneades to AFs: we call the corresponding system **CarneadesToDung**.

3.7 Summary

In this section we have given an overview of several algorithmic approaches to structured argumentation and their respective systems. Formalisms developed for structured argumentation and their implementations draw a quite heterogeneous picture. In particular, algorithms and systems range from query evaluation on the given structure to reasoning on an abstract representation where structural information is abstracted away. In Table 2 we see a summary of the presented approaches that have implementations and how they can be classified. Systems implementing structural reasoning typically solve queries in the form of deciding acceptance of a given claim and constructing arguments for this claim and counterarguments against the claim in a recursive fashion. Abstract reasoning involves construction of an abstract representation, i.e., an AF, and performing reasoning on this

---

[26] Main ASP encoding available under http://sourceforge.net/p/diamond-adf/code/ci/master/tree/lib/theorybase.lp (on 27/04/2017).

[27] http://www.cs.nott.ac.uk/~bmv/COMMA/ (on 27/04/2017).

representation resulting typically in sets of extensions. For reduction-based approaches, the column "language" refers to the target formalism of the approach. These systems typically also include parsers or compilers written in an imperative language that translate or reduce the given input to the formalism. In this table, ASP stands for answer-set programming, SAT for satisfiability solvers, and MUS for solvers capable of solving problems related to minimal unsatisfiable subsets of formulas.

The **Tweety** libraries [Thimm, 2014] implement several reasoning tasks from multiple formalisms for structured argumentation. We name the respective approaches in parenthesis for **Tweety**. We note that not all tools mentioned in Table 2 provide reasoning support themselves, i. e., some tools focus on argument construction and delegate evaluation to other systems. The tools **BA** [Besnard et al., 2010] and **vispartix** [Charwat et al., 2012] handle argument construction for deductive argumentation without evaluation, in particular, **BA** generates arguments and **vispartix** an AF. One of **Tweety**'s algorithms translates a given DeLP to an AF and leaves the choice for an AF reasoner to the user. **CarneadesToDung** [van Gijzel and Nilsson, 2014] translates input as specified in the Carneades model to a Dung AF. **TOAST** [Snaith and Reed, 2012] incorporates Dung-O-Matic [Snaith et al., 2010] for evaluation.

4 Other Implementation Approaches

This chapter would not be complete without a description of implemented systems that provide a general purpose gateway to formal structures of argumentation. They are, for instance, systems supporting text annotation for producing corpora that can be exploited by argument mining algorithms as well as systems for supporting critical thinking by the means of formal models of argumentation thus reusing elements discussed in previous sections. Our aim here is to summarize the most notable examples with some guidance for the reader interested in using—or reusing—existing implementations.

In particular, we analyse 34 promising implementations chosen among those that are active projects. Since it is beyond the scope of this chapter to provide a comprehensive description for each of those, we briefly review them in Section 4.1. Moreover, there are four additional projects that, although they appear to have been discontinued, have been relevant from an academic perspective, and we believe they should be mentioned in order to provide the reader with a complete background. Those are reviewed in Section 4.2, while in Section 4.3 we provide a comparative analysis of the active projects. Finally, the excellent review of Schneider et al. [Schneider et al., 2013] mentions other interesting projects—mostly online platforms—

	Direct	Reduction	Language	(Query-based) Structural Reasoning	Reasoning on Abstract Representation
ASPIC+					
TOAST	Yes		Java		Yes
ASPIC Inference Engine	Yes		Java	Yes	
EPR	Yes		Java	Yes	
Argue tuProlog		Yes	Prolog	Yes	
ABA					
CaSAPI		Yes	Prolog	Yes	
proxdd		Yes	Prolog	Yes	
abagraph		Yes	Prolog	Yes	
grapharg		Yes	Prolog	Yes	
ABAplus		Yes	ASP		Yes
ABAToAF		Yes	ASP		Yes
DeLP					
DeLP client		Yes	Prolog	Yes	
Tweety (DeLP)	Yes		Java	Yes	
Tweety (DeLP to AF)	Yes		Java		Yes
ASP-RP-DeLP		Yes	ASP	Yes	
SAT-R-DeLP		Yes	SAT	Yes	
Deductive					
JArgue	Yes		Java	Yes	
Tweety (deductive)		Yes	Java/MUS	Yes	
vispartix		Yes	ASP		Yes
BA		Yes	MUS		
Carneades					
Carneades		Yes	Prolog	Yes	
CarneadesToDung			Haskell		Yes

Table 2. Summary table for structured implementations.

that are briefly discussed in Section 4.4, even if they do not implement any evident formal model of argumentation.

4.1 Active Projects

The following 34 systems are representative among active projects incorporating some argumentation techniques.

AGORA [Hoffmann, 2005; Hoffmann, 2007] is a Computer-Supported Collaborative Argument Visualization (CSCAV) tool. An argument is defined here as a set of statements—claim and one or more reasons—where the reasons jointly provide support for the claim, or are at least meant to support the claim.

AIFdb [Lawrence et al., 2012b] is a database solution for the Argument Web thus implementing the AIF model of arguments [Bex et al., 2013; Rahwan et al., 2011; Chesñevar et al., 2006]. AIFdb offers an array of web service interfaces allowing a wide range of software to interact with the same argument data. Various dataset are available as part of the Argument Corpora [Reed, 2013].

AnalysisWall [Bex et al., 2013] is a collaborative workspace, a touchscreen measuring 11 feet by 7 feet, located at the University of Dundee.

Arg&Dec [Aurisicchio et al., 2015] is a web application for collaborative decision-making, encompassing the quantitative argumentation-based framework QuAD, and its decision matrix model, assisting their comparison through automated transformation.

ArgTeach [Dauphin and Schulz, 2014] is an interactive tutor that facilitates the learning of different labelling semantics in abstract argumentation. It now exists both as a standalone desktop application and as a web application.[28]

ArgTrust [Tang et al., 2012] relates the grounds of an argument to the agent that supplied the information, and can be used as the basis to compute acceptability statuses of arguments that take trust into account.

ArgueApply [Pührer, 2017] is a Java app for mobile phones, with a graphical interface, that lets users put forward arguments, and positive or negative links between arguments, in a fragment of the GRAPPA [Brewka and Woltran, 2014] language.[29]

ArgMed [Hunter and Williams, 2012; Williams et al., 2015] is a project investigating the use of computational argumentation for analysing and aggregating clinical evidence for making recommendations. In addition to the

---

[28]http://www-argteach.doc.ic.ac.uk/ (on 27/04/2017).
[29]http://www.informatik.uni-leipzig.de/~puehrer/ArgueApply/ (on 27/04/2017).

theoretical framework, it also has a public website.[30]

ArguMed [Verheij, 1998] introduces **ARGUE!**, based on the logical system CUMULA that abstractly models defeasible argumentation [Verheij, 1996a]. The development of **ARGUE!** was soon followed by the **ArguMed** family [Verheij, 2003a] based on the DefLog system [Verheij, 2003b], where dialectical arguments consist of statements that can have two types of connections between them: a statement can support another, or a statement can attack another. Dialectical arguments can be evaluated with respect to a set of prima facie justified assumptions.

Argument Blogging [Bex et al., 2014] allows users to construct debate and discussions across blogs, linking existing and new online resources to form distributed, structured conversations. Arguments and counterarguments can be posed by giving opinions on one's own blog and replying to other bloggers' posts. The resulting argument structure is connected to the Argument Web [Bex et al., 2013], in which argumentative structures are made semantically explicit and machine-processable.

Argunet [Schneider et al., 2007] is a desktop tool coupled with an open source federation system for sharing argument maps.

Arvina [Bex and Reed, 2012; Lawrence et al., 2012a] is a dialogical support system that allows for the structured execution of a reasoning process by implementing dialogue protocols and then allowing users to play the dialogue game against virtual agents and against each other in an instant-messaging environment.

ASPARTIXWeb [Egly et al., 2010b] is a web-based interface to the ASPARTIX system for computing extensions for various semantics of abstract argumentation.[31]

bCisive is a professional argument mapping and critical thinking support system.[32]

CISpaces [Toniolo et al., 2014; Toniolo et al., 2015] is an agent-based tool to help intelligence analysts in acquiring, evaluating, and interpreting information in collaboration. Agents assist analysts in reasoning with different types of evidence to identify what happened and why, what is credible, and how to obtain further evidence. Argument schemes lie at the heart of the tool, and sensemaking agents assist analysts in structuring evidence and identifying plausible hypotheses. A crowdsourcing agent is used to reason about structured information explicitly obtained from groups of contributors, and provenance is used to assess the credibility of hypotheses based

---

[30]http://www0.cs.ucl.ac.uk/staff/a.hunter/projects/argmed/ (on 27/04/2017).
[31]http://rull.dbai.tuwien.ac.at:8080/ASPARTIX/index.faces (on 27/04/2017).
[32]https://www.bcisiveonline.com/ (on 27/04/2017).

on the origin of the supporting information.

Cohere/Compendium [De Liddo and Buckingham Shum, 2010; Shum, 2008] is an open source software for sensemaking using argumentation maps and annotation.

ConargWeb is a web-based interface to the Conarg system for computing extensions of Dung's argumentation frameworks.[33]

CoPe_it! [Tzagarakis et al., 2009] is a tool to support synchronous and asynchronous argumentative collaboration in a Web environment. It introduces the notion of incremental formalization of argumentative collaboration. The tool permits a stepwise evolution of the argumentation space, through which formalization is not imposed by the system but is at the user's control. By permitting the users to formalize the discussion as the collaboration proceeds, more advanced services can be made available. Once the collaboration has been formalized to a certain point, CoPe_it! can exhibit an active behavior facilitating the decision making process.

D-BAS [Krauthoff et al., 2016] is a web and dialogue-based system to facilitate online argumentation, with the aim to guide users through statements, their pro-arguments and counterarguments, and adding new arguments as well as conflicts between these arguments.[34]

Debategraph [Macintosh, 2009] is a collaborative debate visualisation tool.

GERD [Dvořák et al., 2015] is a web-based interface of an ASP-based system for enumerating extensions of various semantics of the framework from [Modgil, 2009], which extends Dung's abstract argumentation framework with preferences among arguments.[35]

Gorgias [Kakas and Moraitis, 2003] is a general argumentation framework that combines preference reasoning and abduction. It can form the basis for reasoning about adaptable preference policies in the face of incomplete information from dynamic and evolving environments [Kakas et al., 1994].

Gorgias-B [Spanoudakis et al., 2016] supports the development of applications of argumentation under **Gorgias**. **Gorgias-B** guides the developer to structure their knowledge at several levels. The first level serves for enumerating the possible decisions and arguments that can support these options under some conditions, while each higher level serves for resolving conflicts at the previous level by taking into account default or contextual knowledge.

---

[33]http://www.dmi.unipg.it/conarg/ (on 27/04/2017).
[34]https://dbas.cs.uni-duesseldorf.de/ (on 27/04/2017).
[35]http://gerd.dbai.tuwien.ac.at/index.php (on 27/04/2017).

Grafix [Cayrol et al., 2014] is a graphical tool for handling abstract argumentation frameworks and bipolar frameworks. Grafix allows editing and drawing of argumentation graphs (or sets of graphs), and the execution of some "predefined treatments" (called "server treatments") on the current graph(s), such as, e.g., computing various acceptability semantics, or computing the strength of arguments.

GrappaVis is a Java graphical tool to specify GRAPPA [Brewka and Woltran, 2014] and ADF [Brewka et al., 2013] frameworks, evaluate them, and visualize the results of the evaluation. In particular, GRAPPA is a general semantical framework for assigning a precise meaning to graphical models of arguments or labelled argument graphs, which makes them suitable for automatic evaluation. GRAPPA rests on the notion of explicit acceptance conditions, as discussed in ADF [Brewka et al., 2013].[36]

MARFs (Markov Argumentation Random Fields) [Tang et al., 2016] is a system combining elements of formal argumentation theory and probabilistic graphical models. In doing so it provides a principled technique for the merger of probabilistic graphical models and non-monotonic reasoning.

Opinion Space [Faridani et al., 2010] is an online interface incorporating ideas from deliberative polling, dimensionality reduction, and collaborative filtering that allows participants to visualize and navigate through a diversity of comments.

OVA+ [Janier et al., 2014] provides a drag-and-drop interface for analysing textual arguments. It is designed to work with web pages It is available as a web interface and does not require a local installation. It also natively handles AIF structures, and supports real-time collaborative analysis.

Parmenides [Cartwright and Atkinson, 2008; Cartwright et al., 2009; Cartwright and Atikinson, 2009] is primarily a forum by which government bodies can present policy proposals to the public so that users can submit their opinions on the justification presented for a particular policy. Within Parmenides, the justification for action is structured to exploit a specific representation of persuasive argument based on the use of argumentation schemes and critical questions.

PIRIKA (PIlot for the RIght Knowledge and Argument) [Oomidou et al., 2014] is an argument-based communication tool for humans and agents, which supplements current communication systems such as Twitter. It allows for asynchronous argumentation for anyone, anytime, anywhere on any issues, as well as synchronous argumentation and stand-alone argumentation.

---

[36]http://www.dbai.tuwien.ac.at/proj/adf/grappavis/ (on 27/04/2017).

Quaestio-it [Evripidou and Toni, 2014] is based on a framework for modelling and analysing social discussions. It offers debating infrastructure for opinion exchanges between users and providing support for extracting intelligent answers to user-posed questions.

Rationale is a professional argument mapping and critical thinking support system.[37]

Reason [Introne, 2009] is a platform for supporting group decisions by leveraging the argumentative structure of deliberative conversation to drive a decision support algorithm. The platform uses argument visualization to mediate the collaborators' conversation.

Truthmapping is a professional, collaborative argument mapping tool.[38]

4.2 Discontinued Projects

In addition to the 34 systems discussed in Section 4.1, we briefly mention the following four as well. Although discontinued at the time of writing, those works have significantly impacted the research field and are still inspirational.

Avicenna [Rahwan et al., 2011] is an OWL-based argumentation system that consists of three main tiers: the data tier, the middle tier, and the client tier. The argumentation ontology is stored in the form of RDF statements (triples) in the back-end database, which constitutes the data tier. The middle tier is responsible for reasoning based on description logics and the interface to the web, through which applications in the client tier connect.

Dispute Finder [Ennals et al., 2010] is a browser extension that alerts a user when information they read online is disputed by a source that they might trust. Dispute Finder examines the text on the page that the user is browsing and highlights any phrases that resemble known disputed claims. If a user clicks on a highlighted phrase then Dispute Finder shows her a list of articles that support other points of view.

SEAS [Lowrance et al., 2008] is a collaborative, semi-automatic approach to evidential reasoning that uses template-based structured argumentation. Graphical depictions of arguments readily convey lines of reasoning, from evidence through to conclusions, making it easy to compare and contrast alternative lines of reasoning.

Trellis [Chklovski et al., 2003] allows users to add their observations, viewpoints, and conclusions as they analyze information by making semantic annotations to documents and other on-line resources. Users can associate specific claims with particular locations in documents used as "sources" for

---

[37]http://rationale.austhink.com/ (on 27/04/2017).
[38]https://www.truthmapping.com/ (on 27/04/2017).

analysis, and then structure these statements into an argument detailing pros and cons on a certain issue.

4.3 Comparative Analysis

To provide a concise overview over the *active* systems discussed in Section 4.1, we identified seven features that characterize the commonalities and differences among those systems, namely whether a system

(F1) is able to handle some form of structured argumentation;

(F2) gives the ability to manipulate arguments;

(F3) is collaborative;

(F4) enables a dialogue between different parties involved in its usage; and, in particular, if it

(F5) enables a dialogue based on speech acts;

(F6) includes a reasoner based on Dung's theory of abstract argumentation; or if it

(F7) includes a reasoner not based on Dung's theory of abstract argumentation.

It is evident that F5 is a specific case of F4: if a system offers speech acts, by definition it also offers a dialogue system. Moreover, F6 and F7 only apparently are mutually exclusive: indeed, a system can offer multiple choices of reasoners—the case of **CISpaces**—or it can encompass Dung's theory of abstract argumentation as a special case—e. g. **MARFs**.

Table 3 provides a comparative overview of the 34 active projects from Section 4.1 with respect to the seven features identified. This list of features is clearly far from being complete or unquestionable. However, it is sufficient for describing a large variety of possible usages of the systems.

Indeed, if a system supports F1 and F6, it is evident that it can be used in the *conventional* meaning of structured argumentation and perhaps it implements a specific approach for structured argumentation [Besnard *et al.*, 2014]. This is, for instance, the case of **OVA+**, which allows to represent and reason about ASPIC+ knowledge bases. Moreover, since **OVA+** also possesses the feature F2, it is evident that it can be used interactively; and since it possesses F3 as well, it can used in a distributed fashion.

It is worth noticing that there is only one system exhibiting all the seven features, CISpaces, which is unfortunately not (yet) available as an open-source implementation. Differently from OVA, CISpaces implements a subset of ASPIC, notably the ability to express only defeasible rules, and it

	F1	F2	F3	F4	F5	F6	F7
AGORA	Yes	Yes	Yes				
AIFdb	Yes		Yes			Yes	
AnalysisWall	Yes	Yes	Yes			Yes	
Arg&Dec		Yes	Yes	Yes			Yes
ArgTeach						Yes	
ArgTrust	Yes	Yes					Yes
ArgueApply		Yes	Yes	Yes		Yes	Yes
ArgMed	Yes	Yes				Yes	
ArguMed	Yes						Yes
Argument Blogging	Yes	Yes	Yes				
Argunet	Yes	Yes	Yes				
Arvina	Yes			Yes	Yes	Yes	
ASPARTIXWeb						Yes	
bCisive	Yes	Yes					
CISpaces	Yes	Yes	Yes	Yes	Yes	Yes	Yes
Cohere/Compendium	Yes	Yes	Yes				
ConargWeb		Yes				Yes	
CoPe_it!	Yes	Yes	Yes				
D-BAS	Yes		Yes	Yes			
Debategraph	Yes	Yes	Yes	Yes			
GERD						Yes	Yes
Grafix						Yes	Yes
GrappaVis	Yes					Yes	Yes
Gorgias	Yes	Yes					Yes
Gorgias-B	Yes	Yes					
MARFs	Yes					Yes	Yes
Opinion Space			Yes				
OVA+	Yes	Yes	Yes	Yes		Yes	
Parmenides	Yes			Yes			
PIRIKA		Yes	Yes			Yes	
Quaestio-it	Yes	Yes	Yes	Yes			Yes
Rationale	Yes	Yes					
Reason	Yes	Yes	Yes				
Truthmapping	Yes	Yes	Yes	Yes			

Table 3. Comparative overview of systems (discontinued systems are omitted) using some form of formal argumentation. F1: structured argumentation; F2: argument manipulation; F3: collaborative; F4: enables dialogues, F5: based on speech acts; F6: Dung's reasoner, or F7: non-Dung's reasoner.

follows a customised methodology for handling preferences, similar to ASPIC+ but using AFRA [Baroni et al., 2011b] as the meta-representation system. However, it also encompasses both the ability to use an evolution of **ArgTrust** as a web-service, as well as models of probabilistic reasoning based on [Li et al., 2012].

To conclude this analysis, it is worth showing the chronological evolution of all 38 systems reviewed in this survey, depicted in Figure 11. It is evident that 2014 has been the most prolific year, as also testified by the significant number (19) of demo submissions to COMMA 2014.

4.4 Projects for Informal Argumentation

Following the review of Schneider et al. [2013], there are further systems worth mentioning that make use of "informal" argumentation techniques. Indeed, they tend to be closer to user experience and they generally have a low entry barrier. At the same time, they do not offer much support for structuring arguments in a formal fashion, nor automated reasoning capabilities.

There is a large number of social networking debating systems such as Arguehow,[39] Climate CoLab [Gürkan et al., 2010], ConsiderIt [Kriplean et al., 2011], ConvinceMe,[40] CreateDebate,[41] Debate.org,[42] Debatepidia,[43] Debatewise,[44] Hypernews,[45] and LivingVote.[46] Further systems worth mentioning are, e.g., Belvedere,[47] an open-source critical thinking support system; the Cabanac's annotation system[48] for investigating social validation of arguments in comments; and DiscourseDB,[49] that is used to collaboratively collect policy-related commentary.

5 Challenges

In this section we discuss current challenges in devising and implementing algorithms for solving problems related to formal argumentation. In particular, for abstract argumentation problems we discuss parallel algorithms (Section 5.1), approximation algorithms (Section 5.2), and dynamic selection of algorithms depending on graph features (Section 5.3). We also have

---

[39] http://arguehow.com/ (on 27/04/2017).
[40] http://hamschank.com/convinceme/index.html (on 27/04/2017).
[41] http://www.createdebate.com/ (on 27/04/2017).
[42] http://debate.org (on 27/04/2017).
[43] http://www.debatepedia.com/ (on 27/04/2017).
[44] http://debatewise.org/ (on 27/04/2017).
[45] http://sourceforge.net/projects/hypernews/ (on 27/04/2017).
[46] http://www.livingvote.org/ (on 27/04/2017).
[47] http://belvedere.sourceforge.net/ (on 27/04/2017).
[48] http://www.irit.fr/~Guillaume.Cabanac/expe/ (on 27/04/2017).
[49] http://www.discoursedb.org/ (on 27/04/2017).

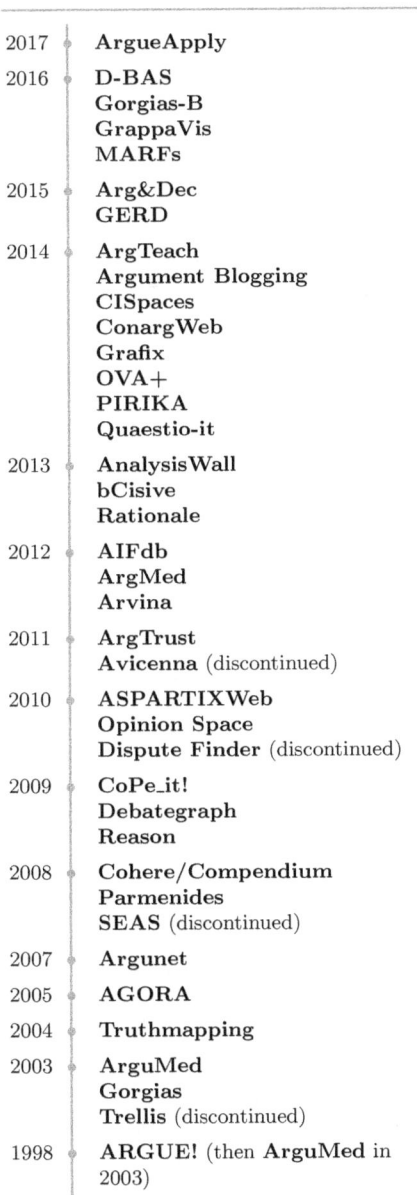

Figure 11. History of systems from Section 4, both active and discontinued. The year refers to the first tracked publication or to the first time the system appears online.

a brief look at advanced techniques and the related challenges for some structured argumentation approaches (Section 5.4).

5.1 Parallelization

As discussed in Chapter 13, reasoning tasks related to computational models of argumentation in general, and abstract argumentation in particular, are usually hard from the perspective of computational complexity. In order to make systems applicable to real-world scenarios, specific measures have to be taken in order to overcome the NP-complexity barrier—or even higher. One such measure is to use *parallelization*. Modern computing systems usually provide many CPU cores that allow for multiple threads to be executed in parallel. Moreover, grid- or cluster-based systems collect the computational capacity of many single machines and provide an abstraction with access to many computing cores. In order to exploit the computational power of such parallel systems, algorithms have to be devised that allow for the decomposition of complex problems, independent solving of the individual sub-problems, and an effective aggregation of the partial results into a global solution. While not every computational problem allows for such a parallelization—or at least does not allow for parallelization with a significant gain in performance—parallelization has been applied to many NP-complete (or harder) problems in the past with some success, most notably to the problem SAT [Hölldobler et al., 2011] allowing for considerable speed-ups on certain subclasses of instances.

For abstract argumentation, a natural feature to exploit for devising parallel algorithms is SCC-recursiveness [Baroni et al., 2005]. A semantics is SCC-recursive if the problem of enumerating the extensions for the graph as a whole can be be decomposed in computing the extensions of its *strongly connected components*[50] (SCC). Once SCCs have been identified, extensions can be computed on each SCC separately and the resulting sub-extensions can be joined in order to obtain the extensions of the whole graph paying attention to the inter-dependencies among different SCCs.[51] This basic approach is followed by the algorithm presented in [Cerutti et al., 2015], which itself is an enhancement to the previously published algorithm from [Cerutti et al., 2014e].

The approach for parallelizing the computation of extensions in abstract argumentation outlined in [Cerutti et al., 2015] is effective as long as the number of SCCs is "relatively" large in comparison to the size of the argumentation framework. Computing the SCCs of a graph can be done in

---

[50] A subgraph of a directed graph is a strongly connected component, if there is a directed path from every vertex to each vertex and the subgraph is maximal.

[51] Other decomposition methods might take advantages of I/O-multipoles [Baroni et al., 2014], but no approaches have been yet proposed.

polynomial time (see e.g. Tarjan's algorithm [Tarjan, 1972]) and, thus, the computational overhead of decomposing the problem is negligible in comparison to the computational effort of computing extensions, which is, as discussed before, often NP-hard or harder, depending on the chosen semantics. The computational effort required for the aggregation step is highly dependent on the actual instance of the problem and may be exponential in the worst case, as a sub-graph may possess an exponential number of extensions [Baumann and Strass, 2014] that need to be aggregated. However, for "reasonable" instances, this step is also negligible in comparison to the effort of computing extensions. As the empirical evaluation in [Cerutti et al., 2015] suggests, exploiting SCC-recursiveness for parallelization may yield a speedup (up to 280%) when increasing the number of cores from 1 to 4.

Another approach to parallelization is not based on decomposing a problem into sub-problems, but on parallel execution of different algorithms for the whole problem. For many computationally hard problems there is usually a limited number of algorithms that can solve "most" of the instances in reasonable time, and the core problem is to determine *which* algorithm should be selected to solve a particular instance. This problem is called the *Algorithm selection* problem and will be discussed in more detail in Section 5.3. It is worth noticing that [Vallati et al., 2017] proposes a first parallel algorithm selection approach. A straightforward solution to this problem is to devise a meta-algorithm that runs several algorithms on the original problem in parallel. As soon as the first algorithm terminates, the meta-algorithm terminates as well and the result of the meta-algorithm is the result of this algorithm. This approach, also called *variant-based parallel computation*, has been implemented in [Craven et al., 2012] for the problem of deciding acceptance of arguments in *assumption-based argumentation* (ABA)[52] and has been applied in the medical domain. More specifically, the approach of [Craven et al., 2012] is based on discussion games (see also Chapter 10) and different algorithms for solving acceptance use different expansion strategies in advancing the game.

The two approaches from above are complementary in the way how parallelization is realized. While the first approach uses a single algorithm and decomposes the problem instance into a parallel execution, the second approach uses multiple algorithms on the whole problem. Of course, combinations of the paradigms are imaginable.

---

[52] While ABA is actually an approach to structured argumentation, we discuss it here as it is the only known parallel approach to structured argumentation.

5.2 Approximation Techniques

Parallelization offers an approach to overcome complexity barriers while maintaining soundness and completeness. A different and also often applied approach is to give up soundness and/or completeness and devise *approximation algorithms*, see e.g. [Vazirani, 2002; Cormen et al., 2009]. Roughly, an approximation algorithm is not expected to solve the problem correctly but only within a certain margin of error. On the other hand, an approximation algorithm is expected to be more efficient than a correct algorithm.

In general, an algorithm A is said to be an ϵ-approximation algorithm for an optimization problem P (with $\epsilon > 0$), if for every instance the output of A is in the interval $[(1-\epsilon)C, (1+\epsilon)C]$, where C is the optimal solution, and ϵ thus represents the relative error in the approximation. Usually, one is interested in polynomial-time ϵ-approximation algorithms with ϵ being as small as possible. In case the algorithm returns more refined solutions—i.e. it decreases the ϵ-approximation further—if provided with additional runtime, it belongs to the class of *anytime algorithms*.

Approximation techniques for problems of abstract argumentation have not been investigated in-depth yet, with only very few exceptions. For example, the equational approach to abstract argumentation [Gabbay, 2012; Gabbay and Rodriguez, 2014] views an argumentation framework as a generator of equations for value assignments V such that $V(X) = 1$ indicates that X is in; $V(X) = 0$ indicates that X is out; and $V(X) \in (0,1)$ that X is undecided. In [Gabbay and Rodriguez, 2014] the authors introduce an iteration schema for computing complete extensions, starting from an arbitrary assignment V_0 and then, by use of a specific update rule, generating a sequence of assignments $V_0, V_1, \ldots$. In [Gabbay and Rodriguez, 2014] it is shown that this sequence will eventually converge and form a complete extension. This algorithm can therefore be interpreted as an anytime algorithm for computing complete extensions, but a thorough analysis of this algorithm in terms of approximation quality has not been done yet.

In the area of *probabilistic abstract argumentation* [Li et al., 2012; Thimm, 2012; Hunter, 2014], which is concerned with combining abstract argumentation frameworks with probabilistic reasoning, approximation techniques from probabilistic reasoning have been applied to overcome the additional complexity necessary to deal with quantitative uncertainty [Hadoux et al., 2015; Li et al., 2012]. As probabilistic abstract argumentation is a topic that will be covered in later volumes of this handbook, we omit discussing these techniques here.

In summary, approximation techniques for computational models of arguments are still underdeveloped, but may gain attention in the near future.

5.3 Algorithm Selection

In Section 5.1 we already discussed the variant-based parallel computation approach of [Craven et al., 2012] which is a specific solution for solving the *Algorithm Selection* problem by running different algorithms for the same problem in parallel. If parallelization is not possible for devising an algorithm, another solution is given by the *algorithm portfolio* approach [Rice, 1976; Leyton-Brown et al., 2003; Xu et al., 2008]. A portfolio is a meta-algorithm that has access to several specific algorithms for solving the same problem. When presented with a problem instance, the meta-algorithm selects one of those specific algorithms. In the case of *dynamic portfolios*, the meta-algorithm first extracts some *features* of the problem instance and then selects an algorithm that has, in a preprocessing step, proven to be the best algorithm for instances with the given features. This approach has been proven quite successful in solving many hard problems, such as SAT [Xu et al., 2008].

The crucial step in developing a dynamic portfolio algorithm is to define which features are relevant both to assess the quality of the algorithms in the preprocessing step and to select the appropriate algorithm during runtime. Furthermore, it is important that the overhead introduced for computing features of the problem instance during runtime is "reasonably" small. In [Vallati et al., 2014b; Cerutti et al., 2014b] the authors presented 50 features of abstract argumentation frameworks and derived empirical performance models (EPMs) to determine the "best" implementation for enumerating preferred extensions, given CPU-time as evaluation criterion and a limited set of solvers. The features considered there were basic graph theory-based measures such as size of the graph, average degree of arguments, flow hierarchy, and so on. The two EPMs presented in [Cerutti et al., 2014b] show an overall accuracy of 80% (classification) and, depending on the implementation, the ability to predict quite accurately the runtime required by a solver to enumerate the preferred extensions (regression). Unsurprisingly, the set of most informative features—according to a greedy forward search-based on the Correlation-based Feature Selection attribute evaluator [Hall, 1998] and with respect to the experimental setting used by the authors—includes the density of the argumentation graph, as well as number of SCCs and the size of the maximum SCC. When the computed EPMs have been applied to the problem of algorithm selection, both of them perform significantly well: in 78% of cases (resp. 75%) the classification-based EPM (the regression-based EPM) selects the best implementation. In most of the cases, 83%, both EPMS select the same algorithm, which is the correct one in 82% of cases.

Complete static and dynamic portfolios have been proposed in [Cerutti *et al.*, 2016d], and parallel portfolios are proposed and discussed in [Vallati *et al.*, 2017]. However, it is still unclear whether there may be better features to use for the selection problem or whether a combination of different techniques discussed in this section may yield improved performance. In [Brochenin *et al.*, 2015], *abstract solvers* [Nieuwenhuis *et al.*, 2006] are used as a formal machinery to formally specify different algorithms addressing extension-enumeration problems. By using these formalizations, algorithms could be combined and extended to more effective algorithms. Hence, using this machinery to also include the concepts discussed in this section may be a fruitful endeavor.

5.4 Advanced Techniques for Structured Argumentation

In structured argumentation, further computational problems than argument evaluation may occur. Many approaches to structured argumentation (see also Chapters 6, 7, 8, and 9) consider a knowledge base formalized in some logical formalism, and then derive arguments and conflicts between them on top of that, cf. Figure 4. Therefore, additional computational effort is required to construct arguments and to discover the conflict relationship between them. In general, computational approaches to structured argumentation can be categorized in two classes: those that use abstract argumentation frameworks as the underlying argument evaluation mechanism and those that provide proprietary evaluation mechanisms.

For the class of approaches providing proprietary evaluation mechanisms—such as Defeasible Logic Programming (see Chapter 8) and earlier versions of Deductive Argumentation (see Chapter 9)—the processes of argument construction, defeat discovery, and argument evaluation are usually intertwined, but each step still imposes some challenges.[53]

For argument construction, an important issue is relevance of arguments. In particular, for approaches building on classical logics—such as Deductive Argumentation (Chapter 9)—the number of arguments that can be derived from knowledge base may be potentially infinite. Given a specific query to the knowledge base, usually only those arguments are constructed that are relevant to the query and possess a certain normal form (in Deductive Argumentation these are the *maximally conservative undercuts*, cf. Chapter 9). In [Besnard and Hunter, 2006] an effective method for constructing both arguments and the defeat relation for a certain query is presented. This method relies on a preprocessing step that generates a so-called *compilation* from a knowledge base, which is an undirected graph with vertices being

---

[53] For those approaches relying on abstract argumentation for argument evaluation, similar sophisticated techniques as outlined in this and the previous sections apply, but will not be discussed separately.

the minimal inconsistent subsets of the knowledge base and two vertices are connected if they have a non-empty intersection. Given a specific query, a traversal algorithm allows the complete construction of an argument tree from this compilation. Considering only *approximate arguments* [Hunter, 2006a]—e.g. arguments which are not necessarily minimal—also allows to gain efficiency by trading-off completeness or soundness (to some extent).

Another advanced technique for structured argumentation is *pruning of dialectical trees* in, e.g., Defeasible Logic Programming (Chapter 8) [Chesñevar et al., 2000; Chesñevar and Simari, 2007; Rotstein et al., 2011]. This technique also offers a solution to refrain from considering all arguments for evaluating a query. This is realized by only expanding the dialectical tree so far until the evaluation status of the query is decided. For example, if an argument possesses multiple attackers, and it can already be decided that the first attacker is ultimately accepted and defeats the argument, then there is no need to evaluate the acceptance status of the remaining attackers as it can already be decided that the argument under consideration is not acceptable. Yet another approach to address the very same issue is to evaluate different argumentation lines in a dialectical tree in parallel [García and Simari, 2000].

6 Evaluation of Implementations

While theoretical approaches to computational models of argumentation are usually analytically evaluated using rationality postulates or comparison of behavior on toy examples—see e.g. [Gorogiannis and Hunter, 2011; Caminada and Amgoud, 2005; Amgoud, 2014]—the evaluation of algorithms and implementations focuses on the three aspects of *correctness*, *performance*, and *usability*. The correctness of algorithms and implementations is usually shown in an analytical way and involves showing that the algorithmic representation corresponds to the formal definition, e.g. that the result of performing an algorithm indeed returns the grounded extension of a given abstract argumentation framework. In order to evaluate an *algorithm* with respect to performance, one usually conducts an analytical runtime or complexity analysis. For the performance evaluation of *implementations* an empirical evaluation on either artificial or real-world benchmarks and runtime measurement on the corresponding computational problems is essential for obtaining a comparative analysis of different approaches. Finally, in order to evaluate the usability of implementations, user studies have to be performed.

For the remainder of this section, we will focus on the problem of empirical performance evaluation of implementations of computational models of argumentation. In particular, we will focus on evaluations of implementations that solve problems for abstract argumentation frameworks, cf. Section 2. The computational problems arising in abstract argumentation have already been discussed in detail in Chapter 13. Those problems are an important aspect of any evaluation of implementations as well, as they provide clear formalizations of what are the expected outcomes of computational tasks. Another important aspect of such evaluations is the identification of suitable benchmarks, i. e. abstract argumentation graphs, that can be used to compare the performance of different implementations, which we discuss in Section 6.1. We discuss existing comparative analyses, in particular the *International Competition on Computational Models of Argumentation* (ICCMA),[54] in Section 6.2.

6.1 Benchmark Examples

A crucial issue in setting up an evaluation of an implementation of abstract argumentation problems is the identification of argument graphs that are used as benchmark examples. Ideally, real-world applications would provide these kind of benchmark graphs in order to test implementations on actually existing problems. Unfortunately, the availability of real-world benchmarks for argumentation problems is quite limited, some few exceptions are [Cabrio *et al.*, 2013; Cabrio and Villata, 2014b; Cabrio and Villata, 2014a] and AIFdb.[55] Moreover, these benchmarks are tailored towards problems of argument mining [Wells, 2014] and their representation as abstract argumentation frameworks usually lead to topologically simple graphs, such as cycle-free graphs, which are unsuitable for comparing abstract argumentation solvers: all classical semantics coincide with grounded semantics on cycle-free graphs [Dung, 1995]. In order to compare solvers for—among others—preferred and stable semantics, artificially-generated argumentation graphs have been used so far.

Generating graphs for testing computational approaches or hypotheses on physical or social phenomena has already some tradition in network theory [Erdös and Rényi, 1959; Albert and Barabási, 2002; III *et al.*, 2012; Tabourier *et al.*, 2011; Barabasi and Albert, 1999]. However, it is questionable whether these graph models are suitable to model argumentation problems. For instance, the Barabási-Albert model [Barabasi and Albert, 1999] generates networks based on *preferential attachment*. The concept *preferential attachment* refers to the tendency of nodes that have already many connections to other nodes, to receive even more connections in the evolution of the network: an example of this phenomenon is the saying "the rich get richer, while the poor get poorer." To the best of our knowledge,

---

[54]http://argumentationcompetition.org (on 27/04/2017).
[55]http://corpora.aifdb.org (on 27/04/2017).

there is no evidence that real-world argumentation adheres to this concept. Another concept from network theory often (indirectly) implemented in graph models is that of *triangle closure*, i.e., the tendency of nodes directly connecting to the neighbors of its neighbors (as in the saying "the friend of my friend is also my friend"). This concept is hardly applicable to argumentation graphs as this would imply that *defense* (an argument attacking the attacker of another argument) tends also to be a *direct attack* (the first argument attacking the argument it also defends).

Graph models from network theory also usually generate undirected graphs. Adapting a model to generate directed edges is of course trivial, but it is questionable whether the resulting graphs have any interpretation with respect to the original intention of the model.

Finally, from the perspective of challenging benchmarks for abstract argumentation, the graphs generated by such models are usually also not adequate. Initial experiments for ICCMA'15 [Thimm et al., 2016] (see also below and the next section) suggest that those generated graphs usually contain an empty or a very small grounded extensions, usually no stable extensions (also due to the triangle closure property), and very few and small complete and preferred extensions. The latter observation is due to the fact that these graph models aim at modeling the "small world" property of many real-world graphs.[56] This leads to many arguments directly or indirectly being in conflict with each other. However, these models have been used for benchmark generation in earlier evaluations of implementations of abstract argumentation solvers [Bistarelli et al., 2013; Bistarelli et al., 2014].

In order to provide challenging benchmarks, ICCMA'15 used proprietary graph generators, each addressing different aspects of computationally hard graphs for specific semantics. For example, the `StableGenerator` aims at generating graphs with many stable extensions, and thus also many complete and preferred extensions. Graphs generated by this generator pose substantial combinatorial challenges for solvers addressing the computational tasks of determining (skeptical or credulous) acceptance of arguments and of enumerating extensions. For a given number of arguments, this generator first identifies a subset of these arguments to form an acyclic subgraph which will contain the grounded extension. Afterwards, another subset of arguments is randomly selected and attacks are randomly added from some arguments within this set to all arguments outside the set (ex-

---

[56]This property basically states that there are always "relatively short" paths from any node to every other node [Watts and Strogatz, 1998], provided that the network is connected and not too complete. For example the theory of "six degrees of separation" suggests that in the social network of the known world the longest shortest path between any two persons is six.

cept to the arguments identified in the first step). This process is repeated until a number of desired stable extensions is reached. The source code for this and other generators can be found in the source code repository[57] of *probo* [Cerutti *et al.*, 2014f], the benchmark suite used to run the competition. Another general tool for generating argumentation frameworks from a set given graph features is given by AFBenchGen[58] [Cerutti *et al.*, 2014d; Cerutti *et al.*, 2016a].

6.2 Comparative Analysis

The first systematic evaluations of implementations of abstract argumentation solvers have been conducted in [Bistarelli *et al.*, 2013; Bistarelli *et al.*, 2014]. In these evaluations a small number of implementations have been evaluated with respect to runtime on graphs generated by different graph models from social networking theory such as the Barabási-Albert model (see above). A similar performance evaluation is provided in [Vallati *et al.*, 2014a; Cerutti *et al.*, 2016d]. In addition, in [Cerutti *et al.*, 2016c] the authors discuss the effect of solver and instances configuration on performance.

A large-scale and systematic comparison of different implementations of computational models of argumentation is offered by the *International Competition on Computational Models of Argumentation* (ICCMA)[59], which has already been mentioned before and is an international event established in 2014. The first instance of the competition took place in 2015 and focused on comparing implementations for various decision and enumeration problems in abstract argumentation.

The competition in 2015 received 18 solvers from research groups in Austria, China, Cyprus, Finland, France, Germany, Italy, Romania, and UK. It was conducted using the benchmark framework *probo* [Cerutti *et al.*, 2014f], which provides the possibility to run the instances on the individual solvers, verify the results, measure the runtime, and log the results accordingly. The software *probo* is written in Java and requires the implementation of a simple command line interface from the participating solvers.[60] All benchmark graphs—generated using proprietary generation algorithms, see previous section—were made available in two file formats. The *trivial graph format*[61] (TGF) is a simple representation of a directed graph which simply

---

[57]http://sourceforge.net/p/probo/code/HEAD/tree/trunk/src/net/sf/probo/generators/ (on 27/04/2017).
[58]https://sourceforge.net/projects/afbenchgen/ (on 27/04/2017).
[59]http://argumentationcompetition.org (on 27/04/2017).
[60]See http://argumentationcompetition.org/2015/iccma15notes_v3.pdf (on 27/04/2017) for the formal interface description.
[61]http://en.wikipedia.org/wiki/Trivial_Graph_Format (on 27/04/2017).

lists all appearing vertices and edges. The *Aspartix format* (APX) [Egly et al., 2008] is an abstract argumentation-specific format which represents an argumentation framework as facts in a logic programming-like way. In order to verify the answers of solvers, the solutions for all instances were computed in advance using the *Tweety libraries for logical aspects of artificial intelligence and knowledge representation*[62] [Thimm, 2014]. Tweety contains naïve algorithms for all considered semantics that implement the formal definitions of all semantics in a straightforward manner and thus provides verified reference implementations for all considered problems. Besides serving as the benchmark framework for executing the competition, *probo* also contains several abstract classes and interfaces for solver specification that can be used by participants in order to easily comply with the solver interface specification.

The competition in 2015 evaluated the runtime performance of the solvers for four different semantics and four different computational tasks, yielding a total of 16 tracks. Among the best solvers throughout all tracks were CoQuiAAS, ArgSemSAT, and LabSATSolver (see also Section 2). For detailed performance comparisons and current competitions see the webpage of ICCMA.[63]

7 Discussion

In this chapter we discussed (1) approaches for addressing reasoning problems in abstract argumentation frameworks; (2) approaches for handling structured argumentation frameworks; and (3) other approaches that might be relevant to the argumentation community although they do not belong to the previous two classes.

As per approaches for abstract argumentation frameworks, it is beyond doubt that currently the majority of proposals adopt a reduction-based approach (Section 2.1), thus relying on SAT-solvers, or CSP-solvers, or ASP-solvers. However, we have covered the few direct implementations as discussed in Section 2.2.

Coming to approaches for structured argumentation frameworks, we considered the four large families developed in some 20 years of studies, viz. (in alphabetical order) ABA, ASPIC+, Deductive argumentation, and DeLP. We also considered the case of Carneades, which is both a formal model of argument structure and evaluation, and a system implementing the model.

Then, we reviewed 34 implemented systems that provide a general purpose gateway to formal structures of argumentation. They can be systems for producing corpora that can be exploited by argument mining algorithms

---

[62]http://tweetyproject.org (on 27/04/2017).
[63]http://argumentationcompetition.org (on 27/04/2017).

as well as system for supporting critical thinking by the means of formal models of argumentation.

This touches one of the main topic of discussion still open in the community, namely applying machine learning techniques for automatic argument elicitation from natural language text, or *argument mining*, see [Budzynska *et al.*, 2014; Wells, 2014] and Chapter 12. This is a fast growing research field, but at the same time, it encompasses a large variety of topics, from mining legal arguments, to mining tweets, and it is unlikely to have a *one-size-fits-all* approach. At the same time, this is an extremely young research field and best practices did not yet emerge in the community.

While we did not devote space to argument mining techniques, we instead discussed what are the main challenges we envisage for implementation of formal argumentation, as well as what are sensible ways for comparing different implementations. In particular, we reviewed (Section 5) the few approaches for making systems applicable to real-world scenarios, and thus overcoming the NP-complexity barrier, namely parallelization and approximation techniques. Moreover, machine learning techniques might also play an important role in selecting the right solver for a specific problem. There are, indeed, some embryonic approaches for automatic algorithm selection on the basis of abstract argumentation frameworks features. However, most—if not all—of the reviewed approaches consider abstract argumentation frameworks only.

This leads us to the last element of discussion we touched in this chapter (Section 6), namely how to compare different systems by the means of benchmarks and competitions. Although the community already made a move in the context of abstract argumentation, with the first edition of the International Competition of Computational Models of Argumentation, we still have a long way ahead for addressing questions related to structured argumentation. Comparative studies on different formalisms, i. e. [Schulz and Caminada, 2015] and [Heyninck and Straßer, 2016], might shed some light on common grounds, thus allowing for a fair comparison.

Acknowledgements

This work has been funded by the Austrian Science Fund (FWF): P25521, I2854 and P30168, and by Academy of Finland under grants 251170 COIN and 284591.

BIBLIOGRAPHY

[Albert and Barabási, 2002] Réka Albert and Albert-Laszló Barabási. Statistical mechanics of complex networks. *Reviews of Modern Physics*, 74(1):47–97, 2002.

[Alsinet *et al.*, 2010] Teresa Alsinet, Ramón Béjar, and Lluis Godo. A characterization of collective conflict for defeasible argumentation. In Pietro Baroni, Federico Cerutti,

Massimiliano Giacomin, and Guillermo R. Simari, editors, *Proceedings of the 3rd International Conference on Computational Models of Argument (COMMA 2010)*, volume 216 of *Frontiers in Artificial Intelligence and Applications*, pages 27–38. IOS Press, 2010.

[Alsinet et al., 2011] Teresa Alsinet, Ramón Béjar, Lluis Godo, and Francesc Guitart. Maximal ideal recursive semantics for defeasible argumentation. In Salem Benferhat and John Grant, editors, *Proceedings of the 5th International Conference on Scalable Uncertainty Management (SUM 2011)*, volume 6929 of *Lecture Notes in Computer Science*, pages 96–109. Springer, 2011.

[Alsinet et al., 2012] Teresa Alsinet, Ramón Béjar, Lluis Godo, and Francesc Guitart. Using answer set programming for an scalable implementation of defeasible argumentation. In *Proceedings of the IEEE 24th International Conference on Tools with Artificial Intelligence (ICTAI 2012)*, pages 1016–1021. IEEE, 2012.

[Alsinet et al., 2013] Teresa Alsinet, Ramón Béjar, Lluis Godo, and Francesc Guitart. On the implementation of a multiple output algorithm for defeasible argumentation. In Weiru Liu, V. S. Subrahmanian, and Jef Wijsen, editors, *Proceedings of the 7th International Conference on Scalable Uncertainty Management (SUM 2013)*, volume 8078 of *Lecture Notes in Computer Science*, pages 71–77. Springer, 2013.

[Ambroz et al., 2013] Thomas Ambroz, Günther Charwat, Andreas Jusits, Johannes P. Wallner, and Stefan Woltran. ARVis: Visualizing relations between answer sets. In Pedro Cabalar and Tran Cao Son, editors, *Proceedings of the 12th International Conference on Logic Programming and Nonmonotonic Reasoning, (LPNMR 2013)*, volume 8148 of *Lecture Notes in Artificial Intelligence*, pages 73–78. Springer, 2013.

[Amgoud and Devred, 2011] Leila Amgoud and Caroline Devred. Argumentation frameworks as constraint satisfaction problems. In Salem Benferhat and John Grant, editors, *Proceedings of the 5th International Conference on Scalable Uncertainty Management (SUM 2011)*, volume 6929 of *Lecture Notes in Computer Science*, pages 110–122. Springer, 2011.

[Amgoud, 2014] Leila Amgoud. Postulates for logic-based argumentation systems. *International Journal of Approximate Reasoning*, 55(9):2028–2048, 2014.

[Arieli and Caminada, 2013] Ofer Arieli and Martin W.A. Caminada. A QBF-based formalization of abstract argumentation semantics. *Journal of Applied Logic*, 11(2):229–252, 2013.

[Aurisicchio et al., 2015] Marco Aurisicchio, Pietro Baroni, Dario Pellegrini, and Francesca Toni. Comparing and integrating argumentation-based with matrix-based decision support in Arg&Dec. In Elizabeth Black, Sanjay Modgil, and Nir Oren, editors, *Proceedings of the 3rd International Workshop on Theory and Applications of Formal Argumentation (TAFA 2015), Revised Selected Papers*, volume 9524 of *Lecture Notes in Computer Science*, pages 1–20. Springer, 2015.

[Barabasi and Albert, 1999] Albert-Laszlo Barabasi and Reka Albert. Emergence of scaling in random networks. *Science*, 286(5439):11, 1999.

[Baroni et al., 2005] Pietro Baroni, Massimiliano Giacomin, and Giovanni Guida. SCC-recursiveness: a general schema for argumentation semantics. *Artificial Intelligence*, 168(1-2):165–210, 2005.

[Baroni et al., 2011a] Pietro Baroni, Martin Caminada, and Massimiliano Giacomin. An introduction to argumentation semantics. *The Knowledge Engineering Review*, 26(4):365–410, 2011.

[Baroni et al., 2011b] Pietro Baroni, Federico Cerutti, Massimiliano Giacomin, and Giovanni Guida. AFRA: argumentation framework with recursive attacks. *International Journal of Approximate Reasoning*, 52(1):19–37, 2011.

[Baroni et al., 2014] Pietro Baroni, Guido Boella, Federico Cerutti, Massimiliano Giacomin, Leendert van der Torre, and Serena Villata. On the Input/Output behavior of argumentation frameworks. *Artificial Intelligence*, 217(0):144–197, 2014.

[Baumann and Strass, 2014] Ringo Baumann and Hannes Strass. On the Maximal and Average Numbers of Stable Extensions. In Elizabeth Black, Sanjay Modgil, and Nir Oren, editors, *Proceedings of the 2nd International Workshop on Theory and Applications of Formal Argumentation (TAFA 2013)*, volume 8306 of *Lecture Notes in Computer Science*, pages 111–126. Springer, 2014.

[Beierle et al., 2015] Christoph Beierle, Florian Brons, and Nico Potyka. A software system using a SAT solver for reasoning under complete, stable, preferred, and grounded argumentation semantics. In Steffen Hölldobler, Markus Krötzsch, Rafael Peñaloza, and Sebastian Rudolph, editors, *Proceedings of the 38th Annual German Conference on AI (KI 2015)*, volume 9324 of *Lecture Notes in Computer Science*, pages 241–248. Springer, 2015.

[Bench-Capon, 2003] T J M Bench-Capon. Persuasion in Practical Argument Using Value Based Argumentation Frameworks. *Journal of Logic and Computation*, 13(3):429–448, 2003.

[Besnard and Doutre, 2004] Philippe Besnard and Sylvie Doutre. Checking the acceptability of a set of arguments. In James P. Delgrande and Torsten Schaub, editors, *Proceedings of the 10th International Workshop on Non-Monotonic Reasoning (NMR 2004)*, pages 59–64, 2004.

[Besnard and Hunter, 2006] Philippe Besnard and Anthony Hunter. Knowledgebase compilation for efficient logical argumentation. In Patrick Doherty, John Mylopoulos, and Christopher A. Welty, editors, *Proceedings of the 10th International Conference on Principles of Knowledge Representation and Reasoning (KR 2006)*, pages 123–133. AAAI Press, 2006.

[Besnard and Hunter, 2008] Philippe Besnard and Anthony Hunter. *Elements of Argumentation*. MIT Press, 2008.

[Besnard and Hunter, 2014] Philippe Besnard and Anthony Hunter. Constructing argument graphs with deductive arguments: a tutorial. *Argument & Computation*, 5(1):5–30, 2014.

[Besnard et al., 2009] Philippe Besnard, Anthony Hunter, and Stefan Woltran. Encoding deductive argumentation in quantified Boolean formulae. *Artificial Intelligence*, 173(15):1406–1423, 2009.

[Besnard et al., 2010] Philippe Besnard, Éric Grégoire, Cédric Piette, and Badran Raddaoui. MUS-based generation of arguments and counter-arguments. In Reda Alhajj, James B. D. Joshi, and Mei-Ling Shyu, editors, *Proceedings of the IEEE International Conference on Information Reuse and Integration, (IRI 2010)*, pages 239–244. IEEE Systems, Man, and Cybernetics Society, 2010.

[Besnard et al., 2014] Philippe Besnard, Alejandro Javier García, Anthony Hunter, Sanjay Modgil, Henry Prakken, Guillermo R. Simari, and Francesca Toni. Introduction to structured argumentation. *Argument & Computation*, 5(1):1–4, 2014.

[Bex and Reed, 2012] Floris Bex and Chris Reed. Dialogue Templates for Automatic Argument Processing. In Bart Verheij, Stefan Szeider, and Stefan Woltran, editors, *Proceedings of the 4th International Conference on Computational Models of Argument (COMMA 2012)*, volume 245 of *Frontiers in Artificial Intelligence and Applications*, pages 366–377. IOS Press, 2012.

[Bex et al., 2013] Floris Bex, John Lawrence, Mark Snaith, and Chris Reed. Implementing the argument web. *Communications of the ACM*, 56(10):66, 2013.

[Bex et al., 2014] Floris Bex, Mark Snaith, John Lawrence, and Chris Reed. Argublogging: An application for the argument web. *Journal of Web Semantics*, 25:9–15, 2014.

[Biere et al., 2009] Armin Biere, Marijn Heule, Hans van Maaren, and Toby Walsh, editors. *Handbook of Satisfiability*, volume 185 of *Frontiers in Artificial Intelligence and Applications*. IOS Press, 2009.

[Bistarelli and Santini, 2010] Stefano Bistarelli and Francesco Santini. A common computational framework for semiring-based argumentation systems. In Helder Coelho,

Rudi Studer, and Michael Wooldridge, editors, *Proceedings of the 19th European Conference on Artificial Intelligence (ECAI 2010)*, volume 215 of *Frontiers in Artificial Intelligence and Applications*, pages 131–136. IOS Press, 2010.

[Bistarelli and Santini, 2011] Stefano Bistarelli and Francesco Santini. ConArg: A constraint-based computational framework for argumentation systems. In Taghi M. Khoshgoftaar and Xingquan (Hill) Zhu, editors, *Proceedings of the 23rd IEEE International Conference on Tools with Artificial Intelligence (ICTAI 2011)*, pages 605–612. IEEE Computer Society Press, 2011.

[Bistarelli and Santini, 2012a] Stefano Bistarelli and Francesco Santini. ConArg: a tool to solve (weighted) abstract argumentation frameworks with (soft) constraints. *CoRR*, abs/1212.2857, 2012.

[Bistarelli and Santini, 2012b] Stefano Bistarelli and Francesco Santini. Modeling and solving AFs with a constraint-based tool: ConArg. In Sanjay Modgil, Nir Oren, and Francesca Toni, editors, *Proceedings of the 1st International Workshop on Theory and Applications of Formal Argumentation (TAFA 2011)*, volume 7132 of *Lecture Notes in Computer Science*, pages 99–116. Springer, 2012.

[Bistarelli et al., 2009] Stefano Bistarelli, Daniele Pirolandi, and Francesco Santini. Solving weighted argumentation frameworks with soft constraints. In Javier Larrosa and Barry O'Sullivan, editors, *Proceedings of the 14th Annual ERCIM International Workshop on Constraint Solving and Constraint Logic Programming (CSCLP 2009), Revised Selected Papers*, volume 6384 of *Lecture Notes in Computer Science*, pages 1–18. Springer, 2009.

[Bistarelli et al., 2013] Stefano Bistarelli, Fabio Rossi, and Francesco Santini. A first comparison of abstract argumentation systems: A computational perspective. In Domenico Cantone and Marianna Nicolosi Asmundo, editors, *Proceedings of the 28th Italian Conference on Computational Logic*, volume 1068 of *CEUR Workshop Proceedings*, pages 241–245. CEUR-WS.org, 2013.

[Bistarelli et al., 2014] Stefano Bistarelli, Fabio Rossi, and Francesco Santini. Benchmarking hard problems in random abstract AFs: The stable semantics. In Simon Parsons, Nir Oren, Chris Reed, and Federico Cerutti, editors, *Proceedings of the 5th International Conference on Computational Models of Argument (COMMA 2014)*, volume 266 of *Frontiers in Artificial Intelligence and Applications*. IOS Press, 2014.

[Bistarelli et al., 2015] Stefano Bistarelli, Fabio Rossi, and Francesco Santini. A comparative test on the enumeration of extensions in abstract argumentation. *Fundamenta Informaticae*, 140(3-4):263–278, 2015.

[Bliem et al., 2012] Bernhard Bliem, Michael Morak, and Stefan Woltran. D-FLAT: Declarative problem solving using tree decompositions and answer-set programming. *TPLP*, 12(4-5):445–464, 2012.

[Bliem, 2012] Bernhard Bliem. Decompose, Guess & Check - Declarative problem solving on tree decompositions. Master's thesis, Vienna University of Technology, 2012. http://permalink.obvsg.at/AC07814499.

[Bondarenko et al., 1997] Andrei Bondarenko, Phan Minh Dung, Robert A. Kowalski, and Francesca Toni. An abstract, argumentation-theoretic approach to default reasoning. *Artificial Intelligence*, 93:63–101, 1997.

[Brachman and Levesque, 2004] Ronald J. Brachman and Hector J. Levesque. *Knowledge Representation and Reasoning*. The Morgan Kaufmann Series in Artificial Intelligence. Morgan Kaufmann Publishers, 2004.

[Brewka and Woltran, 2014] Gerhard Brewka and Stefan Woltran. GRAPPA: A semantical framework for graph-based argument processing. In Torsten Schaub, Gerhard Friedrich, and Barry O'Sullivan, editors, *Proceedings of the 21st European Conference on Artificial Intelligence (ECAI 2014) - Including Prestigious Applications of Intelligent Systems (PAIS 2014)*, volume 263 of *Frontiers in Artificial Intelligence and Applications*, pages 153–158. IOS Press, 2014.

[Brewka et al., 2011] Gerhard Brewka, Thomas Eiter, and Mirosław Truszczyński. Answer set programming at a glance. *Commun. ACM*, 54(12):92–103, 2011.

[Brewka et al., 2013] Gerhard Brewka, Stefan Ellmauthaler, Hannes Strass, Johannes Peter Wallner, and Stefan Woltran. Abstract dialectical frameworks revisited. In *Proceedings of the 23rd International Joint Conference on Artificial Intelligence IJCAI 2013*, pages 803–809. AAAI Press, 2013.

[Brochenin et al., 2015] Remi Brochenin, Thomas Linsbichler, Marco Maratea, Johannes Peter Wallner, and Stefan Woltran. Abstract solvers for dung's argumentation frameworks. In Elizabeth Black, Sanjay Modgil, and Nir Oren, editors, *Proceedings of the 3rd International Workshop on Theory and Applications of Formal Argument (TAFA 2015)*, volume 9524 of *Lecture Notes in Computer Science*. Springer, 2015.

[Bryant and Krause, 2008] Daniel Bryant and Paul J. Krause. A review of current defeasible reasoning implementations. *The Knowledge Engineering Review*, 23(3):227–260, 2008.

[Bryant et al., 2006] Daniel Bryant, Paul J. Krause, and Gerard Vreeswijk. Argue tuProlog: A lightweight argumentation engine for agent applications. In Paul E. Dunne and Trevor J. M. Bench-Capon, editors, *Proceedings of the 1st International Conference on Computational Models of Argument (COMMA 2006)*, volume 144 of *Frontiers in Artificial Intelligence and Applications*, pages 27–32. IOS Press, 2006.

[Budzynska et al., 2014] Katarzyna Budzynska, Mathilde Janier, Juyeon Kang, Chris Reed, Patrick Saint-Dizier, Manfred Stede, and Olena Yaskorska. Towards Argument Mining from Dialogue. In Simon Parsons, Nir Oren, Chris Reed, and Federico Cerutti, editors, *Proceedings of the 5th International Conference on Computational Models of Argument (COMMA 2014)*, volume 266 of *Frontiers in Artificial Intelligence and Applications*, pages 185–196. IOS Press, 2014.

[Cabrio and Villata, 2014a] Elena Cabrio and Serena Villata. Node: A benchmark of natural language arguments. In Simon Parsons, Nir Oren, Chris Reed, and Federico Cerutti, editors, *Proceedings of the 5th International Conference on Computational Models of Argument (COMMA 2014)*, volume 266 of *Frontiers in Artificial Intelligence and Applications*, pages 449–450. IOS Press, 2014.

[Cabrio and Villata, 2014b] Elena Cabrio and Serena Villata. Towards a benchmark of natural language arguments. In Sébastien Konieczny and Hans Tompits, editors, *Proceedings of the 15th International Workshop on Non-Monotonic Reasoning (NMR 2014)*, 2014.

[Cabrio et al., 2013] Elena Cabrio, Serena Villata, and Fabien Gandon. A support framework for argumentative discussions management in the web. In Philipp Cimiano, Óscar Corcho, Valentina Presutti, Laura Hollink, and Sebastian Rudolph, editors, *Proceedings of the 10th Extended Semantic Web Conference (ESWC 2013)*, volume 7882 of *Lecture Notes in Computer Science*. Springer, 2013.

[Caminada and Amgoud, 2005] Martin Caminada and Leila Amgoud. An Axiomatic Account of Formal Argumentation. In Manuela M. Veloso and Subbarao Kambhampati, editors, *Proceedings of the 20th National Conference on Artificial Intelligence and the 17th Innovative Applications of Artificial Intelligence Conference (AAAI 2005)*, pages 608–613. AAAI Press / The MIT Press, 2005.

[Caminada and Amgoud, 2007] Martin Caminada and Leila Amgoud. On the evaluation of argumentation formalisms. *Artificial Intelligence*, 171(5-6):286–310, 2007.

[Caminada, 2007] Martin Caminada. An algorithm for computing semi-stable semantics. In Khaled Mellouli, editor, *Proceedings of the 9th European Conference on Symbolic and Quantitative Approaches to Reasoning with Uncertainty (ECSQARU 2007)*, volume 4724 of *Lecture Notes in Computer Science*, pages 222–234. Springer, 2007.

[Caminada, 2010] Martin Caminada. An algorithm for stage semantics. In Pietro Baroni, Federico Cerutti, Massimiliano Giacomin, and Guillermo R. Simari, editors, *Proceedings of the 3rd International Conference on Computational Models of Argument (COMMA 2010)*, volume 216 of *Frontiers in Artificial Intelligence and Applications*, pages 147–158. IOS Press, 2010.

[Capobianco et al., 2004] Marcela Capobianco, Carlos Iván Chesñevar, and Guillermo R. Simari. An argument-based framework to model an agent's beliefs in a dynamic environment. In Iyad Rahwan, Pavlos Moraitis, and Chris Reed, editors, *Proceedings of the 1st International Workshop on Argumentation in Multi-Agent Systems (ArgMAS 2004), Revised Selected and Invited Papers*, volume 3366 of *Lecture Notes in Computer Science*, pages 95–110. Springer, 2004.

[Cartwright and Atikinson, 2009] Dan Cartwright and Katie Atikinson. Using Computational Argumentation to Support E-participation. *IEEE Intelligent Systems*, pages 42–52, 2009.

[Cartwright and Atkinson, 2008] Dan Cartwright and Katie Atkinson. Political Engagement Through Tools for Argumentation. In Philippe Besnard, Sylvie Doutre, and Anthony Hunter, editors, *Proceedings of the 2nd International Conference on Computational Models of Argument (COMMA 2008)*, volume 172 of *Frontiers in Artificial Intelligence and Applications*, pages 116–127. IOS Press, 2008.

[Cartwright et al., 2009] Dan Cartwright, Katie Atkinson, and Trevor Bench-Capon. Supporting argument in e-democracy. In Alexander Prosser and Peter Parycek, editors, *Proceedings of the 3rd Conference on Electronic Democracy (EDEM 2009)*, pages 151–160, 2009.

[Cayrol et al., 2014] Claudette Cayrol, Sylvie Doutre, and Marie-Christine Lagasquie-Schiex. GRAFIX: a Tool for Abstract Argumentation. In Simon Parsons, Nir Oren, Chris Reed, and Federico Cerutti, editors, *Proceedings of the 5th International Conference on Computational Models of Argument (COMMA 2014)*, volume 266 of *Frontiers in Artificial Intelligence and Applications*, pages 453–454. IOS Press, 2014.

[Cecchi et al., 2006] Laura A. Cecchi, Pablo R. Fillottrani, and Guillermo R. Simari. On the complexity of DeLP through game semantics. In Jürgen Dix and Anthony Hunter, editors, *Proceedings of the 11th International Workshop on Nonmonotonic Reasoning (NMR 2006)*, pages 390–398, 2006.

[Cerutti et al., 2014a] Federico Cerutti, Paul E. Dunne, Massimiliano Giacomin, and Mauro Vallati. Computing preferred extensions in abstract argumentation: a SAT-based approach. In Elizabeth Black, Sanjay Modgil, and Nir Oren, editors, *Proceedings of the 2nd International Workshop on Theory and Applications of Formal Argumentation, Revised Selected papers (TAFA 2013)*, volume 8306 of *Lecture Notes in Computer Science*, pages 176–193. Springer, 2014.

[Cerutti et al., 2014b] Federico Cerutti, Massimiliano Giacomin, and Mauro Vallati. Algorithm Selection for Preferred Extensions Enumeration. In Simon Parsons, Nir Oren, Chris Reed, and Federico Cerutti, editors, *Proceedings of the 5th International Conference on Computational Models of Argument (COMMA 2014)*, volume 266 of *Frontiers in Artificial Intelligence and Applications*, pages 221–232. IOS Press, 2014.

[Cerutti et al., 2014c] Federico Cerutti, Massimiliano Giacomin, and Mauro Vallati. ArgSemSAT: solving argumentation problems using SAT. In Simon Parsons, Nir Oren, Chris Reed, and Federico Cerutti, editors, *Proceedings of the 5th International Conference on Computational Models of Argument (COMMA 2014)*, volume 266 of *Frontiers in Artificial Intelligence and Applications*, pages 455–456. IOS Press, 2014.

[Cerutti et al., 2014d] Federico Cerutti, Massimiliano Giacomin, and Mauro Vallati. Generating challenging benchmark AFs. In Simon Parsons, Nir Oren, Chris Reed, and Federico Cerutti, editors, *Proceedings of the 5th International Conference on Computational Models of Argument (COMMA 2014)*, volume 266 of *Frontiers in Artificial Intelligence and Applications*. IOS Press, 2014.

[Cerutti et al., 2014e] Federico Cerutti, Massimiliano Giacomin, Mauro Vallati, and Marina Zanella. A SCC Recursive Meta-Algorithm for Computing Preferred Labellings in Abstract Argumentation. In Chitta Baral and Giuseppe De Giacomo, editors, *Proceedings of the 14th International Conference on Principles of Knowledge Representation and Reasoning (KR 2014)*, pages 42–51, 2014.

[Cerutti et al., 2014f] Federico Cerutti, Nir Oren, Hannes Strass, Matthias Thimm, and Mauro Vallati. A benchmark framework for a computational argumentation competition (demo paper). In Simon Parsons, Nir Oren, Chris Reed, and Federico Cerutti, editors, *Proceedings of the 5th International Conference on Computational Models of Argumentation (COMMA 2014)*, volume 266 of *Frontiers in Artificial Intelligence and Applications*. IOS Press, 2014.

[Cerutti et al., 2015] Federico Cerutti, Ilias Tachmazidis, Sotirios Batsakis, Mauro Vallati, Massimiliano Giacomin, and Grigoris Antoniou. Exploiting Parallelism for Hard Problems in Abstract Argumentation. In Blai Bonet and Sven Koenig, editors, *Proceedings of the 29th AAAI Conference on Artificial Intelligence (AAAI 2015)*, pages 1475–1481. AAAI Press, 2015.

[Cerutti et al., 2016a] Federico Cerutti, Mauro Vallati, and Massimiliano Giacomin. Generating Structured Argumentation Frameworks: AFBenchGen2. In Pietro Baroni, Thomas F. Gordon, Tatjana Scheffler, and Manfred Stede, editors, *Proceedings of the 6th International Conference on Computational Models of Argument (COMMA 2016)*, pages 467–468, 2016.

[Cerutti et al., 2016b] Federico Cerutti, Mauro Vallati, and Massimiliano Giacomin. jArgSemSAT: an efficient off-the-shelf solver for abstract argumentation frameworks. In James P. Delgrande and Frank Wolter, editors, *15th International Conference on Principles of Knowledge Representation and Reasoning (KR2016)*, pages 541–544, 2016.

[Cerutti et al., 2016c] Federico Cerutti, Mauro Vallati, and Massimiliano Giacomin. On the Effectiveness of Automated Configuration in Abstract Argumentation Reasoning. In Pietro Baroni, Thomas F. Gordon, Tatjana Scheffler, and Manfred Stede, editors, *Proceedings of the 6th International Conference on Computational Models of Argument (COMMA 2016)*, pages 199–206. IOS Press, 2016.

[Cerutti et al., 2016d] Federico Cerutti, Mauro Vallati, and Massimiliano Giacomin. Where Are We Now? State of the Art and Future Trends of Solvers for Hard Argumentation Problems. In Pietro Baroni, Thomas F. Gordon, Tatjana Scheffler, and Manfred Stede, editors, *Proceedings of the 6th International Conference on Computational Models of Argument (COMMA 2016)*, pages 207–218. IOS Press, 2016.

[Cerutti et al., 2017] Federico Cerutti, Mauro Vallati, and Massimiliano Giacomin. An Efficient Java-Based Solver for Abstract Argumentation Frameworks: jArgSemSAT. *International Journal on Artificial Intelligence Tools*, 26(02):1750002–1–26, 2017.

[Charwat et al., 2012] Günther Charwat, Johannes P. Wallner, and Stefan Woltran. Utilizing ASP for Generating and Visualizing Argumentation Frameworks. In Michael Fink and Yuliya Lierler, editors, *Proceedings of the 5th Workshop on Answer Set Programming and Other Computing Paradigms (ASPOCP 2012)*, pages 51–65, 2012.

[Charwat et al., 2015] Günther Charwat, Wolfgang Dvořák, Sarah Alice Gaggl, Johannes Peter Wallner, and Stefan Woltran. Methods for solving reasoning problems in abstract argumentation - A survey. *Artificial Intelligence*, 220:28–63, 2015.

[Charwat, 2012] Günther Charwat. Tree-decomposition based algorithms for abstract argumentation frameworks. Master's thesis, Vienna University of Technology, 2012. http://permalink.obvsg.at/AC07812654.

[Chesñevar and Simari, 2007] Carlos Iván Chesñevar and Guillermo R. Simari. A lattice-based approach to computing warranted beliefs in skeptical argumentation frameworks. In Manuela M. Veloso, editor, *Proceedings of the 20th International Joint Conference on Artificial Intelligence (IJCAI 2007)*, pages 280–285, 2007.

[Chesñevar et al., 2000] Carlos Iván Chesñevar, Guillermo R. Simari, and Alejandro Javier García. Pruning search space in defeasible argumentation. In *Proceedings of the Workshop on Advances and Trends in AI. International Conferences of the Chilean Computer Science Society*, pages 46–55, 2000.

[Chesñevar et al., 2006] Carlos Iván Chesñevar, Jarred McGinnis, Sanjay Modgil, Iyad Rahwan, Chris Reed, Guillermo R. Simari, Matthew South, Gerard A. W. Vreeswijk, and Steven Willmot. Towards an argument interchange format. *The Knowledge Engineering Review*, 21(04):293, 2006.

[Chklovski et al., 2003] Timothy Chklovski, Yolanda Gil, Varun Ratnakar, and John Lee. Trellis: Supporting decision making via argumentation in the semantic web. In Katia Sycara and John Mylopoulos, editors, *Proceedings of the 2nd International Semantic Web Conference (ISWC 2003)*. Citeseer, 2003.

[Cormen et al., 2009] Thomas H. Cormen, Charles E. Leiserson, Ronald L. Rivest, and Clifford Stein. *Introduction to Algorithms*. MIT Press, 2009.

[Courcelle, 1989] B Courcelle. The monadic second-order logic of graphs, II: infinite graphs of bounded width. *Math. Systems Theory*, 21:187–221, 1989.

[Craven and Toni, 2016] Robert Craven and Francesca Toni. Argument graphs and assumption-based argumentation. *Artificial Intelligence*, 233:1–59, 2016.

[Craven et al., 2012] Robert Craven, Francesca Toni, Cristian Cadar, Adrian Hadad, and Matthew Williams. Efficient argumentation for medical decision-making. In Gerhard Brewka, Thomas Eiter, and Sheila A. McIlraith, editors, *Proceedings of the 13th International Conference on Principles of Knowledge Representation and Reasoning (KR 2012)*. AAAI Press, 2012.

[Craven et al., 2013] Robert Craven, Francesca Toni, and Matthew Williams. Graph-based dispute derivations in assumption-based argumentation. In Elizabeth Black, Sanjay Modgil, and Nir Oren, editors, *Proceedings of the 2nd International Workshop on Theory and Applications of Formal Argumentation (TAFA 2013), Revised Selected papers*, volume 8306 of *Lecture Notes in Computer Science*, pages 46–62. Springer, 2013.

[Creignou et al., 2011] Nadia Creignou, Johannes Schmidt, Michael Thomas, and Stefan Woltran. Complexity of logic-based argumentation in post's framework. *Argument & Computation*, 2(2-3):107–129, 2011.

[Cyras and Toni, 2016a] Kristijonas Cyras and Francesca Toni. ABA+: assumption-based argumentation with preferences. In James P. Delgrande and Frank Wolter, editors, *15th International Conference on Principles of Knowledge Representation and Reasoning (KR2016)*, pages 553–556. AAAI Press, 2016.

[Cyras and Toni, 2016b] Kristijonas Cyras and Francesca Toni. ABA+: assumption-based argumentation with preferences. *CoRR*, abs/1610.03024, 2016.

[Darwiche and Marquis, 2002] Adnan Darwiche and Pierre Marquis. A knowledge compilation map. *Journal of Artificial Intelligence Research*, 17:229–264, 2002.

[Dauphin and Schulz, 2014] Jeremie Dauphin and Claudia Schulz. Arg Teach - A Learning Tool for Argumentation Theory. In *Proceedings of the 26th IEEE International Conference on Tools with Artificial Intelligence (ICTAI 2014)*, pages 776–783. IEEE Computer Society, 2014.

[De Liddo and Buckingham Shum, 2010] Anna De Liddo and Simon Buckingham Shum. Cohere: A prototype for contested collective intelligence. In Anna De Liddo, Simon Buckingham Shum, Gregorio Convertino, Ágnes Sándor, and Mark Klein, editors, *ACM Computer Supported Cooperative Work (CSCW 2010) - Workshop: Collective Intelligence In Organizations - Toward a Research Agenda*, 2010.

[Diller et al., 2015] Martin Diller, Johannes Peter Wallner, and Stefan Woltran. Reasoning in abstract dialectical frameworks using quantified boolean formulas. *Argument & Computation*, 6(2):149–177, 2015.

[Doutre and Mengin, 2001] Sylvie Doutre and Jérôme Mengin. Preferred extensions of argumentation frameworks: Query answering and computation. In Rajeev Goré, Alexander Leitsch, and Tobias Nipkow, editors, *Proceedings of the 1st International Joint Conference on Automated Reasoning (IJCAR 2001)*, volume 2083 of *Lecture Notes in Computer Science*, pages 272–288. Springer, 2001.

[Dung et al., 2006] Phan Minh Dung, Robert A. Kowalski, and Francesca Toni. Dialectic proof procedures for assumption-based, admissible argumentation. *Artificial Intelligence*, 170(2):114–159, 2006.

[Dung et al., 2007] Phan Minh Dung, Paolo Mancarella, and Francesca Toni. Computing ideal sceptical argumentation. *Artificial Intelligence*, 171(10-15):642–674, 2007.

[Dung, 1995] Phan Minh Dung. On the Acceptability of Arguments and its Fundamental Role in Nonmonotonic Reasoning, Logic Programming and n-Person Games. *Artificial Intelligence*, 77(2):321–358, 1995.

[Dunne and Bench-Capon, 2002] P E Dunne and T J M Bench-Capon. Coherence in finite argument systems. *Artificial Intelligence*, 141:187–203, 2002.

[Dunne and Bench-Capon, 2003] Paul E Dunne and T J M Bench-Capon. Two party immediate response disputes: properties and efficiency. *Artificial Intelligence*, 149(2):221–250, 2003.

[Dunne et al., 2009] P E Dunne, A Hunter, P McBurney, S Parsons, and M Wooldridge. Inconsistency tolerance in weighted argument systems. In *AAMAS '09: Proceedings of The 8th International Conference on Autonomous Agents and Multiagent Systems*, pages 851–858, Richland, SC, 2009. International Foundation for Autonomous Agents and Multiagent Systems.

[Dunne, 2007] Paul E. Dunne. Computational properties of argument systems satisfying graph-theoretic constraints. *Artificial Intelligence*, 171(10-15):701–729, 2007.

[Dvořák et al., 2010] Wolfgang Dvořák, Stefan Szeider, and Stefan Woltran. Reasoning in argumentation frameworks of bounded clique-width. In Pietro Baroni, Federico Cerutti, Massimiliano Giacomin, and Guillermo R. Simari, editors, *Proceedings of the 3rd Conference on Computational Models of Argument (COMMA 2010)*, Frontiers in Artificial Intelligence and Applications, pages 219–230. IOS Press, 2010.

[Dvořák et al., 2012a] Wolfgang Dvořák, Sebastian Ordyniak, and Stefan Szeider. Augmenting tractable fragments of abstract argumentation. *Artificial Intelligence*, 186:157–173, 2012.

[Dvořák et al., 2012b] Wolfgang Dvořák, Reinhard Pichler, and Stefan Woltran. Towards fixed-parameter tractable algorithms for abstract argumentation. *Artificial Intelligence*, 186:1–37, 2012.

[Dvořák et al., 2012c] Wolfgang Dvořák, Stefan Szeider, and Stefan Woltran. Abstract argumentation via monadic second order logic. In Eyke Hüllermeier, Sebastian Link, Thomas Fober, and Bernhard Seeger, editors, *Proceedings of the 6th International Conference on Scalable Uncertainty Management (SUM 2012)*, volume 7520 of *Lecture Notes in Computer Science*, pages 85–98. Springer, 2012.

[Dvořák et al., 2013a] Wolfgang Dvořák, Sarah A. Gaggl, Johannes P. Wallner, and Stefan Woltran. Making use of advances in answer-set programming for abstract argumentation systems. In Hans Tompits, Salvador Abreu, Johannes Oetsch, Jörg Pührer, Dietmar Seipel, Masanobu Umeda, and Armin Wolf, editors, *Proceedings of the 19th International Conference on Applications of Declarative Programming and Knowledge Management (INAP 2011), Revised Selected Papers*, volume 7773 of *Lecture Notes in Artificial Intelligence*, pages 114–133. Springer, 2013.

[Dvořák et al., 2013b] Wolfgang Dvořák, Michael Morak, Clemens Nopp, and Stefan Woltran. dynPARTIX - A dynamic programming reasoner for abstract argumentation. In Hans Tompits, Salvador Abreu, Johannes Oetsch, Jörg Pührer, Dietmar Seipel, Masanobu Umeda, and Armin Wolf, editors, *Proceedings of the 19th International Conference on Applications of Declarative Programming and Knowledge Management (INAP 2011), Revised Selected Papers*, volume 7773 of *Lecture Notes in Artificial Intelligence*, pages 259–268. Springer, 2013.

[Dvořák et al., 2014] Wolfgang Dvořák, Matti Järvisalo, Johannes Peter Wallner, and Stefan Woltran. Complexity-sensitive decision procedures for abstract argumentation. *Artificial Intelligence*, 206:53–78, 2014.

[Dvořák et al., 2015] Wolfgang Dvořák, Sarah Alice Gaggl, Thomas Linsbichler, and Johannes Peter Wallner. Reduction-Based Approaches to Implement Modgil's Extended Argumentation Frameworks. In Thomas Eiter, Hannes Strass, Miroslaw Truszczynski, and Stefan Woltran, editors, *Advances in Knowledge Representation, Logic Programming, and Abstract Argumentation - Essays Dedicated to Gerhard Brewka on the*

Occasion of His 60th Birthday, volume 9060 of *Lecture Notes in Computer Science*, pages 249–264. Springer, 2015.

[Efstathiou and Hunter, 2008] Vasiliki Efstathiou and Anthony Hunter. Algorithms for effective argumentation in classical propositional logic: A connection graph approach. In Sven Hartmann and Gabriele Kern-Isberner, editors, *Proceedings of the 5th International Symposium on Foundations of Information and Knowledge Systems (FoIKS 2008)*, volume 4932 of *Lecture Notes in Computer Science*, pages 272–290. Springer, 2008.

[Efstathiou and Hunter, 2011] Vasiliki Efstathiou and Anthony Hunter. Algorithms for generating arguments and counterarguments in propositional logic. *International Journal of Approximate Reasoning*, 52(6):672–704, 2011.

[Egly and Woltran, 2006] Uwe Egly and Stefan Woltran. Reasoning in argumentation frameworks using quantified boolean formulas. In Paul E. Dunne and Trevor J. M. Bench-Capon, editors, *Proceedings of the 1st International Conference on Computational Models of Argument (COMMA 2006)*, volume 144 of *Frontiers in Artificial Intelligence and Applications*, pages 133–144. IOS Press, 2006.

[Egly et al., 2008] Uwe Egly, Sarah A. Gaggl, and Stefan Woltran. ASPARTIX: Implementing Argumentation Frameworks Using Answer-Set Programming. In Maria G. de la Banda and Enrico Pontelli, editors, *Proceedings of the 24th International Conference on Logic Programming (ICLP 2008)*, volume 5366 of *Lecture Notes in Computer Science*, pages 734–738. Springer, 2008.

[Egly et al., 2010a] Uwe Egly, Sarah A. Gaggl, and Stefan Woltran. Answer-set programming encodings for argumentation frameworks. *Argument & Computation*, 1(2):147–177, 2010.

[Egly et al., 2010b] Uwe Egly, Sarah Alice Gaggl, Paul Wandl, and Stefan Woltran. ASPARTIX Conquers the Web. Software demonstration at the 3rd International Conference on Computational Models of Argument (COMMA 2010), http://comma2010.unibs.it/demos/Egly_etal.pdf, 2010.

[Eiter and Gottlob, 1995] Thomas Eiter and Georg Gottlob. The Complexity of Logic-Based Abduction. *Journal of the ACM*, 42(1):3–42, 1995.

[Eiter et al., 1997] Thomas Eiter, Georg Gottlob, and Heikki Mannila. Disjunctive datalog. *ACM Transactions on Database Systems*, 22(3):364–418, 1997.

[Ellmauthaler and Strass, 2014] Stefan Ellmauthaler and Hannes Strass. The DIAMOND system for computing with abstract dialectical frameworks. In Simon Parsons, Nir Oren, Chris Reed, and Federico Cerutti, editors, *Proceedings of the 5th International Conference on Computational Models of Argument (COMMA 2014)*, volume 266 of *Frontiers in Artificial Intelligence and Applications*, pages 233–240. IOS Press, 2014.

[Ennals et al., 2010] Rob Ennals, Beth Trushkowsky, and John Mark Agosta. Highlighting disputed claims on the web. In Michael Rappa, Paul Jones, Juliana Freire, and Soumen Chakrabarti, editors, *Proceedings of the 19th International Conference on World Wide Web (WWW 2010)*, pages 341–350, New York, New York, USA, 2010. ACM Press.

[Erdös and Rényi, 1959] Paul Erdös and Alfréd Rényi. On random graphs I. *Publ. Math. Debrecen*, 6:290–297, 1959.

[Evripidou and Toni, 2014] Valentinos Evripidou and Francesca Toni. Quaestio-it.com: a social intelligent debating platform. *Journal of Decision Systems*, 23(3):333–349, 2014.

[Faber et al., 2016] Wolfgang Faber, Mauro Vallati, Federico Cerutti, and Massimiliano Giacomin. Solving set optimization problems by cardinality optimization with an application to argumentation. In Gal A. Kaminka, Maria Fox, Paolo Bouquet, Eyke Hüllermeier, Virginia Dignum, Frank Dignum, and Frank van Harmelen, editors, *22nd European Conference on Artificial Intelligence (ECAI 2016) – Including Prestigious Applications of Artificial Intelligence (PAIS 2016)*, pages 966–973, 2016.

[Faridani et al., 2010] Siamak Faridani, Ephrat Bitton, Kimiko Ryokai, and Ken Goldberg. Opinion space. In Elizabeth D. Mynatt, Don Schoner, Geraldine Fitzpatrick, Scott E. Hudson, W. Keith Edwards, and Tom Rodden, editors, *Proceedings of the 28th International Conference on Human Factors in Computing Systems (CHI 2010)*, pages 1175–1184, New York, New York, USA, 2010. ACM Press.

[Freeman and Farley, 1996] Kathleen Freeman and Arthur M. Farley. A model of argumentation and its application to legal reasoning. *Artificial Intelligence and Law*, 4(3-4):163–197, 1996.

[Gabbay and Rodrigues, 2015] Dov M. Gabbay and Odinaldo Rodrigues. Equilibrium states in numerical argumentation networks. *Logica Universalis*, 9(4):411–473, 2015.

[Gabbay and Rodriguez, 2014] Dov M. Gabbay and Odinaldo Rodriguez. A self-correcting iteration schema for argumentation networks. In Simon Parsons, Nir Oren, Chris Reed, and Federico Cerutti, editors, *Proceedings of the 5th International Conference on Computational Models of Argument (COMMA 2014)*, volume 266 of *Frontiers in Artificial Intelligence and Applications*, pages 377–384. IOS Press, 2014.

[Gabbay, 2012] Dov M. Gabbay. Equational approach to argumentation networks. *Argument & Computation*, 3(2-3):87–142, 2012.

[Gaertner and Toni, 2007a] Dorian Gaertner and Francesca Toni. CaSAPI: a system for credulous and sceptical argumentation. In Guillermo R. Simari and Paolo Torroni, editors, *Workshop on Argumentation for Non-Monotonic Reasoning (NMR 2007)*, pages 80–95, 2007.

[Gaertner and Toni, 2007b] Dorian Gaertner and Francesca Toni. Computing arguments and attacks in assumption-based argumentation. *IEEE Intelligent Systems*, 22(6):24–33, 2007.

[Gaertner and Toni, 2008] Dorian Gaertner and Francesca Toni. Hybrid argumentation and its properties. In Philippe Besnard, Sylvie Doutre, and Anthony Hunter, editors, *Proceedings of the 2nd International Conference on Computational Models of Argument (COMMA 2008)*, volume 172 of *Frontiers in Artificial Intelligence and Applications*, pages 183–195. IOS Press, 2008.

[Gaggl and Manthey, 2015] Sarah Alice Gaggl and Norbert Manthey. ASPARTIX-D: ASP Argumentation Reasoning Tool – Dresden. In Matthias Thimm and Serena Villata, editors, *System Descriptions of the First International Competition on Computational Models of Argumentation (ICCMA'15)*, pages 29–32, 2015.

[Gaggl et al., 2015] Sarah Alice Gaggl, Norbert Manthey, Alessandro Ronca, Johannes Peter Wallner, and Stefan Woltran. Improved answer-set programming encodings for abstract argumentation. *TPLP*, 15(4-5):434–448, 2015.

[García and Simari, 2000] Alejandro Javier García and Guillermo R. Simari. Parallel defeasible argumentation. *Journal of Computer Science & Technology*, 1(2):37–51, 2000.

[García and Simari, 2004] Alejandro Javier García and Guillermo R. Simari. Defeasible logic programming: An argumentative approach. *TPLP*, 4(1-2):95–138, 2004.

[García and Simari, 2014] Alejandro Javier García and Guillermo R. Simari. Defeasible logic programming: DeLP-servers, contextual queries, and explanations for answers. *Argument & Computation*, 5(1):63–88, 2014.

[García, 1997] Alejandro Javier García. Defeasible logic programming: Definition and implementation. Master's thesis, Universidad Nacional del Sur, Computer Science Department, Bahía Blanca, Argentina, 1997.

[Gelfond and Lifschitz, 1991] Michael Gelfond and Vladimir Lifschitz. Classical negation in logic programs and disjunctive databases. *New Generation Computing*, 9(3-4):365–385, 1991.

[Gordon and Walton, 2016] Thomas F. Gordon and Douglas Walton. Formalizing balancing arguments. In Pietro Baroni, Thomas F. Gordon, Tatjana Scheffler, and Manfred Stede, editors, *Proceedings of the 6th International Conference on Computational Models of Argument (COMMA 2016)*, volume 287 of *Frontiers in Artificial*

Intelligence and Applications, pages 327–338. IOS Press, 2016. Revised version under http://www.tfgordon.de/publications/.
[Gordon et al., 2007] Thomas F. Gordon, Henry Prakken, and Douglas N. Walton. The Carneades model of argument and burden of proof. *Artificial Intelligence*, 171(10-15):875–896, 2007.
[Gordon, 2012] Thomas F. Gordon. The Carneades web service. In Bart Verheij, Stefan Szeider, and Stefan Woltran, editors, *Proceedings of the 4th International Conference on Computational Models of Argument (COMMA 2012)*, volume 245 of *Frontiers in Artificial Intelligence and Applications*, pages 517–518. IOS Press, 2012.
[Gordon, 2013] Thomas F. Gordon. Introducing the Carneades web application. In Enrico Francesconi and Bart Verheij, editors, *Proceedings of the International Conference on Artificial Intelligence and Law (ICAIL 2013)*, pages 243–244. ACM, 2013.
[Gorogiannis and Hunter, 2011] Nikos Gorogiannis and Anthony Hunter. Instantiating abstract argumentation with classical logic arguments: Postulates and properties. *Artificial Intelligence*, 175:1479–1497, 2011.
[Grégoire et al., 2009] Éric Grégoire, Bertrand Mazure, and Cédric Piette. Using local search to find msses and muses. *European Journal of Operational Research*, 199(3):640–646, 2009.
[Groza and Groza, 2015] Serban Groza and Adrian Groza. ProGraph: towards enacting bipartite graphs for abstract argumentation frameworks. In Matthias Thimm and Serena Villata, editors, *System Descriptions of the First International Competition on Computational Models of Argumentation (ICCMA'15)*, pages 29–32, 2015.
[Gürkan et al., 2010] Ali Gürkan, Luca Iandoli, Mark Klein, and Giuseppe Zollo. Mediating debate through on-line large-scale argumentation: Evidence from the field. *Information Sciences*, 180(19):3686–3702, 2010.
[Hadjisoteriou, 2015] Evgenios Hadjisoteriou. Computing Argumentation with Matrices. In Claudia Schulz and Daniel Liew, editors, *2015 Imperial College Computing Student Workshop (ICCSW 2015)*, volume 49 of *OpenAccess Series in Informatics (OASIcs)*, pages 29–36, Dagstuhl, Germany, 2015. Schloss Dagstuhl–Leibniz-Zentrum fuer Informatik.
[Hadoux et al., 2015] Emmanuel Hadoux, Aurélie Beynier, Nicolas Maudet, Paul Weng, and Anthony Hunter. Optimization of probabilistic argumentation with markov decision models. In Qiang Yang and Michael Wooldridge, editors, *Proceedings of the 24th International Joint Conference on Artificial Intelligence (IJCAI 2015)*, pages 2004–2010. AAAI Press, 2015.
[Hall, 1998] Mark A. Hall. *Correlation-based Feature Subset Selection for Machine Learning*. PhD thesis, University of Waikato, Hamilton, New Zealand, 1998.
[Heyninck and Straßer, 2016] Jesse Heyninck and Christian Straßer. Relations between assumption-based approaches in nonmonotonic logic and formal argumentation. In Gabriele Kern-Isberner and Renata Wassermann, editors, *Proceedings of the 16th International Workshop on Non-Monotonic Reasoning (NMR 2016)*, 2016.
[Hirsch and Gorogiannis, 2010] Robin Hirsch and Nikos Gorogiannis. The complexity of the warranted formula problem in propositional argumentation. *Journal of Logic and Computation*, 20(2):481–499, 2010.
[Hoffmann, 2005] Michael H. G. Hoffmann. Logical argument mapping: A method for overcoming cognitive problems of conflict management. *International Journal of Conflict Management*, 16(4):304–334, 2005.
[Hoffmann, 2007] Michael H. G. Hoffmann. Logical argument mapping. In Simon Buckingham Shum, Mikael Lind, and Hans Weigand, editors, *Proceedings of the 2nd International Conference on Pragmatic Web (ICPW 2007)*, volume 280 of *ACM International Conference Proceeding Series*, pages 41–47, New York, New York, USA, 2007. ACM.
[Hölldobler et al., 2011] Steffen Hölldobler, Norbert Manthey, Van Hau Nguyen, Julian Stecklina, and Peter Steinke. A short overview on modern parallel SAT-solvers. In

Proceedings of the International Conference on Advanced Computer Science and Information Systems, pages 201–206, 2011.

[Hunter and Williams, 2012] Anthony Hunter and Matthew Williams. Aggregating evidence about the positive and negative effects of treatments. *Artificial intelligence in medicine*, 56(3):173–90, 2012.

[Hunter, 2006a] Anthony Hunter. Approximate arguments for efficiency in logical argumentation. In Jürgen Dix and Anthony Hunter, editors, *Proceedings of the 11th International Workshop on Non-Monotonic Reasoning (NMR 2006)*, pages 383–388, 2006.

[Hunter, 2006b] Anthony Hunter. Contouring of knowledge for intelligent searching for arguments. In Gerhard Brewka, Silvia Coradeschi, Anna Perini, and Paolo Traverso, editors, *Proceedings of the 17th European Conference on Artificial Intelligence (ECAI 2006), Including Prestigious Applications of Intelligent Systems (PAIS 2006)*, volume 141 of *Frontiers in Artificial Intelligence and Applications*, pages 250–254. IOS Press, 2006.

[Hunter, 2014] Anthony Hunter. Probabilistic qualification of attack in abstract argumentation. *International Journal of Approximate Reasoning*, 55(2):607–638, 2014.

[III et al., 2012] Joseph J. Pfeiffer III, Timothy La Fond, Sebastián Moreno, and Jennifer Neville. Fast generation of large scale social networks with clustering. *CoRR*, abs/1202.4805, 2012.

[Introne, 2009] Joshua E. Introne. Supporting group decisions by mediating deliberation to improve information pooling. In Stephanie Teasley, Erling Havn, Wolfgang Prinz, and Wayne Lutters, editors, *Proceedings of the ACM 2009 International Conference on Supporting Group Work*, pages 189–198. ACM, 2009.

[Janier et al., 2014] Mathilde Janier, John Lawrence, and Chris Reed. OVA+: an Argument Analysis Interface. In Simon Parsons, Nir Oren, Chris Reed, and Federico Cerutti, editors, *Proceedings of the 5th International Conference on Computational Models of Argument (COMMA 2014)*, volume 266 of *Frontiers in Artificial Intelligence and Applications*, pages 463–464. IOS Press, 2014.

[Kakas and Moraitis, 2003] Antonis Kakas and P Moraitis. Argumentation based decision making for autonomous agents. In Jeffrey S. Rosenschein, Tuomas Sandholm, Michael Wooldridge, and Makoto Yokoo, editors, *Proceedings of the second international joint conference on Autonomous Agents and Multiagent Systems*, pages 883–890, 2003.

[Kakas et al., 1994] Antonis C. Kakas, Paolo Mancarella, and Phan Minh Dung. The Acceptability Semantics for Logic Programs. In Pascal Van Hentenryck, editor, *Proceedings of the Eleventh International Conference on Logic Programming*, pages 504–519, Cambridge, MA, USA, 1994. MIT Press.

[Krauthoff et al., 2016] Tobias Krauthoff, Michael Baurmann, Gregor Betz, and Martin Mauve. Dialog-based online argumentation. In Pietro Baroni, Thomas F. Gordon, Tatjana Scheffler, and Manfred Stede, editors, *Proceedings of the 6th International Conference on Computational Models of Argument (COMMA 2016)*, volume 287 of *Frontiers in Artificial Intelligence and Applications*, pages 33–40. IOS Press, 2016.

[Kriplean et al., 2011] Travis Kriplean, Jonathan T. Morgan, Deen Freelon, Alan Borning, and Lance Bennett. ConsiderIt: Improving structured public deliberation. In Desney S. Tan, Saleema Amershi, Bo Begole, Wendy A. Kellogg, and Manas Tungare, editors, *Proceedings of the International Conference on Human Factors in Computing Systems (CHI 2011), Extended Abstracts Volume*, pages 1831–1836. ACM, 2011.

[Lagniez et al., 2015] Jean-Marie Lagniez, Emmanuel Lonca, and Jean-Guy Mailly. Coquiaas: A constraint-based quick abstract argumentation solver. In *Proceedings of the 27th IEEE International Conference on Tools with Artificial Intelligence (ICTAI 2015)*, pages 928–935. IEEE Computer Society, 2015.

[Lamatz, 2015] Nico Lamatz. LamatzSolver-v0.1: A grounded extension finder based on the Java-Collection-Framework. In Matthias Thimm and Serena Villata, editors,

System Descriptions of the First International Competition on Computational Models of Argumentation (ICCMA'15), pages 29–32, 2015.

[Lawrence et al., 2012a] John Lawrence, Floris Bex, and Chris Reed. Dialogues on the Argument Web: Mixed Initiative Argumentation with Arvina. In Bart Verheij, Stefan Szeider, and Stefan Woltran, editors, *Proceedings of the 4th International Conference on Computational Models of Argument (COMMA 2012)*, volume 245 of *Frontiers in Artificial Intelligence and Applications*, pages 513–514. IOS Press, 2012.

[Lawrence et al., 2012b] John Lawrence, Floris Bex, Chris Reed, and Mark Snaith. Aifdb: Infrastructure for the argument web. In Bart Verheij, Stefan Szeider, and Stefan Woltran, editors, *Proceedings of the 4th International Conference on Computational Models of Argument (COMMA 2012)*, volume 245 of *Frontiers in Artificial Intelligence and Applications*, pages 515–516. IOS Press, 2012.

[Lehtonen et al., 2017] Tuomo Lehtonen, Johannes P. Wallner, and Matti Järvisalo. From structured to abstract argumentation: Assumption-based acceptance via AF reasoning. In Alessandro Antonucci, Laurence Cholvy, and Odile Papini, editors, *14th European Conference, ECSQARU 2017*, 2017. To appear.

[Leyton-Brown et al., 2003] Kevin Leyton-Brown, Eugene Nudelman, Galen Andrew, Jim McFadden, and Yoav Shoham. A portfolio approach to algorithm selection. In Georg Gottlob and Toby Walsh, editors, *Proceedings of the 18th International Joint Conference on Artificial Intelligence (IJCAI 2003)*, pages 1542–1542. Morgan Kaufmann, 2003.

[Li et al., 2012] Hengfei Li, Nir Oren, and Timothy J. Norman. Probabilistic Argumentation Frameworks. In Sanjay Modgil, Nir Oren, and Francesca Toni, editors, *Proceedings of the 1st International Workshop on Theorie and Applications of Formal Argumentation (TAFA 2011), Revised Selected Papers*, volume 7132 of *Lecture Notes in Computer Science*. Springer, 2012.

[Liao et al., 2013] Beishui Liao, Liyun Lei, and Jianhua Dai. Computing preferred labellings by exploiting sccs and most sceptically rejected arguments. In Elizabeth Black, Sanjay Modgil, and Nir Oren, editors, *Proceedings of the 2nd International Workshop on Theory and Applications of Formal Argumentation (TAFA 2013), Revised Selected papers*, volume 8306 of *LNCS*, pages 194–208. Springer, 2013.

[Liffiton et al., 2016] Mark H. Liffiton, Alessandro Previti, Ammar Malik, and Joao Marques-Silva. Fast, flexible MUS enumeration. *Constraints*, 21(2):223–250, 2016.

[Lowrance et al., 2008] John Lowrance, Ian Harrison, Andres Rodriguez, Eric Yeh, Tom Boyce, Janet Murdock, Jerome Thomere, and Ken Murray. Template-based structured argumentation. In *Knowledge Cartography*, pages 307–333. Springer, 2008.

[Macintosh, 2009] Ann Macintosh. Moving Toward Intelligent Policy Development? *IEEE Intelligent Systems*, 24(5):79–82, 2009.

[McAreavey et al., 2014] Kevin McAreavey, Weiru Liu, and Paul C. Miller. Computational approaches to finding and measuring inconsistency in arbitrary knowledge bases. *International Journal of Approximate Reasoning*, 55(8):1659–1693, 2014.

[Modgil and Caminada, 2009] Sanjay Modgil and Martin Caminada. Proof theories and algorithms for abstract argumentation frameworks. In Iyad Rahwan and Guillermo R. Simari, editors, *Argumentation in Artificial Intelligence*, pages 105–132. Springer, 2009.

[Modgil and Prakken, 2014] Sanjay Modgil and Henry Prakken. The $ASPIC^+$ framework for structured argumentation: a tutorial. *Argument & Computation*, 5(1):31–62, 2014.

[Modgil, 2009] Sanjay Modgil. Reasoning about preferences in argumentation frameworks. *Artificial Intelligence*, 173(9-10):901–934, 2009.

[Nieuwenhuis et al., 2006] R. Nieuwenhuis, A. Oliveras, and C. Tinelli. Solving SAT and SAT modulo theories: From an abstract Davis-Putnam-Logemann-Loveland procedure to DPLL(T). *Journal of the ACM*, 53(6):937–977, 2006.

[Nieves et al., 2008] Juan C. Nieves, Mauricio Osorio, and Ulises Cortés. Preferred extensions as stable models. *TPLP*, 8(4):527–543, 2008.

[Nofal et al., 2012] Samer Nofal, Paul E. Dunne, and Katie Atkinson. On preferred extension enumeration in abstract argumentation. In Bart Verheij, Stefan Szeider, and Stefan Woltran, editors, *Proceedings of the 4th Conference on Computational Models of Argument (COMMA 2012)*, volume 245 of *Frontiers in Artificial Intelligence and Applications*, pages 205–216. IOS Press, 2012.

[Nofal et al., 2014a] Samer Nofal, Katie Atkinson, and Paul E. Dunne. Algorithms for argumentation semantics: Labeling attacks as a generalization of labeling arguments. *Journal of Artificial Intelligence Research*, 49:635–668, 2014.

[Nofal et al., 2014b] Samer Nofal, Katie Atkinson, and Paul E. Dunne. Algorithms for decision problems in argument systems under preferred semantics. *Artificial Intelligence*, 207:23–51, 2014.

[Nofal, 2013] Samer Nofal. *Algorithms for Argument Systems*. PhD thesis, University of Liverpool, 2013. http://research-archive.liv.ac.uk/12173/.

[Oomidou et al., 2014] Yutaka Oomidou, Yuki Katsura, Hajime Sawamura, Jacques Riche, and Takeshi Hagiwara. Asynchronous Argumentation System PIRIKA for Anyone, Anytime, Anywhere, with the Balanced Semantics. In Simon Parsons, Nir Oren, Chris Reed, and Federico Cerutti, editors, *Proceedings of the 5th International Conference on Computational Models of Argument (COMMA 2014)*, volume 266 of *Frontiers in Artificial Intelligence and Applications*, pages 471–472. IOS Press, 2014.

[Parsons et al., 2003] Simon Parsons, Michael Wooldridge, and Leila Amgoud. Properties and complexity of some formal inter-agent dialogues. *Journal of Logic and Computation*, 13(3):347–376, 2003.

[Podlaszewski et al., 2011] Mikolaj Podlaszewski, Martin Caminada, and Gabriella Pigozzi. An implementation of basic argumentation components. In Liz Sonenberg, Peter Stone, Kagan Tumer, and Pinar Yolum, editors, *Proceedings of the 10th International Conference on Autonomous Agents and Multiagent Systems (AAMAS 2011)*, pages 1307–1308. IFAAMAS, 2011.

[Prakken, 2006] Henry Prakken. Combining sceptical epistemic reasoning with credulous practical reasoning. In Paul E. Dunne and Trevor J. M. Bench-Capon, editors, *Proceedings of the 1st International Conference on Computational Models of Argument (COMMA 2006)*, volume 144 of *Frontiers in Artificial Intelligence and Applications*, pages 311–322. IOS Press, 2006.

[Prakken, 2010] Henry Prakken. An abstract framework for argumentation with structured arguments. *Argument & Computation*, 1(2):93–124, 2010.

[Pührer, 2017] Jörg Pührer. ArgueApply: A mobile app for argumentation. In Marcello Balduccini and Tomi Janhunen, editors, *Proceedings of the 14th International Conference on Logic Programming and Nonmonotonic Reasoning (LPNMR 2017)*. Springer, 2017. To appear.

[Rahwan et al., 2011] Iyad Rahwan, Bita Banihashemi, Chris Reed, Douglas Walton, and Sherief Abdallah. Representing and classifying arguments on the Semantic Web. *The Knowledge Engineering Review*, 26(04):487–511, 2011.

[Reed, 2013] Chris Reed. Argument Corpora. Technical report, University of Dundee, 2013.

[Rice, 1976] John R. Rice. The algorithm selection problem. *Advances in Computers*, 15:65–118, 1976.

[Rodrigues, 2016] Odinaldo Rodrigues. Introducing eqargsolver: An argumentation solver using equational semantics. In Matthias Thimm, Federico Cerutti, Hannes Strass, and Mauro Vallati, editors, *Proceedings of the First International Workshop on Systems and Algorithms for Formal Argumentation (SAFA) co-located with the 6th International Conference on Computational Models of Argument (COMMA 2016)*, volume 1672 of *CEUR Workshop Proceedings*, pages 22–33. CEUR-WS.org, 2016.

[Rossi et al., 2006] Francesca Rossi, Peter van Beek, and Toby Walsh. *Handbook of Constraint Programming (Foundations of Artificial Intelligence)*. Elsevier Science Inc., New York, NY, USA, 2006.

[Rotstein et al., 2011] Nicolás D. Rotstein, Sebastian Gottifredi, Alejandro Javier García, and Guillermo R. Simari. A heuristics-based pruning technique for argumentation trees. In Salem Benferhat and John Grant, editors, *Proceedings of the 5th International Conference on Scalable Uncertainty Management (SUM 2011)*, volume 6929 of *Lecture Notes in Computer Science*, pages 177–190. Springer, 2011.

[Russell and Norvig, 2003] Stuart Russell and Peter Norvig. *Artificial Intelligence: A Modern Approach*. Pearson Education, second edition, 2003.

[Schneider et al., 2007] David C Schneider, Christian Voigt, and Gregor Betz. Argunet-a software tool for collaborative argumentation analysis and research. In *7th Workshop on Computational Models of Natural Argument (CMNA VII)*, 2007.

[Schneider et al., 2013] Jodi Schneider, Tudor Groza, and Alexandre Passant. A review of argumentation for the social semantic web. *Semantic Web*, 4(2):159–218, 2013.

[Schulz and Caminada, 2015] Claudia Schulz and Martin W. A. Caminada. On the equivalence between assumption-based argumentation and logic programming. In Sarah Gaggl, Juan Carlos Nieves, and Hannes Strass, editors, *1st International Workshop on Argumentation and Logic Programming (ArgLP 2015)*, 2015.

[Shum, 2008] Simon Buckingham Shum. Cohere: Towards web 2.0 argumentation. In Philippe Besnard, Sylvie Doutre, and Anthony Hunter, editors, *Proceedings of the 2nd International Conference on Computational Models of Argument (COMMA 2008)*, volume 172 of *Frontiers in Artificial Intelligence and Applications*, pages 97–108. IOS Press, 2008.

[Simari, 2011] Guillermo R. Simari. A brief overview of research in argumentation systems. In Salem Benferhat and John Grant, editors, *Proceedings of the 5th International Conference on Scalable Uncertainty Management (SUM 2011)*, volume 6929 of *Lecture Notes in Computer Science*, pages 81–95. Springer, 2011.

[Snaith and Reed, 2012] Mark Snaith and Chris Reed. TOAST: online aspic$^+$ implementation. In Bart Verheij, Stefan Szeider, and Stefan Woltran, editors, *Proceedings of the 4th International Conference on Computational Models of Argument (COMMA 2012)*, volume 245 of *Frontiers in Artificial Intelligence and Applications*, pages 509–510. IOS Press, 2012.

[Snaith et al., 2010] Mark Snaith, Joseph Devereux, John Lawrence, and Chris Reed. Pipelining argumentation technologies. In Pietro Baroni, Federico Cerutti, Massimiliano Giacomin, and Guillermo R. Simari, editors, *Proceedings of the 3rd International Conference on Computational Models of Argument (COMMA 2010)*, volume 216 of *Frontiers in Artificial Intelligence*. IOS Press, 2010.

[Spanoudakis et al., 2016] Nikolaos I. Spanoudakis, Antonis C. Kakas, and Pavlos Moraitis. Gorgias-B: Argumentation in Practice. In Pietro Baroni, Thomas F. Gordon, Tatjana Scheffler, and Manfred Stede, editors, *Proceedings of the 6th International Conference on Computational Models of Argument (COMMA 2016)*, pages 477–478, 2016.

[Sprotte, 2015] Kilian Sprotte. ASGL: Argumentation Semantics in Gecode and Lisp. In Matthias Thimm and Serena Villata, editors, *System Descriptions of the First International Competition on Computational Models of Argumentation (ICCMA'15)*, pages 41–44, 2015.

[Stolzenburg et al., 2003] Frieder Stolzenburg, Alejandro Javier García, Carlos Iván Chesñevar, and Guillermo R. Simari. Computing generalized specificity. *Journal of Applied Non-Classical Logics*, 13(1):87–113, 2003.

[Strass, 2014] Hannes Strass. Implementing instantiation of knowledge bases in argumentation frameworks. In Simon Parsons, Nir Oren, Chris Reed, and Federico Cerutti, editors, *Proceedings of the 5th International Conference on Computational Models of Argument (COMMA 2014)*, volume 266 of *Frontiers in Artificial Intelligence and Applications*, pages 475–476. IOS Press, 2014.

[Tabourier et al., 2011] Lionel Tabourier, Camille Roth, and Jean-Philippe Cointet. Generating constrained random graphs using multiple edge switches. *Journal of Experimental Algorithmics*, 16:1–15, 2011.

[Tang et al., 2012] Yuqing Tang, Kai Cai, Peter McBurney, Elizabeth Sklar, and Simon Parsons. Using argumentation to reason about trust and belief. *Journal of Logic and Computation*, 22(5):979–1018, 2012.

[Tang et al., 2016] Yuqing Tang, Nir Oren, and Katia P. Sycara. Markov Argumentation Random Fields. In Dale Schuurmans and Michael P. Wellman, editors, *Proceedings of the 30th AAAI Conference on Artificial Intelligence (AAAI 2016)*. AAAI Press, 2016.

[Tarjan, 1972] Robert E. Tarjan. Depth-first search and linear graph algorithms. *SIAM Journal on Computing*, 1(2):146–160, 1972.

[Thimm and García, 2010] Matthias Thimm and Alejandro Javier García. Classification and strategical issues of argumentation games on structured argumentation frameworks. In Wiebe van der Hoek, Gal A. Kaminka, Yves Lespérance, Michael Luck, and Sandip Sen, editors, *Proceedings of the 9th International Conference on Autonomous Agents and Multiagent Systems (AAMAS 2010)*, pages 1247–1254. IFAAMAS, 2010.

[Thimm and Kern-Isberner, 2008] Matthias Thimm and Gabriele Kern-Isberner. On the relationship of defeasible argumentation and answer set programming. In Philippe Besnard, Sylvie Doutre, and Anthony Hunter, editors, *Proceedings of the 2nd International Conference on Computational Models of Argument (COMMA 2008)*, volume 172 of *Frontiers in Artificial Intelligence and Applications*, pages 393–404. IOS Press, 2008.

[Thimm et al., 2016] Matthias Thimm, Serena Villata, Federico Cerutti, Nir Oren, Hannes Strass, and Mauro Vallati. Summary report of the first international competition on computational models of argumentation. *AI Magazine*, 37(1):102, 2016.

[Thimm, 2012] Matthias Thimm. A Probabilistic Semantics for Abstract Argumentation. In Luc De Raedt, Christian Bessière, Didier Dubois, Patrick Doherty, Paolo Frasconi, Fredrik Heintz, and Peter J. F. Lucas, editors, *Proceedings of the 20th European Conference on Artificial Intelligence (ECAI 2012)*, volume 242 of *Frontiers in Artificial Intelligence and Applications*. IOS Press, 2012.

[Thimm, 2014] Matthias Thimm. Tweety: A comprehensive collection of java libraries for logical aspects of artificial intelligence and knowledge representation. In Chitta Baral, Giuseppe De Giacomo, and Thomas Eiter, editors, *Proceedings of the Fourteenth International Conference on Principles of Knowledge Representation and Reasoning (KR 2014)*. AAAI Press, 2014.

[Toni and Sergot, 2011] Francesca Toni and Marek Sergot. Argumentation and answer set programming. In Marcello Balduccini and Tran C. Son, editors, *Logic Programming, Knowledge Representation, and Nonmonotonic Reasoning: Essays in Honor of Michael Gelfond*, volume 6565 of *Lecture Notes in Computer Science*, pages 164–180. Springer, 2011.

[Toni, 2013] Francesca Toni. A generalised framework for dispute derivations in assumption-based argumentation. *Artificial Intelligence*, 195:1–43, 2013.

[Toni, 2014] Francesca Toni. A tutorial on assumption-based argumentation. *Argument & Computation*, 5(1):89–117, 2014.

[Toniolo et al., 2014] Alice Toniolo, Timothy Dropps, Robin W. Ouyang, John A. Allen, Timothy J. Norman, Nir Oren, Mani B. Srivastava, and Paul Sullivan. Argumentation-based collaborative intelligence analysis in CISpaces. In Simon Parsons, Nir Oren, Chris Reed, and Federico Cerutti, editors, *Proceedings of the 5th International Conference on Computational Models of Argument (COMMA 2014)*, volume 266 of *Frontiers in Artificial Intelligence and Applications*, pages 481–482. IOS Press, 2014.

[Toniolo et al., 2015] Alice Toniolo, Timothy J. Norman, Anthony Etuk, Federico Cerutti, Robin Wentao Ouyang, Mani Srivastava, Nir Oren, Timothy Dropps, John A. Allen, and Paul Sullivan. Agent Support to Reasoning with Different Types of Evidence in Intelligence Analysis. In Rafael H. Bordini, Edith Elkind, Gerhard Weiss, and Pinar Yolum, editors, *Proceedings of the 14th International Conference on Autonomous Agents and Multiagent Systems (AAMAS 2015)*, pages 781–789, 2015.

[Tzagarakis et al., 2009] Manolis Tzagarakis, Nikos Karousos, Nikos Karacapilidis, and Dora Nousia. Unleashing argumentation support systems on the Web: The case of CoPe_it! In *Proceedings of the Web Science (WebSci'09)*, 2009.

[Vallati et al., 2014a] Mauro Vallati, Federico Cerutti, and Massimiliano Giacomin. Argumentation extensions enumeration as a constraint satisfaction problem: a performance overview. In Richard Booth, Giovanni Casini, Szymon Klarman, Gilles Richard, and Ivan José Varzinczak, editors, *Proceedings of the International Workshop on Defeasible and Ampliative Reasoning (DARe@ECAI 2014)*, volume 1212 of *CEUR Workshop Proceedings*. CEUR-WS.org, 2014.

[Vallati et al., 2014b] Mauro Vallati, Federico Cerutti, and Massimiliano Giacomin. Argumentation frameworks features: an initial study. In Torsten Schaub, Gerhard Friedrich, and Barry O'Sullivan, editors, *Proceedings of the 21st European Conference on Artificial Intelligence (ECAI 2014)*, volume 263 of *Frontiers in Artificial Intelligence and Applications*, pages 1117–1118. IOS Press, 2014.

[Vallati et al., 2015] Mauro Vallati, Federico Cerutti, Wolfgang Faber, and Massimiliano Giacomin. prefMaxSAT: Exploiting MaxSAT for Enumerating Preferred Extensions. In Matthias Thimm and Serena Villata, editors, *System Descriptions of the First International Competition on Computational Models of Argumentation (ICCMA'15)*, pages 58–61, 2015.

[Vallati et al., 2017] Mauro Vallati, Federico Cerutti, and Massimiliano Giacomin. On the combination of argumentation solvers into parallel portfolios. In *Proceedings of the 30th Australasian Joint Conference on Artificial Intelligence*, 2017. To appear.

[van Eemeren et al., 2014] Frans H. van Eemeren, Bart Garssen, Erik C. W. Krabbe, Francisca A. Snoeck Henkemans, Bart Verheij, and Jean H. M. Wagemans. *Handbook of Argumentation Theory*. Springer Netherlands, 2014.

[van Gijzel and Nilsson, 2013] Bas van Gijzel and Henrik Nilsson. Towards a framework for the implementation and verification of translations between argumentation models. In Rinus Plasmeijer, editor, *Proceedings of the 25th Symposium on Implementation and Application of Functional Languages (IFL 2013)*, page 93. ACM, 2013.

[van Gijzel and Nilsson, 2014] Bas van Gijzel and Henrik Nilsson. A principled approach to the implementation of argumentation models. In Simon Parsons, Nir Oren, Chris Reed, and Federico Cerutti, editors, *Proceedings of the 5th International Conference on Computational Models of Argument (COMMA 2014)*, volume 266 of *Frontiers in Artificial Intelligence and Applications*, pages 293–300. IOS Press, 2014.

[Vazirani, 2002] Vijay V. Vazirani. *Approximation Algorithms*. Springer, 2002.

[Verheij, 1996a] Bart Verheij. *Formal studies of argumentation and defeat*. PhD thesis, Universiteit Maastricht, 1996.

[Verheij, 1996b] Bart Verheij. Two approaches to dialectical argumentation:admissible sets and argumentation stages. In John-Jules Ch. Meyer and Linda C. van der Gaag, editors, *Proceedings of the Eighth Dutch Conference on Artificial Intelligence (NAIC'96)*, pages 357–368, Utrecht, NL, 1996.

[Verheij, 1998] Bart Verheij. Argue! - an implemented system for computer-mediated defeasible argumentation. In Han La Poutre and Jaap van den Herik, editors, *NAIC '98. Proceedings of the Tenth Netherlands/Belgium Conference on Artificial Intelligence*, pages 57–66, 1998.

[Verheij, 2003a] Bart Verheij. Artificial argument assistants for defeasible argumentation. *Artificial Intelligence and Law*, 150(1-2):291–324, 2003.

[Verheij, 2003b] Bart Verheij. DefLog: on the Logical Interpretation of Prima Facie Justified Assumptions. *Journal of Computational Logic*, 13(3):319–346, 2003.

[Verheij, 2007] Bart Verheij. A labeling approach to the computation of credulous acceptance in argumentation. In Manuela M. Veloso, editor, *Proceedings of the 20th International Joint Conference on Artificial Intelligence (IJCAI 2007)*, pages 623–628, 2007.

[Visser, 2008] Wietske Visser. Implementation of argument-based practical reasoning. Master's thesis, Utrecht University, 2008.

[Vreeswijk, 1993] Gerard A. Vreeswijk. *Studies in Defeasible Argumentation*. PhD thesis, Department of Computer Science, Free University Amsterdam, 1993.
[Wakaki and Nitta, 2008] Toshiko Wakaki and Katsumi Nitta. Computing argumentation semantics in answer set programming. In Hiromitsu Hattori, Takahiro Kawamura, and Tsuyoshi Ide, editors, *New Frontiers in Artificial Intelligence, (JSAI 2008) Conference and Workshops, Revised Selected Papers*, volume 5447 of *Lecture Notes in Computer Science*, pages 254–269, 2008.
[Walsh, 2000] Toby Walsh. SAT v CSP. In Rina Dechter, editor, *Proceedings of the 6th International Conference on Principles and Practice of Constraint Programming (CP 2000)*, volume 1894 of *Lecture Notes in Computer Science*, pages 441–456. Springer, 2000.
[Walton et al., 2014] Douglas Walton, Christopher W. Tindale, and Thomas F. Gordon. Applying recent argumentation methods to some ancient examples of plausible reasoning. *Argumentation*, 28(1):85–119, 2014.
[Watts and Strogatz, 1998] Duncan J. Watts and Steven H. Strogatz. Collective dynamics of small-world networks. *Nature*, 393(6684):440–442, 1998.
[Wells, 2014] Simon Wells. Argument mining: Was ist das? In *Proceedings of the 14th International Workshop on Computational Models of Natural Argument (CMNA 2014)*, 2014.
[Williams et al., 2015] Matt Williams, Zi Wei Liu, Anthony Hunter, and Fergus Macbeth. An updated systematic review of lung chemo-radiotherapy using a new evidence aggregation method. *Lung cancer (Amsterdam, Netherlands)*, 87(3):290–5, 2015.
[Wooldridge et al., 2006] Michael Wooldridge, Paul E. Dunne, and Simon Parsons. On the complexity of linking deductive and abstract argument systems. In Anthony Cohn, editor, *Proceedings of the 21st National Conference on Artificial Intelligence (AAAI 2006*, pages 299–304. AAAI Press, 2006.
[Wyner et al., 2013] Adam Wyner, Trevor J. M. Bench-Capon, and Paul E. Dunne. On the instantiation of knowledge bases in abstract argumentation frameworks. In João Leite, Tran Cao Son, Paolo Torroni, Leon van der Torre, and Stefan Woltran, editors, *Proceedings of the 14th International Workshop on Computational Logic in Multi-Agent Systems (CLIMA XIV)*, volume 8143 of *Lecture Notes in Computer Science*, pages 34–50. Springer, 2013.
[Xu et al., 2008] Lin Xu, Frank Hutter, Holger H. Hoos, and Kevin Leyton-Brown. Satzilla: Portfolio-based algorithm selection for sat. *Journal of Artificial Intelligence Research*, 32:565–606, 2008.

Federico Cerutti
Cardiff University
United Kingdom
E-mail: CeruttiF@cardiff.ac.uk

Sarah A. Gaggl
Technische Universität Dresden
Germany
E-mail: sarah.gaggl@tu-dresden.de

Matthias Thimm
Universität Koblenz-Landau
Germany
E-mail: thimm@uni-koblenz.de

Johannes P. Wallner
Technische Universität Wien
Austria
E-mail: wallner@dbai.tuwien.ac.at

PART E

ANALYSIS

15
Rationality Postulates: Applying Argumentation Theory for Non-monotonic Reasoning

MARTIN CAMINADA

ABSTRACT. The current book chapter examines how to apply Dung's theory of abstract argumentation to define meaningful forms of non-monotonic inference. The idea is that arguments are constructed using strict and defeasible inference rules, and that it is then examined how these arguments attack (or defeat) each other. The thus defined argumentation framework provides the basis for applying Dung-style semantics, yielding a number of extensions of arguments. As each of the constructed arguments has a conclusion, an extension of arguments has an associated extension of conclusions. It are these extensions of conclusions that we are interested in. In particular, we ask ourselves whether each of these extensions is (1) consistent, (2) closed under the strict inference rules and (3) free from undesired interference. We examine the current generation of techniques to satisfy these properties, and identify some research issues that are yet to be dealt with.

1 Introduction

Argumentation, as it takes place in everyday life, is never completely abstract. Commonly, arguments are exchanged in order to determine what to do or what to believe. These arguments tend to be composed of reasons, some of which are strict and some of which are defeasible. Strict reasons (like rules of logic) provide conclusive evidence for a claim (like "Socrates is a man. All men are mortal. Therefore, Socrates is mortal.") whereas defeasible reasons (like rules of thumb) provide evidence for their claim that is only valid in the absence of counter evidence (like "Tux is a bird. Therefore Tux can fly."). The existence of defeasible reasons illustrates that for commonsense reasoning, classical logic is often not sufficient, and that some form of nonmonotonic reasoning (as for instance provided by formal argumentation theory) is necessary.

Whereas defeasible reasons (formally represented as defeasible rules) provide a basis for nonmonotonic reasoning, strict reasons (formally represented as strict rules) provide the ability to model hard constraints (like "given our budget, if we acquire both product X and Y, then we cannot acquire product Z anymore"). By doing so, strict rules provide an important aspect of commonsense reasoning: the ability to reason about an outside world that has particular constraints (for instance of physical or financial nature) that are not subject to discussion.[1]

Suppose one would like to apply Dung's theory in the presence of strict and defeasible rules. That is, the idea is to apply the strict and defeasible rules to construct the arguments of the argumentation framework.[2] How can one be sure that the outcome makes sense from a logical perspective? Suppose there exists a rule representing the reason "given the current budget, if we acquire both product X and Y, then we cannot acquire product Z anymore", together with various other rules. In that case, what one would like to avoid is arguments for buying product X, Y and Z becoming justified (perhaps even in the same extension) because this would mean the constraint is violated. In principle, we could of course look inside of the arguments to check that what we select does not violate any constraint. However, the whole idea of Dung's abstract argumentation theory[3] is *not* to look at the internal structure of the arguments, and to select them based purely on their position in the graph. However, if one cannot look inside of the arguments when selecting them, then how does one make sure that the overall outcome (regarding conclusions on, say, what to do or what to believe) makes any sense?

In the current chapter, we examine the question of how to apply Dung's theory of abstract argumentation for the purpose of non-monotonic reasoning with strict and defeasible rules. That is, we examine how to apply abstract argumentation semantics while making sure the overall outcome (in terms of justified conclusions) still makes sense. The remaining part of this chapter is structured as follows. First, we will state some formal preliminaries on rule-based argumentation in Section 2. Then, in Section 3 we examine three desirable properties of the overall outcome (direct consistency, indirect consistency and closure) and examine various ways of satisfying these properties. Then, in Section 4 we examine two additional desirable properties (non-interference and crash resistance) that are particularly relevant when the strict rules are derived from classical logic, and again examine various ways of satisfying these properties. We round off with a summary and discussion in Section 5.

---

[1] Some argumentation researchers have claimed (personal communication) that if one digs deep enough, even strict rules start to have exceptions, and that therefore only defeasible rules exist. While this may be true from a philosophical perspective, one often wants to restrict the domain of reasoning and not take the more esoteric exceptions into account. The rule "given the current budget, if we acquire both product X and Y, we cannot acquire product Z anymore" may have exceptions if one is willing to steal, but this exception is of little relevance when the setting is a meeting at work. Also, the very idea of modelling information (be it by means of rules or by any other means) is that one limits oneself to a particular Universe of Discourse. Hence, strict rules can be seen as defeasible rules whose exceptions are beyond our current Universe of Discourse.

[2] Basically, this is done by chaining the rules together into inference trees, like is for instance done in [Modgil and Prakken, 2014; Toni, 2014; Caminada et al., 2014b; Caminada et al., 2015].

[3] Keep in mind that in Dung's theory, arguments are abstract, not atomic. Atomic would mean that arguments have no internal structure at all. Abstract means that arguments do have an internal structure, but that one does not take this structure into account (that is, one has *abstracted* from the internal structure).

2 Formal Preliminaries

In the current section, we outline the process of constructing an argumentation framework from a set of strict and defeasible rules. For current purposes, we base our approach on the work of Caminada *et al.* [2014b].[4]

Definition 1 *Given a logical language that is closed under negation* ($\neg$), *an argumentation system is a tuple* $AS = (\mathcal{R}_s, \mathcal{R}_d, \mathrm{n}, \leq)$ *where:*

- $\mathcal{R}_s$ *is a finite set of strict inference rules of the form* $\varphi_1, \ldots, \varphi_n \to \varphi$ *(where* φ_i, φ *are meta-variables ranging over* $\mathcal{L}$ *and* $n \geq 0$)
- $\mathcal{R}_d$ *is a finite set of defeasible inference rules of the form* $\varphi_1, \ldots, \varphi_n \Rightarrow \varphi$ *(where* φ_i, φ *are meta-variables ranging over* $\mathcal{L}$ *and* $n \geq 0$)
- n *is a partial function such that* $\mathrm{n} : \mathcal{R}_d \longrightarrow \mathcal{L}$
- $\leq$ *is a partial pre-order on* $\mathcal{R}_d$

We write $\psi = -\varphi$ in case $\psi = \neg\varphi$ or $\varphi = \neg\psi$ (we will sometimes informally say that formulas φ and $-\varphi$ are each other's negation).

To keep things simple, we assume that the logical language $\mathcal{L}$ consists of literals only.[5]

In the following definition, arguments are constructed from strict and defeasible rules in an inductive way. This process starts from the strict and defeasible rules with empty antecedents (so where $n = 0$).

Definition 2 *An argument* A *on the basis of an argumentation system* $AS = (\mathcal{R}_s, \mathcal{R}_d, \mathrm{n}, \leq)$ *is defined as:*

1. $A_1, \ldots, A_n \to \psi$ if $A_1 \ldots A_n$ ($n \geq 0$) are arguments, and there is a strict rule $Conc(A_1), \ldots, Conc(A_n) \to \psi$ in $\mathcal{R}_s$. In that case we define
$Conc(A) = \psi$,
$Sub(A) = Sub(A_1) \cup \ldots \cup Sub(A_n) \cup \{A\}$.
$DefRules(A) = DefRules(A_1) \cup \ldots \cup DefRules(A_n)$,
$TopRule(A) = Conc(A_1), \ldots, Conc(A_n) \to \psi$

2. $A_1, \ldots, A_n \Rightarrow \psi$ if $A_1 \ldots A_n$ ($n \geq 0$) are arguments, and there is a defeasible rule $Conc(A_1), \ldots, Conc(A_n) \Rightarrow \psi$ in $\mathcal{R}_d$. In that case we define
$Conc(A) = \psi$,
$Sub(A) = Sub(A_1) \cup \ldots \cup Sub(A_n) \cup \{A\}$,
$DefRules(A) = DefRules(A_1) \cup \ldots \cup DefRules(A_n) \cup \{Conc(A_1), \ldots, Conc(A_n) \Rightarrow \psi\}$,
$TopRule(A) = Conc(A_1), \ldots, Conc(A_n) \Rightarrow \psi$.

---

[4]As such, we will for instance not consider the notion of contraries [Modgil and Prakken, 2014] or any other notions in ASPIC+ that are not relevant for current purposes.
[5]In Section 4 we generalise things by having $\mathcal{L}$ be the language of propositional logic.

Furthermore, for any argument A and set of arguments E:

- *A is strict iff DefRules(A) = ∅; defeasible iff DefRules(A) ≠ ∅;*

- *If DefRules(A) = ∅, then LastDefRules(A) = ∅, else;
  if $A = A_1, \ldots, A_n \Rightarrow \phi$ then LastDefRules(A) = {Conc(A_1), ..., Conc(A_n) $\Rightarrow \phi$},
  otherwise LastDefRules(A) = LastDefRules(A_1)∪...∪LastDefRules(A_n).*

- *Concs(E) = {Conc(A) | A ∈ E}*

- *The closure under strict rules of E, denoted $Cl_S(E)$ is the smallest set containing Concs(E) and the consequent of any strict rule in $\mathcal{R}_s$ whose antecedent is contained in $Cl_S(E)$.*

For current purposes (as well as is done in [Caminada and Amgoud, 2007; Prakken, 2010; Caminada et al., 2014b]) we assume that the set of strict rules is consistent in the following way.

Definition 3 *Let $AS = (\mathcal{R}_s, \mathcal{R}_d, \mathbf{n}, \leq)$ be an argumentation system. We say that AS and $\mathcal{R}_s$ are consistent iff no strict arguments A and B exist such that* $\text{Conc}(A) = -\text{Conc}(B)$

Definition 4 *Let A and B be arguments. We say that*

- *A undercuts B (on B') iff Conc(A) = $-\mathbf{n}(r)$ for some $B' \in \text{Sub}(B)$ with TopRule(B') = r and $r \in \mathcal{R}_d$*

- *A restrictively rebuts B (on B') iff Conc(A) = $-\text{Conc}(B')$ for some $B' \in \text{Sub}(B)$ with TopRule(B') $\in \mathcal{R}_d$*

- *A unrestrictively rebuts B (on B') iff Conc(A) = $-\text{Conc}(B')$ for some $B' \in \text{Sub}(B)$ with B' being a defeasible argument*

To illustrate the difference between restricted rebut and unrestricted rebut, first consider the example of an argumentation system AS_1 with $\mathcal{R}_s = \emptyset$ and $\mathcal{R}_d = \{\Rightarrow a;\ a \Rightarrow b;\ \Rightarrow c;\ c \Rightarrow \neg b\}$. Here, the argument $(\Rightarrow a) \Rightarrow b$ restrictively and unrestrictively rebuts the argument $(\Rightarrow c) \Rightarrow \neg b$, and vice versa. In the argumentation system AS_2 with $\mathcal{R}_s = \{\rightarrow a;\ a \rightarrow b\}$ and $\mathcal{R}_d = \{\Rightarrow c;\ c \Rightarrow \neg b\}$, the argument $(\rightarrow a) \rightarrow b$ restrictively and unrestrictively rebuts the argument $(\Rightarrow c) \Rightarrow \neg b$, but the argument $(\Rightarrow c) \Rightarrow \neg b$ does not restrictively or unrestrictively rebut the argument $(\rightarrow a) \rightarrow b$. In the argumentation system AS_3 with $\mathcal{R}_s = \{a \rightarrow b;\ \rightarrow c\}$ and $\mathcal{R}_d = \{\Rightarrow a;\ c \Rightarrow \neg b\}$ the argument $(\Rightarrow a) \rightarrow b$ restrictively and unrestrictively rebuts the argument $(\rightarrow c) \Rightarrow \neg b$, and the argument $(\rightarrow c) \Rightarrow \neg b$ unrestrictively (but not restrictively) rebuts the argument $(\Rightarrow a) \rightarrow b$. To sum up, with restrictive rebut one needs to check whether the *last* rule of the attacked conclusion[6] is defeasible whereas with unrestricted rebut one needs to check whether *any previous* rule of the attacked conclusion is defeasible.

The intuition behind unrestricted rebut is that a conclusion is defeasible iff it has been derived using at least one defeasible rule. If the conclusion has been derived using strict rules only, then the conclusion is strict and cannot be argued against. The intuition behind restricted rebut, on the other hand, is that (like in classical logic) in order to argue against a particular derivation, one has to argue against its premises. So instead of attacking the consequent of a strict rule, one has to attack its antecedent, unless this antecedent itself consists of the consequents of strict rules, in which case one has to keep on going backwards until finding a defeasible rule. It holds that if A restrictively rebuts B, then A also unrestrictively rebuts B, but not vice versa.

One last subtle aspect of the definition of restricted and unrestricted rebut (Definition 4) is that one only looks at the subargument B' that yields the conclusion that one is arguing against. So in the argumentation system AS_4 with $\mathcal{R}_s = \{\rightarrow c;\ c \rightarrow \neg b\}$ and $\mathcal{R}_d = \{\Rightarrow a;\ a \Rightarrow b;\ \neg b \Rightarrow d\}$ the argument $(\Rightarrow a) \Rightarrow b$ does *not* (restrictively or unrestrictively) rebut the argument $((\rightarrow c) \rightarrow \neg b) \Rightarrow d$, even though the latter argument is defeasible, because the subargument that yields the attacked conclusion $\neg b$ is strict.

The difference between restricted and unrestricted rebut is relevant not just because they are based on different intuitions, but also because choosing to implement either restricted or unrestricted rebut has consequences for how one should define the rest of the argumentation formalism if the aim is to yield some kind of reasonable output in terms of justified conclusions. Details will follow further on in the current chapter.

Apart from (restrictive and unrestrictive) rebutting, Definition 4 also introduces the concept of undercutting. Whereas with rebutting, one argues against the conclusion of an argument (or against the conclusion of a subargument), with undercutting one argues against the applicability of a particular defeasible rule. A classical example of undercutting has been given by Pollock [1995]: "If an object looks red, then it actually is red, unless it is illuminated by a red light". Formally, this can be modelled using argumentation system AS_5 with $\mathcal{R}_s = \{\rightarrow looksred;\ \rightarrow redlight\}$, $\mathcal{R}_d = \{looksred \Rightarrow isred;\ redlight \Rightarrow \neg lris\}$ and $\mathbf{n}(looksred \Rightarrow isred) = lris$. Here, the argument $(\rightarrow looksred) \Rightarrow isred$ is undercut by the argument $(\rightarrow redlight) \Rightarrow \neg lris$. Although undercutting does not play a major role in the remaining part of the current chapter, we have still chosen to introduce it, as it is a piece of functionality that can be implemented while still warranting an overall reasonable outcome regarding the justified conclusions.

Another piece of functionality that some formalisms have implemented is that of argument strength.[7] Argument strength is often defined based on an

---

[6] meaning: of the conclusion one argues against by providing an argument for its contrary

[7] Argument strength is sometimes referred to as *argument preferences* in the work of

ordering of the defeasible rules. However, as arguments can be constructed using more than one defeasible rule, one needs a way of applying the strength ordering between *individual* rules to determine a strength ordering between *sets* of rules. Two principles for doing so have been defined in the literature: the elitist and the democratic set ordering [Modgil and Prakken, 2014; Caminada et al., 2014b].

Definition 5 *Let* $\leq \subseteq (\mathcal{R}_d \times \mathcal{R}_d)$ *be a total pre-ordering on the defeasible inference rules, where as usual,* $r < r'$ *iff* $r \leq r'$ *and* $r \not\leq r'$, *and* $r \equiv r'$ *iff* $r \leq r'$ *and* $r' \leq r$. *Then for any* $\mathcal{E}, \mathcal{E}' \subseteq \mathcal{R}_d \unlhd_{\mathtt{s}}$ ($\mathtt{s} \in \{\mathtt{Eli}, \mathtt{Dem}\}$) *is defined as follows:*

1. *If* $\mathcal{E} = \emptyset$ *then* $\mathcal{E} \not\unlhd_{\mathtt{s}} \mathcal{E}'$;

2. *If* $\mathcal{E}' = \emptyset$ *and* $\mathcal{E} \neq \emptyset$ *then* $\mathcal{E} \unlhd_{\mathtt{s}} \mathcal{E}'$; *else:*

3. *if* $\mathtt{s} = \mathtt{Eli}$: $\mathcal{E} \unlhd_{\mathtt{Eli}} \mathcal{E}'$ *if* $\exists r_1 \in \mathcal{E}$ *s.t.* $\forall r_2 \in \mathcal{E}'$, $r_1 \leq r_2$; *else:*

4. *if* $\mathtt{s} = \mathtt{Dem}$: $\mathcal{E} \unlhd_{\mathtt{Dem}} \mathcal{E}'$ *if* $\forall r_1 \in \mathcal{E}$, $\exists r_2 \in \mathcal{E}'$, $r_1 \leq r_2$.

As usual $\mathcal{E} \lhd_{\mathtt{s}} \mathcal{E}'$ iff $\mathcal{E} \unlhd_{\mathtt{s}} \mathcal{E}'$ and $\mathcal{E}' \not\unlhd_{\mathtt{s}} \mathcal{E}$

The elitist and democratic set ordering principles assume the presence of sets of defeasible rules. This leads to the question of how to determine the relevant sets of defeasible rules when one argument rebuts another. Again, two principles have been formulated in the literature, called *weakest link* and *last link*. With weakest link, one takes into account *all* defeasible rules (of both the rebutting argument and the rebutted (sub)argument), whereas with last link, one takes into account only the *last* defeasible rule(s). Given the weakest link and the last link principles for determining the sets of relevant defeasible rules, as well as the elitist and democratic set ordering principles for evaluating these sets of defeasible rules, one can identify four different principles for determining argument strength.

Definition 6 *Let Ar be the set of arguments that can be constructed using argumentation system* $(\mathcal{R}_s, \mathcal{R}_d, \mathtt{n}, \leq)$. *Then* $\forall A, B \in Ar$:

1. $A \preceq_{\mathtt{Ewl}} B$ *iff* $\mathtt{DefRules}(A) \unlhd_{\mathtt{Eli}} \mathtt{DefRules}(B)$

2. $A \preceq_{\mathtt{Ell}} B$ *iff* $\mathtt{LastDefRules}(A) \unlhd_{\mathtt{Eli}} \mathtt{LastDefRules}(B)$

3. $A \preceq_{\mathtt{Dwl}} B$ *iff* $\mathtt{DefRules}(A) \unlhd_{\mathtt{Dem}} \mathtt{DefRules}(B)$

4. $A \preceq_{\mathtt{Dll}} B$ *iff* $\mathtt{LastDefRules}(A) \unlhd_{\mathtt{Dem}} \mathtt{LastDefRules}(B)$

where $\mathtt{Ewl}, \mathtt{Ell}, \mathtt{Dwl}$ *and* $\mathtt{Dll}$ *respectively denote 'Elitist weakest link', 'Elitist last link', 'Democratic weakest link' and 'Democratic last link'. We may write* $A \prec_p B$ *iff* $A \preceq_p B$ *and* $B \not\preceq_p A$, *and write* $A \approx_p B$ *iff*

Prakken [2010], Modgil and Prakken [2014] and of Caminada *et al.* [2014b].

$A \preceq_p B, B \preceq_p A$ (where $p \in \{\texttt{Ewl}, \texttt{Ell}, \texttt{Dwl}, \texttt{Dll}\}$). It is straightforward to show that $\prec_p$ is a strict partial ordering (irreflexive, transitive and asymmetric).

We are now ready to define the overall notion of defeat. For this, we follow the approach of formalisms like ASPIC+ [Modgil and Prakken, 2014] and ASPIC- [Caminada et al., 2014b], where the notion of defeat stands for attack after argument strength has been taken into account. It is defeat, not attack, that is then used to define the argumentation framework.

Definition 7 *Let Ar be the set of arguments that can be constructed using argumentation system $AS = (\mathcal{R}_s, \mathcal{R}_d, \texttt{n}, \leq)$. Let $\preceq_p$ be the associated argument strength order on Ar as defined in Definition 6. Then $def_{ur} \subseteq Ar \times Ar$ is defined as $(A, B) \in def_{ur}$ iff A undercuts B or A unrestrictively rebuts B on B' and $A \not\prec_p B'$, and $def_{rr} \subseteq Ar \times Ar$ is defined as $(A, B) \in def_{rr}$ iff A undercuts B or A restrictively rebuts B on B' and $A \not\prec_p B'$.*

We observe that the set of arguments Ar, together with the associated defeat relation (either def_{ur} or def_{rr}) defines a Dung-style argumentation framework. On this argumentation framework, one can then apply the standard argumentation semantics, as described in Chapter 4 ("Abstract argumentation frameworks and their semantics") of this volume.

3 Direct Consistency, Indirect Consistency and Closure

To illustrate the issue of rationality postulates, consider the following example.

Example 1 ([Caminada and Amgoud, 2007]) *Consider an argumentation system $AS = (\mathcal{R}_s, \mathcal{R}_d, \texttt{n}, \leq)$ with $\mathcal{R}_s = \{\rightarrow r;\ \rightarrow n;\ m \rightarrow hs;\ b \rightarrow \neg hs\}$, $\mathcal{R}_d = \{r \Rightarrow m;\ n \Rightarrow b\}$, $\texttt{n} = \emptyset$ and $\leq = \emptyset$.*

An intuitive interpretation of this example is the following:
"John wears a ring (r) on his finger. John is also a regular nightclubber (n). Someone who wears a ring on his finger is usually married (m). Someone who is a regular nightclubber is usually bachelor (b). Someone who's married by definition has a spouse (hs). Someone who's bachelor by definition does not have a spouse ($\neg hs$)."

We can construct the following arguments.
$A_1 : \rightarrow r \qquad A_3 : A_1 \Rightarrow m \qquad A_5 : A_3 \rightarrow hs$
$A_2 : \rightarrow n \qquad A_4 : A_2 \Rightarrow b \qquad A_6 : A_4 \rightarrow \neg hs$

If one were to apply unrestricted rebut, the only defeat would be between A_5 and A_6. That is, $def_{ur} = \{(A_5, A_6), (A_6, A_5)\}$. This then implies that for instance the grounded extension is $\{A_1, A_2, A_3, A_4\}$, yielding the associated set of (grounded) justified conclusions $\{r, n, m, b\}$. The problem with these conclusions, however, is that they do not take into account the meaning of the strict rules of the argumentation system: that if one holds the antecedent of a strict rule to be the case, one must also hold what deductively follows from it (the consequent of the rule). For instance, from the fact that we obtain m, together with the strict rule $m \rightarrow hs$ we should also have obtained hs, as a married person by definition has a spouse, so by John being married we cannot escape the conclusion that he has a spouse. Yet, the fact that John has a spouse is not represented in the set of justified conclusions (that is, $hs \notin \{r, n, m, b\}$). This brings us to the first problem: the set of justified conclusions is not closed under the strict rules.

Another problem appears when also applying the strict rule $b \to \neg hs$. After all, John is also considered to be a bachelor, so we cannot escape the conclusion that he does not have a spouse ($\neg hs$). However, when we also apply the rule $m \to hs$, as we did earlier, then we derive that John both has a spouse and does not have a spouse. So not only is our set $\{r, n, m, b\}$ of justified conclusions not closed under the strict rules, if we do try to compute its closure, this closure turns out to be inconsistent!

So far, we examined what happens regarding the justified conclusions in case we apply unrestricted rebut. However, if we were to base the defeat relation on restricted rebut instead, then the outcome would even be worse, as the defeat relation would become empty (that is, $def_{rr} = \emptyset$) which means that (when still applying grounded semantics) one obtains $\{A_1, A_2, A_3, A_4, A_5, A_6\}$ as the grounded extension and $\{r, n, m, b, hs, \neg hs\}$ as the associated justified conclusions. So here, we don't even need to close the justified conclusions under the strict rules in order to obtain an inconsistent outcome, as the set of justified conclusions is already inconsistent by itself.

From Example 1 we observe that there are at least three desirable properties a set of conclusions should satisfy.

Postulate 1 *Let $S \subseteq \mathcal{L}$ be a set of justified conclusions yielded by an argumentation system. S should satisfy:*

- *direct consistency, meaning that $\neg \exists x : x, -x \in S$*

- *closure, meaning that $Cl_{\mathcal{R}_s}(S) = S$*

- *indirect consistency, meaning that $\neg \exists x : x, -x \in Cl_{\mathcal{R}_s}(S)$*

Early formalisations of argumentation theory tried to avoid problems like those illustrated in Example 1 by tinkering with the definition of defeat. However, as explained by Caminada and Amgoud [2007], this does not actually lead to the properties of Postulate 1 being satisfied. Clearly, some more fundamental solutions are needed. In the following two subsections, we examine some of the solutions that have been described in the literature, distinguishing between solutions that have been obtained for restricted rebut and solutions that have been obtained for unrestricted rebut.

3.1 Restricted Rebut Solutions

In the current section, we examine some of the solutions that have been described in the literature for satisfying direct consistency, indirect consistency and closure when the defeat relation is based on restricted rebut.

We recall that, when applying restricted rebut to Example 1 this results in the empty defeat relation, that is $def_{rr} = \emptyset$. One could argue that this is because something is wrong with the information encoded in the argumentation system AS, in particular with the set of strict rule $\mathcal{R}_s$. If one were for instance to add the additional strict rules $\neg hs \to \neg m$ and $hs \to \neg b$ then the problem would be solved. This is because one could then construct additional arguments $A_7 : A_5 \to \neg b$ and $A_8 : A_6 \to \neg m$. It holds that A_7 restrictively rebuts A_4 (as well as each argument that contains A_4, so also A_6 and A_8) and that A_8 restrictively rebuts A_3 (as well as each argument that contains A_3, so also A_5 and A_7). So overall we obtain the argumentation framework shown in Figure 1. This argumentation framework yields the grounded extension $\{A_1, A_2\}$ (with associated conclusions $\{r, n\}$) and preferred extensions $\{A_1, A_2, A_3, A_5, A_7\}$ (with associated conclusions $\{r, n, m, hs, \neg b\}$) and $\{A_1, A_2, A_4, A_6, A_8\}$ (with associated conclusions $\{r, n, b, \neg hs, \neg m\}$). As we can see, each set of conclusions yielded under grounded or preferred semantics satisfies the postulates of direct consistency, closure and indirect consistency.

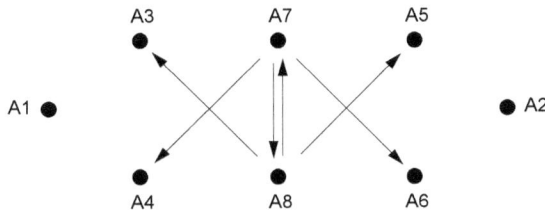

Figure 1. Argumentation framework of Example 1 after adding the rules $\neg hs \to \neg m$ and $hs \to \neg b$.

Adding the rules $\neg hs \to \neg m$ and $hs \to \neg b$ can be seen as a reasonable thing to do. After all, $\mathcal{R}_s$ already contains a rule $m \to hs$, meaning that without possible exception, someone who is married by definition has a spouse. This implies that someone who does not have a spouse cannot be married. Hence, $\neg hs \to \neg m$. Using similar reasoning, one can use the rule $b \to \neg hs$ to derive $hs \to \neg b$. Hence, the rules $\neg hs \to \neg m$ and $hs \to \neg b$ were already "implicitly" contained in $\mathcal{R}_s$. Adding them explicitly can therefore be seen as doing justice to $\mathcal{R}_s$, and has as a side effect that the postulates of direct consistency, closure and indirect consistency become satisfied.

Adding the "contraposed" version of a strict rule is relatively straightforward when the antecedent of the rule consists just of a single formula (as is for instance the case for $m \to hs$ and $b \to \neg hs$) but gets more complicated when the antecedent consists of multiple formulas. For this, a generalised version of contraposition is needed, which is referred to as *transposition* [Caminada and Amgoud, 2007].

Definition 8 ([Caminada and Amgoud, 2007]) Let $\varphi_1, \ldots, \varphi_n \to \varphi$ $(n \geq 0)$ be a strict rule. A transposed version of this rule is of the form $\varphi_1, \ldots, \varphi_{i-1}, -\varphi, \varphi_{i+1}, \ldots, \varphi_n \to -\varphi$ (for some $i \in \{1 \ldots n\}$). We say that a set of strict rules $\mathcal{R}_s$ is closed under transposition when for each strict rule in $\mathcal{R}_s$, each of its transposed versions is also in $\mathcal{R}_s$.

As an example, the strict rule $a, \neg b, c \to d$ has three transposed versions: $\neg d, \neg b, c \to \neg a$, $a, \neg d, c \to b$ and $a, \neg b, \neg d \to \neg c$.

An example of an argumentation formalism that applies transposition to satisfy direct consistency, closure and indirect consistency is ASPIC+ [Modgil and Prakken, 2014]. In ASPIC+ the following design choices have been made:

- the set of strict rules $\mathcal{R}_s$ is consistent and closed under transposition

- restricted rebut is applied

- argument strength is based on a partial pre-order on the defeasible rules, together with either the last-link or weakest link selection principle and either the elitist or democratic set ordering principle[8]

- the argumentation semantics is complete-based, meaning that it selects one or more complete extensions (examples of complete-based semantics are grounded, preferred, complete, semi-stable, ideal and eager semantics)

It is shown that under these choices, the overall outcome of the formalism satisfies direct consistency, closure and indirect consistency.

To understand why transposition plays an important role in satisfying the properties of direct consistency, closure and indirect consistency, it can be useful to give a sketch of proof. We start with the property of direct consistency. Suppose, towards a contradiction, that there exists a complete extension yielding conclusions that are directly inconsistent. This means there exists an argument A for conclusion c and an argument B for conclusion $-c$ (see Figure 2). As the set of strict rules $\mathcal{R}_s$ is consistent, at least one of these arguments must be defeasible. Assume without loss of generality that argument A is defeasible. Then A must contain at least one defeasible rule. Now, identify a defeasible rule r that is "as high as possible" in A (that is, whose distance to the conclusion c is minimal). Let e be the consequent of r and let A_i be the subargument of A that has r as its top rule (so $Conc(A_i) = e$). Let $A_1, \ldots, A_n$ be the subarguments of A that have the same "depth" as A_i (that is, whose respective top-rules have the same distance to conclusion c). It turns out to be possible to build an argument D' that defeats A_i by deriving conclusion $-e$. Recall that "above" each A_i there are only strict rules in A (after all, r was the "highest" defeasible rule in A). In case these strict rules consist of only one layer, there exists a single strict rule $Conc(A_1), \ldots, Conc(A_n) \to c$ with transposed version

---

[8]More precisely, argument strength has to be based on a *reasonable argument ordering* [Modgil and Prakken, 2014], which is satisfied by applying either the weakest link or the last link selection principle, in combination with applying either the democratic or the elitist set ordering principle.

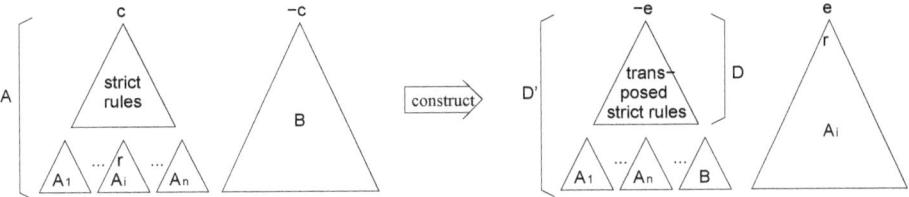

Figure 2. Sketch of proof direct consistency (restricted rebut)

$Conc(A_1), \ldots, Conc(A_{i-1}), -c, Conc(A_{i+1}), \ldots, Conc(A_n) \to -Conc(A_i)$, so $Conc(A_1), \ldots, Conc(A_{i-1}), Conc(B), Conc(A_{i+1}), \ldots, Conc(A_n) \to -c$, which implies we can use $A_1, \ldots A_{i-1}, B$ and $A_{i+1}, \ldots, A_n$ to construct an argument that restrictively rebuts A_i. In case the strict rules above each A_i consist of more than one layer, then one can still use transposition to construct an argument that restrictively rebuts A_i (basically by induction over the number of layers of strict rules). Let D' be the thus constructed argument that restrictively rebuts A_i. As A_i is a subargument of A, it follows that D' also restrictively rebuts A. From the fact that we are considering a complete extension, it follows that the extension has to contain an argument (say C) that defeats D'. However, as each defeasible rule of D' also occurs in A or B, it follows that C also defeats A or B.[9] Hence, the complete extension is not conflict-free. Contradiction.

It is important to observe that the above sketch of proof uses the facts that (1) $\mathcal{R}_s$ is consistent, (2) $\mathcal{R}_s$ is closed under transposition, (3) restricted rebut is being applied, and (4) we are considering a complete extension (or at least an admissible set).[10]

As for the property of closure, suppose there exists a strict rule $\varphi_1, \ldots, \varphi_n \to \varphi$ and that the conclusions $\varphi_1, \ldots, \varphi_n$ are yielded by our complete extension. We need to show that conclusion φ is also yielded by the complete extension. From the fact that conclusions $\varphi_1, \ldots, \varphi_n$ are yielded, it follows that the complete extension contains arguments $A_1, \ldots, A_n$ with conclusions $\varphi_1, \ldots, \varphi_n$ respectively. Now consider the argument $A : A_1, \ldots, A_n \to \varphi$. Let B be an arbitrary argument that defeats A. Then from the definition of defeat, it follows that B also defeats at least one of $A_1, \ldots, A_n$. From the fact that our extension is complete (and therefore also admissible) it follows that it contains an argument (say C) that defeats B. This means that A is defended by the complete extension, and must therefore also be contained in the complete ex-

---

[9]This is straightforward to see when the strength ordering between the rules is empty, but also holds when the strength ordering is non-empty. See the work of Modgil and Prakken [2013] for details.

[10]There are also some requirements regarding argument strength. These are such that $\preceq_{Ewl}$, $\preceq_{Ell}$, $\preceq_{Dwl}$, and $\preceq_{Dll}$ (Definition 6) satisfy them. We refer to the work of Modgil and Prakken [2013; 2014] for details.

tension.[11] This then implies that the complete extension also yields conclusion $Conc(A) = \varphi$.

Given that we have obtained both direct consistency and closure, the property of indirect consistency is trivially satisfied.

As was mentioned above, the property of transposition plays an important role for satisfying direct consistency, closure and indirect consistency. However, if one takes a closer look at the above sketch of proof, what is actually applied is a property that is more general than transposition. Going back to Figure 2 then what is actually needed is that if from $Conc(A_1), \ldots, Conc(A_n)$ one can apply strict rules to derive c, then from $Conc(A_1), \ldots, Conc(A_{i-1}), -c, Conc(A_{i+1}), \ldots, Conc(A_n)$ one can also apply strict rules to derive $-Conc(A_i)$. This property is called *contraposition* by Modgil and Prakken [2013; 2014], who show that direct consistency, closure and indirect consistency are satisfied when the set of strict rules is closed under contraposition.

One can ask the question of whether it is possible to derive even more general conditions than transposition and contraposition, under which direct consistency, closure and indirect consistency are still satisfied. This question is answered positively by Dung and Thang [2014] who present a semi-abstract approach that abstracts away from most aspects of argument structure (making explicit only the notions of a conclusion and that of a subargument). However, their approach does rely on particular constraints on the defeat relation, and it can be observed that these constraints can only be satisfied under restricted (and not unrestricted) rebut.[12]

3.2 Unrestricted Rebut Solutions

Although restricted rebut has become the most popular principle for defining the overall defeat relationship (as is for instance evidenced by the various versions of the ASPIC+ formalism [Prakken, 2010; Modgil and Prakken, 2013; Modgil and Prakken, 2014]) it does have some disadvantages, especially when applied in a dialectical context. Consider for instance the following discussion taken from [Caminada et al., 2014b].

John: *"Bob will attend both AAMAS and IJCAI this year, as he has papers accepted at each of these conferences."*

Mary: *"That won't be possible, as his budget of £1000 only allows for one foreign trip."*

Formally, this discussion can be modelled using the argumentation system $(\mathcal{R}_s, \mathcal{R}_d, \mathfrak{n}, \leq)$ with $\mathcal{R}_d = \{accA \Rightarrow attA;\ accI \Rightarrow attI;\ budget \Rightarrow \neg allboth\}$ and $\mathcal{R}_s = \{\to accA;\ \to accI;\ \to budget;\ attA, attI \to attboth;\ \neg attboth, attI \to \neg attA;\ attA, \neg attboth \to \neg attI\}$.[13]

---

[11]Notice that for this reasoning step, a complete extension is really needed; an admissible set is not sufficient.

[12]More precisely, unrestricted rebut trivialises the notion of a *base* [Dung and Thang, 2014], which prevents the results of Dung and Thang [2014] from being applied in the context of unrestricted rebut.

[13]We observe that $\mathcal{R}_s$ is consistent and closed under transposition.

John: $((\to accA) \Rightarrow attA), ((\to accI) \Rightarrow attI) \to attboth$
Mary: $(\to budget) \Rightarrow \neg attboth$

The problem is that when applying restricted rebut, Mary's argument does not defeat John's argument. This is because the conclusion that Mary wants to attack ($attboth$) is the consequent of a strict rule. If Mary wants to restrictively rebut John's argument, she can only do so by attacking the consequent of a defeasible rule. That is, she would be forced to choose to defeat either $attA$ or $attI$, meaning that she essentially has to utter one of the following statements.

Mary': *Bob won't attend AAMAS because he will already attend IJCAI, and his budget doesn't allow him to attend both.*
Mary'': *Bob won't attend IJCAI because he will already attend AAMAS, and his budget doesn't allow him to attend both.*

The associated formal counterarguments are as follows.
Mary': $((\to budget) \Rightarrow \neg attboth), ((\to accI) \Rightarrow attI) \to \neg attA$
Mary'': $((\to accA) \Rightarrow attA), ((\to budget) \Rightarrow \neg attboth) \to \neg attI$

Critically, Mary does not *know* which of the two conferences Bob will attend, yet the principle of restricted rebut *forces* her to make concrete statements on this. From the perspective of commitment in dialogue [Walton and Krabbe, 1995], this is unnatural. One should not be forced to commit to things one has insufficient reasons to believe in.

It should be stressed that the problem outlined above is particularly relevant in dialectical contexts, where different agents make commitments during the exchange of arguments. This contrasts with a formalism like ASPIC+, which is more monolithic in nature, in that from the given rules and premises, one constructs a graph of each other defeating arguments and simply *computes* which arguments (and associated conclusions) are justified. Concepts like different agents, communication steps or commitment stores do not play a role in AS-PIC+, and hence restricted rebut *seems* acceptable. However, if one wants to add dialectical aspects to formal argumentation (c.f., [Caminada and Wu, 2009; Caminada and Podlaszewski, 2012; Caminada et al., 2014a]) then one is forced to take the limitations of restricted rebut seriously.

The obvious way to deal with problems like sketched above would be to simply replace restricted rebut by unrestricted rebut (thus replacing def_{rr} by def_{ur}). Unfortunately, doing so also has far reaching consequences regarding the ability to satisfy the postulates of indirect consistency and closure. This is illustrated by the following example, taken from [Caminada and Wu, 2011].

Example 2 *Consider the argumentation system* $(\mathcal{R}_s, \mathcal{R}_d, \mathbf{n}, \leq)$ *with* $\mathcal{R}_s = \{\to jw; \to mw; \to sw; mt, st \to \neg jt; jt, st \to \neg mt; jt, mt \to \neg st\}$ *and* $\mathcal{R}_d = \{jw \Rightarrow jt; mw \Rightarrow mt; sw \Rightarrow st\}$. *This example can be interpreted as follows. John, Mary and Suzy want to go cycling in the countryside* ($\to jw; \to mw; \to sw$). *They have a tandem bicycle that each of them would like to be on* ($jw \Rightarrow jt; mw \Rightarrow mt; sw \Rightarrow st$). *However, as the tandem only has two seats, if two of them are on it, the third one cannot be on it* ($mt, st \to \neg jt; jt, st \to \neg mt; jt, mt \to \neg st$). *Using this argumentation system, we can construct the*

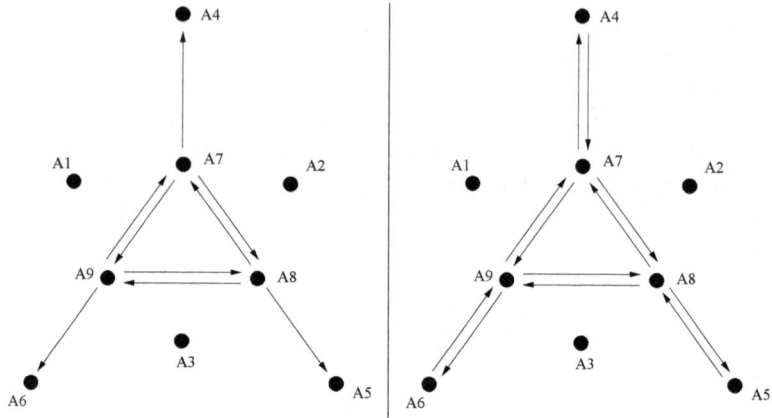

Figure 3. restricted rebut versus unrestricted rebut

following arguments.

$A_1 : \rightarrow jw \quad A_4 : A_1 \Rightarrow jt \quad A_7 : A_5, A_6 \rightarrow \neg jt$
$A_2 : \rightarrow mw \quad A_5 : A_2 \Rightarrow mt \quad A_8 : A_4, A_6 \rightarrow \neg mt$
$A_3 : \rightarrow sw \quad A_6 : A_3 \Rightarrow st \quad A_9 : A_4, A_5 \rightarrow \neg st$

When applying restricted rebut (and assuming the empty rule strength ordering) argument A_7 defeats A_4 (as well as A_8 and A_9, which contain A_4), argument A_8 defeats A_5 (as well as A_7 and A_9, which contain A_5) and argument A_9 defeats A_6 (as well as A_7 and A_8, which contain A_6). This yields the argumentation framework at the left hand side of Figure 3, which we will refer to as AF_{rr}.

AF_{rr} has four complete extensions: $\{A_1, A_2, A_3, A_5, A_6, A_7\}$ (yielding conclusions $\{jw, mw, sw, \neg jt, mt, st\}$), $\{A_1, A_2, A_3, A_4, A_6, A_8\}$ (yielding conclusions $\{jw, mw, sw, jt, \neg mt, st\}$), $\{A_1, A_2, A_3, A_4, A_5, A_9\}$ (yielding conclusions $\{jw, mw, sw, jt, mt, \neg st\}$), and finally $\{A_1, A_2, A_3\}$ (yielding conclusions $\{jw, mw, sw\}$). The first three complete extensions are also preferred (as well as stable and semi-stable). The last one is also grounded. We observe that the conclusions of each complete extension satisfy direct consistency, closure and indirectly consistency.

Now, let us consider what happens if we were to replace restricted rebut by unrestricted rebut. In that case, A_7 would still defeat A_4 (as well as A_8 and A_9), A_8 would still defeat A_5 (as well as A_7 and A_9) and A_9 would still defeat A_6 (as well as A_7 and A_8). However, additionally A_4 would defeat A_7, A_5 would defeat A_8 and A_6 would defeat A_9. This is because A_7, A_8 and A_9 are defeasible arguments, as their subarguments contain defeasible rules. So with unrestricted rebut, the arguments A_4, A_5 and A_6 are able to "strike back" against their respective defeaters. This yields the argumentation framework at the right hand side of Figure 3, which we will refer to as AF_{ur}. AF_{ur} has five complete extensions. The first four are the same as those of AF_{rr}. The fifth one is $\{A_1, A_2, A_3, A_4, A_5, A_6\}$ yielding conclusions $\{jw, mw, sw, jt, mt, st\}$, hence violating closure and indirect consistency. As this fifth complete extension is also preferred, stable and semi-stable, we have a counterexample against applying unrestricted rebut under each of these semantics.

Example 2 illustrates a fundamental difference between restricted and unrestricted rebut. Whereas under restricted rebut (in combination with $\mathcal{R}_s$ being consistent and closed under transposition or contraposition) any admissible set of arguments will yield conclusions that are indirectly consistent, under unrestricted rebut admissibility alone is not sufficient (the set $\{A_1, A_2, A_3, A_4, A_5, A_6\}$ being the counter example). It turns out that what is needed is a property that is stronger than admissibility: strong admissibility [Baroni and Giacomin, 2009; Caminada, 2014].[14] We observe that although the set $\{A_1, A_2, A_3, A_4, A_5, A_6\}$ is admissible, it is not strongly admissible. Furthermore, we observe that the set $\{A_1, A_2, A_3\}$ is both admissible and strongly admissible and yields conclusions $\{jw, mw, sw\}$ that are closed and indirectly consistent.

As the grounded extension is the unique biggest strongly admissible set [Baroni and Giacomin, 2009; Caminada, 2014], grounded semantics is a natural starting point for proving the properties of direct consistency, indirect consistency and closure when applying unrestricted rebut. Proving the property of direct consistency is relatively straightforward. After all, if the grounded extension was to yield conclusions that are directly inconsistent, it would have to contain two arguments A and B with opposite conclusions. As $\mathcal{R}_s$ is consistent, at least one of them has to be defeasible, which means that one would defeat (unrestrictedly rebut) the other, which would implies that the grounded extension is not conflict-free. Contradiction.

Proving the property of closure is a bit more complex, as it is done by induction using the inductive definition of the grounded extension. We refer to the work of Caminada and Amgoud [2007] and of Caminada et al. [2014b] for details. Indirect consistency then follows trivially from direct consistency and closure.

As for argument strength, two possibilities have been observed when it comes to satisfying closure and indirect consistency under unrestricted rebut. The first approach, of Caminada and Amgoud [2007], is to essentially have the empty ordering on the defeasible rules. A later approach, by Caminada et al. [2014b] is to have a total (!) pre-order among the defeasible rules.

An overall overview of approaches to satisfy direct consistency, closure and indirect consistency is provided in Table 1.

---

[14]We recall that a set of arguments $\mathcal{A}rgs$ is strongly admissible iff each $A \in \mathcal{A}rgs$ is defended by some $\mathcal{A}rgs' \subseteq \mathcal{A}rgs \setminus \{A\}$ which in its turn is again strongly admissible. Informally, the idea of strong admissibility is that each argument should be defended without going around in circles.

Table 1. Approaches for satisfying closure and direct/indirect consistency

defeat based on	argument strength	semantics	other conditions	example formalism
restricted rebut	empty	any complete-based semantics	$\mathcal{R}_s$ consistent and closed under transposition	ASPIC [Caminada and Amgoud, 2007]
unrestricted rebut	empty	grounded semantics	$\mathcal{R}_s$ consistent and closed under transposition	ASPIC [Caminada and Amgoud, 2007]
restricted rebut	partial pre-order $\mathcal{R}_d$, last link or weakest link, elitist or democratic	any complete-based semantics	$\mathcal{R}_s$ consistent and closed under transposition/ contraposition	ASPIC+ [Modgil and Prakken, 2014]
unrestricted rebut	total pre-order $\mathcal{R}_d$, last link or weakest link, elitist or democratic	grounded semantics	$\mathcal{R}_s$ consistent and closed under transposition	ASPIC− [Caminada et al., 2014b]

4 Non-Interference and Crash Resistance

One of the issues to decide when formulating an argumentation system is whether the (strict and defeasible) rules should be domain dependent or domain independent. An example of a domain dependent strict rule would be $cow \rightarrow mammal$. An example of a domain independent strict rule would be modus ponens, so $cow, cow \supset mammal \rightarrow mammal$. When the aim is to implement domain independent reasoning, the most obvious thing to do would be to base the strict rules on some form of classical logic. For current purposes, we examine what happens if one were to base the set of strict rules on propositional logic.

Definition 9 *Given the language $\mathcal{L}$ of propositional logic, a* defeasible theory *is a tuple $(P, \mathcal{R}_d, \mathrm{n}, \leq)$ where*

- *P is a consistent set of propositions (called* premises*)*
- *$\mathcal{R}_d$ is a set of defeasible rules of the form $\varphi_1, \ldots, \varphi_n \Rightarrow \varphi$ (where φ_i, φ are meta-variables ranging over $\mathcal{L}$)*
- *n is a function such that $\mathrm{n} : \mathcal{R}_d \longrightarrow \mathcal{L}$*

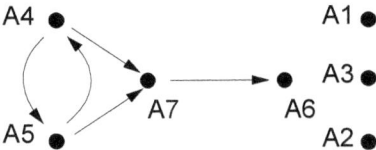

Figure 4. Strict rules as classical logic can have side effects (simple example)

Given a defeasible theory $(P, \mathcal{R}_d, \mathbf{n}, \leq)$, we define the associated argumentation system as $(\mathcal{R}_s, \mathcal{R}_d, \mathbf{n}, \leq)$ with $\mathcal{R}_s = \{\to \varphi \mid \varphi \in P\} \cup \{\varphi_1, \ldots, \varphi_n \to \varphi \mid \varphi_1, \ldots, \varphi_n \vdash \varphi\}$

As P is a consistent set of formulas, $\mathcal{R}_s$ will be consistent. Moreover, $\mathcal{R}_s$ is also closed under transposition. This is because the set $\{\to \varphi \mid \varphi \in P\}$ is trivially closed under transposition (as a rule with an empty antecedent does not have any transposed versions) and the set $\{\varphi_1, \ldots, \varphi_n \to \varphi \mid \varphi_1, \ldots, \varphi_n \vdash \varphi\}$ is closed under transposition as $\varphi_1, \ldots, \varphi_n \vdash \varphi$ implies $\varphi_1, \ldots, \varphi_{i-1}, -\varphi, \varphi_{i+1}, \ldots, \varphi_n \vdash -\varphi$. However, basing strict rules on classical logic also brings an additional type of problems. Consider the following example.

Example 3 *Consider the defeasible theory $(P, \mathcal{R}_d, \mathbf{n}, \leq)$ with $P = \{js, mns\}$, $\mathcal{R}_d = \{js \Rightarrow s;\ mns \Rightarrow \neg s;\ wfr \Rightarrow r\}$ and $\mathbf{n}$ and $\leq$ being the empty ordering. This example can be interpreted as follows. John says the cup of coffee contains sugar, so it probably contains sugar ($\to js;\ js \Rightarrow s$). Mary says the cup of coffee does not contain sugar ($\to mns;\ mns \Rightarrow \neg s$). The weather forecaster predicts rain tomorrow, so it will rain tomorrow ($\to wfr;\ wfr \Rightarrow r$). Hence, although we're not sure about whether the cup of coffee contains sugar, at least we should believe that it will rain tomorrow. Using this argumentation system, at least the following arguments can be constructed.*

$A_1 : \to js$ $A_4 : A_1 \Rightarrow s$
$A_2 : \to mns$ $A_5 : A_2 \Rightarrow \neg s$
$A_3 : \to wfr$ $A_6 : A_3 \Rightarrow r$

However, classical logic also yields the strict rule $s, \neg s \to \neg r$, as $s, \neg s \vdash \neg r$ (ex falso quodlibet). With this rule, we can construct the following argument.
$A_7 : A_4, A_5 \to \neg r$
This yields the argumentation framework of Figure 4.[15]

If one were to apply for instance grounded semantics, the grounded extension $\{A_1, A_2, A_3\}$ would yield conclusions $\{j, m, wf\}$. Thus, the weather forecast is not believed because John and Mary are having a disagreement about a cup of coffee.

The first thing to observe about Example 3 is that the underlying problem

---

[15]Notice that we are applying restricted rebut, but similar problems also occur when applying unrestricted rebut.

cannot be solved simply by removing rules with an inconsistent antecedent. This is because the effects of the rule $s, \neg s \rightarrow \neg r$ can be simulated by the rules $s \rightarrow s \vee \neg r$ and $s \vee \neg r, \neg s \rightarrow \neg r$, which still allow us to construct an argument for $\neg r$ from A_4 and A_5.

One approach that has been proposed in the literature [Prakken, 2010] is to change the semantics. If one were to apply for instance not grounded but preferred semantics to the argumentation framework of Figure 4, then two extensions would result: $\{A_1, A_2, A_3, A_4, A_6\}$ (yielding conclusions $\{j, m, wf, s, r\}$) and $\{A_1, A_2, A_3, A_5, A_6\}$ (yielding conclusions $\{j, m, wf, \neg s, r\}$). We observe that each set of conclusions contains r, so r is a justified conclusion under preferred semantics.

Although changing grounded semantics to preferred semantics seems to yield the desired outcome in Example 3, there exists a slightly more complex example where preferred semantics does *not* yield the desired outcome.

Example 4 *Consider the defeasible theory* $(P, \mathcal{R}_d, \mathrm{n}, \leq)$ *with* $P = \{js, mns,$ $junrel, munrel, wfr\}$, $\mathcal{R}_d = \{js \Rightarrow s;\ mns \Rightarrow \neg s;\ wfr \Rightarrow r;\ junrel \Rightarrow \neg jrel;\ munrel \rightarrow \neg mrel\}$, $\mathrm{n}(js \Rightarrow s) = \mathrm{n}(junrel \Rightarrow \neg jrel) = jrel$, $\mathrm{n}(mns \Rightarrow \neg s) = \mathrm{n}(munrel \Rightarrow \neg mrel) = mrel$ *and* $\leq$ *being the empty ordering. So now, in addition to John saying that the cup of coffee contains sugar, he also says that he is unreliable, so John is probably unreliable* ($junrel \Rightarrow \neg jrel$). *However, if John is unreliable, then the fact that he says something is no longer a reason to believe it. Hence the rule* ($js \Rightarrow s$) *is undercut, just like the rule* ($junrel \Rightarrow \neg jrel$). *Similarly, in addition to Mary saying that the cup of coffee does not contain sugar, she also says that she is unreliable, so Mary is probably unreliable* ($munrel \Rightarrow \neg mrel$). *However, if Mary is unreliable, then the fact that she says something is no longer a reason to believe it. Hence the rule* ($mns \Rightarrow \neg s$) *is undercut, just like the rule* ($munrel \Rightarrow \neg mrel$). *Overall, we can construct at least the following arguments.*

$A_1 : \rightarrow js$ $\qquad A_4 : A_1 \Rightarrow s$
$A_2 : \rightarrow mns$ $\qquad A_5 : A_2 \Rightarrow \neg s$
$A_3 : \rightarrow wfr$ $\qquad A_6 : A_3 \Rightarrow r$
$A_8 : \rightarrow junrel$ $\qquad A_{10} : A_8 \Rightarrow \neg jrel$
$A_9 : \rightarrow munrel$ $\qquad A_{11} : A_9 \Rightarrow \neg mrel$

Classical logic again yields the strict rule $s, \neg s \rightarrow \neg r$, *which allows the construction of the following argument.*
$A_7 : A_4, A_5 \rightarrow \neg r$
This yields the argumentation framework of Figure 5.[16]

In the argumentation framework of Figure 5 there exists just a single complete extension (that is also grounded, preferred, ideal and semi-stable): $\{A_1, A_2, A_3, A_8, A_9\}$ yielding conclusions $\{js, mns, wfr, junrel, munrel\}$. So again, we have that the weather forecast is not believed (under any admissibility-based

---

[16]Notice that we are again applying restricted rebut, although similar problems also occur when applying unrestricted rebut.

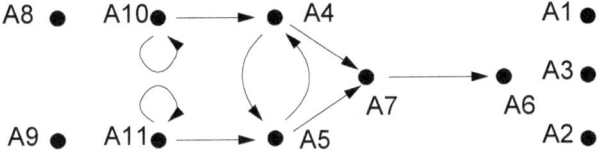

Figure 5. Strict rules as classical logic can have side effects (complex example)

semantics) because John and Mary are having a disagreement about a cup of coffee.

Before continuing to discuss some solutions that have been proposed in the literature, it can be useful to first define what precisely is it that we are trying to satisfy. Or, to put it in other words, what is the property that is actually being violated in Example 3 and Example 4? For this, we follow the approach of Caminada *et al.* [2012].

First of all, if $DT = (P, \mathcal{R}_d, \mathbf{n}, \leq)$ is a defeasible theory, then we write $\texttt{Atoms}(DT)$ for the set of all propositional atoms occurring in DT. We say that defeasible theories DT_1 and DT_2 are syntactically disjoint iff $\texttt{Atoms}(DT_1) \cap \texttt{Atoms}(DT_2) = \emptyset$. For syntactically disjoint defeasible theories $DT_1 = (P_1, \mathcal{R}_{d1}, \mathbf{n}_1, \leq_1)$ and $DT_2 = (P_2, \mathcal{R}_{d2}, \mathbf{n}_2, \leq_2)$ we define the union $DT_1 \cup DT_2$ as $(P_1 \cup P_2, \mathcal{R}_{d1} \cup \mathcal{R}_{d2}, \mathbf{n}_1 \cup \mathbf{n}_2, \leq_1 \cup \leq_2)$. Also, given a defeasible theory DT, we define its consequences $Cn_\sigma(DT)$ as $\{Concs(Args_1), \ldots, Concs(Args_n)\}$ where $Args_1, \ldots, Args_n$ are the extensions of arguments (under semantics σ) of the argumentation framework yielded by defeasible theory DT. Given a set of propositions S and a set of propositional atoms $\mathcal{A}$, we define $S_{|\mathcal{A}}$ as $\{\varphi \in S \mid \text{each atom in } \varphi \text{ is an element of } \mathcal{A}\}$. Similarly, given a set $\mathcal{S} = \{S_1, \ldots, S_n\}$ where each S_i ($i \in \{1 \ldots n\}$) is a set of propositions, we define $\mathcal{S}_{|\mathcal{A}}$ as $\{S_{1|\mathcal{A}}, \ldots, S_{n|\mathcal{A}}\}$.

Definition 10 *An argumentation formalism (applying semantics σ) satisfies non-interference iff for every pair of syntactically disjoint defeasible theories DT_1 and DT_2 it holds that $Cn_\sigma(DT_1)_{|\texttt{Atoms}(DT_1)} = Cn_\sigma(DT_1 \cup DT_2)_{|\texttt{Atoms}(DT_1)}$.*

To see how non-interference can be violated, consider again Example 3. In essence, the defeasible theory of this example can be seen as the union of two syntactically disjoint defeasible theories $DT_1 = (P_1, \mathcal{R}_{d1}, \mathbf{n}_1, \leq_1)$ and $DT_2 = (P_2, \mathcal{R}_{d2}, \mathbf{n}_2, \leq_2)$ with $P_1 = \{wfr\}$, $\mathcal{R}_{d1} = \{wfr \Rightarrow r\}$, $P_2 = \{js, mns\}$, $\mathcal{R}_{d2} = \{js \Rightarrow s;\ mns \Rightarrow \neg s\}$, $\mathbf{n}_1 = \mathbf{n}_2 = \emptyset$ and $\leq_1 = \leq_2 = \emptyset$. When applying grounded semantics, it holds that $Cn_{gr}(DT_1)_{|\texttt{Atoms}(DT_1)} = \{\{wfr, r\}\}$ whereas $Cn_{gr}(DT_1 \cup DT_2)_{|\texttt{Atoms}(DT_1)} = \{\{wfr\}\}$. So merging DT_1 with the completely unrelated defeasible theory DT_2 affects the outcome that is relevant w.r.t. DT_1. Hence, non-interference is violated.

An even stronger property is that of crash resistance.

Definition 11 *A defeasible theory $DT_1 = (P_1, \mathcal{R}_{d1}, \mathbf{n}_1, \leq_1)$ (with $\texttt{Atoms}(DT_1)$ $\subsetneq \texttt{Atoms}(\mathcal{L})$) is called contaminating (under semantics σ) iff for each syntactically disjoint defeasible theory DT_2 it holds that $Cn_\sigma(DT_1) = Cn_\sigma(DT_1 \cup DT_2)$. An argumentation formalism satisfies crash resistance iff there exists no defeasible theory that is contaminating.*

To see how crash resistance can be violated, consider Example 4. Again, the defeasible theory of this example can be seen as the union of two syntactically disjoint defeasible theories $DT_1 = (P_1, \mathcal{R}_{d1}, \mathbf{n}_1, \leq_1)$ and $DT_2 = (P_2, \mathcal{R}_{d2}, \mathbf{n}_2, \leq_2)$ with $P_1 = \{js, mns, junrel, munrel\}$, $\mathcal{R}_{d1} = \{js \Rightarrow s; mns \Rightarrow \neg s; junrel \Rightarrow \neg jrel; munrel \Rightarrow \neg mrel\}$, $\mathbf{n}_1(js \Rightarrow s) = \mathbf{n}_1(junrel \Rightarrow \neg jrel) = jrel$, $\mathbf{n}_1(mns \Rightarrow \neg s) = \mathbf{n}_1(munrel \Rightarrow \neg mrel) = mrel$, $\leq_1 = \emptyset$, $P_2 = \{wfr\}$, $\mathcal{R}_{d2} = \{wfr \Rightarrow r\}$, $\mathbf{n}_2 = \emptyset$ and $\leq_2 = \emptyset$. When applying stable semantics, it holds that $Cn_{st}(DT_1) = \emptyset$, just like $Cn_{st}(DT_1 \cup DT_2) = \emptyset$. Moreover, it can be verified that for *any* DT_2' that is syntactically disjoint with DT_1, it holds that $Cn_{st}(DT_1 \cup DT_2') = \emptyset$, hence violating crash resistance under stable semantics.

Conceptually, the difference between non-interference and crash resistance is as follows. A violation of non-interference means that a defeasible theory somehow influences the entailment of a completely unrelated (syntactically disjoint) defeasible theory when being merged to it. A violation of crash resistance is more severe, as this means that a defeasible theory influences the entailment of a completely unrelated (syntactically disjoint) defeasible theory to such an extent that the actual contents of this other defeasible theory become totally irrelevant. An argumentation formalism that satisfies non-interference also satisfies crash resistance.[17]

Now that the relevant properties have been identified, we proceed to examine some of the approaches in the literature for satisfying these. The first approach to be discussed is that of Wu and Podlaszewski [2015]. Their main idea is simply to erase inconsistent arguments[18] from the argumentation framework before applying argumentation semantics.

Definition 12 *Let (Ar, def) be the argumentation framework constructed from defeasible theory DT (by applying restricted rebut). Let Ar_c be $\{A \in Ar \mid A$ is consistent $\}$ and let def_c be $def \cap (Ar_c \times Ar_c)$. (Ar_c, def_c) is defined as the inconsistency cleaned argumentation framework of DT.*

As an example of how Definition 12 is used, in Example 3 and Example 4 argument A_7 would be removed, as well as all attacks from and to A_7. The resulting inconsistency cleaned argumentation framework is such that r is a conclusion of each complete extension.

One of the main results proved by Wu and Podlaszewski [2015] is that removing inconsistent arguments from the argumentation framework does not

---

[17]That is, as long as the argumentation formalism is *non-trivial* in the sense of [Caminada et al., 2012].

[18]An argument A is called inconsistent iff $\{Conc(A') \mid A' \in Sub(A)\}$ is inconsistent.

lead to any violations of direct consistency, closure and indirect consistency.[19] They also prove that the properties of non-interference and crash resistance are satisfied. However, the work of Wu and Podlaszewski [2015] assumes that the strength ordering among the defeasible rules is the empty one, and they provide an example of how their approach of erasing inconsistent arguments violates consistency and closure when applying non-empty rule strengths in combination with the last link principle.

The second approach to be discussed is that of Grooters and Prakken [2016]. Here, one of the basic ideas is to change the way strict rules are generated from propositional logic. Instead of generating a strict rule $\varphi_1, \ldots, \varphi_n \to \varphi$ whenever $\varphi_1, \ldots, \varphi_n \vdash \varphi$, they are generating such a strict rule only when from some consistent set $\Phi \subseteq \{\varphi_1, \ldots, \varphi_n\}$ it holds that $\Phi \vdash \varphi$. So instead of the strict rules coinciding with *all* propositional entailment, the idea is to have the strict rules coinciding with *consistent* propositional entailment.

However, ruling out inconsistent inferences alone is not sufficient, as the problem of *ex falso quodlibet* can also occur when successively applying several strict inference steps, as was for instance observed earlier, using the rules $s \to s \vee r$ and $s \vee r, \neg s \to \neg r$. The solution proposed by Grooters and Prakken [2016] is simple: when constructing arguments, disallow the application of a strict rule after the application of another strict rule.

It has to be mentioned that the approach of Grooters and Prakken [2016] has not been proven to satisfy any of the properties of direct consistency, closure, indirect consistency, non-interference and crash-resistance. Weaker properties have been proven instead. We refer to [Grooters and Prakken, 2016] for details.

5 Discussion

It is important to observe that the properties examined in the current chapter (sometimes called "rationality postulates" in the literature) are not specific to argumentation theory. In fact, they are general properties that can be applied to each formalism for non-monotonic reasoning that aims to encapsulate some form of strict reasoning. This is why the notion of an argument is not mentioned in the postulates of direct consistency, closure, indirect consistency, non-interference and crash-resistance. Instead, these postulates are defined purely based on the *outcome* (in terms of conclusions) of the argumentation

---

[19]This is unlike what for instance would happen when removing self-defeating (self-undercutting) arguments, which can lead to violations of closure. As an example (free after [Pollock, 1995]) take the argumentation system $(\mathcal{R}_s, \mathcal{R}_d, \mathfrak{n}, \leq)$ with $\mathcal{R}_s = \{\to a;\ b \to \neg c;\ c \to \neg b\}$, $\mathcal{R}_d = \{a \Rightarrow b\}$, $\mathfrak{n}(a \Rightarrow b) = c$ and $\leq = \emptyset$. Here, we can construct arguments $A_1 :\to a$, $A_2 : A_1 \Rightarrow b$ and $A_3 : A_2 \to \neg c$. It holds that A_3 defeats (undercuts) both itself and A_2. This yields a unique complete extension $\{A_1\}$ whose set of conclusions $\{a\}$ satisfies direct consistency, closure and indirect consistency. However, if one were to remove the self-defeating argument A_3, then this would yield a unique complete extension $\{A_1, A_2\}$, whose set of conclusions $\{a, b\}$ violates closure, as it contains b but not $\neg c$. The key point is that whenever one removes a particular class of arguments from the argumentation framework (be it inconsistent or self-attacking arguments) one has to examine whether this results in any violations of direct consistency, indirect consistency and closure.

formalism. That is, the postulates abstract from the notion of an argument.

This is not to say that no postulates have been formulated specifically about the arguments yielded (instead of about the conclusions yielded). An example of such a postulate would be subargument closure [Caminada and Amgoud, 2007]. This postulate says that if a particular extension contains argument A, then it should also contain all subarguments of A (so each $A' \in Sub(A)$). Satisfying subargument closure is not difficult. From the definition of defeat (under either restricted or unrestricted rebut) it follows that each argument that attacks A' also attacks A. So from A being in, say, a complete extension it follows that A is defended against these attackers, so A' is also being defended. Therefore, A' is also part of the complete extension (which contains everything it defends).

In the current chapter, we have mainly focused on rule-based argumentation formalisms, like ASPIC+. However, similar issues also play a role in classical logic based argumentation [Gorogiannis and Hunter, 2011]. Here, the idea is, given a set of propositions Δ (called the *knowledge base*), to construct arguments as pairs $\langle \Phi, \varphi \rangle$ where φ is a proposition (called the *conclusion*) and Φ is a set of propositions (called the *assumptions*) such that $\Phi \vdash \varphi$, $\Phi \nvdash \bot$ and $\neg \exists \phi \in \Phi \colon \Phi \setminus \{\phi\} \vdash \varphi$. Given this argument form, various ways of defining the notion of defeat (or *attack*, as no strength order is taken into account) are examined, especially for their ability to yield a consistent outcome. We refer to the work of Gorogiannis and Hunter [2011] for details. While Gorogiannis and Hunter [2011] do not consider use of preferences, a recent alternative formalisation of classical logic argumentation of D'Agostino and Modgil [2016] satisfies the consistency and non-contamination postulates while supporting the use of preferences. Moreover, this is done without the requirement that an argument's premises need to be checked for consistency and subset minimality, and with the resulting argumentation frameworks only including finite subsets of the arguments defined by a set of classical well-formed formulas. As such, their theory provides a rational account that is suitable for resource bounded agents.

One key point that we want to emphasise is that the satisfaction of rationality postulates is *not* just a matter of theoretical elegance. If we were to apply argumentation theory for practical purposes, to determine what should be the actions to take, and our formalism tells us to put three people on a tandem bicycle, then this advice will be of little use, as the actions to implement it will fail. If we believe the world to be such that there exist some hard (inviolable) constraints, then it makes sense to model these using nondefeasible (strict) rules and expect the argumentation formalism to deal with them in a proper way. Similarly, if one were for instance to build a robot that uses argumentation theory for its internal reasoning, what we would like to avoid is the situation where after being fed some specific snippets of input (like John whispering in its ear "The cup of coffee contains sugar, and I'm unreliable", and Mary whispering in its ear "The cup of coffee contains no sugar, and I'm unreliable") all inference will come to a grinding halt, and the robot essentially stops functioning. Hence, satisfaction of the rationality postulates is important not just for theoretical elegance, but also to make the theory suitable for actual applications.

Given the important role of rationality postulates when it comes to applications of argumentation theory, we observe that the current state of affairs (at the time of writing) is somewhat unsatisfying. As for the postulates of direct consistency, closure and indirect consistency, there seems to be a dilemma. If, on one hand, one chooses to implement restricted rebut then these postulates can be satisfied under any complete-based semantics. The disadvantage, however, is that restricted rebut can be seen as unintuitive, especially in a dialectical context. If, on the other hand, one chooses to implement unrestricted rebut, then the notion of defeat becomes more in line with natural discussion. The disadvantage, however, is that one can only apply grounded semantics, which tends to yield a very sceptical result. Moreover, satisfaction of the rationality postulates is only guaranteed if the strength order on the defeasible rules is either empty or total (hence ruling out a proper partial oder).

As for the postulates of non-inference and crash resistance, the situation is even more troublesome. First of all, all the approaches that we are aware of [Wu, 2012; Wu and Podlaszewski, 2015; Podlaszewski, 2015; Grooters and Prakken, 2016] work only with restricted rebut. Moreover, the approach of Wu and Podlaszewski [2015] requires the empty ordering regarding rule strength, whereas in many application domains different rules can have different strengths. The work of Grooters and Prakken [2016], does allow for a non-empty rule strength ordering, but fails to prove any of the forementioned postulates, opting to prove much weaker properties instead.

Overall, when it comes to the development of formal argumentation theory, one can observe that the topic of pure abstract argumentation tends to receive quite some more research attention than the topic of instantiated argumentation. Much work has for instance been done on how to select nodes from a graph. However, the real challenge is how to select nodes from a graph *in a meaningful way*, that is, such that the overall outcome makes sense from a logical perspective so the conclusions could be relied upon regarding what to do or what to believe. If formal argumentation is to be applied in situations that matter, some proper solutions to the issue of rationality postulates would be highly desirable.

Afterword

At the time the current chapter went to print, a paper of Heyninck and Straßer [2017] has just been accepted to be presented at IJCAI 2017. The authors' main idea is to allow for arguments to be attacked on several of its (sub)conclusions (that is, on the conclusions of one or more of its subarguments). This is done by an attacker with a disjunctive conclusion, such that each disjunct is the contrary (negation) of one of the (sub)conclusions of the attacked argument. As far as we know, this yields the first ever instantiation of Dung's argumentation theory that (1) works with a combination of classical logic and defeasible inference rules, (2) satisfies *all* the rationality postulates and (3) implements argument preferences.

BIBLIOGRAPHY

[Baroni and Giacomin, 2009] P. Baroni and M. Giacomin. Skepticism relations for comparing argumentation semantics. *Int. J. Approx. Reasoning*, 50(6):854–866, 2009.

[Caminada and Amgoud, 2007] M.W.A. Caminada and L. Amgoud. On the evaluation of argumentation formalisms. *Artificial Intelligence*, 171(5-6):286–310, 2007.

[Caminada and Podlaszewski, 2012] M.W.A. Caminada and M. Podlaszewski. Grounded semantics as persuasion dialogue. In Bart Verheij, Stefan Szeider, and Stefan Woltran, editors, *Computational Models of Argument - Proceedings of COMMA 2012*, pages 478–485, 2012.

[Caminada and Wu, 2009] M.W.A. Caminada and Y. Wu. An argument game of stable semantics. *Logic Journal of IGPL*, 17(1):77–90, 2009.

[Caminada and Wu, 2011] M.W.A. Caminada and Y. Wu. On the limitations of abstract argumentation. In Patrick de Causmaecker, Joris Maervoet, Tommy Messelis, Katja Verbeeck, and Tim Vermeulen, editors, *Proceedings of the 23rd Benelux Conference on Artificial Intelligence (BNAIC 2011)*, pages 59–66, 2011.

[Caminada et al., 2012] M.W.A. Caminada, W.A. Carnielli, and P.E. Dunne. Semi-stable semantics. *Journal of Logic and Computation*, 22(5):1207–1254, 2012.

[Caminada et al., 2014a] M.W.A. Caminada, W. Dvořák, and S. Vesic. Preferred semantics as socratic discussion. *Journal of Logic and Computation*, 26:1257–1292, 2014.

[Caminada et al., 2014b] M.W.A. Caminada, S. Modgil, and N. Oren. Preferences and unrestricted rebut. In Simon Parsons, Nir Oren, Chris Reed, and Drederico Cerutti, editors, *Computational Models of Argument; Proceedings of COMMA 2014*, pages 209–220. IOS Press, 2014.

[Caminada et al., 2015] M.W.A. Caminada, S. Sá, J. Alcântara, and W. Dvořák. On the equivalence between logic programming semantics and argumentation semantics. *International Journal of Approximate Reasoning*, 58:87–111, 2015.

[Caminada, 2014] M.W.A. Caminada. Strong admissibility revisited. In Simon Parsons, Nir Oren, Chris Reed, and Frederico Cerutti, editors, *Computational Models of Argument; Proceedings of COMMA 2014*, pages 197–208. IOS Press, 2014.

[D'Agostino and Modgil, 2016] M. D'Agostino and S. Modgil. A rational account of classical logic argumentation for real-world agents. In *European Conference on Artificial Intelligence (ECAI 2016)*, pages 141–149, 2016.

[Dung and Thang, 2014] P.M. Dung and P.M. Thang. Closure and consistency in logic-associated argumentation. *Journal of Artificial Intelligence Research*, 49:79–109, 2014.

[Gorogiannis and Hunter, 2011] N. Gorogiannis and A. Hunter. Instantiating abstract argumentation with classical logic arguments: Postulates and properties. *Artificial Intelligence*, 175(9-10):1479–1497, 2011.

[Grooters and Prakken, 2016] D. Grooters and H. Prakken. Two aspects of relevance in structured argumentation: Minimality and paraconsistency. *Journal of Artificial Intelligence Research*, 56:197–245, 2016.

[Heyninck and Straßer, 2017] J. Heyninck and Ch. Straßer. Revisiting unrestricted rebut and preferences in structured argumentation. In *Proceedings of the 26th International Joint Conference on Artificial Intelligence*, 2017. in print.

[Modgil and Prakken, 2013] S. Modgil and H. Prakken. A general account of argumentation with preferences. *Artificial Intellligence*, 195:361–397, 2013.

[Modgil and Prakken, 2014] S. Modgil and H. Prakken. The ASPIC+ framework for structured argumentation: a tutorial. *Argument & Computation*, 5:31–62, 2014. Special Issue: Tutorials on Structured Argumentation.

[Podlaszewski, 2015] M. Podlaszewski. *Poles Apart: Navigating the Space of Opinions in Argumentation*. PhD thesis, Université du Luxembourg, 2015.

[Pollock, 1995] J.L. Pollock. *Cognitive Carpentry. A Blueprint for How to Build a Person*. MIT Press, Cambridge, MA, 1995.

[Prakken, 2010] H. Prakken. An abstract framework for argumentation with structured arguments. *Argument and Computation*, 1(2):93–124, 2010.

[Toni, 2014] F. Toni. A tutorial on assumption-based argumentation. *Argument & Computation*, 5:89–117, 2014. Special Issue: Tutorials on Structured Argumentation.

[Walton and Krabbe, 1995] D. N. Walton and E. C. W. Krabbe. *Commitment in Dialogue: Basic Concepts of Interpersonal Reasoning*. SUNY Series in Logic and Language. State University of New York Press, Albany, NY, USA, 1995.

[Wu and Podlaszewski, 2015] Y. Wu and M. Podlaszewski. Implementing crash-resistance and non-interference in logic-based argumentation. *Journal of Logic and Computation*, 25(2):303–333, 2015.

[Wu, 2012] Y. Wu. *Between Argument and Conclusion; argument-based approaches to discussion, inference and uncertainty*. PhD thesis, Université du Luxembourg, 2012.

Martin Caminada
Cardiff University
Email: CaminadaM@cardiff.ac.uk

16
The Principle-Based Approach to Abstract Argumentation Semantics

LEENDERT VAN DER TORRE, SRDJAN VESIC

ABSTRACT. The principle-based or axiomatic approach is a methodology to choose an argumentation semantics for a particular application, and to guide the search for new argumentation semantics. This chapter gives a complete classification of the fifteen main alternatives for argumentation semantics using the twenty-seven main principles discussed in the literature on abstract argumentation, extending Baroni and Giacomin's original classification with other semantics and principles proposed in the literature. It also lays the foundations for a study of representation and (im)possibility results for abstract argumentation, and for a principle-based approach for extended argumentation such as bipolar frameworks, preference-based frameworks, abstract dialectical frameworks, weighted frameworks, and input/output frameworks.

1 The principle-based approach

A considerable number of semantics exists in the argumentation literature. Whereas examining the behaviour of semantics on examples can certainly be insightful, a need for more systematic study and comparison of semantics has arisen. Baroni and Giacomin [2007] present a classification of argumentation semantics based on a set of principles. In this chapter, we extend their analysis with other principles and semantics proposed in the literature over the past decade.

The principle-based approach is a methodology that is also successfully applied in many other scientific disciplines. It can be used once a unique universal method is replaced by a variety of alternative methods, for example, once a variety of modal logics is used to represent knowledge instead of unique first order logic. The principle-based approach is also called the axiomatic approach, or the postulate based approach (for example in AGM theory change by Alchourrón *et al.* [1985]).

Maybe the best known example of the principle-based approach is concerned with the variety of voting rules, a core challenge in democratic societies, see, e.g., Tennenholtz and Zohar [2016]. It is difficult to find two countries that elect their governments in the same way, or two committees that decide using exactly the same procedure. Over the past two centuries many voting rules have been proposed, and researchers were wondering how we can know that the currently considered set of voting rules is sufficient or complete, and whether there is no better voting rule that has not been discovered yet. Voting theory addresses what we call the choice and search problems inherent to diversity:

Choice problem: If there are many voting rules, then how to choose one voting rule from this set of alternatives in a particular situation?

Search problem: How to guide the search for new and hopefully better voting rules?

In voting theory, the principle-based approach was introduced by Nobel prize winner Kenneth Arrow. The principle-based approach classifies existing approaches based on axiomatic principles, such that we can select a voting rule based on the set of requirements in an area. Moreover, there may be sets of principles for which no voting rule exist yet. Beyond voting theory, the principle-based approach has been applied in a large variety of domains, including abstract argumentation.

Formal argumentation theory, following the methodology in non-monotonic logic, logic programming and belief revision, defines a diversity of semantics. This immediately raises the same questions that were raised before for voting rules, and in many other areas. How do we know that the currently considered set of semantics is sufficient or complete? May there be a better semantics that has not been discovered yet? Moreover, the same choice and search problems of voting theory can be identified for argumentation theory as well:

Choice problem: If there are many semantics, then how to choose one semantics from this set of alternatives in a particular application?

Search problem: How to guide the search for new and hopefully better argumentation semantics?

The principle-based approach again addresses both problems. For example, if one needs to exclude the possibility of multiple extensions, one may choose the grounded or ideal semantics. If it is important that at least some extension is available, then stable semantics should not be used. As another common example, consider the admissibility principle that if an argument in an extension is attacked, then it is defended against this attack by another argument in the extension. If one needs a semantics that is admissible, then for example CF2 or stage2 cannot be chosen.

Principles have also been used to guide the search for new semantics. For example, the principle of resolution was defined by Baroni and Giacomin [2007], well before resolution based semantics were defined and studied by Baroni et al. [2011b]. Likewise it may be expected that the existing and new principles will guide the further search for suitable argumentation semantics. For example, consider the conflict-freeness principle that says that an extension does not contain arguments attacking each other. All semantics studied in this chapter satisfy this property. If one needs to define new argumentation semantics that are para-consistent in the sense that its extensions are not necessarily conflict free [Arieli, 2015], then one can still adopt other principles such as admissibility in the search for such para-consistent semantics.

The principle-based approach consists of three steps.

The first step in the principle-based approach is to define a general function, which will be the object of study. Kenneth Arrow defined social welfare functions from preference profiles to aggregated preference orders. For abstract argumentation, the obvious candidate is a function from graphs to sets of sets of nodes of the graph. Following Dung's terminology, we call the nodes of the graph arguments, we call sets of nodes extensions, we call the edges attacks, and we call the graphs themselves argumentation frameworks. Moreover, we call the function an argumentation semantics. Obviously nothing hinges on this terminology, and in principle the developed theory could be used for other applications of graph theory as well.

We call this function from argumentation frameworks to sets of extensions a two valued function, as a node is either in the extension, or not. Also multi valued functions are commonly used, in particular three valued functions conventionally called labelings. For three valued labelings, the values are usually called in, out and undecided. Other more general functions have been considered in abstract argumentation, for example in value based argumentation, bipolar argumentation, abstract dialectical frameworks, input/output frameworks, ranked semantics, and more. The principle-based approach can be applied to all of them, but in this chapter we will not consider such generalisations.

The second step of the principle-based approach is to define the principles. The central relation of the principle-based approach is the relation between semantics and principles. In abstract argumentation a two valued relation is used, such that every semantics either satisfies a given property or not. In this case, principles can be defined also as sets of semantics, and they can be represented by a constraint on the function from argumentation frameworks to sets of extensions. An alternative approach used in some other areas gives a numerical value to represent to which degree a semantics satisfies a principle.

The third step of the principle-based approach is to classify and study sets of principles. For example, a set of principles may imply another one, or a set of principles may be satisfiable in the sense that there is a semantics that satisfies all of them. A particular useful challenge is to find a set of principles that characterises a semantics, in the sense that the semantics is the only one that satisfies all the principles. Such characterisations are sometimes called representation theorems.

The principles used in a search problem are typically desirable, and desirable properties are sometimes called postulates. For the mathematical development of a principle-based theory, it may be less relevant whether principles are desirable or not.

Before we continue, we address two common misunderstandings about the principle-based approach, which are sometimes put forward as objections against it.

The first point is that not every function from argumentation frameworks to sets of extensions is an argumentation semantics. In other words, the objection is sometimes raised against the axiomatic approach that it allows for counterintuitive or even absurd argumentation semantics, just like the objection may be raised that not every function from preference profile to candidates is a voting rule. However, in the principle-based approach, such counterintuitive alternatives are excluded by the principles, they are not excluded a priori.

It may be observed that in formal argumentation, this objection is not restricted to principle-based abstract argumentation. A general framework for structured argumentation like ASPIC+ also allows for many counterintuitive or even absurd argumentation theories. However, from the perspective of the principle-based approach, the generality of the ASPIC+ approach can be used to study which combinations of definitions lead to argumentation theories satisfying desired principles [Caminada, 2018].

The second point is that a semantics is fundamentally different from a principle. In general a semantics is a function from argumentation frameworks to sets of extensions, and principles can be defined as sets of such functions and represented by a constraint on such functions. This misunderstanding arises because there are examples where a property can be represented as a semantics. For example, the completeness principle may be defined to state that each extension is complete, and the complete semantics may be defined such that the set of extensions of an argumentation framework are *all* its complete extensions. Likewise, some authors transform the admissibility principle into a "semantics" that associates with a framework all the admissible extensions. In this chapter we do not consider an admissibility semantics defined in this sense, only the admissibility principle.

Finally, we end this introduction with two methodological observations. First, we note that both argumentation semantics and argumentation principles can be organised and clustered in various ways. For example, sometimes a distinction is made between the set of admissibility based semantics and the set of naive based semantics, which are semantics satisfying the admissibility principle and the maximal conflict free principle respectively. In this chapter we have organised the semantics and principles in a way that seemed reasonable to us, but we did not use a systematic approach and we expect that some readers might have preferred an alternative organisation.

Second, while writing the chapter, several readers and reviewers have suggested additional semantics and principles to us. For example, we did not systematically study all resolution based semantics. The reason is pragmatic: this chapter has been growing while we were writing, and at some moment we needed to finish it. Moreover, we excluded several semantics proposed in the literature, such as AD1, AD2, CF1 introduced by Baroni *et al.* [2005], because they have not been further discussed or applied in the formal argumentation literature. However, if some of them will become more popular in the future, then the principle-based study in this chapter has to be extended to them as well. Also various principles are defined and analysed in other chapters, so we decided not to include them. For example, dynamic principles are discussed in other chapters of this volume [Baroni *et al.*, 2018b; Baumann, 2018], including the dynamic principles studied by Baroni *et al.* [2014]. Finally, other dynamic principles have been discussed by Rienstra *et al.* [2015].

The layout of this chapter is as follows. Section 2 introduces the setting and notation, Section 3 introduces the argumentation semantics we study in the rest of the chapter, and Section 4 introduces the principles and presents the table detailing which principles are satisfied by each semantics.

2 Setting and notations

The current section introduces the setting and notations.

Definition 2.1 (Argumentation framework, [Dung, 1995]) *An argumentation framework is a couple $\mathcal{F} = (\mathcal{A}, \mathcal{R})$ where $\mathcal{A}$ is a finite set and $\mathcal{R} \subseteq \mathcal{A} \times \mathcal{A}$. The elements of $\mathcal{A}$ are called* arguments *and $\mathcal{R}$ is called* attack relation*. We say that a attacks b if $(a, b) \in \mathcal{R}$; in that case we also write $a\mathcal{R}b$. For a set $S \subseteq \mathcal{A}$ and an argument $a \in \mathcal{A}$, we say that S attacks a if there exists $b \in S$ such that $b\mathcal{R}a$; we say that a attacks S if there exists $b \in S$ such that $a\mathcal{R}b$. We say that S attacks a set P if there exist $a \in S$, $b \in P$ such that a attacks b.*

We define $S^+ = \{a \in \mathcal{A} \mid S \text{ attacks } a\}$ and $S^- = \{a \in \mathcal{A} \mid a \text{ attacks } S\}$. Moreover, for an argument a, we define $a^+ = \{b \in \mathcal{A} \mid a \text{ attacks } b\}$ and $a^- = \{b \in \mathcal{A} \mid b \text{ attacks } a\}$. We define $S^-_{out} = \{a \in \mathcal{A} \mid a \notin S \text{ and } a \text{ attacks } S\}$. The set of all argumentation frameworks is denoted by $\mathcal{AF}$.

We can observe that an argumentation framework is just a finite graph. In the rest of the chapter, $\mathcal{F} = (\mathcal{A}, \mathcal{R})$ stands for an argumentation framework.

Definition 2.2 (Projection, union, subset) *For an argumentation framework $\mathcal{F} = (\mathcal{A}, \mathcal{R})$ and a set $S \subseteq \mathcal{A}$, we define: $\mathcal{F} \downarrow_S = (S, \mathcal{R} \cap (S \times S))$. Let $\mathcal{F}_1 = (\mathcal{A}_1, \mathcal{R}_1)$ and $\mathcal{F}_2 = (\mathcal{A}_2, \mathcal{R}_2)$ be two argumentation frameworks. We define $\mathcal{F}_1 \cup \mathcal{F}_2 = (\mathcal{A}_1 \cup \mathcal{A}_2, \mathcal{R}_1 \cup \mathcal{R}_2)$. We write $\mathcal{F}_1 \subseteq \mathcal{F}_2$ if and only if $\mathcal{A}_1 \subseteq \mathcal{A}_2$ and $\mathcal{R}_1 \subseteq \mathcal{R}_2$.*

For a set S, we denote its powerset by 2^S. Now we define the notion of semantics. It is a function that, given an argumentation framework $(\mathcal{A}, \mathcal{R})$, returns a set of subsets of $\mathcal{A}$.

Definition 2.3 (Semantics) *An extension-based semantics is a function σ such that for every argumentation framework $\mathcal{F} = (\mathcal{A}, \mathcal{R})$, we have $\sigma(\mathcal{F}) \in 2^{2^{\mathcal{A}}}$. The elements of $\sigma(\mathcal{F})$ are called* extensions*.*

Our definition requires a semantics to satisfy universal domain, i.e. to be defined for every argumentation framework. We could give a more general definition, thus allowing a semantics to be defined only for some argumentation frameworks. We do not do that in order to simplify the setting, since all the semantics of interest for our study are defined for all argumentation frameworks.

3 Semantics

This section introduces different argumentation semantics we study in the rest of the chapter. Note that most of the properties from the literature, which we study in Section 4, can appear in two variants: extension-based and labelling-based. In this chapter, we present their versions for extension-based approach. See the chapter of this volume on abstract argumentation frameworks and their semantics [Baroni et al., 2018a] for further discussion and examples.

We start by introducing the notions of conflict-freeness and admissibility.

Definition 3.1 (Conflict-freeness, admissibility, strong admissibility)
Let $\mathcal{F} = (\mathcal{A}, \mathcal{R})$ and $S \subseteq \mathcal{A}$. Set S is conflict-free *in $\mathcal{F}$ if and only if for every* $a, b \in S$, $(a, b) \notin \mathcal{R}$.

Argument $a \in \mathcal{A}$ is defended *by set S if and only if for every $b \in \mathcal{A}$ such that $b\mathcal{R}a$ there exists $c \in S$ such that $c\mathcal{R}b$. Argument $a \in \mathcal{A}$ is* strongly defended *by set S if and only if for every $b \in \mathcal{A}$ such that $b\mathcal{R}a$ there exists $c \in S \setminus \{a\}$ such that $c\mathcal{R}b$ and c is strongly defended by $S \setminus \{a\}$. S is* admissible *in $\mathcal{F}$ if and only if it is conflict-free and it defends all its arguments. S is* strongly admissible *in $\mathcal{F}$ if and only if it is conflict-free and it strongly defends all its arguments.*

Stable, complete, preferred and grounded semantics were introduced by Dung [1995]:

Definition 3.2 (Complete, stable, grounded, preferred semantics) *Let* $\mathcal{F} = (\mathcal{A}, \mathcal{R})$ *and* $S \subseteq \mathcal{A}$.

- *Set S is a* complete extension *of $\mathcal{F}$ if and only if it is conflict-free, it defends all its arguments and it contains all the arguments it defends.*

- *Set S is a* stable extension *of $\mathcal{F}$ if and only if it is conflict-free and it attacks all the arguments of $\mathcal{A} \setminus S$.*

- *S is the* grounded extension *of $\mathcal{F}$ if and only if it is a minimal with respect to set inclusion complete extension of $\mathcal{F}$.*

- *S is a* preferred extension *of $\mathcal{F}$ if and only if it is a maximal with respect to set inclusion admissible set of $\mathcal{F}$.*

Dung [1995] shows that each argumentation framework has a unique grounded extension. Stable extensions do not always exist, i.e. there exist argumentation frameworks whose set of stable extensions is empty. Semi-stable semantics [Verheij, 1996; Caminada, 2006b] guarantees that every argumentation framework has an extension. Furthermore, semi-stable semantics coincides with stable semantics on argumentation frameworks that have at least one stable extension.

Definition 3.3 (Semi-stable semantics) *Let $\mathcal{F} = (\mathcal{A}, \mathcal{R})$ and $S \subseteq \mathcal{A}$. Set S is a semi-stable extension of $\mathcal{F}$ if and only if it is a complete extension and $S \cup S^+$ is maximal with respect to set inclusion among complete extensions, i.e. there exists no complete extension S_1 such that $S \cup S^+ \subset S_1 \cup S_1^+$.*

Ideal semantics [Dung et al., 2007] is an alternative to grounded semantics. Like grounded semantics, ideal semantics always returns a unique extension, which is also a complete extension [Dung et al., 2007]. From the definition of the grounded semantics, we conclude that the ideal extension is a superset of the grounded extension. Ideal semantics is thus less sceptical than grounded semantics.

Definition 3.4 (Ideal semantics) *Let $\mathcal{F} = (\mathcal{A}, \mathcal{R})$ and $S \subseteq \mathcal{A}$. Set S is the ideal extension of $\mathcal{F}$ if and only if it is a maximal with respect to set inclusion admissible subset of every preferred extension.*

We now introduce eager semantics [Caminada, 2007].

Definition 3.5 (Eager semantics) *Let $\mathcal{F} = (\mathcal{A}, \mathcal{R})$ and $S \subseteq \mathcal{A}$. Set S is the eager extension of $\mathcal{F}$ if and only if it is the maximal with respect to set inclusion admissible subset of every semi-stable extension.*

Caminada [2007] shows that each argumentation framework has a unique eager extension and that the eager extension is also a complete extension. Note that eager semantics is similar to ideal semantics: the ideal extension is the unique biggest admissible subset of every preferred extension; the eager extension is the unique biggest admissible subset of each semi-stable extension. Since each semi-stable extension is a preferred extension [Caminada, 2006a], the eager extension is a superset of the ideal extension.

In our chapter, we want to conduct an exhaustive investigation of properties of extension-based semantics. Thus, for the sake of completeness, we introduce even the semantics that are not very commonly used or studied in the literature, like stage semantics, naive semantics and prudent variants of grounded, complete, stable and preferred semantics.

Stage semantics [Verheij, 1996] was defined in a slightly different setting than ours; we provide an alternative but equivalent definition [Verheij, 1996; Baroni et al., 2011a].

Definition 3.6 (Stage semantics) *Let $\mathcal{F} = (\mathcal{A}, \mathcal{R})$ and $S \subseteq \mathcal{A}$. Set S is a stage extension of $\mathcal{F}$ if and only if S is a conflict-free set and $S \cup S^+$ is maximal with respect to set inclusion, i.e. S is conflict-free, and there exists no conflict-free set S_1 such that $S \cup S^+ \subset S_1 \cup S_1^+$.*

Note the difference between semi-stable and stage semantics: semi-stable extension is a complete extension whereas stage extension is a conflict-free set; stage extension is not necessarily an admissible set.

Definition 3.7 (Naive semantics) *Let $\mathcal{F} = (\mathcal{A}, \mathcal{R})$ and $S \subseteq \mathcal{A}$. Set S is a naive extension of $\mathcal{F}$ if and only if S is a maximal conflict-free set.*

Prudent semantics [Coste-Marquis et al., 2005] is based on the idea that an extension should not contain arguments a and b if a indirectly attacks b. An indirect attack is an odd length attack chain.

Definition 3.8 (Indirect conflict) *Let $\mathcal{F} = (\mathcal{A}, \mathcal{R})$, $S \subseteq \mathcal{A}$ and $a, b \in \mathcal{A}$. We say that a indirectly attacks b if and only if there is an odd-length path from a to b with respect to the attack relation. We say that S is without indirect conflicts and we write $\texttt{wic}(S)$ if and only if there exist no $x, y \in S$ such that x indirectly attacks y.*

The semantics introduced by Dung (grounded, complete, stable, preferred) is based on admissibility; prudent semantics is based on p-admissibility. Prudent semantics is called grounded prudent, complete prudent, stable prudent and preferred prudent by Coste-Marquis et al. [2005]. In order to make the names shorter, we call them p-grounded, p-complete, p-stable and p-preferred.

Definition 3.9 (p-admissible sets) *Let $\mathcal{F} = (\mathcal{A}, \mathcal{R})$ and $S \subseteq \mathcal{A}$. Set S is a p-admissible set in $\mathcal{F}$ if and only if every $a \in A$ is defended by S and S is without indirect conflicts.*

Definition 3.10 (p-complete semantics) *Let $\mathcal{F} = (\mathcal{A}, \mathcal{R})$ and $S \subseteq \mathcal{A}$. Set S is a p-complete extension in $\mathcal{F}$ if and only if S is a p-admissible set and for every argument $a \in \mathcal{A}$ we have: if a defended by S and $S \cup \{a\}$ is without indirect conflicts, then $a \in S$.*

We now introduce p-characteristic function, which is needed to define p-grounded semantics. Note that grounded semantics can be defined using characteristic function, but we preferred to provide an alternative equivalent definition.

Definition 3.11 (p-characteristic function) *The p-characteristic function of an argumentation framework $\mathcal{F} = (\mathcal{A}, \mathcal{R})$ is defined as follows:*

- $\mathcal{CF}^p_\mathcal{F} : 2^\mathcal{A} \to 2^\mathcal{A}$
- $\mathcal{CF}^p_\mathcal{F}(S) = \{a \in \mathcal{A} \mid S \text{ defends } a \text{ and } \texttt{wic}(S \cup \{a\})\}$

Definition 3.12 (p-grounded semantics) *Let $\mathcal{F} = (\mathcal{A}, \mathcal{R})$. Let j be the lowest integer such that*

$$\underbrace{\mathcal{CF}^p_\mathcal{F}(\mathcal{CF}^p_\mathcal{F}(\ldots \mathcal{CF}^p_\mathcal{F}(\emptyset)\ldots))}_{j \text{ times}} = \underbrace{\mathcal{CF}^p_\mathcal{F}(\mathcal{CF}^p_\mathcal{F}(\ldots \mathcal{CF}^p_\mathcal{F}(\emptyset)\ldots))}_{j+1 \text{ times}} = S.$$

The p-grounded extension is the set S.

The p-grounded extension is a p-complete extension [Coste-Marquis et al., 2005]. Note that it is not the case in general that the p-grounded extension is included into every p-preferred extension [Coste-Marquis et al., 2005].

Definition 3.13 (p-stable semantics) *Let $\mathcal{F} = (\mathcal{A}, \mathcal{R})$ and $S \subseteq \mathcal{A}$. Set S is a p-stable extension in $\mathcal{F}$ if and only if S is without indirect conflicts and S attacks (in a direct way) each argument in $\mathcal{A} \setminus S$.*

Definition 3.14 (p-preferred semantics) *Let $\mathcal{F} = (\mathcal{A}, \mathcal{R})$ and $S \subseteq \mathcal{A}$. Set S is a p-preferred extension if and only if S is a maximal for set inclusion p-admissible set.*

Evert p-stable extension is a p-preferred extension [Coste-Marquis et al., 2005].

We now introduce CF2 semantics [Baroni et al., 2005]. For more explanations and examples, the reader is referred to the original paper. The definition of this semantics is complicated; we must introduce several auxiliary definitions in order to present it.

Let us first introduce the notion of strongly connected component (SCC) introduced by Baroni et al. [2005].

Definition 3.15 (Strongly Connected Component) *Let $\mathcal{F} = (\mathcal{A}, \mathcal{R})$. The binary relation of path-equivalence between nodes, denoted as $PE_\mathcal{F} \subseteq \mathcal{A} \times \mathcal{A}$, is defined as follows:*

- *for every $a \in \mathcal{A}$, $(a, a) \in PE_\mathcal{F}$*

- *given two distinct arguments $a, b \in \mathcal{A}$, we say that $(a, b) \in PE_\mathcal{F}$ if and only if and only if there is a path from a to b and a path from b to a.*

The strongly connected components of $\mathcal{F}$ are the equivalence classes of arguments under the relation of path-equivalence. The set of strongly connected components is denoted by $SCCS_\mathcal{F}$. Given an argument $a \in \mathcal{A}$, notation $SCC_\mathcal{F}(a)$ stands for the strongly connected component that contains a.

In the particular case when the argumentation framework is empty, i.e. $\mathcal{F} = (\emptyset, \emptyset)$, we assume that $SCCS_\mathcal{F} = \{\emptyset\}$. The choices in the antecedent strongly connected components determine a partition of the nodes of S into three subsets: defeated, provisionally defeated and undefeated. D stands for defeated, P for provisionally defeated and U for undefeated.

Definition 3.16 (D,P,U [Baroni et al., 2005]) *Given an argumentation framework $\mathcal{F} = (\mathcal{A}, \mathcal{R})$, a set $\mathcal{E} \subseteq \mathcal{A}$ and a strongly connected component $S \in SCCS_\mathcal{F}$, we define:*

- $D_\mathcal{F}(S, \mathcal{E}) = \{a \in S \mid (\mathcal{E} \cap S_{out}^-) \text{ attacks } a\}$

- $P_\mathcal{F}(S, \mathcal{E}) = \{a \in S \mid (\mathcal{E} \cap S_{out}^-) \text{ does not attack } a \text{ and } \exists b \in (S_{out}^- \cap a^-) \text{ such that } \mathcal{E} \text{ does not attack } b\}$

- $U_{\mathcal{F}}(S,\mathcal{E}) = S \setminus (D_{\mathcal{F}}(S,\mathcal{E}) \cup D_{\mathcal{F}}(S,\mathcal{E}))$

We define $UP_{\mathcal{F}}(S,\mathcal{E}) = U_{\mathcal{F}}(S,\mathcal{E}) \cup P_{\mathcal{F}}(S,\mathcal{E})$.

Definition 3.17 (CF2 semantics) Let $\mathcal{F} = (\mathcal{A}, \mathcal{R})$ and $\mathcal{E} \subseteq \mathcal{A}$. Set $\mathcal{E}$ is an extension of CF2 semantics if and only if

- $\mathcal{E}$ is a naive extension of $\mathcal{F}$ if $|SCCS_{\mathcal{F}}| = 1$
- for every $S \in SCCS_{\mathcal{F}}$, $(\mathcal{E} \cap S)$ is a CF2 extension of $\mathcal{F} \downarrow_{UP_{\mathcal{F}}(S,\mathcal{E})}$ otherwise

Observe that $\mathcal{F} \downarrow_{UP_{\mathcal{F}}(S,\mathcal{E})} = \{a \in S \mid \text{there exists no } b \in \mathcal{E} \setminus S \text{ s.t. } (b,a) \in \mathcal{R}\}$.

We now introduce stage2 semantics [Dvorák and Gaggl, 2016].

Definition 3.18 Let $\mathcal{F} = (\mathcal{A}, \mathcal{R})$ and $\mathcal{E} \subseteq \mathcal{A}$. Set $\mathcal{E}$ is a stage2 extension if and only if

- $\mathcal{E}$ is a stage extension of $\mathcal{F}$ if $|SCCS_{\mathcal{F}}| = 1$
- for every $S \in SCCS_{\mathcal{F}}$, $(\mathcal{E} \cap S)$ is a stage2 extension of $\mathcal{F} \downarrow_{UP_{\mathcal{F}}(S,\mathcal{E})}$ otherwise

Dvorák and Gaggl [2016] showed that every stage2 extension is a CF2 extension and that every stable extension is a stage2 extension.

This ends the discussion on extension based semantics of abstract argumentaton. There exist additional proposals for argumentation semantics in the literature, such as for example resolution based semantics of Baroni et al. [2011b], but we do not consider them in this chapter.

In this chapter, we focus on the extension-based approach, which means that each semantics is defined by specifying the extensions it returns for a given argumentation framework. There exists an alternative, labelling-based approach. Instead of calculating extensions, it provides labellings, one labelling being a function that attaches to every argument a label *in*, *out* or *undec* (which stands for "undecided").

Definition 3.19 (Labelling-based semantics) Let $\Lambda = \{in, out, undec\}$. Let $\mathcal{F} = (\mathcal{A}, \mathcal{R})$ be an argumentation framework. A labelling on $\mathcal{F}$ is a total function $\mathcal{L}ab : \mathcal{A} \to \Lambda$. A labelling-based semantics is a function λ defined for every element of $\mathcal{AF}$ such that for every argumentation framework $\mathcal{F}$, we have that $\lambda(\mathcal{F})$ is a set of labellings on $\mathcal{F}$.

To illustrate, let us provide a labelling-based definition of complete semantics.

Definition 3.20 (Complete labelling) Let $\mathcal{F} = (\mathcal{A}, \mathcal{R})$ and $\mathcal{L}ab$ a labelling on $\mathcal{F}$. We say that $\mathcal{L}ab$ is a complete labelling if and only if for every $a \in \mathcal{A}$:

- if a is labelled in then all its attackers are labelled out
- if a is labelled out then none of its attackers is labelled in
- if a is labelled undec then not all its attackers are labelled out and none of its attackers is labelled in.

We denote by $in(\mathcal{L}ab)$ (resp. $out(\mathcal{L}ab)$, $und(\mathcal{L}ab)$) the set of arguments labelled in (resp. out, und).

For every $\mathcal{F} = (\mathcal{A}, \mathcal{R})$, the set of complete extensions under σ is exactly the set $\{in(\mathcal{L}ab) \mid \mathcal{L}ab \text{ is a complete labelling}\}$.

Moreover, there exists a general way that allows to obtain a labelling-based definition of a semantics given its extension-based definition, under the condition that the semantics returns conflict-free sets.

Definition 3.21 (Extension to labelling) *Given an extension $\mathcal{E}$, labelling $\mathcal{L}ab_\mathcal{E}$ is defined as follows: $\mathcal{L}ab_\mathcal{E}(a) = in$ if $a \in \mathcal{E}$, $\mathcal{L}ab_\mathcal{E}(a) = out$ if $a \in \mathcal{E}^+$, $\mathcal{L}ab_\mathcal{E}(a) = und$ otherwise. Then, given a semantics σ, we say that $\mathcal{L}ab$ is a σ labelling of $\mathcal{F}$ if and only if there exists $\mathcal{E} \in \sigma(\mathcal{F})$ such that $\mathcal{L}ab = \mathcal{L}ab_\mathcal{E}$.*

Other ways to obtain a labelling from an extension are possible, for example we could say that an argument is *out* if it is attacked by an argument in the extension, or it attacks an argument in the extension. This would make the definition of *out* more symmetric and more in line with naive based semantics. However, it seems such alternatives have not been explored systematically in the literature. Moreover, even if extension and labelling based semantics are inter-translatable, it may affect other definitions such as equivalence of frameworks. Finally, using Definition 3.21, every principle defined in terms of extension based semantics can be translated into labelings and vice versa, though one of the definitions may be more compact or intuitive than the other.

We saw an intuitive way to define complete labellings in Definition 3.20. Intuitive labelling-based definitions of other semantics also exist in the literature. For example: a grounded labelling is a complete labelling such that the set of arguments labelled *in* is minimal with respect to set inclusion among all complete labellings; a stable labelling is a complete labelling such that the set of undecided arguments is empty; a preferred labelling is a complete labelling such that the set of arguments labelled *in* is maximal with respect to set inclusion among all complete labellings. The reader interested in more details about the labelling-based approach is referred to the paper by Baroni *et al.* [2011a].

4 List of Principles

This section presents the properties from the literature and reviews all the semantics with respect to the properties.

Definition 4.1 (Isomorphic argumentation frameworks) *Two argumentation frameworks $\mathcal{F}_1 = (\mathcal{A}_1, \mathcal{R}_1)$ and $\mathcal{F}_2 = (\mathcal{A}_2, \mathcal{R}_2)$ are isomorphic if and only*

	Defence	Admiss.	Strong adm.	Naivety	Ind. CF	Reinst.	Weak reinst.	CF--reinst.
complete	✓	✓	✗	✗	✗	✓	✓	✓
grounded	✓	✓	✓	✗	✗	✓	✓	✓
preferred	✓	✓	✗	✗	✗	✓	✓	✓
stable	✓	✓	✗	✓	✗	✓	✓	✓
semi-stable	✓	✓	✗	✗	✗	✓	✓	✓
ideal	✓	✓	✗	✗	✗	✓	✓	✓
eager	✓	✓	✗	✗	✗	✓	✓	✓
p-complete	✓	✓	✗	✗	✓	✗	✗	✗
p-grounded	✓	✓	✓	✗	✓	✗	✗	✗
p-preferred	✓	✓	✗	✗	✓	✗	✗	✗
p-stable	✓	✓	✗	✓	✓	✓	✓	✓
naive	✗	✗	✗	✓	✗	✗	✗	✓
CF2	✗	✗	✗	✓	✗	✗	✓	✓
stage	✗	✗	✗	✓	✗	✗	✗	✓
stage2	✗	✗	✗	✓	✗	✗	✓	✓

Table 1. Properties of semantics: basic properties, admissibility and reinstatement

if there exists a bijective function $m : \mathcal{A}_1 \to \mathcal{A}_2$, such that $(a,b) \in \mathcal{R}_1$ if and only if $(m(a), m(b)) \in \mathcal{R}_2$. This is denoted by $\mathcal{F}_1 \doteq_m \mathcal{F}_2$.

The first property, called "language independence" by Baroni and Giacomin [2007] is an obvious requirement for argumentation semantics. It is sometimes called abstraction [Amgoud and Besnard, 2013; Bonzon et al., 2016a] or anonymity [Amgoud et al., 2016].

Principle 1 (Language independence) *A semantics σ satisfies the* language independence *principle if and only if for every two argumentation frameworks $\mathcal{F}_1$ and $\mathcal{F}_2$, if $\mathcal{F}_1 \doteq_m \mathcal{F}_2$ then $\sigma(\mathcal{F}_2) = \{m(\mathcal{E}) \mid \mathcal{E} \in \sigma(\mathcal{F}_1)\}$.*

It is immediate to see that all the semantics satisfy language independence, since the definitions of semantics take into account only the topology of the graph, and not the arguments' names.

Conflict-freeness is one of the basic principles. Introduced by Dung [1995] and stated as a principle by Baroni and Giacomin [2007], it is satisfied by all argumentation semantics studied in this chapter. Note that one can define a non conflict-free semantics [Arieli, 2015]. As another example of relaxing conflict-freeness consider the work by Dunne et al. [2011], who introduce a framework where each attack is associated a weight; given an inconsistency budget β, they accept to disregard the set of attacks up to total weight of β.

Principle 2 (Conflict-freeness) *A semantics σ satisfies the* conflict-freeness *principle if and only if for every argumentation framework $\mathcal{F}$, for every $\mathcal{E} \in \sigma(\mathcal{F})$, $\mathcal{E}$ is conflict-free set in $\mathcal{F}$.*

Defence is a well-known property introduced by Dung [1995].

Principle 3 (Defence) *A semantics σ satisfies the* defence *principle if and only if for every argumentation framework $\mathcal{F}$, for every $\mathcal{E} \in \sigma(\mathcal{F})$, for every $a \in \mathcal{E}$, $\mathcal{E}$ defends a.*

Baroni and Giacomin [2007] show that complete, grounded, preferred, stable, semi-stable, ideal, p-complete, p-grounded, p-preferred, p-stable satisfy defence and that CF2 does not satisfy defence. Let us consider the four remaining semantics: stage, stage2, eager and naive. The argumentation framework from Figure 1 shows that stage, stage2 and naive semantics violate defence since they all return three extensions: $\{a\}$, $\{b\}$ and $\{c\}$. Eager semantics satisfies defence (this follows directly from its definition).

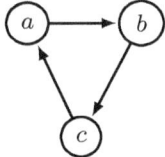

Figure 1. Stage, stage2, naive and CF2 semantics violate admissibility, defence and reinstatement, since they return three extensions: $\{a\}$, $\{b\}$ and $\{c\}$.

Baroni and Giacomin [2007] suppose that every extension is conflict-free. Thus an extension defends all it arguments if and only if it is admissible. However, if conflict-freeness is seen as an optional criterion, we can distinguish between the principles of admissibility and defence.

Principle 4 (Admissibility) *A semantics σ satisfies the* admissibility *principle if and only if for every argumentation framework $\mathcal{F}$, every $\mathcal{E} \in \sigma(\mathcal{F})$ is admissible in $\mathcal{F}$.*

Observation 1 *If a semantics σ satisfies admissibility it also satisfies conflict-freeness and defence.*

We now study the notion of strong admissibility [Baroni and Giacomin, 2007].

Principle 5 (Strong admissibility) *A semantics σ satisfies the* strong admissibility *principle if and only if for every argumentation framework $\mathcal{F}$, for every $\mathcal{E} \in \sigma(\mathcal{F})$ it holds that $a \in \mathcal{E}$ implies that $\mathcal{E}$ strongly defends a.*

Observation 2 *If a semantics σ satisfies strong admissibility then it satisfies admissibility.*

To understand the notion of strong admissibility, consider the example from Figure 2. Set $\{a, d\}$ is admissible but is not strongly admissible. Informally speaking, this is because a is defended by d whereas d is defended by a. The intuition behind strong admissibility is that this kind of defence is not strong enough because it is cyclic, i.e. arguments defend each other. However, argument e is not attacked, thus $\{e\}$ is strongly admissible. Furthermore, $\{e\}$ strongly defends a, so $\{a,e\}$ is strongly admissible. Also, $\{a,e\}$ strongly defends d. Thus $\{a,d,e\}$ is strongly admissible.

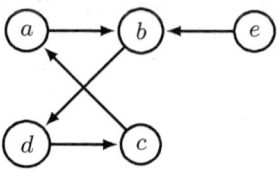

Figure 2. Set $\{a,d\}$ is admissible but is not strongly admissible. Set $\{a,d,e\}$ is admissible and strongly admissible.

Baroni and Giacomin [2007] show that grounded and p-grounded semantics satisfy strong admissibility and that complete, preferred, stable, semi-stable, ideal, p-complete, p-preferred, p-stable and CF2 do not satisfy this principle. Let us consider stage, stage2, eager and naive semantics. Since stage, stage2 and naive semantics violate admissibility, they also violate strong admissibility. To see that eager semantics violates strong admissibility too, consider the example from Figure 3, suggested by Caminada [2007]. The eager extension is $\{b,d\}$; this set is not strongly admissible since it does not strongly defend b.

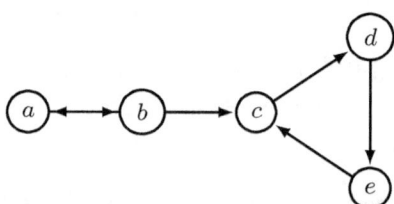

Figure 3. Eager semantics violates strong admissibility because eager extension $\{b,d\}$ does not strongly defend b. The same example shows that eager semantics violates directionality. Observe that $U = \{a,b\}$ is an unattacked set. Denote the whole framework by $\mathcal{F} = (\mathcal{A}, \mathcal{R})$. The eager extension of $\mathcal{F}$ is the set $\{b,d\}$ whereas the eager extension of $\mathcal{F} \downarrow_U$ is the empty set.

Another principle, which we call *naivety*, says that every extension under semantics σ is a naive extension.

Principle 6 (Naivety) *A semantics σ satisfies the naivety principle if and only if for every argumentation framework $\mathcal{F}$, for every $\mathcal{E} \in \sigma(\mathcal{F})$, $\mathcal{E}$ is maximal for set inclusion conflict-free set in $\mathcal{F}$.*

We see directly from the definitions of stable, stage, naive, p-stable and CF2 semantics that they satisfy naivety. Since every stage2 extension is also a CF2 extension [Dvořák and Gaggl, 2016], naivety is also satisfied by stage2 semantics. It is easy to see that the other semantics violate this principle.

Coste-Marquis et al. [2005] introduced prudent semantics, which are based on the notion of indirect conflict-freeness.

Principle 7 (Indirect conflict-freeness) *A semantics σ satisfies the indirect conflict-freeness principle if and only if for every argumentation framework $\mathcal{F}$, for every $\mathcal{E} \in \sigma(\mathcal{F})$, $\mathcal{E}$ is without indirect conflicts in $\mathcal{F}$.*

Observation 3 *If a semantics σ satisfies indirect conflict-freeness then it satisfies conflict-freeness.*

By examining the definitions of prudent semantics, we see that they all satisfy indirect conflict-freeness, since this concept is built in through the use of p-admissibility and p-characteristic function.

The other semantics do not satisfy indirect conflict-freeness. To show this, consider the argumentation framework depicted in Figure 4, suggested by Coste-Marquis et al. [2005]. All the semantics except prudent ones have an extension containing both a and e. Hence, they violate indirect conflict-freeness since e indirectly attacks a.

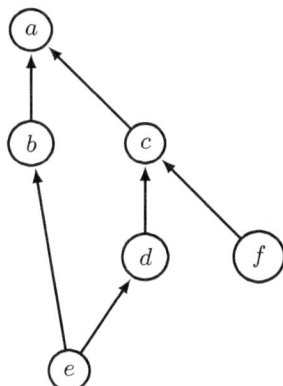

Figure 4. All semantics except prudent semantics violate indirect conflict-freeness. They all yield an extension containing both a and e, even if e indirectly attacks a.

Defence says that an extension must defend all the arguments it contains. Reinstatement can be seen as its counterpart, since it says that an extension must contain all the arguments it defends. This principle was first studied in a systematic way by Baroni and Giacomin [2007].

Principle 8 (Reinstatement) *A semantics σ satisfies the* reinstatement *principle if and only if for every argumentation framework $\mathcal{F}$, for every $\mathcal{E} \in \sigma(\mathcal{F})$, for every $a \in \mathcal{A}$ it holds that if $\mathcal{E}$ defends a then $a \in \mathcal{E}$.*

The results in Table 1 concerning complete, grounded, preferred, stable, semi-stable, ideal, p-complete, p-grounded, p-preferred, p-stable and CF2 semantics were proved by Baroni and Giacomin [2007]. To summarise, all the semantics they study satisfy reinstatement except p-grounded, p-complete, p-preferred and CF2. Let us consider eager, stage, stage2 and naive semantics.

Regarding eager semantics, suppose that $\mathcal{E}$ is an eager extension and that a is defended by $\mathcal{E}$. The eager extension is a complete extension [Caminada, 2007], and complete semantics satisfies reinstatement. Thus, $a \in \mathcal{E}$, which means that eager semantics satisfies reinstatement.

Stage, stage2 and naive semantics violate reinstatement, as proved by Dvořák and Gaggl [2016]. Another way to see this is to consider the counter-example from Figure 1.

Baroni and Giacomin [2007] study another property called weak reinstatement.

Principle 9 (Weak reinstatement) *A semantics σ satisfies the* weak reinstatement *principle if and only if for every argumentation framework $\mathcal{F}$, for every $\mathcal{E} \in \sigma(\mathcal{F})$ it holds that*

$$\mathcal{E} \text{ strongly defends } a \text{ implies } a \in \mathcal{E}.$$

Observation 4 *If a semantics σ satisfies reinstatement then it satisfies weak reinstatement.*

The results in Table 1 concerning complete, grounded, preferred, stable, semi-stable, ideal, p-complete, p-grounded, p-preferred, p-stable and CF2 semantics were proved by Baroni and Giacomin [2007]. From Observation 4 we conclude that eager semantics satisfies weak reinstatement.

Stage and naive semantics violate weak reinstatement as can be seen from Figure 5. This was also shown by Dvořák and Gaggl [2016]. Namely, $\{b\}$ is a stage and a naive extension that strongly defends a but does not contain it. Stage2 semantics does satisfy weak reinstatement [Dvořák and Gaggl, 2016].

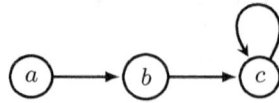

Figure 5. Stage and naive semantics violate weak reinstatement, since $\mathcal{E} = \{b\}$ is an extension that strongly defends a, but $\mathcal{E}$ does not contain a.

The reinstatement principle makes sure that as soon as an argument a is defended by an extension $\mathcal{E}$, a should belong to $\mathcal{E}$—without specifying that a is not in conflict with arguments of $\mathcal{E}$. To take this into account, another principle was defined by Baroni and Giacomin [2007].

Principle 10 ($\mathcal{CF}$-reinstatement) *A semantics σ satisfies the $\mathcal{CF}$-reinstatement principle if and only if for every argumentation framework $\mathcal{F}$, for every $\mathcal{E} \in \sigma(\mathcal{F})$, for every $a \in \mathcal{A}$ it holds that if $\mathcal{E}$ defends a and $\mathcal{E} \cup \{a\}$ is conflict-free then $a \in \mathcal{E}$.*

Observation 5 *If a semantics σ satisfies reinstatement then it satisfies $\mathcal{CF}$-reinstatement.*

The results in Table 1 concerning complete, grounded, preferred, stable, semi-stable, ideal, p-complete, p-grounded, p-preferred, p-stable and CF2 semantics were proved by Baroni and Giacomin [2007].

If $\mathcal{E}$ is a naive extension and a an argument such that $\mathcal{E}$ defends a and $\mathcal{E} \cup \{a\}$ is conflict-free, then $a \in \mathcal{E}$ since $\mathcal{E}$ is a maximal conflict-free set. This means that naive semantics satisfies $\mathcal{CF}$-reinstatement.

Observation 5 implies that eager semantics satisfies $\mathcal{CF}$-reinstatement.

Stage and stage2 semantics satisfy $\mathcal{CF}$-reinstatement, as shown Dvorák and Gaggl [2016].

	I-max.	Allowing abstention	Crash resistance	Non--interference	Direct.	Weak--direct.	Semi--direct.
complete	×	✓	✓	✓	✓	✓	✓
grounded	✓	✓	✓	✓	✓	✓	✓
preferred	✓	×	✓	✓	✓	✓	✓
stable	✓	×	×	×	×	✓	×
semi-stable	✓	×	✓	✓	×	×	×
ideal	✓	✓	✓	✓	✓	✓	✓
eager	✓	✓	✓	✓	✓	✓	✓
p-complete	×	✓	✓	✓	×	×	✓
p-grounded	✓	✓	✓	✓	✓	✓	✓
p-preferred	✓	×	✓	✓	×	×	✓
p-stable	✓	×	×	×	×	✓	×
naive	✓	×	✓	✓	×	×	✓
CF2	✓	×	✓	✓	✓	✓	✓
stage	✓	×	✓	✓	×	×	×
stage2	✓	×	✓	✓	✓	✓	✓

Table 2. Properties of semantics, part 2

The next principle was first considered by Baroni and Giacomin [2007]. It says that an extension cannot contain another extension.

Principle 11 (I-maximality) *A semantics σ satisfies the I-maximality principle if and only if for every argumentation framework $\mathcal{F}$, for every $\mathcal{E}_1, \mathcal{E}_2 \in$*

$\sigma(\mathcal{F})$, if $\mathcal{E}_1 \subseteq \mathcal{E}_2$ then $\mathcal{E}_1 = \mathcal{E}_2$.

I-maximality is trivially satisfied by single extension semantics. It is thus satisfied by eager semantics. We see directly from the definitions of naive and stage semantics that they satisfy I-maximality. Dvorák and Gaggl [2016] show that stage2 semantics satisfies I-maximality. Baroni and Giacomin [2007] show that I-maximality is satisfied by all other semantics except complete and p-complete semantics.

Baroni et al. [2011a] define a principle called rejection, which says that if an argument a is labelled in and a attacks b, then b should be labelled out. If we use the translation from extension to a labelling we mentioned in Definition 3.21, we see that all the labellings satisfy this property. However, it would be possible to be more general by defining a labelling-based semantics that does not satisfy this property. Let us define a semantics σ that always returns a unique labelling such that an argument is labelled in if it is not attacked, it is labelled undec if it is attacked by exactly one argument and it is labelled out otherwise. Consider the example from Figure 5: argument a will be labelled in, argument b undec and argument c out, which violates the rejection principle.

We next consider the allowing abstention principle [Baroni et al., 2011a].

Principle 12 (Allowing abstention) *A semantics σ satisfies the* allowing abstention *principle if and only if for every argumentation framework $\mathcal{F}$, for every $a \in \mathcal{A}$, if there exist two extensions $\mathcal{E}_1, \mathcal{E}_2 \in \sigma(\mathcal{F})$ such that $a \in \mathcal{E}_1$ and $a \in \mathcal{E}_2^+$ then there exists an extension $\mathcal{E}_3 \in \sigma(\mathcal{F})$ such that $a \notin (\mathcal{E}_3 \cup \mathcal{E}_3^+)$.*

Baroni et al. [2011a] show that complete semantics satisfies the previous principle and that preferred, stable, semi-stable, stage and CF2 semantics falsify it. Observe that unique status semantics trivially satisfy this principle. Allowing abstention is thus satisfied by grounded, ideal, eager and p-grounded semantics.[1]

Let us now consider the remaining semantics, namely: naive, p-stable, p-preferred, p-complete and stage2 semantics.

We first prove that p-complete semantics satisfies allowing abstention. We start with a lemma.

Lemma 4.2 *Let $\mathcal{F} = (\mathcal{A}, \mathcal{R})$ be an argumentation framework, GE_p its p-grounded extension and $\mathcal{E} \subseteq \mathcal{A}$ be a set that defends all its arguments. Then, $\mathcal{E}$ does not attack GE_p.*

Proof. Let $\mathcal{CF}^p$ be the p-characteristic function. Denote $\mathrm{GE}_p{}^0 = \emptyset$, $\mathrm{GE}_p{}^1 = \mathcal{CF}^p(\emptyset)$, $\mathrm{GE}_p{}^2 = \mathcal{CF}^p(\mathcal{CF}^p(\emptyset))$, ... and denote by GE_p the p-grounded extension of $\mathcal{F}$. Let $\mathcal{E}$ be a set that defends all its arguments. By means of contradiction, suppose that there exist $x \in \mathcal{E}$, $y \in \mathrm{GE}_p$ such that $x\mathcal{R}y$. Let $k \in \mathbb{N}$ be the

---

[1]Note that Table 2 by Baroni et al. [2011a] specifies that grounded semantics does not satisfy dilemma abstaining. The reason is that Baroni et al. consider the property as being "non-applicable" to unique status semantics (personal communication, 2016).

minimal number such that $y \in \mathrm{GE}_p{}^k$. From the definition of function $\mathcal{CF}^p$, there exists $l < k$ such that there exists $y_1 \in \mathrm{GE}_p{}^l$ such that $y_1 \mathcal{R} x$. Since $\mathcal{E}$ defends all its arguments, there exists $x_1 \in \mathcal{E}$ such that $x_1 \mathcal{R} y_1$. Again, there exists $m < l$ such that there exists $y_2 \in \mathrm{GE}_p{}^m$ such that $y_2 \mathcal{R} x_1$. By continuing this process, we conclude that there exists $y_s \in \mathrm{GE}_p{}^1$ such that there exists $x_s \in \mathcal{E}$ such that $x_S \mathcal{R} y_s$. This is impossible, since the arguments of $\mathrm{GE}_p{}^1$ are not attacked. Contradiction. ∎

Proposition 4.3 *p-complete semantics satisfies allowing abstention.*

Proof. Let $\mathcal{F} = (\mathcal{A}, \mathcal{R})$, let $a, b \in \mathcal{A}$, let $b\mathcal{R}a$ and let $\mathcal{E}_1$ and $\mathcal{E}_2$ be p-complete extensions such that $a \in \mathcal{E}_1$ and $b \in \mathcal{E}_2$. Denote by GE_p the p-grounded extension of $\mathcal{F}$. Let us prove that $a \notin \mathrm{GE}_p$ and that GE_p does not attack a. First, since $b\mathcal{R}a$ and b belongs to a p-complete extension (and every p-complete extension defends all its arguments), Lemma 4.2 implies that $a \notin \mathrm{GE}_p$. Let us now show that GE_p does not attack a. By means of contradiction, suppose the contrary. Let $b \in \mathrm{GE}_p$ be an argument such that $b\mathcal{R}a$. Since $a \in \mathcal{E}_1$, and $\mathcal{E}_1$ defends all its arguments, then there exists $c \in \mathcal{E}_1$ such that $c\mathcal{R}b$. Contradiction with Lemma 4.2. Thus, it must be that GE_p does not attack a. It is known that the p-grounded extension is a p-complete extension [Coste-Marquis et al., 2005]. Thus, we showed that there exists a p-complete extension that neither contains nor attacks argument a. ∎

To see why naive, p-stable, p-preferred and stage2 semantics violate allowing abstention, consider the argumentation framework depicted in Figure 6. The principle is violated since all those semantics return two extensions, $\{a\}$ and $\{b\}$.

Figure 6. Several semantics violate allowing abstention principle.

To define crash resistance [Caminada et al., 2012], we first need to introduce the following two definitions.

Definition 4.4 (Disjoint argumentation frameworks) *Two argumentation frameworks $\mathcal{F}_1 = (\mathcal{A}_1, \mathcal{R}_1)$ and $\mathcal{F}_2 = (\mathcal{A}_2, \mathcal{R}_2)$ are disjoint if and only if $\mathcal{A}_1 \cap \mathcal{A}_2 = \emptyset$.*

A framework $\mathcal{F}^\star$ is contaminating if joining $\mathcal{F}^\star$ with an arbitrary disjoint framework $\mathcal{F}$ results in a framework $\mathcal{F} \cup \mathcal{F}^\star$ having the same extensions as $\mathcal{F}^\star$. The intuition behind this definition is that $\mathcal{F}^\star$ contaminates every framework.

Definition 4.5 (Contaminating) *An argumentation framework $\mathcal{F}^\star$ is contaminating for a semantics σ if and only if for every argumentation framework $\mathcal{F}$ disjoint from $\mathcal{F}^\star$ it holds that $\sigma(\mathcal{F}^\star \cup \mathcal{F}) = \sigma(\mathcal{F}^\star)$.*

A semantics is crash resistant if and only if there are no contaminating frameworks. The intuition behind this name is that a contaminating framework causes the system to crash.

Principle 13 (Crash resistance) *A semantics σ satisfies the crash resistance principle if and only if there are no contaminating argumentation frameworks for σ.*

Crash resistance forbids only the most extreme form of interferences between disjoint subgraphs. A stronger property, non-interference, was defined by Caminada et al. [2012]. We first need to define a notion of isolated set, i.e. a set that neither attacks outside arguments nor is attacked by them.

Definition 4.6 (Isolated set of arguments) *Let $\mathcal{F} = (\mathcal{A}, \mathcal{R})$ be an argumentation framework. A set $S \subseteq \mathcal{A}$ is isolated in $\mathcal{F}$ if and only if*

$$((S \times (\mathcal{A} \setminus S)) \cup ((\mathcal{A} \setminus S) \times S)) \cap \mathcal{R} = \emptyset.$$

A semantics satisfies non-interference principle if for every isolated set S, the intersections of the extensions with set S coincide with the extensions of the restriction of the framework on S.

Principle 14 (Non-interference) *A semantics σ satisfies the non-interference principle if and only if for every argumentation framework $\mathcal{F}$, for every set of arguments S isolated in $\mathcal{F}$ it holds that $\sigma(\mathcal{F} \downarrow_S) = \{\mathcal{E} \cap S \mid \mathcal{E} \in \sigma(\mathcal{F})\}$.*

The previous principle can be made even stronger by considering the case when the set S is not attacked by the rest of the framework, but can attack the rest of the framework. Let us formalize the notion of an unattacked set.

Definition 4.7 (Unattacked arguments) *Given an argumentation framework $\mathcal{F} = (\mathcal{A}, \mathcal{R})$, a set U is unattacked if and only if there exists no $a \in \mathcal{A} \setminus U$ such that a attacks U. The set of unattacked sets in $\mathcal{F}$ is denoted $\mathcal{US}(\mathcal{F})$.*

We can now define the principle of directionality, introduced by Baroni and Giacomin [2007].

Principle 15 (Directionality) *A semantics σ satisfies the directionality principle if and only if for every argumentation framework $\mathcal{F}$, for every $U \in \mathcal{US}(\mathcal{F})$, it holds that $\sigma(\mathcal{F} \downarrow_U) = \{\mathcal{E} \cap U \mid \mathcal{E} \in \sigma(\mathcal{F})\}$.*

Baroni et al. [2011a] show the following dependencies between directionality, interference and crash resistance.

Observation 6 *Directionality implies non interference, and non interference implies crash resistance.*

Let us see which semantics satisfy directionality. Baroni and Giacomin [2007] proved that complete, grounded, preferred, ideal, p-grounded and CF2 semantics satisfy directionality. They also showed that stable, semi-stable, p-complete, p-stable and p-preferred semantics violate this principle. Baroni et al. [2011a] show that stage semantics does not satisfy directionality; however, Dvorák and Gaggl [2016] show that stage2 semantics does satisfy directionality.

The only remaining semantics are eager and naive. The argumentation framework from Figure 7 shows that naive semantics does not satisfy directionality. The argumentation framework from Figure 3 shows that eager semantics

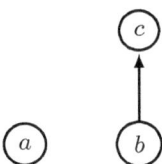

Figure 7. Naive semantics violates directionality and weak directionality. Denote the whole framework by $\mathcal{F} = (\mathcal{A}, \mathcal{R})$. Let $U = \{a, b\}$. Observe that $\{a, c\}$ is a naive extension of $\mathcal{F}$ but that $\{a\}$ is not a naive extension of $\mathcal{F} \downarrow_U$.

does not satisfy directionality.

Let us now consider non-interference. Baroni et al. [2011a] showed that non-interference is satisfied by complete, grounded, preferred, semi-stable, ideal, stage and CF2 semantics. Eager semantics satisfies non-interference since it satisfies directionality. From the definition of non-interference we see that this principle is satisfied by naive semantics. Since p-grounded semantics satisfies directionality, it also satisfies non-interference.

Proposition 4.8 *p-complete, p-preferred semantics satisfy non-interference.*

Proof. We present the proof for p-complete semantics, the one for p-preferred semantics is similar. Let $\mathcal{F} = (\mathcal{A}, \mathcal{R})$ and $\mathcal{A}' \subseteq \mathcal{A}$ be an isolated set in $\mathcal{F}$. Denote by $\mathcal{F}' = (\mathcal{A}', \mathcal{R}')$ the restriction of $\mathcal{F}$ on $\mathcal{A}'$. Let us first suppose that $\mathcal{E}$ is a complete prudent extension of $\mathcal{F}$. Denote $\mathcal{E}' = \mathcal{E} \cap \mathcal{A}'$. We have $icf(\mathcal{E}')$. It is easy to see that every $\alpha \in \mathcal{E}'$ is defended by $\mathcal{E}'$ from all attacks from $\mathcal{A}'$. Also, for an $\alpha \in \mathcal{A}' \setminus \mathcal{E}'$, we can easily see that either $\mathcal{E}' \cup \{\alpha\}$ is not without indirect conflicts or α is attacked by some argument and not defended by $\mathcal{E}'$. Suppose now that $\mathcal{E}'$ is a complete prudent extension of $\mathcal{F}'$. Then $\mathcal{E}'$ is p-admissible in $\mathcal{F}$, so there must be a complete prudent extension $\mathcal{E}''$ of $\mathcal{F}$ such that $\mathcal{E}' \subseteq \mathcal{E}''$. ∎

Stage2 semantics satisfies non-interference since it satisfies directionality. Finally, p-stable semantics violates non-interference. Indeed, as we will soon see, p-stable semantics violates crash resistance. Since non-interference implies crash resistance, we conclude that p-stable semantics violates non-interference.

Let us now consider crash resistance. Baroni et al. [2011a] showed that non-interference is satisfied by complete, grounded, preferred, semi-stable, ideal, stage and CF2 semantics. Eager, naive, p-grounded, p-complete, p-preferred and stage2 semantics satisfy crash resistance since they satisfy non-interference. To see that stable semantics and p-stable semantics violate crash resistance, consider the framework $\mathcal{F}^* = (\{a\}, \{(a,a)\})$. We see that $\mathcal{F}^*$ is contaminating for stable and p-stable semantics. Thus, they both violate crash resistance.

Let us now consider two variants of directionality, called weak directionality and semi-directionality suggested by M. Giacomin (personal communication, 2016).

Principle 16 (Weak directionality) *A semantics σ satisfies the* weak directionality *principle if and only if for every argumentation framework $\mathcal{F}$, for every $U \in \mathcal{US}(\mathcal{F})$, it holds that $\sigma(\mathcal{F} \downarrow_U) \supseteq \{\mathcal{E} \cap U \mid \mathcal{E} \in \sigma(\mathcal{F})\}$.*

Principle 17 (Semi-directionality) *A semantics σ satisfies the* semi-directionality *principle if and only if for every argumentation framework $\mathcal{F}$, for every $U \in \mathcal{US}(\mathcal{F})$, it holds that $\sigma(\mathcal{F} \downarrow_U) \subseteq \{\mathcal{E} \cap U \mid \mathcal{E} \in \sigma(\mathcal{F})\}$.*

Observation 7 *A semantics σ satisfies directionality if and only if σ satisfies both weak directionality and semi-directionality.*

Thus, grounded, complete, preferred, ideal, eager, p-grounded, stage2 and CF2 semantics satisfy both weak directionality and semi-directionality. It is immediate from the definition that stable semantics satisfies weak directionality. Since stable semantics does not satisfy directionality, it does not satisfy semi-directionality.

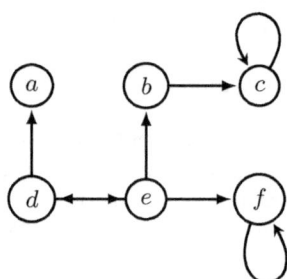

Figure 8. Semi-stable and stage semantics violate weak directionality. Let $U = \{d, e, f\}$. Set $\{b, d\}$ is an extension of this argumentation framework, but $\{b\}$ is not an extension of the restriction of this framework on U.

Example from Figure 8 shows that semi-stable semantics does not satisfy weak directionality. To see that semi-stable semantics does not satisfy semi-directionality, consider the example from Figure 9, suggested by M. Giacomin.

Stage semantics violates weak directionality, the same counter-example as for

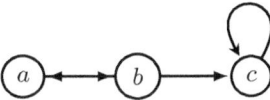

Figure 9. Semi-stable and stage semantics violate semi-directionality. Let $U = \{a,b\}$. Set $\{a\}$ is an extension of the restriction of the framework on U, but there is no extension $\mathcal{E}$ of the whole framework such that $\mathcal{E} \cap U = \{a\}$.

semi-stable semantics (Figure 8) can be used. Stage semantics also violates semi-directionality, and we can again use the same counter-example as for semi-stable semantics (Figure 9).

Directly from the definition of naive semantics we see that it satisfies semi-directionality. Since it does not satisfy directionality, we conclude from Observation 7 that it does not satisfy weak directionality.

Proposition 4.9 *p-complete and p-preferred semantics satisfy semi-directionality.*

Proof. We present the proof for p-complete semantics, the proof for p-preferred semantics is similar. Let $\mathcal{F} = (\mathcal{A}, \mathcal{R})$ be an argumentation framework, $U \subseteq \mathcal{A}$ an unattacked set and $\mathcal{F}' = \mathcal{F} \downarrow_U$ the restriction of $\mathcal{F}$ on U. Let $\mathcal{E}'$ be a p-complete extension of $\mathcal{F}'$. Then $\mathcal{E}'$ is without indirect conflicts and is p-admissible in $\mathcal{F}'$. It is immediate to see that $\mathcal{E}'$ is also p-admissible in $\mathcal{F}$. It is clear that there exists no $x \in U \setminus \mathcal{E}'$ such that x is defended by $\mathcal{E}'$ and $\mathcal{E}' \cup \{x\}$ is without indirect conflicts. Thus, there exists a (possibly empty) set $\mathcal{E} \subset (\mathcal{A} \setminus U)$ such that $\mathcal{E} \cup \mathcal{E}'$ is a p-complete extension. ∎

Since both p-complete and p-preferred semantics violate directionality, the previous proposition and Observation 7 imply that they both violate weak directionality.

Directly from the definition of p-stable semantics, we see that this semantics satisfies weak directionality. From Observation 7 we conclude that it does not satisfy semi-directionality.

We now consider the six properties related to skepticism and resolution adequacy [Baroni and Giacomin, 2007].

The first definition says that a set of extensions Ext$_1$ is more skeptical than Ext$_2$ if the set of skeptically accepted arguments with respect to Ext$_1$ is a subset of the set of skeptically accepted arguments with respect to Ext$_2$.

Definition 4.10 ($\preceq_\cap^E$) *Let* Ext$_1$ *and* Ext$_2$ *be two sets of sets of arguments. We say that* Ext$_1$ $\preceq_\cap^E$ Ext$_2$ *if and only if*

$$\bigcap_{\mathcal{E}_1 \in \text{Ext}_1} \mathcal{E}_1 \subseteq \bigcap_{\mathcal{E}_2 \in \text{Ext}_2} \mathcal{E}_2.$$

The previous definition compares only the intersections of extensions. A finer criterion was introduced by Baroni et al. [2004].

Definition 4.11 ($\preceq_W^E$) *Let* Ext_1 *and* Ext_2 *be two sets of sets of arguments. We say that* $\text{Ext}_1 \preceq_W^E \text{Ext}_2$ *if and only if*

for every $\mathcal{E}_2 \in \text{Ext}_2$, there exists $\mathcal{E}_1 \in \text{Ext}_1$ such that $\mathcal{E}_1 \subseteq \mathcal{E}_2$.

Baroni and Giacomin [2007] refine the previous relation by introducing the following definition.

Definition 4.12 ($\preceq_S^E$) *Let* Ext_1 *and* Ext_2 *be two sets of sets of arguments. We say that* $\text{Ext}_1 \preceq_S^E \text{Ext}_2$ *if and only if* $\text{Ext}_1 \preceq_W^E \text{Ext}_2$ *and*

for every $\mathcal{E}_1 \in \text{Ext}_1$, there exists $\mathcal{E}_2 \in \text{Ext}_2$ such that $\mathcal{E}_1 \subseteq \mathcal{E}_2$.

Letters W and S in the previous definitions stand for *weak* and *strong*. Baroni and Giacomin [2007] showed that the three relations are reflexive and transitive and that they are also in strict order of implication. Namely, given two sets of sets of arguments Ext_1 and Ext_2, we have

Observation 8

$$\text{Ext}_1 \preceq_S^E \text{Ext}_2 \text{ implies } \text{Ext}_1 \preceq_W^E \text{Ext}_2$$

$$\text{Ext}_1 \preceq_W^E \text{Ext}_2 \text{ implies } \text{Ext}_1 \preceq_\cap^E \text{Ext}_2$$

We now define a skepticism relation $\preceq^A$ between argumentation frameworks. It says that $\mathcal{F}_1 \preceq^A \mathcal{F}_2$ if $\mathcal{F}_1$ may have some symmetric attacks where $\mathcal{F}_2$ has a directed attack.

Definition 4.13 ($\preceq^A$) *Given an argumentation framework* $\mathcal{F} = (\mathcal{A}, \mathcal{R})$, *the conflict set is defined as* $\text{CONF}(\mathcal{F}) = \{(a,b) \in \mathcal{A} \times \mathcal{A} \mid (a,b) \in \mathcal{R} \text{ or } (b,a) \in \mathcal{R}\}$. *Given two argumentation frameworks* $\mathcal{F}_1 = (\mathcal{A}_1, \mathcal{R}_1)$ *and* $\mathcal{F}_2 = (\mathcal{A}_2, \mathcal{R}_2)$, *we say that* $\mathcal{F}_1 \preceq^A \mathcal{F}_2$ *if and only if* $\text{CONF}(\mathcal{F}_1) = \text{CONF}(\mathcal{F}_2)$ *and* $\mathcal{R}_2 \subseteq \mathcal{R}_1$.

Observe that $\preceq^A$ is a partial order, as it consists of an equality and a set inclusion relation [Baroni and Giacomin, 2007]. Note that within the set of argumentation frameworks comparable with a given argumentation framework $\mathcal{F}$, there might be several maximal elements with respect to $\preceq^A$, since there might be several ways to replace all symmetric attacks by asymmetric ones.

We can now introduce the skepticism adequacy principle. Its idea is that if $\mathcal{F}$ is more skeptical than $\mathcal{F}'$ then the set of extensions of $\mathcal{F}$ is more skeptical than that of $\mathcal{F}'$.

Principle 18 (Skepticism adequacy) *Given a skepticism relation $\prec^E$ between sets of sets of arguments, a semantics σ satisfies the $\preceq^E$-skepticism adequacy principle if and only if for every two argumentation frameworks $\mathcal{F}$ and $\mathcal{F}'$ such that $\mathcal{F} \preceq^A \mathcal{F}'$ it holds that $\sigma(\mathcal{F}) \preceq^E \sigma(\mathcal{F}')$.*

For example if $\mathcal{F}$ consists of two arguments a and b attacking each other and $\mathcal{F}'$ has only an attack from a to b, then the intersection of the extensions of $\mathcal{F}$ ($\emptyset$ for all semantics) is a subset of extensions of $\mathcal{F}'$, typically $\{a\}$. Roughly speaking: the more symmetric attacks we replace, the more we know, but we do not loose any accepted arguments.

Observation 9

- If σ satisfies $\preceq^E_S$-sk. adequacy then it satisfies $\preceq^E_W$-sk. adequacy
- If σ satisfies $\preceq^E_W$-sk. adequacy then it satisfies $\preceq^E_\cap$-sk. adequacy

	$\preceq^E_\cap$-sk. ad.	$\preceq^E_W$-sk. ad.	$\preceq^E_S$-sk. ad.	$\preceq^E_\cap$-res. ad.	$\preceq^E_W$-res. ad.	$\preceq^E_S$-res. ad.
complete	✓	✓	✗	✗	✗	✗
grounded	✓	✓	✓	✗	✗	✗
preferred	✗	✗	✗	✓	✓	✓
stable	✓	✓	✗	✓	✓	✓
semi-stable	✗	✗	✗	✓	✓	✗
ideal	✗	✗	✗	✗	✗	✗
eager	✗	✗	✗	✗	✗	✗
p-complete	✗	✗	✗	✗	✗	✗
p-grounded	✗	✗	✗	✓	✗	✗
p-preferred	✗	✗	✗	✗	✗	✗
p-stable	✗	✗	✗	✓	✓	✗
naive	✓	✓	✓	✓	✓	✓
CF2	✓	✓	✗	✗	✗	✗
stage	✗	✗	✗	✓	✓	✗
stage2	✗	✗	✗	✗	✗	✗

Table 3. Properties of semantics, skepticism and resolution adequacy

Let us see which semantics satisfy skepticism adequacy. Baroni and Giacomin [2007] proved all the results for grounded, complete, stable, preferred, semi-stable, ideal, all four prudent and CF2 semantics.

Eager semantics does not satisfy $\preceq^E_\cap$-skepticism adequacy, as illustrated by the example depicted in Figure 10. From Observation 9, we conclude that eager semantics violates $\preceq^E_W$-skepticism adequacy and $\preceq^E_S$-skepticism adequacy.

Naive semantics satisfies all three variants of skepticism adequacy since $\mathtt{CONF}(\mathcal{F}_1) = \mathtt{CONF}(\mathcal{F}_2)$ implies $\sigma(\mathcal{F}_1) = \sigma(\mathcal{F}_2)$.

Stage semantics does not satisfy $\preceq^E_\cap$-skepticism adequacy, as illustrated by the example from Figure 11. From Observation 9, we conclude that stage semantics violates $\preceq^E_W$-skepticism adequacy and $\preceq^E_S$-skepticism adequacy.

Finally, stage2 semantics does not satisfy $\preceq^E_\cap$-skepticism adequacy, as illustrated by the example from Figure 12. From Observation 9, we conclude

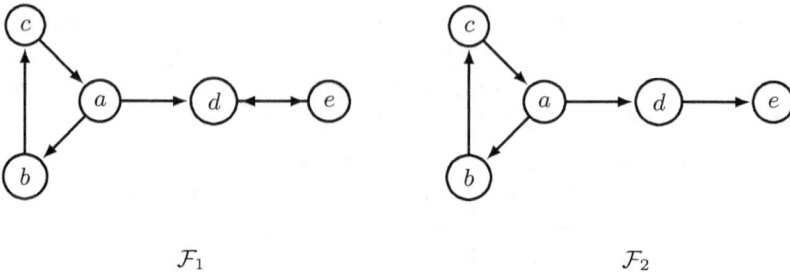

Figure 10. Eager semantics does not satisfy $\preceq_\cap^E$-skepticism adequacy. We have $\mathcal{F}_1 \preceq^A \mathcal{F}_2$. The eager extension of $\mathcal{F}_1$ is $\{e\}$ and the eager extension of $\mathcal{F}_2$ is $\emptyset$. Thus the set of skeptically accepted arguments of $\mathcal{F}_1$ equals $\{e\}$ is not a subset of the set of skeptically accepted arguments of $\mathcal{F}_2$.

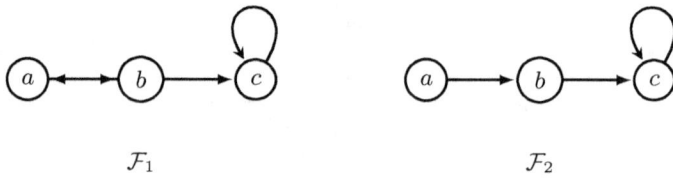

Figure 11. Stage semantics does not satisfy $\preceq_\cap^E$-skepticism adequacy. We have $\mathcal{F}_1 \preceq^A \mathcal{F}_2$. Framework $\mathcal{F}_1$ has a unique stage extension $\{a\}$ and framework $\mathcal{F}_2$ has two stage extensions $\{a\}$ and $\{b\}$. Thus the set of skeptically accepted arguments of $\mathcal{F}_1$ equals $\{a\}$ is not a subset of the set of skeptically accepted arguments of $\mathcal{F}_2$, which is the empty set.

that stage2 semantics violates $\preceq_W^E$-skepticism adequacy and $\preceq_S^E$-skepticism adequacy.

Let us now consider resolution adequacy [Baroni and Giacomin, 2007].

Definition 4.14 (RES) *We denote by $RES(\mathcal{F})$ the set of all argumentation frameworks comparable with $\mathcal{F}$ and maximal with respect to $\preceq^A$.*

Definition 4.15 (UR) *Given an argumentation framework $\mathcal{F}$ and a semantics σ, we define $UR(\mathcal{F}, \sigma) = \bigcup_{\mathcal{F}' \in RES(\mathcal{F})} \sigma(\mathcal{F}')$.*

Principle 19 (Resolution adequacy, [Baroni and Giacomin, 2007])
Given a skepticism relation $\preceq^E$ between sets of sets of arguments, a semantics σ satisfies the $\preceq^E$-resolution adequacy principle if and only if for every argumentation framework $\mathcal{F}$ we have $UR(\mathcal{F}, \sigma) \preceq^E \sigma(\mathcal{F})$.

Figure 12. Stage2 semantics does not satisfy $\preceq_\cap^E$-skepticism adequacy. We have $\mathcal{F}_1 \preceq^A \mathcal{F}_2$. Framework $\mathcal{F}_1$ has a unique stage2 extension $\{a\}$ and framework $\mathcal{F}_2$ has three stage2 extensions $\{a\}$, $\{b\}$ and $\{c\}$. Thus the set of skeptically accepted arguments of $\mathcal{F}_1$ equals $\{a\}$ is not a subset of the set of skeptically accepted arguments of $\mathcal{F}_2$, which is the empty set.

We consider three variants of the resolution adequacy principle: $\preceq_\cap^E$-resolution adequacy, $\preceq_W^E$-resolution adequacy and $\preceq_S^E$-resolution adequacy.

Observation 10

- If σ satisfies $\preceq_S^E$-res. adequacy then it satisfies $\preceq_W^E$-res. adequacy

- If σ satisfies $\preceq_W^E$-res. adequacy then it satisfies $\preceq_\cap^E$-res. adequacy

The results regarding grounded, complete, stable, preferred, semi-stable, ideal, all four prudent and CF2 semantics were shown by Baroni and Giacomin [2007].

Eager semantics violates $\preceq_\cap^E$-resolution adequacy, as illustrated by the example from Figure 13. Consequently, it does not satisfy the other two forms of resolution adequacy. Consider naive semantics; from its definition we see that for every argumentation framework $\mathcal{F}$, for every $\mathcal{F}' \in RES(\mathcal{F})$, we have $\sigma(\mathcal{F}) = \sigma(\mathcal{F}')$. Thus, naive semantics satisfies all three forms of resolution adequacy.

Proposition 4.16 *Stage semantics satisfies $\preceq_W^E$-resolution adequacy.*

Proof. To show this, it is sufficient to show the following claim: for every argumentation framework $\mathcal{F} = (\mathcal{A}, \mathcal{R})$, for every stage extension $\mathcal{E}$ of $\mathcal{F}$, there exists $\mathcal{F}' \in RES(\mathcal{F})$ such that $\mathcal{E}$ is a stage extension of $\mathcal{F}'$. Let $\mathcal{E}$ be a stage extension of $\mathcal{F}$. Let $\mathcal{F}' = (\mathcal{A}, \mathcal{R}') \in RES(\mathcal{F})$ be such that for every $a, b \in \mathcal{A}$ if $a \in \mathcal{E}$ then $(a, b) \in \mathcal{R}'$. (In other words, all attacks from $\mathcal{E}$ are preserved.) $\mathcal{E}$ is conflict-free in $\mathcal{F}'$, and all the attacks from $\mathcal{E}$ are preserved. Observe that the set of conflict-free sets of $\mathcal{F}$ and the set of conflict-free sets of $\mathcal{F}'$ coincide. Also, no conflict-free set attacks more arguments in $\mathcal{F}'$ than it attacks in $\mathcal{F}$. Thus, since $\mathcal{E}$ is a stage extension in $\mathcal{F}$, it is also a stage extension in $\mathcal{F}'$.

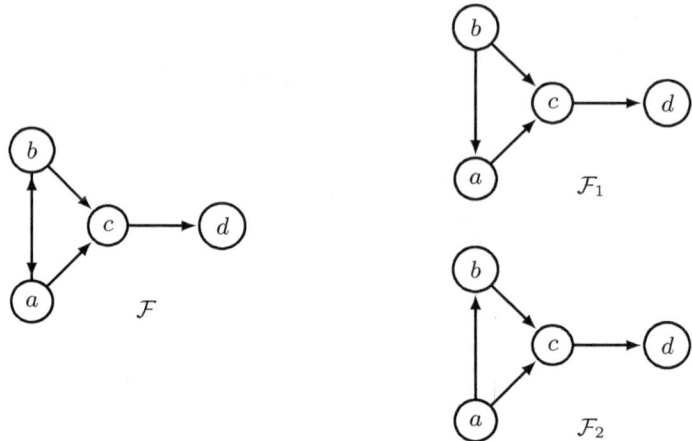

Figure 13. Eager semantics does not satisfy $\preceq_\cap^E$-resolution adequacy. We have $RES(\mathcal{F}) = \{\mathcal{F}_1, \mathcal{F}_2\}$. Namely, the eager extension of $\mathcal{F}_1$ is $\{b, d\}$ and the eager extension of $\mathcal{F}_2$ is $\{a, d\}$. Since the eager extension of $\mathcal{F}$ is the empty set, and $\{a, d\} \cap \{b, d\} = \{d\} \not\subseteq \emptyset$, the criterion is not satisfied.

From the fact that for every argumentation framework $\mathcal{F} = (\mathcal{A}, \mathcal{R})$, for every stage extension $\mathcal{E}$ of $\mathcal{F}$, there exists $\mathcal{F}' \in RES(\mathcal{F})$ such that $\mathcal{E}$ is a stage extension of $\mathcal{F}'$, we conclude that stage semantics satisfies $\preceq_W^E$-resolution adequacy. ∎

Since stage semantics satisfies $\preceq_W^E$-resolution adequacy, then it satisfies $\preceq_\cap^E$-resolution adequacy. The example from Figure 14 shows that stage semantics does not satisfy $\preceq_S^E$-resolution adequacy.

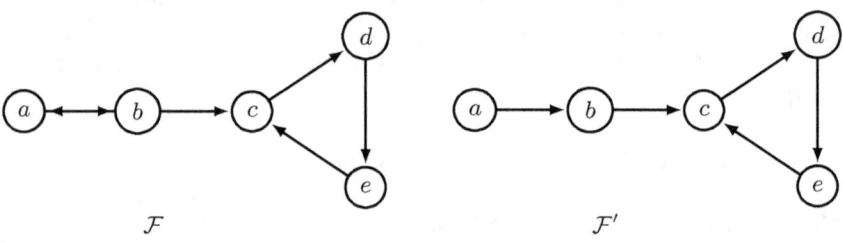

Figure 14. Stage semantics does not satisfy $\preceq_S^E$-resolution adequacy. We have $\mathcal{F}' \in RES(\mathcal{F})$, set $\mathcal{E}' = \{a, c\}$ is a stage extension of $\mathcal{F}'$, but there exists no stage extension $\mathcal{E}$ of $\mathcal{F}$ such that $\mathcal{E}' \subseteq \mathcal{E}$.

Stage2 semantics violates $\preceq_\cap^E$-resolution adequacy, as illustrated by the example from Figure 15. Consequently, it does not satisfy the other two forms of resolution adequacy.

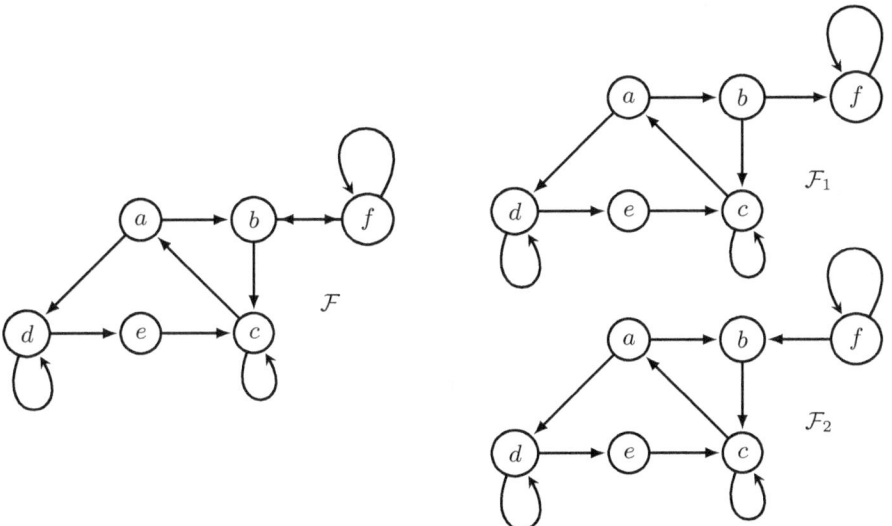

Figure 15. (Example provided by Wolfgang Dvorak, personal communication) Stage2 semantics does not satisfy $\preceq_\cap^E$-resolution adequacy. We have $RES(\mathcal{F}) = \{\mathcal{F}_1, \mathcal{F}_2\}$. Namely, the stage2 extensions of $\mathcal{F}$ are $\{a, e\}$ and $\{b, e\}$, and the stage2 extension of $\mathcal{F}_1$ and $\mathcal{F}_2$ is $\{a, e\}$. Since $\{a, e\} \not\subseteq \{a, e\} \cap \{b, e\} = \{e\}$, the criterion is not satisfied. The intuitive reason for the different behaviour from stage is that resolutions can break up a SCC into several SCCS and arguments that are not in the same SCC are not considered for range maximality.

Baroni et al. [2011b] introduce resolution-based family of semantics, which are developed to satisfy the resolution properties.

Let us now consider the last group of properties listed in Table 4. We first need to define the notion of strong equivalence [Oikarinen and Woltran, 2010]. Two frameworks $\mathcal{F}_1$ and $\mathcal{F}_2$ are strongly equivalent if for every argumentation framework $\mathcal{F}_3$, we have that $\mathcal{F}_1 \cup \mathcal{F}_3$ has the same extensions as $\mathcal{F}_2 \cup \mathcal{F}_3$.

Definition 4.17 (Strong equivalence) *Two argumentation frameworks $\mathcal{F}_1$ and $\mathcal{F}_2$ are strongly equivalent with respect to semantics σ, in symbols $\mathcal{F}_1 \equiv_s^\sigma \mathcal{F}_2$ if and only if for each argumentation framework $\mathcal{F}_3$, $\sigma(\mathcal{F}_1 \cup \mathcal{F}_3) = \sigma(\mathcal{F}_2 \cup \mathcal{F}_3)$.*

An attack is redundant in $\mathcal{F}$ if removing it does not change the extensions of any $\mathcal{F}'$ that contains $\mathcal{F}$.

Definition 4.18 (Redundant attack) *Let $\mathcal{F} = (\mathcal{A}, \mathcal{R})$ be an argumentation framework and σ and semantics. Attack $(a, b) \in \mathcal{R}$ is said to be redundant in $\mathcal{F}$*

	Succinctness	Tightness	Conflict--sensitiveness	Com--closure	SCC--recursiveness	Cardinality
complete	✗	✗	✗	✓	✓	1+
grounded	✗	✓	✓	✓	✓	1
preferred	✗	✗	✓	✓	✓	1+
stable	✗	✓	✓	✓	✓	0+
semi-stable	✗	✗	✓	✓	✗	1+
ideal	✗	✓	✓	✓	✗	1
eager	✗	✓	✓	✓	✗	1+
p-complete	✗	✗	✗	✗	✗	1+
p-grounded	✗	✓	✓	✓	✗	1
p-preferred	✗	✓	✓	✓	✗	1+
p-stable	✗	✓	✓	✓	✗	0+
naive	✗	✓	✓	✓	✗	1+
CF2	✓	✓	✓	✓	✓	1+
stage	✗	✓	✓	✓	✗	1+
stage2	✓	✓	✓	✓	✓	1+

Table 4. Properties of semantics, part 4

with respect to σ if and only if for all argumentation frameworks $\mathcal{F}'$ such that $\mathcal{F} \subseteq \mathcal{F}'$ we have $\sigma(\mathcal{F}') = \sigma(\mathcal{F}' \setminus (a,b))$.

We can now define the succinctness principle [Gaggl and Woltran, 2013].

Principle 20 (Succinctness) *A semantics σ satisfies the* succinctness *principle if and only if no argumentation framework contains a redundant attack with respect to σ.*

Gaggl and Woltran [2013] show that a semantics σ satisfies succinctness if and only if for every two argumentation frameworks $\mathcal{F}_1$ and $\mathcal{F}_2$ strong equivalence under σ coincides with $\mathcal{F}_1 = \mathcal{F}_2$.

Only CF2 and stage2 semantics satisfy succinctness. Namely, Oikarinen and Woltran [2010] showed that the notions of strong equivalence and syntactic equivalence do not coincide under complete, grounded, preferred, stable, semi-stable and ideal semantics. Gaggl and Woltran [2013] show that strong equivalence and syntactic equivalence do not coincide under stage and naive semantics. They also show that strong equivalence coincides with syntactic equivalence under CF2 semantics. Dvořák and Gaggl [2016] show that the same is true under stage2 semantics, which means that it also satisfies succinctness.

Consider eager semantics. Using Theorem 2 by Oikarinen and Woltran [2010], we can see that $\mathcal{F}_1$ and $\mathcal{F}_2$ from Figure 16 are strongly equivalent under semi-stable semantics. Since the eager semantics is uniquely determined by the set of semi-stable extensions, this means that $\mathcal{F}_1$ and $\mathcal{F}_2$ are strongly equivalent under eager semantics. Hence, eager semantics does not satisfy succinctness. Let us now show that all four prudent semantics violate succinctness.

Let $\mathcal{F}_1 = (\mathcal{A}, \mathcal{R}_1)$ and $\mathcal{F}_2 = (\mathcal{A}, \mathcal{R}_2)$ be the two argumentation frameworks

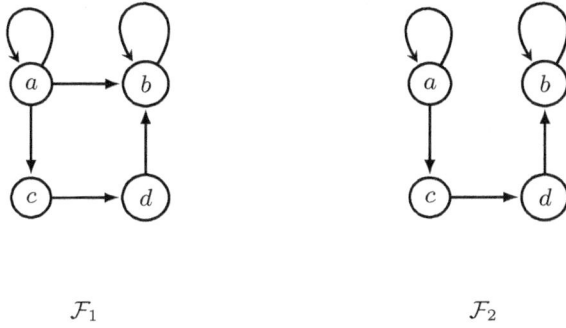

Figure 16. Several semantics violate succinctness

from Figure 16. Let $\mathcal{F} = (\mathcal{A}', \mathcal{R}')$ be an arbitrary argumentation framework. Denote $\mathcal{F}_1' = \mathcal{F}_1 \cup \mathcal{F}$ and $\mathcal{F}_2' = \mathcal{F}_2 \cup \mathcal{F}$. Let us prove that the sets without indirect conflicts of $\mathcal{F}_1'$ and $\mathcal{F}_2'$ coincide. It is immediate to see that if $\mathcal{E} \subseteq \mathcal{A} \cup \mathcal{A}'$ is not without indirect conflicts in $\mathcal{F}_2'$, it is also not without indirect conflicts in $\mathcal{F}_1'$, since $\mathcal{R}_2 \subseteq \mathcal{R}_1$. Let $\mathcal{E} \subseteq \mathcal{A} \cup \mathcal{A}'$ and let us prove that if $\mathcal{E}$ is not without indirect conflicts in $\mathcal{F}_1'$ then it is not without indirect conflicts in $\mathcal{F}_2'$. Let $\{(x_1, x_2), (x_2, x_3), \ldots, (x_{n-1}, x_n)\} \subseteq \mathcal{R}_1 \cup \mathcal{R}'$ with n being even and $x_1, x_n \in \mathcal{E}$. If $\{(x_1, x_2), (x_2, x_3), \ldots, (x_{n-1}, x_n)\} \subseteq \mathcal{R}_2 \cup \mathcal{R}'$ then $\mathcal{E}$ clearly has an indirect conflict in $\mathcal{F}_2'$. Else, it must be that for some $i \in \{1, \ldots, n-1\}$ we have $x_i = a$ and $x_{i+1} = b$. Then $\{(x_1, x_2), \ldots, (x_i, c), (c, d), (d, x_{i+1}), \ldots, (x_{n-1}, x_n)\} \subseteq \mathcal{R}_2 \cup \mathcal{R}'$, thus $\mathcal{E}$ is not without indirect conflicts in $\mathcal{F}_2'$. Hence, the sets without indirect conflicts of $\mathcal{F}_1'$ and $\mathcal{F}_2'$ coincide. It is immediate to see that $\mathcal{E} \subseteq \mathcal{A} \cup \mathcal{A}'$ defends all it arguments in $\mathcal{F}_1'$ if and only if it defends all its arguments in $\mathcal{F}_2'$. Thus, the sets of p-complete extensions of $\mathcal{F}_1'$ and $\mathcal{F}_2'$ coincide. Also, the p-grounded extension of $\mathcal{F}_1'$ is exactly the p-grounded extension of $\mathcal{F}_2'$. Since every $\mathcal{E}$ without indirect conflicts attacks an argument x in $\mathcal{F}_1'$ if and only if $\mathcal{E}$ attacks x in $\mathcal{F}_2'$, p-stable extensions of $\mathcal{F}_1'$ and $\mathcal{F}_2'$ coincide. Since the sets without indirect conflicts coincide, then maximal sets without indirect conflict coincide. Thus, p-preferred extensions of $\mathcal{F}_1'$ and $\mathcal{F}_2'$ coincide. We conclude that all variants of prudent semantics violate succinctness.

The next principle we consider is tightness. Let us first define the notion of pairs. A couple (a, b) is in $\mathcal{P}airs$ if there is an extension containing both a and b.

Definition 4.19 (Pairs) *Given a set of extensions $\mathcal{S} = \{\mathcal{E}_1, \ldots, \mathcal{E}_n\}$, we define*

$$\mathcal{P}airs(\mathcal{S}) = \{(a, b) \mid \text{there exists } \mathcal{E}_i \in \mathcal{S} \text{ such that } \{a, b\} \subseteq \mathcal{E}_i\}.$$

Tightness was introduced by Dunne et al. [2015]. Roughly speaking, it says that if argument a does not belong to extension $\mathcal{E}$, then there must be argument $b \in \mathcal{E}$ which is somehow incompatible with a.

Principle 21 (Tightness) *A set of extensions $\mathcal{S} = \{\mathcal{E}_1, \ldots, \mathcal{E}_n\}$ is tight if and only if for every extension $\mathcal{E}_i$ and for every $a \in \mathcal{A}$ that appears in at least one extension from $\mathcal{S}$ it holds that if $\mathcal{E}_i \cup \{a\} \notin \mathcal{S}$ then there exists $b \in \mathcal{E}_i$ such that $(a, b) \notin \mathcal{P}airs(\mathcal{S})$.*

A semantics σ satisfies the tightness *principle if and only if for every argumentation framework $\mathcal{F}$, $\sigma(\mathcal{F})$ is tight.*

Dunne et al. [2015] show that stable, stage and naive semantics satisfy tightness. Example 4 from their paper shows an argumentation framework $\mathcal{F}$ such that $\sigma(\mathcal{F}) = \{\mathcal{E}_1, \mathcal{E}_2, \mathcal{E}_3\}$ with $\mathcal{E}_1 = \{a, b\}$, $\mathcal{E}_2 = \{a, d, e\}$, $\mathcal{E}_3 = \{b, c, e\}$, under preferred and semi-stable semantics. This example shows that those two semantics violate tightness since $\{a, b, e\}$ is not an extension.

Directly from the definition of tightness, we conclude that unique status semantics satisfy this principle.

Observation 11 *If σ is a semantics that returns exactly one extension for every argumentation framework then σ satisfies tightness.*

Hence, grounded, p-grounded, ideal and eager semantics satisfy tightness. The example from Figure 17 shows that complete and p-complete semantics violate tightness.

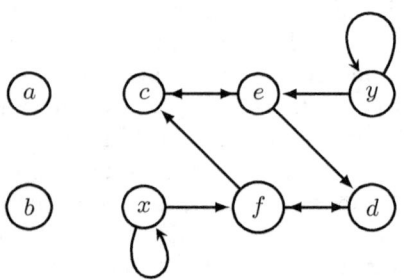

Figure 17. Complete and p-complete semantics violate tightness. There are two extensions $\mathcal{E}_1 = \{a, b\}$, $\mathcal{E}_2 = \{a, b, c, d\}$. Tightness is not satisfied since set $\mathcal{E}_1 \cup \{c\}$ is not an extension.

From Proposition 1 by Dunne et al. [2015], we have that the set of naive extensions is tight for every argumentation framework. Note that when σ is naive semantics and $\mathcal{F}$ an argumentation framework, all the elements of $\sigma(\mathcal{F})$ are pairwise incomparable with respect to $\subseteq$ (i.e. for each S, S', $S \subseteq S'$ implies $S = S'$). Hence, we can apply Lemma 2 by Dunne et al. [2015] and obtain

Observation 12 *If every extension under σ is a maximal conflict-free set, σ satisfies tightness.*

As an immediate consequence, p-stable, CF2 and stage2 semantics satisfy tightness. We now show that p-preferred semantics also satisfies this principle.

Proposition 4.20 *p-preferred semantics satisfies tightness.*

Proof. We use the proof by reductio ad absurdum. Let $\mathcal{E}$ be a p-preferred extension and let a be a credulously accepted argument such that

(1)   for every $b \in \mathcal{E}$ there is a preferred p-extension $\mathcal{E}''$ s.t. $\{a,b\} \subseteq \mathcal{E}''$

By means of contradiction, let us suppose that $\mathcal{E}' = \mathcal{E} \cup \{a\}$ is not a p-preferred extension. From (1), we conclude that $\mathcal{E}'$ is without indirect conflicts. Set $\mathcal{E}'$ is not p-admissible, since that would mean that $\mathcal{E}$ is not a maximal p-admissible set. Since $\mathcal{E}'$ is without indirect conflicts and $\mathcal{E}$ is p-admissible, there exists an argument b_1 such that $b_1 \mathcal{R} a$ and there is no $b' \in \mathcal{E}'$ such that $b' \mathcal{R} b_1$. Denote $B_1 = \{b \mid b\mathcal{R}a\}$.

Note that $\mathcal{E} \neq \emptyset$, since $\mathcal{E} = \emptyset$ would imply that there are no other p-preferred extensions and, consequently, a would not be credulously accepted. Thus, $\mathcal{E} \neq \emptyset$. Let $b \in \mathcal{E}$. From (1), there exists a p-preferred extension $\mathcal{E}_1$ such that $b \in \mathcal{E}_1$ and $a \in \mathcal{E}_1$. Since $a \in \mathcal{E}_1$ then for every $b_1^i \in B_1$ there exists $b_2^i \in \mathcal{E}_1$ such that $b_2^i \mathcal{R}$. Let $B_2 = \{b' \in \mathcal{E}_1 \mid \text{ there exists } b'' \in B_1 \text{ s.t. } b'\mathcal{R}b''\}$. In words, B_2 is the set of arguments from $\mathcal{E}_1$ that attack B_1 (they defend a from B_1).

Let us show that $\mathcal{E} \cup B_2$ is without indirect conflicts. By means of contradiction, suppose $\mathcal{E}$ indirectly attacks B_2. Then $\mathcal{E}$ indirectly attacks a, contradiction. Suppose now that B_2 indirectly attacks $\mathcal{E}$. Since $\mathcal{E}$ is p-admissible, then $\mathcal{E}$ attacks B_2, and thus (like in the previous case) $\mathcal{E}$ indirectly attacks a. Contradiction. So it must be that $\mathcal{E} \cup B_2$ is without indirect conflicts. Note also that since $B_2 \subseteq \mathcal{E}_1$ and $a \in \mathcal{E}_1$, we have that $\mathcal{E}_2 = \mathcal{E} \cup \{a\} \cup B_2$ is without indirect conflicts.

Note that $\mathcal{E}_2$ is not p-admissible, since it is a strict superset of a p-preferred extension. Set $\mathcal{E}$ is p-admissible and B_2 defends a so it must be that some argument(s) of B_2 are not defended by $\mathcal{E}_2$.

Let $B_3 = \{b \mid b\mathcal{R}B_2\}$. It must be that $B_3 \setminus B_2 \neq \emptyset$. Since $B_2 \subseteq \mathcal{E}_1$, and $\mathcal{E}_1$ is p-admissible, there exists $B_4 \subseteq \mathcal{E}_1$ such that B_4 defends B_2. Let $B_4 = \{b' \in \mathcal{E}_1 \mid \text{ there exists } b'' \in B_3 \text{ such that } b'\mathcal{R}b''\}$.

Note that $\mathcal{E}_4 = \mathcal{E} \cup \{a\} \cup B_2 \cup B_4$ is without indirect conflicts. By using the similar reasoning as in the case of $\mathcal{E}_2$, we conclude that $\mathcal{E}_4$ is not p-admissible. Let $B_5 = \{b \mid b\mathcal{R}B_4\}$. We have $B_5 \setminus (B_1 \cup B_3) \neq \emptyset$. By continuing this process, we construct an infinite sequence of different arguments $(b_1, b_3, \ldots, b_{i+1}, \ldots)$ such that $b_1 \in B_1$, $b_3 \in B_3 \setminus B_1$, ..., $b_{i+1} \in B_{i+1} \setminus (B_1 \cup \ldots \cup B_{i-1})$, ..., which is impossible, since the set of arguments is finite. ∎

We now study the notion of conflict-sensitiveness [Dunne et al., 2015]. Note that an equivalent principle was called adm-closure in some papers.

Principle 22 (Conflict-sensitiveness) *A set of extensions* $\mathcal{S} = \{\mathcal{E}_1, \ldots, \mathcal{E}_n\}$ *is conflict-sensitive if and only if for every two extensions* $\mathcal{E}_i$, $\mathcal{E}_j$ *such that* $\mathcal{E}_i \cup \mathcal{E}_j \notin \mathcal{S}$ *it holds that there exist* $a, b \in \mathcal{E}_i \cup \mathcal{E}_j$ *such that* $(a, b) \notin \mathcal{P}airs_\mathcal{S}$.

A semantics σ *satisfies the* conflict-sensitiveness *principle if and only if for every argumentation framework* $\mathcal{F}$, $\sigma(\mathcal{F})$ *is conflict-sensitive.*

This principle checks whether the fact that $\mathcal{E}_i \cup \mathcal{E}_j$ is not an extension is justified by existence of $a \in \mathcal{E}_i$ and $b \in \mathcal{E}_j$ that cannot be taken together. Dunne et al. [2015] show that every tight set is also conflict-sensitive. Thus, grounded, stable, ideal, stage, eager, naive, p-grounded, p-stable, p-preferred, stage2 and CF2 semantics satisfy conflict-sensitiveness. Proposition 2 by Dunne et al. [2015] shows that preferred and semi-stable semantics satisfy conflict-sensitiveness. Our example from Figure 18 shows that complete and p-complete semantics violate this principle. As for tightness, it does not seem that violating this principle is a necessarily a bad thing. It can be rational to ask for both a and b in order to defend e. There is no conflict between a and e, it is just that e needs to be defended from both c and d.

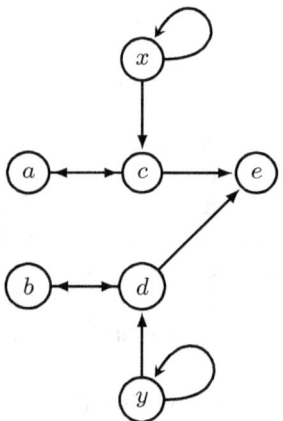

Figure 18. Complete and p-complete semantics violate conflict-sensitiveness. There are four extensions $\mathcal{E}_1 = \emptyset$, $\mathcal{E}_2 = \{a\}$, $\mathcal{E}_3 = \{b\}$, $\mathcal{E}_4 = \{a, b, e\}$. Conflict-sensitiveness is not satisfied since set $\{a, b\}$ is not an extension.

Let us now turn to com-closure [Dunne et al., 2015]. To define this principle, we first need to introduce the notion of completion set. Completion sets are the smallest extensions that contain a given set.

Definition 4.21 (Completion set) *Given a set of extensions* $\mathcal{S} = \{\mathcal{E}_1, \ldots, \mathcal{E}_n\}$ *and a set of arguments* $\mathcal{E}$, *set* $\mathcal{E}'$ *is a completion set of* $\mathcal{E}$ *in* $\mathcal{S}$ *if and only if* $\mathcal{E}'$ *is a minimal for* $\subseteq$ *set such that* $\mathcal{E}' \in \mathcal{S}$ *and* $\mathcal{E} \subseteq \mathcal{E}'$.

Roughly speaking, com-closure says that, given a set of extensions $\mathcal{S}$, if for every $\mathcal{T} \subseteq \mathcal{S}$ each two arguments from sets of $\mathcal{T}$ appear in some extension of $\mathcal{S}$, then $\mathcal{T}$ can be extended to an extension in a unique way.

Principle 23 (Com-closure) *A set of extensions* $\mathcal{S} = \{\mathcal{E}_1, \ldots, \mathcal{E}_n\}$ *is com-closed if and only if for every* $\mathcal{T} \subseteq \mathcal{S}$ *the following holds: if* $(a,b) \in \mathcal{P}airs_\mathcal{S}$ *for each* $a, b \in \cup_{\mathcal{E}_i \in \mathcal{T}} \mathcal{E}_i$, *then* $\cup_{\mathcal{E}_i \in \mathcal{T}} \mathcal{E}_i$ *has a unique completion set in* $\mathcal{S}$.

A semantics σ *satisfies the* com-closure *principle if and only if for every argumentation framework* $\mathcal{F}$, $\sigma(\mathcal{F})$ *is com-closed.*

Dunne et al. [2015] show that each conflict-sensitive set of extensions is com-closed. Thus, all the semantics that satisfy conflict-sensitiveness also satisfy com-closure. Their Proposition 4 shows that complete semantics is com-closed. To see that p-complete semantics does not satisfy com-closure, consider the graph from Figure 19.

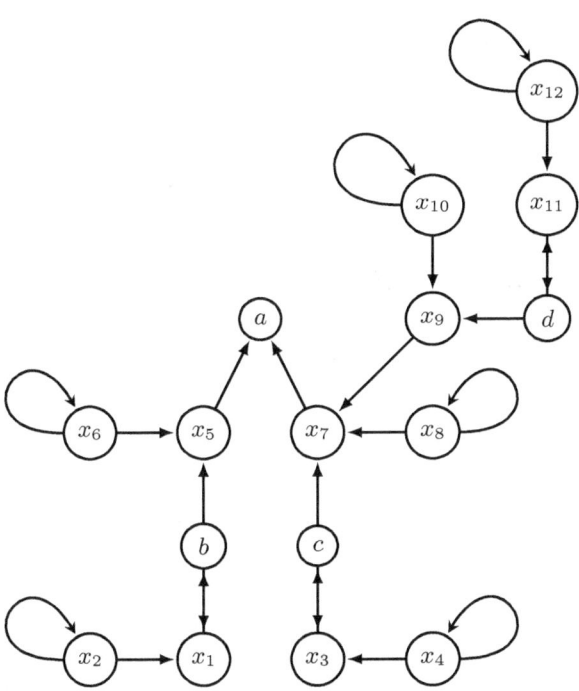

Figure 19. p-complete semantics is not com-closed. There are eight p-complete extensions: $\mathcal{E}_1 = \emptyset$, $\mathcal{E}_2 = \{b\}$, $\mathcal{E}_3 = \{c\}$, $\mathcal{E}_4 = \{d\}$, $\mathcal{E}_5 = \{b,d\}$, $\mathcal{E}_6 = \{c,d\}$, $\mathcal{E}_7 = \{b,c,d\}$, $\mathcal{E}_8 = \{b,c,a\}$. Let $\mathcal{T} = \{\mathcal{E}_2, \mathcal{E}_3\}$. Com-closure is not satisfied since set $\{b,c\}$ has two competition sets, namely $\mathcal{E}_7$ and $\mathcal{E}_8$.

We now study the notion of SCC-recursiveness, which was introduced by Baroni et al. [2005].

Principle 24 (SCC-recursiveness) *A semantics σ satisfies the SCC-recursiveness principle if and only if for every argumentation framework $\mathcal{F} = (\mathcal{A}, \mathcal{R})$ we have $\sigma(\mathcal{F}) = \mathcal{GF}(\mathcal{F}, \mathcal{A})$, where for every $\mathcal{F} = (\mathcal{A}, \mathcal{R})$ and for every set $C \subseteq \mathcal{A}$, the function $\mathcal{GF}(\mathcal{F}, C) \subseteq 2^{\mathcal{A}}$ is defined as follows: for every $\mathcal{E} \subseteq \mathcal{A}$, $\mathcal{E} \in \mathcal{GF}(\mathcal{F}, C)$ if and only if*

- *in case $|SCCS_{\mathcal{F}}| = 1$, $\mathcal{E} \in \mathcal{BF}_S(\mathcal{F}, C)$,*

- *otherwise, $\forall S \in SCCS_{\mathcal{F}}, (\mathcal{E} \cap S) \in \mathcal{GF}(\mathcal{F} \downarrow_{UP_{\mathcal{F}}(S, \mathcal{E})}, U_{\mathcal{F}}(S, \mathcal{E}) \cap C)$,*

where $\mathcal{BF}_S(\mathcal{F}, C)$ is a function, called base function, *that, given an argumentation framework $\mathcal{F} = (\mathcal{A}, \mathcal{R})$, such that $|SCCS(\mathcal{F})| = 1$ and a set $C \subseteq \mathcal{A}$, gives a subset of $2^{\mathcal{A}}$.*

Baroni et al. [2005] proved that grounded, complete, stable and preferred semantics satisfy SCC-recursiveness. CF2 and stage2 semantics also satisfy this principle, since they are defined by using SCC recursive schema. None of the remaining semantics satisfies SCC-recursiveness. To show that ideal, semi-stable, stage and eager semantics does not satisfy SCC-recursiveness, consider the examples from Figures 20 and 21, which are both due to M. Giacomin (personal communication, 2016). Naive semantics does not satisfy SCC-recursiveness

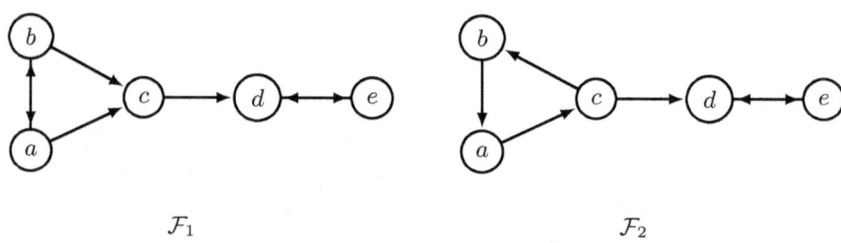

Figure 20. Ideal semantics is not SCC-recursive. Both in $\mathcal{F}_1$ and in $\mathcal{F}_2$, there are two SCCs: $S_1 = \{a, b, c\}$ and $S_2 = \{d, e\}$. Suppose ideal semantics is SCC-recursive. Then, we can calculate the ideal extension of an argumentation framework by starting from S_1 and then continuing to S_2. Denote by $\mathcal{F}_1^1$ the restriction of $\mathcal{F}_1$ on S_1 and by $\mathcal{F}_2^1$ the restriction of $\mathcal{F}_2$ on S_1. The ideal extension of $\mathcal{F}_1^1$ is the empty set. The ideal extension of $\mathcal{F}_2^1$ is also the empty set. So the exact same information is transferred to the next SCC, S_2. The second SCC, S_2 is the same for both frameworks, so given the same information from S_1, both frameworks should have the same ideal extension. However, $\sigma(\mathcal{F}_1) = \emptyset$ whereas $\sigma(\mathcal{F}_2) = \{e\}$. Thus, ideal semantics does not satisfy SCC-recursiveness.

since it ignores the direction of attacks. Consider the example from Figure 22.

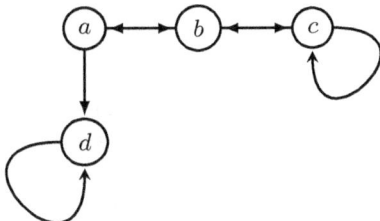

Figure 21. Semi-stable, stage and eager semantics violate SCC-recursiveness. Let σ be stage, semi-stable or eager semantics. Consider the first SCC, $S_1 = \{a, b, c\}$. If we restrict the argumentation framework to S_1, the only extension under σ is $\{b\}$. If σ satisfied SCC-recursiveness, each extension of this framework would contain b, which is not the case, since $\{a\}$ is an extension of this framework under σ.

Figure 22. Naive semantics does not satisfy SCC-recursiveness. Note that the first SCC is $S_1 = \{a\}$. If naive semantics satisfied SCC-recursiveness, every naive extension of the whole framework would contain a, which is not the case since $\{b\}$ is a naive extension of this framework too.

All four prudent semantics violate SCC-recursiveness. Consider the argumentation framework from Figure 4. Let σ be any of the four prudent semantics. In this example, every argument forms an SCC. Thus, each extension must contain both e and f. Furthermore, no extension can contain neither of b, c, d, since they are all attacked by e of f. Finally, if σ satisfied SCC-recursiveness, each extension would contain a, which is not the case.

The results considering cardinality are easy to obtain.

We do not include several properties that are not satisfied by any of the studied semantics. Let us mention three such properties. Downward closure [Dunne et al., 2015] basically says that each subset of each extension is an extension. Non-triviality [Dunne et al., 2012] says that it is not the case that $\sigma(\mathcal{F}) = \{\emptyset\}$; in words, the empty set is not the only extension. Decisiveness [Dunne et al., 2012] is a stronger principle that asks that every framework has exactly one extension $\mathcal{E}$ and that $\mathcal{E}$ is not empty.

5 Summary and outlook

The principle-based approach has developed over the past ten years into a cornerstone of formal argumentation theory, because it allows for a more systematic study and comparison of argumentation semantics. In this chapter we give a complete analysis of the fifteen main alternatives for argumentation semantics using the twenty-seven main principles discussed in the literature on abstract argumentation. Moreover, Caminada [2018] discusses the principles used in structured argumentation, which he calls rationality postulates, and Dung [2016] analyses *prioritised* argumentation using a principle-based or axiomatic approach.

The principle-based approach has also been used to provide a more systematic study and analysis of the semantics of extended argumentation frameworks, of the aggregation of argumentation frameworks, and of the dynamics of argumentation frameworks. For example, principles of ranking-based semantics have been proposed [Amgoud and Ben-Naim, 2016; Amgoud et al., 2017; Bonzon et al., 2016b], where the output is not a set of extensions but a ranking on the set of arguments, and principles have been developed for bipolar argumentation [Cayrol and Lagasquie-Schiex, 2015]. Likewise we expect a further systematic study of weighted argumentation frameworks, preference-based argumentation frameworks, input/output frameworks, abstract dialectical frameworks, and so on. These topics will be discussed in more detail in the second volume of the handbook on formal argumentation.

It may be expected that the principle-based approach will play an even more prominent role in the future of formal argumentation, as the number of alternatives for argumentation semantics increases, new argumentation principles are introduced, and more requirements of actual applications are expressed in terms of such principles. Moreover, in the future applications and principles concerned with infinite frameworks may become more prominent. For example, when the set of arguments becomes infinite, it may be that there are no semi-stable extensions. However, Baumann [2018] illustrates how a meaningful version of eager semantics can be defined, which no longer has the property that it always returns exactly one extension.

Finally, the principle-based approach to formal argumentation may lead to the study of impossibility and possibility results, as well as the development of representation theorems characterising sets of argumentation semantics. The use of the principle-based approach in other areas of reasoning, such as voting theory or AGM theory change, may inspire such further formal investigations.

Acknowledgements

The authors thank Ringo Baumann, Pietro Baroni, Liao Beishui, Martin Caminada, Wolfgang Dvořák and Massimiliano Giacomin for their useful comments and suggestions.

BIBLIOGRAPHY

[Alchourrón et al., 1985] Carlos E. Alchourrón, Peter Gärdenfors, and David Makinson. On the logic of theory change: Partial meet contraction and revision functions. *J. Symb. Log.*, 50(2):510–530, 1985.

[Amgoud and Ben-Naim, 2016] Leila Amgoud and Jonathan Ben-Naim. Evaluation of arguments from support relations: Axioms and semantics. In Subbarao Kambhampati, editor, *Proceedings of the Twenty-Fifth International Joint Conference on Artificial Intelligence, IJCAI 2016, New York, NY, USA, 9-15 July 2016*, pages 900–906. IJCAI/AAAI Press, 2016.

[Amgoud and Besnard, 2013] Leila Amgoud and Philippe Besnard. Logical limits of abstract argumentation frameworks. *Journal of Applied Non-Classical Logics*, 23(3):229–267, 2013.

[Amgoud et al., 2016] Leila Amgoud, Jonathan Ben-Naim, Dragan Doder, and Srdjan Vesic. Ranking arguments with compensation-based semantics. In *Proceedings of the 15th International Conference on Principles of Knowledge Representation and Reasoning, (KR'16)*, pages 12–21, 2016.

[Amgoud et al., 2017] Leila Amgoud, Jonathan Ben-Naim, Dragan Doder, and Srdjan Vesic. Acceptability semantics for weighted argumentation frameworks. In *Proceedings of the Twenty-Sixth International Joint Conference on Artificial Intelligence (IJCAI'2017)*, 2017.

[Arieli, 2015] Ofer Arieli. Conflict-free and conflict-tolerant semantics for constrained argumentation frameworks. *J. Applied Logic*, 13(4):582–604, 2015.

[Baroni and Giacomin, 2007] P. Baroni and M. Giacomin. On principle-based evaluation of extension-based argumentation semantics. *Artificial Intelligence*, 171:675–700, 2007.

[Baroni et al., 2004] Pietro Baroni, Massimiliano Giacomin, and Giovanni Guida. Towards a formalization of skepticism in extension-based argumentation semantics. In *Proceedings of the 4th Workshop on Computational Models of Natural Argument (CMNA'04)*, pages 47–52, 2004.

[Baroni et al., 2005] Pietro Baroni, Massimiliano Giacomin, and Giovanni Guida. SCC-recursiveness: a general schema for argumentation semantics. *Artificial Intelligence*, 168:162–210, 2005.

[Baroni et al., 2011a] Pietro Baroni, Martin Caminada, and Massimiliano Giacomin. An introduction to argumentation semantics. *Knowledge Eng. Review*, 26(4):365–410, 2011.

[Baroni et al., 2011b] Pietro Baroni, Paul E. Dunne, and Massimiliano Giacomin. On the resolution-based family of abstract argumentation semantics and its grounded instance. *Artificial Intelligence*, 175(3-4):791–813, 2011.

[Baroni et al., 2014] Pietro Baroni, Guido Boella, Federico Cerutti, Massimiliano Giacomin, Leendert van der Torre, and Serena Villata. On the input/output behavior of argumentation frameworks. *Artificial Intelligence*, 217:144–197, 2014.

[Baroni et al., 2018a] Pietro Baroni, Martin Caminada, and Massimiliano Giacomin. Abstract argumentation frameworks and their semantics. In Pietro Baroni, Dov Gabbay, Massimilano Giacomin, and Leendert van der Torre, editors, *Handbook of formal argumentation*. College Publications, 2018.

[Baroni et al., 2018b] Pietro Baroni, Massimiliano Giacomin, and Beishui Liao. Locality and modularity in abstract argumentation. In Pietro Baroni, Dov Gabbay, Massimilano Giacomin, and Leendert van der Torre, editors, *Handbook of formal argumentation*. College Publications, 2018.

[Baumann, 2018] Ringo Baumann. On the nature of argumentation semantics: Existence and uniqueness, expressibility, and replaceability. In Pietro Baroni, Dov Gabbay, Massimilano Giacomin, and Leendert van der Torre, editors, *Handbook of formal argumentation*. College Publications, 2018.

[Bonzon et al., 2016a] Elise Bonzon, Jérôme Delobelle, Sébastien Konieczny, and Nicolas Maudet. A comparative study of ranking-based semantics for abstract argumentation. In *AAAI'2016*, 2016.

[Bonzon et al., 2016b] Elise Bonzon, Jérôme Delobelle, Sébastien Konieczny, and Nicolas Maudet. A comparative study of ranking-based semantics for abstract argumentation. In Dale Schuurmans and Michael P. Wellman, editors, *Proceedings of the Thirtieth AAAI Conference on Artificial Intelligence, February 12-17, 2016, Phoenix, Arizona, USA.*, pages 914–920. AAAI Press, 2016.

[Caminada et al., 2012] Martin Caminada, Walter Carnielli, and Paul Dunne. Semi-stable semantics. *Journal of Logic and Computation*, 22(5):1207–1254, 2012.

[Caminada, 2006a] Martin Caminada. On the issue of reinstatement in argumentation. In *Proceedings of the 10th European Conference on Logics in Artificial Intelligence (JELIA'06)*, pages 111–123. Springer, 2006.

[Caminada, 2006b] Martin Caminada. Semi-stable semantics. In *Proceedings of the 1st International Conference on Computational Models of Argument (COMMA'06)*, pages 121–130. IOS Press, 2006.

[Caminada, 2007] Martin Caminada. Comparing two unique extension semantics for formal argumentation: Ideal and eager. In *The 19th Belgian-Dutch Conference on Artificial Intelligence, BNAIC'07*, pages 81–87, 2007.

[Caminada, 2018] Martin Caminada. Rationality postulates: applying argumentation theory for non-monotonic reasoning. In Pietro Baroni, Dov Gabbay, Massimilano Giacomin, and Leendert van der Torre, editors, *Handbook of formal argumentation*. College Publications, 2018.

[Cayrol and Lagasquie-Schiex, 2015] Claudette Cayrol and Marie-Christine Lagasquie-Schiex. An axiomatic approach to support in argumentation. In *Theory and Applications of Formal Argumentation - Third International Workshop, TAFA 2015, Buenos Aires, Argentina, July 25-26, 2015, Revised Selected Papers*, pages 74–91, 2015.

[Coste-Marquis et al., 2005] Sylvie Coste-Marquis, Caroline Devred, and Pierre Marquis. Prudent semantics for argumentation frameworks. In *Proceedings of the 17th International Conference on Tools with Artificial Intelligence (ICTAI'05)*, pages 568–572. IEEE, 2005.

[Dung et al., 2007] Phan Minh Dung, Paolo Mancarella, and Francesca Toni. Computing ideal skeptical argumentation. *Artificial Intelligence*, 171:642–674, 2007.

[Dung, 1995] Phan Minh Dung. On the acceptability of arguments and its fundamental role in nonmonotonic reasoning, logic programming and n-person games. *Artificial Intelligence*, 77:321–357, 1995.

[Dung, 2016] Phan Minh Dung. An axiomatic analysis of structured argumentation with priorities. *Artificial Intelligence*, 231:107–150, 2016.

[Dunne et al., 2011] Paul E. Dunne, Anthony Hunter, Peter McBurney, Simon Parsons, and Michael Wooldridge. Weighted argument systems: Basic definitions, algorithms, and complexity results. *Artificial Intelligence*, 175(2):457–486, 2011.

[Dunne et al., 2012] Paul Dunne, Pierre Marquis, and Michael Wooldridge. Argument aggregation: Basic axioms and complexity results. In *4th International Conference on Computational Models of Argument (COMMA'12)*, pages 129–140, 2012.

[Dunne et al., 2015] Paul E. Dunne, Wolfgang Dvorák, Thomas Linsbichler, and Stefan Woltran. Characteristics of multiple viewpoints in abstract argumentation. *Artificial Intelligence*, 228:153–178, 2015.

[Dvorák and Gaggl, 2016] Wolfgang Dvorák and Sarah Alice Gaggl. Stage semantics and the SCC-recursive schema for argumentation semantics. *Journal of Logic and Computation*, 26(4):1149–1202, 2016.

[Gaggl and Woltran, 2013] Sarah Alice Gaggl and Stefan Woltran. The CF2 argumentation semantics revisited. *Journal of Logic and Comutation*, 23(5):925–949, 2013.

[Oikarinen and Woltran, 2010] E. Oikarinen and S. Woltran. Characterizing strong equivalence for argumentation frameworks. In *Proceedings of the 12th International Conference on Principles of Knowledge Representation and Reasoning, (KR'10)*, pages 123–133, 2010.

[Rienstra et al., 2015] Tjitze Rienstra, Chiaki Sakama, and Leendert van der Torre. Persistence and monotony properties of argumentation semantics. In *Theory and Applications of Formal Argumentation - Third International Workshop, TAFA 2015, Buenos Aires, Argentina, July 25-26, 2015, Revised Selected Papers*, pages 211–225, 2015.

[Tennenholtz and Zohar, 2016] Moshe Tennenholtz and Aviv Zohar. The axiomatic approach and the internet. In Felix Brandt, Vincent Conitzer, Ulle Endriss, Jérôme Lang, and Ariel D. Procaccia, editors, *Handbook of Computational Social Choice*, pages 427–452. Cambridge University Press, 2016.

[Verheij, 1996] Bart Verheij. Two approaches to dialectical argumentation: admissible sets and argumentation stages. In *Proceedings of the Eighth Dutch Conference on Artificial Intelligence (NAIC 1996)*, pages 357–368, 1996.

Leendert van der Torre
University of Luxembourg
Luxembourg
Email: leon.vandertorre@uni.lu

Srdjan Vesic
CRIL
CNRS and University of Artois
Lens, France
Email: vesic@cril.fr

17
On the Nature of Argumentation Semantics: Existence and Uniqueness, Expressibility, and Replaceability

RINGO BAUMANN

ABSTRACT. This chapter is devoted to argumentation semantics which play the flagship role in Dung's abstract argumentation theory. Almost all of them are motivated by an easily understandable intuition of what should be acceptable in the light of conflicts. However, although these intuitions equip us with short and comprehensible formal definitions it turned out that their intrinsic properties such as *existence and uniqueness*, *expressibility*, and *replaceability* are not that easily accessible. The chapter reviews the mentioned properties for almost all semantics available in the literature. In doing so we include two main axes: namely first, the distinction between extension-based and labelling-based versions and secondly, the distinction of different kind of argumentation frameworks such as finite or unrestricted ones.

1 Introduction

Given the large variety of existing logical formalisms it is of utmost importance to select the most adequate one for a specific purpose, e.g. for representing the knowledge relevant for a particular application or for using the formalism as a modeling tool for problem solving. Awareness of the nature of a logical formalism, in other words, of its fundamental intrinsic properties, is indispensable and provides the basis of an informed choice. Apart from the deeper understanding of the considered formalism, the study of such intrinsic properties can help to identify interesting fragments or to develop useful extensions of a formalism. Moreover, the obtained insights can be used to refine existing algorithms, or even give rise to new ones.

Presumably, the best-known intrinsic property of logics is *monotonicity*. *Monotonic logics* like first order logic are perfectly suitable for the formalization of universal truths since in these logics, whenever a formula ϕ is a logical consequence of a set of axioms Σ, it remains true forever and without exception even if we add new axioms to Σ. Formalisms which do not satisfy monotonicity, commonly referred to as *nonmonotonic logics*, allow for defeasible reasoning, i.e. it is possible to withdraw former conclusions (cf. [Brewka, 1992; Gabbay *et al.*, 1994] for excellent overviews). Both kinds of logics have their traditional application domains and apart from this fundamental choice there are many other comparison criteria influencing the decision which logic or which specific semantics of a logic to use in a certain context.

One of the first intrinsic properties which comes to mind is *computational complexity*, i.e. how expensive is it to solve typical decision problems in the candidate formalism. A further related issue is *modularity* which is, among other things, engaged with the question whether it is possible to divide a given theory in subtheories, s.t. the formal semantics of the entire theory can be obtained by constructing the semantics of the subtheories. Both topics were studied in-depth for mainstream nonmonotonic formalisms like default logic [Gottlob, 1992; Turner, 1996], logic programming under certain semantics [Dantsin et al., 1997; Lifschitz and Turner, 1994] as well as abstract argumentation frameworks under various argumentation semantics (cf. Chapters 13 and 18).

In this chapter we give an overview of three further intrinsic properties of abstract argumentation semantics.

1. *existence and uniqueness* Is it possible, and if so how, to *guarantee* the existence of at least one or exactly one extension/labelling by considering the structure of a given AF F only? (cf. Section 2)

2. *expressibility* Is it possible, and if so how, to *realize* a given candidate set of extensions/labellings within a single AF F? (cf. Section 3)

3. *replaceability* Is it possible, and if so how, to *simplify* parts of a given AF F, s.t. the modified version F' and F cannot be semantically distinguished by further information which might be added later to both simultaneously? (cf. Section 4)

The question whether a certain formalism always provides one with a formal meaning or even with a uniquely determined semantical answer is a crucial factor for its suitability for the application in mind. For instance, in contrast to problem solving where a plurality of solutions may possibly be desired, in decision making one might be interested in guaranteeing a single answer provided by a logical formalism. It is well-known that a given theory in propositional logic neither has to possess a model nor, in case of existence, has there to be exactly one. The same applies to logic programs under stable model semantics. In contrast, a propositional theory of positive formulae is always satisfiable and definite logic programs constitute a subclass of logic programs where even uniqueness is guaranteed. In Section 2 we will see that Dung's abstract argumentation semantics behave in a similar way, i.e. the existence or uniqueness of extensions/labellings depend on structural restrictions of argumentation frameworks.

Expressibility is concerned with the expressive power of logical formalisms. The question here is which kinds of model sets are realizable, that is, can be the set of models of a single knowledge base of the formalism. This is a decisive property from an application angle since potential necessary or sufficient properties of model sets may rule out a logic or make it perfectly appropriate for representing certain solutions. For instance, it is well-known that in case of propositional logic any finite set of two-valued interpretations is realizable. This means, given such a finite set $\mathcal{I}$, we always find a set of formulae T, s.t. $Mod(T) = \mathcal{I}$. In case of normal logic programs it is obvious that not all model sets can be expressed, since any set of stable models forms a $\subseteq$-antichain. Remarkably, being such an antichain is not only necessary but even sufficient for realizability w.r.t. stable model semantics [Eiter et al., 2013; Strass, 2015]. In case of abstract argumentation we are equipped with a high number of semantics and in Section 3 we will see that characterizing properties are not that easy. Moreover, as expected, representational limits highly depend on the chosen semantics.

In case of propositional logic we have that – in contrast to all non-monotonic logics available in the literature – standard equivalence, i.e. sharing the same models, even guarantees intersubstitutability in any logical context without loss of information. As an aside, it is not the monotonicity of a certain logic but rather the so-called *intersection property* which guarantees this behavior (cf. [Baumann and Strass, 2016]). Substitutability is of great importance for dynamically evolving scenarios since it allows to simplify parts of a theory without looking at the rest. For this reason, much effort has been devoted to characterizing *strong equivalence* for nonmonotonic formalisms, such as logic programs [Lifschitz et al., 2001], causal theories [Turner, 2004], default logic [Turner, 2001] and nonmonotonic logics in general [Truszczynski, 2006; Baumann and Strass, 2016]. In Section 4 we will see that characterization theorems in case of abstract argumentation are quite different from those for the aforementioned formalisms since being strongly equivalent can be decided by looking at the syntax only.

2 Existence and Uniqueness

Given a certain logical formalism $\mathcal{L}$ together with its semantics $\sigma_\mathcal{L}$. One central question is whether the semantics provides any $\mathcal{L}$-theory T with a formal meaning, i.e. $|\sigma_\mathcal{L}(T)| \geq 1$. A more demanding property than *existence* is *uniqueness*, i.e. $|\sigma_\mathcal{L}(T)| = 1$ for any $\mathcal{L}$-theory T. Clearly, these properties are interesting from several perspectives. For instance, in case of uniqueness, we observe a coincidence of sceptical and credulous reasoning modes. More precisely, if $\sigma_\mathcal{L}(T) = \{E\}$, then $\bigcap \sigma_\mathcal{L}(T) = \bigcup \sigma_\mathcal{L}(T) = E$. Furthermore, if a theory T is interpreted as *meaningful* if and only if $\sigma_\mathcal{L}(T) \neq \emptyset$, then existence might be a desired property. If the latter has to be neglected in the general case, then one further challenge is to identify sufficient properties of $\mathcal{L}$-theories guaranteeing their meaningfulness.

Let us turn to abstract argumentation frameworks [Dung, 1995]. Due to the practical nature of argumentation most work in the literature restricts itself to the case of finite AFs, i.e. any considered AF consists of finitely many arguments and attacks only. For this class of AFs a proof or disproof of existence or uniqueness is mostly straightforward. In the general infinite case however conducting such proofs is more intricate. It usually involves the proper use of set theoretic axioms, like the *axiom of choice* or equivalent statements. Dung already proposed the existence of preferred extensions in the case of infinite argumentation frameworks. It has later on (e.g. [Caminada and Verheij, 2010]) been pointed out that Dung has not been precise with respect to the use of principles. The existence of semi-stable extensions for finitary[1] argumentation frameworks was first shown by Weydert with the use of model-theoretic techniques [Weydert, 2011]. Later on, Baumann and Spanring presented a first comprehensive overview of results regarding existence and uniqueness for a whole bunch of semantics considered in the literature [Baumann and Spanring, 2015]. They provided complete or alternative proofs of already known results and contributed missing results for the infinite or finitary case. We mention two interesting results: Firstly, eager semantics is exceptional among the universally defined semantics since either there is exactly one or there are infinitely many eager extensions. Secondly, stage semantics behaves similarly to semi-stable in the sense that extensions are guaranteed as long as finitary AFs are considered. A further step forward in the systematic analysis of argumentation semantics in the infinite case was presented in [Spanring, 2015]. Spanring studied the relation between non-existence of extensions and the number of non-finitary arguments. It was shown that there are AFs where one single non-finitary argument causes a collapse[2] of semi-stable semantics. Interestingly, all known AFs which do not provide any stage extension possesses infinitely many non-finitary arguments. It is an open question whether this observation applies in general [Spanring, 2015, Conjecture 14].

2.1 Basic Definitions in Dung's Abstract Argumentation Theory

For the sake of self-containedness we review all relevant definitions (for more introductory comments we refer the reader to Chapter 4 in this handbook). The standard way of defining argumentation frameworks is to introduce a certain reference set $\mathcal{U}$, so-called *universe of arguments* and to require, that all arguments used in AFs are elements of this set. More formally, for any AF $F = (A, R)$ we have $A \subseteq \mathcal{U}$ and $R \subseteq A \times A$. In order to be able to consider AFs possessing an arbitrary finite number of arguments or even infinitely many we have to request that $|\mathcal{U}| \geq \aleph_0 = |\mathbb{N}|$. No further conditions are imposed. In the following we use $\mathcal{F}$ as an abbreviation for the set of all AFs (induced by $\mathcal{U}$). An AF F is called *finite* if it possesses finitely many arguments only. Furthermore, we say that F is *finitary* if every argument has only finitely many attackers.

Definition 2.1 *An AF $F = (A, R)$ is called*

1. *finite if $|A| \in \mathbb{N}$,*

---

[1] An argument is called *finitary* if it receives finitely many attacks only. Moreover, an AF is said to be *finitary* if and only if it consists of finitary arguments only (cf. Definition 2.1).

[2] The term *collapse* was firstly introduced in [Spanring, 2015] and it refers to a semantics not providing any extension/labelling for a given AF.

2. *finitary* if for any $a \in A$, $|\{b \in A \mid (b,a) \in R\}| \in \mathbb{N}$ and

3. *arbitrary* or *unrestricted* if $F \in \mathcal{F}$.

In order to formalize the notions of *existence* and *uniqueness* in the context of abstract argumentation theory we have to clarify what we precisely mean by a *semantics*. In the literature two main approaches to argumentation semantics can be found, namely so-called *extension-based* and *labelling-based* versions. The main difference is that extension-based versions return a set of sets of arguments (so-called *extensions*) for any given AF in contrast to a set of sets of n-tuples (so-called *labellings*) as in case of labelling-based approaches. However, from a mathematical point of view both kinds of semantics are instances of Definition 2.2. More precisely, extension-based versions are covered by $n = 1$ and labelling-based approaches can be obtained by setting $n \geq 2$. We use $(2^{\mathcal{U}})^n$ to denote the n-ary cartesian power of $2^{\mathcal{U}}$, i.e. $(2^{\mathcal{U}})^n = \underbrace{2^{\mathcal{U}} \times \cdots \times 2^{\mathcal{U}}}_{n-\text{times}}$.

Definition 2.2 *A semantics is a function* $\sigma : \mathcal{F} \to 2^{(2^{\mathcal{U}})^n}$ *for some* $n \in \mathbb{N}$, *s.t.* $F = (A, R) \mapsto \sigma(F) \subseteq (2^A)^n$.

We now introduce the two different definedness statuses of argumentation semantics which capture the notions of existence and uniqueness, namely so-called *universal* and *unique definedness*. Both versions are relativized to a certain set of AFs. If clear from context, unimportant or if $\mathcal{C} = \mathcal{F}$ we will not mention explicitly the considered set of AFs.

Definition 2.3 *Given a semantics* σ *and a set* $\mathcal{C}$ *of AFs. We say that* σ *is*

1. *universally defined w.r.t.* $\mathcal{C}$ *if* $\forall F \in \mathcal{C}, |\sigma(F)| \geq 1$ *and*

2. *uniquely defined w.r.t.* $\mathcal{C}$ *if* $\forall F \in \mathcal{C}, |\sigma(F)| = 1$.

In this section we are interested in definedness statuses w.r.t. finite, finitary and arbitrary frameworks. Besides conflict-free and admissible sets (abbreviated by *cf* and *ad*) we consider a large number of mature semantics, namely naive, stage, stable, semi-stable, complete, preferred, grounded, ideal, eager semantics as well as the more exotic cf2 and stage2 semantics (abbreviated by $na, stg, stb, ss, co, pr, gr, il, eg, cf2$ and $stg2$ respectively). In the following we introduce the extension-based versions of these semantics (indicated by $\mathcal{E}_\sigma$). Any considered semantics possesses a 3-valued labelling-based version (denoted as $\mathcal{L}_\sigma$). It is important to note that for all considered semantics we do not observe any differences between the definedness statuses of their labelling-based and extension-based versions. For the mature semantics this is due the fact that there is a one-to-one correspondence between σ-extensions and σ-labellings implying that $|\mathcal{E}_\sigma(F)| = |\mathcal{L}_\sigma(F)|$ for any AF F (for more details confer Paragraph *Basic Properties and a Fundamental Relation* in Section 4 as well as Chapter 4 of this handbook).

Before presenting the definitions we have to introduce some notational conventions. Given an AF $F = (A, R)$ and a set $E \subseteq A$. We use E_F^+ or simply, E^+ for $\{b \mid (a, b) \in R, a \in E\}$. Moreover, $E_F^\oplus$ or simply, $E^\oplus$ is called the *range* of E and stands for $E^+ \cup E$. We say a attacks b (in F) if $(a, b) \in R$. An argument a is *defended* by E (in F) if for each $b \in A$ with $(b, a) \in R$, b is attacked by some $c \in E$. Finally, $\Gamma_F : 2^A \to 2^A$ with $I \mapsto \{a \in A \mid a \text{ is defended by } I\}$ denotes the so-called *characteristic function* (of F) [Dung, 1995].

Definition 2.4 *Let $F = (A, R)$ be an AF and $E \subseteq A$.*

1. $E \in \mathcal{E}_{cf}(F)$ iff for no $a, b \in E$, $(a, b) \in R$,
2. $E \in \mathcal{E}_{na}(F)$ iff $E \in \mathcal{E}_{cf}(F)$ and for no $I \in \mathcal{E}_{cf}(F)$, $E \subset I$,
3. $E \in \mathcal{E}_{stg}(F)$ iff $E \in \mathcal{E}_{cf}(F)$ and there is no $I \in \mathcal{E}_{cf}(F)$, s.t. $E^\oplus \subset I^\oplus$,
4. $E \in \mathcal{E}_{stb}(F)$ iff $E \in \mathcal{E}_{cf}(F)$ and $E^\oplus = A$,
5. $E \in \mathcal{E}_{ad}(F)$ iff $E \in \mathcal{E}_{cf}(F)$ and E defends all its elements,
6. $E \in \mathcal{E}_{ss}(F)$ iff $E \in \mathcal{E}_{ad}(F)$ and there is no $I \in \mathcal{E}_{ad}(F)$, s.t. $E^\oplus \subset I^\oplus$,
7. $E \in \mathcal{E}_{co}(F)$ iff $E \in \mathcal{E}_{ad}(F)$ and for any $a \in A$ defended by E in F, $a \in E$,
8. $E \in \mathcal{E}_{pr}(F)$ iff $E \in \mathcal{E}_{ad}(F)$ and for no $I \in \mathcal{E}_{co}(F)$, $E \subset I$,
9. $E \in \mathcal{E}_{gr}(F)$ iff E is the $\subseteq$-least fixpoint of Γ_F,
10. $E \in \mathcal{E}_{il}(F)$ iff $E \in \mathcal{E}_{ad}(F)$, $E \subseteq \bigcap \mathcal{E}_{pr}(F)$ and there is no $I \in \mathcal{E}_{co}(F)$ satisfying $E \subset I \subseteq \bigcap \mathcal{E}_{pr}(F)$,
11. $E \in \mathcal{E}_{eg}(F)$ iff $E \in \mathcal{E}_{ad}(F)$, $E \subseteq \bigcap \mathcal{E}_{ss}(F)$ and there is no $I \in \mathcal{E}_{co}(F)$ satisfying $E \subset I \subseteq \bigcap \mathcal{E}_{ss}(F)$.

Finally, we introduce the recursively defined cf2 and stage2 semantics [Baroni et al., 2005; Dvořák and Gaggl, 2012].

Definition 2.5 *Let $F = (A, R)$ be an AF and $E \subseteq A$.*

1. $E \in \mathcal{E}_{cf2}(F)$ iff
    - $E \in \mathcal{E}_{na}(F)$ if $|SCCs_F = 1|$ and
    - $\forall S \in SCCs_F(E \cap S) \in \mathcal{E}_{cf2}\left(F|_{UP_F(S,E)}\right)$,

2. $E \in \mathcal{E}_{stg2}(F)$ iff
    - $E \in \mathcal{E}_{stg}(F)$ if $|SCCs_F = 1|$ and
    - $\forall S \in SCCs_F(E \cap S) \in \mathcal{E}_{stg2}\left(F|_{UP_F(S,E)}\right)$.

Here $SCCs_F$ denotes the set of all strongly connected components of F, and for any $E, S \subseteq A$, $UP_F(S, E) = \{a \in S \mid \nexists b \in E \setminus S : (b, a) \in R\}$.

The following proposition summarizes well-known subset relations between the considered semantics. For two semantics σ, τ and a certain set of AFs $\mathcal{C}$ we use $\sigma \subseteq_\mathcal{C} \tau$ as a shorthand for $\sigma(F) \subseteq \tau(F)$ for any AF $F \in \mathcal{C}$. The presented relations hold for both extension-based as well as labelling-based versions of the considered semantics. In the interest of readability we present the relations graphically.

Proposition 2.6 *For semantics σ and τ, $\sigma \subseteq_\mathcal{F} \tau$ iff there is a path of solid arrows from σ to τ in Figure 1. A dotted arrow indicates that the corresponding subset relation is guaranteed for finite frameworks only.*

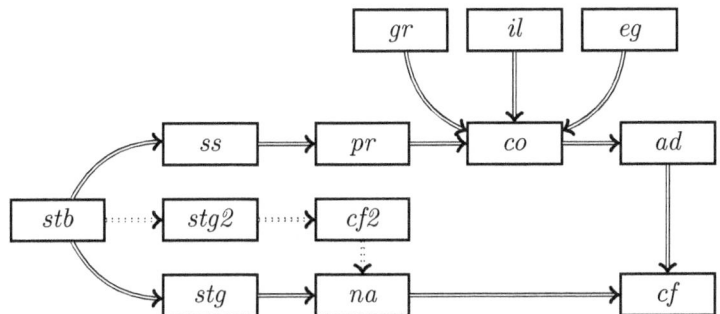

Figure 1: Subset Relations between Semantics

Detailed proofs can be found in [Baumann, 2014b, Proposition 2.7] as well as [Gaggl and Dvořák, 2016, Section 3.1]. Note that the shorthand $\sigma \subseteq_\mathcal{C} \tau$ requires that both semantics are total functions on $\mathcal{C}$ since a framework to which one of these semantics is undefined renders the subset shorthand undefined itself. The following simple example shows that Definition 2.5 does not always provide a definite answer on whether a certain candidate set is an *cf2*-extension or *stg2*-extension, respectively. This is due to the fact that the defined recursion does not terminate necessarily in case of non-finite AFs.[3] Consequently, *stg2* and *cf2* are not total functions regarding arbitrary frameworks.

Example 2.7 (Infinite Recursion [Baumann and Spanring, 2017]) *Consider the following AF $F = (A \cup B, R)$ where*

- $A = \{a_i \mid i \in \mathbb{N}\}$, $B = \{b_i \mid i \in \mathbb{N}\}$ *and*
- $R = \{(b_i, a_i), (a_{i+1}, a_i), (a_i, b_{i+1}) \mid i \in \mathbb{N}\}$

---

[3] We mention that the inventors of both semantics considered finite AFs only [Baroni et al., 2005; Dvořák and Gaggl, 2012]. In case of finite AFs any recursion will terminate no matter which candidate set is considered.

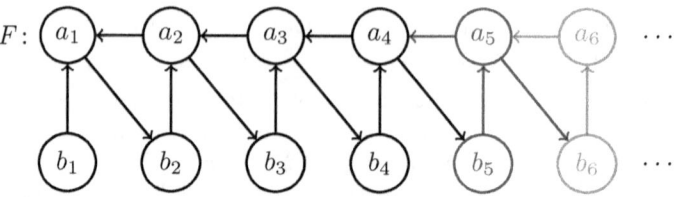

Let $\sigma \in \{cf2, stg2\}$. We want to check whether the candidate set $E = \{b_i \mid i \in \mathbb{N}\}$ is a σ-extension. Observe that the AF F possesses two SCCs, namely one consisting of the single argument b_1 and the other containing the remaining arguments, i.e. $S_1 = \{b_1\}$ and $S_2 = (A \cup B) \setminus \{b_1\}$. For S_1 we end up with the base case returning a positive answer. For S_2 we have to consider the AF $F' = F|_{U P_F(S_2, E)} = F|_{(A \cup B) \setminus \{a_1, b_1\}}$ (since a_1 is attacked by $b_1 \in E \setminus S_2$) and the set $S' = E \cap S_2 = \{b_i \mid i \in \mathbb{N}, i \geq 2\}$. Obviously, determining whether S' is an σ-extension w.r.t. F' is equivalent to decide whether S is an σ-extension w.r.t. F. This means, the consideration of the candidate set E leads to infinite recursion.

2.2 Finite AFs

As a matter of fact, in order to show that a certain semantics σ is not universally defined w.r.t. a certain set $\mathcal{C}$ it suffices to present an AF $F \in \mathcal{C}$, s.t. $\sigma(F) = \emptyset$. Contrastingly, an affirmative answer w.r.t. universal definedness requires a proof involving all AFs in $\mathcal{C}$. Let us consider finite AFs first. It is well-known that stable semantics does not warrant the existence of extensions/labellings even in the case of finite AFs. Witnessing examples are given by odd-cycles (cf. Example 2.8). Interestingly, in case of finite AFs we have that being odd-cycle free is sufficient for warranting at least one stable extension/labelling.[4]

Example 2.8 *The following minimalistic AFs cause a collapse of stable semantics, i.e.* $stb(F_1) = stb(F_3) = \emptyset$.

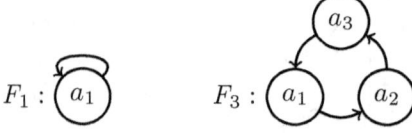

Observe that both frameworks do possess semi-stable, stage2 as well as stage extensions/labellings. The extensions are as follows: For any $\sigma \in \{ss, stg2, stg\}$, $\tau \in \{stg2, stg\}$, $\mathcal{E}_\sigma(F_1) = \{\emptyset\} = \mathcal{E}_{ss}(F_3)$ *and* $\mathcal{E}_\tau(F_3) = \{\{a_1\}, \{a_2\}, \{a_3\}\}$.

Let us consider now semi-stable semantics. Example 2.8 shows that AFs may possess semi-stable extensions even in the absence of stable extensions. Are semi-stable extensions possibly guaranteed in case of finite AFs? Consider

---

[4]This is due to the fact that firstly, in case of finite AFs, being odd-cycle free coincides with being *limited controversial* [Dung, 1995, Definition 32] and secondly, any limited controversial AFs warrants the existence of at least one stable extensions [Dung, 1995, Corollary 36].

the following explanations about the existence of semi-stable extensions taken from [Caminada, 2006]:

> For every argumentation framework there exists at least one semi-stable extension. This is because there exists at least one complete extension, and a semi-stable extension is simply a complete extension in which some property (the union of itself and the arguments it defeats) is maximal.

We would like to point out two issues. Firstly, the presented explanation should not be understood as: Since any semi-stable extension is a complete one and complete semantics is universally defined we conclude that semi-stable semantics is universally defined. Accepting this kind of (false) argumentation would imply the universal definedness of stable semantics since also any stable extension is a complete one. The second issue is that the presented explanation is not precise about *why* it is guaranteed that the non-empty set of complete extensions possesses at least one range-maximal member. The following statement gives a more precise explanation [Caminada et al., 2012]:

> For every (finite) argumentation framework, there exists at least one semi-stable extension. This is because there exists at least one complete extension (the grounded) and the fact that the argumentation framework is finite implies that there exist at most a finite number of complete extensions. The semi-stable extensions are then simply those complete extensions in which some property (its range) is maximal.

This means, the additional argument that we have to compare finitely many complete extensions only justifies the universal definedness of semi-stable extensions in case of finite AFs. Obviously, in case of infinite AFs we cannot expect to have finitely many complete extensions implying that this kind of argumentation is no longer valid for finitary as well as infinite AFs in general.

In the rest of this subsection we want to argue why all considered semantics except the stable one are universally defined in case of finite AFs.[5] Remember that many semantics are looking for certain $\subseteq$-maximal elements. The main advantage in case of finiteness is that it is simply impossible to have infinite $\subseteq$-chains which guarantees the existence of $\subseteq$-maximal elements. Consider the following more detailed explanations. Given a finite AF $F = (A, R)$, i.e. $|A| = n \in \mathbb{N}$. Consequently, $1 \leq |2^A| = 2^n \in \mathbb{N}$. By definition of any extension-based semantics σ we derive $0 \leq |\mathcal{E}_\sigma(F)| \leq 2^n$ since $\mathcal{E}_\sigma(F) \subseteq 2^A$ (cf. Definition 2.2). This means, for any finite F and any semantics σ we have at least one candidate set for being a σ-extension (namely, the empty set) and at most finitely many σ-extensions. In any case, the empty set is conflict-free as well as admissible, i.e. $|\mathcal{E}_{cf}(F)|, |\mathcal{E}_{ad}(F)| \geq 1$. Furthermore, naive and

---

[5] We mention that grounded, ideal and eager semantics are even uniquely defined w.r.t. finite AFs. This will be a by-product of Theorem 2.23, Corollary 2.22 as well as Theorem 2.25.

preferred semantics are looking for $\subseteq$-maximal conflict-free or admissible sets, respectively. Since we have finitely many conflict-free as well as admissible sets only we derive the universal definedness of naive and preferred semantics in case of finite AFs. Combining $\mathcal{E}_{pr} \subseteq \mathcal{E}_{co}$ and $|\mathcal{E}_{pr}(F)| \geq 1$ yields the universal definedness of complete semantics in case of finite AFs. Moreover, since $1 \leq |\mathcal{E}_{cf}(F)|, |\mathcal{E}_{ad}(F)| \leq 2^n$ is given we obtain the universal definedness of stage and semi-stable semantics in case of finite AFs because the existence of $\subseteq$-range-maximal is guaranteed. Let us consider ideal and eager semantics. Candidate sets of both semantics are admissible sets being in the intersection of all preferred or semi-stable extensions, respectively. Note that there is at least one admissible set satisfying this property, namely the empty one since definitely $\emptyset \subseteq \bigcap \mathcal{E}_{pr}(F) \subseteq \mathcal{U}$ as well as $\emptyset \subseteq \bigcap \mathcal{E}_{ss}(F) \subseteq \mathcal{U}$. This means, the sets of candidates are non-empty and finite which guarantees the existence of $\subseteq$-maximal elements implying the universal definedness of ideal and eager semantics in case of finite AFs. The grounded extension, i.e. the $\subseteq$-least fixpoint of the characteristic function Γ_F, is guaranteed due to the monotonicity of Γ_F and the famous Knaster-Tarski theorem [Tarski, 1955]. Finally, even the more exotic stage2 as well as cf2 semantics are universally defined w.r.t. finite AFs. This can be seen as follows: Obviously, finitely many as well as initial SCCs are guaranteed due to finiteness. Consequently, one may start with computing stage/naive extension on these initial components and "propagate" the resulting extensions to the subsequent SCCs and so on. This procedure will definitely terminate and ends up with stage2/cf2 extensions. Apart from stable semantics we have argued that the extension-based versions of all considered semantics are universally defined w.r.t. finite AFs. In case of mature semantics, the result carry over to their labelling-based versions since any of these semantics possesses a one-to-one-correspondence between extensions and labellings. This property does not hold in case of admissible as well as conflict-free sets. However, since any admissible/conflict-free set induce at least one admissible/conflict-free labelling the result applies to their labelling versions too (cf. Chapter 4).

2.3 Arbitrary AFs

Non-well-defined Semantics

In contrast to all other semantics available in the literature, cf2 as well as stage2 semantics were originally defined recursively. The recursive schema is based on the decomposition of AFs along their strongly connected components (SCCs). Roughly speaking, the schema takes a base semantics σ and proceeds along the induced partial ordering and evaluates the SCCs according to σ while propagating relevant results to subsequent SCCs. This procedure defines a $\sigma 2$ semantics.[6] Given so-called *SCC-recursiveness* (cf. Chapter 18) we have to face some difficulties in drawing conclusions with respect to infinite AFs. Firstly,

---

[6]Following this terminology we have to rename *cf2* semantics to *na2* semantics since its base semantics is the naive semantics and not conflict-free sets.

arbitrary AFs need not to possess initial SCCs which is granted for finite AFs. This makes checking whether a certain set is an $\sigma 2$-extension more complicated and in particular, especially due to the recursive definitions not that easy to handle. Secondly, even worse, even if an AF as well as subsequent subframeworks of it possess initial SCCs there is no guarantee that any recursion will stop in finitely many steps. More precisely, as shown in Example 2.7 there might be candidate sets which lead to infinite recursion, i.e. the base case will never be considered. In [Gaggl and Dvořák, 2016, Propositions 2.12 and 3.2] the authors considered alternative non-recursive definitions of cf2 as well as stage2 semantics in case of finite AFs. It is an open question whether these definitions overcome the problem of undefinedness for arbitrary frameworks.

Collapsing Semantics

Dealing with finite AFs is a common as well as attractive and reasonable restriction, due to their computational nature. In the subsection before we have argued that apart from stable semantics all considered semantics are universally defined w.r.t. finite AFs. It is an important observation that warranting the existence of σ-extensions/labellings in case of finite AFs does not necessarily carry over to the infinite case, i.e. the semantics σ does not need to be universally defined w.r.t. arbitrary AFs. Take for instance semi-stable and stage semantics. To the best of our knowledge the first example showing that semi-stable as well as stage semantics does not guarantee extensions/labellings in case of non-finite AFs was given in [Verheij, 2003, Example 5.8.] and is picked up in the following example.

Example 2.9 (Collapse of Stage and Semi-stable Semantics) *Consider the following AF $F = (A \cup B \cup C, R)$ where*

- $A = \{a_i \mid i \in \mathbb{N}\}$, $B = \{b_i \mid i \in \mathbb{N}\}$, $C = \{c_i \mid i \in \mathbb{N}\}$ *and*
- $R = \{(a_i, b_i), (b_i, a_i), (b_i, c_i), (c_i, c_i) \mid i \in \mathbb{N}\} \cup \{(b_i, b_j), (b_i, c_j) \mid i, j \in \mathbb{N}, j < i\}$

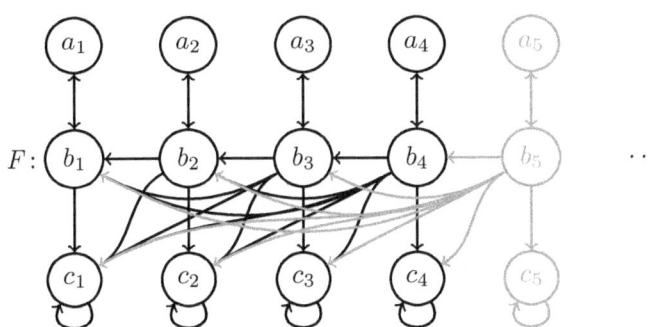

The set of preferred and naive extensions coincide, in particular $\mathcal{E}_{pr}(F) = \mathcal{E}_{na}(F) = \{A\} \cup \{E_i \mid i \in \mathbb{N}\}$ where $E_i = (A \setminus \{a_i\}) \cup \{b_i\}$. Furthermore, none

of these extensions is $\subseteq$-range-maximal since $A^\oplus \subsetneq E_i^\oplus \subsetneq E_{i+1}^\oplus$ for any $i \in \mathbb{N}$. In consideration of ss $\subseteq$ pr and stg $\subseteq$ na (cf. Figure 1) we conclude that this framework possesses neither semi-stable nor stage extensions/labellings.

In Example 2.7 we have seen that cf2 as well as stage2 semantics are not well-defined in general. This means, there are infinite AFs and candidate sets leading to an infinite recursion implying that there is no definite answer on whether such a set is an extension. However, the following example shows that even if for any candidate set a definitive decision is possible there need not to be an extension in contrast to finite AFs.

Example 2.10 (Collapse of Cf2 and Stage2 Semantics) *Taking into account the AF $F = (A \cup B \cup C, R)$ from Example 2.9. Consider the AF $G = F|_B$, i.e. the restriction of F to B.*

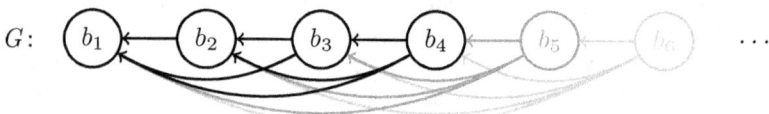

Let $\sigma \in \{cf2, stg2\}$. Obviously, any argument b_i constitutes a SCC $\{b_i\}$ which is evaluated as $\{b_i\}$ by the base semantics of σ. Consequently, $\emptyset$ cannot be a σ-extension. Furthermore, a singleton $\{b_j\}$ cannot be a σ-extension either. The b_i's for $i > j$ are not affected by $\{b_j\}$ and thus, the evaluation of $G|_{UP_G(\{b_i\},\{b_j\})} = G|_{\{b_i\}} = (\{b_i\}, \emptyset)$ do not return $\emptyset$ as required. Finally, any set containing more than two arguments would rule out at least one of them and thus, cannot be a σ-extension. Hence, $|\mathcal{E}_\sigma(G)| = |\mathcal{L}_\sigma(G)| = 0$.

In Example 2.9 we have seen an AF F without any semi-stable and stage extensions/labellings. In [Baumann and Spanring, 2015] the authors studied the question of existence-dependency between both semantics in case of infinite AFs. More precisely, they studied whether it is possible that some AF does have semi-stable but no stage extensions or vice versa, there are stage but no semi-stable extensions. The following Example 2.11 shows that stage extensions might exist even if semi-stable semantics collapses.[7]

Example 2.11 (No Semi-Stable but Stage Extensions/Labellings) *Consider again the AF F depicted in Example 2.9. Using the components of F we define $G = (A \cup B \cup C \cup D \cup E, R \cup R')$ where*

- *$D = \{d_i \mid i \in \mathbb{N}\}$ and $E = \{e_i \mid i \in \mathbb{N}\}$ and*

- *$R' = \{(a_i, d_i), (d_i, a_i), (b_i, d_i), (d_i, b_i), (d_i, c_i), (e_i, d_i), (e_i, e_i) \mid i \in \mathbb{N}\}$*

---

[7]The AF $G = F|_B$ depicted in Example 2.10 witnesses the reverse case. It can be checked that $\mathcal{E}_{ss}(G) = \{\emptyset\}$ and $\mathcal{E}_{stg}(G) = \emptyset$ (cf. [Baumann and Spanring, 2015, Example 2] for further explanations).

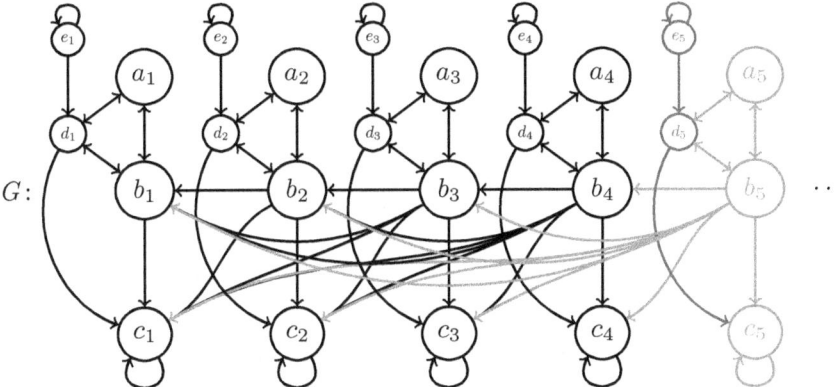

In comparison to Example 2.9 we do not observe any changes as far as preferred and semi-stable semantics are concerned. In particular, $\mathcal{E}_{pr}(G) = \{A\} \cup \{E_i \mid i \in \mathbb{N}\}$ where $E_i = (A \setminus \{a_i\}) \cup \{b_i\}$ and again, none of these extensions is $\subseteq$-range-maximal. Hence, $\mathcal{E}_{ss}(G) = \emptyset$. Observe that we do have additional conflict-free as well as naive sets, especially the set D. Since any $e \in E$ is self-defeating and unattacked and furthermore, $D^{\oplus} = A \cup B \cup C \cup D$ we conclude, $\mathcal{E}_{stg}(G) = \{D\}$. Due to the one-to-one correspondence the collapse or non-collapse transfer to their labelling-based versions.

Universally Defined Semantics

We now turn to semantics which are universally defined w.r.t. the whole class of AFs. The first non-trivial result in this line was already proven by Dung himself, namely the universal definedness of the extension-based version of preferred semantics [Dung, 1995, Corollary 12]. He argued that the Fundamental Lemma (cf. [Dung, 1995, Lemma 10] or Chapter 4 in this handbook) immediately implies that the set of all admissible sets is a complete partial order which means that any $\subseteq$-chain possesses a least upper bound. Then (and this was not explicitly stated in [Dung, 1995]), due to the famous *Zorn's lemma* [Zorn, 1935] the existence of $\subseteq$-maximal admissible sets, i.e. preferred extensions, is guaranteed.

In order to get an idea how things work in the general case we illustrate some proofs in more detail. We will see that a proof of universal definedness w.r.t. arbitrary AFs is completely different to the argumentation in case of finite ones. In order to keep this section self-contained we start with Zorn's lemma and an equivalent version of it.

Lemma 2.12 ([Zorn, 1935]) *Given a partially ordered set $(P, \leq)$. If any $\leq$-chain possesses an upper bound, then $(P, \leq)$ has a maximal element.*

Lemma 2.13 *Given a partially ordered set $(P, \leq)$. If any $\leq$-chain possesses an upper bound, then for any $p \in P$ there exists a maximal element $m \in P$, s.t. $p \leq m$.*

Having Lemma 2.13 at hand we may easily argue that any conflict-free/admissible set is bounded by a naive/preferred extension.

Lemma 2.14 *Given $F = (A, R)$ and $E \subseteq A$,*

1. *if $E \in \mathcal{E}_{cf}(F)$, then there exists $E' \in \mathcal{E}_{na}(F)$ s.t. $E \subseteq E'$ and*
2. *if $E \in \mathcal{E}_{ad}(F)$, then there exists $E' \in \mathcal{E}_{pr}(F)$ s.t. $E \subseteq E'$.*

Proof. For $F = (A, R)$ we have the associated power set lattice $(2^A, \subseteq)$. Consider now the partially ordered fragments $\mathcal{C} = (\mathcal{E}_{cf}(F), \subseteq)$ as well as $\mathcal{A} = (\mathcal{E}_{ad}(F), \subseteq)$. In accordance with Lemma 2.13 the existence of naive and preferred supersets is guaranteed if any $\subseteq$-chain possesses an upper bound in $\mathcal{C}$ or $\mathcal{A}$, respectively. Given a $\subseteq$-chain $\mathcal{E} \subseteq \mathcal{E}_{cf}(F)$ or $\mathcal{E} \subseteq \mathcal{E}_{ad}(F)$, respectively. Consider now $\bar{E} = \bigcup \mathcal{E}$. Obviously, $\bar{E}$ is an upper bound of $\mathcal{E}$, i.e. for any $E \in \mathcal{E}$, $E \subseteq \bar{E}$. It remains to show that $\bar{E}$ is conflict-free or admissible, respectively. Conflict-freeness is a finite condition. This means, if there were conflicting arguments $a, b \in \bar{E}$ there would have to be some conflict-free sets $E_a, E_b \in \bar{E}$, s.t. $a \in E_a$ and $b \in E_b$. Since $\mathcal{E}$ is a $\subseteq$-chain we have $E_a \subseteq E_b$ or $E_b \subseteq E_a$ which contradicts the conflict-freeness of at least one of them. Assume now $\bar{E}$ is not admissible. Consequently, there is some $a \in \bar{E}$ that is not defended by $\bar{E}$. Furthermore, there has to be an $E_a \in \mathcal{E}$, s.t. $a \in E_a$ contradicting the admissibility of $E_a \in \mathcal{E}_{ad}(F)$. ∎

According to the last lemma, we may deduce the universal definedness of the extension-based versions of preferred as well as naive semantics as long as, for any AF F, the existence of at least one conflict-free or admissible set is guaranteed. This is an easy task since the empty set is conflict-free as well as admissible even in the case of arbitrary AFs. Consequently, universal definedness of both extension-based semantics is given and the same applies to their labelling-based versions due to their one-to-one correspondence.

Theorem 2.15 *Let $\sigma \in \{pr, na\}$. The semantics σ is universally defined.*

Remember that no matter which cardinality a considered AF possesses, we have that any preferred extension/labelling is a complete extension/labelling (Proposition 2.6). Thus, having the universal definedness of preferred semantics at hand we deduce that even complete semantics is universally defined w.r.t. the whole class of AFs.

Theorem 2.16 *The semantics co is universally defined.*

Let us consider now eager and ideal semantics. An eager extension is defined as the $\subseteq$-maximal admissible set that is a subset of each semi-stable extension. This is very similar to the definition of an ideal extension where the role of semi-stable extensions is taken over by preferred ones. On a more abstract level, both semantics are instantiations of the following schema.

Definition 2.17 *Let σ be a semantics (so-called base semantics). We define the σ-parametrized semantics ad^σ as follows. For any AF F,*

$$\mathcal{E}_{ad^\sigma} = \max_{\subseteq} \left\{ E \in \mathcal{E}_{ad}(F) \,\bigg|\, E \subseteq \bigcap_{S \in \mathcal{E}_\sigma(F)} S \right\}.$$

These kind of semantics were firstly introduced in [Dvorák et al., 2011]. The authors studied general properties of these semantics in case of finite AFs with the additional restriction that the base semantics σ has to be universally defined. The following general theorem requires neither finiteness of AFs, nor any assumption on the base semantics.

Theorem 2.18 *Any σ-parametrized semantics is universally defined.*

Proof. Given an AF $F = (A, R)$ and a σ-parametrized semantics ad^σ. Consider the set $\Sigma = \left\{ E \in \mathcal{E}_{ad}(F) \mid E \subseteq \bigcap_{S \in \mathcal{E}_\sigma(F)} S \right\}$. Note that in the collapsing case, i.e. $\mathcal{E}_\sigma(F) = \emptyset$, we have: $\bigcap_{S \in \mathcal{E}_\sigma(F)} S = \{x \in \mathcal{U} \mid \forall S \in \mathcal{E}_\sigma(F) : x \in S\} = \mathcal{U}$. However, in any case $\Sigma \neq \emptyset$ since for any F, $\emptyset \in \mathcal{E}_{ad}(F)$ and obviously, $\emptyset \subseteq \bigcap_{S \in \mathcal{E}_\sigma(F)} S \subseteq \mathcal{U}$. In order to show that $\mathcal{E}_{ad^\sigma}(F) \neq \emptyset$ it suffices to prove that $(\Sigma, \subseteq)$ possesses maximal elements. We will use Zorn's lemma. Given a $\subseteq$-chain $\mathcal{E} \in 2^\Sigma$. Consider now $\bar{E} = \bigcup \mathcal{E}$. Analogously to the proof of Lemma 2.14 we may easily show that $\bar{E}$ is conflict-free and even admissible. Moreover, since for any $E \in \mathcal{E}$, $E \subseteq \bigcap_{S \in \mathcal{E}_\sigma(F)} S$ we deduce $\bar{E} \subseteq \bigcap_{S \in \mathcal{E}_\sigma(F)} S$ guaranteeing $\bar{E} \in \Sigma$. Now, applying Lemma 2.12, we deduce the existence of $\subseteq$-maximal elements in Σ, i.e. $|\mathcal{E}_{ad^\sigma}(F)| \geq 1$ concluding the proof. ∎

In particular, we obtain the result for the extension-based versions of eager and ideal semantics and thus, due to the one-to-one correspondence for both labelling-based versions too.

Corollary 2.19 *Let $\sigma \in \{eg, il\}$. The semantics σ is universally defined.*

One obvious question is whether the statement above can be strengthened in the sense that both semantics are even uniquely defined w.r.t. the whole class of AFs. The following proposition, in particular the second item, shows that the unique definedness of eager semantics w.r.t. finite frameworks does not carry over to the general unrestricted case.

Proposition 2.20 *For any F we have:*

1. $ss(F) = \emptyset \Rightarrow eg(F) = pr(F)$ *and*
2. $ss(F) = \emptyset \Rightarrow |eg(F)| \geq \aleph_0 = |\mathbb{N}|$.

Proof. We show both assertions for the extension-based versions.
1.) Given $F = (A, R)$ and let $\mathcal{E}_{ss}(F) = \emptyset$. Hence, $\bigcap_{S \in \mathcal{E}_{ss}(F)} S = \mathcal{U}$. Consequently, $\mathcal{E}_{ss}(F) = \max_{\subseteq} \{E \in \mathcal{E}_{ad}(F) \,|\, E \subseteq \mathcal{U}\}$. This means, $\mathcal{E}_{ss}(F) = \mathcal{E}_{pr}(F)$.
2.) We show the contrapositive. Assume $|\mathcal{E}_{eg}(F)| = n$ for some finite cardinal $n \in \mathbb{N}$. Due to the first statement we derive, $|\mathcal{E}_{pr}(F)| = n$. Since $ss \subseteq pr$ (cf. Proposition 2.6) we have finitely many candidates only. Furthermore, among these preferred extensions has to be at least one $\subseteq$-range-maximal set implying $\mathcal{E}_{ss}(F) \neq \emptyset$. ∎

In a nutshell, if we observe a collapse of semi-stable semantics, then eager and preferred semantics coincide and moreover, we necessarily have infinitely many eager extensions/labellings. An AF witnessing such a behaviour can be found in Example 2.9.

Uniquely Defined Semantics

Although eager and ideal semantics are instances of σ-parametrized semantics we have shown the non-unique definedness (Proposition 2.20) for eager semantics only. This is no coincidence since preferred semantics, the base semantics of ideal semantics is universally defined in contrast to semi-stable semantics, the base semantics of the eager semantics. Moreover, the following theorem shows that any σ-parametrized semantics warrants the existence of exactly one extension if σ-extensions are conflict-free as well as guaranteed ([Dvořák et al., 2011, Proposition 1]).

Theorem 2.21 *Given a σ-parametrized semantics ad^σ, s.t. $\sigma \subseteq cf$ and σ is universally defined w.r.t. a class $\mathcal{C}$, then ad^σ is uniquely defined w.r.t. $\mathcal{C}$.*

Proof. Given an AF $F = (A, R)$. We already know $|\mathcal{E}_{ad^\sigma}(F)| \geq 1$ (Theorem 2.18). Hence, it suffices to show $|\mathcal{E}_{ad^\sigma}(F)| \leq 1$. Suppose, to derive a contradiction, that for some $I_1 \neq I_2$ we have $I_1, I_2 \in \mathcal{E}_{ad^\sigma}(F)$. Consequently, by Definition 2.17, $I_1, I_2 \in \mathcal{E}_{ad}(F)$ and $I_1, I_2 \subseteq \bigcap_{S \in \mathcal{E}_\sigma(F)} S$ as well as neither $I_1 \subseteq I_2$, nor $I_2 \subseteq I_1$. Obviously, $I_1 \cup I_2 \subseteq \bigcap_{S \in \mathcal{E}_\sigma(F)} S$. Since $\mathcal{E}_\sigma(F) \neq \emptyset$ and I_1 as well as I_2 has to be subsets of any σ-extension (which are conflict-free by assumption) we deduce $I_1, I_2 \in \mathcal{E}_{cf}(F)$ and thus, $I_1 \cup I_2 \in \mathcal{E}_{cf}(F)$. Furthermore, since both sets are admissible in F we derive $I_1 \cup I_2 \in \mathcal{E}_{ad}(F)$ contradicting the $\subseteq$-maximality of at least one of the sets I_1 and I_2. ∎

Corollary 2.22 *The semantics il is uniquely defined.*

A further prominent representative of uniquely defined semantics w.r.t. the whole class of AFs is the grounded semantics. Its unique definedness was already implicitly given in [Dung, 1995]. Unfortunately, this result was not explicitly stated in the paper. Nevertheless, in [Dung, 1995, Theorem 25] it was shown that firstly, the set of all complete extensions form a complete semi-lattice w.r.t. subset relation, i.e. the existence of a $\subseteq$-greatest lower bound for any non-empty subset S is implied. Secondly, it was proven that the grounded extension is the ⊆-least complete extension. Consequently, the existence of such a ⊆-least extension is justified via setting $S = \mathcal{E}_{co}(F)$ for any given F. Alternatively, one may stick to the original definition of the grounded extension, namely as ⊆-least fixpoint of the characteristic function Γ_F and argue that the monotonicity of Γ_F as well as the Knaster-Tarski theorem [Tarski, 1955] imply its existence.

Theorem 2.23 *The semantics gr is uniquely defined.*

2.4 Finitary AFs

Let us consider now finitary AFs, i.e. AFs where each argument receives finitely many attacks only. It was already observed by Dung itself that finitary AFs possess useful properties. More precisely, if an AF is finitary, then the characteristic function Γ is not only monotonic, but even ω-continuous [Dung, 1995, Lemma 28] (which does not hold in case of arbitrary AFs [Baumann and Spanring, 2017, Example 1]). This implies that the least fixed point of Γ, i.e. the unique grounded extension, can be "computed" in at most ω steps by iterating Γ on the empty set (cf. [Rudin, 1976] for more details). A further advantage of finitary AFs is that for some semantics σ, the existence or even uniqueness of σ-extension is guaranteed which cannot be shown in general.

Consider again the AF F depicted in Example 2.9. In contrast to finite AFs where the existence of semi-stable as well stage extensions is guaranteed we observed a collapse of both semantics. Not that F is not finitary since, for example, the argument b_1 receives infinitely many attacks. A positive answer in case of semi-stable semantics, i.e. universal definedness w.r.t. finitary AFs was conjectured in [Caminada and Verheij, 2010, Conjecture 1] and firstly proven by Emil Weydert in [Weydert, 2011, Theorem 5.1]. Weydert proved his result in a first order logic setup using generalized argumentation frameworks. Later on, Baumann and Spanring provided an alternative proof using transfinite induction. Moreover, they showed that even stage semantics warrants the existence of at least one extension in case of finitary AFs [Baumann and Spanring, 2015, Theorem 14]. For detailed proofs we refer the reader to the mentioned scientific papers.

Theorem 2.24 *Let $\sigma \in \{ss, stg\}$. The semantics σ is universally defined w.r.t. finitary AFs.*

Applying Theorem 2.21 we derive that exactly one eager extension/labelling is guaranteed as long as the AF in question is finitary.

Theorem 2.25 *The semantics eg is uniquely defined w.r.t. finitary AFs.*

2.5 Summary of Results and Conclusion

In this section we gave an overview on the question whether certain semantics guarantee the existence or even unique determination of extensions/labellings. We have seen that these properties may vary from subclass to subclass. The following table gives a comprehensive overview over results presented in this section. The entry "∃" ("∃!") in row *certain* and column σ indicates that the semantics σ is universally (uniquely) defined w.r.t. the class of *certain* frameworks. No entry reflects the situation that a *certain* AF can be found which do not provide any σ-extension/labelling, i.e. σ collapses. The two question marks represent open problems. Note that we already observed that cf2 as well as stage2 semantics are not well-defined in case of finitary as well arbitrary AFs. This means, there are infinite AFs and candidate sets leading to an infinite recursion implying that there is no definite answer on whether such a set is an extension (Example 2.7). Nevertheless, even if for any candidate set a definitive decision is possible there are infinite (but non-finitary) AFs where both semantics collapse (Example 2.10). In [Baumann and Spanring, 2015, Conjecture 1] it is conjectured that this is impossible in case of finitary frameworks.

	stb	ss	stg	cf2	stg2	pr	ad	co	gr	il	eg	na	cf
finite	∃	∃	∃	∃	∃	∃	∃	∃!	∃!	∃!	∃	∃	∃
finitary		∃	∃	?	?	∃	∃	∃	∃!	∃!	∃!	∃	∃
arbitrary						∃	∃	∃	∃!	∃!	∃	∃	∃

Table 1: Definedness Statuses of Semantics

For a detailed complexity analysis of the associated decision problems, i.e. *Given an AF F. Is $|\sigma(F)| \geq 1$ or even, $|\sigma(F)| = 1$?* we refer the reader to the complexity chapter of this handbook (Chapter 13, Table 1). The mentioned decisions problems are considered for finite AFs only since the input-length, i.e. the length of the formal encoding of an AF has to be finite (for finite representations of infinite AFs we refer the reader to [Baroni et al., 2013]). Due to the table above some complexity results are immediately clear. For instance, the existence problem is trivial for all considered semantics except the stable one. An upper bound for the complexity of the uniqueness problem can be obtained via the complexity of the corresponding verification problem, i.e. *Given an AF F and a set E. Is $E \in \mathcal{E}_\sigma(F)$?*. More precisely, an algorithm which decides the uniqueness problem is the following two-step procedure: first, guessing a certain set E non-deterministically and second, verifying whether this set is an σ-extension.

As already mentioned, most of the literature concentrate on finite AFs for several reasons, especially due to their computational nature. However, allowing an infinite number of arguments is essential in applications where upper bounds on the number of available arguments cannot be established a priori, such as for example in dialogues [Belardinelli et al., 2015] or modeling approaches including time or action sequences [Baumann and Strass, 2012].

Moreover, even actual infinite AFs frequently occur in the instantiation-based context. More precisely, the semantics of so-called *rule-based argumentation formalisms* (cf. Chapter 6 as well as [Prakken, 2010]) is given via the evaluation of induced Dung-style AFs. In this context, even a finite set of rules may lead to an infinite set of arguments as observed in (cf. [Caminada and Oren, 2014; Strass, 2015]).

In 2011, Baroni et al. wrote "As a matter of fact, we are not aware of any systematic literature analysis of argumentation semantics properties in the infinite case." [Baroni *et al.*, 2011, Section 4.4]. Since then only few works have contributed to a better understanding of infinite AFs. In [Baroni *et al.*, 2013] the authors studied to which extent infinite AFs can be finitely represented via formal languages and considered several decision problems within this context. In [Baumann and Spanring, 2015] a detailed study of the central properties of existence and uniqueness as presented in this chapter was given. Recently, the same authors addressed several central issues like *expressibility, intertranslatability* or *replaceability* (cf. Sections 3 and 4) in the general unrestricted case [Baumann and Spanring, 2017].

3 Expressibility

Given a certain logical formalism $\mathcal{L}$ used as knowledge representation language or modelling tool in general. Depending on the application in mind, it might be interesting to know which kinds of model sets are actually expressible in $\mathcal{L}$? More formally, if $\sigma_\mathcal{L}$ denotes the semantics of $\mathcal{L}$, we are interested in determining the set $\mathcal{R}_\mathcal{L} = \{\sigma_\mathcal{L}(T) \mid T \text{ is an } \mathcal{L}\text{-theory}\}$. This task, also known as *realizability* or *defineability*, highly depends on the considered formalism $\mathcal{L}$. Clearly, potential necessary or sufficient properties for being in $\mathcal{R}_\mathcal{L}$, i.e. being $\sigma_\mathcal{L}$-*realizable*, may rule out a logic or make it perfectly appropriate for a certain application. For instance, it is well-known that in case of propositional logic any finite set of two-valued interpretations is realizable. This means, given such a finite set $\mathcal{I}$, we always find a set of formulae T, s.t. $Mod(T) = \mathcal{I}$. Differently, in case of normal logic programs under stable model semantics we have that any finite candidate set is realizable if and only if it forms a $\subseteq$-antichain, i.e. any two sets of the candidate set have to be incomparable with respect to the subset relation. Remarkably, being such an $\subseteq$-antichain is not only necessary but even sufficient for realizability w.r.t. stable model semantics [Eiter *et al.*, 2013; Strass, 2015]. One major application of realizability issues are dynamic evolvements of $\mathcal{L}$-theories like in case of belief revision (cf. [Alchourrón *et al.*, 1985; Williams and Antoniou, 1998; Qi and Yang, 2008; Delgrande and Peppas, 2015; Delgrande *et al.*, 2008; Delgrande *et al.*, 2013; Baumann and Brewka, 2015a; Diller *et al.*, 2015] for several knowledge representation formalisms). Roughly speaking, belief revision deals with the problem of integrating new pieces of information to a current knowledge base which is represented by a certain $\mathcal{L}$-theory T. To this end, you are typically faced with the problem of modifying the given theory T in such a way that the revised version S satisfies $\sigma_\mathcal{L}(S) = M$ for some model set M. Now, before trying to do this revision in a certain minimal way it is essential to know whether M is realizable at all, i.e. $M \in \mathcal{R}_\mathcal{L}$.

The first formal treatment of realizability issues w.r.t. extension-based argumentation semantics was recently given by Dunne et al. [Dunne et al., 2013; Dunne et al., 2015]. They coined the term *signature* for the set of all realizable sets of extensions. The authors provided simple criteria for several mature semantics deciding whether a set of extensions is contained in the corresponding signature. For instance, two obvious necessary conditions in case of preferred semantics (as well as many other semantics) is that a candidate set $\mathbb{S}$ has to be non-empty, due to universal definedness of preferred semantics and second, $\mathbb{S}$ has to be a $\subseteq$-antichain, also known as *I-maximality criterion* [Baroni and Giacomin, 2007]. However, these conditions are not sufficient implying that further requirements has to hold. In case of preferred semantics it turned out that adding the requirement of so-called *conflict-sensitivity* indeed yield a set of characterizing properties. A $\subseteq$-antichain $\mathbb{S}$ is conflict-sensitive if for each pair of distinct sets A and B from $\mathbb{S}$ there are at least one $a \in A$ and one $b \in B$, s.t. a and b do not occur together in any set of $\mathbb{S}$. This implies that there exists an AF F in which the set of its preferred extension coincides with $\mathbb{S} = \{\{a,b\},\{a,c\},\{b,d\},\{c,d\}\}$. Furthermore, since $\{a,b\}$ and $\{b,d\}$ are already contained in $\mathbb{S}$ it is impossible to realize the set $\mathbb{T} = \mathbb{S} \cup \{\{a,d\}\}$ under preferred semantics. From a practical point of view, such realizability insights can be used to limit the search space when enumerating preferred extensions. More precisely, applying the mentioned characterization result we obtain that not only $\{a,d\}$, but also any other set $A \subseteq \{a,b,c,d\}$ can not be a further preferred extension of a certain AF given that we already computed all sets contained in $\mathbb{S}$. As a matter of fact, knowing that a certain set is realizable does not provide one automatically with a witnessing AF. Fortunately, there exist *canonical frameworks* showing realizability in a constructive fashion as shown in [Dunne et al., 2013; Dunne et al., 2015].

Later on, restricted versions of realizability were considered, namely *compact* as well as *analytic realizability* in case of extension-based semantics [Baumann et al., 2014a; Baumann et al., 2014b; Linsbichler et al., 2015; Baumann et al., 2016a]. Both versions are motivated by typical phenomena that can be observed for several semantics. First, there potentially exist arguments in a given AF that do not appear in any extension, so-called *rejected* arguments. Second, most of the argumentation semantics possess the feature of allowing *implicit conflicts*. An implicit conflict arises when two arguments are never jointly accepted although they do not attack each other. In order to understand in which way rejected arguments and implicit conflicts contribute to the expressive power of a certain semantics the notions of compact AFs as well as analytic AFs were introduced. The former kind disallows rejected arguments whereas the latter is free of implicit conflicts. It turned out that for many universally defined semantics the full range of expressiveness indeed relies on the use rejected arguments and implicit conflicts. This means, there are plenty of AFs which do not possess an equivalent AF which is in addition compact or analytic, respectively.

Recently, a first study of extension-based realizability w.r.t. arbitrary frameworks was presented in [Baumann and Spanring, 2017]. The authors compared the expressive power of several mature semantics in the unrestricted setting. Interestingly, the results reveal an intimate connection between arbitrary and finitely compact AFs in terms of expressiveness. Nevertheless, an in-depth analysis of realizability in the unrestricted setting is still missing. For instance, necessary and sufficient properties for being realizable are not considered so far.

There are only few works which have dealt with labelling-based realizability in the context of Dung-style argumentation frameworks. Dyrkolbotn showed that, as long as additional arguments are allowed any finite set of labellings is *realizable under projection* in case of preferred or semi-stable semantics [Dyrkolbotn, 2014]. In order to realize a set of labellings $\mathbb{S}$ under projection it suffices to come up with an AF F, s.t. its set of labellings modulo additional arguments coincide with $\mathbb{S}$. The second work by Linsbichler et al. deals with the standard notion of realizability adapted to labelling-based semantics [Linsbichler et al., 2016]. The authors presented an algorithm which returns either "No" in case of non-realizability or a witnessing AF F in the positive case. Remarkably, the algorithm is not restricted to the formalism of abstract argumentation frameworks only. In fact, it can also be used to decide realizability in case of the more general abstract dialectical frameworks as well as various of its sub-classes [Brewka and Woltran, 2010; Brewka et al., 2013].

3.1 Realizability and Signatures

Let us start with the two central concepts of this section, namely *realizability* as well as *signature*. In a nutshell, we say that a certain set $\mathbb{S}$ is realizable under the semantics σ, if there is an AF F such that its set of σ-labellings/σ-extensions coincides with $\mathbb{S}$. Collecting all realizable sets defines the concept of a signature. In accordance with the existing literature the main part of this section is devoted to finite realizability for extension-based semantics, i.e. signatures which contain set of σ-extensions of finite AFs only. Realizability w.r.t. labelling-based semantics as well as the consideration of infinite AFs will be briefly outlined only. Consider the following general definition of realizability in the context of abstract argumentation.

Definition 3.1 *Given a semantics* $\sigma : \mathcal{F} \to 2^{(2^{\mathcal{U}})^n}$ *and a set* $\mathcal{C} \subseteq \mathcal{F}$. *A set* $\mathbb{S} \subseteq (2^{\mathcal{U}})^n$ *is σ-realizable w.r.t. $\mathcal{C}$ if there is an AF $F \in \mathcal{C}$, s.t. $\sigma(F) = \mathbb{S}$.*

Definition 3.2 *Given a semantics σ and a set $\mathcal{C} \subseteq \mathcal{F}$. The σ-signature w.r.t. $\mathcal{C}$ is defined as* $\Sigma_\sigma^\mathcal{C} = \{\sigma(F) \mid F \in \mathcal{C}\}$.

If clear from context or unimportant we simply speak of *signatures* and write Σ without mentioning a semantics σ or set of AFs $\mathcal{C}$. Similarly, we say that a certain set is *realizable* instead of σ-realizable w.r.t. $\mathcal{C}$. Please observe that both concepts are intimately connected via the following relation: for any set $\mathbb{S}$ we have, $\mathbb{S}$ is realizable if and only if $\mathbb{S} \in \Sigma$. Consequently, if $\mathbb{S}$ is not contained in Σ, then there is no framework whose extensions/labellings are exactly $\mathbb{S}$. Hence, instead of searching for witnessing AFs (which might not exist) it is very attractive to find necessary as well as sufficient properties for the containment of a set $\mathbb{S}$ to a certain signature locally, i.e. by properties of $\mathbb{S}$ itself.

3.2 Signatures w.r.t. Finite AFs

We start with finite realizability. Instantiating Definitions 3.1 and 3.2 with $\mathcal{C} = \{F \in \mathcal{F} \mid F \text{ finite}\}$ formally capture the notions of realizability as well as signatures relativised to finite AFs. Consider the following definitions.

Definition 3.3 *Given a semantics* $\sigma : \mathcal{F} \to 2^{(2^{\mathcal{U}})^n}$. *A set* $\mathbb{S} \subseteq \left(2^{\mathcal{U}}\right)^n$ *is finitely σ-realizable if there is an AF* $F \in \{F \in \mathcal{F} \mid F \text{ finite}\}$, *s.t.* $\sigma(F) = \mathbb{S}$.

Definition 3.4 *Given a semantics* σ. *The finite σ-signature is defined as* $\{\sigma(F) \mid F \in \mathcal{F}, F \text{ finite}\}$ *abbreviated by* Σ_σ^f.

We proceed with further notational shorthands (adjusted to the extension-based approach) which will be used throughout the whole section.

Definition 3.5 ([Dunne et al., 2015]) *Given* $\mathbb{S} \subseteq 2^{\mathcal{U}}$, *we use*

- $Args_\mathbb{S}$ *to denote* $\bigcup_{S \in \mathbb{S}} S$ *and* $\|\mathbb{S}\|$ *for* $|Args_\mathbb{S}|$,
- $Pairs_\mathbb{S}$ *to denote* $\{(a,b) \mid \exists S \in \mathbb{S} : \{a,b\} \subseteq S\}$ *and*
- $dcl(\mathbb{S})$ *to denote (the so-called downward-closure)* $\{S' \subseteq S \mid S \in \mathbb{S}\}$

Furthermore, we say that $\mathbb{S}$ *is an extension-set if* $\|\mathbb{S}\|$ *is a finite cardinal.*

In order to familiarize the reader with the introduced definitions we give the following example.

Example 3.6 *Let* $\mathbb{S} = \{\{a\}, \{a,c\}, \{a,b,d\}\}$. *Then*

- $Args_\mathbb{S} = \{a,b,c,d\}$ *and* $\|\mathbb{S}\| = 4$. *This means,* $\mathbb{S}$ *is an extension-set.*
- $Pairs_\mathbb{S} = \{(a,a),(b,b),(c,c),(d,d),(a,b),(a,c),(a,d),(b,d)\} \cup \{(b,a),(c,a),(d,a),(d,b)\}$
- $dcl(\mathbb{S}) = \{\emptyset, \{a\}, \{b\}, \{c\}, \{d\}, \{a,b\}, \{a,c\}, \{a,d\}, \{b,d\}, \{a,b,d\}\}$

Furthermore, since naive extensions are defined as $\subseteq$-maximal sets and obviously, $\{a\} \subset \{a,c\}$ we deduce that $\mathbb{S}$ is not na-realizable, i.e. $\mathbb{S} \notin \Sigma_{\mathcal{E}_{na}}^f$. Regarding complete semantics we obtain $\mathbb{S} \in \Sigma_{\mathcal{E}_{co}}^f$ witnessed by the following AF F.

$F:$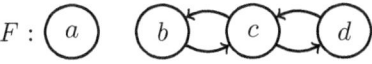

In the following we consider the signatures of the extension-based versions of stable, semi-stable, stage, naive, preferred, complete as well as grounded semantics [Dunne et al., 2013; Dunne et al., 2015]. We provide a bunch of properties where certain subsets of them exactly matches the containment conditions for certain signatures. All properties can be decided by looking on the set in question only.

Semantics based on Conflict-freeness

Our starting point are semantics based on conflict-free sets. Conflict-free sets by themselves inherited their conflict-freeness to any subset of them. More formally, the downward-closure does not vary the set of conflict-free sets for a given AF. A set possessing this property is called *downward-closed*. Clearly, downward-closedness does not hold in case of admissible sets as well as any other reasonable semantics σ where conflict-freeness is just one requirement among others for being a σ-extension. Take for instance naive semantics. Naive extension are defined as $\subseteq$-maximal conflict-free sets. Consequently, the set of all naive extensions is a $\subseteq$-antichain, i.e. any two naive extensions are *incomparable* w.r.t. subset relation. This property also applies to many other semantics, such as stable and stage semantics as well as any uniquely defined semantics. However, although incomparability is a necessary condition for many considered semantics it is certainly not sufficient. Consider therefore the following example taken from [Dunne et al., 2015, Example 1].

Example 3.7 *Consider the $\subseteq$-antichain $\mathbb{S} = \{\{a,b\}, \{a,c\}, \{b,c\}\}$ and a semantics σ which selects its reasonable positions among the conflict-free sets, i.e. $\mathcal{E}_\sigma(F) \subseteq \mathcal{E}_{cf}(F)$ for any AF F. Now suppose there exists an AF F with $\mathcal{E}_\sigma(F) = \mathbb{S}$. Then F must not contain attacks between a and b, a and c, and respectively b and c. This means, $\{a,b,c\} \in \mathcal{E}_{cf}(F)$. But then $\mathcal{E}_\sigma(F)$ typically contains $\{a,b,c\}$.*

There are several ways to define the required property which excludes sets like $\mathbb{S}$ from above. It turned out that in order to characterize conflict-free based semantics like stable, stage and naive semantics a rather strong condition is required, so-called *tightness*. Roughly speaking, if an incomparable set is not tight, then there is a set $S \in \mathbb{S}$ and an argument a not belonging to S, s.t. for any $s \in S$ we find an other $S' \in \mathbb{S}$ with a and s being members of it. The idea behind the notion of being tight is simply that if an argument a does not occur in some extension S there must be a reason for that. The most simple reason one can think of is that there is a conflict between a and some $s \in S$, i.e. a and s do not occur jointly in any extension-set of $\mathbb{S}$ or, in other words, $(a,s) \notin \text{Pairs}_\mathbb{S}$. In a way, this limits the multitude of incomparable elements of an extension-set.

We proceed with the formal definitions.

Definition 3.8 ([Dunne et al., 2013]) *Given $\mathbb{S} \subseteq 2^{\mathcal{U}}$. We call $\mathbb{S}$*

- *downward-closed if $\mathbb{S} = dcl(\mathbb{S})$,*

- *incomparable if $\mathbb{S}$ is a $\subseteq$-antichain and*

- *tight if for all $S \in \mathbb{S}$ and $a \in Args_{\mathbb{S}}$ it holds that if $S \cup \{a\} \notin \mathbb{S}$ then there exists an $s \in S$ such that $(a, s) \notin Pairs_{\mathbb{S}}$.*

Please observe that for incomparable $\mathbb{S}$, the premise of the tightness condition, i.e. $S \cup \{a\} \notin \mathbb{S}$, is always fulfilled. However, tightness and incomparability are independent of each other, i.e. neither tightness implies incomparability or comparability, nor incomparability implies tightness or non-tightness.

Example 3.9 *Consider again the extension-set $\mathbb{S} = \{\{a, b\}, \{a, c\}, \{b, c\}\}$ from Example 3.7. The set $\mathbb{S}$ is incomparable but not tight which can be seen as follows. If setting $S = \{a, b\}$ we observe $S \cup \{c\} \notin \mathbb{S}$. Moreover, for any $s \in S$ we find an $S' \in \mathbb{S}$, s.t. $\{s, c\} = S'$ implying that $(s, c) \in Pairs_{\mathbb{S}}$. More precisely, if $s = a$, then we have $S' = \{a, c\}$ and similarly, if $s = b$ we find $S' = \{b, c\}$.*

Furthermore, it can be checked that $\mathbb{S}' = \{\{a, b\}, \{a, c\}, \{b, d\}, \{c, d\}\}$ or $\mathbb{S}'' = \mathbb{S} \cup \{\{a, b, c\}\}$ are witnessing examples for incomparability and tightness or tightness and comparability, respectively.

Clearly, subsets of incomparable sets are incomparable. Such a kind of inheritance does not hold in case of tight sets (cf. $\mathbb{S}$ and $\mathbb{S}''$ as defined in Example 3.9). Nevertheless, there are non-trivial tight subsets of any tight set. For instance, in any case the set of all $\subseteq$-maximal elements is tight. Furthermore, if a tight set is even incomparable, then any subset of it is tight too.

In the following we present the main statements only. However, in many cases we provide some short comments indicating how to prove the statement in question. For full proofs we refer the reader to the referenced papers.

Lemma 3.10 ([Dunne et al., 2015]) *For a tight extension-set $\mathbb{S} \subseteq 2^{\mathcal{U}}$ we have:*

1. *the $\subseteq$-maximal elements in $\mathbb{S}$ form a tight set, and*

2. *if $\mathbb{S}$ is incomparable then each $\mathbb{S}' \subseteq \mathbb{S}$ is tight.*

Note that the second statement of Lemma 3.10 implies that if the downward-closure of an incomparable extension-set $\mathbb{S}$ is tight, then $\mathbb{S}$ itself has to be tight too.

We proceed with a specific AF and check which properties apply to its different sets of extensions.

Example 3.11 *Consider the following AF F.*

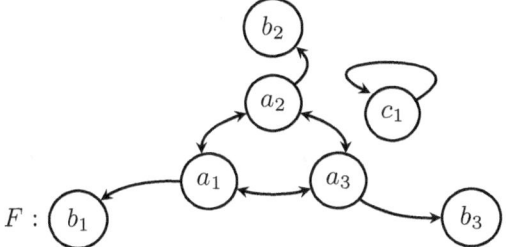

Since c_1 is self-defeating as well as unattacked we obtain $\mathcal{E}_{stb}(F) = \emptyset$. Furthermore, $\mathcal{E}_{stg}(F) = \{\{a_1, b_2, b_3\}, \{a_2, b_1, b_3\}, \{a_3, b_1, b_2\}\}$ and $\mathcal{E}_{na}(F) = \mathcal{E}_{stg}(F) \cup \{\{b_1, b_2, b_3\}\}$. We observe,

1. $\mathcal{E}_{stb}(F), \mathcal{E}_{stg}(F)$ as well as $\mathcal{E}_{na}(F)$ are incomparable,

2. $\mathcal{E}_{stb}(F), \mathcal{E}_{stg}(F)$ as well as $\mathcal{E}_{na}(F)$ are tight and additionally,

3. $dcl(\mathcal{E}_{na}(F))$ and $dcl(\mathcal{E}_{stb}(F))$ are tight and obviously,

4. $\mathcal{E}_{stg}(F)$ and $\mathcal{E}_{na}(F)$ are non-empty.

The first and the last items are not surprising since firstly, all considered semantics satisfy the I-maximality criterion which is just another name for incomparability and secondly, in Section 2 we have already seen that stage extensions are guaranteed for finitary (hence, for finite) frameworks and naive semantics is even universally defined w.r.t. the whole class of AFs. This means, incomparability or non-emptiness of the mentioned sets of σ-extensions do not depend on the specific AF F, but rather apply to any finite AF. Consequently, these properties represent necessary properties regarding realizability. The tightness statements of the second and third items can be checked in a straightforward manner. We now examine that $dcl(\mathcal{E}_{stg}(F))$ is non-tight. This can be seen as follows: Firstly, $\{b_2, b_3\} \in dcl(\mathcal{E}_{stg}(F))$. Now, for b_1 the premise of Definition 3.8 is satisfied, i.e. $\{b_1, b_2, b_3\} \notin dcl(\mathcal{E}_{stg}(F))$. Consequently, since $\{b_1, b_2\}, \{b_1, b_3\} \in dcl(\mathcal{E}_{stg}(F))$ and therefore, $(b_1, b_2), (b_1, b_3) \in Pairs_{dcl(\mathcal{E}_{stg}(F))}$ we deduce the non-tightness of $dcl(\mathcal{E}_{stg}(F))$. This means, tightness of the downward-closure of a given set can not be a necessary criterion for belonging to the stage signature.

We now present the characterization theorems for conflict-free, naive, stable as well as stage signatures. It is somehow surprising that only a few simple properties are sufficient to characterize these different signatures.

Theorem 3.12 ([Dunne et al., 2015]) *Given a set* $\mathbb{S} \subseteq 2^{\mathcal{U}}$, *then*

1. $\mathbb{S} \in \Sigma^f_{\mathcal{E}_{cf}} \Leftrightarrow \mathbb{S}$ *is a non-empty, downward-closed, and tight extension-set,*

2. $\mathbb{S} \in \Sigma^f_{\mathcal{E}_{na}} \Leftrightarrow \mathbb{S}$ *is a non-empty, incomparable extension-set and* $dcl(\mathbb{S})$ *is tight,*

3. $\mathbb{S} \in \Sigma^f_{\mathcal{E}_{stb}} \Leftrightarrow \mathbb{S}$ *is a incomparable and tight extension-set,*

4. $\mathbb{S} \in \Sigma^f_{\mathcal{E}_{stg}} \Leftrightarrow \mathbb{S}$ *is a non-empty, incomparable and tight extension-set.*

We mention that a proof of the characterization theorem above requires two directions. Let us fix a certain semantics $\sigma \in \{cf, na, stb, stg\}$. The first part is to show that for any finite AF F, $\mathcal{E}_\sigma(F)$ satisfies the mentioned properties. Now, for the second part, if a certain extension-set $\mathbb{S}$ satisfies the properties in question, then we have to find a finite AF F, s.t. $\mathcal{E}_\sigma(F) = \mathbb{S}$.

Let us start with the first part. It suffices to consider tightness only since downward-closedness, non-emptiness and incomparability are clear (cf. some explanations given in Example 3.11). It is easy to see that $\mathcal{E}_{cf}(F)$ is tight because if augmenting a conflict-free set S with a non-conflicting argument a yields a conflicting set, then obviously there has to be at least one element in $s \in S$, s.t. $\{a, s\}$ is conflicting. In order to prove that $dcl(\mathcal{E}_{na}(F))$ is tight, it suffice to see that $dcl(\mathcal{E}_{na}(F)) = \mathcal{E}_{cf}(F)$. Consequently, applying Lemma 3.10 we obtain the tightness of $\mathcal{E}_{na}(F)$. Furthermore, with the same lemma, we get that every $\mathbb{S} \subseteq \mathcal{E}_{na}(F)$ is tight. In consideration of $stb \subseteq stg \subseteq na$ (Proposition 2.6) it follows that $\mathcal{E}_{stb}(F)$ as well as $\mathcal{E}_{stg}(F)$ are tight.

In order to show that the mentioned properties are not only necessary but even sufficient we have to come up with witnessing AFs. Consider therefore the following prototype.

Definition 3.13 ([Dunne et al., 2015]) *Given an extension-set $\mathbb{S}$, we define the canonical argumentation framework for $\mathbb{S}$ as*

$$F^{cf}_\mathbb{S} = (Args_\mathbb{S}, (Args_\mathbb{S} \times Args_\mathbb{S}) \setminus Pairs_\mathbb{S}).$$

The idea behind the framework is simple: we draw a relation between two arguments iff they do not occur jointly in any set $S \in \mathbb{S}$. Consequently, for any $\mathbb{S}$, $F^{cf}_\mathbb{S}$ is symmetric. Moreover, in any case, it is self-loop-free since $a \in Args_\mathbb{S}$ implies $(a, a) \in Pairs_\mathbb{S}$. Let us consider the following example.

Example 3.14 *Let* $\mathbb{S} = \{\{a_1, b_2, b_3\}, \{a_2, b_1, b_3\}, \{a_3, b_1, b_2\}, \{b_1, b_2, b_3\}\}$ *and consider the corresponding canonical framework* $F^{cf}_\mathbb{S}$.

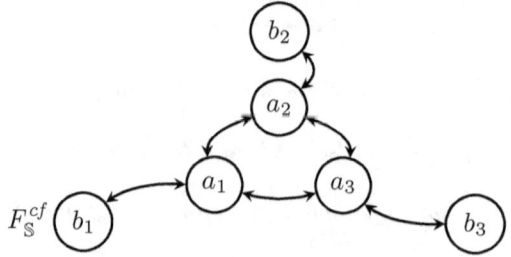

Please note that $\mathbb{S}$ *is non-empty, incomparable as well as possesses a tight downward-closure (cf. Example 3.11). Furthermore,* $F_{\mathbb{S}}^{cf}$ *realizes* $\mathbb{S}$ *under the naive semantics, i.e.* $\mathcal{E}_{na}\left(F_{\mathbb{S}}^{cf}\right) = \mathbb{S}$.

The following proposition shows that this is no coincidence.

Proposition 3.15 ([Dunne et al., 2015]) *For each non-empty, incomparable extension-set* $\mathbb{S}$*, where dcl(*$\mathbb{S}$*) is tight,* $\mathcal{E}_{na}\left(F_{\mathbb{S}}^{cf}\right) = \mathbb{S}$.

Moreover, the canonical framework can also be used as witnessing framework in case of conflict-free sets as stated in the following proposition.

Proposition 3.16 ([Dunne et al., 2015]) *For each non-empty, downward-closed and tight extension-set* $\mathbb{S}$*,* $\mathcal{E}_{cf}\left(F_{\mathbb{S}}^{cf}\right) = \mathbb{S}$.

We proceed with stable and stage semantics. In Theorem 3.12 the only difference between the characterizations of stable and stage signatures is the non-empty requirement for stage semantics. Remember that we are dealing with finite AFs and indeed in case of this restriction stable semantics is the only semantics which does not warrant the existence of extensions (cf. Table 1).[8] This means, stable semantics is the only semantics which may realize the empty extension-set (which is incomparable and tight too). The final step towards concluding Theorem 3.12 is to find witnessing frameworks for any non-empty, incomparable and tight extension-sets. At first we will show that the canonical framework does not do the job in case of these semantics. More precisely, given a non-empty, incomparable as well as tight extension-set $\mathbb{S}$, then the sets of stable as well as stage extensions of the canonical framework $F_{\mathbb{S}}^{cf}$ do not necessarily coincide with $\mathbb{S}$.

Example 3.17 *Consider again Example 3.14. We define* $\mathbb{T} = \mathbb{S}\setminus\{\{b_1, b_2, b_3\}\}$*. Please note that* $F_{\mathbb{T}}^{cf}$ *and* $F_{\mathbb{S}}^{cf}$ *are identical since* $Args_{\mathbb{S}} = Args_{\mathbb{T}}$ *and* $Pairs_{\mathbb{S}} = Pairs_{\mathbb{T}}$*. Furthermore, according to Example 3.11 we have that* $\mathbb{T}$ *is non-empty, incomparable and tight, but* $\mathcal{E}_{stb}\left(F_{\mathbb{T}}^{cf}\right) = \mathcal{E}_{stg}\left(F_{\mathbb{T}}^{cf}\right) = \mathcal{E}_{na}\left(F_{\mathbb{S}}^{cf}\right) = \mathbb{S} \neq \mathbb{T}$*. In order to get rid of the undesired stable as well as stage extension* $E = \{b_1, b_2, b_3\}$ *we may simply add a new self-defeating argument* e *to* $F_{\mathbb{S}}^{cf}$*, s.t.* e *is attacked by all other arguments excepting those stemming from* E*. The following framework* $F_{\mathbb{T}}^{stb}$ *illustrates this idea. Convince yourself that* $\mathcal{E}_{stb}\left(F_{\mathbb{T}}^{stb}\right) = \mathcal{E}_{stg}\left(F_{\mathbb{T}}^{stb}\right) = \mathbb{T}$.

---

[8] For instance, $F = (\{a\}, \{(a,a)\})$ yields $\mathcal{E}_{stb}(F) = \emptyset$.

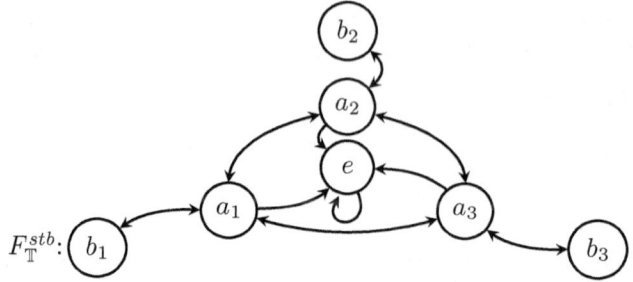

The following definition generalizes the construction idea from above to arbitrary many undesired sets. The subsequent proposition states that we have indeed found witnessing examples for non-empty, incomparable and tight extension-sets as required for Theorem 3.12.

Definition 3.18 ([Dunne et al., 2015]) *Given an extension-set $\mathbb{S}$ and its canonical framework $F_\mathbb{S}^{cf} = (A_\mathbb{S}^{cf}, R_\mathbb{S}^{cf})$. Let $\mathbb{X} = \mathcal{E}_{stb}\left(F_\mathbb{S}^{cf}\right) \setminus \mathbb{S}$ we define*
$$F_\mathbb{S}^{stb} = \left(A_\mathbb{S}^{cf} \cup \{\bar{E} \mid E \in \mathbb{X}\}, R_\mathbb{S}^{cf} \cup \{(\bar{E}, \bar{E}), (a, \bar{E}) \mid E \in \mathbb{X}, a \in Args_\mathbb{S} \setminus E\}\right).$$

Proposition 3.19 ([Dunne et al., 2015]) *For each non-empty, incomparable and tight extension-set $\mathbb{S}$, $\mathcal{E}_{stb}\left(F_\mathbb{S}^{stb}\right) = \mathcal{E}_{stg}\left(F_\mathbb{S}^{stb}\right) = \mathbb{S}$.*

Semantics based on Admissibility

Let us turn now to semantics based on admissible sets. In particular, we provide characterization theorems for the finite signatures w.r.t. admissible sets as well as preferred and semi-stable semantics. In contrast to semantics based on conflict-free sets where the notion of tightness played a decisive role (cf. Theorem 3.12) we have to introduce a new concept, so-called *conflict-sensitivity*. Conflict-sensitivity is a very basic property in the sense that it is fulfilled by almost all semantics σ (or rather, their corresponding sets of σ-extensions) available in the literature. Furthermore, it is strictly weaker than tightness, i.e. tight extension-sets are always conflict-sensitive, but not necessarily vice versa. To explain the difference between these two notions let us consider the following example taken from [Dunne et al., 2015].

Example 3.20 *Consider the following framework F.*

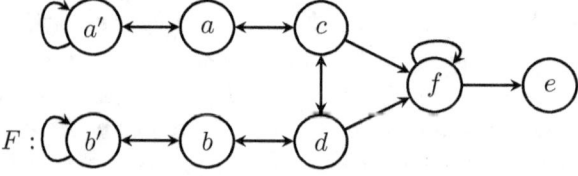

We have $\mathcal{E}_{pr}(F) = \mathcal{E}_{ss}(F) = \mathbb{S} = \{A, B, C\} = \{\{a, b\}, \{a, d, e\}, \{b, c, e\}\}$. First, observe that $\mathbb{S}$ is not tight. This can be seen as follows: Obviously,

$A \cup \{e\} \notin \mathbb{S}$, but both (a,e) and (b,e) are contained in $Pairs_\mathbb{S}$ since $\{a,e\} \subseteq B$ and $\{b,e\} \subseteq C$. This means, although $A \cup \{e\}$ is not a reasonable position w.r.t. preferred and semi-stable semantics we find witnessing extensions, namely B and C, showing that any argument in A is compatible with e, i.e. they can be accepted together. Please observe that this is not true for any two arguments in A and B or A and C, respectively. For instance, $b,d \in A \cup B$, but $(b,d) \notin Pairs_\mathbb{S}$ as well as $a,c \in A \cup C$, but $(a,c) \notin Pairs_\mathbb{S}$. Furthermore, the same applies to B and C, since $c,d \in B \cup C$ and $(c,d) \notin Pairs_\mathbb{S}$.

The following definition precisely formalizes the observed property of the AF F presented in the example above.

Definition 3.21 ([Dunne et al., 2015]) *A set $\mathbb{S} \subseteq 2^\mathcal{U}$ is called conflict-sensitive if for each $A, B \in \mathbb{S}$ such that $A \cup B \notin \mathbb{S}$ it holds that $\exists a, b \in A \cup B :$ $(a,b) \notin Pairs_\mathbb{S}$.*

As the name suggests, the property checks whether the absence of the union of any pair of extensions in an extension-set $\mathbb{S}$ is justified by a conflict indicated by $\mathbb{S}$. Note that for $a, b \in A$ (likewise $a, b \in B$), $(a,b) \in Pairs_\mathbb{S}$ holds by definition. Thus the property of conflict-sensitivity is determined by arguments $a \in A \setminus B$, $b \in B \setminus A$, for $A, B \in \mathbb{S}$. As already indicated tightness implies conflict-sensitivity as stated in the following lemma.

Lemma 3.22 ([Dunne et al., 2015])*Every tight extension-set is also conflict-sensitive.*

Similarly to Lemma 3.10 one may show that the set of all $\subseteq$-maximal elements of a conflict-sensitive set is conflict-sensitive too. Moreover, if the initial set is incomparable in addition, then even any subset of it is conflict-sensitive. Furthermore, in contrast to tight extension-sets it is possible to add the empty set to a conflict-sensitive set without loosing conflict-sensitivity.[9]

Lemma 3.23 ([Dunne et al., 2015])*For a conflict-sensitive ext.-set $\mathbb{S} \subseteq 2^\mathcal{U}$,*

1. *the $\subseteq$-maximal elements in $\mathbb{S}$ form a conflict-sensitive set,*

2. *if $\mathbb{S}$ is incomparable then each $\mathbb{S}' \subseteq \mathbb{S}$ is conflict-sensitive, and*

3. *$\mathbb{S} \cup \{\emptyset\}$ is conflict-sensitive.*

Having conflict-sensitivity at hand, we are now ready to present characterization theorems for the signatures w.r.t. admissible sets as well as preferred and semi-stable semantics. Interestingly, it turns out that preferred and semi-stable semantics are equally expressive in case of finite AFs, i.e. $\Sigma^f_{\mathcal{E}_{pr}} = \Sigma^f_{\mathcal{E}_{ss}}$.

Theorem 3.24 ([Dunne et al., 2015]) *Given a set $\mathbb{S} \subseteq 2^\mathcal{U}$, then*

---
[9]Note that any one-element extension-set $\mathbb{S} \neq \{\emptyset\}$ is tight, whereas $\mathbb{S} \cup \{\emptyset\}$ is not.

1. $\mathbb{S} \in \Sigma^f_{\mathcal{E}_{ad}} \Leftrightarrow \mathbb{S}$ is a conflict-sensitive ext.-set containing $\emptyset$,

2. $\mathbb{S} \in \Sigma^f_{\mathcal{E}_{pr}} \Leftrightarrow \mathbb{S}$ is a non-empty, incomparable and conflict-sensitive ext.-set,

3. $\mathbb{S} \in \Sigma^f_{\mathcal{E}_{ss}} \Leftrightarrow \mathbb{S}$ is a non-empty, incomparable and conflict-sensitive ext.-set.

Let us first argue that the mentioned properties are necessary conditions for being in the corresponding signature. For admissible sets it suffices to recall the following two facts: First, the empty set is admissible by definition; and second, if the union of two admissible sets is conflict-free, then the union is admissible too. In other words, if the union fails to be admissible, then there has to be a conflict proving the conflict-sensitivity of any set of admissible sets. Now, for preferred and semi-stable semantics. Non-emptiness is due to the already shown universal definedness of both semantics in case of finite AFs (cf. Table 1). Moreover, incomparability is clear since both semantics satisfy the I-maximality criterion (cf. Chapter 16). Finally, conflict-sensitivity of sets of admissible sets transfer to sets of preferred extensions via statement 1 of Lemma 3.23 and therefore also to sets of semi-stable extensions via statement 2 of Lemma 3.23 and the fact that $ss \subseteq pr$ (Proposition 2.6).

In order to show that the mentioned properties are not only necessary but even sufficient we have to come up with witnessing AFs. In contrast to conflict-free based semantics we have to find AFs which encode the central notion of admissibility. Please note that the already introduced canonical frameworks $F^{cf}_\mathbb{S}$ as well as $F^{stb}_\mathbb{S}$ (cf. Definitions 3.13 and 3.18) do not comply with the requirements. Consider therefore the following example.

Example 3.25 Let us consider again the non-empty, incomparable as well as tight set $\mathbb{T} = \{\{a_1, b_2, b_3\}, \{a_2, b_1, b_3\}, \{a_3, b_1, b_2\}\}$ together with its corresponding canonical framework $F^{stb}_\mathbb{T}$ as presented in Example 3.17. Due to Lemma 3.22 we have that any tight extension-set is even conflict-sensitive and thus, $\mathbb{T}$ satisfies the necessary requirements of Theorem 3.24. Inspecting the canonical framework reveals that $\mathcal{E}_{pr}\left(F^{stb}_\mathbb{T}\right) = \mathbb{T} \cup \{\{b_1, b_2, b_3\}\} \neq \mathbb{T}$. Although, $\mathcal{E}_{ss}\left(F^{stb}_\mathbb{T}\right) = \mathbb{T}$ one may easily check that non-empty, incomparable as well as conflict-sensitive set $\mathbb{S} = \{\{a, b\}, \{a, d, e\}, \{b, c, e\}\}$ mentioned in Example 3.20 shows that this equality does not hold in general. Likewise, one may prove that the framework $F^{cf}_\mathbb{S}$ is not appropriated as a witnessing prototype for semi-stable as well as preferred semantics.

It turned out that suitable canonical AFs can be built by means of so-called *defense-formulae* as introduced in the following definition.

Definition 3.26 ([Dunne et al., 2015]) Given an extension-set $\mathbb{S}$, the defense-formula $\mathcal{D}^\mathbb{S}_a$ of an argument $a \in Args_\mathbb{S}$ in $\mathbb{S}$ is defined as:

$$\mathcal{D}^\mathbb{S}_a = \bigvee_{S \in \mathbb{S}, s.t. a \in S} \bigwedge_{s \in S \setminus \{a\}} s.$$

$\mathcal{D}_a^\mathbb{S}$ given as (a logically equivalent) CNF is called CNF-defense-formula $\mathcal{CD}_a^\mathbb{S}$ of a in $\mathbb{S}$.

The main idea of the formula $\mathcal{D}_a^\mathbb{S}$ is to describe the conditions for the argument a being in an extension. Note that the variables coincide with the arguments. If $\mathbb{S}$ amounts to a set of admissible extensions, then each disjunct represents a set of arguments A which allows a to join in the sense that $A \cup \{a\}$ is a reasonable position w.r.t. admissible semantics. Put it differently, propositional models of $\mathcal{D}_a^\mathbb{S} \wedge a$ represent (if considered as set of atoms) supersets of certain reasonable position. Please not that a defense-formula $\mathcal{D}_a^\mathbb{S}$ is tautological if and only if $\{a\} \in \mathbb{S}$. We proceed with an example.

Example 3.27 *Consider again the non-empty, incomparable as well as conflict-sensitive set $\mathbb{S} = \{\{a,b\},\{a,d,e\},\{b,c,e\}\}$ stemming from Example 3.20. We obtain the following defense-formulae together with their corresponding CNF-defense-formulae (written in clause form).*

- $\mathcal{D}_a^\mathbb{S} = b \vee (d \wedge e) \equiv (b \vee d) \wedge (b \vee e)$ and $\mathcal{CD}_a^\mathbb{S} = \{\{b,d\},\{b,e\}\}$
- $\mathcal{D}_b^\mathbb{S} = a \vee (c \wedge e) \equiv (a \vee c) \wedge (a \vee e)$ and $\mathcal{CD}_b^\mathbb{S} = \{\{a,c\},\{a,e\}\}$
- $\mathcal{D}_c^\mathbb{S} = b \wedge e$ and $\mathcal{CD}_c^\mathbb{S} = \{\{b,e\}\}$
- $\mathcal{D}_d^\mathbb{S} = a \wedge e$ and $\mathcal{CD}_d^\mathbb{S} = \{\{a,e\}\}$
- $\mathcal{D}_e^\mathbb{S} = (a \wedge d) \vee (b \wedge c) \equiv (a \vee b) \wedge (d \vee b) \wedge (a \vee c) \wedge (d \vee c)$ and $\mathcal{CD}_d^\mathbb{S} = \{\{a,b\},\{a,c\},\{b,d\},\{c,d\}\}$

One simple idea for the realization of a certain set $\mathbb{S}$ under admissible semantics is the following two-step procedure. In the first step, we construct a framework F which maintains all elements of $\mathbb{S}$ as conflict-free sets. This can be done via the the canonical framework $F_\mathbb{S}^{cf}$. In the second step, we augment the initial framework $F_\mathbb{S}^{cf}$, s.t. only elements in $\mathbb{S}$ become admissible. The second step can be realized via adding a certain amount of additional arguments. More precisely, for any argument $a \in Args_\mathbb{S}$ we add n self-conflicting arguments $\alpha_{aC_1}, ..., \alpha_{aC_n}$ if $\left|\mathcal{CD}_a^\mathbb{S}\right| = |\{C_1,...,C_n\}| = n$. Then, for any $i \in \{1,..,n\}$, α_{aC_i} attacks a and is in turn attacked by any argument in C_i. Consider therefore the following example.

Example 3.28 *Again consider the extension-set $\mathbb{S} = \{\{a,b\},\{a,d,e\},\{b,c,e\}\}$ and its corresponding CNF-defense-formulae as presented in Example 3.27. In accordance with the above mentioned two-step procedure we obtain the dashed AF $F_\mathbb{S}^{cf}$ first. Then, in view of the CNF-defense-formulae we have to add 10 additional self-defeating arguments which attacks their corresponding argument. This intermediate step is depicted below.*

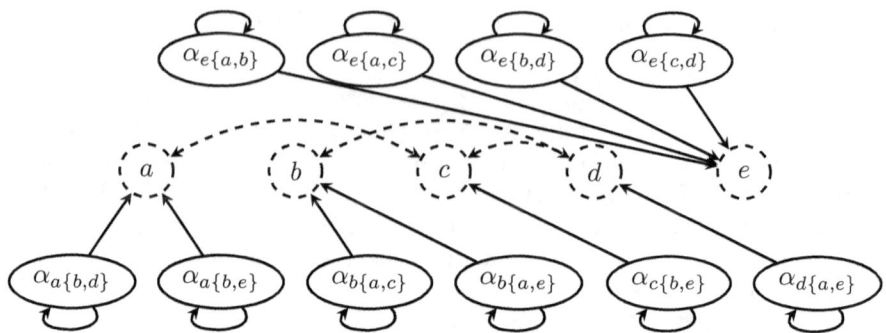

Let us consider the set $\{a,b\} \in \mathbb{S}$. In order for $\{a,b\}$ to be admissible we have to add counter-attacks for the arguments $\alpha_{a\{b,d\}}$, $\alpha_{a\{b,e\}}$, $\alpha_{b\{a,c\}}$ and $\alpha_{b\{a,e\}}$. For instance, $\alpha_{a\{b,d\}}$ is attacked by b and d and so forth. The following figure (built on top of the previous one) depicts resulting counter-attacks for the mentioned 4 arguments highlighted as densely dotted edges. For the sake of clarity we do not perform this construction for the remaining arguments.

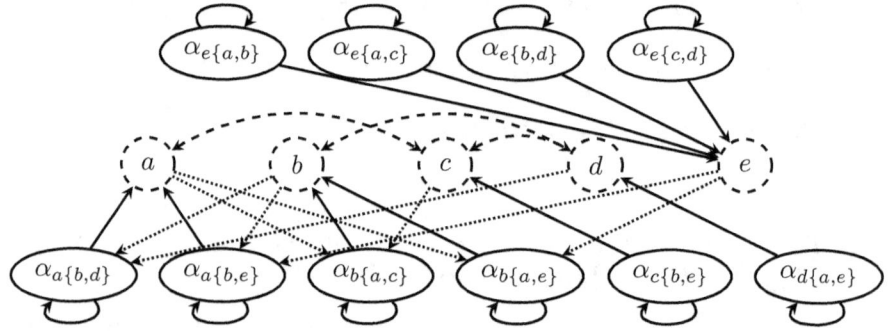

The following definition precisely formalizes the mentioned two-step procedure.

Definition 3.29 ([Dunne et al., 2015]) *Given an extension-set $\mathbb{S}$, the canonical defense-argumentation-framework $F_\mathbb{S}^{def} = (A_\mathbb{S}^{def}, R_\mathbb{S}^{def})$ extends the canonical AF $F_\mathbb{S}^{cf} = (Args_\mathbb{S}, R_\mathbb{S}^{cf})$ as follows:*

$$A_\mathbb{S}^{def} = Args_\mathbb{S} \cup \bigcup_{a \in Args_\mathbb{S}} \{\alpha_{a\gamma} \mid \gamma \in \mathcal{CD}_a^\mathbb{S}\}, \text{ and}$$

$$R_\mathbb{S}^{def} = R_\mathbb{S}^{cf} \cup \bigcup_{a \in Args_\mathbb{S}} \{(b, \alpha_{a\gamma}), (\alpha_{a\gamma}, \alpha_{a\gamma}), (\alpha_{a\gamma}, a) \mid \gamma \in \mathcal{CD}_a^\mathbb{S}, b \in \gamma\}.$$

The subsequent proposition shows that not only all elements in $\mathbb{S}$ become admissible in the constructed AF $F_\mathbb{S}^{def}$, but rather that the set of admissible sets of $F_\mathbb{S}^{def}$ exactly coincides with $\mathbb{S}$ given that $\mathbb{S}$ is conflict-sensitive as well as contains the empty set.

Proposition 3.30 ([Dunne et al., 2015]) *For each conflict-sensitive ext.-set $\mathbb{S}$ where $\emptyset \in \mathbb{S}$, it holds that $\mathcal{E}_{ad}\left(F_{\mathbb{S}}^{def}\right) = \mathbb{S}$.*

Interestingly, we may even use the canonical defense-AF to show that any non-empty, incomparable and conflict-sensitive extension-set $\mathbb{S}$ can be realized under the preferred semantics. This can be seen as follows: First, via Lemma 3.23 we obtain the conflict-sensitivity of $\mathbb{S} \cup \{\emptyset\}$ since $\mathbb{S}$ is assumed to be conflict-sensitive. Consequently, using Proposition 3.31 we obtain $\mathcal{E}_{ad}\left(F_{\mathbb{S} \cup \{\emptyset\}}^{def}\right) = \mathbb{S} \cup \{\emptyset\}$. Since $F_{\mathbb{S}}^{def} = F_{\mathbb{S} \cup \{\emptyset\}}^{def}$ and due to the incomparability of $\mathbb{S}$, we have $\mathcal{E}_{pr}\left(F_{\mathbb{S}}^{def}\right) = \mathbb{S}$ as stated in the following proposition.

Proposition 3.31 ([Dunne et al., 2015]) *For each non-empty, incomparable and conflict-sensitive extension-set $\mathbb{S}$, it holds that $\mathcal{E}_{pr}\left(F_{\mathbb{S}}^{def}\right) = \mathbb{S}$.*

Furthermore, due to a translation result by Dvořák and Woltran we obtain that any non-empty, incomparable and conflict-sensitive extension-set $\mathbb{S}$ can be realized under semi-stable semantics too. More precisely, in [Dvořák and Woltran, 2011] it is shown that for any AF F exists an AF F', s.t. $\mathcal{E}_{pr}(F) = \mathcal{E}_{ss}(F')$.

Proposition 3.32 ([Dunne et al., 2015]) *Each non-empty, incomparable and conflict-sensitive extension-set $\mathbb{S}$ is ss-realizable.*

Uniquely Defined Semantics

Let us finally turn to grounded, ideal and eager semantics. Remember that all mentioned semantics warrants the existence of exactly one extension given that the frameworks in question are finite (cf. Table 1). Furthermore, it is hardly surprising that this property is even sufficient for being in the corresponding signature, since any one-element extension-set $\mathbb{S} = \{E\}$ can be realized via $F_E = (E, \emptyset)$. In particular, we obtain that all three semantics are equally expressive.

Theorem 3.33 ([Dunne et al., 2016]) *Given a set $\mathbb{S} \subseteq 2^{\mathcal{U}}$, then*

1. $\mathbb{S} \in \Sigma_{\mathcal{E}_{gr}}^{f} \Leftrightarrow \mathbb{S}$ *is an extension-set with* $|\mathbb{S}| = 1$,

2. $\mathbb{S} \in \Sigma_{\mathcal{E}_{il}}^{f} \Leftrightarrow \mathbb{S}$ *is an extension-set with* $|\mathbb{S}| = 1$ *and*

3. $\mathbb{S} \in \Sigma_{\mathcal{E}_{eg}}^{f} \Leftrightarrow \mathbb{S}$ *is an extension-set with* $|\mathbb{S}| = 1$.

Summary of Results and Further Remarks

In this subsection we provide a comprehensive overview of characterization results w.r.t. extension-based realizability in case of finite AFs. The following table collect and combine the results of the previous three subsections. The table has to be interpreted as follows: Consider a certain column σ. Then, the entries "×" in rows $r_1,...,r_n$ indicate that for any extension-set $\mathbb{S}$, $\mathbb{S} \in \Sigma^f_{\mathcal{E}_\sigma} \Leftrightarrow r_1, ..., r_n$. Moreover, an entry "→" in row r reflects the fact that the collection of the properties $r_1, ..., r_n$ imply property r.

	cf	na	stb	stg	ad	pr	ss	gr	il	eg
$\mathbb{S} \neq \emptyset$	×	×		×	→	×	×	→	→	→
$\emptyset \in \mathbb{S}$	→				×					
$|\mathbb{S}| = 1$								×	×	×
$dcl(\mathbb{S})$ is tight		×						→	→	→
$\mathbb{S}$ is incomparable		×	×	×		×	×	→	→	→
$\mathbb{S}$ is tight	×	→	×	×				→	→	→
$\mathbb{S}$ is conflict-sensitive	→	→	→	→	×	×	×	→	→	→
$dcl(\mathbb{S}) = \mathbb{S}$	×									

Table 2: Characterizing Properties for Realizable Extension-sets

Remember that the decision whether a certain extension-set $\mathbb{S}$ is realizable can not be done via brute force (i.e., enumerating AFs and checking whether their extensions coincide with $\mathbb{S}$) since there are no a priori bounds on the number of required arguments. Consequently, the results depicted in Table 2 put us in a very good position since now, the question of realizability can be decided locally, i.e. by inspecting the set in question itself. Moreover, all mentioned properties can checked in polynomial time w.r.t. the size of the extensions as shown in [Dunne et al., 2015, Theorem 6]. For the majority of the properties tractability is immediately apparent. The only exception is tightness of the downward-closure of a given extension-set $\mathbb{S}$ since its size is not polynomially bounded in the size of $\mathbb{S}$ (cf. [Dunne et al., 2015, Proposition 12] for a way out of this problem).

By inspecting the respective properties as depicted in Table 2, we can immediately put the signatures of different semantics in relation to each other. The following theorem includes the signature w.r.t. complete semantics in addition. The reason why we did not included complete semantics in our considerations is simply that a precise characterization of the complete signature is still an open problem. Nevertheless, certain necessary properties are already found [Dunne et al., 2015, Proposition 4] justifying items 3 and 4 of the following theorem.

Theorem 3.34 ([Dunne et al., 2015]) *The following relations hold*

1. $\Sigma^f_{\mathcal{E}_{na}} \subset \Sigma^f_{\mathcal{E}_{stg}} \subset \Sigma^f_{\mathcal{E}_{ss}} = \Sigma^f_{\mathcal{E}_{pr}}$,

2. $\Sigma^f_{\mathcal{E}_{stb}} = \Sigma^f_{\mathcal{E}_{stg}} \cup \{\emptyset\}$,

3. $\Sigma^f_{\mathcal{E}_{cf}} \subset \Sigma^f_{\mathcal{E}_{ad}} \subset \Sigma^f_{\mathcal{E}_{co}}$,

4. $\Sigma^f_{\mathcal{E}_\sigma} \subset \Sigma^f_{\mathcal{E}_\tau}$ *where* $\sigma \in \{gr, il, eg\}$, $\tau \in \{na, stb, stg, pr, ss, co\}$ *and*

5. $\left\{ \mathbb{S} \cup \{\emptyset\} \mid \mathbb{S} \in \Sigma^f_{\mathcal{E}_{pr}} \right\} \subset \Sigma^f_{\mathcal{E}_{ad}}$.

The following Venn-diagram provides a compact overview of subset relations between the considered signatures. A bordered area represents a set of extension-sets. The outer ellipse $\mathcal{ES} = \{ \mathbb{S} \subseteq 2^{\mathcal{U}} \mid \mathbb{S}$ is an ext.-set$\}$ stands for the set of all extension-sets over $\mathcal{U}$. Clearly, all other signatures are subsets of $\mathcal{ES}$ by definition. Furthermore, we use $\{\{\emptyset\}\}$ or $\{\emptyset\}$ the set consisting of the single extension-set $\{\emptyset\}$ (realizable by all considered semantics) or the set containing the empty extension-set (realizable by stable semantics only), respectively. The right side of Figure 2 shows signatures of semantics providing only incomparable extension-sets. The intersection of these signatures with $\Sigma^f_{\mathcal{E}_{co}}$ exactly coincides with $\Sigma^f_{\mathcal{E}_{gr}}$ as well as $\Sigma^f_{\mathcal{E}_{il}}$ and $\Sigma^f_{\mathcal{E}_{eg}}$ which contain all extension-sets $\mathbb{S}$ with $|\mathbb{S}| = 1$. Moreover, the only extension-set they have in common with the signatures of conflict-free and admissible sets is the extension-set containing the empty extension. This fact causes the "missing" intersection in the middle of Figure 2.

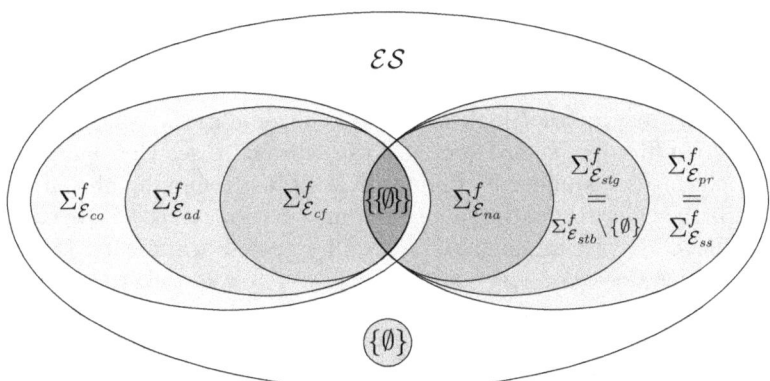

Figure 2: Subset Relations between Finite Signatures

Finally, we want to mention that all considered finite signatures, apart from the complete signature, are closed under non-empty intersections. More precisely, if two finitely σ-realizable sets $\mathbb{S}$ and $\mathbb{T}$ possess a non-empty intersection, then $\mathbb{S} \cap \mathbb{T}$ is finitely σ-realizable too. This feature is mainly due to the fact that subsets of incomparable and tight as well as incomparable and conflict-sensitive sets maintain these properties (cf. Lemmas 3.10 and 3.23).

Theorem 3.35 ([Dunne et al., 2015]) *Let $\sigma \in \{cf, ad, na, stb, stg, pr, ss\}$. For any two finite AFs F_1, F_2 exists an finite AF F, s.t. $\mathcal{E}_\sigma(F) = \mathcal{E}_\sigma(F_1) \cap \mathcal{E}_\sigma(F_2)$ given that $\mathcal{E}_\sigma(F_1) \cap \mathcal{E}_\sigma(F_2) \neq \emptyset$.*[10]

3.3 Signatures w.r.t. Finite, Compact AFs

So far we considered realizibility without any restriction (apart from finiteness) for witnessing AFs. This means, realizing AFs may contain *rejected* arguments, i.e. arguments which do not appear in any extension. Rejected arguments are natural ingredients in typical argumentation scenarios and it is a priori completely unclear in which ways rejected arguments contribute to the expressibility of a particular semantics. In order to have a handle for analyzing the effect of rejected arguments, the class of *compact* AFs and its induced signatures were introduced and studied [Baumann et al., 2014a; Baumann et al., 2014b; Baumann et al., 2016a]. An AF is compact with respect to a semantics σ, if it does not contain rejected arguments, i.e. each of its arguments appears in at least one σ-extension. Now, the main question is whether it is possible to get rid of rejected arguments without changing the outcome? or, in other words: Under which circumstances can AFs be transformed into equivalent compact ones? Note that studying compactness is far from being an academic exercise since there is a fundamental computational significance: When searching for extensions, arguments span the search space, since extensions are to be found among the subsets of the set of all arguments. Hence the more arguments, the larger the search space. Compact AFs are argument-minimal since none of the arguments can be removed without changing the outcome, thus leading to a minimal search space.

Let us first have a brief look on the naive semantics, which is defined as $\subseteq$-maximal conflict-free sets: Here, it is rather easy to see that any AF can be transformed into an equivalent compact AF by just removing all self-defeating arguments. In other words, the same outcome (in terms of the naive extensions) can be achieved by a simplified AF without rejected arguments. This means, naive semantics does not lose expressive power if we stick to compact AFs. However, it is not hard to find semantics where this coincidence does not hold implying that for such semantics the full range of expressiveness indeed relies on the concepts of rejected arguments. Consider therefore the following non-compact AF F.

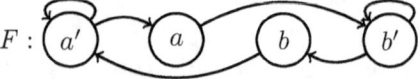

---

[10]The prerequisite of a non-empty intersection can be dropped in case of stable semantics.

Let us consider admissible sets. We obtain $\mathbb{S} = \mathcal{E}_{ad}(F) = \{\emptyset, \{a,b\}\}$. Obviously, any attempt of realizing $\mathbb{S}$ with a compact AF $G = (\{a,b\}, R)$ is doomed to failure since if $\{a,b\}$ is admissible in G we necessarily obtain the admissibility of $\{a\}$ as well as $\{b\}$ proving $\mathbb{S} \neq \mathcal{E}_{ad}(G)$. It was one main result in [Baumann et al., 2014a] to show that the finite, compact signatures w.r.t. stable, preferred, semi-stable, and stage semantics are strict subsets of their corresponding finite signatures. This means, in case of those semantics, sticking to finite, compact AFs implies a loss of expressive power.

Central Definitions and Preliminary Observations

In the following we formally introduce the central notions of *compact argumentation frameworks*, *compact realizibility* as well as *compact signatures*. As already stated, the main idea behind compact AFs is the absence of rejected arguments. For labelling-based semantics σ (i.e., a semantics returning n-tuples) we assume that the first component of their associated σ-labellings are interpreted as *acceptable sets of arguments* in analogy to σ-extensions in case of extension-based semantics. This means, if a certain argument occur in no first component of given σ-labellings we classify it as *rejected*. For a given labelling L we use L^{I} to refer to its first component.

Definition 3.36 *Given a semantics $\sigma : \mathcal{F} \to 2^{\left(2^{\mathcal{U}}\right)^n}$. An AF $F = (A, R)$ is compact for σ (or simply, σ-compact) if $\mathrm{Args}_{\mathcal{E}_\sigma(F)} = A$ (in case of $n = 1$) or $\mathrm{Args}_{\{L^{\mathrm{I}} | L \in \mathcal{L}_\sigma(F)\}} = A$ (for $n \geq 2$), respectively.*

Although extension-based and labelling-based semantics are formally different semantics (according to Definition 2.2) we often speak of the extension-based version or labelling-based version of a certain semantics. This can be formally justified for the considered semantics since there is a close relationship between both versions (cf. Section 2 in Chapter 4 for more explanatory comments as well as Facts 4.38 and 4.39 for some formal relations). The following fact shows that for all considered semantics σ there is no need to distinguish between σ-compactness w.r.t. the extension-based version of σ and σ-compactness w.r.t. the labelling-based version of σ. As an aside, such a coincidence does not require a one-to-one correspondence between the extension-based and labelling-based version of a semantics σ. It suffices that any σ-extension induces a σ-labelling and vice versa in such a way that accepted arguments are preserved (cf. statements 1 and 2 of Fact 4.38).

Fact 3.37 *For any $\sigma \in \{stb, ss, stg, cf2, stg2, pr, ad, co, gr, il, eg, na, cf\}$ and any[11] AF F we have: F is compact for $\mathcal{E}_\sigma$ iff F is compact for $\mathcal{L}_\sigma$.*

In the following we use CAF_σ for AFs compact for σ. Moreover, we use CAF_σ^f to indicate that the considered frameworks are finite in addition. It is intuitively clear that there are AFs F being σ-compact without being τ-compact for two different semantics σ and τ. The following example firstly

---

[11] Indeed, no finiteness restriction is required here.

presented in [Baumann et al., 2014a, Figure 1] provides us with a witnessing framework.

Example 3.38 *Consider the following AF F.*[12]

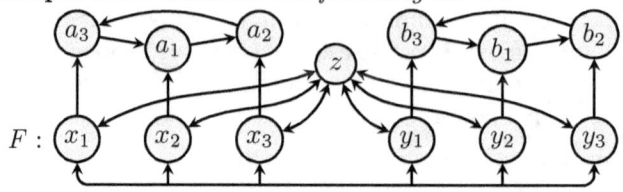

The preferred extensions of F are $\mathcal{E}_{pr}(F) = \{\{z\}, \{x_1, a_1\}, \{x_2, a_2\}, \{x_3, a_3\}, \{y_1, b_1\}, \{y_2, b_2\}, \{y_3, b_3\}\}$, *meaning that F is pr-compact ($F \in CAF_{pr}^f$) since each argument occurs in at least one preferred extension. On the other hand observe that* $\mathcal{E}_{ss}(F) = \mathcal{E}_{pr}(F) \setminus \{\{z\}\}$ *and* $\mathcal{E}_{stg}(F) = \{\{x_i, a_i, b_j\}, \{y_i, b_i, a_j\} \mid 1 \leq i, j \leq 3\}$, *i.e. z is not contained in any semi-stable or stage extension. Therefore F is neither compact for semi-stable nor compact for stage semantics (i.e.* $F \notin CAF_{ss}^f$ *and* $F \notin CAF_{stg}^f$).

How are the different sets of compact AFs related? We start with an easy observation.

Lemma 3.39 ([Baumann et al., 2016a]) *For any two semantics σ and τ such that for each AF F, for every $S \in \mathcal{E}_\sigma(F)$ there is some $S' \in \mathcal{E}_\tau(F)$ with $S \subseteq S'$, we have $CAF_\sigma \subseteq CAF_\tau$.*

Note that $\sigma \subseteq \tau$ is a special case of the premise of Lemma 3.39. Thus, $CAF_\sigma \subseteq CAF_\tau$, whenever $\sigma \subseteq \tau$ (see Figure 1 for an overview). Strict subset relations have to be proven by providing a witnessing AF as presented in Example 3.38. Moreover, $CAF_{pr} = CAF_{co} = CAF_{ad}$ as well as $CAF_{na} = CAF_{cf}$ is justified by Lemma 2.14 and the fact that $pr \subseteq co \subseteq ad$ and $na \subseteq cf$. Finally, in case of the uniquely defined grounded and ideal semantics we have, $F = (A, R)$ is compact if and only if $R = \emptyset$. This in turn implies that F is compact for stable semantics. This means, $CAF_{gr} = CAF_{il} \subset CAF_{stb}$. Remember that eager semantics is uniquely defined w.r.t. finitary AFs only (Theorem 2.25, Example 2.9). Consequently, we may conclude $CAF_{gr}^f = CAF_{eg}^f$ only. Although, the majority of the results do not require the finiteness restriction we present the following theorem in terms of finite AFs. Detailed proofs for the relations between stable, semi-stable, preferred, stage and naive semantics can be found in [Baumann et al., 2016a, Theorem 2].

Theorem 3.40 *The following relations hold:*

1. $CAF_{gr}^f = CAF_{il}^f = CAF_{eg}^f$,

---

[12] The construct in the lower part of the figure represents symmetric attacks between each pair of distinct arguments. We will make use of this style in illustrations throughout the whole section.

2. $CAF^f_{pr} = CAF^f_{co} = CAF^f_{ad}$,

3. $CAF^f_{na} = CAF^f_{cf}$,

4. $CAF^f_{gr} \subset CAF^f_{stb} \subset CAF^f_{ss} \subset CAF^f_{pr} \subset CAF^f_{na}$,

5. $CAF^f_{stb} \subset CAF^f_{stg} \subset CAF^f_{na}$ and

6. $CAF^f_{stg} \not\subseteq CAF^f_{\sigma}$ as well as $CAF^f_{\sigma} \not\subseteq CAF^f_{stg}$ for any $\sigma \in \{pr, ss\}$.

The following figure concisely summarizes all relations mentioned in the theorem above. Directed arrows between two boxes have to be interpreted as strict subset relations between the mentioned sets of compact AFs in these boxes.

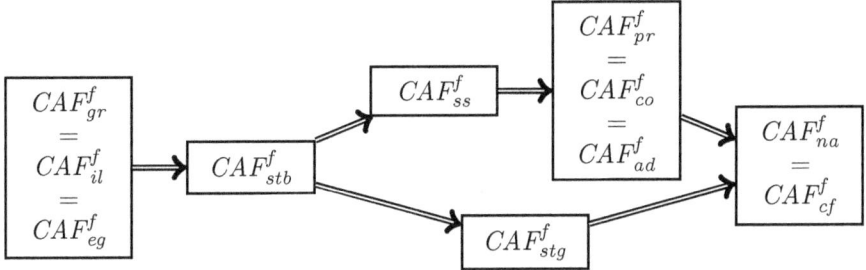

Figure 3: Subset Relations between Finite, Compact AFs

Instantiating Definitions 3.1 and 3.2 with $\mathcal{C} = CAF^f_{\sigma}$ formalize the notions of realizability as well as signatures relativised to finite, compact AFs. Consider the following definitions.

Definition 3.41 *Given a semantics* $\sigma : \mathcal{F} \to 2^{(2^{\mathcal{U}})^n}$. *A set* $\mathbb{S} \subseteq (2^{\mathcal{U}})^n$ *is finitely, compactly σ-realizable if there is an AF* $F \in CAF^f_{\sigma}$, *s.t.* $\sigma(F) = \mathbb{S}$.

Definition 3.42 *Given a semantics* σ. *The finite, compact σ-signature is defined as* $\{\sigma(F) \mid F \in CAF^f_{\sigma}\}$ *abbreviated by* $\Sigma^{f,c}_{\sigma}$.

It is clear that $\Sigma^{c,f}_{\sigma} \subseteq \Sigma^{f}_{\sigma}$ holds for any semantics σ, i.e. finite, compact realizability implies finite realizability. In the following we shed light on the question whether the mentioned subset relation is strict for a given semantics? In other words, we answer the question whether we indeed lose expressive power if sticking to compact AFs.

The Loss or Stability of Expressive Power

Let us consider the uniquely defined grounded, ideal and eager semantics first. We already stated that a set $\mathbb{S}$ is realizable w.r.t. these semantics if and if only if $\mathbb{S}$ is an one-element extension-set if considering finite AFs (Theorem 3.33).

Furthermore, it is immediate that an extension-set $\mathbb{S} = \{E\}$ can be compactly realized via $F_E = (E, \emptyset)$. This means, these semantics do not lose expressive power if we restrict ourselves to compact AFs. Furthermore, the attentive reader may have noticed that the canonical argumentation framework $F_\mathbb{S}^{cf}$, which was used as a witnessing framework for conflict-free sets and naive semantics (cf. Definition 3.13 as well as Propositions 3.15 and 3.16), does not involve further artificial arguments. Thus, it verifies finite, compact realizibility and shows that there is no expressive loss in case of conflict-free sets and naive semantics. For the other considered semantics, namely admissible, stable, stage, semi-stable, preferred as well as complete semantics we have to accept a strict weaker expressibility if we stick to compact AFs. In order to prove that in case of these semantics the full range of expressiveness indeed relies on the concept of rejected arguments we have to come up with witnessing extension-sets. Consider therefore the following example.

Example 3.43 *The extension-set $\mathbb{S} = \{\{a,b\}, \{a,d,e\}, \{b,c,e\}\}$ is realizable under preferred as well as semi-stable semantics (cf. Example 3.20 for a realizing non-compact framework). Let $\sigma \in \{pr, ss\}$. Now suppose there exists an AF $F = (\{a,b,c,d,e\}, R)$, s.t. $\mathcal{E}_\sigma(F) = \mathbb{S}$. Since $\{a,d,e\}, \{b,c,e\} \in \mathbb{S}$ and $\sigma \subseteq cf$ we conclude that there is no attack in R involving e, i.e. e is an isolated argument in F. But then, e is contained in each σ-extension of F contradicting $\{a,b\} \in \mathbb{S}$. In Summary, $\mathbb{S} \in \Sigma_{\mathcal{E}_\sigma}^{f} \setminus \Sigma_{\mathcal{E}_\sigma}^{f,c}$.*

For further witnessing extension-sets we refer the reader to [Baumann et al., 2016a, Propositions 35 and 57] and proceed with the main theorem.

Theorem 3.44 *It holds that*

1. $\Sigma_{\mathcal{E}_\sigma}^{f,c} = \Sigma_{\mathcal{E}_\sigma}^{f}$ *for $\sigma \in \{cf, na, gr, il, eg\}$, and*

2. $\Sigma_{\mathcal{E}_\sigma}^{f,c} \subset \Sigma_{\mathcal{E}_\sigma}^{f}$ *for $\sigma \in \{ad, stb, stg, ss, pr, co\}$.*

In both cases we may benefit of characterization theorems for finite signatures (cf. Theorems 3.12, 3.24 and 3.33). If both signatures are identical (first item), then necessary and sufficient properties for being finitely σ-realizable immediately carry over to finite, compact σ-realizibility. If we observe a strict subset relation (second item), then we obtain at least necessary properties for being in the finite, compact σ-signature.

Theorem 3.45 *Given a set $\mathbb{S} \subseteq 2^\mathcal{U}$, then*

1. $\mathbb{S} \in \Sigma_{\mathcal{E}_{cf}}^{f,c} \Leftrightarrow \mathbb{S}$ *is a non-empty, downward-closed and tight ext.-set,*

2. $\mathbb{S} \in \Sigma_{\mathcal{E}_{na}}^{f,c} \Leftrightarrow \mathbb{S}$ *is a non-empty, incomparable ext.-set and $dcl(\mathbb{S})$ is tight,*

3. $\mathbb{S} \in \Sigma_{\mathcal{E}_{gr}}^{f,c} \Leftrightarrow \mathbb{S}$ *is an ext.-set with $|\mathbb{S}| = 1$,*

4. $\mathbb{S} \in \Sigma^{f,c}_{\mathcal{E}_{il}} \Leftrightarrow \mathbb{S}$ *is an ext.-set with* $|\mathbb{S}| = 1$,

5. $\mathbb{S} \in \Sigma^{f,c}_{\mathcal{E}_{eg}} \Leftrightarrow \mathbb{S}$ *is an ext.-set with* $|\mathbb{S}| = 1$ *and*

6. $\mathbb{S} \in \Sigma^{f,c}_{\mathcal{E}_{stb}} \Rightarrow \mathbb{S}$ *is an incomparable and tight ext.-set,*

7. $\mathbb{S} \in \Sigma^{f,c}_{\mathcal{E}_{stg}} \Rightarrow \mathbb{S}$ *is a non-empty, incomparable and tight ext.-set,*

8. $\mathbb{S} \in \Sigma^{f,c}_{\mathcal{E}_{ad}} \Rightarrow \mathbb{S}$ *is a conflict-sensitive ext.-set containing* $\emptyset$,

9. $\mathbb{S} \in \Sigma^{f,c}_{\mathcal{E}_{pr}} \Rightarrow \mathbb{S}$ *is a non-empty, incomparable and conflict-sensitive ext.-set,*

10. $\mathbb{S} \in \Sigma^{f,c}_{\mathcal{E}_{ss}} \Rightarrow \mathbb{S}$ *is a non-empty, incomparable and conflict-sensitive ext.-set.*

Comparing Finite, Compact Signatures and Final Remarks

In the following we relate the finite, compact signatures of the semantics under consideration to each other. Recall that for finite signatures it holds that $\Sigma^{f}_{\mathcal{E}_{na}} \subset \Sigma^{f}_{\mathcal{E}_{stg}} = \left(\Sigma^{f}_{\mathcal{E}_{stb}} \setminus \{\emptyset\}\right) \subset \Sigma^{f}_{\mathcal{E}_{ss}} = \Sigma^{f}_{\mathcal{E}_{pr}}$ (cf. Figure 2). This picture changes dramatically when considering the relationships between finite, compact signatures as depicted in Figure 4 (incomparable semantics only) and formally stated in Theorem 3.46. The dashed areas represent particular intersections for which the question of existence of extension-sets is still an open question.

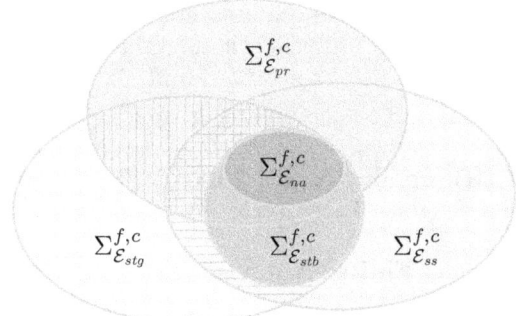

Figure 4: Subset Relations between Finite, Compact Signatures

We proceed with an enumeration of relationships between finite, compact signature including further semantics like conflict-free and admissible sets as well as grounded, ideal, eager and complete semantics. For formal proofs we refer the interested reader to [Baumann et al., 2016a, Theorem 36, Proposition 58].

Theorem 3.46 *The following relations hold:*

1. $\Sigma^{f,c}_{\mathcal{E}_\sigma} \subset \Sigma^{f,c}_{\mathcal{E}_{na}} \subset \Sigma^{f,c}_{\mathcal{E}_\tau}$ *for* $\sigma \in \{gr, il, eg\}$ *and* $\tau \in \{stb, stg, ss, pr\}$,

2. $\Sigma^{f,c}_{\mathcal{E}_{stb}} \subset \Sigma^{f,c}_{\mathcal{E}_{\sigma}}$ for $\sigma \in \{stg, ss\}$,

3. $\Sigma^{f,c}_{\mathcal{E}_{cf}} \subset \Sigma^{f,c}_{\mathcal{E}_{ad}}$,

4. $\Sigma^{f,c}_{\mathcal{E}_{co}} \setminus \Sigma^{f,c}_{\mathcal{E}_{\sigma}} \neq \emptyset$ and $\Sigma^{f,c}_{\mathcal{E}_{\sigma}} \setminus \Sigma^{f,c}_{\mathcal{E}_{co}} \neq \emptyset$ for $\sigma \in \{cf, ad\}$,

5. $\Sigma^{f,c}_{\mathcal{E}_{pr}} \setminus (\Sigma^{f,c}_{\mathcal{E}_{stb}} \cup \Sigma^{f,c}_{\mathcal{E}_{ss}} \cup \Sigma^{f,c}_{\mathcal{E}_{stg}}) \neq \emptyset$,

6. $\Sigma^{f,c}_{\mathcal{E}_{stg}} \setminus (\Sigma^{f,c}_{\mathcal{E}_{stb}} \cup \Sigma^{f,c}_{\mathcal{E}_{pr}} \cup \Sigma^{f,c}_{\mathcal{E}_{ss}}) \neq \emptyset$,

7. $\Sigma^{f,c}_{\mathcal{E}_{stb}} \setminus \Sigma^{f,c}_{\mathcal{E}_{pr}} \neq \emptyset$,

8. $(\Sigma^{f,c}_{\mathcal{E}_{pr}} \cap \Sigma^{f,c}_{\mathcal{E}_{ss}}) \setminus (\Sigma^{f,c}_{\mathcal{E}_{stb}} \cup \Sigma^{f,c}_{\mathcal{E}_{stg}}) \neq \emptyset$ and

9. $\Sigma^{f,c}_{\mathcal{E}_{ss}} \setminus (\Sigma^{f,c}_{\mathcal{E}_{stb}} \cup \Sigma^{f,c}_{\mathcal{E}_{pr}} \cup \Sigma^{f,c}_{\mathcal{E}_{stg}}) \neq \emptyset$.

Comparing the results on expressiveness of the considered semantics as stated in Theorems 3.34 and 3.46 we observe notable differences. When allowing rejected arguments, preferred and semi-stable semantics are equally expressive and at the same time strictly more expressive than stable and stage semantics. As we have seen, this property does not carry over to the compact setting (with the exceptions $\Sigma^{f,c}_{\mathcal{E}_{stb}} \subset \Sigma^{f,c}_{\mathcal{E}_{ss}}$ and $\Sigma^{f,c}_{\mathcal{E}_{stb}} \subset \Sigma^{f,c}_{\mathcal{E}_{stg}}$) where signatures become incomparable.

Finally, regarding the open issues represented as dashed areas in Figure 4. More precisely, it is an open problem whether there are extension-sets lying in the intersection between $\Sigma^{f,c}_{\mathcal{E}_{pr}}$ (resp. $\Sigma^{f,c}_{\mathcal{E}_{ss}}$) and $\Sigma^{f,c}_{\mathcal{E}_{stg}}$ but outside of $\Sigma^{f,c}_{\mathcal{E}_{stb}}$. In [Baumann et al., 2016a] it is conjectured that such extension-sets do not exist.

Conjecture 3.47 ([Baumann et al., 2016a]) It holds that $\Sigma^{f,c}_{\mathcal{E}_{pr}} \cap \Sigma^{f,c}_{\mathcal{E}_{stg}} \subset \Sigma^{f,c}_{\mathcal{E}_{stb}}$ and $\Sigma^{f,c}_{\mathcal{E}_{ss}} \cap \Sigma^{f,c}_{\mathcal{E}_{stg}} = \Sigma^{f,c}_{\mathcal{E}_{stb}}$.

3.4 Signatures w.r.t. Finite, Analytic AFs

We now turn to a further phenomenon, so-called *implicit conflicts* which can be frequently observed in typical argumentation scenarios. Consider therefore the following AF F.

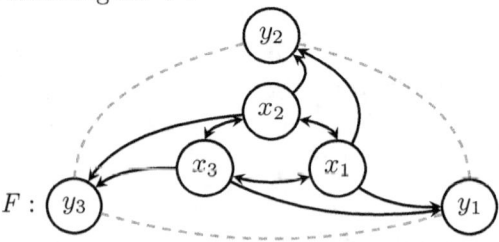

Let us consider stable semantics. Please note that any x_i is jointly acceptable with one specific y_j. More precisely, $\mathcal{E}_{stb}(F) = \{\{x_1, y_3\}, \{x_2, y_1\}, \{x_3, y_2\}\}$

implying that we do not have any rejected arguments, i.e. F is stable compact. What can be said about the two pairs of arguments x_1 and x_2 as well as y_1 and y_2? First of all, both pairs represent a semantical conflict in F since neither of those pairs occur together in any stable extension. In case of x_1 and x_2, the conflict is even a syntactical one since both arguments attack each other in contrast to the pair consisting of y_1 and y_2. This difference leads to the distinction between syntactically underlined *explicit conflicts* and syntactically unfounded *implicit* ones (indicated by dashed lines). In order to understand how implicit conflicts contribute to the expressiveness of a certain semantics, the set of *analytic* AFs and its induced signatures were introduced and studied [Linsbichler et al., 2015; Baumann et al., 2016a]. An analytic framework, i.e. a framework which is free of implicit conflicts maximizes the information on conflicts. One main question is: under which circumstances an arbitrary framework can be transformed into an equivalent analytic one? This question is interesting from a theoretical as well as practical point of view. On the one hand, analytic frameworks are natural candidates for normal forms of AFs, and on the other maximizing the number of explicit conflicts might help argumentation systems to evaluate AFs more efficiently.

Let us consider again the extension-set $\mathbb{S} = \{\{x_1, y_3\}, \{x_2, y_1\}, \{x_3, y_2\}\}$ stemming from the AF F depicted above. Replacing the dashed arrows with symmetric attacks in F shows that $\mathbb{S}$ can be analytically realized under stable semantics. Interestingly, this is no coincidence, since it was shown that in case of stable semantics any AF can be transformed into an equivalent analytical one. However, in general it is not that easy to make implicit conflicts explicit since there are frameworks where any suitable transformation requires the use of additional arguments as shown in [Linsbichler et al., 2015].

Central Definitions and Preliminary Observations

In this section we consider the central notions of *analytic argumentation frameworks*, *analytic realizability* as well as *analytic signatures*. In order to define analytic AF we have to differentiate between the concept of an attack (as a syntactical element) and the concept of a conflict (with respect to the evaluation under a given semantics). More precisely, if two arguments cannot be accepted together, i.e. no reasonable position contain them jointly as elements, we say that these arguments are in conflict. If this conflict is syntactically underlined by an attack between them, we call this conflict explicit, otherwise implicit. Now, an analytic framework is an AF which simply does not contain any implicit conflicts. Consider the following definition.

Definition 3.48 *Given a semantics* $\sigma : \mathcal{F} \to 2^{\left(2^\mathcal{U}\right)^n}$, *an AF* $F = (A, R)$ *and two arguments* $a, b \in A$. *We say that*

1. *a and b are in conflict for* σ *if* $(a, b) \notin \text{Pairs}_{\mathcal{E}_\sigma(F)}$ *(in case of $n = 1$) or* $(a, b) \notin \text{Pairs}_{\{L'| L \in \mathcal{L}_\sigma(F)\}}$ *(for $n \geq 2$), respectively,*

2. *the conflict is explicit w.r.t.* σ *if* $(a, b) \in R$ *or* $(b, a) \in R$, *otherwise implicit,*

3. the AF F is analytic for σ (or σ-analytic) if all conflicts are explicit.

Please notice that Definition 3.48 does not require a and b to be different arguments. In particular, an argument that is not contained in any reasonable position is in conflict with itself. This conflict is explicit if the argument is self-attacking and implicit otherwise. Furthermore, for all considered semantics σ we observe there is no need to distinguish between σ-analyticality w.r.t. the extension-based version of σ and σ-analyticality w.r.t. the labelling-based version of σ (similarly as in case of σ-compactness as stated in Fact 3.37). Please note that this coincidence is justified for any semantics σ whenever $\mathcal{E}_\sigma(F) = \{L^{\mathrm{I}} \mid L \in \mathcal{L}_\sigma(F)\}$ is guaranteed.

Fact 3.49 *For any $\sigma \in \{stb, ss, stg, cf2, stg2, pr, ad, co, gr, il, eg, na, cf\}$ and any AF F we have: F is analytic for $\mathcal{E}_\sigma$ iff F is analytic for $\mathcal{L}_\sigma$.*

In the following we denote the set of all σ-analytic AFs as XAF_σ. To indicate that the frameworks under consideration are finite we use XAF_σ^f. We proceed with an example illustrating the new definitions.

Example 3.50 *As a simple example consider the following AF F depicted below.*

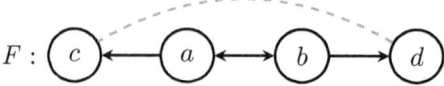

For $\sigma \in \{stb, pr, ss, stg\}$ we have $\mathcal{E}_\sigma(F) = \{\{a,d\}, \{b,c\}\}$. Observe that there is only one implicit conflict, namely the conflict between the arguments c and d, denoted by a dashed line. Hence, F is not σ-analytic, i.e. $F \notin XAF_\sigma^f$. However, since $\mathcal{E}_{na}(F) = \mathcal{E}_\sigma(F) \cup \{\{c,d\}\}$ we have that F is na-analytic, i.e. $F \in XAF_{na}^f$.

As indicated in Example 3.50 the sets of analytic AFs can differ for different semantics. Just like in case of compact AFs (cf. Lemma 3.39) one may easily verify the following lemma which allows to obtain a plenty of subset relations between sets of analytic AFs.

Lemma 3.51 ([Baumann et al., 2016a]) *For any two semantics σ and τ such that for each AF F, for every $S \in \mathcal{E}_\sigma(F)$ there is some $S' \in \mathcal{E}_\tau(F)$ with $S \subseteq S'$, we have $XAF_\sigma \subseteq XAF_\tau$.*

In line with the existing literature we restrict our considerations to finite AFs. Regarding universal (but not uniquely) defined semantics we obtain the same relations as in case of compact AFs (see explanations below Lemma 3.39). In any case we have $XAF_{gr}^f \subseteq XAF_{il}^f \subseteq XAF_{eg}^f$ since ideal semantics accepts more arguments than grounded semantics and eager semantics is even more credulous than ideal semantics. Furthermore, $XAF_{eg}^f \subseteq XAF_{ss}^f$ because the unique eager extension is contained in all semi-stable extension by definition and moreover, semi-stable semantics guarantees reasonable positions in case of finite AFs. Now, let $F = (A, R)$ be analytic w.r.t. eager semantics and $\mathcal{E}_{eg}(F) = \{E\}$. We deduce that all arguments in $A \setminus E$ have to be self-defeating. Consequently, its corresponding (conflict-free) base semantics (cf. Definition 2.17) warrants exactly one extension for F. More precisely, $\mathcal{E}_{ss}(F) = \{E\}$. Finally, due to the self-conflicting arguments and the admissibility of E we obtain $\mathcal{E}_{pr}(F) = \{E\}$ and thus, $\mathcal{E}_{il}(F) = \{E\}$ showing that F is even analytic w.r.t. ideal semantics, i.e. $XAF_{il}^f = XAF_{eg}^f$. The AF $F = (\{a,b\}, \{(a,b), (b,a), (b,b)\})$ proves that a similar result in case of grounded and ideal semantics does not hold. Detailed proofs for the relations between stable, semi-stable, preferred, stage and naive semantics can be found in [Baumann et al., 2016a, Theorem 4].

Theorem 3.52 *The following relations hold:*

1. $XAF_{gr}^f \subset XAF_{il}^f = XAF_{eg}^f \subset XAF_{ss}^f$,

2. $XAF_{pr}^f = XAF_{co}^f = XAF_{ad}^f$,

3. $XAF_{na}^f = XAF_{cf}^f$,

4. $XAF_{stb}^f \subset XAF_{ss}^f \subset XAF_{pr}^f \subset XAF_{na}^f$,

5. $XAF_{stb}^f \subset XAF_{stg}^f \subset XAF_{na}^f$,

6. $XAF_{stg}^f \not\subseteq XAF_{\sigma}^f$ and $XAF_{\sigma}^f \not\subseteq XAF_{stg}^f$ for any $\sigma \in \{pr, ss\}$,

7. $XAF_{\sigma}^f \not\subseteq XAF_{\tau}^f$ and $XAF_{\tau}^f \not\subseteq XAF_{\sigma}^f$ for any $\sigma \in \{gr, il, eg\}$, $\tau \in \{stb, stg\}$.

The following figure summarizes all relation in a compact way. Similarly to Figure 3, a directed arrow between two boxes has to be interpreted as strict subset relation between the mentioned sets of analytic AFs therein.

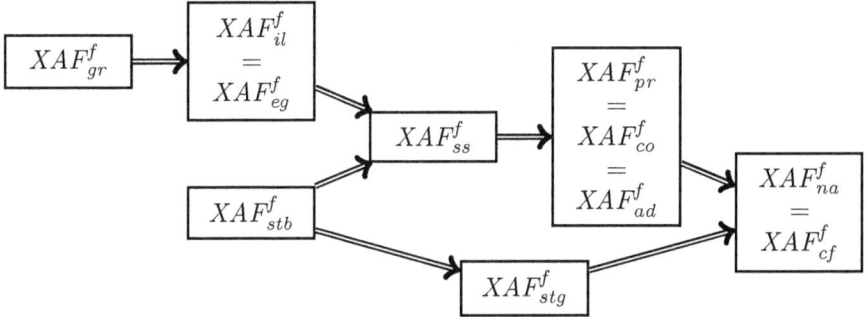

Figure 5: Subset Relations between Finite, Analytic AFs

At this point we want to mention that although Figures 3 and 5 look very similar we have that compactness and analyticality are sufficiently distinct properties. More precisely, apart from the uniquely defined semantics as well as naive semantics and conflict-free sets no subset relations between the sets of compact and analytic frameworks can be stated in general. Sticking to self-loop-free AFs allows one to draw further relations such as analyticality implies compactness for any considered semantics. The main reason for this general relation is that rejected arguments has to be self-defeating in case of analytic frameworks. A selection of proofs of relations listed below can be found in [Baumann et al., 2016a, Proposition 5-8].

Proposition 3.53 *Given an AF F, then*

1. $CAF_\sigma^f \subset XAF_\sigma^f$ for $\sigma \in \{gr, il, eg, na, cf\}$,
2. $CAF_\sigma^f \not\subseteq XAF_\sigma^f$ and $XAF_\sigma^f \not\subseteq CAF_\sigma^f$ for $\sigma \in \{ad, stb, ss, pr, stg, co\}$.

If F is self-loop-free in addition, then

3. $F \in XAF_\sigma^f$ and $F \in CAF_\sigma^f$ for $\sigma \in \{na, cf\}$,
4. $F \in XAF_\sigma^f \Leftrightarrow F \in CAF_\sigma^f$ for $\sigma \in \{gr, il, eg\}$ and
5. $F \in XAF_\sigma^f \Rightarrow F \in CAF_\sigma^f$ for $\sigma \in \{ad, stb, ss, pr, stg, co\}$.

We now precisely formalize the notions of realizibility as well as signatures relativised to finite, analytic AFs. This can be formally done via instantiating Definitions 3.1 and 3.2 with $\mathcal{C} = XAF_\sigma^f$.

Definition 3.54 *Given a semantics $\sigma : \mathcal{F} \to 2^{(2^\mathcal{U})^n}$. A set $\mathbb{S} \subseteq (2^\mathcal{U})^n$ is finitely, analytically σ-realizable if there is an AF $F \in XAF_\sigma^f$, s.t. $\sigma(F) = \mathbb{S}$.*

Definition 3.55 *Given a semantics σ. The finite, analytic σ-signature is defined as $\{\sigma(F) \mid F \in XAF_\sigma^f\}$ abbreviated by $\Sigma_\sigma^{f,x}$.*

The Loss or Stability of Expressive Power

Clearly, every set in the finite, analytic signature of a semantics is also contained in the finite signature. Remember that in case of compact AFs we do not lose any expressive power if considering the uniquely defined grounded, ideal and eager semantics as well as naive semantics and conflict-free sets (Theorem 3.44). These equal expressiveness results carry over to analytic AFs and moreover, even stable and stage semantics may realize the same sets. For instance, consider again the non-analytic AF F as introduced in Example 3.50. One may easily verify that adding an attack from c to d or vice versa yields an AF F' analytic for stable semantics which does not change the set of stable extensions. However, in general it is not that easy to make implicit conflicts explicit but it was shown that the use of additional arguments indeed allows one to turn any finite framework in an analytical one without changing the set of stable or stage extensions, respectively [Baumann et al., 2016a, Proposition

28, Theorem 29]. For the sake of completeness, we mention that it was an open question for a while, known as *Explicit Conflict Conjecture* [Baumann et al., 2014a], whether it is possible, under stable semantics, to translate a given AF into an equivalent analytic one without adding further arguments. In [Baumann et al., 2016a] the conjecture was refuted for stable and even stage semantics. For the remaining semantics, i.e. admissible, semi-stable, preferred and complete semantics the conjecture does not hold either since in case of these semantics we even have that the finite, analytic signature is a strict subset of the corresponding finite one. This means, the full range of expressiveness indeed relies on the use of implicit conflicts. Consider the following example firstly presented in [Baumann et al., 2016a, Example 6].

Example 3.56 *Take into account the AF $F = (A, R)$ as depicted below.*

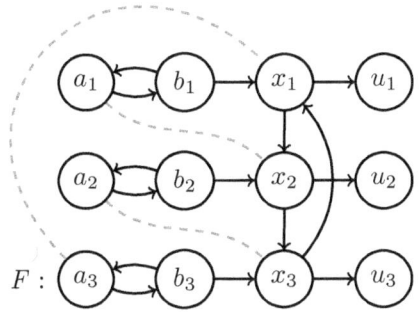

Formally, we have

$A = \{a_i, b_i, x_i, u_i \mid i \in \{1, 2, 3\}\}$ *and*
$R = \{(a_i, b_i), (b_i, a_i), (b_i, x_i), (x_i, u_i) \mid i \in \{1, 2, 3\}\} \cup \{(x_1, x_2), (x_2, x_3), (x_3, x_1)\}$.

Regarding the extension-based version of preferred semantics we obtain the set $\mathbb{S} = \mathcal{E}_{pr}(F) = \{S_a, S_b, A_1, A_2, A_3, B_1, B_2, B_3\}$ *with*

$S_a = \{a_1, a_2, a_3\}$ $\qquad\qquad S_b = \{b_1, b_2, b_3, u_1, u_2, u_3\}$
$A_1 = \{a_2, a_3, b_1, x_2, u_1, u_3\}$ $\qquad B_1 = \{a_1, b_2, b_3, x_1, u_2, u_3\}$
$A_2 = \{a_1, a_3, b_2, x_3, u_1, u_2\}$ $\qquad B_2 = \{a_2, b_1, b_3, x_2, u_1, u_3\}$
$A_3 = \{a_1, a_2, b_3, x_1, u_2, u_3\}$ $\qquad B_3 = \{a_3, b_1, b_2, x_3, u_1, u_2\}$

We observe three implicit conflicts indicated by dashed lines. Consequently, F is not analytic w.r.t. preferred semantics. Moreover, we claim that $\mathbb{S}$ is not analytically pr-realizable at all. For a contradiction we assume that there exists an AF $G \in \text{XAF}_{pr}^f$, s.t. $\mathcal{E}_{pr}(G) = \mathbb{S}$. We now investigate this hypothetical AF G. The main idea is to show that if the conflict between a_1 and x_2 is made explicit, then $\mathbb{S} \neq \mathcal{E}_{pr}(G)$. First, note that G contains at least all arguments in A since $\text{Args}_{\mathbb{S}} = A$. Due to A_3 and B_3 we deduce that $S_a \cup \{u_2\}$ is conflict-free in G. Furthermore, due to A_1, the admissibility of S_a in G and the assumption that all conflicts has to be explicit, we infer that a_1 attacks x_2. Moreover, in consideration of $\mathbb{S}$, it is easy to see that x_2 is the only possible attacker of u_2 among $Args_{\mathbb{S}}$. This implies that S_a defends u_2 against all arguments in $Args_{\mathbb{S}}$. Finally, any additional argument $z \notin Args_{\mathbb{S}}$ in G must be attacked by S_a since G is analytic w.r.t. preferred semantics and S_a must be admissible. This causes $S_a \cup \{u_2\}$ to be admissible in G and hence, S_a cannot be preferred in G. In summary, any AF realizing $\mathbb{S}$ has to be non-analytic for preferred semantics, i.e. $\mathbb{S} \in \Sigma^{f}_{\mathcal{E}_{pr}} \setminus \Sigma^{f,x}_{\mathcal{E}_{pr}}$.

We proceed with the main theorem comparing finite signatures with their corresponding analytical ones.

Theorem 3.57 ([Baumann et al., 2016a]) *It holds that*

1. $\Sigma^{f,x}_{\mathcal{E}_\sigma} = \Sigma^{f}_{\mathcal{E}_\sigma}$ for $\sigma \in \{cf, na, gr, il, eg, stb, stg\}$, and

2. $\Sigma^{f,x}_{\mathcal{E}_\sigma} \subset \Sigma^{f}_{\mathcal{E}_\sigma}$ for $\sigma \in \{ad, ss, pr, co\}$.

In the following we present characterization theorems for finite, analytic signatures or at least necessary properties for being finitely, analytically realizable. All results can be verified via combining the main theorem above as well as the already presented characterization theorems for finite signatures, namely Theorems 3.12, 3.24 and 3.33.

Theorem 3.58 *Given a set* $\mathbb{S} \subseteq 2^{\mathcal{U}}$, *then*

1. $\mathbb{S} \in \Sigma^{f,x}_{\mathcal{E}_{cf}} \Leftrightarrow \mathbb{S}$ is a non-empty, downward-closed and tight ext.-set,

2. $\mathbb{S} \in \Sigma^{f,x}_{\mathcal{E}_{na}} \Leftrightarrow \mathbb{S}$ is a non-empty, incomparable ext.-set and $dcl(\mathbb{S})$ is tight,

3. $\mathbb{S} \in \Sigma^{f,x}_{\mathcal{E}_{gr}} \Leftrightarrow \mathbb{S}$ is an ext.-set with $|\mathbb{S}| = 1$,

4. $\mathbb{S} \in \Sigma^{f,x}_{\mathcal{E}_{il}} \Leftrightarrow \mathbb{S}$ is an ext.-set with $|\mathbb{S}| = 1$,

5. $\mathbb{S} \in \Sigma^{f,x}_{\mathcal{E}_{eg}} \Leftrightarrow \mathbb{S}$ is an ext.-set with $|\mathbb{S}| = 1$,

6. $\mathbb{S} \in \Sigma^{f,x}_{\mathcal{E}_{stb}} \Leftrightarrow \mathbb{S}$ is a incomparable and tight ext.-set,

7. $\mathbb{S} \in \Sigma^{f,x}_{\mathcal{E}_{stg}} \Leftrightarrow \mathbb{S}$ is a non-empty, incomparable and tight ext.-set and

8. $\mathbb{S} \in \Sigma^{f,x}_{\mathcal{E}_{ad}} \Rightarrow \mathbb{S}$ is a conflict-sensitive ext.-set containing $\emptyset$,

9. $\mathbb{S} \in \Sigma^{f,x}_{\mathcal{E}_{pr}} \Rightarrow \mathbb{S}$ is a non-empty, incomparable and conflict-sensitive ext.-set,

10. $\mathbb{S} \in \Sigma^{f,x}_{\mathcal{E}_{ss}} \Rightarrow \mathbb{S}$ is a non-empty, incomparable and conflict-sensitive ext.-set.

Comparing Finite, Analytic Signatures and Final Remarks

So far we have compared finite signatures and finite, analytic signatures for the semantics under consideration. We have seen, for example, that preferred and semi-stable semantics can realize strictly more when allowing the use of implicit conflicts, while this is not the case for stable and stage semantics. In the following we relate the finite, analytic signatures of all considered semantics. Remember that we observed a considerable variety in the relations between incomparable semantics if sticking from finite to finite, compact signatures (cf. Figures 2 and 4). However, in the analytic case we have slight differences only as illustrated in Figure 8 (for incomparable semantics) and formally stated in Theorem 3.60. For instance, preferred and semi-stable signatures do not coincide anymore as shown by the following example taken from [Baumann et al., 2016a, Figure 9, Proof of Theorem 34].

Example 3.59 *Consider the following AF F as depicted below.*

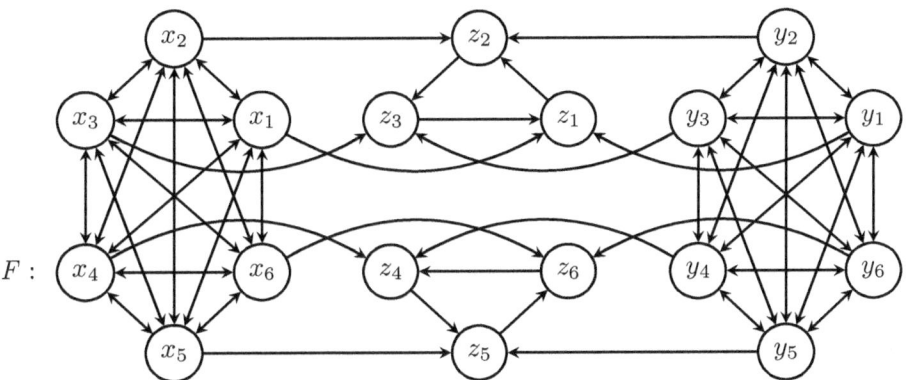

The preferred extension of F can be compactly presented via a cyclic successor functions. More precisely, if $s(1) = 2, s(2) = 3, s(3) = 1$ and $s(4) = 5, s(5) = 6, s(6) = 4$, then $\mathcal{E}_{pr}(F) = \mathbb{S} = \mathbb{S}_0 \cup \mathbb{S}_1 \cup \mathbb{S}_2$ with

$\mathbb{S}_0 = \{\{x_i, y_j, z_{s(i)}, z_{s(j)}\} \mid i \in \{1,2,3\}, j \in \{4,5,6\} \text{ or } i \in \{4,5,6\}, j \in \{1,2,3\}\}$,
$\mathbb{S}_1 = \{\{x_i, y_i, z_{s(i)}\} \mid i \in \{1,2,3,4,5,6\}\}$ *and*
$\mathbb{S}_2 = \{\{x_i, y_{s(i)}, z_{s(s(i))}\}, \{x_{s(i)}, y_i, z_{s(s(i))}\} \mid i \in \{1,2,3,4,5,6\}\}$.

This means, F is pr-analytic and therefore, $\mathbb{S} \in \Sigma_{\mathcal{E}_{pr}}^{f,x}$. We show now that $\mathbb{S} \notin \Sigma_{\mathcal{E}_{ss}}^{f,x}$. Assume that there is some $G = (B, S) \in XAF_{ss}^f$ with $\mathcal{E}_{pr}(G) = \mathbb{S}$. We take a look at $\mathbb{S}_1$ and more specifically $\{x_1, y_1, z_2\} \in \mathbb{S}_1$. Now we need an explicit conflict between x_1 and x_4, but in the selected set only x_1 can possibly defend against this attack, hence $(x_1, x_4) \in S$. The same argument works for x_1 and x_3 as well as z_2 and z_3, meaning that also $(x_1, x_3), (z_2, z_3) \in S$. For symmetry reasons $\{(x_i, x_j), (x_j, x_i), (y_i, y_j), (y_j, y_i) \mid i \in \{1,2,3\}, j \in \{4,5,6\}\} \subseteq S$ and $\{(x_{s(i)}, x_i), (z_i, z_{s(i)}) \mid i \in \{1,2\ldots 6\}\} \subseteq S$.

We take a look at $\mathbb{S}_2$ and more specifically $\{x_1, y_2, z_3\} \in \mathbb{S}_2$. As there should be an explicit conflict between x_1 and x_2 with only x_1 possibly defending this extension against x_2 we need $(x_1, x_2) \in S$. Further as in this set only y_2 and z_3 can possibly attack z_2 we have the set $\{y_2, z_3\}$ attacking z_2. For symmetry reasons $\{(x_i, x_{\mathrm{s}(i)}), (y_i, y_{\mathrm{s}(i)}) \mid i \in \{1, 2 \ldots 6\}\} \subseteq S$ and each set $\{x_i, z_{\mathrm{s}(i)}\}, \{y_i, z_{\mathrm{s}(i)}\}$ for $i \in \{1, 2 \ldots 6\}$ attacks z_i.

Finally we take a look at $\mathbb{S}_0$ and specifically the set $I = \{x_1, y_4, z_2, z_5\} \in \mathbb{S}_0$. Since I necessarily is an admissible extension in an analytic AF we have that I attacks all rejected arguments. By the above observations we now have that I even attacks all arguments not being member of I in G, which means that I is a stable extension and stable semantics and semi-stable semantics thus coincide on G. But then, with $J = \{x_1, y_1, z_2\} \in \mathbb{S}_1$ not being in conflict with for instance z_4 we have that J can not be a stable or semi-stable extension in G concluding $\mathbb{S} \notin \Sigma^{f,x}_{\mathcal{E}_{ss}}$.

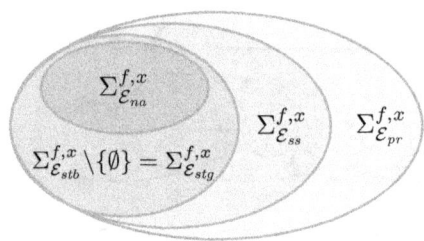

Figure 6: Subset Relations between Finite, Analytic Signatures

Theorem 3.60 ([Dunne et al., 2015]) *The following relations hold:*

1. $\Sigma^{f,x}_{\mathcal{E}_\sigma} \subset \Sigma^{f,x}_{\mathcal{E}_{na}} \subset \Sigma^{f,x}_{\mathcal{E}_{stg}} \subset \Sigma^{f,x}_{\mathcal{E}_{ss}} \subset \Sigma^{f,x}_{\mathcal{E}_{pr}}$ for $\sigma \in \{gr, il, eg\}$,

2. $\Sigma^{f,x}_{\mathcal{E}_{stb}} = \Sigma^{f,x}_{\mathcal{E}_{stg}} \cup \{\emptyset\}$,

3. $\Sigma^{f,x}_{\mathcal{E}_{cf}} \subset \Sigma^{f,x}_{\mathcal{E}_{ad}}$ and

4. $\Sigma^{f,x}_{\mathcal{E}_{co}} \setminus \Sigma^{f,x}_{\mathcal{E}_\sigma} \neq \emptyset$ and $\Sigma^{f,x}_{\mathcal{E}_\sigma} \setminus \Sigma^{f,x}_{\mathcal{E}_{co}} \neq \emptyset$ for $\sigma \in \{cf, ad\}$.

3.5 Remarks on Unrestricted AFs and Intertranslatability

Recently, some first results regarding expressibility w.r.t. unrestricted frameworks were presented in [Baumann and Spanring, 2017]. Remember that the set of unrestricted frameworks, abbreviated by $\mathcal{F}$, contains all AFs $F = (A, R)$, s.t. $A \subseteq \mathcal{U}$ (cf. Section 2.1 for further information). This means, $\mathcal{F}$ contains finite as well as infinite AFs and especially, AFs possessing all available arguments. It is obvious that signatures w.r.t. unrestricted frameworks contain more realizable sets then their finite counterparts since finite AFs may realize finite as well as finitely many extensions only. The following definition formally captures all considered types of signatures (cf. Definitions 3.4, 3.42 and 3.55) without any finite assumption.

Definition 3.61 *Given a semantics σ. We call the set S the*

1. *(unrestricted) σ-signature if $S = \{\sigma(F) \mid F \in \mathcal{F}\}$ abbreviated by Σ_σ,*
2. *compact σ-signature if $S = \{\sigma(F) \mid F \in CAF\}$ abbreviated by Σ_σ^c and*
3. *analytic σ-signature if $S = \{\sigma(F) \mid F \in XAF\}$ abbreviated by Σ_σ^x.*

In [Baumann and Spanring, 2017] the authors were interested in a comparison of the expressive power of several mature semantics in the unrestricted setting. The following result shows that the relation between unrestricted signatures is intimately connected to their relation in case of finite, compact signatures. More precisely, non-empty relative complements in case of finite, compact signatures between two semantics carry over to their unrestricted versions. The main reason for this relation is the fact that unrestricted frameworks may contain any available argument of the universe $\mathcal{U}$.

Theorem 3.62 ([Baumann and Spanring, 2017]) *Given two semantics $\sigma, \tau \in \{na, stb, stg, ss, pr, co, gr, il, eg, cf, ad\}$ we have:*

1. *If $\Sigma_{\mathcal{E}_\sigma}^{f,c} \setminus \Sigma_{\mathcal{E}_\tau}^{f,c} \neq \emptyset$, then $\Sigma_{\mathcal{E}_\sigma}^{c} \setminus \Sigma_{\mathcal{E}_\tau}^{c} \neq \emptyset$ and*
2. *If $\Sigma_{\mathcal{E}_\sigma}^{c} \setminus \Sigma_{\mathcal{E}_\tau}^{c} \neq \emptyset$, then $\Sigma_{\mathcal{E}_\sigma} \setminus \Sigma_{\mathcal{E}_\tau} \neq \emptyset$.*

The following example illustrates the main proof idea.

Example 3.63 *Let $\mathcal{E} \in \Sigma_{\mathcal{E}_{pr}}^{f,c} \setminus \Sigma_{\mathcal{E}_{stb}}^{f,c}$ (cf. Figure 4) and $F = (A, R)$ a witnessing framework. This means, F is finite, $\mathcal{E}_{pr}(F) = \mathcal{E}$ and pr-compact, i.e. $\bigcup \mathcal{E} = A$. Consider now $H = (\mathcal{U}, R)$. Obviously, $\mathcal{E}' = \mathcal{E}_{pr}(H) = \{E \cup (\mathcal{U} \setminus A) : E \in \mathcal{E}\}$ and $\bigcup \mathcal{E}' = \mathcal{U}$ showing the σ-compactness of H. In particular, $\mathcal{E}' \in \Sigma_{\mathcal{E}_{pr}}^c$. Note that any stb-realization of $\mathcal{E}'$ has to be compact too since there are no additional arguments available. Assume $\mathcal{E}' \in \Sigma_{\mathcal{E}_{stb}}^c$, i.e. there is an AF $G' = (\mathcal{U}, R')$, s.t. $\mathcal{E}_{stb}(G') = \mathcal{E}'$. We observe that due to conflict-freeness there can not be attacks in G' between arguments from A and $\mathcal{U} \setminus A$ nor between any of the arguments from $\mathcal{U} \setminus A$. Consequently, $G = (A, R')$ is finite, $\mathcal{E}_{stb}(G) = \mathcal{E}$ and stb-compact implying that $\mathcal{E} \in \Sigma_{\mathcal{E}_{stb}}^{f,c}$ in contradiction to the initial assumption.*

Now we are prepared for a comparison in case of unrestricted frameworks. Ignoring the superscripts in Figure 4 provides you with a graphical representation for selected semantics.

Theorem 3.64 *For unrestricted signatures the following hold:*

1. *$\{\{E\} \mid E \subseteq \mathcal{U}\} = \Sigma_{\mathcal{E}_\sigma} \subset \Sigma_{\mathcal{E}_{na}} \subset \Sigma_{\mathcal{E}_\tau}$ for $\sigma \in \{gr, il\}, \tau \in \{stb, stg, ss, pr\}$,*

2. $\Sigma_{\mathcal{E}_{eg}} \subset \Sigma_{\mathcal{E}_{pr}}$,

3. $\Sigma_{\mathcal{E}_{stb}} \subset \Sigma_{\mathcal{E}_{\sigma}}$ for $\sigma \in \{stg, ss\}$,

4. $\Sigma_{\mathcal{E}_{pr}} \setminus (\Sigma_{\mathcal{E}_{stb}} \cup \Sigma_{\mathcal{E}_{ss}} \cup \Sigma_{\mathcal{E}_{stg}}) \neq \emptyset$,

5. $\Sigma_{\mathcal{E}_{stg}} \setminus (\Sigma_{\mathcal{E}_{stb}} \cup \Sigma_{\mathcal{E}_{pr}} \cup \Sigma_{\mathcal{E}_{ss}}) \neq \emptyset$,

6. $\Sigma_{\mathcal{E}_{stb}} \setminus \Sigma_{\mathcal{E}_{pr}} \neq \emptyset$,

7. $\Sigma_{\mathcal{E}_{ss}} \setminus (\Sigma_{\mathcal{E}_{stb}} \cup \Sigma_{\mathcal{E}_{pr}} \cup \Sigma_{\mathcal{E}_{stg}}) \neq \emptyset$,

8. $\Sigma_{\mathcal{E}_{co}} \setminus \Sigma_{\mathcal{E}_{\sigma}} \neq \emptyset$ and $\Sigma_{\mathcal{E}_{\sigma}} \setminus \Sigma_{\mathcal{E}_{co}} \neq \emptyset$ for $\sigma \in \{cf, ad\}$,

9. $\Sigma_{\mathcal{E}_{cf}} \subset \Sigma_{\mathcal{E}_{ad}}$.

Finally, we briefly consider the closely related topic of *intertranslatability*. Intertranslatability revolves around the idea of mapping one semantics to another. One main motivation for studying this issue is the possibility to reuse a solver for one semantics for another [Dvořák and Woltran, 2011]. The main tool for this endeavour are functions mapping AFs to AFs, so-called *translations* formally defined as follows.

Definition 3.65 *[Dvořák and Woltran, 2011] Given two semantics σ, τ. A function $f : \mathcal{F} \to \mathcal{F}$ is called an exact translation for $\sigma \to \tau$, if $\sigma(F) = \tau(f(F))$ for any AF F. It is called a faithful translation if for any AF F first $|\sigma(F)| = |\tau(f(F))|$ and second $\sigma(F) = \{S \cap A(F) \mid S \in \tau(f(F))\}$.*

Please note that for some semantics there are no exact translations available due to reasons inherent to those semantics. For instance, preferred semantics satisfies *I-maximality*, i.e. for any AF F, $\mathcal{E}_{pr}(F)$ forms a $\subseteq$-antichain [Baroni and Giacomin, 2007]. This implies that an exact translation $\mathcal{E}_{ad} \to \mathcal{E}_{pr}$ can not exist since for $F = (\{a\}, \emptyset)$ we observe $\{\emptyset, \{a\}\} = \mathcal{E}_{ad}(F)$. Sticking to faithful translations provides us with a positive answer if we consider finite AFs only [Spanring, 2012, Translation 3.1.85]. Interestingly, the considered translation does not serve in the general unrestricted case and interestingly, it was shown that a search for a suitable translation will never succeed (cf. [Baumann and Spanring, 2017, Example 6]).

The following theorem (a generalization of the finite version from [Dvořák and Spanring, 2016, Section 6.1]) establishes a close relation between realizability and intertranslatability as promised, namely: if τ is not less expressive than σ, then σ can be exactly translated to τ and vice versa.

Theorem 3.66 ([Baumann and Spanring, 2017]) *Given semantics σ, τ. We have: $\Sigma_\sigma \subseteq \Sigma_\tau$ if and only if there is an exact translation for $\sigma \to \tau$.*

The following Figure 7 illustrates translational (im)possibilities in an eye-catching way. Figure 7b summarizes known results regarding faithful translations in the finite case [Dvořák and Woltran, 2011; Spanring, 2012; Dvořák and

Spanring, 2016], augmented with obvious insights for unique status semantics il and eg. For semantics σ, τ, encirclement in the same component indicates bidirectional translations. An arrow from σ to τ means directional translations. If there is no directed path (for instance for na to cf, or for cf to gr), then there is no translation. Figure 7a features the same visualization for unrestricted AFs. Dropping the finiteness restriction has some further consequences for the considered semantics, namely exact and faithful intertranslatability coincide. It is an open question whether both forms of translations are essentially the same in the general unrestricted setting. In consideration of Theorem 3.66 we may interpret Figure 7a as a comparison of the expressiveness of the considered semantics. That is, $\Sigma_\sigma \subset \Sigma_\tau$ if and only if there is a directed path from σ to τ.

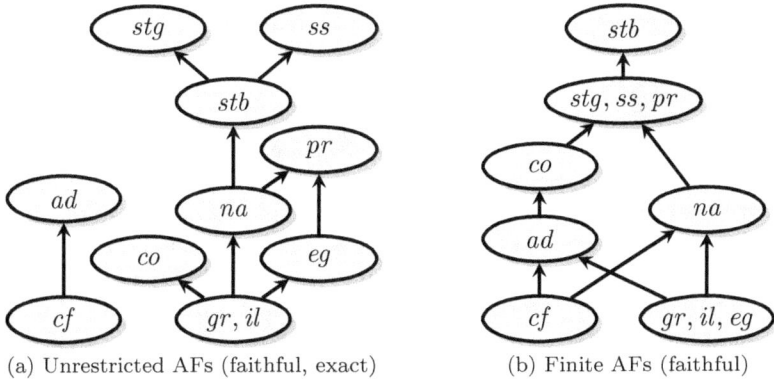

(a) Unrestricted AFs (faithful, exact)   (b) Finite AFs (faithful)

Figure 7: Translational (Im)Possibilities

As a final note, in contrast to the unrestricted setting Baumann and Spanring observed that for slightly restricted AFs $F = (A, R)$, s.t. $|A| \leq |\mathcal{U} \setminus A|$ it is possible to provide exact and efficiently computable translations from preferred to semi-stable semantics via $f(F) = F' = (A', R')$ with $A' = A \cup \{a' \mid a \in A\}$ and $R' = R \cup \{(a, a'), (a', a') \mid a \in A\}$. It is an interesting question whether this restriction allows for similar translational possibilities as in case of finite AFs.

3.6 Realizability and Signatures for Labelling-based Versions

Although any considered semantics σ possesses an extension-based version (indicated by $\mathcal{E}_\sigma$) as well as a closely related 3-valued labelling-based version (indicated by $\mathcal{L}_\sigma$) we formally have that both versions are different semantics (or more precisely, functions) in the sense of Definition 2.2. This formal difference has some impact on realizability as well as signatures. Let us consider realizability in the realm of finiteness. As a matter of fact, for any considered 3-valued labelling-based version $\mathcal{L}_\sigma$ we have: if $F = (A, R)$ and $L = (L^{\mathrm{I}}, L^{\mathrm{O}}, L^{\mathrm{U}}) \in \mathcal{L}_\sigma(F)$, then $A = L^{\mathrm{I}} \cup L^{\mathrm{O}} \cup L^{\mathrm{U}}$. This means, σ-labellings assign a status to any argument in F. Now, in case of finite AFs we know that potentially realizable sets of labellings have to involve finitely many arguments only. Moreover, these finitely many arguments precisely determine the set A of witnessing AFs $F = (A, R)$.[13] Consider therefore the following example.

Example 3.67 *Consider the following set of 3-valued labellings* $\mathbb{S} = \{(\{a\}, \emptyset, \{b, c\}), (\{a, b\}, \{c\}, \emptyset)\}$. *Is* $\mathbb{S}$ *co-realizable? Since* $\{a\} \cup \emptyset \cup \{b, c\} = \{a, b, c\}$ *we deduce that candidates have to be members of* $\mathcal{C} = \{F = (A, R) \mid A = \{a, b, c\}\}$. *Note that* $|\mathcal{C}| = 2^{|\{a,b,c\}|^2} = 2^9 = 512$. *Clearly, this is a huge number, but it is a finite one. Consequently, the question of realizability can be decided by computing the σ-labellings of all AFs in* $\mathcal{C}$. *Of course, any intelligent search algorithm would involve further information like $\{a, b\}$ has to be conflict-free in a witnessing AF. Such an observation would decrease the number of candidates to* $2^5 = 32$. *However, in both cases one would find the unique witnessing framework F, i.e.* $\mathcal{L}_{co}(F) = \mathbb{S}$, *as depicted below.*

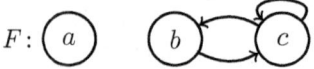

The example above shows that the search space can be very large even in case of small numbers of arguments. Consequently, locally verifiable necessary as well as sufficient properties for realizability just like in case of extension-based semantics are of high interest too. To the best of our knowledge only two papers have dealt with labelling-based realizability in the context of AFs. The first study was presented by Dyrkolbotn [Dyrkolbotn, 2014]. The author showed that, as long as additional arguments are allowed any finite set of labellings is realizable under preferred and semi-stable semantics. It is important to emphasize that Dyrkolbotn uses a more relaxed notion of realizability, namely *realizibility under projection* (cf. Definition 3.72). The other work [Linsbichler et al., 2016] deals with the standard notion of finite realizability (Definition 3.3). The authors presented an algorithm which returns either "No" in case of non-realizibility or a witnessing AF F in the positive case. The algorithm is not purely a guess-and-check method since it also includes a propagation step where certain necessary properties of witnessing AFs are processed. Remarkably, the algorithm is not restricted to the formalism of abstract argumentation frameworks only. In fact, it can also be used to decide realizability in case of the more general abstract dialectical frameworks as well as various of its sub-classes [Brewka and Woltran, 2010; Brewka et al., 2013].

Preliminary Results for Labelling-based Signatures

In the following we shed light on general relations between the labelling-based and extension-based signatures of the considered semantics. Fortunately, due to former characterization results we will even achieve characterizing or at least necessary properties for finite realizability regarding labelling-based versions. We proceed with the definition of an *labelling-set* which is the n-valued analogon (for $n \geq 2$) to an extension-set as introduced in Definition 3.5. A labelling-set

---

[13]This is exactly the point which does not carry over to finite realizability in case of extension-based semantics (cf. statement 2 of Theorem 3.44).

is a finite set of n-tuples which are dealing with the same set of arguments and moreover, any n-tuple assigns exactly one status to each argument in question.

Definition 3.68 *Given* $\mathbb{S} \subseteq (2^\mathcal{U})^n$. *$Args_\mathbb{S}$ denotes $\bigcup_{L=(L_1,\ldots,L_n)\in\mathbb{S}} \bigcup_{i=1}^n L_i$ and $\|\mathbb{S}\|$ stands for $|Args_\mathbb{S}|$. We say that $\mathbb{S}$ is a labelling-set if*

1. *$\|\mathbb{S}\|$ is a finite cardinal,*
2. *for any $L = (L_1,\ldots,L_n) \in \mathbb{S}$, $Args_\mathbb{S} = \bigcup_{i=1}^n L_i$ and*
3. *for any $L = (L_1,\ldots,L_n) \in \mathbb{S}$, $L_1,\ldots,L_n$ are pairwise disjoint.*

The following proposition establishes a connection between extension-based and labelling-based realizibility for any considered semantics. Roughly speaking, it states that labelling-based realizability requires extension-based realizability of the corresponding sets of in-labelled arguments. For a 3-tuple $L = (L_1, L_2, L_3)$ we also write $(L^\mathtt{I}, L^\circ, L^\mathtt{U})$ as usual.

Proposition 3.69 *Given a set of 3-tuples $\mathbb{S} \subseteq (2^\mathcal{U})^3$. For any semantics $\sigma \in \{stb, ss, stg, cf2, stg2, pr, ad, co, gr, il, eg, na, cf\}$ we have,*

1. $\mathbb{S} \in \Sigma_{\mathcal{L}_\sigma} \Rightarrow \{L^\mathtt{I} \mid L \in \mathbb{S}\} \in \Sigma_{\mathcal{E}_\sigma}$ \hfill *(unrestricted realizability)*
2. $\mathbb{S} \in \Sigma_{\mathcal{L}_\sigma}^c \Rightarrow \{L^\mathtt{I} \mid L \in \mathbb{S}\} \in \Sigma_{\mathcal{E}_\sigma}^c$ \hfill *(compact realizability)*
3. $\mathbb{S} \in \Sigma_{\mathcal{L}_\sigma}^x \Rightarrow \{L^\mathtt{I} \mid L \in \mathbb{S}\} \in \Sigma_{\mathcal{E}_\sigma}^x$ \hfill *(analytic realizability)*
4. $\mathbb{S} \in \Sigma_{\mathcal{L}_\sigma}^f \Rightarrow \{L^\mathtt{I} \mid L \in \mathbb{S}\} \in \Sigma_{\mathcal{E}_\sigma}^f$ \hfill *(finite realizability)*
5. $\mathbb{S} \in \Sigma_{\mathcal{L}_\sigma}^{f,c} \Rightarrow \{L^\mathtt{I} \mid L \in \mathbb{S}\} \in \Sigma_{\mathcal{E}_\sigma}^{f,c}$ \hfill *(finite, compact realizability)*
6. $\mathbb{S} \in \Sigma_{\mathcal{L}_\sigma}^{f,x} \Rightarrow \{L^\mathtt{I} \mid L \in \mathbb{S}\} \in \Sigma_{\mathcal{E}_\sigma}^{f,x}$ \hfill *(finite, analytic realizability)*

Please note that the implications above are justified for any semantics σ whenever the different versions of it satisfy $\mathcal{E}_\sigma(F) = \{L^\mathtt{I} \mid L \in \mathcal{L}_\sigma(F)\}$ for any relevant AF F. In the former sections we already presented characterization theorems or at least necessary properties for being finitely realizable regarding extension-based versions (cf. Theorems 3.12, 3.24 and 3.33). Combining these results with the proposition above yields the following necessary properties for finite realizability in the labelling-based case. Note that the mentioned implications apply to finite, compact as well as finite, analytic signatures too since $\Sigma_{\mathcal{L}_\sigma}^{f,c} \subseteq \Sigma_{\mathcal{L}_\sigma}^f$ as well as $\Sigma_{\mathcal{L}_\sigma}^{f,x} \subseteq \Sigma_{\mathcal{L}_\sigma}^f$ by definition. In case of grounded, ideal and eager semantics we have that being an one-element labelling-set is necessary and even sufficient for being finitely realizable. One may easily verify that the only-if-directions of these semantics are justified by the witnessing framework $F_L = (L^\mathtt{I} \cup L^\circ \cup L^\mathtt{U}, \{(i,o) \mid i \in L^\mathtt{I}, o \in L^\circ\} \cup \{(u,u) \mid u \in L^\mathtt{U}\})$ given that $\mathbb{S} = \{L\}$.

Theorem 3.70 *Given a set of 3-tuples* $\mathbb{S} \subseteq (2^{\mathcal{U}})^3$, *then*

1. $\mathbb{S} \in \Sigma^f_{\mathcal{L}_{cf}} \Rightarrow \{L^I \mid L \in \mathbb{S}\}$ *is a non-empty, downward-closed and tight extension-set,*

2. $\mathbb{S} \in \Sigma^f_{\mathcal{L}_{na}} \Rightarrow \{L^I \mid L \in \mathbb{S}\}$ *is a non-empty, incomparable extension-set and* $dcl(\mathbb{S})$ *is tight,*

3. $\mathbb{S} \in \Sigma^f_{\mathcal{L}_{gr}} \Leftrightarrow \mathbb{S}$ *is a labelling-set with* $|\mathbb{S}| = 1$,

4. $\mathbb{S} \in \Sigma^f_{\mathcal{L}_{il}} \Leftrightarrow \mathbb{S}$ *is a labelling-set with* $|\mathbb{S}| = 1$,

5. $\mathbb{S} \in \Sigma^f_{\mathcal{L}_{eg}} \Leftrightarrow \mathbb{S}$ *is a labelling-set with* $|\mathbb{S}| = 1$,

6. $\mathbb{S} \in \Sigma^f_{\mathcal{L}_{stb}} \Rightarrow \{L^I \mid L \in \mathbb{S}\}$ *is a incomparable and tight extension-set,*

7. $\mathbb{S} \in \Sigma^f_{\mathcal{L}_{stg}} \Rightarrow \{L^I \mid L \in \mathbb{S}\}$ *is a non-empty, incomparable and tight extension-set,*

8. $\mathbb{S} \in \Sigma^f_{\mathcal{L}_{ad}} \Rightarrow \{L^I \mid L \in \mathbb{S}\}$ *is a conflict-sensitive ext.-set containing* $\emptyset$,

9. $\mathbb{S} \in \Sigma^f_{\mathcal{L}_{pr}} \Rightarrow \{L^I \mid L \in \mathbb{S}\}$ *is a non-empty, incomparable and conflict-sensitive extension-set,*

10. $\mathbb{S} \in \Sigma^f_{\mathcal{L}_{ss}} \Rightarrow \{L^I \mid L \in \mathbb{S}\}$ *is a non-empty, incomparable and conflict-sensitive extension-set.*

Realizibility under Projection

We turn now to *realizability under projection* which was firstly considered in [Dyrkolbotn, 2014]. In order to realize a set of labellings $\mathbb{S}$ under projection it suffices to come up with an AF F, s.t. its set of labellings restricted to the relevant arguments coincide with $\mathbb{S}$. Consider therefore the following illustrating example.

Example 3.71 *Given* $\mathbb{S} = \{(\{a\}, \{b\}, \emptyset), (\{b\}, \{a\}, \emptyset), (\emptyset, \{a, b\}, \emptyset)\}$. *We note that the corresponding set of sets of in-labelled arguments* $\mathbb{S}^I = \{\emptyset, \{a\}, \{b\}\}$ *violates incomparability. Thus, applying statement 9 of Theorem 3.70 we derive that* $\mathbb{S}$ *is not finitely pr-realizable. Consider now the following AF F.*

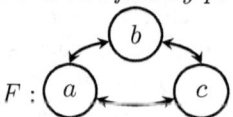

We obtain $\mathcal{L}_{pr}(F) = \{(\{a\}, \{b, c\}, \emptyset), (\{b\}, \{a, c\}, \emptyset), (\{c\}, \{a, b\}, \emptyset)\}$. *Now, if we restrict any labelling* $L = (L^I, L^O, L^U) \in \mathcal{L}_{pr}(F)$ *to the arguments a and b, i.e.* $L|_{\{a,b\}} = (L^I \cap \{a, b\}, L^O \cap \{a, b\}, L^U \cap \{a, b\})$ *we obtain exactly all labellings in* $\mathbb{S}$. *In this sense,* $\mathbb{S}$ *is pr-realizable under projection.*

We proceed with the formal definitions. For the sake of completeness we introduce realizability under projection and its corresponding signatures w.r.t. any kind of semantics as defined in Definition 2.2.

Definition 3.72 *Given a semantics* $\sigma : \mathcal{F} \to 2^{(2^{\mathcal{U}})^n}$. *A set* $\mathbb{S} \subseteq (2^{\mathcal{U}})^n$ *is σ-realizable under projection if there is an AF F, s.t.* $\sigma(F)|_{Args_{\mathbb{S}}} = \{E|_{Args_{\mathbb{S}}} \mid E \in \mathcal{E}_\sigma(F)\} = \mathbb{S}$ *(in case of $n=1$) or* $\sigma(F)|_{Args_{\mathbb{S}}} = \{L|_{Args_{\mathbb{S}}} \mid L \in \mathcal{L}_\sigma(F)\} = \mathbb{S}$ *(for $n \geq 2$), respectively.*

Definition 3.73 *Given a semantics σ. The unrestricted as well as finite σ-projection-signatures are defined as follows:*

1. $\Sigma^p_\sigma = \{\sigma(F)|_B \mid F = (A, R) \in \mathcal{F}, B \subseteq A\}$ *and*

2. $\Sigma^{f,p}_\sigma = \{\sigma(F)|_B \mid F = (A, R) \in \mathcal{F}, F \text{ is finite}, B \subseteq A\}$

Analogously to Proposition 3.69 we state the following relation between labelling-based and extension-based versions of the considered semantics.

Proposition 3.74 *Given a set of 3-tuples $\mathbb{S} \subseteq (2^{\mathcal{U}})^3$. For any semantics $\sigma \in \{stb, ss, stg, cf2, stg2, pr, ad, co, gr, il, eg, na, cf\}$ we have,*

1. $\mathbb{S} \in \Sigma^p_{\mathcal{L}_\sigma} \Rightarrow \{L^I \mid L \in \mathbb{S}\} \in \Sigma^p_{\mathcal{E}_\sigma}$ *(unrestr. realizability under projection)*

2. $\mathbb{S} \in \Sigma^{f,p}_{\mathcal{L}_\sigma} \Rightarrow \{L^I \mid L \in \mathbb{S}\} \in \Sigma^{f,p}_{\mathcal{E}_\sigma}$ *(finite realizability under projection)*

As a matter of fact, any projection signature is a superset of the corresponding signature. The following question then arises naturally: how much more sets can be generated if we stick to realizability under projection? For instance, we have already seen that even comparable sets are realizable under projection by semantics satisfying incomparability (Example 3.71). It was the main result in [Dyrkolbotn, 2014, Theorem 3.1] that in case of semi-stable and preferred semantics indeed any 3-valued labelling-set is finitely realizable under projection. The proof relies on two basic constructions. The first step *generates* an AF, consisting of so-called *circuits*, s.t. its set of preferred as well as semi-stable labellings restricted to the relevant arguments contains any possible labelling. The second construction *eliminates* undesired labellings step by step. Combining this realizability result with statement 2 of Proposition 3.74 yields the following theorem.

Theorem 3.75 *Let $\sigma \in \{pr, ss\}$. We have,*

1. $\Sigma^{f,p}_{\mathcal{L}_\sigma} = \left\{\mathbb{S} \subseteq (2^{\mathcal{U}})^3 \mid \mathbb{S} \text{ is a labelling-set}\right\}$ *and*

2. $\Sigma^{f,p}_{\mathcal{E}_\sigma} = \{\mathbb{S} \subseteq 2^{\mathcal{U}} \mid \mathbb{S} \text{ is an extension-set}\}$.

3.7 Final Remarks and Conclusion

We have dealt with different forms of realizability in the context of abstract argumentation frameworks. In accordance with the existing literature the main part of this section was devoted to finite realizability for extension-based semantics. However, for any semantics σ we may state the following general subset relations depicted as Venn-diagram.

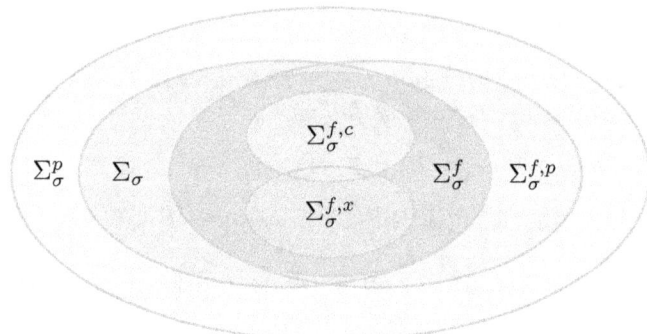

Figure 8: Subset Relations between Different Kinds of Signatures

In case of the extension-based versions of naive, grounded, ideal, eager, stable, stage, preferred and semi-stable semantics as well as conflict-free and admissible sets we provided exact characterizations for their corresponding general signatures. We have seen that for some semantics we do not lose any expressive power if sticking to compact or analytic AFs, i.e. $\Sigma_\sigma^f = \Sigma_\sigma^{f,c}$ or $\Sigma_\sigma^f = \Sigma_\sigma^{f,x}$, respectively. However, for certain prominent semantics, e.g. preferred semantics we have that the expressive power indeed relies on the use of rejected arguments or implicit conflicts. For such semantics, it remains an open problem to present exact characterizations for finite, compact or finite, analytic realizability, respectively. In case of labelling-based versions of semi-stable and preferred semantics we have seen that any labelling-set is realizable under projection. In [Dyrkolbotn, 2014] it was already noted that this equality does not hold for any semantics. For instance, the empty labelling is admissible for any AF F. Hence, in case of admissible semantics, no labelling-set is realizable under projection if it fails to include the empty labelling.

Finally, let us mention some computational issues not considered so far. It can be said that on the one hand, the classes of finite, compact and finite, analytic provide computational benefits both in practice and in terms of theoretical worst-case analysis. On the other hand testing for membership in one of the classes is, for most of the semantics, of rather high complexity and thus these classes cannot be directly used to improve systems. We refer the interested reader to [Baumann et al., 2016a] for more details. Moreover, in general, given an extension-set $\mathbb{S}$, deciding whether $\mathbb{S}$ is compactly realizable is a hard problem, that is, by definition of the decision problem there are no good reasons to believe that we can do any better than guessing a compact AF and checking whether its extension-set coincides with $\mathbb{S}$. Nevertheless, for some semantics we have seen that finite, compact realizability can be characterized locally, i.e. by properties of $\mathbb{S}$ itself (as shown in Theorem 3.45). In this case, finite, compact realizability can be checked in polynomial time as for standard finite realizability [Dunne et al., 2015, Theorem 6]. Moreover, in [Baumann et al., 2016a] a huge number of shortcuts to detect non-compactness are provided. By shortcut we mean a property of the given extension-set $\mathbb{S}$ that is easily computable (preferably in polynomial time) which (sometimes) provides us with a definitive answer to the decision problem. These shortcuts are related to numerical aspects of argumentation frameworks like results concerning maximal number of extensions [Baumann and Strass, 2013].

4 Replaceability

Given a certain logical formalism $\mathcal{L}$ and two syntactically different $\mathcal{L}$-theories T_1 and T_2. One central question is whether, and if so, how to decide that these $\mathcal{L}$-theories represent the same information? Of course, in order to answer this question we have to clarify what we exactly mean by sharing the same information first. Note that there is neither a uniquely determined, nor a certain preferred interpretation by the formalism $\mathcal{L}$ itself. For instance, equating information with possessing the same semantics yields to the well-known notion of *ordinary* or *standard equivalence*. This means, assuming that $\sigma_\mathcal{L}$ is the semantics of $\mathcal{L}$ we might answer that T_1 and T_2 are equivalent if and only if $\sigma_\mathcal{L}(T_1) = \sigma_\mathcal{L}(T_2)$. A more demanding interpretation of sharing the same information is to require that T_1 and T_2 are semantically indistinguishable even if further $\mathcal{L}$-theories T are added to both simultaneously. More formally, we may state: T_1 and T_2 are considered to be equivalent if and only if $\sigma_\mathcal{L}(T_1 \cup T) = \sigma_\mathcal{L}(T_2 \cup T)$ for any theory T. This notion is known as *strong equivalence* and is of high interest for any logical formalism since it allows one to locally replace, and thus give rise for simplification, parts of a given theory without changing the semantics of the latter. In contrast to classical (monotone) logics where standard and strong equivalence coincide (cf. [Baumann and Strass, 2016] for more detailed information on this issue), it is possible to find ordinary but not strongly equivalent objects for any nonmonotonic formalism available in the literature. Consequently, much effort has been devoted to characterizing strong equivalence for nonmonotonic formalisms such as logic programs [Lifschitz et al., 2001], causal theories [Turner, 2004], default logic [Turner, 2001] as well as nonmonotonic logics in general [Truszczynski, 2006].

In [Oikarinen and Woltran, 2011] the authors introduced the notion of strong equivalence for abstract AFs. They provided a series of characterization theorems for deciding strong equivalence of two AFs with respect to several semantics. In view of the fact that strong equivalence is defined semantically it is the main and quite surprisingly insight that being strongly equivalent can be decided syntactically. More precisely, they introduced the notion of a *kernel* of an AF F which is (informally speaking) a subgraph of F where certain attacks are deleted and showed that syntactical identity of suitably chosen kernels characterizes strong equivalence w.r.t. the considered semantics. Strong equivalence is, as its name suggests, a very (and often unnecessarily to) strong notion of equivalence if dynamic evolvements are considered. In many argumentation scenarios the type of modification which may potentially occur can be anticipated and furthermore, more importantly, does not range over *arbitrary expansions* as required for strong equivalence. Let us consider the instantiation-based context where AFs are built from an underlying knowledge base. Here, we typically observe that older arguments and their corresponding attacks survive and only new arguments which may interact with the previous ones arise given that a new piece of information is added to the underlying knowledge base. This type of dynamic evolution is a so-called *normal expansion* and its corresponding equivalence notion were firstly studied in [Baumann, 2012a]. Over the last five years several equivalence notions taking into account specific types of evolvements reflecting the nature of various argumentation scenarios were defined and characterized. The considered dynamic scenarios range from the most general form, so-called *updates* [Baumann, 2014a] where arguments and attacks can be deleted and added to different types of *expansions* [Oikarinen and Woltran, 2011; Baumann, 2012a; Baumann and Brewka, 2010] and *deletions* [Baumann, 2014a] where arguments and/or attacks are allowed to be added or deleted in a certain way only.

Into the year 2015 all characterization theorems were stated in terms of extension-based semantics. Recently, Baumann presents their labelling-based counterparts and showed that, although labelling-based semantics contain more information then there extension-based counterpart, there is a majority of equivalence relations where labelling-based and extension-based versions coincide [Baumann, 2016]. Even more recently, a first consideration of strong equivalence regarding unrestricted frameworks were presented in [Baumann and Spanring, 2017]. It turned out that there are no characterizational differences compared to the finite case as long as the AFs in question are *jointly expandable*, i.e. that the existence of fresh arguments is guaranteed.

Another approach somehow complementary to the ones mentioned before is presented in [Baroni et al., 2014] where sharing the same information is interpreted as possessing the same Input/Output behavior. Roughly speaking, the main idea is to consider an argumentation framework as a kind of black box which receives some input from the external world (i.e, a set of external arguments) via incoming attacks and produces an output to the external world via outgoing attacks. Such an interacting module is called an *argumentation multipole*. Two multipoles connected with the same external world are considered as *Input/Output equivalent* if the effects, i.e. the produced labellings for external arguments are the same for any reasonable input-labelling. This notion yields the possibility of replacing a multipole with another one embedded in a larger framework without affecting the labellings of the unmodified part of the initial framework. The interested reader is referred to Chapter 18 for further information. In the following we shed light on equivalence notions induced by certain dynamic scenarios.

4.1 Dynamic Scenarios and Corresponding Equivalence Notions

There are two main classes of dynamic scenarios, namely *expansions* and *deletions*. Both of them can be further divided in *normal* and *local* versions. These scenarios are motivated by real-world argumentation as well as instantiation-based argumentation [Caminada and Amgoud, 2007]. For instance, let us consider the dynamics of a discussion or dispute illustrated by the following citation [Besnard and Hunter, 2009]:

> How does argumentation usually take place? Argumentation starts when an initial argument is put forward, making some claim. An objection is raised, in the form of a counterargument. The latter is addressed in turn, eventually giving rise to a counter-counterargument, if any. And so on.

This means, in order to strengthen the own point of view or to rebut the opponents arguments it is natural that one tries to come up with *stronger* arguments, i.e. new arguments which are not attacked by the former arguments. This type of dynamics is formally captured by so-called *strong expansions* [Baumann and Brewka, 2010]. The formal counterpart of it, so-called *weak expansions* [Baumann and Brewka, 2010], where the new arguments do not attack (but may be attacked by) the old ones seem to be more an academic exercise than a task with practical relevance with regard to real-world argumentation.[14] Let us turn to instantiation-based argumentation where arguments and attacks stem from an underlying knowledge base (cf. Chapter 6 for detailed information as well as Figure 2 in Chapter 4 for an illustration). What happens on the abstract level if a new piece of information is added? It turns out that in almost all deductive argumentation systems older arguments and their corresponding attacks survive and only new arguments which may interact with the previous ones arise. This type of dynamic evolution is formally captured by so-called *normal expansions*. *Local expansions* in contrast, i.e. expansions where new attacks are added only correspond to re-instantiations if we change to a less restrictive notion of attack (cf. [Besnard and Hunter, 2001] for different attack notions).

We start with the definition of the different types of expansions together with some introducing examples.

Definition 4.1 ([Baumann and Brewka, 2010]) *An AF G is an expansion of AF $F = (A, R)$ (for short, $F \preceq_E G$) iff $G = (A \stackrel{.}{\cup} B, R \stackrel{.}{\cup} S)$ for some (maybe empty) sets B and S. An expansion is called*

---

[14]We mention that they do play a decisive role w.r.t. computational issues, so-called *splitting methods* (cf. [Baumann, 2011; Baumann et al., 2011; Baumann et al., 2012]).

1. *normal* $(F \preceq_N G)$ *iff* $\forall ab\ ((a,b) \in S \rightarrow a \in B \vee b \in B)$,
2. *strong* $(\mathcal{F} \preceq_S G)$ *iff* $\mathcal{F} \preceq_N G$ *and* $\forall ab\ ((a,b) \in S \rightarrow \neg(a \in A \wedge b \in B))$,
3. *weak* $(\mathcal{F} \preceq_W G)$ *iff* $\mathcal{F} \preceq_N G$ *and* $\forall ab\ ((a,b) \in S \rightarrow \neg(a \in B \wedge b \in A))$,
4. *local* $(F \preceq_L G)$ *iff* $B = \emptyset$.

For short, being a normal expansion means that new attacks must involve at least one new argument in contrast to local expansions where new attacks involve old arguments only. Moreover, strong and weak expansions are normal and their names refer to properties of the additional arguments, namely arguments which are never attacked by former arguments (so-called *strong* arguments) and arguments which do not attack former arguments (so-called *weak* arguments).

Observe that any arbitrary expansion can be splitted up in a normal and a local part. This can be nicely seen in the following example.

Example 4.2 *The AF F is the initial framework. An arbitrary, normal, strong, weak or local expansion of it are given by F_E, F_N, F_S, F_W and F_L, respectively. Grey-highlighted arguments or attacks represent added information.*

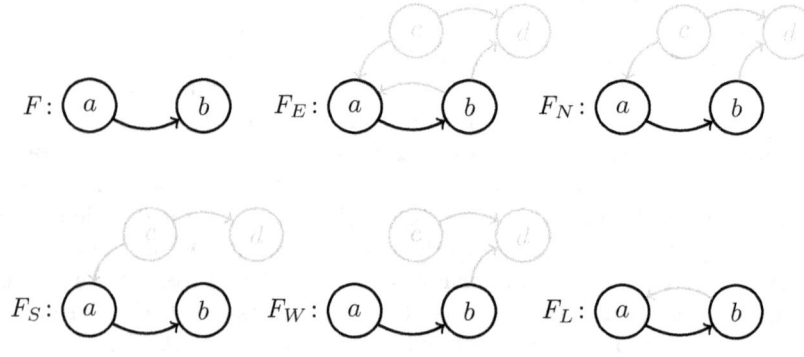

Figure 9: Different Kinds of Expansions

In 2014 the natural counter-parts (or more precisely, inverse operations) to arbitrary, normal and local expansions, so-called *deletions* were introduced [Baumann, 2014a]. Furthermore, the most general form of a dynamic scenario (where expansion and deletion can be combined) a so-called *update* were considered too. Analogously to expansions, any arbitrary deletion can be splitted in a normal and a local part. This means, a *normal deletion* retract arguments and their corresponding attacks. *Local deletions* in contrast delete attacks only.[15]

---

[15]We mention that *strong* as well as *weak deletions* are not introduced/considered so far. They could be easily defined as inverse operations of their expansion counterparts. Before doing so, it would be interesting to identify real-world situations or instantiation-based dynamics were such kind of evolvements naturally occur.

The main motivation behind these notions stems from instantiation-based context. More precisely, a normal deletion on the abstract level correspond to deleting information of a given knowledge base. Changing to a more restrictive notion of attack correspond to a local deletion and a combination of both of them give rise to an arbitrary deletion on the abstract level. We proceed with the formal definitions as well as introductory examples.

Definition 4.3 ([Baumann, 2014a]) *Given an AF $F = (A, R)$, a set of arguments B and a set of attacks S as well as a further AF H. The AF*

$$G = (F \setminus [B, S]) \cup H := ((A, R \setminus S)|_{A \setminus B}) \cup H$$

is called an update of F (for short, $F \asymp_U G$). An update is called a

1. *deletion ($F \succeq_D G$) iff $H = (\emptyset, \emptyset)$,*

2. *normal deletion ($F \succeq_{ND} G$) iff ($F \succeq_D G$) and $S = \emptyset$,*

3. *local deletion ($F \succeq_{LD} G$) iff $F \succeq_D G$ and $B = \emptyset$.*

Let us take a closer look at the definition of $G = (F \setminus [B, S]) \cup H$. The AF H plays the role of added information, i.e. it contains new arguments and attacks. Consequently, for all kind of deletions we have $H = (\emptyset, \emptyset)$ which leaves us with $G = F \setminus [B, S]$. The set B contains arguments which have to deleted. Since attacks depend on arguments we have to delete the attacks which involve arguments from B too. This operation is formally captured by the restriction of F to $A \setminus B$. Furthermore, the set S contains particular attacks which have to be deleted. This means, the pair $[B, S]$ does not necessarily have to be an AF. Therefore we use $[B, S]$ instead of (B, S). If clear from context we use B and S instead of $[B, \emptyset]$ or $[\emptyset, S]$, i.e. we simply write $F \setminus B$ as well as $F \setminus S$ for normal or local deletions, respectively.

Example 4.4 *The AF F represents the initial situation. An update as well as arbitrary, normal or local deletion of it are given by F_U, F_D, F_{ND} and F_{LD}. Grey-highlighted arguments or attacks represent added information in contrast to dotted arguments and attacks which represent deleted objects.[16] More formally, in accordance with Definition 4.3 we have that $F_U = (F \setminus [B, S]) \cup H$, $F_D = F \setminus [B, S]$, $F_{ND} = F \setminus B$, $F_{LD} = F \setminus S$ where the set of arguments $B = \{c\}$, the set of attacks $S = \{(b, a)\}$ and the AF $H = (\{b, d, e, f\}, \{(d, b), (e, f), (f, d)\})$.*

---

[16]This convention will be used throughout the whole section.

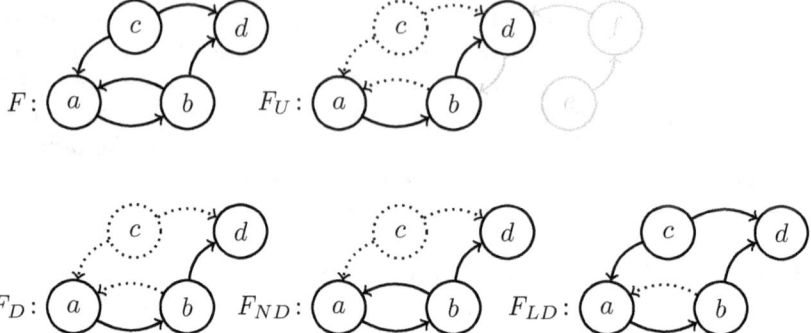

Figure 10: An Update and Different Kinds of Deletions

We now turn to the corresponding equivalence notions (cf. [Baumann and Strass, 2015, Section 3.8] for chronological order). Two AFs F and G are said to be *ordinarily equivalent* w.r.t. a semantics σ if they possess the same σ-extensions/labellings. In this case, we say that F and G possess the same *explicit* information. In contrast, sharing the same *implicit* information, i.e. being semantically indistinguishable w.r.t. any suitable future scenario is a much more demanding property which allows to replace F and G by each other without loss of semantical information.

Example 4.5 *Consider the following AFs F and G. We have $\mathcal{E}_{pr}(F) = \mathcal{E}_{pr}(G) = \{\{a\}\}$. This means, F and G possess the same explicit information w.r.t. preferred semantics or in other words, they are ordinarily equivalent.*

Assume that expansions as well deletions are the dynamic scenarios of interest. This means, we ask whether the AFs F and G even possess the same implicit information w.r.t. expansions or deletions, respectively? In order to give a negative answer one has to come up with one single dynamic scenario were the revised versions possess different preferred extensions. A positive answer in contrast is a statement about infinitely many dynamic scenarios (even in case of finite AFs). In this example, we give a negative answer for both modification types.

In case of expansions, we conjoin to both the AF $H = (\{a,b\},\{(b,a)\})$. Consider the resulting frameworks below. We have $\mathcal{E}_{pr}(F \cup H) = \{\{a\},\{b\}\}$ and since $G \cup H = G$ we obtain $\mathcal{E}_{pr}(G \cup H) = \{\{a\}\}$ without re-computing.

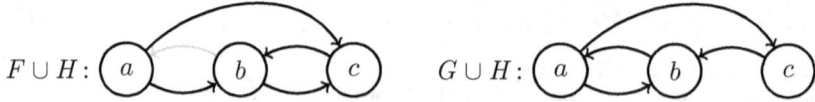

To reveal the inherent difference between F and G in case of deletions we may retract with the argument c. Consider the resulting (normal) deletions $F \setminus \{c\}$ and $G \setminus \{c\}$ of F or G, respectively. Now, $\{b\}$ becomes a preferred extension in $F \setminus \{c\}$ but still not in $G \setminus \{c\}$.

We now formally define what we precisely mean by possessing the same implicit information. As already stated, the first paper in this line of work was [Oikarinen and Woltran, 2011] engaged with characterizing *strong equivalence*. For the sake of clarity and comprehensibility we use the term *expansion equivalence* since strong equivalence [Oikarinen and Woltran, 2011, Definition 2] corresponds to semantical indistinguishability w.r.t. arbitrary expansions.

Definition 4.6 *Given a semantics σ. Two AFs F and G are*

1. *ordinarily equivalent w.r.t. σ (for short, $F \equiv^\sigma G$) iff $\sigma(F) = \sigma(G)$,*

2. *expansion equivalent w.r.t. σ (for short, $F \equiv^\sigma_E G$) iff for each AF H we have, $F \cup H \equiv^\sigma G \cup H$,*

3. *normal expansion equivalent w.r.t. σ (for short, $F \equiv^\sigma_N G$) iff for each AF H, such that $F \preceq_N F \cup H$ and $G \preceq_N G \cup H$ we have, $F \cup H \equiv^\sigma G \cup H$,*

4. *strong expansion equivalent w.r.t. σ (for short, $F \equiv^\sigma_S G$) iff for each AF H, such that $F \preceq_S F \cup H$ and $G \preceq_S G \cup H$ we have, $F \cup H \equiv^\sigma G \cup H$,*

5. *weak expansion equivalent w.r.t. σ (for short, $F \equiv^\sigma_W G$) iff for each AF H, such that $F \preceq_W F \cup H$ and $G \preceq_W G \cup H$ we have, $F \cup H \equiv^\sigma G \cup H$,*

6. *local expansion equivalent[17] w.r.t. σ (for short, $F \equiv^\sigma_L G$) iff for each AF H, such that $A(H) \subseteq A(F \cup G)$ we have, $F \cup H \equiv^\sigma G \cup H$.*

7. *update equivalent w.r.t. σ (for short, $F \equiv^\sigma_U G$) iff for any pair $[B, S]$ and any AF H we have, $(F \setminus [B, S]) \cup H \equiv^\sigma (G \setminus [B, S]) \cup H$,*

8. *deletion equivalent w.r.t. σ (for short, $F \equiv^\sigma_D G$) iff for any pair $[B, S]$ we have, $F \setminus [B, S] \equiv^\sigma G \setminus [B, S]$,*

9. *normal deletion equivalent w.r.t. σ (for short, $F \equiv^\sigma_{ND} G$) iff for any set of arguments B we have, $F \setminus B \equiv^\sigma G \setminus B$,*

10. *local deletion equivalent w.r.t. σ (for short, $F \equiv^\sigma_{LD} G$) iff for any set of attacks S we have, $F \setminus S \equiv^\sigma G \setminus S$,*

---

[17]Note that a suitable AF H is not necessarily a local expansion of F and G in the sense of Definition 4.1. Nevertheless, we may loosely speak about local expansions.

Remember that there are several relations between the considered dynamic scenarios. For instance, in accordance with Definitions 4.1 and 4.3, any normal expansion (deletion) is an arbitrary expansion (deletion). Furthermore, in the light of Definition 4.6, we certainly affirm that expansion equivalence is much more demanding then local expansion equivalence. In other words, local expansion equivalence of two AFs is an immediate and unavoidable consequence of being expansion equivalent. Finally, any considered equivalence notion is at least as demanding then ordinary equivalence.[18] Please note that these relations do not depend on certain properties of a considered semantics. Consequently, Figure 11 gives a preliminary overview for such interrelations (arising from the definitions) between the introduced equivalence notions for any possible semantics. For reasons, which will become clearer later, we also consider the identity relation. For two equivalence notion Φ and Ψ we have $\Phi \subseteq \Psi$ iff there is a link from Φ to Ψ.

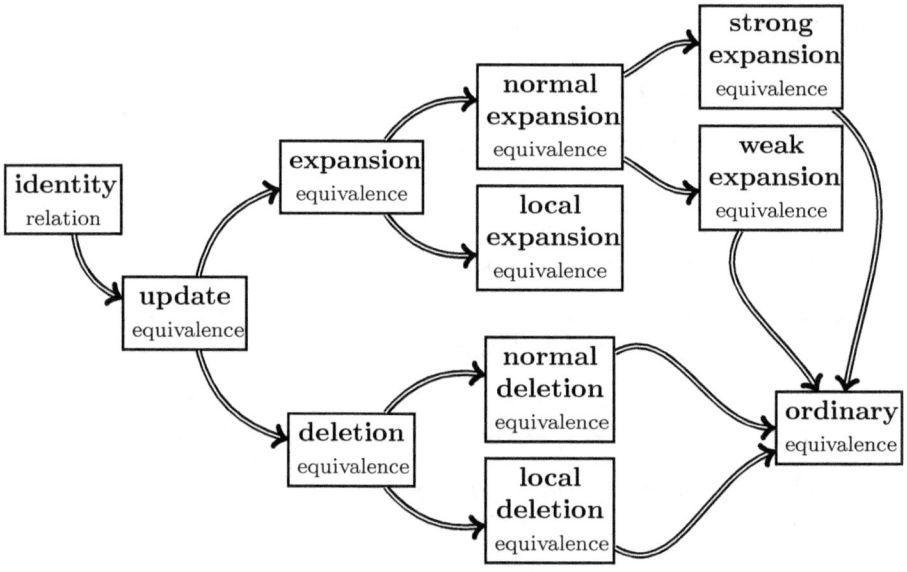

Figure 11: Preliminary Subset Relations between Equivalence Notions

In the remainder of this section we shed light on the question of *how* to determine whether two AFs are equivalent w.r.t. certain scenarios? As a by-product of these characterization results we will see that for many semantics the preliminary relations between the introduced equivalence notions depicted above can be delineated in a much more compact way. The majority of the presented characterization results is devoted to finite AFs as well as extension-

---

[18]The empty framework $(\emptyset, \emptyset)$ as well as the empty pair $[\emptyset, \emptyset]$ justifies this assertion for any type of expansions or deletions, respectively.

based semantics. We will see that there are some differences if sticking to unrestricted frameworks or the corresponding labelling-based versions.

4.2 Characterization Theorems for Extension-based Semantics

The Central Notion of Expansion Equivalence

In order to get an idea of how to find a characterization we start with some reflections. For this purpose we consider the most restrictive semantics, namely the stable one as well as the most prominent type of equivalence, namely expansion equivalence. What are necessary features of expansion equivalence w.r.t. stable semantics, i.e. which properties are implied if two AFs F and G are expansion equivalent? In consideration of Figure 11 we deduce their ordinary equivalence, i.e. $\mathcal{E}_{stb}(F) = \mathcal{E}_{stb}(G)$. Note that possessing the same set of extensions neither imply sharing the same arguments nor sharing the same self-loops as shown in the following example.

Example 4.7 *Consider the AFs F, G and H. Each two of them are ordinarily equivalent since $\mathcal{E}_{stb}(F) = \mathcal{E}_{stb}(G) = \mathcal{E}_{stb}(H) = \{\{a\}\}$.*

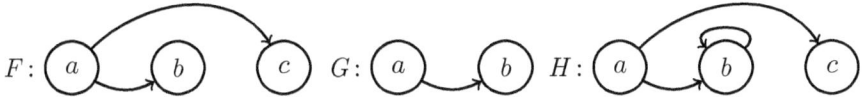

The AFs $I_1 = (\{c\}, \emptyset)$ and $I_2 = (\{a, b, c\}, \{(b, a), (b, c)\})$ witness that neither F and G, nor F and H are expansion equivalent w.r.t. stable semantics. Convince yourself that $\mathcal{E}_{stb}(F \cup I_1) = \{\{a\}\} \neq \{\{a, c\}\} = \mathcal{E}_{stb}(G \cup I_1)$ and $\mathcal{E}_{stb}(F \cup I_2) = \{\{a\}, \{b\}\} \neq \{\{a\}\} = \mathcal{E}_{stb}(G \cup I_2)$.

Restricting ourselves to finite AFs, it is not difficult to see that in case of expansion equivalence w.r.t. stable semantics the observed relation between non-sharing the same arguments/loops and non-equivalence does hold in general. In other words, possessing the same arguments as well as possessing the same loops are indeed necessary conditions for being expansion equivalent in the finite setting.

Let us summarize our observations in the following fact.

Fact 4.8 *Given two finite AFs F and G. If $F \equiv_E^{\mathcal{E}_{stb}} G$, then*

1. *$\mathcal{E}_{stb}(F) = \mathcal{E}_{stb}(G)$,*

2. *$A(F) = A(G)$ and*

3. *$L(F) = L(G)$.*

As already stated in Figure 11, being identical (i.e. $A(F) = A(G)$ and $R(F) = R(G)$) is sufficient for being expansion equivalent. Combining this undeniable fact together with the second and third items of Fact 4.8 encourages one to search for syntactical properties sufficient as well as necessary for being expansion equivalent. In order to guarantee the first item of Fact 4.8

we have to identify attacks which do not contribute anything when computing stable extensions. Moreover, these attacks which do not affect the evaluation of a given AF F have to be *redundant*, no matter how F is extended. Remember that being a stable extension can be simply verified by checking whether the set in question is conflict-free and possesses a full range.[19] This means, good candidates for "useless" attacks w.r.t. stable semantics should fulfill the following two properties: firstly, having or not having such an attack does not change the status of a set from being conflict-free to conflicting or vice versa and secondly, having or not having such an attack does not affect the range of a conflict-free set. Certainly, an attack (a, b) stemming from a self-defeating argument a does not change the conflict status of a certain set E. This can be seen as follows: If $a \in E$, then E was conflicting as well as remains conflicting after deleting or adding (a, b). Furthermore, if $a \notin E$, then E might be conflicting or not. In either case the conflict status of E does not change if (a,b) is added or removed since $\{a, b\} \not\subseteq E$. Finally, such an attack (a, b) might have an influence on the range of conflicting sets but it definitely has not in case of conflict-free sets since $a \notin E$ can not be questioned.

Example 4.9 *Consider the following AF F. We have, $\mathcal{E}_{stb}(F) = \{\{a\}\}$.*

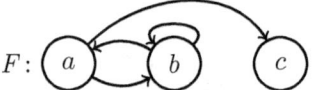

According to our considerations above adding or deleting an attack stemming from the self-defeating argument b does not change the semantics. Consider therefore the following three possible "manipulations".

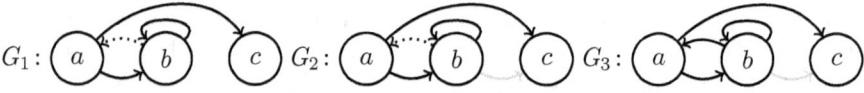

Indeed, $\mathcal{E}_{stb}(F) = \mathcal{E}_{stb}(G_1) = \mathcal{E}_{stb}(G_2) = \mathcal{E}_{stb}(G_3) = \{\{a\}\}$ support our claims for the static case. We encourage the reader to try to do the impossible, namely semantically distinguish the AFs F and its manipulations by an arbitrary expansion.

It was the main result in [Oikarinen and Woltran, 2011] that expansion equivalence can be indeed decided by looking at the syntax only. The authors introduced so-called *kernels* which are simply functions mapping each AF F to its redundancy-free version. This means, the kernel of an AF F does not possess any redundant attack. Put it differently, for any surviving attack exist at least one dynamic scenario were deleting this attack would cause a semantical difference. We proceed with the formal definition of the very first kernels

---

[19]The topic of *verifiability* of argumentation semantics σ was studied in [Baumann et al., 2016b]. The main question is which (minimal amount of) information on top of conflict-free sets is exactly needed to determine whether a certain set is a σ-extension.

already introduced in [Oikarinen and Woltran, 2011]. We sometimes call them *classical*.

Definition 4.10 *Let $\sigma \in \{stb, ad, gr, co\}$. The σ-kernel $k(\sigma) : \mathcal{F} \to \mathcal{F}$ with $k(\sigma)(F) = F^{k(\sigma)} = (A, R^{k(\sigma)})$ for a given AF $F = (A, R)$ is defined as:*

$$R^{k(stb)} = R \setminus \{(a,b) \mid a \neq b, (a,a) \in R\},$$
$$R^{k(ad)} = R \setminus \{(a,b) \mid a \neq b, (a,a) \in R, \{(b,a),(b,b)\} \cap R \neq \emptyset\},$$
$$R^{k(gr)} = R \setminus \{(a,b) \mid a \neq b, (b,b) \in R, \{(a,a),(b,a)\} \cap R \neq \emptyset\},$$
$$R^{k(co)} = R \setminus \{(a,b) \mid a \neq b, (a,a), (b,b) \in R\}.$$

In order to get an idea of how the classical kernels work we proceed with an example.

Example 4.11 *Consider again the AF G_3 depicted in Example 4.9. We apply now all classical kernels.*

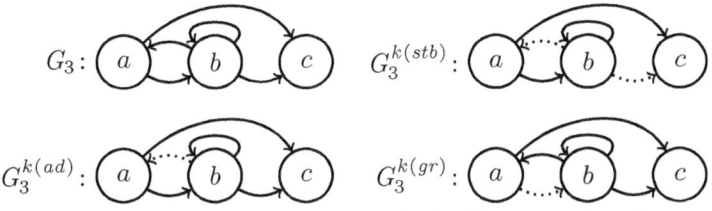

The stable kernel deletes all attacks (a,b) stemming from a self-defeating argument a. A deletion of (a,b) in case of the grounded kernel additionally requires that a is counter-attacked by b or b is self-defeating or both. Interchanging a and b yields the condition for deletion in case of the grounded kernel. Finally, $G_3^{k(co)} = G_3$ since deleting an attack (a,b) w.r.t. the complete kernel requires that both arguments a and b are self-defeating.

Before turning to characterization theorems, we collect some useful properties of the introduced kernels. The following fact contains intrinsic properties of the classical kernels.[20] More precisely, any classical kernel k is *node-preserving* and *loop-preserving*, i.e. the sets of arguments and self-defeating arguments do not change if applying k. Moreover, in the absence of self-loops, each AF coincides with its classical kernels. Furthermore, the decision whether an attack (a,b) has to be deleted does not depend on further arguments than a and b. Put differently, the reason of being redundant is *context-free*, i.e. it stems from the arguments themselves. The last two properties claim that equality of kernels is *robust* w.r.t. further compositions as well as deleting arguments and corresponding attacks. For a given AF $F = (B, S)$ we use $A(F), R(F)$ and $L(F)$

---

[20]Although most of the properties are immediately clear even in case of unrestricted frameworks we will state all of them for finite AFs only as done in the existing literature. The same applies to Fact 4.18. Some results regarding unrestricted frameworks can be found in Section 4.3.

to refer to its arguments, attacks and self-defeating arguments, i.e. $A(F) = B$, $R(F) = S$ and $L(F) = \{a \in A(F) \mid (a,a) \in R(F)\}$.

Fact 4.12 (cf. [Oikarinen and Woltran, 2011; Baumann, 2014a]) *Given $k \in \{k(stb), k(ad), k(gr), k(co)\}$. For any finite AF F we have:*

1. $A(F) = A\left(F^k\right)$, *(node-preserving)*

2. $L(F) = L\left(F^k\right)$, *(loop-preserving)*

3. $L(F) = \emptyset \Rightarrow F = F^k$ *and (sufficient condition for identity)*

4. $(a,b) \in R\left(F^k\right) \Leftrightarrow (a,b) \in R\left((F|_{\{a,b\}})^k\right)$. *(context-freeness)*

Furthermore, for finite AFs F and G we have:

4. *If $F^k = G^k$, then $(F \cup H)^k = (G \cup H)^k$ for any finite AF H and* ($\cup$-*robustness*)

5. *If $F^k = G^k$, then $(F \setminus B)^k = (G \setminus B)^k$ for any finite set of args B.* ($\setminus$-*robustness*)

We proceed with extrinsic properties, i.e. features of kernels in presence of semantics. More precisely, stable, admissible, grounded and complete semantics are insensitive w.r.t. the application of their corresponding classical σ-kernel, i.e. the set of σ-extensions remains unchanged. Furthermore, the admissible kernel neither effects semi-stable, eager, preferred and ideal semantics. Similarly in case of stable kernel and stage semantics.

Fact 4.13 ([Oikarinen and Woltran, 2011; Gaggl and Woltran, 2013]) *For any finite AF F we have:*

1. $\mathcal{E}_\sigma(F) = \mathcal{E}_\sigma\left(F^{k(\sigma)}\right)$ for $\sigma \in \{stb, ad, gr, co\}$,

2. $\mathcal{E}_\sigma(F) = \mathcal{E}_\sigma\left(F^{k(ad)}\right)$ for $\sigma \in \{ss, eg, pr, il\}$ and

3. $\mathcal{E}_{stg}(F) = \mathcal{E}_{stg}\left(F^{k(stb)}\right)$.

As already mentioned, kernels play a decisive role in deciding expansion equivalence. In general, we say that an equivalence notion $\equiv$ *is characterizable through k* or simply, k *is a characterizing kernel (of $\equiv$)* if for any two AFs F and G, $F \equiv G$ iff $F^k = G^k$. This means, proving whether two frameworks are equivalent can be done by simply checking whether the corresponding kernels are identical. Note that all classical kernels can be efficiently constructed from a given AF. The following main theorem states that for all nine considered semantics σ there is a certain classical kernel k, s.t. expansion equivalence w.r.t. σ is characterizable through k in the finite setting. This is a very remarkable result since expansion equivalence is defined semantically. For instance, two finite

AFs F and G are expansion equivalent w.r.t. stable semantics if and only if the associated stable kernels $F^{k(stb)}$ and $G^{k(stb)}$ are syntactically equal. Observe that there is no need to introduce further kernels since one single kernel may serve for different semantics.

Theorem 4.14 *[Oikarinen and Woltran, 2011; Gaggl and Woltran, 2013] For finite AFs F and G we have:*

1. $F \equiv_E^{\mathcal{E}_\sigma} G \Leftrightarrow F^{k(\sigma)} = G^{k(\sigma)}$ *for any* $\sigma \in \{stb, ad, co, gr\}$,
2. $F \equiv_E^{\mathcal{E}_\sigma} G \Leftrightarrow F^{k(ad)} = G^{k(ad)}$ *for any* $\sigma \in \{pr, il, ss, eg\}$ *and*
3. $F \equiv_E^{\mathcal{E}_{stg}} G \Leftrightarrow F^{k(stb)} = G^{k(stb)}$.

Having Theorem 4.14 at hand we can now formally verify that all AFs depicted in Example 4.9 are expansion equivalent w.r.t. stable semantics. This means, the recommended search for arbitrary expansions revealing semantical difference between them will never succeed. As an aside, one might get the impression that the syntactical characterization presented in Theorem 4.14 is somehow unique. This is not true. Consider therefore the equivalence class $[F]_E^{\mathcal{E}_{stb}} = \{G \mid F \equiv_E^{\mathcal{E}_{stb}} G\}$ induced by F. Mathematically speaking, the stable kernel $F^{k(stb)}$ represents the least (w.r.t. subgraph-relation) element in $[F]_E^{\mathcal{E}_{stb}}$. It is not difficult to prove that $[F]_E^{\mathcal{E}_{stb}}$ even possesses a greatest element, namely $F^{k'(stb)} = (A(F), R(F) \cup \{(a,b) \mid a \neq b, (a,a) \in R(F)\})$, i.e. the framework resulting from F by adding (instead of deleting) all redundant attacks. In case of finite AFs it can be shown with reasonable effort that expansion equivalence w.r.t. stable semantics is characterizable through $k'(stb)$ too. In the same manner, all other semantics considered in Theorem 4.14 possess alternative "greatest elements" characterizations. We will see that the so-called *naive kernel* (compare Definition 4.17) provides such a kind of characterization for naive semantics. The reason for this "choice" is simply that the induced equivalence classes do not necessarily possess a least element in case of naive semantics.

Finally, let us turn to the more exotic cf2 as well as stage2 semantics which are defined via a recursive schema based on the decomposition of AFs along their strongly connected components (SCCs). These semantics are exceptional regarding expansion equivalence since in contrast to all other semantics considered in this section we have that even attacks between two self-attacking arguments are *meaningful*. This means, the presence or absence of such attacks may change the outcome of an AF. Moreover, it turned out that any attack is non-redundant. In summary, expansion equivalence coincides with syntactical identity or more formally, for any finite AF F, $\left|[F]_E^{\mathcal{E}_{cf2}}\right| = \left|[F]_E^{\mathcal{E}_{stg2}}\right| = |\{F\}| = 1$.

Theorem 4.15 *[Gaggl and Woltran, 2013; Gaggl and Dvořák, 2016] Given $\sigma \in \{cf2, stg2\}$. For finite AFs F and G we have,*

$$F \equiv_E^{\mathcal{E}_\sigma} G \Leftrightarrow F = G.$$

Further Equivalence Notions Characterizable through Kernels

Let us turn to the remaining equivalence notions? Are there similar syntax-based characterization results?

Weaker Notions of Expansion Equivalence Let us consider less demanding notions than expansion equivalence, e.g. normal and local expansion equivalence. In consideration of Definition 4.1 we do not have good reasons to believe that two AFs could be semantically distinguished by normal or local expansions, given that we only have a witnessing arbitrary expansion showing their non-equivalence. It was one surprising result in this line of research, that for many semantics, expansion equivalence coincide with definitorially weaker notions of it. This implies that weaker notions than expansion equivalence can be characterized by classical kernels too. The first results in this respect were already given in [Oikarinen and Woltran, 2011, Theorem 8]. The authors showed that for some semantics expansion equivalence and local expansion equivalence coincide if considering finite AFs. It is worthwhile to gain a thorough understanding of this relation since it actually means that if there is an arbitrary expansion which semantically distinguish two finite AFs, than there has to be a local expansion doing likewise. Later it was shown that even normal expansion equivalence coincides with expansion equivalence for a whole bunch of semantics [Baumann, 2012a]. Interestingly, in contrast to local expansion equivalence, there are (to the best of our knowledge) no semantics together with witnessing AFs known which show that this coincidence does not hold in general.

Example 4.16 *Consider the following AFs F and G. According to Theorem 4.14 they are not expansion equivalent w.r.t. preferred semantics since $F^{k(ad)} = F \neq G = G^{k(ad)}$.*

As already stated (up to now) normal expansion equivalence coincides with expansion equivalence for any considered semantics. One possible scenario which makes the predicted different behaviour explicit is the following.

Formally, we define $H = (\{b, c, d\}, [(b, d), (d, c)\})$ and we obtain $\{\{a, d\}\} = \mathcal{E}_{pr}(F \cup H) \neq \{\emptyset\} = \mathcal{E}_{pr}(G \cup H)$. We encourage the reader to try to find a witnessing example showing that F and G are not local expansion equivalent w.r.t. preferred semantics. Due to Theorem 4.20 there has to be at least one distinguishing local expansion.

How do the semantics behave in case of strong expansion equivalence? Remember, a special feature of strong expansions is that a former attack between old arguments will never become a counterattack to an added attack. In this sense, former attacks do not play a role with respect to being a potential defender of an added argument. Hence, in contrast to arbitrary expansions where such attacks might be relevant, we may delete them without changing the behavior with respect to further evaluations. To make this point clearer consider again the AF $F \cup H$ depicted in Example 4.16. Note that the already existing attack (a, b) in F becomes a defending attack of the newly added argument d. This means, such attacks in fact play an important role with respect to further evaluation in case of arbitrary expansions. It was one main result in [Baumann, 2012a] that for some semantics attacks like (a, b) in F are indeed redundant w.r.t. strong expansions. Even more surprising, strong expansion equivalence is characterizable through kernels. Therefore, more involved kernel definitions, so-called σ-*-kernels had to be introduced. These kernels allow more deletions than their classical counterparts for expansion equivalence. In contrast to them, σ-*-kernels are *context-sensitive*, i.e. the question whether an attack (a, b) is redundant can not be answered by considering the arguments a and b only [Baumann, 2014a].

The first three kernels presented in the definition below were firstly introduced in [Baumann, 2012a] with the objective to characterize strong expansion equivalence with respect to certain semantics. For the sake of completeness we also present the so-called *stg-*-kernel* as well as *na-kernel* [Baumann and Woltran, 2016; Baumann et al., 2016b].[21]

Definition 4.17 *Let* $\sigma \in \{ad, gr, co, stg\}$. *The σ-*-kernel* $k^*(\sigma) : \mathcal{F} \to \mathcal{F}$ *with* $k^*(\sigma)(F) = F^{k^*(\sigma)} = \left(A, R^{k^*(\sigma)}\right)$ *for a given AF* $F = (A, R)$ *is defined as:*

$$R^{k^*(ad)} = R \setminus \{(a, b) \mid a \neq b, ((a, a) \in R \land \{(b, a), (b, b)\} \cap R \neq \emptyset)$$
$$\lor ((b, b) \in R \land \forall c \ ((b, c) \in R \to \{(a, c), (c, a), (c, c), (c, b)\} \cap R \neq \emptyset))\},$$

$$R^{k^*(gr)} = R \setminus \{(a, b) \mid a \neq b, ((b, b) \in R \land \{(a, a), (b, a)\} \cap R \neq \emptyset)$$
$$\lor ((b, b) \in R \land \forall c \ ((b, c) \in R \to \{(a, c), (c, a), (c, c)\} \cap R \neq \emptyset))\},$$

$$R^{k^*(co)} = R \setminus \{(a, b) \mid a \neq b, ((a, a), (b, b) \in R) \lor ((b, b) \in R \land (b, a) \notin R$$
$$\land \forall c \ ((b, c) \in R \to \{(a, c), (c, a), (c, c), (c, b)\} \cap R \neq \emptyset))\},$$

$$R^{k^*(stg)} = R \setminus \{(a, b) \mid a \neq b, (a, a) \in R \lor \forall c \ (c \neq a \to (c, c) \in R)\}$$

$$R^{k(na)} = R \cup \{(a, b) \mid a \neq b, \{(a, a), (b, a), (b, b)\} \cap R \neq \emptyset\}.$$

The latter represents the so-called na-kernel $F^{k(na)} = \left(A, R^{k(na)}\right)$.

For an illustrating example we refer the reader to Example 4.19. Analogously to Fact 4.12 we collect some properties of the newly introduced kernels. The

---

[21] As an aside, we use the supplement "*", whenever the kernel in question is non-classical and expansion equivalence is already characterized by another kernel.

first three properties are immediately clear by definition.[22] The robustness w.r.t. deletions and corresponding attacks is less obvious but it is already shown for all considered kernels (except the stg-*-kernel) in case of finite AFs (cf. [Baumann, 2014a, Theorems 6 and 14]).

Fact 4.18 *Given* $k \in \{k^*(ad), k^*(gr), k^*(co), k^*(stg), k(na)\}$ *as well as* $k^* \in \{k^*(ad), k^*(gr), k^*(co), k^*(stg)\}$. *For two finite AFs F and G we have:*

1. $A(F) = A(F^k)$, *(node-preserving)*

2. $L(F) = L(F^k)$, *(loop-preserving)*

3. $L(F) = \emptyset \Rightarrow F = F^{k^*}$ *and* *(sufficient condition for identity)*

4. *If* $F^k = G^k$, *then* $(F \setminus B)^k = (G \setminus B)^k$ *for any finite set of arguments B.* *($\setminus$-robustness)*

Let us consider the ad-*-kernel (which, as we shall see, characterizes strong expansion equivalence for preferred semantics) in more detail. Consider the first disjunct. This first condition is exactly the same as in case of the ad-kernel (compare Definition 4.10), i.e. an attack (a, b) has to be deleted if a is self-attacking and at least one of the attacks (b, a) or (b, b) exist. The second disjunct provides one with further options to delete an attack (a, b), namely if b is self-defeating and furthermore, for all arguments c which are attacked by b at least one of the following conditions has to be fulfilled:

1. a attacks c,

2. c attacks a,

3. c attacks c,

4. c attacks b.

The motivation for the second disjunct is the following: At first observe that b cannot be an element of any conflict-free set. Consequently, in case of strong expansions the attack (a, b) may only be relevant with respect to the defense of c. In the first three cases this relevance becomes unimportant since $\{a, c\}$ is conflicting. In the fourth case the redundancy of (a, b) with respect to the defense of c is given by the fact that c already defends itself against b. Please note that the consideration of $c = a$ or $c = b$ is not excluded by Definition 4.17. The following frameworks exemplify different cases.

---

[22]The AF $F = (\{a, b\}, \{(a, b)\})$ shows that the naive kernel has to be excluded from item 3 of Fact 4.18 since $F^{k(na)} = (\{a, b\}, \{(a, b), (b, a)\}) \neq F$.

Example 4.19 *The following graphs show six frameworks and their corresponding ad-*-kernels. The dotted attacks represent initial attacks which have to be deleted if applying the ad-*-kernel.*

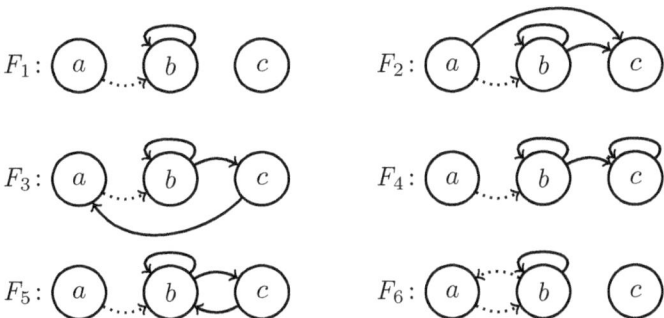

Consider the formal description of $R^{k^*(ad)}$ as given in Definition 4.17. The AF F_1 is somehow the base case since the only argument c, s.t. $(b,c) \in R(F_1)$ is b itself. Since $(b,b) \in R(F_1)$ we deduce that the considered intersection is non-empty and thus, the deletion of (a,b) is justified. The subsequent four frameworks F_2, F_3, F_4 and F_5 are the base case plus one further argument c different from a and b, s.t. for any $i \in \{2,3,4,5\}$, $(b,c) \in R(F_i)$. The last framework F_6 illustrates the case b counterattacks a. Note that the reason to delete (a,b) is somehow self-referential since (additionally to the base case) it is justified by $(a,b) \in R(F_6)$. Due to the first disjunct (i.e. just like in case of the classical ad-kernel) even the attack (b,a) has to be deleted.

We proceed with further characterization theorems.[23] An comprehensive overview of equivalence notion and their characterizing kernels in case of finite AFs and extension-based semantics is presented in Figure 12.

Theorem 4.20 *[Oikarinen and Woltran, 2011; Baumann, 2012a; Baumann and Woltran, 2016; Baumann et al., 2016b] For finite AFs F and G we have the following coincidences.*

1. $F \equiv_E^{\mathcal{E}_\sigma} G \Leftrightarrow F \equiv_N^{\mathcal{E}_\sigma} G$ for $\sigma \in \{stg, stb, ss, eg, ad, pr, il, gr, co, na, cf2, stg2\}$,

2. $F \equiv_E^{\mathcal{E}_\sigma} G \Leftrightarrow F \equiv_L^{\mathcal{E}_\sigma} G$ for $\sigma \in \{ss, eg, ad, pr, il, na\}$ and

3. $F \equiv_E^{\mathcal{E}_\sigma} G \Leftrightarrow F \equiv_S^{\mathcal{E}_\sigma} G$ for $\sigma \in \{stg, stb, ss, eg, na\}$.

Furthermore, for any two finite AFs F and G we have the following non-classical characterizations.

4. $F \equiv_L^{\mathcal{E}_{stg}} G \Leftrightarrow F^{k^*(stg)} = G^{k^*(stg)}$,

5. $F \equiv_S^{\mathcal{E}_\sigma} G \Leftrightarrow F^{k^*(ad)} = G^{k^*(ad)}$ for $\sigma \in \{ad, pr, il\}$,

---
[23] Please note that the results in case of cf2 and stage2 semantics have never been published before.

6. $F \equiv_S^{\mathcal{E}_\sigma} G \Leftrightarrow F^{k^*(\sigma)} = G^{k^*(\sigma)}$ for $\sigma \in \{co, gr\}$ and

7. $F \equiv_E^{\mathcal{E}_{na}} G \Leftrightarrow F^{k(na)} = G^{k(na)}$.

At this point we want to highlight a very surprising relation. Remember that normal expansion equivalence and normal deletion equivalence are completely unrelated in the general picture (cf. Figure 11). The observation that the characterizing kernels (including the identity map in case of cf2 and stage2 semantics) of normal expansion equivalence w.r.t. all considered semantics in this section satisfy \-robustness (cf. Facts 4.12 and 4.18) reveals that normal expansion equivalence implies normal deletion equivalence for these semantics.

Corollary 4.21 *Given $\sigma \in \{stg, stb, ss, eg, ad, pr, il, gr, co, na, cf2, stg2\}$ and two finite AFs F and G. We have: $F \equiv_N^{\mathcal{E}_\sigma} G \Rightarrow F \equiv_{ND}^{\mathcal{E}_\sigma} G$.*

The attentive reader may have noticed that we do not have characterized local expansion equivalence w.r.t. stable, complete as well as grounded extension-based semantics. We mention that all three equivalence notions are already characterized but the characterization theorems are not purely kernel-based (cf. [Oikarinen and Woltran, 2011, Theorems 9,10,11]). Furthermore, it can be checked that none of the kernels presented in Definitions 4.10 and 4.17 serve as a characterizing kernel. Consider therefore the following example [Oikarinen and Woltran, 2011, Example 15].

Example 4.22 *The AFs F and G are local expansion equivalent w.r.t. stable semantics. This can be seen as follows. Given an AF H, s.t. $A(H) \subseteq \{a, b\}$. If $(a, b) \in R(H)$ and $(a, a) \notin R(H)$, we obtain $\mathcal{E}_{stb}(F \cup H) = \mathcal{E}_{stb}(G \cup H) = \{\{a\}\}$. Otherwise, $\mathcal{E}_{stb}(F \cup H) = \mathcal{E}_{stb}(G \cup H) = \emptyset$.*

$F:$ (a) (b)　　$G:$ (b)

Remember that all introduced kernels are node-preserving (Facts 4.12 and 4.18). Consequently, none of them may serve as a characterizing kernel for local expansion equivalence w.r.t. stable semantics.

We mention that weak expansion equivalence is already characterized in case of stable semantics [Baumann, 2012a, Proposition 3] as well as admissible, preferred and complete semantics [Baumann and Brewka, 2015b, Theorem 1]. All characterization results are not kernel-based. For instance, two AFs are weak expansion equivalent w.r.t. stable semantics iff both do not possess stable extensions at all or if they share the same arguments and at the same time possess the same stable extensions. Consequently, $F = (\{a\}, \{(a, a)\})$ and $G = (\{a, b, c\}, \{(a, b), (b, c), (c, a)\})$ are weak expansion equivalent w.r.t. stable semantics. Both frameworks witness that any potential characterizing kernel k is necessarily neither node- nor loop-preserving.

As a final note, we are not aware of any study of weaker notions of expansion equivalence in case of cf2 as well as stage2 semantics.

Notions of Deletion Equivalence and Update Equivalence We start with local deletion equivalence. Remember that local deletion equivalent AFs cannot be semantically distinguished by deleting a certain set of attacks in both simultaneously. How "strong" is this notion? Are there redundant attacks or even redundant arguments?

Example 4.23 *Consider the following AFs F, G and H.*

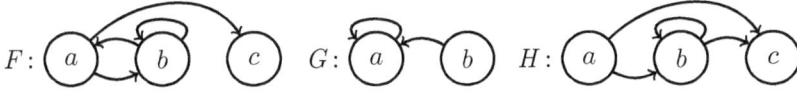

The AFs F and G do not possess the same arguments. Let us delete all occurring attacks, i.e. $S_A = R(F) \cup R(G)$. We obtain the following local deletions where $\{a,b,c\} \in \mathcal{E}_\sigma(F \setminus S_A) \setminus \mathcal{E}_\sigma(G \setminus S_A)$ for all semantics σ considered in this section.

The AFs F and H possess the same arguments but differ in their attack-relation, e.g. $(b,c) \in R(H) \setminus R(F)$. This difference can be made more explicit if defining $S_R = (R(F) \cup R(H)) \setminus \{(b,c)\}$. Consider the resulting local deletions.

$F \setminus S_R:$ (a) (b) (c) $H \setminus S_R:$ (a) (b) (c)

Once again we have $\{a,b,c\} \in \mathcal{E}_\sigma(F \setminus S_R)$ for all known semantics σ and $\{a,b,c\} \notin \mathcal{E}_\sigma(H \setminus S_R)$ if assuming conflict-freeness of the considered semantics.

The observations above indicate that there is not much space for redundancy in case of local expansion equivalence and indeed, it was one main result in [Baumann, 2014a] that local expansion equivalence collapse to identity for all semantics considered in this section. Moreover, instead of proving this one by one for any semantics the author followed the line in [Baroni and Giacomin, 2007] and provide abstract criteria guaranteeing the coincidence with syntactical identity. These criteria are very weak requirements, namely *conflict-freeness* ($\mathcal{CF}$) and the principle of *isolate-inclusion* ($\mathcal{II}$). The latter is fulfilled by a semantics σ iff for any AF F, the set of all isolated arguments is contained in at least one σ-extension. Observe that any considered semantics apart from stable semantics satisfy $\mathcal{II}$.[24]

---

[24] Note that only universally defined semantics σ, i.e. semantics which warrants the existence of at least one σ-extension (cf. Definition 2.3), may satisfy isolate-inclusion. A counterexample in case of stable semantics is given by $F = (\{a,b\}, \{(b,b)\})$. Obviously, a is isolated but $\mathcal{E}_{stb}(F) = \emptyset$. Nevertheless, local expansion equivalence in case of stable semantics collapse to identity too.

Theorem 4.24 ([Baumann, 2014a]) *Given a semantics σ satisfying $\mathcal{CF}$ and $\mathcal{II}$. For two finite AFs F and G we have:*

$$F \equiv_{LD}^{\mathcal{E}_\sigma} G \Leftrightarrow F = G.$$

Since being identical implies local deletion equivalence we deduce that all equivalence notion "inbetween" them collapse to identity too (cf. Figure 11).

Proposition 4.25 *Given a semantics σ satisfying $\mathcal{CF}$ and $\mathcal{II}$. For any two finite AFs F and G we have:*

$$F \equiv_{U}^{\mathcal{E}_\sigma} G \Leftrightarrow F \equiv_{D}^{\mathcal{E}_\sigma} G \Leftrightarrow F = G.$$

This means, for semantics satisfying conflict-freeness and isolate-inclusion any argument/attack may play a crucial role with respect to further evaluations if updates, deletions or local deletions are considered. Note that the results may apply to future semantics. In order to refine the general picture (as depicted in Figure 11) for the semantics considered in this section we state the following relations.[25]

Corollary 4.26 *Let $\sigma \in \{stg, stb, ss, eg, ad, pr, il, gr, co, na, cf2, stg2\}$. For any two finite AFs F and G we have:*

1. $F \equiv_{U}^{\mathcal{E}_\sigma} G \Leftrightarrow F \equiv_{D}^{\mathcal{E}_\sigma} G \Leftrightarrow F \equiv_{LD}^{\mathcal{E}_\sigma} G \Leftrightarrow F = G,$ $\hfill (k = id)$

2. $F \equiv_{D}^{\mathcal{E}_\sigma} G \Rightarrow F \equiv_{E}^{\mathcal{E}_\sigma} G,$ $\hfill$ *(deletion vs. expansion)*

3. $F \equiv_{LD}^{\mathcal{E}_\sigma} G \Rightarrow F \equiv_{L}^{\mathcal{E}_\sigma} G.$ $\hfill$ *(local versions)*

The Exceptional Case of Normal Deletion Equivalence

Normal deletion equivalence, where the retraction of arguments and corresponding attacks is considered, is exceptional in several regards. Firstly, the characterization theorems for admissible, complete and grounded semantics partially rely on σ-*-kernels. Remember that these kernels were originally introduced to characterize strong expansion (cf. Theorem 4.20). Secondly, normal deletion equivalent AFs do not even have to share the same arguments and thus give space for simplifications.

Example 4.27 *Consider the following AFs F and G. We have $\mathcal{E}_{ad}(F) = \mathcal{E}_{ad}(G) = \{\emptyset, \{a\}\}$. Even more, for any set of arguments B, $\mathcal{E}_{ad}(F \setminus B) = \mathcal{E}_{ad}(G \setminus B)$ showing their normal deletion equivalence, i.e. $F \equiv_{ND}^{\mathcal{E}_{ad}} G$.*

---

[25]The results in case of cf2 and stage2 semantics have never been published before. It can be checked that both semantics satisfy the preconditions of Theorem 4.25.

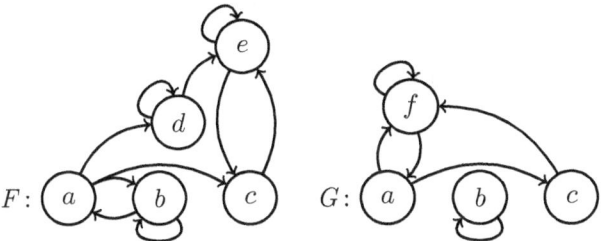

Observe that the non-shared arguments d, e and f do not play a role for the evaluation w.r.t. admissible semantics since firstly, they are self-defeating and thus cannot be part of an admissible set; and secondly, if they attack a non-looping argument shared by both arguments, e.g. e attacks c in F or f attacks a in G, then they are counter-attacked by the same argument, i.e. c attacks e in F and a attacks f in G. Consequently, they cannot influence potential admissible sets being a subset of $\{a,b\}$. Finally, let us consider the ad-*-kernel of both frameworks (cf. Example 4.19 and the comments above for more details).

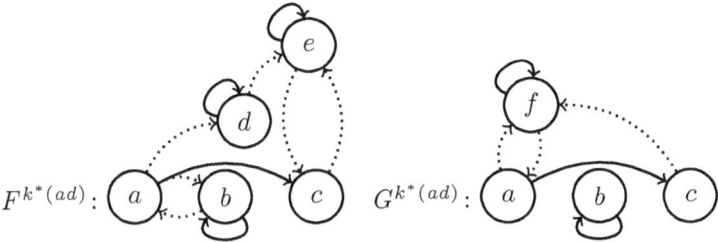

Obviously, F and G do not possess the same kernels but note that their restrictions to the shared arguments do, i.e. $\left(F|_{\{a,b,c\}}\right)^{k^*(ad)} = \left(G|_{\{a,b,c\}}\right)^{k^*(ad)}$.

It turned out that the issues raised in Example 4.27 are essential to characterize normal deletion equivalence w.r.t. admissible semantics. In case of complete and grounded semantics slightly different conditions have to be fulfilled, namely w.r.t. the non-shared arguments we have "it is forbidden to be attacked" instead of "counter-attack if attacked" like in case of admissible semantics and furthermore, instead of the ad-*-kernel the corresponding σ-*-kernels are used. Consider therefore the following definition and the characterization theorem. We use Δ to denote the symmetric difference, i.e. $A \Delta A' = A \setminus A' \cup A' \setminus A$. Moreover, $NL(F) = A(F) \setminus L(F)$, i.e. $NL(F)$ contains all arguments of F which are not self-defeating.

Definition 4.28 Given $F = (A, R)$ and $G = (A', R')$ and let $\sigma \in \{co, gr\}$.

1. $Loop(F,G) \Leftrightarrow_{def} L(F \cup G|_{A \Delta A'}) = A \Delta A'$,
    ("non-shared args are self-defeating")

2. $Att^{ad}(F,G) \Leftrightarrow_{def} \forall b \in A \setminus A' \ \forall a \in NL(F|_{A \cap A'}) : ((b,a) \in R \to (a,b) \in R)$
    $\land \ \forall b \in A' \setminus A \ \forall a \in NL(G|_{A \cap A'}) : ((b,a) \in R' \to (a,b) \in R')$,

("counter-attack if attacked")

3. $Att^\sigma(F,G) \Leftrightarrow_{def} \forall b \in A \setminus A' \; \forall a \in NL(F|_{A \cap A'}) : (b,a) \notin R$
   $\wedge \; \forall b \in A' \setminus A \; \forall a \in NL(G|_{A \cap A'}) : (b,a) \notin R'$.
   ("it is forbidden to be attacked")

Theorem 4.29 ([Baumann, 2014a]) *Let $\sigma \in \{ad, co, gr\}$. Given two finite AFs $F = (A, R)$ and $G = (A', R')$ and let $I = A \cap A'$,*

$$F \equiv_{ND}^{\mathcal{E}_\sigma} G \Leftrightarrow Loop(F,G), \; Att^\sigma(F,G), (F|_I)^{k^*(\sigma)} = (G|_I)^{k^*(\sigma)}.$$

In contrast to admissible, complete and grounded semantics where normal deletion equivalence is indeed weaker than normal expansion equivalence we observe that these notions coincide in case of stable semantics. This means, normal deletion equivalence w.r.t. stable semantics is characterized by the classical stable kernel too.

The following theorem corrects the corresponding result in [Baumann, 2014a, Theorem 10] which did not take into account that an empty framework possess a stable extension, namely the empty one.[26]

Theorem 4.30 *For finite AFs F and G we have:*

$$F \equiv_{ND}^{\mathcal{E}_{stb}} G \Leftrightarrow F^{k(stb)} = G^{k(stb)}.$$

Proof. ($\Rightarrow$) We show the contrapositive, i.e. $F^{k(stb)} \neq G^{k(stb)} \Rightarrow F \not\equiv_{ND}^{\mathcal{E}_{stb}} G$.
1st case: Assume $A\left(F^{k(stb)}\right) \neq A\left(G^{k(stb)}\right)$ and w.l.o.g. let $a \in A\left(F^{k(stb)}\right) \setminus A\left(G^{k(stb)}\right)$. Since the stable kernel is node-preserving (Fact 4.12) we obtain $G \setminus B = (\emptyset, \emptyset)$ and $F \setminus B \in \{(\{a\}, \emptyset), (\{a\}, \{(a,a)\})\}$ if $B = (A(F) \cup A(G)) \setminus \{a\}$. In either case, $\emptyset \in \mathcal{E}_{stb}(G) \setminus \mathcal{E}_{stb}(F)$ since $\mathcal{E}_{stb}(F) \in \{\emptyset, \{\{a\}\}\}$. From now on we assume $A\left(F^{k(stb)}\right) = A\left(G^{k(stb)}\right)$.
2nd case: Consider $R\left(F^{k(stb)}\right) \neq R\left(G^{k(stb)}\right)$ and w.l.o.g. let $(a,b) \in R\left(F^{k(stb)}\right) \setminus R\left(G^{k(stb)}\right)$. Let $a = b$. Remember that the stable kernel is loop-preserving (Fact 4.12). Therefore, $(a,a) \in R(F) \setminus R(G)$. We obtain $G \setminus B = (\{a\}, \emptyset)$ and $F \setminus B = (\{a\}, \{(a,a)\})$ if $B = (A(F) \cup A(G)) \setminus \{a\}$. Hence, $\emptyset = \mathcal{E}_{stb}(F) \neq \mathcal{E}_{stb}(G) = \{\{a\}\}$. From now on we assume $L\left(F^{k(stb)}\right) = L\left(G^{k(stb)}\right)$. Consider now $a \neq b$. Consequently, $(a,b) \in R(F)$ and $(a,a) \notin R(F)$. Hence, $(a,a) \notin R(G)$ and furthermore, $(a,b) \notin R(G)$. Define $B = (A(F) \cup A(G)) \setminus \{a,b\}$. In any case, $\{a\} \in \mathcal{E}_{stb}(F \setminus B) \setminus \mathcal{E}_{stb}(G \setminus B)$ concluding the if-direction.
($\Leftarrow$) Given $F^{k(stb)} = G^{k(stb)}$. Applying Theorems 4.14 and 4.20 one after the other yields $F \equiv_E^{\mathcal{E}_{stb}} G$ and then $F \equiv_N^{\mathcal{E}_{stb}} G$. Finally, Corollary 4.21 justifies $F \equiv_{ND}^{\mathcal{E}_{stb}} G$ concluding the proof. ∎

---

[26]We mention that Theorem 10 in [Baumann, 2014a] hold, given that resulting AFs have to be non-empty. The claimed normal deletion equivalence of the AFs F and G depicted in [Baumann, 2014a, Example 4] can be disproved by setting $B = \{a, b, c, f\}$.

Characterization Theorems in Case of Self-loop-free AFs

We already observed that apart from naive kernel any mentioned kernel k does not change anything if the considered AF F is self-loop-free, i.e. $F = F^k$ (cf. Facts 4.12 and 4.18). Consequently, any equivalence relation characterizable through such a kernel collapses to identity if we restrict ourselves to self-loop-free AFs. This is stated in the following theorem.

Theorem 4.31 *Given a relation* $\equiv\ \subseteq \mathcal{F} \times \mathcal{F}$ *characterizable through* k *where* $k \in \{k(stb), k(ad), k(gr), k(co), k^*(ad), k^*(gr), k^*(co), k^*(stg)\}$. *For any self-loop-free AFs F and G,*
$$F \equiv G \Leftrightarrow F = G.$$

We will refrain from listing all combinations of semantics and equivalence notions characterizable through a kernel mentioned in the theorem above. Please confer Figures 12 and 15 for compact overviews. For all such combinations, self-loop-free AFs are redundancy-free, i.e. all attacks as well as arguments may play a crucial role w.r.t. further evaluations and thus, there is no space for simplification. In the introductory part of this section we noted that many equivalence notions, e.g. normal and local expansion equivalence are motivated by the instantiation-based context where AFs are built from an underlying knowledge base. However, we want to mention that there are some formalisms like classical logic-based argumentation where self-attacking arguments do not occur [Besnard and Hunter, 2001, Theorem 4.13], while for other systems, e.g. ASPIC self-defeating arguments indeed may arise [Prakken, 2010, Section 7].

Summary of Results and Conclusion

In the presented results the notion of a kernel played a crucial role. Indeed, kernels are interesting from several perspectives: First, they allow to decide the corresponding notion of equivalence by a simple check for topological (i.e. syntactical) equality. Moreover, all kernels we have obtained so far can be efficiently constructed from a given argumentation framework. This means, if a certain equivalence notion is characterizable through such a kernel, then we have tractability of the associated decision problem.

The following Figure 12 provides a comprehensive overview of the state of the art in case of extension-based semantics. The entry "k" in row M and column σ indicates that $\equiv_M^{\mathcal{E}_\sigma}$ is characterizable through k. The abbreviation "id" stands for identity map and the question mark represents an open problem. Further abbreviations like "L" and "Att^σ" refer to additional conditions relevant in case of normal deletion equivalence (cf. Theorem 4.29). The entry "$[m,n]$" indicates three facts. First, the characterization problem is already solved in Theorem/Proposition n in m.[27] Second, the characterization result is not (purely) kernel-based and third, it can be checked that none of the introduced kernels serve as a characterization.

---

[27]For m we use the following assignments: 1 = [Baumann, 2011], 2 =[Baumann and Brewka, 2015b], 3 =[Baumann and Brewka, 2013] and 4 =[Oikarinen and Woltran, 2011]

	stg	stb	ss	eg	ad	pr	il	gr	co	na	cf2	stg2
W	?	[1,3]	?	?	[2,1]	[3,1]	?	?	[2,1]	?	?	?
L	$k^*(stg)$	[4,9]	$k(ad)$	$k(ad)$	$k(ad)$	$k(ad)$	$k(ad)$	[4,10]	[4,11]	$k(na)$	?	?
E	$k(stb)$	$k(stb)$	$k(ad)$	$k(ad)$	$k(ad)$	$k(ad)$	$k(ad)$	$k(gr)$	$k(co)$	$k(na)$	id	id
N	$k(stb)$	$k(stb)$	$k(ad)$	$k(ad)$	$k(ad)$	$k(ad)$	$k(ad)$	$k(gr)$	$k(co)$	$k(na)$	id	id
S	$k(stb)$	$k(stb)$	$k(ad)$	$k(ad)$	$k^*(ad)$	$k^*(ad)$	$k^*(ad)$	$k^*(gr)$	$k^*(co)$	$k(na)$	?	?
ND	?	$k(stb)$	?	?	$k^*(ad)$, L, Att^{ad}	?	?	$k^*(gr)$, L, Att^{gr}	$k^*(co)$, L, Att^{co}	?	?	?
D	id	id	id	id	id	id	id	id	id	id	id	id
LD	id	id	id	id	id	id	id	id	id	id	id	id
U	id	id	id	id	id	id	id	id	id	id	id	id

Figure 12: Extension-based Characterizations for Finite AFs

Remember that any arbitrary expansion (deletion) can be split into a normal and local part. So one natural conjecture is that normal and local expansion (deletion) equivalence jointly imply expansion (deletion) equivalence. Using the results presented in this section we can not only verify the addressed conjecture but even give a significantly stronger result. In fact, the main and quite surprisingly relations for the considered semantics can be briefly and concisely stated in the following two equations, namely "normal expansion equivalence = expansion equivalence" and "local deletion equivalence = deletion equivalence".

The fact that different notions of equivalence might or might not coincide is interesting from a conceptual point of view. To illustrate this let us have a look at normal and strong expansion equivalence. Recall that normal expansions add new arguments and possibly new attacks which involve at least one of the fresh arguments, while strong expansions (a subclass of normal expansions) restrict the possible attacks between the new arguments and the old ones to a single direction. In dynamic settings, both concepts can be justified in the sense that new arguments might be raised but this will not influence the relation between already existing arguments. For strong expansions, only strong arguments will be raised, i.e. arguments which cannot be attacked by existing ones. The corresponding equivalence notions now check whether two AFs are "equally robust" to such new arguments, and indeed, normal expansion equivalence always implies strong expansion equivalence but the other direction is only true for some of the semantics, namely stage, stable, semi-stable, eager and naive semantics. One interpretation is that when two AFs are not normal expansion equivalent, then this can be made explicit by only posing strong arguments (not attacked by existing ones), while for the other semantics this is not the case. For this particular example, it seems that the notion of admissibility which is more "explicit" in the admissible, preferred, ideal, grounded and complete semantics is responsible for the fact that frameworks might be strong expansion equivalent but not normal expansion equivalent.

In Figure 11 we presented preliminary relations between several notions of equivalence which hold for any semantics. The refinement depicted in Figure 13 applies to any extension-based semantics considered in this section.

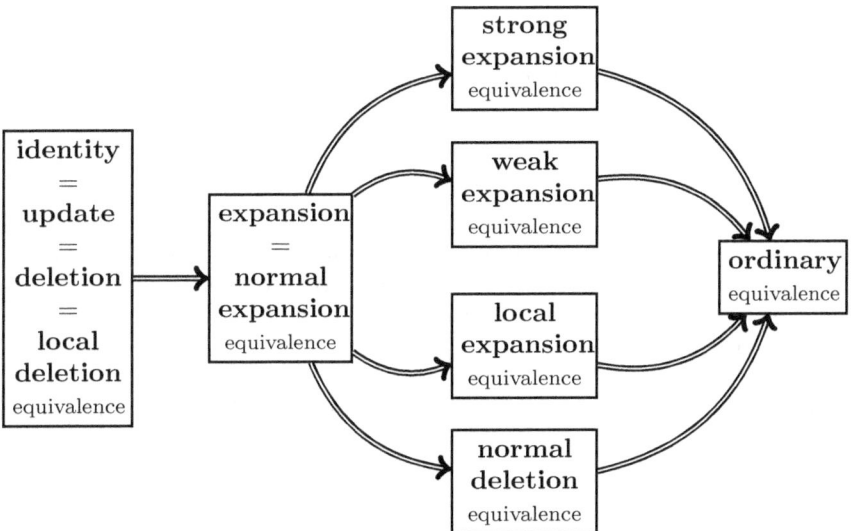

Figure 13: Relations for $\sigma \in \{stg, stb, ss, eg, ad, pr, il, gr, co, na, cf2, stg2\}$ - Extension-based Versions and Finite AFs

Finally, we present the overall picture for the most prominent semantics, namely the stable one. Interestingly, in contrast to Figure 13 all equivalence notions are comparable, i.e. they are totally ordered w.r.t. $\subseteq$. Comprehensive overviews for single semantics can be found in [Baumann, 2014b, Section 5.5.2] or [Baumann and Brewka, 2015b]. The latter also contains a comparison to different notions of *minimal change equivalence* firstly introduced in [Baumann, 2012b]. As an aside, very recently the authors of [Baumann et al., 2017] introduced so-called *C-relativized equivalence* that subsumes ordinary and expansions equivalence as its extreme corner cases. The set C represents so-called *core* arguments which will not be directly touched by the possible expansions. This means, for any set C we obtain a further intermediate notion between expansion and ordinary equivalence. However, due to its recency further relations are not studied so far.

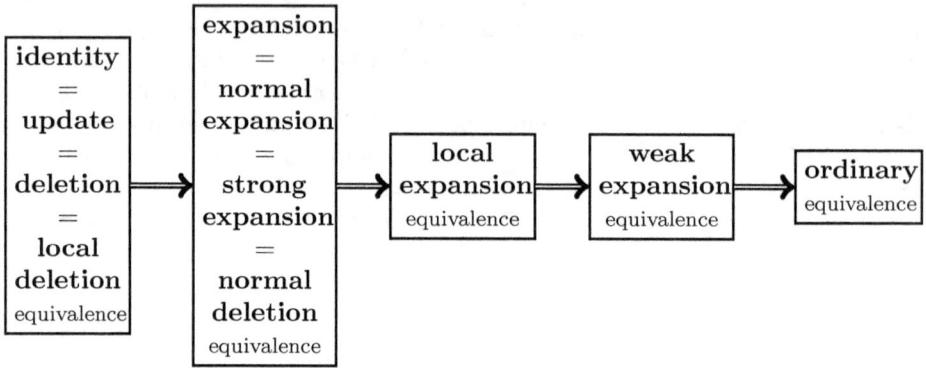

Figure 14: Stable Semantics - Extension-based Version and Finite AFs

4.3 Equivalence in the Light of Unrestricted Frameworks

Recently, a first study of several abstract properties in the unrestricted setting were presented in [Baumann and Spanring, 2017]. The main result regarding expansion equivalence can be summarized as follows: All characterization results carry over to the unrestricted setting as long as the AFs in question are *jointly expandable* (w.r.t. $\mathcal{U}$). Consider therefore the following definition and the corresponding characterization theorem.

Definition 4.32 F and G are jointly expandable if $\mathcal{U} \setminus (A(F) \cup A(G)) \neq \emptyset$.

Theorem 4.33 *[Baumann and Spanring, 2017]* For jointly expandable AFs F and G we have:

1. $F \equiv_E^{\mathcal{E}_\sigma} G \Leftrightarrow F^{k(\sigma)} = G^{k(\sigma)}$ for any $\sigma \in \{stb, ad, co, gr, na\}$,

2. $F \equiv_E^{\mathcal{E}_\sigma} G \Leftrightarrow F^{k(ad)} = G^{k(ad)}$ for any $\sigma \in \{pr, il, ss, eg\}$ and

3. $F \equiv_E^{\mathcal{E}_{stg}} G \Leftrightarrow F^{k(stb)} = G^{k(stb)}$.

The main proof strategies are straightforward extensions of those presented in [Oikarinen and Woltran, 2011]. However, finiteness assumptions are often used implicitly and one has to pay attention whether a certain reasoning step (e.g. subset relation between semantics, definedness statuses of semantics, finitely many extensions etc.) carry over to the infinite setting.

Interestingly, in case of the admissible as well as naive kernel we may even drop the restriction of joint expandability as stated in the following theorem.

Theorem 4.34 *[Baumann and Spanring, 2017]* For unrestricted AFs F and G we have:

1. $F \equiv_E^{\mathcal{E}_{na}} G \Leftrightarrow F^{k(na)} = G^{k(na)}$ and

2. $F \equiv_E^{\mathcal{E}_\sigma} G \Leftrightarrow F^{k(ad)} = G^{k(ad)}$ for any $\sigma \in \{ad, pr, il, ss, eg\}$.

The following two examples taken from [Baumann and Spanring, 2017] show that this assertion does not hold for all kernels considered in this section. The main reason for this different behaviour is that for some semantics it plays a decisive role whether AFs can be expanded by "fresh" arguments which is not given for unrestricted frameworks in general but guaranteed for jointly expandable AFs (cf. Definition 4.32).

Example 4.35 Given $c \in \mathcal{U}$ and define $F = (\mathcal{U} \setminus \{c\}, \{(a,a) \mid a \in \mathcal{U} \setminus \{c\}\})$ and $G = (\mathcal{U}, \{(a,a) \mid a \in \mathcal{U} \setminus \{c\}\})$. For any H we observe $\mathcal{E}_{stb}(F \cup H) = \mathcal{E}_{stb}(G \cup H)$. In particular,

$$\mathcal{E}_{stb}(F \cup H) = \begin{cases} \{\{c\}\}, & \text{if } \{(c,a) \mid a \in \mathcal{U} \setminus \{c\}\} \subseteq R(H) \text{ and } (c,c) \notin R(H) \\ \emptyset, & \text{otherwise} \end{cases}$$

Consequently, $F \equiv_E^{\mathcal{E}_{stb}} G$ although $A(F) \neq A(G)$ (and thus, $F^{k(stb)} \neq G^{k(stb)}$).

Example 4.36 Consider the AFs $F = (\mathcal{U}, \{(a,a) \mid a \in \mathcal{U}\})$ as well as $G = (\mathcal{U}, \{(a,b) \mid a,b \in \mathcal{U}, a \neq b\})$. Applying the grounded kernel does not change anything for either framework, i.e. $F^{k(gr)} = F$ and $G = G^{k(gr)}$. Due to the absence of unattacked arguments we deduce $\mathcal{E}_{gr}(F \cup H) = \mathcal{E}_{gr}(G \cup H) = \{\emptyset\}$ for any AF H. Consequently, $F \equiv_E^{\mathcal{E}_{gr}} G$ although $F^{k(gr)} \neq G^{k(gr)}$.

4.4 Characterization Theorems for Labelling-Based Semantics

We now return to the finite setting and consider the second main approach used for evaluating argumentation scenarios, namely labelling-based semantics. As a matter of fact, the labelling-based versions of all considered semantics provides one with more information than their extension-based counter-parts. More precisely, the defined 3-valued labellings assign a status to any argument of the considered AF F, i.e. in addition to the information which arguments are *accepted* we also have labels for the remaining arguments indicating that they are either *rejected* or *undecided* with respect to F (cf. Chapter 4 for more background and precise definitions). It is well known that many semantics establish a one-to-one correspondence between their extension-based and labelling-based versions. This means, any labelling is associated with exactly one extension and vice versa. It is not immediately apparent whether this property guarantees that there is a coincidence of the extension-based and labelling-based equivalence notions. In [Baumann, 2016] a negative answer was given. The main reason for the invalidity is that AFs may possess the same extensions without sharing the same arguments which is impossible in case of labellings since any argument has to be labelled. Furthermore, even sharing the same arguments does not ensure the validity of the converse direction. Consider therefore the following example.

Example 4.37 *Consider the AFs F and G as depicted below. Although both frameworks possess the same unique preferred extension, they do not share the same preferred labellings. More precisely, $\mathcal{E}_{pr}(F) = \mathcal{E}_{pr}(G) = \{\{a\}\}$ but $\{(\{a\}, \{b\}, \emptyset)\} = \mathcal{L}_{pr}(F) \neq \mathcal{L}_{pr}(G) = \{(\{a\}, \emptyset, \{b\})\}$.*

$F : \; a \quad\quad b \quad\quad\quad G : \; a \quad\quad b$

Moreover, observe that $F^{k^*(ad)} = G = G^{k^*(ad)}$. Consequently, both frameworks are even strong expansion equivalent w.r.t. preferred extension-based semantics (Theorem 4.20). This means, equivalence notions may differ considerably if considered under the extension-based or labelling-based approach.

In contrast to extension-based semantics where characterization results are spread over a high number of publications there is only one reference, namely [Baumann, 2016] concerned with labelling-based semantics. The author considered 8 different equivalence notions w.r.t. 8 prominent labelling-based semantics in the finite setting. In effect, similarly to extension-based semantics, almost all labelling-based equivalence notions can be decided syntactically. Differently from the extension-based approach we observe a much more homogeneous picture. For instance, there is no need for the more sophisticated σ-*-kernels as we will see.

Basic Properties and a Fundamental Relation

Before turning to the main results we start with some preliminary facts relating σ-extensions and σ-labellings. In the following we restrict ourselves to the semantics considered in [Baumann, 2016]. For any 3-valued labelling $L = (L_1, L_2, L_3)$ we use $L = (L^{\text{I}}, L^{\circ}, L^{\text{U}})$ as usual.

Fact 4.38 *Given a finite AF $F = (A, R)$ and $E \subseteq A$. We write $E^{\mathcal{L}}$ for $(E, E^+, A \setminus E^{\oplus})$. For all $\sigma \in \{stb, ss, eg, ad, pr, il, gr, co\}$ we have,*

1. *If $L \in \mathcal{L}_\sigma(F)$, then $L^{\text{I}} \in \mathcal{E}_\sigma(F)$,*   *(extension induced by labelling)*

2. *If $E \in \mathcal{E}_\sigma(F)$, then $E^{\mathcal{L}} \in \mathcal{L}_\sigma(F)$ and*   *(labelling induced by extension)*

3. *Obviously, $(E^{\mathcal{L}})^{\text{I}} = E$.*   *($I \circ \mathcal{L} = id$)*

We point out that the first two properties mentioned in Fact 4.38 do not ensure that there is a one-to-one correspondence between σ-labellings and σ-extensions. This desirable feature (which would indeed justify the terms σ-labellings and σ-extensions) is given if additionally, labellings are uniquely determined by their in-labelled arguments.

Fact 4.39 *Given a finite AF $F = (A, R)$ and a set $E \subseteq A$. For all semantics $\sigma \in \{stb, ss, eg, pr, il, gr, co\}$ we have,*

1. *For any $L, M \in \mathcal{L}_\sigma(F), L^{\text{I}} = M^{\text{I}}$ iff $L = M$,*
     (uniquely determined by in-labels)

2. Given $L \in \mathcal{L}_\sigma(F)$, then $(L^I)^\mathcal{L} = L$ and $\qquad$ ($\mathcal{L} \circ I = id$)

3. $|\mathcal{L}_\sigma(F)| = |\mathcal{E}_\sigma(F)|$. $\qquad$ (same cardinality)

As an aside, we mention that (although not immediately apparent) the first two items of Fact 4.39 are equivalent independently of any semantics definition. Please note that admissible labellings are excluded from Fact 4.39. The AF F depicted in Example 4.37 shows that this is no coincidence. It possesses two admissible labellings associated with one admissible extension. More precisely, the admissible labellings $(\{a\}, \{b\}, \emptyset)$ as well as $(\{a\}, \emptyset, \{b\})$ refer to the same admissible extension $\{a\}$.

We proceed with a general relation between labelling-based and extension-based versions of certain equivalence notion. More precisely, for any considered semantics and any equivalence notion presented in Definition 4.6 we have that being equivalent w.r.t. labellings implies being equivalent w.r.t. extensions. The main reason for this fundamental relation is the following lemma stating that possessing the same labellings implies sharing the same extensions. We mention that this property is already guaranteed if the semantics σ in question satisfies that any σ-extension induces an σ-labelling and vice versa (cf. statements 1 and 2 of Fact 4.38).

Lemma 4.40 ([Baumann, 2016]) *Given two finite AFs F and G. For any semantics $\sigma \in \{stb, ss, eg, ad, pr, il, gr, co\}$ we have,*

$$\mathcal{L}_\sigma(F) = \mathcal{L}_\sigma(G) \Rightarrow \mathcal{E}_\sigma(F) = \mathcal{E}_\sigma(G).$$

Proof. Reductio ad absurdum. Assume $\mathcal{E}_\sigma(F) \neq \mathcal{E}_\sigma(G)$. Then, w.l.o.g. exists $E \in \mathcal{E}_\sigma(F) \setminus \mathcal{E}_\sigma(G)$. Consequently, $E^\mathcal{L} \in \mathcal{L}_\sigma(F)$ (item 2 of Fact 4.38). Thus, $E^\mathcal{L} \in \mathcal{L}_\sigma(G)$ (assumption). Hence, $(E^\mathcal{L})^I \in \mathcal{E}_\sigma(G)$ (item 1 of Fact 4.38). Furthermore, $(E^\mathcal{L})^I = E \in \mathcal{E}_\sigma(G)$ (item 3 of Fact 4.38). Contradiction! ∎

We now present the fundamental relation between labelling-based and extension-based equivalence notion.

Theorem 4.41 ([Baumann, 2016]) *Given two finite AFs F and G. For any $\sigma \in \{stb, ss, eg, ad, pr, il, gr, co\}$ and any $M \in \{W, L, E, N, S, ND, D, LD, U\}$ we have,*

$$F \equiv_M^{\mathcal{L}_\sigma} G \Rightarrow F \equiv_M^{\mathcal{E}_\sigma} G.$$

Proof. We show the contrapositive. Assume $F \not\equiv_M^{\mathcal{E}_\sigma} G$. This means, there is a certain scenario S according to M, s.t. $\mathcal{E}_\sigma(S(F)) \neq \mathcal{E}_\sigma(S(G))$.[28] Consequently, $\mathcal{L}_\sigma(S(F)) \neq \mathcal{L}_\sigma(S(G))$ (Lemma 4.40) proving $F \not\equiv_M^{\mathcal{L}_\sigma} G$. ∎

---

[28] For instance, in case of expansion equivalence (i.e. M = E) a scenario S is simply the union with a further AF H, i.e. $S(F) = F \cup H$ and $S(G) = G \cup H$.

In Example 4.37 we have seen that the converse direction does not hold in general. Nevertheless, there is huge number of equivalence notions where labelling-based and extension-based versions do indeed coincide (cf. Figure 15 for an overview).

Coincidences of Extension-based and Labelling-based Versions

Remember that the identity relation is the finest equivalence relation. Furthermore, it is already shown that deletion, local deletion as well as update equivalence w.r.t. $\mathcal{E}_\sigma$ collapse to identity (see Figure 13). Consequently, applying the fundamental relation stated in Theorem 4.41 we obtain the identical characterization results w.r.t. labelling-based semantics.

Theorem 4.42 ([Baumann, 2016]) *For finite AFs F and G, a scenario $M \in \{D, LD, U\}$ and a semantics $\sigma \in \{stb, ss, eg, ad, pr, il, gr, co\}$ we have,*

$$F \equiv_M^{\mathcal{L}_\sigma} G \Leftrightarrow F = G.$$

Analogously to extension-based semantics (cf. Fact 4.13) we have that there are combinations of kernels and semantics σ, s.t. the application of a kernel does not vary the set of σ-labellings.

Fact 4.43 *For any finite AF F,*

1. $\mathcal{L}_\sigma(F) = \mathcal{L}_\sigma\left(F^{k(\sigma)}\right)$ *for $\sigma \in \{co, stb, gr\}$ and*

2. $\mathcal{L}_\tau(F) = \mathcal{L}_\tau\left(F^{k(ad)}\right)$ *for $\tau \in \{ss, eg, pr, il\}$.*

The fact above is the decisive property which allows one to carry over further kernel-based characterization results for extension-based semantics to their labelling-based version. In order to show this result it was necessary to find a condition for equality of two complete labellings of different AFs. Remember that two complete labellings of the same framework are identical if and only if they possess the same in-labelled arguments (Fact 4.39). In case of different AFs we have to require additionally that both frameworks share the same arguments and the same range w.r.t. the set of in-labelled arguments.

Fact 4.44 *Given two finite AFs F and G as well as $L \in \mathcal{L}_{co}(F)$ and $M \in \mathcal{L}_{co}(G)$. We have $L = M$ iff simultaneously $A(F) = A(G)$, $L^I = M^I$ and $R_F^+(L^I) = R_G^+(M^I)$.*

Please observe that admissible labellings do not fulfill Fact 4.44. Consider for instance again the AF F depicted in Example 4.37 and its two admissible labellings $(\{a\}, \{b\}, \emptyset)$ and $(\{a\}, \emptyset, \{b\})$.

We proceed with the main coincidence theorem. It stipulates that several expansion equivalence relations as well as weaker notions do not distinguish between their labelling-based and extension-based version. This means, kernel-based characterization results (depicted in Figure 11) carry over to labelling-based semantics. Similarly to extension-based semantics we present an overview of characterizing kernels at the end of this section (cf. Figure 15).

Theorem 4.45 ([Baumann, 2016]) *Given finite AFs F and G. We have,*

1. $F \equiv_M^{\mathcal{E}_\sigma} G \Leftrightarrow F \equiv_M^{\mathcal{L}_\sigma} G$ *for* $\sigma \in \{stb, ss, eg, pr, il, gr, co\}, M \in \{E, N\}$,

2. $F \equiv_L^{\mathcal{E}_\sigma} G \Leftrightarrow F \equiv_L^{\mathcal{L}_\sigma} G$ *for* $\sigma \in \{ss, eg, pr, il\}$ *and*

3. $F \equiv_S^{\mathcal{E}_\sigma} G \Leftrightarrow F \equiv_S^{\mathcal{L}_\sigma} G$ *for* $\sigma \in \{stb, ss, eg\}$.

Non-Coincidence of Extension-based and Labelling-based Versions
We now leave the realm of uniformity of extension-based and labelling-based characterizations. This section is divided into three parts. We start with characterization theorems for admissible labellings. In particular, we will see that the admissible kernel (originally introduced to characterize equivalence notions w.r.t. admissible extension-based semantics) does not serve as characterizing kernel for admissible labellings. We then proceed with strong expansion equivalence w.r.t. labellings. We will see that the remaining notions are characterizable via traditional kernels instead of σ-*-kernels. In the third part we consider normal deletion equivalence w.r.t. labelling-based semantics. In contrast to their extension-based versions where many notions has defied any attempt of solving, we present characterization theorems based on traditional kernels for all eight considered semantics.

Expansion Equivalence w.r.t. Admissible Labellings Expansion equivalence as well as its local, normal and strong versions w.r.t. admissible extensions are characterizable through the admissible kernel. The following example shows that this assertion does not hold in case of admissible labellings.

Example 4.46 *The following two AFs possess the same admissible kernels, namely $F^{k(ad)} = G^{k(ad)} = F$. Consequently, applying characterization theorems for extension-based semantics we obtain $F \equiv_M^{\mathcal{E}_{ad}} G$ for $M \in \{L, E, N\}$ (cf. Figure 12).*

$F : \quad \widehat{a} \rightleftarrows \widehat{b} G : \quad \widehat{a} \rightarrow \widehat{b}$

Observe that $(\{b\}, \emptyset, \{a\}) \in \mathcal{L}_{ad}(G) \setminus \mathcal{L}_{ad}(F)$ because the argument a cannot be undecided in F since it attacks the in-labelled argument b. Thus $F \not\equiv_M^{\mathcal{L}_{ad}} G$ for $M \in \{L, E, N, S\}$.

Let us assume that the equivalence notions considered in the example above are characterizable through a certain kernel k. Due to the fundamental relation (Theorem 4.41) and the characterization results w.r.t. admissible extensions (Figure 12), we already know that the kernel k has to satisfy the following implication: $F^k = G^k \Rightarrow F^{k(ad)} = G^{k(ad)}$ for any two AFs F and G. This means, we are looking for a weaker kernel than the admissible one in the sense that first, everything which is redundant w.r.t. k has to be redundant w.r.t. the admissible kernel too; and second, an attack from a to b has to survive even if a is self-defeating and b counterattacks a. One candidate for k is the complete kernel since redundancy w.r.t. the complete kernel implies redundancy w.r.t. to the admissible one, and furthermore, it deletes an attack between two arguments if and only if both are self-defeating. And indeed, it was shown that expansion equivalence as well as its local, normal and strong variant w.r.t. admissible labellings are characterizable through the complete kernel as stated by the following theorem.

Theorem 4.47 ([Baumann, 2016]) *Given finite AFs F and G. We have,*

$$F \equiv_M^{\mathcal{L}_{ad}} G \Leftrightarrow F^{k(co)} = G^{k(co)} \text{ with } M \in \{L, E, N, S\}.$$

Strong Expansion Equivalence for Preferred, Ideal, Grounded and Complete Labellings In this subsection we present characterization theorems for strong expansion equivalence w.r.t. labelling-based preferred, ideal, grounded and complete semantics. Remember that in case of strong expansions a former attack between old arguments will never become a counterattack to an added attack. Consequently, in contrast to arbitrary expansions former attacks do not play a role with respect to being a potential defender of an added argument. The context-sensitive σ-*-kernels took these considerations into account and allow for more deletions than their classical counterparts.

Example 4.48 *According to Definition 4.17 we have, $F^{k^*(\sigma)} = G^{k^*(\sigma)}$ for any semantics $\sigma \in \{ad, gr, co\}$. More precisely, the attacks (a,b) in F as well as (c,b) in G are redundant w.r.t. all three σ-*-kernels. This means, in consideration of Figure 12 both frameworks are strong expansion equivalent w.r.t. the extension-based versions of preferred, ideal, grounded and complete semantics.*

Consider the following dynamic scenario where a stronger argument than the former ones is added. Formally, we conjoin the AF $H = (\{c, d\}, \{(d, c)\})$ to both frameworks F and G.

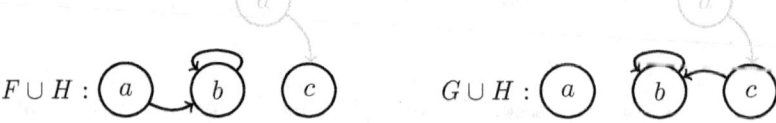

Note that both frameworks has to possess the same σ-extension since $G \equiv_S^{\mathcal{E}_\sigma} H$ for $\sigma \in \{pr, il, gr, co\}$ is already ensured. Furthermore, we observe $(\{a, d\}, \{b, c\},$

$\emptyset) \in \mathcal{L}_\sigma(F \cup H) \setminus \mathcal{L}_\sigma(G \cup H)$ since b cannot be out-labelled in $G \cup H$ because there is no in-labelled attacker. This means, $F \not\equiv_S^{\mathcal{L}_\sigma} G$ for $\sigma \in \{pr, il, gr, co\}$.

Analogously to the previous section let us assume that strong expansion equivalence w.r.t. the considered labelling-based semantics are characterizable through a certain kernel k. We immediately obtain, $F^k = G^k \Rightarrow F^{k^*(\sigma)} = G^{k^*(\sigma)}$ for any two AFs F and G. Possible candidates are the classical counterparts of the σ-*-kernels and indeed it was shown that these kernels guarantee the desired outcome. This means, in case of strong expansion equivalence w.r.t. preferred, ideal, grounded and complete semantics we have that the labelling-based version is characterizable through a classical σ-kernel if and only if the extension-based version is characterizable through the corresponding σ-*-kernel.

Theorem 4.49 ([Baumann, 2016]) *Given finite AFs F and G. We have,*

1. $F \equiv_S^{\mathcal{L}_\sigma} G \Leftrightarrow F^{k(ad)} = G^{k(ad)}$ for $\sigma \in \{pr, il\}$,

2. $F \equiv_S^{\mathcal{L}_{gr}} G \Leftrightarrow F^{k(gr)} = G^{k(gr)}$ and

3. $F \equiv_S^{\mathcal{L}_{co}} G \Leftrightarrow F^{k(co)} = G^{k(co)}$.

Normal Deletion Equivalence Characterizing normal deletion equivalence in case of extension-based semantics is exceptional in several regards. Remember that normal deletions retract arguments and their corresponding attacks. Firstly, only a few characterization results are achieved (cf. Figure 12). Furthermore, apart from stable semantics, none of the characterization results is purely kernel-based, i.e. beside the equality of kernels on certain parts of the frameworks further loop- as well as attack-conditions have to be satisfied. Finally, quite surprisingly, normal deletion equivalent AFs do not even have to share the same arguments enabling equivalence classes with an infinite number of elements. Being equivalent w.r.t. labellings and possessing different arguments at the same time is impossible in case of labellings since any argument has to be labelled. It turned out that any considered labelling-based semantics is characterizable through traditional kernels and thus, do not share any of the features mentioned above. Consider the following main theorem.

Theorem 4.50 ([Baumann, 2016]) *Given finite AFs F and G. We have,*

1. $F \equiv_{ND}^{\mathcal{L}_{stb}} G \Leftrightarrow F^{k(stb)} = G^{k(stb)}$,

2. $F \equiv_{ND}^{\mathcal{L}_\sigma} G \Leftrightarrow F^{k(ad)} = G^{k(ad)}$ for $\sigma \in \{ss, eg, pr, il\}$,

3. $F \equiv_{ND}^{\mathcal{L}_\sigma} G \Leftrightarrow F^{k(co)} = G^{k(co)}$ for $\sigma \in \{ad, co\}$ and

4. $F \equiv_{ND}^{\mathcal{L}_{gr}} G \Leftrightarrow F^{k(gr)} = G^{k(gr)}$.

Summary of Results and Conclusion

The following Figure 15 presents a comprehensive overview of the state of the art in case of labelling-based semantics. Analogously to Figure 12 the entry "k" in row M and column σ indicates that $\equiv_M^{\mathcal{L}_\sigma}$ is characterizable through k given the finiteness restriction. The abbreviation "id" stands for identity map and the question mark represents an open problem.[29] A grey-highlighted entry reflects the situation that extension-based and labelling-based version do not coincide.

	stb	ss	eg	ad	pr	il	gr	co
L	?	$k(ad)$	$k(ad)$	$k(co)$	$k(ad)$	$k(ad)$		?
E	$k(stb)$	$k(ad)$	$k(ad)$	$k(co)$	$k(ad)$	$k(ad)$	$k(gr)$	$k(co)$
N	$k(stb)$	$k(ad)$	$k(ad)$	$k(co)$	$k(ad)$	$k(ad)$	$k(gr)$	$k(co)$
S	$k(stb)$	$k(ad)$	$k(ad)$	$k(co)$	$k(ad)$	$k(ad)$	$k(gr)$	$k(co)$
ND	$k(stb)$	$k(ad)$	$k(ad)$	$k(co)$	$k(ad)$	$k(ad)$	$k(gr)$	$k(co)$
D	id	id	id	id	id	id	id	id
LD	id	id	id	id	id	id	id	id
U	id	id	id	id	id	id	id	id

Figure 15: Labelling-based Characterizations for Finite AFs

In contrast to extension-based semantics we observe a much more homogeneous picture. Firstly, there is no need for the more sophisticated σ-*-kernels. Secondly, normal deletion equivalence w.r.t. labelling-based semantics is naturally incorporated in the overall picture in the sense that it coincides with its corresponding expansion, normal expansion and strong expansion equivalence notions.

The following Figure 16 applies to each one of the eight labelling-based semantics considered in this section. In comparison to Figure 11 where pre-

---

[29] In contrast to extension-based semantics the labelling versions of conflict-free-based semantics like stage, naive, cf2 as well as stage2 semantics (cf. [Caminada, 2011; Gaggl and Dvořák, 2016]) as well as weak expansion equivalence at all were not considered so far and thus, represent open problems too.

liminary relations are depicted it illustrates (to a certain extent) a collapse of the diversity of the introduced equivalence notions in case of labelling-based semantics.

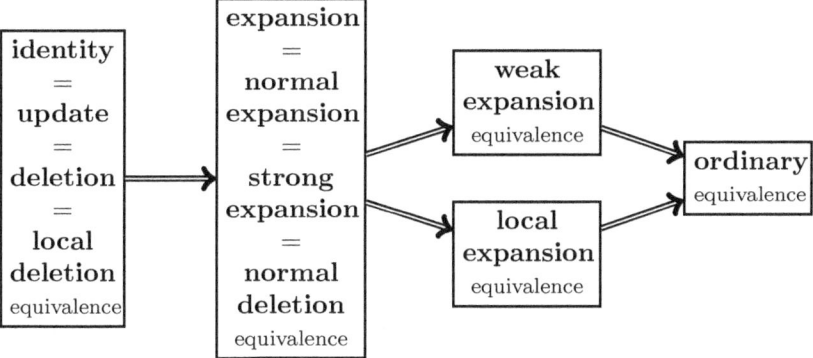

Figure 16: Relations for $\sigma \in \{stb, ss, eg, ad, pr, il, gr, co\}$ - Labelling-based Versions and Finite AFs

4.5 Final Remarks

In this section we motivated and discussed several notions of equivalence in the context of abstract argumentation and provided an exhaustive number of characterization theorems for extension-based as well as labelling-based semantics. In general we may state that Dung's abstract argumentation frameworks are a very compact formalism since the majority of the considered equivalence notion possess only little space for redundancy. Moreover, most of these notions collapse to identity if self-loop-free AFs are considered. This means, in this case any subframework of the AF in question may play a decisive role w.r.t. further evaluations and thus, cannot be locally replaced by another. This insight is sometimes used as an argument against the usefulness of the study of equivalence notions in the context of abstract argumentation. Obviously, we agree that if you are expecting much space for simplification, then the results are somehow disappointing but let us not lose sight of the fact that this is only clear *after* it has been proved. Furthermore, as already stated, the results underline that in case of abstract argumentation (almost) everything is meaningful similar to other non-monotonic formalisms available in the literature (cf. [Lifschitz et al., 2001] for logic programs, [Turner, 2004] for causal theories, [Turner, 2001] for default logic and [Truszczynski, 2006] for nonmonotonic logics in general). However, one decisive difference to these formalisms is that equivalence notions in case of abstract argumentation can be decided syntactically. Indeed, kernels are interesting from several perspectives: First, they allow to decide the corresponding notion of equivalence by a simple check for topological equality and second, all kernels we have obtained so far can be efficiently constructed from a given argumentation framework. This means, if a certain equivalence notion is characterizable through such a kernel, then we have tractability of the associated decision problem.

Acknowledgements

We would like to thank Wolfgang Dvořák for fruitful discussions which helped to improve the article.

BIBLIOGRAPHY

[Alchourrón et al., 1985] C. E. Alchourrón, P. Gärdenfors, and D. Makinson. On the logic of theory change: Partial meet contraction and revision functions. *Journal of Symbolic Logic*, 50:510–530, 1985.

[Baroni and Giacomin, 2007] Pietro Baroni and Massimiliano Giacomin. On principle-based evaluation of extension-based argumentation semantics. *Artificial Intelligence*, 171:675–700, 2007.

[Baroni et al., 2005] Pietro Baroni, Massimiliano Giacomin, and Giovanni Guida. SCC-recursiveness: a general schema for argumentation semantics. *Artificial Intelligence*, 168:162–210, 2005.

[Baroni et al., 2011] Pietro Baroni, Martin Caminada, and Massimiliano Giacomin. An introduction to argumentation semantics. *The Knowledge Engineering Review*, 26:365–410, 2011.

[Baroni et al., 2013] Pietro Baroni, Federico Cerutti, Paul E. Dunne, and Massimiliano Giacomin. Automata for infinite argumentation structures. *Artificial Intelligence*, 203:104–150, 2013.

[Baroni et al., 2014] Pietro Baroni, Guido Boella, Federico Cerutti, Massimiliano Giacomin, Leendert W. N. van der Torre, and Serena Villata. On the input/output behavior of argumentation frameworks. *Artificial Intelligence*, 217:144–197, 2014.

[Baumann and Brewka, 2010] Ringo Baumann and Gerhard Brewka. Expanding argumentation frameworks: Enforcing and monotonicity results. In *Computational Models of Argument: Proceedings of COMMA 2010, Desenzano del Garda, Italy, September 8-10, 2010.*, pages 75–86, 2010.

[Baumann and Brewka, 2013] Ringo Baumann and Gerhard Brewka. Analyzing the equivalence zoo in abstract argumentation. In *Computational Logic in Multi-Agent Systems - 14th International Workshop, CLIMA XIV, Corunna, Spain, September 16-18, 2013. Proceedings*, pages 18–33, 2013.

[Baumann and Brewka, 2015a] Ringo Baumann and Gerhard Brewka. AGM meets abstract argumentation: Expansion and revision for dung frameworks. In *IJCAI, Proceedings of the Twenty-Fourth International Joint Conference on Artificial Intelligence, 2015, Buenos Aires, Argentina, July 25-31, 2015*, pages 2734–2740, 2015.

[Baumann and Brewka, 2015b] Ringo Baumann and Gerhard Brewka. The equivalence zoo for Dung-style semantics. *Journal of Logic and Computation*, 2015.

[Baumann and Spanring, 2015] Ringo Baumann and Christof Spanring. Infinite argumentation frameworks - On the existence and uniqueness of extensions. In *Advances in Knowledge Representation, Logic Programming, and Abstract Argumentation - Essays Dedicated to Gerhard Brewka on the Occasion of His 60th Birthday*, pages 281–295, 2015.

[Baumann and Spanring, 2017] Ringo Baumann and Christof Spanring. A study of unrestricted abstract argumentation frameworks. In *IJCAI, Proceedings of the 26th International Joint Conference on Artificial Intelligence, Melbourne, China, Australia, August 19-25 ,2017*, page to appear, 2017.

[Baumann and Strass, 2012] Ringo Baumann and Hannes Strass. Default reasoning about actions via abstract argumentation. In *Computational Models of Argument: Proceedings of COMMA 2012,, September 10-12, 2012, Vienna, Austria*, pages 297–309, 2012.

[Baumann and Strass, 2013] Ringo Baumann and Hannes Strass. On the maximal and average numbers of stable extensions. In *Theory and Applications of Formal Argumentation - Second International Workshop, TAFA 2013, Beijing, China, August 3-5, 2013, Revised Selected papers*, pages 111–126, 2013.

[Baumann and Strass, 2015] Ringo Baumann and Hannes Strass. Open problems in abstract argumentation. In *Advances in Knowledge Representation, Logic Programming, and Abstract Argumentation - Essays Dedicated to Gerhard Brewka on the Occasion of His 60th Birthday*, pages 325–339, 2015.

[Baumann and Strass, 2016] Ringo Baumann and Hannes Strass. An abstract logical approach to characterizing strong equivalence in logic-based knowledge representation formalisms. In *Principles of Knowledge Representation and Reasoning: Proceedings of the Fifteenth International Conference, KR 2016, Cape Town, South Africa, April 25-29, 2016.*, pages 525–528, 2016.

[Baumann and Woltran, 2016] Ringo Baumann and Stefan Woltran. The role of self-attacking arguments in characterizations of equivalence notions. *Journal of Logic and Computation*, 26(4):1293–1313, 2016.

[Baumann et al., 2011] Ringo Baumann, Gerhard Brewka, and Renata Wong. Splitting argumentation frameworks: An empirical evaluation. In *Theorie and Applications of Formal Argumentation - First International Workshop, TAFA 2011. Barcelona, Spain, July 16-17, 2011, Revised Selected Papers*, pages 17–31, 2011.

[Baumann et al., 2012] Ringo Baumann, Gerhard Brewka, Wolfgang Dvorák, and Stefan Woltran. Parameterized splitting: A simple modification-based approach. In *Correct Reasoning - Essays on Logic-Based AI in Honour of Vladimir Lifschitz*, pages 57–71, 2012.

[Baumann et al., 2014a] Ringo Baumann, Wolfgang Dvorák, Thomas Linsbichler, Hannes Strass, and Stefan Woltran. Compact argumentation frameworks. In *ECAI 2014 - 21st European Conference on Artificial Intelligence, 18-22 August 2014, Prague, Czech Republic - Including Prestigious Applications of Intelligent Systems (PAIS 2014)*, pages 69–74, 2014.

[Baumann et al., 2014b] Ringo Baumann, Wolfgang Dvorák, Thomas Linsbichler, Hannes Strass, and Stefan Woltran. Compact argumentation frameworks. *CoRR*, abs/1404.7734, 2014.

[Baumann et al., 2016a] Ringo Baumann, Wolfgang Dvorák, Thomas Linsbichler, Christof Spanring, Hannes Strass, and Stefan Woltran. On rejected arguments and implicit conflicts: The hidden power of argumentation semantics. *Artificial Intelligence*, 241:244–284, 2016.

[Baumann et al., 2016b] Ringo Baumann, Thomas Linsbichler, and Stefan Woltran. Verifiability of argumentation semantics. *CoRR*, abs/1603.09502, 2016.

[Baumann et al., 2017] Ringo Baumann, Wolfgang Dvořák, Thomas Linsbichler, and Stefan Woltran. A general notion of equivalence for abstract argumentation. In *IJCAI, Proceedings of the 26th International Joint Conference on Artificial Intelligence, Melbourne, China, Australia, August 19-25 ,2017*, page to appear, 2017.

[Baumann, 2011] Ringo Baumann. Splitting an argumentation framework. In *Logic Programming and Nonmonotonic Reasoning - 11th International Conference, LPNMR 2011, Vancouver, Canada, May 16-19, 2011. Proceedings*, pages 40–53, 2011.

[Baumann, 2012a] Ringo Baumann. Normal and strong expansion equivalence for argumentation frameworks. *Artificial Intelligence*, 193:18–44, 2012.

[Baumann, 2012b] Ringo Baumann. What does it take to enforce an argument? minimal change in abstract argumentation. In *ECAI 2012 - 20th European Conference on Artificial Intelligence. Including Prestigious Applications of Artificial Intelligence (PAIS-2012) System Demonstrations Track, Montpellier, France, August 27-31 , 2012*, pages 127–132, 2012.

[Baumann, 2014a] Ringo Baumann. Context-free and context-sensitive kernels: Update and deletion equivalence in abstract argumentation. In *ECAI 2014 - 21st European Conference on Artificial Intelligence, 18-22 August 2014, Prague, Czech Republic - Including Prestigious Applications of Intelligent Systems (PAIS 2014)*, pages 63–68, 2014.

[Baumann, 2014b] Ringo Baumann. *Metalogical Contributions to the Nonmonotonic Theory of Abstract Argumentation*. College Publications - Studies in Logic, 2014.

[Baumann, 2016] Ringo Baumann. Characterizing equivalence notions for labelling-based semantics. In *Principles of Knowledge Representation and Reasoning: Proceedings of the Fifteenth International Conference, KR 2016, Cape Town, South Africa, April 25-29, 2016.*, pages 22–32, 2016.

[Belardinelli et al., 2015] Francesco Belardinelli, Davide Grossi, and Nicolas Maudet. Formal analysis of dialogues on infinite argumentation frameworks. In *IJCAI, Proceedings of the 24th International Joint Conference on Artificial Intelligence, Buenos Aires, Argentina, July 25-31 ,2015*, pages 861–867, 2015.

[Besnard and Hunter, 2001] Philippe Besnard and Anthony Hunter. A logic-based theory of deductive arguments. *Artificial Intelligence*, 128:203–235, 2001.

[Besnard and Hunter, 2009] Philippe Besnard and Anthony Hunter. Argumentation based on classical logic. In *Argumentation in Artificial Intelligence*, pages 133–152. Springer, 2009.

[Brewka and Woltran, 2010] Gerhard Brewka and Stefan Woltran. Abstract dialectical frameworks. In *Principles of Knowledge Representation and Reasoning: Proceedings of the Twelfth International Conference, KR 2010, Toronto, Ontario, Canada, May 9-13, 2010*, 2010.

[Brewka et al., 2013] Gerhard Brewka, Hannes Strass, Stefan Ellmauthaler, Johannes Peter Wallner, and Stefan Woltran. Abstract dialectical frameworks revisited. In *IJCAI, Proceedings of the 23rd International Joint Conference on Artificial Intelligence, Beijing, China, August 3-9, 2013*, pages 803–809, 2013.

[Brewka, 1992] Gerhard Brewka. Nonmonotonic reasoning: Logical foundations of commonsense. *SIGART Bull.*, 3(2):28–29, April 1992. Reviewer-Marek, V. W.

[Caminada and Amgoud, 2007] Martin Caminada and Leila Amgoud. On the evaluation of argumentation formalisms. *Artificial Intelligence*, 171:286–310, 2007.

[Caminada and Oren, 2014] Martin W. A. Caminada and Nir Oren. Grounded semantics and infinitary argumentation frameworks. In *Benelux Conference on Artificial Intelligence*, 2014.

[Caminada and Verheij, 2010] Martin W.A. Caminada and Bart Verheij. On the existence of semi-stable extensions. In *Benelux Conference on Artificial Intelligence*, 2010.

[Caminada et al., 2012] Martin W. Caminada, Walter A. Carnielli, and Paul E. Dunne. Semi-stable semantics. *Journal of Logic and Computation*, 22:1207–1254, 2012.

[Caminada, 2006] Martin Caminada. Semi-stable semantics. In *Computational Models of Argument: Proceedings of COMMA 2006, September 11-12, 2006, Liverpool, UK*, pages 121–130, 2006.

[Caminada, 2011] Martin Caminada. A labelling approach for ideal and stage semantics. *Argument and Computation*, 2:1–21, 2011.

[Dantsin et al., 1997] Evgeny Dantsin, Thomas Eiter, Georg Gottlob, and Andrei Voronkov. Complexity and expressive power of logic programming. In *Proceedings of the Twelfth Annual IEEE Conference on Computational Complexity, Ulm, Germany, June 24-27, 1997*, pages 82–101, 1997.

[Delgrande and Peppas, 2015] James P. Delgrande and Pavlos Peppas. Belief revision in horn theories. *Artificial Intelligence*, 218:1–22, 2015.

[Delgrande et al., 2008] James P. Delgrande, Torsten Schaub, Hans Tompits, and Stefan Woltran. Belief revision of logic programs under answer set semantics. In *Principles of Knowledge Representation and Reasoning: Proceedings of the Eleventh International Conference, KR 2008, Sydney, Australia, September 16-19, 2008*, pages 411–421, 2008.

[Delgrande et al., 2013] James P. Delgrande, Pavlos Peppas, and Stefan Woltran. Agm-style belief revision of logic programs under answer set semantics. In *Logic Programming and Nonmonotonic Reasoning, 12th International Conference, LPNMR 2013, Corunna, Spain, September 15-19, 2013. Proceedings*, pages 264–276, 2013.

[Diller et al., 2015] Martin Diller, Adrian Haret, Thomas Linsbichler, Stefan Rümmele, and Stefan Woltran. An extension-based approach to belief revision in abstract argumentation. In *IJCAI, Proceedings of the Twenty-Fourth International Joint Conference on Artificial Intelligence, 2015, Buenos Aires, Argentina, July 25-31, 2015*, pages 2926–2932, 2015.

[Dung, 1995] Phan Minh Dung. On the acceptability of arguments and its fundamental role in nonmonotonic reasoning, logic programming and n-person games. *Artificial Intelligence*, 77:321–357, 1995.

[Dunne et al., 2013] Paul E. Dunne, Wolfgang Dvořák, Thomas Linsbichler, and Stefan Woltran. Characteristics of multiple viewpoints in abstract argumentation. In *4th Workshop on Dynamics of Knowledge and Belief*, pages 16–30, 2013.

[Dunne et al., 2015] Paul E. Dunne, Wolfgang Dvořák, Thomas Linsbichler, and Stefan Woltran. Characteristics of multiple viewpoints in abstract argumentation. *Artificial Intelligence*, 228:153–178, 2015.
[Dunne et al., 2016] Paul E. Dunne, Christof Spanring, Thomas Linsbichler, and Stefan Woltran. Investigating the relationship between argumentation semantics via signatures. In *IJCAI, Proceedings of the Twenty-Fifth International Joint Conference on Artificial Intelligence, 2016, New York, NY, USA, 9-15 July 2016*, pages 1051–1057, 2016.
[Dvořák and Spanring, 2016] Wolfgang Dvořák and Christof Spanring. Comparing the expressiveness of argumentation semantics. *Journal of Logic and Computation*, 2016.
[Dvořák et al., 2011] Wolfgang Dvořák, Paul E. Dunne, and Stefan Woltran. Parametric properties of ideal semantics. In *IJCAI, Proceedings of the 22nd International Joint Conference on Artificial Intelligence, Barcelona, Catalonia, Spain, July 16-22, 2011*, pages 851–856, 2011.
[Dvořák and Gaggl, 2012] Wolfgang Dvořák and Sarah Alice Gaggl. Incorporating stage semantics in the scc-recursive schema for argumentation semantics. In *Proceedings of the 14th International Workshop on Non-Monotonic Reasoning (NMR 2012)*, 2012.
[Dvořák and Woltran, 2011] Wolfgang Dvořák and Stefan Woltran. On the intertranslatability of argumentation semantics. *Journal of Artificial Intelligence Research*, 41:445–475, 2011.
[Dyrkolbotn, 2014] Sjur Kristoffer Dyrkolbotn. How to argue for anything: Enforcing arbitrary sets of labellings using afs. In *Principles of Knowledge Representation and Reasoning: Proceedings of the Fourteenth International Conference, KR 2014, Vienna, Austria, July 20-24, 2014*, 2014.
[Eiter et al., 2013] Thomas Eiter, Michael Fink, Jörg Pührer, Hans Tompits, and Stefan Woltran. Model-based recasting in answer-set programming. *Journal of Applied Non-Classical Logics*, 23(1-2):75–104, 2013.
[Gabbay et al., 1994] Dov M. Gabbay, C. J. Hogger, and J. A. Robinson, editors. *Handbook of Logic in Artificial Intelligence and Logic Programming: Nonmonotonic Reasoning and Uncertain Reasoning*, volume 3. Oxford University Press, Inc., New York, NY, USA, 1994.
[Gaggl and Dvořák, 2016] Sarah Alice Gaggl and Wolfgang Dvořák. Stage semantics and the SCC-recursive schema for argumentation semantics. *Journal of Logic and Computation*, 26(4):1149–1202, February 2016.
[Gaggl and Woltran, 2013] Sarah Alice Gaggl and Stefan Woltran. The cf2 argumentation semantics revisited. *Journal of Logic and Computation*, 23:925–949, 2013.
[Gottlob, 1992] Georg Gottlob. Complexity results for nonmonotonic logics. *Journal of Logic and Computation*, 2(3):397–425, 1992.
[Lifschitz and Turner, 1994] Vladimir Lifschitz and Hudson Turner. Splitting a logic program. In Pascal Van Hentenryck, editor, *International Conference on Logic Programming*, pages 23–37. MIT Press, 1994.
[Lifschitz et al., 2001] Vladimir Lifschitz, David Pearce, and Agustín Valverde. Strongly equivalent logic programs. *ACM Transactions on Computational Logic*, 2:526–541, 2001.
[Linsbichler et al., 2015] Thomas Linsbichler, Christof Spanring, and Stefan Woltran. The hidden power of abstract argumentation semantics. In *Theory and Applications of Formal Argumentation - Third International Workshop, TAFA 2015, Buenos Aires, Argentina, July 25-26, 2015, Revised Selected Papers*, pages 146–162, 2015.
[Linsbichler et al., 2016] Thomas Linsbichler, Jörg Pührer, and Hannes Strass. A uniform account of realizability in abstract argumentation. In *ECAI 2016 - 22nd European Conference on Artificial Intelligence, 29 August-2 September 2016, The Hague, The Netherlands - Including Prestigious Applications of Artificial Intelligence (PAIS 2016)*, pages 252–260, 2016.
[Oikarinen and Woltran, 2011] Emilia Oikarinen and Stefan Woltran. Characterizing strong equivalence for argumentation frameworks. *Artificial Intelligence*, 175:1985–2009, 2011.
[Prakken, 2010] Henry Prakken. An abstract framework for argumentation with structured arguments. *Argument and Computation*, 1:93–124, 2010.
[Qi and Yang, 2008] Guilin Qi and Fangkai Yang. A survey of revision approaches in description logics. In *Proceedings of the 21st International Workshop on Description Logics (DL2008), Dresden, Germany, May 13-16, 2008*, 2008.

[Rudin, 1976] Walter Rudin. *Principles of Mathematical Analysis.* International series in pure and applied mathematics. McGraw-Hill, 1976.

[Spanring, 2012] Christof Spanring. Intertranslatability results for abstract argumentation semantics. Master's thesis, Uni Wien, 2012.

[Spanring, 2015] Christof Spanring. Hunt for the collapse of semantics in infinite abstract argumentation frameworks. In *2015 Imperial College Computing Student Workshop, ICCSW 2015, September 24-25, 2015, London, United Kingdom,* pages 70–77, 2015.

[Strass, 2015] Hannes Strass. The relative expressiveness of abstract argumentation and logic programming. In *Proceedings of the Twenty-Ninth AAAI Conference on Artificial Intelligence, January 25-30, 2015, Austin, Texas, USA.*, pages 1625–1631, 2015.

[Tarski, 1955] Alfred Tarski. A lattice-theoretical fixpoint theorem and its applications. *Pacific Journal of Mathematics,* 5(2):285–309, 1955.

[Truszczynski, 2006] Miroslaw Truszczynski. Strong and uniform equivalence of nonmonotonic theories - an algebraic approach. *Annals of Mathematics and Artificial Intelligence,* 48:245–265, 2006.

[Turner, 1996] Hudson Turner. Splitting a default theory. In William J. Clancey and Daniel S. Weld, editors, *Conference on Artificial Intelligence and Innovative Applications of Artificial Intelligence Conference,* pages 645–651. AAAI Press / The MIT Press, 1996.

[Turner, 2001] Hudson Turner. Strong equivalence for logic programs and default theories (made easy). In *Logic Programming and Nonmonotonic Reasoning, 6th International Conference, LPNMR 2001, Vienna, Austria, September 17-19, 2001, Proceedings,* pages 81–92, 2001.

[Turner, 2004] Hudson Turner. Strong equivalence for causal theories. In *Logic Programming and Nonmonotonic Reasoning, 7th International Conference, LPNMR 2004, Fort Lauderdale, FL, USA, January 6-8, 2004, Proceedings,* pages 289–301, 2004.

[Verheij, 2003] Bart Verheij. Deflog: On the logical interpretation of prima facie justified assumptions. *Journal of Logic and Computation,* 13(3):319–346, 2003.

[Weydert, 2011] Emil Weydert. Semi-stable extensions for infinite frameworks. In *Benelux Conference on Artificial Intelligence,* pages 336–343, 2011.

[Williams and Antoniou, 1998] Mary-Anne Williams and Grigoris Antoniou. A strategy for revising default theory Extensions. In *Principles of Knowledge Representation and Reasoning: Proceedings of the Sixth International Conference, KR 1998, Trento, Italy, June 2-5, 1998.*, pages 24–35, 1998.

[Zorn, 1935] Max Zorn. A remark on method in transfinite algebra. *Bulletin of the American Mathematical Society,* 41:667–670, 1935.

Ringo Baumann
Computer Science Institute, Intelligent Systems Group
Leipzig University, Germany
Email: baumann@informatik.uni-leipzig.de

18
Locality and Modularity in Abstract Argumentation

PIETRO BARONI, MASSIMILIANO GIACOMIN, BEISHUI LIAO

ABSTRACT. This chapter presents the main results achieved by the research activities aimed at overcoming the traditional "monolithic" view of abstract argumentation by investigating locality and modularity properties in this context. Three basic locality properties of argumentation semantics, namely directionality, SCC-recursiveness, and decomposability, are introduced and their relationships analyzed. On this basis, the relevant notions of argumentation modules, namely various kinds of interacting subframeworks within an argumentation framework, are discussed. To exemplify their practical applicability, the use of locality and modularity properties in the study of divide-and-conquer algorithms and of interchangeability and summarization methods for argumentation frameworks is discussed.

1 Introduction

In their original formulation, abstract argumentation frameworks [Dung, 1995] are regarded as monolithic entities and the relevant argumentation semantics are defined referring to an argumentation framework as a whole. While this *single block* vision is sufficient to introduce these fundamental notions and to investigate many of their properties, it reveals its limitations, both from a theoretical and a practical perspective, whenever one is interested in focusing attention on a specific portion of a framework, regarded in some way as an autonomous subframework, and on its interactions with the remaining part(s) of the framework.

This *local focus* can be motivated by various reasons.

In many contexts some arguments (and in particular their conclusions, though not represented explicitly at the abstract level) may be regarded as more important than others. For instance in a legal reasoning scenario involving a dispute, the arguments whose conclusions concern the final decision may attract more attention than the many other ones that are built around the case.

In a dynamic scenario, the arguments and their relationships may change: it may happen that changes occur, and hence require attention and proper management, only in a subpart of the whole framework, for instance because the effect of changes propagates through the existing argument relationships and, given the structure of the framework, some arguments are not reached by this propagation.

Further, given its abstract nature, an argumentation framework may contain arguments of different nature and in particular produced by different kinds of reasoning (e.g. practical vs. epistemic reasoning). Then one might argue that different kinds of semantics (e.g. a more credulous vs a more skeptical one) should be applied to different parts of the framework, an option not encompassed by traditional semantics notions.

These context-dependent aspects call for several tasks which are not encompassed by the monolithic view of abstract argumentation, such as: assessing the status of a selected subframework only, summarizing a complex framework into a simpler "equivalent" one which provides a synthetic view about the most important arguments, upgrading the status of arguments in a subframework after some change, analyzing the interactions among distinct subframeworks.

Even in absence of context dependent factors, like importance or change, attention on subframeworks may be motivated by the interest in the application of divide-and-conquer strategies for computational purposes. In fact, as in many other areas of computer science, one may consider tackling a problem by dividing it into subproblems of smaller size to obtain, under appropriate conditions, an efficiency gain.

For instance the task of computing the set of extensions prescribed by a given semantics could be faced by computing separately local extensions at the level of some subframeworks and then properly combining these local extensions together. Similarly the task of determining the justification status of a given argument might, under appropriate conditions, be solved without considering the whole framework, since considering a suitable neighborhood of the argument could be enough.

Considerations of this kind are not unique to abstract argumentation and it is worth mentioning that similar issues have been investigated also in the context of other formalisms for nonmonotonic reasoning. For instance, as pointed out in [Baumann, 2014], the basic idea of splitting, before being developed for abstract argumentation [Baumann, 2011], has been investigated for logic programs under answer set semantics [Lifschitz and Turner, 1994], for default logic [Turner, 1996], and for auto-epistemic logic [Geffner and Pearl, 1992].

Altogether, these motivations led, over the years, to an increasing interest in the investigation of locality and modularity notions in abstract argumentation, which are the subject of this chapter.

The presentation of this matter starts in Section 2 with the description of three properties of argumentation semantics that are inherently related to some kind of local focus within an argumentation framework, namely directionality, SCC-recursiveness, and decomposability. Then, in Section 3 the relationships between these three properties are analyzed, pointing out some intuitive commonalities but also some essential distinctions which are worth noting. Building on the above concepts, Section 4 turns to the issue of modularity in abstract argumentation, by discussing the different kinds of subframeworks which can be regarded as modular units with respect to the three basic properties previously

	$\mathcal{AD}$	$\mathcal{CO}$	$\mathcal{ST}$	$\mathcal{GR}$	$\mathcal{PR}$	$\mathcal{ID}$	$\mathcal{EAG}$ $\mathcal{SST}$
Directionality (Defs. 2.2 - 2.3)	Yes	Yes	No	Yes	Yes	Yes	No
SCC-recursiveness (Def. 2.10)	Yes	Yes	Yes	Yes	Yes	No	No
Full decomposability (Def. 2.23)	Yes	Yes	Yes	No	No	No	No
Top-down decomp. (Def. 2.26)	-	Yes	Yes	Yes	Yes	No	No
Bottom-up decomp. (Def. 2.27)	-	Yes	Yes	No	No	No	No
Transparency (Def. 6.7)	Yes	Yes	Yes	Yes	No	No	No
Strong transparency (Def. 6.7)	Yes	Yes	Yes	Yes	No	No	No

Table 1. Locality and transparency properties of admissibility-based argumentation semantics.

introduced. A first area of application of locality and modularity, namely the investigation of algorithms based on the divide-and-conquer paradigm, is discussed in Section 5, while Section 6 deals with the problem of interchangeability of subframeworks within a global framework and its use for summarization purposes. Finally, Section 7 concludes the chapter.

2 Locality properties in abstract argumentation

In this section we describe the underlying intuitions and introduce the formal definitions of the locality properties we consider in this chapter, namely directionality, SCC-recursiveness, and decomposability. We assume as background the contents of Chapter 4 of this volume and the relevant notation as far as the definitions of abstract argumentation frameworks and their semantics are concerned, while, regarding the three main properties we focus on, we allow some redundancy with Chapter 4 and with other chapters where some of them are also mentioned, in order to limit the dependencies of the present chapter. Also, in this chapter we restrict to finite argumentation frameworks since some notions, like SCC-recursiveness, are not immediately extendable to infinite frameworks.

An overview of the satisfaction of the considered locality properties by the considered semantics is given in tables 1 and 2 including also, for convenience of presentation, the transparency properties introduced in Section 6.3. In the tables, *Yes* and *No* have the obvious meanings, *?* indicates an open question, and - that the property is not applicable to the considered semantics.

2.1 Directionality

The directionality property corresponds to the idea that the relation of attack encodes a form of dependency and that arguments can affect each other only following the direction of attacks.

Intuitively, the justification state of an argument a should be affected only by the justification state of the attackers of a (which in turn are affected by their attackers and so on), while the arguments which only receive an attack

	$\mathcal{NA}$	$\mathcal{STG}$	$\mathcal{CF}2$	$\mathcal{STG}2$
Directionality (Defs. 2.2 - 2.3)	No	No	Yes	Yes
SCC-recursiveness (Def. 2.10)	No	No	Yes	Yes
Full decomposability (Def. 2.23)	No	No	?	?
Top-down decomp. (Def. 2.26)	-	-	?	?
Bottom-up decomp. (Def. 2.27)	-	-	?	?
Transparency (Def. 6.7)	No	No	?	?
Strong transparency (Def. 6.7)	No	No	?	?

Table 2. Locality and transparency properties of non admissibility-based argumentation semantics.

from a (and in turn those which are attacked by them and so on) or those which are simply not connected with a through the attack relation should not have any effect on the state of a.

This consideration can be extended from individual arguments to sets of arguments: if a set of arguments is *unattacked* (i.e. it does not receive attacks from arguments outside the set) it should be unaffected by the remaining part of the argumentation framework. In more formal terms, given an argumentation semantics σ, the evaluation produced (either in terms of extensions or of labellings) by σ on the global framework, when projected on an unattacked set should coincide with the evaluation produced by σ on the argumentation framework consisting only of the same unattacked set. This is the *directionality* property which is formalized in the following definitions[1].

Definition 2.1 *Given an argumentation framework $AF = \langle Ar, att \rangle$, a set $Args \subseteq Ar$ is* unattacked *in AF if and only if $\nexists (a,b) \in att$ such that $a \in (Ar \setminus Args), b \in Args$.*

Definition 2.2 *An extension-based semantics σ satisfies the* directionality *property if and only if for any argumentation framework $AF = \langle Ar, att \rangle$, for any set of arguments $Args$ unattacked in AF it holds that $\mathcal{AE}_\sigma(AF, Args) = \mathcal{E}_\sigma(AF\downarrow_{Args})$ where $\mathcal{AE}_\sigma(AF, Args) \triangleq \{(E \cap Args) \mid E \in \mathcal{E}_\sigma(AF)\}$.*

Definition 2.3 *A labelling-based semantics σ satisfies the* directionality *property if and only if for any argumentation framework $AF = \langle Ar, att \rangle$, for any set of arguments $Args$ unattacked in AF it holds that $\mathcal{AL}_\sigma(AF, Args) = \mathcal{L}_\sigma(AF\downarrow_{Args})$ where $\mathcal{AL}_\sigma(AF, Args) \triangleq \{Lab \cap (Args \times \Lambda) \mid Lab \in \mathcal{L}_\sigma(AF)\}$.*

---
[1] We provide distinct definitions since, in principle, there might exist labelling-based semantics which have no extension-based counterpart. In practice, for all semantics considered in this chapter, both an extension-based and a labelling-based version exist and are in close correspondence, as explained in chapter 4. In this context, the property of directionality can be regarded as being basically the same in the two versions.

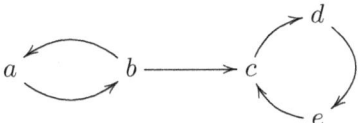

Figure 1. The leading example used in this chapter.

It can be proved (see in particular [Baroni and Giacomin, 2007]) that admissible, complete, grounded, preferred, ideal, CF2 and stage2 semantics satisfy the directionality property.

Example 2.4 *Consider the argumentation framework $AF = \langle \{a,b,c,d,e\}, \{(a,b),(b,a),(b,c),(c,d),(d,e),(e,c)\}\rangle$ shown in Figure 1 and complete semantics (denoted as $\mathcal{CO}$). It holds that $\mathcal{E}_{\mathcal{CO}}(AF\downarrow_{\{a,b\}}) = \{\emptyset, \{a\}, \{b\}\}$. Globally $\mathcal{E}_{\mathcal{CO}}(AF) = \{\emptyset, \{a\}, \{b,d\}\}$, from which $\mathcal{AE}_{\mathcal{CO}}(AF, \{a,b\}) = \{\emptyset, \{a\}, \{b\}\} = \mathcal{E}_{\mathcal{CO}}(AF\downarrow_{\{a,b\}})$.*

It is also known that naïve, stable, semi-stable, eager, and stage semantics are not directional. Intuitively this is motivated by the fact that these semantics share the property of maximizing the arguments attacked by (or conflicting with, in the case of naïve semantics) an extension, which implies that changing the attacked arguments may have a "backward" (i.e. non directional) effect.

Example 2.5 *Consider again the example of Figure 1. The set $\{a,b\}$ is unattacked and it holds $\mathcal{E}_\sigma(AF\downarrow_{\{a,b\}}) = \{\{a\},\{b\}\}$ for stable, semi-stable and stage semantics while for the same semantics $\mathcal{E}_\sigma(AF) = \{\{b,d\}\}$ and hence $\mathcal{AE}_\sigma(AF, \{a,b\}) = \{\{b\}\} \neq \mathcal{E}_\sigma(AF\downarrow_{\{a,b\}})$. For eager semantics (denoted as $\mathcal{EAG}$) $\mathcal{E}_{\mathcal{EAG}}(AF\downarrow_{\{a,b\}}) = \{\emptyset\}$, while $\mathcal{E}_{\mathcal{EAG}}(AF) = \{\{b,d\}\}$. It holds therefore that $\mathcal{AE}_{\mathcal{EAG}}(AF, \{a,b\}) = \{\{b\}\} \neq \mathcal{E}_{\mathcal{EAG}}(AF\downarrow_{\{a,b\}})$.*

The counter-example runs in a fully analogous way in the labelling-based approach. Figure 1 does not provide a counterexample to directionality of naïve semantics (denoted as $\mathcal{NA}$). This is given trivially by a framework like $AF_1 = \langle\{a,b\},\{(a,b)\}\rangle$. Here $\{a\}$ is unattacked and $\mathcal{E}_{\mathcal{NA}}(AF\downarrow_{\{a\}}) = \{\{a\}\}$, while $\mathcal{E}_{\mathcal{NA}}(AF_1) = \{\{a\},\{b\}\}$ yielding $\mathcal{AE}_{\mathcal{NA}}(AF, \{a\}) = \{\emptyset, \{a\}\} \neq \mathcal{E}_{\mathcal{NA}}(AF\downarrow_{\{a\}})$.

2.2 SCC-recursiveness

The property of *SCC-recursiveness* [Baroni et al., 2005] is based on the graph theoretical notion of strongly connected component (SCC). Briefly, strongly connected components provide a unique partition of a directed graph into disjoint parts where all nodes are mutually reachable (it is assumed that reachability is a reflexive relation). Formally, strongly connected components are the equivalence classes induced by the path equivalence (i.e. mutual reachability) relation between nodes as specified by the following definitions.

Definition 2.6 *Given an argumentation framework $AF = \langle Ar, att \rangle$ and two distinct arguments $a, b \in Ar$ a path from a to b is a sequence $a_0, a_1, \ldots, a_n$ with $n \geq 1$ such that $a_0 = a$, $a_n = b$, and for every $1 \leq i \leq n$, $a_{i-1} \in a_i^-$.*

Definition 2.7 *Given an argumentation framework $AF = \langle Ar, att \rangle$, the binary relation of path-equivalence between arguments, denoted as $PE_{AF} \subseteq (Ar \times Ar)$, is defined as follows:*

- $\forall a \in Ar, (a, a) \in PE_{AF}$

- *given two distinct arguments $a, b \in Ar$, $(a, b) \in PE_{AF}$ if and only if there is a path from a to b and a path from b to a.*

Definition 2.8 *The strongly connected components of AF are the equivalence classes of arguments under the relation of path-equivalence. The set of the strongly connected components of AF is denoted as SCCS_{AF}.*

An important property of the SCC decomposition is that the graph obtained considering SCCs as single nodes is acyclic, i.e. the attack relation induces a partial order between the SCCs.

Example 2.9 *In the argumentation framework of Figure 1 it is easy to see that there are two groups of mutually reachable arguments, namely two SCCs: $S_1 = \{a, b\}$ and $S_2 = \{c, d, e\}$ where S_1 precedes S_2 in the partial order induced by the attack relation.*

As introduced in Chapter 4, the SCC recursive scheme exploits this property and can be intuitively regarded as a constructive procedure to incrementally build extensions following the partial order between SCCs.

Formally, a semantics satisfies the SCC-recursiveness property if each of its extensions in any argumentation framework fits the constructive scheme sketched above, as specified by Definition 2.10.

Definition 2.10 *An argumentation semantics σ is SCC-recursive if and only if for any argumentation framework $AF = \langle Ar, att \rangle$, $\mathcal{E}_\sigma(AF) = \mathcal{GF}(AF, Ar)$, where for any $AF = \langle Ar, att \rangle$ and any set $C \subseteq Ar$, the function $\mathcal{GF}(AF, C) \subseteq 2^{Ar}$ is defined as follows:*
for any $E \subseteq Ar$, $E \in \mathcal{GF}(AF, C)$ if and only if

- *in case $|\text{SCCS}_{AF}| = 1$, $E \in \mathcal{BF}_\sigma(AF, C)$*

- *otherwise, $\forall S \in \text{SCCS}_{AF}$ $(E \cap S) \in \mathcal{GF}(AF\!\downarrow_{UP_{AF}(S,E)}, U_{AF}(S, E) \cap C)$*

where

- *$\mathcal{BF}_\sigma(AF, C)$ is a (semantics specific) function, called base function, that, given an argumentation framework $AF = \langle Ar, att \rangle$ such that $|\text{SCCS}_{AF}| = 1$ and a set $C \subseteq Ar$, returns a subset of 2^{Ar};*

- for $S \in \text{SCCS}_{AF}$, $E \subseteq Ar$, $U_{AF}(S, E) = \{a \in S \mid ((a^- \setminus S) \cap E = \emptyset) \land ((a^- \setminus S) \subseteq E^+)\}$;

- for $S \in \text{SCCS}_{AF}$, $E \subseteq Ar$, $UP_{AF}(S, E) = \{a \in S \mid (a^- \setminus S) \cap E = \emptyset\}$.

Partly due to its recursive nature, Definition 2.10 and its detailed application may appear complex. However the underlying basic ideas are relatively simple and can be summarized as follows:

1. the argumentation framework is partitioned into its strongly connected components; they form a partial order which encodes the dependencies existing among them (in line with the directionality principle);

2. the possible starting points for extension construction (in other words, the candidate partial extensions) within each initial strongly connected component are determined using a semantic-specific base function $\mathcal{BF}_\sigma$, which returns the extensions of argumentation frameworks consisting of a single strongly connected component. Note in particular that for every initial strongly connected component S it holds that $\forall E \subseteq Ar$ $UP_{AF}(S, E) = U_{AF}(S, E) = S$. If the AF consists of exactly one SCC S (which is of course also initial) then the first branch of Definition 2.10 applies, yielding $\mathcal{GF}(AF, Ar) = \mathcal{BF}_\sigma(AF, Ar) = \mathcal{BF}_\sigma(AF, S) = \mathcal{BF}_\sigma(AF\downarrow_S, S)$. Otherwise, if there is more than one SCC, for every initial SCC S the second branch of Definition 2.10 gives rise to $E \cap S \in \mathcal{GF}(AF\downarrow_{UP_{AF}(S,E)}, U_{AF}(S, E) \cap Ar) = \mathcal{GF}(AF\downarrow_S, S) = \mathcal{BF}_\sigma(AF\downarrow_S, S)$ where the last equality follows from the recursive application of first branch of Definition 2.10 (illustrated above) given the fact that $AF\downarrow_S$ consists of a unique SCC.

3. for each candidate partial extension, the nodes directly attacked within subsequent strongly connected components are suppressed (this corresponds to the fact that in the recursive step $AF\downarrow_{UP_{AF}(S,E)}$ is considered) and the distinction between arguments defended or not defended by the arguments in the candidate extension is (possibly) taken into account (in particular defended arguments correspond to the second parameter of $\mathcal{GF}$ in the recursive step, namely $U_{AF}(S, E) \cap C$, the set C "propagates" this information through the recursive steps);

4. the steps 1-3 above are applied recursively on the restricted argumentation frameworks obtained at step 3.

Example 2.11 *To exemplify, consider again the argumentation framework in Figure 1 using preferred semantics (denoted as $\mathcal{PR}$) which is known to be SCC-recursive.*

Step 1 above corresponds to determine that $\text{SCCS}_{AF} = \{S_1, S_2\}$, *where* $S_1 = \{a, b\}$ *and* $S_2 = \{c, d, e\}$ *and* S_1 *is initial, as easily seen.*

Step 2 involves applying a semantics specific base function $\mathcal{BF}_\sigma(AF, C)$ *on the restricted argumentation frameworks corresponding to the initial SCCs* $(AF\downarrow_{S_1}$ *only, in our example), where the parameter* C *is initially set to the whole set of arguments, namely* $\{a, b, c, d, e\}$. *The base function* $\mathcal{BF}_{\mathcal{PR}}(AF, C)$ *of preferred semantics has been identified in [Baroni et al., 2005] and basically it returns the "standard" preferred extensions of* AF *which are included in* C. *In our example* $\mathcal{BF}_{\mathcal{PR}}(AF\downarrow_{S_1}, S_1 \cap \{a, b, c, d, e\}) = \{\{a\}, \{b\}\}$, *yielding two partial candidate extensions each playing the role of* $E \cap S_1$ *for the next steps.*

Step 3 (and then 4) has therefore to be considered separately for the two candidate partial extensions.

Starting with $\{a\}$, *and considering its effects on* S_2 *it can be seen that* $U_{AF}(S_2, E) = U_{AF}(S_2, E \cap S_1) = U_{AF}(S_2, \{a\}) = \{c, d, e\}$. *In particular,* c *belongs to* $U_{AF}(S_2, E)$ *since its only attacker outside* S_2, *namely* b, *is attacked by* $E \cap S_1$, *while* d *and* e *belong to* $U_{AF}(S_2, E)$ *since they do not receive any attack from outside* S_2 *(and in particular from* $E \cap S_1$). *Analogously it can be seen that* $UP_{AF}(S_2, E) = UP_{AF}(S_2, E \cap S_1) = UP_{AF}(S_2, \{a\}) = \{c, d, e\}$. *Then, according to the recursive branch of Definition 2.10, one can obtain the various possible values of* $E \cap S_2$ *as the outcomes of* $\mathcal{GF}(AF\downarrow_{UP_{AF}(S,E)}, U_{AF}(S, E) \cap C)$ *which in the example means* $\mathcal{GF}(AF\downarrow_{\{c,d,e\}}, \{c, d, e\} \cap \{a, b, c, d, e\})$. *Steps 1-3 are now applied to the restricted framework. In this case* $AF\downarrow_{\{c,d,e\}}$ *consists of a single SCC which is of course initial (Step 1), therefore the non-recursive branch of Definition 2.10 applies and* $\mathcal{BF}_{\mathcal{PR}}$ *is used (Step 2). In particular it turns out that* $\mathcal{BF}_{\mathcal{PR}}(AF\downarrow_{\{c,d,e\}}, \{c, d, e\}) = \{\emptyset\}$. *Since there are no other SCCs, step 3 is void and the process terminates: each of the outcomes (in this case just one, namely* $\emptyset$) *obtained in the last step, combined with the candidate partial extension* $\{a\}$, *gives rise to an extension of* AF, *yielding* $\{a\} \cup \emptyset = \{a\} \in \mathcal{E}_{\mathcal{PR}}(AF)$.

Turning to the effects of set $\{b\}$, *now playing the role of* $E \cap S_1$, *on* S_2, *it can be seen that* $U_{AF}(S_2, \{b\}) = \{d, e\}$: c *does not belong to* $U_{AF}(S_2, E)$ *since it has an attacker, namely* b, *in* $E \cap S_2$, *while* d *and* e *belong to* $U_{AF}(S_2, E)$ *since they do not receive any attack from outside* S_2 *(and in particular from* $E \cap S_1$). *Analogolousy it can be seen that* $UP_{AF}(S_2, \{b\}) = \{d, e\}$. *Again, according to the recursive branch of Definition 2.10, one can obtain the various possible values of* $E \cap S_2$ *as the outcomes of* $\mathcal{GF}(AF\downarrow_{UP_{AF}(S,E)}, U_{AF}(S, E) \cap C)$ *which in the example means* $\mathcal{GF}(AF\downarrow_{\{d,e\}}, \{d, e\} \cap \{a, b, c, d, e\})$. *Again steps 1-3 are now applied to the restricted framework:* $AF' = AF\downarrow_{\{d,e\}}$ *consists of two SCCs (Step 1), namely* $S'_1 = \{d\}$ *and* $S'_2 = \{e\}$, *thus the recursive branch of Definition 2.10 has to be applied again. The initial SCC of* AF' *is* S'_1 *and, rather obviously,* $\mathcal{BF}_{\mathcal{PR}}(AF'\downarrow_{S'_1}, S'_1 \cap \{d, e\}) = \{\{d\}\}$ *(Step 2), i.e.* d *belongs to the candidate extension. Then, since* e *is attacked by* d, $UP_{AF'}(S'_2, \{d\}) = \emptyset$ *(Step 3). This yields an empty argumentation framework for the last (substantially void but formally important) recursive step. Clearly an empty argumentation framework admits only an empty (partial) extension. Summing up we get* $\{b\} \cup \{d\} \cup \emptyset = \{b, d\} \in \mathcal{E}_{\mathcal{PR}}(AF)$ *and, given that the constructive procedure is completed, altogether* $\{\{a\},\{b,d\}\} = \mathcal{E}_{\mathcal{PR}}(AF)$.

It is instructive to compare the application of the same procedure on the same argumentation framework for stable semantics (denoted as $\mathcal{ST}$*), which is also known to be SCC-recursive. Skipping many details, the behavior of stable semantics on the initial SCC* S_1 *of AF is the same as preferred semantics, i.e* $\mathcal{BF}_{\mathcal{ST}}(AF\!\downarrow_{S_1},\{a,b\}) = \{\{a\},\{b\}\}$*, yielding the same partial candidate extensions playing the role of* $E \cap S_1$ *for the next steps. Now, of course, the effect of the two candidate extensions on* S_2 *is the same as above and, for the candidate* $\{a\}$ *we get to consider the restricted framework* $AF\!\downarrow_{\{c,d,e\}}$*, consisting of a single SCC. Stable semantics fails to produce any extension (not even the empty one) for this framework, i.e.* $\mathcal{BF}_{\mathcal{ST}}(AF\!\downarrow_{\{c,d,e\}},\{c,d,e\}) = \emptyset$*. This corresponds to a failure of this branch of the SCC-recursive procedure: nothing can be combined with the candidate extension* $\{a\}$ *which, in a sense, has to be discarded and does not give rise to a global extension in this case. The treatment of the candidate* $\{b\}$ *is instead the same as for preferred semantics, yielding finally* $\{\{b,d\}\} = \mathcal{E}_{\mathcal{ST}}(AF)$.

It is shown in [Baroni et al., 2005] that the admissible, complete, grounded, preferred, and stable semantics are SCC-recursive. CF2 and stage2 semantics are SCC-recursive by construction, while ideal, naïve, stage, eager, and semi-stable semantics fail to satisfy this property (see also [Baroni et al., 2014b]). These results are summarised by the following proposition.

Proposition 2.12 *The semantics* $\mathcal{AD}$, $\mathcal{CO}$, $\mathcal{GR}$, $\mathcal{PR}$, $\mathcal{ST}$, $\mathcal{CF}2$, $\mathcal{STG}2$ *are SCC-recursive. The semantics* $\mathcal{ID}$, $\mathcal{NA}$, $\mathcal{STG}$, $\mathcal{EAG}$, $\mathcal{SST}$ *are not SCC-recursive.*

2.3 Decomposability

The definitions of the properties of directionality and SCC-recursiveness are based on specific partitions of an argumentation framework. The former considers a simple twofold partition consisting of an unattacked set of arguments and the rest of the framework, while the latter refers to a (recursively applied) partition into strongly connected components.

Considering arbitrary partitions leads to the notion of semantics decomposability. Intuitively, a full notion of decomposability amounts to require that, given an arbitrary partition of an argumentation framework into a set of subframeworks, the outcomes produced by a given semantics σ can be obtained as a combination of the outcomes produced by a local counterpart of σ applied separately on each subframework and vice versa.

Making this intuition formal requires developing a quite articulated treatment (see [Baroni et al., 2014a]). To this purpose, in the following we use the labelling-based approach to semantics definition since it allows a more compact and convenient presentation. As far as the set of labels Λ is concerned, we assume for all semantics but stable $\Lambda = \{\text{in}, \text{out}, \text{undec}\}$ while for stable semantics we assume $\Lambda = \{\text{in}, \text{out}\}$, given that the undec label is prevented by its definition.

As a first step, given that unconstrained partitions have to be considered, one needs to formally describe the interactions between the subframeworks corresponding to arbitrary sets of arguments.

Intuitively, given an argumentation framework $AF = \langle Ar, att \rangle$ and a subset $Args$ of its arguments, the elements affecting $AF\!\downarrow_{Args}$ include the arguments attacking $Args$ from the outside, called *input* arguments, and the attack relation from the input arguments to $Args$, called *conditioning relation*.

Definition 2.13 Given $AF = \langle Ar, att \rangle$ and a set $Args \subseteq Ar$, the input of $Args$, denoted as $Args^{inp}$, is the set $\{b \in Ar \setminus Args \mid \exists a \in Args, (b,a) \in att\}$, the conditioning relation of $Args$, denoted as $Args^R$, is defined as $att \cap (Args^{inp} \times Args)$.

Example 2.14 Letting in Figure 1 $Args = \{c,d\}$, we get $Args^{inp} = \{b,e\}$ and $Args^R = \{(b,c),(e,c)\}$.

It is then necessary to consider each subframework as an individual entity interacting with the other ones: this gives rise to the notion of *argumentation framework with input*, consisting of an argumentation framework $AF = \langle Ar, att \rangle$ (playing the role of a partial argumentation framework), a set of external input arguments $\mathcal{I}$, a labelling $\mathcal{L}ab_{\mathcal{I}}$ assigned to them and an attack relation $att_{\mathcal{I}}$ from $\mathcal{I}$ to Ar. In particular the labelling $\mathcal{L}ab_{\mathcal{I}}$ represents a generic label assignment to the input arguments, to be taken into account in the semantics evaluation inside the partial framework.

Definition 2.15 An *argumentation framework with input* is a tuple $(AF, \mathcal{I}, \mathcal{L}ab_{\mathcal{I}}, att_{\mathcal{I}})$, including an argumentation framework[2] $AF = \langle Ar, att \rangle$, a set of arguments $\mathcal{I}$ such that $\mathcal{I} \cap Ar = \emptyset$, a labelling $\mathcal{L}ab_{\mathcal{I}} \in \mathfrak{L}_{\mathcal{I}}$ and a relation $att_{\mathcal{I}} \subseteq \mathcal{I} \times Ar$, where $\mathfrak{L}_{\mathcal{I}}$ is the set of all possible labellings of the set of arguments $\mathcal{I}$.

Example 2.16 Continuing the running example, one may consider an argumentation framework with input $AFI = (AF, \mathcal{I}, \mathcal{L}ab_{\mathcal{I}}, att_{\mathcal{I}})$, where $AF = \langle \{c,d\}, \{(c,d)\}\rangle$, $\mathcal{I} = \{b,e\}$, $\mathcal{L}ab_{\mathcal{I}} = (\emptyset, \{b\}, \{e\})$, and $att_{\mathcal{I}} = \{(b,c),(e,c)\}$. Note that here we use the notation where each labelling is represented as a triple, where the first element is the set of arguments labelled **in**, the second one includes those labelled **out**, and the third one those labelled **undec**.

A *local function* is, at a general level, the counterpart of an argumentation semantics for argumentation frameworks with input: it specifies a set of labellings (produced by the semantics evaluation) for the arguments of the framework, taking into account the influence of the input.

---

[2] In the following, unless otherwise specified, we will implicitly assume $AF = \langle Ar, att \rangle$.

Definition 2.17 *A local function F assigns to any argumentation framework with input a (possibly empty) set of labellings of AF, i.e. $F(AF, \mathcal{I}, \mathcal{L}ab_\mathcal{I}, att_\mathcal{I}) \in 2^{\mathfrak{L}(\Lambda, AF)}$, where $\mathfrak{L}(\Lambda, AF)$ is the set of all Λ-labellings of AF (see chapter 4).*

Given a semantics σ, the point is then to identify the corresponding local function for argumentation frameworks with input. This is achieved by transforming an argumentation framework with input into a "standard" argumentation framework through a suitable replacement of input arguments.

Definition 2.18 *Given an argumentation framework with input $(AF, \mathcal{I}, \mathcal{L}ab_\mathcal{I}, att_\mathcal{I})$, the standard argumentation framework w.r.t. $(AF, \mathcal{I}, \mathcal{L}ab_\mathcal{I}, att_\mathcal{I})$ is defined as $AF' = (Ar \cup \mathcal{I}', att \cup R'_\mathcal{I})$, where $\mathcal{I}' = \mathcal{I} \cup \{a' \mid a \in \text{out}(\mathcal{L}ab_\mathcal{I})\}$ and $R'_\mathcal{I} = att_\mathcal{I} \cup \{(a', a) \mid a \in \text{out}(\mathcal{L}ab_\mathcal{I})\} \cup \{(a, a) \mid a \in \text{undec}(\mathcal{L}ab_\mathcal{I})\}$.*

Roughly, the standard argumentation framework puts AF under the influence of $(\mathcal{I}, \mathcal{L}ab_\mathcal{I}, att_\mathcal{I})$, by adding $\mathcal{I}$ to Ar and $att_\mathcal{I}$ to att, and by enforcing the label $\mathcal{L}ab_\mathcal{I}$ for the arguments of $\mathcal{I}$ in this way:

- for each argument $a \in \mathcal{I}$ such that $\mathcal{L}ab_\mathcal{I}(a) = \text{out}$, an unattacked argument a' is included which attacks a, in order to get a labelled out by all labellings of AF';

- for each argument $a \in \mathcal{I}$ such that $\mathcal{L}ab_\mathcal{I}(a) = \text{undec}$, a self-attack is added to a in order to get it labelled undec by all labellings of AF';

- each argument $a \in \mathcal{I}$ such that $\mathcal{L}ab_\mathcal{I}(a) = \text{in}$ is left unattacked, so that it is labelled in by all labellings of AF'.

Example 2.19 *The standard argumentation framework AF' for the argumentation framework with input AFI previously introduces is defined as $AF' = \langle \{b, c, d, e, b'\}, \{(c, d), (b, c), (e, c), (b', b), (e, e)\} \rangle$.*

Given a semantics σ its *canonical local function* for an argumentation framework with input is then easily defined on the basis of the corresponding standard argumentation framework.

Definition 2.20 *Given a semantics σ, the canonical local function of σ (also called local function of σ) is defined as $F_\sigma(AF, \mathcal{I}, \mathcal{L}ab_\mathcal{I}, att_\mathcal{I}) = \{\mathcal{L}ab\!\downarrow_{Ar} \mid \mathcal{L}ab \in \mathcal{L}_\sigma(AF')\}$, where $AF = \langle Ar, att \rangle$ and AF' is the standard argumentation framework w.r.t. $(AF, \mathcal{I}, \mathcal{L}ab_\mathcal{I}, att_\mathcal{I})$.*

In words, σ is applied on the standard argumentation framework and the resulting labellings are restricted on the arguments of the argumentation framework with input.

Example 2.21 *For $\sigma \in \{\mathcal{AD}, \mathcal{CO}, \mathcal{GR}, \mathcal{PR}, \mathcal{ID}, \mathcal{SST}\}$ we get $F_\sigma(AFI) = \{(\emptyset, \emptyset, \{c, d\})\}$. For $\sigma \in \{\mathcal{STG}, \mathcal{CF}2, \mathcal{STG}2\}$ we get $F_\sigma(AFI) = \{(\{c\}, \{d\}, \emptyset)\}$. For $\sigma = \mathcal{NA}$ we get $F_\sigma(AFI) = \{(\{c\}, \{d\}, \emptyset), (\{d\}, \{c\}, \emptyset)\}$.*

Note that since the argumentation framework with input AFI considered in the example uses the label undec, it is out of the scope of stable semantics. For stable semantics the definition of standard argumentation framework is the same but the fragment concerning $\text{undec}(\mathcal{L}ab_{\mathcal{I}})$ never applies.

While Definition 2.20 identifies a canonical local function for any semantics σ, it is shown in [Baroni et al., 2014a] that this notion is useful, in the context of the study of decomposability, only for semantics satisfying a set of basic properties, recalled below, expressing compatibility with complete semantics.

Definition 2.22 *A semantics σ is* complete-compatible *iff the following conditions hold:*

1. *For any argumentation framework $AF = \langle Ar, att \rangle$, every labelling $\mathcal{L} \in \mathcal{L}_\sigma(AF)$ satisfies the following conditions:*

    - *if $a \in Ar$ is unattacked, i.e. $a^- = \emptyset$, then $\mathcal{L}(a) = \text{in}$*
    - *if $b \in Ar$ and there is an unattacked argument a which attacks b, then $\mathcal{L}(b) = \text{out}$*
    - *if $c \in Ar$ is self-attacking, and there are no attackers of c besides c itself, then $\mathcal{L}(c) = \text{undec}$*

2. *for any set of arguments $\mathcal{I}$ and any labelling $\mathcal{L}ab_{\mathcal{I}} \in \mathfrak{L}_{\mathcal{I}}$, the argumentation framework $AF' = \langle \mathcal{I}', att' \rangle$, where $\mathcal{I}' = \mathcal{I} \cup \{a' \mid a \in \text{out}(\mathcal{L}ab_{\mathcal{I}})\}$ and $att' = \{(a', a) \mid a \in \text{out}(\mathcal{L}ab_{\mathcal{I}})\} \cup \{(a, a) \mid a \in \text{undec}(\mathcal{L}ab_{\mathcal{I}})\}$, admits a (unique) labelling, i.e. $|\mathcal{L}_\sigma(AF')| = 1$.*

Beyond complete semantics itself, *complete-compatible* semantics include grounded, preferred, stable, semi-stable, ideal, eager, $\mathcal{CF}2$, and stage2 semantics. Note that, in this respect, for stable semantics the assumption $\Lambda = \{\text{in}, \text{out}\}$ is crucial.

On the above bases, a formal notion of semantics decomposability can be introduced. To this purpose, consider a generic argumentation framework $AF = \langle Ar, att \rangle$ and an arbitrary partition of Ar, i.e. a set $\{P_1, \ldots, P_n\}$ such that $\forall i \in \{1, \ldots, n\}$ $P_i \subseteq Ar$ and $P_i \neq \emptyset$, $\bigcup_{i=1\ldots n} P_i = Ar$ and $P_i \cap P_j = \emptyset$ for $i \neq j$. Such a partition identifies the restricted argumentation frameworks $AF\!\downarrow_{P_1}, \ldots, AF\!\downarrow_{P_n}$, that affect each other with the relevant input arguments and conditioning relations as stated in Definition 2.13. Intuititively a semantics σ is decomposable if σ can be put in correspondence with a local function F such that:

- every labelling prescribed by σ on AF, namely every element of $\mathcal{L}_\sigma(AF)$, corresponds to the union of n "compatible" labellings $\mathcal{L}ab_{P_1}, \ldots, \mathcal{L}ab_{P_n}$ of the restricted argumentation frameworks, each labelling being obtained applying F;

- in turn, each union of n "compatible" labellings $\mathcal{L}ab_{P_1}, \ldots, \mathcal{L}ab_{P_n}$ obtained applying F to the restricted frameworks gives rise to a labelling of AF.

The compatibility constraint mentioned above reflects the fact that any labelling of a restricted framework is used by F for computing the other ones: $\mathcal{L}ab_{P_i}$ plays a role in determining $\mathcal{L}ab_{P_1}, \ldots, \mathcal{L}ab_{P_{i-1}}, \mathcal{L}ab_{P_{i+1}}, \ldots, \mathcal{L}ab_{P_n}$ and vice versa. This means that $\mathcal{L}ab_{P_1}, \ldots, \mathcal{L}ab_{P_n}$ are compatible if each $\mathcal{L}ab_{P_i}$ is produced by F for $AF\downarrow_{P_i}$ with the input arguments P_i^{inp} labelled according to $\mathcal{L}ab_{P_1}, \ldots, \mathcal{L}ab_{P_{i-1}}, \mathcal{L}ab_{P_{i+1}}, \ldots, \mathcal{L}ab_{P_n}$. Definition 2.23 synthesizes all these considerations.

Definition 2.23 *A semantics σ is* fully decomposable *(or simply decomposable) iff there is a local function F such that for every argumentation framework $AF = \langle Ar, att \rangle$ and every partition $\mathcal{P} = \{P_1, \ldots, P_n\}$ of Ar it holds that $\mathcal{L}_\sigma(AF) = \mathcal{U}(\mathcal{P}, AF, F)$ where $\mathcal{U}(\mathcal{P}, AF, F) \triangleq \{\mathcal{L}ab_{P_1} \cup \ldots \cup \mathcal{L}ab_{P_n} \mid \mathcal{L}ab_{P_i} \in F(AF\downarrow_{P_i}, P_i^{inp}, (\bigcup_{j=1\ldots n, j \neq i} \mathcal{L}ab_{P_j})\downarrow_{P_i inp}, P_i^R)\}$.*

It is proved in [Baroni et al., 2014a] that, if a complete-compatible semantics σ is fully decomposable, then the local function appearing in Definition 2.23 coincides with the canonical local function F_σ.

Proposition 2.24 *Given a complete-compatible semantics σ, if σ is fully decomposable then there is a unique local function satisfying the conditions of Definition 2.23, coinciding with the canonical local function F_σ.*

Example 2.25 *Continuing the example of Figure 1 and considering the partition $\{\{c, d\}, \{a, b, e\}\}$, full decomposability of a complete-compatible semantics σ requires that the canonical local function F_σ is such that the labellings of AF are exactly those obtained by the union of the compatible labellings of $AF\downarrow_{\{c,d\}}$ and $AF\downarrow_{\{a,b,e\}}$ given by the local function itself. For the sake of illustration, let us consider complete semantics and its canonical local function $F_{\mathcal{CO}}$.*

The labelling $(\{d\}, \{c\}, \emptyset)$ is compatible with $(\{b\}, \{a, e\}, \emptyset)$, since the first is obtained by $F_{\mathcal{CO}}$ with b labelled in *and e labelled* out*, and the latter is obtained by $F_{\mathcal{CO}}$ with d labelled* in*. On the other hand, for instance the labelling $(\{a\}, \{b\}, \{e\})$ is not compatible with $(\{d\}, \{c\}, \emptyset)$, while it is compatible with $(\emptyset, \emptyset, \{d, c\})$ and it can be noted that also $(\emptyset, \emptyset, \{a, b, e\})$ is compatible with $(\emptyset, \emptyset, \{d, c\})$. Overall, three global labellings arise from the combinations of the compatible outcomes of $F_{\mathcal{CO}}$, namely $(\{b, d\}, \{a, c, e\}, \emptyset)$, $(\{a\}, \{b\}, \{c, d, e\})$, $(\emptyset, \emptyset, \{a, b, c, d, e\})$, corresponding to the complete labellings of AF.*

As evidenced by the two-bullet description given before Definition 2.23, full decomposability can be viewed as the conjunction of two partial decomposability properties, namely *top-down decomposability* and *bottom-up decomposability*.

In words, a semantics is top-down decomposable if the procedure to compute the global labellings identified by Definition 2.23 is complete, i.e. all of the global labellings can be obtained by combining the labellings prescribed by F_σ for the restricted subframeworks, even if putting together labellings of the restricted subframeworks may give rise to some "spurious" labellings besides the correct ones. The following definition formalizes this intuition.

Definition 2.26 *A complete-compatible semantics σ is* top-down decomposable *iff for any argumentation framework $AF = \langle Ar, att\rangle$ and any partition $\mathcal{P} = \{P_1, \ldots, P_n\}$ of Ar, it holds that $\mathcal{L}_\sigma(AF) \subseteq \mathcal{U}(\mathcal{P}, AF, F_\sigma)$.*

While top-down decomposability corresponds to completeness of the procedure identified by Definition 2.23, bottom-up decomposability requires its soundness, i.e. that any combination of local labellings is a global labelling, while it is not guaranteed that all global labellings can be obtained in this way.

Definition 2.27 *A complete-compatible semantics σ is* bottom-up decomposable *iff for any argumentation framework $AF = \langle Ar, att\rangle$ and any partition $\mathcal{P} = \{P_1, \ldots, P_n\}$ of Ar, it holds that $\mathcal{L}_\sigma(AF) \supseteq \mathcal{U}(\mathcal{P}, AF, F_\sigma)$.*

A comment on the two definitions above is in order. While the definition of full decomposability applies to any kind of semantics and requires the existence of a local function satisfying the decomposability property, Definitions 2.26 and 2.27 are restricted to complete-compatible semantics and refer to the canonical local function F_σ to avoid triviality: the local function returning all the possible labellings of AF trivially satisfies the inclusion condition of Definition 2.26 for any semantics, while the local function always returning the empty set trivially satisfies the condition of Definition 2.27. This is the reason why both definitions refer to the specific canonical local function, which makes sense for complete-compatible semantics in the light of Proposition 2.24. If a semantics is not complete-compatible (examples are naïve and stage semantics) then the notion of canonical local function is meaningless, since the labelling $\mathcal{L}ab_\mathcal{I}$ would not be in general enforced for the arguments of $\mathcal{I}$ in the standard argumentation framework w.r.t. $(AF, \mathcal{I}, \mathcal{L}ab_\mathcal{I}, att_\mathcal{I})$.

Concerning complete-compatible semantics, it is proved in [Baroni et al., 2014a] that complete and stable[3] semantics are fully decomposable, grounded and preferred semantics are only top-down decomposable, while ideal and semi-stable semantics do not satisfy any of the decomposability properties and decomposability of CF2 and stage2 semantics is an open question. The negative example provided for ideal semantics in [Baroni et al., 2014a] directly applies to eager semantics too. As to semantics which are not complete-compatible, only full decomposability can be considered. Admissible semantics is fully decomposable [Baroni et al., 2014a], while it is relatively easy to see that naïve

---

[3]A more restricted notion of decomposability for stable semantics, called *generalized splitting*, was previously considered in [Baumann et al., 2012; Baumann, 2014].

and stage semantics are not. These results are summarised by the following proposition.

Proposition 2.28 *Among complete-compatible semantics, $\mathcal{CO}$ and $\mathcal{ST}$ are fully decomposable (hence also top-down and bottom-up decomposable); the semantics $\mathcal{GR}$ and $\mathcal{PR}$ are top-down decomposable but not bottom-up (nor fully) decomposable; the semantics $\mathcal{ID}$, $\mathcal{EAG}$, and $\mathcal{SST}$ are neither top-down nor bottom-up (nor fully) decomposable. Among non complete-compatible semantics, $\mathcal{AD}$ is fully decomposable, while $\mathcal{NA}$ and $\mathcal{STG}$ are not.*

3 Relationships among locality properties

The three locality properties reviewed in Section 2 share the very basic idea of considering semantics evaluation on some kind of subframework of an argumentation framework interacting with the rest of the framework. Attack direction plays a crucial role in the definition of the subframeworks considered by both directionality and SCC-recursiveness, while subframeworks are arbitrary for decomposability. Given the basic intuitive commonalities among these properties, it is important to discuss their relationships, which turn out to be less simple and immediate than one may expect at first glance.

Let's start by considering directionality and SCC-recursiveness, as they share the key role ascribed to attack direction. It is important to remark here that though the definitions of these two properties have some basic structural similarity (in particular they both focus on properties of the intersections of the global extensions with specific parts of the framework) they reflect rather different intuitions. On the one hand, directionality can be essentially regarded as getting the same extensions when one obtains the same framework from different ones by suppressing some parts that are topologically uninfluential with respect to the remaining common one. On the other hand, SCC-recursiveness requires that all the intersections of a global extension with the different SCCs satisfy some (articulated) constraints, within the same framework.

To appreciate the difference, the following two points can be evidenced:

- directionality does not involve a notion of incremental construction of extensions, since it deals with what one finds out when cancelling (rather than adding) arguments, while SCC-recursiveness has an embedded notion of extension construction, starting from the initial SCCs and then following their partial order;

- SCC-recursiveness does not require uniform outcomes on equal restrictions of distinct frameworks, since the constraints of the SCC-recursive scheme need to be satisfied for the set of extensions of each framework individually. On the contrary, directionality requires a sort of regularity across globally different, but partially equal, frameworks.

The differences pointed out above are made apparent by the fact that there are non directional but SCC-recursive semantics, like stable semantics, as well as directional but non SCC-recursive semantics, like ideal semantics (see in particular [Baroni et al., 2014b]). Moreover, these differences are crucial with respect to the role that these properties may play in supporting the study of efficient computational methods for abstract argumentation, discussed in Section 5.

While the points above make evident that, in general, directionality and SCC-recursiveness are independent properties, the perceptive reader might feel that this fact is somehow counterintuitive: one might expect that SCC-recursiveness, being a stronger and more articulated property, implies the weaker and simpler property of directionality.

This intuition is indeed correct if one adds the requirement that the semantics prescribes a nonempty set of extensions (or labellings) for each argumentation framework: this is, in fact, not the case for stable semantics while it holds for all the other semantics in the literature we are aware of (apart possible variations of the notion of stable semantics itself). The fact that, under this additional condition, SCC-recursiveness implies directionality is formally stated in the following proposition, proved in [Baroni et al., 2005].

Proposition 3.1 *Any SCC-recursive semantics σ such that $\forall AF = \langle Ar, att \rangle$ $\forall C \subseteq Ar$ $\mathcal{BF}_\sigma(AF, C) \neq \emptyset$, satisfies the directionality criterion.*

Turning to the relationships between decomposability and the other locality properties, it can be remarked first of all that the fact that stable semantics is fully decomposable shows, also in this case, that directionality is not a necessary condition for decomposability. It is also evident that it is neither a sufficient one, given that, in particular, ideal semantics, being directional, fails to satisfy any form of decomposability. If one adds the requirement that a semantics is complete-compatible and always prescribes a nonempty set of extensions, it can be seen that directionality becomes a necessary condition for full decomposability. Investigating further detailed relationships is an open research direction.

Proposition 3.2 *Any complete-compatible fully decomposable semantics σ such that $\forall AF = \langle Ar, att \rangle$ $\mathcal{E}_\sigma(AF) \neq \emptyset$ satisfies the directionality criterion.*

It is worth recalling that in [Baumann, 2014] a result bearing some analogy with Propositions 3.1 and 3.2 has been proved, showing a relationship between the satisfaction of the splitting property[4] [Baumann, 2011; Baumann, 2014] and directionality (again under the condition that the semantics prescribes a nonempty set of extensions for each argumentation framework).

Proposition 3.3 *Any semantics σ satisfying the splitting property and such that $\forall AF = \langle Ar, att \rangle$ $\mathcal{E}_\sigma(AF) \neq \emptyset$ satisfies the directionality criterion.*

---

[4]The splitting property will be recalled in the final part of Section 5.3 and basically corresponds to a special form of decomposability, where an argumentation framework is partitioned into two parts, one of which is an unattacked set.

Turning to the relationships between SCC-recursiveness and decomposability, they are largely an open research question, however some basic observations can be drawn. First, the fact that some SCC-recursive semantics like grounded and preferred, are not fully decomposable, shows that SCC-recursiveness is not sufficient for full decomposability, as one may expect. It is however worth noting that complete, stable, grounded and preferred semantics are SCC-recursive and satisfy top-down decomposability, while ideal and semi-stable semantics, which are not SCC-recursive, are not decomposable in any sense. On this basis, one might conjecture that top-down decomposability has some significant relationship with SCC-recursiveness.

Conjectures apart, the notion of strongly connected components has been used in [Baroni et al., 2014a] to identify classes of partitions of argumentation frameworks where argumentation semantics feature stronger decomposability properties. More precisely, if one restricts to partitions $\{P_1, \ldots, P_n\}$ where each P_i coincides with the union of a set of strongly connected components, it turns out that also grounded and preferred semantics are fully decomposable (see Theorem 7 of [Baroni et al., 2014a]).

Restricting the class of considered partitions does not help ideal and semi-stable semantics instead, which still fail to satisfy any form of decomposability even in the case where the only considered partition for each argumentation framework is exactly the one corresponding to its strongly connected components.

These results, though obtained specifically for the semantics mentioned above, suggest again that it is likely that there are quite some interesting relationships between decomposability properties and SCC-recursiveness: generalizing them to any SCC-recursive semantics represents an interesting research problem.

4 Modularity in abstract argumentation

The different locality properties previously reviewed identify different kind of subframeworks that can be regarded as "modules" in the context of semantics evaluation. In this section we recap and compare these different notions of "module" altogether and introduce further relevant modularity notions.

Starting from the simplest concept, the property of directionality identifies unattacked sets as basic useful modules, whose semantics evaluation is unaffected by the other parts of the framework and is guaranteed not to change if those other parts are modified, provided that the condition of being unattacked is preserved. As mentioned in the previous section, this involves a sort of regularity across globally different, but partially equal, frameworks and hence a notion of "reuse": the results obtained on a partial framework corresponding to an unattacked set can be reused in every argumentation framework where such partial framework appears (still unattacked). This suggests that another basic modular notion, complementary to the one of unattacked set, involves the remaining part of the framework, which is (possibly) affected by the unattacked set.

This complementary notion corresponds in fact to the formal definition of conditioned argumentation framework adopted in particular in [Liao et al., 2011] as a basis of the division-based method, to be recalled in Section 5. Here, an argumentation framework is regarded as divided into two parts. The first one, called the unaffected part, corresponds, by definition, to an unattacked set. The second part, namely the rest of the framework, may receive attacks from the first one and hence be "conditioned" by it: this is formalised by the notion of conditioned argumentation framework.

Definition 4.1 *Given an argumentation framework* $AF_1 = \langle Ar_1, att_1 \rangle$, *a conditioned argumentation framework with respect to* AF_1 *is a tuple* $\mathcal{CAF} = (\langle Ar_2, att_2 \rangle, (\mathcal{C}(Ar_1), \mathcal{I}_{(\mathcal{C}(Ar_1), Ar_2)}))$ *in which*

- $\langle Ar_2, att_2 \rangle$ *is an argumentation framework that is conditioned by* $\mathcal{C}(Ar_1)$ *in which* $Ar_2 \cap Ar_1 = \emptyset$;

- $\mathcal{C}(Ar_1) \subseteq Ar_1$ *is a nonempty set of arguments (called conditioning arguments) that have interactions with arguments in* Ar_2, *i.e.,* $\forall a \in \mathcal{C}(Ar_1)$ $\exists b \in Ar_2$ *such that* $(a, b) \in \mathcal{I}_{(\mathcal{C}(Ar_1), Ar_2)}$;

- $\mathcal{I}_{(\mathcal{C}(Ar_1), Ar_2)} \subseteq \mathcal{C}(Ar_1) \times Ar_2$ *is the set of interactions from the arguments in* $\mathcal{C}(Ar_1)$ *to the arguments in* Ar_2

Example 4.2 *It is possible to consider the argumentation framework presented in Figure 1, as consisting of an argumentation framework* $AF_1 = \langle \{a, b\}, \{(a, b), (b, a)\} \rangle$ *and of a conditioned argumentation framework with respect to* AF_1 $\mathcal{CAF} = ((\{c, d, e\}, \{(c, d), (d, e), (e, c)\}), (\{b\}, \{(b, c)\}))$ *where in* AF_1 *there is a single conditioning argument, namely b and the interaction from* AF_1 *to* AF_2 *consists of the attack from b to c.*

A closely related notion is the one of *splitting* of an argumentation framework [Baumann, 2011; Baumann, 2014], recalled below.

Definition 4.3 *Given* $AF_1 = \langle Ar_1, att_1 \rangle$ *and* $AF_2 = \langle Ar_2, att_2 \rangle$ *with* $Ar_1 \cap Ar_2 = \emptyset$ *and* $att_3 \subseteq Ar_1 \times Ar_2$, *the tuple* (AF_1, AF_2, att_3) *is called a splitting of the framework* $AF = \langle Ar_1 \cup Ar_2, att_1 \cup att_2 \cup att_3 \rangle$.

In words, AF_1 corresponds to the restriction of AF to its unattacked set Ar_1, which can be regarded as a conditioning framework in terms of Definition 4.1, while AF_2 corresponds essentially to the conditioned framework in Definition 4.1 and att_3 to $\mathcal{I}_{(\mathcal{C}(Ar_1), Ar_2)}$.

Example 4.4 *A splitting of the argumentation framework presented in Figure 1 is given by the tuple* (AF_1, AF_2, att_3) *where* $AF_1 = \langle \{a, b\}, \{(a, b), (b, a)\} \rangle$, $AF_2 = \langle \{c, d, e\}, \{(c, d), (d, e), (e, c)\} \rangle$, *and* $att_3 = \{(b, c)\}$.

The reader may have noted that conditioned argumentation frameworks and splittings are very close and also bear some similarity with argumentation frameworks with input, as will be discussed later. For this reason we will speak of conditioning and conditioned frameworks also for the case of splittings, even if this terminology was not used in [Baumann, 2011; Baumann, 2014].

Turning to SCC-recursiveness, clearly it identifies strongly connected components as modular entities. At an intuitive level, this can be regarded as a refinement of the twofold partition corresponding to directionality. In particular, it is easy to see that any (nonempty) unattacked set necessarily coincides with the union of a set of strongly connected components which do not receive attacks from other strongly connected components. In particular, initial strongly connected components correspond to the minimal, with respect to set inclusion, unattacked sets of a framework while larger unattacked sets necessarily include at least an initial strongly connected component together with other SCCs. Dually, the second part of a framework, which receives attacks from an unattacked set and corresponds to a conditioned argumentation framework as specified by Definition 4.1, consists of a set of SCCs of the original framework too, accompanied by the information about the attacks they receive. In this sense, SCCs may be regarded as the finest modular notion related to the idea of capturing a dependence ordering based on the direction of attacks. Note however that no direct counterpart of the notion of conditioned argumentation framework is present in the SCC-recursive scheme: this is due to the fact that the effect of the preceding (according to the direction of attacks) SCCs on the subsequent ones is embedded in the recursive branch of Definition 2.10, and in particular in the elements $UP_{AF}(S, E)$ and $U_{AF}(S, E)$.

Decomposability, referring to arbitrary partitions, does not rely on directionality and uses argumentation frameworks with input as modular entities. A comparison with conditioned argumentation frameworks and splittings is worth drawing. All notions focus on a (partial) argumentation framework (AF in Definition 2.15, $\langle Ar_2, att_2 \rangle$ in Definition 4.1, AF_2 in Definition 4.3), receiving some influence from some other arguments (explicitly denoted as $\mathcal{I}$ in Definition 2.15 and as $\mathcal{C}(Ar_1)$ in Definition 4.1, left implicit in Definition 4.3), through an attack relation ($att_\mathcal{I}$ in Definition 2.15, $\mathcal{I}_{(\mathcal{C}(Ar_1), Ar_2)}$ in Definition 4.1, att_3 in Definition 4.3). Thus argumentation frameworks with input, conditioned argumentation frameworks and splittings are basically the same entity from a structural point of view. The main difference in their definition, reflecting a difference in their intended use, concerns the representation of the actual "input" they receive from outside: the definition of argumentation framework with input includes a labelling $\mathcal{L}ab_\mathcal{I}$ of the arguments attacking from outside, which is the actual "input" in a given context, while a similar entity is not present in Definitions 4.1 and 4.3. This is because in Definitions 4.1 and 4.3 the input is assumed to come from a whole unaffected framework ($AF_1 = \langle Ar_1, att_1 \rangle$), through the set of conditioning arguments $\mathcal{C}(Ar_1)$ and the relevant interactions $\mathcal{I}_{(\mathcal{C}(Ar_1), Ar_2)}$.

In particular, in [Liao et al., 2011] the input is determined by the extensions (or labellings) of AF_1 according to a given semantics, which then have an effect on AF_2 through $\mathcal{C}(Ar_1)$ and $\mathcal{I}_{(\mathcal{C}(Ar_1),Ar_2)}$. To put it in other words, the idea is that extension computation in AF_2 depends on AF_1 but not vice versa, hence one can use the extensions of AF_1 as fixed conditions to determine the extensions of AF_2. This is formally expressed by the notion of assigned conditioned argumentation framework [Liao et al., 2011].

Definition 4.5 Let $\mathcal{CAF} = (\langle Ar_2, att_2 \rangle, (\mathcal{C}(Ar_1), \mathcal{I}_{(\mathcal{C}(Ar_1),Ar_2)}))$ be a conditioned argumentation framework with respect to $AF_1 = \langle Ar_1, att_1 \rangle$ and let $E_1 \in \mathcal{E}_\sigma(AF_1)$. The assigned $\mathcal{CAF}$ with respect to E_1 is defined as $\mathcal{CAF}[E_1] = (\langle Ar_2, att_2 \rangle, (\mathcal{C}(Ar_1)[E_1], \mathcal{I}_{(\mathcal{C}(Ar_1),Ar_2)}))$ in which $\mathcal{C}(Ar_1)[E_1]$ is a triple which partitions the elements of $\mathcal{C}(Ar_1)$ according to their status with respect to the extension E_1: $\mathcal{C}(Ar_1)[E_1] \triangleq (\mathcal{C}(Ar_1) \cap E_1, \mathcal{C}(Ar_1) \cap E_1^+, \mathcal{C}(Ar_1) \setminus (E_1 \cup E_1^+))$.

Example 4.6 Continuing from example 4.2, consider the three complete extensions of AF_1, namely $E_1 = \emptyset$ $E'_1 = \{a\}$ and $E''_1 = \{b\}$. Then we have $\mathcal{C}(Ar_1)[E_1] = (\emptyset, \emptyset, \{b\})$, $\mathcal{C}(Ar_1)[E'_1] = (\emptyset, \{b\}, \emptyset)$, $\mathcal{C}(Ar_1)[E''_1] = (\{b\}, \emptyset, \emptyset)$ given that the only member of $\mathcal{C}(Ar_1)$ is respectively not included in nor attacked by E_1, attacked by E'_1, and included in E''_1.

In the context of splittings [Baumann, 2011; Baumann, 2014], the formalization of the influence of an extension of a framework on the arguments of a conditioned framework is more articulated and is expressed through structural changes to the conditioned framework itself. First, the explicit suppression of the arguments belonging to the conditioned framework which are attacked by the members of the extension is captured by the notion of *reduct*.

Definition 4.7 Given $AF = \langle Ar, att \rangle$, a set of arguments $Args$ such that $Args \cap Ar = \emptyset$, and $att' \subseteq Args \times Ar$ the $(Args, att')$-reduct of AF denoted as $red_{Args,att'}(AF)$ is the argumentation framework $red_{Args,att'}(AF) = \langle \overline{Ar}, \overline{att} \rangle$ where $\overline{Ar} = \{a \in Ar \mid \nexists b \in Args \text{ such that } (b,a) \in att'\}$ and $\overline{att} = \{(a,b) \in att \mid a, b \in \overline{Ar}\}$.

Note that in Definition 4.7, the set $Args$ is meant to coincide with an extension E of the conditioning framework AF_1.

Second, a self-attack is added to those arguments in the conditioned framework which have to be regarded as undecided according to the influence of the conditioning framework. This is captured by the notion of *modification*.

Definition 4.8 Given $AF = \langle Ar, att \rangle$, a set of arguments $Args$ such that $Args \cap Ar = \emptyset$, and $att' \subseteq Args \times Ar$ the $(Args, att')$-modification of AF denoted as $mod_{Args,att'}(AF)$ is the argumentation framework $mod_{Args,att'}(AF) = \langle Ar, att \cup \{(a,a) \mid \exists b \in Args \text{ such that } (b,a) \in att'\} \rangle$.

Note that in Definition 4.8, the set $Args$ is meant to include the arguments of the conditioning framework AF_1 which are not included in nor attacked by a given extension E of AF_1.

As mentioned above, an assigned $\mathcal{CAF}$ and the notions of reduct and modification in splittings are meant to capture a fixed effect coming from an extension prescribed by a given semantics on an unaffected framework. In the context of argumentation frameworks with input, an unaffected part is not considered since this notion is meant to represent various "modules" mutually influencing each other.

Nevertheless, a basic similarity can be identified in the way the external influence is captured in conditioned argumentation frameworks and argumentation frameworks with input: in both cases the conditioning arguments are partitioned into three sets, the difference being that in argumentation frameworks with input this is achieved directly by the labelling $\mathcal{L}ab_{\mathcal{I}}$ while in an assigned conditioned argumentation frameworks the partition is induced by the extension E_1. More precisely

- the set of arguments labelled in by $\mathcal{L}ab_{\mathcal{I}}$ corresponds to $\mathcal{C}(Ar_1) \cap E_1$ in Definition 4.5;

- the set of arguments labelled out by $\mathcal{L}ab_{\mathcal{I}}$ corresponds to $\mathcal{C}(Ar_1) \cap E_1^+$ in Definition 4.5;

- the set of arguments labelled undec by $\mathcal{L}ab_{\mathcal{I}}$ corresponds to $\mathcal{C}(Ar_1) \setminus (E_1 \cup E_1^+)$ in Definition 4.5

Though more implicitly, this is also the partition of the arguments considered in splittings, through definitions 4.7 and 4.8 and in the SCC-recursive scheme through Definition 2.10. As already mentioned, in the formalization of splittings the external arguments belonging to E (first item above) are meant to play the role of $Args$ in Definition 4.7, the external arguments undecided with respect to E (third item above) are meant to play the role of $Args$ in Definition 4.8, while the external arguments attacked by E (second item above) are considered implicitly as having no effect.

Similarly, in the SCC-recursive scheme the external arguments belonging to E (used implicitly to determine the set $UP_{AF}(S, E)$) correspond to the first item above, the external arguments attacked by E (used to determine the set $U_{AF}(S, E)$) correspond to the second item above, and the remaining external arguments correspond to the third item.

These commonalities have a formal counterpart in the notion of *effect* introduced in [Baroni et al., 2014a]. Intuitively the effect of a set of arguments $\mathcal{I}$ with an assigned labelling $\mathcal{L}ab_{\mathcal{I}}$ on another set of arguments $Args$ can be modelled as the labelling that would be induced on $Args$ by considering only the attacks from $\mathcal{I}$ to $Args$. For instance, if all arguments in $\mathcal{I}$ attacking an argument $a \in Args$ are labelled out according to $\mathcal{L}ab_{\mathcal{I}}$ then, considering only these attacks, a would be in. The following definition formalizes this intuition.

Definition 4.9 *Given a set of arguments $\mathcal{I}$, a labelling $\mathcal{L}ab_\mathcal{I} \in \mathfrak{L}_\mathcal{I}$, a set of arguments $\mathcal{A}rgs$ such that $\mathcal{I} \cap \mathcal{A}rgs = \emptyset$ and a relation $R_{INP} \subseteq \mathcal{I} \times \mathcal{A}rgs$, the effect of $(\mathcal{I}, \mathcal{L}ab_\mathcal{I}, R_{INP})$ on $\mathcal{A}rgs$, denoted as $\text{eff}_{\mathcal{A}rgs}(\mathcal{I}, \mathcal{L}ab_\mathcal{I}, R_{INP})$, is defined as*

$$\{(a, \text{out}) \mid a \in \mathcal{A}rgs, \exists b \in \mathcal{I} : (b, a) \in R_{INP} \wedge \mathcal{L}ab_\mathcal{I}(b) = \text{in}\} \cup$$
$$\{(a, \text{undec}) \mid a \in \mathcal{A}rgs, \exists b \in \mathcal{I} : (b, a) \in R_{INP} \wedge \mathcal{L}ab_\mathcal{I}(b) = \text{undec},$$
$$\nexists c \in \mathcal{I} : (c, a) \in R_{INP} \wedge \mathcal{L}ab_\mathcal{I}(c) = \text{in}\} \cup$$
$$\{(a, \text{in}) \mid a \in \mathcal{A}rgs, \nexists b \in \mathcal{I} : (b, a) \in R_{INP} \wedge \mathcal{L}ab_\mathcal{I}(b) \in \{\text{in}, \text{undec}\}\}$$

Example 4.10 *Referring again to the example of Figure 1, let $\mathcal{I} = \{a, e\}$, $\mathcal{L}ab_\mathcal{I} = \{(\{a\}, \text{undec}), (\{e\}, \text{in})\}$, $\mathcal{A}rgs = \{b, c, d\}$, and $R_{INP} = \{(a,b), (e,c)\}$. Then $\text{eff}_{\mathcal{A}rgs}(\mathcal{I}, \mathcal{L}ab_\mathcal{I}, R_{INP}) = \{(b, \text{undec}), (c, \text{out}), (d, \text{in})\}$.*

By definition, $\text{eff}_{\mathcal{A}rgs}(\mathcal{I}, \mathcal{L}ab_\mathcal{I}, R_{INP})$ only depends on the labelling of the arguments in $\mathcal{I}$ that attack $\mathcal{A}rgs$ through R_{INP}. Moreover each argument in $\mathcal{A}rgs$ not receiving attacks from $\mathcal{I}$ is labelled in according to $\text{eff}_{\mathcal{A}rgs}(\mathcal{I}, \mathcal{L}ab_\mathcal{I}, R_{INP})$. Thus, in the particular case where $\mathcal{I} = \emptyset$ (thus also $\mathcal{L}ab_\mathcal{I}$ and R_{INP} are empty), it turns out that $\text{eff}_{\mathcal{A}rgs}(\emptyset, \emptyset, \emptyset) = \{(a, \text{in}) \mid a \in \mathcal{A}rgs\}$.

The notion of effect is important because it can be proved that, for many semantics, it is only and exactly the effect to determine the behavior of the canonical local function on an argumentation framework with input. More precisely, a semantics σ is said to be *effect-dictated* if, given $AF = \langle Ar, att \rangle$, $F_\sigma(AF, \mathcal{I}, \mathcal{L}ab_\mathcal{I}, att_\mathcal{I})$ only depends on $\text{eff}_{Ar}(\mathcal{I}, \mathcal{L}ab_\mathcal{I}, att_\mathcal{I})$, rather than on the whole labelling $\mathcal{L}ab_\mathcal{I}$ and the specific relation $att_\mathcal{I}$, as formalized by the following definition.

Definition 4.11 *A semantics σ is effect-dictated if $(\text{eff}_{Ar}(\mathcal{I}_1, \mathcal{L}ab_{\mathcal{I}_1}, att_{\mathcal{I}_1}) = \text{eff}_{Ar}(\mathcal{I}_2, \mathcal{L}ab_{\mathcal{I}_2}, att_{\mathcal{I}_2})) \Rightarrow F_\sigma(AF, \mathcal{I}_1, \mathcal{L}ab_{\mathcal{I}_1}, att_{\mathcal{I}_1}) = F_\sigma(AF, \mathcal{I}_2, \mathcal{L}ab_{\mathcal{I}_2}, att_{\mathcal{I}_2})$ for every AF, $\mathcal{I}_1$, $\mathcal{I}_2$, $\mathcal{L}ab_{\mathcal{I}_1}$, $\mathcal{L}ab_{\mathcal{I}_2}$, $att_{\mathcal{I}_1}$ and $att_{\mathcal{I}_2}$, where $AF = \langle Ar, att \rangle$ is an argumentation framework, $\mathcal{I}_1$ and $\mathcal{I}_2$ are two sets of arguments such that $\mathcal{I}_1 \cap Ar = \emptyset$ and $\mathcal{I}_2 \cap Ar = \emptyset$, $\mathcal{L}ab_{\mathcal{I}_1} \in \mathfrak{L}_{\mathcal{I}_1}$ and $\mathcal{L}ab_{\mathcal{I}_2} \in \mathfrak{L}_{\mathcal{I}_2}$ two labellings of $\mathcal{I}_1$ and $\mathcal{I}_2$ respectively, and $att_{\mathcal{I}_1} \subseteq \mathcal{I}_1 \times Ar$ and $att_{\mathcal{I}_2} \subseteq \mathcal{I}_2 \times Ar$ two attack relations.*

Many of the semantics considered in this chapter are effect-dictated as shown by the following proposition [Baroni et al., 2014a].

Proposition 4.12 *The semantics $\mathcal{AD}, \mathcal{CO}, \mathcal{ST}, \mathcal{GR}, \mathcal{PR}, \mathcal{ID}, \mathcal{SST}, \mathcal{CF}2, \mathcal{STG}2$ are effect-dictated.*

While the discussion above points out a basic commonality concerning the influence of external arguments in the context of different modular notions in abstract argumentation, the perceptive reader might observe that something is "missing" in these notions: they all deal with unidirectional influences from the outside to the considered "module", while in general a "module" may receive and give influences at the same time.

Indeed, while unidirectional influences are sufficient to support the definition of decomposability, as illustrated in Section 2.3, they are not enough if one wants to make a step further to consider properties like module interchangeability (developed in Section 6) which inherently involve bidirectional interactions.

To address this requirement, a richer modular notion, called *argumentation multipole* [Baroni et al., 2014a] needs to be considered. Briefly, the idea is that a multipole is an argumentation framework interacting with an external set of arguments E by both receiving and launching attacks, as specified in the following definition.

Definition 4.13 *An* Argumentation Multipole *(or, briefly, multipole)* $\mathcal{M}$ *w.r.t. a set* E *is a tuple* (AF, R_{INP}, R_{OUTP}), *where letting* $AF = \langle Ar, att \rangle$ *it holds that* $Ar \cap E = \emptyset$, $R_{INP} \subseteq E \times Ar$, *and* $R_{OUTP} \subseteq Ar \times E$. *Extending the notation introduced in Definition 2.13, we denote as* $\mathcal{M}^{inp}$ *the set* $\{a \in E \mid \exists b \in Ar, (a, b) \in R_{INP}\}$, *i.e. including the arguments of* E *which attack* Ar *through* R_{INP}. *Moreover, we denote as* $\mathcal{M}^{outp}$ *the set* $\{a \in Ar \mid \exists b \in E, (a, b) \in R_{OUTP}\}$, *i.e. including the arguments of* AF *attacking* E *through* R_{OUTP}.

Example 4.14 *Referring again to Figure 1 and letting* $E = \{a, d\}$, *we have that* $\mathcal{M} = (AF, R_{INP}, R_{OUTP})$ *is an argumentation multipole w.r.t.* E *with* $AF = (\{b, c, e\}, \{(b, c), (e, c)\})$, $R_{INP} = \{(a, b), (d, e)\}$, $R_{OUTP} = \{(b, a), (c, d)\}$. *Then* $\mathcal{M}^{inp} = \{a, d\}$ *and* $\mathcal{M}^{outp} = \{b, c\}$.

Argumentation multipoles provide a natural extension and generalization of argumentation frameworks with input in some respects. First, and quite obviously, while in argumentation frameworks with input the external set $\mathcal{I}$ is meant only to provide attacks, in argumentation multipoles the external set E has a twofold role and, while R_{INP} in Definition 4.13 corresponds to $att_\mathcal{I}$ in Definition 2.15, Definition 4.13 includes the additional output relation R_{OUTP}. Moreover, in Definition 4.13 there is no counterpart of the labelling $\mathcal{L}ab_\mathcal{I}$ of the external arguments. This is because in the study of interchangeability properties of argumentation multipoles the effect of multiple configurations of inputs (and outputs) needs to be considered, as will be evidenced in Section 6, while embedding a single input configuration within the definition of argumentation frameworks with input was enough to deal with decomposability properties in an appropriate manner.

Having provided an overview of the basic locality properties in abstract argumentation and of the relevant modular notions, in the next two sections we describe two application areas for these concepts. First in Section 5 we discuss how they can be exploited for the sake of improving efficiency in computational problems related to abstract argumentation, while in Section 6 we deal with the issues of interchangeability and summarization.

5 Efficient computation based on locality and modularity

As discussed in chapter 13, in general many natural questions regarding arguments acceptability are computationally intractable. For this reason, in recent years, an increasing amount of work has been focused on identifying tractable fragments of argumentation frameworks and developing more efficient algorithms, as surveyed in chapter 14. Note that while these approaches have made great achievements, since they treat each argumentation framework as a monolithic entity, their efficiency can be further improved if the properties of locality and modularity are exploited. In this section, we review three locality-based approaches, and show that they can be used to improve the efficiency of computing the semantics of a dynamic, a static and a partial argumentation framework, respectively [Liao et al., 2010; Liao et al., 2011; Baumann et al., 2011; Liao, 2013; Baroni et al., 2014b; Liao, 2014].

5.1 Computing partial semantics of argumentation

When querying the status of some arguments in an abstract argumentation framework, the computation may only take into consideration a subset of arguments that are *relevant* to these arguments. This phenomenon appears in many situations. Let us consider the following two examples.

First, when developing proof theories and algorithms for argumentation frameworks [Modgil and Caminada, 2009], for a given semantics, there are some "local" questions with respect to a subset $Args$ of arguments, such as:

(a) Is $Args$ contained in an extension? (credulous membership question)

(b) Is $Args$ contained in all extensions? (skeptical membership question)

(c) Is $Args$ attacked by an extension?

(d) Is $Args$ attacked by all extensions?

When answering these "local" questions, one might not have to figure out the extensions of a whole framework, but only the extensions of a part of the framework.

Second, when agents engage in dialogues, they have goals to make some arguments (say, a set $Args$ of arguments) acceptable or unacceptable [Boella et al., 2012]. At each turn, an agent may choose some arguments from a set of available arguments, and add them to the framework. For each addition, the status of arguments of the framework may change accordingly. However, since it is necessary to figure out only the status evolution of arguments in $Args$, other arguments that are irrelevant to $Args$ may not be taken into consideration.

Motivated by the above examples, an approach for evaluating the status of a subset of arguments in an argumentation frameworks is presented in [Liao and Huang, 2013]. The basic idea is as follows.

Given an argumentation framework and a subset $Args$ of arguments within it, the *minimal* set of arguments that are *relevant* to the arguments in $Args$

(called the set of *relevant arguments* of $\mathcal{A}rgs$) is identified. Intuitively, an argument a which is not already in $\mathcal{A}rgs$ is relevant to $\mathcal{A}rgs$ if there is an attack path from a to an element of $\mathcal{A}rgs$.

Definition 5.1 *Given $AF = \langle Ar, att \rangle$ and $\mathcal{A}rgs \subseteq Ar$, the set of relevant arguments of $\mathcal{A}rgs$ is defined as follows:*

$$rlvt_{AF}(\mathcal{A}rgs) = \mathcal{A}rgs \cup$$

$$\left(\bigcup_{a \in \mathcal{A}rgs} \{b \in Ar \setminus \mathcal{A}rgs : \text{ there is a path from } b \text{ to } a \text{ w.r.t. } att\} \right)$$

Then, under a given argumentation semantics σ, the set of extensions of the sub-framework that is induced by the set of relevant arguments of $\mathcal{A}rgs$ (called a *partial semantics* of the argumentation framework with respect to $\mathcal{A}rgs$) is defined as follows.

Definition 5.2 *Let $AF = \langle Ar, att \rangle$ be an argumentation framework, $\mathcal{A}rgs \subseteq Ar$ a set of arguments, and $rlvt_{AF}(\mathcal{A}rgs)$ the set of relevant arguments of $\mathcal{A}rgs$. Under a semantics σ, the partial semantics of AF with respect to $\mathcal{A}rgs$ is defined as $\mathcal{E}_\sigma(\langle rlvt_{AF}(\mathcal{A}rgs), R_{rlvt_{AF}(\mathcal{A}rgs)} \rangle)$, where $R_{rlvt_{AF}(\mathcal{A}rgs)} = (rlvt_{AF}(\mathcal{A}rgs) \times rlvt_{AF}(\mathcal{A}rgs)) \cap att$.*

Proposition 5.3 *Let $AF = \langle Ar, att \rangle$ be an argumentation framework, and $\mathcal{A}rgs \subseteq Ar$ a set of arguments. Under a semantics σ satisfying directionality property, for every argument $a \in \mathcal{A}rgs$, a is skeptically justified (respectively, credulously justified and indefensible[5]) with respect to $\mathcal{E}_\sigma(AF)$, if and only if a is skeptically justified (respectively, credulously justified and indefensible) with respect to $\mathcal{E}_\sigma(\langle rlvt_{AF}(\mathcal{A}rgs), R_{rlvt_{AF}(\mathcal{A}rgs)} \rangle)$.*

According to Proposition 5.3, given $AF = \langle Ar, att \rangle$ and $\mathcal{A}rgs \subseteq Ar$, we may evaluate the status of arguments in $rlvt_{AF}(\mathcal{A}rgs)$ independently, without computing the status of other arguments that are irrelevant to $\mathcal{A}rgs$. Furthermore, it is desirable that after the partial semantics of AF with respect to some sets of arguments are obtained, they can be reused in the subsequent computation. This vision is embodied by the following three basic properties of partial semantics of argumentation: *monotonicity, combinability,* and *extensibility*.

First, given $AF = \langle Ar, att \rangle$ and $\mathcal{A}rgs, \mathcal{A}rgs' \subseteq Ar$, the *monotonicity* of partial semantics of argumentation can be informally expressed as follows: if $rlvt_{AF}(\mathcal{A}rgs) \subseteq rlvt_{AF}(\mathcal{A}rgs')$, then the justification status of each argument in $\mathcal{A}rgs$ evaluated with respect to $\mathcal{E}_\sigma(\langle rlvt_{AF}(\mathcal{A}rgs), R_{rlvt_{AF}(\mathcal{A}rgs)} \rangle)$ is the same as the one evaluated with respect to $\mathcal{E}_\sigma(\langle rlvt_{AF}(\mathcal{A}rgs'), R_{rlvt_{AF}(\mathcal{A}rgs')} \rangle)$. In other words, after we get the partial semantics of AF with respect to $\mathcal{A}rgs'$ (i.e.,

---
[5] An argument which is not skeptically nor credulously justified is called indefensible in [Liao and Huang, 2013].

$\mathcal{E}_\sigma(\langle rlvt_{AF}(\mathcal{A}rgs'), R_{rlvt_{AF}(\mathcal{A}rgs')}\rangle))$, we may evaluate the justification status of arguments in $\mathcal{A}rgs$ with respect to $\mathcal{E}_\sigma(\langle rlvt_{AF}(\mathcal{A}rgs'), R_{rlvt_{AF}(\mathcal{A}rgs')}\rangle)$ directly, and do not need to compute the partial semantics of AF with respect to $\mathcal{A}rgs$.

Second, given $AF = \langle Ar, att\rangle$ and $\mathcal{A}rgs, \mathcal{A}rgs' \subseteq Ar$, the *extensibility* of partial semantics of argumentation can be informally expressed as follows: If $rlvt_{AF}(\mathcal{A}rgs) \subseteq rlvt_{AF}(\mathcal{A}rgs')$, then the partial semantics of AF with respect to $\mathcal{A}rgs$ can be *extended* to form the partial semantics of AF with respect to $\mathcal{A}rgs'$. Let $Comp = rlvt_{AF}(\mathcal{A}rgs') \setminus rlvt_{AF}(\mathcal{A}rgs)$ denote the *complement* of $rlvt_{AF}(\mathcal{A}rgs)$ in $rlvt_{AF}(\mathcal{A}rgs')$. Let $(\langle Comp, R_{Comp}\rangle, (Comp^-, \mathcal{I}_{Comp}))$ be a conditioned sub-framework with respect to $\langle rlvt_{AF}(\mathcal{A}rgs), R_{rlvt_{AF}(\mathcal{A}rgs)}\rangle$, where $R_{Comp} = att \cap (Comp \times Comp)$ and $\mathcal{I}_{Comp} = att \cap (Comp^- \times Comp)$. According to [Liao, 2014], the following proposition holds.

Proposition 5.4 *For all* $\sigma \in \{\mathcal{AD}, \mathcal{CO}, \mathcal{GR}, \mathcal{PR}\}$, *it holds that*

$$\mathcal{E}_\sigma(\langle rlvt_{AF}(\mathcal{A}rgs'), R_{rlvt_{AF}(\mathcal{A}rgs')}\rangle) = \{E_1 \cup E_2 \mid$$
$$(E_1 \in \mathcal{E}_\sigma(\langle rlvt_{AF}(\mathcal{A}rgs), R_{rlvt_{AF}(\mathcal{A}rgs)}\rangle)) \wedge$$
$$(E_2 \in \mathcal{E}_\sigma((\langle Comp, R_{Comp}\rangle, (Comp^-[E_1], \mathcal{I}_{Comp}))))\}$$

Proposition 5.4 indicates that the partial semantics of AF with respect to $\mathcal{A}rgs'$ is obtained by extending the partial semantics of AF with respect to $\mathcal{A}rgs$.

Third, given $AF = \langle Ar, att\rangle$ and $B, C \subseteq Ar$, the *combinability* of partial semantics of argumentation can be informally expressed as follows: The combination of partial semantics of AF with respect to B and the one with respect to C is equal to the partial semantics of AF with respect to $B \cup C$. Formally, we have the following definition.

Definition 5.5 *Given* $AF = \langle Ar, att\rangle$, $B, C \subseteq Ar$, *let* $Int = rlvt_{AF}(B) \cap rlvt_{AF}(C)$ *denote the* intersection *of* $rlvt_{AF}(B)$ *and* $rlvt_{AF}(C)$. *The combination of partial semantics of* AF *with respect to* B *and the one with respect to* C *under a given semantics* σ, *denoted as* $CombExt_\sigma(\langle rlvt_{AF}(B \cup C), R_{rlvt_{AF}(B \cup C)}\rangle)$, *is defined as follows:*

$$CombExt_\sigma(\langle rlvt_{AF}(B \cup C), R_{rlvt_{AF}(B \cup C)}\rangle) \triangleq \{E_1 \cup E_2 \mid$$
$$(E_1 \in \mathcal{E}_\sigma(\langle rlvt_{AF}(B), R_{rlvt_{AF}(B)}\rangle)) \wedge$$
$$(E_2 \in \mathcal{E}_\sigma(\langle rlvt_{AF}(C), R_{rlvt_{AF}(C)}\rangle)) \wedge$$
$$(E_1 \cap Int = E_2 \cap Int)\}$$

Proposition 5.6 *For all* $\sigma \in \{\mathcal{AD}, \mathcal{CO}, \mathcal{GR}, \mathcal{PR}\}$, *it holds that* $\mathcal{E}_\sigma(\langle rlvt_{AF}(B \cup C), R_{rlvt_{AF}(B \cup C)}\rangle) = CombExt_\sigma(\langle rlvt_{AF}(B \cup C), R_{rlvt_{AF}(B \cup C)}\rangle)$.

According to this property, after we get the partial semantics of AF with respect to some subsets of Ar respectively, we may get the partial semantics of a larger subset of arguments by means of semantics combination.

The ideas of partially evaluating the status of arguments and the combinability of partial semantics are illustrated by the following example.

Example 5.7 *Given an argumentation framework $AF = \langle Ar, att \rangle$ illustrated in Figure 2, let $Args_1 = \{a_9, a_{10}\}$, according to the directionality of argumentation semantics, only the status of a_1 and a_2 may affect the status of a_9 and a_{10}. In other words, when computing the status of arguments in $Args_1$, we may only consider the set $\{a_1, a_2, a_9, a_{10}\}$ of arguments.*
Let $Args_2 = \{a_{11}, a_{12}\}$, and $Int_1 = rlvt_{AF}(Args_1) \cap rlvt_{AF}(Args_2)$. $Int_1 = \{a_1, a_2\}$. Under preferred semantics, $\mathcal{E}_{\mathcal{PR}}(\langle rlvt_{AF}(Args_1), R_{rlvt_{AF}(Args_1)}\rangle) = \{E_{1,1}, E_{1,2}, E_{1,3}\}$, in which $E_{1,1} = \{a_1, a_9\}$, $E_{1,2} = \{a_1, a_{10}\}$, $E_{1,3} = \{a_2, a_{10}\}$, $\mathcal{E}_{\mathcal{PR}}(\langle rlvt_{AF}(Args_2), R_{rlvt_{AF}(Args_2)}\rangle) = \{E_{2,1}, E_{2,2}, E_{2,3}\}$, in which $E_{2,1} = \{a_1, a_3, a_{12}\}$, $E_{2,2} = \{a_2, a_{11}\}$, and $E_{2,3} = \{a_2, a_{12}\}$. Then, the partial semantics of AF with respect to $Args_1 \cup Args_2$ can be obtained by means of semantics combination: $\mathcal{E}_{\mathcal{PR}}(\langle rlvt_{AF_1}(Args_1 \cup Args_2), R_{rlvt_{AF_1}(Args_1 \cup Args_2)}\rangle) = \{E_{3,1}, E_{3,2}, E_{3,3}, E_{3,4}\}$, in which $E_{3,1} = E_{1,1} \cup E_{2,1} = \{a_1, a_3, a_9, a_{12}\}$, $E_{3,2} = E_{1,2} \cup E_{2,1} = \{a_1, a_3, a_{10}, a_{12}\}$, $E_{3,3} = E_{1,3} \cup E_{2,2} = \{a_2, a_{10}, a_{11}\}$, $E_{3,4} = E_{1,3} \cup E_{2,3} = \{a_2, a_{10}, a_{12}\}$.

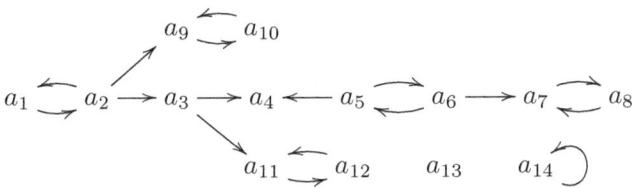

Figure 2. The relevant set of arguments of $Args_1$

The efficiency of computing partial semantics of argumentation has been evidenced by the empirical results presented in [Liao, 2014]. In this empirical investigation, the partial semantics of argumentation is computed by using an efficient ASP solver DLV.

Figure 3 shows the execution time when the number of nodes is from 50 to 100 and the edge density of argumentation frameworks is 1%, 2%, 3% and 20% respectively, where nodes are arguments, edges are attacks, and the density of an argument graph is defined as $\frac{|R|}{|A|*(|A|-1)}*100\%$. In Figure 3, the following notations are used: "$di\%$-whole", denoting the time for computing the extensions of the (whole) argumentation framework whose edge density is $i\%$; "$di\%$-part-q1"(respectively "$di\%$-part-q10%", and "$di\%$-part-q20%"), denoting the time for computing the partial semantics of the argumentation framework whose edge density is $i\%$, when the number of arguments to be queried is 1 (respectively, 10 percent of the number of nodes, and 20 percent of the number of nodes). The results show that when the edge density of an argumentation framework is 1%, which means that when the size of the argumentation framework is 100, the ratio of nodes to edges is 1:1, the average time for computing the partial semantics is much lower than that of computing the semantics of the whole argumentation framework. With the increase of edge density, the average time for computing the partial semantics becomes closer and closer to that of computing the semantics of the whole argumentation framework.

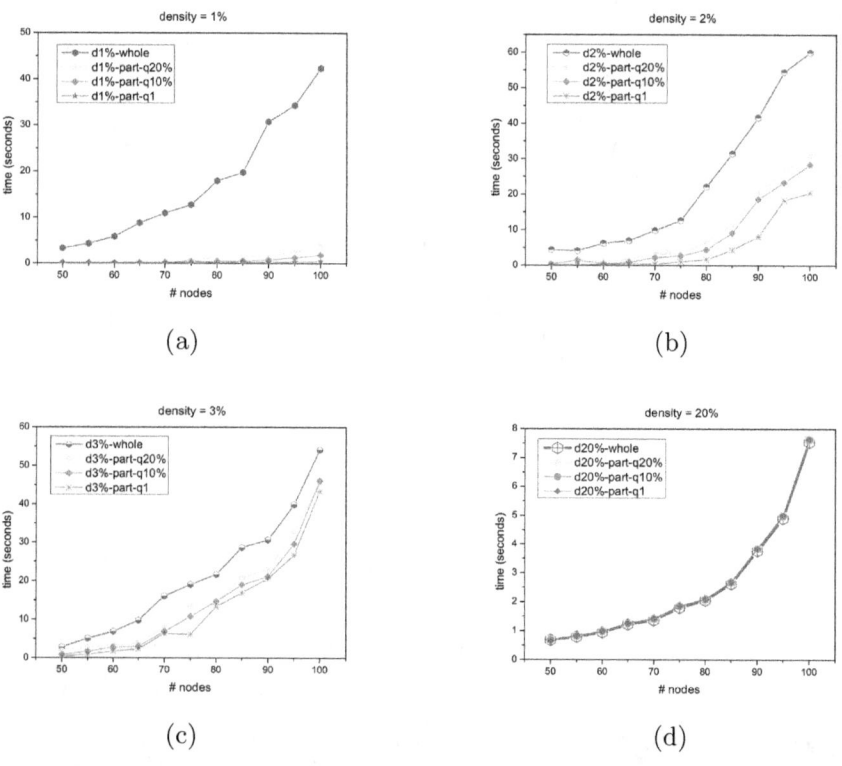

Figure 3. Plots showing the average execution time (revised by dropping some cases where the execution time of computing the extensions of the whole argumentation framework is 3 times more than the average execution time in remaining cases) when the edge density of argumentation frameworks is 1%, 2%, 3% and 20% respectively.

5.2 Computing semantics of a static argumentation framework

A given argumentation framework can be viewed as a snapshot of a dynamic argumentation system, and therefore is called "static argumentation framework" [Liao, 2014]. Since the tractability of an argumentation framework is closely related to its structure [Dunne, 2007] and its size, it is intuitively feasible to improve the efficiency of computation by taking advantage of the special topologies of argumentation frameworks, and by reducing their sizes. Indeed, some tractable classes of argumentation frameworks have been identified so far, including acyclic argumentation frameworks [Dung, 1995], symmetric argumentation frameworks [Coste-Marquis et al., 2005], and bounded tree-width argumentation frameworks [Dunne, 2007; Dvořák et al., 2010], etc. Moreover, even if an argumentation framework globally does not belong to any tractable class, it is possible to exploit a decomposition of the framework to achieve tractability at a local level.

Recent proposals in this important research direction include the approach based on splitting argumentation frameworks [Baumann et al., 2011], the approach of decomposition-based incremental computation [Liao and Huang, 2012; Liao, 2013] and the SCC-recursive meta-algorithm presented in [Cerutti et al., 2014], etc. All these approaches are based on the SCC-decomposition of the framework. We describe the basic idea of the decomposition-based approach in the following.

Given a generic argumentation framework, it is firstly decomposed into a set of sub-frameworks (each of which is a single SCC or the union of a set of SCCs), among which some may be acyclic sub-frameworks. Then, the extensions of the argumentation framework are computed incrementally, following a partial order of the set of SCCs. In each iteration, the extensions of each sub-framework are computed locally. As a result, the tractability of acyclic subframeworks can be exploited.

When decomposing an argumentation framework, it is necessary to consider how to organize the sub-frameworks, such that their semantics can be computed incrementally. According to the partial order over the set of SCCs, it is natural to adopt a *layered approach*. The basic idea is that the sub-frameworks in the same layer can be handled in parallel, while the sub-frameworks in a given layer are only dependent on the sub-frameworks in the lower layers and, in the lowest layer, all sub-frameworks are independent.

In order to identify acyclic subframeworks the basic notion of trivial SCC is needed.

Definition 5.8 *Let* $SCCS_{AF}$ *be the set of SCCs of an argumentation framework* $AF = \langle Ar, att \rangle$. *For all* $S \in SCCS_{AF}$, *if there is only one argument (say a) in S and the only argument does not self-attack, i.e.,* $(a, a) \notin att$, *then S is called a* trivial *SCC, else a* nontrivial *SCC.*

Based on trivial SCCs, a decomposition into sub-frameworks is generated such that:

(i) each non-trivial SCC is used to induce a sub-framework, and

(ii) the union of a set of trivial SCCs is used to induce an acyclic sub-framework, taking into account the partial order of dependence relation among subframeworks induced by the attack relation.

In order to realize a layered decomposition approach, given the set of SCCs of an argumentation framework, a natural number is then assigned to each SCC, called the *level* of the SCC, indicating the order according to which the SCC can be processed. Formally, the notion of the *level* of a SCC is defined as follows.

Definition 5.9 *Let* $SCCS_{AF}$ *be the set of SCCs of an argumentation framework* $AF = \langle Ar, att \rangle$. *For all* $S \in SCCS_{AF}$, *the level of* S *is a function* $\rho : SCCS_{AF} \to \mathbb{N}$. *Recursively,* $\rho(S)$ *is defined as follows:*
1. *If* S *has no parent, then* $\rho(S) = 0$;
2. *else:*
   (a) *when* S *is nontrivial,* $\rho(S) = \max\{\rho(P) + 1 : P \text{ is a parent of } S\}$;
   (b) *when* S *is trivial,* $\rho(S) = \max\{\rho(P) + 1, \rho(Q) : P \text{ is a nontrivial parent of } S, Q \text{ is a trivial parent of } S\}$.

After all SCCs of an argumentation framework are assigned a level, layers can be easily identified.

Definition 5.10 *Let* $SCCS_{AF}$ *be the set of SCCs of an argumentation framework* $AF = \langle Ar, att \rangle$, *and* $l_{max} = \max\{\rho(S) : S \in SCCS_{AF}\}$. *Then,* $SCCS_{AF}$ *can be decomposed into* $l_{max} + 1$ *layers, denoted as* $SCCS_{AF}^0, \ldots,$ *and* $SCCS_{AF}^{l_{max}}$, *respectively, in which*

(1) $$SCCS_{AF}^i = \{S \in SCCS_{AF} \mid \rho(S) = i\}$$

Furthermore, Trv_i *and* Non_i ($0 \leq i \leq l_{max}$) *will denote the set of trivial, and nontrivial SCCs of layer* i, *respectively.*

It obviously holds that:

(2) $$SCCS_{AF}^i = Trv_i \cup Non_i$$

In each layer i, $0 \leq i \leq l_{max}$, every nontrivial SCC is used to induce a sub-framework that contains cycles, and the union of all trivial SCCs is used to induce an acyclic sub-framework. Formally, let $U_i = \cup_{S_i \in Trv_i} S_i$ be the union of the set of trivial SCCs in Layer i. For every $P \in Non_i \cup \{U_i\}$, P is used to induce a (conditioned) sub-framework $(\langle P, \mathcal{R}_P \rangle, (P^-, \mathcal{I}_P))$, where $\mathcal{R}_P = att \cap (P \times P)$, $P^- = \{A \in Ar \setminus P \mid \exists B \in P, \text{ s.t. } (A, B) \in att\}$, and $\mathcal{I}_P = att \cap (P^- \times P)$.

This yields a decomposition of an argumentation framework into a set of (conditioned) sub-frameworks located in a number of layers.

Definition 5.11 *Given an argumentation framework* $AF = \langle Ar, att \rangle$, *a decomposition of* AF, *denoted as* $\text{decomp}(AF)$, *is defined as a tuple:*

(3) $$\text{decomp}(AF) = (\text{decomp}_0(AF), \ldots, \text{decomp}_{l_{max}}(AF))$$

where $\text{decomp}_i(AF) = \{(\langle P, \mathcal{R}_P \rangle, (P^-, \mathcal{I}_P)) \mid P \in \mathcal{N}on_i \cup \{U_i\}\}$, $0 \leq i \leq l_{max}$

After an argumentation framework is decomposed according to Definition 5.11, the computation of extensions proceeds incrementally from layer 0 to layer l_{max}. First, the computation of a layer may include the following steps:

- constructing assigned conditioned sub-frameworks in Layer i ($i \geq 1$),
- computing the extensions of each sub-framework in Layer i ($i \geq 0$),
- horizontally combining the extensions of Layer i ($i \geq 0$), and
- vertically combining the extensions of layers from 0 to i ($i \geq 1$).

For every $\sigma \in \{\mathcal{AD}, \mathcal{CO}, \mathcal{PR}, \mathcal{GR}\}$, it holds that the extensions of the original argumentation framework coincide with the ones obtained by combining the extensions of all sub-frameworks.

To evaluate the efficiency of the decomposition-based approach, an experiment and its results can be found in [Liao, 2014]. In this experiment, the decomposition-based algorithm (or briefly, Decomp. Alg.) is compared to Modgil and Caminada's labelling-based algorithms (or briefly, MC Alg.) [Modgil and Caminada, 2009]. The configuration of the experiment is as follows. For each assignment of the number of nodes and the ratio of the number of edges to the number of nodes, the program was executed 20 times. In each time, an argumentation framework was generated at random, and then its preferred labellings were computed by using the MC algorithm and the decomposition-based algorithm, respectively. Table 3 shows the average results of the number of SCCs and the execution time of the two approaches.

In this table, "#nodes", "#SCCS" and "#timeout" denote, respectively, the number of nodes of an argumentation framework, the number of SCCs, and the number of timeouts (when the time for computing the preferred labellings of an argumentation framework is over 30 minutes) among the 20 times of execution[6]. When the average time shown in Table 3 was computed, only the cases without timeouts were considered. The results of this table show that when the ratio of the number of edges to the number of nodes is around 1.5:1, the decomposition-based algorithm is much more efficient. With the increase of the ratios of the number of edges to the number of nodes, the execution time of the two algorithms increase, and the computational advantage of the decomposition-based algorithm decreases.

5.3 Improving efficiency in dynamic argumentation frameworks

In many contexts argumentation frameworks are used to represent dynamic scenarios (e.g. multi-agent interactions) where arguments and their attack relations change [Parsons et al., 1998; Kraus et al., 1998; Rahwan et al., 2003;

---

[6]Since in many cases, the execution time might last very long, to make the test easier, when the time for computing the preferred labellings of an argumentation framework is over 30 minutes, we stopped the execution by setting a break in the program.

ratio	#nodes	#SCCS	Decomp. Alg.		MC Alg.	
			time (seconds)	#timeout	time (seconds)	#timeout
	100	95	0.013	0	1.444	0
	120	114	0.008	0	0.022	3
1:1	140	135	0.012	0	0.055	3
	160	156	0.015	0	5.033	1
	180	171	0.017	0	1.656	4
	200	192	0.019	0	0.328	1
	15	10	0.451	0	9.818	0
	17	11	0.015	0	54.236	2
1.5:1	19	11	0.001	0	57.379	2
	21	12	0.002	1	1.775	3
	23	16	4.496	0	40.126	2
	25	18	0.002	0	2.933	3
	15	7	2.133	0	95.911	4
	17	7	1.501	1	12.971	8
2:1	19	9	0	3	0.576	5
	21	10	0.001	1	0.59	6
	23	9	0.451	3	24.94	7
	25	10	0.119	3	34.87	4

Table 3. Average results of the two algorithms

Amgoud et al., 2008]. When an argumentation framework is modified, recomputing the status of all arguments from scratch may be redundant and inefficient. As a matter of fact, for each modification to an argumentation framework, it may be the case that only a small part of the arguments are affected. So, if it is feasible to evaluate the status of those affected arguments locally, then the status of the remaining (unaffected) arguments can be reused [Liao et al., 2011; Baumann, 2011]. In the following we will explain the division-based approach [Liao et al., 2011] in more detail.

This basic idea of the division-based approach was introduced in [Liao et al., 2010] and developed in [Liao et al., 2011; Baroni et al., 2014b]. Briefly, to efficiently deal with a modification of an original framework $AF = \langle Ar, att \rangle$, the resulting framework $AF' = \langle Ar', att' \rangle$ is partitioned into an *affected* and an *unaffected* subframework (denoted respectively as $AF_a = \langle Ar_a, att_a \rangle$ and $AF_u = \langle Ar_u, att_u \rangle$). Existing outcomes of extension computation in the unaffected subframework can then be reused and only the affected subframework needs some recomputing.

The identification of the unaffected part corresponds to the property of directionality. In fact, the unaffected part consists of the arguments which are unreachable from the arguments or attacks that have been added or deleted, i.e. the set Ar_u is by definition an unattacked set.

By directionality, the set of extensions of $AF\downarrow_{Ar_u}$ can be derived simply by restriction from the extensions of the original argumentation framework, which are assumed to be available. Formally for a semantics σ satisfying the directionality property it holds that:

$$(4) \qquad \mathcal{E}_\sigma(AF\downarrow_{Ar_u}) = \{E \cap Ar_u \mid E \in \mathcal{E}_\sigma(AF)\}.$$

The process of partial recomputing of extensions is then formalized by using the notion of conditioned argumentation framework, the idea being that, after a modification, the unaffected subframework corresponds to the conditioning part while the affected part, which receives attacks from the unaffected one, is the conditioned framework.

For a given semantics σ, the incremental computation presented in [Liao et al., 2011] considers each extension E_1 of the unaffected subframework, namely $AF\downarrow_{Ar_u}$, which gives rise to an assigned $\mathcal{CAF}$.

Traditional Dung's semantics notions can be extended to assigned conditioned frameworks in a relatively straightforward way (see [Liao et al., 2011; Baroni et al., 2014b] for details): $\mathcal{E}_\sigma(\mathcal{CAF}[E_1])$ will denote the extensions prescribed by semantics σ for the assigned conditioned framework $\mathcal{CAF}[E_1]$. On this basis the "conditioned" extensions prescribed by σ for this assigned $\mathcal{CAF}$ can be computed. Combining E_1 with a conditioned extension gives rise to a global extension of the modified framework according to σ. The combination process envisaged is synthesised by the following definition.

Definition 5.12 *Given an argumentation framework $AF = \langle Ar, att \rangle$ and an unaffected part $AF_u = \langle Ar_u, att_u \rangle$, let $E_1 \in \mathcal{E}_\sigma(AF_u)$ an extension of AF_u with respect to a semantics σ, and $\mathcal{CAF}[E_1]$ be the relevant assigned conditioned subframework, i.e. $\mathcal{CAF}[E_1] \triangleq (\langle Ar_a, att_a \rangle, (\mathcal{C}(Ar_u)[E_1], \mathcal{I}_{(\mathcal{C}(Ar_u), Ar_a)}))$ with $\mathcal{I}_{(\mathcal{C}(Ar_u), Ar_a)} = att' \cap (Ar_u \times Ar_a)$. The set of resulting combined extensions of AF' is defined as follows:*

$$(5) \qquad \mathcal{CE}_\sigma(AF') = \{E_1 \cup E_2 \mid E_1 \in \mathcal{E}_\sigma(AF_u) \wedge E_2 \in \mathcal{E}_\sigma(\mathcal{CAF}[E_1])\}$$

Then for suitable semantics σ, it holds that

$$(6) \qquad \mathcal{CE}_\sigma(AF') = \mathcal{E}_\sigma(AF')$$

i.e. that the global extensions of the framework AF' coincide with the combinations of Definition 5.12.

In particular it is proved in [Liao et al., 2011] that equation (6) holds for complete, grounded and preferred semantics. Variants of the division-based method for stable and ideal semantics (including a correction to [Liao et al., 2011]) are presented in [Baroni et al., 2014b].

A result similar to (6) has been considered in the context of splittings [Baumann, 2011; Baumann, 2014] and is recalled below. First, we say that a semantics σ satisfies the splitting property if the two relationships stated in Definition 5.13 hold. Note that they bear some similarity with bottom-up and top-down decomposability (indeed they are a special case of these notions, restricted to the specific way of partitioning the framework corresponding to splitting) and that they use the concepts of reduct and modification introduced in Section 4.

Definition 5.13 *A semantics σ satisfies the splitting property if for every argumentation framework $AF = \langle Ar, att \rangle$ admitting a splitting (AF_1, AF_2, att_3) with $AF_1 = \langle Ar_1, att_1 \rangle$ and $AF_2 = \langle Ar_2, att_2 \rangle$ it holds that:*

1. *If $E_1 \in \mathcal{E}_\sigma(AF_1)$ and $E_2 \in \mathcal{E}_\sigma(mod_{U_{E_1}, att_3}(red_{E_1, att_3}(AF_2)))$ then $E_1 \cup E_2 \in \mathcal{E}_\sigma(AF)$*

2. *If $E \in \mathcal{E}_\sigma(AF)$ then $E \cap Ar_1 \in \mathcal{E}_\sigma(AF_1)$ and $E \cap Ar_2 \in \mathcal{E}_\sigma(mod_{U_{E \cap Ar_1}, att_3}(red_{E \cap Ar_1, att_3}(AF_2)))$*

where, given an extension E of $AF_1 = \langle Ar_1, att_1 \rangle$, U_E is defined as follows: $U_E = \{a \in Ar_1 \mid a \notin E, a^- \cap E = \emptyset\}$.

In words, if E_1 is a σ-extension of the conditioning framework AF_1 and E_2 is a σ-extension of the framework obtained from the conditioned framework AF_2 by applying first the reduct and then the modification on the basis of E_1, then combining E_1 and E_2 a global extension of the whole framework is obtained. Vice versa if E is an extension of the whole framework, then it can be decomposed into a σ-extension of the conditioning framework AF_1 and a σ-extension of the modified framework obtained from the framework AF_2 as described above.

It has been proved that admissible, grounded, complete, preferred and stable semantics satisfy the splitting property.

Proposition 5.14 *The semantics $\mathcal{AD}, \mathcal{GR}, \mathcal{CO}, \mathcal{PR},$ and $\mathcal{ST}$ satisfy the splitting property.*

Similarly to the approaches previously described, splitting can be used as a basis for the design of divide-and-conquer algorithms. In particular, an empirical evaluation of the efficiency gains that can be achieved by taking into account splitting in the computation of grounded, preferred and stable extensions has been reported in [Baumann et al., 2011].

6 Interchangeability and summarization

This section deals with the issue of interchangeability of "argumentation modules", specifically of argumentation multipoles, in the context of a global argumentation framework: this property is obviously useful in several respects and in particular as far as the summarization of argumentation frameworks is concerned, as it will be discussed later.

Interchangeability of two modules requires first of all that they are compatible as far as the connections with the rest of the framework are concerned, i.e. that the interfaces they expose are such that wherever one of the modules can be "plugged in", the other can too. Besides this *plug-level* interchangeability, it is of great interest to characterize the *behavior-level* interchangeability of modules, namely to identify the situations where internally different modules can be freely interchanged without affecting the global behavior of the framework they belong to, since their Input/Output behavior is equivalent in this respect.

As to the *plug-level* interchangeability, a very loose notion is sufficient: recalling from Definition 4.13 that a multipole is defined with respect to an external set of arguments E, two multipoles are regarded as interchangeable if they refer to the same set E.

Behavior-level interchangeability is definitely more complex (and more interesting): it involves characterizing first the "black-box behavior" of the multipole, as a basis for defining the Input/Output equivalence between multipoles and then investigating the relevant interchangeability properties. Of course, since the behavior of a multipole is determined by the specific argumentation semantics σ adopted, all considerations and definitions will be semantics-dependent.

6.1 Multipole effect and Input/Output equivalence

The first step consists in characterizing the effect of a multipole on the set of external arguments E, for a given semantics σ, taking into account that the multipole also receives some input from E. The idea is that, given a multipole $\mathcal{M}$ w.r.t. a set E, for any "input" labelling $\mathcal{L}_E \in \mathfrak{L}_E$, an argumentation framework with input corresponding to the multipole is induced and the local function F_σ prescribes a set of labellings for it. Each of these labellings has its own effect on E (Definition 4.9), therefore the global effect of the multipole receiving an input $\mathcal{L}_E$ is a set of labellings of E whose members are all the single effects.

This is formalised by the following definition.

Definition 6.1 *Let* $\mathcal{M} = (AF, R_{INP}, R_{OUTP})$ *a multipole w.r.t. a set E and σ an argumentation semantics. Given a labelling $\mathcal{L}_E \in \mathfrak{L}_E$, the σ-effect of* $(\mathcal{M}, \mathcal{L}_E)$ *on E, denoted as σ-$\mathtt{eff}_E(\mathcal{M}, \mathcal{L}_E)$, is defined as σ-$\mathtt{eff}_E(\mathcal{M}, \mathcal{L}_E) = \{\mathtt{eff}_E(\mathcal{M}^{outp}, \mathcal{L}\downarrow_{\mathcal{M}^{outp}}, R_{OUTP}) \mid \mathcal{L} \in F_\sigma(AF, \mathcal{M}^{inp}, \mathcal{L}_E\downarrow_{\mathcal{M}^{inp}}, R_{INP})\}$.*

Note that if $F_\sigma(AF, \mathcal{M}^{inp}, \mathcal{L}_E\downarrow_{\mathcal{M}^{inp}}, R_{INP}) = \emptyset$, i.e. the local function prescribes no labelling, then σ-$\mathtt{eff}_E(\mathcal{M}, \mathcal{L}_E) = \emptyset$.

Example 6.2 *Continuing with the example of Figure 1, consider the multipole $\mathcal{M}_1 = (AF\downarrow_{\{a,b\}}, \emptyset, \{(b,c)\})$ with respect to the set $E = \{c, d, e\}$. $\mathcal{M}_1$ (whose only possible labelling $\mathcal{L}_E\downarrow_{\mathcal{M}_1^{inp}}$ is the empty set, since $\mathcal{M}_1$ receives no attacks from outside) has two preferred labellings, one where b is* $\mathtt{in}$ *and another where b*

is out, hence for any $\mathcal{L}_E \in \mathfrak{L}_E$, $\mathcal{PR}\text{-eff}_E(\mathcal{M}_1, \mathcal{L}_E) = \{(\{d,e\}, \{c\}, \emptyset), (\{c,d,e\}, \emptyset, \emptyset)\}$.

Consider now replacing $\mathcal{M}_1$ with the multipole $\mathcal{M}_2 = (AF_2, \emptyset, \{(b_2, c)\})$ where $AF_2 = \langle \{a_1, b_1, a_2, b_2\}, \{(a_1, b_1), (b_1, a_2), (a_2, b_2), (b_2, a_1)\} \rangle$. Also $\mathcal{M}_2$ has two preferred labellings, one where b_2 is in and another where b_2 is out, leading to $\mathcal{PR}\text{-eff}_E(\mathcal{M}_1, \mathcal{L}_E) = \mathcal{PR}\text{-eff}_E(\mathcal{M}_2, \mathcal{L}_E)$ for every $\mathcal{L}_E \in \mathfrak{L}_E$.

In the example above, intuitively, one can then say that $\mathcal{M}_1$ and $\mathcal{M}_2$ are equivalent with respect to E and also that $\mathcal{M}_1$, being "more compact", may be regarded as a summarization of $\mathcal{M}_2$.

Generalizing this intuition, two multipoles $\mathcal{M}_1$ and $\mathcal{M}_2$ w.r.t. E can be considered σ-equivalent if, for any possible labelling $\mathcal{L}_E \in \mathfrak{L}_E$, i.e. for any possible input they receive from E, the effect on E is the same, namely if $\sigma\text{-eff}_E(\mathcal{M}_1, \mathcal{L}_E) = \sigma\text{-eff}_E(\mathcal{M}_2, \mathcal{L}_E)$. For reasons that will be clear later, it is also useful to identify multipoles that have the same effect only for a subset of input labellings: in order to capture this possibility, it is useful to define equivalence under a set of labellings of E.

Definition 6.3 *Two multipoles $\mathcal{M}_1$ and $\mathcal{M}_2$ w.r.t. a set E are Input/Output σ-equivalent (or simply σ-equivalent) under a set of labellings $\mathfrak{L}' \subseteq \mathfrak{L}_E$ iff for any labelling $\mathcal{L}_E \in \mathfrak{L}'$ it holds that $\sigma\text{-eff}_E(\mathcal{M}_1, \mathcal{L}_E) = \sigma\text{-eff}_E(\mathcal{M}_2, \mathcal{L}_E)$. The multipoles $\mathcal{M}_1$ and $\mathcal{M}_2$ are σ-equivalent iff they are σ-equivalent under $\mathfrak{L}_E$.*

Obviously if two multipoles w.r.t. E are σ-equivalent then they are σ-equivalent under any set $\mathfrak{L}' \subseteq \mathfrak{L}_E$.

6.2 Multipole replacement

As anticipated in the example in the previous subsection, an argumentation multipole can be viewed as a component of an argumentation framework that can be *replaced* with another multipole giving rise to a (possibly) different argumentation framework. In particular, given an argumentation framework $AF = \langle Ar, att \rangle$, one may partition the set of arguments $\mathcal{A}rgs$ into two sets, i.e. a set E which is not involved in the replacement and the set $D_1 = Ar \setminus E$ which is replaced along with the relevant attacks: the set D_1 identifies the multipole $\mathcal{M}_1 = (AF\downarrow_{D_1}, att \cap (E \times D_1), att \cap (D_1 \times E))$ w.r.t. E, which can be replaced with another multipole $\mathcal{M}_2$ w.r.t. the same set E.

Definition 6.4 *Let $AF = \langle Ar, att \rangle$ be an argumentation framework, and $E \subseteq Ar$ be a subset of its arguments. Let $D_1 = Ar \setminus E$, $R^1_{INP} = att \cap (E \times D_1)$ and $R^1_{OUTP} = att \cap (D_1 \times E)$. A replacement $\mathcal{R}$ is a tuple $(AF, \mathcal{M}_1, \mathcal{M}_2)$ where $\mathcal{M}_1 = (AF\downarrow_{D_1}, R^1_{INP}, R^1_{OUTP})$ and $\mathcal{M}_2$ is an argumentation multipole w.r.t. E. The set E is called the* invariant set *of the replacement $\mathcal{R}$. Assuming $\mathcal{M}_2 = ((D_2, R_{D_2}), R^2_{INP}, R^2_{OUTP})$, the result of the replacement $\mathcal{R}$, denoted as $T(\mathcal{R})$, is the argumentation framework $AF_2 = (E \cup D_2, (att \cap E \times E) \cup R^2_{INP} \cup R_{D_2} \cup R^2_{OUTP})$.*

It is easy to see that, for every AF, $T(AF, \mathcal{M}_1, \mathcal{M}_1) = AF$. Moreover, letting $AF_2 = T(AF, \mathcal{M}_1, \mathcal{M}_2)$ it holds that $T(AF_2, \mathcal{M}_2, \mathcal{M}_1) = AF$.

While Definition 6.4 leaves room for any possible replacement, not all of them can be considered legitimate. In particular, one has to focus on replacements involving multipoles having the same Input/Output behavior, otherwise in most cases the labellings of the resulting frameworks would be different in the invariant set E, leading to changes in the status assignment of its arguments. For instance, in the example of the previous section, replacing $\mathcal{M}_1$ with a multipole $\mathcal{M}_2$ including a single argument attacking c would imply having only one preferred labelling and changing completely the relevant behavior.

Thus a natural requirement for legitimate replacements is that the involved multipoles are σ-equivalent. However it may be observed that this requirement may turn out to be too strong: some of the input labellings for a multipole with respect to E may turn out to be irrelevant, given that they can not be generated by the semantics σ on E.

On the basis of this observation, a replacement may be regarded as legitimate even if the involved multipoles are not equivalent under all labellings, provided that they are equivalent under the *possible* ones (in a sense, input labellings that never occur are neglected as the "don't care terms" in digital logic). Of course replacing a multipole with an equivalent one gives more guarantees, since in this case equivalence holds independently of the context (in particular, the multipoles would remain equivalent even modifying the attack relations between arguments of the invariant set E).

In order to distinguish between the two cases, a replacement is called *contextually legitimate* in the first case, and simply *legitimate* in the latter. Independently of its legitimacy properties, a replacement is called *safe* if it does not yield modifications of the labellings in the invariant set E.

Definition 6.5 Let σ be an argumentation semantics and $AF = \langle Ar, att \rangle$ be an argumentation framework. A replacement $\mathcal{R} = (AF, \mathcal{M}_1, \mathcal{M}_2)$ with invariant set E is σ-legitimate if $\mathcal{M}_1$ and $\mathcal{M}_2$ are σ-equivalent, it is contextually σ-legitimate if $\mathcal{M}_1$ and $\mathcal{M}_2$ are σ-equivalent under $\mathfrak{L}_\mathcal{R}^\sigma$, where $\mathfrak{L}_\mathcal{R}^\sigma \triangleq \{F_\sigma(AF\downarrow_E, \mathcal{M}_1{}^{outp}, \mathcal{L}_1, R^1_{OUTP}) \mid \mathcal{L}_1 \in \mathfrak{L}_{\mathcal{M}_1 outp}\} \cup \{F_\sigma(AF\downarrow_E, \mathcal{M}_2{}^{outp}, \mathcal{L}_2, R^2_{OUTP}) \mid \mathcal{L}_2 \in \mathfrak{L}_{\mathcal{M}_2 outp}\}$. Moreover, $\mathcal{R}$ is σ-safe if $\{\mathcal{L}\downarrow_E \mid \mathcal{L} \in \mathcal{L}_\sigma(AF)\} = \{\mathcal{L}\downarrow_E \mid \mathcal{L} \in \mathcal{L}_\sigma(T(AF, \mathcal{M}_1, \mathcal{M}_2))\}$.

Example 6.6 Continuing the example, it can be easily seen that the replacement $(AF, \mathcal{M}_1, \mathcal{M}_2)$ is legitimate and safe for all the admissibility-based semantics considered in this chapter.

Consider instead the following multipole with respect to E: $\mathcal{M}'_2 = (AF'_2, \emptyset, \{(b', c)\})$ where $AF'_2 = \langle \{b'\}, (b', b') \rangle$.

Then considering grounded semantics we have that $\mathcal{GR}\text{-eff}_{\{E\}}(\mathcal{M}_1, \mathcal{L}_E) = \mathcal{GR}\text{-eff}_E(\mathcal{M}'_2, \mathcal{L}_E) = \{(\{d, e\}, \emptyset, \{c\})\}$, for every $\mathcal{L}_E \in \mathfrak{L}_E$, hence the replacement $(AF, \mathcal{M}_1, \mathcal{M}'_2)$ is legitimate (and also safe, as easily seen) for grounded semantics. The same does not hold for preferred semantics (and for all other semantics considered in this chapter). In particular $\mathcal{PR}\text{-eff}_E(\mathcal{M}'_2, \mathcal{L}_E) = \{(\{d,e\}, \emptyset, \{c\})\}$ differing from $\mathcal{PR}\text{-eff}_E(\mathcal{M}_1, \mathcal{L}_E) = \{(\{d,e\}, \{c\}, \emptyset), (\{c,d,e\}, \emptyset, \emptyset)\}$ hence the replacement $(AF, \mathcal{M}_1, \mathcal{M}'_2)$ is not legitimate (nor safe) for preferred semantics.

6.3 Semantics transparency

Given a semantics σ, one may then wonder whether a (possibly contextually) σ-legitimate replacement is always σ-safe, i.e. replacing a multipole with an equivalent multipole preserves the labellings in the invariant set of the replacement. This property actually does not hold for all semantics: a semantics such that legitimate replacements are always safe is called *transparent*, while a semantics such that contextually legitimate replacements are always safe is called *strongly transparent*.

Definition 6.7 *A semantics σ is* transparent *if any σ-legitimate replacement is σ-safe, it is* strongly transparent *if any contextually σ-legitimate replacement is σ-safe.*

Since any legitimate replacement is also contextually legitimate, any strongly transparent semantics is also transparent. Note that the notions of replacement and transparent semantics refer to partitions of argumentation frameworks into just two subframeworks, i.e. one corresponding to the replaced multipole $\mathcal{M}_1$ (or the replacing one $\mathcal{M}_2$) and the other identified by the invariant set E. This is not restrictive, since one can treat a multiple replacement of several multipoles as a sequence of replacements each involving just one multipole. It is proved in [Baroni et al., 2014a] that safeness is preserved by a sequence of safe replacements, and the same holds for skeptical and credulous justification of those arguments that are not replaced.

One can now observe that intuitively there is a close relationship between decomposability and transparency: if a semantics is decomposable, i.e. the labellings prescribed for an argumentation framework are completely determined by applying the canonical local function to the elements of a partition, then one may expect that replacing a multipole with another one having the same Input/Output behavior has no impact on the invariant set of the replacement. This intuition is confirmed by a formal result (Theorem 8 of [Baroni et al., 2014a], reported below as Proposition 6.8), showing that decomposability of an effect-dictated semantics σ is a sufficient condition for strong transparency under the rather obvious condition that the semantics prescribes the empty extension (or equivalently the empty labelling) for the empty argumentation framework.

Proposition 6.8 *Let $AF_\emptyset = \langle \emptyset, \emptyset \rangle$ and σ be an effect-dictated semantics such that $\mathcal{E}_\sigma(AF_\emptyset) = \{\emptyset\}$. If σ is decomposable then σ is strongly transparent.*

This in particular entails that admissible, complete and stable semantics are strongly transparent.

Decomposability is however not necessary: it is proved in [Baroni et al., 2014a] that also grounded semantics, while not being decomposable, is strongly transparent. Preferred, semi-stable, ideal, eager, naïve, and stage semantics instead fail to satisfy the transparency property.

Like in the case of decomposability, some semantics recover some transparency property if one restricts the kind of partitions/multipoles considered. In particular preferred semantics is strongly transparent with respect to partitions consisting of sets of strongly connected components and ideal semantics is strongly transparent for acyclic multipoles (see [Baroni et al., 2014a] for details).

6.4 From transparency to summarization

The results on semantics transparency are a key element for the study of summarization of argumentation frameworks. The idea is that an argumentation framework can be transformed, for the sake of easier presentation or analysis, into a *summarized* form where attention can be focused on a few key arguments, while other parts of the framework are replaced by simpler, but semantically equivalent, versions.

Given the adopted semantics, the results provided in this section show which replacements (under which conditions) are allowable. Other interesting issues, like defining a notion of (minimal) canonical form for equivalent multipoles or synthesising an argumentation multipole with a desired Input/Output behavior under some semantics and/or structural constraints are currently being investigated [Giacomin et al., 2016].

It is however worth noticing that the basic results already available are sufficient to provide examples of summarization on significant real cases. For instance, in [Baroni et al., 2014a] the summarization of two different argument-based representations of the famous Popov vs. Hayashi legal case is presented: we give a brief account here, referring the reader to the original source for details.

The first representation, taken from [Wyner and Bench-Capon, 2007], consists of 17 arguments related by a (mostly acyclic) attack relation, while the second, in a sense finer grained, representation taken from [Prakken, 2012] consists of 65 arguments related by both the attack and the subargument (which can be ignored for our purposes) relations.

Applying iteratively the notion of safe replacement the first representation has been summarized with an argumentation framework consisting of 4 arguments, while the second one has been summarized with an argumentation framework consisting of 5 arguments.

In both cases two key conflicting arguments, corresponding to the resolution of the dispute in favour of Popov and Hayashi respectively, are preserved in the summarization. As a matter of fact the dispute was not resolved in favour of either of them, since an equal share of the contended good (actually a valuable baseball ball) was decided. The summarized versions enable a simple direct comparison of the two reconstructions. In particular 4 arguments are quite similar in both reconstructions: there are two arguments corresponding to the resolution of the dispute in favour of either side, each being attacked by another unattacked argument. As a consequence, both one-sided options are labelled out by (the only labelling prescribed by) every semantics in both summarized frameworks. The difference is then that in the second reconstruction the option of equal share corresponds to an explicit argument (the additional fifth one), which is labelled in by (the only labelling prescribed by) every semantics in the summarized framework, while in the first reconstruction the option of equal share emerges implicitly from the rejection of the other ones. Of course, due to the safeness of the applied replacements, the above observations were somehow applicable to the original versions of the frameworks too, but they were much less evident at eye inspection, being cluttered by the presence of many other arguments and relations, providing a detailed account of the case but actually not necessary for a synthetic analysis.

Summarization is a quite open and fertile field of investigation: in the example described above an ad hoc summarization was applied, while the issue of identifying generally applicable summarization methods, defining their desirable properties, and analyzing their computational complexity is largely unexplored in the literature. The study of the interplay between the notions of summarization and equivalence is also an interesting research topic.

7 Conclusions

While not explicitly considered in the original definition of the formalism of argumentation frameworks, locality and modularity properties are receiving an increasing attention by the research community in abstract argumentation, as they feature a significant theoretical interest and at the same time show direct applicability for practical purposes (like the study of efficient divide-and-conquer algorithms or of sound summarization methods).

This chapter is meant to provide a reasonably complete overview of the basic definitions and results available in the literature about the three key properties of directionality, SCC-recursiveness and decomposability, and to give an illustration of their applicability in the two areas mentioned above.

Retrospectively, the fact that Dung's original semantics all satisfy SCC-recursiveness and top-down decomposability, and all but stable semantics satisfy directionality, may suggest that these properties are intimately related to the basic principles underlying the definition of (most) abstract argumentation semantics. For this reason the study of locality and modularity properties, though initiated relatively recently, can be regarded as part of the foundational aspects of the area.

Several more advanced topics and directions of ongoing research can be identified in the literature: we conclude by giving some hints in this respect.

The study of applying heterogeneous semantics to an argumentation framework, considered first in [Rienstra et al., 2011], appears a very interesting research direction to pursue. Briefly, the idea is that different semantics are applied to different parts of an argumentation framework and that these heterogeneous local semantics assessments are then combined to yield a global extension or labelling.

Furthermore, while the notion of equivalence between multipoles is essentially different with respect to the notions of equivalence between argumentation frameworks, considered for instance in [Oikarinen and Woltran, 2011; Baumann and Brewka, 2013] (see also chapter 17), it would be interesting to develop a full analysis of the relationships of Input/Output multipole equivalence with other equivalence notions considered both inside the abstract argumentation field and outside it. In particular the notion of modular equivalence of logic programs appears relevant in this respect [Oikarinen and Woltran, 2006].

Finally, the generic idea of argumentation module (and in particular the notion of argumentation multipole) suggests the study of argumentation synthesis problems: given a desired Input/Output behavior, generating an argumentation framework which produces it, possibly under some constraints concerning its structure and/or the semantics to be adopted.

Last but not least, the study of local properties and decomposition methods is not limited to Dung's argumentation frameworks and the relevant investigation needs to be extended to other argumentation formalisms too. In particular, the reader may refer to [Linsbichler, 2014; Gaggl and Strass, 2014] for results concerning Abstract Dialectical Frameworks.

These and the further open problems mentioned along the chapter suggest that the study of locality and modularity issues in abstract argumentation is only at its beginning and will witness a steady growth in the years to come.

Acknowledgments

The authors are grateful to the reviewers for their careful analysis of the chapter and their helpful comments.

BIBLIOGRAPHY

[Amgoud et al., 2008] Leila Amgoud, Yannis Dimopoulos, and Pavlos Moraitis. A general framework for argumentation-based negotiation. In I. Rahwan, S. Parsons, and C. Reed, editors, *Argumentation in Multi-Agent Systems - 4th Int. Workshop (ArgMAS 2007). Revised Selected and Invited Papers*, volume 4946 of *Lecture Notes in Computer Science*. Springer, 2008.

[Baroni and Giacomin, 2007] Pietro Baroni and Massimiliano Giacomin. On principle-based evaluation of extension-based argumentation semantics. *Artificial Intelligence*, 171(10-15):675–700, 2007.

[Baroni et al., 2005] Pietro Baroni, Massimiliano Giacomin, and Giovanni Guida. SCC-recursiveness: a general schema for argumentation semantics. *Artificial Intelligence*, 168(1-2):165–210, 2005.

[Baroni et al., 2014a] Pietro Baroni, Guido Boella, Federico Cerutti, Massimiliano Giacomin, Leendert W. N. van der Torre, and Serena Villata. On the input/output behavior of argumentation frameworks. *Artificial Intelligence*, 217:144–197, 2014.

[Baroni et al., 2014b] Pietro Baroni, Massimiliano Giacomin, and Beishui Liao. On topology-related properties of abstract argumentation semantics. A correction and extension to dynamics of argumentation systems: A division-based method. *Artificial Intelligence*, 212:104–115, 2014.

[Baumann and Brewka, 2013] Ringo Baumann and Gerhard Brewka. Analyzing the equivalence zoo in abstract argumentation. In J. Leite, T. Cao Son, P. Torroni, L. van der Torre, and S. Woltran, editors, *Proc. of the 14th Int. Workshop on Computational Logic in Multi-Agent Systems (CLIMA XIV)*, volume 8143 of *Lecture Notes in Computer Science*, pages 18–33. Springer, 2013.

[Baumann et al., 2011] Ringo Baumann, Gerhard Brewka, and Renata Wong. Splitting argumentation frameworks: An empirical evaluation. In S. Modgil, N. Oren, and F. Toni, editors, *Theory and Applications of Formal Argumentation - First Int. Workshop (TAFA 2011). Revised Selected Papers*, volume 7132 of *Lecture Notes in Computer Science*, pages 17–31. Springer, 2011.

[Baumann et al., 2012] Ringo Baumann, Gerhard Brewka, Wolfgang Dvořák, and Stefan Woltran. Parameterized splitting: A simple modification-based approach. In E. Erdem, J. Lee, Y. Lierler, and D. Pearce, editors, *Correct Reasoning - Essays on Logic-Based AI in Honour of Vladimir Lifschitz*, volume 7265 of *Lecture Notes in Computer Science*, pages 57–71. Springer, 2012.

[Baumann, 2011] Ringo Baumann. Splitting an argumentation framework. In J. P. Delgrande and W. Faber, editors, *Proc. of the 11th Inf. Conf. on Logic Programming and Nonmonotonic Reasoning (LPNMR 2011)*, volume 6645 of *Lecture Notes in Computer Science*, pages 40–53. Springer, 2011.

[Baumann, 2014] Ringo Baumann. *Metalogical Contributions to the Nonmonotonic Theory of Abstract Argumentation*. College Publications - Studies in Logic, 2014.

[Boella et al., 2012] Guido Boella, Dov M. Gabbay, Alan Perotti, Leendert van der Torre, and Serena Villata. Conditional labelling for abstract argumentation. In S. Modgil, N. Oren, and F. Toni, editors, *Theory and Applications of Formal Argumentation - 1st Int. Workshop (TAFA 2011), Revised Selected Papers*, volume 7132 of *Lecture Notes in Computer Science*, pages 232–248. Springer, 2012.

[Cerutti et al., 2014] Federico Cerutti, Massimiliano Giacomin, Mauro Vallati, and Marina Zanella. An SCC recursive meta-algorithm for computing preferred labellings in abstract argumentation. In C. Baral, G. De Giacomo, and T. Eiter, editors, *Proc. of the 14th Int. Conf. on Principles of Knowledge Representation and Reasoning (KR 2014)*, pages 42–51. AAAI Press, 2014.

[Coste-Marquis et al., 2005] Sylvie Coste-Marquis, Caroline Devred, and Pierre Marquis. Symmetric argumentation frameworks. In L. Godo, editor, *Proc. of the 8th European Conf. on Symbolic and Quantitative Approaches to Reasoning with Uncertainty ECSQARU 2005*, volume 3571 of *Lecture Notes in Computer Science*, pages 317–328. Springer, 2005.

[Dung, 1995] Phan Minh Dung. On the acceptability of arguments and its fundamental role in nonmonotonic reasoning, logic programming and n-person games. *Artificial Intelligence*, 77:321–357, 1995.

[Dunne, 2007] Paul E. Dunne. Computational properties of argument systems satisfying graph-theoretic constraints. *Artificial Intelligence*, 171:10–15, 2007.

[Dvořák et al., 2010] Wolfgang Dvořák, Reinhard Pichler, and Stefan Woltran. Towards fixed-parameter tractable algorithms for argumentation. In F. Lin, U. Sattler, and M. Truszczynski, editors, *Proc. of the 12th Int. Conf. on Principles of Knowledge Representation and Reasoning (KR 2010)*, pages 112–122. AAAI Press, 2010.

[Gaggl and Strass, 2014] Sarah Alice Gaggl and Hannes Strass. Decomposing abstract dialectical frameworks. In S. Parsons, N. Oren, C. Reed, and F. Cerutti, editors, *Proc. of the 5th Int. Conf. on Computational Models of Argument (COMMA 2014)*, pages 281–292. IOS Press, 2014.

[Geffner and Pearl, 1992] Hector Geffner and Judea Pearl. Conditional entailment: bridging two approaches to default reasoning. *Artificial Intelligence*, 53:209–244, 1992.

[Giacomin et al., 2016] Massimiliano Giacomin, Thomas Linsbichler, and Stefan Woltran. On the functional completeness of argumentation semantics. In C. Baral, J. P. Delgrande, and F. Wolter, editors, *Proc. of the 15th Int. Conf. on Principles of Knowledge Representation and Reasoning (KR 2016)*, pages 43–52. AAAI Press, 2016.

[Kraus et al., 1998] S. Kraus, K. Sycara, and A. Evenchik. Reaching agreements through argumentation: a logical model and implementation. *Artificial Intelligence*, 104(1-2):1–69, 1998.

[Liao and Huang, 2012] Beishui Liao and Huaxin Huang. Computing the extensions of an argumentation framework based on its strongly connected components. In *Proc. of the 24th IEEE Int. Conf. on Tools with Artificial Intelligence (ICTAI 2012)*, pages 1047–1052. IEEE Computer Society, 2012.

[Liao and Huang, 2013] Beishui Liao and Huaxin Huang. Partial semantics of argumentation: Basic properties and empirical results. *Journal of Logic and Computation*, 23(3):541–562, 2013.

[Liao et al., 2010] Beishui Liao, Li Jin, and Robert C. Koons. Dynamics of argumentation systems: a basic theory. In A. Voronkov, G. Sutcliffe, M. Baaz, and C. G. Fermüller, editors, *Proc. of the 17th Int. Conf. on Logic for Programming, Artificial Intelligence, and Reasoning - Short Papers (LPAR-17-short)*, pages 47–51, 2010.

[Liao et al., 2011] Beishui Liao, Li Jin, and Robert C. Koons. Dynamics of argumentation systems: A division-based method. *Artificial Intelligence*, 175(11):1790–1814, 2011.

[Liao, 2013] Beishui Liao. Toward incremental computation of argumentation semantics: A decomposition-based approach. *Annals of Mathematics and Artificial Intelligence*, 67:319–358, 2013.

[Liao, 2014] Beishui Liao. *Efficient Computation of Argumentation Semantics*. Intelligent systems series. Academic Press, 2014.

[Lifschitz and Turner, 1994] Vladimir Lifschitz and Hudson Turner. Splitting a logic program. In P. Van Hentenryck, editor, *Proc. of the 11th Int. Conf. on Logic Programming (ICLP 1994)*, pages 23–37. MIT Press, 1994.

[Linsbichler, 2014] Thomas Linsbichler. Splitting abstract dialectical frameworks. In S. Parsons, N. Oren, C. Reed, and F. Cerutti, editors, *Proc. of the 5th Int. Conf. on Computational Models of Argument (COMMA 2014)*, pages 357–368. IOS Press, 2014.

[Modgil and Caminada, 2009] Sanjay Modgil and Martin W. A. Caminada. Proof theories and algorithms for abstract argumentation frameworks. In I. Rahwan and G.R. Simari, editors, *Argumentation in Artificial Intelligence*, pages 105–129. Springer, 2009.

[Oikarinen and Woltran, 2006] Emilia Oikarinen and Stefan Woltran. Modular equivalence for normal logic programs. In G. Brewka, S. Coradeschi, A. Perini, and P. Traverso, editors, *Proc. of the 17th European Conf. on Artificial Intelligence (ECAI 2006)*, volume 141 of *Frontiers in Artificial Intelligence and Applications*, pages 412–416. IOS Press, 2006.

[Oikarinen and Woltran, 2011] Emilia Oikarinen and Stefan Woltran. Characterizing strong equivalence for argumentation frameworks. *Artificial Intelligence*, 175:1985–2009, 2011.

[Parsons et al., 1998] S. Parsons, C.A. Sierra, and N.R. Jennings. Agents that reason and negotiate by arguing. *Journal of Logic and Computation*, 8(3):261–292, 1998.

[Prakken, 2012] Henry Prakken. Reconstructing Popov v. Hayashi in a framework for argumentation with structured arguments and Dungean semantics. *Artificial Intelligence and Law*, 20(1):57–82, 2012.

[Rahwan et al., 2003] Iyad Rahwan, Sarvapali D. Ramchurn, Nicholas R. Jennings, Peter McBurney, Simon Parsons, and Liz Sonenberg. Argumentation-based negotiation. *The Knowledge Engineering Review*, 18(4):343–375, 2003.

[Rienstra et al., 2011] Tjitze Rienstra, Alan Perotti, Serena Villata, Dov M. Gabbay, and Leendert van der Torre. Multi-sorted argumentation frameworks. In S. Modgil, N. Oren, and F. Toni, editors, *Theory and Applications of Formal Argumentation - First Int. Workshop (TAFA 2011). Revised Selected Papers*, volume 7132 of *Lecture Notes in Computer Science*, pages 231–245. Springer, 2011.

[Turner, 1996] Hudson Turner. Splitting a default theory. In W. J. Clancey and D. S. Weld, editors, *Proc. of the 13th National Conf. on Artificial Intelligence (AAAI 96) Volume 1.*, pages 645–651. AAAI Press / The MIT Press, 1996.

[Wyner and Bench-Capon, 2007] Adam Z. Wyner and Trevor J. M. Bench-Capon. Argument schemes for legal case-based reasoning. In A. R. Lodder and L. Mommers, editors, *Proc. of the 20th Annual Conf. on Legal Knowledge and Information Systems (JURIX 2007)*, volume 165 of *Frontiers in Artificial Intelligence and Applications*, pages 139–149. IOS Press, 2007.

Pietro Baroni
Dip. Ingegneria dell'Informazione
University of Brescia, Italy
Email: pietro.baroni@unibs.it

Massimiliano Giacomin
Dip. Ingegneria dell'Informazione
University of Brescia, Italy
Email: massimiliano.giacomin@unibs.it

Beishui Liao
Center for the Study of Language and Cognition
Zhejiang University, Hangzhou, China
Email: baiseliao@zju.edu.cn

19
Argumentation, Nonmonotonic Reasoning and Logic
ALEXANDER BOCHMAN

ABSTRACT. In this chapter, we will explore the respective roles of logic and nonmonotonic reasoning in argumentation. As a first step, we introduce the notion of collective argumentation as a logical basis of argumentation frameworks, and provide it with a natural (four-valued) logical semantics. This will allows us, in particular, to augment the underlying language with appropriate logical connectives that will transform abstract argumentation frameworks into a reasoning system with full-fledged logical capabilities. On the way, we will show not only that argumentation and logic are important for nonmonotonic reasoning, but also the other way round, namely that the main nonmonotonic formalisms and argumentation systems constitute actually primary instantiations of Dung's abstract argumentation in appropriately extended logical languages.

1 The new stage of argumentation theory

Dung's argumentation frameworks are viewed today as a general formal basis of the argumentation theory, but they have deep roots in nonmonotonic reasoning, and it is due to these roots that they constitute a new stage in the development of the theory of argumentation.

Traditional formal argumentation theory (see, e.g., [Hamblin, 1971]) has been based, implicitly or explicitly, on the standard deductive paradigm, according to which our corpus of knowledge and beliefs comprises a set of propositional (factual or epistemic) assertions, coupled with a set of strict, universal deductive rules (= the Logic) that govern their acceptance. These rules allow us, in particular, to derive (support) further propositional claims, as well as to reveal possible inconsistencies among them. This underlying logic can also be given a precise argumentative (dialectical) formulation in the form of allowable attack and defense moves in argumentation games ([Lorenzen and Lorenz, 1978]). In this way, the traditional formal argumentation theory could be largely viewed as a 'human-friendly' instantiation of standard deductive reasoning.

The above deductive paradigm has been challenged, however, with the advent of nonmonotonic reasoning in AI. Studies in the latter (including related areas such as the AGM theory of belief revision), as well as contemporaneous studies of defeasible reasoning in philosophical logic (see [Pollock, 1987]), have shown that epistemic states underlying our reasoning are much more complex and structured than plain sets of beliefs governed by logic. They have shown, in particular, an important role of *default assumptions* in our reasoning, assumptions that we normally accept in the absence of evidence to the contrary. These assumptions usually appear in the conditional form *"If A, then normally B"*, and it can even be argued that such normality conditionals constitute one of the central ingredients of our commonsense epistemic states.

Despite their presumptive acceptability status, default assumptions are *defeasible*, that is, they can be attacked, and even eventually refuted, due to other assumptions and available evidence. The corresponding adjudication process, however, already cannot be represented as a deductive inference or proof in some logic, primarily because it is in general *non-local* and *non-monotonic*. The eventual acceptability of such assumptions depends on other assumptions present, and it can change from acceptance to rejection and vice versa with addition of new assumptions or facts. This reasoning process displays, however, distinctive features of genuine argumentation.

As a matter of fact, the intimate connections between argumentation and nonmonotonic reasoning has been noticed at the very beginning of the studies in nonmonotonic reasoning. The starting point of this understanding can be found already in the Truth Maintenance System (TMS) of [Doyle, 1979] and assumption-based truth maintenance (ATMS) of [de Kleer, 1986]. This understanding has even led to a general view of NMR as a theory of the reasoned use of assumptions in [Doyle, 1994]. In fact, even before Dung, a significant argumentation-based representation of default logic and other nonmonotonic formalisms has been suggested in [Lin and Shoham, 1989].

Developing further this line of research, Dung has shown a fundamental and unifying role of argumentation in logic programming and general nonmonotonic formalisms such as default and modal nonmonotonic logics. More precisely, he has shown that all these formalisms can be viewed as particular instantiations of a uniform argumentation scheme that implements the principle of default acceptability for arguments in an abstract framework based solely on a single relation of attack among them.

Dung's argumentation frameworks have had two crucial novel features. First, they implemented the basic principle of default acceptance for arguments[1]. On Dung's vision, however, this principle admits a number of different interpretations, which lead to different possible *nonmonotonic semantics* for the argumentation frameworks.

The second, more formal, novel feature of the Dung's formalism was the asymmetric (directional) character of the attack relation; it was this 'degree of freedom' that has allowed to provide an adequate representation of the above-mentioned nonmonotonic formalisms. This feature has also marked an important formal difference with the traditional, deductive argumentation that has been based primarily on symmetric inconsistency relations.

It could even be argued that the main contribution of Dung's theory has

---

[1] Dung himself has called it the basic principle of argumentation and described it in [Dung, 1995b] as the principle *"The one who has the last word laughs best"*.

consisted in incorporating these two novel features as central conceptual ingredients of argumentation. It is this conceptual advancement that has given the argumentation theory its current impetus.

2 Logic in argumentation

One of the fundamental questions that have been re-opened, however, with the advent of Dung's argumentation frameworks was the question of the relation between argumentation and logic. Indeed, on the face of it, Dung's abstract frameworks do not include, or even require, any explicit logical components. On the other hand, it has been shown in subsequent argumentation literature that arbitrary, unrestricted combinations of argumentation frameworks with deductive rules may lead to patently inappropriate results, so such compositions should be constrained by some (more or less) reasonable 'rationality postulates' - see, e.g., [Caminada and Amgoud, 2007; Amgoud and Besnard, 2013; Dung and Thang, 2014].

As we are going to show in this Chapter, though the relations between argumentation and logic have irrevocably changed, logic still plays (or, better, should play) an important role in argumentation. However, we contend that a crucial prerequisite for a proper understanding of this role amounts to a clear separation of the logical and non-monotonic aspects of argumentation.[2] In fact, the latter objective is not specific to argumentation theory, but pertains to all nonmonotonic formalisms.

In nonmonotonic formalisms, logic no longer 'pervades the world' (using the famous Wittgenstein's phrase). Namely, the logic, taken by itself, cannot provide the final 'output' of these formalisms; this latter task is relegated to the associated nonmonotonic semantics.

Nevertheless, logic still plays a distinctive and even crucial role in these formalisms. First of all, logic and its associated (monotonic) semantics should still provide a formal interpretation and meaning for the very syntax of a nonmonotonic formalism. Note that a nonmonotonic semantics is usually defined as a distinguished *subset* of the corresponding logical semantics, so it cannot be used for interpreting the source language. In addition, the logic provides deductive inferences that are 'safe' with respect to the nonmonotonic semantics, so it can be used to facilitate proofs and computations of the latter.

However, an even more profound benefit of the separation between logical and nonmonotonic aspects of a reasoning formalism emerges from the fact that, once the separation is made, many of these formalisms can be reconstructed as instantiations of the same nonmonotonic semantics in different logical languages.

The field of formal argumentation is abundant with different formalisms, which creates a fertile ground for extensive and rapid development. But there

---

[2]In this respect, our construction below will be distinct from a large number of other suggested ways of combining logic with argumentation - see, e.g., [Boella et al., 2005; Caminada and Gabbay, 2009; Gabbay, 2011; Strasser and Seselja, 2010].

is also a lot of conceptual affinity among these argumentation formalisms, as well as between the latter and the major knowledge representation languages in AI. It is this affinity that allows us to use many of them for basically the same reasoning tasks. This situation creates, however, an obvious incentive for unification, namely for constructing a general theory of argumentation and reasoning where these formalisms could find their proper and hospitable place.

As we are going to show in this study, the logic appropriate for Dung's argumentation frameworks can be constructed on the basis of a four-valued logical semantics that can be found, in effect, already in [Jakobovits and Vermeir, 1999]. In that paper, the authors described a general semantic framework based on acceptance and rejection of arguments. This semantics was essentially four-valued, because assignments of acceptance and rejection to arguments were primarily viewed as mutually independent, which permitted valuations in which arguments can be both accepted and rejected, or neither accepted, nor rejected. The semantics suggested in [Jakobovits and Vermeir, 1999] were designed, however, to be generalizations of existing *nonmonotonic* argumentation semantics, so they incorporated also some non-logical, nonmonotonic features (see below). Still, we will show that a (properly generalized) four-valued semantics can be used as a *logical* basis of Dung's argumentation frameworks. It will allows us, in particular, to augment the underlying language with appropriate logical connectives that will transform this abstract argumentation to real argumentation reasoning with full-fledged logical capabilities.

As we are going to see in what follows, the assumption-based frameworks of [Bondarenko *et al.*, 1997] (see also Chapter 7 of this Handbook) could be viewed as a 'focal point' of this logical development. On our reconstruction of the latter, assumption-based frameworks can be obtained from abstract Dung's frameworks just by adding a particular negation connective to the underlying language of arguments. This connective will also allow us to establish straightforward relations between attack and inference, as well as between (non-propositional) arguments and (propositional) assumptions that can be viewed as their reified counterparts in the object language.

Further stages of this logical development will allow us to provide a more systematic description of many other argumentation and general nonmonotonic formalisms, such as logic programming, default logic, abstract dialectical frameworks and the causal calculus.

The general picture that will emerge from this formal development is not only that argumentation is important for nonmonotonic reasoning, but also the other way round, namely that the main nonmonotonic formalisms and argumentation systems constitute actually primary instantiations of Dung's abstract argumentation in appropriately extended logical languages.

3 Abstract Collective Argumentation

As a general formal basis of argumentation theory, we will use the formalism of collective argumentation suggested in [Bochman, 2003a] as a 'disjunctive' generalization of Dung's argumentation theory. In this formalism, a primitive attack relation holds between *sets* of arguments[3]: in the notation introduced below, $a \hookrightarrow b$ says that a set a of arguments attacks a set of arguments b. This fact implies, of course, that these two sets arguments are incompatible. $a \hookrightarrow b$ says, however, more than that, namely that the set a of arguments, being accepted, provides a reason, or explanation, for rejection of the set of arguments b. Accordingly, the attack relation will not in general be symmetric, since in this situation acceptance of b need not give reasons for rejection of a. In addition, the attack relation is not reducible to attacks between individual arguments. For instance, we can disprove some conclusion jointly supported by a disputed set of arguments, though no particular argument in the set, taken alone, could be held responsible for this.

In what follows, $a, b, c, \ldots$ will denote finite sets of arguments, while $u, v, w, \ldots$ will denote arbitrary such sets. We will use the same agreements for the attack relation as for usual consequence relations. Thus, $a, a_1 \hookrightarrow b, B$ will have the same meaning as $a \cup a_1 \hookrightarrow b \cup \{B\}$, etc.

In what follows, proofs will be provided only for the main theorems. Proofs of all the other claims mentioned in this and the next section can be found in [Bochman, 2005].

Definition 3.1 *Let $\mathcal{A}$ be a set of arguments. A (collective) attack relation is a relation $\hookrightarrow$ on finite sets of arguments satisfying the following postulate:*

Monotonicity *If $a \hookrightarrow b$, then $a, a_1 \hookrightarrow b, b_1$.*

As we will see below, the above Monotonicity postulate turns out to be sufficient to characterize the primary logic behind the attack relation.

Though defined initially on finite sets of arguments, the attack relation can be extended to arbitrary such sets by imposing the compactness requirement: for any $u, v \subseteq \mathcal{A}$,

(Compactness) $u \hookrightarrow v$ iff there exist finite $a \subseteq u$ and $b \subseteq v$ such that $a \hookrightarrow b$.

The original Dung's argumentation frameworks can be seen as a special case of collective argumentation that satisfies additional properties (cf. [Kakas and Toni, 1999]):

Definition 3.2 *An attack relation will be called*

- affirmative *if no argument set attacks the empty set $\emptyset$;*

- local *if it satisfies the following condition:*

  *(**Locality**)   If $a \hookrightarrow b, b_1$, then either $a \hookrightarrow b$, or $a \hookrightarrow b_1$.*

---
[3] A similar idea has been suggested in [Nielsen S.H., 2007], though only for attacking sets, not attacked ones.

- normal *if it is both affirmative and local.*

Then the following facts can be easily verified:

Lemma 3.3
- *If $\hookrightarrow$ is a normal attack relation, then $a \hookrightarrow b$ holds if and only if $a \hookrightarrow A$, for some $A \in b$.*

- *If $\hookrightarrow$ is a local attack relation, then $a \hookrightarrow b$ holds iff either $a \hookrightarrow \emptyset$, or $a \hookrightarrow A$, for some $A \in b$.*

Thus, the normal attack relation is reducible to the relation $a \hookrightarrow A$ between argument sets and single arguments, and the resulting theory will coincide, in effect, with that given in [Dung, 1995a]. A slightly more general local attack relation admits also constraints of the form $a \hookrightarrow$. Such a constraint says that the argument set a is *unacceptable* (due to Monotonicity, it attacks any argument whatsoever).

By an *argument theory* we will mean an arbitrary set of attacks $a \hookrightarrow b$ between finite argument sets. Now, since the Monotonicity condition is a 'Horn' one, any argument theory Δ generates a unique least attack relation that we will denote by $\hookrightarrow_\Delta$. The latter is obtained from Δ just by closing it with respect to the Monotonicity rule. Accordingly, $\hookrightarrow_\Delta$ can be described directly as follows:

$$u \hookrightarrow_\Delta v \quad \text{iff} \quad a \hookrightarrow b \in \Delta, \text{ for some } a \subseteq u, b \subseteq v.$$

An argument theory will be called *definite*, if it consists of attack rules of the form $a \hookrightarrow A$, where A is a single argument, and *singular*, if it has only attacks of the form $a \hookrightarrow b$, where b contains no more than one argument. Then the preceding lemma, coupled with the above representation, immediately implies the following simple observation:

Lemma 3.4 *An attack relation is local (respectively, normal) if and only if it is generated by a singular (resp., definite) argument theory.*

Thus, the differences between general, local and normal argumentation are reducible to the differences between corresponding generating argument theories.

3.1 Four-valued logical semantics

Collective argumentation can be given a four-valued semantics that can be seen as describing the (abstract) *meaning* of the attack relation. This formal meaning stems from the following understanding of an attack $a \hookrightarrow b$:

> *If all arguments in a are accepted, then at least one of the arguments in b should be rejected.*

The argumentation theory does not impose, however, the classical constraints on acceptance and rejection of arguments, so an argument can be both accepted and rejected, or neither accepted, nor rejected. Such an understanding can be captured formally by assigning any argument A a *subset* $\nu(A) \subseteq \{t, f\}$, where t denotes acceptance (truth), while f denotes rejection (falsity). This is nothing other than the well-known *Belnap's interpretation* of four-valued logic (see [Belnap, 1977]). On this understanding, $t \in \nu(A)$ means that an argument A is accepted, while $f \in \nu(A)$ means that A is rejected. In accordance with this, collective argumentation acquires a four-valued logical semantics described below.

Definition 3.5 *An attack* $a \hookrightarrow b$ *will be said to* hold *in a four-valued interpretation* ν *of arguments, if either* $t \notin \nu(A)$, *for some* $A \in a$, *or* $f \in \nu(B)$, *for some* $B \in b$.
An interpretation ν *will be called a* model *of an argument theory* Δ *if every attack from* Δ *holds in* ν.

Since an attack relation can be seen as a special kind of an argument theory, the above definition determines also the notion of a model for an attack relation.

Any pair (u, v) of argument sets determines a four-valued interpretation in which u is the set of true (i.e., accepted) arguments, while v is the set of arguments that are not false (non-rejected), and vice versa, any four-valued interpretation corresponds to such a pair of argument sets. In fact, we will often identify in what follows four-valued interpretations with their associated pairs of argument sets.

In this sense canonical models of an attack relation or an argument theory can be identified with bitheories described in the next definition.

Definition 3.6 *A pair* (u, v) *of arguments will be called a* bitheory *of an argument theory* Δ, *if* $u \not\hookrightarrow_\Delta v^4$.

It can be easily verified that any bitheory of Δ corresponds to a four-valued model of the latter.

Definition 3.7 • *A bitheory* (u, v) *of an argument theory will be called* consistent, *if* $u \subseteq v$, *and* complete, *if* $v \subseteq u$.

• *An argument set* u *will be called* consistent, *if* $u \not\hookrightarrow u$.

Consistent bitheories correspond to consistent four-valued interpretations, namely to interpretations in which no argument is both accepted and rejected. Similarly, complete bitheories correspond to complete four-valued interpretations in which every argument is either accepted, or rejected. Such constrained interpretations will play an important role in what follows. Finally, consistent argument sets correspond exactly to bitheories that are both consistent and complete. This notion of consistency provides an appropriate generalization of the notion of a conflict-free argument set in Dung's argumentation theory.

---

[4] where $\not\hookrightarrow$ means that the attack relation does not hold.

For a set I of four-valued interpretations, we will denote by $\hookrightarrow_I$ the set of all attacks that hold in each interpretation from I. Then the following result is actually a basic representation theorem showing that the four-valued semantics is adequate for collective argumentation.

Theorem 3.8 $\hookrightarrow$ *is an attack relation iff it coincides with* $\hookrightarrow_I$, *for some set of four-valued interpretations* I.

Proof. For any set I of interpretations, the attack relation $\hookrightarrow_I$ satisfies Monotonicity, so it determines a collective attack relation. Now, if $\hookrightarrow$ is a collective attack relation, let I denote the set of all interpretations corresponding to bitheories of $\hookrightarrow$. We will show that $\hookrightarrow_I = \hookrightarrow$.

Since bitheories correspond to models of an argumentation theory, any attack $u \hookrightarrow v$ will hold in every interpretation from $\hookrightarrow_I$. Consequently $\hookrightarrow \;\subseteq\; \hookrightarrow_I$. Now, if $u \not\hookrightarrow v$, then (u,v) is a bitheory of $\hookrightarrow$, and hence its corresponding interpretation, say ν, belongs to I. But $u \hookrightarrow v$ clearly does not hold in ν, so $u \not\hookrightarrow_I v$. This completes the proof. ∎

3.2 Logical kinds of argumentation

We will describe now three special kinds of collective argumentation called, respectively, classical, negative and positive argumentation. On the semantic level, these kinds of argumentation will correspond to restrictions of four-valued reasoning to two- and three-valued reasoning. For all these kinds of argumentation, the attack relation will be defined by 'borrowing' arguments of the opposite side in order to disprove the latter. Classical argumentation will give an abstract description of classical consistency-based reasoning. Negative argumentation will be shown to be especially appropriate for describing the stable nonmonotonic semantics.

In ordinary disputation and argumentation the parties can provisionally accept some of the arguments defended by their adversaries in order to disprove the latter. Three basic cases of such an 'argument sharing' in attacking the opponents are described in the following definition (see also [Bondarenko *et al.*, 1997]).

Definition 3.9 *Given an attack relation* $\hookrightarrow$, *we will say that*

- *a classically attacks b (notation* $a \hookrightarrow^\circ b$) *if* $a, b \hookrightarrow a, b$;
- *a negatively attacks b (notation* $a \hookrightarrow^- b$) *if* $a \hookrightarrow a, b$;
- *a positively attacks b (notation* $a \hookrightarrow^+ b$) *if* $a, b \hookrightarrow b$.

In a classical attack, the proponent shows, in effect, that her arguments are incompatible with that of the opponent. In a positive attack, the proponent temporarily accepts opponent's arguments in order to disprove the latter, while in a negative attack she shows that her arguments are sufficient for challenging an addition of the opponent's arguments. Clearly, if a attacks b directly, then it attacks the latter classically, positively and negatively, though not vice versa.

As can be seen, $\hookrightarrow^\circ$, $\hookrightarrow^-$ and $\hookrightarrow^+$ are also attack relations. Moreover, it turns out that all of them can be given an invariant structural characterization in terms of additional rules imposed on the attack relation. For explanatory reasons, we will begin below with a simplest such kind, namely the classical argumentation.

3.2.1 Classical argumentation

Classical argumentation can be seen as an 'upper bound' of collective argumentation; it is a simplest kind of argumentation which amounts, in effect, to classical consistency reasoning.

Definition 3.10 *An attack relation will be called* classical *if $a, b \hookrightarrow a, b$ always implies $a \hookrightarrow b$.*

It can be easily verified that an argument theory based on a classical attack $\hookrightarrow^\circ$ will be classical. Moreover, the latter determines a least classical 'closure' of the source attack relation:

Lemma 3.11 $\hookrightarrow^\circ$ *is a least classical attack relation containing* $\hookrightarrow$.

An immediate consequence of the above lemma is that classical attack relations are precisely attack relations of the form $\hookrightarrow^\circ$. In other words, classical attack relations provide a canonical description of argumentation based on a classical attack.

An attack relation is classical if and only if it satisfies:

(**Symmetry**) $a \hookrightarrow b, c$ iff $a, b \hookrightarrow c$.

As a special case of Symmetry, we have

$$u \hookrightarrow v \text{ iff } \emptyset \hookrightarrow u, v \text{ iff } u, v \hookrightarrow \emptyset.$$

This shows that a classical attack amounts to inconsistency in the full classical sense. It should be noted in this respect that the classical closure $\hookrightarrow^\circ$ of an attack relation $\hookrightarrow$ preserves consistent argument sets. Namely, the definition of $\hookrightarrow^\circ$ immediately implies that it has the same consistent argument sets as $\hookrightarrow$.

As could be anticipated, classical argumentation can be characterized semantically by restricting the set of four-valued interpretations to classical two-valued ones, namely to interpretations that assign only **t** or only **f** to the arguments. This means that any argument is either accepted or rejected in an interpretation, but not both.

Theorem 3.12 *An attack relation is classical if and only if it is determined by a set of classical interpretations.*

Proof. As can be verified, any set of classical interpretations determines a classical attack relation. Now, if $\hookrightarrow$ is a classical attack relation, let I_c denote the set of (classical) interpretations corresponding to bitheories of the form (u,u). If $u \not\hookrightarrow v$, then $u, v \not\hookrightarrow u, v$, and therefore the interpretation ν corresponding to a bitheory $(u \cup v, u \cup v)$ belongs to I_c. But $u \hookrightarrow v$ clearly does not hold in ν, so $u \not\hookrightarrow_{I_c} v$. This shows that $\hookrightarrow$ is determined by the set of classical interpretations I_c. ∎

Finally, classical argumentation can be seen as a combination of positive and negative argumentation. The proof follows immediately from the definitions of the respective attack relations.

Lemma 3.13 *For any attack relation* $\hookrightarrow$, $(\hookrightarrow^-)^+ = (\hookrightarrow^+)^- = \hookrightarrow^\circ$.

The above result says, in particular, that positive and negative argumentation are incompatible on pain of collapsing to classical reasoning.

3.2.2 Negative argumentation

The definition below provides a general description of collective argumentation based on a negative attack.

Definition 3.14 *An attack relation will be called* negative *if* $a \hookrightarrow a, b$ *always implies* $a \hookrightarrow b$.

To begin with, it can be easily verified that any attack relation of the form $\hookrightarrow^-$ will be negative. Moreover, the latter determines a least negative closure of the source attack relation:

Lemma 3.15 $\hookrightarrow^-$ *is a least negative attack relation containing* $\hookrightarrow$.

An immediate consequence of the above lemma is that negative attack relations are precisely relations of the form $\hookrightarrow^-$. In other words, negative attack relations provide a canonical description of argumentation based on negative attacks.

The following result gives an important alternative characterization of negative argumentation.

Lemma 3.16 *An attack relation is negative iff it satisfies:*

(Import)   *If* $a \hookrightarrow b, c$, *then* $a, b \hookrightarrow c$.

As a special case of Import, we have that if $a \hookrightarrow b$, then $a, b \hookrightarrow \emptyset$. Thus, any negative attack relation is bound to be non-affirmative. Furthermore, this implies that inconsistent argument sets attack any argument:

$$\text{If } v \hookrightarrow v, \text{ then } v \hookrightarrow u.$$

This feature is responsible for the fact that only stable sets constitute a reasonable nonmonotonic semantics for negative argumentation (see below).

Negative argumentation can be characterized semantically by restricting the set of possible four-valued interpretations to *consistent* ones that do not assign the set $\{t, f\}$ to arguments. This means that no argument can be both accepted and rejected in an interpretation.

Theorem 3.17 *An attack relation is negative if and only if it is determined by a set of consistent interpretations.*

Proof. If an attack $a \hookrightarrow b$ does not hold in a consistent interpretation ν, then all arguments in a are accepted, and no argument from b is rejected in ν. Since ν is consistent, no assumption in $a \cup b$ is rejected in ν, so the attack $a \hookrightarrow a, b$ also does not hold in ν. This shows that an attack relation determined by consistent interpretations is bound to be negative. In the other direction, if an attack relation $\hookrightarrow$ is negative, and $u \not\hookrightarrow v$, then $(u, u \cup v)$ is a consistent bitheory of $\hookrightarrow$ (since $u \not\hookrightarrow u, v$). Moreover, $u \hookrightarrow v$ does not hold in the (consistent) interpretation corresponding to this bitheory. As follows from the proof of the main representation theorem, this means that any negative attack relation is determined by a set of consistent interpretations. ∎

3.2.3 Positive argumentation

The definition below provides a structural description of positive argumentation.

Definition 3.18 *An attack relation will be called* positive *if $a, b \hookrightarrow b$ always implies $a \hookrightarrow b$.*

Any attack relation $\hookrightarrow^+$ will be positive. Moreover, the latter determines a least positive extension of the source attack relation.

Lemma 3.19 $\hookrightarrow^+$ *is a least positive argument theory containing* $\hookrightarrow$.

The lemma implies that positive attack relations are precisely relations of the form $\hookrightarrow^+$, and hence they give a canonical description of argumentation based on positive attacks.

Similarly to negative argumentation, positive argumentation can be characterized by the 'exportation' property described in the lemma below:

Lemma 3.20 *An attack relation is positive iff it satisfies:*

(Export)   *If $a, b \hookrightarrow c$, then $a \hookrightarrow b, c$.*

Positive argumentation can also be characterized semantically by restricting the set of possible four-valued interpretations to *complete* ones, namely to interpretations that do not assign $\emptyset$ to arguments. This means that every argument is either accepted or rejected in an interpretation (or both).

Theorem 3.21 *An attack relation is positive if and only if it is determined by a set of complete interpretations.*

The proof of the above theorem is perfectly similar to the case of negative argumentation. It turns out, however, that the positive argumentation, taken in its full generality, is not appropriate for the main nonmonotonic semantics, namely the stable semantics. Still, the reader can find in the literature a number of weaker argumentation systems that incorporate some of the features of positive argumentation. We will mention below only one important logical principle of this kind.

Consistent argumentation. A number of argumentation systems suggested in the literature (see, e.g., [Kakas *et al.*, 1994]) are based on the idea that inconsistent arguments should not form a legitimate attack on other arguments. A simplest way to incorporate this idea into an argumentation theory consists in using the following modification of an attack relation:

Definition 3.22 *Given an attack relation $\hookrightarrow$, we will say that a consistently attacks b (notation $a \hookrightarrow^c b$) if either $a \hookrightarrow b$, or $b \hookrightarrow b$.*

It turns out that this kind of an attack relation can also be given a logical description.

Definition 3.23 *An attack relation will be called* consistent *if $b \hookrightarrow b$ implies $\emptyset \hookrightarrow b$.*

As can be seen, consistent attack relations embody the most significant feature of positive argumentation, namely that inconsistent arguments are attacked by any argument. Still, consistency in the above sense is a weaker property than positivity (Export).

As before, it can be shown that consistent attack relations are precisely relations of the form $\hookrightarrow^c$. Finally, a semantic characterization of consistent argumentation can be obtained by requiring that any inconsistent argument set is rejected in at least one four-valued interpretation. This requirement is met by restricting the set of possible four-valued interpretations to *quasi-reflexive* ones, namely to sets of interpretations I such that if $(u,v) \in I$, then the corresponding classical interpretation (v,v) also belongs to I. As a matter of fact, this semantic constraint plays a prominent role in describing the logics appropriate for the stable semantics of logic programs (see [Bochman and Lifschitz, 2011]).

4 Nonmonotonic semantics

In the preceding section we have described a structural logical basis of argumentation. As we have argued in the introduction, however, the argumentation theory should be viewed as a two-layered formalism which has both logical and nonmonotonic components. This means that, in addition to the logical semantics, an argumentation formalism should be assigned also a *nonmonotonic* semantics that will determine the actual acceptance and rejection of arguments in each reasoning context. As one of its main objectives, the latter semantics should incorporate and thoroughly implement the basic principle of default acceptance for arguments.

Partly due to historical reasons (primarily, the logic programming origins), there is a bewildering number of nonmonotonic semantics that are actively investigated in the current argumentation literature (see, e.g., Chapters 2 and 4 of the Handbook). There have been a number of attempts to systematize these semantics (see, e.g., [Baroni and Giacomin, 2007]), though no uniform picture has been emerged.

In some sense, an attempt to systematize the various nonmonotonic semantics of argumentation is similar to an attempt to systematize logics in general, and as for the latter, it appears to be doomed from the very beginning. Worse still, the modern formal argumentation theory is still too young to provide substantive evidence for (or against) specific semantics, and thereby implicitly preserves hopes that they could be found useful in the future.

We will attempt to provide below a rough sketch of the basic principles and desiderata for constructing the nonmonotonic semantics of argumentation, which will also implicitly single out certain preferences, or priorities, between them. Our main underlying idea is that we should always try to apply the best (rather than the most general) semantics that is consistent with the constraints of the application in question. Of course, our position on this issue is not uncontroversial, but we contend that it is a reasonable and defensible position.

As a starting point, we will formulate the main *principle of argumentation* as the claim that arguments (in sharp distinction with factual assertions) bear with them the presumption of acceptance:

An argument is accepted unless there is a reason for its rejection.

One of the important ways of interpreting the above principle amounts to viewing arguments as *abducibles* in the framework of an argumentation theory. This understanding can serve as a guidance in determining the associated nonmonotonic semantics.

Now, in the framework of the formal argumentation theory, the reasons for rejection of arguments come only in the form of attacks by other arguments. Thus, our logical interpretation of the attack relation immediately sanctions that if an argument A attacks an argument B, and A is a accepted, then B should be rejected. In what follows, we will say that an argument is *refuted*, if it is attacked by an accepted argument set. Then our main principle of argumentation implies that an argument should be accepted whenever all its attacking arguments are not accepted. In other words, it evolves to

An argument is accepted if and only if it is not refuted.

Now, if we combine the above principle with the natural 'classical' requirement that any argument should be either accepted, or rejected, but not both, we will immediately obtain the primary nonmonotonic semantics of argumentation, the *stable semantics*[5]. According to this semantics, acceptable sets of arguments are conflict-free sets that attack any argument outside them.

In the general correspondence between Dung's argumentation theory and other nonmonotonic formalisms, the stable semantics corresponds to the main nonmonotonic semantics of the latter. This, as well as many other facts (some of which will be detailed later in this study), make the stable semantics a proper candidate on the role of the *standard nonmonotonic semantics* for argumentation, much in the same sense as the classical logic can be viewed as the standard logic for our reasoning (whatever the objections one could possibly have against this logic).

Despite its naturalness and simplicity, however, there are also quite simple argumentation frameworks where the stable semantics fails to determine an acceptable set of arguments[6]. Such situations create an obvious incentive for trying alternative, more tolerant, nonmonotonic semantics[7].

It turns out that the general four-valued logical semantics of acceptance and rejection of arguments provides all the necessary 'degrees of freedom' for defining such alternative nonmonotonic semantics, and the way to do this amounts to adopting different 'partial' generalizations of the main argumentation principle in the four-valued setting.

Retaining our earlier definition of refutation, a first such relaxed argumentation principle can be formulated as follows:

An argument is rejected if and only if it is refuted.

Note that the above principle is not equivalent to our original main argumentation principle, since the assignments of acceptance and rejection are logically independent. Instead, combined with our logical characterization of the attack relation, this principle will give us precisely the notion of labeling from [Jakobovits and Vermeir, 1999].

An even stronger general constraint on nonmonotonic semantics can be obtained by adding the following alternative generalization of the main argumentation principle:

An argument is accepted if and only if all its attackers are rejected.

Now, if we will restrict the set of valuations to consistent ones, we will obtain exactly the Caminada labellings (see [Caminada and Gabbay, 2009]). These labellings have been shown to encompass the main nonmonotonic semantics of Dung's argumentation frameworks.

---

[5] see also [Pollock, 1987].

[6] A simplest such framework comprises a single argument that attacks itself.

[7] As a side remark, it is important to bear in mind, however, that the nonmonotonic semantics is not intended to replace the logical semantics in all its functions and capacities. In particular, the nonmonotonic semantics *should not* be required to deliver a consistent extension in any situation (just as a definite description in classical first-order logic cannot be required to always determine a unique referent).

In the rest of this section we are going to provide a more detailed description of the nonmonotonic semantics of argumentation.

4.1 Normal (Dung) argumentation

As a convenient starting point, we will describe now a range of nonmonotonic semantics for normal argument theories suggested in [Dung, 1995b]. All these semantics can be defined in terms of the following two notions:

Definition 4.1 *Given an attack relation* $\hookrightarrow$, *an argument A will be called* allowable *for a set of arguments u, if $u \not\hookrightarrow A$, and* acceptable *for u, if u attacks any argument set that attacks A.*

$[u]$ will denote the set of all assumptions allowable by u:

$$[u] = \{A \mid u \not\hookrightarrow A\}.$$

The origins of this operator can be found already in [Pollock, 1987], and it has been extensively used in [Dung, 1995b].

Note that $[\ ]$ is an anti-monotonic operator on argument sets, that is, $u \subseteq v$ implies $[v] \subseteq [u]$. Moreover, the set of arguments that are acceptable for u coincides with $[[u]]$, where $[[\ ]]$ is obviously a monotonic operator.

Using the above notions, we can give a rather simple characterization of the basic nonmonotonic models of a normal argumentation.

Definition 4.2 *An argument set u is*

- conflict-free *if $u \subseteq [u]$;*
- admissible *if it is conflict-free and $u \subseteq [[u]]$;*
- a complete extension *if it is conflict-free and $u = [[u]]$;*
- a preferred extension *if it is a maximal complete extension;*
- a stable extension *if $u = [u]$.*

As has been shown in [Dung, 1995a], the above models correspond to well-known semantics for normal logic programs.

4.2 Stable and partial stable semantics

Though the above notions and models of Dung's argumentation theory have been defined for arbitrary collective attack relations, it should be clear that they are adequate only for normal argumentation, since they are based only on singular attacks. Still, we will see that in some important cases the more general models defined below will coincide with their normal counterparts. Unfortunately, it will turn out that only a small part of the nice and well-organized structure of nonmonotonic models for normal argumentation can be transferred into a general framework of collective argumentation.

If (u,v) is a bitheory of an attack relation (that is, $u \not\hookrightarrow v$), then v is always included in a maximal set v_1 such (u,v_1) is a bitheory. The corresponding four-valued model contains a minimal set of rejected arguments for a given set u of accepted ones.

For an argument set u, we will denote by $\langle u \rangle$ the set of all maximal argument sets v such that $u \not\hookrightarrow v$. The operator $\langle u \rangle$ will play in what follows the same role as the operator $[u]$ in normal argumentation. In particular, using this operator, we can give the following rather simple description of stable and partial stable models of collective argumentation.

Definition 4.3
- *An argument set u will be called* stable *if $u \in \langle u \rangle$.*
- *A bitheory (u,v) will be called* partial stable *if $u \in \langle v \rangle$, and $v \in \langle u \rangle$.*

As has been shown in [Bochman, 2003a], under a general correspondence between collective argumentation and *disjunctive* logic programs, the above models correspond precisely to the well-known semantics for such logic programs, given in the literature.

Stable argument sets correspond precisely to partial stable bitheories of the form (u,u). Note also that if (u,v) is a partial stable bitheory, then (v,u) will also be partial stable, and vice versa. A usual additional condition imposed on partial stable models amounts, however, to requiring that (u,v) should be a consistent bitheory (that is, $u \subseteq v$). The corresponding models will be called *consistent* partial stable bitheories. Note, however, that the bitheories (u,v) and (v,u) provide the same information about 'classical' acceptance and rejection of assumptions, namely they single out the same assumptions that are accepted without being rejected, and same rejected assumptions that are not also accepted.

The following lemmas give more direct, and often more convenient, descriptions of the above models.

Lemma 4.4 *An argument set u is stable iff* $u = \{A \mid u \not\hookrightarrow u, A\}$.

The above equation says that a stable argument u consists of all arguments A such that u does not attack $u \cup \{A\}$. A similar description can be given for partial stable models.

Lemma 4.5 (u,v) *is a partial stable bitheory if and only if*

$$v = \{A \mid u \not\hookrightarrow v, A\} \quad \text{and} \quad u = \{A \mid v \not\hookrightarrow u, A\}.$$

The next result shows that stable argument sets and consistent partial stable bitheories of a normal attack relation coincide, respectively, with stable and complete extensions.

Lemma 4.6 *If $\hookrightarrow$ is a normal attack relation, then*

- *stable argument sets coincide with stable extensions;*
- *(u, v) is a consistent partial stable bitheory iff u is a complete extension, and $v = [u]$.*

Furthermore, it can be shown that partial stable bitheories are representable as stable argument sets of a certain 'doubled' attack relation.

Let $\hookrightarrow$ be an attack relation on a set $\mathcal{A}$ of arguments. For each $A \in \mathcal{A}$, we introduce a new argument A'. For any subset u of $\mathcal{A}$, we will denote by u' the set $\{A' \mid A \in u\}$. Now we define a new attack relation $\hookrightarrow_\circ$ on $\mathcal{A} \cup \mathcal{A}'$ as follows:

$$a, b' \hookrightarrow_\circ c, d' \equiv a \hookrightarrow d \text{ or } b \hookrightarrow c.$$

Then we have

Theorem 4.7 *A bitheory (u, v) is partial stable in $\hookrightarrow$ if and only if $u \cup v'$ is a stable argument set in $\hookrightarrow_\circ$.*

The above theorem shows, in effect, that partial stable models are essentially stable argument sets 'in disguise'. Unfortunately, in the case of collective argumentation they lack most of the structural properties they had in the normal case of Dung's theory. Most importantly, they do not form a lower semilattice (under the standard information order over four-valued models -see, e.g., [Fitting, 1991]) and, in particular, there may be no least partial stable model. In fact, unlike the normal case, collective argument theories may have no partial stable models at all.

Example 4.8 *(1) It can be verified that the argument theory*

$$\{\hookrightarrow A, B, C;\ A \hookrightarrow B;\ B \hookrightarrow C;\ C \hookrightarrow A\}$$

does not have any partial stable bitheories.

(2) the argument theory $\{\hookrightarrow A, B;\ A \hookrightarrow A;\ B \hookrightarrow B\}$ does not have consistent partial stable models, though $(\{A\}, \{B\})$ and $(\{B\}, \{A\})$ are its partial stable bitheories.

4.3 Admissibility semantics

Finally we will briefly consider collective counterparts of admissible argument sets in normal argumentation. Recall that the latter have been defined as conflict-free argument sets that counterattack any argument against them. This definition can be naturally generalized to collective argumentation as follows:

Definition 4.9 *A consistent argument set u will be called* admissible *if, for any v, if $v \hookrightarrow u$, then $u \hookrightarrow v$.*

Admissible argument sets can also be described in terms of the $\langle\ \rangle$ operator. Namely, u is admissible if no argument set from $\langle u \rangle$ attacks u. Clearly, admissibility reduces to Dung-admissibility for normal attack relations. Unfortunately, in the context of collective argumentation the notion of admissibility behaves in a much less ordered fashion than in the Dung's theory. Note, in particular, that even stable argument sets need not be admissible in this sense:

Example 4.10 *Let us consider an argument theory* $\{A \hookrightarrow B \hookrightarrow A, B\}$. *As can be seen,* $\{B\}$ *is a stable argument set of this theory, but it is not admissible: we have* $A \hookrightarrow B$, *though* $B \not\hookrightarrow A$.

It turns out, however, that consistent stable extensions are always both admissible and stable.

Lemma 4.11 *Any consistent stable extension of an attack relation is both an admissible and stable argument set.*

4.4 Underlying argumentation logics

Any nonmonotonic semantics implicitly determines an appropriate underlying logic, a logic that preserves this semantics under all expansions of the associated nonmonotonic theory. In this section, we describe the effects of imposing logical constraints, described earlier, on the nonmonotonic argumentation semantics.

As an 'upper' limiting case, the classical argumentation theory drastically simplifies the whole range of nonmonotonic semantics:

Lemma 4.12 *If* $\hookrightarrow$ *is a classical attack relation, then*

- *Admissible sets coincide with consistent sets;*
- *Stable sets coincide with maximal consistent sets;*
- *Partial stable bitheories coincide with stable ones.*

A more discriminate look reveals that it is the 'positive ingredient' of classical argumentation that could be held responsible for this trivializing effect. More precisely, already the consistency property produces the same result:

Lemma 4.13 *If* $\hookrightarrow$ *is a consistent attack relation, then*

- *stable argument sets coincide with maximal consistent sets;*
- *consistent partial stable bitheories are bitheories of the form* (u, u), *where* u *is a stable argument set.*

On the other hand, the results below will show that negative argumentation constitutes an adequate and very convenient framework for studying the stable semantics.

As a first step, the next result shows that stable argument sets of an attack relation are precisely stable extensions of its negative closure.

Lemma 4.14 *Stable argument sets of an attack relation $\hookrightarrow$ coincide with the stable extensions of $\hookrightarrow^-$.*

The above result shows that Dung's stable extensions and stable argument sets of collective argumentation are indeed close relatives. As a consequence, we immediately obtain

Corollary 4.15
- *Stable argument sets of a negative attack relation coincide with its stable extensions.*
- *Any attack relation $\hookrightarrow$ has the same stable argument sets as $\hookrightarrow^-$.*

The second claim above implies that Import is an argumentation rule that preserves stable argument sets, and hence negative argumentation turns out to be appropriate for the stable nonmonotonic semantics. An additional consequence of the above results is the eventual reduction of stable argument sets to stable extensions of Dung's argumentation theory. Lemma 4.14 says, in effect, that, after extending a given attack relation to a negative one (by closing it with respect to Import), we can restrict ourselves to its normal sub-relation; stable extensions of the resulting normal attack relation will coincide with stable argument sets of the original attack relation.

Our next result shows, however, that negative argumentation trivializes partial stable semantics.

Lemma 4.16 *Partial stable bitheories of a negative attack relation are bitheories of the form (u, u), where u is a stable argument set.*

Thus, negative argumentation reduces partial stable models to stable ones. In fact, negative argumentation seems to exclude all nonmonotonic semantics other than the stable one.

Our final result shows that admissible argument sets still play an important role in negative argumentation.

Theorem 4.17 *Let $\hookrightarrow$ be a negative attack relation.*

- *If u is an admissible argument set in $\hookrightarrow$, and v a consistent argument set that includes u, then v is also admissible in $\hookrightarrow$.*

- *Stable argument sets of $\hookrightarrow$ coincide with maximal admissible argument sets.*

The above theorem demonstrates that the structure of admissible and stable argument sets in negative argumentation is very simple. Namely, they behave much like logically consistent sets. There is, however, a crucial difference: the empty set $\emptyset$ is not, in general, admissible. Moreover, a negative attack relation may have no admissible arguments at all; this happens precisely when it has no stable argument sets.

5 Negation, deduction and assumptions

The notion of an argument is often taken as primitive in argumentation theory, which, among other theoretical advantages, allows for a possibility of considering arguments that are non-propositional in character (e.g., arguments as inference rules, or derivations). Still, there exists a natural, direct connection between abstract argumentation frameworks and traditional deductive argumentation; it has been established, in effect, already in [Bondarenko et al., 1997][8]. In this formalism of *assumption-based argumentation* arguments were constructed as plain deductive arguments that may involve, however, auxiliary propositional *assumptions*. Moreover, the attack relation can already be defined in this framework, so the assumption-based argumentation can be viewed as a special case of Dung's abstract argumentation. Nevertheless, it has been shown in [Bondarenko et al., 1997] that this special kind of argumentation still provides a natural and powerful generalization of the main nonmonotonic formalisms and various semantics for logic programming.

As we are going to show in this section, the entire formalism of assumption-based argumentation can be obtained just by adding a single negation connective to the logical system of abstract argumentation, a connective that is actually implicit in the formalism of [Bondarenko et al., 1997] in the form of the contrary mapping on assumptions. This move will also constitute a first, and most important, step towards a full-fledged theory of *propositional argumentation* that will be described subsequently.

Let us extend our underlying language with a negation connective $\sim$ having the following precise (four-valued) semantic definition[9]:

$\sim A$ is accepted iff A is rejected

$\sim A$ is rejected iff A is accepted.

The above definition makes $\sim$ a particular four-valued connective; it will be called a *global negation*, since it switches the evaluation contexts between acceptance and rejection.

An axiomatization of this negation in abstract argumentation theory can be obtained by imposing the following rules on the attack relation:

$$A \hookrightarrow \sim A \qquad \sim A \hookrightarrow A$$

(AN) If $a \hookrightarrow A, b$ and $a, \sim A \hookrightarrow b$, then $a \hookrightarrow b$

If $a, A \hookrightarrow b$ and $a \hookrightarrow b, \sim A$, then $a \hookrightarrow b$

Attack relations satisfying the above postulates will be called *N-attack relations*. It turns out that the latter are inter-definable with a particular kind of consequence relations.

---

[8] See also [Kowalski and Toni, 1996].

[9] This negation connective played a prominent role in Belnap's information lattices [Belnap, 1977].

Recall that a *Scott consequence relation*, known also as a multiple-conclusion consequence relation [Shoesmith and Smiley, 1978; Gabbay, 1981; Segerberg, 1982; Wojcicki, 1988], is a binary relation between *sets* of propositions that is required to satisfy the following will-known postulates:

(**Reflexivity**) $A \Vdash A$;

(**Monotonicity**)   If $a \Vdash b$ and $a \subseteq a'$, $b \subseteq b'$, then $a' \Vdash b'$;

(**Cut**)   If $a \Vdash b, A$ and $a, A \Vdash b$, then $a \Vdash b$,

In this logical framework, our target consequence relations can be described as follows:

Definition 5.1 *A* Belnap consequence relation *in a propositional language with a global negation $\sim$ is a Scott consequence relation satisfying the following two Double Negation rules for the global negation:*

$$A \Vdash \sim\sim A \qquad \sim\sim A \Vdash A.$$

For any set u of propositions, we will denote by $\sim u$ the set $\{\sim A \mid A \in u\}$. Now, for a given N-attack relation, we can define the following consequence relation:

(CA) $\qquad\qquad a \Vdash b \equiv a \hookrightarrow \sim b$

Similarly, for any Belnap consequence relation we can define the corresponding attack relation as follows:

(AC) $\qquad\qquad a \hookrightarrow b \equiv a \Vdash \sim b$

As has been shown in [Bochman, 2003a], the above definitions establish an exact equivalence between N-attack relations and Belnap consequence relations. This correspondence allows us to represent an assumption-based argumentation framework from [Bondarenko *et al.*, 1997] entirely in the framework of attack relations (see below).

N-attack relations allow to provide simpler alternative descriptions of negative and positive argumentation. Thus, the rule Import of negative argumentation for such attack relations is equivalent to the condition

$$A, \sim A \hookrightarrow \emptyset.$$

whereas the rule Export of positive argumentation is equivalent to

$$\emptyset \hookrightarrow A, \sim A.$$

The above condition explicitly says that any argument should be either rejected or accepted. As could be expected, it is equivalent also to the principle of *reasoning by cases*:

(**Factoring**)   If $a, A \hookrightarrow b$ and $a, \sim A \hookrightarrow b$, then $a \hookrightarrow b$.

Assumptions versus factual propositions. Though the global negation $\sim$ is a logically well-defined connective, it implicitly interferes with the main principle of argumentation that presupposes an asymmetric treatment of acceptance and rejection for arguments. Indeed, if A is an argument, then $\sim A$ cannot already be viewed as an argument, since otherwise presumptive acceptance of $\sim A$ would directly imply presumptive *rejection* of A itself!

The emerging problem immediately reminds us, however, that our commonsense epistemic states are not homogeneous: in addition to normality assumptions (that can be viewed as primitive arguments), they contain also ordinary factual claims. Furthermore, the latter have in a sense an opposite nature as compared to arguments. Namely, they are presumably rejected unless we have reasons for their acceptance.

A simplest and perhaps the most natural way of resolving the above issues consists in a clear separation between assumptions and factual propositions; it has been actually implemented in the assumption-based argumentation of [Bondarenko et al., 1997].

Assumption-based argumentation (ABA). Slightly changing the formulation of [Bondarenko et al., 1997], an assumption-based argumentation framework can be defined as a triple consisting of an underlying deductive system (including the current set of beliefs), a distinguished subset of propositions Ab called *assumptions*, and a mapping from Ab to the set of all propositions of the language that determines the *contrary* $\overline{A}$ of any assumption A.

Now, the underlying deductive system can be expressed directly in the framework of N-attack relations by identifying deductive rules $a \vdash A$ with attacks of the form $a \hookrightarrow \sim A$. Furthermore, the global negation $\sim$ can also serve as a faithful logical formalization of the operation of taking the contrary. More precisely, given an arbitrary underlying language $\mathcal{L}$ that does not contain $\sim$, we can *define* assumptions as propositions of the form $\sim A$, where $A \in \mathcal{L}$. Then, since $\sim$ satisfies double negation, a negation of an assumption will be a proposition from $\mathcal{L}$. Accordingly, N-attack relations can be seen as a proper generalization of the assumption-based framework.

Remark 5.2 *It should be mentioned that our representation assign a bit more structure and properties to assumptions and the contrary mapping than it was originally assumed in ABA. Still, it can be verified that this extended representation is fully conservative with respect to the applications of this argumentation theory to other nonmonotonic formalisms, described in [Bondarenko et al., 1997].*

In [Bondarenko et al., 1997], the connection between (assumption-based) argumentation and main nonmonotonic formalisms has been established by showing that these nonmonotonic systems can be viewed as assumption-based frameworks just by defining assumptions and their contraries. As a partial converse of these results, we are going to show below that many of these formalisms constitute actually primary instantiations of propositional argumentation in appropriately chosen logical languages.

6 Default argumentation

Taking seriously the idea of propositional argumentation, it is only natural to make further steps toward extending the underlying language of arguments to the usual classical propositional language. These steps should be coordinated, however, with the inherently four-valued nature of the attack relation. And the way to do this amounts to requiring that the relevant classical connectives should behave in a usual classical way with respect to both acceptance and rejection of arguments.

As a first such connective, we introduce the *conjunction* $\wedge$ of arguments that is determined by the following familiar semantic conditions:

$A \wedge B$ is accepted iff A is accepted and B is accepted

$A \wedge B$ is rejected iff A is rejected or B is rejected

As can be seen, $\wedge$ behaves as an ordinary classical conjunction with respect to acceptance and rejection of arguments. On the other hand, it is a four-valued connective, since the above conditions determine a four-valued truth-table for conjunction in the Belnap's interpretation of four-valued logic (see [Belnap, 1977]). The following postulates provide a simple syntactic characterization of this connective for attack relations:

$(A_\wedge)$
$$a, A \wedge B \hookrightarrow b \text{ iff } a, A, B \hookrightarrow b$$
$$a \hookrightarrow A \wedge B, b \text{ iff } a \hookrightarrow A, B, b$$

Collective attack relations satisfying these postulates will be called *conjunctive*. The next result shows that they give a complete description of the four-valued conjunction.

Corollary 6.1 *An attack relation is conjunctive if and only if it coincides with $\hookrightarrow_I$, for some set of four-valued interpretations I in a language with the four-valued conjunction $\wedge$.*

An immediate benefit of introducing conjunction into the language of argumentation is that any finite set of arguments a becomes reducible to a single argument $\bigwedge a$:

$$a \hookrightarrow b \text{ iff } \bigwedge a \hookrightarrow \bigwedge b.$$

As a result, the collective attack relation in this language is reducible to an attack relation between individual arguments, just as it has been assumed in [Dung, 1995b].

Having a conjunction at our disposal, we only have to add a classical negation $\neg$ in order to obtain a full classical language. Moreover, since sets of arguments are reducible to their conjunctions, we can represent the resulting argumentation theory using just a binary attack relation on classical formulas.

As a basic condition on argumentation in the classical propositional language, we will require only that the attack relation should respect the classical entailment $\vDash$ in the precise sense of being monotonic with respect to $\vDash$ on both sides.

Definition 6.2 *A propositional attack relation is a relation $\hookrightarrow$ on the set of classical propositions satisfying the following postulates:*

(Left Strengthening) *If $A \vDash B$ and $B \hookrightarrow C$, then $A \hookrightarrow C$;*

(Right Strengthening) *If $A \hookrightarrow B$ and $C \vDash B$, then $A \hookrightarrow C$;*

(Truth) $\mathbf{t} \hookrightarrow \mathbf{f}$;

(Falsity) $\mathbf{f} \hookrightarrow \mathbf{t}$.

Left Strengthening says that logically stronger arguments should attack any argument that is attacked already by a logically weaker argument, and similarly for Right Strengthening. Truth and Falsity postulates characterize the limit cases of argumentation by stipulating that any tautological argument attacks any contradictory one, and vice versa.

There exists a simple definitional way of extending the above attack relation to a collective attack relation between arbitrary sets of propositions. Namely, for any sets u, v of propositions, we can define $u \hookrightarrow v$ as follows:

$$u \hookrightarrow v \equiv \text{ there exist finite } a \subseteq u, b \subseteq v \text{ such that } \bigwedge a \hookrightarrow \bigwedge b$$

The resulting attack relation will satisfy the properties of collective argumentation, as well as the postulates $(A_\wedge)$ for conjunction.

Finally, in order to acquire full expressive capabilities of the argumentation theory, we can add the global negation $\sim$ to the language. Actually, a rather simple characterization of the resulting collective argumentation theory can be obtained by accepting the basic postulates AN for $\sim$, plus the following rule that permits the use of classical entailment in attacks:

Classicality If $a \vDash A$, then $a \hookrightarrow \sim A$ and $\sim A \hookrightarrow a$.

It can be verified that the resulting system satisfies all the postulates for propositional argumentation. The system will be used later for a direct representation of default logic.

6.1 Logical semantics

A semantic interpretation of propositional attack relations can be obtained by generalizing four-valued interpretations to pairs (u, v) of deductively closed theories, where u is the set of accepted propositions, while v the set of propositions that are not rejected. Such pairs will be called *bimodels*, while a set of bimodels will be called a *binary semantics*.

Definition 6.3 *An attack $A \hookrightarrow B$ will be said to be valid in a binary semantics $\mathcal{B}$ if there is no bimodel (u,v) from $\mathcal{B}$ such that $A \in u$ and $B \in v$.*

We will denote by $\hookrightarrow_\mathcal{B}$ the set of attacks that are valid in a semantics $\mathcal{B}$. This set forms a propositional attack relation. Moreover, the following result shows that propositional attack relations are actually complete for the binary semantics.

Theorem 6.4 $\hookrightarrow$ *is a propositional attack relation if and only if it coincides with $\hookrightarrow_\mathcal{B}$, for some binary semantics $\mathcal{B}$.*

6.2 Default logic

Now we will show that propositional argumentation provides a direct representation of Reiter's default logic [Reiter, 1980].

Given a system of propositional argumentation in the classical language augmented with the global negation $\sim$, we will interpret Reiter's default rule $a{:}b/A$ as an attack[10]

$$a, \sim \neg b \hookrightarrow \sim A,$$

or, equivalently, as a rule $a, \sim \neg b \Vdash A$ of the associated Belnap consequence relation. Similarly, an axiom A of a default theory will be interpreted as an attack $\mathbf{t} \hookrightarrow \sim A$. For a default theory Δ, we will denote by $tr(\Delta)$ the corresponding argument theory obtained by this translation.

By our general agreement, by *assumptions* we will mean propositions of the form $\sim A$, where A is a classical proposition. For a set u of classical propositions, we will denote by $\tilde{u}$ the set of assumptions $\{\sim A \mid A \notin u\}$. Finally, a set w of assumptions will be called *stable* in an argument theory Δ if, for any assumption A, $A \in w$ iff $w \not\hookrightarrow_\Delta A$, where $\hookrightarrow_\Delta$ is the least propositional attack relation containing Δ. Then we have

Theorem 6.5 *A set u of classical propositions is an extension of a default theory Δ if and only if $\tilde{u}$ is a stable set of assumptions in $tr(\Delta)$.*

The above result is similar to the corresponding representation result in [Bondarenko et al., 1997, Theorem 3.10], but it is much simpler, and is formulated entirely in the framework of propositional attack relations. The simpler representation was made possible due to the fact that propositional attack relations already embody the deductive capabilities treated as an additional ingredient in assumption-based frameworks.

7 Probative and causal argumentation

We will introduce now some stronger propositional attack relations that satisfy further reasonable postulates:

(Left Or) If $A \hookrightarrow C$ and $B \hookrightarrow C$, then $A \vee B \hookrightarrow C$;

---
[10]As before, we use set notation according to which $\neg b$ denotes the set $\{\neg B \mid B \in b\}$.

(**Right Or**)   If $A \hookrightarrow B$ and $A \hookrightarrow C$, then $A \hookrightarrow B \vee C$;

(**Self-Defeat**)   If $A \hookrightarrow A$, then $\mathbf{t} \hookrightarrow A$.

Definition 7.1 *A propositional attack relation will be called* probative *if it satisfies Left Or,* basic, *if it also satisfies Right Or, and* causal, *if it is basic and satisfies Self-Defeat.*

Probative argumentation allows for reasoning by cases. Its semantic interpretation can be obtained by restricting bimodels to pairs (α, v), where α is a world (maximal classically consistent set). The corresponding binary semantics will also be called *probative*. Similarly, the semantics for basic argumentation is obtained by restricting bimodels to world pairs (α, β); such a binary semantics will be called *basic*. Finally, the *causal* binary semantics is obtained from the basic semantics by requiring further that (α, β) is a bimodel only if (β, β) is also a bimodel.

Corollary 7.2 *A propositional attack relation is probative [basic, causal] iff it is determined by a probative [resp. basic, causal] binary semantics.*

Basic propositional argumentation can already be given a purely four-valued semantic interpretation, in which the classical negation $\neg$ has the following semantic description:

$\neg A$ is accepted iff A is not accepted

$\neg A$ is rejected iff A is not rejected

A syntactic characterization of this connective in collective argumentation can be obtained by imposing the rules

$$A, \neg A \hookrightarrow \qquad \hookrightarrow A, \neg A$$

($A_\neg$)   If $a, A \hookrightarrow b$ and $a, \neg A \hookrightarrow b$ then $a \hookrightarrow b$

If $a \hookrightarrow b, A$ and $a \hookrightarrow b, \neg A$ then $a \hookrightarrow b$

Then a basic propositional attack relation can be alternatively described as a collective attack relation satisfying the rules $(A_\wedge)$ and $(A_\neg)$. Moreover, the global negation $\sim$ can be added to this system just by adding the corresponding postulates (AN). It turns out, however, that the global negation is *eliminable* in this setting via to the following reductions:

($R_\sim$) $a, \sim A \hookrightarrow b \equiv a \hookrightarrow b, \neg A \qquad a \hookrightarrow \sim A, b \equiv a, \neg A \hookrightarrow b$

$a, \neg \sim A \hookrightarrow b \equiv a \hookrightarrow b, A \qquad a \hookrightarrow \neg \sim A, b \equiv a, A \hookrightarrow b$

As a result, the basic attack relation can be safely restricted to an attack relation in a classical language.

Finally, the rule Self-Defeat of causal argumentation gives a formal representation for an often expressed desideratum that self-conflicting arguments should not participate in defeating other arguments (see, e.g., [Bondarenko et al., 1997]). This aim is achieved in our setting by requiring that such arguments are attacked even by tautologies, and hence by any argument whatsoever.

7.1 Argumentation vs. causal reasoning

Probative attack relations turn out to be equivalent to general production inference relations from [Bochman, 2003b; Bochman, 2004a], a variant of input-output logics from [Makinson and van der Torre, 2000].

A *production inference relation* is a relation $\Rightarrow$ on the set of classical propositions satisfying the following rules:

(**Strengthening**)   If $A \vDash B$ and $B \Rightarrow C$, then $A \Rightarrow C$;

(**Weakening**)   If $A \Rightarrow B$ and $B \vDash C$, then $A \Rightarrow C$;

(**And**)   If $A \Rightarrow B$ and $A \Rightarrow C$, then $A \Rightarrow B \wedge C$;

(**Truth**) $\mathbf{t} \Rightarrow \mathbf{t}$;

(**Falsity**) $\mathbf{f} \Rightarrow \mathbf{f}$.

A production rule $A \Rightarrow B$ can be informally interpreted as saying that A *causes*, or *explains* B. A characteristic property of production inference is that reflexivity $A \Rightarrow A$ does not hold for it. Production rules are extended to rules with sets of propositions in premises by requiring that $u \Rightarrow A$ holds for a set u of propositions iff $\bigwedge a \Rightarrow A$, for some finite $a \subseteq u$. $\mathcal{C}(u)$ will denote the set of propositions produced by u:

$$\mathcal{C}(u) = \{A \mid u \Rightarrow A\}$$

The production operator $\mathcal{C}$ plays much the same role as the usual derivability operator for consequence relations.

A production inference relation is called *basic*, if it satisfies

(**Or**)   If $A \Rightarrow C$ and $B \Rightarrow C$, then $A \vee B \Rightarrow C$.

and *causal*, if it is basic and satisfies, in addition

(**Coherence**)   If $A \Rightarrow \neg A$, then $A \Rightarrow \mathbf{f}$.

It has been shown in [Bochman, 2003b] that causal inference relations provide a complete description of the underlying logic of causal theories from [McCain and Turner, 1997] (see also [Giunchiglia et al., 2004]).

It turns out that the binary semantics, introduced earlier, is appropriate also for interpreting production inference:

Definition 7.3 *A rule $A \Rightarrow B$ is valid in a binary semantics $\mathcal{B}$ if, for any bimodel $(u, v) \in \mathcal{B}$, $A \in u$ only if $B \in v$.*

As has been shown in [Bochman, 2004a], the above semantics is adequate for production inference relations. Moreover, the semantics for basic production inference can be obtained by restricting bimodels to world pairs (α, β), while the semantics for causal inference is obtained by requiring, in addition, that (α, β) is a bimodel only if (α, α) is also a bimodel.

Now, the correspondence between probative argumentation and production inference can be established directly on the syntactic level using the following definitions:

(PA) $\qquad\qquad A \Rightarrow B \equiv \neg B \hookrightarrow A;$

(AP) $\qquad\qquad A \hookrightarrow B \equiv B \Rightarrow \neg A.$

Under these correspondences, the rules of a probative attack relation correspond precisely to the postulates for production relations. Moreover, the correspondence extends also to a correspondence between basic and causal argumentation, on the one hand, and basic and causal production inference, on the other. Hence the following result is straightforward.

Lemma 7.4 *If $\hookrightarrow$ is a probative [basic, causal] attack relation, then (PA) determines a [basic, causal] production inference relation, and vice versa, if $\Rightarrow$ is a [basic, causal] production inference relation, then (AP) determines a probative [basic, causal] attack relation.*

Remark 7.5 *A seemingly more natural correspondence between propositional argumentation and production inference can be obtained using the following definitions:*

$$A \Rightarrow B \equiv A \hookrightarrow \neg B \qquad A \hookrightarrow B \equiv A \Rightarrow \neg B.$$

By these definitions, A explains B if it attacks $\neg B$, and vice versa, A attacks B if it explains $\neg B$. Unfortunately, this correspondence, though plausible by itself, does not take into account the intended understanding of arguments as (negative) assumptions. As a result, it cannot be extended directly to the correspondence between the associated nonmonotonic semantics, described below.

As our next result, we will establish a correspondence between the nonmonotonic semantics of causal inference relations and that of causal argumentation.

The nonmonotonic semantics of a causal inference relation is a set of its *exact worlds*, namely worlds α such that $\alpha = \mathcal{C}(\alpha)$ (see [Bochman, 2004a]). Such a world satisfies the rules of the causal relation, and any proposition that holds in it is explained by the causal rules.

A *causal theory* is an arbitrary set of production rules. By a nonmonotonic semantics of a causal theory Δ we will mean the exact worlds of the least causal relation containing Δ.

The correspondence between exact worlds and stable sets of assumptions is established in the next theorem.

Theorem 7.6 *If Δ is a causal theory, and Δ_a its corresponding argument theory given by (AP), then a world α is an exact world of Δ iff $\tilde{\alpha}$ is a stable set of assumptions in Δ_a.*

The above result shows, in effect, that propositional argumentation subsumes causal reasoning as a special case. Moreover, it can be shown that causal attack relations constitute a strongest argumentation system suitable for this kind of nonmonotonic semantics.

7.2 Abstract dialectical frameworks (ADFs)

As we are going to show in this section, Abstract Dialectical Frameworks [Brewka and Woltran, 2010; Brewka et al., 2013] can be viewed, in effect, as yet another bridge between argumentation and causal reasoning.

A detailed description of ADFs can be found in Chapter 5 of this Handbook. Accordingly, we will restrict our descriptions below only to the features of the latter that are relevant for our exposition.

Abstract Dialectical Frameworks have been introduced as an abstract argumentation formalism purported to capture more general forms of argument interaction than just attacks among arguments. To achieve this, each argument (or statement) in an ADF is associated with an *acceptance condition*, which is some propositional function determined by arguments that are linked to it. Using such acceptance conditions, ADFs allow to express that arguments may jointly support another argument, or that two arguments may jointly attack a third one, and so on. Dung's argumentation frameworks are recovered in this setting by acceptance condition saying that an argument is accepted if none of its parents is.

Formally, an abstract dialectical framework is a directed graph whose nodes represent statements or positions which can be accepted or not. The links represent dependencies: the status of a node s only depends on the status of its parents (denoted $par(s)$), that is, the nodes with a direct link to s. In addition, each node s has an associated acceptance condition C_s specifying the exact conditions under which s is accepted. C_s is a function assigning to each subset of $par(s)$ one of the truth values $\mathbf{t}, \mathbf{f}$. Intuitively, if for some $R \subseteq par(s)$ we have $C_s(R) = \mathbf{t}$, then s will be accepted provided the nodes in R are accepted and those in $par(s) \setminus R$ are not accepted.

Definition 7.7 *An abstract dialectical framework is a tuple $D = (S, L, C)$ where*

- *S is a set of statements (positions, nodes),*

- *$L \subseteq S \times S$ is a set of links,*

- *$C = \{C_s\}_{s \in S}$ is a set of total functions $C_s : 2^{par(s)} \to \{\mathbf{t}, \mathbf{f}\}$, one for each statement s. C_s is called acceptance condition of s.*

A more 'logical' representation of ADFs can be obtained simply by assigning each node s a *classical* propositional formula corresponding to its acceptance condition C_s (see [Ellmauthaler, 2012]). In this case we can tacitly assume that the acceptance formulas implicitly specify the parents a node depends on. It is then not necessary to give the links L, so an ADF D amounts to a tuple (S, C) where S is a set of statements, and C is a set of propositional formulas, one for each statement from S. The notation $s[C_s]$ has been used by the authors to denote the fact that C_s is the acceptance condition of s.

A two-valued interpretation v is a (two-valued) *model* of an ADF (S, C) whenever for all statements $s \in S$ we have $v(s) = v(\varphi_s)$, that is, v maps exactly those statements to true whose acceptance conditions are satisfied under v. This notion of a model provides a natural semantics for ADFs. In addition to this semantics, however, the authors define appropriate generalizations for all the major semantics of Dung's argumentation frameworks. Following the 'revised' description in [Brewka *et al.*, 2013], all these semantics are defined by generalizing the two-valued interpretations to three-valued ones. All of them are formulated using the basic operator Γ_D over three-valued interpretations that was introduced, in effect, already in [Brewka and Woltran, 2010]. For an ADF D and a three-valued interpretation v, the interpretation $\Gamma_D(v)$ is given by the mapping

$$s \mapsto \prod \{w(\varphi_s) \mid w \in [v]_2\},$$

where $[v]_2$ is the set of all two-valued interpretations that extend v.

For each statement s, the operator Γ_D returns the consensus truth value for its acceptance formula φ_s, where the consensus takes into account all possible two-valued interpretations w that extend the input valuation v. Then two-valued models of D are precisely those classical interpretations that are fixed points of Γ_D.

Taken in its full generality, however, the operator Γ_D allows to define generalizations of all the major Dung's argumentation semantics as follows.

A three-valued interpretation v for an ADF D is

- admissible iff $v \leq_i \Gamma_D(v)$;

- complete iff $\Gamma_D(v) = v$;

- preferred iff it is $\leq_i$-maximal admissible.

- grounded iff it is the least fixpoint of Γ_D.

As has been shown, the above definitions provide proper generalizations of the corresponding semantics for Dung's argumentation frameworks and, moreover, preserve much of the properties and relations of the latter.

7.2.1 The causal representation

We will describe now a uniform and modular translation of ADFs into the causal calculus. Actually, the key to this translation can be found in the striking similarity between the official definition of an ADF and the notion of a *causal model*, used by Judea Pearl in [Pearl, 2000]. Causal models are defined as triples $M = \langle U, V, F \rangle$, where U is a set of *exogenous* variables, V is a finite set of *endogenous* variables, while F is a set of functions that determine the values of each endogenous variable in terms of other variables.

Symbolically, F is represented as a set of *structural* equations

$$v_i = f_i(pa_i, u_i) \quad i = 1, \ldots, n$$

where pa_i is any realization of the unique minimal set of variables PA_i in $V \backslash \{V_i\}$ (parents) sufficient for representing f_i, and similarly for $U_i \subseteq U$.

In Pearl's account, every instantiation $U = u$ of the exogenous variables determines a particular "causal world" of the causal model. Such worlds stand in one-to-one correspondence with the solutions to the above equations in the ordinary mathematical sense. However, structural equations also encode causal information in their very syntax by treating the variable on the left-hand side of $=$ as the effect and treating those on the right as causes. Accordingly, the equality signs in structural equations convey the asymmetrical relation of "is determined by".

Being restricted to the classical propositional language, Pearl's notion of a causal model can be reduced to the following notion of a Boolean causal model that has been used in [Bochman and Lifschitz, 2015]:

Definition 7.8 *Assume that the set of propositional atoms is partitioned into a set of* exogenous *atoms and a finite set of* endogenous *atoms.*

- *A Boolean structural equation is an expression of the form $p = F$, where p is an endogenous atom and F is a propositional formula in which p does not appear.*

- *A Boolean causal model is a set of Boolean structural equations $p = F$, one for each endogenous atom p.*

As can be seen, the above definition is much similar to the logical reformulation of ADFs, with structural equations $p = F$ playing essentially the same role as the acceptance conditions $p[F]$. The differences are that only endogenous atoms are determined by their associated conditions in causal models, but on the other hand, there are no restrictions on appearances of atoms on both sides in ADF's acceptance conditions. Furthermore, plain (two-valued) models of ADFs correspond precisely to causal worlds of the causal model, as defined in [Bochman and Lifschitz, 2015]:

Definition 7.9 *A* solution *(or a* causal world*) of a Boolean causal model M is any propositional interpretation satisfying the equivalences $p \leftrightarrow F$ for all equations $p = F$ in M.*

Now, a modular representation of Boolean causal models as causal theories of the causal calculus has been given in [Bochman and Lifschitz, 2015], and it can now be seamlessly transformed into the following causal representation of ADFs:

Definition 7.10 (*Causal representation of an ADF*) *For any ADF D, Δ_D is the causal theory consisting of the rules*

$$F \Rightarrow p \quad \text{and} \quad \neg F \Rightarrow \neg p$$

for all acceptance conditions $p[F]$ in D.

The above representation is fully modular, and it will be taken as a uniform basis for the correspondences described in this section.

To begin with, the correspondence results from [Bochman and Lifschitz, 2015] immediately imply

Theorem 7.11 *The two-valued semantics of an ADF D corresponds precisely to the causal nonmonotonic semantics of Δ_D.*

As a consequence, the full system of causal inference provides a precise logical basis for this nonmonotonic semantics.

Furthermore, it can be shown that the above causal representation also survives the transition to three-valued models of ADFs. An essential precondition of this causal representation, however, amounts to transforming the underlying semantic interpretations of ADFs in terms of three-valued models into ordinary classical logical descriptions. In fact, the very possibility of such a classical reformulation stems from the crucial fact that the basic operator Γ of an ADF, described earlier, is defined, ultimately, in terms of ordinary classical interpretations extending a given three-valued one.

Any three-valued interpretation v on the set of statements S can be faithfully encoded using an associated set of literals $[v] = S_0 \cup \neg S_1$ such that $S_0 = \{p \in S \mid v(p) = \mathbf{t}\}$ and $S_1 = \{p \in S \mid v(p) = \mathbf{f}\}$. Moreover, this set of literals generates a unique deductively closed theory $\text{Th}([v])$ that corresponds in this sense to the source three-valued interpretation v. Conversely, let us say that a deductively closed set u is a *literal theory*, if it is a deductive (classical) closure of some set of literals. Then the latter set of literals will correspond to a unique three-valued interpretation v such that $u = \text{Th}([v])$. These simple facts establish a precise bi-directional correspondence between three-valued interpretations and classical literal theories.

Now, a broader correspondence between various semantics of ADFs and general nonmonotonic semantics of the causal calculus arises from the fact that the operator Γ of an ADF naturally corresponds to a particular causal operator of the associated causal theory.

Let L denote the set of classical literals of the underlying language. We will denote by $\mathcal{C}^L$ the restriction of a causal operator $\mathcal{C}$ to literals, that is, $\mathcal{C}^L(u) = \mathcal{C}(u) \cap L$. Now, it can be shown that the operator Γ of ADFs corresponds precisely to this 'literal restriction' of the causal operator associated with a basic production inference:

Lemma 7.12 *For any three-valued interpretation v,*

$$[\Gamma_D(v)] = \mathcal{C}_D^L([v]),$$

where $\mathcal{C}_D$ is a basic production operator corresponding to Δ_D.

The above equation has immediate consequences for the broad correspondence between the semantics of ADFs that are defined in terms of the operator Γ_D and natural sets of propositions definable wrt associated causal theory. Thus, we have

Theorem 7.13 *Complete models of an ADF D correspond precisely to the fixed points of $\mathcal{C}_D^L$:*

$$v = \Gamma_D(v) \quad \text{iff} \quad [v] = \mathcal{C}_D^L([v])$$

As a result, we immediately conclude that preferred models of an ADF correspond to maximal fixpoints of $\mathcal{C}_D^L$ (with respect to set inclusion), while the grounded model corresponds to the least fixpoint of $\mathcal{C}_D^L$.

Further details about these correspondences are discussed in [Bochman, 2016].

7.3 Logic programming

To complete the circle of representations, described in this study, we will show in this section that the formalism of logic programming itself, which could be seen as one of the main sources of Dung's argumentation theory, can also be viewed as a very specific kind of propositional argumentation.

A *general logic program* Π is a set of rules of the form[11]

(*) $\qquad\qquad\qquad\qquad$ **not** $d, c \leftarrow a,$ **not** b

where a, b, c, d are finite sets of propositional atoms. These are program rules of a most general kind that contain disjunctions and negations as failure **not** in their heads. As has been shown in [Bochman, 2004b], general logic programs are representable as causal theories obtained by translating program rules (*) as causal rules

$$d, \neg b \Rightarrow \bigwedge a \to \bigvee c,$$

and adding a formalization of the Closed World Assumption:

(Default Negation) $\neg p \Rightarrow \neg p$, for any propositional atom p.

---
[11]As before, **not** a denotes the set $\{\textbf{not}\ A \mid A \in a\}$.

Now, due to the correspondence between causal reasoning and argumentation, this causal theory can be transformed (using (PA)) into an argument theory that consists of attacks

(AL) $\qquad a, \neg c \hookrightarrow \neg b, d$

plus the 'argumentative' Closed World Assumption:

(Default Assumption) $\quad p \hookrightarrow \neg p$, for any atom p.

Let $tr(\Pi)$ denote the argument theory obtained by this translation from a logic program Π. Then we obtain

Theorem 7.14 *A set u of propositional atoms is a stable model of a logic program Π iff $\tilde{u}$ is a stable set of assumptions in $tr(\Pi)$.*

It is interesting to note that, due to the reduction rules $(R_\sim)$ for the global negation $\sim$, described earlier, the above representation (AL) of the program rules is equivalent to $a, \sim b \hookrightarrow \sim c, d$, and therefore to the inference rules

$$a, \sim b \Vdash c, \sim d$$

of the associated Belnap consequence relation. For normal logic programs (single atoms in heads), this latter representation coincides with that given in [Bondarenko et al., 1997].

8 Conclusions

The main objective of this study consisted in showing that both logic and nonmonotonic reasoning constitute two distinct, but essential, components of argumentation. On the way, we have shown that propositional argumentation suggests a viable and useful extension of the abstract argumentation theory that allows us to endow argumentation with full-fledged logical capabilities. The resulted theory has allowed us, in particular, to provide a systematic description of a large number of argumentation and nonmonotonic formalisms, such as assumption-based argumentation, abstract dialectical frameworks, default logic, logic programming and causal reasoning. It is natural to expect that further development of this approach to argumentation may bring additional theoretical and practical benefits.

One of the basic tasks that still need to be resolved with respect to the suggested formalism of propositional argumentation is a *systematic* connection of the latter with the more 'standard' approach of *structural argumentation* (see Chapter 6 of this Handbook) in which arguments are represented directly as derivations (proofs) constructed from strict and defeasible inference rules. Recall that both kinds of formalisms, propositional and proof theoretic ones, are peacefully coexisting in the majority of traditional logical systems, so its only natural to expect that the same kind of correspondence could be established also for the formal argumentation theory at its present, essentially nonmonotonic stage of development.

BIBLIOGRAPHY

[Amgoud and Besnard, 2013] L. Amgoud and P. Besnard. Logical limits of abstract argumentation frameworks. *Journal of Applied Non-Classical Logics*, 23(3):229–267, 2013.

[Baroni and Giacomin, 2007] P. Baroni and M. Giacomin. On principle-based evaluation of extension-based argumentation semantics. *Artif. Intell.*, 171(10-15):675–700, 2007.

[Belnap, 1977] N. D. Belnap, Jr. A useful four-valued logic. In M. Dunn and G. Epstein, editors, *Modern Uses of Multiple-Valued Logic*, pages 8–41. D. Reidel, 1977.

[Bochman and Lifschitz, 2011] A. Bochman and V. Lifschitz. Yet another characterization of strong equivalence. In M. Hermenegildo and T. Schaub, editors, *Technical Communications of the 26th Int'l. Conference on Logic Programming (ICLP'10)*, volume 7 of *Leibniz International Proceedings in Informatics (LIPIcs)*, pages 281–290, Dagstuhl, Germany, 2011. Schloss Dagstuhl–Leibniz-Zentrum fuer Informatik.

[Bochman and Lifschitz, 2015] A. Bochman and V. Lifschitz. Pearl's causality in a logical setting. In *Proceedings of the Twenty-Ninth AAAI Conference on Artificial Intelligence, January 25-30, 2015, Austin, Texas, USA.*, pages 1446–1452. AAAI Press, 2015.

[Bochman, 2003a] A. Bochman. Collective argumentation and disjunctive logic programming. *Journal of Logic and Computation*, 9:55–56, 2003.

[Bochman, 2003b] A. Bochman. A logic for causal reasoning. In *IJCAI-03, Proceedings of the Eighteenth International Joint Conference on Artificial Intelligence, Acapulco, Mexico, August 9-15, 2003*, pages 141–146, Acapulco, 2003. Morgan Kaufmann.

[Bochman, 2004a] A. Bochman. A causal approach to nonmonotonic reasoning. *Artificial Intelligence*, 160:105–143, 2004.

[Bochman, 2004b] A. Bochman. A causal logic of logic programming. In D. Dubois, C. Welty, and M.-A. Williams, editors, *Proc. Ninth Conference on Principles of Knowledge Representation and Reasoning, KR'04*, pages 427–437, Whistler, 2004.

[Bochman, 2005] A. Bochman. *Explanatory Nonmonotonic Reasoning*. World Scientific, 2005.

[Bochman, 2016] A. Bochman. Abstract dialectical argumentation among close relatives. In P. Baroni, T. F. Gordon, T. Scheffler, and M. Stede, editors, *Computational Models of Argument - Proceedings of COMMA 2016, Potsdam, Germany, 12-16 September, 2016.*, volume 287 of *Frontiers in Artificial Intelligence and Applications*, pages 127–138. IOS Press, 2016.

[Boella et al., 2005] G. Boella, J. Hulstijn, and L. W. N. v. d. Torre. A logic of abstract argumentation. In S. Parsons, N. Maudet, P. Moraitis, and I. Rahwan, editors, *Argumentation in Multi-Agent Systems*, page 2941. Springer, 2005.

[Bondarenko et al., 1997] A. Bondarenko, P. M. Dung, R. A. Kowalski, and F. Toni. An abstract, argumentation-theoretic framework for default reasoning. *Artificial Intelligence*, 93:63–101, 1997.

[Brewka and Woltran, 2010] G. Brewka and S. Woltran. Abstract dialectical frameworks. In *Principles of Knowledge Representation and Reasoning: Proceedings of the Twelfth International Conference, KR 2010, Toronto, Ontario, Canada, May 9-13, 2010*, 2010.

[Brewka et al., 2013] G. Brewka, H. Strass, S. Ellmauthaler, J. P. Wallner, and S. Woltran. Abstract dialectical frameworks revisited. In *IJCAI 2013, Proceedings of the 23rd International Joint Conference on Artificial Intelligence, Beijing, China, August 3-9, 2013*, 2013.

[Caminada and Amgoud, 2007] M. Caminada and L. Amgoud. On the evaluation of argumentation formalisms. *Artif. Intell.*, 171(5-6):286–310, 2007.

[Caminada and Gabbay, 2009] M. W. A. Caminada and D. M. Gabbay. A logical account of formal argumentation. *Studia Logica*, 93(2-3):109–145, 2009.

[de Kleer, 1986] J. de Kleer. An assumption-based TMS. *Artificial Intelligence*, 28:127–162, 1986.

[Doyle, 1979] J. Doyle. A truth maintenance system. *Artif. Intell.*, 12:231–272, 1979.

[Doyle, 1994] J. Doyle. Reasoned assumptions and rational psychology. *Fundamenta Informaticae*, 20:3573, 1994.

[Dung and Thang, 2014] P. M. Dung and P. M. Thang. Closure and consistency in logic-associated argumentation. *J. Artif. Intell. Res. (JAIR)*, 49:79–109, 2014.

[Dung, 1995a] P. M. Dung. An argumentation-theoretic foundation for logic programming. *J. of Logic Programming*, 22:151–177, 1995.

[Dung, 1995b] P. M. Dung. On the acceptability of arguments and its fundamental role in non-monotonic reasoning, logic programming and n-persons games. *Artificial Intelligence*, 76:321–358, 1995.
[Ellmauthaler, 2012] S. Ellmauthaler. Abstract Dialectical Frameworks: Properties, Complexity, and Implementation. Master's thesis, Technische Universität Wien, Institut für Informationssysteme, 2012.
[Fitting, 1991] M. C. Fitting. Bilattices and the semantics of logic programming. *Journal of Logic Programming*, 11:91–116, 1991.
[Gabbay, 1981] D. M. Gabbay. *Semantical Investigations in Heyting's Intuitionistic Logic*. D. Reidel, 1981.
[Gabbay, 2011] D. M. Gabbay. Dungs argumentation is essentially equivalent to classical propositional logic with the Peirce-Quine dagger. *Logica Universalis*, 5(2):255318, 2011.
[Giunchiglia et al., 2004] E. Giunchiglia, J. Lee, V. Lifschitz, N. McCain, and H. Turner. Nonmonotonic causal theories. *Artificial Intelligence*, 153:49–104, 2004.
[Hamblin, 1971] C. L. Hamblin. Mathematical models of dialogue. *Theoria*, 37:130–155, 1971.
[Jakobovits and Vermeir, 1999] H. Jakobovits and D. Vermeir. Robust semantics for argumentation frameworks. *Journal of Logic and Computation*, 9:215–261, 1999.
[Kakas and Toni, 1999] A. C. Kakas and F. Toni. Computing argumentation in logic programming. *Journal of Logic and Computation*, 9:515–562, 1999.
[Kakas et al., 1994] A. C. Kakas, P. Mancarella, and P. M. Dung. The acceptability semantics for logic programs. In P. Van Hentenryck, editor, *Proceedings Int. Conf. on Logic Programming, ICLP-94*, pages 504–519, Camb., MA, 1994. MIT Press.
[Kowalski and Toni, 1996] R. A. Kowalski and F. Toni. Abstract argumentation. *Artificial Intelligence and Law*, 4:275–296, 1996.
[Lin and Shoham, 1989] F. Lin and Y. Shoham. Argument systems: A uniform basis for nonmonotonic reasoning. In *Proceedings of 1st Intl. Conference on Principles of Knowledge Representation and Reasoning*, pages 245–255, Stanford, CA, 1989.
[Lorenzen and Lorenz, 1978] P. Lorenzen and K. Lorenz. *Dialogische Logik*. Wissenschaftliche Buchgesellschaft, Darmstadt, 1978.
[Makinson and van der Torre, 2000] D. Makinson and L. van der Torre. Input/Output logics. *Journal of Philosophical Logic*, 29:383–408, 2000.
[McCain and Turner, 1997] N. McCain and H. Turner. Causal theories of action and change. In *Proceedings AAAI-97*, pages 460–465, 1997.
[Nielsen S.H., 2007] Parsons S. Nielsen S.H. A generalization of dungs abstract framework for argumentation: Arguing with sets of attacking arguments. In Rahwan I. Maudet N., Parsons S., editor, *Argumentation in Multi-Agent Systems. ArgMAS 2006*, volume 4766 of *Lecture Notes in Computer Science*. Springer, 2007.
[Pearl, 2000] J. Pearl. *Causality: Models, Reasoninig and Inference*. Cambridge UP, 2000. 2nd ed., 2009.
[Pollock, 1987] J. L. Pollock. Defeasible reasoning. *Cognitive Science*, 11(4):481518, 1987.
[Reiter, 1980] R. Reiter. A logic for default reasoning. *Artificial Intelligence*, 13:81–132, 1980.
[Segerberg, 1982] K. Segerberg. *Classical Propositional Operators*. Clarendon Press, 1982.
[Shoesmith and Smiley, 1978] D. J. Shoesmith and T. J. Smiley. *Multiple-Conclusion Logic*. Cambridge University Press, 1978.
[Strasser and Seselja, 2010] C. Strasser and D. Seselja. Towards the proof-theoretic unification of Dungs argumentation framework: an adaptive logic approach. *Journal of Logic and Computation*, 21:133156, 2010.
[Wojcicki, 1988] R. Wojcicki. *Theory of Logical Calculi*, volume 199 of *Synthese Library*. Kluwer Ac. Publ., 1988.

Alexander Bochman
Computer Science Department
Holon Institute of Technology (HIT), Israel
Email: bochmana@hit.ac.il

www.ingramcontent.com/pod-product-compliance
Lightning Source LLC
Chambersburg PA
CBHW071147230426
43668CB00009B/864